Nutrient Metabolism
Structures, Functions, and Genes

Nutrient Metabolism
Structures, Functions, and Genes

Second Edition

Martin Kohlmeier
Research Professor,
Department of Nutrition,
UNC Gillings School of Global Public Health,
and UNC Nutrition Research Institute,
University of North Carolina,
USA

AMSTERDAM • BOSTON • HEIDELBERG • LONDON
NEW YORK • OXFORD • PARIS • SAN DIEGO
SAN FRANCISCO • SINGAPORE • SYDNEY • TOKYO
Academic Press is an imprint of Elsevier

Academic Press is an imprint of Elsevier
125, London Wall, EC2Y 5AS.
525 B Street, Suite 1800, San Diego, CA 92101-4495, USA
225 Wyman Street, Waltham, MA 02451, USA
The Boulevard, Langford Lane, Kidlington, Oxford OX5 1GB, UK

Second edition 2015
First edition 2003
Reprinted 2006

ISBN: 978-0-12-387784-0

British Library Cataloguing-in-Publication Data
A catalogue record for this book is available from the British Library.

Library of Congress Cataloging-in-Publication Data
A catalog record for this book is available from the Library of Congress.

For Information on all Academic Press publications
visit our website at http://store.elsevier.com/

Working together
to grow libraries in
developing countries

www.elsevier.com • www.bookaid.org

Publisher: Cathleen Sether
Acquisition Editor: Andrea Dierna
Editorial Project Manager: Marisa LaFleur
Production Project Manager: Lucía Pérez
Designer: Mark Rogers

Typeset by MPS Limited, Chennai, India
www.adi-mps.com

*To the advancement of nutrition science and the
continuous improvement of the education of health
care providers around the world.*

Contents

Preface..xi
Acknowledgments... xiii
List of Abbreviations...xv
Introduction...xxiii

CHAPTER 1	**Chemical Senses** 1	
	Smell 1	
	Taste 4	
	Intestinal Sensing 16	
	Physical Sensing and Chemesthesis....................................20	
CHAPTER 2	**Intake Regulation** 25	
	Appetite 25	
	Thirst 32	
CHAPTER 3	**Absorption, Transport, and Retention** 37	
	Digestion and Absorption..37	
	Microbiome 61	
	Renal Processing 69	
	The Blood–Brain Barrier..81	
	Materno-Fetal Nutrient Transport 87	
CHAPTER 4	**Xenobiotics** 95	
	Caffeine 95	
	Heterocyclic Amines 100	
	Nitrite/Nitrate 107	
CHAPTER 5	**Fatty Acids** 111	
	Structure and Function of Fatty Acids 111	
	Overfeeding 143	
	Acetate...147	
	Myristic Acid..153	
	Conjugated Linoleic Acid 157	
	Docosahexaenoic Acid 164	
	Trans-Fatty Acids 174	
	Chlorophyll/Phytol/Phytanic Acid 179	
CHAPTER 6	**Carbohydrates, Alcohols, and Organic Acids** 187	
	Carbohydrates 187	
	Glucose 191	

Fructose 207

Galactose 213

Xylitol 219

Pyruvate 223

Oxalic Acid...228

Ethanol 231

Methanol 238

CHAPTER 7   Nonnutrients and Bioactives 243

Indigestible Carbohydrates...243

Flavonoids and Isoflavones 247

Garlic Bioactives 260

CHAPTER 8   Amino Acids and Nitrogen Compounds 265

Structure and Function of Amino Acids...266

Starvation 291

Glutamate 294

Glutamine 302

Glycine 309

Threonine 316

Serine..321

Alanine 329

Phenylalanine 336

Tyrosine 342

Tryptophan 349

Methionine 359

Cysteine 368

Lysine 376

Leucine 382

Valine..389

Isoleucine 396

Aspartate 401

Asparagine..408

Arginine..413

Proline 423

Histidine 431

Citrulline 440

Taurine..443

Creatine 450

Carnitine 454

Melatonin 461

Choline 468

CHAPTER 9   Fat-Soluble Vitamins and Nonnutrients **479**
Free Radicals and Antioxidants 479
Vitamin A...486
Vitamin D...501
Vitamin E...514
Vitamin K...526
Cholesterol...539
Lipoic Acid...553
Ubiquinone...560

CHAPTER 10   Water-Soluble Vitamins and Nonnutrients **567**
Methylation..567
Vitamin C...570
Thiamin..580
Riboflavin...589
Niacin..599
Vitamin B6..610
Folate..620
Vitamin B12...632
Biotin...642
Pantothenate..647
Queuine...654
Biopterin..657
Inositol..663

CHAPTER 11   Minerals and Trace Elements **673**
Water...673
Sodium..679
Potassium..685
Chlorine...690
Iron...697
Copper...709
Zinc...716
Manganese...725
Calcium..730
Phosphorus...737
Magnesium...744
Iodine..748
Fluorine...755
Sulfur...758
Selenium..766
Molybdenum...773

Cobalt...778
Chromium 782
Boron...785
Silicon...789
Bromine...793
Arsenic 797
Vanadium 802
Nickel..805

CHAPTER 12 Applications...809
Genetic Variation...809
Nutrient Adequacy and Supplementation 816
Nutrient Interactions 819
GRAS Database 822
Using Molecular Databases 822

Index 825

Preface

It has been twelve years since the first edition of *Nutrient Metabolism* was published. That was also the year when a first read of our genome was completed and the blueprint for all human existence was deciphered. Since then many of the formerly white areas in the vast expanses of the nutrient metabolism map have been colored in. We now know much more about the metabolic fates of nutrients and about the molecular actors and mechanisms responsible for the underlying processes. A major reason for the explosive advances in the understanding of nutrient metabolism has been the massive investigative use of all kinds of "*-omics*".

One of these high-resolution technologies is ***genomics***. We now know most genes. A few may still play hard to get, but there are not going to be a lot of surprises. The attention, therefore, has turned increasingly to the multiple products individual genes can generate, how their production is regulated, and what the gene products actually do. High-throughput analyzers, which can read millions of DNA or RNA bases in days, have made it possible to study genes and their expression in detail in many different species of model animals and in human populations. Who would have thought that the blood group secretor status is a major determinant of vitamin B12 sufficiency in older adults? A wave of genome-wide association studies has yielded this and many other new and often unexpected insights into the inner workings of nutrient metabolism.

We can also use ***epigenomics*** to monitor in detail many chemical modifications of particular DNA sequences. The most widely known kind, the methylation of cytosine bases in promoter regions, will usually silence expression of the associated gene. Each cell type has its own set of characteristic epigenetic modifications that determine the nature of the cells and what they do. While many methylation patterns are tissue specific, others are shared more globally and can be read from blood cells. Several nutrients are important for the epigenetic status at numerous DNA sites of interest because they are needed for a steady supply of methyl groups. Research in recent years has greatly increased our understanding of the relevant pathways, variation of the involved genes, and the role of nutrients, phytochemicals, and feeding status in setting epigenetic patterns.

The advances in ***proteomics*** also had their share in deepening the understanding of nutrient metabolism. Proteomic analyses, which resolve proteins and peptides from different tissues and cell lines with high-resolution techniques, have concluded that fewer than 20,000 genes are coding for proteins. The approach has helped to clarify in which tissues particular proteins are present and contribute to nutrient metabolism there.

Many of the new insights have come from ***metabolomic*** analyses. This approach measures many different small metabolites in fluids and tissues. Current technologies already can resolve several thousand different compounds and this number is rapidly growing. The distinct compounds in foods and the metabolic products derived from them most likely are more than one hundred thousand in number. For example, the consumption of a particular fruit juice leads to a signature increase in urine of urolithin A glucuronide, pyrogallol sulfate, and ascorbic acid sulfate, which can be linked to their specific precursors, ellagitannins and ascorbic acid. Being able to make sense of food- and nutrient-specific metabolite patterns is becoming increasingly important. An understanding of the nature of food constituents, their properties, and typical metabolic conversions helps with the interpretation of complex chemical signatures in biological fluids.

The role of gut bacteria explored by **microbiomics**, for nutrient metabolism did not receive much attention when the first edition was put together, but this aspect absolutely had to be added now. The human intestines contain about ten times as many microorganisms as there are cells in the human body taken together. The collection of bacteria, yeasts, and viruses in the intestinal tract constitute the human gut microbiome. Many microbiota are metabolically very active. They partake of the food that comes their way, convert some of the dietary fiber into fuels for their host, and provide essential vitamins. They also chemically alter bile acids and other compounds in the intestinal lumen before they are absorbed. Thus, the gut microbiome functions like a chemical factory that is normally well integrated into overall human metabolism and critically important for good health. The shifts in the balance of many nutrients due to disruption of the microbiome by dietary factors, infections, or antibiotics are increasingly recognized, opening up new approaches for prevention and treatment of diseases.

All those advances in nutrition science notwithstanding, most of what we have known at the time of the first edition has not changed. On the positive side, there have been few fundamental surprises and much of the emerging new information is consistent with previously established science. On the negative side, the paradigms and dogmas governing practice and policies are notoriously slow to change and that is particularly true for nutrition. For example, every word in the first edition about setting the Dietary Recommended Intakes (DRI) is as applicable now as it was then. Despite our rapidly growing recognition of extensive genomic and metabolic diversity even within seemingly homogenous populations, current intake recommendations still assume that everybody has more or less the same metabolism. This is certainly contradicted by the known genomic, proteomic, metabolomic, and microbiomic variations that make each of us different, sometimes very different. The smallest structural difference in an enzyme, transporter, or receptor can profoundly change its function and alter how much of a nutrient an individual needs. Similarly, common epigenetic alterations of our genome and subtle shifts in microbiome patterns have important nutritional implications. The information in this book is meant to help put the nutritional consequences of such functional differences into perspective.

Martin Kohlmeier

Chapel Hill, April 2015

Acknowledgments

I am indebted to many people for their support, encouragement, and helpful discussions while working on this new edition. The support and collaborative environment of the UNC Nutrition Research Institute has been particularly helpful. I also could not have done this without the consistent and caring support of the Nutrition in Medicine team at the University of North Carolina at Chapel Hill.

List of Abbreviations

1,25-(OH)2-D	1,25-dihydroxyvitamin D
1,25-D	1α,25-dihydroxyvitamin D
1,7-X	1,7-dimethylxanthine
24,25-D	24R,25-dihydroxyvitamin D
25-D	25-hydroxyvitamin D
25-OH-D	25-trihydroxyvitamin D
5-HT	5-hydroxytryptamine
5-HTOL	5-hydroxytryptophol
6-SMT	6-sulfatoxymelatonin
A2S	ascorbate-2-sulfate
ABC	ATP-binding cassette (ABCA1, ABCG5, ABCG8)
ADH	alcohol dehydrogenase (Chapters 6, 9, and 12)
ADH	antidiuretic hormone (Chapters 1, 3, and 11)
ADPR	ADP ribose
AgRP	agouti-related protein
AI	Adequate Intake
AI-2	autoinducer 2
AK	adenylate kinase
Ala	L-alanine
ALDH	aldehyde dehydrogenase
AMP	adenosine monophosphate
AMPK	AMP-activated protein kinase
ANP	atrial natriuretic peptide
apBGlu	p-acetamidobenzoylglutamate
apoAI	apolipoprotein AI
apoAII	apolipoprotein AII
apoAIV	apolipoprotein AIV
apoB	apolipoprotein B
apoB48	apolipoprotein B48
apoB100	apolipoprotein B100
apoC-I	apolipoprotein C-I
apoC-II	apolipoprotein C-II
apoC-III	apolipoprotein C-III
apoE	apolipoprotein E
APS	adenosine 5'-phosphosulfate
AQP	aquaporin
Arg	L-arginine
As	arsenic
ASC	L-ascorbic acid
Asn	L-asparagine
Asp	L-aspartate
ATP7A	copper-transporting ATPase (ATP7A, ATP7B)

B	boron
B6	vitamin B6
B12	vitamin B12
BBB	blood–brain barrier
BCAA	branched chain amino acids
BGT1	sodium chloride–dependent betaine transporter
BH4	5,6,7,8-tetrahydrobiopterin
BM	basal membrane
Br	bromine
CACT	carnitine-acylcarnitine translocase
cADPR	cyclic ADP-ribose
cAMP	cyclic adenosine monophosphate
Carbs	carbohydrates
CaSR	calcium-sensing receptor
CAT-2	cationic amino acid transporter 2
CCK	cholecystokinin
CD	Celiac disease
CFTR	cystic fibrosis transmembrane regulator
Chol	cholesterol
Chylos	chylomicrons
CLA	conjugated linoleic acid
CLD	congenital diarrhea gene
Co	cobalt
CoA	coenzyme A
Cr	chromium
CRBP2	cellular retinol–binding protein 2
CSF	cerebrospinal fluid
CssC	L-cystine
CTR	copper transporter (CTR1, CTR2)
CYP	cytochrome P450
Cys	L-cysteine
D2	vitamin D2
D3	vitamin D3
DADS	diallyl disulfide
DADSO	diallyl thiosulfinate
DATS	diallyl trisulfide
DBP	vitamin D–binding protein
DHA	dehydro-L-ascorbic acid (Chapter 10)
DHA	docosahexaenoic acid (Chapters 3, 5, 12)
DHF	dihydrofolate
DiMeIQx	2-amino-3,4,8-trimethylimidazo[4,5-*f*]quinoxaline
DL-AP4	DL(+)-2-amino-phosphono-butyric acid
DMA(III)	dimethylarsinous acid
DMA(V)	dimethylarsinic acid
DMT1	divalent metal ion transporter 1

DPDS	dipropyl disulfide
DRA	down-regulated in adenoma gene
DRI	Dietary Reference Intake
DTDST	diastrophic dysplasia sulfate transporter
EAR	Estimated Average Requirement
ECaC1	epithelial calcium channel 1
ECI	enoyl-CoA isomerase
EMS	eosinophilia-myalgia syndrome
EPA	eicosapentaenoic acid
ETF	electron-transfer flavoprotein
F	fluorine
F6P	fructose 6-phosphate
FABPpm	plasma membrane fatty acid binding protein
FAD	flavin adenine dinucleotide
FATP-1	fatty acid transport protein 1
FIGLU	formiminoglutarnate
FMN	flavin mononucleotide
FOLR	folate receptor (FOLR1, FOLR2, FOLR3)
Fru	D-fructose
GABA	gamma-aminobutyric acid
Gal	D-α-galactose
GalNAc	Gal N-acetylglucosamine
GAP	glyceraldehyde 3-phosphate
GFR	Glomerular Filtration Rate
GIc	D-glucose
Gln	L-glutamine
GLP-1	glucagon-like peptide 1
Glu	L-glutamate
GLUT	glucose transporter (GLUT1, GLUT2, GLUT3, GLUT4, GLUT5)
Glvr-1	gibbon ape leukemia virus receptor
Gly	glycine
GMP	guanosine-5′-monophosphate
Grp120	G-protein coupled receptor 120
GSH	glutathione
HA	heterocyclic amine
Hcys	homocysteine
HDL	high-density lipoprotein
HFE	hemochromatosis gene
HIF-1	hypoxia-inducible factor 1
His	L-histidine
HMGCoA	hydroxymethylglutaryl-CoA
HNE	4-hydroxy-2,3-trans-nonenal
Hypro	4-hydroxyproline

IL-6	interleukin-6
Ile	L-isoleucine
IMP	inosine-5′-monophosphate
IOM	Institute of Medicine
IP3	inositol triphosphate
IP5	inositol pentaphosphate
IP6	inositol hexaphosphate
IQ	2-amino-3-methylimidazo[4,5-*f*]quinoline
IRE	iron-responsive element
IRE-BP	iron-responsive element–binding protein
IRP	iron regulatory protein
K1	phylloquinone
KAT	kynurenine-oxoglutarate aminotransferase
LAT	L-type amino acid transporter (LAT1, LAT2)
LDH	lactate dehydrogenase isoform
LDL	low-density lipoprotein
LDL-R	LDL receptor
Leu	L-leucine
Lp	lipoprotein
LPL	lipoprotein lipase
LRP	LDL receptor-related protein (LRP1, LRP2)
Lys	L-lysine
Man	D-mannose
MC	monocarboxylate transporter
MCH	melanin-concentrating hormone
MCR4	melanocortin receptor 4
MCT	proton/monocarboxylic acid cotransporter (MCT1, MCT2)
MDR	multidrug resistance gene (MDR1, MDR2, MDR3)
MECI	mitochondrial enoyl-CoA isomerase
MelQx	2-amino-3,8-dimethylimidazo[4,5-*f*]quinoxaline
MEOS	microsomal ethanol oxidizing system
Met	L-methionine
MFP2	peroxisomal multifunctional protein 2
MG	macroglycogen
Mg	magnesium
MK4	menaquinone-4
MMA(III)	monomethylarsonous acid
MMA(V)	monomethylarsonic acid
Mn	manganese
Mo	molybdenum
MRP	multidrug resistance protein (MRP1, MRP2)
MSG	monosodium glutamate
MSH	alpha-melanocyte stimulating hormone
MTA	5′-methylthioadenosine

MTF1	metal response element-binding transcription factor 1
MTHFD	methylenetetrahydrofolate dehydrogenase
MTHFR	methylenetetrahydrofolate reductase
mTORC1	mammalian target of rapamycin 1
MTP	microsomal triglyceride transfer protein
MTR	5-methyltetrahydrofolate-homocysteine *S*-methyltransferase
MVM	microvillous membrane
NaCl	sodium chloride
NAD	nicotinamide adenine dinucleotide
NADP	nicotinamide phosphate adenine dinucleotide
NAT	arylamine *N*-acetyltransferase
NBCl	sodium/bicarbonate cotransporter 1
NCX1	Na^+/Ca^{2+} exchanger
NE	niacin equivalent
NHE	sodium/hydrogen exchanger (NHE1, NHE2, NHE3)
NKCC	sodium-potassium-chloride transporter (NKCC1, NKCC2)
NMDA	N-methyl D-aspartate
NMN	nicotinamide mononucleotide
NPY	neuropeptide Y
OAT	ornithine-oxo-acid aminotransferase
OTC	ornithine carbamoyltransferase
OCT	organic cation transporter (OCT1, OCT2, OCT2)
P5C	delta-1-pyrroline-5-carboxylate
pABG	para-aminobenzoic acid
PANK	pantotheine kinase
PAPS	3'-phosphoadenosine 5'-phosphosulfate
PAT1	putative anion ion transporter 1
P-cADPR	2'-phospho-cyclic ADP-ribose
PDS	Pendred syndrome gene
PEITC	beta-phenylethyl isothiocyanate
PEM	protein-energy malnutrition
PEMT	phosphatidylethanolamine-N-methyltransferase
PepT	hydrogen ion/peptide cotransporter (PepT1, PepT2)
PET	positron emission tomography
PG	proglycogen
Phe	L-phenylalanine
PhIP	2-amino-1-methyl-6-phenylimidazo[4,5-β]pyridine
PKU	phenylketonuria
PLP	pyridoxal 5'-phosphate
PMP	pyridoxamine 5'-phosphate
PNP	pyridoxine 5'-phosphate
Pro	L-proline
PROP	6-*n*-propyl-2-thiouracil
PTH	parathyroid hormone

PTHrP	PTH-related protein
PTP	6-pyruvoyltetrahydro pterin
PYY	peptide tyrosine-tyrosine
Q10	ubiquinone-10
RAE	retinol activity equivalent
RALDH	retinal dehydrogenase (RALDH1, RALDH2)
RAP	receptor-associated protein
RBP	retinol–binding protein (RBP1, RBP2, RBP3, RBP4)
RCP	riboflavin carrier protein
RDA	recommended dietary allowance
RFC1	reduced folate carrier
ROS	reactive oxygen species
SAC	*S*-allyl cysteine
SAH	*S*-adenosyl-L-homocysteine
SAM	*S*-adenosyl-methionine
SCPx	sterol carrier protein X
SDA	semidehydroascorbate
Se	selenium
SEC	*S*-ethyl *cysteine*
Ser	L-serine
SGLT	sodium-glucose cotransporter (SGLT1, SGLT2)
Si	silicon
SMVT	sodiurn-dependent multivitamin transporter
SPC	S-propyl cysteine
SR-B1	scavenger receptor class B type 1
SREBP	sterol regulatory element–binding protein
SVCT	sodium-ascorbate transporter (SVCT1, SVCT2)
TAP	tocopherol-associated protein
TAT1	T-type amino acid transporter 1
TAUT	taurine transporter
TBP	tocopherol–binding protein
TCP	thiamin carrier protein
TF	transferrin
TfR	transferrin receptor
TGT	RNA-guanine transglycosylase
THF	5,6,7,8-tetrahydrofolate
Thr	L-threonine
ThTr2	thiamin transporter 2
TMP	thiamin monophosphate
TmP	tubular maximum for phosphate
TPP	thiamin pyrophosphate
TRL	triglyceride-rich lipoprotein
Trp	L-tryptophan
Trp-P-1	3-amino-1,4-dimethyl-SH-pyrido[4,3-β]indole

Trp-P-2	3-amino-1-methyl-5H-pyrido[4,3-β]indole
TRPA1	transient receptor potential ankyrin 1
TRPV1	transient receptor potential vanilloid 1
TSH	thyroxine-stimulating hormone
TTP	alpha-tocopherol transfer protein
TTP	thiamin triphosphate
TVA	trans-vaccenic acid
Tyr	L-tyrosine
UGT	UDP-glucuronosyltransferase
UL	tolerable upper intake level
UV-B	ultraviolet B light
Val	L-valine
vitA	vitamin A
vitD	vitamin D
vitE	vitamin E
vitK	vitamin K
VLDL	very-low-density lipoprotein
VNO	vomeronasal organ
XSP	xylulose 5-phosphate
ZIP1	zinc- and iron-regulated protein 1
ZnT1	zinc transporter 1

Introduction

NUTRIENTS
WHAT NUTRIENTS ARE

Humans depend for sustenance on what we consume, but fortunately we are not what we eat. The foods that we eat and drink are broken down by grinding and digestion, sorted by selective absorption, and changed in metabolic reactions. The ultimate fate of an absorbed compound in the body depends on how quickly it is metabolized for energy production or excreted via bile, feces, or urine. Most ingested compounds leave again after a short time—which is for the best. Some of those that linger (such as pesticides, toxic heavy metals, or too much fat) are unwelcome guests that make us wish for more short-term visitors like vitamin C (ascorbic acid), which cheerfully do their job and leave again in a timely manner.

Nutrients are compounds from food or bacteria in the gut that the body uses for its normal physiological functions. This broad definition includes compounds that are utilized directly for energy production (ethanol), to aid in metabolism (coenzymes), to build body structures (cholesterol), or to serve in a specific cellular function (bromine, for the oxidative burst of eosinophils). A nutrient has been considered essential in short-lived organisms if its lack prevents the organisms from completing their life cycle and if the compound is directly involved in the function of the organisms (Epstein, 1994). This definition has some obvious shortcomings for humans because many of the current concerns are about health long beyond reproductive age. Indeed, more than half of healthy people's life spans occur after reproduction ceases (Blurton Jones et al., 2002). Prevention of cardiovascular disease, cancer, and dementia has come to the forefront of today's healthcare efforts. The significance of an individual's exposure to particular food compounds for a span of nearly 100 years is just beginning to come into focus.

ESSENTIAL BUILDING BLOCKS

At least 26 different elements are used to build the human body and keep it functioning; 14 of them are usually found in quantities of at least 1 g or more in a young 70-kg person. Fluorine (as fluoride) strengthens and protects teeth (and possibly also bones), but it is not considered essential. Additional elements, including tin, rubidium, germanium, and lithium, are consumed in small but significant amounts with food and are regularly present in the body. It is uncertain, however, whether these are truly needed at any stage in life or provide any health benefit for some or all humans (Nielsen, 2001).

The body can use many elements only when they are ingested in specific chemical configurations. Carbon has to be supplied mainly as digestible carbohydrate, alcohol, fat, or protein. Nitrogen has to be consumed as protein or amino acids; small amounts of ammonia and some nitrogen compounds can also be utilized. Hydrogen comes with water and smaller quantities from the oxidation and other metabolic reactions of nutrients. Sulfur has to come with methionine, cysteine, and sulfate. Phosphorus can be utilized only as phosphate or pyrophosphate. Trivalent chromium supports the regulation of glucose metabolism, while hexavalent chromium is a very toxic allergen and carcinogen. Therefore, the physiological properties of such elements really depend on their specific chemical forms (Table I.1).

Table I.1 Elements of the Body

	Intake (g/day)	Body Content (g/70 kg)	
Oxygen	2000	45,000	Estimated
Carbon	305	16,000	Estimated
Hydrogen	275	4000	Estimated
Nitrogen	10	1850	Cohn et al. (1983)
Calcium	1	1150	Cohn et al. (1983)
Phosphorus	1	450	Cohn et al. (1976), Aloia et al. (1984)
Potassium	3	140	Cohn et al. (1983); Larson et al. (2003)
Sulfur	0.3	100	Estimated
Sodium	3	82	Brennan et al. (1980)
Chloride	6	75	Cohn et al. (1983)
Magnesium	0.3	25	Elin (1987), Rude and Gruber (2004)
Fluorine	0.002	6	Hać et al. (1997)
Iron	0.008	3	Puliyel et al. (2013)
Zinc	0.011	2	King and Keen (1994)
Copper	0.0009	0.08	Uauy et al. (2008)
Selenium	0.000055	0.017	Schroeder et al. (1970)
Iodine	0.00015	0.015	Hays (2001)
Manganese	0.0002	0.015	Versieck (1985)
Silicon	0.04	0.003	Lugowski et al. (2000)
Boron	0.0008	0.003	Meacham and Hunt (1998)
Bromine	0.003		Shiraishi et al. (2009)
Nickel	0.0003		Health Canada (2007)
Molybdenum	0.0002		Barceloux (1999)
Cobalt	0.000015		Health Canada (2007)
Arsenic	0.00007		Health Canada (2007)
Chromium	0.00002		
Vanadium			
Tin			
Rubidium	0.0032		Health Canada (2007)
Germanium			
Lithium			
Aluminum	0.0076		Health Canada (2007)
Cadmium	0.000015		Health Canada (2007)

ESSENTIAL NUTRIENTS

Only 24 complex nutrients are absolutely essential since they cannot be adequately produced from precursors. Precursors can provide a small percentage of the required amounts of water (oxidation of macronutrients) and niacin (metabolism of tryptophan), but most has to come from the outside.

Some of these nutrients are essential only for humans and possibly a few other species. Thus, most mammals can synthesize ascorbic acid from glucose and produce daily amounts equivalent to several grams per 70 kg body weight. Humans, like the other primates, have lost their ability to synthesize vitamin C from L-gulonate because their gene for L-gulonolactone oxidase (EC1.1.3.8) is riddled with crippling mutations. Other nutrients that cannot be synthesized at all by humans include 9 amino acids, the omega-3 and omega-6 fatty acids, 11 vitamins, and queuine. Queuine appears to be important since a dedicated human enzyme inserts this nucleotide-like compound into specific DNA sequences that promote colonocyte stability and function. Vitamin D is not on this list of essential nutrients since it is synthesized in the body from 7-dehydrocholesterol (also the immediate precursor of newly synthesized cholesterol) and people can thrive without any intake. If anything, the ultraviolet (UV) light that is needed for vitamin D production in skin might be considered essential, rather than the vitamin itself. Under some conditions, especially in children and older people living at high latitudes and staying indoors too much, dietary intake becomes important for optimal health.

This is no different from arginine, which is also critical for health and needs to be partially supplied with the diet if vitamin B6 and niacin status is inadequate, especially during times of great need. Food compounds such as phytate (Heim et al., 2002) still may turn out to be important, if not essential, as new evidence sheds light on their role in human metabolism.

The list of essential nutrients may serve as a reminder that only a few of them are actually nutrients of concern in an affluent society. Many essential nutrients are harmful if consumed in excess. The bottom line is that for every nutrient, optimal intake levels need to be determined, and that too much is often as problematic as too little (Tables I.2 and I.3).

Table I.2 Essential Nutrients

Water	Generates aqueous environment, protons, hydroxide ions
Sugars	Energy production, synthesis of most organics
Amino groups	Constituent of proteins, mediator synthesis
Leucine	Energy source, protein and beta-leucine synthesis
Valine	Energy source, protein synthesis
Isoleucine	Energy source, protein synthesis
Lysine	Energy source, protein synthesis
Tryptophan	Energy source, protein, niacin, and mediator synthesis
Phenylalanine	Energy source, protein, tyrosine, mediator, and pigment synthesis
Methionine	Energy source, protein and cysteine synthesis, methyl donor
Threonine	Energy source, protein and glycine synthesis
Histidine	Energy source, protein synthesis
Omega-3 fatty acids	Energy source, mediator synthesis
Omega-6 fatty acids	Energy source, mediator synthesis
Vitamin A	Synthesis of vision pigment, regulator of numerous genes
Vitamin E	Antioxidant
Vitamin C	Antioxidant, cofactor of numerous enzymes
Riboflavin	Cofactor of numerous enzymes

(*Continued*)

Table I.2 (Continued)

Niacin	Cofactor of several hundred enzymes
Pantothenate	Cofactor of numerous enzymes
Folic acid	Cofactor of numerous enzymes
Vitamin B6	Cofactor of several hundred enzymes
Vitamin K	Cofactor of one enzyme that modifies many proteins
Thiamin	Cofactor of 5 enzymes, neuronal action
Biotin	Cofactor of 4 enzymes, stabilizing DNA, additional actions
Vitamin B12	Cofactor of 3 enzymes
Queuine	Stabilizes specific tRNAs in the colon
Sodium	Osmolyte, enzyme cofactor, cotransport
Potassium	Signal transduction, enzyme cofactor
Chloride	Osmolyte, cotransport, digestion, immune defense
Iron	Cofactor of numerous enzymes and proteins
Zinc	Cofactor of numerous enzymes and transcription factors
Copper	Cofactor of numerous enzymes and proteins
Manganese	Cofactor of numerous enzymes
Iodine	Constituent of thyroid hormones
Selenium	Cofactor of 13 enzymes and proteins
Molybdenum	Cofactor of 4 enzymes, additional actions
Chromium	Constituent of chromomodulin, interaction with DNA
Cobalt	Constituent of vitamin B12, cofactor of methionine aminopeptidase
Bromine	Halogenating oxidant of eosinophils
Boron	Unknown
Silicon	Unknown
Arsenic	Unknown
Vanadium	Unknown
Nickel	Unknown
Tin	Unknown
Rubidium	Unknown
Germanium	Unknown
Lithium	Unknown .
Aluminum	Unknown
Cadmium	Unknown
Lead	Unknown

NONESSENTIAL ORGANIC MICRONUTRIENTS

Some compounds can be synthesized by humans, but production may not cover needs at all times, especially at certain times in the life cycle. Thus, food sources have to augment endogenous synthesis of arginine, cysteine, taurine, and docosahexaenoic acid (an omega-3 fatty acid) to meet the needs of infants. Severe injury, infections, chronic diseases, or other temporary circumstances also may increase needs

Table I.3 Conditionally Essential Nutrients

Vitamin D	Regulates numerous genes
Choline	Phospholipid synthesis, methyl group donor
Arginine	Constituent of protein, creatine, nitrous oxide synthesis
Tyrosine	Constituent of protein, mediator, pigment synthesis
Cysteine	Glutathione synthesis, conjugation, signaling
Taurine	Constituent of bile acids, osmolyte
Choline	Constituent of phospholipids, methyl group donor, neurotransmitter
Lipoic acid	Antioxidant, cofactor of 4 enzymes
Ubiquinone	Cofactor of oxidative phosphorylation
Carnitine	Cofactor for fatty acid translocation
DHA	Constituent of structural phospholipids, mediator

Table I.4 Nonessential Nutrients Conferring Health Benefits

Fluorine	Stabilizes tooth minerals as fluoride
Flavonoids	Antioxidants, phytoestrogens
Carotenoids	Antioxidants, some are vitamin A precursors

beyond the capacity of endogenous synthesis. On the other hand, dietary intakes of specific nonessential nutrients may become more important when genetic variants create a bottleneck in the synthesis of a particular compound, such as carnitine. Vitamin D provides another illustration of an endogenously synthesized compound that can be conditionally essential. Humans can produce large amounts of cholecalciferol (vitamin D_3), so long as their skin is exposed to sufficiently intense sunlight for long enough periods of time. Only life at higher latitudes (especially during winter months) or indoors, or prevention of skin exposure to sun by clothing or sunscreen, makes any dietary intake of this misnamed nutrient necessary (cholecalciferol is not even an amine, much less a vitamin). Decreased availability (for instance, of lipoic acid) due to disruption of their production by intestinal bacteria should also be mentioned.

Undoubtedly, there are numerous other dietary compounds, particularly those of plant origin, that affect health, sometimes to a very significant extent. Thousands of these compounds and their metabolites are detectable in blood and act on tissues (Wishart et al., 2013).

Examples of such substances include a wide range of polysaccharides, flavonoids, phytosterols, saponins, and other constituents of plant-derived foods that have shown some promise for the prevention of atherosclerosis, cancer, or other debilitating diseases. The real question about such nutrientlike compounds is how much can be safely consumed and whether higher than typical consumption with food provides any worthwhile health benefit (Table I.4).

REFERENCES

Aloia, J.F., Vaswani, A.N.M., Yeh, J.K., Ellis, K., Cohn, S.H., 1984. Total body phosphorus in postmenopausal women. Mineral Electrolyte Metab. 10, 73–76.
Barceloux, D.G., 1999. Molybdenum. J. Toxicol. Clin. Toxicol. 37, 231–237.

Blurton Jones, N.G., Hawkes, K., O'Connell, J.E., 2002. Antiquity of postreproductive life: are there modern impacts on hunter-gatherer postreproductive life spans? Am. J. Hum. Biol. 14, 184–205.

Brennan, B.L., Yasumura, S., Letteri, J.M., Cohn, S.H., 1980. Total body electrolyte composition and distribution of body water in uremia. Kidney Int. 17, 364–371.

Cohn, S.H., Brennan, B.L., Yasumura, S., Vartsky, D., Vaswani, A.N., Ellis, K.J., 1983. Evaluation of body composition and nitrogen content of renal patients on chronic dialysis as determined by total body neutron activation. Am. J. Clin. Nutr. 38, 52–58.

Cohn, S.H., Vaswani, A., Zanzi, I., Aloia, J.F., Roginsky, M.S., Ellis, K.J., 1976. Changes in body chemical composition with age measured by total-body neutron activation. Metabolism 25, 85–95.

Elin, R., 1987. Assessment of magnesium status. Clin. Chem. 33, 1965–1970.

Epstein, E., 1994. The anomaly of silicon in plant biology. Proc. Natl. Acad. Sci. USA 91, 11–17.

Haćt, E., Czarnowski, W., Gos, T., Krechniak, J., 1997. Lead and fluoride content in human bone and hair in the Gdańsk region. Sci. Total Environ. 206, 249–254.

Hays, M.T., 2001. Estimation of total body iodine content in normal young men. Thyroid 11, 671–675.

Health Canada. 2007. Canadian total diet study. Dietary intakes of contaminants and other chemicals for different age–sex groups of Canadians. Vancouver. Health 2011;26:81–92. Available at: <http://www.hc-sc.gc.ca/fn-an/surveill/total-diet/intake-apport/chem_age-sex_chim_2007-eng.php>.

Heim, M., Johnson, J., Boess, E., Bendik, I., Weber, P., Hunziker, W., et al., 2002. Phytanic acid, a natural peroxisome proliferator-activated receptor (PPAR) agonist, regulates glucose metabolism in rat primary hepatocytes. FASEB J. 16, 718–720.

King, J.C., Keen, C.L., 1994. Zinc. In: Shils, M.E., Olson, J.A., Shike, M. (Eds.), Modern Nutrition in Health and Disease, eighth ed. Lea & Febiger, Philadelphia, PA, pp. 214–230.

Larsson, I., Lindroos, A.K., Peltonen, M., Sjostrom, L., 2003. Potassium per kilogram fat-free mass and total body potassium: predictions from sex, age, and anthropometry. Am. J. Physiol. Endocrinol. Metab. 284, E416–E423.

Lugowski, S.J., Smith, D.C., Bonek, H., Lugowski, J., Peters, W., Semple, J., 2000. Analysis of silicon in human tissues with special reference to silicone breast implants. J. Trace Elem. Med. Biol. 14, 31–42.

Meacham, S.L., Hunt, C.D., 1998. Dietary boron intakes of selected populations in the United States. Biol. Trace Elem. Res. 66, 65–78.

Nielsen, F.H., 2001. Boron, manganese, molybdenum, and other trace elements. In: Bowman, B.A., Russell, R.M. (Eds.), Present Knowledge in Nutrition, eighth ed. ILSI Press, Washington, DC, pp. 384–400.

Puliyel, M., Sposto, R., Berdoukas, V.A., Hofstra, T.C., Nord, A., Carson, S., et al., 2013. Ferritin trends do not predict changes in total body iron in patients with transfusional iron overload. Am. J. Hematol. 89, 391–394.

Rude, RK, Gruber, HE., 2004. Magnesium deficiency and osteoporosis: animal and human observations. J. Nutr. Biochem. 15, 710–716.

Schroeder, H.A., Frost, D.V., Balassa, J.J., 1970. Essential trace metals in man: selenium. J. Chron. Dis. 23, 227–243.

Shiraishi, K., Ko, S., Muramatsu, Y., Zamostyan, P.V., Tsigankov, N.Y., 2009. Dietary iodine and bromine intakes in Ukrainian subjects. Health Phys. 96, 5–12.

Uauy, R., Maass, A., Araya, M., 2008. Estimating risk from copper excess in human populations. Am. J. Clin. Nutr. 88, 867S–871S.

Versieck, J., 1985. Trace elements in human body fluids and tissues. Crit. Rev. Clin. Lab. Sci. 22, 97–184.

Wishart, D.S., Jewison, T., Guo, A.C., Wilson, M., Knox, C., Liu, Y., et al., 2013. HMDB 3.0—The Human Metabolome Database in 2013. Nucleic Acids Res. 41 (Database issue), D801–D807.

CHEMICAL SENSES

CHAPTER OUTLINE

Smell 1
Taste 4
Intestinal Sensing 16
Physical Sensing and Chemesthesis 20

SMELL

The ability to detect odors from foods is an important appetite stimulant and often initiates the secretion of digestive juices. Aversive reactions to foul or otherwise disagreeable odors provide some protection against ingestion of unsafe foods. Impaired ability to smell, which becomes more common with advancing age, increases the risk of inadequate food intake and food poisoning.

ABBREVIATION

VNO   vomeronasal organ

ANATOMICAL STRUCTURE

Humans detect odors by olfactory epithelia covering about $10\,cm^2$ at the top of the nasal passage and transmitted through perforations in the skull along cranial nerve 1 (the olfactory nerve) to the limbic, cortical, hippocampal, and hypothalamic regions. A layer of mucus from specialized secretory cells, which constitute the Bowman's glands, separates the lumen of the nasal passage from the epithelial surface (Purves et al., 2001). The actual detecting units are receptor neurons (bipolar nerve cells), and supporting cells surround them. Basal cells give rise to new receptor neurons that reach maturity within 30–120 days (Costanzo and Graziadei, 1983). The receptor neurons extend cilia (microvilli) into the mucus on the epithelial surface. Odors can reach this surface both from the nostrils (sniffing) and from the pharynx (retronasal space, during chewing and swallowing). Odorants present in the epithelial mucus layer can bind to specific receptors on the cilia, trigger a cyclic adenosine monophosphate (cAMP)–mediated signaling cascade inside the receptor cell and thereby initiate depolarization. Action potentials propagate along the olfactory cell axons to mitral and tufted cells in the glomeruli of the olfactory bulb.

Each glomerulus appears to have a characteristic, limited molecular receptive range and thereby contributes to odor discrimination. Such odor-specific coordination may be related, in part, to input from lateral dendritic connections, which modulate the output of mitral/tufted cells to higher brain regions. Olfactory receptor cells with evoked activity survive better than nonstimulated cells, pointing to use-based selection as an important principle in the organization and maintenance of the olfactory system. The vomeronasal organ (VNO, Jacobson's organ) is a walled-off cluster of chemosensory cells at the anterior septum of the nose with a narrow opening that is easily missed (Smith et al., 2001). Many adults have only one identifiable VNO, and many have none at all. The irregular presence, the high proportion of nonfunctional pseudogenes coding for associated receptors, and other reasons have led some to believe that the VNO is a vestigial organ without functional significance (Kouros-Mehr et al., 2001). Others still consider the possibility of functional importance (Meredith, 2001). The VNO in rodents detects pheromones, a group of odorant molecules released by the animals to signal sexual and social states. Nasal irritation by odorants such as acetic acid (vinegar) is transmitted by stimulation of the trigeminus nerve (cranial nerve V; Rauchfuss et al., 1987).

Impairment of the sense of smell becomes more common with advancing age. Partially this reflects a general decline in neuronal survival and function. Indeed, diminished odor recognition both correlates with cognitive decline in nondemented older people (Swan and Carmelli, 2002) and may indicate the progression of Alzheimer's dementia. Trauma is another possible cause for a diminished sense of smell. The olfactory nerve is the cranial nerve that is most often damaged by a fracture of the skull base (Kruse and Awasthi, 1998).

MOLECULAR MECHANISMS OF OLFACTION

Healthy individuals can detect several hundred specific odors (and possibly thousands). Humans are able to distinguish aliphatic ketones (2-butanone to 2-decanone) and acetic esters (ethyl acetate to *n*-octyl acetate) based on carbon chain length (Laska and Hubener, 2001). However, as part of mixtures, only a maximum of three or four individual components can be identified. For some compounds, the addition of a second odorant is sufficient to obliterate detection of the first (Jinks and Laing, 2001).

An extensive family of transmembrane receptors recognizes volatile compounds. Some of these genes are related to major histocompatibility complex (MHC) genes and share their ability to bind and respond adaptively to diverse molecules. The human genome contains more than a thousand genes related to olfactory receptors. However, most of these are pseudogenes and not expressed (Younger et al., 2001; Zhang and Firestein, 2002). So far, only a few odors have been linked to particular receptors (Touhara, 2001).

Several gene products, other than receptors, have been linked to olfaction, but a comprehensive understanding of the entire process is still lacking. Both catecholamines and cholecystokinin (CCK) are involved in the olfactory signaling cascade. The odorant-binding protein is a member of the alpha-2-microglobulin superfamily in nasal epithelium. Additional odorant-binding proteins are expressed in olfactory neurons (Raming et al., 1993).

It may be of interest to note that cyanide, which is both a strong odorant and a potent inhalatory toxin, is specifically detoxified in nasal mucosa by the enzyme rhodanese (Lewis et al., 1991). Since this enzyme is otherwise mainly expressed in the liver, its presence in the nasal mucosa may confer some protection while sniffing at unknown food sources.

VARIATION IN SMELL SENSITIVITY

The ability of individuals to detect low concentrations of particular odorants varies greatly between individuals and tends to decline with age. The inability to detect a particular odor, such as butylmercaptan (the overpowering foul odorant in skunk secretion), has been referred to as *smell-blindness* (analogous to color blindness, where sensitivity to one color is lost, but others are still detected).

Androstenone: About 50% of people are not able to detect the odor of androstenone, even at high concentrations; 15% are moderately sensitive, and 35% are very sensitive (0.2 ppb in air). The sensitivity appears to be inducible by prior exposure.

Asparagus: One in 10 adults detects with high sensitivity a particular odor associated with asparagus consumption (Lison et al., 1980). The odorant substance may be methanethiol. Originally, it had been thought that some people excreted this compound more effectively than others, but there seems to be little variation in that respect.

Musk: As many as 7% of Caucasians appear to be unable to detect musklike odors.

REFERENCES

Costanzo, R.M., Graziadei, P.P., 1983. A quantitative analysis of changes in the olfactory epithelium following bulbectomy in hamster. J. Comp. Neurol. 215, 370–381.

Jinks, A., Laing, D.G., 2001. The analysis of odor mixtures by humans: evidence for a configurational process. Physiol. Behav. 72, 51–63.

Kouros-Mehr, H., Pintchovski, S., Melnyk, J., Chen, Y.J., Friedman, C., Trask, B., et al., 2001. Identification of non-functional human VNO receptor genes provides evidence for vestigiality of the human VNO. Chem. Senses 26, 1167–1174.

Kruse, J.J., Awasthi, D., 1998. Skull-base trauma: neurosurgical perspective. J. Cranio-Maxillofac Trauma 4, 8–14.

Laska, M., Hubener, E., 2001. Olfactory discrimination ability for homologous series of aliphatic ketones and acetic esters. Behav. Brain Res. 119, 193–201.

Lewis, J.L., Rhoades, C.E., Gervasi, P.G., Griffith, W.C., Dahl, A.R., 1991. The cyanide-metabolizing enzyme rhodanese in human nasal respiratory mucosa. Toxicol. Appl. Pharmacol. 108, 114–120.

Lison, M., Blondheim, S.H., Melmed, R.N., 1980. A polymorphism of the ability to smell urinary metabolites of asparagus. Br. Med. J. 281, 1676–1678.

Meredith, M., 2001. Human vomeronasal organ function: a critical review of best and worst cases. Chem. Senses 26, 433–445.

Purves, D., Augustine, G.J., Fitzpatrick, D., Katz, L.C., LaMantia, A.S., McNamara, J.O. et al., (Eds.), 2001. Neuroscience, second ed. Sinauer Associates, Sunderland, MA.

Raming, K., Krieger, J., Strotmann, J., Boekhoff, I., Kubick, S., Baumstark, C., et al., 1993. Cloning and expression of odorant receptors. Nature 361, 353–356.

Rauchfuss, A., Hiller, E., Leitner, H., Wollmer, W., Reaktion des, M., 1987. Tensor tympani—ausgelöst durch nasal applizierte Trigeminusreizstoffe. Laryngol. Rhinol. Otol. 66, 131–132.

Smith, T.D., Buttery, T.A., Bhatnagar, K.E., Burrows, A.M., Mooney, M.E., Siegel, M.I., 2001. Anatomical position of the vomeronasal organ in postnatal humans. Ann. Anat. 183, 475–479.

Swan, G.E., Carmelli, D., 2002. Impaired olfaction predicts cognitive decline in nondemented older adults. Neuroepidemiology 21, 58–67.

Touhara, K., 2001. Functional cloning and reconstitution of vertebrate odorant receptors. Life Sci. 68, 2199–2206.

Younger, R.M., Amadou, C., Bethel, G., Ehlers, A., Lindahl, K.E., Forbes, S., et al., 2001. Characterization of clustered MHC-linked olfactory receptor genes in human and mouse. Genome. Res. 11, 519–530.

Zhang, X., Firestein, S., 2002. The olfactory receptor gene superfamily of the mouse. Nat. Neurosci. 5, 124–133.

Zhao, H., Reed, R.R., 2001. X inactivation of the OCNCI channel gene reveals a role for activity-dependent competition in the olfactory system. Cell 104, 651–660.

TASTE

The ability of healthy people to distinguish a wide range of tastes (gustation) is not only essential for their enjoyment of food and drink, but it also promotes digestion and protects against harmful compounds. We now have to recognize that delineation of the four basic taste qualities (salty, sweet, bitter, and sour) so familiar to European and American audiences is an overly narrow cultural construct that leaves people ignorant of their additional sensory abilities. For instance, Japanese culture has given a specific name (umami) to the savory flavor typical of meats and seafood. People can also learn to taste calcium in solutions, an ability that they did not know about before and for which a suitable name still does not exist (Tordoff et al., 2012). Again, Japanese connoisseurs come to the rescue and offer the descriptive term *kokumi* (heartiness). Our senses also detect metallic, fatty and starchy tastes, temperature, spiciness, and surface structure (hardness, size, roughness, and fluidity). Odors contribute as much to the gustatory experience as the flavors detected through receptors in taste buds (Purves et al., 2001).

ABBREVIATIONS

cAMP	cyclic adenosine monophosphate
DL-AP4	DL(+)-2-amino-phosphono-butyric acid
GMP	guanosine 5′-monophosphate
IMP	inosine-5′-monophosphate
IP3	inositol triphosphate
MSG	monosodium glutamate
PET	positron emission tomography
PROP	6-*n*-propyl-2-thiouracil

TASTE BUDS

Perception of most taste qualities is mediated by about 4000 taste buds. These onion-shaped structures are about 50 μm wide and 80 μm deep. Nearly half of them are embedded adjacent to either side of the trenches around the nine circumvallate papillae (Purves et al., 2001). Fungiform papillae, which are located in large numbers at the tip and the foremost two-thirds of the tongue, carry about three taste buds each (Figure 1.1). The foliate papillae, which resemble parallel ridges on either side of the root of the tongue, contain a total of about 600 taste buds (Figure 1.2). A smaller number of them reside on the epiglottis. Taste buds were also found on the soft palate by some investigators (Kikuchi et al., 1988; Imfeld and Schroeder, 1992), but not by others (Cleaton-Jones, 1975).

All taste buds carry receptors at least for all the basic taste qualities, though the quantitative mixture varies among the different locations of the oral cavity. Taste buds that have a high propensity for detecting bitter tastes are especially well represented in the foliate papillae, whereas those detecting sweet taste are more common at the front of the tongue. The sides of the back half of the tongue are especially salt-sensitive, and the sides of the middle portion of the tongue detect sour flavors especially well.

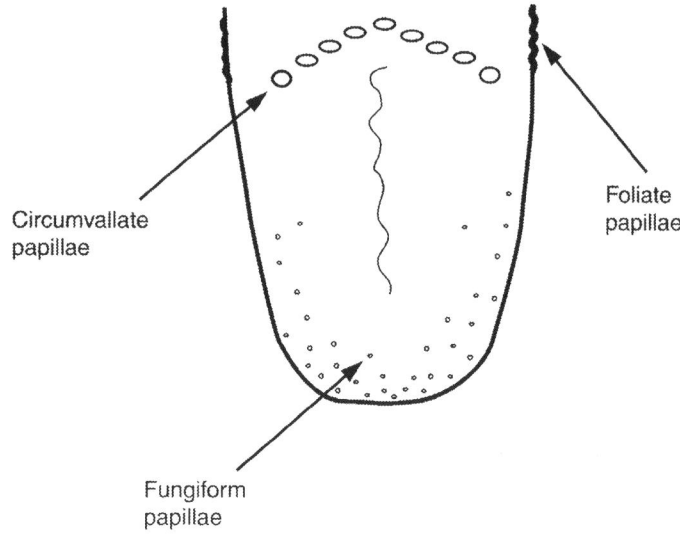

FIGURE 1.1

The papillae on the tongue contain 75% of all taste buds.

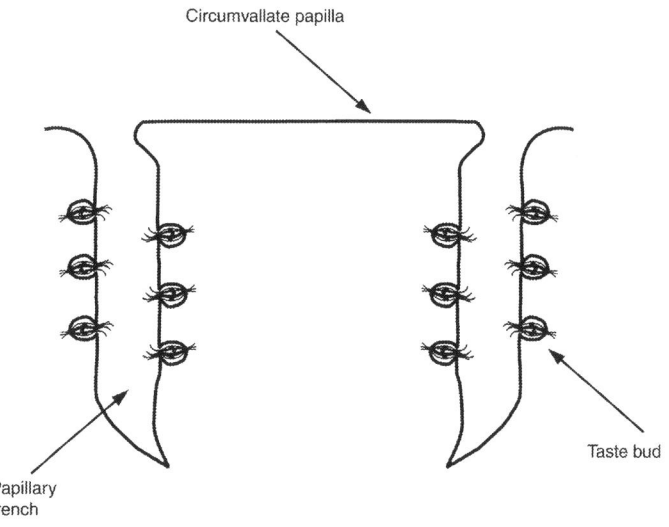

FIGURE 1.2

Taste buds adjacent to the *trenches* of a circumvallate papilla.

Each taste bud is populated by 30–100 cells (Roper, 2013). These cells fall into three distinct groups. The receptor (type II) cells have microvilli that extend through a taste pore into the trench around the papilla and detect mainly sweet, umami, and bitter tastes. Type I cells are like glia cells and wrap around the receptor cells. Presynaptic (type III) cells have synapses; release serotonin, norepinephrine, and gamma-aminobutyric acid (GABA); detect sour taste directly; and amplify the other tastes (Roper, 2013). Whether each individual cell detects only one taste quality is still debated (Gilbertson et al., 2001; Wilson, 2001; Roper, 2013). The cells in the taste buds respond when sapid molecules (those are the chemicals that carry flavors in foods) act on ion channels and specific receptors on their apical surface. The receptor cells are associated with G-proteins, which initiate second messenger cascades. Eventually, this leads to the depolarization of the cell and release of a transmitter (serotonin) into the synaptic cleft at the base of the bipolar cell (Roper, 2013). Serotonin binds to a serotonin receptor on the adjacent primary sensory neuron and triggers its depolarization.

The receptor cells have a limited life span (about a month) and must be replaced by replicating basal cells at the base of the taste buds. Carbonic anhydrase VI (gustin), a zinc-containing glycoprotein that is specifically expressed in salivary glands, is critical for maintenance of taste bud growth (possibly by acting on bud stem cells) and function (Henkin et al., 1999). Adequate zinc intake appears to promote taste acuity.

The taste buds have additional sensors, which do not convey a taste sensation but participate directly in the regulation of nutrient metabolism. The circumvallate papillae contain CD36 (Degrace-Passilly and Besnard, 2012) and the G-protein-coupled receptor 120 (GPR120) (Ichimura et al., 2014), which respond to unsaturated fatty acids with medium and long chain lengths. Carriers of variants with reduced function tend to eat more fat and have increased obesity risk (Waguri et al., 2013). Glucagon-like peptide 1 (GLP-1) is expressed together with GPR120. This very short-lived incretin is encoded by a transcript variant of the GCG gene (which also encodes glucagon and several other signaling peptides) and may translate the CD36/GPR120 fatty acid–sensing information into regulatory actions on energy homeostasis. It promotes glucose-stimulated insulin secretion from the pancreas and suppresses the release of glucagon. The detection of nutrients in the mouth thus sets the hormonal stage for their metabolic processing and disposition before they even reach the intestines.

INNERVATION

The bipolar cells of the taste buds connect to the synapses of afferent gustatory axons. Individual primary sensory neurons carry information on more than one taste quality (Gilbertson et al., 2001). The corda tympani of cranial nerve VII (facial nerve) carries the gustatory axons from the fungiform and anterior foliate papillae, and the lingual branch of cranial nerve IX (glossopharyngeal nerve) carries the axons from the posterior foliate papillae and circumvallate papillae. The superficial petrosal branch of cranial nerve VII is responsible for innervation of the taste buds on the soft palate, and the superior branch of cranial nerve X (vagal nerve) links to the taste buds on the epiglottis. Some taste categories are predominantly linked to particular nerves. Sweet and salty tastes are largely associated with the corda tympani of cranial nerve VII, while sour and bitter tastes are predominantly (but not exclusively) associated with cranial nerve IX. The chorda tympani of cranial nerve VII also is important for the maintenance of actin filaments in taste pore cells.

The gustatory axons travel within their cranial nerves to the nucleus of the solitary tract in the medulla, where they link to neurons leading to the medial half of the ventral posterior medial nucleus

Table 1.1 Cranial Nerves Carrying Afferent Gustatory Neurons	
Cranial n.V	**Mechanoreceptors (size/texture)**
	Temperature
	Irritants and pungent compounds
Corda tympani n.VII	Fungiform and anterior foliate papillae
Superficial petrosal n.VII	Taste buds of soft palate
Lingual branch n.IX	Posterior foliate and circumvallate papillae
Superior branch n.X	Taste buds of epiglottis

in the thalamus. Another set of neurons then connects to the gustatory cortex of the insula and frontal operculum. The stimulatory effect of sapid compounds can be detected in these cortical areas by positron emission tomography (PET) scans (Gautier et al., 1999). Interneuron connections also exist within the solitary tract to sympathetic and parasympathetic neurons of the visceral organs. Thus, gustatory senses have direct input on the motility of stomach and diaphragm and on the secretion of digestive juices and hormones (Table 1.1).

TASTE-ENHANCING SECRETIONS

Most sapid compounds are water soluble and need to be washed into the trenches around the taste buds to reach the microvillous filaments of bipolar taste cells. Some of the necessary fluids come from the ingested foods or drinks themselves. Saliva from sublingual, submandibular, and parotid glands augments these ingested liquids. Healthy adults secrete about 1500 ml/day; chemical stimuli, as well as odors, tastes, and food images, increase the flow. Normal saliva usually has a pH of 5–7 and contains water (99%), electrolytes and minerals, proteins, and numerous other active ingredients (Neyraud et al., 2012).

Additional small secretory glands adjacent to the foliate and circumvallate papillae, the Von Ebner glands, secrete lipocalin 1 (Von Ebner gland protein 1), and other molecules (Spielman et al., 1993). Whether some facilitate and modify taste perception, possibly by engulfing sapid molecules, is under dispute (Creuzenet and Mangroo, 1998). The peptide YY (peptide tyrosine-tyrosine, PYY) is among those proteins secreted by taste buds (Acosta et al., 2011). It binds to its receptor Y2R in the basal layer of the tongue epithelia and von Ebner's glands, stimulates stomach motility, and inhibits appetite.

MOLECULAR BASIS OF TASTE SIGNALING

Diverse structures form the proximal end of the taste-sensing cascade. Best known are receptors that trigger G-protein-mediated signals when specific molecules bind to them, as detailed in the following text. Understanding of the molecular machinery connecting these receptor-mediated effects downstream to the nerve endings is still a work in progress. The CALHM1 ion channel mediates further downstream the detection of bitter, sweet, or savory flavor molecules (Taruno et al., 2013).

Other detection modes use structures that carry the sapient molecules themselves into the cell, such as the CD36 fatty acid transporter, or allow passage across the cell membrane via specialized ion channels.

SALTY TASTE

The relatively small number of compounds with a salty taste are small cations, including sodium chloride, potassium chloride, lithium chloride, and cesium chloride. Some also have a distinct bitter taste, which limits their acceptance as salt substitutes. Habit and body status strongly influence the perception of salty flavors. While small amounts of salty compounds tend to be desirable, large doses or combinations with sweet-tasting foods can cause aversion.

Salty-tasting compounds are detected by sodium channels on the apical surface of bipolar cells (Lin and Kinnamon, 1999). The more sodium flows into a taste cell the more it is likely to depolarize. Antidiuretic hormone (ADH), aldosterone, and other water- and electrolyte-regulating hormones regulate the apical sodium channels. Amiloride and other diuretics similarly act on the channels and diminish salt-tasting sensitivity. Protons permeate these channels more slowly than sodium. Therefore, acids act as competitive inhibitors of salt tasting.

SWEET TASTE

Sweet taste is generally perceived as pleasant and satisfying. The large spectrum of sweet-tasting molecules is chemically diverse and includes carbohydrates, proteins, amino acids, peptides and complex proteins, heterocyclic compounds, terpenoids, flavonoids, and steroidal compounds from plants. Their sweet taste has hundreds or even thousands of times more intensity than sugar in some instances. Only a few of the intensely sweet compounds are approved in the United States for consumption (e.g., aspartame, acesulfame potassium, saccharin, sucralose, ammoniated glycyrrhizin, and thaumatin) (Figure 1.3), and a few others are approved in other countries (Kinghorn et al., 1998). Sweet-tasting carbohydrates include the oligosaccharides sucrose and lactose and the monosaccharides glucose, fructose, and galactose. The potency of carbohydrates differs considerably; fructose and sucrose taste sweeter than glucose (Stone and Oliver, 1969; Schiffman et al., 2000). The sugar alcohols xylitol, sorbitol, lactitol, and maltitol are slightly less sweet than glucose.

Some amino acids, peptides, and proteins also have a sweet taste. Examples include D-amino acids and the artificial sweetener aspartame (N-L-α-aspartyl-L-phenylalanine 1-methyl ester). Lysozyme (EC3.2.1.17) from chicken and goose eggs tastes sweet, possibly due to an enzymic property that is independent of its ability to cleave peptidoglycan heteropolymers (Masuda et al., 2001). Of greater practical importance may be the plant-derived complex proteins monellin, thaumatin, mabinlin,

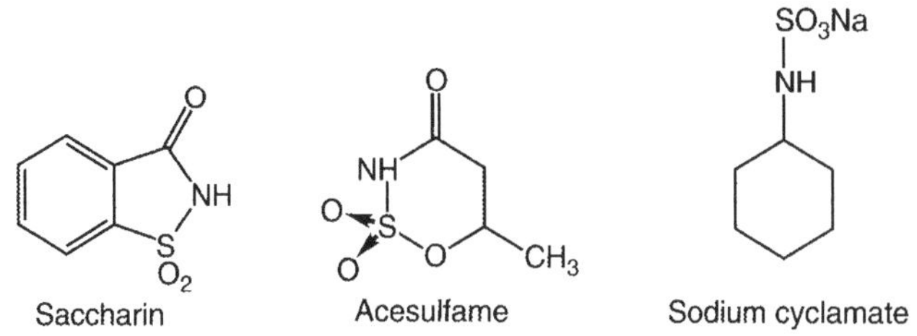

Saccharin   Acesulfame   Sodium cyclamate

FIGURE 1.3

Artificial sweeteners.

pentadin, brazzein, and curculin, which are several thousand times sweeter than glucose (Faus, 2000). Water and even sour substances still taste sweet after exposure to curculin has stopped (Kurihara, 1992). Miraculin is another taste-modifying glycoprotein that makes sour-tasting foods taste sweet (Theerasilp and Kurihara, 1998).

Saccharin and other heterocyclic compounds also have an intense sweet taste, though with some lingering bitterness (Home et al., 2002). This is an important example of overlapping taste qualities in a single compound. The nitrogen-containing sweetener cyclamate has similar characteristics. Concerns about the potential carcinogenicity of saccharine and cyclamate need to be mentioned.

The roots and rhizome of licorice contain 6–14% of the sweet-tasting terpenoid glycyrrhizin. Several related compounds also contribute to the sweet taste of the plant (Liu et al., 2000). Numerous plant glycosides, including monoterpenoids, diterpenoids, triterpenoids, saponins, and proanthocyanidins, have an intense sweet taste (Kinghorn et al., 1998). Another example of a plant-derived molecule with high potency is telosmoside A15, a polyoxypregnane glycoside (Huan et al., 2001). Sweet taste is inhibited, on the other hand, by the triterpene glycosides gymnemic acid and ziziphin, and the peptide gurmarin (Kurihara, 1992).

Sweet taste is perceived through the interaction of ingested molecules with one or more specific receptors. Sugars bind to G-protein-coupled receptors that activate adenylate cyclase. The binding of synthetic sweeteners to G-protein-coupled receptors, in contrast, may trigger the release of inositol triphosphate (IP3) by phospholipase C. Both IP3 and cAMP increase intracellular calcium ion concentration by promoting the release of calcium ions from intracellular stores and the flow of calcium ions across the basolateral membrane into the cell. The rise in calcium ion concentration can then trigger the secretion of the neurotransmitter serotonin into the synaptic cleft.

Heterodimers of TAS1R3, a member of the T1R family of taste receptors (Max et al., 2001), with either TAS1R1 or TAS1R2 respond to sweet-tasting substances, including sucrose and saccharin (Nelson et al., 2001). The TAS1R1/TAS1R3 dimer is present only in the front of the tongue; the TAS1R2/TAS1R3 dimer is present on the back of the tongue and on the palate. Intracellular signaling elicited by stimulation of these receptor dimers is mediated by the G-protein alpha-gustducin, a protein closely related to transducin of the rods in the retina of the eye.

BITTER TASTE

A broad spectrum of structurally diverse compounds has a bitter taste, including caffeine, theobromine (the principal alkaloid in cocoa and chocolate), quinine (in tonic water), nicotine, L-amino acids (with the exception of L-glutamate), urea, magnesium and cesium salts, phenols and polyphenols, flavonoids (such as naringin in grapefruit and methylnaringenin in hops), catechins (in tea), and local anesthetics (Figure 1.4). The bitter-tasting compound 6-n-propyl-2-thiouracil (PROP) is commonly used in taste tests (Drewnowski et al., 2001). Use of phenylthiocarbamide (PTC) for this purpose in humans has been largely discontinued because of its suspected carcinogenicity.

A large family of taste receptors, comprising more than 40 distinct molecular species, sense bitter tastes (Adler et al., 2000; Matsunami et al., 2000). The receptors activate an intracellular signaling cascade via alpha-gustducin, analogous to receptors responding to sweet-tasting compounds. One of these receptors, TASRB7 (Matsunami et al., 2000), responds to the test substance PROP (Bartoshuk et al., 1994). Another one, TAS2R9, senses cycloheximide. While individual taste cells express several different receptors (Adler et al., 2000; Matsunami et al., 2000), humans can still discriminate between

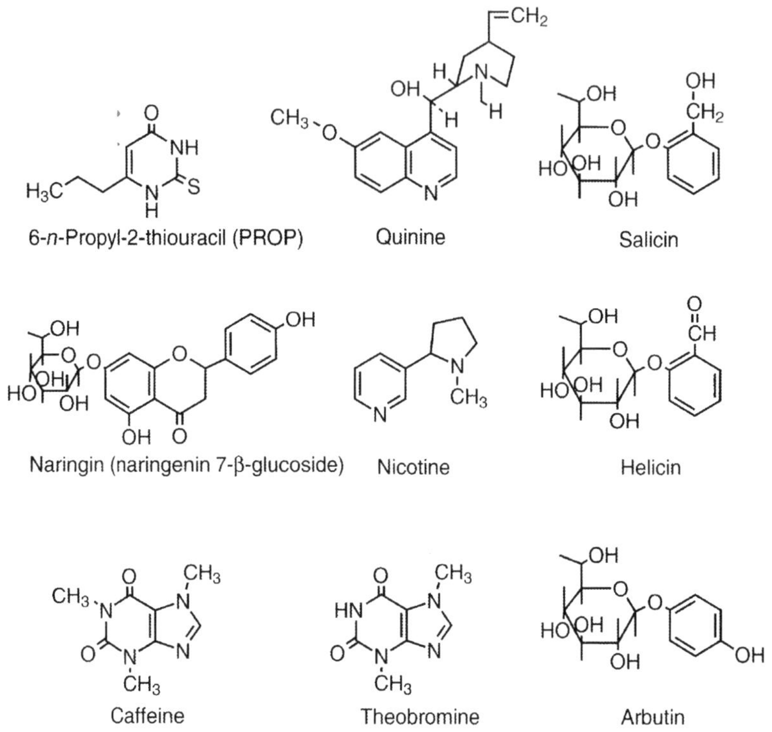

FIGURE 1.4

Bitter-tasting compounds.

several bitter-tasting compounds. It has been suggested that individual bitter-tasting substances preferentially activate distinct subpopulations of taste cells and thereby generate a compound-specific pattern of neuronal signals (Caicedo and Roper, 2001).

Additional mechanisms for bitter taste sensing have been described but are less well characterized. Thus, the bitter-tasting compound denatonium appears to inhibit outward potassium flow and possibly induces signaling through the release of intracellular calcium stores (Yah et al., 2001).

SOUR TASTE

The sour taste of foods and beverages is largely due to organic acids, including acetic, citric, malic, and fumaric acids in fruits and vegetables and tartaric acid in wine. Dilute solutions of many inorganic acids also taste sour. One example is the phosphoric acid in cola beverages. This taste is elicited when protons or acids move into the taste cells. Several different pathways allow the entry of molecules associated with sour taste, including amiloride-sensitive sodium channels, proton-sensitive cation channels, and carriers of weak organic acids. Carbon dioxide permeates the cell membrane directly and is then dissociated through the action of zinc-dependent carbonic anhydrase (EC4.2.1.1). An additional mode of taste cell stimulation by sour-tasting compounds is the direct activation of hyperpolarization-activated and cyclic-nucleotide-gated (HCN) channels by extracellular protons (Stevens et al., 2001).

The perceived intensity of a compound or mixture in solution correlates only loosely with its pH. The likely reason is that the drop in intracellular pH induced by exposure of taste cells to the solution is a direct cause of cell depolarization and signaling (Lyall et al., 2001). Since weak acids enter in their neutral form and then equilibrate with the intracellular medium, the perceived acidity may not correlate with the pH of a test solution applied to taste cells. Carbon dioxide buffered with sodium bicarbonate to a neutral pH thus retains its sour taste. Similarly, buffered acetic acid elicits similar taste intensity as the free acid despite the considerable difference in pH. The perceived acidity of equimolar solutions also differs considerably between compounds. Acetic acid, for example, is a more potent stimulus than citric acid or hydrochloric acid at an identical pH of the test solution.

MEATY TASTE (UMAMI)

The taste quality known as *umami* is distinct from other taste qualities (particularly salty), and might best be described as the savory taste in a wide variety of foods, including vegetables and meats. Umami is elicited by foods or solutions containing monosodium glutamate (MSG), guanosine 5′-monophosphate (GMP), inosine-5′-monophosphate (IMP), ibotenate, DL(+)-2-amino-phosphono-butyric acid (DL-AP4), or trans-(±)-1-aminocyclopentone-trans-1,3-dicarboxylic acid (Kurihara and Kashiwayanagi, 2000).

MSG and GMP strongly potentiate each other; the taste threshold is also lowered by the presence of salt (Kurihara and Kashiwayanagi, 2000). Specific L-glutamate-containing oligopeptides also enhance the umami taste when combined with IMP, including Glu–Glu, Glu–Val, Ala–Asp–Glu, Ala–Glu–Asp, Asp–Glu–Glu, Ser–Pro–Glu, Glu–Glu–Asn (Maehashi et al., 1999).

On a molecular level, the umami taste is mediated by a truncated form of the metabotropic glutamate receptor 4 (mGluR4), whose G-protein generates cAMP as a second messenger (San Gabriel et al., 2009). L-glutamate also acts on ionotropic receptors that activate ion channels (Lin and Kinnamon, 1999).

HEARTY TASTE (*KOKUMI*)

It will be a surprise for most people brought up on European or American standard foods that there is still another distinct savory flavor, which is described by a Japanese term, *kokumi* (which means "hearty" or "mouthful"). The flavor is usually elicited by gamma-glutamyl peptides with one or two aliphatic amino acids. These kokumi substances do not have a strong flavor by themselves but are most effective when enhancing other savory substances. The paradigmatic kokumi peptide is gamma-glutamyl-valyl-glycine, which is found in fish sauces, soy sauces, strong fermented cheeses, and even in processed avocado. Another peptide with distinct kokumi properties is glutathione, gamma-glutamyl-cysteinyl-glycine (Ohsu et al., 2010).

Surprisingly, the sensor for typical kokumi substances is the calcium-sensing receptor CaSR (Maruyama et al., 2012), possibly in conjunction with TAS1R2 receptors (Ohsu et al., 2010; Brennan et al., 2014).

FAT TASTE

Humans can taste fat in food through the action of the fatty acid translocase CD36 in circumvallate taste buds on the tongue (Degrace-Passilly and Besnard, 2012). Since CD36 can only sense free fatty acids, the fat (triglycerides) in foods has to be cleaved first. There is a special lingual lipase (EC3.1.1.3) in saliva that does not need bile acids and phospholipids unlike the pancreas lipase working on triglycerides in the small intestine. Individuals can differ considerably in their ability to sense fat in foods and in beverages such as milk (Pepino et al., 2012).

CALCIUM TASTE

Humans can specifically recognize calcium in solutions, but they require training before they can put a name to the taste. This is a reminder that the association of a sensory perception with a categorical taste quality is a learned response and the concept of a few basic tastes is a cultural construct (Tordoff et al., 2012). With some training, most people can distinguish between foods that have a calcium-rich taste and others that don't. Once this largely conceptual barrier was crossed, it became clear that calcium sensing is mediated by the receptor TAS1R3, which contains a calcium-binding motif resembling that in the calcium receptor (Tordoff et al., 2012). This is a surprising finding since TAS1R3 mediates both sweet and umami tastes. Each of these three taste qualities (calcium-rich, sweet, and savory) are perceived distinctly and without significant overlap. It is not clear how one and the same receptor can have such versatility. The taste pertains only to calcium and does not respond to magnesium. The calcium receptor itself and additional receptors may also contribute to the sensory detection of calcium.

OTHER TASTE QUALITIES

Recent findings provide strong support that additional taste qualities are sensed by some mammals and that this may extend to humans. Further research will have to determine the extent to which such mechanisms contribute to the human gustatory experience.

There have been intermittent reports on the ability of humans to discern the fat content of foods (Perkins et al., 1990; Mattes, 1996; Yackinous and Guinard, 2001). It remains an open question, however, whether fat content is deduced from the texture of foods or whether humans have a specific sense for fat perception as rodents do. The modification of ion flux through potassium channels by polyunsaturated fatty acids may by itself confer fat-sensing ability to rats (Gilbertson et al., 1997). An alternative mechanism may be triggered by fat uptake into taste cells via the fatty acid transporter 1 (Fukuwatari et al., 1997).

L-Arginine stimulates specific reconstituted channels, which have been localized to the pores of guinea-pig taste buds (Grosvenor et al., 1998).

Some foods or food additives (e.g., saccharin) have a distinct metallic aftertaste. Metallic tastes also may be induced by a wide spectrum of compounds acting directly on the central nervous system. This includes medications (e.g., anesthetics, antitumor agents, chemoprotectants, H1/H2-blockers, and antibiotics), as well as neurotoxin contaminants in some fish (Pearn, 2001). Smoking causes a strongly aversive metallic taste when the oral cavity has been exposed to silver acetate. The use of silver acetate in gums or lozenges as a smoking deterrent appears to be of limited use in cessation programs, however (Hymowitz and Eckholdt, 1996).

VARIATION IN TASTE SENSITIVITY

About a quarter of American women are able to detect particularly low concentrations of PROP, about half have intermediate sensitivity, and a quarter cannot taste the compound at all (Bartoshuk et al., 1994).

The most sensitive group, the supertasters, also appears to be more sensitive to most tastes than others. High PROP sensitivity may even extend to enhanced perception of fat in foods (Tepper and Nurse, 1998). Supertasters may have a higher than average number of fungiform papillae with an especially high density of taste buds (Tepper and Nurse, 1998). Additional variation, and in particular the inability to taste PROP at all, may be due to a polymorphism of the TRB7 gene (Matsunami et al., 2000). Numerous common genetic variants are now known to affect all taste qualities (Kohlmeier, 2012). Genetic differences in taste sensitivity matter since they can affect health factors such as avoidance of broccoli and other cruciferous vegetables (Duffy et al., 2010), the ability to tell the difference between skim and full-fat milk (Pepino et al., 2012), or the risk of nicotine addiction (Enoch et al., 2001).

Some tastes may be perceived only by a minority of people to begin with. About 8% of Americans report a soapy flavor in fresh coriander leaves (cilantro), while the vast majority of their friends and family are oblivious to this taste quality (Erikksson et al., 2012).

Taste sensitivity also varies with age. Young infants acquire the ability to taste salt only after about 4 months. At the other extreme of the age spectrum, taste acuity declines in many elderly people (Duffy, 1999).

Physical status and medications also significantly affect taste perception. Diuretics are known to blunt salt perception (Lin et al., 1999). Physical exercise appears to increase the preference for sucrose and citric acid, but not sodium chloride, caffeine, and MSG (Horio and Kawamura, 1998). And then there are many foods or ingested compounds that modify the gustatory experience. Eating artichokes makes water taste sweet in most but not all people. Gourmets have long used this effect to improve the palatability of tannin-rich red wines. The bitter-tasting antiseptic mouthwash ingredient chlorhexidine (a bis-cationic biguanide) reduces perceived intensity of sodium chloride and quinine solutions and induces the perception of bitterness in salts that otherwise do not have a bitter taste (Frank et al., 2001).

REFERENCES

Acosta, A., Hurtado, M.D., Gorbatyuk, O., La Sala, M., Duncan, D., Aslanidi, G., et al., 2011. Salivary PYY: a putative bypass to satiety. PLoS One 6 (10), e26137.

Adler, E., Hoon, M.A., Mueller, K.L., Chandrashekar, J., Ryba, N.J., Zuker, C.S., 2000. A novel family of mammalian taste receptors. Cell 100, 693–702.

Bartoshuk, L.M., Duffy, V.B., Miller, I.J., 1994. PTC/PROP tasting: anatomy, psychophysics, and sex effects. Physiol. Behav. 56, 1165–1171.

Brennan, S.C., Davies, T.S., Schepelmann, M., Riccardi, D., 2014. Emerging roles of the extracellular calcium-sensing receptor in nutrient sensing: control of taste modulation and intestinal hormone secretion. Br. J. Nutr., 1–7.

Caicedo, A., Roper, S.D., 2001. Taste receptor cells that discriminate between bitter stimuli. Science 291, 1557–1560.

Cleaton-Jones, P., 1975. Normal histology of the human soft palate. J. Biol. Bucc. 3, 265–276.

Creuzenet, C., Mangroo, D., 1998. Physico-chemical characterization of human von Ebner gland protein expressed in *Escherichia coli:* implications for its physiological role. Protein Expr. Purif. 14, 254–260.

Degrace-Passilly, P., Besnard, P., 2012. CD36 and taste of fat. Curr. Opin. Clin. Nutr. Metab. Care 15, 107–111.

Drewnowski, A., Henderson, S.A., Barratt-Fortnell, A., 2001. Genetic taste markers and food preferences. Drug Metab. Disp. 29, 535–538.

Duffy, V.B., Hayes, J.E., Davidson, A.C., Kidd, J.R., Kidd, K.K., Bartoshuk, L.M., 2010. Vegetable intake in college-aged adults is explained by oral sensory phenotypes and TAS2R38 Genotype. Chemosens. Percept. 3, 137–148.

Duffy, V.B., 1999. Smell, taste, and somatosensation in the elderly. In: Chernoff, R. (Ed.), Geriatric Nutrition Aspen Publishers, Gaithersburg, MD, pp. 170–211.

Enoch, M.A., Harris, C.R., Goldman, D., 2001. Does a reduced sensitivity to bitter taste increase the risk of becoming nicotine addicted? Addict. Behav. 26, 399–404.

Erikkson, N., Wu, S., Chuong, B.D., Kiefer, A.K., Tung, J.Y., Mountain, J.L., et al., 2012. A genetic variant near olfactory receptor genes influences cilantro preference. <http://arxiv.org/pdf/1209.2096v1.pdf>.

Faus, I., 2000. Recent developments in the characterization and biotechnological production of sweet-tasting proteins. Appl. Microbiol. Biotech. 53, 145–151.

Frank, M.E., Gent, J.F., Hettinger, T.P., 2001. Effects of chlorhexidine on human taste perception. Physiol. Behav. 74, 85–99.

Fukuwatari, T., Kawada, T., Tsuruta, M., Hiraoka, T., Iwanaga, T., Sugimoto, E., et al., 1997. Expression of the putative membrane fatty acid transporter (FAT) in taste buds of the circumvallate papillae in rats. FEBS Lett. 414, 461–464.

Gautier, J.E., Chen, K., Uecker, A., Bandy, D., Frost, J., Salbe, A.D., et al., 1999. Regions of the human brain affected during a liquid-meal taste perception in the fasting state: a positron emission tomography study. Am. J. Clin. Nutr. 70, 806–810.

Gilbertson, T.A., Boughter Jr., J.D., Zhang, H., Smith, D.V., 2001. Distribution of gustatory sensitivities in rat taste cells: whole-cell responses to apical chemical stimulation. J. Neurosci. 21, 4931–4941.

Gilbertson, T.A., Fontenot, D.T., Liu, L., Zhang, H., Monroe, W.T., 1997. Fatty acid modulation of K^+ channels in taste receptor cells: gustatory cues for dietary fat. Am. J. Physiol. 272, C1203–C1210.

Grosvenor, W., Feigin, A.M., Spielman, A.I., Finger, T.E., Wood, M.R., Hansen, A., et al., 1998. The arginine taste receptor. Physiology, biochemistry, and immunohistochemistry. Ann. N.Y. Acad. Sci. 855, 134–142.

Henkin, R.I., Martin, B.M., Agarwal, R.E., 1999. Decreased parotid saliva gustin/carbonic anhydrase VI secretion: an enzyme disorder manifested by gustatory and olfactory dysfunction. Am. J. Med. Sci. 318, 380–391.

Horio, T., Kawamura, Y., 1998. Influence of physical exercise on human preferences for various taste solutions. Chem. Senses 23, 417–421.

Horne, J., Lawless, H.T., Speirs, W., Sposato, D., 2002. Bitter taste of saccharin and acesulfame-K. Chem. Senses 27, 31–38.

Huan, V.D., Ohtani, K., Kasai, R., Yamasaki, K., Tuu, N.V., 2001. Sweet pregnane glycosides from *Telosma procumbens*. Chem. Pharmaceut. Bull. 49, 453–460.

Hymowitz, N., Eckholdt, H., 1996. Effects of a 2.5-mg silver acetate lozenge on initial and long-term smoking cessation. Prev. Med. 25, 537–546.

Ichimura, A., Hara, T., Hirasawa, A., 2014. Regulation of energy homeostasis via GPR120. Front. Endocrinol. (Lausanne) 5, 111.

Imfeld, T.N., Schroeder, H.E., 1992. Palatal taste buds in man: topographical arrangement in islands of keratinized epithelium. Anat. Embryol. 185, 259–269.

Kikuchi, T., Kusakari, J., Kawase, T., Takasaka, T., 1988. Electrogustometry of the soft palate as a topographic diagnostic method for facial paralysis. Acta OtoLaryngol. 458, 134–138.

Kinghorn, A.D., Kaneda, N., Baek, N.I., Kennelly, E.J., Soejarto, D.D., 1998. Noncariogenic intense natural sweeteners. Med. Res. Rev. 18, 347–360.

Kohlmeier, M., 2012. The genetics of taste and smell. In: Kohlmeier, M. (Ed.), Nutrigenetics: Applying the Science of Personal Nutrition Academic Press, San Diego, CA, pp. 82–92.

Kurihara, Y., 1992. Characteristics of antisweet substances, sweet proteins, and sweetness-inducing proteins. Crit. Rev. Food Sci. Nutr. 32, 231–252.

Kurihara, K., Kashiwayanagi, M., 2000. Physiological studies on umami taste. J. Nutr. 130, 931S–934S.

Lin, W., Finger, T.E., Rossier, B.C., Kinnamon, S.C., 1999. Epithelial Na^+ channel subunits in rat taste cells: localization and regulation by aldosterone. J. Comp. Neurol. 405, 406–420.

Lin, W., Kinnamon, S.C., 1999. Physiological evidence for ionotropic and metabotropic glutamate receptors in rat taste cells. J. Neurophysiol. 82, 2061–2069.

Liu, H.M., Sugimoto, N., Akiyama, T., Maitani, T., 2000. Constituents and their sweetness of food additive enzymatically modified licorice extract. J. Agric. Food Chem. 48, 6044–6047.

Lyall, V., Alam, R.I., Phan, D.Q., Ereso, G.L., Phan, T.H., Malik, S.A., et al., 2001. Decrease in rat taste receptor cell intracellular pH is the proximate stimulus in sour taste transduction. Am. J. Physiol. Cell Physiol. 281, C1005–C1013.

Maehashi, K., Matsuzaki, M., Yamamoto, Y., Udaka, S., 1999. Isolation of peptides from an enzymatic hydrolysate of food proteins and characterization of their taste properties. Biosci. Biotech. Biochem. 63, 555–559.

Maruyama, Y., Yasuda, R., Kuroda, M., Eto, Y., 2012. Kokumi substances, enhancers of basic tastes, induce responses in calcium-sensing receptor expressing taste cells. PLoS One 7 (4), e34489.

Masuda, T., Ueno, Y., Kitabatake, N., 2001. Sweetness and enzymatic activity of lysozyme. J. Agric. Food Chem. 49, 4937–4941.

Matsunami, H., Montmayeur, J.E., Buck, L.B., 2000. A family of candidate taste receptors in human and mouse. Nature 404, 601–604.

Mattes, R.D., 1996. Oral fat exposure alters postprandial lipid metabolism in humans. Am. J. Clin. Nutr. 63, 911–917.

Max, H., Shanker, Y.G., Huang, L., Rong, M., Liu, Z., Campagne, F., et al., 2001. A candidate taste receptor gene near a sweet taste locus. Nat. Neurosci. 4, 492–498.

Nelson, G., Hoon, M.A., Chandrashekar, J., Zhang, Y., Ryba, N.J., Zuker, C.S., 2001. Mammalian sweet taste receptors. Cell 106, 381–390.

Neyraud, E., Palicki, O., Schwartz, C., Nicklaus, S., Feron, G., 2012. Variability of human saliva composition: possible relationships with fat perception and liking. Arch. Oral Biol. 57, 556–566.

Ohsu, T., Amino, Y., Nagasaki, H., Yamanaka, T., Takeshita, S., Hatanaka, T., et al., 2010. Involvement of the calcium-sensing receptor in human taste perception. J. Biol. Chem. 285, 1016–1022.

Pearn, J., 2001. Neurology of ciguatera. J. Neurol. Neurosurg. Psychiat. 70, 4–8.

Pepino, M.Y., Love-Gregory, L., Klein, S., Abumrad, N.A., 2012. The fatty acid translocase gene CD36 and lingual lipase influence oral sensitivity to fat in obese subjects. J. Lipid. Res. 53, 561–566.

Perkins, K.A., Epstein, L.H., Stiller, R.E., Fernstrom, M.H., Sexton, J.E., Jacob, R.G., 1990. Perception and hedonics of sweet and fat taste in smokers and nonsmokers following nicotine intake. Pharmacol. Biochem. Behav. 35, 671–676.

Purves, D., Augustine, G.J., 2001. Fitzpatrick 19. In: Katz, L.C., LaMantian, A.S., McNamara, J.O., Williams, S.M. (Eds.), Neuroscience Sinauer Associates, Sunderland, MA, pp. 317–344.

Roper, S.D., 2013. Taste buds as peripheral chemosensory processors. Semin. Cell Dev. Biol. 24, 71–79.

San Gabriel, A., Maekawa, T., Uneyama, H., Torii, K., 2009. Metabotropic glutamate receptor type 1 in taste tissue. Am. J. Clin. Nutr. 90, 743S–746S.

Schiffman, S.S., Sattely-Miller, E.A., Graham, B.G., Booth, B.J., Gibes, K.M., 2000. Synergism among ternary mixtures of fourteen sweeteners. Chem. Senses 25, 131–140.

Spielman, A.I., D'Abundo, S., Field, R.B., Schmale, H., 1993. Protein analysis of human von Ebner saliva and a method for its collection from the foliate papillae. J. Dent. Res. 72, 1331–1335.

Stevens, D.R., Seifert, R., Bufe, B., Muller, E., Kremmer, E., Gauss, R., et al., 2001. Hyperpolarization-activated channels HCNI and HCN4 mediate responses to sour stimuli. Nature 413, 631–635.

Stone, H., Oliver, S.M., 1969. Measurement of the relative sweetness of selected sweeteners and sweetener mixtures. J. Food Sci. 34, 215–222.

Taruno, A., Vingtdeux, V., Ohmoto, M., Ma, Z., Dvoryanchikov, G., Li, A., et al., 2013. CALHM1 ion channel mediates purinergic neurotransmission of sweet, bitter and umami tastes. Nature 495 (7440), 223–226.

Tepper, B.J., Nurse, R.J., 1998. PROP taster status is related to fat perception and preference. Ann. N.Y. Acad. Sci. 855, 802–804.

Theerasilp, S., Kurihara, Y., 1998. Complete purification and characterization of the taste-modifying protein, miraculin, from miracle fruit. J. Biol. Chem. 263, 11536–11539.

Tordoff, M.G., Alarcón, L.K., Valmeki, S., Jiang, P., 2012. T1R3: a human calcium taste receptor. Sci. Rep. 2, 496.

Waguri, T., Goda, T., Kasezawa, N., Yamakawa-Kobayashi, K., 2013. The combined effects of genetic variations in the GPR120 gene and dietary fat intake on obesity risk. Biomed. Res. 34, 69–74.

Wilson, J.E., 2001. Bitter-sweet research. Scientist 15, 18–19.

Yackinous, C., Guinard, J.X., 2001. Relation between PROP taster status and fat perception, touch, and olfaction. Physiol. Behav. 72, 427–437.

Yah, W., Sunavala, G., Rosenzweig, S., Dasso, M., Brand, J.G., Spielman, A.I., 2001. Bitter taste transduced by PLC-beta(2)-dependent rise in IP(3) and alpha-gustducin-dependent fall in cyclic nucleotides. Am. J. Physiol. Cell. Physiol. 280, C742–C751.

INTESTINAL SENSING

A small part of the intestinal wall consists of a range of specialized endocrine cells and other related cells (Reimann et al., 2012; van der Wielen et al., 2014). More than 20 different cell types carry nutrient sensors that provide critical input for the orchestration of secretory, absorptive, and metabolic processes, as well as the control of systemically acting endocrine signals. Information from various types of enteroendocrine cells is also carried to the brain through the vagal nerve.

Some of the intestinal sensors recapitulate taste receptors of the mouth, in several instances using the same genes. Others are not identical but respond to the same chemical characteristics that taste receptors detect. This means that the flavors in foods do not just taste good while we eat them, but they also have significant digestive, metabolic, and behavioral effects. Such effects include release of digestive enzymes and chemicals, regulation of appetite to prevent overeating, and support of intestinal wall renewal. The use of noncaloric flavoring agents may thus have unexpected and unintended consequences.

ABBREVIATIONS

CCK	cholecystokinin
GLP-1	glucagon-like peptide 1
GPR120	G-protein-coupled receptor 120
PYY	peptide YY (tyrosine-tyrosine)

NUTRIENT-SENSING CELL TYPES

Enteral hormones: The diverse types and lineages of enteral cells produce many different signaling compounds and release them to nearby neighboring cells or into blood circulation. Major enteral hormones linked to nutrient sensors are CCK, secretin, glucose-dependent insulinotropic peptide (GIP), GLP-1, GLP-2, PYY, and neurotensin. The properties of individual nutrient-sensing cells depend on their location in the intestinal region, their position in the intestinal villi (base, tip, or crypts), and their combination of active nutrient sensors (TAS1R1/TAS1R3, TAS1R2/TAS1R2, mGluR5, GPRC6A, CaSR, and others). Ghrelin-secreting cells are mainly in the antrum of the stomach. Enteroendocrine cells located at the base of the villi and near crypts often produce a range of hormones (specifically CCK, secretin, GIP, and neurotensin), while expression of one of them predominates in most cells migrating toward the villus tips and depending on the intestinal segment (Egerod, et al., 2012; Engelstoft et al., 2013a).

Ghrelin-secreting cells: The ghrelin produced by cells in the lower part of the stomach (antrum), and to a lesser extent in the small intestine, is secreted most actively in the fasting state, and hardly after a meal. The fatty acid receptor GPR120 is colocalized with ghrelin in the duodenum and thus forms a sensor unit there (Ichimura et al., 2014). A second hormone, obestatin, is produced by alternative cleavage from the same preproghrelin/obestatin precursor encoded by the GHRL gene. The feeding state is monitored on these cells by a broad complement of receptors, including FFAR2 (short-chain fatty acids), FFAR4 (long-chain fatty acids), CasR (calcium and amino acids), GPR81 (lactate), a β1-adrenergic receptor, receptors for GIP, secretin, melanocortin 4, and CGRP, as well as several somatostatin-responsive receptors (Engelstoft et al., 2013b).

Enterocytes: The bulk of the luminal side of the small intestine consists of the enterocytes, which are responsible for absorption of molecules from the intestinal lumen. Their exquisite nutrient-sensing properties modify expression and translocation events in the responding cell, but they do not communicate specifically to other cells or tissues what nutrients they detected in the intestinal lumen.

I cells: This type of enteroendocrine cell is found particularly in the duodenum. It secretes cholecystokinin (CCK) when its nutrient sensors detect amino acids in the intestinal lumen. CCK in circulation triggers gallbladder contraction and signals satiety in the brain.

S cells: These cells in the crypts of the duodenum release secretin, which inhibits acid production in the stomach and promotes bicarbonate release from the pancreas. Thus, when a food bolus leaves the stomach, this hormone orchestrates secretions toward a higher pH that is optimal for the activity of the digestive enzymes in the duodenum. Secretin also binds to a specific receptor in ghrelin-secreting cells in the stomach (Engelstoft et al., 2013b).

K cells: These cells respond mostly to sweet-tasting substances. They secrete gastric inhibitory polypeptides, which promote insulin release from pancreatic beta cells.

L cells: These cells, located mostly in the ileum, respond to glucose (Reimann et al., 2012) and other luminal nutrients by releasing glucagon-like proteins 1 (GLP-1) and 2 (GLP-2), which are both derived by selective cleavage from the same proglucagon precursor encoded by the GCG gene. GLP-1 sensitizes beta cells in pancreatic islets to stimulation by glucose. GLP-2 stimulates proliferation of cells in the intestines, bones, brain, and other tissues.

Tuft cells: These are open-type cells that have microfilaments with nutrient-sensing properties directed toward the intestinal lumen (Tolhurst et al., 2012). They are also sometimes called *brush cells* or *caveolated cells*. They express the taste receptor–associated G-protein alpha-gustducin and the taste cell–associated cation-channel TRPM5, which suggests chemosensor properties. It remains to be seen what the other components of this putative sensing apparatus are. The cells also express the signaling peptides beta-endorphin, met-enkephalin, and uroguanylin (regulating water and electrolyte balance in the small intestine and the kidneys), presumably in response to the detected composition of the intestinal contents.

Paneth cells: These crypt cells have important immune functions for local and systemic protection. They are finely tuned to sense the presence of pathogenic viruses (Hirao et al., 2014) and secrete immunoglobulin A (IgA) into their venous blood supply.

SUGARS

For coordination of carbohydrate absorption, the small intestine uses the same taste receptor system that detects sweet taste in the mouth, consisting of heterodimers with TAS1R2 and TAS1R3, signaling through G-protein-mediated cAMP release to alpha-gustducin and transducin. In particular, the receptors in ileal L cells sense unabsorbed sugars. Activation of this sweet taste sensor by glucose, fructose, sucrose, and other sweet-tasting compounds promotes the rapid insertion of the high-capacity sugar transporter GLUT2 into the apical enterocyte membrane (Mace et al., 2009; Reimann et al., 2012). The sensors also respond to some nonnutritive sweeteners, such as sucralose.

AMINO ACIDS

Ghrelin-secreting cells in the stomach and small intestine and several other specialized cells in the small intestine use the calcium-sensing receptor CaSR to sense L-phenylalanine, L-tyrosine, L-tryptophan, and L-histidine. Several other amino acids and small peptides are detected with lower sensitivity (Brennan et al., 2014).

Most important for detecting amino acids in the lumen of the small intestine is CaSR. Cells with CaSR respond by releasing CCK when they sense the presence of their kind of amino acid in the small intestinal lumen. K cells and L cells also carry CaSR and respond to stimulation by L-phenylalanine, L-tryptophan, L-arginine, L-asparagine, and L-glutamine with the release of gluco-indulinotropic peptide, glucagon-like peptide 1 and PYY (Mace et al., 2012; Reimann et al., 2012).

The TAS1R1–TAS1R3 dimer that detects savory (umami) flavors in the mouth is also present in the small intestine. Located on the luminal side of enterocytes, Paneth cells, and specialized enteroendocrine cells, the taste receptor dimers respond specifically strongly to L-alanine and L-serine and less well to a few other amino acids, but not to L-tryptophan or D-amino acids (Brennan et al., 2014; Daly et al., 2013). Just as in the mouth, the effect of the amino acids is potentiated by inosine-5'-monophosphate (IMP). The presence of L-glutamate in the small intestinal lumen also rapidly recruits GLUT2 and amino acid transporters such as EAAC1 to enterocyte membranes that enhance the absorption of both glucose and amino acids (Mace et al., 2009). Luminal amino acids trigger in enteroendocrine cells the release of the signaling peptide CCK, with strong effects on the gallbladder (causing its contraction and the release of bile), neighboring enteroendocrine cells, and, not least, the brain.

The metabotropic glutamate receptor 5 (mGluR5, GRM5) is a G-protein-linked receptor that responds specifically to L-glutamate (Brennan et al., 2014). GPR93 senses protein and proteolytic degradation products (Choi et al., 2007). The GPRC6A receptor is yet another G-protein-coupled receptor in the small intestinal wall. It senses L-lysine, L-arginine, and, to a lesser degree, glycine, L-alanine, L-serine, and L-cysteine (Brennan et al., 2014). It is interesting to note that in other tissues, GPRC6A responds to undercarboxylated osteocalcin.

FAT

Oleic acid directly activates afferent neurons in the vagus nerve and thus signals to the brain the arrival of digested fat in the small intestine (Webster and Beyak, 2013). Additional fat sensors in the small intestine are the fatty acid transporter CD36 and the G-protein-coupled receptor 120 (GPR120). CD36 senses long-chain fatty acids, and GPR120 responds best to alpha-linolenic and other omega-3 fatty acids (Paulsen et al., 2014). GPR120 triggers the release of several incretins, including GLP-1 (Mace et al., 2014) and CCK (Ichimura et al., 2014). Genetically dysfunctional GPR120 sensors tend to promote excessive abdominal fat accumulation both in a mouse knock-out model and in people with loss-of-function variants (Ichimura et al., 2012).

FFA1 (GPR40) activation by medium- and long-chain fatty acids triggers the release of GLP-1 from intestinal L cells, GI from K cells, and CCK from I cells (Hara et al., 2014). Free fatty acid receptor 2 (FFA2, GPR43) is most responsive to the short-chain fatty acids acetate and proprionate, and also to a lesser degree to slightly longer fatty acids. Activation of FFA2 in the small intestine triggers the release of PYY, GLP-1, and GLP-2 from L cells and promotes intestinal motility (Hara et al., 2014). Free fatty acid receptor 3 (FFA3, GPR41) responds most to short-chain fatty acids and triggers the release of GLP-1 and GIP secretion. Cells in the colon with the short-chain fatty acid receptors 2 (GPR43/FFAR2) and 3 (GPR41/FFAR3) detect acetic acid, propionic acid, and butyric acid from bacterial fermentation of dietary fiber, resistant starch, and other undigested polysaccharides (Nøhr et al., 2013). FFA3 and FFA2 detect short-chain fatty acids in the colon and link this information through enteroendocrine cells and sympathetic pathways with appetite regulation (López Soto et al., 2014) and immune response (Ulven, 2012).

Another mode of nutrient-intake sensing responds to the presence of bile acids in the ileal lumen, which triggers the release of the signaling peptides PYY and GLP-1 (Ullmer et al., 2013).

CALCIUM

The CaSR in the small intestinal endothelium detects the presence of ingested calcium in the intestinal lumen, and its signals are likely to contribute significantly to the maintenance of calcium homeostasis in the body (Garg and Mahalle, 2013). High calcium intake might thus directly decrease fractional calcium absorption and possibly also affect other regulators of calcium disposition.

REFERENCES

Brennan, S.C., Davies, T.S., Schepelmann, M., Riccardi, D., 2014. Emerging roles of the extracellular calcium-sensing receptor in nutrient sensing: control of taste modulation and intestinal hormone secretion. Br. J. Nutr., 1–7.

Choi, S., Lee, M., Shiu, A.L., Yo, S.J., Halldén, G., Aponte, G.W., 2007. GPR93 activation by protein hydrolysate induces CCK transcription and secretion in STC-1 cells. Am. J. Physiol. Gastrointest. Liver Physiol. 292, G1366–G1375.

Daly, K., Al-Rammahi, M., Moran, A., Marcello, M., Ninomiya, Y., Shirazi-Beechey, S.P., 2013. Sensing of amino acids by the gut-expressed taste receptor T1R1-T1R3 stimulates CCK secretion. Am. J. Physiol. Gastrointest. Liver Physiol. 304, G271–G282.

Egerod, K.L., Engelstoft, M.S., Grunddal, K.V., Nøhr, M.K., Secher, A., Sakata, I., et al., 2012. A major lineage of enteroendocrine cells coexpress CCK, secretin, GIP, GLP-1, PYY, and neurotensin but not somatostatin. Endocrinology 153, 5782–5795.

Engelstoft, M.S., Egerod, K.L., Lund, M.L., Schwartz, T.W., 2013a. Enteroendocrine cell types revisited. Curr. Opin. Pharmacol. 13, 912–921.

Engelstoft, M.S., Park, W.M., Sakata, I., Kristensen, L.V., Husted, A.S., Osborne-Lawrence, S., et al., 2013b. Seven transmembrane G protein-coupled receptor repertoire of gastric ghrelin cells. Mol. Metab. 2, 376–392.

Garg, M.K., Mahalle, N., 2013. Calcium homeostasis, and clinical or subclinical vitamin D deficiency—Can a hypothesis of "intestinal calcistat" explain it all? Med. Hypotheses 81, 253–258.

Hara, T., Kashihara, D., Ichimura, A., Kimura, I., Tsujimoto, G., Hirasawa, A., 2014. Role of free fatty acid receptors in the regulation of energy metabolism. Biochim. Biophys. Acta 1841, 1292–1300.

Hirao, L.A., Grishina, I., Bourry, O., Hu, W.K., Somrit, M., Sankaran-Walters, S., et al., 2014. Early mucosal sensing of SIV infection by Paneth cells induces IL-1β production and initiates gut epithelial disruption. PLoS Pathog. 10, e1004311.

Ichimura, A., Hara, T., Hirasawa, A., 2014. Regulation of energy homeostasis via GPR120. Front. Endocrinol. (Lausanne) 5, 111.

Ichimura, A., Hirasawa, A., Poulain-Godefroy, O., Bonnefond, A., Hara, T., Yengo, L., et al., 2012. Dysfunction of lipid sensor GPR120 leads to obesity in both mouse and human. Nature 483 (7389), 350–354.

López Soto, E.J., Gambino, L.O., Mustafá, E.R., 2014. Free fatty acid receptor 3 is a key target of short chain fatty acid: what is the impact on the sympathetic nervous system? Channels (Austin) 8 (3).

Mace, O.J., Schindler, M., Patel, S., 2012. The regulation of K- and L-cell activity by GLUT2 and the calcium-sensing receptor CasR in rat small intestine. J. Physiol. 590, 2917–2936.

Mace, O.J., Lister, N., Morgan, E., Shepherd, E., Affleck, J., Helliwell, P., et al., 2014. Expression of the fatty acid receptor GPR120 in the Gut of Diet-induced-obese rats and its role in GLP-1 secretion. PLoS One 9 (2), e88227.

Mace, O.J., Lister, N., Morgan, E., Shepherd, E., Affleck, J., Helliwell, P., et al., 2009. An energy supply network of nutrient absorption coordinated by calcium and T1R taste receptors in rat small intestine. J. Physiol. 587 (Pt 1), 195–210.

Nøhr, M.K., Pedersen, M.H., Gille, A., Egerod, K.L., Engelstoft, M.S., Husted, A.S., et al., 2013. GPR41/FFAR3 and GPR43/FFAR2 as cosensors for short-chain fatty acids in enteroendocrine cells vs FFAR3 in enteric neurons and FFAR2 in enteric leukocytes. Endocrinology 154, 3552–3564.

Paulsen, S.J., Larsen, L.K., Hansen, G., Chelur, S., Larsen, P.J., Vrang, N., 2014. Expression of the fatty acid receptor GPR120 in the gut of diet-induced-obese rats and its role in GLP-1 secretion. PLoS One 9 (2), e88227.

Reimann, F., Tolhurst, G., Gribble, F.M., 2012. G-protein-coupled receptors in intestinal chemosensation. Cell Metab. 15, 421–431.

Tolhurst, G., Reimann, F., Gribble, F.M., 2012. Intestinal sensing of nutrients. Handb. Exp. Pharmacol. 209, 309–335.

Ullmer, C., Alvarez Sanchez, R., Sprecher, U., Raab, S., Mattei, P., Dehmlow, H., et al., 2013. Systemic bile acid sensing by G protein-coupled bile acid receptor 1 (GPBAR1) promotes PYY and GLP-1 release. Br. J. Pharmacol. 169, 671–684.

Ulven, T., 2012. Short-chain free fatty acid receptors FFA2/GPR43 and FFA3/GPR41 as new potential therapeutic targets. Front. Endocrinol. (Lausanne) 3, 111.

van der Wielen, N., van Avesaat, M., de Wit, N.J., Vogels, J.T., Troost, F., Masclee, A., et al., 2014. Cross-species comparison of genes related to nutrient sensing mechanisms expressed along the intestine. PLoS One 9 (9), e107531.

Webster, W.A., Beyak, M.J., 2013. The long chain fatty acid oleate activates mouse intestinal afferent nerves *in vitro*. Can. J. Physiol. Pharmacol. 9, 375–379.

PHYSICAL SENSING AND CHEMESTHESIS

The oral and pharyngeal cavities constitute a very vulnerable access route to the lower air passages and digestive tract. Many different harmful solids, fluids, and gases are potentially lethal threats and have to be guarded against at a moment's notice. This includes exposure to overly cold or hot items, abrasive or cutting items, and corrosive chemicals. Diverse sensors guard the portal to the sensitive passages beyond and prevent harmful structures from passing (or punish humans painfully if they do). Many plants have evolved mechanisms to defend against being consumed and deploy mechanical, chemical, and psychological warfare. The latter is particularly interesting because it is not all that it seems. Sharp, pungent tastes, for instance, can be fear inducing, but even the hottest jalapeño only leverages human responses without possessing any chemically corrosive properties.

ABBREVIATIONS

TRPV1 transient receptor potential vanilloid 1
TRPA1 transient receptor potential ankyrin 1

MECHANORECEPTORS AND NOCICEPTORS

The oral cavity, like most of the body surface, contains numerous and fairly evenly distributed mechano-receptors (Purves et al., 2001). These receptors have a crucial protective function by imposing a strong aversion against ingesting dangerously hot items, caustic chemicals, or mechanically destructive objects. However, these receptors also contribute importantly to the gustatory experience by sensing the temperature and texture of foods and conveying hot and pungent tastes. Excessive sensitivity to minor irritants in foods, which is not rare in elderly people (Duffy, 1999), can become a difficult barrier to normal eating.

A wide range of irritants trigger polymodal nociceptive neurons that extend mainly from the ganglia of cranial nerve V (trigeminus nerve) and, to a lesser extent, from those of cranial nerves IX (glossopharyngeal nerve) and X (vagus nerve). The nerve endings do not have a specific sensory structure but respond to stimuli along their length. The conduction of the information is relatively slow because nociceptive neurons are not myelinated.

Some of the irritants that trigger nociceptor neurons are sensed at much lower levels by specific olfactory and gustatory cells. Commonly ingested compounds that can trigger nociceptors at some level of exposure include hot spices (e.g., capsaicin in peppers), ethanol, acetic acid (in vinegar), carbon dioxide (in carbonated beverages), and menthol (in cough drops). Even salt at high concentration (i.e., greater than 1 mol/l) activates these nociceptors and elicits an irritant sensation distinct from its salty taste. Activation of nociceptors elicits a burning sensation; in extreme instances, pain may be felt. Independent of higher central nervous function, nociception also tends to trigger autonomous responses, such as sweating, salivation, and lacrimation.

HOT, SPICY TASTE

Capsaicin (in chili peppers), piperine (in black pepper), allyl isothiocyanate (mustard oil in mustard, wasabi, and other cruciferous vegetables), and anandamide act on the capsaicin receptor (transient receptor potential vanilloid 1, TRPV1) on low-threshold heat-sensing neurons with small to medium diameters, which are present throughout the oral cavity. TRPV1 is a cation channel, which otherwise would be activated by exposure to temperatures over $43°C$. Capsaicin does not traverse the channel itself; rather, it binds to an extracellular moiety and opens the channel for cations. The ensuing influx of sodium and calcium ions then leads to the more or less persistent activation of the nociceptor neuron (Tominaga and Julius, 2000). Acids can activate the receptor directly and also increase its responsiveness to capsaicin and heat (Jordt et al., 2000). It is of note that capsaicin does not directly damage cells, the considerable potential for perceived discomfort notwithstanding. It is only the response to neural stimulation that induces considerable self-harm. Because they lack a capsaicin-responsive TRPV1, birds can tolerate large doses of capsaicin, while mammals cannot. This means that birds can happily and safely feed on seeds laced with capsaicin, while squirrels cannot.

While capsaicin has no significant effect on gastrointestinal motility, piperine reduces it through a mechanism that does not appear to involve TRPV1 (Izzo et al., 2001). The two spicy compounds also appear to affect intestinal barrier function differently: capsaicin, but not piperine, increases cell permeability for ions and medium-sized macromolecules (Jensen-Jarolim et al., 1998). Piperine has also been reported to affect the metabolism of drugs and other xenobiotics by decreasing some phase I enzymes (P450 2E1), but increasing other phase I enzymes (P450 2B and P450 1A) and phase II enzymes (glucuronidation) (Kang et al., 1994; Shoba et al., 1998).

A second type of nociceptor cation channel, transient receptor potential ankyrin 1 (TRPA1), similarly responds to capsaicin, allyl isothiocyanate, and other pungent compounds in foods by inducing pain, swelling, and increased heat sensitivity (Alpizar et al., 2014) (Figure 1.5).

COLD PERCEPTION

Menthol ([1α,2β,5a]-5-methyl-2-[1-methylethyl]cyclohexanol), 1-menthone ([2β,5α]-5-methyl-2-[1-methylethyl]cyclohexanone), cinnamaldehyde (3-phenyl-2-propenal), and similar phenolic compounds (Figure 1.6) raise the temperature threshold of the cold-sensing sodium- and calcium-channel TRPM8 (Clapham, 2002). Upon exposure to such compounds, the channel becomes active at 27–31°C (instead of the normal trigger temperature, which is 5°C less) and projects to the brain the sensation of coolness via the trigeminus nerve.

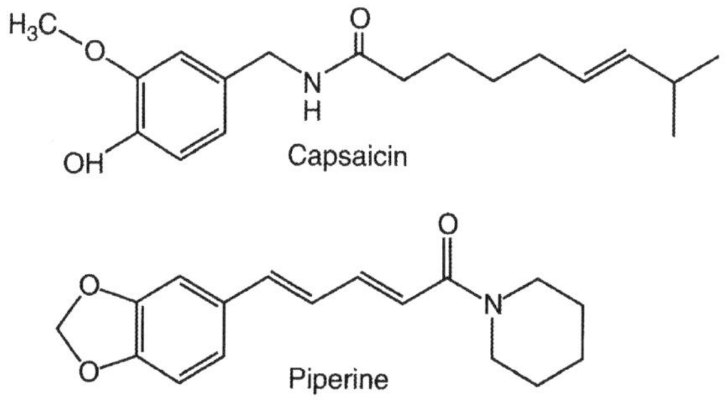

FIGURE 1.5

The active compounds in pungent spices.

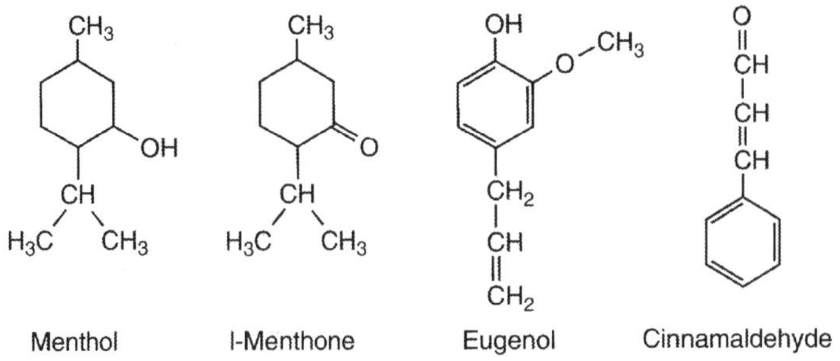

FIGURE 1.6

Several phenolic compounds induce a cold sensation in the mouth.

REFERENCES

Alpizar, Y.A., Boonen, B., Gees, M., Sanchez, A., Nilius, B., Voets, T., et al., 2014. Allyl isothiocyanate sensitizes TRPV1 to heat stimulation. Pflügers Arch. 466, 507–515.

Clapham, D.E., 2002. Signal transduction. Hot and cold TRP ion channels. Science 295, 2228–2229.

Duffy, V.B., 1999. Smell, taste, and somatosensation in the elderly. In: Chernoff, R. (Ed.), Geriatric Nutrition Aspen Publishers, Gaithersburg, MD, pp. 170–211.

Izzo, A.A., Capasso, R., Pinto, L., Di Carlo, G., Mascolo, N., Capasso, E., 2001. Effect of vanilloid drugs on gastrointestinal transit in mice. Br. J. Pharmacol. 132, 1411–1416.

Jensen-Jarolim, E., Gajdzik, L., Haberl, I., Kraft, D., Scheiner, O., Graf, J., 1998. Hot spices influence permeability of human intestinal epithelial monolayers. J. Nutr. 128, 577–581.

Jordt, S.E., Tominaga, M., Julius, D., 2000. Acid potentiation of the capsaicin receptor determined by a key extracellular site. Proc. Natl. Acad. Sci. USA 97, 8134–8139.

Kang, M.H., Won, S.M., Park, S.S., Kim, S.G., Novak, R.E., Kim, N.D., 1994. Piperine effects on the expression of P4502E1, P4502B and P4501A in rat. Xenobiotica 24, 1195–1204.

Purves, D., Augustine, G.J., Fitzpatrick, D., Katz, L.C., LaMantian, A.S., McNamara, J.O. (Eds.), 2001. Neuroscience Sinauer Associates, Sunderland, MA.

Shoba, G., Joy, D., Joseph, T., Majeed, M., Rajendran, R., Srinivas, P.S., 1998. Influence of piperine on the pharmacokinetics of curcumin in animals and human volunteers. Planta Med. 64, 353–356.

Tominaga, M., Julius, D., 2000. Capsaicin receptor in the pain pathway. Jap. J. Pharmacol. 83, 20–24.

INTAKE REGULATION

2

CHAPTER OUTLINE

Appetite 25
Thirst 32

APPETITE

The ability to detect odors from foods is an important appetite stimulant and often initiates the secretion of digestive juices. Aversive reactions to foul or otherwise disagreeable odors provide some protection against ingestion of unsafe foods. Impaired ability to smell, which becomes more common with advancing age, increases the risk of inadequate food intake and of food poisoning.

ABBREVIATIONS

AgRP	agouti-related protein
α-MSH	alpha-melanocyte-stimulating hormone
AMPK	AMP-activated protein kinase (EC2.7.11.1)
apoAII	apolipoprotein AII
apoAIV	apolipoprotein AIV
CCK	cholecystokinin
MCH	melanin-concentrating hormone
MCR4	melanocortin receptor 4mTORC1 mammalian target of rapamycin 1
NPY	neuropeptide Y

Appetite is a feeling that is largely related to neuronal activities in the hypothalamus in response to nutritional status, recent intakes, health, pregnancy status, habituation, moods, social cues, and other complex inputs. Mechanisms that promote dietary intake are called *orexigenic. Anorexia* is a state with suppressed intake behavior. Energy balance is a major determinant, but lack of a few nutrients (e.g., sodium) also induces more or less specific cravings. Excess of some nutrients can decrease appetite, on the other hand. Foods with an imbalance of certain nutrients (e.g., isolated lack of a branched-chain amino acid) can promote conditioned taste aversion (Gietzen and Magrum, 2001). Since this means that humans make some of their food choices based on nutritional needs, some feel inclined to speak of innate "nutritional wisdom." The high prevalence of obesity and micronutrient deficiencies in affluent populations highlights the limitations of such regulatory mechanisms.

Regulation of intake behavior is closely integrated with both other brain functions and the control of energy metabolism. Hormones, such as insulin, that were originally associated with the regulation of glucose utilization and other metabolic functions are now believed to directly affect appetite. Mediators like leptin, which were discovered because of their function in intake regulation, now reveal unrelated functions outside the brain.

The control of water-intake behavior is discussed later in this chapter in the section on thirst.

CENTRAL APPETITE REGULATION

Much of the activity related to appetite perception and intake regulation has been located in the arcuate nucleus of the hypothalamus near the third ventricle, the nearby ventromedial nucleus, and lateral areas of the hypothalamus. Increased expression of the oncogene c-los in some of these brain regions correlates with appetite perception.

NPY/melanocortin neurons: Two types of neurons, which are located mainly in the arcuate nucleus, constitute a significant portion of appetite perception circuitry (Schwartz and Morton, 2002). One type is characterized by its production of neuropeptide Y (NPY) and agouti-related protein (AgRP). Production of melanocortin characterizes the second type of neurons. The action of NPY/AgRP neurons on other neurons nearby stimulates the perception of appetite. The binding of melanocortin to melanocortin receptor 4 (MC4R) on hypothalamic neurons inhibits appetite. AgRP specifically blocks the binding of melanocortin to MC4R and other melanocortin receptors (MC1R and MC3R).

The (appetite-promoting) NPY/AgRP neurons are stimulated by ghrelin (signaling an empty stomach and duodenum) and inhibited by leptin (signaling fat-filled adipose tissue), insulin (signaling plentiful carbohydrate supplies), and PYY3-36 (signaling a filled distal intestine). They are also inhibited by their own activity and by melanocortin neuron input. PYY3-36 action is mediated by a type of NPY receptor called Y2R, which responds to both intact NPY and fragments such as PYY3-36. The mammalian target of rapamycin complex 1 (mTORC1) functions as a molecular integrator of the diverse incoming appetite signals (Watterson et al., 2013). Insulin and leptin increase mTORC1 activity by acting through phosphatidylinositol 3-kinase (phosphoinositide-3-kinase, PI3K, EC2.7.1.137) and RAC-alpha serine/threonine-protein kinase (protein kinase B, PKB, EC2.7.11.1). Ghrelin, in contrast, stimulates the expression of AMP-activated protein kinase (AMPK, EC2.7.11.1), which then decreases mTORC1 activity.

The (appetite-decreasing) melanocortin neurons are inhibited by leptin and insulin, and by input from NPY/AgRP neurons. Melanocortin is produced by posttranslational processing (Pritchard et al., 2002) of the much larger peptide proopiomelanocortin (POMC), which is also the precursor of alpha-, beta-, and gamma-MSH (melanocyte-stimulating hormones) and adrenocorticotrophic hormone (ACTH).

Neurotransmitters: Several types of neurons that are intimately involved in the regulation of appetite (and also in thirst regulation) use serotonin (5-hydroxytryptamin) as a neurotransmitter. Food intake stimulates the release of serotonin by such neurons and thereby induces a feeling of satiation. Various drugs (e.g., dexfenfluramine) that increase the availability of serotonin induce weight loss, and also serious side effects like pulmonary hypertension and heart valve damage (Michelakis and Weir, 2001).

Two distinct receptors respond to dopamine binding. The D1 receptors in the hypothalamus promote food intake, while D2 receptors inhibit it. Norepinephrin appears to slow eating behavior by inhibiting motor neurons in the medulla involved in the involuntary control of ingestion and swallowing (Nasse and Travers, 2014).

Some peripheral mediators of food intake regulation, such as amylin, act on dopamine receptors. Catecholamine- and histamine-producing neurons also contribute to appetite suppression. Beta-adrenergic agonists (e.g., sibutramine) are used to induce weight loss.

Other central mediators: Galanin, opioid peptides, growth hormone-releasing hormone, and other hormones increase food intake as part of their broad activity profiles. Galanin in the brain stimulates appetite, especially for fat. Specialized neuroendocrine cells in the hypothalamus, other brain regions, and the gastrointestinal tract secrete this peptide together with prolactin. Three galanin receptors in the brain and many other tissues respond to galanin.

Melanin-concentrating hormone (MCH) is a cyclic small neuropeptide from the lateral hypothalamic area and arcuate nucleus that promotes food intake. It acts on target neurons by activating the G-protein-linked MCH receptor 1 (Marsh et al., 2002), and possibly also the NPY Y1 receptor (Chaffer and Morris, 2002).

Neurons of the lateral, dorsomedial, and perifornical areas of the hypothalamus near the base of the stalk of the pituitary produce the appetite-promoting peptides orexin A (hypocretin 1) and orexin B (hypocretin 2) from the same precursor. The orexin-producing neurons respond directly to low glucose availability. Orexins stimulate appetite by activating opioid receptors (Clegg et al., 2002). These peptides are also involved in the regulation of wakefulness and other behaviors (Shirasaka et al., 2002). Further food-intake-inhibiting central mediators include cocaine- and amphetamine-regulated transcript (CART) peptide (a regulator of pituitary function) and corticotrophin-releasing factor (CRF).

Cannabinoids: The G-protein-coupled cannabinoid receptors CB1 (mainly in the brain) and CB2 (mainly in immune cells) respond to both endocannabinoids and several phytochemicals from the cannabis plant (Onaivi et al., 2002). Putative endogenous ligands for these receptors include 2-arachidoylglycerol, 2-arachidonyl glyceryl ether, and anandamide. An important function of this signaling mechanism may be the inhibition of neurotransmitter release by target neurons. Cannabis-derived compounds, including tetrahydrocannabinol, cannabidiol (nonpsychotropic), and delta-9-tetrahydrocannabinol promote appetite. A distinct mechanism (such as delta-9-tetrahydrocannabinol) achieves the suppression of nausea and vomiting (in cancer and AIDS patients) and thereby an improvement of intake desire (Mechoulam and Hanu, 2001).

SENSORY INPUT

Visual and acoustic cues, odors, textures, and tastes can trigger a sequence of centrally mediated events (cephalic phase response) that result in greater vagal activity and the stimulation of exocrine (saliva and gastric juice) and endocrine (insulin) secretion (Nederkoorn et al., 2000; Smeets et al., 2010). Cephalic (brain-related) responses also cause a general arousal, as reflected by more frequent gastric contractions, higher variability of heart rate, and higher blood pressure. Just the thought of a favored food can trigger salivation and other typical cephalic phase responses. A fasting state, approaching habitual mealtimes, and social context contribute to the stimulation. Many of the raw signals are interpreted in the context of previous experiences. Preferences for many foods are learned responses based on cultural norms or associations with strongly positive or negative events.

Impairment of the senses for taste and smell, which becomes more marked with advancing age, can diminish adequate appetite and often becomes a significant barrier for normal intake behavior (Schiffman and Graham, 2000).

ENTERAL INPUT

Neural connections: The stomach and intestine contain stretch receptors and chemoreceptors that signal satiation both locally and to the brain. These entities are not well characterized yet. Afferent vagus fibers can convey signals from the gastrointestinal tract to the brain.

Intake stimulation: After half a day or so of fasting, the sense of hunger may become strong and the stomach may start growling audibly. The physiological basis for this response is not entirely understood, but some correlates have been uncovered. A fall in glucose, and possibly an increase in circulating insulin levels, induces appetite and hunger (Bray, 2000). Hormones from the digestive tract can signal the need for feeding to the brain. Such hunger-signaling compounds are called *orexigenic*. Anorexigenic mediators are those with the opposite effect, slowing or stopping the desire to eat. Ghrelin is the principal orexigenic peptide hormone, and it is secreted in the blood by X/A-like cells, mainly in the stomach but fewer in the duodenum and pancreas, in response to several hours of fasting (Perret et al., 2014). Secretion promptly stops upon eating. The diurnal variation in ghrelin release is thought to set meal patterns in humans. Average circulating levels may contribute to long-term intake regulation. Removal of large parts of ghrelin-producing tissue during gastric bypass surgery may be an important mechanism for the sustained weight loss following such intervention. Ghrelin *O*-acyltransferase (GOAT, membrane-bound *O*-acyltransferase domain-containing protein 4, MBOAT4, EC2.3.1.-) attaches a fatty acid to the third serine in the newly secreted protein and thereby transforms ghrelin into the predominantly active form, which can then bind and activate a specific ghrelin receptor (growth hormone secretagogue receptor, GHSR-1a) in many different segments of the brain (including the hypothalamus, hippocampus, ventral tegmental area, nucleus accumbens, and amygdala) and in other tissues. Ghrelin is thought to increase the activity of AMPK in the brain, which in turn decreases mTORC1 activity and increases hypothalamic AgRP messenger RNA (mRNA) levels (Watterson et al., 2013). Ghrelin not only promotes appetite directly, but it contributes to reward signaling after consumption of a tasty food (Schellekens et al., 2013). It has additional very diverse biological activities that differ between specific peripheral target tissues. The ghrelin receptor can form heterodimers with various other receptors (including dopamine 1 receptor, 5-HT2C receptor, and MC3 receptor) and thereby lose some of its activity (Schellekens et al., 2013). Acylated ghrelin is rapidly deacylated by acyl-protein thioesterase 1 (APT1, EC3.1.2.-) in blood (Satou and Sugimoto, 2012); most of the protein circulates as des-acyl-ghrelin in peripheral blood. Both the acyl and des-acyl forms are biologically active, but their qualitative and quantitative effects differ (Heppner et al., 2014).

Intake inhibition: Cholecystokinin (CCK) plays a major role in appetite regulation by limiting meal size and providing a sense of fullness. This proteohormone is secreted by I endocrine cells of the small intestine in response to the presence of fat and other nutrients. CCK-A receptors in the small intestine terminate intake through as-yet-unclear mechanisms (Moran, 2000). While it seems quite certain that CCK does not act directly on CCK-B receptors in the brain, the level of CCK secretion modulates long-term feeding behavior in addition to the short-term local effects. Several pancreatic hormones, including glucagon, amylin, and pancreatic polypeptide, also reduce appetite (Bray, 2000).

Other gastrointestinal mediators that inhibit appetite include gastrin inhibitory peptide (GIP), gastrin-releasing peptide (GRP; related to bombesin), and glucagon-like peptides 1 (GLP-1) and 2 (GLP-2).

The normal digestive process generates products that exert some feedback regulation by limiting intake. The proteolysis of pancreatic colipase generates the pentapeptide enterostatin (Val–Pro–Asp–Pro–Arg), which suppresses appetite and increases insulin secretion (Bouras et al., 1995). Pyruvate and lactate from breakdown of carbohydrates and proteins highlight metabolites that slow intake. The extent of fat absorption is coupled with the production of apolipoprotein AIV (a normal chylomicron constituent), which has appetite-suppressing characteristics (Tso et al., 2001).

INDIVIDUAL NUTRIENTS

Carbohydrates: The predominant factor that makes people feel hungry is a decrease in blood glucose concentrations. Specific glucose-sensing neurons in the ventromedial hypothalamus provide the brain with information on blood concentrations (Ngarmukos et al., 2001). The inhibition of NPY neurons by insulin provides a more long-term signal. When less insulin gets to the brain, the NPY neurons become more active and stimulate the appetite center. The hormone amylin, which is secreted jointly with insulin, acutely induces satiety on its own by acting on dopamine D2 receptors (Lutz et al., 2001). Decreased secretion with declining glucose levels is another reason why people feel hungry when they miss a meal.

Protein: An imbalance with inadequacy of one of the branched-chain amino acids (i.e., valine, leucine, and isoleucine) leads to the development of conditioned taste aversion (Gietzen and Magrum, 2001). Expression of several genes changes in response to inadequacy, but the mechanism responsible for learning to balance essential amino acid intake remains elusive.

Fat: Long-term regulation of energy intake is closely coupled to the status of fat stores. Adipocytes release leptin in proportion to their fat content. When fat mass declines even by a small fraction, the NPY cells in brain increase their activity and induce appetite in the hypothalamic intake regulation center. The adipose tissue also secretes agouti protein (Kim and Moustaid-Moussa, 2000), which may promote appetite by blocking the action of melanocortin on MCR4 in the brain (Voisey and van Daal, 2002), as outlined previously for the AgRP from NPY neurons. This could indicate that adipose tissue in obese people promotes appetite whatever they do: when their fat mass declines, the decreasing leptin secretion promotes appetite, and when the fat mass stays constant, the agouti protein still does its work.

The following example may illustrate links between recent fat intake and appetite at the level of gastrointestinal metabolism. Fat consumption triggers CCK release, which increases the secretion of digestive compounds from pancreas. Among them are lipase and colipase, which are crucial for the normal digestion of dietary triglycerides and cholesterol esters. Eventually, the pancreatic proteases digest colipase and generate the pentapeptide enterostatin, among other fragments. Enterostatin is a potent inhibitor of appetite, especially for fat, and it also induces the secretion of insulin. The hormone trigger (CCK) itself also limits fat intake (Moran, 2000). As a result of fat absorption, triglyceride-rich chylomicrons enter the circulation. Apolipoprotein AIV (apoAIV) is one of the normal protein constituents of chylomicrons. ApoAIV in circulation acts through as-yet-unknown mechanisms to suppress appetite, especially for fatty foods (Tso et al., 2001).

Apolipoprotein AII (apoAII) is another chylomicron constituent that mediates the satiating effects of fat, particularly the unsaturated kind, as it is absorbed in the small intestine (Smith et al., 2012). ApoAII may well act through ghrelin. A common genetic variant of the APOA2 promoter demonstrates the importance of this link: carriers of the variant gain more body weight than noncarriers when they eat a lot of saturated fat because this signaling axis is not sufficiently effective (Smith et al., 2013).

Salt: Sodium depletion can induce a craving for salty foods. This feeling is at least partially mediated by angiotensin I, which is synthesized in the brain (McKinley et al., 2001). Most food choices involving salty foods, however, are unlikely to be related to actual depletion and arise more from a pre-set preference for such items. Excessive salt intake leads to salt aversion. Volume depletion and thirst also suppress salt appetite (Thunhorst and Johnson, 2001).

Other nutrients: There are some indications of an innate preference for calcium-containing foods that may become more acute during pregnancy (Tordoff, 2001·). The compulsive consumption of gritty nonfood items, such as clay, by pregnant women is usually called *pica* (Rose et al., 2000). Unmet calcium requirements in conjunction with hormonal fluctuations may induce indiscriminate cravings in women, which decrease with calcium supplementation (Thys-Jacobs, 2000). The extent to which pica behavior and other cravings are linked to actual calcium deficiency remains a matter of dispute. Pica behavior has also been linked to iron deficiency and the cravings may subside with improvement of iron status (Rose et al., 2000). Iron deficiency may induce such cravings for nonfood items also in children and nonpregnant adults. A third of US patients with sickle cell anemia reported pica behavior (Ivascu et al., 2001), and it was most often those with the most severe nutritional deficiencies.

A preference for magnesium-rich foods was observed in magnesium-depleted rats (McCaughey and Tordoff, 2002). Similarly, mice can sense calcium in solutions. This calcium-tasting sense of mice depends on the taste receptor Tas1r3 (Tordoff et al., 2008). More investigation is needed to determine whether such need-based preferences also exist in humans.

PHYSICAL EXERCISE

Intense physical activity can suppress appetite, possibly through increased serotonin concentration in the hypothalamus (Avraham et al., 2001). Physical exercise appears to increase the preference for sucrose and citric acid (Horio and Kawamura, 1998).

REFERENCES

Avraham, Y., Hao, S., Mendelson, S., Berry, E.M., 2001. Tyrosine improves appetite, cognition, and exercise tolerance in activity anorexia. Med. Sci. Sports Exerc. 33, 2104–2110.

Bouras, M., Huneau, J.E., Luengo, C., Erlanson-Albertsson, C., Tome, D., 1995. Metabolism of enterostatin in rat intestine, brain membranes, and serum: differential involvement of proline-specific peptidases. Peptides 16, 399–405.

Bray, G.A., 2000. Afferent signals regulating food intake. Proc. Nutr. Soc. 59, 373–384.

Chaffer, C.L., Morris, M.J., 2002. The feeding response to melanin-concentrating hormone is attenuated by antagonism of the NPYY(1)-receptor in the rat. Endocrinology 143, 191–197.

Clegg, D.J., Air, E.L., Woods, S.C., Seeley, R.J., 2002. Eating elicited by orexin-α, but not melanin-concentrating hormone, is opioid mediated. Endocrinology 143, 2995–3000.

Gietzen, D.W., Magrum, L.J., 2001. Molecular mechanisms in the brain involved in the anorexia of branched-chain amino acid deficiency. J. Nutr., 851S–855S.

Heppner, K.M., Piechowski, C.L., Müller, A., Ottaway, N., Sisley, S., Smiley, D.L., et al., 2014. Both acyl and des-acyl ghrelin regulate adiposity and glucose metabolism via central nervous system ghrelin receptors. Diabetes 63, 122–131.

Horio, T., Kawamura, Y., 1998. Influence of physical exercise on human preferences for various taste solutions. Chem. Senses 23, 417–421.

Ivascu, N.S., Sarnaik, S., McCrae, J., Whitten-Shurney, W., Thomas, R., Bond, S., 2001. Characterization of pica prevalence among patients with sickle cell disease. Arch. Ped. Adolescent. Med. 155, 1243–1247.

Kim, S., Moustaid-Moussa, N., 2000. Secretory, endocrine and autocrine/paracrine function of the adipocyte. J. Nutr. 130, 3110S–3115S.

Lutz, T.A., Tschudy, S., Mollet, A., Geary, N., Scharrer, E., 2001. Dopamine D(2) receptors mediate amylin's acute satiety effect. Am. J. Physiol. Reg. Integr. Comp. Physiol. 280, R1697–R1703.

Marsh, D.J., Weingarth, D.T., Novi, D.E., Chen, H.Y., Trumbauer, M.E., Chen, A.S., et al., 2002. Melanin-concentrating hormone 1 receptor-deficient mice are lean, hyperactive, and hyperphagic and have altered metabolism. Proc. Natl. Acad. Sci. USA 99, 3240–3245.

McCaughey, S.A., Tordoff, M.G., 2002. Magnesium appetite in the rat. Appetite 38, 29–38.

McKinley, M.J., Allen, A.M., Mathai, M.L., May, C., McAllen, R.M., Oldfield, B.J., et al., 2001. Brain angiotensin and body fluid homeostasis. Jap. J. Physiol. 51, 281–289.

Mechoulam, R., Hanu, K., 2001. The cannabinoids: an overview. Therapeutic implications in vomiting and nausea after cancer chemotherapy, in appetite promotion, in multiple sclerosis and in neuroprotection. Pain Res. Manag. 6, 67–73.

Michelakis, E.D., Weir, E.K., 2001. Anorectic drugs and pulmonary hypertension from the bedside to the bench. Ant. J. Med. Sci. 321, 292–299.

Moran, T.H., 2000. Cholecystokinin and satiety: current perspectives. Nutrition 16, 858–865.

Nasse, J.S., Travers, J.B., 2014. Adrenoreceptor modulation of oromotor pathways in the rat medulla. J. Neurophysiol. [Epub ahead of print].

Nederkoorn, C., Smulders, F.T., Jansen, A., 2000. Cephalic phase responses, craving and food intake in normal subjects. Appetite 35, 45–55.

Ngarmukos, C., Baur, E.L., Kumagai, A.K., 2001. Co-localization of GLUT1 and GLUT4 in the blood-brain barrier of the rat ventromedial hypothalamus. Brain Res. 900, 1–8.

Onaivi, E.S., Leonard, C.M., Ishiguro, H., Zhang, P.W., Lin, Z., Akinshola, B.E., et al., 2002. Endocannabinoids and cannabinoid receptor genetics. Progr. Neurobiol. 66, 307–344.

Perret, J., De Vriese, C., Delporte, C., 2014. Polymorphisms for ghrelin with consequences on satiety and metabolic alterations. Curr. Opin. Clin. Nutr. Metab. Care 17, 306–311.

Pritchard, L.E., Turnbull, A.V., White, A., 2002. Pro-opiomelanocortin processing in the hypothalamus: impact on melanocortin signalling and obesity. J. Endocrinol. 172, 411–421.

Rose, E.A., Porcerelli, J.H., Neale, A.V., 2000. Pica: common but commonly missed. J. Am. Board Fam. Pract. 13, 353–358.

Satou, M., Sugimoto, H., 2012. The study of ghrelin deacylation enzymes. Methods Enzymol. 514, 165–179.

Schellekens, H., van Oeffelen, W.E., Dinan, T.G., Cryan, J.F., 2013. Promiscuous dimerization of the growth hormone secretagogue receptor (GHS-R1a) attenuates ghrelin-mediated signaling. J Biol. Chem. 288, 181–191.

Schiffman, S.S., Graham, B.G., 2000. Taste and smell perception affect appetite and immunity in the elderly. Eur. J. Clin. Nutr. 54, S54–S63.

Schwartz, M.W., Morton, G.J., 2002. Keeping hunger at bay. Nature 418, 595–597.

Shirasaka, T., Kunitake, T., Takasaki, M., Kannan, H., 2002. Neuronal effects of orexins: relevant to sympathetic and cardiovascular functions. Reg. Peptides. 104, 91–95.

Smeets, P.A., Erkner, A., de Graaf, C., 2010. Cephalic phase responses and appetite. Nutr. Rev. 68, 643–655.

Smith, C.E., Ordovás, J.M., Sánchez-Moreno, C., Lee, Y.C., Garaulet, M., 2012. Apolipoprotein AII polymorphism: relationships to behavioural and hormonal mediators of obesity. Int. J. Obes. (Lond) 36, 130–136.

Smith, C.E., Tucker, K.L., Arnett, D.K., Noel, S.E., Corella, D., Borecki, I.B., et al., 2013. Apolipoprotein A2 polymorphism interacts with intakes of dairy foods to influence body weight in 2 U.S. populations. J. Nutr. 143, 1865–1871.

Thunhorst, R.L., Johnson, A.K., 2001. Effects of hypotension and fluid depletion on central angiotensin-induced thirst and salt appetite. Am. J. Physiol. Reg. Integr. Comp. Physiol. 281, R1726–R1733.

Thys-Jacobs, S., 2000. Micronutrients and the premenstrual syndrome: the case for calcium. J. Am. Coll. Nutr. 19, 220–227.

Tordoff, M.G., 2001. Calcium: taste, intake, and appetite. Physiol. Rev. 81, 1567–1597.

Tordoff, M.G., Shao, H., Alarcón, L.K., Margolskee, R.F., Mosinger, B., Bachmanov, A.A., et al., 2008. Involvement of T1R3 in calcium-magnesium taste. Physiol. Genomics 34, 338–348.

Tso, P., Liu, M., Kalogeris, T.L., Thomson, A.B., 2001. The role of apolipoprotein A-IV in the regulation of food intake. Ann. Rev. Nutr. 21, 231–254.

Voisey, J., van Daal, A., 2002. Agouti: from mouse to man, from skin to fat. Pigment Cell Res. 15, 10–18.

Watterson, K.R., Bestow, D., Gallagher, J., Hamilton, D.L., Ashford, F.B., Meakin, P.J., et al., 2013. Anorexigenic and orexigenic hormone modulation of mammalian target of rapamycin complex 1 activity and the regulation of hypothalamic agouti-related protein mRNA expression. Neurosignals 21, 28–41.

THIRST

ABBREVIATION

ADH   antidiuretic hormone (vasopressin)

Water intake is the main protective response against increased solute and sodium concentration in the blood (McKinley et al., 2006). Thirst is a compelling central nervous stimulus for the intake of water-rich fluids with low solute content. It appears to be experienced even before birth (Caston-Balderrama et al., 2001). Water deprivation may lead to the perception of water where there is none (Changizi and Hall, 2001), and there is measurable relief when water is imbibed (Hallschmid et al., 2001). Dilute aqueous drinks are preferred to quench thirst, but drinks with higher solute concentrations (even saline) may also be accepted (De Luca et al., 2002).

The sensation perceived as thirst results from the integration of stimulatory and inhibitory signals from peripheral sensors and from within the brain itself. The physiological state that is sensed is predominantly volume depletion and increased concentration of osmotically active plasma constituents (osmolality). At around 140 mmol/l, sodium is the main solute in blood, and an increase in it is known to elicit thirst. Glucose and other small molecules can also increase osmolality enough under some circumstances (e.g., hyperglycemia in diabetes mellitus) to increase the urge to drink. Centrally or peripherally acting stimulants can also affect thirst. A common example is caffeine, which increases thirst, especially in nonhabituated consumers (Smit and Rogers, 2000). Aging tends to decrease thirst perception and blunt the urge to replenish volume as needed (Stachenfeld et al., 1997).

PERIPHERAL SENSORS

Thirst-stimulating input: Specialized cells embedded in the vascular endothelium at the branch point of the carotid arteries (carotid sinus) respond to pressure and thus detect volume changes. While it is clear that these signals promote thirst, the pathways and mechanisms remain obscure (Mann et al., 1987). Parasympathetic neurons carry the signals to the brain, where they eventually reach the vasomotor center of the medulla and the paraventricular nuclei. When blood pressure in the sinus bulb decreases, stimulation of the sinus baroreceptors slows and secretion of an antidiuretic hormone (ADH) called vasopressin from the paraventricular nuclei into circulation diminishes. Low blood volume is likely to generate a distinct and probably weaker thirst-inducing signal. The thirst signal in response to low blood volume is elicited by angiotensin II acting on angiotensin II receptors in the subfornical organ, a brain region near the ventricles with high vascularization, and lack of separation of the brain tissue from blood circulation by a blood–brain barrier (Stricker and Hoffmann, 2007).

Thirst-inhibiting input: Otopharyngeal mechanoreceptors inhibit thirst in anticipation of the effect of drinking, even before the ingested water has reached circulation and had a chance to normalize osmolality (Thompson et al., 1987). Unabated drinking without the input from such a delay circuit would favor abrupt excessive intake in response to acute volume contraction (for instance, after sweating during a workout). While the intestines and kidneys can deal with large fluid loads, excessive water intake exacts an avoidable energy cost (for absorption and excretion) and may interfere with the sodium-driven absorption of many nutrients from the small intestine. Similarly, an increased water load may be detected by sodium receptors in the stomach (Stricker and Sved, 2000). As arterial baroreceptors detect increased blood pressure, thirst sensation tends to decrease (Stricker and Sved, 2000).

SENSORS IN THE BRAIN

Blood hyperosmolality (primarily increased sodium concentration) is sensed by osmoreceptors in the hypothalamus (vascular organ of the lamina terminalis). The vascular endothelial cells do not maintain tight seals (blood–brain barrier) in this region, and diffusion of water and small solutes proceeds rapidly and relatively unimpeded. An increase of as little as 1% from basal osmolality of 280–290 mosmol/kg induces the sensation of thirst (Robertson et al., 1982).

Both sodium-sensitive and solute-density (osmolality-) sensitive neurons in the brain are known to participate in the regulation of water-intake behavior (Denton et al., 1996; Bourque and Oliet, 1997). For each quality, there are distinct excitatory and inhibitory signals (Stricker and Sved, 2000). One of the sensors in sodium-sensitive neurons of the mouse brain appears to be the voltage-gated sodium channel Na_v2/NaG (Watanabe et al., 2000). These and other thirst-related molecular events may be triggered by expression of the early oncogene c-fos (De Luca et al., 2002).

Circumventricular cells detect the concentration of ADH, angiotensin II (Sewards and Sewards, 2000), and relaxin (Sunn et al., 2002). This means that peripheral events, such as diminished renal perfusion, are linked to the thirst centers of the brain.

The renal peptide angiotensin II provides another important link between the kidneys and thirst perception in the brain (Sewards and Sewards, 2000). Angiotensin II is secreted by iuxtaglomerular cells of the renal afferent arterioles in response to low volume and sympathetic stimulation. Doubts have been expressed, however, whether angiotensin II acts directly on the brain or whether it acts through its effect on sodium balance (Gordon et al., 1997). Evidence from direct stimulation experiments in mice support the concept that angiotensin II induces thirst by acting directly on brain structures (Thunhorst and Johnson, 2001). Adaptation to high salt intake (4.6 g sodium/day) greatly raises the plasma sodium threshold at which thirst is perceived, presumably through a decrease of baseline angiotensin II release (Gordon et al., 1997).

The peptide hormone relaxin is secreted by the corpus luteum during pregnancy. Relaxin induces water drinking through direct action on the brain. Brain regions that are known to be related to thirst perception respond to relaxin stimulation with typical molecular events, such as c-fos expression (Sunn et al., 2002).

COMPLEX CENTRAL NERVOUS SYSTEM INPUT

Quenching of thirst is a pleasurable experience for most people. A measurable change in excitability underscores the role of the frontocortical brain region in thirst perception (McKinley et al., 2006).

It is not rare that mental disorders are associated with inappropriate water intake. Both insufficient and excessive water intake can occur. Excessive water intake (up to 20 l/day) is most often attributable to schizophrenia (Assouly-Besse et al., 1996). Hypodipsic hypernatremia (elevated plasma sodium concentration with a lack of urge to drink) may be due to brain injury (Nguyen et al., 2001), viral destruction (Keuneke et al., 1999), or vascular damage (Schaad et al., 1979).

Thirst perception is further modulated by input from cortical areas that respond to mood and social stimuli (Mann et al., 1987). Drinking is often a social activity and as much subject to habit as to conscious decisions and involuntary drives.

REFERENCES

Assouly-Besse, F., Seletti, B., Lamarque, I., Elghozi, D., Petitjean, F., 1996. Le syndrome "polydipsie, hypo-natremie intermittence et psychoses": diagnostic et conduite therapeutique á propos d'un cas. Ann. Med. Psychol. 154, 259–263.

Bourque, C.W., Oliet, S.H., 1997. Osmoreceptors in the central nervous system. Annu. Rev. Physiol. 59, 601–619.

Caston-Balderrama, A., Nijland, M.J., McDonald, T.J., Ross, M.G., 2001. Intact osmoregulatory centers in the preterm ovine fetus: Fos induction after an osmotic challenge. Am. J. Physiol. Heart Circ. Physiol. 281, H2626–H2635.

Changizi, M.A., Hall, W.G., 2001. Thirst modulates a perception. Perception 30, 1489–1497.

De Luca Jr., L.A., Xu, Z., Schoorlemmer, G.H.M., Thunhorst, R.E., Beltz, T.G., Menani, J.V., et al., 2002. Water deprivation-induced sodium appetite: humoral and cardiovascular mediators and immediate early genes. Am. J. Physiol. Regul. Integr. Comp. Physiol. 282, R552–R559.

Denton, D.A., McKinley, M.J., Weisinger, R.S., 1996. Hypothalamic integration of body fluid regulation. Proc. Natl. Acad. Sci. USA 93, 7397–7404.

Gordon, M.S., Majzoub, J.A., Williams, G.H., Gordon, M.B., 1997. Sodium balance modulates thirst in normal man. Endocrinol. Res. 23, 377–392.

Hallschmid, M., Molle, M., Wagner, U., Fehm, H.E., Born, J., 2001. Drinking related direct current positive potential shift in the human EEG depends on thirst. Neurosci. Lett. 311, 173–176.

Keuneke, C., Anders, H.J., Schlöndorff, D., 1999. Adipsic hypernatremia in two patients with AIDS and cytomegalovirus encephalitis. Am. J. Kidney Dis. 33, 379–382.

Mann, J.F., Johnson, A.K., Ganten, D., Ritz, E., 1987. Thirst and the renin–angiotensin system. Kidney Int. 21, S27–S34.

McKinley, M.J., Denton, D.A., Oldfield, B.J., De Oliveira, L.B., Mathai, M.L., 2006. Water intake and the neural correlates of the consciousness of thirst. Semin. Nephrol. 26, 249–257.

Nguyen, B.N., Yablon, S.A., Chert, C.Y., 2001. Hypodispic hypernatremia and diabetes insipidus following anterior communicating artery aneurysm clipping: diagnostic and therapeutic challenges in the amnestic rehabilitation patient. Brain Inj. 15, 975–980.

Robertson, G.L., Aycinena, E., Zerbe, R.L., 1982. Neurogenic disorders of osmoregulation. Am. J. Med. 72, 339–353.

Schaad, U., Vassella, E., Zuppinger, K., Oetliker, O., 1979. Hypodipsia–hypernatremia syndrome. Helv. Paed. Acta 34, 63–76.

Sewards, T.V., Sewards, M.A., 2000. The awareness of thirst: proposed neural correlates. Conscious. Cogn. 9, 463–487.

Smit, H.J., Rogers, P.J., 2000. Effects of low doses of caffeine on cognitive performance, mood and thirst in low and higher caffeine consumers. Psychopharmacology 152, 167–173.

Stachenfeld, N.S., DiPietro, L., Nadel, E.R., Mack, G.W., 1997. Mechanism of attenuated thirst in aging: role of central volume receptors. Am. J. Physiol. 272, R148–R157.

Stricker, E.M., Hoffmann, M.L., 2007. Presystemic signals in the control of thirst, salt appetite, and vasopressin secretion. Physiol. Behav. 91, 404–412.

Stricker, E.M., Sved, A.E., 2000. Thirst. Nutrition 16, 821–826.

Sunn, N., Egli, M., Burazin, T.C., Burns, E., Colvill, L., Davern, E., et al., 2002. Circulating relaxin acts on subfornical organ neurons to stimulate water drinking in the rat. Proc. Natl. Acad. Sci. USA 99, 1701–1706.

Thompson, C.J., Burd, J.M., Baylis, P.H., 1987. Acute suppression of plasma vasopressin and thirst after drinking in hypernatremic humans. Am. J. Physiol. 252, R1138–R1142.

Thunhorst, R.L., Johnson, A.K., 2001. Effects of hypotension and fluid depletion on central angiotensin-induced thirst and salt appetite. Am. J. Physiol. Reg. Integr. Comp. Physiol. 281, R1726–R1733.

Watanabe, E., Fujikawa, A., Matsunaga, H., Yasoshima, Y., Sako, N., Yamamoto, T., et al., 2000. Nav2/NaG channel is involved in control of salt-intake behavior in the CNS. J. Neurosci. 20, 7743–7751.

ABSORPTION, TRANSPORT, AND RETENTION

3

CHAPTER OUTLINE

Digestion and Absorption.. 37
Microbiome 61
Renal Processing.. 69
The Blood–Brain Barrier 81
Materno-Fetal Nutrient Transport 87

DIGESTION AND ABSORPTION

With the exception of inhaled oxygen, all of the atoms and molecules for building, maintaining, and powering the body normally have to be taken in by mouth. Food and beverages then have to be processed in the digestive tract to become useful. The various organs in the digestive tract help to break down and modify ingested food constituents and help with their transfer to other organs and tissues. It is important to recognize that many components support function and health without ever being absorbed. They may provide bulk for a steady flow of residual matter through the intestines and feed the commensal microbes that keep us healthy. The intestinal tract also is an important organ for removing metabolic waste products and excess nutrients that come from other parts of the body.

The absorbed compounds then have to get to a wide range of tissues and organs, which involves a wide range of transport mechanisms and anatomical arrangements. Transfer of molecules into and out of the brain and the materno-fetal exchange of nutrients and waste products involve particularly complex processes. On the other side of nutrient homeostasis, removal of metabolic end products and undesirable excess is as important as intake. Harmful substances and metabolites would accumulate without selective excretion and eventually make life impossible. The kidneys are critical for that function, but bile secretion and a number of additional minor functions contribute significantly as well. Most of the underlying processes are highly selective and tightly regulated, ensuring the right balance between elimination and retention.

Note: The term *secretion* usually refers to events that move substances into glandular ducts, the gastrointestinal tract, other compartments, or fluid streams. The term *excretion* indicates that substances are eliminated from the body, such as with feces or urine.

ABBREVIATIONS

CCK	cholecystokinin
FAD	flavin adenine dinucleotide
FMN	flavin mononucleotide
GLUT1	glucose transporter 1 (SLC2A1)
GLUT2	glucose transporter 2 (SLC2A2)
GLUT5	glucose transporter 5 (SLC2A5)
NHE3	sodium/hydrogen exchanger 3 (SLC9A3)
PAT1	putative anion ion transporter 1 (SLC26A6)
SGLT1	sodium-glucose cotransporter 1 (SLC5A1)

THE GASTROINTESTINAL TRACT

The various segments of the gastrointestinal tract cooperate to acquire and modify food, extract nutrients from it, and prepare the residue for excretion. This makes it clear that the organs assisting in digestion, such as pancreas and liver, are as important as the intestines.

Oral cavity and esophagus glands

Solid foods are chewed, which increases the surface area for digestion and facilitates mixing with saliva. Several glands (i.e., sublingual, submandibular, parotid, von Ebner's, and other glands) emptying into the oral cavity normally produce about 1500 ml of saliva per day. Salivation is almost exclusively under neural control and can be stimulated by odors, tastes, and tactile and chemical manipulation. Images of foods can also stimulate salivation very potently (Drummond, 1995).

Saliva is a thin secretory fluid that contains minerals, enzymes, antibacterial compounds (such as thiocyanate, hydrogen peroxide, and secretory immunoglobulin A), and numerous other substances (Table 3.1). Actual concentrations strongly depend on the flow rate (Neyraud et al., 2012).

Tannins bind salivary proline-rich proteins and thereby reduce viscosity, increase lubricant characteristics of saliva, and convey the perception of astringency. Tannin-rich beverages (such as tea, coffee, and red wine) reduce adhesion of food particles to the oral mucosa, allowing their rapid oral clearance (Prinz and Lucas, 2000).

Stomach

Anatomical structure: The luminal surface of the stomach folds into deep glandular recesses. The mucosa consists mainly of parietal (oxyntic) cells (secretion of hydrochloric acid and intrinsic factor), chief (peptic, zymogenic) cells (secretion of enzymes), and mucus-secreting cells. The neck area near the opening of the gastric gland ducts contains stem cells that are the source for replacement cells. The much smaller number of interspersed enteroendocrine cells (Bordi et al., 2000) includes enterochromaffine-like (ECL) cells (histamine production), D cells (somatostatin production), enterochromaffine (EC) cells (serotonin production), A cells (glucagon), G cells (gastrin production), and cells that release motilin-related peptide.

Secretory function: The stomach adds hydrochloric acid, digestive enzymes, and specific binding proteins (e.g., intrinsic factor for the binding of vitamin B12) to ingested food (Table 3.2). About 1000–3000 ml

Table 3.1 Typical Composition of Saliva

Volume 1500 ml	pH 5–7
Sodium 2–21 mmol/l	Chloride 5–40 mmol/l
Potassium 10–36 mmol/l	Bicarbonate 2–13 mmol/l
Calcium 1.2–2.8 mmol/l	Phosphate 1.4–39 mmol/l
Magnesium 0.08–0.5 mmol/l	Thiocyanate 0.4–1.2 mmol/l
Cortisol	Hydrogen peroxide
Immunoglobulin A	Lactoferrin
Alpha-amylase (EC3.2.1.1)	Lingual lipase (EC3.1.1.3)
Salivary acid phosphatases A + B (EC3.1.3.2)	Lysozyme (EC3.2.1.17)
N-Acetylmuramyl-L-alanine amidase (EC3.5.1.28)	Salivary lactoperoxidase (EC1.11.1.7)
NAD(P)H dehydrogenase-quinone (EC1.6.99.2)	Superoxide dismutase (EC1.15.1.1)
Glutathione transferases α, μ, π (EC2.5.1.I 8)	Tissue kallikrein (EC3.4.21.35)
Class 3 aldehyde dehydrogenase (EC1.2.1.3)	
Glucose-6-phosphate isomerase (EC5.3.1.9)	

Table 3.2 Gastric Secretions for the Support of Digestion and Absorption

Hydrochloric acid	0.01–0.1 M
Pepsinogen A	EC3.4.23.3
Gastricsin (pepsinogen C)	EC3.4.23.3
Gastric lipase	EC3.1.1.3
Carboxyl ester hydrolase	EC3.1.1.1
Pancreatic lipase-related protein 2	EC3.1.1.26
Mucus	
Intrinsic factor	
Haptocorrins (R proteins)	
Gastrin	
Cholecystokinin	
Somatostatin	
Histamine	
Endorphins	
Serotonin	

is secreted per day, containing about 0.1 mol/l hydrochloric acid (pH 1–1.5). The presence of food in the stomach, and even the sight or thought of appetizing foods, get the juices flowing. Low intragastric pH and fullness of the terminal ileum and colon inhibit secretion. The latter effect, often called the *ileal brake*, is at least partially mediated by peptide YY (PYY; Yang, 2002). The acid-stimulating effect of several hormones, including gastrin and cholecystokinin (CCK), is mediated by histamine release from ECL cells. Stimulation of histamine-2 (H2) receptors, gastrin receptors, cholinergic receptors (parasympathetic innervation), and the divalent ion-sensing receptor known as stomach calcium-sensing receptor (SCAR) (Geibel et al., 2001) on parietal cells stimulates gastric acid output directly. Hormones including gastrin and CCK induce histamine release by ECL cells. Stimulation of G cells by the presence of protein-rich foods and alcoholic beverages in the stomach increases gastrin production.

The hydrochloric acid inactivates potential pathogens, denatures dietary proteins, and provides the optimal pH for protein digestion by pepsin (Figure 3.1). The hydration of carbon dioxide by isoforms I and II of carbonate dehydratase (carbonic anhydrase, EC4.2.1.1, zinc-dependent) generates bicarbonate, a prolific source of protons, in parietal cells of the stomach. The protons are then pumped across the secretory canalicular membrane in exchange for potassium ions into the glandular lumen by hydrogen/potassium-exchanging ATPase (EC3.6.3.10, magnesium-dependent). The parallel export of chloride ions via the chloride channel 2 (ClC2) completes gastric acid synthesis (Sherry et al., 2001). The basolateral sodium-potassium-chloride cotransporter (NKCC1, SLC12A2) and the bicarbonate/chloride exchanger 2 (SLC4A2) provide chloride from pericapillary fluid to parietal cells. This cell type also produces intrinsic factor, whose production diminishes whenever acid output decreases.

The chief cells produce pepsinogen A (EC3.4.23.1) and gastricsin (pepsinogen C, EC3.4.23.3), which are activated when they come in contact with gastric acid. Both enzymes cleave with broad specificity and break down most proteins from small- to medium-sized peptides. Other products include gastric lipase (EC3.1.1.3) and vitamin B12-binding haptocorrins (R proteins).

Motility: The stomach retains solids for some time. Segmental contraction of its muscles (peristalsis), especially of the distal portions, contributes to the mixing and grinding of ingested foods. Some nutrients (including alcohol, molybdenum, nicotinate, and nicotinamide) can be absorbed from the stomach. Small portions of the stomach contents are propelled into the small intestine by peristalsis while the distal (antral) lumen is open. Hormonal signaling stimulated by acidity, sugars, specific amino acids, and monoglycerides in the duodenum regulates such periodic release of gastric contents (Meyer, 1994).

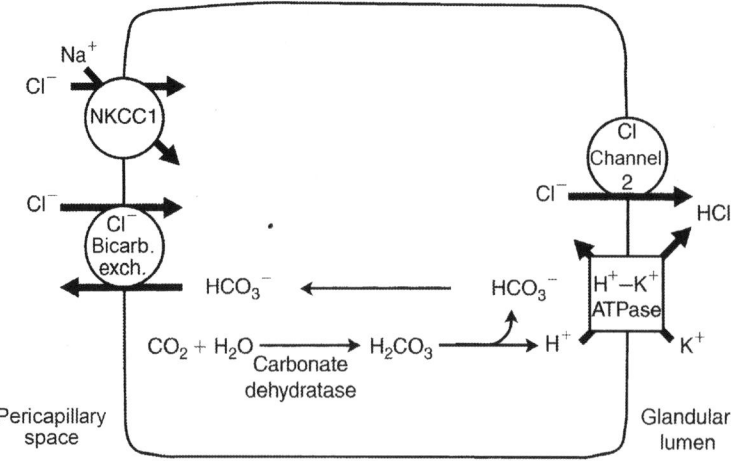

FIGURE 3.1

The production of gastric acid depends on an abundant supply of chloride.

Small intestine

Gross anatomical structure: The distal (pyloric) end of the stomach connects to a tubular organ of about 500–600 cm in length—the small intestine. The initial 30-cm-long segment is called the *duodenum*. The next segment, the *jejunum*, is approximately 200 cm long. There is no clear transition to the ileum, which is the last 300 cm of the small intestine The intestines are attached on one side to a mesenteric stem that carries blood vessels and lymph ducts.

The mesentery is covered with folds of adipose tissue that may account for a third or more of total body fat in obese people (abdominal obesity). The main function of the small intestine is the digestion and absorption of most nutrients from the food mixture coming from the stomach.

The small intestine is characterized by the numerous folds that protrude into the lumen and greatly expand the surface area available for digesting and absorbing compounds from the intestinal contents (chyme). The luminal surface is interspersed with small (1 mm diameter) accumulations of lymphatic tissue (Peyer's plaques). Another type of anatomical feature in the submucosa of the proximal duodenum is Brunner's glands. These glands produce mucin, bicarbonate, protective peptides, growth factors, and other compounds that help to protect the lining of the small intestine where it is most exposed to the corrosive effects of gastric acid, pancreatic enzymes, and bile constituents (Krause, 2000). Epithelial cells lining Brunner's glands also produce duodenase (preliminary EC3.4.21.B3), which is the critical initiator of pancreatic enzyme activation (Zamolodchikova et al., 2013).

Microanatomical features: The small intestine is composed of discernible layers throughout. The mucosa is the layer of epithelial cells that form the luminal surface and are directly responsible for local digestion and absorption. The major cell types of the small intestinal mucosa are enterocytes (95%), goblet cells, enteroendocrine cells, Paneth cells, cup cells, tuft cells, and microfold (M) cells (Dalal and Radhi, 2013). The luminal surface of the small intestine is arranged in a folded pattern. At the bottom of the folds are the crypts of Lieberkühn, and the tips of the folds are the villi. New (nascent) cells continuously arise from the dividing multipotent stem cells at the base of the crypts. A new cell rises toward the tip of its villus and is shed after 3–5 days. There are as many as 15 crypts, with several hundred cells surrounding each villus, which in turn consists of about 7000 cells (Potten and Loeffler, 1990; Dalal and Radhi, 2013).

It is of the utmost importance to recognize that functions and events that have been loosely attributed to the intestinal mucosa in general may in fact occur exclusively in specific intestinal segments, cell types, villus locations, and degrees of cell maturity. Even modest changes in a particular cell's population numbers, maturation speed, or expression pattern may profoundly affect an individual's response to a food or nutrient. Many earlier assumptions have to be revisited and research on the regulation of intestinal cells is actively proceeding. Particularly important are new concepts about the close interaction of the community of intestinal organisms (gut microbiota) on mucosal cell proliferation and function (Kurashima et al., 2013).

Small intestinal enterocytes are the major site of terminal digestion, uptake, processing, and transport into circulation for most nutrients. The luminal (apical) side of individual cells is covered with numerous protrusions that greatly increase the surface area. This luminal surface is called the *brush border membrane* because of its bristlelike appearance. The individual protrusions are called *microvilli*. These ultramicroscopic structures should not be confused with the villi, which are much larger structures comprising several thousand cells and are barely visible with the naked eye.

The mucosa also contains a significant number of goblet cells, which are interspersed between the enterocytes. These cells secrete mucins and other proteins for facilitating the smooth passage of intestinal contents. They release additional factors that block entry of pathogens (barrier function).

The enteroendocrine cells of the small intestine include at least 10 distinct types, including secretin-producing S cells, cholecystokinin-producing I cells, L cells (PYY, glucagon, glucagon-like peptides GLP-1 and GLP-2), D cells (somatostatin), K cells (glucose-dependent insulinotropic peptide), and EC cells (serotonin).

Paneth cells are located at the bottom of the crypts. Paneth cell secretions are critical to maintaining gut wall integrity and serve as the first line of antimicrobial defense in the small intestine (Zamolodchikova et al., 2013). They serve these functions partly by secreting lysozyme (EC3.2.1.17), phospholipase A2 (EC3.1.1.4), cathepsin G, and other peptides with antimicrobial, trophic, and paracrine properties into the intestinal lumen. The secretory granules of the Paneth cells contain alpha-defensins (cryptidins) and other peptides packaged together with matrilysin (EC3.4.24.23 requires zinc and calcium), a matrix metalloproteinase that cleaves and thereby activates the propeptides upon release of the granules into the crypts. Paneth cells also produce synaptophysin, a neuroendocrine peptide (Rubio, 2012) and release guanylin (guanylate cyclase activating peptide-1), an intestinal natriuretic peptide (Sindic, 2013) that responds to salt. Guanylin is secreted into the intestinal lumen, where it inhibits sodium absorption and water secretion by slowing sodium/hydrogen exchange and increases chloride and bicarbonate secretion. The peptide and its analog, uroguanylin, are also secreted in the blood and act on the kidneys, where they increase sodium and potassium secretion.

Cup cells are a cell type in the ileum distinguished from enterocytes by a shorter brush border membrane, the expression of vimentine, lower alkaline phosphatase activity, and other characteristics. Cup cells make up more than 1% of the cells in the luminal layer of the ileal mucosa (Ramirez and Gebert, 2003). It has been surmised that the surface glycoproteins on this cell type may serve to capture and process intestinal bacteria for initiation of an immune response (Ramirez and Gebert, 2003). These authors also suggested that cup cells may represent the cell type that expresses the cystic fibrosis transmembrane conductance regulator (CFTR), which might be important for the typical immune pathology in people with cystic fibrosis.

Tuft cells structurally resemble taste cells (Gerbe et al., 2012). They express the TAS1R1–TAS1R3 dimer, which in taste cells responds to amino acids (particularly glutamate) and nucleotides (inosine monophosphate and guanosine monophosphate). These cells also express alpha-gustducin and other proteins that are typically involved in transduction of taste signals. In the aggregate, the presence of these taste-related elements strongly suggests that tuft cells have chemosensory functions, but significant uncertainty remains. They also appear to secrete bicarbonate ions in response to acid exposure.

Microfold or membraneous cells (M cells) constitute the interface of the mucosa with the intestinal lymphoid follicles, the so-called Peyer's patches (Neutra, 1998). They take up molecules and particles from the intestine and transfer them into the extracellular space of the lymphoid follicles, where lymphocytes can interact with these potential pathogens and stage an immune response.

Counteracting stomach acidity: The food chyme coming from the stomach into the small intestine is strongly acidic. Secretions from the intestinal wall and pancreas rapidly raise the pH to 6. Much of the neutralizing action is attributable to the putative anion transporter (PAT1, SLC26A6) that moves bicarbonate into the proximal intestinal lumen (Wang et al., 2002). Since chloride ions are taken up in return, this exchanger (which actually is not at all putative—the name just stuck after it had been used for a number of years, even after the mechanism was confirmed) neutralizes the acidity and recovers chloride at the same time and in a tightly coordinated fashion. The sodium/hydrogen exchangers 2 (NHE2, SLC9A2) and 3 (NHE3, SLC9A3) also contribute to bicarbonate transport and pH adjustment (Repishti et al., 2001). The duodenal mucosa typically produces 1–21 of fluid with about 250 mmol

Table 3.3 Typical Composition of Bile

Volume	500–1000 ml/day	pH 7.3–7.7
Sodium		Chloride
Potassium		Bicarbonate
Calcium		Copper
Phospholipids		Cholesterol
Bile acids		Bilirubin
Vitamin B12		Corrinoids
Ceruloplasmin		Haptoglobin
Hemopexin		Mucin glycoproteins
Alkaline phosphatase		Lactate dehydrogenase
Lysosomal enzymes		Alcohol dehydrogenase

bicarbonate (Gullo et al., 1987). A firmly adhering layer of mucus gel from the goblet cells provides additional protection against the corrosive effects of stomach acidity and food contents (Atuma et al., 2001). This protective layer is present throughout the intestines.

Bile: Liver cells secrete water, electrolytes, metabolites, and xenobiotic compounds (Table 3.3) into the biliary ducts (Reshetnyak, 2013). This pathway is of particular importance for the excretion of all kinds of lipophilic and otherwise poorly water-soluble metabolites. The ducts empty bile into larger branches and finally through the bile duct into the small intestine just a few centimeters below the pylorus. A duct from the pancreas joins the bile duct about 6–8 cm before the common duct finally reaches the small intestine. This last duct segment carries both bile and pancreatic secretions. A ring of smooth muscles (sphincter of Oddi) encircles the common duct just downstream of the confluences of the biliary and pancreatic segments and thereby controls the rate of flow from both sources. The liver produces about 0.5–1 l of bile per day.

Bile is an alkaline liquid that is yellow, green, brown, or nearly black (Sutor and Wilkie, 1976). Bile that has been stored in the gallbladder is significantly more concentrated than freshly secreted bile. The major constituents are water, phospholipids (usually around 4%), bile acids (around 12%), and variable amounts of cholesterol. Bile also contains electrolytes, minerals, and trace elements. Some nutrients, (including vitamin B12, choline, copper, manganese, and molybdenum), metabolites (such as bilirubin), and xenobiotics (such as medications, industrial contaminants, some heavy metals, and phytochemicals) are secreted with bile and thus are subject to enterohepatic cycling. Bile also contains enzymes (alkaline phosphatase) and numerous other proteins and peptides (Zhou et al., 2005), whose functional significance is not understood in most instances.

Pancreatic secretions: The *pancreas* is a glandular organ that lies adjacent to the curvature of the duodenum and below the stomach. The secretory ducts join with the bile duct shortly before leading into the duodenum. The opening of the joint duct into the duodenal lumen is called the *papilla of Vater* (papilla Vateri). An additional smaller duct carries only pancreatic secretions. The pancreas is both an endocrine organ (output into the bloodstream) and an exocrine organ (output into the intestines). The endocrine products include insulin, glucagon, and other hormones with direct bearing on nutrient utilization and disposition. The exocrine secretions contain, among other minor ingredients, sodium, bicarbonate, chloride, and a diverse set of digestive enzymes. Trypsin (EC3.4.21.4), chymotrypsin B

Table 3.4 Typical Composition of Pancreatic Secretions

Volume 1000–3000 ml	pH 7.2–7.4
120 mmol/l bicarbonate	140 mmol/l sodium
70 mmol/l chloride	5 mmol/l potassium
	2 mmol/l calcium
Trypsin I–III (EC3.4.21.4)	Pro-chymotrypsin C (EC3.4.21.2)
Pro-alpha-chymotrypsin (EC3.4.21.1)	Tissue kallikrein (EC3.4.21.35)
Pro-carboxypeptidases A1/2 (EC3.4.2.1)	Pro-carboxypeptidase B (EC3.4.17.2)
Endopeptidases EL3A/B (EC3.4.21.70)	Pro-elastases IIA/B (EC3.4.21.71)
Pancreatic prolipase (EC3.1.1.3)	Colipase
Lipase-related proteins 1/2 (EC3.1.1.3)	Prophospholipase B (EC3.1.1.5)
Pro-carboxylesterlipase (EC3.1.1.13)	Deoxyribonuclease I (EC3.1.21.1)
Prophospholipase A2 (EC3.1.1.4)	Alpha-amylase (EC3.2.1.1)
Pancreatic ribonuclease (EC3.1.27.5)	

(EC3.4.21.1), chymotrypsin C (EC3.4.21.2), carboxypeptidases A1 and A2 (EC3.4.2.1), elastases IIA and IIB (EC3.4.21.71), lipase and lipase-related proteins 1 and 2 (EC3.1.1.3), carboxylester lipase (EC3.1.1.13), and phospholipase A2 (EC3.1.1.4) are secreted as proenzymes that have to be cleaved before they become active. The other enzymes are active as secreted. Bicarbonate output of the pancreas is stimulated by the peptide hormone secretin from S cells in the small intestinal wall that respond to low pH and fatty acids. The peptide hormone cholecystokinin (identical to pancreozymin) from I cells (enteroendocrine cells) increases the output of pancreatic enzymes in response to the presence of fat and protein in the small intestine (Table 3.4).

Digestion: At this point, much of the ingested food has been broken down mechanically (by chewing and by the grinding action of the stomach) to small particles, denatured by concentrated acid, and predigested by alpha-amylase, proteases, and lipases from saliva and the stomach. With the movement of small portions of this mixture into the duodenum, digestion can start in earnest.

First, a cascade of proteases must remove the lead sequences from the newly secreted pancreatic proenzymes. Duodenase (preliminary EC3.4.21.B3, Zamolodchikova et al., 2013) from duodenal (Brunner's) glands cleaves and thereby activates the brush border protease enteropeptidase (enterokinase, EC3.4.21.9). Enterokinase then activates trypsin (EC3.4.21.4), and trypsin finally activates the other proenzymes.

Carbohydrates are broken up into oligosaccharides by alpha-amylase (EC3.2.1.1), and other secreted enzymes act on their targets. A particular feature of the digestion and absorption of fat and many fat-soluble compounds is the need to first create an emulsion.

Mixing of fat with the bile acids and phospholipids in bile spontaneously forms mixed micelles. Predigestion of fat by lingual and gastric lipase generates monoglycerides, which act as additional emulsifiers for the ingested fats. Lipase action then gradually cleaves the triglycerides in the micelles and makes them available for absorption. Bile also contains alkaline sphingomyelinase (EC3.1.4.12), which cleaves ceramides (Duan and Nilsson, 1997).

Table 3.5 Enzymes at the Brush Border Membrane That Prepare Nutrient Molecules for Absorption

Leucine aminopeptidase (LAP) (EC3.4.11.1)

Membrane alanine aminopeptidase (aminopeptidase N) (EC3.4.11.2)

Aminopeptidase A (EC3.4.11.7)

X-pro aminopeptidase (aminopeptidase P) (EC3.4.11.9)

Glycylleucyl dipeptidase (EC3.4.13.18)

Dipeptidyl-peptidase IV (EC3.4.14.5)

Angiotensin I-converting enzyme (ACE) (EC3.4.15.1)

Carboxypeptidase P (EC3.4.17.16)

Glutamate carboxypeptidase II (EC3.4.17.21)

Folylpoly-γ-glutamatec arboxypeptidase (glutamatec arboxypeptidase II) (EC3.4.19.8)

Enterokinase (EC3.4.21.9)

Duodenase (EC3.4.21.B3)

Neprilysin (neutral endopeptidase) (EC3.4.24.11)

Membrane dipeptidase (EC3.4.13.19)

Meprin A (*N*-benzoyl-L-tyrosyl-*p*-aminobenzoicacid hydrolase, EC3.4.24.18)

Meprin B (EC3.4.24.63)

γ-Glutamyl transpeptidase (gamma-GT) (EC2.3.2.2)

Sucrase/alpha-glucosidase (EC3.2.1.10/EC3.2.1.48)

Lactase/phlorhizin hydrolase (EC3.2.1.108)

Trehalase (EC3.2.1.28)

Alkaline phosphatase (EC3.1.3.1)

5′-Nucleotide phosphodiesterase (EC3.1.4.1)

Phospholipase B (E 3.1.1.5)

Most of the enzymes that render the half-digested nutrient molecules fit for absorption reside on the brush border membrane of the small intestine (Table 3.5). Some of the enzymes cluster near the channels and transporters for uptake, as found with lactase and the associated sodium-glucose transporter 1 (Mizuma and Awazu, 1998).

Large intestine

The *colon* is a tubular organ of 1200–1500 cm in length that connects to the distal end of the small intestine. A valve known as the *Bauhinian valve* limits the backflow of fecal matter from the colon into the small intestine. The small portion (about 7 cm) of the colon that extends below the valve is called the *cecum;* it ends in a much narrower, worm-shaped portion, the *appendix*, which contains lymphatic tissue. The main function of the colon is the recovery of fluids and electrolytes from the intestinal contents. Some micronutrients, such as biotin, pantothenate, and vitamin K, may be absorbed from bacterial production.

In contrast to the small intestine, the mucosa of the colon is relatively smooth without villi, and the enterocytes of the colon (colonocytes) do not have microvilli. The colon contains a much larger number of mucin-producing goblet cells than the small intestine. The greater number of goblet cells means that the mucous layer covering the colonic mucosa is also much thicker (close to 1 mm). The absence of folds, crypts, and villi, as well as better lubrication, obviously facilitates the movement of feces through the colon even when the excreta become increasingly solid.

MOVEMENT OF FOOD THROUGH THE GI TRACT

Movement of food through the mouth and esophagus proceeds largely by voluntary muscle contractions assisted by reflexes, which prompt swallowing of ingested foods and prevent aspiration. Coordinated contractions (peristalsis) propel small portions of chyme (the slurry of partially digested food and digestive juices) through the pylorus into the duodenum. Segments of the small intestines of a few centimeters in length can contract in a coordinated fashion and thereby generate slow peristaltic waves. The propulsive motion is essential to push the chyme through the small intestine. In healthy subjects, food takes about 290 min after ingestion to reach the colon (Geypens et al., 1999). This interval is often called the *oro-cecal transit time*. Digestion and absorption of most bioavailable compounds in the intestinal content proceed while the chime moves through the small intestine. Fecal matter is moved through the colon by a series of concerted contractions about once a day, usually triggered by food intake. It is important to remember that a large proportion of chyme constituents, such as water and electrolytes, bile acids, lipids, and proteins, are of endogenous origin and need to be recovered. The fecal mass remains for another few hours to several days in the large intestine, where most of the remaining water and electrolytes are extracted.

ENTEROHEPATIC CIRCULATION

Some nutrients and metabolites are subject to repeated cycling between the small intestine and the liver. This happens when a compound is actively secreted into bile and then absorbed from the intestinal lumen and returned to the liver. The compounds may be changed during the cycling, either by bacterial action in the intestines or by metabolism after absorption. Such enterohepatic circulation creates a metabolic communication line between the liver, the intestines, and the microbes in the intestinal lumen (liver–gut crosstalk).

Bile has the benefit of the digestive and absorptive activities of the entire length of the small and large intestines. Recovery of water, minerals, bile acids, and phospholipids from bile is nearly complete. Additional key nutrients, as well as numerous metabolites, nonnutrient bioactives, xenobiotics, and medications, are secreted with bile, absorbed again from the intestines, and returned to the liver for secretion. Relatively common inflammatory (e.g., primary sclerosing cholangitis in patients with ulcerative colitis), anatomical (cholestasis or loss of bowel segments), hormonal (diminished release of intestinal hormones such as glucagon-like peptide and PYY), metabolic (excessive bile acid synthesis), and nutritional (pernicious anemia due to vitamin B12 deficiency) conditions cause significant health problems by disrupting or altering enterohepatic circulation of nutrients and metabolites (Islam and DiBaise, 2012; Navaneethan, 2014).

Bile acids

Anion exchange in the jejunum and sodium-mediated transport in the ileum recover most bile acids and bile acid conjugates (Amelsberg et al., 1999). The taurine that is cleaved in the terminal ileum from bile acid conjugates by bacterial enzymes is also recovered with high efficiency by the chloride-dependent taurine transporter (TAUT, SLC6A6).

The cycling of bile constituents between liver and intestines exposes bile constituents that reach the terminal ileum to bacterial modification. An important example is the intestinal conversion of newly synthesized cholic acid to the secondary bile acids (derived from a primary bile acid by bacterial action in the ileum and colon) deoxycholic acid and lithocholic acid (Kakiyama et al., 2013). Bile acid–CoA ligase (SLC27A5, EC6.2.1.7) and bile acid–CoA:amino acid N-acyltransferase (BAAT, EC2.3.1.65) in the liver work in concert to improve the water-solubility of cholic acid by linking it to glycine or taurine (Setchell et al., 2013). The bile salt export pump (BSEP/ABCB11) or the adenosine triphosphate (ATP)–powered multidrug resistance protein 2 (MRP2, ABCC2) can then push the conjugated bile acid into bile.

A small percentage of bile acid is taken from the primary bile ducts right back into hepatocytes. The apical bile salt transporter ASBT (SLC10A2) can move the bile acid across the luminal membrane of the cholangiocytes (the epithelial cells forming the walls of the bile ducts) and then with MRP3/4 or OSTα/β across the basolateral membrane into the bloodstream. The sodium-taurocholate cotransporting polypeptide (NTCP, SLC10A1) finally takes it into the hepatocytes. This phenomenon is called *cholehepatic shunting* (Anwer and Stieger, 2014).

Most bile acids, however, travel unhindered down the bile duct, possibly staying for a while in the gallbladder (if one still exists), and finally reach the lumen of the proximal small intestine. There, it mixes with the food chime and other intestinal secretions to aid the absorption of fat and fat-soluble nutrients. The apical bile salt transporter ASBT (SLC10A2) uses a sodium cotransport mechanism for bile acid uptake into the enteral cell. The ileal bile acid–binding protein (IBABP) is critical for getting bile acids from the luminal to the basolateral side of the enterocyte. The heterodimeric organic solute transporter alpha-beta (OSTα/OSTβ) then completes transfer across the basolateral membrane.

Since bile acid transport is most active in the ileum, the bacteria there may cleave some of the conjugates and convert some of the cholic acid to deoxycholic acid. Ultimately, more than 95% of the bile acids are then absorbed by passive diffusion and active transport. Both unchanged (primary) and metabolized (secondary) bile acids eventually travel through the portal vein to the liver. The sodium-taurocholate cotransporting polypeptide (NTCP/SLC10A1) or organic anion transporting polypeptide 2 (OATP2/OATP1B1) moves conjugated bile acids into the hepatocyte, and then another cycle can begin.

Bile acids bind to nuclear farnesoid X receptors (FXR) in the liver and gut, providing the signal for feedback inhibition of bile acid synthesis. Bile acids also bind to bile acid receptor 5 (TGR5). These receptors modify release of incretin hormone and fibroblast growth factor 19 (FGF19), providing the liver and other tissues with the signals that regulate cholesterol homeostasis and energy metabolism.

Since different bile acids have very different agonistic strengths, the maturation of primary to secondary bile acid has significant homeostatic consequences. The precise type and extent of bacterial processing modifies the predisposition of gaining weight, developing diabetes, cancer risk, and other health outcomes.

Vitamin B12

The importance of hepatobiliary circulation can be illustrated by the example of vitamin B12. Several micrograms of this vitamin are secreted into bile every day and then nearly completely recovered from the small intestine. The recycling of this nutrient is so efficient under normal circumstances that normal liver stores last for decades. The benefit is that it eliminates irreversibly inactivated cobalamin metabolites and potential harmful corrinoids, which are cobalamin-like compounds without vitamin

B12 activity abundantly produced by gut bacteria (Degnan et al., 2014). However, if reabsorption becomes less effective (usually due to a lack of intrinsic factor), vitamin B12 stores can be depleted within a few months.

Molybdenum

There appears to be significant enterohepatic cycling of molybdenum. Molybdenum stores in the body appear to be regulated through adjustments of the enteral loss rate during cycling. An ATP-binding cassette (ABC) transporter in bacteria has been found to mediate molybdate transport (Neubauer et al., 1999). An analog of this transporter may also operate in humans.

Phytochemicals

Many ingested and absorbed flavonoids and other plant-derived compounds are taken up by the liver, excreted into bile, and absorbed again, creating some enterohepatic circulation. Retention, and therefore the degree of cycling, often depends on the chemical properties of the ingested compound and the extent of metabolic modifications (Terao and Mukai, 2014). Prenylated flavonoids such as 8-prenylquercetin are absorbed to a greater degree than the parent compounds without the prenyl chain (in this case, quercetin), but excreted more slowly into bile (Mukai et al., 2013). Postprandial metabolic modifications of flavonoids (Chen et al., 2014) with impacts on bioavailability and enterohepatic circulation include O-methylation by catechol-O-methyltransferase (COMT, EC2.1.1.6) and conjugation to sulfate by sulfate transferases (in the liver, mainly SULT1A1, EC2.8.2.1) or to glucuronide by uridine diphosphate (UDP)–glucuronosyl-transferase (mainly the isoenzyme UGT1A1, EC2.4.1.17) (Davis and Brodbelt, 2008).

INTESTINAL ABSORPTION OF NUTRIENTS AND NONNUTRIENTS

Anatomical considerations

Lymph system: Long-chain fatty acids, sterols, the fat-soluble vitamins A, D, E, and K, carotenoids and other fat-soluble bioactives, and some fat-soluble drugs and xenobiotics are packaged by enterocytes into chylomicrons and moved into the lymph ducts of the proximal small intestines. The smaller lymph ducts drain into the thoracic duct (alimentary duct) which in turn guides its contents to the large vein (vena cava) near the heart.

Portal vein: The bulk of compounds absorbed from the intestines are moved into the extracellular space on the basolateral side of the enterocytes and from there into the capillaries that eventually drain into the hepatic portal vein. The portal vein branches in the liver into smaller veins and from there into the sinusoids. This arrangement gives hepatocytes, Kupffer cells, stellate cells, and other liver cells privileged access to the freshly absorbed nutrients and other compounds. The resulting first-pass effect means that a large portion of some compounds is removed from circulation before it ever reaches extrahepatic tissues, which is particularly important for extraction and storage of excess retinol by the stellate cells and the processing of gut bacteria, bacterial antigens, and endotoxins by Kupffer cells (Bilzer et al., 2006).

Molecular transport mechanisms

Paracellular diffusion: Water, some inorganic ions, and a few other very small compounds can bypass the intestinal cells altogether by traveling (in either direction) through very narrow (4–8Å) pores of the tight junctions sealing the spaces between enterocytes. The tight junctions of the proximal small intestine consist of fewer strands of sealing proteins with pore sizes of about 8 Å and are therefore more permeable than distal intestinal segments with tight junction pore sizes of only 4 Å.

Active ATP-driven transport: Several ATP-hydrolyzing complexes transport nutrients across luminal, intracellular, and basolateral membranes. Crossing intracellular membranes is often necessary to move molecules out of endosomes/lysosomes, across the inner mitochondrial membrane or into secretory compartments. There is a group of nearly 50 ABC transporters that use the energy from ATP hydrolysis to move medium-sized molecules across membranes or to adjust their shape for regulatory purposes. Several are critical for intestinal absorption, such as CFTR (ABCC7) for the regulation of chloride secretion, the ABC transporters A1 and G5/G8 for the control of cholesterol absorption efficiency, and MRP2 (ABCC2) for folate export across the basolateral membrane. Then there are highly specialized ATPases for the absorption of copper and calcium. Menkes protein (ATP7A, EC3.6.3.4) transports copper into secretory vesicles for export into the blood. The calcium-transporting ATPase Ib (plasma membrane calcium-pumping ATPase 1b, PMCA1b, EC3.6.3.8) moves calcium directly across the basolateral membrane.

The ATPases that pump sodium and potassium do the really heavy work in nutrient absorption (Figure 3.2). The sodium/potassium-exchanging ATPase (EC3.6.3.9) labors at the basolateral membrane

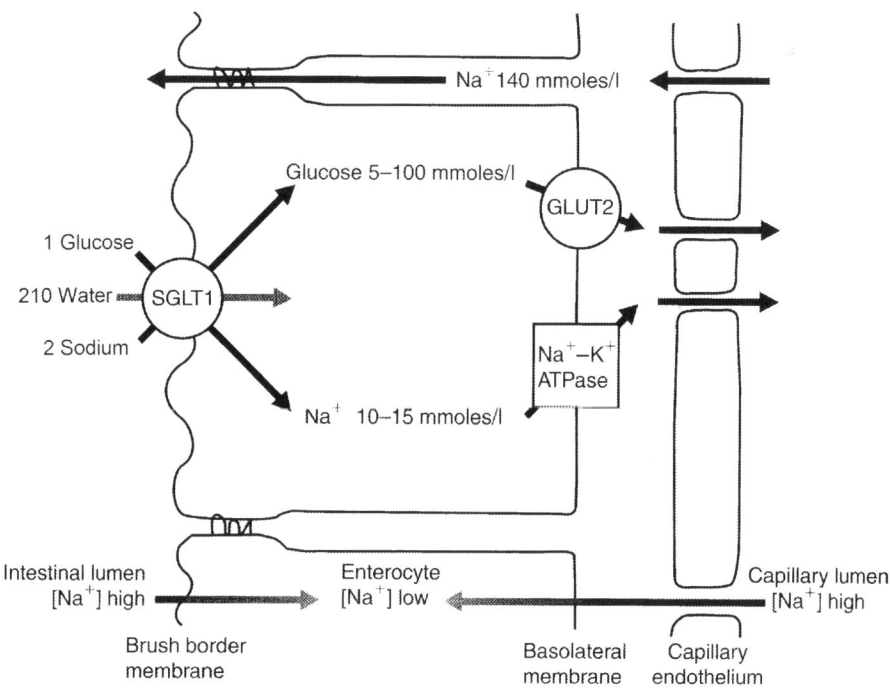

FIGURE 3.2

The sodium gradient established by sodium–potassium ATPase drives nutrient transport from the intestinal lumen into enterocytes.

of all enterocytes throughout the small and large intestines. Each ATP-hydrolyzing cycle pumps three sodium ions out of the cell into basolateral space and pulls in two potassium ions in exchange. This establishes the low intracellular sodium concentration (Zuidema et al., 1986) that is the main driving force for active nutrient transport.

Sodium cotransport: Specific transporters use the electrochemical potential by firmly coupling the movement of the nutrient ligand to that of sodium along the steep sodium gradient. Examples for sodium-driven bulk transporters are the sodium-glucose transporter 1 (SLC2A2) and the amino acid transport system B° (ASCT2, SLC1A5), which every day move several hundred grams of nutrients plus several liters of water out of the small intestinal lumen. It should be noted that similar cotransporters at the basolateral membrane allow postprandial nutrient flux from the enterocytes toward the blood capillaries, and also in the opposite direction during fasting, to supply nutrients for the enterocyte's own considerable needs.

Proton cotransport: The sodium/hydrogen exchanger 3 (NHE3, SLC9A3) at the luminal side and the sodium/hydrogen exchanger 2 (NHE2, SLC9A2) at the basolateral side move protons out of the enterocytes and establish a significant proton gradient. Hydrogen ions (protons) can then drive nutrient cotransport just like sodium ions do. Examples of proton-driven transporters are the proton/monocarboxylic acid cotransporter 1 (MCT1, SLC16A1) for lactate, pyruvate, acetate, propionate, benzoate, and nicotinate (Orsenigo et al., 1999) and the hydrogen ion/peptide cotransporter 1 (PepT1, SLC15A1).

Chloride cotransport: The taurine transporter (TAUT, SLC6A6) uses the chloride gradient in the ileum for the uptake of taurine from bile acids.

Exchangers: Some transporters function in such a way that a mass gradient pushes another type of molecule in the opposite direction. The abovementioned sodium/hydrogen exchangers, for instance, use the gradient-driven inward movement of sodium to push protons out of the cell. Another example is the putative anion transporter 1 (PAT1, SLC26A6) in the proximal small intestine, which very effectively couples the recovery of chloride with the countertransport of bicarbonate for neutralizing gastric (hydrochloric) acid. A group of membrane-anchored glycoproteins at both sides of the enterocytes uses neutral amino acids to move other amino acids in the opposite direction. This means that alanine (or another neutral amino acid consumed in bulk), whose intracellular concentration increases after a meal due to uptake via the sodium-driven transport system B° moves back into the intestinal lumen to drive cystine (CssC, oxidized cysteine) into the cell. Of course, the alanine then is taken up again via system B°.

Facilitated diffusion: Several transporters mediate the selective transfer of nutrients along their concentration gradient. Important examples are the transporter for fructose (GLUT5, SLC2A5) on the luminal side of the proximal small intestine, and the glucose transporter 2 (GLUT2, SLC2A2) at the basolateral side. GLUT2 also serves nutritive functions for enterocytes as mentioned previously for the sodium-driven amino acid transporters. Another example is TAT1 (SLC16A10) at the basolateral membrane. It colocalizes with exchangers and recycles aromatic amino acids to help the efflux of other amino acids through the exchangers (Ramadan et al., 2007).

Intracellular transformation: Phosphorylation or other chemical changes commonly take place after the uptake of nutrients to prevent them from returning into the intestinal lumen by the way they entered. The absorption of vitamin B6 provides an example for such "trapping." The pyridoxin carrier accepts only free pyridoxin, and conversion to pyridoxin phosphate keeps the equilibrium always in favor of the influx direction.

Nutrients are also metabolized to provide energy and material for the needs of the rapidly proliferating intestinal cells. Glutamine, for example, is a major energy fuel as well as a nitrogen source for the intestines. Some glutamine in enterocytes is also used for the synthesis of ornithine and citrulline, which is exported for use in the urea cycle of liver and kidneys and for the arginine synthesis. Thus, only some of the absorbed glutamine reaches blood circulation.

Unmediated transcellular diffusion: Very few compounds can directly cross the formidable barrier of the bilayer membranes on the apical and basolateral sides of the cells lining the intestinal lumen. These are mainly small lipophilic compounds, urea, and gases (i.e., hydrogen, methane, and hydrogen sulfide).

Transcytosis: The brush border membrane folds in and connects to vesicular structures of the endosomal compartment that also provides a secretory pathway on the basolateral side. This endosomal mechanism can transport intestinal peptide hormones as well as small amounts of ingested proteins and peptides (Ziv and Bendayan, 2000). Transcytosis explains how cow milk protein or gluten can evade intracellular catabolism.

Absorption of macronutrients

Proteins, peptides, and amino acids: Foods are broken down by proteases and aminopeptidases from the stomach (pepsin and gastricsin), pancreas (trypsin, alpha-chymotrypsin, chymotrypsin C, elastases IIA and IIB, endopeptidases EA and EB, and carboxypeptidases A1 and A2), and intestinal brush border. Protrypsin has to be activated in the intestinal lumen by the brush border protease enterokinase, and additional trypsin and the other pancreatic proenzymes by trypsin. About half of the digested proteins are taken up by small intestinal enterocytes as free amino acids via specific transporters. The other half is taken up as dipeptides and tripeptides via the hydrogen ion/peptide cotransporters 1 (PepT1, SLC15A1) and 2 (PepT2, SLC15A2). Hydrolysis of these small peptides is completed inside the enterocytes by various cytosolic proteases and peptidases. Transfer of the intracellular amino acids into portal blood is mediated by another set of specific transporters. Minute amounts of larger peptides and proteins may occasionally reach the bloodstream, through transcytotic transport via endosomes (Ziv and Bendayan, 2000), by paracellular passage or transport with immune cells.

Carbohydrates: Digestion of starches and related carbohydrates, initiated by salivary alpha-amylase in the mouth, continues as pancreatic alpha-amylase is added. Four oligosaccharidase complexes, maltase/glucoamylase (EC3.2.1.20/EC3.2.1.3), sucrase/isomaltase (EC3.2.1.48/EC3.2.1.10), lactase/phlorhizin hydrolase (EC3.2.1.108/EC3.2.1.62), and trehalase (EC3.2.1.28), located on the small intestinal brush border, complete carbohydrate digestion (Murray et al., 2000). The resulting monosaccharides are taken up into enterocytes, mainly in the jejunum. D-Glucose and D-galactose are taken up mainly through the sodium-glucose cotransporter 1 (SGLT1, SLC5A1), which transfers 2 sodium and more than 200 water molecules along with each sugar molecule. A specific fructose transporter may exist in the apical enterocyte membrane, but it has not been characterized yet. Glucose and galactose are transported out of the enterocytes into portal blood mainly via the glucose transporter 2 (GLUT2, SLC2A2) and to a lesser extent the uniporter SLC50A1 (Wright, 2013), fructose via GLUT5 (SLC2A5) and SLC5A9 (Tazawa et al., 2005). Other monosaccharides and sugar alcohols may be absorbed via the paracellular pathway following solvent drag.

Lipids: Diglycerides and triglycerides are cleaved by enzymes from the mouth (lingual lipase, EC3.1.1.3), stomach (gastric lipase, EC3.1.1.3), pancreas (pancreatic lipase, EC3.1.1.3, pancreatic lipase-related proteins 1 and 2, carboxylester lipase, EC3.1.1.1), and the small intestine (glycerol-ester hydrolase). These lipases differ greatly in their pH optima and cofactor requirements, reflecting the respective environments where they normally are active. Gastric lipase may be responsible for as much as 10–20% of fat digestion (Carriere et al., 1993), and be particularly important for the generation of monoglycerides and diglycerides needed for micelle formation. Pancreatic lipase acts optimally at alkaline pH on bile acid-coated micelles and requires another protein, colipase, as a cofactor to penetrate the bile acid layer of the micelle. The precursor form, procolipase, has to be converted in the small intestinal lumen by trypsin to become active.

As most of the other pancreatic proenzymes, prophospholipase A2 (EC3.1.1.4) is activated in the small intestinal lumen through trypsin (EC3.4.21.4) cleavage. This enzyme removes the middle fatty acid from phospholipids. The resulting lysophospholipid is hydrolyzed further by pancreatic carboxylesterase (EC3.1.1.1), as are cholesteryl-, tocopheryl-, retinyl-, and cholecalciferyl esters.

The free fatty acids resulting from any of these actions are taken up, mainly from the jejunum and proximal ileum, into enterocytes, where they are esterified again as triglycerides, also as cholesteryl esters and phospholipids. Most fatty acids, especially those with 14 or more carbons, are assembled in the enterocytes as part of chylomicrons that are then secreted into intestinal lymph. Cholesterol, the fat-soluble vitamins A, D, E, and K, carotenoids, ubiquinones, and various other fat-soluble food constituents are brought into circulation through the same route. Understanding of the exact mechanism of uptake for cholesterol and other neutral sterols is still incomplete. A saturable transport protein is likely to facilitate the initial transfer from mixed micelles into the brush border membrane (Detmers et al., 2000).

Short- and medium-chain fatty acids (up to a carbon-chain length of 12) are mainly transferred into portal blood as free acids. Significant quantities of short-chain fatty acids (mainly acetate, propionate, and butyrate) are produced during the bacterial fermentation of undigested carbohydrates in the terminal ileum and colon. These monocarboxylic acids are taken up (at least partly) into colonocytes through proton-linked monocarboxylate transporters (mainly MCT1 and SLC16A1, but other MCTs may also contribute), and secreted again across the basolateral membrane via an anion.

Absorption of water-soluble vitamins

Ascorbate: Ascorbic acid (vitamin C) is oxidized in the small intestinal lumen (possibly by ceruloplasmin) and taken up as dehydroascorbate via GLUT1 (SLC2A1). Once inside the cell, dehydroascorbate is reduced again to ascorbate and transported into portal blood by the sodium-dependent ascorbate transporter 1 (SLC23A1).

Thiamin: Phosphorylated forms are cleaved by brush border alkaline phosphatase (EC3.1.3.1). The free thiamin can be taken up via thiamin transporter 2 (THTR2, SLC19A3), mainly from the proximal small intestine, much less from the colon. Transport across the basolateral membrane into portal blood uses the thiamin transporter 1 (ThTr1, SLC19A2). Significant absorption from the colon has been demonstrated including via a TPP-specific transporter (Said, 2013).

Riboflavin: About 75% is absorbed from the jejunum with the ingestion of a large dose (20 mg). Absorption efficiency declines steeply with higher intakes. Flavin adenine dinucleotide (FAD) is cleaved by 5′-nucleotide phosphodiesterase (EC3.1.4.1), flavin mononucleotide (FMN) by alkaline phosphatase (EC3.1.3.1) from the jejunal brush border. Riboflavin is released from FMN through unknown means (Byrd et al., 1985). Plant-derived riboflavin glycosides are cleaved by as-yet-uncharacterized enzymes (possibly lactase). Export into blood is mainly after conversion to FMN and FAD. Riboflavin is absorbed mainly from the jejunum, and only to a much lesser degree from the large intestine (Said et al., 2005). Uptake proceeds through the proton cotransporter RFVT3 (SLC52A3). Another two transporters, RFVT1 (SLC52A1) and RFVT2 (SLC52A2), mediate efflux across the basolateral membrane through an as-yet-unexplored mechanism (Yonezawa and Inui, 2013).

Vitamin B6: After digestion of phosphorylated species by alkaline phosphatase (EC3.1.3.1) and other phosphatases, and of glycosylated species by mucosal pyridoxinebeta-D-glucoside hydrolase, pyridoxine enters duodenal and jejunal enterocytes by nonsaturable diffusion through unidentified carriers. B6 is trapped metabolically by phosphorylation in the cytosol (pyridoxal kinase, EC2.7.1.35). Some of the pyridoxine glycosides may enter via sodium-glucose cotransporter 1 (SLC5A1) and are then cleaved by cytosolic pyridoxine-beta-D-glucoside hydrolase. Most B6 is converted to pyridoxal and appears to cross the basolateral membrane without a phosphate group attached; the mechanism of this transfer is still uncertain.

Niacin: Absorption from the stomach and proximal small intestine is nearly complete up to doses in the gram range. Nucleotide pyrophosphatase (EC3.6.1.9) cleaves the nicotinamide-containing nucleotides (NAD and NADP). Bilitranslocase in the stomach, and anion antiporter AE2 and the proton cotransporter MCT1 (SLC16A1) in the small intestine take up nicotinic acid across the apical membrane of jejunum (Takanaga et al., 1996). Uptake of nicotinamide is not well understood. Export into the blood is mainly as nicotinamide.

Folate: Most (85%) of a moderate dose of synthetic folate is absorbed, but only about half of the folate comes from foods. The difference lies in the poly-gammaglutamyl side chain of food folate, which must first be shortened to one glutamyl residue by the zinc-dependent brush border enzyme pteroylpoly-gamma-glutamate carboxypeptidase (EC3.4.17.21). Mono-gammaglutamylfolate is taken up through a high-affinity folate transporter (reduced folate carrier 1, SLC19A1) and moved into portal blood by an as-yet-unknown mechanism. The ATP-driven ABC transporter MRP2 (ABCC2) and related proteins may provide the pathways for export into the blood (Wang et al., 2001).

Vitamin B12: Healthy people absorb from the distal small intestine about half of the vitamin B12 in foods, and a greater proportion of synthetic vitamin B12. Efficient absorption of dietary (but not synthetic) vitamin B12 requires proteolytic digestion of binding proteins and the presence of intrinsic factor from chief cells of the stomach. The intrinsic factor receptor (cubilin) takes up the vitamin B12-intrinsic factor complex from the ileal lumen. Inhibition of gastric acid secretion with H2-inhibitors will also decrease intrinsic factor secretion and potentially vitamin B12 absorption.

Pantothenate, biotin, lipoic acid: The sodium-dependent multivitamin transporter (SLC5A6) mediates uptake of pantothenic acid, biotin, and lipoic acid into enterocytes of both the small and large intestine.

Absorption of fat-soluble vitamins

Vitamin A: Absorption of retinol, retinylesters, and retinoic acid is fairly complete (70–90%), whereas less than 3% of the vitamin A–forming carotenoids are absorbed. Both types are fat-soluble, have to be incorporated into mixed micelles, and depend both on intake of some fat and on normal fat digestion. Retinylesters have to be cleaved by lipase (EC3.1.1.3), sterol esterase (EC3.1.1.13) from the pancreas, or an unknown brush border esterase. Cooking, grinding, and protease digestion are essential to release carotenoids from their enclosed granules in plant foods. Without such pretreatment, absorption is minimal. After uptake into enterocytes, retinol tends to be esterified. A small percentage (around 15%) of absorbed β-carotene is cleaved by β-carotene 15,15'-dioxygenase (EC1.14.99.36, requires iron and bile acids). Cleavage of β-carotene generates two retinal molecules, cleavage of alpha-carotene, cryptoxanthin, and a few other pro–vitamin A carotenoids yields one retinal. Eccentric cleavage generates very small amounts of potentially important apocarotenals. Retinol, retinylesters, and carotenoids are exported with chylomicrons.

Vitamin D: Absorption is nearly complete so long as some fat is ingested with the vitamin and fat digestion functions normally. Vitamin D is taken up with mixed micelles and brought into circulation with chylomicrons.

Vitamin E: All dietary forms of vitamin E are completely absorbed with mixed micelles, so long as fat is absorbed. Export from small intestine into lymph uses chylomicrons.

Vitamin K: The phylloquinone, tightly embedded in the chloroplasts of green leafy vegetables, becomes available only if these foods are cooked or very finely ground. Absorption depends on normal fat digestion, just as with the other fat-soluble vitamins, to generate mixed micelles as a vehicle. Under optimal conditions of food preparation and fat intake, less than 20% of the vitamin K in greens is absorbed; otherwise, almost nothing is. Vitamin E or lutein from high-dosed supplements decreases vitamin K absorption. The vitamin K in oils, cheeses, fermented soy products, and synthetic additives is absorbed much better. Transport of absorbed vitamin K into blood uses chylomicrons.

Absorption of water and electrolytes

pH: Food is normally mixed with considerable amounts of hydrochloric acid in the stomach. The admixture of bicarbonate from the duodenal mucosa and pancreatic secretions abruptly brings the pH of the intestinal chyme back up to about 6 in the duodenum, and increases it gradually to 7.4 toward the end of the small intestine (Fallingborg, 1999). The putative anion transporter (PAT1, SLC26A6) is responsible for bicarbonate transport into the duodenal lumen in exchange for chloride. The intestinal content becomes slightly more acidic at the beginning of the colon (pH around 5.7) and eventually drops to pH 6.7 again by the time the feces reach the rectum (Fallingborg, 1999). Sodium/hydrogen exchanger 3 (NHE3, SLC9A3) at the luminal side and sodium/hydrogen exchanger 2 (NHE2, SLC9A2) at the basolateral side are responsible for most of the pH adjustments in jejunum, ileum, and colon. Bacterial production of D-lactic acid from fermentable carbohydrates is an important factor for the slightly acidic pH in colon.

Water: The mucosa of the proximal small intestine has a net water secretion of about 30 ml/h (Knutson et al., 1995). Water from the basolateral and pericapillary space readily flows across the tight junctions, if the lumen contains a high concentration of osmotically active nutrients (e.g., simple sugars). Water is largely taken up again into small intestinal enterocytes by cotransport with sodium and glucose via SGLT1 (Wright and Loo, 2000), and exported across the basolateral membrane both via aquaporin 3 (Purdy et al., 1999) and aquaporin 4 (Wang et al., 2000). Aquaporin 3 expression increases from the stomach and is greatest in the ileum and colon. The large intestine further removes remaining water, potassium, and other electrolytes. Both aquaporin 3 and 4 (Wang et al., 2000) contribute to water transport across the basolateral membrane.

Sodium: Digestive secretions (saliva, stomach secretions, bile, pancreatic juice, and intestinal secretions) move large amounts of sodium into the digestive tract. Most of it is reabsorbed in the small intestine. The sodium/potassium-exchanging ATPase (EC3.6.3.9) at the basolateral side of small and large intestines generates a steep gradient with several-fold lower sodium concentration within the enterocytes than in the intestinal lumen or the basolateral and pericapillary fluid. This gradient drives most other active transcellular absorption mechanisms. Many of these operate as sodium cotransporters and thus explain much of the sodium uptake from the intestinal lumen. The absorption of 100 g of carbohydrate brings along 26 g of sodium! The sodium/hydrogen exchanger NHE3 (SLC9A3) is expressed on apical membranes of intestinal epithelial cells and is thought to play a major role in salt and bicarbonate absorption.

Chloride: The putative anion transporter 1 (PAT1, SLC26A6) in the small intestine recovers most of the large amounts secreted as gastric acid. At least two anion exchangers (including AE2 and AE3) at the basolateral membrane of the ileum and colon move chloride in exchange for bicarbonate toward the bloodstream (Alrefai et al., 2001). Much smaller amounts accompany the uptake of a few nutrients via chloride-dependent transporters, such as system $B^{o,+}$ and the IMINO transporter. The chloride conductance regulator (cystic fibrosis transmembrane regulator, CFTR, ABCC7) regulates chloride secretion into pancreatic secretions and from the mucosa into the small intestinal lumen.

Absorption of minerals and trace elements

Calcium: More than 40% of calcium from food is absorbed by young adults, more by infants, and as little as 20% by older people. Optimal vitamin D status is a prerequisite for high absorption. Phosphate and oxalate reduce fractional absorption by forming poorly soluble complexes. Entry of calcium into the enterocyte is favored energetically by the more than 1000-fold difference between luminal and intracellular concentration (10^{-3} vs. 10^{-6} molar), and the strong electronegativity of the enterocyte compared to the intestinal lumen. At low to moderate intakes, most absorption proceeds via the apical calcium transport protein (CaT1), transcellular transport involving calbindin-D9k and other proteins, and active extrusion by the basolateral membrane calcium-pumping ATPase (isoform K of PMCA1, ATP2B1, EC3.6.3.8). The expression of calbindin-D9k, but not of PMCA1, is upregulated by 1,25-dihydroxyvitamin D (Waiters et al., 1999). At higher concentrations, the less efficient paracellular diffusion pathway becomes significant.

Magnesium: Absorption from distal jejunum and ileum is about 30–60% (Sabatier et al., 2002). Phosphate and phytate decrease absorption. The magnesium channel TRPM6 (transient receptor potential melastatin 6) mediates uptake with high affinity. Solvent drag across the tight junctions between the enterocytes accounts for one-third of the absorption at high intake levels.

Phosphate: Absorption of inorganic phosphate from the small intestine is very efficient (60–70%), depending to a minor degree on 1,25-dihydroxyvitamin D stimulation. Even better is the availability of the phosphate in phospholipids (from food, bile, and shedding of old enterocytes). Absorption of the latter is not well inhibited by typical phosphate binders. Many organophosphates are cleaved prior to absorption of the phosphate ion by alkaline phosphatase (EC3.1.3.1). Both alkaline phosphatase and inorganic pyrophosphatase (EC3.6.1.1) hydrolyze pyrophosphate. Type I (SLC17A1) and type IIb (NaPi3B, SLC34A2) sodium/phosphate cotransporters and possibly an additional sodium-independent transporter take up inorganic phosphate. Phospholipid phosphate enters as lysophospholipid after cleavage by phospholipase A2 (EC3.1.1.4) from pancreas.

Sulfur: Absorption of sulfate is probably very high throughout the intestinal tract and only minimal amounts are lost (Florin et al., 1991). Transport proceeds via members of the SLC26 family, including SLC26A2 (DTDST), SLC26A3 (CLD/DRA), and SLC26A4 (PDS), which also accept various other anions (Markovich, 2001).

Iron: Ferrous iron (Fe^{2+}) moves into enterocytes of the small intestine via the divalent metal ion transporter (DMT1, SLC11A2), ferric iron (Fe^{3+}) does so via a mechanism that involves duodenal cytochrome B.

Heme can be taken up through a second distinct mechanism. A proton-dependent carrier, folate transporter/heme carrier protein 1 (PCFT/HCP1, SLC46A1), moves the intact heme molecule from the intestinal lumen into enterocytes (Shayeghi et al., 2005). The feline leukemia virus subgroup C receptor (FLVCR1, SLC49A1) also is a heme transporter (Khan and Quigley, 2013). Although it is better known for its critical function in erythroid bone marrow cells, it is highly expressed in the intestines. Yet another heme transporter within the enterocytes, encoded by SLC48A1 (heme responsive gene-1, HRG-1), is thought to move heme from the endosome into the cytosol (Khan and Quigley, 2013). Heme oxygenase (EC1.14.99.3) inside the cell can then release the iron from its porphyrine frame and make it available for processing and export across the basolateral membrane into blood just like nonheme iron.

Transferrin, taken up from circulation through the transferrin receptor, is commonly thought to be responsible for iron transport into portal blood. Recently, however, the gene product of Ireg1 has been suggested to mediate ferrous iron export across the basolateral membrane (McKie et al., 2000). The proportion of dietary iron taken up into the enterocytes depends on the constituents in a given meal: heme is much better absorbed than nonheme iron; ascorbate, citrate, and other organic acids promote uptake; phytate, tannins, and calcium inhibit it. Absorption efficiency is tightly regulated through changes in the expression of DMT1, ferritin (which promotes loss during shedding of cells by preventing iron export), the ferroportin (SLC40A1), and the transferrin receptor (which promote export). All of the genes in question are associated with iron-responsive elements (IREs) that bind the multifunctional iron-sulfur protein aconitase and respond to the iron content of this regulator.

Copper: Maximal absorption from stomach and small intestine decreases from more than half of low doses to less than 15% above 6–7 mg (Wapnir, 1998). DMT1 (SLC11A2) is the main luminal transporter. The copper-transporting ATPase 7A (Menkes protein, ATP7A, EC3.6.3.4) exports copper across the basolateral membrane. Ascorbate, zinc, and phytate decrease absorption.

Zinc: Under optimal conditions, more than 70% of a small dose (3 mg) is absorbed from the small intestine, and to a lesser degree from the stomach and colon. Zinc complexes with histidine, cysteine, and nucleotides are absorbed better than zinc alone. Phytate interferes with zinc absorption (Lönnerdal, 2000). DMT1 (SLC11A2) takes up zinc ions. Zinc-peptide complexes are transported by the hydrogen ion/peptide cotransporter (PepT1, SLC15A1). Binding to intestinal metallothionein regulates absorption efficiency. Export from enterocytes across the basolateral membrane uses zinc transporters 1 (ZnT1, SLC30A1) and 2 (ZnT2, SLC30A2) and the zinc-regulated transporter 1 (SLC39A1).

Manganese: Less than 2% of ingested food manganese is absorbed from the small intestine via DMT1 (SLC11A2) and other components of nonheme iron absorption. Availability is higher from human milk (8%) than from other foods.

Chromium: Absorption is inversely related to intake levels and is usually around 0.5–2%. Phytate decreases bioavailability, and ascorbic acid increases it. Absorption of chromium picolinate is higher (2–5%) and proceeds by a different (unknown) mechanism.

Iodine: Iodine, iodide, iodate, and thyroxine are absorbed nearly to completion. Iodide uptake in the stomach, small intestine, and colon uses the sodium/iodide symporter (SLC5A5) and possibly other chloride and anion transporters.

Fluoride: About half of ingested fluoride is absorbed from the mouth, stomach, and small intestine by proton-cotransport and possibly also by anion exchange.

Bromide: Absorption from the entire intestine uses chloride transport mechanisms and ion exchange systems.

Selenium: Most dietary selenium is absorbed in the small intestine as selenocysteine and selenomethionine, or complexed as selenodicysteine and selenodiglutathione via amino acid (SLC3A1 and SLC1A4) and peptide (SLC15A1) transporters (Vendeland et al., 1992).

Molybdenum: Absorption of most forms with the exception of molybdenum sulfide is nearly complete at typical intake levels. Sulfate inhibits absorption.

Nickel: Absorption is limited to about 1% and proceeds mainly via DMT1. Replete iron status and the presence of phytate, tannins, and calcium in the same meal appear to decrease fractional absorption.

Vanadium: A small percentage (1–5%) is absorbed from the small intestine, possibly via nonheme iron pathways.

Arsenic: Arsenate, arsenite, and organic arsenic compounds are nearly completely absorbed from the small intestine. Arsenobetaine and arsenocholine are taken up via amino acid transporters.

Absorption of phytochemicals

Polyphenols: Absorption ranges from minimal to more than 30% between individual compounds (Crozier et al., 2010; Clifford et al., 2013). Uptake is best if any attached sugars are removed by lactase or bacterial β-glucosidases in the intestinal lumen (Johnston et al., 2005). MRP1 and several other transporters move conjugates across the basolateral lumen into enterocytes.

The ATP-binding cassette transporters MDR1 (Hoffmeyer et al., 2000), MRP2 (Walgren et al., 2000), BCRP/ABCG2 (Tan et al., 2013) are intestinal efflux transporters that pump particular flavonoid conjugates and some of their metabolites across the apical enterocyte membrane back into the intestinal lumen.

REFERENCES

Alrefai, W.A., Tyagi, S., Nazir, T.M., Barakat, J., Anwar, S.S., Hadjiagapiou, C., et al., 2001. Human intestinal anion exchanger isoforms: expression, distribution, and membrane localization. Biochim. Biophys. Acta 1511, 17–27.

Amelsberg, A., Jochims, C., Richter, C.P., Nitsche, R., Folsch, U.R., 1999. Evidence for an anion exchange mechanism for uptake of conjugated bile acid from the rat jejunum. Am. J. Physiol. 276 (3 Pt 1), G737–G742.

Anwer, M.S., Stieger, B., 2014. Sodium-dependent bile salt transporters of the SLC10A transporter family: more than solute transporters. Pflügers. Arch. 466, 77–89.

Atuma, C., Strugala, V., Allen, A., Holm, L., 2001. The adherent gastrointestinal mucus gel layer: thickness and physical state *in vivo*. Am. J. Physiol. Gastroint. Liver Physiol. 280, G922–G929.

Bilzer, M., Roggel, F., Gerbes, A.L., 2006. Role of Kupffer cells in host defense and liver disease. Liver Int. 26, 1175–1186.

Bordi, C., D'Adda, T., Azzoni, C., Ferraro, G., 2000. Classification of gastric endocrine cells at the light and electron microscopical levels. Microsc. Res. Tech. 48, 258–271.

Byrd, J.C., Feamey, F.J., Kim, Y.S., 1985. Rat intestinal nucleotide-sugar pyrophosphatase. Localization, partial purification, and substrate specificity. J. Biol. Chem. 260, 7474–7480.

Carriere, E., Barrowman, J.A., Verger, R., Laugier, R., 1993. Secretion and contribution to lipolysis of gastric and pancreatic lipases during a test meal in humans. Gastroenterology 105, 876–888.

Chen, Z., Zheng, S., Li, L., Jiang, H., 2014. Metabolism of flavonoids in human: a comprehensive review. Curr. Drug Metab. 15, 48–61.

Clifford, M.N., van der Hooft, J.J., Crozier, A., 2013. Human studies on the absorption, distribution, metabolism, and excretion of tea polyphenols. Am. J. Clin. Nutr. 98, 1619S–1630S.

Crozier, A., Del Rio, D., Clifford, M.N., 2010. Bioavailability of dietary flavonoids and phenolic compounds. Mol. Aspects Med. 31, 446–467.

Dalal, J., Radhi, M., 2013. Intestinal stem cells: common signal pathways, human disease correlation, and implications for therapies. ISRN Stem Cells Article ID 372068.

Davis, B.D., Brodbelt, J.S., 2008. Regioselectivity of human UDP-glucuronosyl-transferase 1A1 in the synthesis of flavonoid glucuronides determined by metal complexation and tandem mass spectrometry. J. Am. Soc. Mass Spectrom. 19, 246–256.

Degnan, P.H., Barry, N.A., Mok, K.C., Taga, M.E., Goodman, A.L., 2014. Human gut microbes use multiple transporters to distinguish vitamin B_{12} analogs and compete in the gut. Cell Host Microbe 15, 47–57.

Detmers, P.A., Patel, S., Hernandez, M., Montenegro, J., Lisnock, J.M., Pikounis, B., et al., 2000. A target for cholesterol absorption inhibitors in the enterocyte brush border membrane. Biochim. Biophys. Acta 1486, 243–252.

Drummond, P.D., 1995. Effect of imagining and actually tasting a sour taste on one side of the tongue. Physiol. Behav. 57, 373–376.

Duan, R.D., Nilsson, A., 1997. Purification of a newly identified alkaline sphingomyelinase in human bile and effects of bile salts and phosphatidylcholine on enzyme activity. Hepatology 26, 823–830.

Fallingborg, J., 1999. Intraluminal pH of the human gastrointestinal tract. Dan. Med. Bull. 46, 183–196.

Florin, T., Neale, G., Gibson, G.R., Christl, S.U., Cummings, J.H., 1991. Metabolism of dietary sulphate: absorption and excretion in humans. Gut 32, 766–773.

Geibel, J.E., Wagner, C.A., Caroppo, R., Qureshi, I., Gloeckner, J., Manuelidis, L., et al., 2001. The stomach divalent ion-sensing receptor scar is a modulator of gastric acid secretion. J. Biol. Chem. 276, 39549–39552.

Gerbe, F., Legraverend, C., Jay, P., 2012. The intestinal epithelium tuft cells: specification and function. Cell Mol. Life Sci. 69, 2907–2917.

Geypens, B., Bennink, R., Peeters, M., Evenepoel, E., Mortelmans, L., Maes, B., et al., 1999. Validation of the lactose-[^{13}C] ureide breath test for determination of orocecal transit time by scintigraphy. J. Nucl. Med. 40, 1451–1455.

Gullo, L., Pezzilli, R., Priori, E., Baldoni, E., Paparo, E., Mattioli, G., 1987. Pure pancreatic juice collection over 24 consecutive hours. Pancreas 2, 620–623.

Hoffmeyer, S., Burk, O., von Richter, O., Arnold, H.E., Brockmoller, J., Johne, A., et al., 2000. Functional polymorphisms of the human multidrug-resistance gene: multiple sequence variations and correlation of one allele with P-glycoprotein expression and activity *in vivo*. Proc. Natl. Acad. Sci. USA 97, 3473–3478.

Islam, R.S., DiBaise, J.K., 2012. Bile acids: an underrecognized and underappreciated cause of chronic diarrhea. Pract. Gastroent. 110, 32–44.

Johnston, K., Sharp, P., Clifford, M., Morgan, L., 2005. Dietary polyphenols decrease glucose uptake by human intestinal Caco-2 cells. FEBS Lett. 579, 1653–1657.

Kakiyama, G., Pandak, W.M., Gillevet, P.M., Hylemon, P.B., Heuman, D.M., Daita, K., et al., 2013. Modulation of the fecal bile acid profile by gut microbiota in cirrhosis. J. Hepatol. 58, 949–955.

Kassie, E., Rabot, S., Kundi, M., Chabicovsky, M., Qin, H.M., Knasmuller, S., 2001. Intestinal microflora plays a crucial role in the genotoxicity of the cooked food mutagen 2-amino-3-methylimidazo [4,5-φ]quinoline. Carcinogenesis 22, 1721–1725.

Khan, A.A., Quigley, J.G., 2013. Heme and FLVCR-related transporter families SLC48 and SLC49. Mol. Aspects Med. 34, 669–682.

Knutson, T.W., Knutson, L.E., Hogan, D.L., Koss, M.A., Isenberg, J.I., 1995. Human proximal duodenal ion and water transport. Role of enteric nervous system and carbonic anhydrase. Dig. Dis. Sci. 40, 241–246.

Krause, W.J., 2000. Brunner's glands: a structural, histochemical and pathological profile. Progr. Histochem. Cytochem. 35, 259–367.

Kurashima, Y., Goto, Y., Kiyono, H., 2013. Mucosal innate immune cells regulate both gut homeostasis and intestinal inflammation. Eur. J. Immunol. 43, 3108–3115.

Lipsky, J.J., 1994. Nutritional sources of vitamin K. Mayo Clin. Proc. 69, 462–466.

Lönnerdal, B., 2000. Dietary factors influencing zinc absorption. J. Nutr. 130, 1378S–1383S.

Macfarlane, G.Z., Macfarlane, S., 1997. Human colonic microbiota: ecology, physiology and metabolic potential of intestinal bacteria. Stand. J. Gastroenterol. 222, 3–9.

Markovich, D., 2001. Physiological roles and regulation of mammalian sulfate transporters. Physiol. Rev. 81, 1499–1533.

Maroni, L., van de Graaf, S.F., Hohenester, S.D., Oude Elferink, R.P., Beuers, U., 2015. Fucosyltransferase 2: a genetic risk factor for primary sclerosing cholangitis and crohn's disease-a comprehensive review. Clin. Rev. Allergy Immunol. 48 (2–3), 182–191.

McKie, A.T., Marciani, E., Rolfs, A., Brennan, K., Wehr, K., Barrow, D., et al., 2000. A novel duodenal iron-regulated transporter, IREGI, implicated in the basolateral transfer of iron to the circulation. Mol. Cell 5, 299–309.

Meyer, J.H., 1994. The stomach and nutrition. In: Shils, M.E., Olson, J.A., Shike, M. (Eds.), Modern Nutrition in Health and Disease, eighth ed. Lea and Febiger, Philadelphia, PA, pp. 1029–1035.

Mizuma, T., Awazu, S., 1998. Intestinal Na^+/glucose cotransporter-mediated transport of glucose conjugate formed from disaccharide conjugate. Biochim. Biophys. Acta 1379, 1–6.

Morris, R.C., Brown, K.G., Elliott, M.S., 1999. The effect of queuosine on tRNA structure and function. J. Biomol. Struct. Dynam. 16, 757–774.

Mukai, R., Fujikura, Y., Murota, K., Uehara, M., Minekawa, S., Matsui, N., et al., 2013. Prenylation enhances quercetin uptake and reduces efflux in Caco-2 cells and enhances tissue accumulation in mice fed long-term. J. Nutr. 143, 1558–1564.

Murray, I.A., Coupland, K., Smith, J.A., Ansell, I.D., Long, R.G., 2000. Intestinal trehalase activity in a UK population: establishing a normal range and the effect of disease. Br. J. Nutr. 83, 241–245.

Navaneethan, U., 2014. Hepatobiliary manifestations of ulcerative colitis: an example of gut-liver crosstalk. Gastroenterol. Rep. (Oxf) 2, 193–200.

Neubauer, H, Pantel, I, Lindgren, PE, Götz, F., 1999. Characterization of the molybdate transport system ModABC of *Staphylococcus carnosus*. Arch. Microbiol. 172 (2), 109–115.

Neutra, M.R., 1998. Current concepts in mucosal immunity.V. Role of M cells in transepithelial transport of antigens and pathogens to the mucosal immune system. Am. J. Physiol. 274 (5 Pt 1), G785–G791.

Neyraud, E., Palicki, O., Schwartz, C., Nicklaus, S., Feron, G., 2012. Variability of human saliva composition: possible relationships with fat perception and liking. Arch. Oral Biol. 57, 556–566.

Orsenigo, M.N., Tosco, M., Bazzini, C., Laforenza, U., Faelli, A., 1999. A monocarboxylate transporter MCT1 is located at the basolateral pole of rat jejunum. Exp. Physiol. 84, 1033–1042.

Potten, C.S., Loeffler, M., 1990. Stem cells: attributes, cycles, spirals, pitfalls and uncertainties. Lessons for and from the crypt. Development 110, 1001–1020.

Prinz, J.E., Lucas, P.W., 2000. Saliva tannin interactions. J. Oral. Rehab. 27, 991–994.

Purdy, M.J., Cima, R.R., Doble, M.A., Klein, M.A., Zinner, M.J., Soybel, D.I., 1999. Selective decreases in levels of mRNA encoding a water channel (AQP3) in ileal mucosa after ileostomy in the rat. J. Gastroint. Surg. 3, 54–60.

Ramadan, T., Camargo, S.M., Herzog, B., Bordin, M., Pos, K.M., Verrey, F., 2007. Recycling of aromatic amino acids via TAT1 allows efflux of neutral amino acids via LAT2-4F2hc exchanger. Pflügers Arch. 454, 507–516.

Ramirez, C., Gebert, A., 2003. Vimentin-positive cells in the epithelium of rabbit ileal villi represent cup cells but not M-cells. J. Histochem. Cytochem. 51, 1533–1544.

Repishti, M., Hogan, D.E., Pratha, V., Davydova, L., Donowitz, M., Tse, C.M., et al., 2001. Human duodenal mucosal brush border Na(+)/H(+) exchangers NHE2 and NHE3 alter net bicarbonate movement. Am. J. Physiol. Gastroint. Liver Physiol. 281, G159–G163.

Reshetnyak, V.I., 2013. Physiological and molecular biochemical mechanisms of bile formation. World J. Gastroenterol. 19, 7341–7360.

Rubio, C.A., 2012. Paneth cells and goblet cells express the neuroendocrine peptide synaptophysin. I. Normal duodenal mucosa. In Vivo 26, 135–138.

Sabatier, M., Arnaud, M.J., Kastenmayer, E., Rytz, A., Barclay, D.V., 2002. Meal effect on magnesium bioavailability from mineral water in healthy women. Am. J. Clin. Nutr. 75, 65–71.

Said, H.M., Wang, S., Ma, T.Y., 2005. Mechanism of riboflavin uptake by cultured human retinal pigment epithelial ARPE-19 cells: possible regulation by an intracellular Ca^{2+}-calmodulin-mediated pathway. J. Physiol. 566 (Pt 2), 369–377.

Setchell, K.D., Heubi, J.E., Shah, S., Lavine, J.E., Suskind, D., Al-Edreesi, M., et al., 2013. Genetic defects in bile acid conjugation cause fat-soluble vitamin deficiency. Gastroenterology 144, 945–955.

Shayeghi, M., Latunde-Dada, G.O., Oakhill, J.S., Laftah, A.H., Takeuchi, K., Halliday, N., et al., 2005. Identification of an intestinal heme transporter. Cell 122, 789–801.

Sherry, A.M., Malinowska, D.H., Morris, R.E., Ciraolo, G.M., Cuppoletti, J., 2001. Localization of ClC-2 Cl-channels in rabbit gastric mucosa. Am. J. Physiol. Cell Physiol. 280, C1599–C1606.

Sindic, A., 2013. Current understanding of guanylin peptides actions. ISRN Nephrol. 2013, 813648.

Sutor, D.J., Wilkie, L.I., 1976. Diurnal variations in the pH of pathological gallbladder bile. Gut 17, 971–974.

Takanaga, H., Maeda, H., Yabuuchi, H., Tamai, I., Higashida, H., Tsuji, A., 1996. Nicotinic acid transport mediated by pH-dependent anion antiporter and proton cotransporter in rabbit intestinal brush border membrane. J. Pharm. Pharmacol. 48, 1073–1077.

Tan, K.W., Li, Y., Paxton, J.W., Birch, N.P., Scheepens, A., 2013. Identification of novel dietary phytochemicals inhibiting the efflux transporter breast cancer resistance protein (BCRP/ABCG2). Food Chem. 138, 2267–2274.

Tazawa, S., Yamato, T., Fujikura, H., Hiratochi, M., Itoh, F., Tomae, M., et al., 2005. SLC5A9/SGLT4, a new Na^+-dependent glucose transporter, is an essential transporter for mannose, 1,5-anhydro-D-glucitol, and fructose. Life Sci. 76, 1039–1050.

Terao, J., Mukai, R., 2014. Prenylation modulates the bioavailability and bioaccumulation of dietary flavonoids. Arch. Biochem. Biophys. 559C, 12–16.

Vendeland, S.C., Deagen, J.T., Whanger, P.D., 1992. Uptake of selenotrisulfides of glutathione and cysteine by brush border membranes from rat intestines. J. Inorg. Biochem. 47, 131–140.

Waiters, J.R., Howard, A., Lowery, L.L., Mawer, E.B., Legon, S., 1999. Expression of genes involved in calcium absorption in human duodenum. Eur. J. Clin. Invest. 29, 214–219.

Walgren, R.A., Karnaky Jr., K.J., Lindenmayer, G.E., Walle, T., 2000. Efflux of dietary flavonoid quercetin 4′-beta-glucoside across human intestinal Caco-2 cell monolayers by apical multidrug resistance-associated protein-2. J. Pharmacol. Exp. Ther. 294, 830–836.

Wang, K.S., Ma, T., Filiz, E., Verkman, A.S., Bastidas, J.A., 2000. Colon water transport in transgenic mice lacking aquaporin-4 water channels. Am. J. Physiol. Gastroint. Liver Physiol. 279, G463–G470.

Wang, Y., Zhao, R., Russell, R.G., Goldman, I.D., 2001. Localization of the murine reduced folate carrier as assessed by immunohistochemical analysis. Biochim. Biophys. Acta 1513, 49–54.

Wang, Z., Petrovic, S., Mann, E., Soleimani, M., 2002. Identification of an apical Cl(−)/HCO_3(−) exchanger in the small intestine. Am. J. Physiol. Gastroint. Liver Physiol. 282, G573–G579.

Wapnir, R.A., 1998. Copper absorption and bioavailability. Am. J. Clin. Nutr. 67, 1054S–1060S.

Wright, E.M., Loo, D.D., 2000. Coupling between Na^+, sugar, and water transport across the intestine. Ann. NY Acad. Sci. 915, 54–66.

Wright, E.M., 2013. Glucose transport families SLC5 and SLC50. Mol. Aspects Med. 34, 183–196.

Yang, H., 2002. Central and peripheral regulation of gastric acid secretion by peptide YY. Peptides 23, 349–358.

Yonezawa, A., Inui, K., 2013. Novel riboflavin transporter family RFVT/SLC52: identification, nomenclature, functional characterization and genetic diseases of RFVT/SLC52. Mol. Aspects Med. 34, 693–701.

Zamolodchikova, T.S., Scherbakov, I.T., Khrennikov, B.N., Svirshchevskaya, E.V., 2013. Expression of duodenase-like protein in epitheliocytes of Brunner's glands in human duodenal mucosa. Biochemistry (Mosc) 78, 954–957.

Zhou, H., Chen, B., Li, R.X., Sheng, Q.H., Li, S.J., Zhang, L., et al., 2005. Large-scale identification of human biliary proteins from a cholesterol stone patient using a proteomic approach. Rapid. Commun. Mass. Spectrom. 9 (23), 3569–3578.

Ziv, E., Bendayan, M., 2000. Intestinal absorption of peptides through the enterocytes. Microscopy Res. Techn. 49, 346–352.

Zuidema, T., Kamermans, M., Siegenbeek van Heukelom, J., 1986. Influence of glucose absorption on ion activities in cells and submucosal space in goldfish intestine. Pfl. Arch. Eur. J. Physiol. 407, 292–298.

MICROBIOME

The oro-gastro-intestinal tract contains about 10 times as many microorganisms as the human body contains cells. One might be tempted to see humans as a convenient container for its community of microbes. However this may be, the typical microorganisms in and on the human shell have evolved continuously with our ancestors and our lifestyles, and form a functional unit with their human host.

The term *microbiome* refers to the ecological community at a particular location, describing both composition and activities. The most common and best-understood life forms in this community are bacteria, yeasts, and viruses. The largest numbers of these types of microorganisms reside in the mouth, distal ileum, and large intestine, but none of the other segments are sterile. Most of these microorganisms are commensals, meaning that they normally coexist with us without causing harm or benefit. There is definitely some mutualism going on for numerous species because we keep them warm and fed, and they produce some nutrients or provide other benefits for us. Pathogenic organisms creep in every now and then, and sometimes an otherwise harmless or even beneficial type turns bad. The timing, amount, and composition of ingested food are major forces for favoring or inhibiting specific microbiome community members. They depend on what we eat, and we eat depending on what they are. Understanding the interactions between our individual gut microbiomes and our foods helps increasingly to decipher disease processes and develop intervention solutions.

ABBREVIATION

CD   celiac disease

NOMENCLATURE AND CLASSIFICATION

The term *microbiota* conventionally refers to a wide variety of single-cell organisms. Systematic classification (illustrated here for *Lactobacillus acidophilus*) is organized in ever greater detail by separating domain (here, *Bacteria*) into phylum (e.g., *Firmicutes*), and into class (e.g., *Bacilli*), and then into order (e.g., *Lactobacillales*), family (e.g., *Lactobacillaceae*), and genus (e.g., *Lactobacillus*), and further into species (e.g., *L. acidophilus*) and possibly subspecies (e.g., *L. acidophilus La-14*, a commercially available strain used as an experimental probiotic). As shown here, the name of the genus (plural is *genera*) is commonly abbreviated when the full name has been used earlier or if the usage is very common. The word *species* after the genus name (abbreviated as *spp.*) indicates that several different species are included without distinguishing between them (e.g., *Lactobacillus* spp.). Names of microbiota, like those of other organisms, are commonly italicized to indicate that they refer to life forms.

SEGMENT-SPECIFIC MICROBIOME DENSITY

Bacteria: Normally, the upper small intestine contains few, if any, bacteria. Most ingested microorganisms are inactivated by the acidity of the stomach, digestive secretions (containing lysozyme, proteases, lipases, DNAses, RNAses, and many other enzymes), and the intestinal immune defense system (antibodies, lymphocytes, and macrophages from Peyer's plaques, and other intestinal structures). Colonization of the intestine by aerobic and anaerobic bacteria becomes significant only in the terminal ileum and is extensive throughout the colon. In those segments, the intestinal flora shows considerable diversity. Since the introduction of DNA-based typing, it has become clear that previous culture-based assessments were not representative. Instead, many previously unknown types are now understood to be dominant. The main phyla of normal intestinal bacteria are *Firmicutes*, *Bacteroidetes*, *Actinobacteria*, and *Proteobacteria*. Clustering of major groups defines three common enterotypes with strong components of *Bacteroides*, *Prevotella*, or *Ruminococcus* species (Wu et al., 2011). *Escherichia*, *Bifidobacteria*, *Lactobacilli*, and *Clostridia* have importance in particular aspects—just not quantitatively.

Intestinal virome: Viruses constitute another previously unappreciated group of gut microbiota (Minot et al., 2011). These are bacteriophages that infect bacteria and thereby directly influence their ability to thrive. They can change the genome of bacteria by integrating into their DNA. In particular, they can transfer sequences for toxins or drug resistance from one type of bacteria into another. Thus, DNA encoding resistance to antibiotics can be transferred (Wang et al., 2010). The community of viruses (virome) in the intestines tends to be shared by mother and child, is relatively stable over time, and responds to dietary changes (Minot et al., 2011). The specific impact of individual nutrients and foods remains to be seen.

Yeasts: Various yeast species are present in the intestine, but their role in healthy people is not fully understood. They appear to be most important for their interactions with other microbes in the microbiome community. They become mostly relevant in situations of excessive expansion. Yeast overgrowth in the small intestine can be a significant problem in rare cases (Jansson-Nettelbladt et al., 2006).

COLONIZATION

Vaginally born infants receive a first set of microorganisms from their mother (Makino et al., 2013). Colonization of the intestinal tract of newborns then proceeds during the first few months of life in a well-ordered sequence (La Rosa et al., 2014). *Bacilli* tend to settle in earliest. In particular, *Bifidobacterium breve* has a strong presence in the gut microbiome of very young breast-fed infants (Turroni et al., 2014). *Gammaproteobacteria* typically come a little bit later, followed by *Clostridia*, until eventually the normal, complex ecology is established a few months after birth.

Some of the typical gut bacteria are closely adapted to their mammalian host, even down to the intestinal geography and specific microscopic location. Experiments in rodents have demonstrated that *Bacteroides* species express several proteins (commensal colonization factors) that aid them to move through the mucus layer of the colonic lumen and home in on the crypts (Lee et al., 2013). The identified proteins constitute an impressive toolset, get to the target location, stay in place (Sus-like proteins that bind to cell surface oligosaccharides), and feed the microorganisms (polysaccharidases) and the host (glycosidases cleaving otherwise unavailable food compounds). The investigated *Bacteroides fragilis* species even has a chemical sensor that detects *N*-acetyl-D-lactosamine indicative of the mucus layer that newly ingested organisms have to penetrate if they are to colonize the crypts. *B. bifidum* in the intestine of young infants makes a virtue out of this voyage by cleaving specific glycoproteins of the mucus layer and then utilizing the sugars as its main carbon source (Turroni et al., 2014).

It is safe to assume that many other microbial species have systems that are similarly adapted to their individual human hosts and allow them to colonize their very own intestinal niche and outcompete most other microbes there.

LUMINAL SURFACE FACTORS

The nature of the luminal surface of the gastrointestinal walls greatly influences what kind of microbiome settles in. There is now detailed information on some of the factors that make a more or less hospitable environment for particular species. One of them is the extent of fucosylation of type III glycans (containing Gal-beta1,3-GalNAc chains) at the luminal cell surface of enterocytes. These fucosylated ABO and Lewis blood group cell surface antigens are targets for several pathogenic organisms, including *Noro* viruses and *Helicobacter pylori* (Maroni et al., 2015). The fucose of these surface glycans also supports some commensal bacteria such as *Bifidobacteria* and *Bacteroides* species, which metabolize it as an energy source. It is not surprising, therefore, that nonsecretors shed fewer *Bifidobacteria* with feces than do secretors (Maroni et al., 2015).

The surface antigen group HLA-DQ2, which predisposes to the development of celiac disease, also has a significant influence on microbiome composition. Infants with HLA-DQ2 tend to have more *Firmicutes* and *Protobacteria* and fewer *Actinobacteria* than children without this marker (Olivares et al., 2014a,b). The infants also had fewer *Bifidobacteria* species. This is the microbiome environment that will eventually trigger celiac disease (CD) in a small minority (2–5%) of these infants at some time in the future. It may well be that the inherited molecular surface composition favors specific microbiota and those, in turn, drive the development of the autoimmune disease CD.

DISRUPTION OF THE GUT MICROBIOME

Antibiotics: Oral ingestion of antibiotics may severely diminish gut bacteria across the board, only some selected species, or have minimal impact. All of those scenarios are possible, depending on the amount and type of the antibiotic (Preidis and Versalovic, 2009). Generally, treatment effects have been underappreciated until sequencing could document the dramatic loss of species diversity with even a single course of a general-purpose antibiotic like ciprofloxacin. Return of the full original microbiotic spectrum often takes months, and many species never come back. These changes are typically very subtle and do not cause diarrhea or other intestinal dysfunction. Instead, they may cause long-term metabolic shifts with unrecognized health consequences (Keeney et al., 2014). Treatment with clarithromycin, for instance, is associated with weight gain (Lane et al., 2011). Similar associations have been observed with several other types of antibiotics (Keeney et al., 2014).

Infections: The growth of some microbiota disrupts the balance of the community of organisms in the gut. For example, chronic intestinal infection with *Clostridium difficile* causes watery diarrhea, abdominal distress, fever, and other signs of local and systemic illness. This pathological expansion at the expense of beneficial species often happens when treatment with antibiotics wipes out many of the beneficial species. The consequence is often a shift in the spectrum of intestinal bile acids (Weingarden et al., 2014) because the surviving species have different metabolic properties than the eradicated forms. Normally, commensal bacteria in the ileum and colon cleave taurine from taurocholate and dehydroxylate some of the resulting cholic acid to deoxycholic acid. Conjugated chenodeoxycholic acid is similarly metabolized to lithocholic and ursodeoxycholic acid. Since *C. difficile* needs taurocholate for optimal growth and are inhibited by the secondary bile acids (lithodeoxycholic and ursodeoxycholic acid), the normal bile acid–metabolizing bacteria limit expansion of the harmful bacteria. The decline of the normal bacterial population then establishes a kind of metabolic feedback loop that can sustain chronic and difficult to treat *C. difficile* overgrowth.

Transplanting live fecal bacteria from a healthy individual to someone with overgrowth can reestablish the ecological balance in the intestinal lumen and crowd out the pathological *C. difficile* bacteria (Weingarden et al., 2014). This seemingly counterintuitive approach, called *fecal transplantation*, is rapidly getting wider use due to its proven effectiveness.

NUTRITIONAL FACTORS

High-fat, low-fiber: People with habitually high fat intake and low fiber consumption often have a gut microbiome pattern with a significant *Bacteroides* component (Wu et al., 2011).

High-fiber, low-fat: People who eat a lot of fiber-rich foods and little fat-rich foods tend to have a significant number of *Prevotella* species in their colon (Wu et al., 2011). Switching from high fat intake to a low-fat diet does not change the microbiome pattern for at least several weeks, though it will do that in the long term.

Meat consumption: The microbiome differs between omnivores and people following a meat-free diet pattern. Consumption of animal protein and cholesterol is associated with a *Bacteroides*-type pattern (Wu et al., 2011). It should not surprise that vegetarians tend to have more a *Prevotella*-dominated enterotype, as would be expected for a group with usually high fiber intake. Consumption of carnitine, a nutrient particularly present in meat and other animal-derived foods, gives rise to intestinal production of trimethylamine (TMA) production. A microbial bifunctional carnitine oxygenase and reductase (CntAB) with a characteristic Rieske domain adds oxygen to the carnitine molecule and then cleaves it into TMA and malic semialdehyde (Zhu et al., 2014). This is relevant because TMA is the precursor of trimethylamine oxide (TMAO), an atherosclerosis-promoting metabolite (Brown and Hazen, 2014). Carnitine supplementation in a rodent model was found to change the microbiome composition from a low-production type to one that generates large amounts of TMA. The identification of the Rieske domain in the *CntAB* gene responsible for the carnitine oxygenation is helpful because its presence allows for identifying specific bacteria that can generate TMA from carnitine and potentially developing inhibitors to suppress TMA generation. Known TMA producers include *Acinetobacter* species and *Serratia marcescens*. It appears that breakdown of carnitine to TMA occurs mainly in omnivores, but not in vegans (Koeth et al., 2013). Following a meat-free food pattern may not be the only promising approach to minimizing production of TMA. Another strategy may be to add specific methane-producing archebacteria, such as *Methanomassiliicoccus luminyensis* B10 (Brugère et al., 2014), that may deplete excess TMA.

Milk consumption: The microbiome composition depends to some extent on the long-term ingestion of specific macronutrients, such as the sugar in milk. Some microbiota species decline, while others thrive when intake patterns change. For instance, the presence of lactose after weaning promotes the proliferation of *Lactobacilli* species (Daly et al., 2014). The saturated fat in milk, on the other hand, appears to favor the expansion of *Deltaproteobacteria*, an otherwise minor group of microorganisms. These bacteria produce highly toxic hydrogen sulfide than can damage the colonic epithelium and initiate local inflammatory events (Sartor, 2012).

Obesity: People with a great excess of body fat tend to have a higher ratio of *Firmicutes* to *Bacteroides* species than normal-weight individuals (Turnbaugh et al., 2006). At the same time, the number of species declines with high weight gain. This alteration of microbiota spectrum was replicated in rodent models. Bariatric surgery can revert the microbiome within days to both a higher number of *Bacteroides* species and to greater species diversity overall (Aron-Wisnewsky et al., 2014). It is possible that some of the weight loss after surgery is directly attributable to the microbiome change.

PREBIOTICS

Indigestible oligosaccharides and polysaccharides that support the expansion of beneficial gut microbiota are commonly called *prebiotics*. This term is used mostly in the context of dietary supplements that promise benefits mediated by a more favorable and balanced gut ecology. The same compounds are often normal constituents of human milk or other everyday foods, particularly vegetables, fruits, grains, and legumes.

Nondigestible oligosaccharides promote the proliferation of normal intestinal microflora (Walker, 2013; Oozeer et al., 2013). *Bifidobacteria* and *Lactobacilli*, in particular, generate lactate and tend to induce a slightly acidic environment. Other bacteria produce significant amounts of acetate, propionate, and butyrate from undigested carbohydrates, such as incompletely digested starch and different kinds of dietary fiber. Inulin, a storage carbohydrate in plant roots, chicory, and many fruits and vegetables, is a typical enhancer of *Bifidobacteria*. Its partial utilization by these bacteria produces short-chain fatty acids (particularly butyrate), which is an important energy source for the epithelium of the colon wall.

Human milk contains a great variety of indigestible oligosaccharides—probably more than 200 of them (Barile and Rastall, 2013). These milk oligosaccharides are produced in the mammary gland by the addition of glucose, galactose, *N*-acetyl-glucosamine units, and fucose units. Acidic oligosaccharides also contain *N*-acetyl-neuraminic acid. Individual bacterial species such as *Bifidobacterium infantis* have distinct preferences. They thrive best when fed with short-chain neutral oligosaccharides (Ruiz-Moyano et al., 2013).

In regular foods, good sources of prebiotic oligosaccharides (Moshfegh et al., 1999) include leeks (12 g/100 g), onions (5 g/100 g), asparagus (5 g/100 g), and whole-wheat flour (4.8 g/100 g).

PROBIOTICS

The World Health Organization (WHO) has defined probiotics as "live microorganisms, which, when consumed in adequate amounts, confer a health benefit on the host" (FAO/WHO, 2001). Various traditional fermented foods meet this definition. Fermented milk (yogurt, kefir, and others) and vegetables (such as sauerkraut and kimchi) taste sour because *Lactobacilli*, *Bifidobacteria*, and other microorganisms use sugars in the parent food and release short-chain acids, including lactic acid, acetic acid, succinic acid, and propionic acid. In many cases, some of the ingested live cultures survive transit through the stomach and contribute to the microbial ecology of the distal intestines. Some commercial yogurts and other fermented foods are cultured with organisms that have been selected to resist inactivation by gastric acid and therefore are said to be particularly effective probiotics.

As mentioned previously, infants with the HLA genotypes DQ2 or DQ8 have fewer *Bifidobacteria* than noncarriers (Olivares et al., 2014b). However, feeding cultured *Bifidobacteria* can remedy this deficit (Olivares et al., 2014a). Thus, the composition of the gut microbiota can change in response to ingested microbes, particularly if they are consumed in large quantities and survive the sterilizing zone of the stomach.

Probiotics are usually products that contain live cultures of presumably beneficial bacteria or yeasts. Typical bacterial species in probiotics are from the genera *Bifidobacterium*, *Lactobacillus*, *Lactococcus*, and *Streptococcus*, and less often for *Bacillus*, *Bacteroides*, *Enterococcus*, *Escherichia*, *Faecalibacterium*, and *Propionibacterium*. It is important to recognize that even closely related species or subspecies often have very different properties. Commercial probiotics, therefore, need to be well defined and maintained true to type if any health claim is to be made.

MICROBIAL METABOLITES

Nutrients: Many physiological microorganisms produce compounds that contribute to the nutriture of the host. Nutrient production highlights the symbiotic (mutualist) relationship between the gut microbiome and the human host. Some bacteria break down dietary fiber and generate short-chain fatty acids (mainly acetic, propionic, and butyric acid) as important energy fuels for the local enterocytes. These short-chain fatty acids also influence hepatic lipid metabolism.

Bacteria of the lower intestine are also important sources of pantothenate (Said, 2013), biotin (Said, 2013), queuine (Morris et al., 1999), and vitamin K. How much vitamin K is actually available in humans from the terminal ileum and the colon is difficult to assess (Lipsky, 1994).

Toxins: Intestinal microorganisms can release toxic metabolites, particularly in response to particular dietary patterns or ingredients. The production of such toxic bacterial products depends on both the pattern of the microflora and on the presence of microbiological nutrients in the colon. Fermentable carbohydrates, including dietary fiber, tend to suppress the production of various indols, phenols, and heterocyclic amines (Macfarlane and Macfarlane, 1997; Kassie et al., 2001). Expansion of hydrogen sulfide producers such as *Deltaproteobacteria* can be driven by milk fat (Sartor, 2012). It has also been suggested that undigested lactose in the colon is associated in some susceptible individuals with increased release of methylglyoxal, a potent toxin that may explain some of the systemic symptoms of people suffering from an excess of undigested carbohydrates (Campbell et al., 2010).

In rare instances of severe infection, yeasts like *Candida albicans* can convert significant amounts of available carbohydrate to ethanol (autobrewery syndrome). While forensic investigators routinely reject this phenomenon as a defense in drunk driving indictments (Logan and Jones, 2000), yeast-mediated ethanol production in the intestine has been confirmed in cases of severe infections (Jansson-Nettelbladt et al., 2006). Endogenous ethanol production does seem to be a reality in Japanese men with the "Mei-Tei-Sho" syndrome (Picot et al., 1997). A confluence of yeast overgrowth, excessive rice intake, and malabsorption may be responsible.

METABOLIC, DEVELOPMENTAL, AND HEALTH CONSEQUENCES

Intestinal microbiota influences the development of the immune system (Kabat et al., 2014). Changes of the microbiome can increase the risk of infectious diseases, such as otitis media and viral diarrhea (ESPGHAN Committee on Nutrition et al., 2009).

Colonization with specific bacteria promotes early development of the innervation of the intestines (Collins et al., 2014). Others drive angiogenesis in the intestinal wall (Reinhardt et al., 2012).

Some commensal microbiota increase resistance of their local environment against atypical, potentially pathogenic, organisms. One of the defense mechanisms is the release of growth-inhibiting or even bactericidal compounds (Donia et al., 2014). In the process of keeping them away from their territory, these beneficial organisms also protect their human host against infection by pathogenic organisms.

Obesity, diabetes, cardiovascular disease, and allergies are also related to microbiome composition (Keeney et al., 2014).

REFERENCES

Aron-Wisnewsky, J., Clement, K., 2014. The effects of gastrointestinal surgery on gut microbiota: potential contribution to improved insulin sensitivity. Curr. Atheroscler. Rep. 16, 454.

Barile, D., Rastall, R.A., 2013. Human milk and related oligosaccharides as prebiotics. Curr. Opin. Biotechnol. 24, 214–219.

Brown, J.M., Hazen, S.L., 2014. Metaorganismal nutrient metabolism as a basis of cardiovascular disease. Curr. Opin. Lipidol. 25, 48–53.

Brugère, J.F., Borrel, G., Gaci, N., Tottey, W., O'Toole, P.W., Malpuech-Brugère, C., 2014. Archaebiotics: proposed therapeutic use of archaea to prevent trimethylaminuria and cardiovascular disease. Gut Microbes. 5, 5–10.

Campbell, A.K., Matthews, S.B., Vassel, N., Cox, C.D., Naseem, R., Chaichi, J., et al., 2010. Bacterial metabolic "toxins": a new mechanism for lactose and food intolerance, and irritable bowel syndrome. Toxicology 278, 268–276.

Collins, J., Borojevic, R., Verdu, E.F., Huizinga, J.D., Ratcliffe, E.M., 2014. Intestinal microbiota influence the early postnatal development of the enteric nervous system. Neurogastroenterol. Motil. 26, 98–107.

Daly, K., Darby, A.C., Hall, N., Nau, A., Bravo, D., Shirazi-Beechey, S.P., 2014. Dietary supplementation with lactose or artificial sweetener enhances swine gut Lactobacillus population abundance. Br. J. Nutr. 111 (Suppl. 1), S30–S35.

Donia, M.S., Cimermancic, P., Schulze, C.J., Wieland Brown, L.C., Martin, J., Mitreva, M., et al., 2014. A systematic analysis of biosynthetic gene clusters in the human microbiome reveals a common family of antibiotics. Cell 158 (6), 1402–1414.

ESPGHAN Committee on Nutrition, Agostoni, C., Braegger, C., Decsi, T., Kolacek, S., Koletzko, B., Michaelsen, K.F., et al., 2009. Breastfeeding: a commentary by the ESPGHAN Committee on Nutrition. J. Pediatr. Gastroenterol. Nutr. 49, 112–125.

FAO/WHO, 2001. Health and nutritional properties of probiotics in food including powder milk with live lactic acid bacteria. Report of a Joint FAO/WHO Expert Consultation on Evaluation of Health and Nutritional Properties of Probiotics in Food Including Powder Milk with Live Lactic Acid Bacteria. Basel, Switzerland: World Health Organization.

Jansson-Nettelbladt, E., Meurling, S., Petrini, B., Sjölin, J., 2006. Endogenous ethanol fermentation in a child with short bowel syndrome. Acta Paediatr. 95, 502–504.

Kabat, A.M., Srinivasan, N., Maloy, K.J., 2014. Modulation of immune development and function by intestinal microbiota. Trends Immunol. 35, 507–517.

Keeney, K.M., Yurist-Doutsch, S., Arrieta, M.C., Finlay, B.B., 2014. Effects of antibiotics on human microbiota and subsequent disease. Annu. Rev. Microbiol. 68, 217–235.

Koeth, R.A., Wang, Z., Levison, B.S., Buffa, J.A., Org, E., Sheehy, B.T., et al., 2013. Intestinal microbiota metabolism of L-carnitine, a nutrient in red meat, promotes atherosclerosis. Nat. Med. 19, 576–585.

Lane, J.A., Murray, L.J., Harvey, I.M., Donovan, J.L., Nair, P., Harvey, R.F., 2011. Randomised clinical trial: *Helicobacter pylori* eradication is associated with a significantly increased body mass index in a placebo controlled study. Aliment Pharmacol. Ther. 33, 922–929.

La Rosa, P.S., Warner, B.B., Zhou, Y., Weinstock, G.M., Sodergren, E., Hall-Moore, C.M., et al., 2014. Patterned progression of bacterial populations in the premature infant gut. Proc. Natl. Acad. Sci. USA 111, 12522–12527.

Lee, S.M., Donaldson, G.P., Mikulski, Z., Boyajian, S., Ley, K., Mazmanian, S.K., 2013. Bacterial colonization factors control specificity and stability of the gut microbiota. Nature 501 (7467), 426–429.

Logan, B.K., Jones, A.W., 2000. Endogenous ethanol "auto-brewery syndrome" as a drunk-driving defence challenge. Med. Sci. Law 40, 206–215.

Makino, H., Kushiro, A., Ishikawa, E., Kubota, H., Gawad, A., Sakai, T., et al., 2013. Mother-to-infant transmission of intestinal bifidobacterial strains has an impact on the early development of vaginally delivered infant's microbiota. PLoS One 8 (11), e78331.

Minot, S., Sinha, R., Chen, J., Li, H., Keilbaugh, S.A., Wu, G.D., et al., 2011. The human gut virome: inter-individual variation and dynamic response to diet. Genome. Res. 21, 1616–1625.

Moshfegh, A.J., Friday, J.E., Goldman, J.P., Ahuja, J.K., 1999. Presence of inulin and oligofructose in the diets of Americans. J. Nutr. 129, 1407S–1411S.

Olivares, M., Castillejo, G., Varea, V., Sanz, Y., 2014a. Double-blind, randomised, placebo-controlled intervention trial to evaluate the effects of *Bifidobacterium longum* CECT 7347 in children with newly diagnosed coeliac disease. Br. J. Nutr. 112, 30–40.

Olivares, M., Neef, A., Castillejo, G., Palma, G.D., Varea, V., Capilla, A., et al., 2014b. The HLA-DQ2 genotype selects for early intestinal microbiota composition in infants at high risk of developing coeliac disease. Gut 64 (3), 406–417.

Oozeer, R., van Limpt, K., Ludwig, T., Ben Amor, K., Martin, R., Wind, R.D., et al., 2013. Intestinal microbiology in early life: specific prebiotics can have similar functionalities as human-milk oligosaccharides. Am. J. Clin. Nutr. 98, 561S–571S.

Picot, D., Lauvin, R., Hellegouarc'h, R., 1997. Intra-digestive fermentation in intestinal malabsorption syndromes: relations with elevated serum activity of gamma-glutamyl-transpeptidase. Gastroenterol. Clin. Biol. 21, 562–566.

Preidis, G.A., Versalovic, J., 2009. Targeting the human microbiome with antibiotics, probiotics, and prebiotics: gastroenterology enters the metagenomics era. Gastroenterology 136, 2015–2031.

Reinhardt, C., Bergentall, M., Greiner, T.U., Schaffner, F., Ostergren-Lundén, G., Petersen, L.C., et al., 2012. Tissue factor and PAR1 promote microbiota-induced intestinal vascular remodeling. Nature 483 (7391), 627–631.

Ruiz-Moyano, S., Totten, S.M., Garrido, D.A., Smilowitz, J.T., German, J.B., Lebrilla, C.B., et al., 2013. Variation in consumption of human milk oligosaccharides by infant gut-associated strains of *Bifidobacterium breve*. Appl. Environ. Microbiol. 79, 6040–6049.

Said, H.M., 2013. Recent advances in transport of water-soluble vitamins in organs of the digestive system: a focus on the colon and the pancreas. Am. J. Physiol. Gastrointest. Liver Physiol. 305, G601–G610.

Sartor, R.B., 2012. Gut microbiota: Diet promotes dysbiosis and colitis in susceptible hosts. Nat. Rev. Gastroenterol. Hepatol. 9, 561–562.

Turnbaugh, P.J., Ley, R.E., Mahowald, M.A., Magrini, V., Mardis, E.R., Gordon, J.I., 2006. An obesity-associated gut microbiome with increased capacity for energy harvest. Nature 444 (7122), 1027–1031.

Turroni, F., Duranti, S., Bottacini, F., Guglielmetti, S., Van Sinderen, D., Ventura, M., 2014. Bifidobacterium bifidum as an example of a specialized human gut commensal. Front. Microbiol. 5, 437.

Walker, W.A., 2013. Initial intestinal colonization in the human infant and immune homeostasis. Ann. Nutr. Metab. 63 (Suppl. 2), 8–15.

Wang, X., Kim, Y., Ma, Q., Hong, S.H., Pokusaeva, K., Sturino, J.M., et al., 2010. Cryptic prophages help bacteria cope with adverse environments. Nat. Commun. 1, 147.

Weingarden, A.R., Chen, C., Bobr, A., Yao, D., Lu, Y., Nelson, V.M., et al., 2014. Microbiota transplantation restores normal fecal bile acid composition in recurrent *Clostridium difficile* infection. Am. J. Physiol. Gastrointest. Liver Physiol. 306, G310–G319.

Wu, G.D., Chen, J., Hoffmann, C., Bittinger, K., Chen, Y.Y., Keilbaugh, S.A., et al., 2011. Linking long-term dietary patterns with gut microbial enterotypes. Science 334 (6052), 105–108.

Zhu, Y., Jameson, E., Crosatti, M., Schäfer, H., Rajakumar, K., Bugg, T.D., et al., 2014. Carnitine metabolism to trimethylamine by an unusual Rieske-type oxygenase from human microbiota. Proc. Natl. Acad. Sci. USA 111, 4268–4273.

RENAL PROCESSING

The kidneys selectively secrete dead-end products of nutrient metabolism, such as urea and uric acid; metabolize nutrients and other compounds; and regulate blood volume and acid–base balance. The list of nutrients and nutrient like compounds whose availability is at least partly determined by the kidney includes vitamin D, creatine, sodium, chloride, potassium, calcium, phosphate, and magnesium.

ABBREVIATIONS	
ANP	atrial natriuretic peptide
AQP	aquaporin
DBP	vitamin D–binding protein
GFR	glomerular filtration rate
25-OH-D	25-hydroxyvitamin D
1,25-$(OH)_2$-D	1,25-dihydroxyvitamin D
PTH	parathyroid hormone
RBP	retinol–binding protein
TAUT	taurine transporter (SLC6A6)

RENAL ANATOMY

The kidneys are a pair of bean-shaped organs that contain about a million individual microscopic functional units called *nephrons*. Each nephron consists of a tangle of capillaries inside a capsule (Bowman's capsule) connected to a series of tubules (i.e., proximal tubule, loop of Henle, distal tubule, and collecting duct) that eventually drain into the ureter (Figure 3.3).

The filtering layer of the glomerulus is formed by the fenestrated capillary endothelium, an extracellular glomerular basement membrane, and a layer of specialized epithelial cells called *podocytes* (Smoyer and Mundel, 1998). The podocyte foot processes are tightly connected to the basement membrane in a typical interlaced pattern. Small slitlike areas between the foot processes are thinned to a narrow diaphragm formed of rod-shaped units with some similarity to the tight junctions connecting tubular epithelial cells. This arrangement allows the unhindered passage of water and solutes up to a molecular weight of about 5–6 kDa. As their size increases, molecules are less likely to pass through the slits. Only a negligible percentage of albumin and similarly sized proteins crosses into the ultrafiltrate (Rose, 1989). The resulting ultrafiltrate passes into Bowman's space and from there into the proximal tubule. The amount of ultrafiltrate produced per minute is the glomerular filtration rate (GFR). Healthy kidneys in young women produce about 95 ml of ultrafiltrate per minute. Men produce slightly more (120 ml/min) because they are heavier on average. GFR declines with age, both due to nonspecific processes and disease-typical glomerular deterioration.

A healthy, young 70-kg man with a GFR of 125 ml/min produces about 180 l of ultrafiltrate per day, which contains large amounts of small solutes. This corresponds to many times a typical day's intake not only of water, but also of essential nutrients such as sodium (580 g), potassium (>25 g), calcium (>8 g), and vitamin D (>1000 µg). The proximal tubule extends from the glomerulus to the loop of Henle. Characteristics of its epithelial cells change from the early convoluted segment (S1 cells), to the late convoluted segment and early straight (pars recta) segment (S2 cells), and late straight segment (S3 cells). The apical membrane facing the tubular lumen is folded (microvilli) and covered by protrusions (brush border). The basolateral membrane is adjacent to the interstitium and peritubular

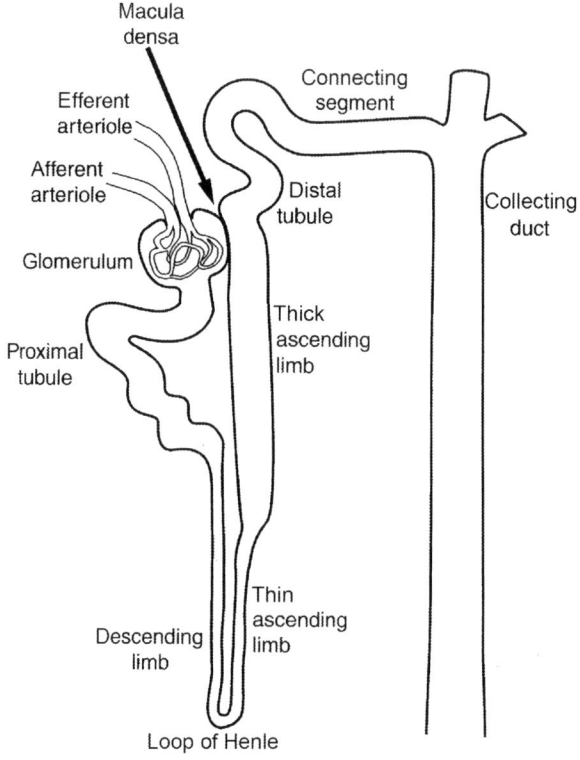

FIGURE 3.3

The anatomical organization of the nephrons.

capillaries. Tight junctions, which are strandlike structures, seal the space between the epithelial cells near the luminal side. The tight junctions of the proximal tubular epithelial cells are more permeable for water and electrolytes than the tight junctions downstream (Gumbiner, 1987). About 120 l/day of water (two-thirds of the amount filtered by the glomeruli) is reabsorbed from the proximal tubule, along with electrolytes, minerals and trace elements, glucose, amino acids, vitamins, and other filtered plasma constituents (Figure 3.4).

The *loop of Henle* is a hairpin arrangement that extends from the renal cortex into the medulla and consists of the narrowbore descending and thin ascending limbs followed by the thick ascending limb. Adjacent to the loops are the vasa recta, which start from glomeruli near the cortex–medulla interface, extend into the medulla, and return to the cortex along with the ascending limb of the loop. The capillaries are fully permeable to small- and medium-sized molecules. This is the site where further sodium chloride and water are reabsorbed by countercurrent exchange, intraluminal osmolality greatly increases toward the tip of the loop, reaching as much as 1200 mosmol/l. High intracellular concentrations of osmolytes, including sorbitol, myo-inositol, taurine, and betaine, protect the Henle loop epithelia from the potentially disastrous effects of an excessive osmolar gradient. The import of myo-inositol and

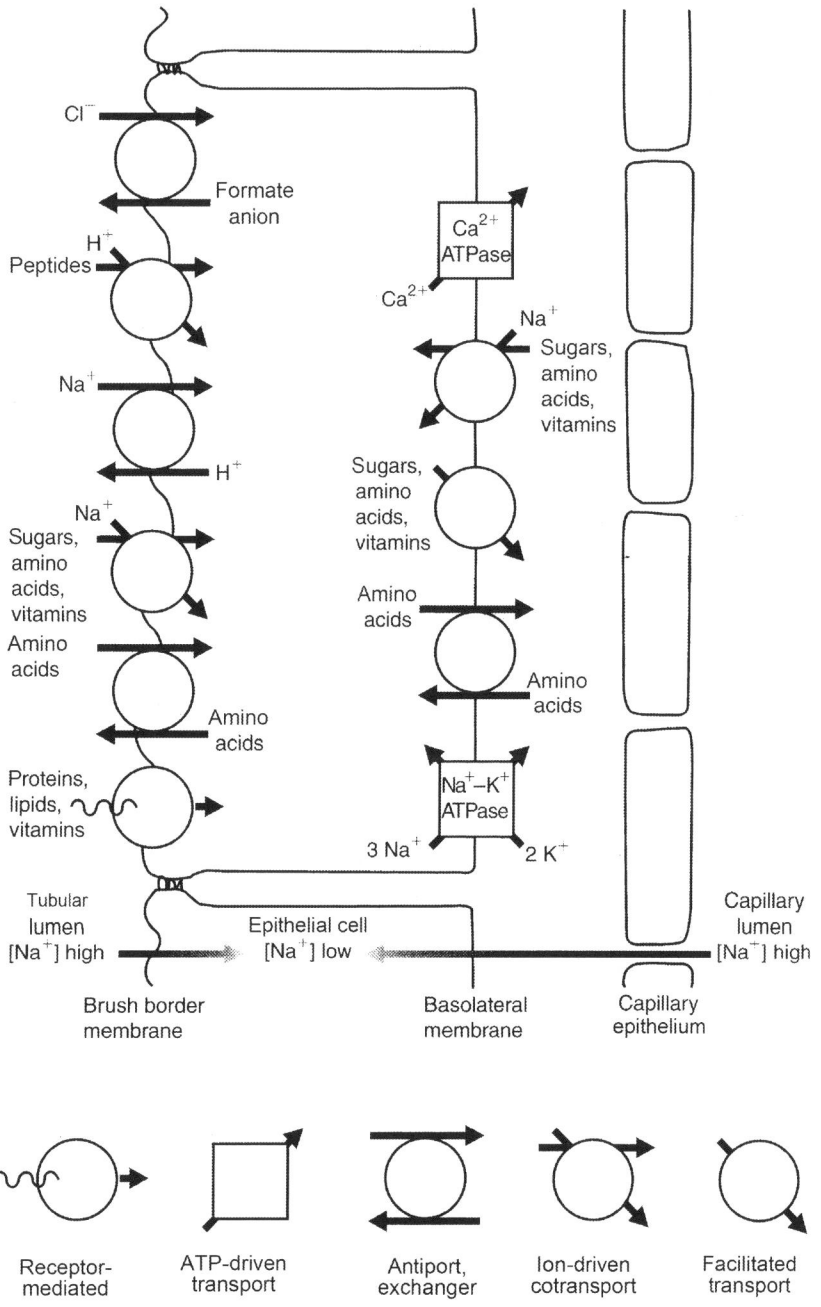

FIGURE 3.4

Diverse mechanisms mediating the recovery of nutrients from the proximal tubular lumen.

betaine is mediated by their respective transporters [namely, sodium-dependent myo-inositol transporter (SLC5A3) and betaine transporter BGT-1 (SLC6A12, sodium chloride-dependent)]; taurine is taken up via the taurine transporter (TAUT, SLC6A6); and sorbitol is produced locally (Bitoun et al., 2001).

The distal tubule extends from the macula densa (an important site for blood volume regulation) to the connecting segment. The tight junctions between its epithelial cells, as those of the following segments, prevent most uncontrolled paracellular movement of electrolytes from the highly concentrated luminal fluid into the blood. Another 5% of the originally filtered sodium chloride is reabsorbed here. Residual amino acids and other complex organic compounds can still be recovered from this nephron segment.

The connecting segment contributes to calcium recovery (driven by the calcium-transporting ATPase, EC3.6.3.8, at the basolateral membrane), absorbs some sodium, and reabsorbs calcium.

About two-thirds of cells forming the cortical collecting tubule are principal cells with the ability to further modify the electrolyte content of the luminal contents. Intercalated cells, the remainder of the cells in the cortical collecting tubule, contribute to sodium-independent acid–base balance. Intercalated cells differ from more proximal tubular cells in that their ATPases (proton and potassium-transporting) are located at the luminal membrane rather than the peritubular membrane.

The medullary-collecting tubule modifies urine composition further by adjusting its water, electrolyte, and proton content as needed. Water recovery from the inner medullar segment is under the control of antidiuretic hormone.

Salvage of complex nutrients

Complex nutrients are recovered very efficiently from ultrafiltrate in the proximal renal lumen so long as intake levels are modest and renal function is normal. Some of this reabsorptive activity continues in parts of the distal tubule. The luminal side of the tubular epithelial cells has a brush border membrane with numerous specific transport systems that mediate the uptake of carbohydrates, proteins and amino acids, vitamins, and most other essential nutrients. In many instances, these are the same systems that also mediate nutrient uptake across the small intestinal brush border membrane. The major driving force for uptake from the lumen into the tubular epithelial cells is the low intracellular sodium concentration that is maintained by sodium/potassium ATPase at the basolateral membrane of both proximal and distal tubular cells. Additional gradients involved in tubular reuptake include protons, formate, and an electric potential difference which favors inflow of cations. Receptor-mediated pinocytosis is another important mechanism of concentrative transport. In most cases, a distinct set of transporters and channels then mediates the transport out of the epithelial cell across the basolateral membrane, this time mainly driven by the concentration gradient of the transported molecules, antiport mechanisms, or active transport. Once they have reached the basolateral intercellular space, the molecules can then move into the luminal space of peritubular blood capillaries by simple diffusion. Neither the basement membrane adjacent to the tubular cell layer nor the fenestrated epithelium of the capillaries constitutes a significant barrier to this last step of solute transfer from tubular lumen to capillary lumen.

Carbohydrates: The sugar content of the ultrafiltrate reflects the composition of plasma since these small molecules are readily filtered. Recovery of D-glucose and D-galactose from the lumen via sodium-glucose cotransporters proceeds with high capacity and low affinity in segments S1 and S2 of the proximal tubule, and with low capacity but high affinity in segment S3. Glucose salvage becomes noticeably incomplete (i.e., glucose appears in urine) when the concentration in blood exceeds about 1800 mg/l. This threshold rises as GFR decreases (Rose, 1989, pp. 102–103). Fructose crosses the brush border membrane via its own transporter, GLUT5 (Mate et al., 2001). D-Mannose uptake across the brush border membrane proceeds via a sodium-dependent transporter distinct from

Table 3.6 Amino Acid Transporters in the Human Kidney

Transporters	In	Out	In
Apical			
pepT1 and pepT2	Na$^+$		Di- and tripeptides
BAT1/b$^{o,+}$ + rBAT		Neutral amino acids	K, H, R, E, D, S, T, F, W, G, A, C, V, I, L, P, M, cystine, ornithine
ASC	Na$^+$		G, A, S, C, T
B^o/B/NBB	Na$^+$		V, I, L, T, F, W, [A, S, C]
B$^{o,+}$	Na$^+$		H, C, R, taurine, beta-alanine, carnitine
TAUT	2 Na$^+$ Cl$^-$		Taurine, beta-alanine
BGT-1	3 Na$^+$		Betaine
GAT-1 and GAT-3	Na$^+$ Cl$^-$		GABA, hypotaurine, beta-alanine
IMINO	Na$^+$		P, OH-P, taurine, beta-alanine
EAAC1/X$^-_{AG}$	3 Na$^+$	K$^+$	D, E
OCTN2	Na$^+$		Carnitine
y$^+$ CAT			R, K, ornithine, choline, polyamines
Basolateral			
y(+)LAT1 + 4F2	Neutral amino acids		K, R, H, Q, N, ornithine, choline, orotate
LAT2 + 4F2	Neutral amino acids		Y, F, W,T, N, I, C, S, L, V, Q, [H, A, M, G]
A	Na$^+$		A,S,Q
ASCT1	Na$^+$		G, A, S, C, T
TAUT	Na$^+$ Cl$^-$		Taurine, beta-alanine
BGT-1/GAT-2	Na$^+$ Cl$^-$		Betaine, hypotaurine, beta-alanine
system T (TAT1)	Neutral amino acids		F, Y, W
asc-1 + 4F2	Neutral amino acids		G, A, S, C, T, D-serine

the sodium-glucose transporters. Its renal recovery is a critical element for the regulation of D-mannose homeostasis (Blasco et al., 2000). All major sugars are transferred across the basolateral membrane by the glucose transporter 2 (GLUT2).

Citrate: The sodium/dicarboxylate cotransporter (NaDC-1, SLC13A2) in the proximal tubule mediates citrate recovery. The efficiency of this process is determined by acid–base balance, increasing with acidosis. Since citrate competes with phosphate and oxalate for binding to calcium, its residual concentration in urine contributes to protection against calcium oxalate and calcium phosphate stone formation (Brardi et al., 2012). Daily citrate excretion typically is several hundred milligrams (Schwille et al., 1979).

Proteins and amino acids: Many different transporters contribute to the recovery of proteins, peptides, and amino acids from ultrafiltrate in the kidneys (Table 3.6). Several specific proteins, including retinol–binding protein (RBP), vitamin D–binding protein (DBP), transcobalamin II, insulin, and lysozyme, are taken up intact by megalin-mediated endocytosis as described in more detail later in this chapter. Most of the smaller proteins are hydrolyzed by various brush border exoenzymes, including membrane Pro-X carboxypeptidase (EC3.4.17.16) and angiotensin I–converting enzyme (ACE; EC3.4.15.1).

Two distinct sodium/peptide cotransporters then mediate the uptake of dipeptides and tripeptides, but not of free amino acids. Sodium/peptide cotransporter 1 (PepT1, SLC15A1) in the S1 segment of the proximal tubule has lower affinity for the oligopeptides than sodium/peptide cotransporter 2 (PepT2, SLC15A2) in the S3 segment (Shen et al., 1999).

Neutral amino acids enter epithelial cells mainly via the sodium-dependent neutral amino acid transporters $B°$ (Avissar et al., 2001), ASCT2, and $B^{o,+}$. Glutamate and aspartate use the EAACI/X Am transport system. The sodium-dependent transporters GAT-1 and GAT-3, which are better known for their role in neurotransmitter recovery in the brain, ferry gamma-aminobutyric acid (GABA), hypotaurine, and beta-alanine across the proximal tubular brush border membrane (Muth et al., 1998). Proline, hydroxyproline, taurine, and beta-alanine are taken up by the sodium-dependent imino transporter (Urdaneta et al., 1998), and betaine enters via the sodium- and chloride-dependent betaine transporter (SLC6A12). Taurine uptake via the taurine transporter (TAUT, SLC6A6) is sodium- and chloride-dependent (Chesney et al., 1990). High concentrations of osmolytes, such as taurine and betaine, protect epithelial cells against the high osmotic pressure in the medulla. Specificity and capacity of the sodium-dependent transporters is expanded considerably by the rBAT (SLC3A1)–linked transporter BAT1 (SLC7A9). This transporter, which accounts for most, if not all, activity of system $b^{o,+}$ shuttles small and large neutral amino acids across the brush border membrane in exchange for other neutral amino acids. Carnitine enters the cell via the organic cation transporter OCTN2 in exchange for tetraethylammonium or other organic cations (Ohashi et al., 2001). Amino acids are utilized to some extent in tubular epithelial cells for protein synthesis, energy production, and other metabolic pathways. The case of hydroxyproline is somewhat special since the kidneys are the main sites of hydroxyproline metabolism, mainly to serine and glycine (Lowry et al., 1985). Hydroxyproline is derived from dietary collagen and from endogenous muscle, connective tissue, and bone turnover. It reaches the mitochondria of the tubular epithelial cells through a translocator that is distinct from that for proline (Atlante et al., 1994). Hydroxyproline is then oxidized by 4-oxoproline reductase (hydroxyproline oxidase; EC1.1.1.104) to 4-oxoproline (Kim et al., 1997). 4-Hydroxy-2-oxoglutarate aldolase (EC4.1.3.16) generates pyruvate and glyoxylate. Glycine is produced when the pyridoxal-phosphate-dependent alanine-glyoxylate aminotransferase (EC2.6.1.44) uses alanine for the amination of glyoxylate.

Transport: The combination of $b^{o,+}$AT (BAT1, SLC7A9) and rBAT (SLC3A1) moves the bulk of amino acids across the apical (luminal) membrane of the proximal tubule and some downstream nephron segments.

The main sodium-dependent amino acid transporters of the basolateral membrane of the proximal tubule are system A (preferentially transporting alanine, serine, and glutamine) and ASCT1 (transporting alanine, serine, cysteine, and threonine). Net transfer of individual amino acids, importantly, depends on their own concentration gradient. As on the luminal side, some transporters operate in exchange mode. Small neutral amino acids are the main counter molecules because their concentration is the highest. Glycoprotein 4F2 (SLC3A2) anchors the amino acid exchangers typical for this side to the basolateral membrane (Verrey et al., 1999). The L-type amino acid transporter LAT2 (SLC7A8) accepts most neutral amino acids for transport in either direction. The facilitated diffusion transporter TAT1 (SLC16A10) colocalizes with LAT2. It recycles aromatic amino acid to help the efflux of other amino acids (Ramadan et al., 2007).

Arginine and other cationic amino acids can pass through related heterodimers; one of these is 4F2 in combination with y(+)LAT1 (SLC7A7), while another one consists of 4F2 and y(+)LAT2 (SLC7A6). These transporters can exchange a cationic amino acid for a neutral amino acid plus a sodium ion. The combination of asc-1 (SLC7A10) and 4F2 mediates the transport of small neutral amino acids, beta-alanine, and some D amino acids at the basolateral membrane of the thin limbs of Henle's loop, distal tubules, and collecting ducts (Pineda et al., 2004).

GAT-2 mediates betaine, beta-alanine, and some taurine transport. The same compounds may also leave via the sodium chloride-dependent taurine transporter (SLC6A6).

Urea: One of the major functions of the kidneys is to eliminate the potentially toxic end products of amino acid utilization. As the tubular fluid is concentrated, a urea concentration gradient builds up that drives the passive diffusion of urea across the tubular epithelium into the peritubular blood capillary. This diffusion is little hindered by cell membranes because these are readily permeable to lipid soluble urea. Due to this reabsorption, only about half of the filtered urea (>50 g/day) is excreted with urine. A much smaller amount of protein-derived nitrogen is excreted as ammonia. The ammonia can be generated from glutamine by glutaminase and is secreted into the distal (S3) part of the proximal tubulus via the sodium/hydrogen ion antiporter (SLC9A1) in a sodium/ammonium ion exchange mode.

Vitamins: The most significant effect of the kidney may be on vitamin D. After it is synthesized in the skin or absorbed from food, vitamin D is converted rapidly to 25-hydroxyvitamin D (25-OH-D) in the liver. 25-OH-D is secreted in the blood, where it circulates in association with DBP. Owing to its relatively small size, a significant percentage of the complex gets into renal ultrafiltrate. Megalin, a member of the lipoprotein receptor family, binds DBP, and mediates its uptake into the epithelial cells of the proximal tubule. 25-OH-D can then be hydroxylated by mitochondrial vitamin D-1-alpha-hydroxylase (P450c1 alpha, CYP27B1) to 1,25-dihydroxyvitamin D (1,25-$(OH)_2$-D). Parathyroid hormone (PTH), calcitonin (Shinki et al., 1999), phosphate concentration in blood (Prince et al., 1988), and other factors tightly control the rate of 1,25-$(OH)_2$-D synthesis. In situations of limited vitamin D availability, however, the supply of precursor taken up from the proximal lumen is important. Diminished filtration in patients with end-stage renal disease severely limits vitamin D activation with all the attendant consequences of 1,25-dihydroxyvitamin D deficiency.

Cobalamin uptake from the proximal lumen is mediated by megalin as well. In blood, and hence in the filtrate, transcobalamin II is the cobalamin carrier protein. Additional transcobalamin II appears to be secreted into the proximal tubular lumen, which would ensure maximal recovery. Contrast this with the mechanism for intestinal absorption, where cobalamin is bound to intrinsic factor and taken up via cubilin. Retinol, which circulates in blood bound to retinol–binding protein (RBP), is another vitamin relying on megalin for salvage from ultrafiltrate. Clara cell secretory protein (CCSP) is a blood transport protein for lipophilic (xenobiotic) substances, including polychlorinated biphenyl metabolites. This versatile carrier, with any ligands that might be attached to it, is extracted from primary filtrate by cubilin. The complex is then targeted toward lysosomes by its coreceptor megalin (Burmeister et al., 2001; Christensen and Birn, 2001). Thiamin pyrophosphate is dephosphorylated and the free thiamin taken up from the tubular lumen by a thiamin/H^+ antiporter with a 1:1 stoichiometric ratio (Gastaldi et al., 2000). Transport across the basolateral membrane uses an as-yet-uncharacterized ATP-driven thiamin carrier. Similarly, a nucleotide pyrophosphatase (EC3.6.1.9) cleaves several vitamin-derived nucleotides, including NAD, NADP, FAD, and coenzyme A. While this enzyme is certainly expressed in the distal tubule, its presence in proximal tubules has not been reported. The free vitamins (riboflavin, niacin, and pantothenate) can be taken up via their respective transport systems. Pantothenate, like biotin and lipoate, is taken up from the proximal tubular lumen via the sodium-dependent multivitamin transporter (SLC5A6). Folate is recovered from the proximal tubular lumen by folate receptors; the reduced folate carrier 1 (SLC19A1) then completes transport across the basolateral membrane in exchange for organic phosphate (Sikka and McMartin, 1998; Wang et al., 2001).

WATER, ELECTROLYTES, MINERALS, AND TRACE ELEMENTS

Water: About two-thirds of the filtered water is reabsorbed from the proximal tubular lumen. Much of this water movement is by transcellular cotransport following the sodium-dependent uptake of organic compounds such as glucose. Water can also pass directly between the lumen and the basolateral space by passive diffusion. The tight junctions that seal the intercellular gaps near the luminal side permit more water movement in either direction than in the more distal segments of the nephron (Gumbiner, 1987). The countercurrent principle is responsible for further concentration of the luminal fluid in the descending limb of the loop of Henle. Water from the descending limb follows the sodium and potassium gradient built up by ATPase-driven electrolyte transport on the adjacent ascending limb. Final urine concentration is adjusted by water diffusion out of the distal and collecting tubules.

Aquaporins (AQPs) are well-defined water channels on the luminal and basolateral side of the tubular epithelium (Nielsen et al., 2002). The kidney expresses at least seven different AQPs. Their presence and regulated permeability determines whether water can follow a concentration gradient from the tubular lumen into the epithelial cells and from there into the pericapillary space. Proximal tubules and the descending limbs of Henle's loop contain AQP7 at the luminal side and AQP1 at the basolateral side. The ascending limb of the loop is largely devoid of AQPs and thus not well water permeable. The principal cells of the connecting and the collecting tubules contain AQP2 at the luminal side, regulated by antidiuretic hormone (arginine vasopressine, ADH), and AQP3 plus AQP4 at the basolateral side. The roles of AQP6 and AQP8 are less well defined.

Sodium, potassium, and chloride: The handling of the three main electrolytes in the kidney is intrinsically linked. The sodium–potassium ATPase at the basolateral membrane of most nephron segments pumps three sodium ions from the cell into the intercellular space in exchange for two potassium ions. This pumping action maintains the intracellular sodium concentration much lower than in the tubular lumen. The resulting sodium concentration gradient drives the cotransport for the recovery of glucose, amino acids, and other nutrients in the proximal tubule. At the same time, a large percentage of the filtered sodium is taken up. The dissociation of carbonic acid by carbonate dehydratase (carbonic acid anhydrase, EC4.2.1.1, zinc-dependent) provides the protons, which are exchanged for luminal sodium by the sodium/hydrogen ion antiporter (SLC9A1). The bicarbonate ions, in turn, drive sodium transport across the basolateral side membrane via the sodium/bicarbonate cotransporter (SLC4A4). Chloride is taken up from the proximal tubular lumen in exchange for formate by the formate/chloride exchanger and carried across the basolateral membrane together with a potassium ion by the potassium chloride cotransporter. A smaller proportion of the filtered chloride reaches the basolateral intercellular space directly by diffusion across the tight junctions. As pointed out previously, the tight junctions of the proximal nephron segments are more permeable than those of the more distal segments and allow some movement of electrolytes in either direction, depending on the direction of the prevailing concentration gradient.

Sodium chloride flows into the interstitium around the thin ascending limb of Henle's loop by passive diffusion. In the segments distal to the thin ascending limb, additional electrolytes are actively recovered. A passive carrier on the luminal side with a sodium-potassium-chloride stoichiometry of 1:1:2 and the sodium–potassium ATPase, in conjunction with potassium channels on the luminal side and chloride channels on the basolateral side, provide for the mechanisms of sodium chloride reabsorption in the thick ascending limb of Henle's loop. Sodium chloride reabsorption in the distal tubule depends on sodium chloride cotransporters on both sides of the epithelial cell and the sodium–potassium

ATPase on the basolateral side. Final adjustments to electrolyte composition occur in the collecting tubules. In most situations, this final nephron segment provides for net secretion of potassium into urine. Aldosterone in the principal cells of the cortical collecting tubules modulates the permeability of the tight junctions (for the paracellular diffusion of chloride) and of potassium and sodium channels. The intercalated cells of cortical-collecting tubules contain a potassium ATPase, which allows for net potassium reabsorption in response to depletion.

Calcium: A calcium-binding protein mediates calcium recovery from the lumen of the distal nephron segments between thick ascending limb and the connecting segment. The calcium-ATPase (EC3.6.3.8) at the basolateral side of the tubular epithelium establishes the sodium-independent calcium gradient that drives calcium transfer from the lumen back into the bloodstream. Typically, most of the filtered calcium (>8 g/day) is recovered through this mechanism, but the efficiency is largely regulated by hormones and other agents. Most important, calcium-ATPase is stimulated when PTH binds to the PTH receptor on the basolateral side of the epithelial cell and triggers a cyclic adenosine monophosphate (cAMP)–mediated signaling cascade.

Magnesium: Free ionized magnesium and magnesium complexed with small anions, which constitutes the bulk of circulating magnesium, is freely filtered by the renal glomerulus. Normally, less than 5% of the filtered magnesium is lost with urine. Some of the luminal magnesium is recovered from the proximal tubule, a much greater proportion from the thick ascending limb of Henle's loop and from the distal tubule.

Phosphate: The type IIa sodium/phosphate cotransporter of the proximal tubular brush border membrane extracts most (80–95%) phosphate from the ultrafiltrate (Muter et al., 2001). PTH modulates fractional phosphate reabsorption by decreasing transporter expression. Much less is known about the mechanisms underlying exit across the basolateral membrane. Several anion exchange pathways are likely to contribute. These pathways can also supply phosphate from the blood if uptake from the tubular lumen becomes insufficient for the cells' own needs.

Iodide: The chloride/iodide transporter pendrin (SLC26A4) mediates iodide uptake across the luminal membranes of proximal tubular cells (Soleimani et al., 2001) and intercalated cells of the cortical collecting ducts (Royaux et al., 2001). Iodide may then be pumped by the sodium/iodide symporter across the basolateral membrane into the perivascular space (Spitzweg et al., 2001) and reenter blood circulation.

Active secretion of food compounds

Many ingested and absorbed food constituents are actively secreted into urine. The organic anion transporters 1 (OAT1) and 3 (OAT3) take up acidic xenobiotics from the pericapillary space in exchange for alpha-ketoglutarate and other dicarboxylic acids (Burckhardt and Wolff, 2000). Similarly, the organic cation transporters 1 (OCT1), 2 (OCT2), and 3 (OCT3) are present in the basolateral membrane of the proximal and distal tubules. The niacin catabolite *N*-methyl-nicotinamide (NMN) is one of the transported cations. Several active transport systems on the luminal side complete secretion. Multidrug resistance protein 2 (MRP2; ABCC2) transports creatinine, urate, hippurate, ketoacids, and salicylates (Berkhin and Humphreys, 2001).

Xenobiotic–cysteine conjugates that have been secreted into the lumen can be cleaved by the PLP-dependent enzyme cysteine conjugate beta-lyase (EC4.4.1.13). The reaction releases ammonia and pyruvate, which can be reabsorbed. The thio-derivatives of the xenobiotics may then react further (e.g., oxidize), or they are excreted unchanged (van Bladeren, 2000).

Hormones affecting renal function

Several key hormones, including ADH, aldosterone, angiotensin II, atrial natriuretic peptide (ANP), prostaglandins, and PTH, control nutrient homeostasis by acting on renal cells. In some instances, hormones act by modulating tubular reabsorption or secretion of nutrients or nutrient metabolites; in others, by determining the rate of metabolic transformation of nutrients. A wide variety of additional hormones and hormone-like factors (which are not mentioned in the following section) may also affect renal disposition of nutrients directly or indirectly.

ADH is formed in supraoptic and ventricular nuclei and secreted into cerebrospinal fluid (CSF), the portal capillaries of the median eminence (which supply the anterior pituitary via the long portal vein), and the posterior pituitary gland. ADH release is governed through osmoreceptors. Through the venous system of the pituitary gland, ADH can then reach systemic circulation and act on kidney, blood vessels, and other tissues. ADH activates cAMP-mediated signaling in epithelial cells of collecting tubules, thereby increasing water permeability of the apical membranes. ADH acts mainly by inducing the insertion of AQ2 into the luminal membrane of the ADH-sensitive collecting tubules. The increase in the number of these water channels facilitates osmotic equilibration and water moves from the lumen into the more concentrated interstitium. Without adequate ADH concentration, urine cannot be concentrated properly in certain situations (e.g., with many old people).

Aldosterone is synthesized from cholesterol in the zona glomerulosa of the adrenal cortex. Secretion of the hormone is promoted via the renin–angiotensin cascade in response to low renal perfusion and sympathetic activation. Renin is produced by the iuxtaglomerular cells located in the afferent arterioles of glomeruli near the macula densa. Renin (EC3.4.23.15) is a peptidase that converts angiotensinogen into angiotensin I (comprised of 10 amino acids). Another peptidase, angiotensin-converting enzyme (ACE, EC3.4.15.1), generates angiotensin II by removing two additional amino acids. Angiotensin II then promotes aldosterone production in the adrenal gland. Angiotensin II also acts directly on the sodium–proton antiporter in the proximal tubule, which increases sodium reabsorption. Aldosterone acts mainly on connecting segments and collecting tubules. After entering the epithelial cells of those nephron segments aldosterone combines with a cytosolic receptor and binds to DNA in the nucleus. Gene activation by the aldosterone-receptor complex promotes sodium–potassium ATPase pumping at the basolateral membrane and increases the number of sodium and potassium channels in the luminal membrane. An important consequence of aldosterone action is the increase in potassium content of urine. ANP is produced by myocardial cells in the atria and portions of the ventricles of the heart in response to volume expansion. ANP acts on a specific membrane receptor, which triggers a signaling cascade via the formation of cGMP. ANP increases glomerular filtration and decreases sodium reabsorption from collecting tubules.

Prostaglandins are produced within and outside the kidney by cyclooxygenase from long-chain polyunsaturated fatty acid precursors. PGE2 is a major form synthesized in the tubular epithelium. PGE2 and prostacyclin are produced in glomeruli and vascular epithelium. Amount and type of synthesized prostaglandin is influenced not only by medical drugs (nonsteroidal anti-inflammatory agents such as acetylsalicylic acid), but also by intake levels of omega-3 fatty acids (e.g., eicosapentaenoic acid in fish oils and flax seed), omega-6 fatty acids (e.g., linoleic acid in many seed oils), and salicylates (in many fruits and vegetables). Prostaglandins limit the activity of ADH and ensure adequate blood flow by easing vasoconstriction of the renal arterioles.

PTH is secreted by the four parathyroid glands in response to low ionized calcium concentration in blood. The stimulating influence of low blood calcium concentration is attenuated by the interaction of 1,25-(OH)2-D with specific receptors in the parathyroid glands (Russell et al., 1986). Binding of PTH to its receptors on tubular epithelial cells triggers a cAMP-initiated signaling cascade. PTH increases calcium reabsorption through activation of calcium-ATPase at the basolateral membrane and it promotes the synthesis of the active form of vitamin D, 1,25-(OH)2-D. At the same time, PTH slows phosphate uptake from the proximal tubular lumen via the sodium/phosphate cotransporter.

The capillary endothelial cells adjacent to the tubular epithelium in response to low oxygen saturation produce more than 90% of the body's erythropoietin. The main action of erythropoietin is the proliferation and differentiation of red blood cell precursors. The attempt to overcome anemia resulting from inadequate erythropoietin production (e.g., in renal disease) with increased iron intakes is likely to cause more harm than benefit.

REFERENCES

Atlante, A., Passarella, S., Quagliariello, E., 1994. Spectroscopic study of hydroxyproline transport in rat kidney mitochondria. Biochem. Biophys. Res. Comm. 202, 58–64.

Avissar, N.E., Ryan, C.K., Ganapathy, V., Sax, H.C., 2001. Na^+-dependent neutral amino acid transporter ATB^0 is a rabbit epithelial cell brush border protein. Am. J. Physiol. Cell Physiol. 281, C963–C971.

Berkhin, E.B., Humphreys, M.H., 2001. Regulation of renal tubular secretion of organic compounds. Kidney Int. 59, 17–30.

Bitoun, M., Levillain, O., Tappaz, M., 2001. Gene expression of the taurine transporter and taurine biosynthetic enzymes in rat kidney after antidiuresis. Pfl. Arch. Eur. J. Physiol. 442, 87–95.

Blasco, T., Aramayona, J.J., Alcalde, A.I., Halaihel, N., Sarasa, M., Sorribas, V., 2000. Expression and molecular characterization of rat renal D-mannose transport in *Xenopus oocytes*. J. Membr. Biol. 178, 127–135.

Brardi, S., Imperiali, P., Cevenini, G., Verdacchi, T., Ponchietti, R., 2012. Effects of the association of potassium citrate and agropyrum repens in renal stone treatment: results of a prospective randomized comparison with potassium citrate. Arch. Ital. Urol. Androl. 84, 61–67.

Burckhardt, G., Wolff, N.A., 2000. Structure of renal organic anion and cation transporters. Am. J. Physiol. Renal. Physiol. 278, F853–F866.

Burmeister, R., Boe, I.M., Nykjaer, A., Jacobsen, C., Moestrup, S.K., Verroust, P., et al., 2001. A two-receptor pathway for catabolism of Clara cell secretory protein in the kidney. J. Biol. Chem. 276, 13295–13301.

Chesney, R.W., Zelikovic, I., Jones, D.E., Budreau, A., Jolly, K., 1990. The renal transport of taurine and the regulation of renal sodium chloride-dependent transporter activity. Ped. Nephrol. 4, 399–407.

Christensen, El, Birn, H., 2001. Megalin and cubilin: synergistic endocytic receptors in renal proximal tubule. Am. J. Physiol. Renal Fluid Electrolyte Physiol. 280, F562–F573.

Gastaldi, G., Cova, E., Verri, A., Laforenza, U., Faelli, A., Rindi, G., 2000. Transport of thiamin in rat renal brush border membrane vesicles. Kidney Int. 57, 2043–2054.

Gumbiner, B., 1987. Structure, biochemistry, and assembly of tight junctions. Am. J. Physiol. 253, C749.

Kim, S.Z., Varvogli, L., Waisbren, S.E., Levy, H.L., 1997. Hydroxyprolinemia: comparison of a patient and her unaffected twin sister. J. Pediat. 130, 437–441.

Lowry, M., Hall, D.E., Brosnan, J.T., 1985. Hydroxyproline metabolism by the rat kidney: distribution of renal enzymes of hydroxyproline catabolism and renal conversion of hydroxyproline to glycine and serine. Metab. Clin. Exp. 34, 955–961.

Mate, A., de la Hermosa, M.A., Barfull, A., Planas, J.M., Vazquez, C.M., 2001. Characterization of D-fructose transport by rat kidney brush border membrane vesicles: changes in hypertensive rats. Cell Mol. Life Sci. 58, 1961–1967.

Muter, H., Hernando, N., Forster, I., Biber, J., 2001. Molecular aspects in the regulation of renal inorganic phosphate reabsorption: the type IIa sodium/inorganic phosphate co-transporter as the key player. Curr. Opin. Nephrol. Hypertens. 10, 555–561.

Muth, T.R., Ahn, J., Caplan, M.J., 1998. Identification of sorting determinants in the C-terminal cytoplasmic tails of the gamma-aminobutyric acid transporters GAT-2 and GAT-3. J. Biol. Chem. 273, 25616–25627.

Nielsen, S., Frøkiæer, J., Marples, D., Kwon, T.H., Agre, P., Knepper, M.A., 2002. Aquaporins in the kidney: from molecules to medicine. Phys. Rev. 82, 205–244.

Ohashi, R., Tamai, I., Nezu Ji, J., Nikaido, H., Hashimoto, N., Oku, A., et al., 2001. Molecular and physiological evidence for multifunctionality of carnitine/organic cation transporter OCTN2. Mol. Pharmacol. 59, 358–366.

Pineda, M., Font, M., Bassi, M.T., Manzoni, M., Borsani, G., Marigo, V., et al., 2004. The amino acid transporter asc-1 is not involved in cystinuria. Kidney Int. 66, 1453–1464.

Prince, R.L., Hutchison, B.G., Kent, J.C., 1988. Calcitriol deficiency with retained synthetic reserve in chronic renal failure. Kidney Int. 33, 722–728.

Ramadan, T., Camargo, S.M., Herzog, B., Bordin, M., Pos, K.M., Verrey, F., 2007. Recycling of aromatic amino acids via TAT1 allows efflux of neutral amino acids via LAT2-4F2hc exchanger. Pflügers Arch. 454, 507–516.

Rose, B.D., 1989. Clinical Physiology of Acid-Base and Electrolyte Disorders. McGraw-Hill, New York, NY.

Royaux, I.E., Wall, S.M., Karniski, L.P., Everett, L.A., Suzuki, K., Knepper, M.A., et al., 2001. Encoded by the Pendred syndrome gene, resides in the apical region of renal intercalated cells and mediates bicarbonate secretion. Proc. Natl. Acad. Sci. USA 98, 4221–4226.

Russell, J., Lettieri, D., Sherwood, L.M., 1986. Suppression by 1,25(OH)2D3 of transcription of pre-proparathyroid hormone gene. Endocrinology 119, 2864–2866.

Schwille, P.O., Scholz, D., Paulus, M., Engelhardt, W., Sigel, A., 1979. Citrate in daily and fasting urine: results of controls, patients with recurrent idiopathic calcium urolithiasis, and primary hyperparathyroidism. Invest. Urol. 16, 457–462.

Shen, H., Smith, D.E., Yang, T., Huang, Y.G., Schnermann, J.B., Brosius III, F.C., 1999. Localization of PEPT1 and PEPT2 proton-coupled oligopeptide transporter mRNA and protein in rat kidney. Am. J. Physiol. 276, F658–F665.

Shinki, T., Ueno, Y., DeLuca, H.E., Suda, T., 1999. Calcitonin is a major regulator for the expression of renal 25-hydroxyvitamin D3-1alpha-hydroxylase gene in normocalcemic rats. Proc. Natl. Acad. Sci. USA 96, 8253–8258.

Sikka, P.K., McMartin, K.E., 1998. Determination of folate transport pathways in cultured rat proximal tubule cells. Chem. Biol. Interact. 114, 15–31.

Smoyer, W.E., Mundel, P., 1998. Regulation of podocyte structure during the development of nephrotic syndrome. J. Mol. Med. 76, 172–183.

Soleimani, M., Greeley, T., Petrovic, S., Wang, Z., Amlal, H., Kopp, P., et al., 2001. Pendrin: an apical $Cl^-/OH^-/HCO_3^-$ exchanger in the kidney cortex. Am. J. Physiol. Renal Fluid Electrolyte Physiol. 280, F356–F364.

Spitzweg, C., Dutton, C.M., Castro, M.R., Bergen, E.R., Goellner, J.R., Heufelder, A.E., et al., 2001. Expression of the sodium iodide symponer in human kidney. Kidney Int. 59, 1013–1023.

Urdaneta, E., Barber, A., Wright, E.M., Lostao, M.E., 1998. Functional expression of the rabbit intestinal Na+/L-proline cotransporter (IMINO system) in *Xenopus laevis* oocytes. J. Physiol. Biochem. 54, 155–160.

Van Bladeren, P.J., 2000. Glutathione conjugation as a bioactivation reaction. Chem. Biol. Interact. 129, 61–76.

Verrey, E., Jack, D.L., Paulsen, I.T., Saier Jr., M.H., Pfeiffer, R., 1999. New glycoprotein-associated amino acid transporters. J. Membrane Biol. 172, 181–192.

Wang, Y., Zhao, R., Russell, R.G., Goldman, I.D., 2001. Localization of the murine reduced folate carrier as assessed by immunohistochemical analysis. Biochim. Biophys. Acta 1513, 49–54.

THE BLOOD–BRAIN BARRIER

The brain needs a steady supply of nutrients and essential metabolites. However, it is also critical that the removal of metabolic waste products and nutrient excess can proceed as needed. The interface between blood and brain is formed by two distinct barrier systems (Takeda, 2000). The blood–brain barrier (BBB) separates the blood in most of the brain from brain tissue. The blood–cerebrospinal fluid barrier is the much smaller barrier in the choroid plexus, where much of the liquid (i.e., CSF) that bathes and shelters the brain is formed. Nutrients and metabolites move through the CSF by diffusion and convection. Individual cells of the brain can obtain specific molecules either from nearby blood vessels (after the molecules crossed the BBB) or from the CSF.

ABBREVIATIONS

BBB	blood–brain barrier
CSF	cerebrospinal fluid
DMT1	divalent metal ion transporter 1 (SLC11A2)
GLUT1	glucose transporter 1 (SLC2A1)
MCT1	proton/monocarboxylic acid cotransporter 1 (SLC16A1)
OCT2	organic cation transporter 2
SMVT	sodium-dependent multivitamin transporter (SLCSA6)

ANATOMICAL BACKGROUND

A web of arteries, capillaries, and veins permeates the brain, like every other large organ, supplies it with a constant stream of oxygen and nutrients, and carries away the waste products. However, the BBB restricts the exchange between the blood and the intercellular space in the brain much more than in other organs (Pardridge, 1998). Unusually effective tight junctions seal the narrow spaces between the endothelial cells of brain capillaries. These tight junction seals are formed by multiple threaded strands of specialized proteins, including claudin-1, claudin-2, occludin, 7H6, ZO-1, ZO-2, and ZO-3 (Kniesel and Wolburg, 2000; Prat et al., 2001). The foot processes (podocytes) of astrocytes (microglia) extend to the basal side of the capillary endothelial cells and cover a significant portion of its abluminal surface. The narrow space adjacent to the endothelial cells is occupied by a basement membrane with collagen type IV, laminin, fibronectin, and proteoglycans (Prat et al., 2001).

A blood–nerve barrier with properties similar to the BBB also separates the major nerves from their blood supply (Allt and Lawrenson, 2000). Another type of interface between the bloodstream and brain space is seen in the choroid plexus. These highly convoluted structures in the ventricles produce CSF, which bathes and internally suspends the brain and the spinal cord and conveys nutrients and waste products around the intracerebral space. The choroidal epithelial cells are polarized cells with a specialized complement of carriers and channels on the luminal and abluminal sides. The spaces between adjacent choroidal cells are sealed much less tightly than those between typical brain capillaries and allow significant movement of water, electrolytes, and small molecules through them. Transport from blood into brain can thus follow the more circuitous route of secretion into CSF and later uptake into brain cells.

A few other specialized regions (including circumventricular organs and the pituitary gland) are exempt from the strict separation of the blood and brain spaces. One particular example is the vascular organ of the lamina terminalis in the hypothalamus. The endothelial cells of the blood capillaries in this region do not maintain tight seals and allow rapid and unimpeded diffusion of water and small solutes (Johnson et al., 1996) to presumed osmosensors on the abluminal side (Stricker and Hoffmann, 2007).

NUTRIENT TRANSPORT ACROSS THE BBB

For all practical purposes, nutrients and their derivatives that are to reach the brain have to be taken up from the capillary lumen into the endothelial cells, traffic to the other side, and exit across the abluminal (basal) membrane. With very few exceptions (such as gases and some lipophilic compounds), efficient transfer therefore requires the mediation by specific carriers, receptors, or channels. Many of those have been identified and extensively characterized, but additional ones are likely to contribute. In some instances, especially at high blood concentrations, a carrier or channel is an important conduit of transfer for a nutrient for which it has much lower affinity than for its dominant substrate. Such seemingly unspecific transfer is likely to account for significant residual transport capacity even when the designated dominant transporter of a nutrient is absent (e.g., GLUT1 deficiency does not completely abolish glucose transport into the brain).

Carbohydrates

The glucose transporter 1 (GLUT1, SLC2A1) mediates the transfer of glucose, which is the predominant energy source in the brain. Some cells lining the third ventricle have GLUT2 (SLC2A2). The insulin-inducible carrier GLUT4 (SLC2A4) is only expressed, along with GLUT1, at the BBB of the ventromedial hypothalamus, where it may link the bloodstream to glucose-sensing neurons (Ngarmukos et al., 2001). In light of the near-exclusive dependence on GLUT1 for glucose transport into the brain, the consequences of inborn absence of this transporter, which include seizures and developmental delay, appear less catastrophic than one might expect (Boles et al., 1999).

Lipids

Lipoproteins, cholesterol, and most fatty acids are effectively excluded from the brain. Long-chain polyunsaturated fatty acids are selectively transported by an unknown lipoprotein receptors and essential fatty acids transferred across the abluminal membrane by specific transporters (Edmond, 2001).

Fatty acid–derived metabolites are an important energy fuel for the brain, particularly with prolonged starvation. The proton/monocarboxylic acid cotransporter 1 (MCT1, SLC16A1) carries ketone bodies (and other carboxylic acids, including acetate, lactate, pyruvate, propionate, and butyrate) across both sides of brain endothelial cells. About 0.5 nmol/ml/min are transported (Blomqvist et al., 1995). Ketosis strongly induces both MCT1 and GLUT1 (Leino et al., 2001). Fasting for a few days, which induces ketosis, increases the efficiency of both glucose and beta-hydroxybutyrate transfer across the BBB (Hasselbalch et al., 1995).

Amino acids

A specific set of amino acid transporters mediates the transfer of amino acids. While most amino acids are preferentially carried into the brain, transport out of the brain is equally important to remove potentially excitotoxic amino acids, such as glutamate (Hosoya et al., 1999). A few specific proteins (e.g., leptin and insulin) can cross the BBB to a limited extent by endocytosis (Tamai et al., 1997).

The L-type transporter 1 (LAT1) is the main conduit for neutral amino acids across both sides of the endothelial cells (Duelli et al., 2000; Killian and Chikhale, 2001).

The closely related transporter LAT2 is also expressed at the BBB (Wagner et al., 2001). The sodium-dependent system A appears to contribute to some degree to the transfer of neutral amino acids from the blood into the brain (Kitazawa et al., 2001). Arginine and other cationic amino acids can cross via the 4F2-anchored exchange complex y^+ LAT2 (Broer et al., 2000). Sodium-linked system N mediates the transfer of glutamate and aspartate into brain, but also the elimination into circulation when there is an excess in the brain (Ennis et al., 1998; Hosoya et al., 1999; Smith, 2000).

Choline, an amine, is an essential precursor for phospholipid structures of the brain and for the neurotransmitter acetylcholine. The organic cation transporter 2 (OCT2, SLC22A2), which is located at the luminal side of the brain capillary endothelial cell, drives choline transport into the brain (Sweet et al., 2001). The mechanism of transport across the abluminal membrane remains unclear. At the same time, the concentration of choline in CSF is kept much lower than in the rest of the interstitial fluid of the brain. This is related to the uptake of choline from CSF via OCT2 and subsequent extrusion into blood circulation by an unknown mechanism.

Vitamins

Like any other tissue, the brain depends on adequate supplies of essential nutrients, and vitamins are no exception. Several transporters have been identified that mediate the selective transfer of specific vitamers into the brain, but often their exact location is not known.

Vitamin C: Oxidized vitamin C (dehydroascorbate) enters the brain capillary endothelium via GLUTI and is trapped inside by reduction to ascorbate. The sodium-ascorbate cotransporter 2 (SLC23A1) then completes transfer into the brain (Liang et al., 2001). This concentrative transport mechanism sustains a roughly 10-fold higher ascorbate concentration in the brain compared to the blood.

Thiamin: Free thiamin and some TMP cross the BBB (Patrini et al., 1988). The thiamin transporter 1 (SLC19A2) is expressed in the brain, but its role at the BBB remains to be elucidated. A choline transport system has been described in the brain that also appears to mediate thiamin uptake (Kang et al., 1990).

Riboflavin: The free form of riboflavin rapidly crosses from blood circulation into the brain or in the reverse direction. Riboflavin is trapped inside the endothelial cell as FMN and FAD (Spector, 1980). No specific carriers for riboflavin across either side of the endothelial barrier have been identified yet.

Vitamin B6: While it is clear that the pyridoxal in blood eventually reaches the brain (Sakurai et al., 1991), little is known about the mechanisms involved in this transfer.

Biotin, pantothenate, and lipoate: A high-affinity sodium-dependent multivitamin transporter (SMVT, SLC5A6) helps to move biotin, pantothenate, and lipoate from blood into the brain (Prasad et al., 1998). So far, however, it is not known whether this carrier suffices for transfer across the BBB, or whether additional mechanisms complement its action.

Niacin: Positron emission tomography (PET) scans have demonstrated the rapid transfer of both nicotinamide and nicotinate into the brain (Hankes et al., 1991), but the underlying mechanisms are not well understood.

Vitamin B12: Megalin, which mediates endocytosis of the vitamin B12/transcobalamin II complex, is present at the luminal side of the BBB (Zlokovic et al., 1996). It seems likely that this member of the LDL receptor family provides the main route of entry for cobalamin. The actual contribution of this pathway to vitamin B12 uptake and the mechanism of export across the abluminal membrane remain to be learned.

Fat-soluble vitamins: Megalin is also likely to be important for the transfer of vitamins A and D because it binds RBP and DBP (Christensen and Birn, 2001). The concentration of vitamin E in CSF is 100-fold lower than in plasma (Pappert et al., 1996), and there is uncertainty how even small amounts get into the brain. High-density lipoprotein (HDL), which is an important carrier of vitamin E in the blood, enters brain capillary endothelium via an HDL-binding receptor (Goti et al., 2000). It is not known how the vitamin components are separated from other lipid constituents (which are not transferred into brain) and how they cross the abluminal membrane. HDL-mediated transfer delivers to the brain all alpha-tocopherol isomers with similar efficiency. One may assume that vitamin K, which is carried by HDL in part (Kohlmeier et al., 1995), also enters the brain through this pathway, but this has not been explored yet. Vitamin E, carotenoids, flavonoids, and similar antioxidants also are important for the protection of the BBB against the damaging effects of oxygen free radicals.

Minerals and trace elements

Iron: Brain is critically dependent on adequate iron supplies but threatened at the same time by the potential toxicity of excessive concentrations. Brain dysfunction in older people (e.g., Parkinson's disease) is often accompanied by iron accumulation in some brain regions, which may add to the damage. The BBB effectively separates brain iron metabolism from whole-body iron metabolism, and all proteins involved in iron metabolism are produced within the brain (Rouault, 2001). A tentative model of iron transfer across the BBB might include the following mechanistic components: iron-carrying transferrin binds to the transferrin receptor 1, the complex enters the endothelial cell through the endocytotic pathway, and the proton-coupled divalent metal ion transporter 1 (DMT1, SLC11A2) then extracts iron from the endosomes (Burdo et al., 2001). A significant amount of iron may be directly handed to DMT1 in the adjacent astrocyte foot processes.

Lactoferrin may be another vehicle to convey iron across brain capillary endothelial cells, possibly mediated by low-density lipoprotein receptor-related protein. A much smaller amount of iron (and other metals) may reach the intracerebral space through nonbarrier sites (Gross and Weindl, 1987).

Redistribution of iron within the brain may use transport within axons as well as with CSE. The choroid plexus recovers iron from CSF through the transferrin-transferrin receptor pathway. Iron export from the brain into the blood also involves the membrane protein ferroportin 1 (metal transport protein 1, MTP1, SLC11A3), but the precise mechanism still needs to be resolved (Burdo et al., 2001). It has been further proposed that melanotransferrin at the luminal side of the BBB, which binds ferric iron, functions as an iron sensor (Rouault, 2001). Excessive iron concentration in blood, particularly of the portion that is not bound to transferrin, can increase oxidative damage to the BBB and increase its permeability.

Other metals: Delivery of copper proceeds is from ceruloplasmin and albumin in blood and involves DMT1. Active transport by Cu-ATPase is likely to promote removal of copper from the brain (Qian et al., 1998). As with iron, an excess of copper can damage the BBB (Stuerenburg, 2000). Copper as a cofactor of copper/zinc-dependent superoxide dismutase (EC1.15.1.1), on the other hand, helps to maintain BBB integrity (Kim et al., 2001).

Zinc is an essential cofactor for numerous enzymes in the brain, coordinates to regulators of gene transcription (zinc finger proteins), and modulates neurotransmission directly (ionic zinc in vesicles) and indirectly [interaction with GABA and *N*-methyl-D-aspartate (NMDA) receptors]. Transport from the blood into and out of CSF via the choroid plexus is a major pathway for the maintenance of brain zinc homeostasis, but the molecular mechanisms at that site are no better resolved than for brain capillary epithelium. Ionic zinc can cross into the endothelial cells of the BBB through zinc transporter 1 (ZnT1, SLC30A1). Some of the zinc taken up into the endothelial cells can be transported by ZnT2 (SLC30A2) to intracellular vesicles (zincosomes) and sequestered there as a complex with metallothionein and available for mobilization during times of need (Bobilya et al., 2008). Uptake is likely to involve mediation of uptake from the blood by a zinc–histidinyl complex and DMT1 (Takeda, 2000).

Manganese rapidly traverses from the blood across the BBB (Rabin et al., 1993), presumably using transferrin, the transferrin receptor 1, DMT1, and other elements of the iron transport system.

Nonnutrient bioactives and xenobiotics

Many phytochemicals in foods are absorbed and circulate via the blood. Common examples include flavonoids (e.g., naringenin in grapefruit), phenolic compounds (e.g., catechins in black tea), and indoles (e.g., indolcarbinol in cabbage). The BBB actively prevents the transfer from blood into brain of many of these compounds (Strazielle and Ghersi-Egea, 1999; Miller et al., 2002). As in other polarized epithelia, xenobiotics are preferentially conjugated to glutathione or glucuronate and then actively pumped across the membrane into the brain capillary lumen. The ATP-driven pumps for xenobiotic extrusion across the luminal membrane may include P-glycoprotein (ABCB1) and Mrp2 (ABCC2).

REFERENCES

Allt, G., Lawrenson, J.G., 2000. The blood–nerve barrier: enzymes, transporters and receptors—a comparison with the blood–brain barrier. Brain Res. Bull. 52, 1–12.

Blomqvist, G., Thorell, J.O., Ingvar, M., Grill, V., Widen, L., Stone-Elander, S., 1995. Use of R-beta-[1-^{11}C] hydroxybutyrate in PET studies of regional cerebral uptake of ketone bodies in humans. Am. J. Physiol. 269, E948–E959.

Bobilya, D.J., Gauthier, N.A., Karki, S., Olley, B.J., Thomas, W.K., 2008. Longitudinal changes in zinc transport kinetics, metallothionein and zinc transporter expression in a blood–brain barrier model in response to a moderately excessive zinc environment. J. Nutr. Biochem. 19, 129–137.

Boles, R.G., Seashore, M.R., Mitchell, W.G., Kollros, P.R., Mofidi, S., Novotny, E.J., 1999. Glucose transporter type 1 deficiency: a study of two cases with video-EEG. Eur. J. Pediatr. 158, 978–983.

Broer, A., Wagner, C.A., Lang, F., Broer, S., 2000. The heterodimeric amino acid transporter 4F2hc/y$^+$LAT2 mediates arginine efflux in exchange with glutamine. Biochem. J. 349, 787–795.

Burdo, J.R., Menzies, S.L., Simpson, I.A., Garrick, L.M., Garrick, M.D., Dolan, K.G., et al., 2001. Distribution of divalent metal transporter 1 and metal transport protein 1 in the normal and Belgrade rat. J. Neurosci. Res. 66, 1198–1207.

Christensen, EI, Birn, H., 2001. Megalin and cubilin: synergistic endocytic receptors in renal proximal tubule. Am. J. Physiol. Renal Fluid Electrolyte Physiol. 280, F562–F573.

Duelli, R., Enerson, B.E., Gerhart, D.Z., Drewes, L.R., 2000. Expression of large amino acid transporter LAT1 in rat brain endothelium. J. Cereb. Blood Flow Metabol. 20, 1557–1562.

Edmond, J., 2001. Essential polyunsaturated fatty acids and the barrier to the brain: the components of a model for transport. J. Mol. Neurosci. 16, 181–193.

Ennis, S.R., Kawai, N., Ren, X.D., Abdelkarim, G.E., Keep, R.E., 1998. Glutamine uptake at the blood–brain barrier is mediated by N-system transport. J. Neurochem. 71, 2565–2573.

Goti, D., Hammer, A., Galla, H.J., Malle, E., Sattler, W., 2000. Uptake of lipoprotein-associated alpha-tocopherol by primary porcine brain capillary endothelial cells. J. Neurochem. 74, 1374–1383.

Gross, P.M., Weindl, A., 1987. Peering through the windows of the brain. J. Cereb. Blood Flow Metabol. 7, 663–672.

Hankes, L.V., Coenen, H.H., Rota, E., Langen, K.J., Herzog, H., Wutz, W., et al., 1991. Effect of Huntington's and Alzheimer's diseases on the transport of nicotinic acid or nicotinamide across the human blood–brain barrier. Adv. Exp. Med. Biol. 294, 675–678.

Hasselbalch, S.G., Knudsen, G.M., Jakobsen, J., Hageman, L.P., Holm, S., Paulson, O.B., 1995. Blood–brain barrier permeability of glucose and ketone bodies during short-term starvation in humans. Am. J. Physiol. 268, E1161–E1166.

Hosoya, K., Sugawara, M., Asaba, H., Terasaki, T., 1999. Blood–brain barrier produces significant efflux of L-asparatic acid but not D-aspartic acid: *in vivo* evidence using the brain efflux index method. J. Neurochem. 73, 1206–1211.

Johnson, A.K., Cunningham, J.T., Thunhorst, R.L., 1996. Integrative role of the lamina terminalis in the regulation of cardiovascular and body fluid homeostasis. Clin. Exp. Pharmacol. Physiol. 23, 183–191.

Kang, Y.S., Terasaki, T., Ohnishi, T., Tsuji, A., 1990. *In vivo* and *in vitro* evidence for a common carrier mediated transport of choline and basic drugs through the blood–brain barrier. J. Pharmacobiol. Dynamics 13, 353–360.

Killian, D.M., Chikhale, P.J., 2001. Predominant functional activity of the large, neutral amino acid transporter (LAT1) isoform at the cerebrovasculature. Neurosci. Lett. 306, 1–4.

Kim, G.W., Lewen, A., Copin, J., Watson, B.D., Chan, P.H., 2001. The cytosolic antioxidant, copper/zinc superoxide dismutase, attenuates blood–brain barrier disruption and oxidative cellular injury after photothrombotic cortical ischemia in mice. Neuroscience 105, 1007–1018.

Kitazawa, T., Hosya, K., Watanabe, M., Takashima, T., Ohtsuki, S., Takanaga, H., et al., 2001. Characterization of the amino acid transport of new immortalized choroid plexus epithelial cell lines: a novel *in vitro* system for investigating transport functions at the blood–cerebrospinal fluid barrier. Pharmaceut. Res. 18, 16–22.

Kniesel, U., Wolburg, H., 2000. Tight junctions of the blood–brain barrier. Cell Mol. Neurobiol. 20, 57–76.

Kohlmeier, M., Saupe, J., Drossel, H.J., Shearer, M.J., 1995. Variation of phylloquinone (vitamin K1) concentrations in hemodialysis patients. Thromb. Haemostasis 74, 1252–1254.

Leino, R.L., Gerhart, D.Z., Duelli, R., Enerson, B.E., Drewes, L.R., 2001. Diet-induced ketosis increases monocarboxylate transporter (MCT1) levels in rat brain. Neurochem. Int. 38, 519–527.

Liang, W.J., Johnson, D., Jarvis, S.M., 2001. Vitamin C transport systems of mammalian cells. Mol. Membrane Biol. 18, 87–95.

Miller, D.S., Graeff, C., Droulle, L., Fricker, S., Fricker, G., 2002. Xenobiotic efflux pumps in isolated fish brain capillaries. Am. J. Physiol. Regul. Integr. Com. Physiol. 282, R191–R198.

Ngarmukos, C., Baur, E.L., Kumagai, A.K., 2001. Co-localization of GLUT1 and GLUT4 in the blood–brain barrier of the rat ventromedial hypothalamus. Brain Res. 900, 1–8.

Pappert, E.J., Tangney, C.C., Goetz, C.G., Ling, Z.D., Lipton, J.W., Stebbins, G.T., et al., 1996. Alpha tocopherol in the ventricular cerebrospinal fluid of Parkinson's disease patients: dose–response study and correlations with plasma levels. Neurology 47, 1037–1042.

Pardridge, W.M., 1998. Blood–brain barrier carrier-mediated transport and brain metabolism of amino acids. Neurochem. Res. 23, 635–644.

Patrini, C., Reggiani, C., Laforenza, U., Rindi, G., 1988. Blood–brain transport of thiamine monophosphate in the rat: a kinetic study *in vivo*. J. Neurochem. 50, 90–93.

Prasad, P.D., Wang, H., Kekuda, R., Fujita, T., Fei, Y.J., Devoe, L.D., et al., 1998. Cloning and functional expression of a cDNA encoding a mammalian sodium dependent vitamin transporter mediating the uptake of pantothenate, biotin, and lipoate. J. Biol. Chem. 273, 7501–7506.

Prat, A., Biernacki, K., Wosik, K., Antel, J.P., 2001. Glial cell influence on the human blood–brain barrier. GLIA 36, 145–155.

Qian, Y., Tiffany-Castiglioni, E., Welsh, J., Harris, E.D., 1998. Copper efflux from murine microvascular cells requires expression of the menkes disease Cu-ATPase. J. Nutr. 128, 1276–1282.

Rabin, O., Hegedus, L., Bourre, J.M., Smith, Q.R., 1993. Rapid brain uptake of manganese (II) across the blood–brain barrier. J. Neurochem. 61, 509–517.

Rouault, T.A., 2001. Systemic iron metabolism: a review and implications for brain iron metabolism. Ped. Neurol. 25, 130–137.

Sakurai, T., Asakura, T., Mizuno, A., Matsuda, M., 1991. Absorption and metabolism of pyridoxamine in mice. I. Pyridoxal as the only form of transport in blood. J. Nutr. Sci. Vitaminol. 37, 341–348.

Smith, Q.R., 2000. Transport of glutamate and other amino acids at the blood–brain barrier. J. Nutr. 130, 1016S–1022S.

Spector, R., 1980. Riboflavin homeostasis in the central nervous system. J. Neurochem. 35, 202–209.

Strazielle, N., Ghersi-Egea, J.E., 1999. Demonstration of a coupled metabolism of brain protection toward xeno-biotics. J. Neurosci. 19, 6275–6289.

Stricker, E.M., Hoffmann, M.L., 2007. Presystemic signals in the control of thirst, salt appetite, and vasopressin secretion. Physiol. Behav. 91, 404–412.

Stuerenburg, H.J., 2000. CSF copper concentrations, blood–brain barrier function, and ceruloplasmin synthesis during the treatment of Wilson's disease. J. Neural. Transm. General Section 107, 321–329.

Sweet, D.H., Miller, D.S., Pritchard, J.B., 2001. Ventricular choline transport: a role for organic cation transporter 2 expressed in choroid plexus. J. Biol. Chem. 276, 41611–41619.

Takeda, A., 2000. Movement of zinc and its functional significance in the brain. Brain Res. Rev. 34, 137–148.

Tamai, I., Sai, Y., Kobayashi, H., Kamata, M., Wakamiya, T., Tsuji, A., 1997. Structure–internalization relation-ship for adsorptive-mediated endocytosis of basic peptides at the blood–brain barrier. J. Pharmacol. Exp. Ther. 280, 410–415.

Wagner, C.A., Lang, F., Broer, S., 2001. Function and structure of heterodimeric amino acid transporters. Am. J. Physiol. Cell Physiol. 281, C1077–C1093.

Zlokovic, B.V., Martel, C.L., Matsubara, E., McComb, J.G., Zheng, G., McCluskey, R.T., et al., 1996. Glycoprotein 330/megalin: probable role in receptor-mediated transport of apolipoprotein J alone and in a complex with Alzheimer disease amyloid beta at the blood–brain and blood–cerebrospinal fluid barriers. Proc. Natl. Acad. Sci. USA 93, 4229–4234.

MATERNO-FETAL NUTRIENT TRANSPORT

Transfer of nutrients to the growing child and the management of waste and metabolites are extremely complex and highly selective events. The amniotic sac and placenta shield the new life against the unfiltered onslaught of environmental chemicals when it is most vulnerable, produce important growth factors, and buffer nutrient excess and deficiency, even at the expense of the mother's own needs.

ABBREVIATIONS	
BM	basal membrane (fetal side)
LAT1	L-type amino acid transporter 1 (SLC7A5)
LAT2	L-type amino acid transporter 2 (SLC7A8)
MVM	microvillous membrane (maternal side)

NUTRITURE OF THE EARLY EMBRYO

During the first few days of embryonic development, nutrients are directly transferred from the mother to the embryo without an intervening structure. Nutrient transfer to the embryo initially proceeds via the amnion fluid of the yolk sac until the placenta forms. The 8-week-old embryo already has a well-developed placenta, which by then is the dominant organ for nutrient transfer from maternal blood to fetal circulation.

THE MATURE PLACENTA

The placenta, which weighs about 500 g at term, provides the interface between maternal and fetal blood circulation. On the maternal side, the placenta is firmly attached to the endometrium and derives its blood supply from the spiral arteries of the uterine wall. The maternal blood flows from arterial openings into the intervillous space of the placenta and returns into the exit openings of maternal veins.

A system of finely branched fingerlike tissue (villi) is suspended in the intervillous space, bathed in the pool of maternal blood. The villi consist of fetal capillaries in the core covered by a contiguous layer of trophoblast cells. The surface of the placental villi covers about $1 \, m^2$ at the end of the first trimester and as much as $11 \, m^2$ at term (Mayhew, 1996). The capillaries are the end branches of an arteriovenous system that comes together in two arteries, and a single vein is embedded in the umbilical cord. These blood vessels connect directly to the fetal vascular system.

THE MATERNO-FETAL BARRIER

Compounds that move from maternal into fetal circulation encounter a layer of syncytiotrophoblast (syntrophoblast), a layer of cytotrophoblasts, and the endothelial cell layer of the fetal capillaries (Figure 3.5). During the later months of pregnancy, the cytotrophoblast layer becomes incomplete and eventually disappears. The endothelium of the fetal capillaries constitutes no significant barrier. Pores in the interendothelial clefts provide for the relatively unrestricted diffusion of small and large molecules (Michel and Curry, 1999). The syntrophoblast, on the other hand, is fused into a contiguous cell layer without gaps, and it blocks the transit of any compound that is not transported either via a carrier or channel or by endocytosis. The discussion of materno-fetal nutrient transfer in this section will focus primarily on the syntrophoblast, since this cell layer constitutes the main materno-fetal barrier and is the main site of placental nutrient metabolism later in pregnancy.

The maternal side (microvillous membrane, MVM) of the syntrophoblast has numerous cell protrusions (microvilli), while the fetal side (basal membrane, BM) is relatively smooth. Due to this structural difference, the ratio of MVM-to-BM surface areas is typically about 6:1 (Teasdale and Jean-Jacques, 1988).

Carbohydrates

The glucose transporter 1 (GLUT1) is active at both sides of the syntrophoblast cell layer (Illsley, 2000). This transporter mediates the transfer of both glucose and galactose, but not of other sugars to any significant extent. Expression of GLUT1 appears to be down-regulated in response to high maternal glucose concentrations only during the first trimester and then stay constant for the remainder of the pregnancy (Jansson et al., 2001). GLUT3 and other additional glucose transporters are also active in placenta, but they may be more important for placenta nutriture than for transfer to the fetus.

Lipids

The placenta preferentially transports essential polyunsaturated fatty acids (Dutta-Roy, 2000), particularly arachidonic acid (omega-6 fatty acid) and docosahexaenoic acid (DHA; omega-3 fatty acid). A wide range of fatty acids is taken up from maternal circulation with lipoproteins. Some metabolic

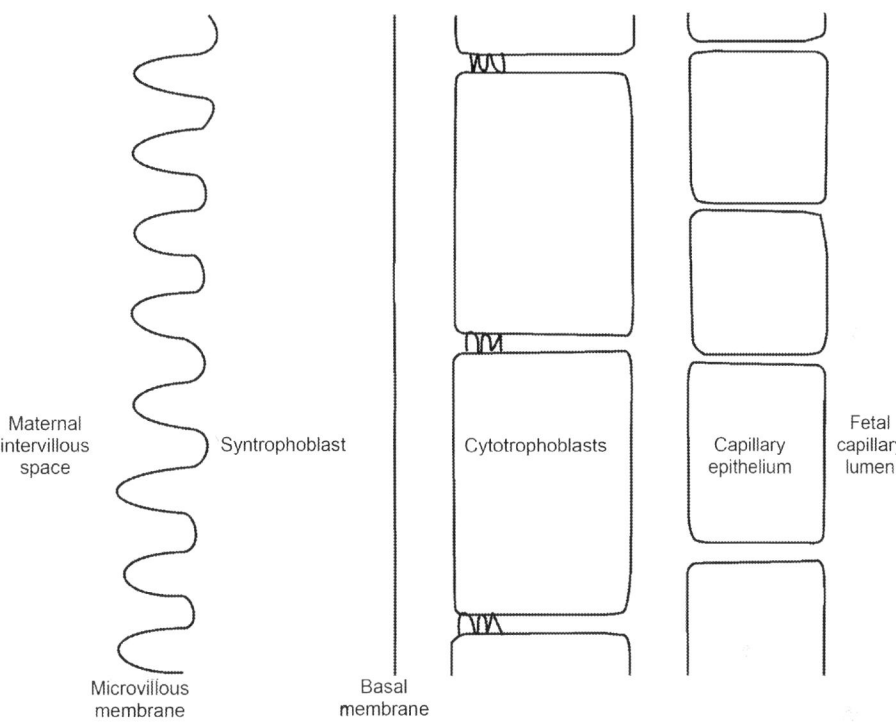

FIGURE 3.5

Structural organization of the human placenta.

processing of essential fatty acids occurs in the syntrophoblast, particularly the conversion of linoleic acid to arachidonic acid and of eicosapentaenic acid to DHA by chain elongation and introduction of additional double bonds. A placenta-specific fatty acid binding protein preferentially binds these long-chain polyunsaturated fatty acids (Dutta-Roy, 2000) and passes them on to other binding proteins and eventually to an unknown transporter.

The placenta takes up cholesterol from maternal blood with HDL and other lipoproteins (Christiansen-Weber et al., 2000). Endocytotic uptake of maternal HDL is mediated by the intrinsic factor receptor (cubilin), a member of the low-density lipoprotein (LDL) receptor protein family (Kozyraki et al., 1999). The placenta also contains the LDL receptor (LDL-R), the acetylated-LDL receptor, the apoE receptor, the very-low-density lipoprotein (VLDL) receptor, the scavenger receptor class B type I, and megalin. Significant additional amounts of cholesterol may enter the syntrophoblast without the mediation of receptors (Wyne and Woollett, 1998). Steroid hormone synthesis in the placenta is the main use for cholesterol taken up from maternal blood. The ATP-driven transporter ABCA1 carries some cholesterol across the BM to the fetus (Christiansen-Weber et al., 2000).

Amino acids

Net transfer: Adequate transfer of amino acids from mother to her unborn child is critical to support the very rapid growth of the fetus. Amino acids are used extensively by the fetus, and by the placenta itself, as building blocks for protein and peptide synthesis and as an important energy source. Leucine, isoleucine, valine, serine, and glutamine constitute the bulk of net amino acid uptake by the placenta from maternal circulation (Cetin, 2001). Other amino acids are taken up at more modest rates. At the other end of this spectrum is glycine, which plays a major role in placental metabolism, but of which there is little net transfer.

Placental metabolism: Amino acid metabolism in the placenta is very active, providing energy and protein precursors for the growth and sustenance of this crucial organ. The placenta is also the major site for synthesis of glycine from serine, both from maternal and fetal circulation (Cetin, 2001).

Transporters: The interior of the syntrophoblast cell is electronegative in relation to the outside. This electric potential difference and the sodium gradient are the driving forces for concentrative amino acid transport from the maternal to the fetal side of the placenta. Sodium-dependent transport systems A (for alanine, serine, cysteine, methionine, praline, asparagine, and glutamine), $B°$ (for valine, iso-leucine, leucine, threonine, phenylalanine, and tryptophan), and, to a slightly lesser extent, ASC (for glycine, alanine, serine, cysteine, and threonine) transport neutral amino acids across both MVM and BM (Jansson, 2001). Two distinct system A transporters have been found in the placenta, ATA1 and ATA2, but their localization is still uncertain. Indeed, ATA2 expression is higher in the placenta than in all other tissues tested (Hatanaka et al., 2000), pointing to the special importance of this transporter as the driving force for all materno-fetal amino acid transport.

Branched-chain (valine, leucine, and isoleucine) and aromatic amino acids (tryptophan, phenylalanine, and tyrosine, as well as the thyroid hormone triiodothyrosine), which these bulk transporters do not accept, cross the MVM via LAT1 (Ritchie and Taylor, 2001), and the BM via LAT2 (Kudo and Boyd, 2001). LAT1 and LAT2 transport their substrates in exchange for other neutral amino acids and are not dependent on sodium or proton gradients. The main driving force is thus the concentration gradient of the neutral bulk amino acids that is established by the sodium-dependent transporters.

Transport system beta imports mainly beta-alanine and taurine. The cationic amino acids arginine and lysine, as well as choline and polyamines, are carried across the MVM by several members of the system y^+ (CAT-1/SLC7AI, CAT-4/SLC7A4, and CAT-2B/SLC7A2). The heterodimer formed by y^+LAT1 (SLC7A7) and 4F2 (SLC3A2) carries them across the BM (Kudo and Boyd, 2001).

Several members of transport system X^-_{AG}, including EAAT 1 (SLC1A3), EAAT2 (SLC1A2), EAAT3 (SLC1A1), and EAAT4 (SLC1A6), mediate glutamate and aspartate uptake from maternal blood. Because these acidic amino acids are metabolized extensively within the placenta, there is no net significant transfer to the fetus (Jansson, 2001).

The placenta also contains TAT1, a transporter with system T characteristics, which mediates con-centration-driven transfer of phenylalanine, tyrosine, and tryptophan (Kim et al., 2001).

Vitamins

Several of the water-soluble vitamins have their own specific transporters that ensure the transfer of these essential nutrients to the fetus, possibly even at the mother's expense. Information on others is still lacking. The fat-soluble vitamins tend to be taken up from maternal blood with lipoproteins and are exported specifically to the fetal side.

Vitamin C: GLUT1 accepts the predominant form of vitamin C in blood (DHA) as a substrate. The reduced form (ascorbate) can enter via sodium-dependent ascorbate transporter 1 (SVCT1, SLC23A1).

Most DHA is reduced inside the syntrophoblast layer, and ascorbate is then exported to the fetal side via SVCT2.

Thiamin: Two specific transporters contribute to concentrative thiamin transfer to the fetus, but the exact locations and mechanisms still must be resolved (Dutta et al., 1999; Rajgopal et al., 2001). Both the thiamin transporter 1 (SLC19A2) and ThTr2 (SLC19A3) are abundantly expressed in the placenta.

Riboflavin: Alkaline phosphatase has to cleave the FAD in maternal blood before the free riboflavin can be taken up into the syntrophoblasts via an as-yet-unknown carrier. The riboflavin carrier protein (RCP) is an intracellular protein that is essential for riboflavin transport across the syntrophoblast (Schneider, 1996). RFVT1 (SLC52A1) is highly expressed in the placenta and clearly constitutes one part of the materno-fetal transport mechanism (Yonezawa and Inui, 2013). The complete details still need to be investigated.

Niacin: The information on transfer of niacin to the fetus is quite limited. Since little maternal nicotinate reaches the fetus (Baker et al., 1981), the transfer of other metabolites must occur.

Vitamin B6: Pyridoxal enters the syntrophoblast by an unknown mechanism. The free pyridoxal is phosphorylated to pyridoxal-5′-phosphate and then exported to the fetal side by another unknown mechanism (Schenker et al., 1992).

Folate: Folate receptors (FOLRs) in the intervillous blood space bind 5-methyltetrahydrofolate and concentrate it on the maternally facing chorionic surface. The resulting concentration gradient drives folates across the placental barrier (Henderson et al., 1995). The folate receptors alpha (FOLR1) and beta (FOLR2), as well as the reduced folate carrier (SLC19A1), are expressed in the placenta.

Biotin, pantothenate, and lipoic acid: The sodium-dependent multivitamin transporter (SMVT, SLC5A6) operates at the MVM as the exclusive carrier for biotin, pantothenate, and lipoate (Prasad et al., 1998). There is still uncertainty about the mechanism responsible for extrusion across the BM.

Minerals and trace metals

Iron: Over the course of a healthy pregnancy, about 500 mg of iron is shuttled from mother to fetus. Transferrin receptors at the MVM bind iron-carrying transferrin from maternal blood and enter the syntrophoblast with endosomes. The divalent metal ion transporter 1 (DMT1) pumps iron out of these endosomes and across the BM (Georgieff et al., 2000). Another pathway uses the integrin ferroportin-1 at the MVM (Donovan et al., 2000), iron oxidation by an incompletely characterized placental copper oxidase (not identical with ceruloplasmin), and ferric iron transfer to fetal transferrin.

Maternal iron deficiency increases the efficiency of iron transfer to the fetus by increasing expression of copper oxidase, transferrin receptor (Gambling et al., 2001), and other elements of the iron transfer apparatus. Another placental iron-transport protein, uteroferrin (ACP5, a lysosomal glycoprotein with tartrate-resistant alkaline phosphatase type 5 properties), can cross the placenta and deliver some of its iron to the fetus (Laurenz et al., 1997).

Calcium: Late in pregnancy, large amounts of calcium are transported across the placenta. Most (i.e., 80%) of total fetal accretion occurs during the last trimester. Maternal ionized calcium enters the syntrophoblast through calcium channels and binds to intracellular proteins, including 9 kDa calcium-binding protein (calbindin, 9CBP). The net flow of calcium is driven by calcium ATPase at the BM. Several components of the calcium transport machinery (including MVM calcium channels, calbindin, and the calcium-ATPase) are induced by 1,25-dihydroxyvitamin D (1,25-$(OH)_2$-D; Hoenderop et al., 1999).

Xenobiotics: The ATP-binding cassette transporter ABCP (MXR, BCRP, and ABCG2) is a xenobiotic transporter that is highly expressed in the placenta (Allikmets et al., 1998).

REFERENCES

Allikmets, R., Schriml, L.M., Hutchinson, A., Romano-Spica, V., Dean, M., 1998. A human placenta-specific ATP-binding cassette gene (ABCP) on chromosome 4q22 that is involved in multidrug resistance. Cancer Res. 58, 5337–5339.

Baker, H., Frank, O., Deangelis, B., Feingold, S., Kaminetzky, H.A., 1981. Role of placenta in maternal-fetal vitamin transfer in humans. Am. J. Obstet. Gynecol. 141, 792–796.

Cetin, I., 2001. Amino acid interconversions in the fetal-placental unit: the animal model and human studies *in vivo*. Pediatr. Res. 49, 148–153.

Christiansen-Weber, T.A., Voland, J.R., Wu, Y., Ngo, K., Roland, B.L., Nguyen, S., et al., 2000. Functional loss of ABCA1 in mice causes severe placental malformation, aberrant lipid distribution, and kidney glomerulone-phritis as well as high-density lipoprotein cholesterol deficiency. Am. J. Pathol. 157, 1017–1029.

Donovan, A., Brownlie, A., Zhou, Y., Shepard, J., Pratt, S.J., Moynihan, J., et al., 2000. Positional cloning of zebrafish ferroportin 1 identifies a conserved vertebrate iron exporter. Nature 403, 776–781.

Dutta, B., Huang, W., Molero, M., Kekuda, R., Leibach, F.H., Devoe, L.D., et al., 1999. Cloning of the human thiamine transporter, a member of the folate transporter family. J. Biol. Chem. 274, 31925–31929.

Dutta-Roy, A.K., 2000. Transport mechanisms for long-chain polyunsaturated fatty acids in the human placenta. Am. J. Clin. Nutr. 71, 315S–322S.

Gambling, L., Danzeisen, R., Gair, S., Lea, R.G., Charania, Z., Solanky, N., et al., 2001. Effect of iron deficiency on placental transfer of iron and expression of iron transport proteins *in vivo* and *in vitro*. Biochem. J. 356, 883–889.

Georgieff, M.K., Wobken, J.K., Weile, J., Burdo, J.R., Connor, J.R., 2000. Identification and localization of divalent metal transporter-1 (DMT-1) in term human placenta. Placenta 21, 799–804.

Hatanaka, T., Huang, W., Wang, H., Sugawara, M., Prasad, P.D., Leibach, F.H., et al., 2000. Primary structure, functional characteristics and tissue expression pattern of human ATA2, a subtype of amino acid transport system A. Biochim. Biophys. Acta 1467, 1–6.

Henderson, G.I., Perez, T., Schenker, S., Mackins, J., Antony, A.C., 1995. Maternal-to-fetal transfer of 5-methyl-tetrahydrofolate by the perfused human placental cotyledon: evidence for a concentrative role by placental folate receptors in fetal folate delivery. J. Lab. Clin. Med. 126, 184–203.

Hoenderop, J.G., van der Kemp, A.W., Hartog, A., van de Graaf, S.K., van Os, C.H., Willems, P.H., et al., 1999. Molecular identification of the apical Ca^{2+} channel in 1,25-dihydroxyvitamin D3-responsive epithelia. J. Biol. Chem. 274, 8375–8378.

Illsley, N.P., 2000. Glucose transporters in the human placenta. Placenta 21, 14–22.

Jansson, T., 2001. Amino acid transporters in the human placenta. Ped. Res. 49, 141–147.

Jansson, T., Ekstrand, Y., Wennergren, M., Powell, T.L., 2001. Placental glucose transport in gestational diabetes mellitus. Am. J. Obstet. Gynecol. 184, 111–116.

Kim, D.K., Kanai, Y., Chairoungdua, A., Matsuo, H., Cha, S.H., Endou, H., 2001. Expression cloning of a Na^+-independent aromatic amino acid transporter with structural similarity to H^+/monocarboxylate transporters. J. Biol. Chem. 276, 17221–17228.

Kozyraki, R., Fyfe, J., Kristiansen, M., Gerdes, C., Jacobsen, C., Cui, S., et al., 1999. The intrinsic factor vitamin B12 receptor, cubilin, is a high-affinity apolipoprotein A-I receptor facilitating endocytosis of high-density lipoprotein. Nat. Med. 5, 656–661.

Kudo, Y., Boyd, C.A., 2001. Characterisation of L-tryptophan transporters in human placenta: a comparison of brush border and basal membrane vesicles. J. Physiol. 531, 405–416.

Laurenz, J.C., Hadjisavas, M., Schuster, D., Bazer, F.W., 1997. The effect of uteroferrin and recombinant GM-CSF on hematopoietic parameters in normal female pigs (*Sus scrofa*. Comp. Biochem. Physiol. Part B 118, 579–586.

Mayhew, T.M., 1996. Patterns of villous and intervilious space growth in human placentas from normal and abnormal pregnancies. Eur. J. Obstet. Gynecol. Reprod. Biol. 68, 75–82.

Michel, C.C., Curry, F.E., 1999. Microvascular permeability. Physiol. Rev. 79, 703–761.

Prasad, P.D., Wang, H., Kekuda, R., Fujita, T., Fei, Y.J., Devoe, L.D., et al., 1998. Cloning and functional expression of a cDNA encoding a mammalian sodium-dependent vitamin transporter mediating the uptake of pantothenate, biotin, and lipoate. J. Biol. Chem. 273, 7501–7506.

Rajgopal, A., Edmondson, A., Goldman, I.D., Zhao, R., 2001. SLC19A3 encodes a second thiamine transporter ThTr2. Biochim. Biophys. Acta 1537, 175–178.

Ritchie, J.W., Taylor, P.M., 2001. Role of the System L permease LAT1 in amino acid and iodothyronine transport in placenta. Biochem. J. 356, 719–725.

Schenker, S., Johnson, R.E., Mahuren, J.D., Henderson, G.I., Coburn, S.P., 1992. Human placental vitamin B6 (pyridoxal) transport: normal characteristics and effects of ethanol. Am. J. Physiol. 262, R966–R974.

Schneider, W.J., 1996. Vitellogenin receptors: oocyte-specific members of the low-density lipoprotein receptor supergene family. Int. Rev. Cytol. 166, 103–137.

Teasdale, E., Jean-Jacques, G., 1988. Intrauterine growth retardation: morphometry of the microvillous membrane of the human placenta. Placenta 9, 47–55.

Wyne, K.L., Woollett, L.A., 1998. Transport of maternal LDL and HDL to the fetal membranes and placenta of the golden Syrian hamster is mediated by receptor-dependent and receptor-independent processes. J. Lipid Res. 39, 518–553.

Yonezawa, A., Inui, K., 2013. Novel riboflavin transporter family RFVT/SLC52: identification, nomenclature, functional characterization and genetic diseases of RFVT/SLC52. Mol. Aspects Med. 34, 693–701.

CHAPTER

XENOBIOTICS

4

CHAPTER OUTLINE

Caffeine ... 95
Heterocyclic Amines ... 100
Nitrite/Nitrate ... 107

CAFFEINE

Caffeine (3,7-dihydro-1,2,7-trimethyl-1H-purine-2,6-dione, 1,3,7-trimethylxanthine, 1,3,7-trimethyl-2,6-dioxopurine; molecular weight 194) is a commonly consumed, bitter-tasting dietary stimulant with a purine-like structure (Figure 4.1).

ABBREVIATIONS	
1,7-X	1,7-dimethylxanthine
FMO3	flavin-containing monooxygenase 3

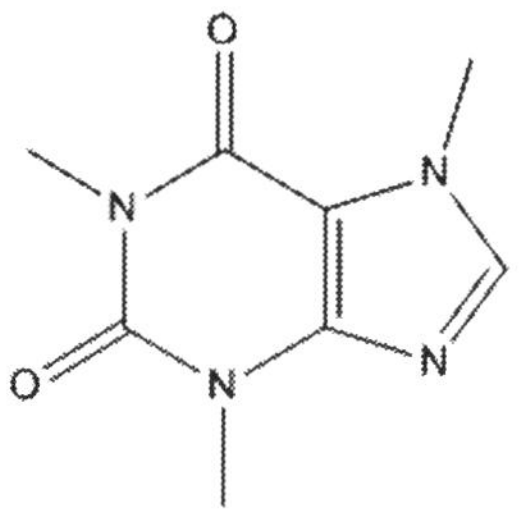

FIGURE 4.1

Caffeine.

NUTRITIONAL SUMMARY

Function: Caffeine and its metabolites increase attentiveness, exercise performance, and affect sleep patterns.

Food sources: Coffee, tea, maté, caffeinated beverages, and stimulant supplements provide pharmacologically significant amounts to the overwhelming majority of adults.

Requirements: No intakes are needed to improve health.

Deficiency: The absence of caffeine intake is not known to adversely affect health.

Excessive intake: Caffeine intake, particularly late in the day, can interfere with sleep. Daily intake of several cups of caffeinated coffee is likely to increase the risk of miscarriage. Even modest consumption levels increase renal calcium loss and may increase the risk of osteoporosis and bone fracture. Abrupt caffeine withdrawal in habituated users can induce severe headaches.

DIETARY SOURCES AND OTHER SOURCES

Some plants, such as *Caffea arabica* L. and *Caffea canephora* L. (coffee), *Camellia sinensis* [green, oolong, white, red, black, and dark (pu-erh) tea], *Ilex paraguayensis* (maté), *Paullinia cupana* (guaraná seeds), synthesize caffeine and related xanthine derivatives from adenosine. The most common usage of caffeine-containing plants is to prepare beverages. Caffeine content of brewed coffees varies widely around a typical value of 90 mg per 240 ml (8 oz) serving, ranging from 29 to 130 mg in one survey (McCusker et al., 2003). Content was found to differ twofold from 1 day to another, even in the same coffee shop.

Methylxanthines, including caffeine, can be removed with chlorinated solvents (dimethyl chloride, chloroform), hot water, steam, or supercritical CO_2 (Saldaña et al., 2002). The caffeine content of decaffeinated brews is still significant though usually well below 10 mg per 240 ml *NB*. Decaffeination removes a large percentage of epigallocatechin gallate (EGCG) and other catechines from green tea (Henning et al., 2003). Many other beverages, particularly sodas and waters, have caffeine added.

In many countries, including the United States and most of Europe, daily consumption of caffeinated beverages is the norm. Typical intakes among consumers are around 300 mg/day.

The Chinese tea (*Camellia assamica*) variety known as *kucha* contains significant amounts of theacrine (1,3,7,9-tetramethyluric acid) in addition to caffeine (Zheng et al., 2002).

INTESTINAL ABSORPTION

Absorption of caffeine (and its metabolites) from the small intestine is rapid and nearly complete. Caffeine appears in the blood within an hour of ingestion (Lang et al., 2013). The transporters and mechanisms responsible for absorption are not well understood. There appears to be considerable overlap with transporters responsible for purine, uric acid, and glucose absorption.

TRANSPORT AND CELLULAR UPTAKE

Materno-fetal transfer: Caffeine and its metabolites cross the placental barrier through as-yet-unidentified mechanisms (Abdi et al., 1993).

Blood–brain barrier (BBB): Caffeine crosses into the brain by unknown means.

METABOLISM

The main caffeine catabolite produced in the liver is likely to be 1,3,7-trimethyluric acid (Roberts et al., 1994).

The initial major step in caffeine (and paraxanthine) metabolism is demethylation at positions 3 and 7 by the microsomal oxidase CYP1A2 (Gu et al., 1992). The expression and activity of this flavin adenine dinucleotide (FAD)–containing microsomal oxidase in liver is highly variable (by a factor of 40 or more) and tends to be higher in men than in women (Schweikl et al., 1993). A common genetic variant is associated with delayed breakdown (Cornelis et al., 2006).

The ethanol-inducible CYP2E1 is quite effective in converting caffeine to theobromine and theophylline (Gu et al., 1992). CYP2C9 also contributes to 7-N-demethylation of caffeine to theophylline (Utoh et al., 2013).

Another metabolite, 1-methylxanthine, is probably generated in the kidneys (Delahunty and Schoendorfer, 1998).

Due to wide variation in catabolic rates, particularly due to polymorphisms of the CYP1A2 gene, the half-life of caffeine in blood after ingestion ranges from 1.5 to more than 9 h. The rate of conversion to theobromine by flavin-containing monooxygenase 3 (FMO3) varies greatly (Park et al., 1999) (Figure 4.2).

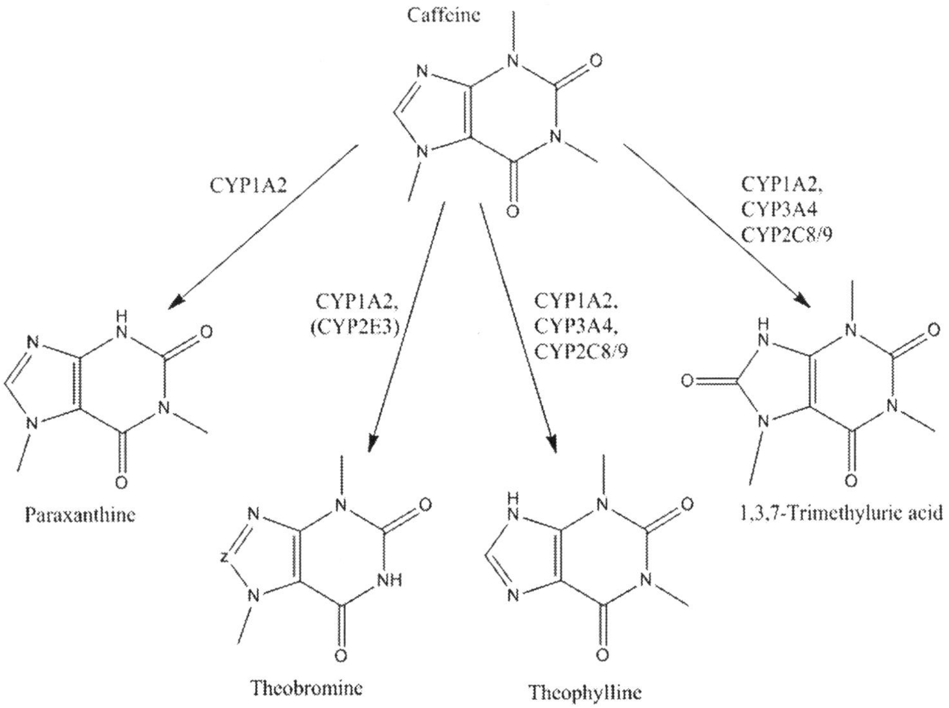

FIGURE 4.2

Caffeine metabolism.

EXCRETION

Typical urinary metabolites include paraxanthine, theobromine, 1-methylxanthine, and 1,3,7-trimethyluric acid (Ullrich et al., 1992; Regal et al., 1998). Excretion of caffeine and its urinary metabolites declines with increasing dose.

Function

Receptor interaction: Caffeine acts on several receptor types, including purinoceptors and purinceptors.

Caffeine stimulates adenosine receptors A1, A2A, A2B, and A3. The metabolite theophylline has much greater affinity to these receptors, while theobromine has much lower affinity.

Most individuals perceive the stimulatory effect of ingested caffeine, though at different threshold levels (Mumford et al., 1994).

Phosphodiesterase inhibition: Caffeine inhibits diverse phosphodiesterases, including PDE1, PDE4, and PDE5 (all EC3.1.4). These enzymes are modulators of cAMP and other signaling compounds.

Addiction: Caffeine consumption is remarkably popular around the world and may extend to more than 90% in many populations. Withdrawal can cause a wide range of symptoms, including irritability, sleepiness, dysphoria, delirium, nausea, vomiting, rhinorrhea, nervousness, restlessness, anxiety, muscle tension, muscle pains, and flushed face (Dews et al., 2002). Regular caffeine users often develop mild to severe migraine-type headaches when they miss out on their usual caffeine dose for a day. Whether all of this adds up to a finding of addictive qualities in caffeine continues to be a matter of debate (Budney and Emond, 2014).

Sleep: Many people find it difficult to fall asleep when they consume too much caffeine, particularly when they are unaccustomed to the consumed amounts. Individuals with the common variant rs5751876 C in the gene coding for the adenosine A2A receptor (ADORA2A) are particularly susceptible to this effect (Byrne et al., 2012). Other common variants associated with the caffeine-induced sleep disturbance suggest the importance of the *PRIMA1* gene (anchoring of acetylcholinesterase to the neuron membrane), *MTUS2* (involved in neuron development), and a melatonin receptor, *MTNR1B*.

Cognitive function: Controlled human studies suggest that modest caffeine doses can improve information processing (Warburton et al., 2001). This may be associated with its propensity to increase irritability, nervousness, and anxiety. Caffeine also acts on orexin-containing neurons that may play a role in the stimulating effect of caffeine (Murphy et al., 2003).

Diuresis: Caffeine has as a very mild diuretic effect, which is usually reinforced by the associated water intake. A 300-mg dose of caffeine will typically increase urine output by about 100 ml (Zhang et al., 2014). Exercise appears to abolish this caffeine-induced diuresis.

Ergogenic effects: Caffeine can improve performance in elite athletes and therefore its permissible intake before competitions is regulated. Competitive rowers, for instance, improved their power by almost 2% with the consumption (45 min before the test) of 3 mg caffeine/kg body weight (Christensen et al., 2014). Caffeine appears to increase pain tolerance during exhaustive exertion (Maridakis et al., 2007).

Calcium homeostasis: Modest caffeine intake increases renal calcium loss by 25% or more (Heaney and Rafferty, 2001). Distal renal tubuli contain both A1 and A2 adenosine receptors, through which caffeine and its metabolites can influence calcium and magnesium recovery (Kang et al., 2001).

Caffeine promotes calcium release from the sarcoplasmic reticulum of skeletal muscle by acting on the ryanodine receptor 1.

Paraxanthine decreases intracellular calcium concentration (Hawke et al., 2000).

Excessive intakes: Daily intake of several cups of caffeinated coffee is likely to increase the risk of miscarriage. In one case–control study, women with caffeine consumption in excess of 300 mg/day were twice as likely to miscarry compared to women with lower intake (Gianelli et al., 2003).

High caffeine intake can precipitate an acute heart attack. About half of all middle-aged men, those with genetically slow caffeine metabolism are particularly vulnerable when they consume at least three large cups of coffee (Cornelis et al., 2006).

Migraine relief: People with mild migraine headaches often respond well to moderate doses of caffeine (30–200 mg), often in combination with other pain medications. Since responsiveness differs greatly, migraine sufferers need to test the effectiveness and the required dose. Caffeine is an adenosine receptor antagonist and might work through overly relaxed meningeal or dural blood vessels (Haanes and Edvinsson, 2014).

Diagnostic use: ^{13}C-labeled caffeine can be used to assess liver function (Park et al., 2003). Since the oxidation of the methyl groups to CO_2 depends almost exclusively on microsomal enzymes in the liver, delayed exhalation of ^{13}C from the labeled test dose indicates liver failure.

REFERENCES

Abdi, F., Pollard, I., Wilkinson, J., 1993. Placental transfer and foetal disposition of caffeine and its immediate metabolites in the 20-day pregnant rat: function of dose. Xenobiotica 23, 449–456.

Budney, A.J., Emond, J.A., 2014. Caffeine addiction? Caffeine for youth? Time to act! Addiction 109, 1771–1772.

Byrne, E.M., Johnson, J., McRae, A.F., Nyholt, D.R., Medland, S.E., Gehrman, P.R., et al., 2012. A genome-wide association study of caffeine-related sleep disturbance: confirmation of a role for a common variant in the adenosine receptor. Sleep 35, 967–975.

Christensen, P.M., Petersen, M.H., Friis, S.N., Bangsbo, J., 2014. Caffeine, but not bicarbonate, improves 6 min maximal performance in elite rowers. Appl. Physiol. Nutr. Metab. 39, 1058–1063.

Cornelis, M.C., El-Sohemy, A., Kabagambe, E.K., Campos, H., 2006. Coffee, CYP1A2 genotype, and risk of myocardial infarction. JAMA 295, 1135–1141.

Delahunty, T., Schoendorfer, D., 1998. Caffeine demethylation monitoring using a transdermal sweat patch. J. Anal. Toxicol. 22, 596–600.

Dews, P.B., O'Brien, C.P., Bergman, J., 2002. Caffeine: behavioral effects of withdrawal and related issues. Food Chem. Toxicol. 40, 1257–1261.

Gianelli, M., Doyle, P., Roman, E., Pelerin, M., Hermon, C., 2003. The effect of caffeine consumption on the risk of miscarriage. Paediatr. Perinat. Epidemiol. 17, 316–323.

Gu, L., Gonzalez, F.J., Kalow, W., Tang, B.K., 1992. Biotransformation of caffeine, paraxanthine, theobromine and theophylline by cDNA-expressed human CYP1A2 and CYP2E1. Pharmacogenetics 2 (2), 73–77.

Haanes, K.A., Edvinsson, L., 2014. Expression and characterization of purinergic receptors in rat middle meningeal artery-potential role in migraine. PLoS One 9 (9), e108782.

Hawke, T.J., Allen, D.G., Lindinger, M.I., 2000. Paraxanthine, a caffeine metabolite, dose dependently increases [Ca(2+)](i) in skeletal muscle. J. Appl. Physiol. 89, 2312–2317.

Heaney, R.P., Rafferty, K., 2001. Carbonated beverages and urinary calcium excretion. Am. J. Clin. Nutr. 74, 343–347.

Henning, S.M., Fajardo-Lira, C., Lee, H.W., Youssefian, A.A., Go, V.L., Heber, D., 2003. Catechin content of 18 teas and a green tea extract supplement correlates with the antioxidant capacity. Nutr. Cancer 45, 226–235.

Kang, H.S., Kerstan, D., Dai, L.J., Ritchie, G., Quamme, G.A., 2001. Adenosine modulates Mg(2+) uptake in distal convoluted cells via A(1) and A(2) purinoceptors. Am. J. Physiol. Renal. Physiol. 281, F1141–F1147.

Lang, R., Dieminger, N., Beusch, A., Lee, Y.M., Dunkel, A., Suess, B., et al., 2013. Bioappearance and pharmacokinetics of bioactives upon coffee consumption. Anal. Bioanal. Chem. 405, 8487–8503.

Maridakis, V., O'Connor, P.J., Dudley, G.A., McCully, K.K., 2007. Caffeine attenuates delayed-onset muscle pain and force loss following eccentric exercise. J. Pain 8, 237–243.

McCusker, R.R., Goldberger, B.A., Cone, E.J., 2003. Caffeine content of specialty coffees. J. Anal. Toxicol. 27, 520–522.

Mumford, G.K., Evans, S.M., Kaminski, B.J., Preston, K.L., Sannerud, C.A., Silverman, K., et al., 1994. Discriminative stimulus and subjective effects of theobromine and caffeine in humans. Psychopharmacology 115, 1–8.

Murphy, J.A., Deurveilher, S., Semba, K., 2003. Stimulant doses of caffeine induce c-FOS activation in orexin/hypocretin-containing neurons. Neuroscience 121, 269–275.

Park, C.S., Chung, W.G., Kang, J.H., Roh, H.K., Lee, K.H., Chah, Y.N., 1999. Phenotyping of flavin-containing monooxygenase using caffeine metabolism and genotyping of FMO3 gene in a Korean population. Pharmacogenetics 9, 155–164.

Park, G.J., Katelaris, P.H., Jones, D.B., Seow, F., Le Couteur, D.G., Ngu, M.C., 2003. Validity of the 13C-caffeine breath test as noninvasive, qualitative test of liver function. Hepatology 38, 1227–1236.

Regal, K.A., Howald, W.N., Peter, R.M., Gartner, C.A., Kunze, K.L., Nelson, S.D., 1998. Subnanomolar quantification of caffeine's *in vitro* metabolites by stable isotope dilution gas chromatography–mass spectrometry. J. Chromatogr. B Biomed. Sci. Appl. 708, 75–85.

Roberts, E.A., Furuya, K.N., Tang, B.K., Kalow, W., 1994. Caffeine biotransformation in human hepatocyte lines derived from normal liver tissue. Biochem. Biophys. Res. Commun. 201 (2), 559–566.

Saldaña, M.D.A., Zetl, C., Mohamed, R.S., Brunner, G., 2002. Extraction of methylxanthines from guaraná seeds, maté leaves, and cooa beans using supercritical carbon dioxide and ethanol. J. Agric. Food Chem. 50, 4820–4826.

Schweikl, H., Taylor, J.A., Kitareewan, S., Linko, P., Nagorney, D., Goldstein, J.A., 1993. Expression of CYP1A1 genes in human liver. Pharmacogenetics 3, 239–249.

Ullrich, D., Compagnone, D., Munch, B., Brandes, A., Hille, H., Bircher, J., 1992. Urinary caffeine metabolites in man. Age-dependent changes and pattern in various clinical situations. Eur. J. Clin. Pharmacol. 43, 167–172.

Utoh, M., Murayama, N., Uno, Y., Onose, Y., Hosaka, S., Fujino, H., et al., 2013. Monkey liver cytochrome P450 2C9 is involved in caffeine 7-N-demethylation to form theophylline. Xenobiotica 43, 1037–1042.

Warburton, D.M., Bersellini, E., Sweeney, E., 2001. An evaluation of a caffeinated taurine drink on mood, memory and information processing in healthy volunteers without caffeine abstinence. Psychopharmacology 158, 322–328.

Zhang, Y., Coca, A., Casa, D.J., Antonio, J., Green, J.M., Bishop, P.A., 2014. Caffeine and diuresis during rest and exercise: a meta-analysis. J. Sci. Med. Sport [Epub ahead of print].

Zheng, X.Q., Ye, C.X., Kato, M., Crozier, A., Ashihara, H., 2002. Theacrine (1,3,7,9-tatramethyluric acid) synthesis in leaves of a Chinese tea, kucha (*Camellia assamica* var. kucha). Phytochemistry 60, 129–134.

HETEROCYCLIC AMINES

Heterocyclic amines (HAs) and heterocyclic aromatic amines (HAAs) include many carcinogenic compounds produced during cooking.

ABBREVIATIONS

NAT	arylamine N-acetyltransferase
CYP	cytochrome P450
DiMeIQx	2-amino-3,4,8-trimethylimidazo[4, 5-*f*]quinoxaline
HA	heterocyclic amine
IQ	2-amino-3-methylimidazo[4, 5-*f*]quinoline

MelQx	2-amino-3,8-dimethylimidazo[4, 5-*f*]quinoxaline
MRP	multidrug resistance protein
PEITC	beta-phenylethyl isothiocyanate
PhIP	2-amino-1-methyl-6-phenylimidazo[4,5-β]pyridine
Trp-P-1	3-amino-1,4-dimethyl-5H-pyrido[4,3-β]indole
Trp-P-2	3-amino-1-methyl-5H-pyrido[4,3-β]indole
UGT	UDP-glucuronosyltransferase

NUTRITIONAL SUMMARY

Function: HAs are potent mutagens and potential carcinogens.

Food sources: Cooking of flesh foods, including beef, pork, poultry, and fish, produces microgram amounts of HAs per serving from the pyrolytic rearrangement of creatine and amino acids.

Requirements: Beneficial effects of intakes at any level are not likely.

Excessive intake: People with even modest HA intake from cooked meat and fish have more DNA adducts in tissues and a greater risk to develop cancers at various sites than individuals with minimal intake (Figures 4.3 and 4.4).

DIETARY SOURCES

Frying, broiling, and other forms of heating meats generate a wide variety of creatine-derived carcinogens (Schut and Snyderwine, 1999; Wyss and Kaddurah-Daouk, 2000). One of these is 2-amino-1-methyl-6-phenylimidazo[4,5-β]pyridine (PhIP), generated by the pyrolysis of creatine in the presence

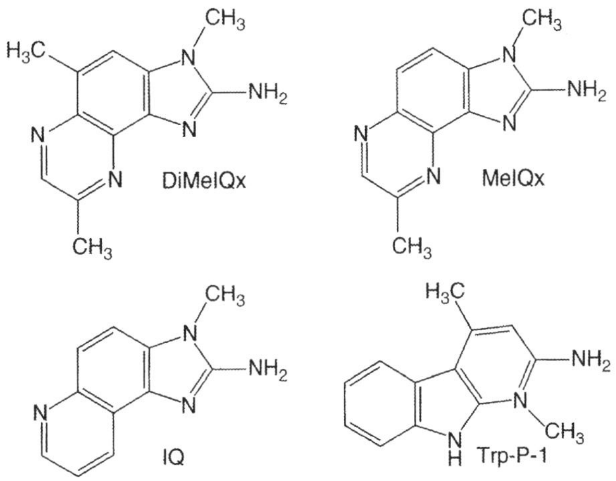

FIGURE 4.3

HAs in cooked meat.

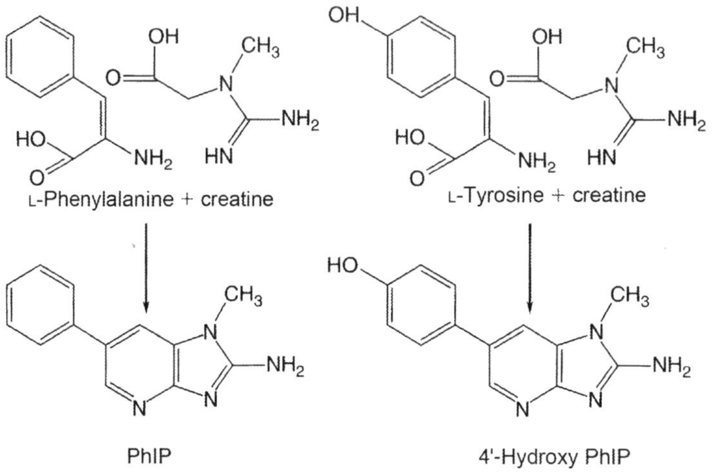

FIGURE 4.4

Cooking of meat generates PhIP.

of phenylalanine, threonine, or tyrosine; another is 2-amino-3,8-dimethylimidazo[4, 5-*f*]quinoxaline (MeIQx) derived from creatine, glycine, and glucose (Oguri et al., 1998). Cooked chicken was found to contain 18–21 µg of PhIP per 200 g serving (Kulp et al., 2000).

Intakes of PhIP have been estimated to range from nanograms to tens of micrograms per day, depending on flesh food consumption and preparation methods (Layton et al., 1995).

DIGESTION AND ABSORPTION

The site, mechanism, and effectiveness of HA uptake from the intestinal lumen is incompletely understood. More is known about several transporters that pump HA back into the lumen, some of it from circulating blood. At equal concentrations, the basolateral to apical efflux is several-fold greater than the apical to basolateral influx (Walle and Walle, 1999). The ABC transporter multidrug resistance protein 1 (MRP1) pumps its substrates across the basolateral lumen into the enterocyte (Walle and Walle, 1999). P-glycoprotein, the product of the *MDR-1* gene (Walle and Walle, 1999; Hoffmeyer et al., 2000), MRP2 (Walgren et al., 2000), and possibly MRP3, pump HAs (and many other xenobiotic compounds) across the apical enterocyte membrane into the intestinal lumen.

METABOLISM

Metabolism of xenobiotics often involves two types of metabolic processes, mainly in the liver, to a lesser extent in the kidney and other organs (Figure 4.5). The first, activating reactions, is referred to as *phase I reactions.* Phase I reactions most often involve hydroxylation or oxidation of the compound; a variety of other reactions are also possible. The second wave, the *phase II reactions*, includes conjugating and other modifying reactions. These molecular modifications tend to make the original compound more polar and enhance their renal excretion. Phase I reactions often increase the reactivity of xenobiotics toward DNA and proteins. Phase II enzymes, on the other hand, tend to decrease the potential

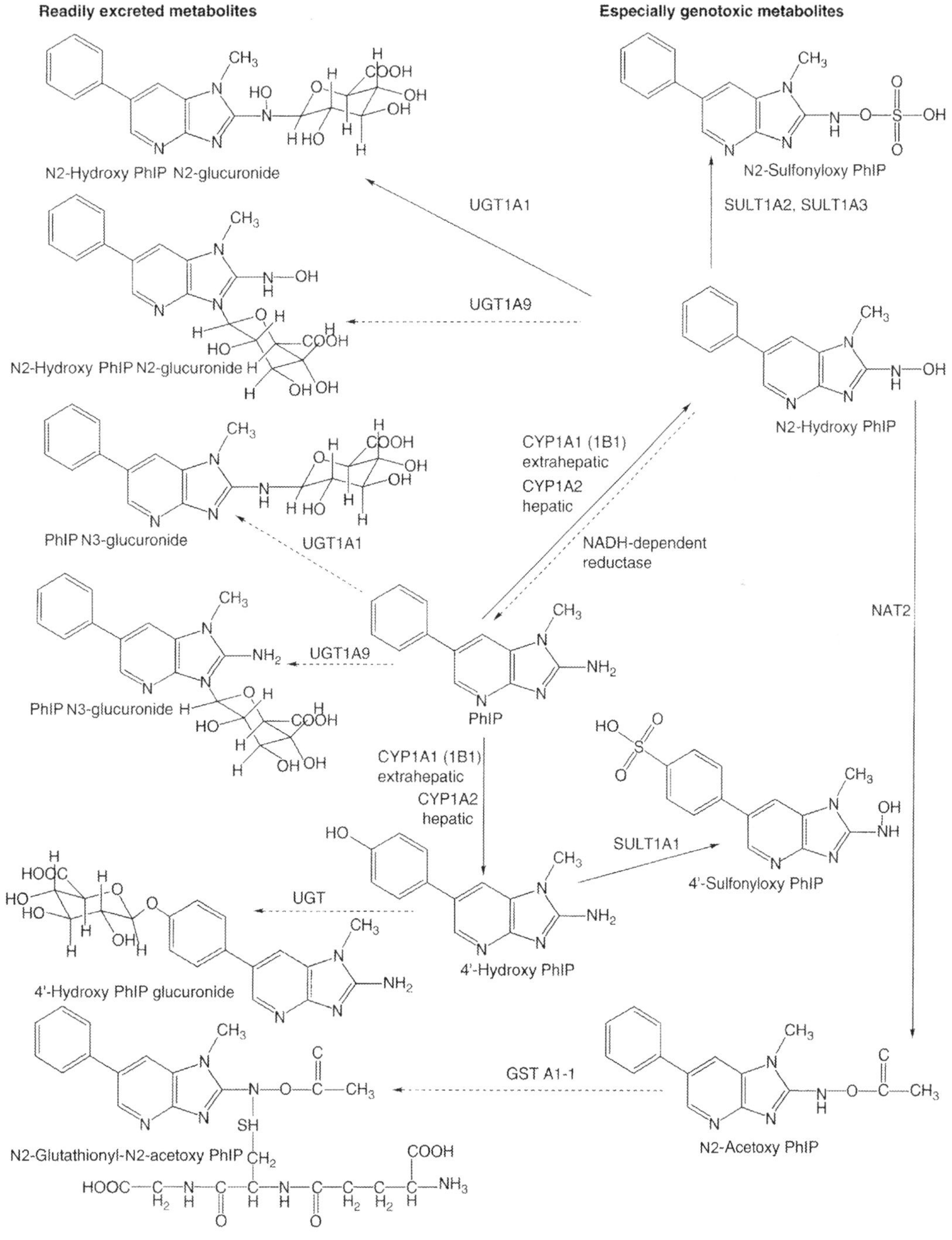

FIGURE 4.5

Metabolism of PhIP.

for harm by speeding up their elimination. However, this is not universally the case. Since phase I enzyme activities are a prerequisite for phase II–dependent modification, the concerted and balanced actions of members from both classes minimize risk. What is more, some phase II–catalyzed reactions (e.g., sulfotransferase-catalyzed generation of highly reactive N-sulfonyloxy HA derivatives) actually increase the potential for harm. Many of the enzymes that catalyze either phase I or phase II enzymes vary greatly in their activity due to differences in genetic makeup, gender, age, or exposure to these or other compounds (Williams, 2001).

Phase I reactions: The initial step of HA metabolism, as of most xenobiotics, usually occurs in the liver and is mainly hydroxylation by microsomal cytochrome P450 isoenzyme 1A2 (CYP1A2, EC1.14.14.1). In peripheral tissues, the isoenzymes CYP1A1 and CYP1B1 catalyze the same reaction (Gooderham et al., 2001). Reactions catalyzed by epoxide hydrolases (EC3.3.2.3, microsomal and cytosolic isoenzymes), numerous peroxidases, and other enzymes may contribute to a lesser degree in the liver, small intestine, or other tissues (Williams and Phillips, 2000). The main hydroxylation product is N2-hydroxy-PhIP and ring-hydroxylated products also arise (Buonarati et al., 1992), but much less are produced in humans than in rodents. CYPI A2, CYPIAI, and CYPIBI are inducible by various xenobiotics, including polycyclic aromatic hydrocarbons (PAHs), dioxins, and polychlorinated biphenyls (PCBs). Their expression is inhibited by some isothiocyanates, including sulforaphane from broccoli (1-isothiocyanate-4-methylsulfinylbutane) and beta-phenylethyl isothiocyanate (PEITC) from water cress.

Phase II reactions: Several UDP-glucuronosyltransferases (UGTs) in the liver, less in other tissues, can conjugate HAs. The isoenzyme UGT1A1 conjugates preferentially at N2, UGT1A9, preferentially at N3 (Malfatti and Felton, 2001). Overall, the predominant product is N2-hydroxy glucuronide (Gooderham et al., 2001). Conjugation also occurs at N2 (without prior hydroxylation), N3, and at hydroxy groups of the phenyl ring. Glucuronides can be excreted into bile. N3-glucuronides, but not N2-glucuronides, can then be cleaved by bacterial beta-glucuronidases in the large intestine (Gooderham et al., 2001).

The sulfotransferases SULTIA2, SULTIA3 (phenol-sulfating phenol sulfotransferase l, EC2.8.2.1) in the liver, and SULT1E1 in mammary tissue can activate N-hydroxylated PhIP (Williams, 2001). SULT1E1 is inducible by progesterone. Since N-sulfonyloxy derivatives of HAs are highly reactive, their metabolic end point tends to be the formation of adducts with DNA and other macromolecules (see later in this chapter). Arylamine N-acetyltransferases (NAT, EC2.3.1.5) are another group of enzymes that rapidly metabolize hydroxylated HAs and thereby make them more reactive. Fast NAT2 acetylators produce more HA/DNA adducts than slow acetylators (Calabrese, 1996). Reduced glutathione can be conjugated with some HAs. Such reactions are catalyzed by a family of microsomal glutathione S-transferases (GST, EC2.5.1.18) with broad specificity (van Bladeren, 2000). In the case of PhIP, only N-acetoxy-PhIP can form such a conjugate, catalyzed by the isoenzyme GST A1-l; the other isoenzymes are either little or not at all reactive (Coles et al., 2001). Conjugation of PhIP to glutathione makes it less likely to bind to DNA (Nelson et al., 2001) and promotes its transport (mediated by MRP2 and other transporters) out of cells (Dietrich et al., 2001) and its excretion with urine. An additional detoxification pathway is the niacin adenine dinucleotide (NADH)–dependent reduction of N2-hydroxy-PhIP to its amine (King et al., 1999).

Many phase II enzymes are inducible by food constituents, including thiocyanates (in broccoli, watercress, and other *Brassica* species), flavonoids, and polyphenols (Wilkinson and Clapper, 1997).

EXCRETION

The bulk of HAs is excreted with urine as glucuronide conjugates that are freely filtered. Additional secretion via MDRs or multidrug resistance proteins (MRPs) may be of importance. Fractional excretion of PhIP and its metabolites as a percentage of intake was between 5% and 50% in healthy subjects given the same meal (Kulp et al., 2000). Much of the variation may be due to differences in digestion and absorption. Broccoli consumption (induction of phase II enzymes) increases fractional excretion. Note that lower fractional excretion of identical oral doses indicates either lower absorption, higher retention due to adduct formation, or a combination of the two. Some excretion also occurs with bile. MDR2 can transport HA across the apical (canalicular) membranes of hepatocytes against high concentration gradients (Elferink et al., 1997).

EFFECTS

The main relevance of HAs in human nutrition is their potential to form DNA adducts and to induce mutations upon replication of the affected DNA strands. Thus, after exposure to a single oral dose of radiolabeled PhIP, about one PhIP-DNA adduct per 10^{10} nucleotides was detected in human mammary tissue (Lightfoot et al., 2000). Binding of HA metabolites to DNA is not random but appears to prefer some bases and sequences. PhIP metabolites, for example, often induce a characteristic G deletion at 5′-GGGA-3′ sites (Okochi et al., 1999).

While PhIP and related HAs themselves are not directly carcinogenic, their N-hydroxylated, N-hydroxy-sulfated, or otherwise N-hydroxy-substituted products induce cancer in many tissues in all investigated mammalian species, including humans (Ito et al., 1991; Sinha et al., 2000; Gooderham et al., 2001). In contrast, the ring-hydroxylated metabolites (which are produced to a much lesser extent in humans than in rodents) have much lower mutagenic potential (Gooderham et al., 2001).

REFERENCES

Buonarati, M.H., Roper, M., Morris, C.J., Happe, J.A., Knize, M.G., Felton, J.S., 1992. Metabolism of 2-amino-l-methyl-6-phenylimidazo[4,5-β] pyridine and effect of dietary fat. Carcinogenesis 15, 2429–2433.

Calabrese, E.J., 1996. Biochemical individuality: the next generation. Reg. Toxicol. Pharmacol. 24, S58–S67.

Coles, B., Nowell, S.A., MacLeod, S.L., Sweeney, C., Lang, N.E., Kadlubar, F.E., 2001. The role of human glutathione S-transferases (hGSTs) in the detoxification of the food-derived carcinogen metabolite N-acetoxy-PhIP and the effect of a polymorphism in hGSTA1 on colorectal cancer risk. Mut. Res. 482, 3–10.

Dietrich, C.G., de Waart, D.R., Ottenhoff, R., Bootsma, A.H., van Gennip, A.H., Elferink, R.P., 2001. Mrp2-deficiency in the rat impairs biliary and intestinal excretion and influences metabolism and disposition of the food-derived carcinogen 2-amino-1-methyl-6- phenylimidazo. Carcinogenesis 22, 805–811.

Elferink, R.E., Tytgat, G.N., Groen, A.K., 1997. Hepatic canalicular membrane 1. The role of mdr2 P-glycoprotein in hepatobiliary lipid transport. FASEB J. 11, 19–28.

Gooderham, N.L., Murray, S., Lynch, A.M., Yadollahi-Farsani, M., Zhao, K., Boobis, A.R., et al., 2001. Food-derived heterocyclic amine mutagens: variable metabolism and significance to humans. Drug Metab. Disp. 29, 529–534.

Hoffmeyer, S., Burk, O., von Richter, O., Arnold, H.E., Brockmoller, J., Johne, A., et al., 2000. Functional polymorphisms of the human multidrug-resistance gene: multiple sequence variations and correlation of one allele with P-glycoprotein expression and activity *de novo*. Proc. Natl. Acad. Sci. USA 97, 3473–3478.

Ito, N., Hasegawa, R., Sano, M., Tamano, S., Esumi, H., Takayama, S., et al., 1991. A new colon and mammary carcinogen in cooked food, 2-amino-1-methyl-6-phenylimidazol[4,5-β] pyridine (PhIP). Carcinogenesis 12, 1503–1506.

King, R.S., Teitel, C.H., Shaddock, J.G., Casciano, D.A., Kadlubar, F.E., 1999. Detoxification of carcinogenic aromatic and heterocyclic amines by enzymatic reduction of N-hydroxyderivatives. Cancer Lett. 143, 167–171.

Kulp, K.S., Knize, M.G., Malfatti, M.A., Salmon, C.P., Felton, J.S., 2000. Identification of urine metabolites of 2-amino-i-methyl-6-phenylimidazo[4,5-β] pyridine (PhIP) following consumption of a single cooked chicken meal in humans. Carcinogenesis 21, 2065–2072.

Layton, D.W., Bogen, K.T., Knize, M.G., Hatch, F.T., Johnson, V.M., Felton, J.S., 1995. Cancer risk of heterocyclic amines in cooked foods: an analysis and implications for research. Carcinogenesis 16, 39–52.

Lightfoot, T.J., Coxhead, J.M., Cupid, B.C., Nicholson, S., Garner, R.C., 2000. Analysis of DNA adducts by accelerator mass spectrometry in human breast tissue after administration of 2-amino-1-methyl-6-phenylimidazo[4,5-β]pyridine and benzo[α]pyrene. Mut. Res. 472, 119–127.

Malfatti, M.A., Felton, J.S., 2001. N-glucuronidation of 2-amino-l-methyl-6-phenylimidazo[4,5-β] pyridine (PhIP) and N-hydroxy-PhIP by specific human UDP-glucuronosyltransferases. Carcinogenesis 22, 1087–1093.

Nelson, C.P., Kidd, L.C., Sauvageot, J., Isaacs, W.B., De Marzo, A.M., Groopman, J.D., et al., 2001. Protection against 2-hydroxyamino-l-methyl-6-phenylimidazo[4,5-β] pyridine cytotoxicity and DNA adduct formation in human prostate by glutathione S-transferase P1. Cancer Res. 61, 103–109.

Oguri, A., Suda, M., Totsuka, Y., Sugimura, T., Wakabayashi, K., 1998. Inhibitory effects of antioxidants on formation of heterocyclic amines. Mut. Res. 402, 237–245.

Okochi, E., Watanabe, N., Shimada, Y., Takahashi, S., Wakazono, K., Shirai, T., et al., 1999. Preferential induction of guanine deletion at 5′-GGGA-3′ in rat mammary glands by 2-amino-l-methyl-6-phenylimidazo[4,5β] pyridine. Carcinogenesis 20, 1933–1938.

Schut, H.A., Snyderwine, E.G., 1999. DNA adducts of heterocyclic amine food mutagens: implications for mutagenesis and carcinogenesis. Carcinogenesis 20, 353–368.

Sinha, R., Gustafson, D.R., Kulldroff, M., Wen, W.Q., Cerhan, J.R., Zheng, W., 2000. 2-amino-l-methyl-6-phenylimidazo[4,5-β] pyridine, a carcinogen in high-temperature-cooked meat, and breast cancer risk. J. Natl. Cancer Inst. 92, 1352–1354.

van Bladeren, P.J., 2000. Glutathione conjugation as a bioactivation reaction. Chem. Biol. Interact. 129, 61–76.

Walgren, R.A., Karnaky Jr., K.J., Lindenmayer, G.E., Walle, T., 2000. Efflux of dietary flavonoid quercetin 4′-beta-glucoside across human intestinal Caco-2 cell monolayers by apical multidrug resistance-associated protein-2. J. Pharmacol. Exp. Ther. 294, 830–836.

Walle, U.K., Walle, T., 1999. Transport of the cooked-food mutagen 2-amino-l-methyl-6-phenylimidazo-[4,5-β] pyridine (PhIP) across the human intestinal Caco-2 cell monolayer: role of efflux pumps. Carcinogenesis 20, 2153–2157.

Wilkinson IV, J., Clapper, M.L., 1997. Detoxication enzymes and chemoprevention. Exp. Biol. Med. 216, 192–200.

Williams, J.A., 2001. Single nucleotide polymorphisms, metabolic activation and environmental carcinogenesis: why molecular epidemiologists should think about enzyme expression. Carcinogenesis 22, 209–214.

Williams, J.A., Phillips, D.H., 2000. Mammary expression of xenobiotic metabolizing enzymes and their potential role in breast cancer. Cancer Res. 60, 4667–4677.

Wyss, M., Kaddurah-Daouk, R., 2000. Creatine and creatinine metabolism. Physiol. Rev. 80, 1107–1213.

Yang, F., Comtois, A.S., Fang, L., Hartman, N.G., Blaise, G., 2002. Nitric oxide-derived nitrate anion contributes to endotoxic shock and multiple organ injury/dysfunction. Crit. Care Med. 30, 650–657.

NITRITE/NITRATE

Nitrite (NO_2^-), molecular weight 46.0055, is a water-soluble anion consisting of one nitrogen and two oxygens. Nitrate (NO_3^-), molecular weight 62.0049, is similarly a water-soluble anion, but with three oxygens and one nitrogen (Figure 4.6).

NUTRITIONAL SUMMARY

Function: Nitrite is used in food management to prevent spoilage and to confer the bright red appearance of meats. It is used as a fertilizer and may be retained, particularly in leafy vegetables. Very small amounts are used to relieve acute chest pain from insufficient blood flow to the heart muscles.

Food sources: Cured meats are the main food sources of nitrite. Other nitrite-cured foods usually contribute less. Spinach and other leafy vegetables may have increased content when grown with nitrate fertilizer.

Requirements: There is no nutritional requirement for nitrite or nitrate.

Deficiency: Low nitrite or nitrate intake is not known to cause specific symptoms or conditions.

Excessive intake: Intake of nitrite increases the risk of carcinogen formation and, with chronic exposure, a heightened cancer risk. Young infants have increased vulnerability to methemoglobin formed upon exposure to nitrite.

ENDOGENOUS PRODUCTION

Significant amounts of nitrate are produced in the body from arginine-derived nitric oxide (Leaf et al., 1989). Infection, inflammation, and injury increase endogenous nitrate production because macrophages and other cells of the immune and tissue repair systems increase their nitric oxide production and release.

Bacteria, including those in the mouth and gastrointestinal tract, can produce biologically relevant amounts of nitrite from nitrate. Orange juice, ascorbic acid, and other organic acids in foods can suppress nitrite production when ingested together with nitrates.

DIETARY SOURCES

Most dietary nitrate comes from vegetables, particularly leafy vegetables (lettuces, spinach, arugula, and others) that have been treated with nitrate fertilizer. Tap water is usually a lesser source. Well water,

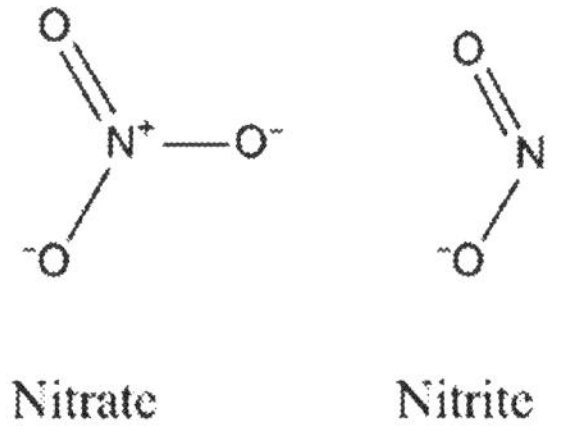

FIGURE 4.6

Nitrite and nitrate.

particularly in areas with significant agricultural runoff from heavily fertlilizer-treated fields and gardens, may be significantly contaminated (Richard et al., 2014). Daily nitrate intake is around 80 mg in the United States (Dellavalle et al., 2013) and around 180 mg in Shanghai (Dellavalle et al., 2014). Daily intakes can easily exceed 300 mg with high consumption of green vegetables (Bondonno et al., 2014). Acceptable daily intake (ADI) level is 3.7 mg/kg in Europe. Vegetarians tend to have several times higher nitrate exposure due to their higher consumption of fresh greens and other nitrate-rich vegetables.

Dietary nitrite comes mainly from sausages and meats cured to prevent spoilage of foods. The curing salts usually contain nitrite (e.g., Prague powder #1, with 6.75% sodium nitrite and 93.25% sodium chloride), sometimes also some nitrate (e.g., Prague powder #2, with 6.75% sodium nitrite, 4.00% sodium nitrate, and 89.25% sodium chloride). Recommended use is 2.2 g of curing salt per 1000 g of meat. Nitrite is classified by the U.S. Food and Drug Administration (FDA) as "generally recognized as safe" (GRAS) as a food additive. Daily nitrite intake is around 1 mg in the United States (Dellavalle et al., 2013).

ABSORPTION

Nitrate and nitrite are taken up from the intestinal lumen with high efficiency by various unspecific mechanisms. Nitrite, but not nitrate, competes with bicarbonate for the proton-coupled folate transporter (SLC46A1). Another example is the chloride anion exchanger SLC26A3 in the luminal membrane of colon cells, which mediates uptake of nitrate in exchange for chloride (Hayashi et al., 2009). SLC4A1 (anion exchanger 1), SLC4A2, and SLC26A6 at their respective expression sites appear to have similar properties.

TRANSPORT AND CELLULAR UPTAKE

Blood circulation: Serum concentration of nitrate in healthy people is around 28 μmol/l (Larsen et al., 2014). Nitrite concentrations can be expected around 230 nmol/l. Bacteria in the mouth or intestines can greatly increase nitrite concentrations in the blood by converting nitrate if it is available (Larsen et al., 2014). Uptake into tissues proceeds through unspecific transporters, such as the anion exchanger SLC4A3 in the membrane of red blood cells.

BBB: Since nitrite and nitrate are intermediary and end products, respectively, of nitric oxide, removal from brain is critically important. With acute stimulation of nitric oxide production by inflammation, infection, or injury, significant amounts have to be handled (Yang et al., 2002). It is unclear to what extent and how brain capillary endothelial cells move nitrate and nitrite into and out of the brain.

Materno-fetal transfer: Nitrate and nitrite from external exposure and endogenous production have to be removed from fetal tissues and circulation to avoid accumulation to toxic levels. There is no detailed information about the mechanisms controlling transfer of nitrate and nitrite to or from the mother.

METABOLISM

Bacteria in the gastrointestinal tract can generate significant amounts of nitrite and nitrous oxide. Patients with bacterial overgrowth in the stomach due to atrophic gastritis can generate enough nitrous oxide that it is detectable in their breath (Mitsui and Kondo, 2004).

Colonocytes can oxidize nitrite to nitrate (Roediger and Radcliffe, 1988).

Ubiquinone-dependent nitrite reductase (EC1.7.2.1) in mitochondria and deoxyhemoglobin in blood convert nitrite into nitric oxide (Kozlov et al., 1999; Demoncheaux et al., 2002; Gladwin et al., 2009).

When nitrite comes in contact with hemoglobin, it forms methemoglobin. This reaction can be reversed by the action of NADH dehydrogenase (NADH-diaphorase, EC1.6.99.3).

Storage
There are no known specific storage sites for nitrate or nitrite.

Excretion
Nitrate and nitrite are very small molecules that stay unhindered with the primary ultrafiltrate. It is not clear how much is reabsorbed across the nephron. Some unspecific recovery is known to occur through various anion transporters. One example is the anion exchanger 1 (SLC4A1) at the basolateral membrane of intercalated cells that can move nitrate into the blood in exchange for chloride.

Regulation
No specific homeostatic mechanisms for regulating nitrite or nitrate concentrations in the body are known.

Function
Vasodilation: Nitrate dilates small arteries. The effect has been demonstrated even with the small amounts in spinach and other leafy vegetables (Lidder and Webb, 2013). Several nitrate derivatives (nitroglycerin, isosorbide mononitrate, isosorbide dinitrate, pentaerythritol tetranitrate, and sodium nitroprusside) are used for the short-term symptomatic relief of angina pectoris (coronary ischemia or heart attack).

It has been suggested that nitrate, which after reduction to nitrite is a potential precursor of the vasodilator nitric oxide, reduces blood pressure, inhibits platelet aggregation, improves arterial perfusion, and reduces arterial stiffness (Lidder and Webb, 2013). A diet with high nitrate content has been observed to increase the blood flow to the brain (Presley et al., 2011).

Energy metabolism: Nitrate intake reduces the resting metabolic rate (RMR) by increasing the efficiency of mitochondria. An added intake of 300 g spinach reduced the RMR by 4% in healthy young men and women (Larsen et al., 2014). The effect is mediated by nitrite, possibly after conversion to nitric oxide.

Antimicrobial action: Nitrite is particularly important to prevent botulism, a particularly dangerous form of food poisoning due to the toxin produced by *Clostridium botulis*. Nitrite salts also inhibit the growth of the much-more-pervasive spoiler of foods, *Clostridium perfringens*, and other microbes causing food poisoning.

Methemoglobin formation: Exposure to nitrite converts hemoglobin to methemoglobin, which has little capacity for transporting oxygen. Conversion of more than 10–15% of total circulating hemoglobin reduces oxygenation of sensitive tissues, in particular of the brain. The condition is called *cyanosis* and is indicated by a bluish tint of the lips. The condition is often fatal when methemoglobin accounts for more than 70% of total hemoglobin (Richard et al., 2014).

Nitrosamine formation: Incubation of proteins and amino acids with nitrate or nitrite in the stomach can generate nitrosamines, which are potent carcinogens. Consumption of ascorbic acid together with the nitrate dose will block most of the nitrosamine formation (Tannenbaum et al., 1991; Krul et al., 2004).

Thyroid function: High nitrate or nitrite intake may slow iodide uptake and interfere with thyroid health by competing for the sodium-iodide cotransporter SLC5A5 (Gatseva and Argirova, 2008).

Diabetes: Nitrate may act as an endocrine disruptor, interfere with beta-cell function, and induce hyperglycemia, at least in rodents (El-Wakf et al., 2014).

REFERENCES

Bondonno, C.P., Liu, A.H., Croft, K.D., Ward, N.C., Yang, X., Considine, M.J., et al., 2014. Short-term effects of nitrate-rich green leafy vegetables on blood pressure and arterial stiffness in individuals with high-normal blood pressure. Free Radic. Biol. Med. 77, 353–362.

Dellavalle, C.T., Daniel, C.R., Aschebrook-Kilfoy, B., Hollenbeck, A.R., Cross, A.J., Sinha, R., et al., 2013. Dietary intake of nitrate and nitrite and risk of renal cell carcinoma in the NIH-AARP Diet and Health Study. Br. J. Cancer 108, 205–212.

Dellavalle, C.T., Xiao, Q., Yang, G., Shu, X.O., Aschebrook-Kilfoy, B., Zheng, W., et al., 2014. Dietary nitrate and nitrite intake and risk of colorectal cancer in the Shanghai Women's Health Study. Int. J. Cancer 134, 2917–2926.

Demoncheaux, E.A., Higenbottam, T.W., Foster, P.J., Borland, C.D., Smith, A.P., Marriott, H.M., et al., 2002. Circulating nitrite anions are a directly acting vasodilator and are donors for nitric oxide. Clin. Sci. 102, 77–83.

El-Wakf, A.M., Hassan, H.A., Mahmoud, A.Z., Habza, M.N., 2014. Fenugreek potent activity against nitrate-induced diabetes in young and adult male rats. Cytotechnology, http://dx.doi.org/doi:10.1007/s10616-014-9702-7.

Gatseva, P.D., Argirova, M.D., 2008. High-nitrate levels in drinking water may be a risk factor for thyroid dysfunction in children and pregnant women living in rural Bulgarian areas. Int. J. Hyg. Environ. Health 211 (5–6), 555–559.

Gladwin, M.T., Grubina, R., Doyle, M.P., 2009. The new chemical biology of nitrite reactions with hemoglobin: R-state catalysis, oxidative denitrosylation, and nitrite reductase/anhydrase. Acc. Chem. Res. 42, 157–167.

Hayashi, H., Suruga, K., Yamashita, Y., 2009. Regulation of intestinal Cl^-/HCO_3^- exchanger SLC26A3 by intracellular pH. Am. J. Physiol. Cell Physiol. 296, C1279–C1290.

Kozlov, A.V., Gille, L., Staniek, K., Nohl, H., 1999. Dihydrolipoic acid maintains ubiquinone in the antioxidant active form by two-electron reduction of ubiquinone and one-electron reduction of ubisemiquinone. Arch. Biochem. Biophys. 363, 148–154.

Krul, C.A., Zeilmaker, M.J., Schothorst, R.C., Havenaar, R., 2004. Intragastric formation and modulation of N-nitrosodimethylamine in a dynamic *in vitro* gastrointestinal model under human physiological conditions. Food Chem. Toxicol. 42, 51–63.

Larsen, F.J., Schiffer, T.A., Ekblom, B., Mattsson, M.P., Checa, A., Wheelock, C.E., et al., 2014. Dietary nitrate reduces resting metabolic rate: a randomized, crossover study in humans. Am. J. Clin. Nutr. 99, 843–850.

Leaf, C.D., Wishnok, J.S., Tannenbaum, S.R., 1989. L-arginine is a precursor for nitrate biosynthesis in humans. Biochem. Biophys. Res. Commun. 163, 1032–1037.

Lidder, S., Webb, A.J., 2013. Vascular effects of dietary nitrate (as found in green leafy vegetables and beetroot) via the nitrate–nitrite–nitric oxide pathway. Br. J. Clin. Pharmacol. 75, 677–696.

Mitsui, T., Kondo, T., 2004. Increased breath nitrous oxide after ingesting nitrate in patients with atrophic gastritis and partial gastrectomy. Clin. Chim. Acta 345, 129–133.

Presley, T.D., Morgan, A.R., Bechtold, E., Clodfelter, W., Dove, R.W., Jennings, J.M., et al., 2011. Acute effect of a high nitrate diet on brain perfusion in older adults. Nitric Oxide 24, 34–42.

Richard, A.M., Diaz, J.H., Kaye, A.D., 2014. Reexamining the risks of drinking-water nitrates on public health. Ochsner J. 14, 392–398.

Roediger, W.E., Radcliffe, B.C., 1988. Role of nitrite and nitrate as a redox couple in the rat colon. Implications for diarrheal conditions. Gastroenterology 94, 915–922.

Tannenbaum, S.R., Wishnok, J.S., Leaf, C.D., 1991. Inhibition of nitrosamine formation by ascorbic acid. Am. J. Clin. Nutr. 53, 247S–250S.

FATTY ACIDS

5

CHAPTER OUTLINE

Structure and Function of Fatty Acids 111
Overfeeding 143
Acetate 147
Myristic Acid 153
Conjugated Linoleic Acid 157
Docosahexaenoic Acid 164
Trans-Fatty Acids 174
Chlorophyll/Phytol/Phytanic Acid 179

STRUCTURE AND FUNCTION OF FATTY ACIDS

Fatty acids are monocarbonic acids with chain lengths between 2 and 26 (and sometimes more). Most fatty acids in plants and animals have an even carbon number and no branching. One or more double bonds may occur, mostly in cis configuration. More than 40 different fatty acids are commonly encountered in foods.

ABBREVIATIONS

apoC-I	apolipoprotein C-I
CoA	coenzyme A
ETF	electron-transfer flavoprotein
FATP-1	fatty acid transport protein 1 (CD36, SLC27A1)
HDL	high-density lipoprotein
HNE	4-hydroxy-2,3-trans-nonenal
LDL	low-density lipoprotein
Lp	lipoprotein
LPL	lipoprotein lipase (EC3.1.1.34)
MFP2	peroxisomal multifunctional protein 2
MTP	microsomal triglyceride transfer protein
SCPx	sterol carrier protein X
TRL	triglyceride-rich lipoprotein
VLDL	very low-density lipoprotein

NUTRITIONAL SUMMARY

Function: Fat (neutral fat, triglyceride) has the highest energy contents of all macronutrients, providing about 9 kcal/g. Complete oxidation depends on thiamin, riboflavin, niacin, vitamin B12, biotin, pantothenate, carnitine, ubiquinone, iron, and magnesium. Specific types of fatty acids are needed for hormonal action and cell signaling (eicosanoids), structural compounds, and the modification of proteins and many other molecules.

Food sources: Meats and oils contribute most of the fat in foods. Oils, hydrogenated plant fats, and animal fats are used extensively in mixed dishes, baked goods, and snacks.

Requirements: Current recommendations suggest limiting saturated and trans-fat intake to <10% of total energy intake, and keeping total fat intake around or below 30%. At least 3 g of omega-3 fatty acids (especially in fish, marine oils, flaxseed, and linseed oil), and 6 g of omega-6 fatty acids (in many plant oils) should be consumed daily. One type of essential fatty acid cannot substitute for the other.

Deficiency: Inadequate intake of omega-6 fatty acids causes scaly skin deterioration and increases susceptibility to the cholesterol-raising effects of saturated and trans-fats. Inadequate omega-3 fatty acid intake impairs immune function and leads to the development of dermatitis (after many weeks).

Excessive intake: High fat consumption is likely to result in excessive total energy intake and eventually increase body weight and fatness. A high percentage of body fat is associated with an increased risk of hyperlipidemia, hyperuricemia, high blood pressure, atherosclerosis, and other health risks. High intake of specific fatty acids may have its own risks distinct from the risk of high total fat intake; examples are the cholesterol-raising action of myristic, palmitic, and trans-fatty acids, the induction of bleeding by polyunsaturated fatty acids, the initiation of heart arrhythmia by stearic acid, and increased risk of some cancers in people with high long-chain omega-3 fatty content in phospholipids (Brasky et al., 2013).

STRUCTURES

The characteristic feature of fatty acids is the carboxy function attached to an aliphatic body. Fatty acids in plants and animals most commonly have an even number of carbons, but a small percentage of odd-numbered fatty acids are also present in many species. Short-chain fatty acids contain 2, 3, or 4 carbons, and medium-chain fatty acids have between 6 and 12 carbons. Long-chain fatty acids are those with 14–18 carbons, but a larger number is possible (very-long-chain fatty acids). While most fatty acids have straight chains, branching does occur. Some fatty acids have double bonds linking carbons.

Their number, type, and position influence physical and metabolic properties. It is characteristic of cis double bonds that the hydrogens at the two involved carbons are on the same side (which is what *cis* means in Latin) and that the acyl chain is bent at a stark angle with some freedom of movement. In trans-fatty acids, the hydrogens are at different sides of the chain and the acyl chain is more or less straight and rigid. Fatty acids without double bonds are called *saturated*. Monounsaturated fatty acids have one double bond, while polyunsaturated fatty acids have more than one. All fatty acids have systematic names that consist of the Greek number term for the chain length followed by the ending *-ic* and the word *acid* or just the ending *-ate*. The saturated fatty acid with 18 carbons is called *octadecanoic acid* or *octadecanoate,* therefore (Table 5.1, Figure 5.1).

If the chain contains a double bond, this is indicated by adding to the Greek number term the number of the first double-bonded carbon (counting from the carboxyl end) and the ending *-enic* or *-enate.* Thus, the 18-carbon fatty acid with a double bond between carbons 9 and 10 has the systematic name of

Table 5.1 Saturated Fatty Acids

Trivial Name	Systematic Name	Carbons: Double Bonds
Short-chain fatty acids		
Acetic acid	Ethanoic acid	C2:0
Propionic acid	Propanoic acid	C3:0
Butyric acid	Butanoic acid	C4:0
Medium-chain fatty acids		
Caproic acid	Hexanoic acid	C6:0
Caprylic acid	Octanoic acid	C8:0
Capric acid	Decanoic acid	C10:0
Lauric acid	Dodecanoic acid	C12:0
Long-chain fatty acids		
Myristic acid	Tetradecanoic acid	C14:0
Pentadecanoic acid	Pentadecanoic acid	C15:0
Palmitic acid	Hexadecanoic acid	C16:0
Margarinic acid	Heptadecanoic acid	C17:0
Stearic acid	Octadecanoic acid	C18:0
Arachidic acid	Eicosanoic acid	C20:0
Behenic acid	Docosanoic acid	C22:0
Lignoceric acid	Tetracosanoic acid	C24:0
Cerotenic acid	Hexacosanoic acid	C26:0
Branched-chain fatty acids		
	Iso-tetradecanoic acid	Iso-C14:0
	Iso-pentadecanoic acid	Iso-C15:0
	Anteiso-pentadecanoic acid	Anteiso-C15:0
	Iso-hexadecanoic acid	Iso-C16:0
	Iso-heptadecanoic acid	Iso-C17:0
	Anteiso-heptadecanoic acid	Anteiso-C17:0
	18-methyleicosanic acid	Anteiso-C21:0
Phytanic acid	3,7,11,15-tetramethylhexadecanoic acid	3,7,11,15-tetramethyl-C16
Pristanic acid	2,6,10,14-tetramethylpentadecanoic acid	2,6,10,14-tetramethyl-C15

octadec-9-enoic acid or *octa-9-enoate*. If there is more than one double bond, all carbon positions are indicated and followed by the Greek number term (*di-, tri-, tetra-, penta-, hexa-*, etc.), referring to the number of double bonds, followed by the ending *-enic* or *-enate*. This would make the 18-carbon fatty acid with double bonds at carbons 9 and 12 an *octa-9,12-dienoic acid* or *octa-9,12-dienoate*.

The usual assumption is that these are cis double bonds unless indicated otherwise. If there are trans-double bonds or the presence of cis double bonds is to be emphasized, the term *cis* or *trans* is inserted before the position number. To abbreviate this lengthy notation, the letters *c* or *t* are often used

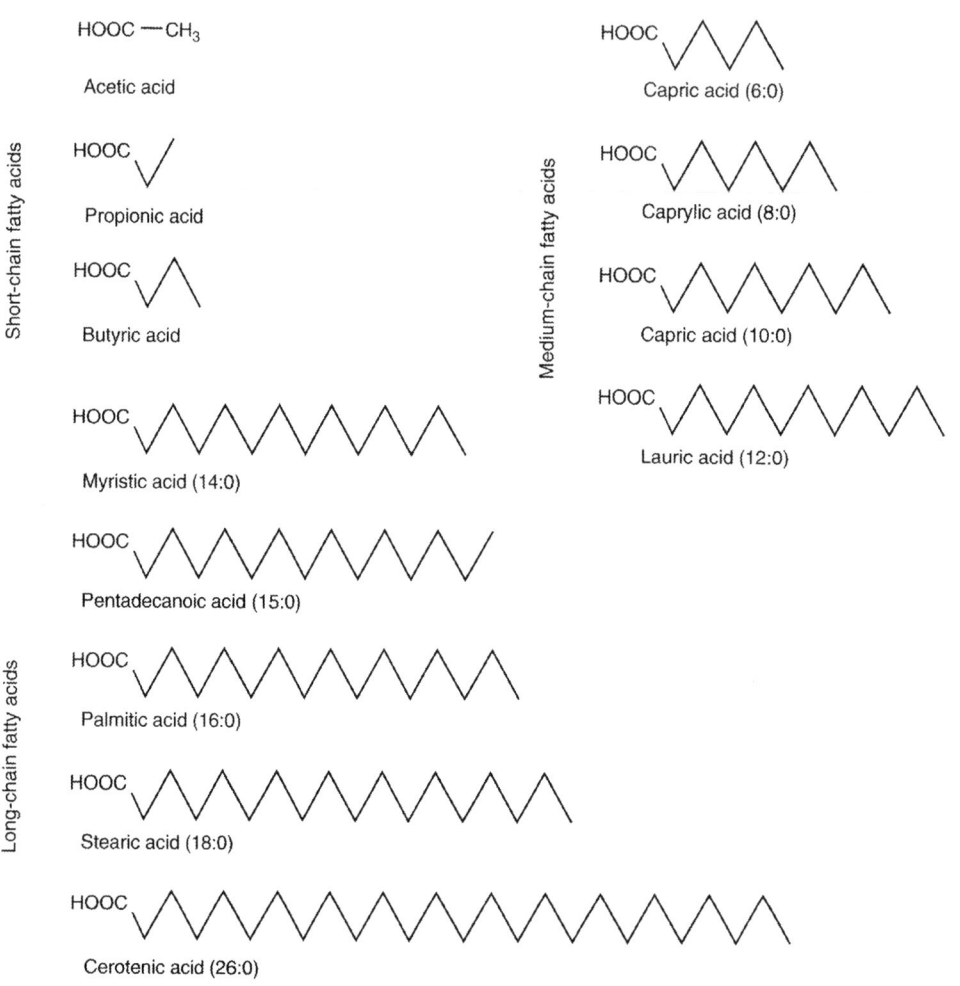

FIGURE 5.1

Common saturated fatty acids.

instead. The common fatty acids have trivial names that avoid the burdensome use of the systematic names (Table 5.2, Figure 5.2).

It is certainly easier to refer to *arachidonic acid,* which unambiguously identifies eicosa-cis6,cis9,cis12,cis15-tetraenoic acid, than to use the systematic name. If there are two or more double bonds in natural fatty acids, they usually recur at three-carbon intervals. Two families of polyunsaturated fatty acids have particular metabolic importance and have to come from the diet. One family, the omega-3-fatty acids, comprises fatty acids with a double bond between the third and fourth carbon, counting from the acyl end, and one or more double bonds at three carbon intervals. The counting,

Table 5.2 Monounsaturated Fatty Acids

Trivial Name	Systematic Name	Carbons: Double Bonds
Myristoleic acid	Tetradec-c7-enoic acid	C14:1 n7
Palmitoleic acid	Hexadec-c7-enoic acid	C16:1 n7
Petroselinic acid	Octadec-c6-enoic acid	C18:1 n12
Petroselaidic acid	Octadec-t6-enoic acid	C18:1 n12
Oleic acid	Octadec-c9-enoic acid	C18:1 n9
Elaidic acid	Octadec-t9-enoic acid	C18:1 n9
Trans-vaccenic acid	Octadec-t11-enoic acid	C18:1 n7
Petroselinic acid	Octadec-c6-enoic acid	C18:1 n12
Gondoic acid	Eicosa-c11-enoic acid	C20:1 n9
Erucic acid	Docosa-c13-enoic acid	C22:1 n9
Cetoleic acid	Docosa-c11-enoic acid	C22:1 n11
Brassidic acid	Docosa-t13-enoic acid	C22:1 n9
Nervonic acid	Tetracosa-c15-enoic acid	C24:1 n9

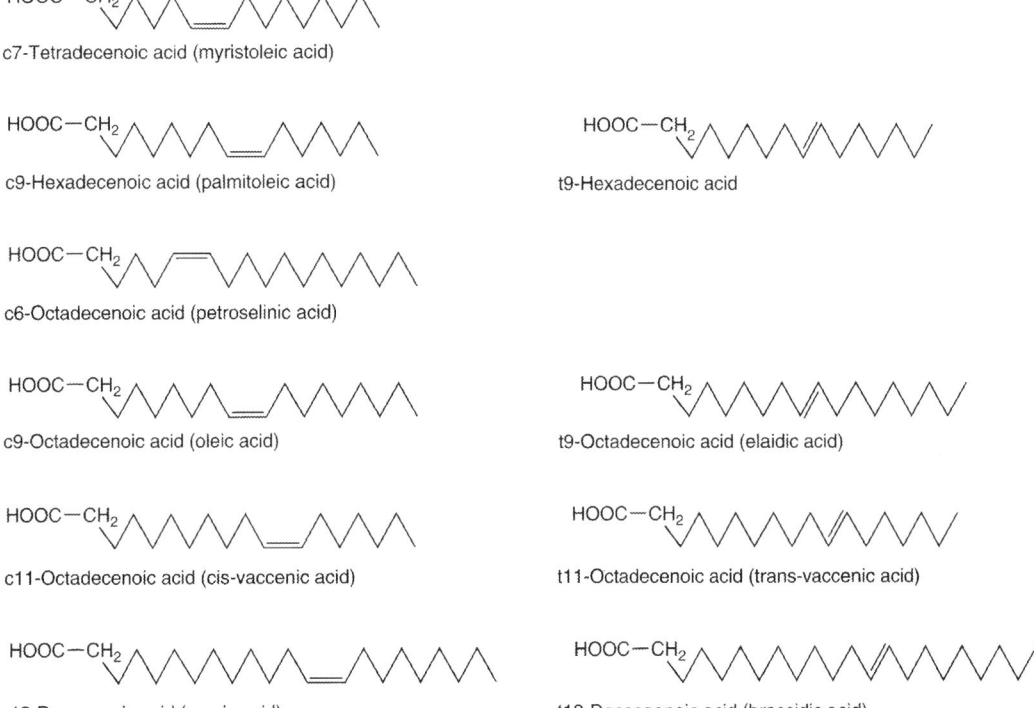

c7-Tetradecenoic acid (myristoleic acid)

c9-Hexadecenoic acid (palmitoleic acid)

t9-Hexadecenoic acid

c6-Octadecenoic acid (petroselinic acid)

c9-Octadecenoic acid (oleic acid)

t9-Octadecenoic acid (elaidic acid)

c11-Octadecenoic acid (cis-vaccenic acid)

t11-Octadecenoic acid (trans-vaccenic acid)

c13-Docosaenic acid (erucic acid)

t13-Docosaenoic acid (brassidic acid)

FIGURE 5.2

Cis- and trans-unsaturated fatty acids.

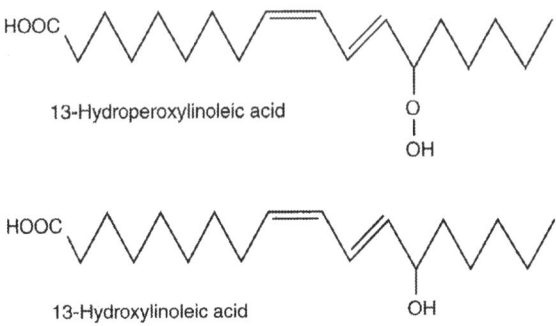

13-Hydroperoxylinoleic acid

13-Hydroxylinoleic acid

FIGURE 5.3

Free radicals can modify unsaturated fatty acids.

according to this nomenclature, starts from the tail end. The common notation uses the omega symbol (ω, sometimes represented as w, or even as n) since omega is the last letter in the Greek alphabet and traditionally has indicated the end. The other family, the omega-6 fatty acids, comprises fatty acids with double bonds between the sixth and seventh carbon, counting from the acyl end, and one or more additional double bonds at three-carbon intervals from there.

Peroxidation by exposure to free radicals can modify unsaturated fatty acids (Figure 5.3) and generate conjugated dienic double bonds adjacent to hydroperoxy or hydroxy groups that dramatically alter their chemical and metabolic properties (e.g., orientation in a membrane) (Table 5.3, Figures 5.4 and 5.5).

While the vast majority of fatty acids in foods have straight chains, there are a few branched-chain fatty acids in certain foods. Iso-fatty acids contain a methyl branch that extends from the second-to-last carbon; anteiso-fatty acids carry their methyl branch at the third-to-last carbon (Figure 5.6). The very short branched-chain fatty acids are typically derived from bacterial sources, while some of the longer ones are produced in mammals (including humans) by the extension of valine metabolites (iso-fatty acids) and isoleucine metabolites (anteiso-fatty acids). Recent evidence also points to a specific role of branched long-chain fatty acids (including 18-methyleicosanic acid), some of them hydroxylated, with double bonds, or both, such as structural components of hair (Jones and Rivett, 1997) and skin and fatty secretions (sebum) from skin (Steward and Downing, 1990). Phytanic acid and pristanic acid are mostly derived from phytol, which provides the side chain of chlorophyll and has multiple methyl branches.

The bulk of dietary fatty acids is bonded to glycerol. Monoglycerides contain one fatty acid, diglycerides contain two, and triglycerides contain three. A large percentage of cholesterol and plant sterols in foods is esterified to a fatty acid, more likely an unsaturated fatty acid than another kind. Phospholipids by definition contain fatty acids (Figure 5.7).

The appearance of dietary fats (triglycerides) provides clues about their fatty acid composition. Fats that are solid (white) at room temperature are composed mainly of saturated fats. Oils, which are triglycerides that are liquid at room temperature, contain significant percentages of unsaturated fats.

Table 5.3 Polyunsaturated Fatty Acids

Trivial Name	Systematic Name	Carbons: Double Bonds
Omega-6 PUFA		
Linoleic acid	Octadeca-c9,c12-dienoic acid	C18:2n6
Gamma-linolenic acid	Octadeca-c6,c9,c12-trienoic acid	C18:3n6
Dihomo-gamma-linolenic acid	Docosa-c8,c11,c14-trienoic acid	C20:3n6
Arachidonic acid	Docosa-c5,c8,c11,c14-tetraenoic acid	C20:4n6
Adrenic acid	Docosa-c7,c10,c13,c16-tetraenoic acid, DTA	C22:4n6
Osbond acid	Docosa-c4,c7,c10,c13,c16-pentaenoic acid	C22:5n6
Omega-3 PUFA		
	Hexadeca-c4,c7,c10,c13-tetraenoic acid	C16:4n3
Alpha-linolenic acid	Octadeca-c9,c12,c15-trienoic acid	C18:3n3
Stearidonic acid	Octadeca-c6,c9,c12,c15-tetraenoic acid	C18:4n3
Timnodonic acid	Eicosa-cS,c8,c11,c14,c17-pentaenoic acid, EPA	C20:5n3
Clupadonic acid	Docosa-c7,c10,c13,c16,c19-pentaenoic acid	C22:5n3
	Docosa-c4,c7,c10,c13,c16,c19-hexaenoic, DHA	C22:6n3
Other PUFA		
Mead acid	Eicosa-c5,c8,c11-trienoic acid	C20:3n9
Conjugated linoleic acid (CLA)	28 different potential isomers	C20:2n9
Rumenic acid (a particular CLA)	Octadeca-c9,t11-dienoic acid	C20:2n9

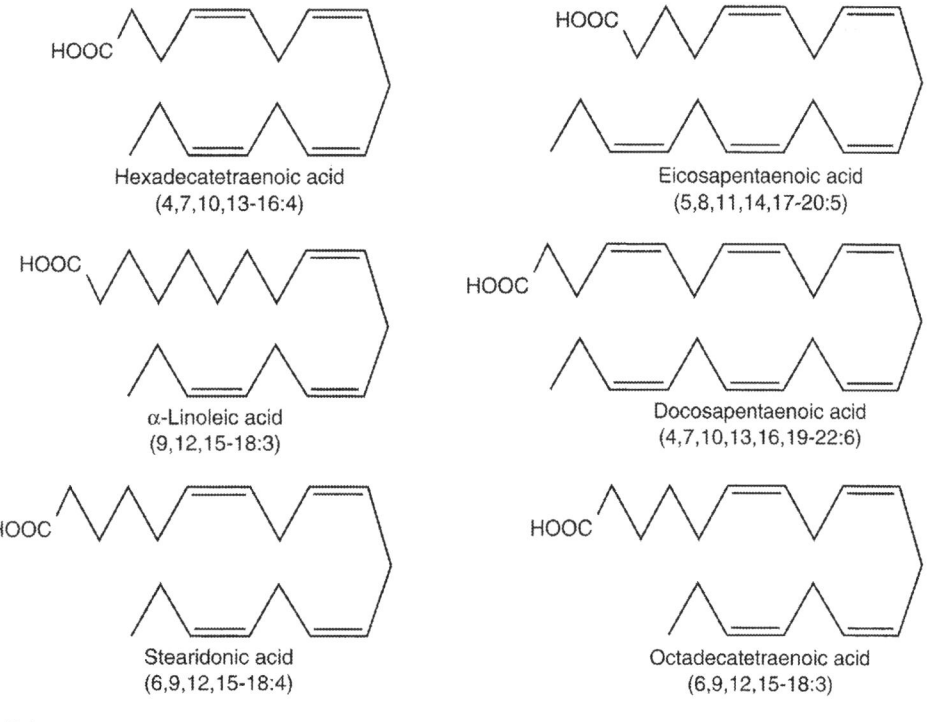

FIGURE 5.4

Omega-3 polyunsaturated fatty acids.

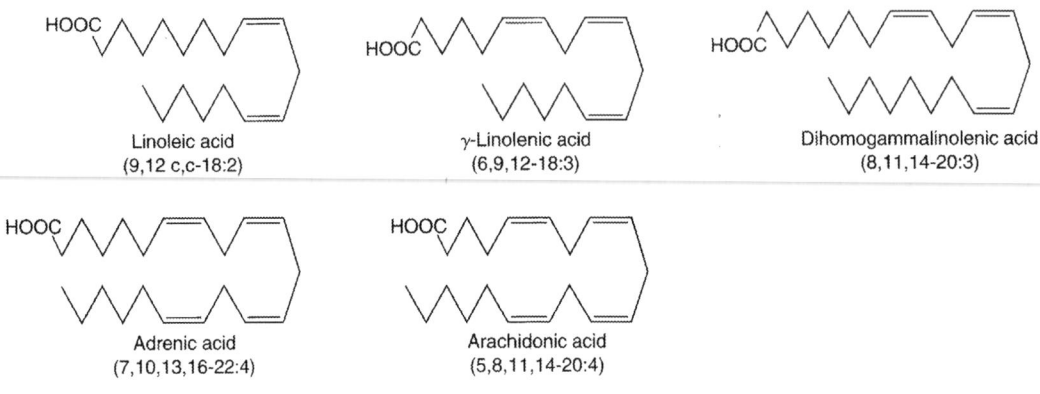

FIGURE 5.5

Omega-6 polyunsaturated fatty acids.

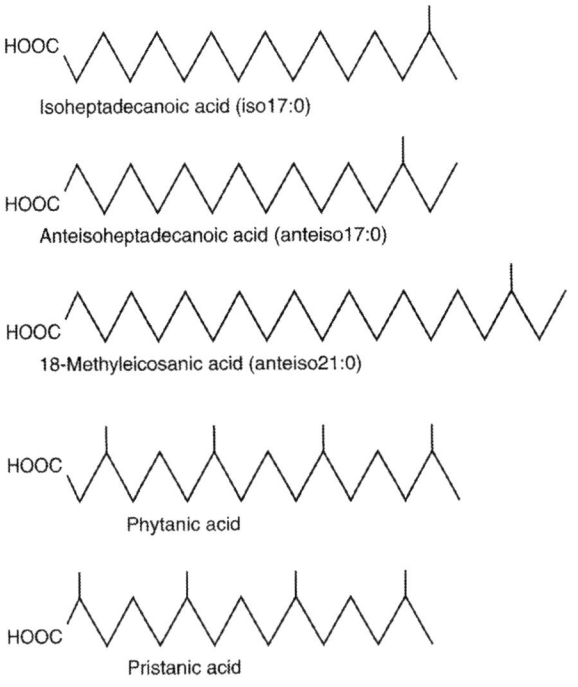

FIGURE 5.6

Branched-chain fatty acids.

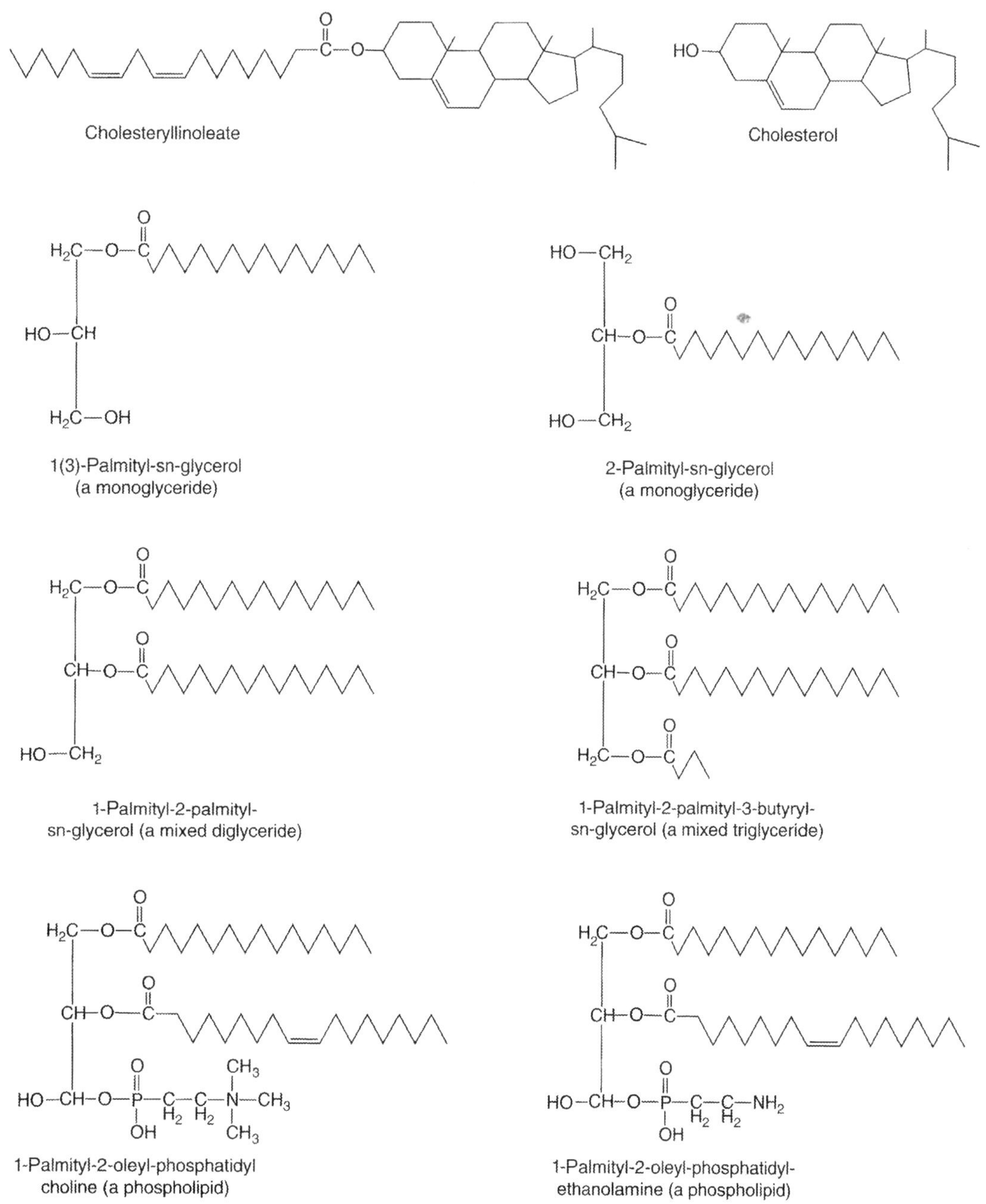

Cholesteryllinoleate

Cholesterol

1(3)-Palmityl-sn-glycerol
(a monoglyceride)

2-Palmityl-sn-glycerol
(a monoglyceride)

1-Palmityl-2-palmityl-
sn-glycerol (a mixed diglyceride)

1-Palmityl-2-palmityl-3-butyryl-
sn-glycerol (a mixed triglyceride)

1-Palmityl-2-oleyl-phosphatidyl
choline (a phospholipid)

1-Palmityl-2-oleyl-phosphatidyl-
ethanolamine (a phospholipid)

FIGURE 5.7

Fatty acids are part of different complex lipids.

ENDOGENOUS SOURCES

Complete de novo *synthesis*: The extent to which carbohydrates other than fructose are converted into fat is still under dispute (Hellerstein, 2001), not least because of great differences between individuals (Hudgins et al., 2011). This much seems certain: the liver produces only a few grams of fat per day from acetyl–coenzyme A (acetyl-CoA), even during periods of massive carbohydrate overfeeding. Adipose tissue may be responsible for the conversion of carbohydrate into fat at a much higher rate (Aarsland et al., 1997). Synthesis, as far as it takes place, tends to proceed to chain lengths of 16 or 18 carbons with little release of the intermediate metabolite myristate.

Synthesis depends on plentiful supplies of acetyl-CoA, which is moved from mitochondria to the cytosol by the citrate shuttle mechanism. Citrate, which is produced from acetyl-CoA and oxaloacetate by citrate synthase (EC4.1.3.7), moves across the inner mitochondrial membrane via the tricarboxylate transporter (SLC25A1) in exchange for malate. ATP-citrate (pro-S-)-lyase (citrate cleavage enzyme, EC4.1.3.8) can then release acetyl-CoA for fatty acid synthesis and malate for transport back into mitochondria and another cycle of acetyl-CoA export. Alternatively, nicotinamide-adenine-dinucleotide-phosphate (NADP)–dependent malic enzyme 1 (EC1.1.1.40) can generate pyruvate, which in turn is shuttled back into mitochondria.

The first key step of fatty acid synthesis and elongation in cytoplasma is malonyl-CoA production by acetyl-CoA carboxylase (EC6.4.1.2). This enzyme is biotinylated by biotin-[acetyl-CoA-carboxylase] ligase (EC6.3.4.15), inactivated by [acetyl-CoA carboxylase] kinase (EC2.7.1.128), and reactivated by [acetyl-CoA carboxylase]-phosphatase (EC3.1.3.44). Two genes encode isoenzymes with different tissue expression patterns. The alpha enzyme is expressed in muscle and many other tissues except liver. The beta enzyme in muscle and liver occurs in short and long isoforms due to alternative splicing.

Fatty acid synthase (EC2.3.1.85) is a multienzyme complex in the acyl-carrier protein, which moves the growing fatty acyl-CoA chain from site to site until the full-length chain is released. A notable feature of acyl-carrier protein is the phosphopantetheine group. Holo-acyl-carrier protein synthase (EC2.7.8.7) transfers this CoA-derived molecule to a specific serine group of the nascent protein. Acyl-carrier protein is the only use of pantothenate in a form other than CoA (Figure 5.8).

The fatty acid synthase complex comprises components that catalyze the reactions of acyl-carrier protein *S*-acetyltransferase (EC2.3.1.38), acyl-carrier protein *S*-malonyltransferase (EC2.3.1.39), 3-oxoacyl-acyl-carrier protein synthase (EC2.3.1.41), 3-oxoacyl-acyl-carrier protein reductase (EC1.1.1.100), 3-hydroxypalmitoyl-acylcarrier protein dehydratase (EC4.2.1.61), enoyl-acyl-carrier protein reductase (NADPH, B-specific, EC1.3.1.10), and oleoyl-acyl-carrier protein hydrolase (EC3.1.2.14).

These activities add in sum with each cycle a malonyl-CoA to the growing chain, oxidize two nicotine adenine dinucleotide phosphate (NADPH) molecules to NADP, and release a carbon dioxide molecule and a free CoA molecule.

Chain elongation: The length of fatty acids with 12, 14, or 16 carbons can be extended in the cytosol by the addition of 2-carbon units (Moon et al., 2001). The first step is the conjugation to CoA by long-chain-fatty-acid-CoA ligase (EC6.2.1.3). A specific NADH-using long-chain fatty acyl elongase (ELOV6, EC2.3.1.199) adds malonyl-CoA to the acyl-CoA and releases carbon dioxide and free CoA (Figure 5.9). This (and presumably additional related enzymes) are active mainly in hepatocytes, adipocytes, and mammary glands during lactation.

Desaturation: Several enzyme systems can introduce double bonds at specific sites of a well-defined range of fatty acyl-CoAs. The ferroenzyme stearyl-CoA desaturase (delta-9 desaturase, EC1.14.99.1),

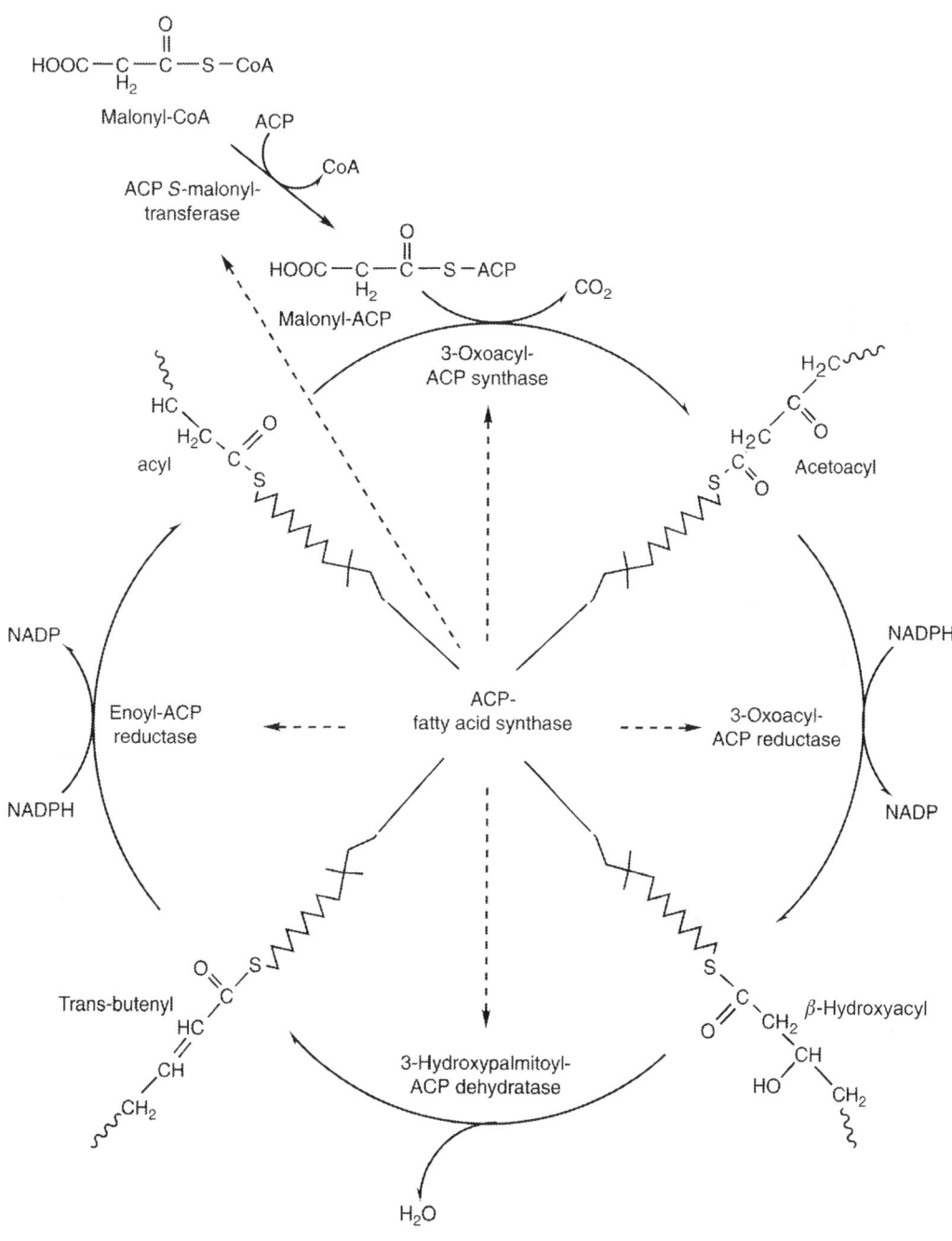

FIGURE 5.8

De novo synthesis is usually a minor source of fatty acids in humans.

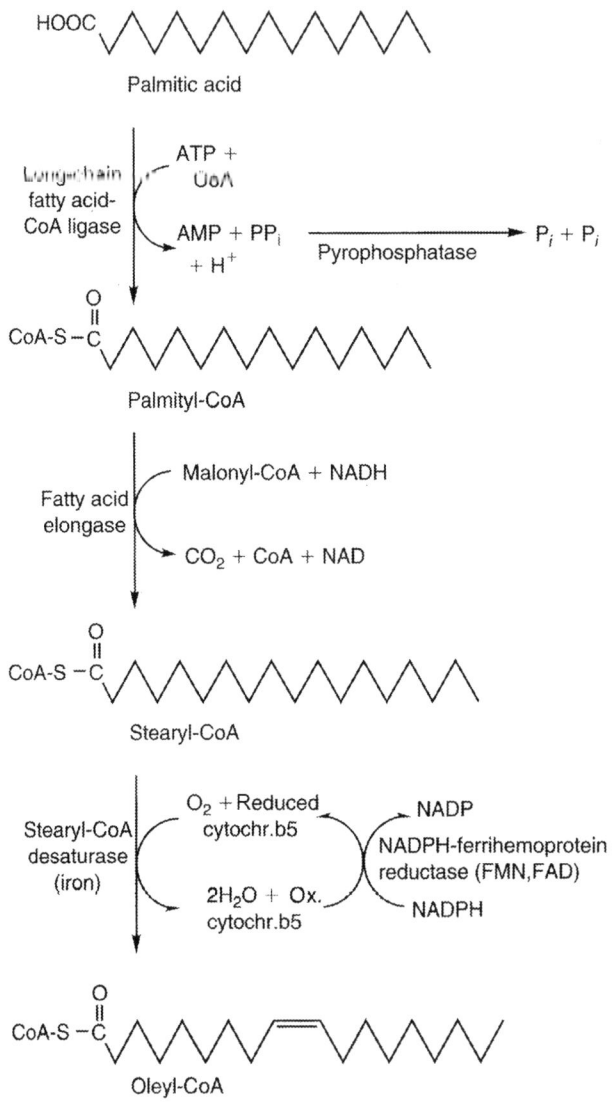

FIGURE 5.9

Chain elongation and desaturation.

which is attached through a myristate anchor to the cytoplasmic side of the endoplasmic reticulum, uses cytochrome b5 and oxygen to introduce a double bond at carbon 9 of acyl-CoAs with chain lengths between 14 and 18. Flavin mononucleotide (FMN)– and flavin adenine dinucleotide (FAD)–containing cytochrome b5 reductase (EC1.6.2.2) then reactivates this electron donor in an NADH-dependent reaction.

A distinct linoleoyl-CoA desaturase (delta-6 desaturase, FADS2, EC1.14.19.-) acts on 18:2n-6, 18:3n-3, 24:4n-6, and 24:5n-3 (de Antueno et al., 2001). This ferroenzyme uses the cytochrome b5 system like other desaturases.

DIETARY SOURCES

The bulk of dietary fatty acids is consumed as triglycerides. Smaller amounts are taken up with diglycerides and monoglycerides, phospholipids, and cholesteryl esters; plasmalogens, glycolipids, and other complex compounds contribute relatively little in a mixed diet.

Fatty acid composition is greatly dependent on the type of food and its origin. Almost all of them have an even number of carbons. Small amounts of pentadecanoic acid come with dairy foods (Wolk et al., 2001) and peanuts. The intake of heptadecanoic acid and other odd-chain fatty acids is even lower. While oleic acid (octadec-c9-enoic acid) is the main isomer in most foods, several other forms are consumed in small amounts. One of these, petroselinic acid (octadec-c6-enoic acid), is even the major form in the seed oil of coriander and other *Umbelliferae* species. Transvaccenic acid (octadec-t11-enoic acid) is a naturally occurring minor constituent of ruminant fat and may contribute some health benefits as a precursor of conjugated linoleic acid (CLA). The other trans-fatty acids in foods, in contrast, arise from partial hydrogenation of polyunsaturated fatty acids and are associated with unfavorable health outcomes. Erucic acid (docosa-c13-enoic acid) is another monoenic fatty acid of ill repute, due to concerns that it might cause cardiac lipidosis and other harm. Genetic plant modification has reduced the content in rapeseed oil (the main source) from more than 40% to only a few percent, and this led to the renaming of the new product as *canola oil.*

Plant foods tend to contain more long-chain monounsaturated and polyunsaturated acids and short- and medium-chain fatty acids. Animal foods usually have more saturated long-chain fatty acids, many exceptions notwithstanding. Dairy foods provide most of the myristic acid, plant oils most of the linoleic acid and linolenic acid, and coldwater fish most of the eicosapentaenoic acid and docosahexaenoic acid (DHA) in the diet. Some edible marine algae, such as *Undaria pinnifida* and *Ulva pertusa*, contain the unusual omega-3 fatty acid hexadeca-4,7,10,13-tetraenoic acid. Blackcurrant seeds are a rare plant source (2–4%) of stearidonic acid (octadeca-6,9,12,15-tetraenoic acid).

Phospholipids and cholesterol esters tend to have a higher proportion of long-chain polyunsaturated fatty acids than triglycerides. Game animals (e.g., boar) living in the wild have much leaner meat than closely related domesticated animals and more polyunsaturated than saturated fatty acids. The heating of unsaturated fats, such as during deep frying, causes oxidation. A conjugated diene content of 4–20% has to be considered normal in such foods and may be higher (Staprans et al., 1999). The smaller fatty acids are volatile and tend to have strong odors, as is the case with butyric acid. For instance, lipases from specific bacteria release odorous fatty acids from triglycerides in cheeses, giving them their characteristic aroma. The compound responsible for the typical odor in goat's milk products is thought to be 4-ethyloctanoate (Alonso et al., 1999). Dairy products are also the major source of a wide range of branched-chain fatty acids such as iso-C15:0 and iso-C17:0 (derived from leucine metabolism), anteiso-C15:0 and anteiso-C17:0 (from isoleucine), and iso-C16:0 (from valine). Human milk contains nearly 40 different fatty acids, including the abovementioned branched-chain fatty acids.

DIGESTION

Diglycerides and triglycerides are hydrolyzed by several lipases in the upper digestive tract. Corresponding to their target environment, they differ in pH optimum and cofactor requirements. Lingual lipase is active at neutral to alkaline pH, while gastric lipase (also secreted with milk) tolerates the acidic environment of the stomach; neither depends on further cofactors. Mixed micelles are generated upon mixing of bile acids and phospholipids from bile with dietary triglycerides, and stabilized by the addition of monoglycerides from ongoing triglyceride hydrolysis. Bile-salt activated lipase (EC3.1.1.3) from the pancreas efficiently hydrolyzes mixed micelles at the alkaline pH typical of small intestine. The enzyme from the pancreas acts in concert with a colipase, which allows the lipase to penetrate the bile acid coating of mixed micelles and interact with the glycerides inside. This lipase also has carboxylester lipase (EC3.1.1.13) activity and cleaves fatty acids off cholesterol esters. Phospholipase A2 (EC3.1.1.4) from pancreas cleaves off the middle fatty acid (position sn-2).

ABSORPTION

Free acids and monoglycerides from mixed micelles transfer into enterocytes of the small intestine by diffusion and facilitated transport (Figure 5.10). A wide range of fatty acids is taken up, including

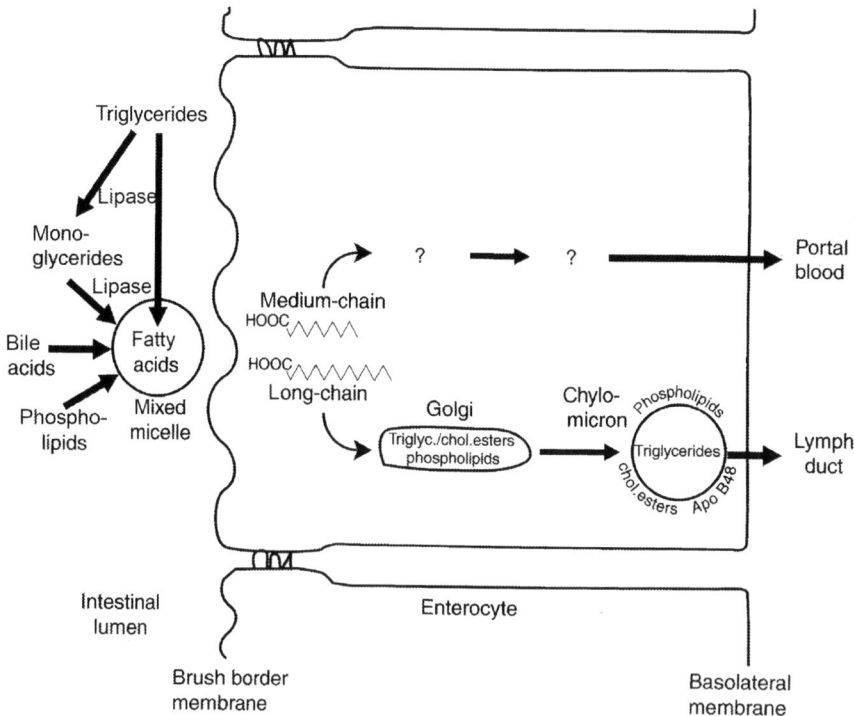

FIGURE 5.10

Intestinal absorption of fat.

branched-chain and oxidized (dienic) fatty acids. Long-chain fatty acids tend to be conjugated to CoA by one of several specific long-chain fatty acid CoA ligases. Most of the acyl-CoA is used for triglyceride synthesis (for pathways see the section "Storage," later in this chapter); a much smaller proportion contributes to the synthesis of cholesterol esters, phospholipids, and other complex lipids. Some acyl-CoA undergoes beta-oxidation and is used as an energy fuel for the enterocytes.

Triglycerides are assembled into chylomicrons with cholesterol esters, phospholipids, and one molecule of apolipoprotein B48 per particle, and secreted into intestinal lymphatic vessels. The fatty acid binding proteins 1 and 2, microsomal triglyceride transfer protein (MTP), and presumably additional proteins are necessary for the intracellular transport of fatty acids and subsequent assembly and secretion of chylomicrons (Dannoura et al., 1999). Ingested fat reaches the thoracic duct, packaged into chylomicrons, within 80 min (Qureshy et al., 2001) and is transported from there with lymph into the vena cava.

Free fatty acids (mainly those with 12 or fewer carbon atoms) also can be transferred directly into portal blood through incompletely understood mechanisms.

A highly specific cytidine deaminase (apolipoprotein B editing catalytic subunit 1, APOBEC1, EC3.5.4.-) is part of a multiprotein complex, which modifies cytidine 6666 of the apoB messenger RNA (mRNA) to uridine and thereby introduces a stop codon into the sequence. Due to this modification, the intestinal protein transcripts are shortened to about 48% of the full-length version (hence B48). In humans, this enzyme is expressed only in the small intestine (Cheng et al., 2001), which means that all intestinal apoB is of the apoB48 variety. Rodents and many other mammals, in contrast, produce in their enterocytes both apoB48 and the full-length version, apoB100.

TRANSPORT AND CELLULAR UPTAKE

Blood circulation: Free fatty acids in blood occur almost exclusively bound noncovalently to albumin. Postabsorptive free fatty acids that are carried in portal blood are taken up into hepatocytes by active transport during their first pass through the liver. Skeletal and cardiac muscle cells and adipocytes get fatty acids from circulating triglyceride-rich lipoproteins (TRLs), chylomicrons, and very low-density lipoproteins (VLDLs). Muscle cells of the heart depend for most of their energy on long-chain fatty acids taken up from circulation. Triglycerides in TRLs are hydrolyzed by luminal lipoprotein lipase. To be accessible for lipase action, TRLs have to contain sufficient amounts of the activating peptide apolipoprotein C-II (apoC-II); TRL also contain apolipoprotein C-III (apoC-III), which inhibits lipoprotein lipase (LPL, EC3.1.1.34) activity, thereby pacing the release of fatty acids from TRLs. Another TRL constituent is apolipoprotein C-I (apoC-I), a small peptide constituent of chylomicron and VLDL remnants. It inhibits the binding of lipoproteins to the low-density-lipoprotein (LDL) receptor, LDL receptor-related protein, and the VLDL receptor; slows lipid exchange mediated by the cholesteryl ester transfer protein; and decreases cellular fatty acid uptake (Shachter, 2001).

Most lipoprotein processing occurs in the blood. Chylomicrons, which carry triglycerides from small intestinal absorption, and VLDLs, which transport excess triglycerides from the liver (Figure 5.11), release much of triglycerides as they pass through small arterioles and arterial capillaries. LPL on the endothelial surface of these vessels cleaves triglycerides within the lipoprotein, and the resulting fatty acids and glycerol are ejected because they are no longer compatible with the lipophilic environment of the lipoproteins. The free fatty acids have a high probability to be released in close proximity to muscle

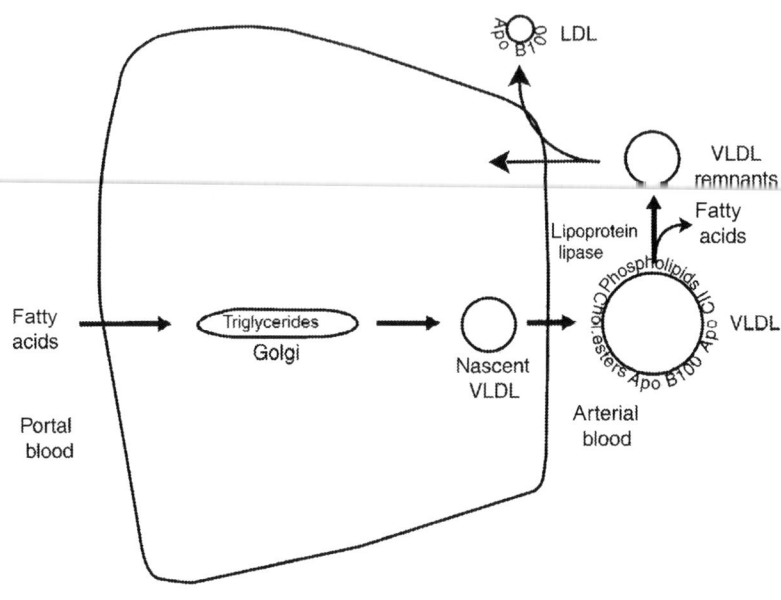

FIGURE 5.11

Transport of fat from the liver.

cells, their primary target. Fatty acids can easily diffuse through the gaps between adjoining endothelial cells of muscle capillaries into the pericapillary space around muscle cells. At least six distinct fatty acid transport proteins are expressed in various tissues. The uptake of the bulk of free fatty acids into muscle cells and adipocytes depends on the fatty acid transport protein 1 (FATP-1, CD36, SLC27A1; Martin et al., 2000), either in conjunction with or augmented by plasma membrane fatty acid–binding protein (FABPpm; Storch and Thumser, 2000) and fatty acid translocase (FAT). Transport across the cell membrane appears to be coupled to acyl-CoA synthase (EC6.2.1.3) activity. Apparently, this activity does not always require a separate protein since FATP4 in adipocytes functions as an acyl-CoA synthetase (Herrmann et al., 2001). Fatty acids can also be taken up into some cells with the entire TRL. Specific receptors recognize chylomicron remnants (Brown et al., 2000) and VLDLs (Tacken et al., 2001). Fatty acids that remain in circulation bind to albumin. Their main fate is uptake in the liver via a proton-driven fatty acid transporter (Elsing et al., 1996).

Blood–brain barrier (BBB): Transport of fatty acids from circulation into the brain is very limited and highly selective (Edmond, 2001). The mechanism for the specific transfer of essential fatty acids, which are needed for the synthesis of many brain-typical structural compounds, remains unclear. Small amounts may be transported with high-density lipoproteins (HDLs), which enter brain capillary endothelium via an HDL-binding receptor (Goti et al., 2000).

Materno-fetal transfer: The growing fetus is especially dependent on a generous supply of long-chain polyunsaturated fatty acids. About 3 g of DHA has to be supplied every day to a third-trimester fetus to ensure normal brain development (Clandinin et al., 1981). Both albumin-bound fatty acids and fatty acids released from VLDLs by LPLs at the maternal face of the placenta are available for transport. Fatty acid uptake is mediated by FATP, FABPpm, and FAT, and export to the fetal side involves mainly FATP (Dutta-Roy, 2000). The exact role and identity of these transporters remain to be clarified. The placental FABPpm, for instance, is not identical with the form in other tissues and transports essential fatty acids more effectively than smaller and saturated fatty acids. While long-chain essential fatty acids are preferentially transported across the placenta, the entire spectrum of fatty acids, including trans-fatty acids, reaches the fetus. Fatty acids in fetal circulation are bound largely to alpha-fetoprotein.

Metabolism

Oxidative metabolism: Fatty acids are broken down through diverse pathways in at least three different organelle compartments. Beta-oxidation in mitochondria is the main fate of straight-chain fatty acids with 18 or fewer carbons. Peroxisomes take care of the more unusual fatty acids, such as those with more than 18 carbons. At least two different alpha-oxidation pathways also operate in peroxisomes, one for phytanic acid and a different one for long-chain hydroxy acids, such as cerebronic acid (2-hydroxy-tetracosanoic acid). Attack of the acyl end of fatty acids in smooth endoplasmic reticulum produces dicarboxylic acids (omega-oxidation). With few exceptions the fatty acid metabolism begins with its conjugation to CoA. Different ligases act on the various fatty acids. Humans have at least six separate isoenzymes of long-chain-fatty-acid-CoA ligase (EC6.2.1.3) for the activation of fatty acids that differ in substrate preference and tissue expression pattern. The reaction is driven by adenosine triphosphate (ATP) and subsequent pyrophosphate hydrolysis (inorganic pyrophosphatase, EC3.6.1.1, magnesium-dependent) (Figure 5.12).

Mitochondrial beta-oxidation: Muscle, heart, and other tissues utilize fatty acids, mainly via beta-oxidation in mitochondria. Carnitine is needed to shuttle medium- and long-chain fatty acids across the inner mitochondrial membrane. Long-chain fatty acid transport protein appears to participate in this transfer in an as-yet-unknown way. First, the fatty acid is activated by conjugation to CoA by one of several ligases that differ with regard to preferred chain length. The resulting acyl-CoA can permeate by diffusion across the porous outer mitochondrial membrane, on the inside of which it encounters palmitoyl-CoA:L-carnitine O-palmitoyltransferase I. This enzyme transfers the fatty acid from CoA to carnitine and can then be taken across the inner mitochondrial membrane by a specific carrier (carnitine-acylcarnitine translocase). Palmitoyl-CoA:L-carnitine O-palmitoyltransferase II moves the fatty acid to CoA again, and finally, exchange of the free carnitine for the next acylcarnitine back across the inner membrane completes the cycle.

During each round of mitochondrial beta-oxidation, the fatty acid is shortened by two carbons, reducing equivalents for oxidative phosphorylation are generated, and one acetyl-CoA is released. These rounds proceed until the final thiolase cleavage generates either two acetyl-CoA (even-number chain length) or an acetyl-CoA and a propionyl-CoA (odd-number chain length). Depending on the fatty acid chain length of even-numbered fatty acids, the initial step is catalyzed by long-chain acyl-CoA dehydrogenase (EC1.3.99.13, for CoA esters of straight fatty acids with 10–19 carbons), acyl-CoA dehydrogenase (EC1.3.99.3, for octanoyl-CoA), butyryl-CoA dehydrogenase (EC1.3.99.2, for butyryl-CoA and hexanoyl-CoA), 2-methyl-CoA dehydrogenase (EC1.3.99.12, for 2-methylbutanoyl-CoA and isobutyryl-CoA), and isovaleryl-CoA dehydrogenase (EC1.3.99.10, for 3-methylbutanoyl CoA and valeryl-CoA). All of these dehydrogenases contain FAD and use the FAD-containing electron-transfer flavoprotein (ETF) as acceptor. The ETF dehydrogenase (EC1.5.5.1), which again contains FAD in addition to its iron-sulfur centers, moves the electrons from reduced ETF to ubiquinol

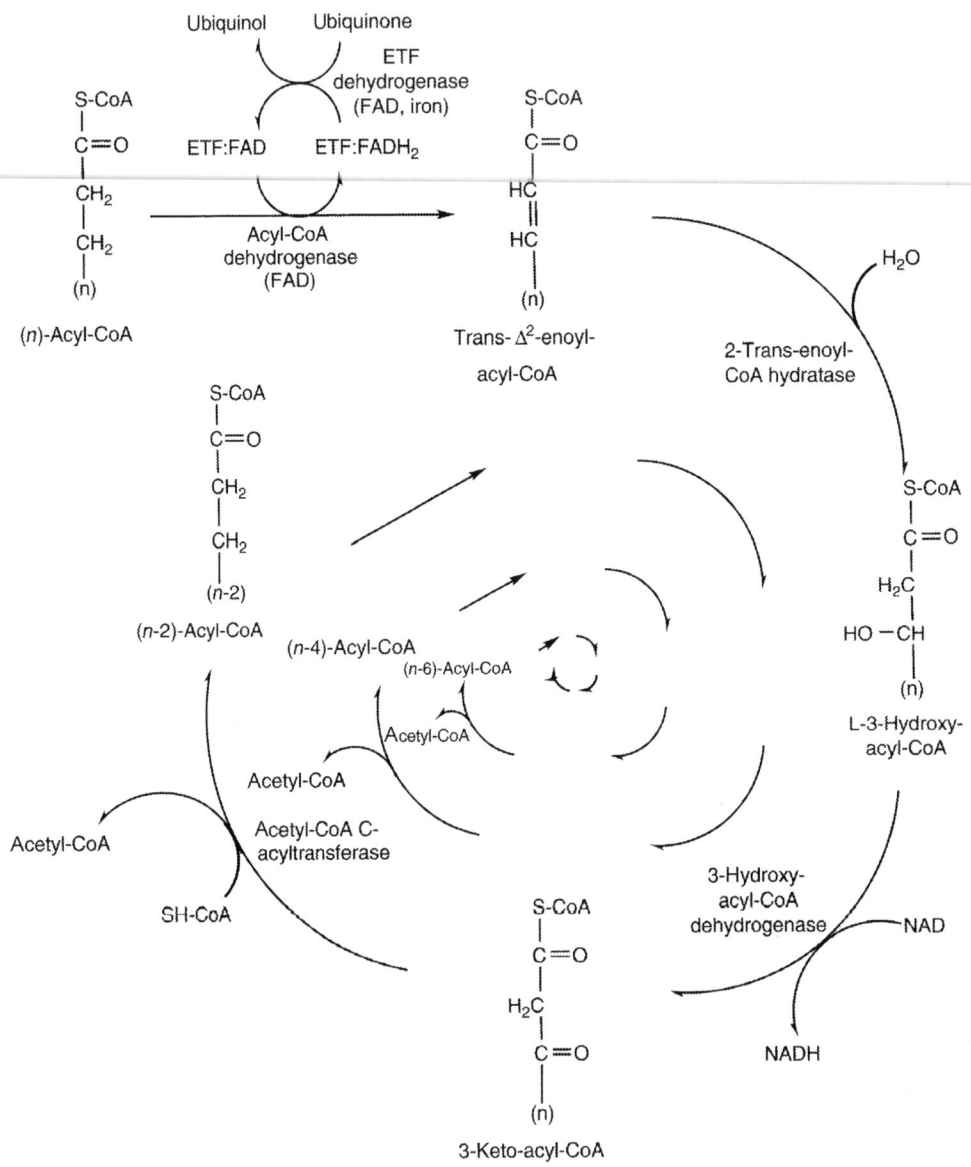

FIGURE 5.12

Mitochondrial beta-oxidation of fatty acids.

for utilization by oxidative phosphorylation. The subsequent enzymes for mitochondrial beta-oxidation are 2-trans-enoyl-CoA hydratase (EC4.2.1.17), long-chain-3-hydroxyacyl-CoA dehydrogenase (EC1.1.1.211), and acetyl-CoA C-acyltransferase (thiolase, EC2.3.1.16). Segments with a double bond require the action of dodecenoyl-CoA delta-isomerase (EC5.3.3.8) instead of long-chain acyl-CoA dehydrogenase. Cis-unsaturated fatty acids produce a cis-delta2 acyl-CoA intermediate at some point. Dodecenoyl-CoA delta-isomerase (EC5.3.3.8) converts this cis-unsaturated compound into the normal trans-isomer substrate of 2-trans-enoyl-CoA hydratase, and beta-oxidation can continue. If one or more double bonds follow, the acyl dehydrogenase reaction two cycles later generates cis-delta4-acyl-CoA. This intermediate is not a suitable substrate for 2-trans-enoyl-CoA hydratase; instead, it requires the action of 2,4-dienoyl-CoA reductase (EC1.3.1.34), and then that of dodecenoyl-CoA delta-isomerase.

These two reactions complete a round of beta-oxidation. Mitochondrial beta-oxidation of branched fatty acids can lead to the generation of a 2-methyl branched-acyl-CoA intermediate, which can then be oxidized by NAD-dependent 2-methyl-3 hydroxybutyryl-CoA dehydrogenase (MHBD, EC1.1.1.178). This pathway parallels the breakdown of isoleucine. Acetyl-CoA acyltransferase (EC2.3.1.16) cleaves the intermediate into propionyl-CoA and acetyl-CoA.

Peroxisomal beta-oxidation of straight-chain fatty acids: The classical peroxisomal beta-oxidation pathway metabolizes saturated acyl-CoA with fatty acid carbons between 8 and 26 and no branching (Figure 5.13). Beta-oxidation becomes possible after linking to CoA. Fatty acids with 20 or more carbons are imported by the peroxisomal ABC half-transporter ALDP (ALD protein, ABCD1, defective in adrenoleukodystrophy) and esterified to CoA by very-long-chain acyl-CoA synthetase (SLC27A2, EC 6.2.1.3). First, the acyl-CoA is oxidized to trans-2-enoyl acyl CoA by FAD-dependent acyl-CoA oxidase (EC1.3.3.6) and then converted to L-3-hydroxyacyl CoA and 3-ketoacyl CoA by peroxisomal multifunctional protein 2 (MFP2, comprising activities enoyl-CoA hydratase, EC4.2.1.17; 3,2-trans-enoyl-CoA isomerase, EC5.3.3.8; and 3-hydroxyacyl-CoA dehydrogenase, EC1.1.1.35). Finally, acetyl-CoA and the shortened acyl-CoA are released by a peroxisome-specific acetyl-CoA C-acyltransferase (3-ketoacyl-CoA thiolase, EC2.3.1.16). Beta-oxidation usually terminates in peroxisomes when a chain length of 10–14 is reached and the resulting medium-chain acyl-CoA is metabolized further in mitochondria.

Peroxisomal beta-oxidation of branched-chain fatty acids: An important example of branched-chain fatty acids is pristanic acid, which results from alpha-oxidation of phytanic acid (Verhoeven et al., 1998). Branched-chain fatty acids are activated by very-long-chain acyl-CoA synthetase (SLC27A2, EC6.2.1.3), either in the peroxisomes themselves or in mitochondria or endoplasmic reticulum. The branched-chain acyl-CoA is then oxidized by 2-methylacyl-CoA dehydrogenase (branched-chain acyl-CoA oxidase, EC1.3.99.12), metabolized to D-3-hydroxyacyl CoA and 3-ketoacyl CoA by MFP2, and cleaved by the peroxisomal protein sterol carrier protein X (SCPx), which is not induced by clofibrate.

Peroxisomal alpha-oxidation: Phytanic acid is metabolized in peroxisomes by phytanoyl-CoA dioxygenase (phytanoyl-CoA hydroxylase (EC1.14.11.18), cofactors iron and ascorbate). Then 2-hydroxy-phytanoyl-CoA lyase (EC4.1.-.-, thiamin pyrophosphate and magnesium-dependent) cleaves 2-hydroxy-phytanoyl-CoA into pristanal and formyl-CoA (Foulon et al., 1999). After fatty aldehyde dehydrogenase (ALDH3A2, EC1.2.1.3) has oxidized pristanal, the resulting pristanic acid undergoes further beta-oxidation in mitochondria.

The alpha-oxidation of cerebronic acid (2-hydroxytetracosanoic acid) from cerebrosides and sulfatides proceeds by a peroxisomal pathway distinct from that for phytanic acid (Sandhir et al., 2000).

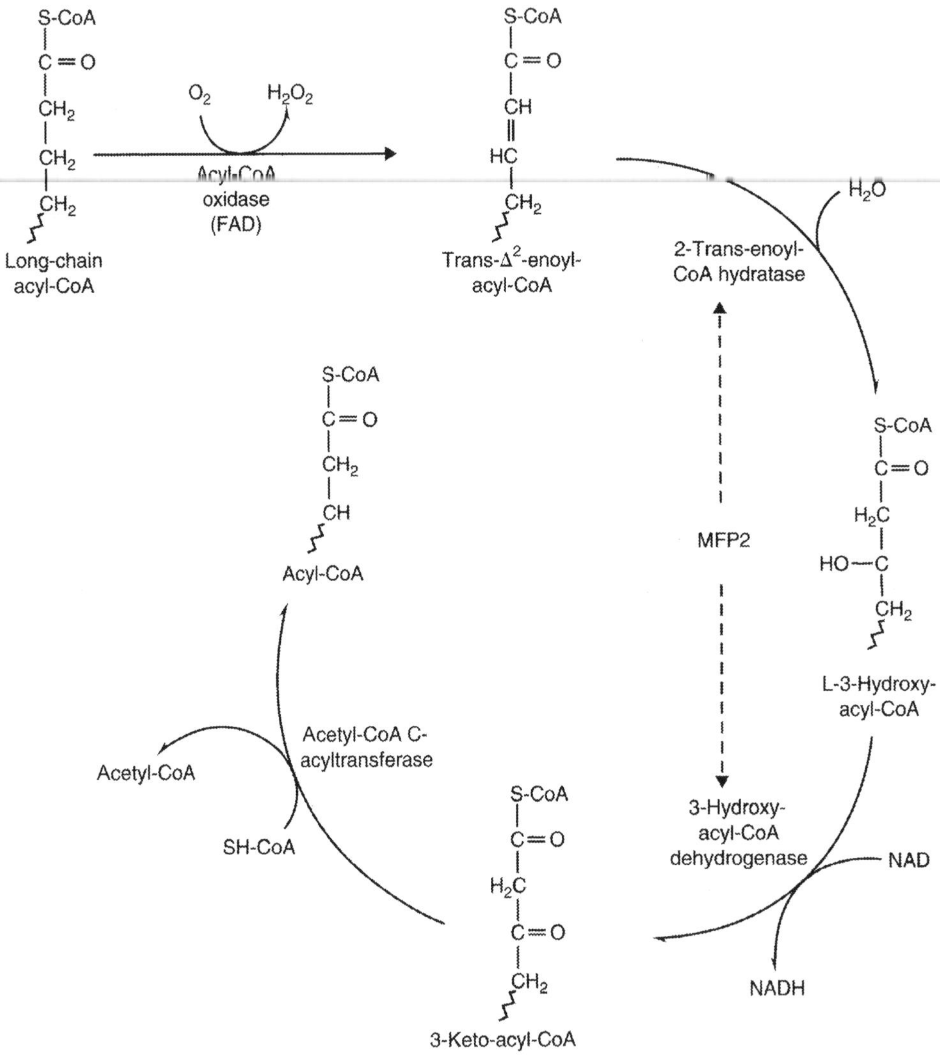

FIGURE 5.13

Peroxisomal beta-oxidation of fatty acids.

Microsomal omega-oxidation: Oxidation starting at the acyl end (Figure 5.14) appears to contribute significantly to fatty acid metabolism in the liver (Rognstad, 1995) and brain (Alexander et al., 1998). It proceeds only to a chain length of 12–14 carbons. The resulting dicarboxylic acids are excreted with urine. Omega-2 oxidation, starting at the second carbon from the acyl end, also has been found to occur (Costa et al., 1996). Omega oxidation may be of particular importance for the metabolism of cytotoxic products of fatty acid peroxidation. The typical product of fatty acid peroxidation, 4-hydroxy-2-nonenal, can thus be oxidized to the less harmful 4-hydroxynonenoic acid (Laurent et al., 2000).

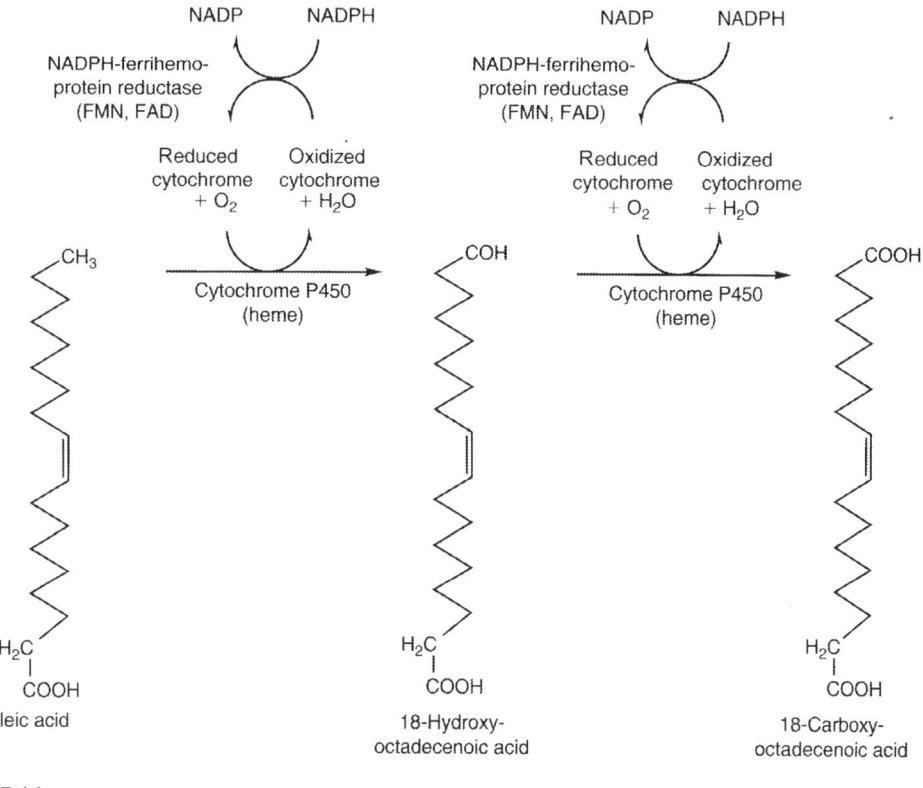

FIGURE 5.14

Microsomal omega-oxidation of fatty acids.

Polyunsaturated fatty acids are oxygenated by microsomal cytochrome P450-enzymes in numerous additional ways that account for only a minute fraction of total metabolism, but may give rise to significant physiological and pathological effects. Epoxidation, allylic, and bis-allylic hydroxylation, as well as hydroxylation with double bond migration, have been described (Oliw, 1994).

Oxidized fatty acids: Stress, exertion, infection, and inflammation increase oxygen free radical generation and, as a consequence, the exposure of unsaturated fatty acids to oxidative reactants. Double bonds in the acyl moieties of phospholipids, cholesterol esters, and other complex fatty acid–derived compounds are highly susceptible to oxidation upon free radical attack (Figure 5.15). Considerable amounts of oxidized fatty acids are also taken up when people consume fried foods (Staprans et al., 1999; Wilson et al., 2002) and may lead to an inflated estimate of fatty acid peroxidation. A typical scenario is exposure of phospholipid acyl double bonds in the inner mitochondrial membrane to oxygen free radicals from oxidative phosphorylation reactions. Ensuing reactions may introduce a hydroxyl group adjacent to a trans-double bond. Alternatively, a fatty aldehyde is generated. The main products of cholesteryllinoleate from LDL exposed to monocytes are 9-hydroperoxyoctadienoic acid and 13-hydroperoxyoctadienoic acid (Folcik and Cathcart, 1994).

The products of oxidized fatty acid metabolism, especially 4-hydroxy-2,3-transnonenal (HNE), crotonaldehyde, and malondialdehyde, are highly reactive and cytotoxic metabolites that damage

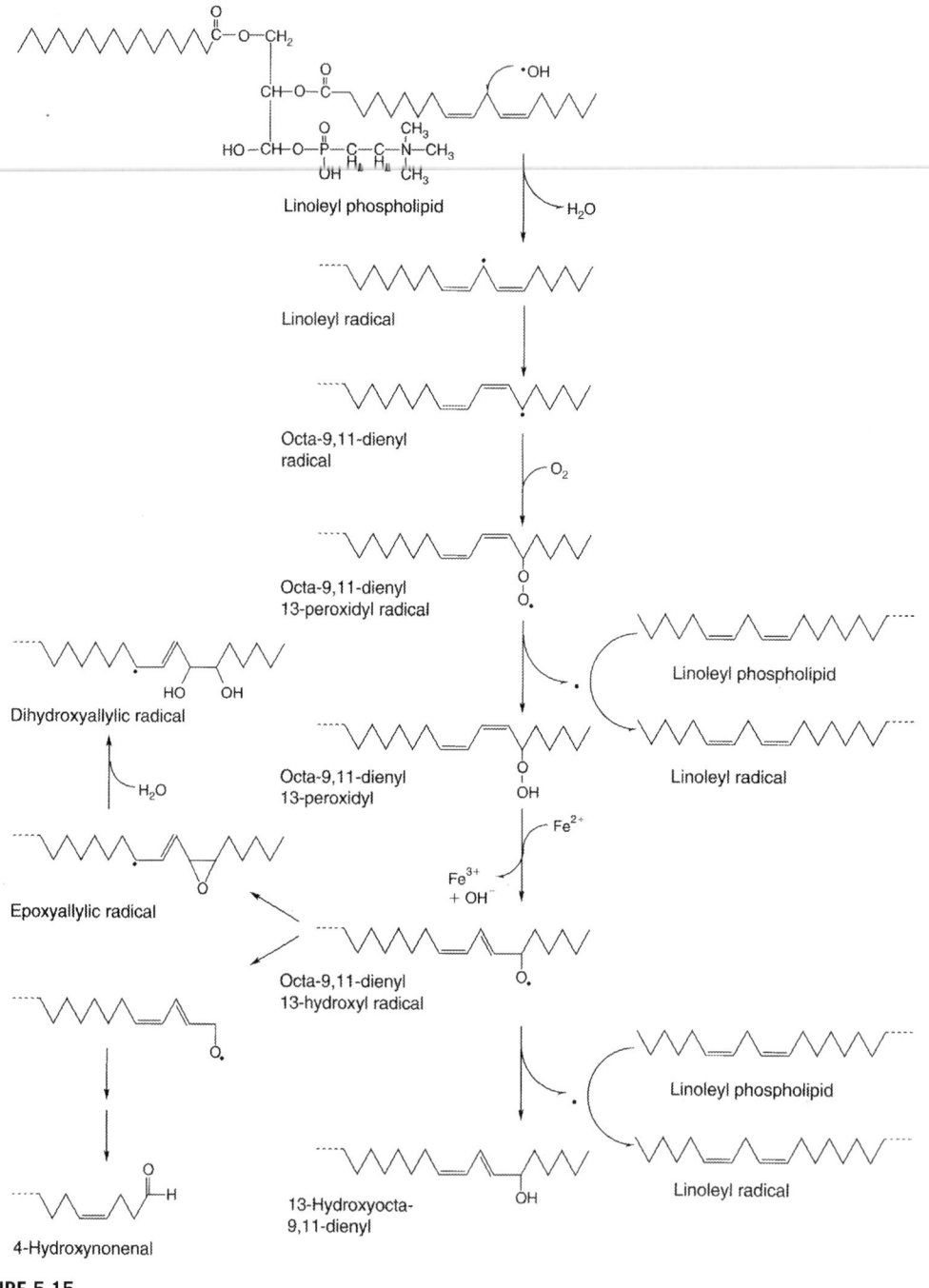

FIGURE 5.15

Oxygen free radicals react with unsaturated fatty acids.

DNA and proteins. Malondialdehyde is also a typical byproduct of prostaglandin synthesis from poly-unsaturated fatty acids. The reaction of malondialdehyde with DNA, identified by the formation of pyrimido[1,2-a]purin-10(3H)-one and other adducts, greatly increases cancer risk (Marnett, 1999). Exposure of LDL to fatty aldehydes interferes with LDL-receptor-mediated clearance and increases lipid deposition in arteries (Tanaga et al., 2002). HNE can be conjugated to glutathione (by glutathione transferases A4-4 and 5.8, EC2.5.1.18) and then pumped out of the cell by ATP-dependent transport that involves RLIP76 (Cheng et al., 2001). Extracellular gamma-glutamyltranspeptidase (EC2.3.2.2) cleaves glutamate from the glutathione moiety and releases the cytotoxic conjugate cysteinylglycine-4-hydroxy-2,3-trans-nonenal (Enoiu et al., 2002).

HDL has a very high content of phospholipids with polyunsaturated fatty acids and is thus very susceptible to oxidative damage. Apolipoprotein AI can terminate fatty acid radical chain reactions by directing the free radicals to core phospholipids and generate fatty aldehydes. Paraoxonase (PON, comprising activities of both arylesterase, EC3.1.1.2; and aryldialkylphosphatase, EC3.1.8.1) cleaves the fatty aldehydes from damaged phospholipids (Figure 5.16) (Ahmed et al., 2001). The free fatty

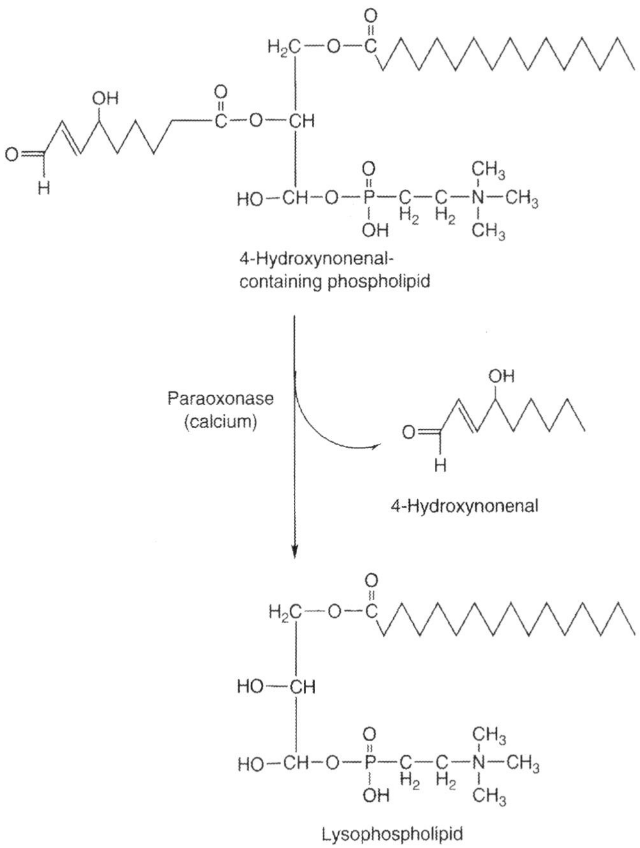

FIGURE 5.16

Metabolism of oxidized fatty acids.

aldehydes can then be metabolized as outlined previously. Two paraoxonases, PON1 and PON3, are primarily associated with apolipoprotein AI in HDL. The more ubiquitously expressed PON2 provides antioxidant protection for LDL and most tissues (Ng et al., 2001). The antioxidant potency of the paraoxonases appears to be distinct from the ability to cleave organophosphates. Paraoxonase activity shows as much as 40-fold variation due to genetic polymorphism and dietary and exogenous factors, including vitamin C and E status (Jarvik et al., 2002).

Acetyl-CoA oxidation: Mitochondria are the major site of acetyl-CoA utilization via the tricarboxylic acid (Krebs) cycle. Citrate synthase (EC4.1.3.7) joins acetyl-CoA to oxaloacetate. The citrate from this reaction can then be metabolized further providing FADH, NADH, and succinate for oxidative phosphorylation and ATP or guanosine triphosphate (GTP) from succinyl-CoA (Figures 5.15–5.17).

Ketogenesis: The production rate of acetyl-CoA from fatty acid beta-oxidation in the liver with prolonged fasting often exceeds the capacity of the Krebs cycle. The CoA for continued beta-oxidation and other functions can be released through the production of acetoacetate in three steps. Enzymic reactions convert excess mitochondrial acetyl-CoA either to acetoacetate or 3-hydroxybutyrate (the main product); acetoacetate can spontaneously decarboxylate to acetone (Figure 5.17). The term *ketone bodies* is misleading when applied to these three products, but it continues to be widely used. The acetyl-CoA condensation sequence frees CoA for continued breakdown of fatty acids, mainly in the liver during extended periods of fasting or in situations of abnormally high lipolysis (as in diabetes due to low insulin concentration). Since the reactions occur in mitochondria, ketogenesis regulation is independent of the same reactions for cholesterol synthesis in cytosol. The typical odor of a fasting individual is partially related to exhaled acetone formed from acetoacetate. The conversion of acetoacetate into beta-hydroxybutyrate taxes the body's acid-buffering capacity and may cause a drop in blood pH (acidosis) in diabetics and similarly susceptible patients. None of these events is related to dietary intake of acetate in typical quantities.

Acetyl-CoA C-acetyltransferase (thiolase, EC2.3.1.9) joins two acetyl-CoA molecules, and hydroxymethylglutaryl-CoA synthase (HMG-CoA synthase, EC4.1.3.5) adds another one. The mitochondrial isoform of HMG-CoA synthase is genetically distinct from the cytosolic one, which generates the precursor for cholesterol synthesis. Hydroxymethylglutaryl-CoA lyase (HMG-CoA lyase, EC4.1.3.4) finally generates acetoacetate by cleaving off acetyl-CoA from the HMG-CoA intermediate. Spontaneous decarboxylation of acetoacetate generates the dead-end product acetone.

Acetoacetate can also be reduced to beta-hydroxybutyrate by NADH-dependent 3-hydroxybutyrate dehydrogenase (EC1.1.1.30). This enzyme is allosterically activated by phosphatidyl choline. The reaction is fully reversible. Net flux depends on substrate concentrations. Acetoacetate and beta-hydroxybutyrate (but not acetone) can become a significant energy fuel for the brain after several days of adaptation to starvation conditions.

Propionyl-CoA metabolism: For its breakdown to continue, propionyl-CoA has to be ferried from peroxisomes into mitochondria, where the biotin-containing enzyme propionyl-CoA-carboxylase (EC6.4.1.3) adds a carbon, methylmalonyl-CoA converts D-methylmalonyl-CoA into the L-form, and methylmalonyl-CoA mutase (EC5.4.99.2, contains 5′-deoxyadenosylcobalamin) finally produces the Krebs cycle metabolite succinyl-CoA (Figure 5.18).

EXCRETION

Normally, *neither* free fatty acids nor fatty acid–containing compounds are excreted via urine. Excretion with bile is minimal, mainly as a component of phospholipids that are readily reabsorbed from the small intestinal lumen. Thus, almost no fatty acid is lost from the body once it has been absorbed.

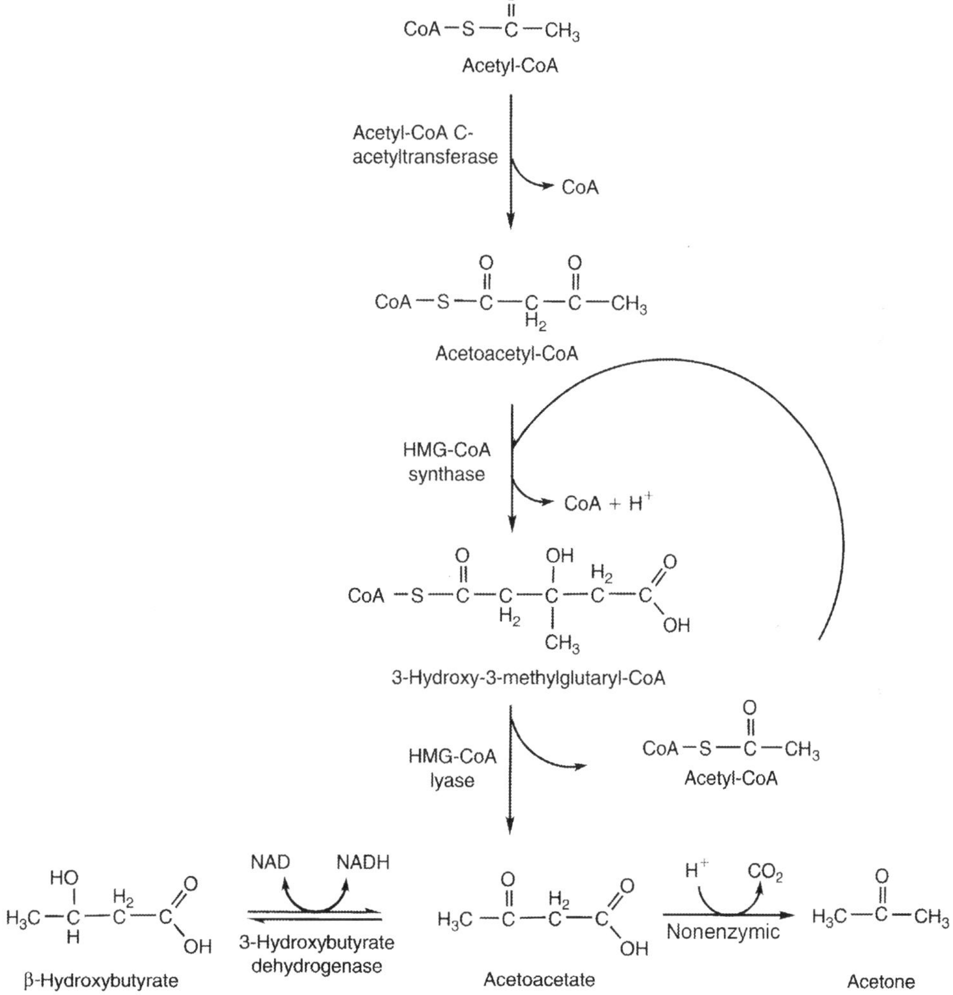

FIGURE 5.17

Metabolism of ketone bodies.

STORAGE

Fat is well suited for storage. It has by far the highest energy density of all nutrients and requires the smallest possible space. The lack of electric charges and insolubility in water means that it does not exert osmotic pressure and is chemically quite inert.

Adipocytes are specialized cells that are able to store large amounts of fatty acids as triglyceride. Fatty acids in adipocytes and other tissues are converted into acyl-CoA by several ligases specific for short-chain, medium-chain, and long-chain fatty acids as outlined previously. Triglyceride synthesis occurs at the cytosolic face of the microsomal membrane in most tissues, most extensively in adipose

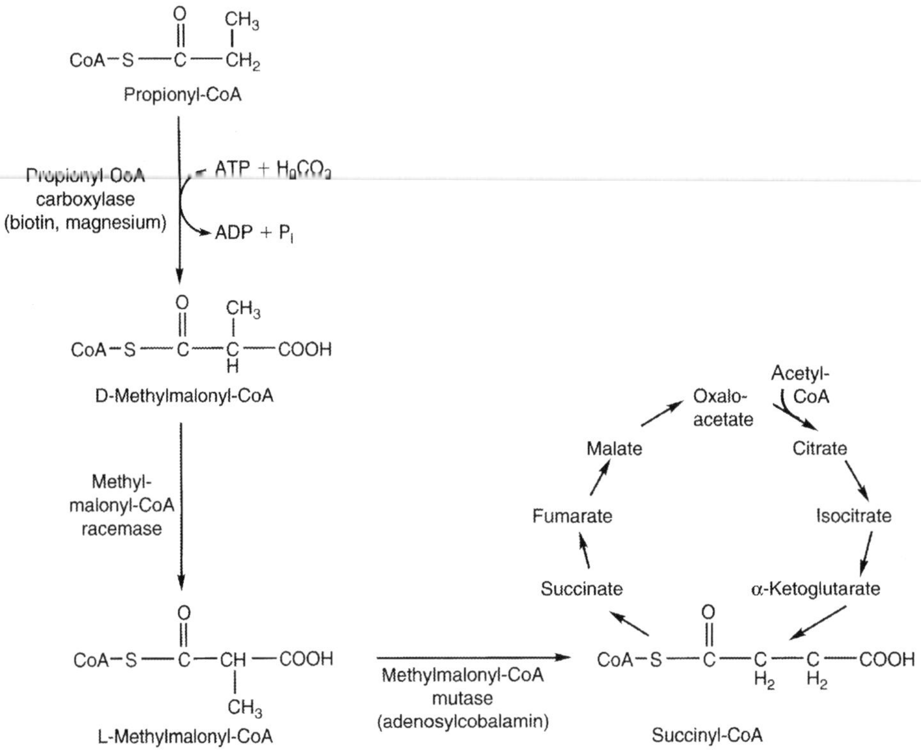

FIGURE 5.18

The metabolism of propionyl-CoA from odd-chain fatty acids.

tissue, liver, and muscle. The initial steps can also lead to phospholipid synthesis. Only the last step, the one that adds a third fatty acid, commits the fatty acids to triglyceride synthesis.

Glycerol-3-phosphate O-acyltransferase (EC2.3.1.15) links the first acyl-CoA to glycerol-3-phosphate (Figure 5.19). An alternative peroxisomal pathway with glycerone-phosphate O-acyltransferase (dihydroxyacetone phosphate acyltransferase, EC2.3.1.42), which supports mainly the synthesis of etherlipids and plasmalogens, uses dihydroxyacetone phosphate as the initial fatty acid acceptor. In this case, the product is converted to 1-acylglycerol-3-phosphate by acylglycerone-phosphate reductase (acyldihydroxyacetone phosphate reductase, EC1.1.1.101). Next, 1-Acylglycerol-3-phosphate acyltransferase (EC2.3.1.51) adds a second activated fatty acid. At least five different genes code for proteins with the same enzyme function, and additional isoforms arise from alternative splicing. Phosphatidate phosphatase (EC3.1.3.4), the regulatory enzyme of triglyceride synthesis, removes the phosphate group. Triglyceride synthesis is completed when diacylglycerol O-acyltransferase (EC2.3.1.20) adds a third fatty acid. A wide range of fatty acids can be incorporated at positions 1 and 3, but palmitate is preferred. The second fatty acid often is unsaturated. Overall, the fatty acid composition reflects long-term fatty acid intake patterns (Kohlmeier and Kohlmeier, 1995).

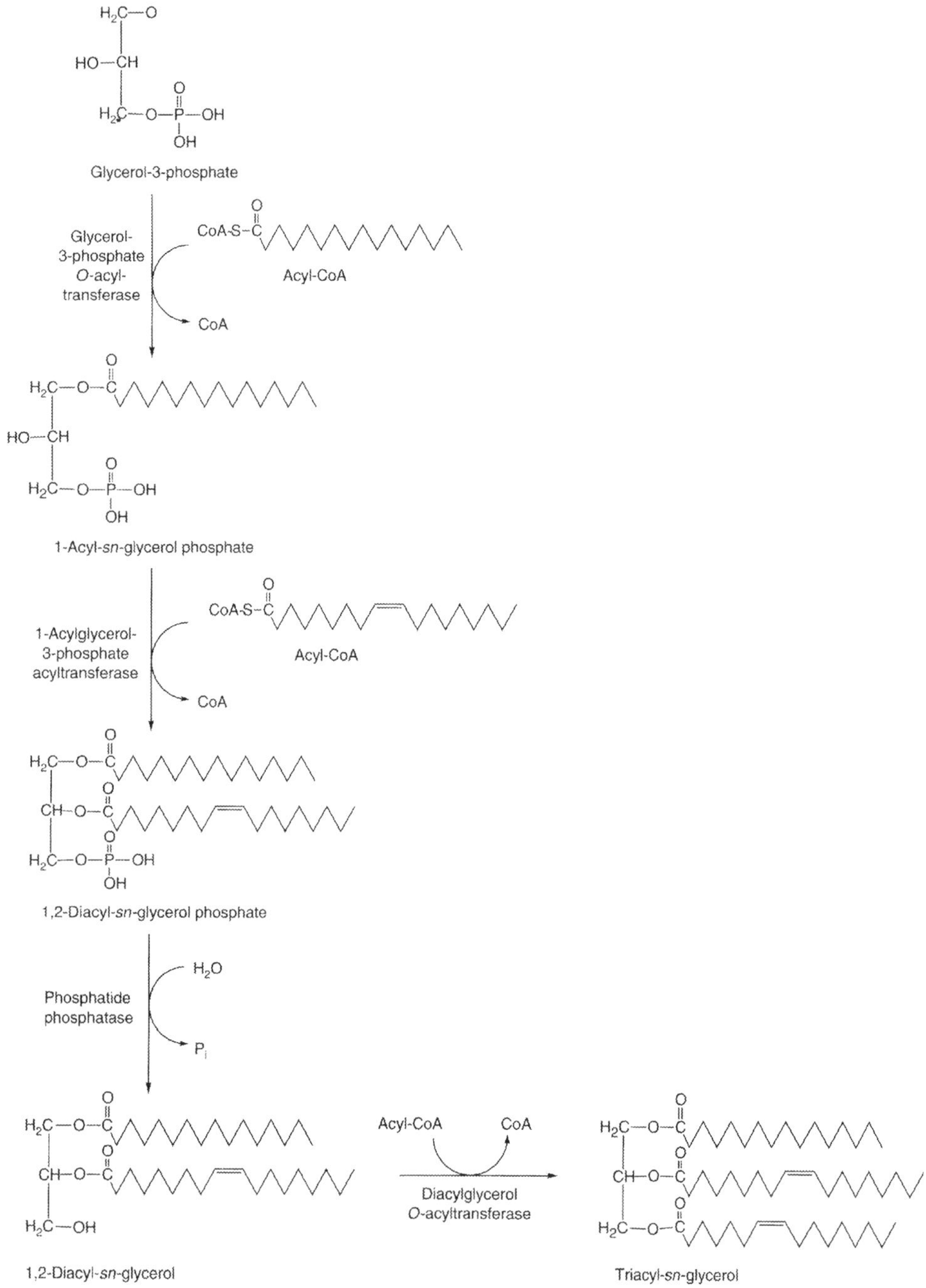

FIGURE 5.19

Fat storage.

Stored triglycerides can be released again into the blood circulation by the combined activity of hormone-sensitive lipase and monoglyceride lipase. The activity of hormone-sensitive lipase is under hormonal and neuronal control through the cyclic adenosine monophosphate (cAMP)–mediated phosphorylation of serine 563; the enzyme is activated by adrenaline (epinephrine) and inactivated by insulin.

REGULATION

Intake regulation: Adipose tissue produces several hormone-like factors and cytokines, some of which signal fat content to the brain and other tissues. Leptin crosses the blood–brain barrier (BBB) and stimulates neuropeptide Y (NPY) secretion in the brain. NPY induces satiety and thereby slows food-seeking behavior. Leptin also affects numerous metabolic processes in other tissues. The rate of fat utilization by muscle depends to a considerable extent on the rate of uptake, which is decreased by leptin (Steinberg et al., 2002).

Fat storage: Energy metabolism depends to a large extent on fat stores to buffer variation in the availability of fuel energy from dietary sources. Storage and mobilization are largely under the control of insulin, glucagon, and adrenaline. Insulin slows the mobilization of fatty acids from adipocytes by inhibiting hormone-sensitive lipase (through cAMP-mediated phosphorylation by protein kinase A). Insulin also promotes fatty acid synthesis, though net production seems to be minor, as pointed out previously. Glucagon and adrenaline slow fatty acid synthesis. Sterol regulatory element-binding proteins (SREBPs) mediate some of the effects by modulating the expression of genes involved in fatty acid synthesis. The more important adrenaline effect is, however, its strong promotion of lipolysis.

Free fatty acid receptors: Several receptors bind and respond to specific fatty acids in diverse tissues (Hara et al., 2014). The individual receptors overlap to a small extent in their response to specific fatty acids, usually differing greatly in responsiveness.

Free fatty acid receptor 1 (FFA1, GPR40) is a G-protein-coupled receptor mainly on the surface of beta cells in the pancreas (Briscoe et al., 2003) and of neurons in the brain. FFA1 responds to many different fatty acids, ranging in chain length from 6 (n-caproic acid) to 22 and possibly more. Omega-3 polyunsaturated fatty acids may be slightly more effective FFA1 ligands than saturated fatty acids. Activation of FFA1 in pancreatic beta cells appears to lower blood glucose concentrations and promote insulin production. FFA1 activation triggers the release of glucagon-like peptide 1 (GLP-1) from intestinal L cells, gastric inhibitory polypeptide (GIP) from K cells, and cholecystokinin (CCK) from I cells.

Free fatty acid receptor 2 (FFA2, GPR43) is most responsive to the short-chain fatty acids acetate and proprionate, and also to a lesser degree to slightly longer fatty acids. Activation of FFA2 in the small intestine triggers the release of PYY, GLP-1, and GLP-2 from L cells and promotes intestinal motility. Binding of acetate to FFA2 in adipocytes leads to the inhibition of lipolysis and the promotion of adipogenesis.

Free fatty acid receptor 3 (FFA3, GPR41) responds most to short-chain fatty acids and triggers release of GLP-1 and GIP secretion. FFA3 and FFA2 detect short-chain fatty acids in the colon and link this information through enteroendocrine cells and sympathetic pathways with appetite regulation (López Soto et al., 2014) and immune response (Ulven, 2012).

Peroxisomal metabolism: Fatty acid oxidation in peroxisomes, which may help to cope with excess, is regulated by the nuclear peroxisome proliferator–activated receptors (PPARs). These receptors are inducible by a wide range of compounds, including lipid-lowering drugs (clofibrate and related fibrates) and glucose-lowering drugs (thiazolidinediones), phthalate plasticizers, leukotriene antagonists, and herbicides. At least three genetically distinct forms exist (alpha, beta, and delta), with different patterns of tissue expression and inducibility by specific compounds. PPAR delta is induced by unsaturated fatty acids.

Fatty acid composition: The ratio of saturated to unsaturated fatty acids in membrane lipids is maintained within a narrow range, since this determines membrane fluidity. Transforming growth factor β and other cytokines contribute to the control of membrane fluidity by increasing stearyl-CoA desaturase (EC1.14.99.5) expression.

FUNCTION

Fuel energy: Its high energy content makes fat (triglycerides) a central player in fuel metabolism. On average, fats provide about 9 kcal/g. Their full oxidation depends on adequate supplies of thiamin, riboflavin, niacin, vitamin B12, biotin, pantothenate, carnitine, ubiquinone, iron, and magnesium. Additional nutrients are needed for the metabolism of some fatty acids, such as vitamin B12 for odd-chain fatty acids, or thiamin for phytanic acid. Since fat contains much less oxygen than carbohydrate, the ratio of carbon dioxide production to oxygen consumption (respiratory quotient) is much lower (0.7). At the same time, slightly more oxygen is needed to produce the same amount of energy from a fatty acid than from sugar.

Complex lipid synthesis: The fatty acid–derived acyl chains in phospholipids and cholesterol esters provide more than half of the lipid mass in membranes. Phosphatidylcholine-sterol *O*-acyltransferase [lecithin-cholesterol acyltransferase (LCAT, EC2.3.1.43)] transfers the middle (sn-2) fatty acid from phosphatidylcholine to cholesterol and other sterols (Figure 5.20). Since the sn-2 position of these phospholipids contains predominantly unsaturated fatty acids, more than half of the fatty acids in cholesterol esters are linoleate, and more than 10% are other highly unsaturated fatty acids (Smedman et al., 1999).

Phospholipids share the triglyceride synthesis pathway to the penultimate or the final step. Phosphatidate cytidylyltransferase (EC2.7.7.41, magnesium-dependent) replaces the phosphate group of phosphatidate with cytidyl diphosphate (CDP). CDP-diacylglycerol-inositol 3-phosphatidyltransferase (EC2.7.8.1 1) then completes the synthesis of phosphatidyl inositol. Phosphatidylglycerol is produced from by sequential action of CDP-diacylglycerol-glycerol-3-phosphate 3-phosphatidyltransferase (glycerophosphate phosphatidyltransferase, EC2.7.8.5) and phosphatidylglycerophosphatase (EC3.1.3.27). The synthesis of other phospholipids starts from 1,2-diacylglycerol.

Cholinephosphate cytidylyltransferase (EC2.7.7.15) generates phosphatidylcholine, and ethanolamine-phosphate cytidylyltransferase (EC2.7.7.14) produces phosphatidylethanolamine. Two CDP-diacylglycerol-serine *O*-phosphatidyltransferase [phosphatidylserine synthase (PSS), base exchange enzyme, EC2.7.8.8] genes, PSS-1 and PSS-2, encode enzymes that can replace ethanolamine with serine to generate phosphatidylserine. PSS-1 can also replace choline with serine. Cardiolipid and other more complex phospholipids are generated from these basic phospholipids.

Protein acylation: Myristic acid or palmitic acid can be attached to specific sites of numerous proteins (Resh, 2012). The hydrophobic side chain often is important for anchoring of receptors, transporters, and enzymes to membranes. Examples for myristoylated proteins are alpha (KAPA) and beta (KAPB) catalytic subunits of cAMP-dependent protein kinase (EC2.7.1.37). A cysteine residue in tumor necrosis factor (TNF) is linked to a palmityl residue (Utsumi et al., 2001).

Eicosanoid synthesis: Polyunsaturated fatty acids are essential precursors for a multimembered group of signaling compounds that affect platelet aggregation, uterine contractions, inflammation, pain

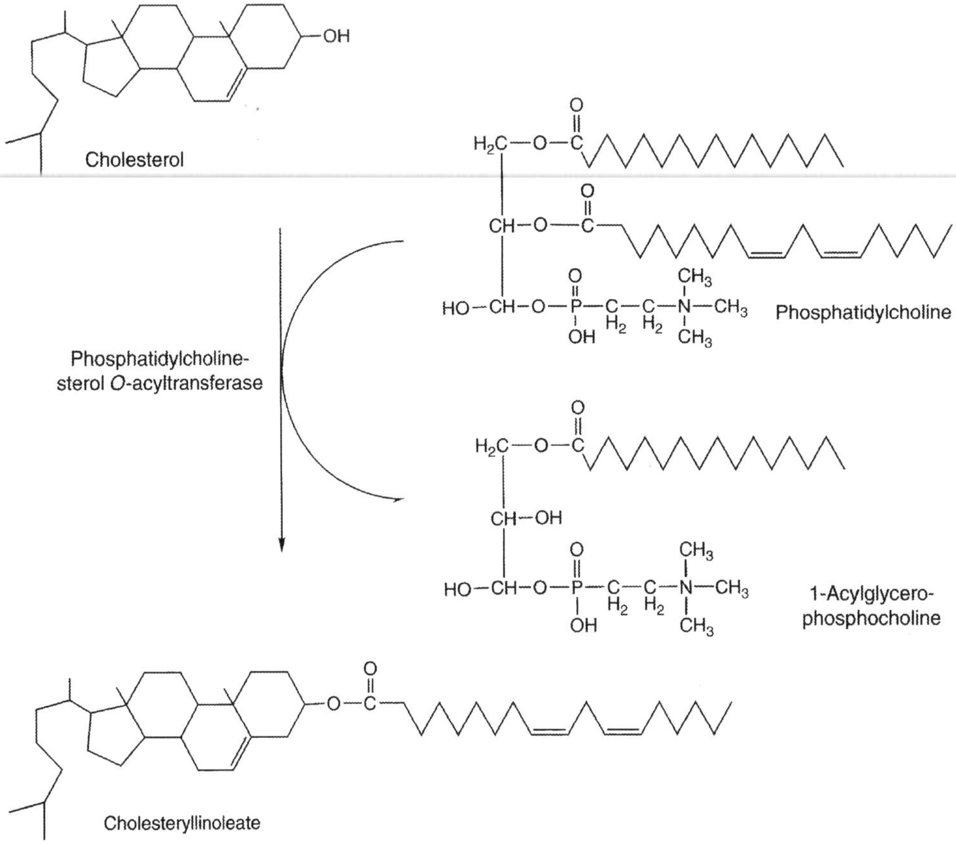

FIGURE 5.20

Synthesis of cholesterol esters.

sensation, blood flow, bone repair, and numerous other effects. Many of these effects are initiated when an eicosanoid binds to one of many specific receptors and starts an intracellular signaling cascade. The term *eicosanoid* is derived from the Greek term for the number of carbons in arachidonic acid, one of the fatty acid precursors. Prostaglandin-endoperoxide synthase (prostaglandin H synthase, Cox-1/ Cox-2, EC1.14.99.1) is responsible for the first step in the synthesis of prostaglandins, prostacyclin, and thromboxanes. Arachidonate 5-1ipoxygenase (EC1.13.11.34, iron-dependent) catalyzes the first step of leukotriene synthesis. Both enzymes accept arachidonic acid, linolenic acid, adrenic acid, gamma-linolenic acid, eicosapentaenoic acid, and other long-chain polyunsaturated fatty acids. The structure of the reaction products depends on the substrate (Larsen et al., 1996). Each precursor fatty acid generates eicosanoids with characteristic activity profiles. The difference in activities of compounds derived from omega-6 fatty acids versus those derived from omega-3 fatty acids is considerable and has been investigated extensively.

REFERENCES

Aarsland, A., Chinkes, D., Wolfe, R.R., 1997. Hepatic and whole-body fat synthesis in humans during carbohydrate overfeeding. Am. J. Clin. Nutr. 65, 1774–1782.

Ahmed, Z., Ravandi, A., Maguire, G.E., Emili, A., Draganov, D., La Du, B.N., et al., 2001. Apolipoprotein A-I promotes the formation of phosphatidylcholine core aldehydes that are hydrolyzed by paraoxonase (PON-1) during high intensity lipoprotein oxidation with a peroxynitrite donor. J. Biol. Chem. 276, 24473–24481.

Alexander, J.J., Snyder, A., Tonsgard, J.H., 1998. Omega-oxidation of monocarboxylic acids in rat brain. Neurochem. Res. 23, 227–233.

Alonso, L., Fontecha, J., Lozada, L., Fraga, M.J., Juarez, M., 1999. Fatty acid composition of caprinemilk: major, branched-chain, and trans fatty acids. J. Dairy. Sci. 82, 878–884.

Brasky, T.M., Darke, A.K., Song, X., Tangen, C.M., Goodman, P.J., Thompson, I.M., et al., 2013. Plasma phospholipid fatty acids and prostate cancer risk in the SELECT trial. J. Natl. Cancer Inst. 105, 1132–1141.

Briscoe, C.P., Tadayyon, M., Andrews, J.L., Benson, W.G., Chambers, J.K., Eilert, M.M., et al., 2003. The orphan G protein-coupled receptor GPR40 is activated by medium and long chain fatty acids. J. Biol. Chem. 278, 11303–11311.

Brown, M.L., Ramprasad, M.E., Umeda, P.K., Tanaka, A., Kobayashi, Y., Watanabe, T., et al., 2000. A macrophage receptor for apolipoprotein B48: cloning, expression, and atherosclerosis. Proc. Natl. Acad. Sci. 97, 7488–7493.

Cheng, J.Z., Sharma, R., Yang, Y., Singhal, S.S., Sharma, A., Saini, M.K., et al., 2001. Accelerated metabolism and exclusion of 4-hydroxynonenal through induction of RLIP76 and hGST5.8 is an early adaptive response of cells to heat and oxidative stress. J. Biol. Chem. 276, 41213–41223.

Clandinin, M.T., Chappell, J.E., Heim, T., Swyer, P.R., Chance, G.W., 1981. Fatty acid accretion in fetal and neonatal liver: implications for fatty acid requirements. Early Hum. Dev. 5, 7–14.

Costa, C.C., Dorland, L., Kroon, M., Tavares de Almeida, I., Jakobs, C., Duran, M., 1996. 3-, 6- and 7- hydroxyoctanoic acids are metabolites of medium-chain triglycerides and excreted in urine as glucuronides. J. Mass Spectrom. 31, 633–638.

Dannoura, A.H., Berriot-Varoqueaux, N., Amati, P., Abadie, V., Verthier, N., Schmitz, J., et al., 1999. Anderson's disease: exclusion of apolipoprotein and intracellular lipid transport genes. Arterioscl. Thromb. Vasc. Biol. 19, 2494–2508.

de Antueno, R.J., Knickle, L.C., Smith, H., Elliot, M.L., Allen, S.L., Nwaka, S., et al., 2001. Activity of human Delta5 and Delta6 desaturases on multiple n-3 and n-6 polyunsaturated fatty acids. FEBS Lett. 509, 77–80.

Dutta-Roy, A.K., 2000. Transport mechanisms for long-chain polyunsaturated fatty acids in the human placenta. Am. J. Clin. Nutr. 71, 315S–322S.

Edmond, J., 2001. Essential polyunsaturated fatty acids and the barrier to the brain: the components of a model for transport. J. Mol. Neurosci. 16, 181–193.

Elsing, C., Kassner, A., Stremmel, W., 1996. Effect of surface and intracellular pH on hepatocellular fatty acid uptake. Am. J. Physiol. 271, G1067–G1073.

Enoiu, M., Herber, R., Wennig, R., Marson, C., Bodaud, H., Leroy, P., et al., 2002. Gamma-glutamyltranspeptidase-dependent metabolism of 4-hydroxynonenal- glutathione conjugate. Arch. Biochem. Biophys. 397, 18–27.

Folcik, V.A., Cathcart, M.K., 1994. Predominance of esterified hydroperoxy-linoleic acid in human monocyte-oxidized LDL. J. Lip. Res. 35, 1570–1582.

Foulon, V., Antonenkov, V.D., Croes, K., Waelkens, E., Mannaerts, G.P., Van Veldhoven, P.P., et al., 1999. Purification, molecular cloning, and expression of 2-hydroxyphytanoyl- CoA lyase, a peroxisomal thiamine pyrophosphate-dependent enzyme that catalyzes the carbon-carbon bond cleavage during alpha-oxidation of 3-methyl-branched fatty acids. Proc. Natl. Acad. Sci. USA 96, 10039–10044.

Goti, D., Hammer, A., Galla, H.J., Malle, E., Sattler, W., 2000. Uptake of lipoprotein-associated alpha-tocopherol by primary porcine brain capillary endothelial cells. J. Neurochem. 74, 1374–1383.

Hara, T., Kashihara, D., Ichimura, A., Kimura, I., Tsujimoto, G., Hirasawa, A., 2014. Role of free fatty acid receptors in the regulation of energy metabolism. Biochim. Biophys. Acta. 1841, 1292–1300.

Hellerstein, M.K., 2001. No common energy currency: *de novo* lipogenesis as the road less traveled. Am. J. Clin. Nutr. 74, 707–708.

Herrmann, T., Buchkremer, E., Gosch, I., Hall, A.M., Bernlohr, D.A., Stremmel, W., 2001. Mouse fatty acid transport protein 4 (FATP4): characterization of the gene and functional assessment as a very long chain acyl-CoA synthetase. Gene 270, 31–40.

Hudgins, L.C., Parker, T.S., Levine, D.M., Hellerstein, M.K., 2011. A dual sugar challenge test for lipogenic sensitivity to dietary fructose. J. Clin. Endocrinol. Metab. 96, 861–868.

Jarvik, G.E., Tsai, N.T., McKinstry, L.A., Wani, R., Brophy, V.H., Richter, R.J., et al., 2002. Vitamin C and E intake is associated with increased paraoxonase activity. Arterioscl. Thromb. Vasc. Biol. 22, 1329–1333.

Jones, L.N., Rivett, D.E., 1997. The role of 18-methyleicosanic acid in the structure and formation of mammalian hair fibres. Micron. 28, 469–485.

Kohlmeier, L., Kohlmeier, M., 1995. Adipose tissue as a medium for epidemiologic exposure assessment. Env. Health Persp. 103, 99–106.

Larsen, L.N., Dahl, E., Bremer, J., 1996. Peroxidative oxidation of leuco-dichlorofluorescein by prostaglandin H synthase in prostaglandin biosynthesis from polyunsaturated fatty acids. Biochim. Biophys. Acta. 1299, 47–53.

Laurent, A., Perdu-Durand, E., Alary, J., Debrauwer, L., Cravedi, J.P., 2000. Metabolism of 4-hydroxynonenal, a cytotoxic product of lipid peroxidation, in rat precision-cut liver slices. Toxicol. Lett. 114, 203–214.

López Soto, E.J., Gambino, L.O., Mustafá, E.R., 2014. Free fatty acid receptor 3 is a key target of short chain fatty acid: what is the impact on the sympathetic nervous system? Channels (Austin) 8 (3), 169–171.

Marnett, L.J., 1999. Chemistry and biology of DNA damage by malondialdehyde. IARC Sci. Publ. 150, 17–27.

Martin, G., Nemoto, M., Gelman, L., Geffroy, S., Najib, J., Fruchart, J.C., et al., 2000. The human fatty acid transport protein-1 (SLC27A1: FATP-1) cDNA and gene: organization, chromosomal localization, and expression. Genomics 66, 296–304.

Moon, Y.A., Shah, N.A., Mohapatra, S., Warrington, J.A., Horton, I.D., 2001. Identification of a mammalian long chain fatty acyl elongase regulated by sterol regulatory element-binding proteins. J. Biol. Chem. 276, 45358–45366.

Ng, C.J., Wadleigh, D.J., Gangopadhyay, A., Hama, S., Grijalva, V.R., Navab, M., et al., 2001. Paraoxonase-2 is a ubiquitously expressed protein with antioxidant properties and is capable of preventing cell-mediated oxidative modification of low density lipoprotein. J. Biol. Chem. 276, 44444–44449.

Oliw, E.H., 1994. Oxygenation of polyunsaturated fatty acids by cytochrome P450 monooxygenases. Progr. Lipid. Res. 33, 329–354.

Qureshy, A., Kubota, K., Ono, S., Sato, T., Fukuda, H., 2001. Thoracic duct scintigraphy by orally administered 1-123 BMIPP: normal findings and a case report. Clin. Nucl. Med. 26, 847–855.

Resh, M.D., 2012. Targeting protein lipidation in disease. Trends Mol. Med. 18, 206–214.

Rognstad, R., 1995. On the estimation of alternative pathways of fatty acid oxidation in the liver *in vivo*. Bull. Math. Biol. 57, 191–203.

Sandhir, R., Khan, M., Singh, I., 2000. Identification of the pathway of alpha-oxidation of cerebronic acid in peroxisomes. Lipids 35, 1127–1133.

Shachter, N.S., 2001. Apolipoproteins C-I and C-III as important modulators of lipoprotein metabolism. Curr. Opin. Lipid. 12, 297–304.

Smedman, A.E., Gustafsson, I.B., Berglund, L.G., Vessby, B.O., 1999. Pentadecanoic acid in serum as a marker for intake of milk fat: relations between intake of milk fat and metabolic risk factors. Am. J. Clin. Nutr. 69, 22–29.

Staprans, I., Hardman, D.A., Pan, X.M., Feingold, K.R., 1999. Effect of oxidized lipids in the diet on oxidized lipid levels in postprandial serum chylomicrons of diabetic patients. Diabetes Care 22, 300–306.

Steinberg, G.R., Dyck, D.J., Calles-Escandon, J., Tandon, N.N., Luiken, J.J., Glatz, J.F., et al., 2002. Chronic leptin administration decreases fatty acid uptake and fatty acid transporters in rat skeletal muscle. J. Biol. Chem. 277, 8854–8860.

Steward, M.E., Downing, D.T., 1990. Unusual cholesterol esters in the sebum of young children. J. Invest. Dermatol. 95, 603–606.

Storch, J., Thumser, A.E., 2000. The fatty acid transport function of fatty acid-binding proteins. Biochim. Biophys. Acta. 1486, 28–44.

Tacken, P.J., Hofker, M.H., Havekes, L.M., van Dijk, K.W., 2001. Living up to a name: the role of the VLDL receptor in lipid metabolism. Curt. Opin. Lipid. 12, 275–279.

Tanaga, K., Bujo, H., Inoue, M., Mikami, K., Kotani, K., Takahashi, K., et al., 2002. Increased circulating malondialdehyde-modified LDL levels in patients with coronary artery diseases and their association with peak sizes of LDL particles. Art. Thromb. Vase. Biol. 22, 662–666.

Ulven, T., 2012. Short-chain free fatty acid receptors FFA2/GPR43 and FFA3/GPR41 as new potential therapeutic targets. Front. Endocrinol. (Lausanne) 3, 111.

Utsumi, T., Takeshige, T., Tanaka, K., Takami, K., Kira, Y., Klostergaard, J., et al., 2001. Transmembrane TNF (pro-TNF) is palmitoylated. FEBS Lett. 500, 1–6.

Verhoeven, N.M., Wanders, R.J.A., Poll-The, B.T., Saudubray, J.M., Jacobs, C., 1998. The metabolism of phytanic acid and pristanic acid in man: a review. J. Inher. Metab. Dis. 21, 697–728.

Wilson, R., Lyall, K., Smyth, L., Fernie, C.E., Riemersma, R.A., 2002. Dietary hydroxy fatty acids are absorbed in humans: implications for the measurement of "oxidative stress" *in vivo*. Free Rad. Biol. Med. 32, 162–168.

Wolk, A., Furuheim, M., Vessby, B., 2001. Fatty acid composition of adipose tissue and serum lipids are valid biological markers of dairy fat intake in men. J. Nutr. 131, 828–833.

OVERFEEDING

ABBREVIATIONS

CoA	coenzyme A
cAMP	cyclic adenosine monophosphate
LDL	low-density lipoprotein
VLDL	very low-density lipoprotein

THE NATURE OF OVERFEEDING

Unhealthful expansion of fat stores due to persistent overfeeding is fast becoming the norm in the United States and many other affluent countries. The major killers of older adults (type 2 diabetes and cardiovascular disease) hold their ground despite large advances in medical technology and now reach into adolescent and even childhood populations.

Common diseases associated with persistent overfeeding include insulin resistance and type 2 diabetes, hyperlipidemia and dyslipidemia, hypertension, hyperuricemia and gout, heart disease, stroke, sleep apnea, cancer, cholesterol gallstone disease, and osteoarthritis (Pi-Sunyer, 1999). Even high levels of energy intake without severe body fatness may be undesirable. Evidence is mounting, for instance, that suggests a lower cancer risk with energy restriction (Thompson et al., 2002).

One of the primary consequences of overfeeding is the expansion of fat stores. Subcutaneous and intraabdominal adipose tissues are the main sites of triglyceride storage, but fat accumulations also occur in other tissues, especially in diabetics and obese people (Ravussin and Smith, 2002). Large intraabdominal deposits (signaled by great abdominal girth) are associated with greater cardiovascular and other health risks than subcutaneous fat (Stevens, 1995). Preferential distribution of excess fat, resulting more or less in applelike shapes for men and pear shapes for women, may explain some of the differences in health consequences between the genders (Bertrais et al., 1999).

THE FATE OF EXCESS CARBOHYDRATE

There has been a spirited debate about the question whether humans convert excess glucose into fat (Hellerstein, 2001). The more recent evidence indicates that such *de novo* fat synthesis is very limited, accounting for <10 g/day under all but the most extraordinary circumstances. This finding should not detract from the obvious fact that *de novo* synthesis from carbohydrate is not needed to explain the storage of fat in overfed humans. The carbohydrate is utilized preferentially, and the fat is left over to be stored. So long as people consume more energy than they expend, there will be enough fat to put into storage and expand the fat mass further.

A separate issue is the impact of overfeeding on lipoproteins in blood. The production rate of triglyceride-containing VLDL is very low in healthy lean subjects. This is important because the main atherogenic (atherosclerosis-promoting) lipoprotein fraction, LDL, derives exclusively from VLDL. Even modest overfeeding increases VLDL production and thereby raises harmful LDL concentration in blood. High sugar consumption is a particularly potent recipe for raising VLDL production. Conversion of carbohydrate to saturated fat is partially responsible for the additional VLDL triglyceride synthesis (Hudgins et al., 2000).

INSULIN RESISTANCE AND HYPERLIPIDEMIA

One of the most persistent effects of overfeeding and obesity is the declining ability of tissues to increase glucose uptake upon stimulation with insulin. Insulin resistance is the conventional term for this phenomenon. The effect is particularly severe in muscle cells since they import glucose largely via the insulin-stimulated transporter GLUT4 (SLC2A4). Glucose transfer into the brain is little affected because it proceeds mainly via GLUT1 (SLC2A1). Several explanations for the high prevalence of increased insulin resistance in overfed and obese people have been put forward. One of these sees the problem in the typically elevated free fatty acid concentrations in the blood (Kraegen et al., 2001). The rise in free fatty acid concentration, which is typical after a fat-rich meal, rapidly decreases insulin effectiveness in skeletal muscle. The same is seen with chronic elevation of free fatty acid concentration due to obesity. Another mechanism may involve the hexosamine nutrient-sensing pathway mentioned later in this chapter.

A third scenario focuses on the depressed production of adiponectin by expanded and overfed adipose tissue (Tsao et al., 2002). Low circulating levels of this peptide hormone, which is produced only in adipose tissue, induce insulin resistance (Maeda et al., 2002). Adiponectin blunts the typical rise of plasma free fatty acid after a fatty meal, which should relieve the free fatty acid–related resistance. A direct adiponectin effect promoting insulin action is also possible.

Insulin serves in adipocytes to slow the release of fatty acids. The binding of circulating insulin to specific receptors activates adenylate cyclase (EC4.6.1.1) and the ensuing rise of intracellular

$3',5'$-cAMP concentration inhibits the activity of hormone-sensitive lipase (EC3.1.1.3). The bottom line is that insulin resistance is associated with accelerated turnover of fat stores. The release of fatty acids from adipose tissue closely correlates with fat mass. The age-typical doubling of fat stores from about 9 kg in a lean young man of average weight (70 kg, BMI 21) to 18 kg 30 years later (77 kg, BMI 25) is likely to increase VLDL production several-fold. Insulin resistance then just adds to the quandary of the obese person by further ballooning VLDL output (Couillard et al., 1998). On top of all this, an increase of fat intake adds to the fat load of the liver and cranks up VLDL synthesis all on its own. Since VLDL is the obligate precursor of LDL, both overfeeding and obesity raise the concentration of atherosclerosis-promoting lipoproteins in blood. Therefore, hyperlipidemia and cardiovascular disease are very common in obese people. An illustration of the role of expanded fat stores in the genesis of hyperlipidemia is the close association between body fatness and LDL cholesterol concentration observed in patients with LDL receptor defects (Gaudet et al., 1998).

NUTRIENT SENSING

Abundance alters the utilization of energy-rich nutrients in characteristic ways and modifies intake behavior. One of the metabolites that serves as fuel sensors is malonyl-CoA. Since the acetyl-CoA for its synthesis comes from carbohydrate, fat, and protein breakdown, malonyl-CoA is well suited to signal fuel nutrient availability. A high malonyl-CoA concentration slows the rate of fat oxidation by inhibiting carnitine palmitoyl transferase-mediated fatty acid import into mitochondria (Chien et al., 2000). At the same time, fat storage increases with high levels of malonyl-CoA (Abu-Elheiga et al., 2001).

The hexosamine pathway provides another nutrient-sensing mechanism. A high level of hexosamine metabolites (particularly UDP-N-acetylglucosamine) increases the glycosylation of regulatory proteins and decreases the expression of mitochondrial genes for oxidative phosphorylation (Obici et al., 2002). The hexosamine metabolites also down-regulate the insulin-dependent uptake of glucose and trigger the release of leptin and other satiety signals. The high free fatty acid concentrations in blood that are typical for overfeeding and obesity might cause insulin resistance through stimulation of the hexosamine pathway.

BLOOD COAGULATION

Large fat stores increase the risk of myocardial infarction and stroke partially due to an increased tendency to form thrombi (blood clots). This is not surprising, since adipose tissue, especially within the abdomen, plays an active role in fibrinolysis. Obese people have higher blood concentrations of coagulation factors VIII and VII, fibrinogen, von Willebrand factor, and plasminogen activator inhibitor (Mertens and Van Gaal, 2002).

HYPERURICEMIA

Elevated blood concentration of uric acid is a cardiovascular risk indicator that is closely related to overfeeding and obesity (Nakanishi et al., 2001). The hyperuricemia of most obese people is due to increased uric acid synthesis and not due to impaired renal elimination. Increased production from dietary nucleic acid precursors contributes to the uric acid burden. Many obese people with severe hyperuricemia suffer from gout. This condition is characterized by the deposition of uric acid crystals in joints and connective tissue at other sites and the ensuing inflammation and pain.

REFERENCES

Abu-Elheiga, L., Matzuk, M.M., Abo-Hashema, K.A., Wakil, S.J., 2001. Continuous fatty acid oxidation and reduced fat storage in mice lacking acetyl-CoA carboxylase 2. Science 291, 2613–2616.

Bertrais, S., Balkau, B., Vol, S., Forhan, A., Calvet, C., Marre, M., et al., 1999. Relationships between abdominal body fat distribution and cardiovascular risk factors: an explanation for women's healthier cardiovascular risk profile. The DESIR Study. Int. J. Obes. Rel. Metab. Dis. 23, 1085–1094.

Chien, D., Dean, D., Saha, A.K., Flatt, J.E., Ruderman, N.B., 2000. Malonyl-CoA content and fatty acid oxidation in rat muscle and liver *in vivo*. Am. J. Physiol. Endocrin. Metab. 279, E259–E265.

Couillard, C., Bergeron, N., Prud'homme, D., Bergeron, J., Tremblay, A., Bouchard, C., et al., 1998. Postprandial triglyceride response in visceral obesity in men. Diabetes 47, 953–960.

Gaudet, D., Vohl, M.C., Perron, P., Tremblay, G., Gagne, C., Lesiege, D., et al., 1998. Relationships of abdominal obesity and hyperinsulinemia to angiographically assessed coronary artery disease in men with known mutations in the LDL receptor gene. Circulation 97, 871–877.

Hellerstein, M.K., 2001. No common energy currency: *de novo* lipogenesis as the road less traveled. Am. J. Clin. Nutr. 74, 707–708.

Hudgins, L.C., Hellerstein, M.K., Seidman, C.E., Neese, R.A., Tremaroli, J.D., Hirsch, J., 2000. Relationship between carbohydrate-induced hypertriglyceridemia and fatty acid synthesis in lean and obese subjects. J. Lipid Res. 41, 595–604.

Kraegen, E.W., Cooney, G.J., Ye, J.M., Thompson, A.L., Furler, S.M., 2001. The role of lipids in the pathogenesis of muscle insulin resistance and beta cell failure in type II diabetes and obesity. Exp. Clin. Endocrinol. Diab. 109, S189–S201.

Maeda, N., Shimomura, I., Kishida, K., Nishizawa, H., Matsuda, M., Nagaretani, H., et al., 2002. Diet-induced insulin resistance in mice lacking adiponectin/ACRP30. Nat. Med. 8, 731–737.

Mertens, I., Van Gaal, L.E., 2002. Obesity, haemostasis and the fibrinolytic system. Obesity Rev. 3, 85–101.

Nakanishi, N., Yoshida, H., Nakamura, K., Suzuki, K., Tatara, K., 2001. Predictors for development of hyperuricemia: an 8-year longitudinal study in middle-aged Japanese men. Metab. Clin. Exp. 50, 621–626.

Obici, S., Wang, J., Chowdury, R., Feng, Z., Siddhanta, U., Morgan, K., et al., 2002. Identification of a biochemical link between energy intake and energy expenditure. J. Clin. Invest. 109, 1599–1605.

Pi-Sunyer, F.X., 1999. Comorbidities of overweight and obesity: current evidence and research issues. Med. Sci. Sports Exerc. 31, S602–S608.

Ravussin, E., Smith, S.R., 2002. Increased fat intake, impaired fat oxidation, and failure of fat cell proliferation result in ectopic fat storage, insulin resistance, and type 2 diabetes mellitus. Ann. NY Acad. Sci. 967, 363–378.

Stevens, J., 1995. Obesity, fat patterning and cardiovascular risk. Adv. Exp. Med. Biol. 369, 21–27.

Thompson, H.L., Zhu, Z., Jiang, W., 2002. Protection against cancer by energy restriction: all experimental approaches are not equal. J. Nutr. 132, 1047–1049.

Tsao, T.S., Lodish, H.E., Fruebis, J., 2002. ACRP30, a new hormone controlling fat and glucose metabolism. Eur. J. Pharmacol. 440, 213–221.

ACETATE

The term *acetate* (acetic acid; molecular weight 60) refers to both the carbonic acid with the acrid odor and its less odorous salts (Figure 5.21).

ABBREVIATIONS

CoA   coenzyme A
MCT1   monocarboxylate transporter 1 (SLC16A1)

NUTRITIONAL SUMMARY

Function: Acetate, and particularly its conjugate with CoA (acetyl-CoA), is a critical intermediary metabolite for the utilization of carbohydrates, some amino acids (lysine, leucine, isoleucine, phenylalanine, tyrosine, and tryptophan), fatty acids, and alcohol. It can be used as a precursor for fatty acid and cholesterol synthesis. Acetate can also be utilized as an energy fuel; its complete oxidation requires thiamin, riboflavin, niacin, pantothenate, lipoate, ubiquinone, iron, and magnesium.

Food sources: Only very small amounts are consumed with foods, mainly vinegar, fruits, and vegetables. Alcohol is converted completely into acetate. Several hundred grams of acetyl-CoA are generated daily from the breakdown of carbohydrates, fat, and protein.

Requirements: No dietary acetate intake is necessary. A beneficial effect of moderate vinegar intake on blood sugar control and chronic inflammatory polyarthritis has been claimed.

Excessive intake: High intakes of acetic acid (more than 10–20 g/day) may induce gastric discomfort, alter pH balance (metabolic acidosis), cause the loss of bone minerals, and increase the risk of dental erosion.

ENDOGENOUS PRODUCTION

The metabolism of carbohydrates, amino acids, and fatty acids generates several hundred grams of acetate per day, mainly as acetyl-CoA (Figure 5.22). Depending on intake, significant amounts of free acetate may also be generated from ethanol (Figure 5.23). Most is utilized within the cells or tissues where the acetate or acetyl-CoA is generated; some is transported to other tissues and utilized there.

Carbohydrates: The amount of acetate generated from glucose depends on the proportion used for glycolysis (as opposed to the smaller fraction metabolized via the pentose phosphate pathway) and the proportion used for the generation of oxaloacetate from pyruvate. Typically, about half a gram of acetate (as acetyl-CoA) is generated per gram of absorbed carbohydrate.

Amino acids: Acetyl-CoA is generated during the catabolism of isoleucine, leucine, and threonine. Lysine and tryptophan each generates two acetyl-CoA molecules. Metabolism of cysteine, alanine, and

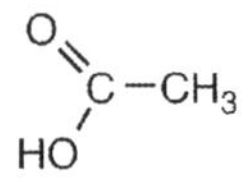

$$\overset{O}{\underset{HO}{\underset{\diagup}{\overset{\diagdown}{C}}}}-CH_3$$

FIGURE 5.21

Acetate.

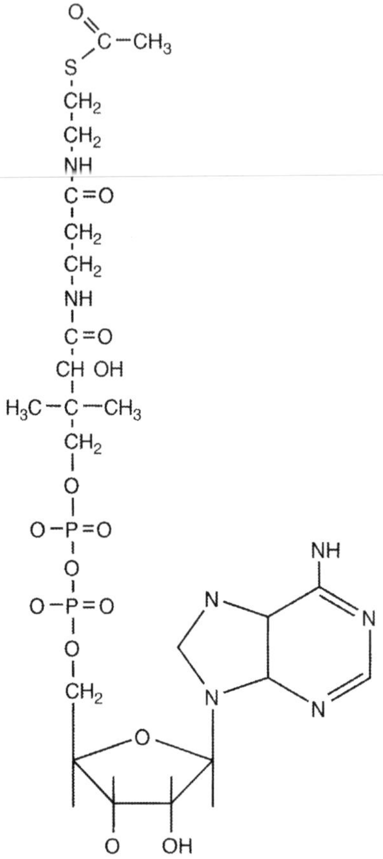

FIGURE 5.22

Acetyl-CoA is a critical intermediate of fuel metabolism.

tryptophan generates pyruvate, which may be converted into acetyl-CoA. Acetoacetate is generated by the catabolism of phenylalanine, tyrosine, and leucine (for the latter in addition to 1 mol of acetyl-CoA). The acetoacetate can be activated by 3-oxoacid CoA-transferase (succinyl-CoA transferase, EC2.8.3.5) and then cleaved by acetyl-CoA C-acetyltransferase (thiolase, EC2.3.1.9) to generate 2 mol of acetyl-CoA. A minor pathway of threonine breakdown generates free acetate.

Fatty acids: 1 mol of acetyl-CoA is released with each cycle of fatty acid beta-oxidation.

Alcohol: Ethanol is oxidized by various alcohol dehydrogenases (EC1.1.1.1) or the microsomal ethanol oxidizing system (MEOS, unspecific monooxygenases of the cytochrome P-450 family, EC1.14.14.1), in conjunction with several types of aldehyde dehydrogenase (EC1.2.1.3, EC1.2.1.4, and EC1.2.1.5) or acetaldehyde oxidase (EC1.2.3.1). Ethanol metabolism occurs mainly in the liver, and most of the resulting acetate is released into the circulation (Siler et al., 1999); 1 g of ethanol generates about 1.3 g of acetate.

Fiber: Normal intestinal bacteria in the colon break down nondigestible carbohydrates and release significant amounts of short-chain fatty acids, including acetate.

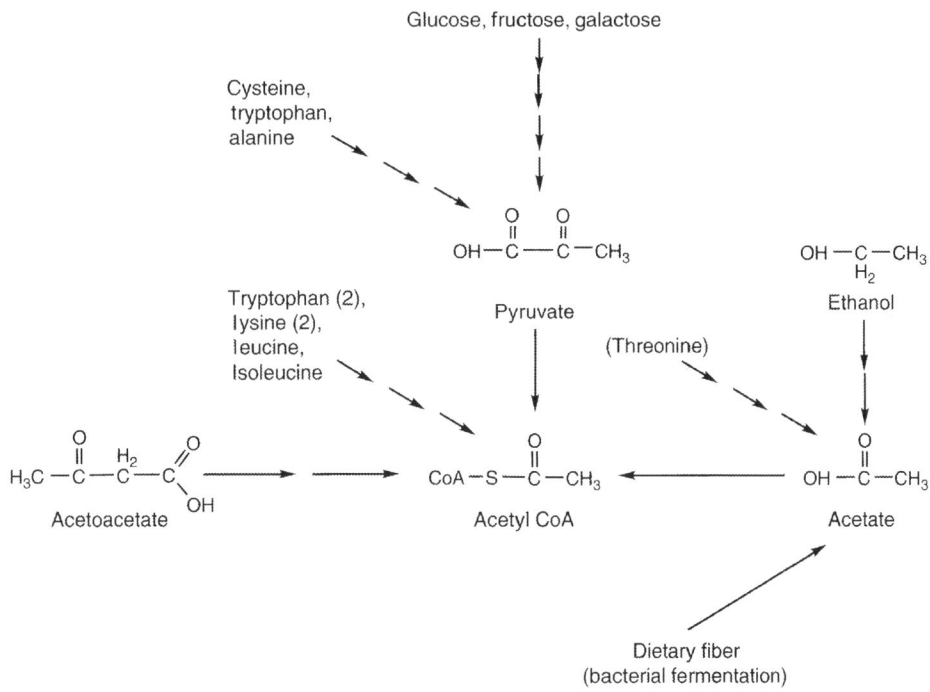

FIGURE 5.23

Endogenous sources of acetate and its metabolites.

DIETARY SOURCES

Acetate is ingested mostly as vinegar (content typically 5–6%) and with pickled, marinated, or fermented foods. Typical intake is likely to be <1 g/day corresponding to about 1 tbsp (15 ml) of vinegar. Much smaller amounts are present in a wide range of plant- and animal-derived foods as acetyl-CoA.

INTESTINAL ABSORPTION

Absorption of acetate from the small intestine (Watson et al., 1991; Tamai et al. 1995), especially the jejunum, appears to proceed mainly via the proton/monocarboxylic acid cotransporter (MCT1, SLC16A1), which is possibly present in the apical and certainly in the basolateral enterocyte membrane (Garcia et al., 1994; Orsenigo et al., 1999). Acetate can also be absorbed from the colon and rectum, which are important sites of bacterial production from dietary fiber (Wolever et al., 1995). MCT1 and possibly the SCFA/HCO_3 antiporter contribute to this uptake (Stein et al., 2000).

The flow of protons across the luminal membrane of the proximal colon via the sodium/hydrogen exchanger also promotes the protonation of the acetate anion and its subsequent passage into the enterocyte by nonionic diffusion (von Engelhardt et al., 1993).

TRANSPORT AND CELLULAR UPTAKE

Blood circulation: The proton/monocarboxylic acid cotransporter (MCT1, SLC16A1) is the main carrier for uptake of acetate, acetoacetate, and beta-hydroxybutyrate by the liver and other critical tissues.

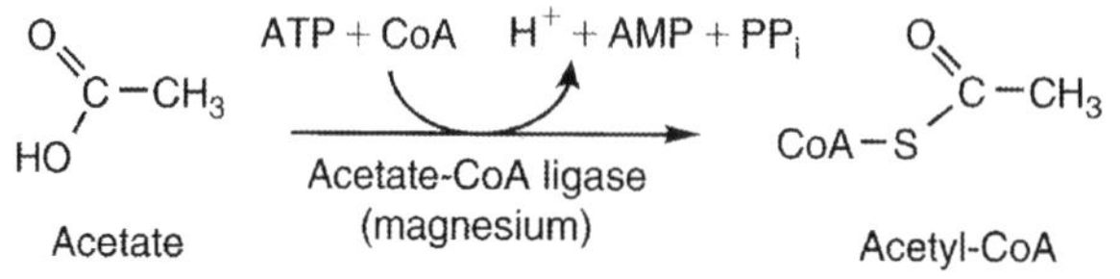

FIGURE 5.24

Acetate must be utilized before it can be utilized.

The related carriers MCT2 (SLC16A7), MCT3 (SLC16A8), and MCT4 (SLC16A3) have much more limited distribution.

BBB: Limited transport of acetate occurs across the epithelial cells of the BBB (Terasaki et al., 1991) via MCT1 on both sides of the brain capillary endothelial cell (Halestrap and Price, 1999). Interestingly, the foot processes of astroglial cells, which form part of the BBB, express MCT2. This carrier has a much greater affinity for monocarboxylates than MCT1 (Halestrap and Price, 1999). Permeability of the BBB increases greatly after several days of starvation and in diabetes mellitus.

METABOLISM

Acetate can be utilized by muscle and other peripheral tissues (Pouteau et al., 1996). Complete oxidation of acetate requires thiamin, riboflavin, niacin, pantothenate, lipoate, ubiquinone, iron, and magnesium.

First, free acetate must be conjugated to CoA by acetate-CoA ligase (thiokinase, EC6.2.1.1) (Figure 5.24). Most acetyl-CoA is utilized in mitochondria via the tricarboxylic acid (Krebs) cycle. Citrate synthase (EC4.1.3.7) joins acetyl CoA to oxaloacetate. The citrate from this reaction can then be metabolized further, providing FADH, NADH, and succinate for oxidative phosphorylation and ATP or GTP from succinyl CoA. The production rate of acetyl-CoA from fatty acid beta-oxidation in the liver with prolonged fasting usually exceeds the capacity of the Krebs cycle. The CoA for continued beta-oxidation and other functions can be released through the production of acetoacetate in three steps. The typical odor of a fasting individual is partially related to exhaled acetone formed from acetoacetate. The conversion of acetoacetate into beta-hydroxybutyrate taxes the body's acid-buffering capacity and may cause a drop in blood pH (acidosis) in diabetics and similarly susceptible patients. None of these events is related to dietary intake of acetate.

Ketogenesis takes place in the mitochondria, where fatty acid catabolism generates acetyl-CoA (Figure 5.25). Acetyl-CoA C-acetyltransferase (thiolase, EC2.3.1.9) joins two acetyl-CoA molecules, and hydroxymethylglutaryl-CoA synthase (HMG-CoA synthase, EC4.1.3.5) adds another one. The mitochondrial isoform of HMG-CoA synthase is genetically distinct from the cytosolic one, which generates the precursor for cholesterol synthesis. Hydroxymethylglutaryl-CoA lyase (HMG-CoA lyase, EC4.1.3.4) finally generates acetoacetate by cleaving off acetyl-CoA from the HMG-CoA intermediate. Spontaneous decarboxylation of acetoacetate generates the dead-end product acetone.

Acetoacetate can also be reduced to beta-hydroxybutyrate by NADH-dependent 3-hydroxybutyrate dehydrogenase (EC1.1.1.30). This enzyme is allosterically activated by phosphatidyl choline. The reaction is fully reversible. Net flux depends on substrate concentrations. Acetoacetate and beta-hydroxybutyrate (but not acetone) can become a significant energy fuel for brain after several days of adaptation to starvation conditions.

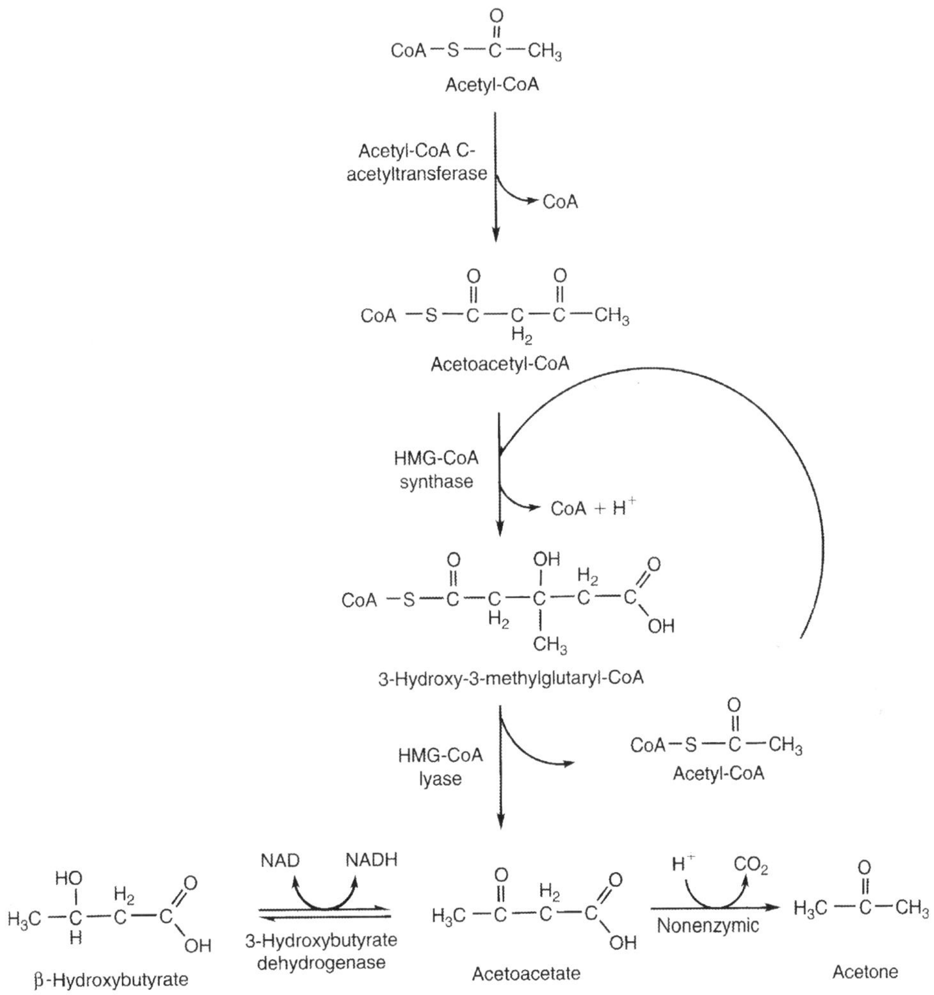

FIGURE 5.25

Ketogenesis frees up CoA from acetyl-CoA.

STORAGE

Other than the rapidly metabolized amounts in cellular cytosol and body fluids, acetate is not stored to a significant extent.

EXCRETION

Owing to its small molecular size, the renal glomerular membrane does not retain acetate. Nearly all of the filtered acetate is recovered from the proximal renal tubular lumen. Much of the uptake from the.tubular lumen is mediated by the proton/monocarboxylic acid cotransporter 2 (MCT2, SLC16A2), which has a several-fold higher affinity for its ligands than MCT1. Additional transporters, including MCT1, are likely to play a role in acetate salvage from renal ultrafiltrate.

REGULATION

Acetyl-CoA activates allosterically the biotin-dependent enzyme pyruvate carboxylase (EC6.4.1.1) and thereby stimulates Krebs cycle throughput.

DIETARY EFFECTS

Acetate inhibits lipolysis and replaces fat in the fuel mixture (Siler et al., 1999). Acetic acid also lowers blood sugar levels (Ogawa et al., 2000), possibly by decreasing the activities of sucrase, maltase, trehalase, and lactase (Ogawa et al., 2000), or by delaying gastric emptying (Liljeberg and Bjorck, 1998).

Dietary vinegar was found to enhance intestinal calcium absorption in rats (Kishi et al., 1999), but at the same time, it may increase urinary mineral loss and cause osteoporosis (Lhotta et al., 1998).

A combination of vinegar and honey has been claimed to be effective for the self-treatment of chronic inflammatory polyarthritis (Camara and Danao-Calnara 1999). But drinking vinegar just once a week appears to be sufficient to increase the risk of dental erosion (Jarvinen et al., 1991).

REFERENCES

Camara, K., Danao-Calnara, T., 1999. Awareness of, use and perception of efficacy of alternative therapies by patients with inflammatory arthropathies. Hawaii Med. J. 58, 329–332.

Garcia, C.K., Goldstein, J.L., Pathak, R.K., Anderson, R.G., Brown, M.S., 1994. Molecular characterization of a membrane transporter for lactate, pyruvate, and other monocarboxylates: implications for the Cori cycle. Cell 76, 865–873.

Halestrap, A.E., Price, N.T., 1999. The proton-linked monocarboxylate transporter (MCT) family: structure, function and regulation. Biochem. J. 343 (Pt 2), 281–299.

Jarvinen, V.K., Rytomaa, I.I., Heinonen, O.P., 1991. Risk factors in dental erosion. J. Dent. Res. 70, 942–947.

Kishi, M., Fukaya, M., Tsukamoto, Y., Nagasawa, T., Takehana, K., Nishizawa, N., 1999. Enhancing effect of dietary vinegar on the intestinal absorption of calcium in ovariectomized rats. Biosci. Biotech. Biochem. 63, 905–910.

Lhotta, K., Hofle, G., Gasser, R., Finkenstedt, G., 1998. Hypokalemia, hyperreninemia and osteoporosis in a patient ingesting large amounts of cider vinegar. Nephron. 80, 242–243.

Liljeberg, H., Bjorck, I., 1998. Delayed gastric emptying rate may explain improved glycaemia in healthy subjects to a starchy meal with added vinegar. Eur. J. Clin. Nutr. 52, 368–371.

Ogawa, N., Satsu, H., Watanabe, H., Fukaya, M., Tsukamoto, Y., Miyamoto, Y., et al., 2000. Acetic acid suppresses the increase in disaccharidase activity that occurs during culture of caco-2 cells. J. Nutr. 130, 507–513.

Orsenigo, M.N., Tosco, M., Bazzini, C., Laforenza, U., Faelli, A., 1999. A monocarboxylate transporter MCT1 is located at the basolateral pole of rat jejunum. Exp. Physiol. 84, 1033–1042.

Pouteau, E., Piloquet, H., Maugeais, P., 1996. Kinetic aspects of acetate metabolism in healthy humans using [1-^{13}C]acetate. Am. J. Physiol. 271, E58–E64.

Siler, S.Q., Neese, R.A., Hellerstein, M.K., 1999. *De novo* lipogenesis, lipid kinetics, and whole-body lipid balances in humans after acute alcohol consumption. Am. J. Clin. Nutr. 70, 928–936.

Stein, J., Zores, M., Schroder, O., 2000. Short-chain fatty acid (SCFA) uptake into Caco-2 cells by a pH-dependent and carrier mediated transport mechanism. Eur. J. Nutr. 39, 121–125.

Tamai, I., Takanaga, H., Maeda, H., Sai, Y., Ogihara, T., Higashida, H., et al., 1995. Participation of a proton-cotransporter, MCT1, in the intestinal transport of monocarboxylic acids. Biochem. Biophys. Res. Comm. 214, 482–489.

Terasaki, T., Takakuwa, S., Moritani, S., Tsuji, A., 1991. Transport of monocarboxylic acids at the blood-brain barrier: studies with monolayers of primary cultured bovine brain capillary endothelial cells. J. Pharmacol. Exp. Ther. 258, 932–937.

von Engelhardt, W., Burmester, M., Hansen, K., Becket, G., Rechkemmer, G., 1993. Effects of amiloride and ouabain on short-chain fatty acid transport in guinea-pig large intestine. J. Physiol. 460, 455–466.

Watson, A.J., Brennan, E.A., Farthing, M.J., Fairclough, P.D., 1991. Acetate uptake by intestinal brush border membrane vesicles. Gut 32, 383–385.

Wolever, T.M., Trinidad, T.E., Thompson, L.U., 1995. Short chain fatty acid absorption from the human distal colon: interactions between acetate, propionate and calcium. J. Am. Coll. Nutr. 14, 393–398.

MYRISTIC ACID

Myristate (myristic acid and tetradecanoic acid; molecular weight 228) is a saturated fatty acid with 14 carbons in a straight chain (Figure 5.26).

ABBREVIATIONS

CoA   coenzyme A
ETF   electron-transfer flavoprotein

NUTRITIONAL SUMMARY

Function: Myristate has one of the highest energy contents of any nutrient, providing about 9 kcal/g. Complete oxidation depends on riboflavin, niacin, pantothenic acid, carnitine, ubiquinone, iron, and magnesium.

Food sources: Milk fat and other animal fats are especially rich sources, but many solid plant fats also contain myristate, especially after hydrogenation.

Requirements: Current recommendations suggest limiting total saturated fat intake to less than 10% of total energy intake.

Deficiency: There is no indication that a lack of myristate intake causes any untoward health consequences.

FIGURE 5.26

Myristic acid.

Excessive intake: Myristate intake strongly raises LDL cholesterol concentrations and increases cardiovascular risk.

ENDOGENOUS SOURCES

The extent of *de novo* fatty acid synthesis, which occurs in cytosol of adipose tissue and the liver, is still disputed (Hellerstein, 2001). Synthesis, so far as it takes place, tends to proceed to chain lengths of 16 or 18 carbons, with little release of the intermediate metabolite myristate.

DIETARY SOURCES

Myristate is a minor, but very characteristic, component of milk (8–12% of total fat) and ruminant fat (about 3%). The amounts in fats from other sources are much smaller. Intakes of healthy Swedish men were around 4 g/day (Wolk et al., 2001).

DIGESTION AND ABSORPTION

Myristate-containing fats are absorbed to near completion from the small intestine. The myristate and other fatty acids from lipase-mediated hydrolysis combine with bile acids, monoglycerides, and phospholipids into mixed micelles. The micellar lipids are taken up into the small intestinal enterocytes through a mechanism that needs further elucidation. Myristate is then used mainly for the synthesis of triglycerides, which are secreted with chylomicrons into intestinal lymph ducts.

TRANSPORT AND CELLULAR UPTAKE

Blood circulation: Myristate in plasma is mainly bound to cholesterol and other complex lipids in lipoproteins and taken up into cells with them. Muscles, liver, and adipose tissue readily take up free fatty acids through an incompletely understood mechanism.

BBB: The transfer of fatty acids in general into the brain is limited and involves largely receptor-mediated endocytosis of lipoproteins.

Materno-fetal transfer: While myristate, like most fatty acids, reaches the fetus, the amounts and responsible mechanisms are not well understood.

METABOLISM

Chain elongation and desaturation: While some chain elongation and desaturation may occur, the extent is likely to be small.

Mitochondrial catabolism: Long-chain fatty acid CoA ligase 1 or 2 (EC6.2.1.3) activates myristate, and the combined action of camitine, palmitoyl-CoA:L-carnitine *O*-palmitoyltransferase I (EC2.3.1.21, on the outside), translocase, and palmitoyl-CoA:L-carnitine *O*-palmitoyltransferase II (EC2.3.1.21, on the inside) shuttles it into mitochondria (Figure 5.27).

The successive actions of long-chain acyl-CoA dehydrogenase (EC1.3.99.13), enoyl-CoA hydratase, L-3-hydroxyacyl-CoA dehydrogenase, and thiolase remove two carbons as acetyl-CoA and generate FAD and NADH. The acyl-CoA dehydrogenase forms a complex in the mitochondrial matrix with the ETF (contains FAD), and the iron-sulfur protein ETF dehydrogenase (EC1.5.5.1, also contains FAD),

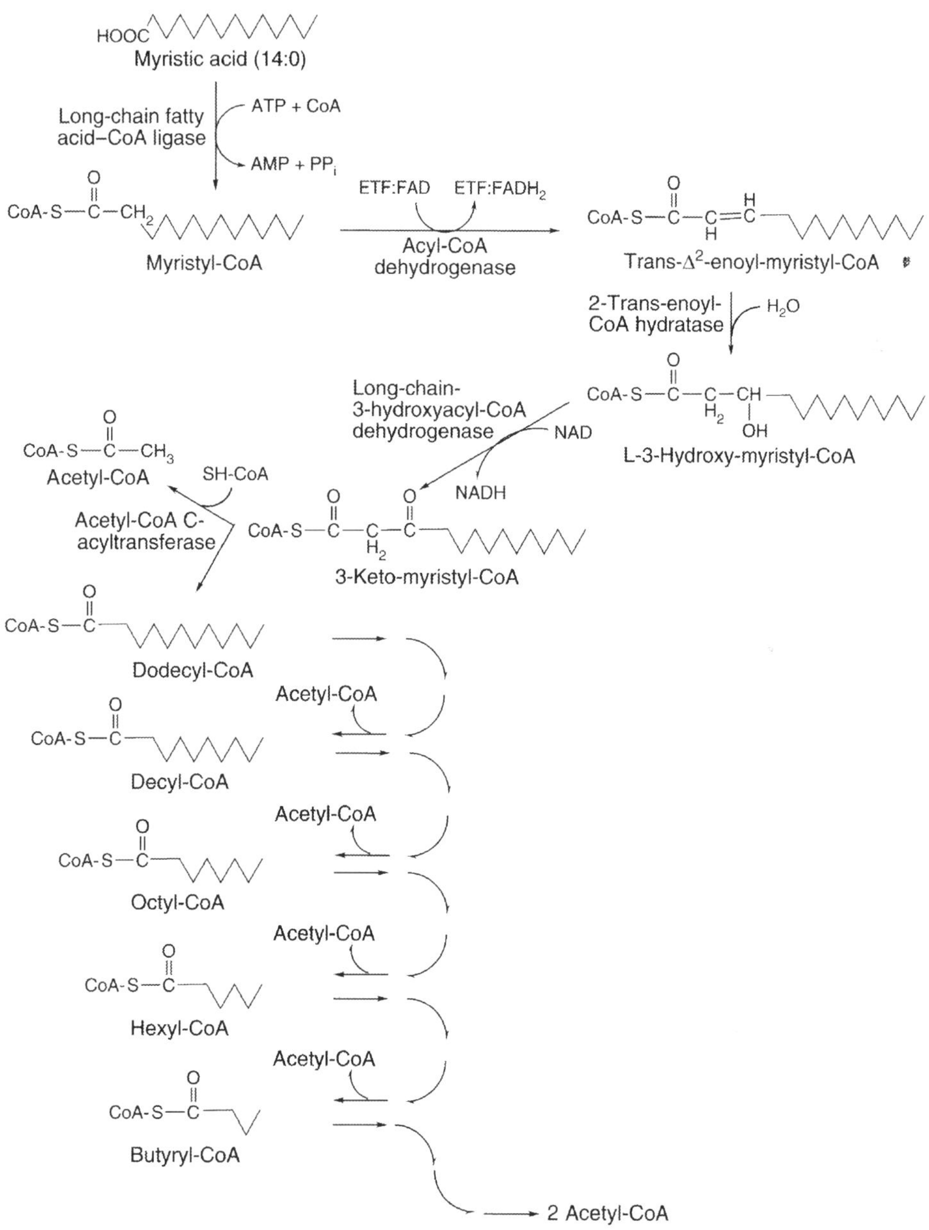

FIGURE 5.27

Breakdown of myristic acid occurs via β-oxidation.

which hands off the reducing equivalents to ubiquinone for oxidative phosphorylation. This sequence is repeated five times. The last cycle releases two acetyl-CoA molecules.

Peroxisomal catabolism: Myristate is less effectively catabolized in peroxisomes than longer fatty acids. If it is taken into peroxisomes at all, it will undergo only one or two beta-oxidation cycles, since medium-chain acyl-CoA molecules tend to leave peroxisomes and metabolism continues in mitochondria. After activation by one of several available long-chain fatty acid CoA ligases (EC6.2.1.3), the beta-oxidation cycle in peroxisomes uses FAD-dependent acyl-CoA oxidase (EC1.3.3.6), peroxisomal multifunctional protein 2 (MFP2, comprising activities EC4.2.1.17, EC5.3.3.8, and EC1.1.1.35), and peroxisome-specific acetyl-CoA C-acyltransferase (3-ketoacyl-CoA thiolase, EC2.3.1.16).

STORAGE

Adipose tissue typically contains 2–5%, depending strongly on dairy fat intake (Garaulet et al., 2001; Wolk et al., 2001). Myristate is released with normal adipose tissue turnover (about 1–2% of body fat per day).

EXCRETION

As with all fatty acids, there is no mechanism that could mediate significant excretion of myristate, even with significant excess.

REGULATION

The total fat content of the body is protected by powerful appetite-inducing mechanisms that include the action of leptin and other humoral mediators. Adipocytes release the proteohormone leptin commensurate to their fat content. Leptin binds to a specific receptor in the brain and decreases appetite through a signaling cascade that involves NPY. If the fat content of adipose tissue decreases, less leptin is sent to the brain and appetite increases.

There is no indication that the amounts of myristate in the body or concentrations in specific tissues or compartments are homeostatically controlled.

FUNCTION

Fuel energy: The oxidation of myristate supports the generation of about 92 ATP'(6×2.5 from NADH, 6×1.5 from FADH 2, about 70 from acetyl-CoA, -2 for ligation to CoA). This corresponds to an energy yield of about 9 kcal/g. Complete oxidation of myristate requires adequate supplies of riboflavin, niacin, pantothenic acid, carnitine, ubiquinone, iron, and magnesium.

Membrane anchor for proteins: Some proteins, especially those with signaling function, are acylated with myristate as a substrate (Resh, 2012). This lipophilic side chain can nestle into membranes and thus anchor the attached proteins. The specific type of fatty acid determines a preference for membrane regions and precise protein positioning with regard to the membrane surface. A typical example for myristylated proteins are kinases of the Src family. Loss of the myristate anchor can contribute to breast cancer phenotypes. With other proteins, such as c-Abl tyrosine kinase, myristate binding alters the conformation of the mature protein.

Hyperlipidemic potential: Myristate raises LDL cholesterol concentrations in plasma to a greater degree than other saturated fatty acids (Mensink et al., 1994), especially when polyunsaturated fatty acids contribute less than 5% of total energy. It is not known to what extent a concurrent increase in HDL cholesterol concentration can offset the known detrimental effect of higher LDL cholesterol levels.

REFERENCES

Garaulet, M., Perez-Llamas, F., Perez-Ayala, M., Martinez, E., de Medina, F.S., Tebar, F.J., et al., 2001. Site-specific differences in the fatty acid composition of abdominal adipose tissue in an obese population from a Mediterranean area: relation with dietary fatty acids, plasma lipid profile, serum insulin, and central obesity. Ant. J. Clin. Nutr. 74, 585–591.

Hellerstein, M.K., 2001. No common energy currency: *de novo* lipogenesis as the road less traveled. Am. J. Clin. Nutr. 74, 707–708.

Mensink, R.P., Temme, E.H., Hornstra, G., 1994. Dietary saturated and trans fatty acids and lipoprotein metabolism. Ann. Med. 26, 461–464.

Resh, M.D., 2012. Targeting protein lipidation in disease. Trends Mol. Med. 18, 206–214.

Wolk, A., Furuheim, M., Vessby, B., 2001. Fatty acid composition of adipose tissue and serum lipids are valid biological markers of dairy fat intake in men. J. Nutr. 131, 828–833.

CONJUGATED LINOLEIC ACID

Conjugated linoleic acid (CLA) is a term comprising 28 isomers of octadecadienoic acid (molecular weight 280) that have two double bonds separated by one single bond (Figure 5.28).

ABBREVIATIONS

CoA	coenzyme A
ECI	enoyl-CoA isomerase (EC5.3.3.8)
ETF	electron-transfer flavoprotein
FABPpm	plasma membrane fatty acid binding protein
FATP-1	fatty acid transport protein 1 (CD36, SLC27A1)
LPL	lipoprotein lipase
MECI	mitochondrial enoyl-CoA isomerase (EC5.3.3.8)
MTP	microsomal triglyceride transfer protein
TVA	trans-vaccenic acid
VLDL	very low-density lipoprotein

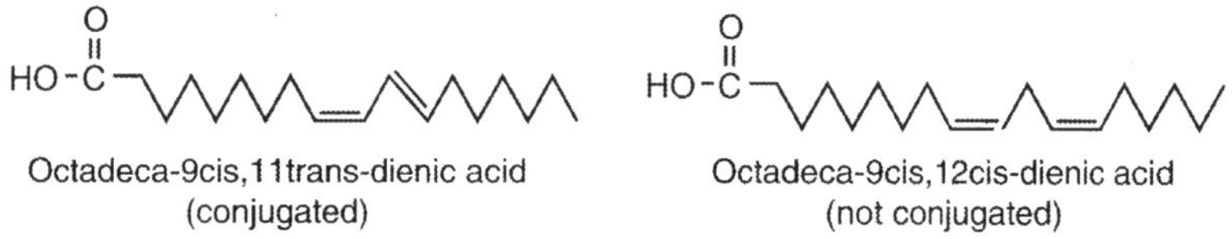

Octadeca-9cis,11trans-dienic acid
(conjugated)

Octadeca-9cis,12cis-dienic acid
(not conjugated)

FIGURE 5.28

The double bonds are separated by one single bond in this CLA isomer, but by two in linoleic acid.

NUTRITIONAL SUMMARY

Function: The nonessential group of fatty acids collectively called *conjugated linoleic acid (CLA)* may have some cancer-preventive potential. CLA is used as an energy fuel like other fatty acids. Complete oxidation depends on thiamin, riboflavin, niacin, pantothenate, carnitine, lipoate, ubiquinone, iron, and magnesium.

Food sources: CLA is a normal component of ruminant fat in dairy products and meat.

Deficiency: There is no indication for harmful effects in people with very low or absent CLA intake.

Excessive intake: The long-term health consequences of high-dose CLA supplement use are not known.

ENDOGENOUS SOURCES

Metabolism of trans11-octadecenoic acid (trans-vaccenic acid, TVA) generates cis9, trans11-linoleic acid in humans. The conversion occurs in the endoplasmic reticulum through the addition of a double bond by the stearoyl-CoA desaturase (delta-9-desaturase, EC1.14.99.5; contains iron) as originally observed in ruminants (Parodi, 1994; Santora et al., 2000). The desaturase uses cytochrome b5, which in turn is reduced by the flavoenzyme cytochrome-b5 reductase (EC1.6.2.2, contains FAD). About one-fifth of a moderate TVA dose (1.5 g/day) is converted to CLA, but individual metabolic capacities differ considerably (Turpeinen et al., 2002). Typical daily intakes of TVA from dairy products are around 1.3–1.8 g (Emken, 1995; Wolff, 1995). High intake of polyunsaturated fatty acids decreases stearoyl-CoA desaturase activity, whereas diets low in fat, high in carbohydrate, or high in cholesterol decrease it (Turpeinen et al., 2002).

Some *Bifidobacterium* species convert linoleic acid and linolenic acid to their conjugated isomers. This raises the possibility that the microbiota in some human individuals have some limited intestinal CLA production (Gorissen et al., 2010).

DIETARY SOURCES

CLA is consumed with fat-containing foods of ruminant origin, such as milk, butter, cheese, and beef. Cis9, trans11-linoleic acid (cis9, trans11-octadecadienoic acid, rumenic acid) accounts for three-quarters or more of the total CLA in ruminant fat. Most of the remainder consists of trans7, cis9-octadienoic acid and trans10, cis12-octadecadienoic acid (Figures 5.29 and 5.30).

Synthetic products (e.g., Clarinol) contain a wider range of isomers. Mean daily CLA intakes, mainly in ruminant fat (including from dairy and beef), have been reported to be between 52 and 310 mg (Ritzenhaler et al., 1998; Salminen et al., 1998).

DIGESTION AND ABSORPTION

Bile-salt activated lipase (EC3.1.1.3) from the pancreas in conjunction with colipase is the major digestive enzyme for the hydrolysis of CLA-containing diglycerides and triglycerides. Micelles form spontaneously from the mixture of fatty acids, monoglycerides, bile acids, and phospholipids. CLA is taken up into the small intestine by diffusion and facilitated transport from mixed micelles and is conjugated to CoA by long-chain fatty acid CoA ligase (EC6.2.1.3). Most of the acyl-CoA is used for the synthesis

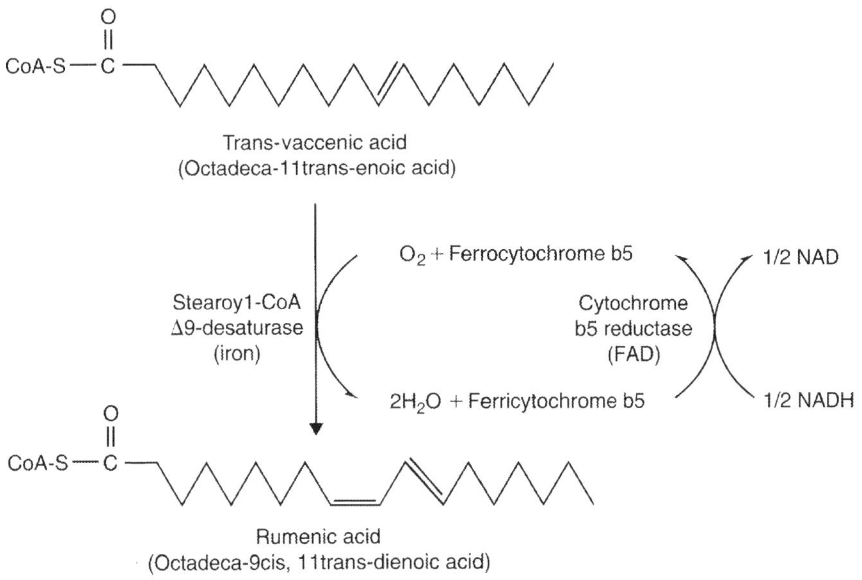

FIGURE 5.29

Dietary TVA is a precursor for endogenous CLA synthesis.

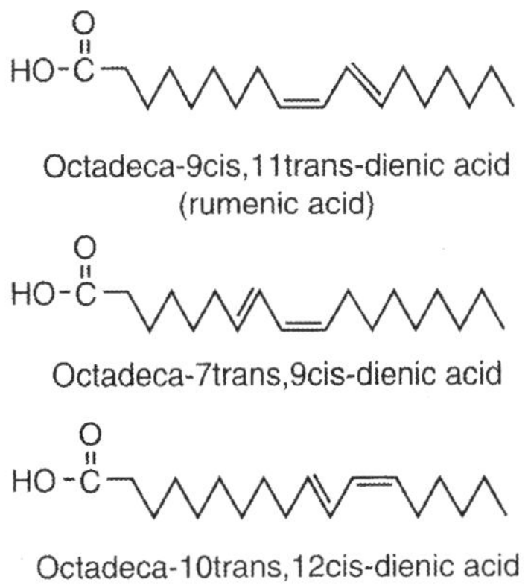

FIGURE 5.30

The main CLA isomers in ruminant fat.

of triglycerides, cholesterol esters, and phospholipids. Triglycerides are assembled into chylomicrons with cholesterol esters, phospholipids, and one molecule of apolipoprotein B48 per particle, and then secreted into intestinal lymphatic vessels. The fatty acid binding proteins 1 and 2, MTP, and presumably additional proteins are necessary for the intracellular transport of fatty acids and subsequent assembly and secretion of chylomicrons.

TRANSPORT AND CELLULAR UPTAKE

Blood circulation: Most CLA in blood is a constituent of triglycerides in chylomicrons, VLDL, and other lipoproteins. The relatively low percentage of cis9, trans11-linoleic acid in serum of people with low ruminant fat intake (about 0.15% of total fatty acids) rapidly increases (to 0.3% or more with typical intakes) with higher consumption (Turpeinen et al., 2002). Lipoprotein lipase (LPL, EC3.1.1.3) on the endothelial surface of small arterioles and capillaries releases CLA from triglycerides within chylomicrons and VLDL. At least six distinct fatty acid transport proteins are expressed in various tissues. The uptake of the bulk of free fatty acids into muscle cells and adipocytes depends on the fatty acid transport protein 1 (FATP-I, CD36, SLC27A1), either in conjunction with or augmented by plasma membrane fatty acid–binding protein (FABPpm; Storch and Thumser, 2000) and FAT. Transport across the cell membrane appears to be coupled to acyl-CoA synthase (EC6.2.1.3) activity. Apparently, this activity does not always require a separate protein since FATP4 in adipocytes functions as an acyl-CoA synthase (Herrmann et al., 2001). Fatty acids that remain in circulation bind to albumin. A proton-driven fatty acid transporter mediates their uptake into liver cells (Elsing et al., 1996).

BBB: It is not clear how extensive the transfer of nonessential fatty acids to the brain is. Lipoprotein-mediated transfer tends to favor long-chain polyunsaturated essential fatty acids but is likely to unspecifically bring along some CLA.

Materno-fetal transfer: Albumin-bound fatty acids and fatty acids released from VLDL by LPL at the maternal face of the placenta can be transported to the fetus. FATP, FABPpm, and FAT on the maternal side, and FATP on the fetal side are likely to be involved (Dutta-Roy, 2000).

METABOLISM

Beta-oxidation, both in mitochondria and in peroxisomes, is the main metabolic fate of CLA, discussed here with the example of rumenic acid (cis9, trans11-linoleic acid) (Figure 5.31). CoA-linked CLA is transferred to carnitine by palmitoyl-CoA:L-carnitine *O*-palmitoyltransferase I (EC2.3.1.21). Carnitine acylcarnitine translocase (CACT, SLC25A20) then moves the conjugate across the inner mitochondrial membrane in exchange for a free carnitine. Palmitoyl-CoA:L-carnitine *O*-palmitoyltransferase II (EC2.3.1.21) links the fatty acid again to CoA. Long-chain acyl-CoA dehydrogenase (EC1.3.99.13) starts the first cycle of beta-oxidation with the transfer of electrons from CLA via FAD and ubiquinone to the electron-transfer system. This FAD-containing enzyme forms a complex with ETF (another FAD-containing protein) and ETF dehydrogenase (EC1.5.5.1, contains both FAD and iron) in the mitochondrial matrix. Trans-enoyl-CoA hydratase (EC4.2.1.17) oxidizes the 2-trans metabolite. Another oxidation step, catalyzed by long-chain-3-hydroxyacyl-CoA dehydrogenase (EC1.1.1.211), generates the 3-keto intermediate. Thiolysis by acetyl-CoA C-acyltransferase (thiolase, EC2.3.1.16) releases acetyl-CoA. Two more rounds of beta-oxidation with the same sequence of reactions follow.

The cis3 double bond of the dodeca-cis3,trans5-dienoyl-CoA is not processed by trans-enoyl-CoA hydratase (EC4.2.1.17), which would normally act at this point of the cycle. It is likely that dodecenoyl-CoA delta-isomerase (enoyl-CoA isomerase, EC5.3.3.8) converts the intermediate dodeca-trans3,5-dienoyl-CoA to its trans2, cis5 isomer. Liver mitochondria contain the genetically distinct enzymes

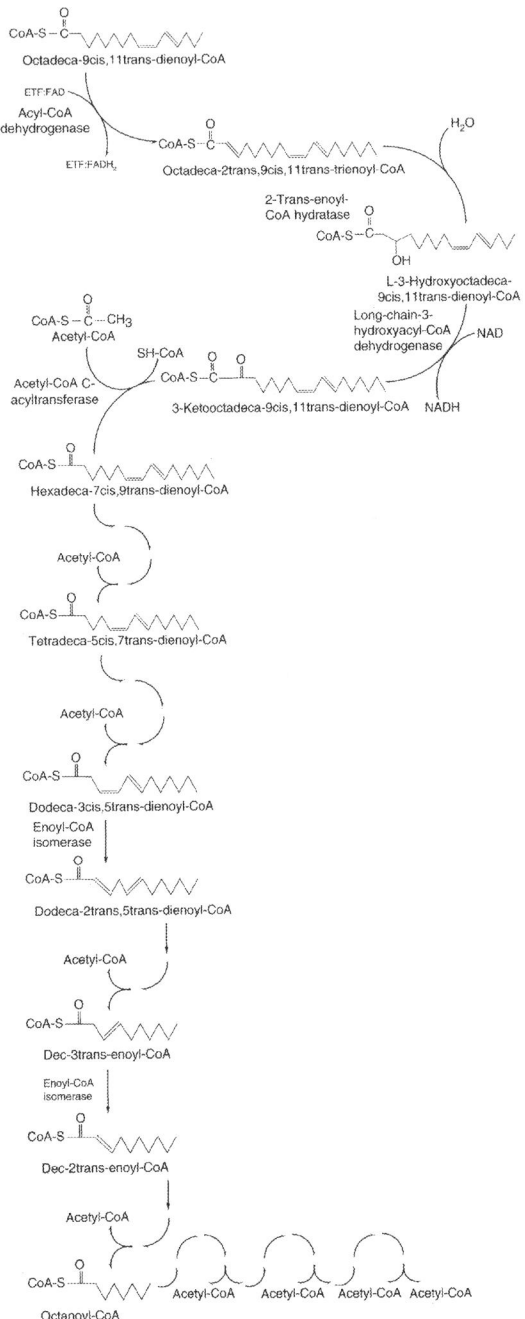

FIGURE 5.31

Putative pathway for the metabolism of rumenic acid.

enoyl-CoA isomerase (ECI) and mitochondrial enoyl-CoA isomerase (MECI) with different catalytic profiles (Zhang et al., 2002). In peroxisomes, both the multifunctional enzyme 1 and ECI have this ECI activity. It is not clear, however, to what extent each of these three enzymes isomerizes dodeca-trans3,5-dienoyl-CoA and analogous intermediates with conjugated double bonds. The successive actions of 2-trans-enoyl-CoA hydratase, long-chain-3-hydroxyacyl-CoA dehydrogenase, and acetyl-CoA C-acyltransferase then shorten the molecule to dec-trans3-enoyl-CoA.

ECI moves the double bond to the 2 position, as discussed previously. Among the three isomerases, ECI has the highest preference for this reaction (Zhang et al., 2002). Completion of this round of beta-oxidation generates the saturated intermediate octanoyl-CoA and another three rounds finish off the breakdown of rumenic acid into nine 2-carbon fragments. The acetyl-CoA moieties released with each cycle are utilized further through the Krebs cycle.

Just like other long-chain fatty acids, CLA may be modified by extension of chain length and addition of double bonds. The 18:3, 20:3, and 20:4 derivatives have been detected in adipose tissue. In particular, the arachidonic acid analogs may be preferentially incorporated into phospholipids (Banni et al., 2001). The 20 carbon metabolites may also give rise to eicosanoidlike compounds with as-yet-unknown properties.

STORAGE

The proportion of triglycerides with conjugated double bonds in adipose tissue correlates with the dietary intake of CLA (Sebedio et al., 2001).

EXCRETION

As is the case with other fatty acids, virtually no CLA is lost with feces or urine.

FUNCTION

Fuel metabolism: CLA can be utilized as an energy fuel providing about 9 kcal/g, the low consumption levels notwithstanding. Complete oxidation requires adequate supplies of thiamin, riboflavin, niacin, pantothenate, carnitine, lipoate, ubiquinone, iron, and magnesium.

Other effects: Numerous metabolic changes are associated with increasing dietary intake of CLA. These include induction of lipid peroxidation (Basu et al., 2000) through acting as a peroxisome pro-liferator (Belury et al., 1997), alteration of prostaglandin synthesis through an action on prostaglandin H synthase (Bulgarella et al., 2001), and promotion of apoptosis (Park et al., 2001). The antioxidant capacity of CLA *in vivo* is less certain.

The presumed health consequences, possibly including anticancer action, cardiovascular disease prevention, and weight loss promotion (Riserus et al., 2001), still require substantiation by further studies.

REFERENCES

Banni, S., Carla, G., Angioni, E., Murru, E., Scanu, P., Melis, M.E., et al., 2001. Distribution of conjugated linoleic acid and metabolites in different lipid fractions in the rat liver. J. Lipid Res. 42, 1056–1061.

Basu, S., Smedman, A., Vessby, B., 2000. Conjugated linoleic acid induces lipid peroxidation in humans. FEBS Lett. 468, 33–36.

Belury, M.A., Moya-Camarena, S.Y., Liu, K.L., Van den Heuvel, J.E., 1997. Dietary conjugated linoleic acid induces peroxisome-specific enzyme accumulation and ornithine decarboxylase activity in mouse liver. J. Nutr. Biochem. 8, 579–584.

Bulgarella, J.A., Patton, D., Bull, A.W., 2001. Modulation ofprostaglandin H synthase activity by conjugated linoleic acid (CLA) and specific CLA isomers. Lipids 36, 407–412.

Dutta-Roy, A.K., 2000. Transport mechanisms for long-chain polyunsaturated fatty acids in the human placenta. Am. J. Clin. Nutr. 71, 315S–322S.

Elsing, C., Kassner, A., Stremmel, W., 1996. Effect of surface and intracellular pH on hepatocellular fatty acid uptake. Am. J. Physiol. 271, G1067–G1073.

Emken, E.A., 1995. Trans fatty acids and coronary heart disease risk. Physicochemical properties, intake and metabolism. Am. J. Clin. Nutr. 62, 659S–669S.

Herrmann, T., Buchkremer, E., Gosch, I., Hall, A.M., Bernlohr, D.A., Stremmel, W., 2001. Mouse fatty acid transport protein 4 (FATP4): characterization of the gene and functional assessment as a very long chain acyl-CoA synthetase. Gene 270, 31–40.

Park, H.S., Ryu, J.H., Ha, Y.L., Park, J.H., 2001. Dietary conjugated linoleic acid (CLA) induces apoptosis of colonic mucosa in 1,2-dimethylhydrazine-treated rats: a possible mechanism of the anticarcinogenic effect by CLA. Br. J. Nutr. 86, 549–555.

Parodi, P.W., 1994. Conjugated linoleic acid: an anticarcinogenic fatty acid present in milk fat. Aust. J. Dairy Technol. 49, 93–97.

Riserus, U., Berglund, L., Vessby, B., 2001. Conjugated linoleic acid (CLA) reduced abdominal adipose tissue in obese middle-aged men with signs of the metabolic syndrome: a randomized controlled trial. Int. J. Obes. 25, 1129–1135.

Ritzenhaler, K., McGuire, M.K., Falen, R., Schultz, T.D., McGuire, M.A., 1998. Estimation of conjugated linoleic acid (CLA) intake. FASEB J. 12, A527.

Salminen, I., Mutanen, M., Jauhiainen Aro, A., 1998. Dietary trans fatty acids increase conjugated linoleic fatty acid levels in human serum. J. Nutr. Biochem. 9, 93–98.

Santora, J.E., Palmquist, D.L., Roehrig, K.L., 2000. Trans-vaccenic acid is desaturated to conjugated linoleic acid in mice. J. Nutr. 130, 208–215.

Sebedio, J.L., Angioni, E., Chardigny, J.M., Gregoire, S., Juaneda, E., Berdeaux, O., 2001. The effect of conjugated linoleic acid isomers on fatty acid profiles of liver and adipose tissues and their conversion to isomers of 16:2 and 18:3 conjugated fatty acids in rats. Lipids 36, 575–582.

Storch, J., Thumser, A.E., 2000. The fatty acid transport function of fatty acid-binding proteins. Biochim. Biophys. Acta 1486, 28–44.

Turpeinen, A.M., Mutanen, M., Aro, A., Salminen, I., Basu, S., Palmquist, D.L., et al., 2002. Bioconversion of vaccenic acid to conjugated linoleic acid in humans. Am. J. Clin. Nutr. 76, 504–510.

Wolff, R.L., 1995. Content and distribution of trans-18:1 acids in ruminant milk and meat fats. Their importance in European diets and their effect on human milk. J. Am. Oil Chem. Soc. 72, 259–272.

Zhang, D., Yu, W., Geisbrecht, B.V., Gould, S.J., Sprecher, H., Schulz, H., 2002. Functional characterization of Delta3,Delta2-enoyl-CoA isomerases from rat liver. J. Biol. Chem. 277, 9127–9132.

DOCOSAHEXAENOIC ACID

Docosahexaenoic acid (DHA; molecular weight 328) is an omega-3 polyunsaturated fatty acid (Figure 5.32).

ABBREVIATIONS

CoA	coenzyme A
DHA	docosahexaenoic acid
EPA	eicosapentaenoic acid
ETF	electron-transfer flavoprotein

NUTRITIONAL SUMMARY

Function: The essential fatty acid docosahexaenoic acid (DHA) becomes a component of complex lipids in membranes (especially of the retina), nerve insulation (myelin in the brain), and other structures; is the precursor for signaling molecules (prostaglandins and other eicosanoids); and provides about 9 kcal/g when used as an energy fuel. Complete oxidation depends on thiamin, riboflavin, niacin, pantothenate, carnitine, ubiquinone, iron, and magnesium.

Food sources: Salmon, herring, mackerel, and trout are good DHA sources. Fish-oil capsules are used by some as a dietary supplement. Some DHA can also be formed from alpha-linolenic acid in flaxseed (linseed) and other plant sources. The synthesis requires adequate supplies of riboflavin, niacin, pantothenate, magnesium, and iron.

Requirements: Canadian recommendations for daily consumption of total omega-3 fatty acids are 1.2–1.6 g. Recommendations in the United Kingdom are 1% of total energy intake as alpha-linolenic acid, and 0.5% as EPA plus DHA. Omega-6 fatty acids cannot substitute for omega-3 fatty acids.

Deficiency: Inadequate omega-3 fatty acid intake impairs immune function and leads to the development (after many weeks) of dermatitis. Low intake increases the risk of sudden death due to heart disease and the risk for some cancers.

Excessive intake: High omega-3 fatty acid consumption prolongs bleeding time (Dyerberg and Bang, 1979).

ENDOGENOUS SOURCES

Alpha-linolenic acid, eicosapentaenoic acid (EPA), and other omega-3 fatty acid can be converted to DHA, mainly in the liver (Figure 5.33). A series of elongation and desaturation steps in the endoplasmic

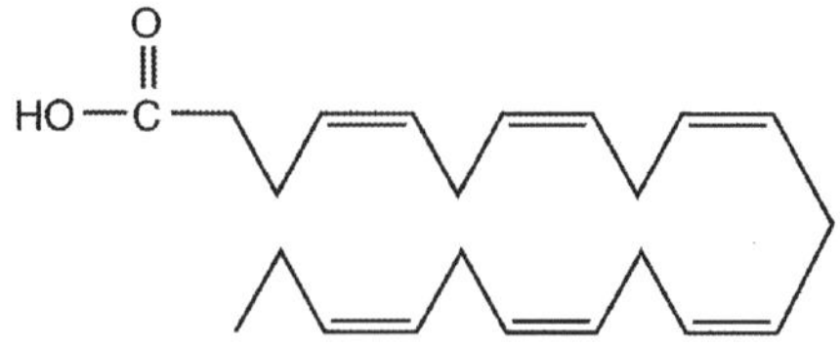

FIGURE 5.32

DHA.

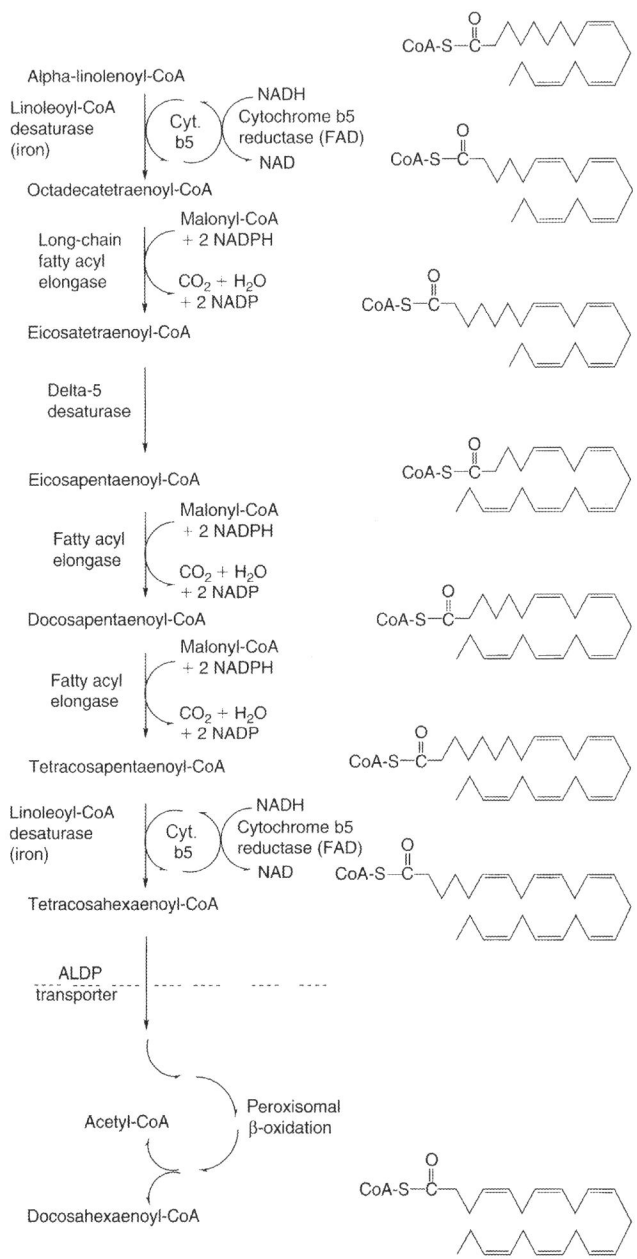

Alpha-linolenoyl-CoA

Linoleoyl-CoA
desaturase
(iron)

Cyt.
b5

NADH
Cytochrome b5
reductase (FAD)

NAD

Octadecatetraenoyl-CoA

Long-chain
fatty acyl
elongase

Malonyl-CoA
+ 2 NADPH

$CO_2 + H_2O$
+ 2 NADP

Eicosatetraenoyl-CoA

Delta-5
desaturase

Eicosapentaenoyl-CoA

Fatty acyl
elongase

Malonyl-CoA
+ 2 NADPH

$CO_2 + H_2O$
+ 2 NADP

Docosapentaenoyl-CoA

Fatty acyl
elongase

Malonyl-CoA
+ 2 NADPH

$CO_2 + H_2O$
+ 2 NADP

Tetracosapentaenoyl-CoA

Linoleoyl-CoA
desaturase
(iron)

Cyt.
b5

NADH
Cytochrome b5
reductase (FAD)

NAD

Tetracosahexaenoyl-CoA

ALDP
transporter

Acetyl-CoA

Peroxisomal
β-oxidation

Docosahexaenoyl-CoA

FIGURE 5.33

A small percentage of alpha-linolenic acid is converted to DHA.

reticulum generates the 24-carbon intermediate tetracosahexaenoic acid, which then has to be shortened to DHA by peroxisomal beta-oxidation (Ferdinandusse et al., 2001). DHA synthesis requires adequate supplies of riboflavin, niacin, pantothenate, biotin, iron, and magnesium. The capacity for such conversion reactions is limited in the very young infant, and adequate amounts of preformed DHA have to be supplied (Clandinin et al., 1981).

The first step is the conjugation of the precursor to CoA by the iron-enzyme long-chain-fatty-acid-CoA ligase (EC6.2.1.3). Linoleoyl-CoA desaturase (delta-6 desaturase, EC1.14.99.25) uses the cytochrome b5 system to add a double bond.

FAD-containing cytochrome b5 reductase (EC1.6.2.2) can reactivate the b5 electron donor in an NADH-dependent reaction. A specific NADH-using long-chain fatty acyl elongase (no EC number assigned) adds malonyl-CoA to the acyl-CoA and releases carbon dioxide and free CoA. The biotin-enzyme acetyl CoA carboxylase (EC6.4.1.2) produces malonyl-CoA by carboxylating acetyl-CoA. Delta-5 desaturase (EC1.14.19.-) introduces another double bond. This enzyme is not yet well characterized.

The next round of elongation generates EPA, which is functionally important in itself. Apparently, there is no enzyme available that can introduce a double bond at position 4 of EPA. This difficulty is sidestepped by one more round of elongation and addition of the final double bond in position 6 by linoleoyl-CoA desaturase. However, the reaction sequence generates a 24-carbon product, tetracosahexaenoic acid. To produce DHA, the intermediate has to be moved into a peroxisome (by the peroxisomal ABC half-transporter ALDP, ABCD1) and shortened to 22-carbon-length by one round of beta-oxidation (Su et al., 2001).

Peroxisomal beta-oxidation depends on the successive action of FAD-dependent acyl-CoA oxidase (EC1.3.3.6) and peroxisomal multifunctional protein 2 (MFP2, comprising activities enoyl-CoA hydratase, EC4.2.1.17; 3,2-trans-enoyl-CoA isomerase, EC5.3.3.8; and 3-hydroxyacyl-CoA dehydrogenase, EC1.1.1.35; and acetyl-CoA C-acyltransferase, EC2.3.1.16). SCPx also can catalyze the final step. Peroxisomes are indispensable for this final activation of DHA, since mitochondria can process only fatty acids with up to 22 carbons (Singh et al., 1984). Individuals with defective or absent peroxisomes can produce EPA, but not DHA, from omega-3 fatty acid precursors (Martinez et al., 2000).

DIETARY SOURCES

The few good dietary sources of DHA include salmon (14.6 mg/g), herring (11.1 mg/g), mackerel (7.0 mg/g), swordfish (6.8 mg/g), trout (5.2 mg/g), and halibut (3.7 mg/g). These fish also contain significant amounts of EPA and other omega-3 fatty acids. DHA is mainly present in the middle position (sn-2) of the fish-oil triglycerides (Yoshida et al., 1999).

Flaxseed (linseed) is a particularly rich source of omega-3 fatty acids because it contains 181 mg/g of alpha-linolenic acid. English walnuts can provide 68 mg/g. Canola oil contains about 92 mg/g and soybean oil about 78 mg/g. Perilla oil has an exceptionally high alpha-linolenic acid content (630 mg/g). Due to its propensity to spoil rapidly by oxidation, it is rarely used, other than in a few Asian dishes and medicinal concoctions. The combined omega-3 fatty acid content of most other animal- and plant-derived foods is well under 10 mg/g.

Fish-oil capsules, used as dietary supplements, typically provide a combination of DHA, EPA, and other minor omega-3 fatty acids. Vitamin D can be a significant and welcome component of such supplements (particularly those containing cod liver oil), but is not likely to increase intake to harmful levels, even in combination with other vitamin D–containing supplements. An excess of retinol is a concern, though.

Typical dietary omega-3 fat intakes depend to a large extent on habitual use of the major sources. The daily dose of combined omega-3 fatty acids for Americans is about 1.6 g, but less than 200 mg of that is DHA (Kris-Etherton et al., 2000). People using a typical Mediterranean diet get about 0.3% of their total fat intake as DHA and about 2.1% as omega-3 fatty acids combined (Garaulet et al., 2001).

DIGESTION AND ABSORPTION

Bile-salt-activated lipase (EC3.1.1.3) from pancreas in conjunction with colipase is the major digestive enzyme for the hydrolysis of DHA-containing diglycerides and triglycerides and cholesterol esters. Phospholipase A2 (EC3.1.1.4) from pancreas cleaves DHA-rich phospholipids. Micelles form spontaneously from the mixture of fatty acids, monoglycerides, bile acids, and phospholipids. DHA enters enterocytes of the small intestine by diffusion and facilitated transport from mixed micelles and is conjugated to CoA by long-chain fatty acid CoA ligase (EC6.2.1.3). Most of the acyl-CoA is used for the synthesis of triglycerides, cholesterol esters, and phospholipids. Triglycerides are assembled into chylomicrons with cholesterol esters, phospholipids, and one molecule of apolipoprotein B48 per particle, and secreted into intestinal lymphatic vessels. The fatty acid binding proteins 1 and 2, MTP, and presumably additional proteins are necessary for the intracellular transport of fatty acids and subsequent assembly and secretion of chylomicrons (Dannoura et al., 1999).

TRANSPORT AND CELLULAR UPTAKE

Blood circulation: DHA in fasting plasma (about 2.4% of total fatty acids or 30–100 mg/l) is mainly a constituent of phospholipids and cholesterol in LDL and other lipoproteins. Erythrocytes contain about 10–15 ng/10^6 cells or 50–75 mg/l (Martinez et al., 2000). Lipoproteins are taken up into muscle, liver, and other tissues via specific receptor-mediated endocytosis. Lipoprotein lipase (LPL, EC3.1.1.3) on the endothelial surface of small arterioles and capillaries releases free DHA from triglycerides within chylomicrons and VLDL. At least six distinct fatty acid transport proteins are expressed in various tissues. The uptake of the bulk of free fatty acids into muscle cells and adipocytes depends on the fatty acid transport protein 1 (FATP-1, CD36, SLC27A1; Martin et al., 2000), either in conjunction with or augmented by plasma membrane fatty acid binding protein (FABPpm; Storch and Thumser, 2000) and FAT. Transport across the cell membrane appears to be coupled to acyl-CoA synthase (EC6.2.1.3) activity. Apparently, this activity does not always require a separate protein, since FATP4 in adipocytes functions as an acyl-CoA synthetase (Herrmann et al., 2001).

Fatty acids can also be taken up into some cells with the entire TRL. Specific receptors recognize chylomicron remnants (Brown et al., 2000) and VLDL (Tacken et al., 2001). Fatty acids that remain in circulation bind to albumin. Their main fate is uptake into the liver via a proton-driven fatty acid transporter (Elsing et al., 1996).

BBB: Transport of fatty acids from circulation into brain, via a poorly understood mechanism, favors DHA and other polyunsaturated fatty acids (Edmond, 2001). The mechanism for the specific transfer of essential fatty acids, which are needed for the synthesis of many brain-typical structural compounds, remains unclear. Small amounts may be transported with HDLs, which enter brain capillary endothelium via a HDL-binding receptor (Goti et al., 2000).

Materno-fetal transfer: A third-trimester fetus requires about 3 g of DHA per day for structures in the growing brain (Clandinin et al., 1981). Both albumin-bound fatty acids and fatty acids released from VLDL by LPL at the maternal face of the placenta are available for transport. FATP, FABPpm, and FAT move DHA into the syntrophoblast. FATP provides the main mechanism for export to the fetal side (Dutta-Roy, 2000). The exact role and identity of these transporters remain to be clarified. The placental FABPpm, which is distinct from related transporters in other tissues, transports DHA more effectively than smaller and saturated fatty acids. Most DHA in fetal circulation is bound to alpha-fetoprotein (Calvo et al., 1988).

METABOLISM

Energy metabolism: DHA-CoA can undergo beta-oxidation both in mitochondria and in peroxisomes. Nearly one-tenth of ingested DHA is shortened to EPA (Conquer and Holub, 1997), which has its own specific functions (Figure 5.34). The enzyme that conjugates DHA to CoA in the brain for the initial activation step has a much lower K_m than other long-chain fatty acid CoA ligases (Bazan, 1990). Palmitoyl-CoA:L-carnitine O-palmitoyltransferase I (EC2.3.1.21) replaces the CoA attached to DHA with carnitine and carnitine acylcarnitine translocase (CACT, SLC25A20) moves the conjugate across the inner mitochondrial membrane in exchange for a free carnitine. Palmitoyl-CoA:L-carnitine O-palmitoyltransferase II (EC2.3.1.21) links DHA again to CoA. The first cycle of DHA-CoA beta-oxidation, catalyzed by long-chain acyl-CoA dehydrogenase (EC1.3.99.13), transfers electrons from the substrate via FAD and ubiquinone to the electron-transfer system. This FAD-containing enzyme forms a complex with ETF (another FAD-containing protein) and ETF dehydrogenase (EC1.5.5.1, contains both FAD and iron) in the mitochondrial matrix. The conjugated delta2, delta4-double bond prevents the usual hydration step. This barrier is sidestepped since NADPH-dependent 2,4-dienoyl-CoA reductase (EC1.3.1.34) saturates the newly introduced double bond and dodecenoyl-CoA delta-isomerase (EC5.3.3.8) converts the delta3-cis double bond into a delta2-trans-double bond. The successive actions of 2-trans-enoyl-CoA hydratase (EC4.2.1.17), long-chain-3-hydroxyacyl-CoA dehydrogenase (EC1.1.1.211), and acetyl-CoA C-acyltransferase (thiolase, EC2.3.1.16) complete cycle 1. The resulting EPA-CoA can either be used for its own specific purposes or continue through beta-oxidation. The breakdown of EPA-CoA takes nine cycles of beta-oxidation (Figure 5.35). Cycle 1 is catalyzed by long-chain acyl-CoA dehydrogenase, 2-trans-enoyl-CoA hydratase, long-chain-3-hydroxyacyl-CoA dehydrogenase, and acetyl-CoA C-acyltransferase. Cycle 2 starts with conversion of the delta-3-cis double bond in 3c,6c,9c,12c,15c-octadecapentaenoate to a delta-2-trans-double bond (thus omitting the FADH2-generating step), 2-transenoyl-CoA hydratase, long-chain-3-hydroxyacyl-CoA dehydrogenase, and acetyl-CoA C-acyltransferase complete this round. Cycle 3 comprises long-chain acyl-CoA dehydrogenase, 2,4-dienoyl-CoA reductase, dodecenoyl-CoA delta-isomerase, 2-trans-enoyl-CoA hydratase, long-chain-3-hydroxyacyl-CoA dehydrogenase, and acetyl-CoA C-acyltransferase, just like the cycle for DHA to EPA conversion. The next three cycles are analogous to 1, 2, and 3. Another two cycles are analogous to 1 and 2, except that acyl-CoA dehydrogenase (EC1.3.99.3) catalyzes the initial oxidation step. The successive actions of butyryl-CoA dehydrogenase (EC1.3.99.2), 2-transenoyl-CoA hydratase, long-chain-3-hydroxyacyl-CoA dehydrogenase, and acetyl-CoA C-acyltransferase complete one final round of beta-oxidation. The acetyl-CoA moieties released with each cycle are utilized further through the Krebs cycle.

Resolvins and protectin D1: We now know three alternative DHA hydroxylation pathways, whose entry points are determined by the enzyme that acts on DHA. The pathway leading to the generation of D-resolvins 1, 2, 3, and 4, and 17-S-Neuroprotectin D1 starts with the production of 17-S-H(p)DHA by

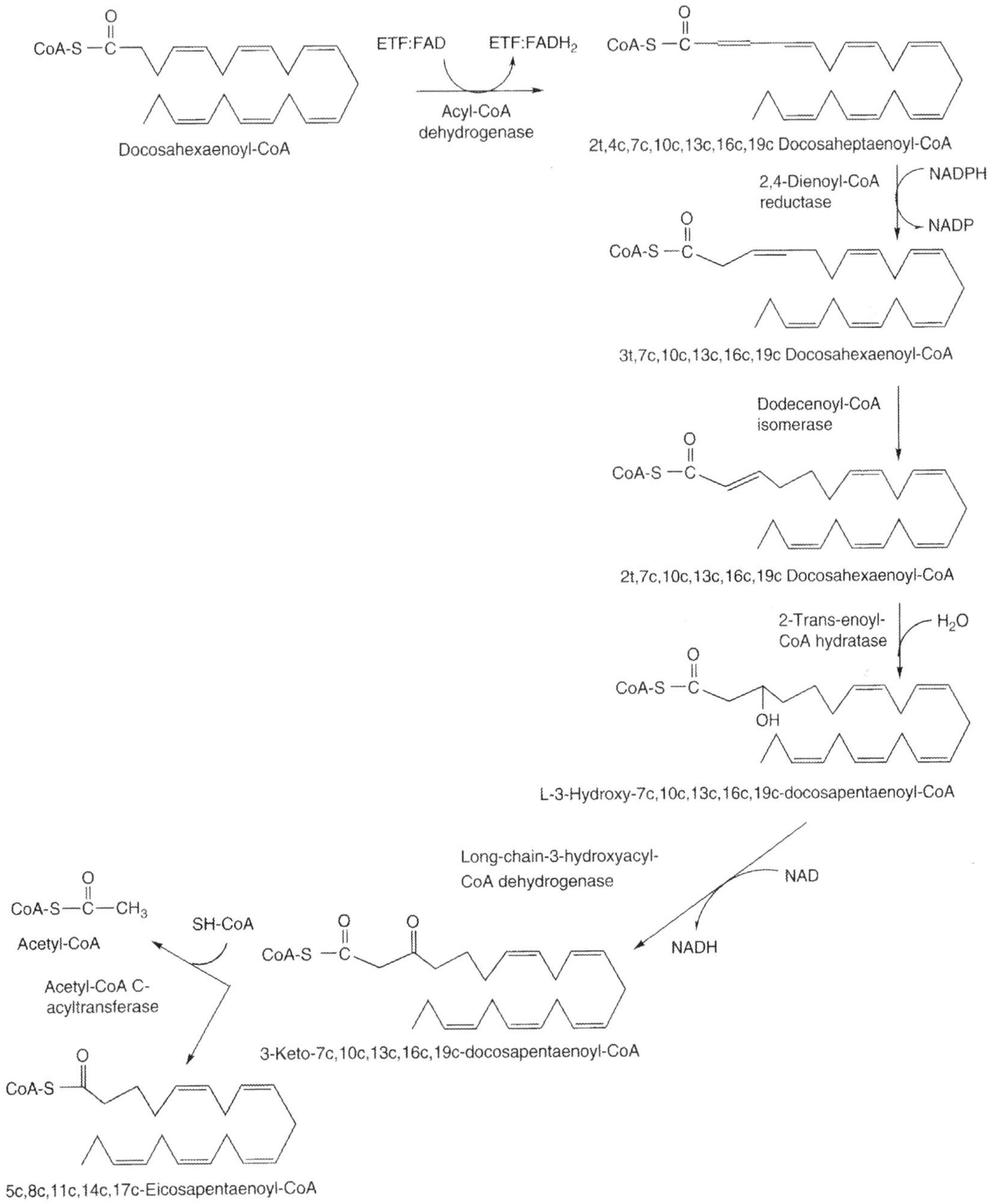

FIGURE 5.34

DHA retroconversion to EPA.

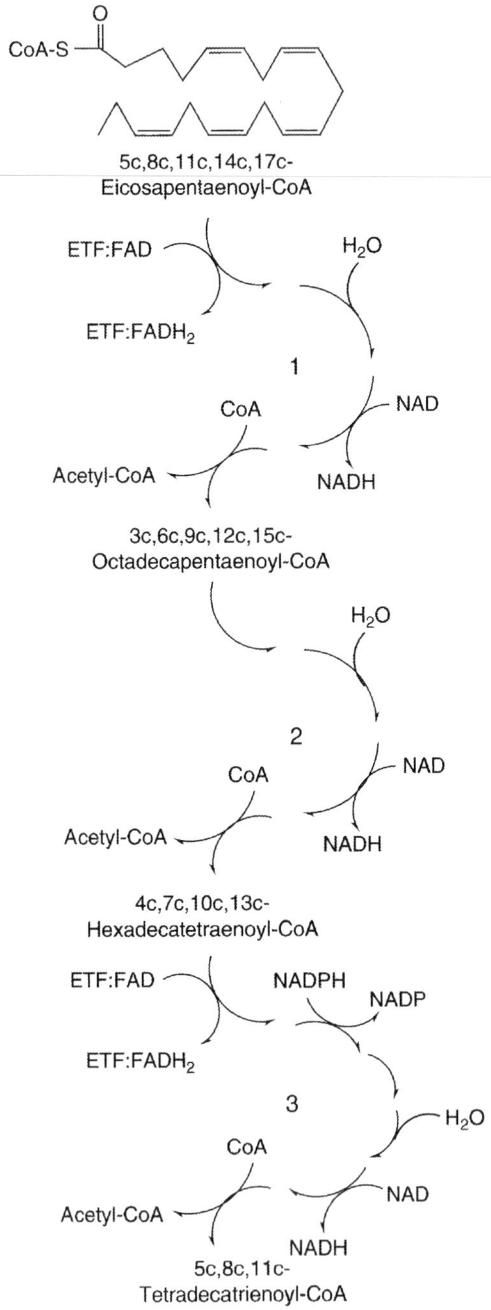

FIGURE 5.35

EPA breakdown.

arachidonate 15-lipoxygenase (EC1.13.11.33). Prostaglandin-endoperoxide synthase 2 (PTGS, COX-2, EC1.14.99.1)—but only the acetylated form—generates 17R-H(p)DHA, which forms the substrate for the synthesis of a different set of resolvins: 17R-resolvin D1, D2, D3, D4, and 17R-protectin D1 (Serhan, 2014).

STORAGE

Blood carries about 400–800 mg DHA, much of which becomes available for other uses upon uptake of lipoproteins and degradation of red cell membranes. Adipose tissue at all sites contains some DHA (about 0.25% in a Mediterranean obese population), though the percentage tends to be lower than in the diet (Garaulet et al., 2001). Adipose tissue near the intestines may contain slightly more than at other sites (0.35%). Assuming 10% total body fat, an average-sized (70-kg) lean man may be expected to carry about 18 g of DHA, of which 1–2% (i.e., 180–360 mg) are mobilized per day. Women tend to have slightly higher DHA stores due to their higher body fat mass. Lipids in other tissues, especially brain, contain a high percentage of DHA, but slow tissue turnover is likely to limit the amounts available for reuse at other sites.

EXCRETION

Losses of DHA, as of any other fatty acid, are minimal and occur mainly via skin, feces, and body fluids.

FUNCTION

Energy fuel: While DHA and other omega-3 fatty acids are not an important energy source, most is eventually utilized through beta-oxidation, providing about 9 kcal/g. Full oxidation depends on adequate supplies of thiamin, riboflavin, niacin, pantothenate, carnitine, ubiquinone, iron, and magnesium.

Precursor for complex brain lipids: DHA is indispensable for the myelinization of neurons in the brain and for the functioning of photoreceptor cells in the eye (Martinez et al., 2000). While DHA comprises one-third of the fatty acids in aminophospholipids of gray matter (Bazan and Scott, 1990), the exact functional significance of these and other complex lipids remains to be elucidated further. In particular, DHA is the precursor of a series of very long hexaenoic acids in the brain (24:6, 26:6, 28:6, 30:6, 32:6, 34:6, 36:6) with elusive functions.

Prostanoid synthesis: DHA is a precursor of EPA, which gives rise to the synthesis of 3-series prostaglandins and 5-series leukotrienes (Belch and Hill, 2000). Both DHA and EPA inhibit the synthesis of omega-6 fatty acid-derived prostaglandins, such as PGE2 and PGF2a (Noguchi et al., 1995). Nonenzymic peroxidation of DHA in the brain produces a large series of prostaglandin-like compounds, neuroprostanes, which may induce neuronal injury, but also have functional significance (Bernoud-Hubac et al., 2001). Another group of DHA-derived signaling compounds are the D-resolvins and protectin D1 (Weylandt et al., 2012).

Cardiovascular disease: DHA and other omega-3 fatty acids slow the development of heart disease and its sequelae (Bucher et al., 2002) through several mechanisms. Important effects include lowering elevated concentrations of both cholesterol- and triglyceride-rich lipoproteins in blood, possibly through preferential utilization (Madsen et al., 1999), decreasing platelet aggregation, and stabilizing heart rhythm (Harper and Jacobson, 2001). The latter is particularly important in older people with significant coronary pathology. A large prospective study found that DHA stores were a strong predictor for sudden death from cardiac causes (Albert et al., 2002). Contrary to some earlier expectations, raising DHA intakes moderately does not appear to increase oxidative stress in humans (Mori et al., 2000).

Cancer: DHA and other omega-3 fatty acids appear to decrease the risk of some cancers. The involved mechanisms are not well understood. Women with relatively high DHA content of their adipose tissue were found to have much lower than average breast cancer risk (odds ratio of 0.31, Maillard et al., 2002). The anticancer potential may be partially explained by the well-established ability of DHA to induce apoptosis (Chert and Istfan, 2000).

Mental health: It has been suggested that low intake of omega-3 fatty acids increases the risk of depression and suicide (Brunner et al., 2002). DHA may also affect appetite and mood through a decrease in leptin production (Reseland et al., 2001).

REFERENCES

Albert, C.M., Campos, H., Stampfer, M.J., Ridker, P.M., Manson, J.E., Willett, W.C., et al., 2002. Blood levels of long-chain n-3 fatty acids and the risk of sudden death. N. Engl. J. Med. 346, 1113–1118.

Bazan, N.G., 1990. Supply of n-3 polyunsaturated fatty acids and their significance in the central nervous system Nutrition and the Brain, vol. 8. Raven Press, New York, NY, pp. 1–24.

Bazan, N.G., Scott, B.L., 1990. Dietary omega-3 fatty acids and accumulation of docosahexaenoic acid in rod photoreceptor cells of the retina and at synapses. Ups. J. Med. Sci. Suppl. 48, 97–107.

Belch, J.J.E., Hill, A., 2000. Evening primrose oil and borage oil in rheumatologic conditions. Am. J. Clin. Nutr. 71, 352S–356S.

Bernoud-Hubac, N., Davies, S.S., Boutaud, O., Montine, T.J., Roberts II, L.J., 2001. Formation of highly reactive gamma-ketoaldehydes (neuroketals) as products of the neuroprostane pathway. J. Biol. Chem. 276, 30964–30970.

Brown, M.L., Ramprasad, M.E., Umeda, P.K., Tanaka, A., Kobayashi, Y., Watanabe, T., et al., 2000. A macrophage receptor for apolipoprotein B48: cloning, expression, and atherosclerosis. Proc. Natl. Acad. Sci. 97, 7488–7493.

Brunner, J., Parhofer, K.G., Schwandt, P., Bronisch, T., 2002. Cholesterol, essential fatty acids, and suicide. Pharmacopsychiatry 35, 1–5.

Bucher, H.C., Hengstler, P., Schindler, C., Meier, G., 2002. N-3 polyunsaturated fatty acids in coronary heart disease: a meta-analysis of randomized controlled trials. Am. J. Med. 112, 298–304.

Calvo, M., Naval, J., Lampreave, E., Uriel, J., Pineiro, A., 1988. Fatty acids bound to alpha-fetoprotein and albumin during rat development. Biochim. Biophys. Acta. 959, 238–246.

Chert, Z.Y., Istfan, N.W., 2000. Docosahexaenoic acid is a potent inducer of apoptosis in HT-29 colon cancer cells. Prostagl Leukotr Essential Fatty Acids 63, 301–308.

Clandinin, M.T., Chappell, J.E., Heim, T., Swyer, P.R., Chance, G.W., 1981. Fatty acid accretion in fetal and neonatal liver: implications for fatty acid requirements. Early Hum. Dev. 5, 7–14.

Conquer, J.A., Holub, B.J., 1997. Dietary docosahexaenoic acid as a source of eicosapentaenoic acid in vegetarians and omnivores. Lipids 32, 341–345.

Dannoura, A.H., Berriot-Varoqueaux, N., Amati, P., Abadie, V., Verthier, N., Schmitz, J., et al., 1999. Anderson's disease: exclusion of apolipoprotein and intracellular lipid transport genes. Arterioscl. Thromb. Vasc. Biol. 19, 2494–2508.

Dutta-Roy, A.K., 2000. Transport mechanisms for long-chain polyunsaturated fatty acids in the human placenta. Am. J. Clin. Nutr. 71, 315S–322S.

Dyerberg, J., Bang, H.O., 1979. Haemostatic function and platelet polyunsaturated fatty acids in Eskimos. Lancet 2 (8140), 433–435.

Edmond, J., 2001. Essential polyunsaturated fatty acids and the barrier to the brain: the components of a model for transport. J. Mol. Neurosci. 16, 181–193.

Elsing, C., Kassner, A., Stremmel, W., 1996. Effect of surface and intracellular pH on hepatocellular fatty acid uptake. Am. J. Physiol. 271, G1067–G1073.

Ferdinandusse, S., Denis, S., Mooijer, P.A., Zhang, Z., Reddy, J.K., Spector, A.A., et al., 2001. Identification of the peroxisomal beta-oxidation enzymes involved in the biosynthesis of docosahexaenoic acid. J. Lipid Res. 42, 1987–1995.

Garaulet, M., Pérez-Llamas, F., Pérez-Ayala, M., Martínez, P., de Medina, F.S., Tebar, F.J., et al., 2001. Site-specific differences in the fatty acid composition of abdominal adipose tissue in an obese population from a Mediterranean area: relation with dietary fatty acids, plasma lipid profile, serum insulin, and central obesity. Am. J. Clin. Nutr. 74, 585–591.

Goti, D., Hammer, A., Galla, H.J., Malle, E., Sattler, W., 2000. Uptake of lipoprotein-associated alpha-tocopherol by primary porcine brain capillary endothelial cells. J. Neurochem. 74, 1374–1383.

Harper, C.R., Jacobson, T.A., 2001. The fats of life: the role of omega-3 fatty acids in the prevention of coronary heart disease. Arch. Intern. Med. 161, 2185–2192.

Herrmann, T., Buchkremer, F., Gosch, I., Hall, A.M., Bernlohr, D.A., Stremmel, W., 2001. Mouse fatty acid transport protein 4 (FATP4): characterization of the gene and functional assessment as a very long chain acyl-CoA synthetase. Gene 270, 31–40.

Kris-Etherton, P.M., Taylor, D.S., Yu-Poth, S., Huth, P., Moriarty, K., Fishell, V., et al., 2000. Polyunsaturated fatty acids in the food chain in the United States. Am. J. Clin. Nutr. 71, 179S–188S.

Madsen, L., Rustan, A.C., Vaagenes, H., Berge, K., Dyroy, E., Berge, R.K., 1999. Eicosapentaenoic and docosahexaenoic acid affect mitochondrial and peroxisomal fatty acid oxidation in relation to substrate preference. Lipids 34, 951–963.

Maillard, V., Bougnoux, P., Ferrari, P., Jourdan, M.L., Pinault, M., Lavillonniere, F., et al., 2002. N-3 and N-6 fatty acids in breast adipose tissue and relative risk of breast cancer in a case–control study in Tours, France. Int. J. Cancer 98, 78–83.

Martin, G., Nernoto, M., Gelman, L., Geffroy, S., Najib, J., Fruchart, J.C., et al., 2000. The human fatty acid transport protein-1 (SLC27A1; FATP-1) cDNA and gene: organization, chromosomal localization, and expression. Genomics 66, 296–304.

Martinez, M., Vázquez, E., Garcia-Silva, M.T., Manzanares, J., Bertran, J.M., Castelló, E., et al., 2000. Therapeutic effects of docosahexaenoic acid ethyl ester in patients with generalized peroxisomal disorders. Am. J. Clin. Nutr. 71, 376S–385S.

Mori, T.A., Puddey, I.B., Burke, V., Croft, K.D., Dunstan, D.W., Rivera, J.H., et al., 2000. Effect of omega 3 fatty acids on oxidative stress in humans: GC–MS measurement of urinary F2-isoprostane excretion. Redox Rep. 5, 45–46.

Noguchi, M., Earashi, M., Minami, M., Kinoshita, K., Miyazaki, I., 1995. Effects ofeicosapentaenoic and docosahexaenoic acid on cell growth and prostaglandin E and leukotriene B production by a human breast cancer cell line (MDA-MB-231). Ontology 52, 458–464.

Reseland, J.E., Haugen, E., Hollung, K., Solvoll, K., Halvorsen, B., Brude, I.R., et al., 2001. Reduction of leptin gene expression by dietary polyunsaturated fatty acids. J. Lipid Res. 42, 743–750.

Serhan, C.N., 2014. Pro-resolving lipid mediators are leads for resolution physiology. Nature 510, 92–101.

Singh, I., Moser, A.E., Goldfischer, S., Moser, H.W., 1984. Lignoceric acid is oxidized in the peroxisome: implications for the Zellweger cerebro-hepato-renal syndrome and adrenoleukodystrophy. Proc. Natl. Acad. Sci. USA 81, 4203–4207.

Storch, J., Thumser, A.E., 2000. The fatty acid transport function of fatty acid-binding proteins. Biochim. Biophys. Acta 1486, 28–44.

Su, H.M., Moser, A.B., Moser, H.W., Watkins, P.A., 2001. Peroxisomal straight-chain acyl-CoA oxidase and D-bifunctional protein are essential for the retroconversion step in docosahexaenoic acid synthesis. J. Biol. Chem. 276, 38115–38120.

Tacken, P.J., Hofker, M.H., Havekes, L.M., van Dijk, K.W., 2001. Living up to a name: the role of the VLDL receptor in lipid metabolism. Curr. Opin. Lipid. 12, 275–279.

Weylandt, K.H., Chiu, C.Y., Gomolka, B., Waechter, S.F., Wiedenmann, B., 2012. Omega-3 fatty acids and their lipid mediators: towards an understanding of resolvin and protectin formation. Prostaglandins Other Lipid Mediat. 97, 73–82.

Yoshida, H., Mawatari, M., Ikeda, I., Imaizumi, K., Seto, A., Tsuji, H., 1999. Effect of dietary seal and fish oils on triacylglycerol metabolism in rats. J. Nutr. Sci. Vitaminol. 45, 411–421.

TRANS-FATTY ACIDS

Trans-unsaturated fatty acids (trans-fatty acids, trans-fats) contain double bonds in trans-configuration, as opposed to the cis-configured double bonds in the vast majority of natural fats and oils (Figure 5.36).

ABBREVIATIONS

CLA	conjugated linoleic acid
CoA	coenzyme A
TVA	trans-vaccenic acid

NUTRITIONAL SUMMARY

Function: Trans-fats raise LDL cholesterol concentrations.

Food sources: Most trans-fats come from industrial hydrogenation of unsaturated fats, particularly for the production of shortening and other products with long shelf life and desired physical properties.

Requirements: All consumption of foods with partially hydrogenated fat should be kept to a minimum. Commercial sales of foods with significant artificial trans-fat content are being phased out.

Deficiency: There is no known requirement or health benefit related to the consumption of trans-fats from partially hydrogenated fats.

Excessive intake: Consumption over the long term of more than immeasurable amounts of trans-fats from partially hydrogenated fats are likely to increase cardiovascular disease and cancer.

ENDOGENOUS SOURCES

Humans are not known to generate endogenously significant amounts of trans-fatty acids.

DIETARY SOURCES

More than half of the fatty acids in margarine, baked goods, and candy may be in the trans form (Enig et al., 1990). The major trans-fatty acid in hydrogenated fat is transelaidic acid (t-18:1n-9); in partially hydrogenated vegetable oils, it is trans, translinoelaidic acid (t,t-18:2n-6). It is of note that complete hydrogenation is not likely to generate trans-fatty acids. Some margarines in other countries are made

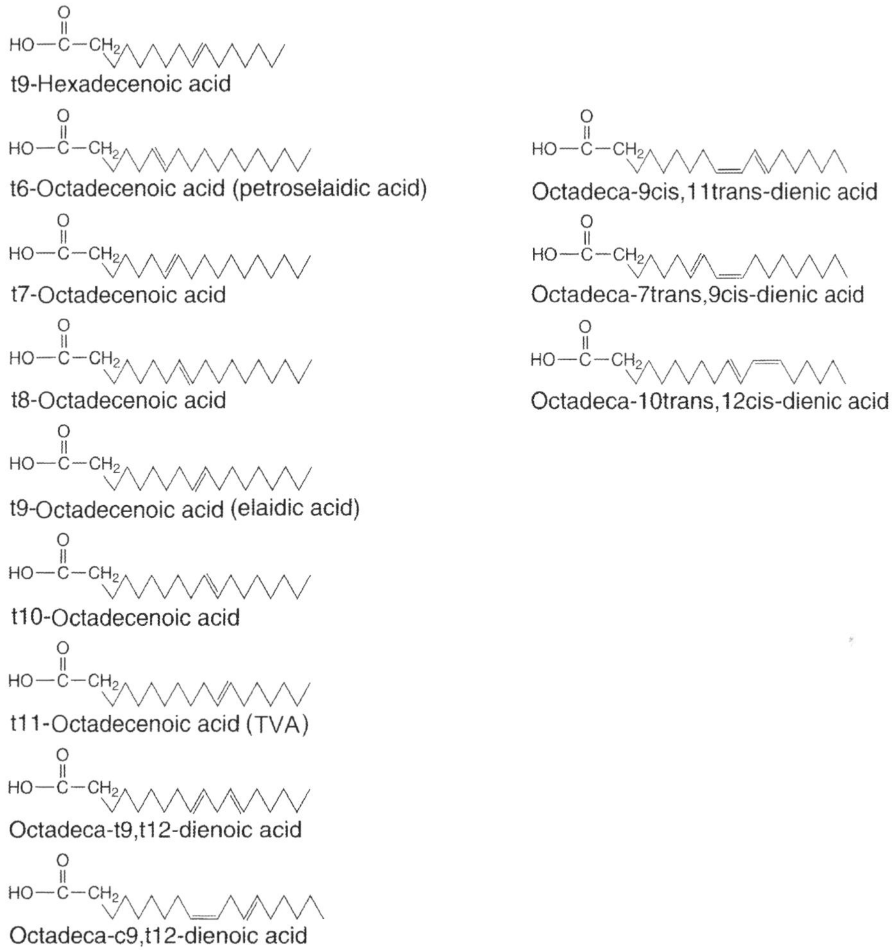

t9-Hexadecenoic acid

t6-Octadecenoic acid (petroselaidic acid)

Octadeca-9cis,11trans-dienic acid

t7-Octadecenoic acid

Octadeca-7trans,9cis-dienic acid

t8-Octadecenoic acid

Octadeca-10trans,12cis-dienic acid

t9-Octadecenoic acid (elaidic acid)

t10-Octadecenoic acid

t11-Octadecenoic acid (TVA)

Octadeca-t9,t12-dienoic acid

Octadeca-c9,t12-dienoic acid

FIGURE 5.36

Trans-fatty acids in foods.

of a mixture of fully hydrogenated and unhydrogenated oils. Typical daily intakes tend to be higher in the United States than in Europe. The Food and Drug Administration (FDA) is reclassifying partially hydrogenated oils, which account for most of the intake of artificially produced trans-fats, from their previous generally recognized as safe (GRAS) status to that of a food additive (Brownell and Pomeranz, 2014). Food additives need approval and are subject to close FDA oversight. This regulatory change can be expected to greatly curb dietary exposure of Americans to trans-fats.

Milk contains about 4% trans-fatty acids (Precht and Molkentin, 1999). Mean daily intake of trans11-octadecenoic acid (18:1t11, trans-vaccenic acid, TVA) with dairy may be around 1.3–1.8 g (Emken, 1995; Wolff, 1995).

Note: Hydrogenation also saturates the naturally occurring phylloquinone (vitamin K) in oils and appears to render it largely inactive (Booth et al., 2001).

TRANSPORT AND CELLULAR UPTAKE

Blood circulation: Trans-fatty acids in plasma are mainly bound to triglycerides and other complex molecules in lipoproteins and taken into cells with them. Muscles, liver, and adipose tissue readily take up free fatty acids through an incompletely understood mechanism.

BBB: The transfer of fatty acids in general into the brain is limited and involves largely receptor-mediated endocytosis of lipoproteins.

Materno-fetal transfer: Trans-fatty acids cross the human placenta and reach the fetus (Koletzko, 1991). The amounts and responsible mechanisms are not well understood).

METABOLISM

Mitochondrial catabolism: The trans-fatty acid (say, in this case, TVA) is activated by long-chain fatty acid CoA ligase 1 or 2 (EC6.2.1.3) and then shuttled across the inner mitochondrial membrane by the combined action of carnitine, palmitoyl-CoA:L-carnitine *O*-palmitoyltransferase I (EC2.3.1.21, on the outside), translocase, and palmitoyl-CoA:L-carnitine *O*-palmitoyltransferase II (EC2.3.1.21, on the inside). The successive actions of long-chain acyl-CoA dehydrogenase (EC1.3.99.13), enoyl-CoA hydratase, L-3-hydroxyacyl-CoA dehydrogenase, and thiolase remove the first two carbons as acetyl-CoA and generates FAD and NADH (Figure 5.37). The acyl-CoA dehydrogenase forms a complex in the mitochondrial matrix with the ETF (contains FAD), and the iron-sulfur protein ETF dehydrogenase (EC1.5.5.1, also contains FAD), which hands off the reducing equivalents to ubiquinone for oxidative phosphorylation.

Another three beta-oxidation cycles follow. At that point, the trans-double bond intermediate released at the end of the cycle (trans-3 decenyl-CoA) is positioned between carbons 3 and 4. Dodecenoyl-CoA delta-isomerase (EC5.3.3.8) moves the trans-double bond and generates an A2-enoyl-CoA, just like the acyl-CoA dehydrogenase would have done in the absence of a double bond. Due to the missed oxidation step, each double bond decreases the energy yield slightly and generates only about 12.5 ATP molecules (2.5 from NADH and 10 from acetyl-CoA) instead of the 14 (2.5 from NADH, 1.5 from FADH 2, and 10 from acetyl-CoA) with a complete cycle.

Peroxisomal catabolism: After activation by one of several available long-chain fatty acid CoA ligases (EC6.2.1.3), the beta-oxidation cycle in peroxisomes uses FAD-dependent acyl-CoA oxidase (EC1.3.3.6), peroxisomal multifunctional protein 2 (MFP2, comprising activities EC4.2.1.17, EC5.3.3.8, and EC1.1.1.35), and peroxisome-specific acetyl- CoA C-acyltransferase (3-ketoacyl-CoA thiolase, EC2.3.1.16). If beta-oxidation is continued down to trans-3 decenyl-CoA, a peroxisome-specific dodecenoyl-CoA delta-isomerase (EC5.3.3.8) is available to deal with the trans-double bond. Shorter acyl-CoA molecules tend to leave peroxisomes, and beta-oxidation is completed in mitochondria.

STORAGE

Adipose tissue contains as much as 5%. The predominant species are 18:1 trans-isomers, including those with double bonds at carbons 6–11. Minor trans-isomers in adipose tissue are those of 16:1, 18:2 (including 18:2c9t11, 18:2c9t12, 18:2c9t13, 18:2t9t12, and 18:2t9t11), and 18:3. Other tissues,

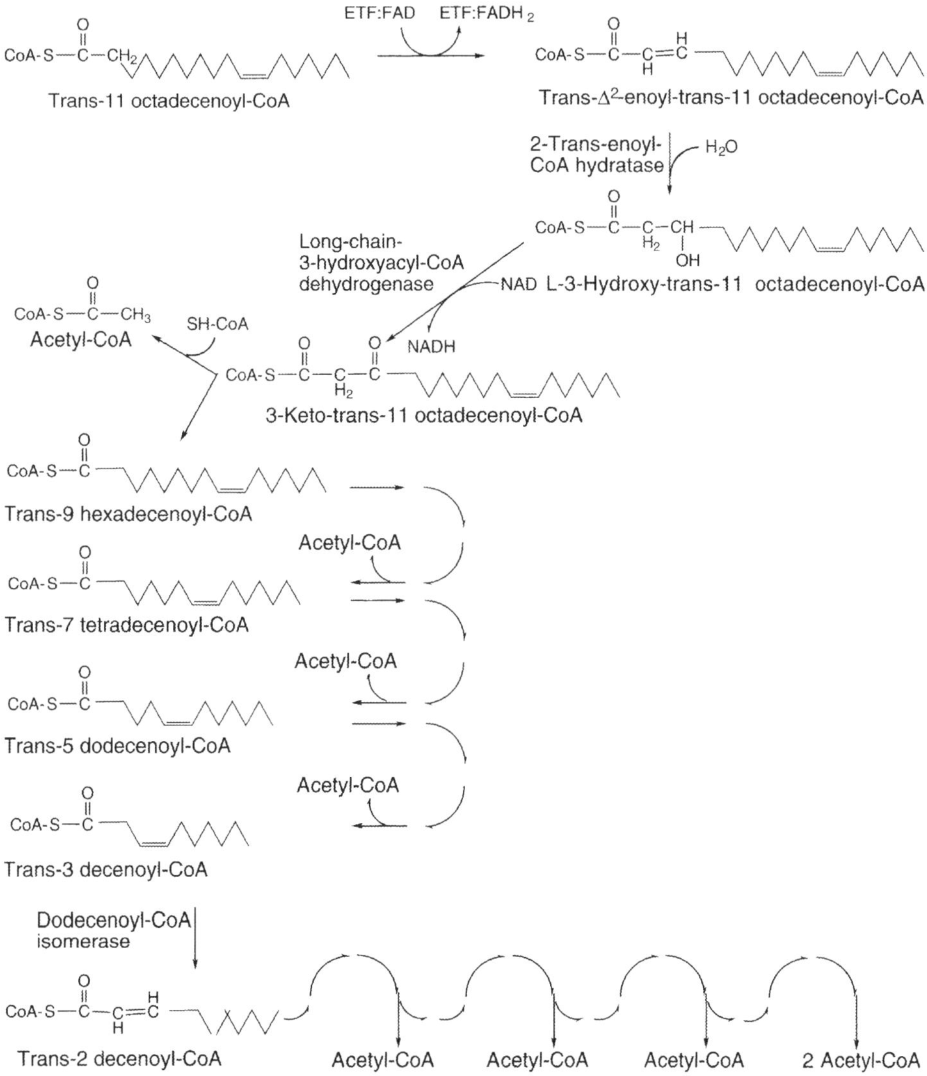

FIGURE 5.37

Mitochondrial metabolism of trans-fatty acids.

including the arteries and heart, may contain as much as 10% trans-fatty acids (Johnston et al., 1957). Trans-fatty acids are released with normal adipose tissue turnover (about 1–2% of body fat per day).

EXCRETION

As with all fatty acids, there is no mechanism that could mediate significant excretion of trans-isomers, even in a situation of great excess.

REGULATION

There is no indication for homeostatic control of trans-fatty acid concentrations in the body.

FUNCTION

Fuel energy: The energy yield of trans-fatty acids may be slightly (about 10%) less than that of saturated fatty acids. Complete oxidation of trans-fatty acids requires adequate supplies of thiamin, riboflavin, niacin, pantothenic acid, carnitine, ubiquinone, iron, and magnesium.

Conjugated linoleic acid: A significant portion of TVA may be converted endogenously to the CLA cis9, trans11-linoleic acid (Parodi, 1994; Santora et al., 2000). However, surmised health benefits of CLA, such as lower cancer risk and strengthened immune function, lack confirmation from human studies (Kelly, 2001).

Insulin resistance: Contrary to various initial observations, a meta-analysis of seven randomized, placebo-controlled clinical trials finds no consistent effect of trans-fat intake on glucose homeostasis (Aronis et al., 2012).

Inflammation: Regular consumption of significant amounts of trans-fats produced by oil hydrogenation promotes systemic inflammation (Mozaffarian et al., 2009). This does not apply to naturally occurring trans-fats, such as CLA and TVA.

Cardiovascular risk: Trans-fats are likely to promote atherosclerosis and risk of myocardial infarction and stroke through increased LDL cholesterol concentration in blood and other incompletely understood mechanisms (Brouwer et al., 2013).

REFERENCES

Aronis, K.N., Khan, S.M., Mantzoros, C.S., 2012. Effects of trans fatty acids on glucose homeostasis: a meta-analysis of randomized, placebo-controlled clinical trials. Am. J. Clin. Nutr. 96, 1093–1099.

Booth, S.L., Lichtenstein, A.H., O'Brien-Morse, M., McKeown, N.M., Wood, R.J., Saltzman, E., et al., 2001. Effects of a hydrogenated form of vitamin K on bone formation and resorption. Am. J. Clin. Nutr. 74, 783–790.

Brouwer, I.A., Wanders, A.J., Katan, M.B., 2013. Trans fatty acids and cardiovascular health: research completed? Eur. J. Clin. Nutr. 67, 541–547.

Brownell, K.D., Pomeranz, J.L., 2014. The trans-fat ban—food regulation and long-term health. N. Engl. J. Med. 370, 1773–1775.

Emken, E.A., 1995. Trans fatty acids and coronary heart disease risk. Physicochemical properties, intake and metabolism. Am. J. Clin. Nutr. 62, 659S–669S.

Enig, M., Atal, S., Keeny, M., Sampugna, J., 1990. Isomeric trans fatty acids in the US diet. J. Am. Coll. Nutr. 9, 471–486.

Gorissen, L., Raes, K., Weckx, S., Dannenberger, D., Leroy, F., De Vuyst, L., et al., 2010. Production of conjugated linoleic acid and conjugated linolenic acid isomers by Bifidobacterium species. Appl. Microbiol. Biotechnol. 87, 2257–2266.

Johnston, P.V., Johnson, O.C., Kummerow, F.A., 1957. Occurrence of trans fatty acids in human tissue. Science 126, 698–699.

Kelly, G.S., 2001. Conjugated linoleic acid: a review. Altern. Med. Rev. 6, 367–382.

Koletzko, B., 1991. Zufuhr, Stoffwechsel und biologische Wirkungen trans-isomerer Fettsäuren bei Säuglingen. Nahrung 35, 229–283.

Mozaffarian, D., Aro, A., Willett, W.C., 2009. Health effects of trans-fatty acids: experimental and observational evidence. Eur. J. Clin. Nutr. 63 (Suppl. 2), S5–S21.

Parodi, P.W., 1994. Conjugated linoleic acid: an anticarcinogenic fatty acid present in milk fat. Aust. J. Dairy Technol. 49, 93–97.

Precht, D., Molkentin, J., 1999. C18:1, C18:2 and CI8:3 trans and cis fatty acid isomers including conjugated cis delta 9, trans delta 11 linoleic acid (CLA) as well as total fat composition of German human milk lipids. Nahrung 43, 233–244.

Santora, J.E., Palmquist, D.L., Roehrig, K.L., 2000. Trans-vaccenic acid is desaturated to conjugated linoleic acid in mice. J. Nutr. 130, 208–215.

Wolff, R.L., 1995. Structural importance of the cis-5 ethylenic bond in the endogenous desaturation product of dietary elaidic acid, cis-5,trans-9 18:2 acid, for the acylation of rat mitochondria phosphatidylinositol. Lipids 30, 893–898.

CHLOROPHYLL/PHYTOL/PHYTANIC ACID

Chlorophyll is the green pigment that enables plants to capture light for photosynthesis. Chlorophyll consists of a porphyrin ring with magnesium and the phytol side chain. Phytol is a diterpene with two potential isomeric forms, (2*E*,7*R*,11*R*)-3,7,11,15-tetramethyl-2-hexadecen-1-ol (E-phytol) and (2*Z*,7*R*,11*R*)-3,7,11,15- tetramethyl-2-hexadecen-1-ol (Z-phytol). Phytanic acid is a phytol metabolite (Figure 5.38).

ABBREVIATIONS

CoA   coenzyme A
MFP2   multifunctional protein 2
SCPx   sterol carrier protein X

NUTRITIONAL SUMMARY

Function: It has been suggested that phytanic acid and other phytol metabolites specifically bind to retinoid X receptor (RXR)–like receptors and participate in the regulation of the cell cycle.

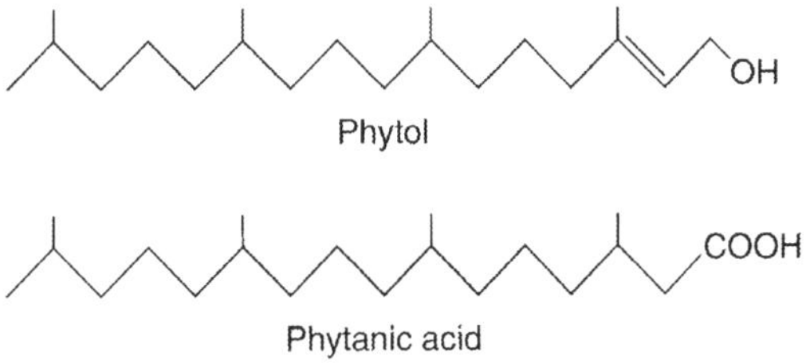

Phytol

Phytanic acid

FIGURE 5.38

Phytol and phytanic acid.

Food sources: Green foods contain varying amounts of chlorophyll; the intensity of the green coloration is a visual guide for the concentration. Plants also may contain significant amounts of phytol or phytanic acid. Dairy fat and meat fat from ruminants contain phytanic acid.

Requirements: Not known.

Deficiency: No specific symptoms or chronic disease risk has been linked to low intake.

Excessive intake: There is no indication that high intake of chlorophyll or phytol with green foods or of phytanic acid with dairy or beef is harmful except in the rare individuals with genetically defective metabolism of phytanic acid (Refsum, Zellweger, and Sjögren-Larson syndromes).

DIETARY SOURCES

Foods contain various forms of chlorophyll and degradation products from which phytol can be released by bacterial action; the percentage of phytol generated during intestinal digestion of these compounds is presumed to be low (Avigan, 1966). Fresh spinach, one of the richest dietary sources of chlorophyll, contains about 7 mg of chlorophyll per gram of dry weight (Anonymous, 1993).

Phytol is present in nuts, spinach, and coffee. Significant amounts (typically about 50–100 mg/day) of the phytol metabolite phytanic acid (3,7,11,15-tetramethyl-hexadecanoic acid; both 3D- and 3L-isomers occur naturally) are consumed with a wide variety of foods (Steinberg, 1995). Sources with particularly high concentrations include tuna in oil (0.57 mg/g dry matter), lamb (0.49 mg/g dry matter), and soya oil (0.14 mg/g) (Masters-Thomas et al., 1980).

A metabolite of phytanic acid, pristanic acid (2,6,10,14-tetramethylpentadecanoic acid; both 2D- and 2L-isomers occur naturally), is present in some foods. Another related compound is pristane (isoprenoid alkane 2,6,10,14-tetramethylpentadecane), which is a significant component (14%) of the unsaponifiable fraction in shark liver oil.

DIGESTION AND ABSORPTION

Food processing and digestion can convert chlorophyll into the pheophytin, pyropheophytin, and pheophorbide derivatives. Phytanic and pristanic acids are nearly completely absorbed from the small intestine. Pancreatic lipase (EC3.1.1.3) and phospholipases A2 (EC3.1.1.4) and B (EC3.1.1.5) in the digestive tract release phytanic and pristanic acids from dietary triglycerides or other lipids, which are then incorporated into mixed micelles. These micelles transfer their content to enterocytes through poorly understood mechanisms. Phytanic and pristanic acids can then be incorporated into triglycerides and other lipids and exported with chylomicrons, just like other long-chain fatty acids.

TRANSPORT AND CELLULAR UPTAKE

Blood circulation: Phytanic acid is transported in all major lipoprotein classes and enters cells through receptor-mediated uptake; reverse transport from tissue deposits is thought to occur via HDL (Wierzbicki et al., 1999). Blood concentrations are strongly influenced by dietary intake of dairy fat and butter (Wright et al., 2014).

BBB: The extent and mechanisms of transport of phytanic acid from blood to brain remain unclear.

Materno-fetal transfer: Very little phytanic acid appears to reach the fetus (Zomer et al., 2000).

METABOLISM

Phytol oxidation: Information on phytol conversion to phytanic acid in humans is limited and is largely extrapolated from animal experiments (Gloerich et al., 2007). In mice, an unknown alcohol dehydrogenase (ADH, EC1.1.1.1) can oxidize both phytol isomers to phytenal and microsomal fatty aldehyde dehydrogenase (FALDH, EC1.2.1.48), and then converts E-phytenal into E-phytenic acid. Acyl-CoA synthetase (EC6.2.1.3) and an unknown reductase then generate phytanoyl-CoA.

Phytanic acid alpha-oxidation: Phytanic acid is metabolized mainly in the liver and kidney, but not to a significant extent in the brain, nerves, or muscles (Figure 5.39). Complete oxidation requires

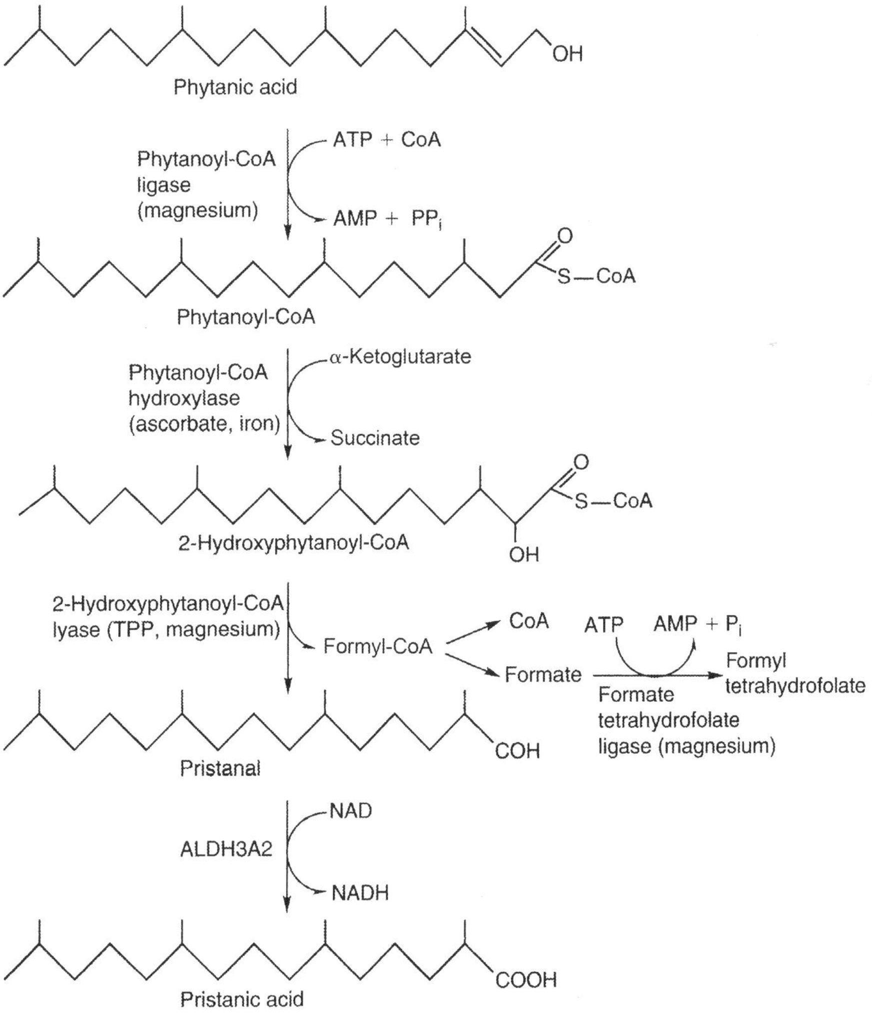

FIGURE 5.39

Oxidation of phytanic acid.

adequate supplies of ascorbate, thiamin, riboflavin, niacin, folate, vitamin B12, pantothenate, biotin, carnitine, iron, and magnesium.

Phytanic acid breakdown is initiated by activation to phytanoyl-CoA, followed by one round of alpha-oxidation (to pristanic acid) and continued by three cycles of peroxisomal beta-oxidation and at least another round of mitochondrial beta-oxidation (Verhoeven et al., 1998). Omega-oxidation of phytanate also occurs, but it is normally less important than alpha-oxidation.

Phytanoyl-CoA ligase on the cytoplasmic side of the peroxisomal membrane conjugates phytanic acid (EC6.2.1.24) in a reaction that requires ATP and magnesium. Phytanoyl-CoA is then imported into the peroxisome by an unknown mechanism.

Owing to the presence of a methyl group in the beta position of phytanic acid, beta-oxidation is blocked and metabolism proceeds by oxidation of the alpha-carbon (the one adjacent to the carbonyl conjugated to CoA) by phytanoyl-CoA dioxygenase (phytanoyl-CoA hydroxylase, EC1.14.11.18). This enzyme uses iron and ascorbate as cofactors and oxygen and alpha-ketoglutarate as cosubstrates. The resulting 2-hydroxyphytanoyl-CoA is then cleaved by 2-hydroxyphytanoyl-CoA lyase (EC4.1.-.-) into pristanal and formyl-CoA. 2-hydroxyphytanoyl-CoA lyase contains thiamin pyrophosphate as a prosthetic group and is magnesium-dependent (Foulon et al., 1999). It is possible that this reaction occurs in a microsomal compartment, and that the precursor and product are shuttled in and out of the peroxisomes. Formyl-CoA disintegrates rapidly into formate and free CoA. Formate-tetrahydrofolate ligase (EC6.3.4.3, an activity of the trifunctional protein C1-THF synthase) links formate to tetrahydrofolate in an ATP-dependent reaction. Finally, pristanal is oxidized to pristanic acid by fatty aldehyde dehydrogenase (ALDH3A2; EC1.2.1.3).

Pristanic acid beta-oxidation: Long-chain-fatty-acid-CoA ligase (EC6.2.1.3) in the cytosol conjugates pristanic acid to CoA (Figure 5.40). The ATP-binding cassette transporter ABCD3 then has to move this bulky conjugate into peroxisomes before beta-oxidation can start (Ferdinandusse et al., 2014).

The following peroxisomal beta-oxidation of pristanic acid generates propionyl-CoA using 2-methylacyl-CoA dehydrogenase (branched-chain acyl-CoA oxidase, EC1.3.99.12) and the 2-enoyl-CoA hydratase (EC4.2.1.17) and 3-hydroxyacyl-CoA dehydrogenase (EC1.1.1.35) activities of the peroxisomal multifunctional protein 2 (MFP2). The acyl CoA oxidase uses FAD as a hydrogen acceptor and is specific for 2-methyl-branched fatty acids and bile acid intermediates in human liver and kidney (Vanhove et al., 1993). 2-Methylacyl-CoA dehydrogenase reacts only with (2S)-pristanic acid; (2R)-pristanic acid has to be isomerized by peroxisomal alpha-methylacyl-CoA racemase (EC5.1.99.4) prior to further catabolism. The peroxisomal protein SCPx also has propionyl-CoA C(2)-trimethyltridecanoyltransferase (EC2.3.1.154), which allows it to release propionyl-CoA from 3-keto-pristanoyl-CoA. The next cycle of beta-oxidation generates acetyl-CoA, and another cycle produces propionyl-CoA again.

Propionyl-CoA metabolism: Prior to its breakdown, propionyl-CoA has to be ferried from peroxisomes into mitochondria where the biotin-containing enzyme propionyl-CoA-carboxylase (EC6.4.1.3) adds a carbon, methylmalonyl-CoA converts D-methylmalonyl-CoA into the L-form, and finally, methylmalonyl-CoA mutase (EC5.4.99.2, contains 5′-deoxyadenosylcobalamin) produces the Krebs cycle metabolite succinyl-CoA.

4,8-dimethylnonanoyl-CoA oxidation: After three cycles of beta-oxidation, the pristanoyl CoA molecule has been shortened to a mid-size acyl CoA that is not effectively cleaved further by the peroxisomal enzymes. 4,8-dimethylnonanoyl-CoA is linked to carnitine, probably by peroxisomai carnitine

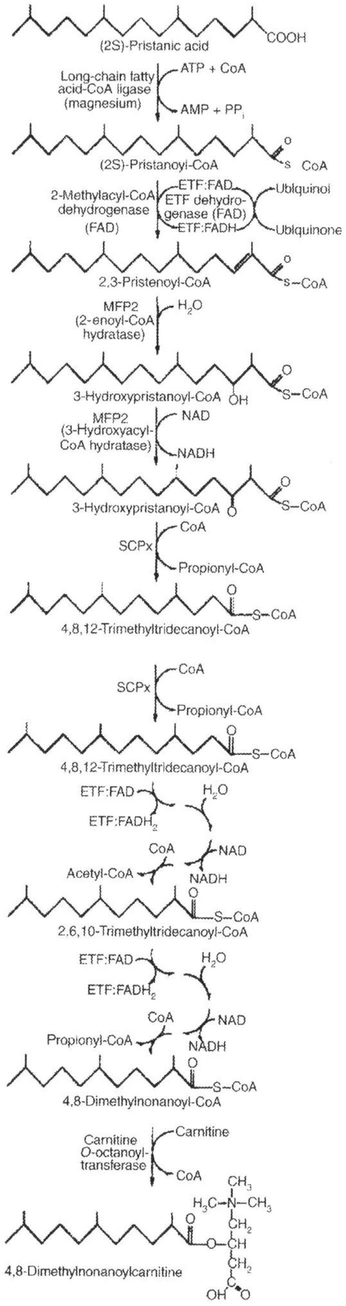

FIGURE 5.40

Oxidation of pristanic acid.

O-octanoyltransferase (EC2.3.1.137), and then exported. The C11-carnitine ester is transported across the mitochondrial inner membrane by carnitine acylcarnitine translocase and reesterified by palmitoyl-CoA:L-carnitine *O*-palmitoyltransferase 11 (EC2.3.1.21). One cycle of beta-oxidation then generates 2,6-dimethylheptanoyl-CoA, the fate of which remains to be elucidated.

STORAGE

Phytanic acid and pristanic acid are minor constituents of adipose tissue triglycerides. Release due to normal adipose tissue turnover is not likely to contribute much to total daily phytanic or pristanic acid input.

EXCRETION

It is unlikely that significant amounts of the fatty acid intermediates are excreted. Little is known about the excretion of phytol.

REGULATION

Peroxisome proliferator–activated receptor alpha (PPARalpha) activates several of the enzymes involved in phytol metabolism (Gloerich et al., 2007). It is not known whether there is any homeostatic regulation of any metabolites in the phytanic acid catabolic pathway. Phytanic acid is a transcriptional activator of liver fatty acid binding protein (L-FABP) expression and this effect is mediated via PPARalpha (Wolfrum et al., 1999). Phytanoyl-CoA dioxygenase (EC1.14.11.18) may be involved in intracellular signaling (Chambraud et al., 1999), since it is the specific target of the immunophilin FKBP52, a protein that binds to the immunosuppressant FK506.

EFFECTS

Chlorophyll and phytol derivatives from food may have some relevance in the prevention of carcinogenesis and teratogenesis, possibly through the binding of mutagens (Chernomorsky et al., 1999) or the induction of apoptosis (Komiya et al., 1999).

Nuclear hormone receptor activation: It has been suggested that phytol, phytanic acid, and related chlorophyll metabolites function as essential nutrients whose function on cellular metabolism is exerted through RXR-like receptors with narrow specificity. Phytol appears to induce RXR-dependent transcription at physiological concentrations (Kitareewan et al., 1996). Phytol has also been shown to activate uncoupling protein-1 gene transcription and brown adipocyte differentiation (Schluter et al., 2002).

One of the consequences of impaired phytanic acid metabolism (e.g., in Refsum's disease) might involve excessive activation of phytanic acid targets in the nucleus. Phytanic acid is a powerful activator of the PPAR alpha and the three retinoid X receptors (RXR) alpha, beta, and gamma.

Controlled studies have found no deodorizing effect of chlorophyll in patients with ileostomy (Christiansen et al., 1989) or urine incontinence (Nahata et al., 1983).

NUTRITIONAL ASSESSMENT

Concentrations of phytanic acid concentrations in blood are important for the management of patients with defective phytanic acid metabolism (due to Refsum disease, Zellweger syndrome, and a few other rare conditions). Concentrations in these patients reflect their dietary intake (Rüether et al., 2010). Since the realization that Refsum disease responds favorably to diet modification and lipoprotein removal continuous assessment of exposure to phytanic acid has become important. The assessment is confounded by mobilization from adipose tissue stores.

REFERENCES

Anonymous, 1993. Chlorophyll. In: Macrae, R., Robinson, R.K., Sadler, M.J. (Eds.), Encyclopedia of Food Science Food Technology and Nutrition Academic Press, London, pp. 904–911.

Avigan, J., 1966. The presence of phytanic acid in normal human and animal plasma. Biochim. Biophys. Acta 116, 391–394.

Chambraud, B., Radanyi, C., Camonis, J.H., Rajkowski, K., Schumacher, M., Baulieu, E.E., 1999. Immunophilins, Refsum disease, and lupus nephritis: the peroxisomal enzyme phytanoyl-CoA alpha-hydroxylase is a new FKBP-associated protein. Proc. Natl. Acad. Sci. 96, 2104–2109.

Chernomorsky, S., Segelman, A., Poretz, R.D., 1999. Effect of dietary chlorophyll derivatives on mutagenesis and tumor cell growth. Teratog. Carcinog. Mutagen. 19, 313–322.

Christiansen, S.B., Byel, S.R., Stromsted, H., Stenderup, J.K., Eickhoff, J.H., 1989. Nedsaetter klorofyl kolostomiopererede patienters lugtgener? Ugeskrift Laeger 151, 1753–1754.

Ferdinandusse, S., Jimenez-Sanchez, G., Koster, J., Denis, S., Van Roermund, C.W., Silva-Zolezzi, I., et al., 2014. A novel bile acid biosynthesis defect due to a deficiency of peroxisomal ABCD3. Hum. Mol. Genet. 24, 361–370.

Foulon, V., Antonenkov, V.D., Croes, K., Waelkens, E., Mannaerts, G.P., Van Veldhoven, P.E., et al., 1999. Purification, molecular cloning, and expression of 2-hydroxyphytanoyl-CoA lyase, a peroxisomal thiamine pyrophosphate-dependent enzyme that catalyzes the carbon–carbon bond cleavage during alpha-oxidation of 3-methyl-branched fatty acids. Proc. Natl. Acad. Sci. USA 96, 10039–10044.

Gloerich, J., van den Brink, D.M., Ruiter, J.P., van Vlies, N., Vaz, F.M., Wanders, R.J., et al., 2007. Metabolism of phytol to phytanic acid in the mouse, and the role of PPARalpha in its regulation. J. Lipid. Res. 48, 77–85.

Kitareewan, S., Burka, L.T., Tomer, K.B., Parker, C.E., Deterding, L.J., Stevens, R.D., et al., 1996. Phytol metabolites are circulating dietary factors that activate the nuclear receptor RXR. Mol. Biol. Cell 7, 1153–1166.

Komiya, T., Kyohkon, M., Ohwaki, S., Eto, J., Katsuzaki, H., Imai, K., et al., 1999. Phytol induces programmed cell death in human lymphoid leukemia Molt 4B cells. Int. J. Mol. Med. 4, 377–380.

Masters-Thomas, A., Bailes, J., Billimoria, J.D., Clemens, M.E., Gibberd, F.B., Page, N.G., 1980. Heredopathia atactica polyneuritiformis (Refsum's disease). 2: Estimation of phytanic acid in foods. J. Hum. Nutr. 34, 251–254.

Nahata, M.C., Slencsak, C.A., Kamp, J., 1983. Effect ofchlorophyllin on urinary odor in incontinent geriatric patients. Drug Intell. Clin. Pharm. 17, 732–734.

Rüether, K., Baldwin, E., Casteels, M., Feher, M.D., Horn, M., Kuranoff, S., et al., 2010. Adult Refsum disease: a form of tapetoretinal dystrophy accessible to therapy. Surv. Ophthalmol. 55, 531–538.

Schluter, A., Barbera, M.J., Iglesias, R., Giralt, M., Villarroya, E., 2002. Phytanic acid, a novel activator of uncoupling protein-1 gene transcription and brown adipocyte differentiation. Biochem. J. 362, 61–69.

Steinberg, D., 1995. Refsum disease. In: Scriver, C.R., Beaudet, A.L., Sly, W.S., Valle, D. (Eds.), The Metabolic and Molecular Bases of Inherited Disorders McGraw-Hill, New York, NY, pp. 2351–2369.

Vanhove, G.E., Van Veldhoven, P.E., Fransen, M., Denis, S., Eyssen, H.J., Wanders, R.J.A., et al., 1993. The CoA esters of 2-methylbranched chain fatty acids and of the bile acid intermediates di- and trihydroxy-coprostanic acids are oxidized by one single peroxisomal branched chain acyl-CoA oxidase in human liver and kidney. J. Biol. Chem. 268, 10335–10344.

Verhoeven, N.M., Wanders, R.J., Poll-The, B.T., Saudubray, J.M., Jakobs, C., 1998. The metabolism of phytanic acid and pristanic acid in man: a review. J. Inher. Metab. Dis. 21, 697–728.

Wierzbicki, A.S., Sankaralingam, A., Lumb, P.J., Hardman, T.C., Sidey, M.C., Gibberd, F.B., 1999. Transport of phytanic acid on lipoproteins in Refsum disease. J. Inher. Metab. Dis. 22, 29–36.

Wolfrum, C., Ellinghaus, P., Fobker, M., Seedorf, U., Assmann, G., Borchers, T., et al., 1999. Phytanic acid is ligand and transcriptional activator of murine liver fatty acid binding protein. J. Lipid Res. 40, 708–714.

Wright, M.E., Albanes, D., Moser, A.B., Weinstein, S.J., Snyder, K., Männistö, S., et al., 2014. Serum phytanic and pristanic acid levels and prostate cancer risk in Finnish smokers. Cancer Med. [Epub ahead of print].

Zomer, A.W., Jansen, G.A., Van Der Burg, B., Verhoeven, N.M., Jakobs, C., Van Der Saag, P.T., et al., 2000. Phytanoyl-CoA hydroxylase activity is induced by phytanic acid. Eur. J. Biochem. 267, 4063–4067.

CARBOHYDRATES, ALCOHOLS, AND ORGANIC ACIDS

CHAPTER OUTLINE

Carbohydrates 187
Glucose 191
Fructose 207
Galactose 213
Xylitol 219
Pyruvate 223
Oxalic Acid 228
Ethanol 231
Methanol 238

CARBOHYDRATES

Carbohydrates are common biological compounds consisting of carbon, oxygen, and hydrogen in ratios of 1:1:2. They form the basic building blocks of plant structures and are the main energy fuel for most people.

ABBREVIATIONS

Carbs	carbohydrates
Fru	D-fructose
Glc	D-glucose
Gal	d-α-galactose
GalNAc	Gal N-acetylglucosamine
Man	D-mannose
UDPGal	UDP-galactose

NUTRITIONAL SUMMARY

Function: Digestible carbohydrates (carbs) are used as an energy fuel, for the synthesis of glycoproteins and glycolipids, and as a general precursor for most complex organic compounds in the body.

Food sources: The carbs in plant foods are starches and sugars. Dietary fiber from plants, which chemically also are carbohydrates, are indigestible by definition and supply energy only indirectly after bacterial fermentation in the colon to short-chain fatty acids. Animal foods contain glycogen and small amounts of glucose (Glc) and other sugars.

Requirements: While there are no distinct requirements, a healthy diet should provide more than half of the energy as carbs. Much of this should come from whole-grain foods and other sources of complex, slowly digestible carbs.

Deficiency: The health risks of low carb intake remain in dispute, and depend to a large extent on the alternative energy fuel sources (protein, saturated fat, monounsaturated fat, etc.). The risk is greatest if the intake of complex carbohydrates is very low. Most often, the cited effects relate to increased atherosclerosis risk. Low intake associated with inadequate total energy intake causes starvation.

Excessive intake: The risk is greatest if high carb consumption is predominantly in the form of simple sugars and occurs in the context of excessive total energy fuel intake. High concentrations promote the nonenzymic formation of protein–sugar adducts that can interfere with protein function. Long-term dangers are increased risk of obesity, diabetes mellitus, atherosclerosis, renal disease, and retinal damage. The frequent use of simple sugars in conjunction with poor oral hygiene increases the risk of dental caries and tooth loss.

DIETARY SOURCES

Carbs are pervasive compounds in plants, where they are used as energy fuel and very extensively for extracellular structures. Only a small fraction of the carbs in nature are fit for consumption by humans. The vast majority is contained in cellulose and other indigestible macromolecules in trees, shrubs, and grasses. Digestible carbs in fruits, grains, tubers, stalks, marrows, and other plant parts are often commingled with toxic or impalatable compounds that protect plants against grazing.

SIMPLE SUGARS

Most edible sugars are based on six-carbon sugar units, all of them D-isomers. A sugar with an aldehyde residue at the end is called an *aldose;* one with a keto group at carbon 2 is called a *ketose.* Usually, the number of carbons is appended to the term indicating the type of sugar. Based on Greek number terms, *hexose* indicates six carbons and *pentose* implies five carbons. Thus, glucose is an aldohexose and fructose is a ketohexose. Among the eight possible D-aldohexoses, only three (glucose, galactose, and mannose) have significance for human nutrition or metabolism. The other five (allose, altrose, gulose, idose, and talose) are rarely or never encountered in human foods, and information on their bioavailability and health impact is limited or absent. Interference of these sugar isomers with transport and metabolism of the main dietary sugars is possible, as shown by the inhibition of glucose transport by allose (Pratt et al., 1994), possibly in its phosphorylated form (Ullrey and Kalckar, 1991). Many of the seven aldoses with shorter chains can be readily metabolized. Glyceraldehyde, after phosphorylation by triokinase (EC2.7.1.28), is readily metabolized via the glycolytic pathway. Phosphorylated erythrose and ribose are common metabolites of the pentose–phosphate pathway. Ribose has only slightly lower bioavailability than glucose (Karimzadegan et al., 1979). A relatively small percentage of ingested xylose (30–40%) is normally absorbed, and most of the absorbed sugar is rapidly excreted again with urine. These characteristics have led to the use of D-xylose as a test substance for small bowel integrity. The bioavailability of the other smaller aldoses (erythrose, threose, arabinose, and lyxose) appears to be even lower (or even minimal), and not enough is known about metabolic and health consequences of consumption. Similarly, there are four ketohexoses, of which fructose is well absorbed. Tagatose is absorbed well (81%) and is increasingly used as a low-energy sweetener because it appears to have much lower energy content than glucose (Normén et al., 2001). The other two ketohexoses, sorbose and psicose, do not appear to have nutritional value, but not enough is known to assess potential health risks. All of the four smaller ketoses can be utilized. Dihydroxyacetone can be used for glycolysis or gluconeogenesis after phosphorylation by glycerone kinase (EC2.7.1.29). Xylulose is a normal metabolic intermediate of D-glucuronate breakdown that can be activated by xylulokinase (EC2.7.1.17) and utilized through the pentose–phosphate pathway. The same goes for ribulose, which is activated by ribulokinase (EC2.7.1.16). Erythrulose is a minor product of the pentose–phosphate pathway (Solov'eva et al., 2001), but its further metabolic fate is not well understood. D-erythrulose reductase (EC1.1.1.162) uses nicotine adenine dinucleotide phosphate (NADPH) or nicotinamide adenine dinucleotide (NADH) for the reduction to erythritol (Maeda et al., 1998). Isomerization and epimerization of D-erythrose 4-phosphate to D-erythrulose 4-phosphate have also been described (Hosomi et al., 1986). The sugar alcohols sorbitol and xylitol have uses as sugar substitutes because their metabolism is not insulin-dependent. Xylitol appears to reduce the incidence of dental caries (Levine, 1998) and inner ear infections (Uhari et al., 1998).

OLIGOSACCHARIDES

Glucose α-[1 > 4] oligomers occur in many foods, especially after germination (malt) or fermentation (industrial products). Maltose contains two glucose molecules, maltotriose three, and so on. Isomaltose (Glc α-[1 > 6] Glc), alpha-limit dextrin (small mixed α-[1 > 4], and α-[1 > 6] Glc polymer) in foods are mainly fermentation products or the result of industrial processing. Trehalose (Glc α-[1 > 1] Glc) is a relatively unusual glucose dimer in human foods. A few grams of it may be ingested with a serving of some wild mushrooms (Arola et al., 1999). Yeasts and other single-cell organisms also contain trehalose. Trehalose is now used increasingly as a food additive. Sucrose (Glc α-[1 > β2] Fru) is the common sugar from sugar beets or sugarcane. Significant amounts are consumed with many sweet fruits. The sucrose in peaches may constitute as much as half of their dry weight. Lactose is a mixed dimer of galactose and glucose (the α-D-galactopyranosyl-[1 > 4] D-glucose) that is present only in milk and milk products.

DIGESTIBLE POLYSACCHARIDES

Starches constitute the bulk of carbs in tubers, grains, and many fruits and vegetables. Starches are present in specialized plant granules as glucose polymers of diverse length and degree of branching. Amylose is a largely linear α-[1 > 4] Glc polymer, amylopectin is a mixed polymer with predominantly α-[1 > 4] linkages and about 1/25 α-[1 > 6] linkages. Because these polymers are essentially locked up in granules with some water-repelling properties, unprocessed starches are not well digested, even with extensive chewing and mastication. Boiling allows water to penetrate and break open the granule envelopes and partially unfolds the starch inside. Briefly cooked (al dente) pasta and grains resist digestion more than the same foods cooked for a long time. Starches that release their glucose over a period of several hours after a meal are often referred to as *resistant starch*. It is important to remember that the processing history of the food is as important for its glucose release kinetics as its nominal glucose content. Animal-derived foods, especially liver, contain some glycogen, a glucose polymer with a mixture of α-[1 > 4] and about a quarter α-[1 > 6] linkages.

GUSTATORY PERCEPTION

Many simple sugars and related compounds have a pleasant sweet taste, which is mediated through taste receptor heterodimers ofTAS1R3 with either TAS1R1or TAS1R2 (Nelson et al., 2001).

Sweet-tasting sugars include the monosaccharides glucose, fructose, and galactose; the disaccharides sucrose and lactose; and the sugar alcohols xylitol, sorbitol, and maltitol. The potency of carbohydrates differs considerably: fructose and sucrose taste considerably sweeter than glucose (Stone and Oliver, 1969).

REFERENCES

Arola, H., Koivula, T., Karvonen, A.L., Jokela, H., Ahola, T., Isokoski, M., 1999. Low trehalase activity is associated with abdominal symptoms caused by edible mushrooms. Scand. J. Gastroent. 34, 898–903.
Hosomi, S., Nakai, N., Kogita, J., Terada, T., Mizoguchi, T., 1986. Mechanism of enzymic isomerization and epimerization of D-erythrose 4-phosphate. Biochem. J. 239, 739–743.
Karimzadegan, E., Clifford, A.J., Hill, F.W., 1979. A rat bioassay for measuring the comparative availability of carbohydrates and its application to legume foods, pure carbohydrates and polyols. J. Nutr. 109, 2247–2259.
Levine, R.S., 1998. Briefing paper: xylitol, caries and plaque. Br. Dent. J. 185, 520.
Maeda, M., Hosomi, S., Mizoguchi, T., Nishihara, T., 1998. D-erythrulose reductase can also reduce diacetyl: further purification and characterization of D-erythrulose reductase from chicken liver. J. Biochem. 123, 602–606.
Nelson, G., Hoon, M.A., Chandrashekar, J., Zhang, Y., Ryba, N.J., Zuker, C.S., 2001. Mammalian sweet taste receptors. Cell 106, 381–390.
Normén, L., Lærke, H.N., Jensen, B.B., Langkilde, A.M., Andersson, H., 2001. Small-bowel absorption of D-tagatose and related effects on carbohydrate digestibility: an ileostomy study. Am. J. Clin. Nutr. 73, 105–110.
Pratt, S.E., Colby-Germinario, S., Manuel, S., Germinario, R.J., 1994. Evidence that modulation of glucose transporter intrinsic activity is the mechanism involved in the allose-mediated depression of hexose transport in mammalian cells. J. Cell Physiol. 161, 580–588.
Solov'eva, O.N., Bykova, I.A., Meshalkina, L.E., Kovina, M.V., Kochetov, G.A., 2001. Cleaving of ketosubstrates by transketolase and the nature of the products formed. Biochemistry 66, 932–936.
Stone, H., Oliver, S.M., 1969. Measurement of the relative sweetness of selected sweeteners and sweetener mixtures. J. Food Sci. 34, 215–222.
Uhari, M., Kontiokari, T., Niemela, M., 1998. A novel use of xylitol sugar in preventing acute otitis media. Pediatrics 102, 879–884.
Ullrey, D.B., Kalckar, H.M., 1991. Search for cellular phosphorylation products of D-allose. Proc. Natl. Acad. Sci. USA 88, 1504–1505.

GLUCOSE

D-Glucose (Glc; molecular weight 180) is an aldohexose, often present as a component of polymeric and other complex structures (Figure 6.1).

ABBREVIATIONS

CoA	coenzyme A
Fru	D-fructose
Glc	D-glucose
Gal	d-α-galactose
GalNAc	Gal N-acetylglucosamine
GLUT1	glucose transporter 1 (SLC2A1)
GLUT2	glucose transporter 2 (SLC2A2)
GLUT4	glucose transporter 4 (SLC2A4)
GLUT5	glucose transporter 5 (SLC2A5)
Man	D-mannose
MCT1	proton/monocarboxylic acid cotransporter 1 (SLC16A1)
MG	macroglycogen
PG	proglycogen
SGLT1	sodium-glucose cotransporter 1 (SLCSA1)
SGLT2	sodium-glucose cotransporter 2 (SLCSA2)

NUTRITIONAL SUMMARY

Function: Glucose (Glc) is used as an energy fuel, for the synthesis of glycoproteins and glycolipids, and as a general precursor for most complex organic compounds in the body.

Food sources: Glc is the exclusive constituent of starches, maltodextrin, maltose, isomaltose, and trehalose from plant foods and glycogen from animal foods. It is combined with other monosaccharides in sucrose and lactose.

Requirements: A healthy diet should provide at least 130 g/day of carbohydrate, and that means some form of glucose (Food and Nutrition Board, 2002).

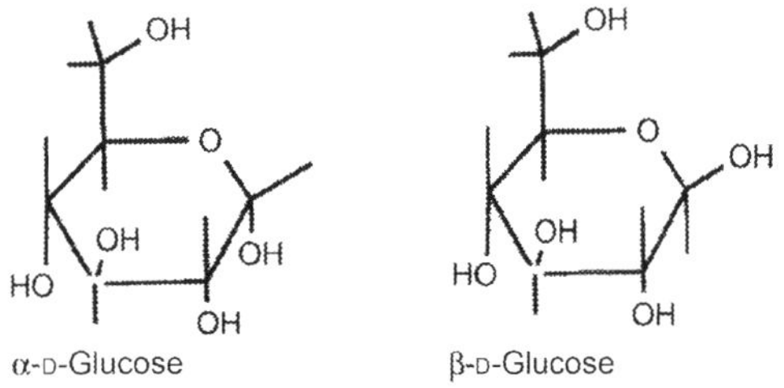

α-D-Glucose   β-D-Glucose

FIGURE 6.1

D-Glucose and L-glucose.

Deficiency: Since Glc constitutes the bulk of carbohydrates in adults, the potential health risk of low intake (though this is disputed) is that of low total carbohydrate intake. One obvious risk is due to over-all low energy consumption (starvation, anorexia). A different situation concerns the long-term effects of an energy-balanced, low-carbohydrate diet with a correspondingly higher proportion of proteins and fats. Concerns with such diets are largely about potential risks of the high protein and/or fat intake, not the low carbohydrate content *per se*. Moderately low blood Glc concentration (hypoglycemia) induces hunger, sweating, tachycardia, and dizziness. Severe hypoglycemia may cause loss of consciousness, coma, and organ damage.

Excessive intake: High Glc consumption is most likely to be harmful, if it causes total energy intake to exceed expenditure and body fatness to increase. The consequences of obesity are well known and include higher risk of hypertension, hyperlipidemia, diabetes mellitus, cardiovascular disease, and shortened life span. Glc intake needs to be most carefully balanced in people with diabetes mellitus because of the high potential for damage from high blood and tissue concentrations and the risks related to metabolic decompensation. Simple sugars promote dental caries and tooth loss, especially if dental hygiene is not meticulous.

ENDOGENOUS SOURCES

Glc can be synthesized from a wide variety of intermediary metabolites in foods including glycolytic metabolites (glycerol, glyceraldehyde-3-phosphate, 3-phosphoglycerate, 2,3-diphosphoglycerate, pyruvate, and lactate), glucogenic amino acids (especially alanine), and tricarboxylic cycle intermediates (oxaloacetate, α-ketoglutarate, citrate, isocitrate, succinate, fumarate, and malate). Fructose (Fru), galactose (Gal), mannose (Man), and other sugars can be converted, mainly in the liver and kidney, into Glc.

Gluconeogenesis: The liver and the kidneys have the largest capacity for glucose synthesis from lactate, protein-derived precursors, or glycerol (from triglyceride hydrolysis). The Krebs cycle intermediate oxaloacetate is the common intermediate for Glc synthesis from lactate and amino acids. Glc synthesis from glycerol joins the shared pathway at the level of dihydroxyacetone phosphate. It should be noted that only about 90 g of Glc could be generated from 1 kg of fat. NAD-dependent L-lactate dehydrogenase (EC1.1.1.27) oxidizes lactate to pyruvate, which can be converted by the biotin-containing enzyme pyruvate carboxylase (EC6.4.1.1) into oxaloacetate. When protein is broken down in muscle and other tissues during times of need, the flux of alanine and glutamine to the liver and kidneys increases. Alanine is transaminated to pyruvate, which gives rise to oxaloacetate as just described. Five of the Krebs cycle reactions convert the glutamate metabolite α-ketoglutarate into oxaloacetate. The rate-limiting step then is the phosphorylating decarboxylation of oxaloacetate by guanosine triphosphate (GTP)–dependent phosphoenolpyruvate carboxykinase (PEPCK, EC4.1.1.32). The isoenzymes in cytosol (PEPCK1) and in mitochondria (PEPCK2) are genetically distinct. The mitochondrial enzyme requires manganese as a cofactor. All but two of the remaining nine steps are the same as for Glc breakdown (glycolysis) and are described in slightly more detail later in this chapter. Synthesis of the Glc precursor glucose 6-phosphate proceeds via reactions catalyzed by phosphopyruvate hydratase (EC4.2.1.11), phosphoglycerate mutase (EC5.4.2.1), phosphoglycerate kinase (EC2.7.2.3), glyceraldehyde 3-phosphate dehydrogenase (EC1.2.1.12), triose isomerase (EC5.3.1.1), fructose-bisphosphate aldolase (EC4.1.2.13), fructose 1,6-bisphosphatase (EC3.1.3.11), and glucose 6-phosphate isomerase (EC5.3.1.9). The key gluconeogenic enzyme fructose 1,6-bisphosphatase is inhibited by fructose 2,6-bisphosphate. A bifunctional regulatory protein comprises both the synthetic activity of 6-phosphofructo-2-kinase (EC2.7.1.105) and the opposite activity of fructose-2,6-bisphosphate 2-phosphatase (EC3.1.3.46). At least four different genes code for isoenzymes in the liver, heart, brain, and testis. Phosphorylation brings out the 6-phosphofructo-2-kinase activity and abolishes fructose 2,6-bisphosphate 2-phosphatase activity. Dephosphorylation switches those activities again. The glucose 6-phosphatase (EC3.1.3.9, zinc-containing) then completes Glc synthesis. Glc synthesis from glycerol, which is of particular importance in fasting and starvation, starts with the rate-limiting step of activation by glycerol kinase (EC2.7.1.30). Three genes encode distinct isoenzymes, and alternative splicing generates additional tissue-specific isoforms. The resulting glycerol 3-phosphate can then be converted into dihydroxyacetone phosphate by NAD-dependent glycerol 3-phosphate dehydrogenase (EC1.1.1.8) in cytosol or by the flavin adenine dinucleotide (FAD)–containing glycerol 3-phosphate dehydrogenase (EC1.1.99.5) complex at the mitochondrial membrane. This latter mitochondrial complex transfers reducing equivalents via ubiquinone directly to the electron-transport chain for oxidative phosphorylation. The last five gluconeogenic enzymes described above can then complete Glc synthesis from dihydroxyacetone phosphate (Figure 6.2).

DIETARY SOURCES

Bioavailable food sources are monomeric Glc, oligosaccharides and polysaccharides containing only Glc (starches, maltodextrin, maltose, isomaltose, and trehalose in plant foods, and glycogen in animal foods), and the mixed disaccharides lactose (α-D-galactopyranosyl-[1 > 4) D-glucose] and sucrose [Glc α-(1 > β2] fructose). About two-thirds of carbohydrate intake in developed countries is Glc. Glc is combined with other monosaccharides in sucrose and lactose. Amylose, the α-(1 > 4) Glc polymer with a molecular weight of around 60 kD, typically comprises about 20% of starch in plants and is thus consumed with many foods. Amylose provides 8–10% of the energy in a mixed diet in the United States. Amylopectin, a mixed α-[1 > 4] and (about 1/25) α-[1 > 6] Glc polymer, typically accounts for 80% of the starch in plant foods and provides nearly half of the food energy in the United States. Glycogen, a mixture of Glc polymers with mainly α-[1 > 4] and some α-[1 > 6] Glc linkages, is consumed with meats (typically <3 mg/g) and liver (about 30 mg/g). While trehalose (Glc α-[1-α] Glc) is ubiquitous in nature, the amounts consumed with common foods are small. A typical serving of mushrooms does not contain more than 6 g (Arola et al., 1999). Yeast and other single-cell foods are other food sources containing more than minimal amounts. Trehalose is also used in some countries as a food additive.

Glc is constituent of many other plant oligosaccharides and polysaccharides, such as cellulose, raffinose, stachyose, and verbascose, which are not at all or only partially hydrolyzed by humans but which have their own distinct effects on the digestive tract, nutritional status, and health. Fermentation of such dietary fiber by intestinal bacteria can generate methane and hydrogen gas and cause abdominal discomfort and flatulence. In appropriate quantities, however, dietary fiber can reduce cancer and cardiovascular risk and promote normal bowel movements.

DIGESTION AND ABSORPTION

Digestion: Alpha-amylase (EC3.2.1.1), both from salivary gland and pancreas, cleaves [1 > 4] Glc bonds in starch, glycogen, and similar polysaccharides and oligosaccharides. Amylose is not well digestible unless cooked, steeped, or thoroughly chewed and wetted because otherwise, the starch

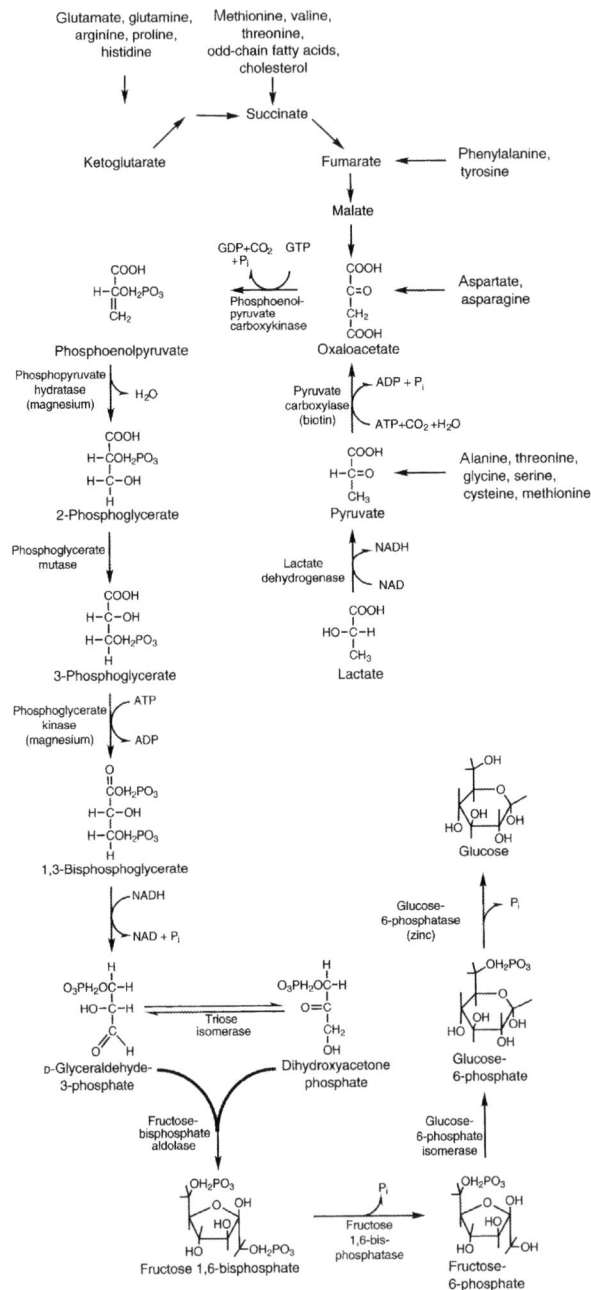

FIGURE 6.2

Gluconeogenesis from endogenous and exogenous precursors.

granules in foods remain inaccessible for digestion by alpha-amylase; most cooked amylose is digested and absorbed. Alpha-amylase produces a mixture of maltotriose, maltose, glucose, and oligomers (isomaltose and alpha-limit dextrins) containing both 1,4- and 1,6-α-D-glucosidic bonds. Maltase-glucoamylase (EC3.2.1.20/EC3.2.1.3) hydrolyzes terminal nonreducing 1,4-linked glucose residues. The brush border sucrase/isomaltase (EC3.2.1.48/EC3.2.1.10) complex finally hydrolyzes the 1,6-α-D-glucosidic bonds of branched oligomers, as well as the 1,6-α-D-glucosidic bonds in maltose and sucrose.

Lactose and cellobiose are cleaved by the brush border enzyme lactase (EC3.2.1.108), a β-glucosidase residing mainly on the microvillar tips of the proximal small intestine. Trehalose from mushrooms, yeast, and other single-cell food sources is cleaved by the α-glucosidase trehalase (EC3.2.1.28), another intestinal brush border enzyme (Murray et al., 2000).

Epidermal growth factor (EGF), which binds to a specific membrane receptor, exerts control over the expression of the intestinal brush border enzymes, promoting lactase expression upon feeding after birth. Expression of the other two enzymes is low during infancy, and all three enzymes are repressed in epithelial cells of the distal small intestine and colon (reference values for mucosal disaccharidase activities in children can be found in Gupta et al., 1999). About 1 in 500 North Americans does not express sucrase. Trehalase deficiency is not common except in Greenland Inuit, 10–15% of whom do not express this enzyme. Nonetheless, some Caucasians suffer mild abdominal discomfort when eating trehalose-rich mushrooms (Arola et al., 1999). Intestinal lactase activity usually declines within a few years to a small percentage of infancy values. Persistently high lactase expression is found in people of North European descent and some North African, Arabian, and Asian populations (Harvey et al., 1998). It may be worth mentioning that smoking appears to decrease the activities of both lactase and trehalase (Kaura et al., 2001). The main biological significance of intestinal trehalase may be the inactivation of trehalose-containing compounds of pathogens (e.g., trehalose-6,6'-dimycolate in *Mycoplasma tuberculosis*). The energy yield of glucose from digested trehalose may be much less important in comparison.

Unabsorbed oligosaccharides in legumes and other vegetables often cause flatulence due to metabolism to methane by intestinal bacteria. Food-grade preparations of α-galactosidase are now commercially available (Ganiats et al., 1994). This enzyme cleaves galactose from raffinose, stachyose, and verbascose, and the residual sucrose can then be digested by brush border sucrase.

Absorption: Glc is actively taken up into enterocytes along with two sodium ions and 210 water molecules by the sodium-glucose cotransporter (SGLT1, SLC5A1) (Figure 6.3). Uptake of 100 g of glucose, galactose, or both via this transporter thus transfers 25 g of sodium ions and 2.1 l of water from the lumen into and across the intestinal mucosa. A regulatory protein, RSC1A1, inhibits SGLT1 activity. The facilitative glucose transporter 2 (GLUT2, SLC5A2) provides an additional (though minor) entry route (Helliwell et al., 2000). Most Glc leaves enterocytes again rapidly and is transferred into portal blood via the facilitative glucose transporter 2 (GLUT2, SLC5A2, Levin, 1994). GLUT1 (SLC2A1) is also present on the basolateral side of intestinal cells and transports Glc, though the extent of its contribution is uncertain (Pascoe et al., 1996). Weak efflux is supported by the uniporter SLC50A1 (Wright, 2013), fructose via GLUT5 (SLC2A5).

Enterocytes use some of the ingested Glc to meet their own energy and growth needs. When the intestinal lumen is empty and intracellular Glc concentration declines, the direction of the flux across the basolateral membrane reverses, and Glc moves from capillary blood into the cells.

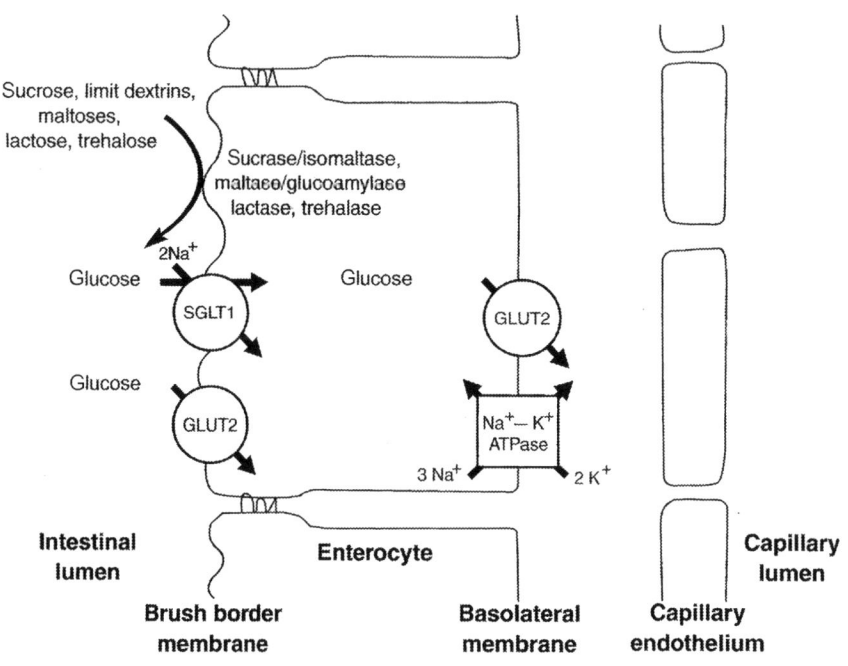

FIGURE 6.3

Intestinal absorption of glucose.

TRANSPORT AND CELLULAR UPTAKE

Blood circulation: Glc is dissolved in plasma in its free form. Typical blood concentrations are around 1000 mg/l and vary normally by 50% or more. Current blood Glc concentration depends on composition of recent meals, time elapsed since recent intakes, and the action of insulin and other hormones. Several facilitative glucose transporters have been identified that mediate Glc uptake into cells (Shepherd and Kahn, 1999), in some instances also in the reverse direction.

GLUT1 is the constitutive conduit for Glc entry into most cells. Both GLUT1 (SLC2A1) and GLUT4 (SLC2A4) are present in muscle and adipose tissue. In these tissues, GLUT1 provides a constant low influx of Glc, while GLUT4 can accommodate much higher transport rates upon stimulation by insulin. Liver imports Glc via GLUT1, but it uses GLUT2 for export in arterial blood (Nordlie et al., 1999). Glc uptake by erythrocytes is mediated at least in part by a chloride/bicarbonate anion exchanger (band 3 of red cell membrane, SLC4A1). Glucose transporters with more limited tissue distribution include the brain and neuron-specific GLUT3, GLUT6 (in the brain, leukocytes, and adipose tissue), GLUT8 (mainly in testicular cells and less in skeletal muscle and the heart, small intestine, and brain), GLUT10 in the human heart, lung, brain, liver, skeletal muscle, pancreas, placenta, and kidney (Dawson et al., 2001), and GLUT11 in skeletal and heart muscle (Sasaki et al., 2001). GLUT7 is the transporter for glucose out of the endoplasmic reticulum after its dephosphorylation by glucose-6-phosphatase (EC3.1.3.9).

Blood–brain barrier (BBB): GLUT1 is the main Glc transporter on both sides of brain capillary barrier epithelial cells. However, the absence of GLUT1 is not fatal (Boles et al., 1999), and some Glc may reach the brain through other carriers or channels. The capillary endothelial cells in a limited area of the ventromedial hypothalamus contain GLUT4 in addition to GLUT1. This transporter is thought to link the bloodstream to glucose-sensing neurons (Ngarmukos et al., 2001) but is not a quantitatively important mechanism of Glc transfer into the brain.

Materno-fetal transfer: GLUT1 is the main Glc carrier on both sides of the syntrophoblast (Illsley, 2000). GLUT3 and additional transporters are present in the placenta but may be more important for nutriture of the placenta itself than for transport to the fetus. Expression of GLUT1 in the placenta is limited for the remainder of the pregnancy when maternal blood Glc concentrations are high in the first trimester (Jansson et al., 2001).

METABOLISM

There are two major metabolic pathways for the utilization of Glc. The main route is the glycolytic pathway, which proceeds via pyruvate and acetyl-CoA to the Krebs cycle or feeds various synthetic pathways through its intermediate metabolites. Complete oxidation of Glc through this route yields 10 NADH, 2 reduced ubiquinones, and 4 ATP/GTP. If there is not enough oxygen (anaerobic conditions) for NADH utilization, Glc metabolism can be terminated at the level of lactate without a net production of NADH. Glc breakdown via glycolysis and the Krebs cycle is the staple of muscles and most other cells. The pentose–phosphate cycle (hexose monophosphate shunt) removes 1 carbon from Glc with each cycle. This pathway is particularly important for rapidly growing cells because it generates 2 NADPH (used for many synthetic pathways) with each cycle and provides ribose for DNA and RNA synthesis. Red blood cells also depend largely on the pentose–phosphate cycle for their fuel metabolism.

Glycolysis: The initial phosphorylation of glucose is catalyzed by hexokinase (HK, EC2.7.1.1) on the outer mitochondrial membrane (Figure 6.4). Four genes encode HK that are present at different levels in most tissues. Alternative splicing of HK 4 (glucokinase) produces two liver-specific isoforms and a pancreas-specific one. Alternative promoters respond selectively to insulin (liver isoforms) or glucose (pancreas isoform). The large number of genes and isoforms and their different characteristics are commensurate to the diverse needs in different tissues that can be met by a finely tuned mixture. The product, glucose 6-phosphate, allosterically inhibits all of these forms. An alternative for Glc phosphorylation in the liver is a nonclassical function of the zinc-enzyme glucose 6-phosphatase (EC3.1.3.9). This is actually a multicomponent complex embedded in the endoplasmic reticulum membrane, which comprises both complex catalytic activities and at least four distinct substrate transport properties. Glucose 6-phosphatase can use both carbamyl-phosphate and pyrophosphate as phosphate donors (Nordlie et al., 1999). The next steps depend on glucose 6-phosphate isomerase (EC5.3.1.9) and phosphofructokinase-1 (EC2.7.1.11). Phosphofructokinase is activated by the regulatory metabolite fructose 2,6-bisphosphate described previously. The resulting fructose 1,6-bisphosphate is cleaved into three-carbon molecules by fructose-bisphosphate aldolase (aldolase, EC4.1.2.13), a key regulatory enzyme for glycolysis that is activated by AMP, adenosine diphosphate (ADP), and fructose bisphosphate and inhibited by the downstream products citrate and ATE. Three different genes code for the main forms of the latter in the muscle (aldolase A), liver (aldolase B), and brain (aldolase C), and additional isoforms are generated by alternative splicing. Triose isomerase (EC5.3.1.1) converts dihydroxyacetone phosphate into glyceraldehyde 3-phosphate in a near-equilibrium reaction. Glyceraldehyde 3-phosphate dehydrogenase/phosphorylating (GAPDH, EC1.2.1.12) for the following oxidizing reaction exists as muscle and liver forms encoded by different genes. Metabolism to pyruvate continues with phosphoglycerate kinase (EC2.7.2.3, ubiquitous and testis specific forms), phosphoglycerate mutase (EC5.4.2.1, three

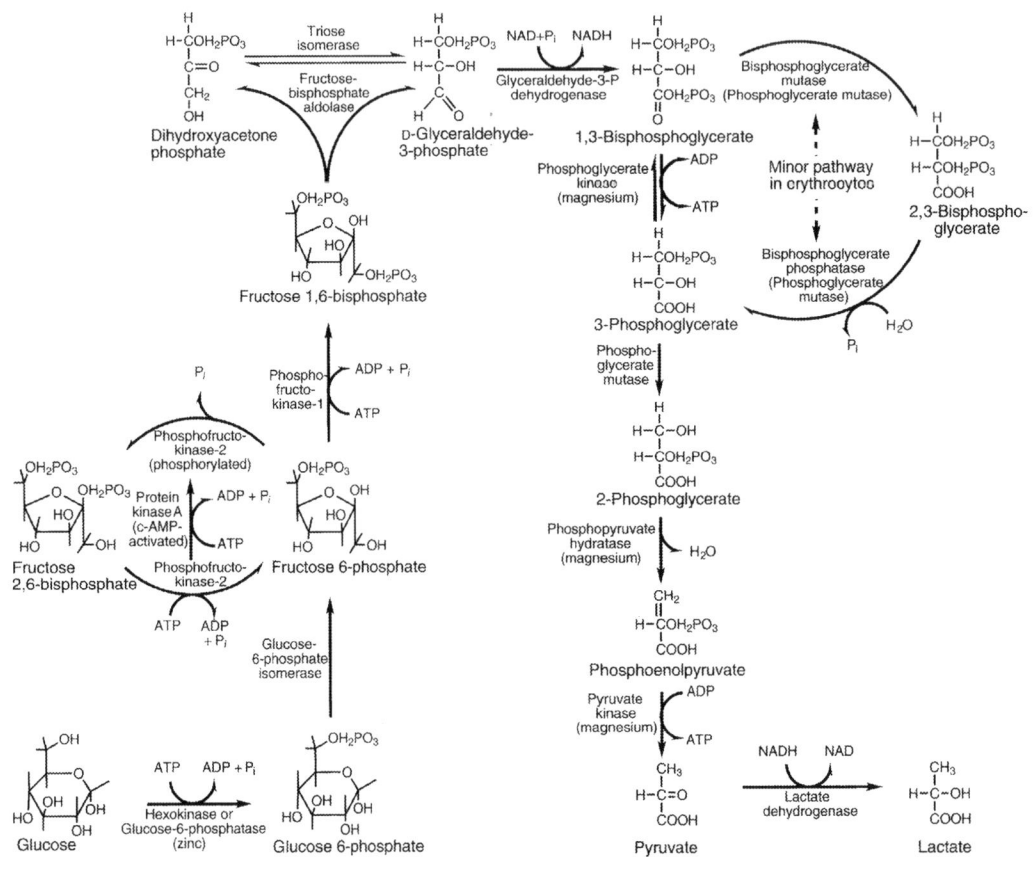

FIGURE 6.4

Glycolysis encompasses the initial anaerobic steps of glucose metabolism.

different isoenzymes for muscle, erythrocytes, and other tissues), phosphopyruvate hydratase (enolase, EC4.2.1.11, magnesium dependent, multiple isoenzymes encoded by at least four genes), and pyruvate kinase (EC2.7.1.40, multiple isoenzymes due to three genes and alternative splicing).

Anaerobic metabolism: The capacity for adenosine triphosphate (ATP) production is more likely to be limited by the availability of oxygen for oxidative phosphorylation than by the availability of oxidizable substrate. This is typical for intense short-term exercise. Muscles can metabolize anaerobically, though with a much smaller energy yield than with aerobic metabolism. In this case, pyruvate is reduced to lactate by L-lactate dehydrogenase (EC1.1.1.27), providing a renewed supply of oxidized nicotinamide adenine dinucleotide (NAD) for continued glycolysis. Two molecules of ATP and two lactates can be produced anaerobically from one glucose molecule. The protons arising from the lactate production increase intracellular acidity and help to push out excess lactate via the proton/monocarboxylic acid cotransporter 1 (MCT1, SLC16A1). Lactate is readily taken up by the liver, used for Glc synthesis, and returned into circulation and muscle again as needed. This shuttling of lactate and Glc

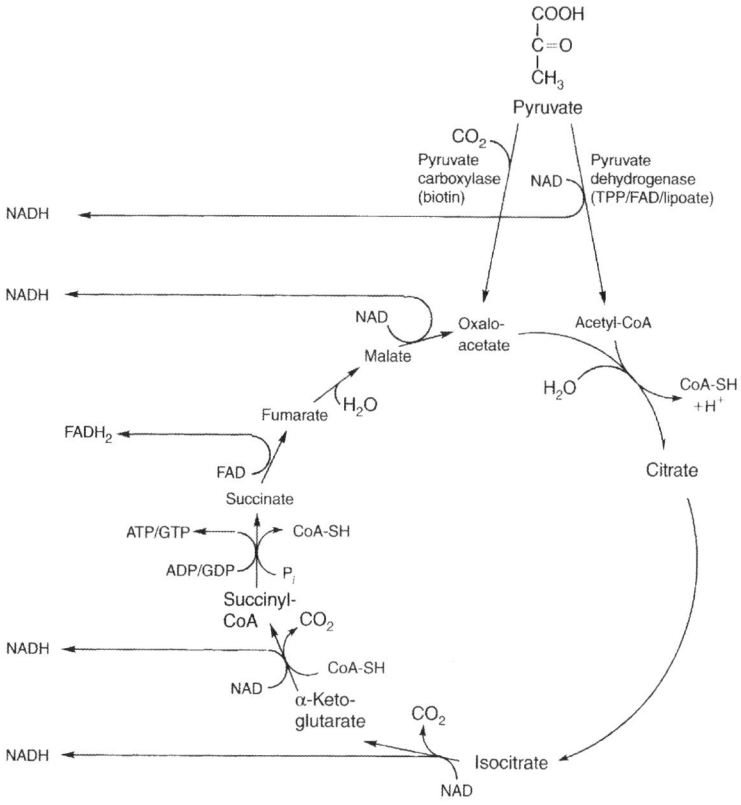

FIGURE 6.5

The aerobic part of glucose metabolism starts with the transfer of pyruvate into mitochondria.

between muscle and liver (the Cori cycle) allows individual muscles to continue working without the burden of metabolic liabilities from previous bouts of exercise.

Aerobic metabolism: Pyruvate is transported across the inner mitochondrial membrane by pyruvate translocase, where the enzymes for oxidative metabolism reside (Figure 6.5). If there is an adequate supply of oxygen, pyruvate is metabolized by the pyruvate dehydrogenase (EC1.2.4.1) complex to acetyl-CoA in an NADH-producing reaction. The multisubunit enzyme complex requires thiamin pyrophosphate (bound to the EI subunits), lipoate (bound to the E2 subunits, dihydrolipoamide *S*-acetyltransferase, EC2.3.1.12), and FAD (bound to the E3 subunits, dihydrolipoamide dehydrogenase, EC1.8.1.4). Phosphorylation of serines in the E1 subunit by [pyruvate dehydrogenase (lipoamide)] kinase (EC2.7.1.99) inactivates the enzyme complex. The dephosphorylation by [pyruvate dehydrogenase (lipoamide)]-phosphatase (EC3.1.3.43) activates it again.

Pentose–phosphate pathway: This alternative pathway for Glc metabolism is especially important for rapidly dividing tissues, because it generates ribose 5-phosphate, which is the sugar precursor for DNA and RNA synthesis, and NADPH, which is used by many biosynthetic pathways (Figure 6.6).

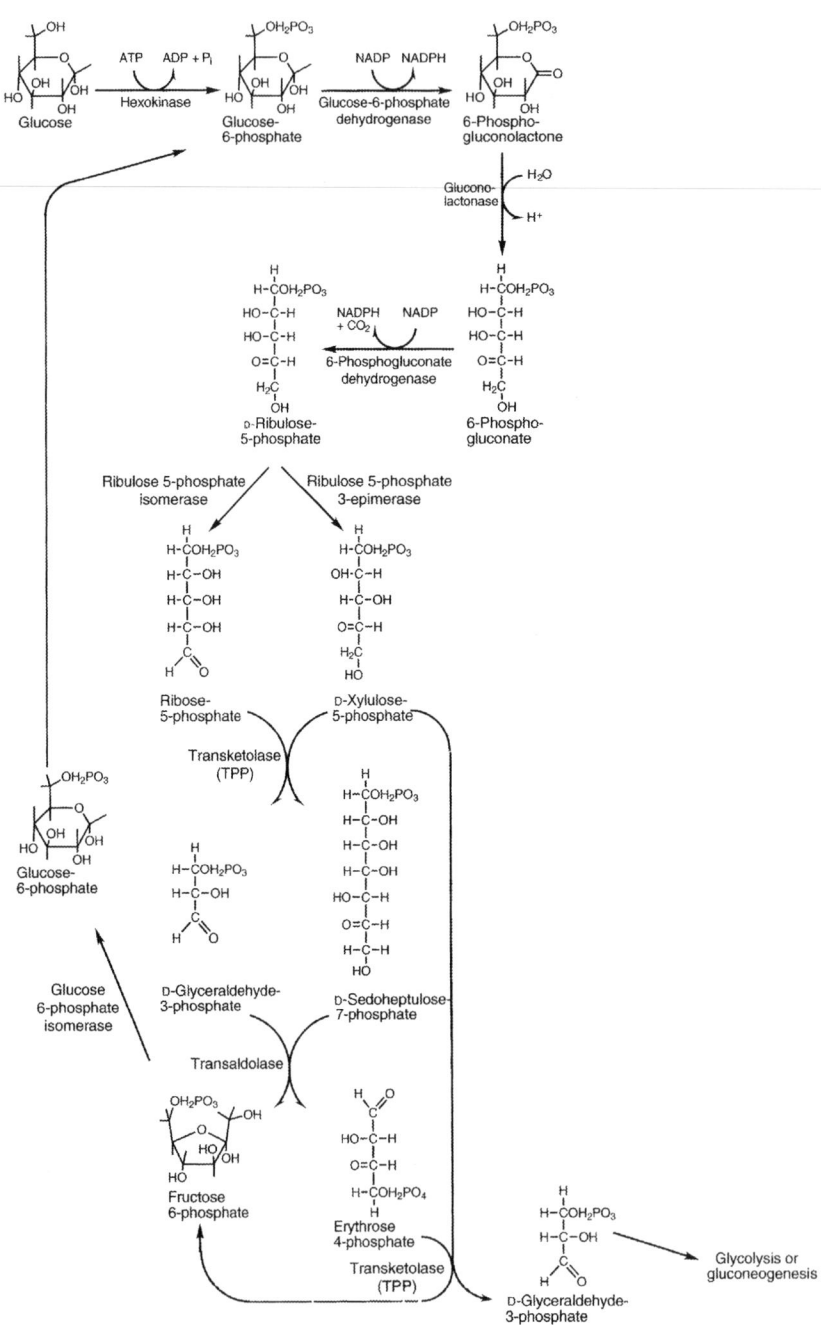

FIGURE 6.6

The pentose–phosphate pathway is a major source of NADPH.

NADPH is essential for the reduction of oxidized glutathione in erythrocytes. Reduced NADPH availability (typical with glucose 6-phosphate dehydrogenase deficiency) increases erythrocyte vulnerability to oxidative stress and tendency for hemolysis. The ingestion of the pyrimidine aglycone divicine with *Vicia fava* beans (or exposure to their pollen) in individuals with glucose 6-phosphate dehydrogenase deficiency induces oxidative modification of hemoglobin and may precipitate an acute hemotoxic crisis in them (McMillan et al., 2001). After phosphorylation of Glc by hexokinase (EC2.7.1.1) as described above, the successive actions of glucose 6-phosphate dehydrogenase (EC1.1.1.49), gluconolactonase (EC3.1.1.17), and 6-phosphogluconate dehydrogenase (EC1.1.1.44) generate the pentose ribulose 5-phosphate. The first and third reactions generate NADPH. Transketolase (EC2.2.1.1) with covalently bound thiamin-pyrophosphate catalyzes two rearrangement reactions. One of these converts two pentose phosphates (X5P and ribose 5-phosphate) into a set of compounds with seven (D-sedoheptulose 7-phosphate) and three (glyceraldehyde 3-phosphate, GAP) carbons. The other one rearranges X5P plus erythrose 4-phosphate into glyceraldehyde 3-phosphate and fructose 6-phosphate (F6P). A third possibility is the rearrangement of two X5P molecules into two GAP molecules and one erythrulose (Bykova et al., 2001). The same reactions are catalyzed by transketolase 2 (Coy et al., 1996), with different isoforms in the brain and heart generated by alternative splicing. Variants of the transketolase 2 gene may be implicated in the pathogenesis of Wernicke–Korsakoff syndrome. Transaldolase (EC2.2.1.2) complements the transketolase-catalyzed rearranging reactions by converting the compounds with seven and three carbons into erythrose 4-phosphate (4 carbons) and F6P (6 carbons). Two additional steps, catalyzed by glucose 6-phosphate isomerase (EC5.3.1.9) and glucose 6-phosphatase (EC3.1.3.9), can then generate glucose. Alternatively, depending on feeding status, 6-phosphofructokinase (phosphofructokinase 1, EC2.7.1.11) can initiate utilization via glycolysis.

STORAGE

Glycogen, a large polymer with predominant α-[1 > 4] links and a smaller number of α-[1 > 6] cross-links, is the storage form of Glc. There are two types of glycogen with different metabolic properties. Proglycogen (PG) is characterized by relatively small size (around 400 kDa) and is the predominant form in muscle (as much as 55 mg/g dry weight). Macroglycogen (MG) can be as big as 10 million kDa. Muscle can contain as much as 55 mg PG/g dry weight, and 22 mg MG/g dry weight (Shearer et al., 2000). Liver contains about 80 g of glycogen (Petersen et al., 2001). Glycogen synthesis from glucose 6-phosphate proceeds in a three-step process catalyzed by phosphoglucomutase (EC5.4.2.2, magnesium-dependent), UTP-glucose 1-phosphate uridylyltransferase (UDP glucose pyrophosphorylase, EC2.7.7.9, two isoenzymes), glycogen synthase (GYS, EC2.4.1.11), and 1,4-α-glucan branching enzyme (EC2.4.1.18). The isoenzyme GYS 1 is the main isoenzyme in muscle, GYS2 is mostly expressed in the liver. Alpha 1–6 cross-links are added by 1,4-α-glucan branching enzyme (EC2.4.1.18). Glycogen synthase can only act on an oligosaccharide (primer) with several α-[1 > 4] linked Glc residues attached to glycogenin-1 or glycogenin-2 (EC2.4.1.186), scaffoldlike proteins with the ability to catalyze the manganese-dependent transfer of uridine diphosphate (UDP)–linked glucosyl residues to itself (autocatalysis). Glycogen synthase and glycogenin constitute an enzyme complex. The glycogenin concentration in muscle is proportional to the number of glycogen molecules (Shearer et al., 2000). Repletion of spent glycogen stores (for instance, after a long-distance run) starts with new primers and newly synthesized glycogenin and may take several days. Glycogen phosphorylase (EC2.4.1.1) cleaves Glc residues one at a time off the nonreducing end of glycogen molecules and releases them as glucose

1-phosphate. There are at least three distinct glycogen phosphorylases (EC2.4.1.1) with tissue-specific expression (liver, muscle, and brain types), all of which require lysine-bound pyridoxal 5′-phosphate (PLP) as a cofactor. AMP activates glycogen phosphorylase, while ATP, ADP, and glucose 6-phosphate inhibit the enzyme. Branched ends of glycogen are not substrates for glycogen phosphorylase; instead, they are cleaved by debranching enzyme. Two activities reside on the same polypeptide. Oligo-1,4-1,4-glucanotransferase (EC2.4.1.25) moves the 1,4 alpha-linked chain segment to another 4-position in the molecule, which then leaves the chain end with the 1,6-alpha-linked glucose exposed. The amylo-1,6-glucosidase (EC3.2.1.33) activity can then cleave off the 1,6-alpha-linked Glc. Alternative splicing of the same gene produces several tissue-specific isoforms. The ATP-yield from glycogen oxidation is slightly higher than from free Glc because the main glycogen cleavage product does not require ATP-dependent phosphorylation.

A cascade of phosphorylating and dephosphorylating enzymes under cyclic adenosine monophosphate (cAMP)–mediated hormonal control modulates the activities of glycogen storing and mobilizing enzymes. By convention, the lowercase letter *a* may be attached to the name of the active forms of these enzymes, and the letter *b* to the name of the inactivated forms. Thus, the active form is glycogen synthase a, and the inactive form glycogen synthase b. Signaling through this system slows glycogen deposition and accelerates glucose release from glycogen. The binding of hormones to G-protein-linked receptors raises the intracellular concentration of cAMP, which in turn activates protein kinase A (EC2.7.1.37). This enzyme near the top of the signaling cascade contains two regulatory chains and two catalytic chains, which respond to calcium ions and other effectors. Phosphorylation by protein kinase A activates glycogen synthase a kinase (EC2.7.1.37), which in turn inactivates glycogen synthase by phosphorylation. Protein kinase A also activates phosphorylase b kinase (EC2.7.1.37), which in turn activates glycogen phosphorylase. Phosphorylase b kinase is a multisubunit complex that includes the calcium-binding protein calmodulin and is exquisitely sensitive to the intracellular calcium ion concentration.

The cAMP-induced actions are constantly opposed by corresponding dephosphorylating enzyme activities. Protein phosphatase 1 (PPI, EC3.1.3.16) reverses the activation of glycogen synthase, phosphorylase kinase, and glycogen phosphorylase. Inhibitor 2 and glycogen synthetase kinase 3 fold the newly synthesized catalytic subunit of protein phosphatase 1 and attach it to a targeting subunit. If it is attached to G(L), the hepatic glycogen-targeting subunit (expressed in both the liver and muscle, despite its name), the complex binds to glycogen and modulates the level of active enzymes involved in glycogen metabolism (Munro et al., 2002). Glycogen phosphorylase phosphatase (EC3.1.3.17) inactivates the phosphorylase by removing the four phosphates that link two dimers in the active tetrameric form. Glycogen-synthase-D-phosphatase (EC3.1.3.42), on the other hand, removes an inactivating phosphate from glycogen synthase and gets it started.

EXCRETION

Glc passes into renal primary filtrate owing to its small molecular size and complete water solubility. SGLT1 actively transports Glc into tubular epithelium, from where it is exported into the blood mainly via the high-capacity transporter GLUT2, and less via GLUT1. So long as the transport capacity of the sodium-glucose cotransporter is not exceeded, little glucose is lost via urine. In young healthy individuals, losses occur only at blood glucose concentrations above 1800 mg/l; this threshold decreases with age and renal insufficiency.

REGULATION

Glc concentrations in tissues and body fluids are stabilized by many diverse mechanisms, many of which involve the action of specific hormones. Overall homeostasis is maintained through directing the flux of Glc to or from glycogen stores, balancing glycolysis versus gluconeogenesis, and promoting protein catabolism in times of need.

Hormonal regulation: Among the many hormones with some effect on particular tissues or metabolic sequences, a few stand out because of their dominant and overriding actions on Glc disposition. Insulin promotes uptake and oxidation of Glc by tissues and favors storage, particularly in the postprandial phase. Glucagon in response to low blood Glc concentration increases Glc release from storage and synthesis from precursors. Adrenaline (epinephrine) mobilizes stores and accelerates utilization.

Insulin is produced in the beta cells of pancreatic islet cells and released in a zinc-dependent process with its companion, amylin. The rate of production and release into circulation are related to Glc-sensing mechanisms in the beta cell. ATP generation from Glc and cytosolic calcium concentration are thought to be critical for Glc sensing. A zinc-containing enzyme, insulysin (EC3.4.24.56), inactivates insulin irreversibly in many tissues (Ding et al., 1992). Insulysin activity is inhibited by high concentrations of both amylin and insulin (Mukherjee et al., 2000). Insulin binds to specific insulin receptors in muscles, adipocytes, and some other insulin-sensitive tissues and triggers a signaling cascade with the receptor kinase activity. The chromium-containing peptide chromodulin binds to the insulin-activated insulin receptor and optimizes its receptor kinase activity (Vincent, 2000). In response to the insulin-initiated signaling cascade, GLUT4 (SLC2A4) moves to the plasma membrane and increases Glc uptake into insulin-stimulated cells several-fold. Another important insulin effect is increased transcription of hepatic hexokinase 4 (glucokinase), which increases the availability of glucose 6-phosphate, the precursor for glycolysis and glycogen synthesis. Glycolysis is further promoted by increased concentrations of the regulatory metabolite fructose 2,6-bisphosphate (due to induction of 6-phosphofructo-2-kinase, EC2.7.1.105, and lower expression of fructose-2,6-bisphosphate 2-phosphatase, EC3.1.3.46). At the same time, gluconeogenesis is blocked by the inhibiting effect of insulin on phosphoenolpyruvate carboxykinase (EC4.1.1.32) and of fructose 2,6-bisphosphate on fructose 1,6-bisphosphatase (EC3.1.3.11). Insulin promotes glycogenesis through increasing the availability of the glucose 6-phosphate precursor and decreasing the phosphorylation of enzymes of glycogen metabolism.

The metabolic functions of the insulin companion amylin, which tend to be in opposition to insulin action, are only beginning to be understood. They include promotion of glycogen breakdown and inhibition of glycogen synthesis. Years of excessive amylin secretion may be responsible for the beta cell decline in obesity and insulin resistance. Amylin may promote the deposition of amyloid plaques (Hayden and Tyagi, 2001) and induce beta cell apoptosis (Saafi et al., 2001). Amylin is also a potent neurotransmitter that helps to align food intake with glycemic status (Mietlicki-Baase et al., 2013).

Glucagon is produced and secreted by the alpha cells of the pancreas in response to low Glc concentration. Glucagon promotes the release of glucose 1-phosphate from glycogen. Adrenaline and the less potently acting noradrenaline stimulate the breakdown of glycogen. These catecholamines also counteract the inhibitory effects of nonglucose fuels on glycolysis.

Appetite and satiety: Low blood Glc concentration induces the feeling of hunger. According to the long-held glucostatic theory, specific areas of the brain, such as paraventricular and supraoptic portions of the hypothalamus, integrate input from peripheral and central Glc-responsive sensors and generate appetite sensation (Briski, 2000). Amylin, on the other hand, is secreted in response to feeding and increased blood Glc concentration and acts on histamine H1 receptors with a significant satiety-inducing and anorectic effect (Mietlicki-Baase and Hayes, 2014). A satiety-inducing effect of insulin has also been reported, but it may be weak or mediated through other effectors such as amylin.

Postprandial metabolism: The influx of newly absorbed Glc and other nutrients alters the balance of hormonal and metabolic activities. As outlined above, the rate of insulin (and amylin) secretion increases and the rate of glucagon decreases in response to the higher blood Glc concentration. Gluconeogenesis is effectively turned off and glycolysis is turned on. Glc utilization occurs in preference to fat oxidation. When high carbohydrate intake is coupled with excessive total energy intake, fat (both from diet and from adipose tissue turnover) is preferentially deposited, and the carbohydrate is used as the near-exclusive energy fuel. In fact, the release of fat from adipose tissue is slowed by the increased action of insulin. This is a reminder that both the timing and quantity of carbohydrate ingestion matter.

The deposition of glycogen in the liver and muscles increases, though with a considerable time lag. Reconstitution of depleted glycogen stores is likely to take 1–2 days (Shearer et al., 2000). Carbohydrate loading for 1 or more days can increase glycogen stores by a third or more (Tarnopolsky et al., 2001). Repleting glycogen stores by carbohydrate feeding on the evening before elective surgery instead of fasting is likely to improve outcome and reduce hospital stays (Nygren et al., 2001).

Exercise: A burst of exertion (such as in a short sprint) taxes the capacity of muscle to generate ATP for contraction. Glycolytic breakdown of Glc to lactate is an inefficient mode of fuel utilization because it generates only 2 ATP per glucose molecule. The advantages are that glycolysis is fast, because only one reactions is needed, and that it does not require oxygen (anaerobic pathway). The resulting lactate moves from the muscle cell into circulation via the monocarboxylate transporter 1 (MCT1, SLC16A1). Due to the cotransport of protons, increasing acidification of the muscle cells will promote lactate export. Lactate is used in the liver for gluconeogenesis, and the resulting Glc returned to muscle for another potential round through this lactate–glucose (Cori) cycle. Another of the many adaptations to muscle exertion is the increased activity of GLUT4, which promotes Glc influx from circulation.

Fasting and starvation: When tissue levels of Glc decline and new supplies from food are not forthcoming, the liver and kidneys begin to release Glc into circulation. This Glc comes initially from glycogen stores and from the use of Glc metabolites (lactate, pyruvate, and others) for gluconeogenesis; later, it comes from tissue protein.

FUNCTION

Fuel energy: Glc, from both dietary and endogenous sources, is the predominant energy source of most tissues. The brain, which normally uses Glc to the near-exclusion of other fuels for its energy metabolism (Wahren et al., 1999), can take up nearly 100 g/day, and more upon intense stimulation and use (Dienel and Hertz, 2001). Muscles rely almost as much on Glc as an energy fuel. Even after 1 h of running and considerable depletion of glycogen reserves, more than 70% of the energy is derived from glucose (Arkinstall et al., 2001). Complete oxidation of Glc requires adequate supplies of thiamin, riboflavin, niacin, pantothenate, ubiquinone, iron, and magnesium, and yields about 4 kcal/g.

Reducing equivalents: The metabolism via glycolysis and Krebs cycle generates 10 reduced NADH per completely oxidized Glc (and 2 reduced ubiquinones directly for oxidative phosphorylation). Metabolism through the pentose–phosphate cycle generates 12 reduced NADPH. These reducing equivalents are important prerequisites for the synthesis of many compounds and are essential for maintaining the appropriate redox state of ascorbate, glutathione, and other components of cellular antioxidant defenses.

Fructose precursor: Glc can provide the synthesis of fructose when intakes become low. Fructose is the precursor for the synthesis of hexosamines, which participate in nutrient sensing by modifying signaling proteins (Hanover, 2001) and are constituents of glycans (chondroitins, keratans, dermatans, hyaluronan, heparans, and heparin) in the extracellular matrix of all tissues. Fructose is the predominant energy fuel of spermatozoa. NADP-dependent aldehyde reductase (aldose reductase, EC1.1.1.21) reduces Glc to sorbitol, which is then converted into fructose by zinc-requiring L-iditol 2-dehydrogenase (EC1.1.1.14).

UDP-galactose precursor: Synthesis of cerebrosides, gangliosides, glucosaminoglycans (chondroitin sulfate, dermatan sulfates, and keratan sulfates), and numerous glycoproteins, as well as lactose for milk production, starts with UDP-galactose. UTP-glucose-1 phosphate uridylyltransferase (EC2.7.7.9) can link Glc to UDP, and UDP-glucose-4'-epimerase (GALE, EC5.1.3.2) then generates UDP-galactose in the next and final step.

Carbon source: Numerous endogenously synthesized compounds originate from intermediates of Glc metabolism, in particular dihydroxyacetone phosphate (glycerol in triglycerides, phospholipids, cerebrosides, and gangliosides), 3-phosphoglycerate (serine and glycine), pyruvate (alanine), acetyl-CoA (cholesterol, bile acids, and steroid hormones), α-ketoglutarate (glutamate, glutamine, proline, and arginine), succinyl-CoA (heme), and oxaloacetate (aspartate and asparagine).

Hexosamines: Glc is a precursor for glucosamine 6-phosphate synthesis by glutamine: fructose-6-phosphate transaminase/isomerizing (GFAT, EC2.6.1.16, contains covalently bound pyridoxal 5'-phosphate). *N*-Acetyl glucosamine and other hexosamines are formed after the initial rate-limiting GFAT reaction. The addition of O-linked *N*-acetylglucosamine to proteins can modify their signaling function and give them roles in nutrient sensing (Hanover, 2001). Glc-derived hexosamines are critical constituents of glycans (chondroitins, keratans, dermatans, hyaluronan, heparans, and heparin) in the extracellular matrix of all tissues.

REFERENCES

Arkinstall, M.J., Bruce, C.R., Nikopoulos, V., Garnham, A.P., Hawley, J.A., 2001. Effect of carbohydrate ingestion on metabolism during running and cycling. J. Appl. Physiol. 91, 2125–2134.

Arola, H., Koivula, T., Karvonen, A.L., Jokela, H., Ahola, T., Isokoski, M., 1999. Low trehalase activity is associated with abdominal symptoms caused by edible mushrooms. Scand. J. Gastroenterol. 34, 898–903.

Boles, R.G., Seashore, M.R., Mitchell, W.G., Kollros, P.R., Mofodi, S., Novotny, E.J., 1999. Glucose transporter type I deficiency: a study of two cases with video-EEG. Eur. J. Pediatr. 158, 978–983.

Briski, K.P., 2000. Intraventricular 2-deoxy-D-glucose induces Fos expression by hypothalamic vasopressin, but not oxytocin neurons. Brain Res. Bull. 51, 275–280.

Bykova, I.A., Solovjeva, O.N., Meshalkina, L.E., Kovina, M.V., Kochetov, G.A., 2001. One-substrate transketolase-catalyzed reaction. Biochem. Biophys. Res. Comm. 280, 845–847.

Coy, J.F., Dubel, S., Kioschis, P., Thomas, K., Micklem, G., Delius, H., et al., 1996. Molecular cloning of tissue-specific transcripts of a transketolase-related gene: implications for the evolution of new vertebrate genes. Genomics 32, 309–316.

Dawson, P.A., Mychaleckyj, J.C., Fossey, S.C., Mihic, S.J., Craddock, A.L., Bowden, D.W., 2001. Sequence and functional analysis of GLUT10: a glucose transporter in the type 2 diabetes-linked region of chromosome 20q 12–13. I. Mol. Genet. Metabol. 74, 186–199.

Dienel, G.A., Hertz, L., 2001. Glucose and lactate metabolism during brain activation. J. Neurosci. Res. 66, 824–838.

Ding, L., Becker, A.B., Suzuki, A., Roth, R.A., 1992. Comparison of the enzymatic and biochemical properties of human insulin-degrading enzyme and *Escherichia coli* protease III. J. Biol. Chem. 267, 2414–2420.

Food and Nutrition Board, Institute of Medicine, 2002. Dietary Reference Intakes for Energy, Carbohydrate, Fiber, Fat, Fatty Acids, Cholesterol, Protein, and Amino Acids (macronutrients). National Academy Press, Washington, DC.

Gupta, S.K., Chong, S.K., Fitzgerald, J.E., 1999. Disaccharidase activities in children: normal values and comparison based on symptoms and histologic changes. J. Pediatr. Gastroenterol. Nutr. 28, 246–251.

Hanover, J.A., 2001. Glycan-dependent signaling: *O*-linked *N*-acetylglucosamine. FASEB J. 15, 1865–1876.

Harvey, C.B., Hollox, E.J., Poulter, M., Wang, Y., Rossi, M., Auricchio, S., et al., 1998. Lactase haplotype frequencies in Caucasians: association with the lactase persistence/non-persistence polymorphism. Ann. Hum. Genet. 62, 215–223.

Hayden, M.R., Tyagi, S.C., 2001. "A" is for amylin and amyloid in type 2 diabetes mellitus. J. Pancreas 2, 124–139.

Helliwell, P.A., Richardson, M., Affleck, J., Kellett, G.L., 2000. Stimulation of fructose transport across the intestinal brush-border membrane by PMA is mediated by GLUT2 and dynamically regulated by protein kinase C. Biochem. J. 350, 149–154.

Illsley, N.P., 2000. Glucose transporters in the human placenta. Placenta 21, 14–22.

Jansson, T., Ekstrand, Y., Wennergren, M., Powell, T.L., 2001. Placental glucose transport in gestational diabetes mellitus. Am. J. Obstet. Gynecol. 184, 111–116.

Kaura, D., Bhasin, D.K., Rana, S.V., Katyal, R., Vaiphei, K., Singh, K., 2001. Alterations in duodenal disaccharidases in chronic smokers. Indian J. Gastroenterol. 20, 62–63.

Levin, R.J., 1994. Digestion and absorption of carbohydrates—from molecules and membranes to humans. Am. J. Clin. Nutr. 59, 690S–698S.

McMillan, D.C., Bolchoz, L.J., Jollow, D.J., 2001. Favism: effect of divicine on rat erythrocyte sulfhydryl status, hexose monophosphate shunt activity, morphology, and membrane skeletal proteins. Toxicol. Sci. 62, 353–359.

Mietlicki-Baase, E.G., Hayes, M.R., 2014. Amylin activates distributed CNS nuclei to control energy balance. Physiol. Behav. 136C, 39–46.

Mietlicki-Baase, E.G., Rupprecht, L.E., Olivos, D.R., Zimmer, D.J., Alter, M.D., Pierce, R.C., et al., 2013. Amylin receptor signaling in the ventral tegmental area is physiologically relevant for the control of food intake. Neuropsychopharmacology 38, 1685–1697.

Mukherjee, A., Song, E., Kihiko-Ehmann, M., Goodman Jr., J.P., Pyrek, J.S., Estus, S., et al., 2000. Insulysin hydrolyzes amyloid beta peptides to products that are neither neurotoxic nor deposit on amyloid plaques. J. Neurosci. 20, 8745–8749.

Munro, S., Cuthbertson, D.J., Cunningham, J., Sales, M., Cohen, P.T., 2002. Human skeletal muscle expresses a glycogen-targeting subunit of PP1 that is identical to the insulin-sensitive glycogen-targeting subunit G(L) of liver. Diabetes 51, 591–598.

Murray, I.A., Coupland, K., Smith, J.A., Ansell, I.D., Long, R.G., 2000. Intestinal trehalase activity in a UK population: establishing a normal range and the effect of disease. Br. J. Nutr. 83, 241–245.

Ngarmukos, C., Baur, E.L., Kumagai, A.K., 2001. Co-localization of GLUT1 and GLUT4 in the blood–brain barrier of the rat ventromedial hypothalamus. Brain Res. 900, 1–8.

Nordlie, R.C., Foster, J.D., Lange, A.J., 1999. Regulation of glucose production by the liver. Annu. Rev. Nutr. 19, 379–406.

Nygren, J., Thorell, A., Ljungqvist, O., 2001. Preoperative oral carbohydrate nutrition: an update. Curr. Opin. Clin. Nutr. Metab. Care 4, 255–259.

Pascoe, W.S., Inukai, K., Oka, Y., Slot, J.W., James, D.E., 1996. Differential targeting of facilitative glucose transporters in polarized epithelial cells. Am. J. Physiol. 271, C547–C554.

Petersen, K.F., Cline, G.W., Gerard, D.E., Magnusson, I., Rothman, D.L., Shulman, G.I., 2001. Contribution of net hepatic glycogen synthesis to disposal of an oral glucose load in humans. Metab. Clin. Exp. 50, 598–601.

Saafi, E.L., Konarkowska, B., Zhang, S., Kistler, J., Cooper, G.J., 2001. Ultrastructural evidence that apoptosis is the mechanism by which human amylin evokes death in RINm5F pancreatic islet beta-cells. Cell Biol. Int. 25, 339–350.

Sasaki, T., Minoshima, S., Shiohama, A., Shintani, A., Shimizu, A., Asakawa, S., et al., 2001. Molecular cloning of a member of the facilitative glucose transporter gene family GLUT11 (SLC2A11) and identification of transcription variants. Biochem. Biophys. Res. Comm. 289, 1218–1224.

Shearer, J., Marchand, I., Sathasivam, E., Tarnopolsky, M.A., Graham, T.E., 2000. Glycogenin activity in human skeletal muscle is proportional to muscle glycogen concentration. Am. J. Physiol. Endocrinol. Metab. 278, E177–E180.

Shepherd, P.R., Kahn, B.B., 1999. Glucose transporters and insulin action: implications for insulin resistance and diabetes mellitus. N. Engl. J. Med. 341, 248–257.

Tarnopolsky, M.A., Zawada, C., Richmond, L.B., Carter, S., Shearer, J., Graham, T., et al., 2001. Gender differences in carbohydrate loading are related to energy intake. J. Appl. Physiol. 91, 225–230.

Vincent, J.B., 2000. The biochemistry of chromium. J. Nutr. 130, 715–718.

Wahren, J., Ekberg, K., Fernqvist-Forbes, E., Nair, S., 1999. Brain substrate utilisation during acute hypoglycaemia. Diabetologia 42, 812–818.

Wright, E.M., 2013. Glucose transport families SLC5 and SLC50. Mol. Aspects Med. 34, 183–196.

FRUCTOSE

Fructose (α-D-fructose, levulose, fruit sugar; molecular weight 180) is a six-carbon ketose (ketohexose) (Figure 6.7).

ABBREVIATIONS

Fru	D-fructose
Glc	D-glucose
GalNAc	Gal N-acetylglucosamine
GLUT2	glucose transporter 2 (SLC2A2)
GLUT5	glucose transporter 5 (SLC2AS)

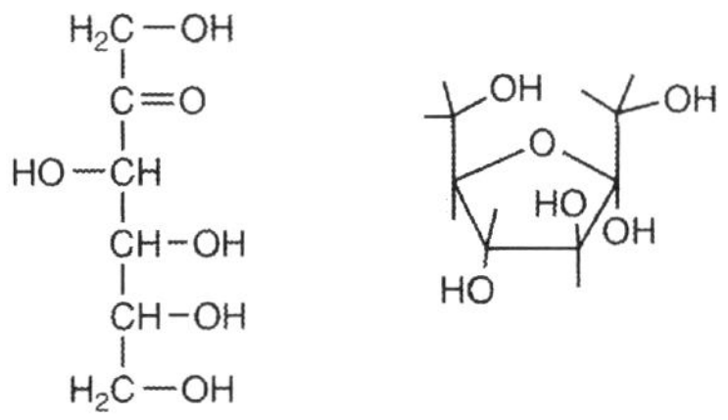

FIGURE 6.7

Alternative structural representations of α-D-fructose.

NUTRITIONAL SUMMARY

Function: Fructose (Fru) is used as an energy fuel and for the synthesis of glycoproteins and glycolipids.

Food sources: Most Fru is consumed with high-fructose corn syrup, refined sugar (sucrose), fruits, and vegetables.

Requirements: No dietary Fru is needed since the required amounts are modest and can easily be produced endogenously from glucose.

Deficiency: There is no information on adverse health effects due to low intake.

Excessive intake: Fructose intolerance is a rare genetic condition that causes hypoglycemia, hypophosphatemia, metabolic acidosis, vomiting, and hyperuricemia in response to intakes of a few grams. Very high Fru intake may cause cataracts and increase oxidative stress.

ENDOGENOUS SOURCES

Some Fru can be produced in the body from glucose (Glc). This pathway is particularly important in the testis, where Fru constitutes the predominant energy fuel of spermatozoa. Aldehyde reductase (aldose reductase, EC1.1.1.21, activated by sulfate) uses NADP for the reduction of glucose to sorbitol. L-Iditol 2-dehydrogenase (EC1.1.1.14, cofactor zinc) can then complete the conversion to Fru. Both steps are reversible (Figure 6.8).

DIETARY SOURCES

High-fructose corn syrup is a major source of Fru in the United States, where this pervasive sweetener is added to many industrially produced foods, including ketchup and bread. The Fru content in this syrup is increased by conversion of its Glc during industrial processing using glucose isomerase/D-xylulose ketol-isomerase (EC5.3.1.5). A more conventional source is the disaccharide sucrose, consisting of Glc α-[1 > β2] Fru.

Mixtures with equal amounts of monomeric Fru and Glc are 1.3 times sweeter than the same amount of sucrose (Stone and Oliver, 1969). Fruits and vegetables also contain significant amounts of monomeric Fru and sucrose. About half of the dry weight of peaches is sucrose.

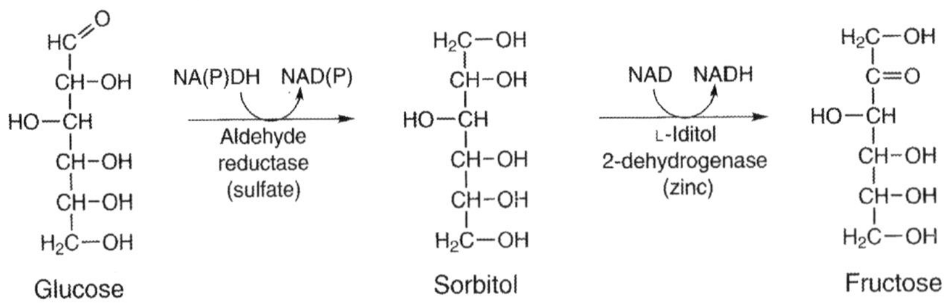

FIGURE 6.8

Fructose synthesis.

Daily Fru intake may be as high as 100 g, especially in populations with high intakes of sucrose and high-fructose corn syrup (Ruxton et al., 1999). Per-capita disappearance of fructose was 81 g/day in the United States for 1997 (Elliott et al., 2002).

DIGESTION AND ABSORPTION

The disaccharide sucrose (Glc α-[1 > β2] Fru) is hydrolyzed by sucrose ot-glucosidase (EC3.2.1.48), a component of the brush border enzyme complex sucrase-isomaltase. The facilitative transporter GLUT5 (SLC2A5), the sodium-dependent transporter SGLT4/SLC5A9 (Tazawa et al., 2005), and to a lesser extent GLUT2 (SLC2A2), mediate Fru uptake from the small-intestinal lumen, mainly the jejunum (Helliwell et al., 2000). Diffusion and paracellular passage via glucose-activated solute drag also may contribute to absorption. Large quantities (25 g) are poorly absorbed and will cause malabsorption symptoms in as many as one-third of healthy subjects (Born et al., 1994). GLUT2 (SLC2A2) facilitates the transport of Fru out of enterocytes into interstitial fluid from where it enters the portal bloodstream (Figure 6.9).

TRANSPORT AND CELLULAR UPTAKE

Blood circulation: Fru is transported in blood as a serum solute. The concentration in plasma of healthy people is around 0.13 mmol/l and increases in response to very high Fru intake (Hallfrisch et al., 1986). Fru is taken up into cells via the facilitative transporters GLUT2 (Colville et al., 1993) and GLUT5. Uptake of Fru into spermatozoa depends on GLUT5.

BBB: There is no evidence that significant amounts of Fru cross from the blood into the brain.
Materno-fetal transfer: The net transfer of Fru to the fetus is unknown but is likely to be small.

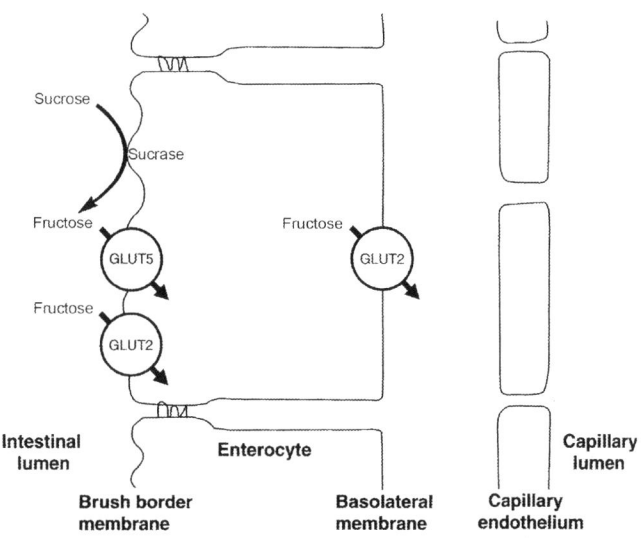

FIGURE 6.9

Intestinal absorption of fructose.

METABOLISM

Most Fru is metabolized in the liver, which explains some of the metabolic differences between Fru and glucose. The dominant metabolic pathway proceeds via fructose 1-phosphate and joins the glycolysis pathway at the level of the trioses glyceraldehyde 3-phosphate and dihydroxyacetone phosphate (Figure 6.10). A smaller proportion joins the glycolysis pathway immediately through the phosphorylation to the glycolysis intermediate fructose 6-phosphate by magnesium-dependent hexokinase (EC2.7.1.1).

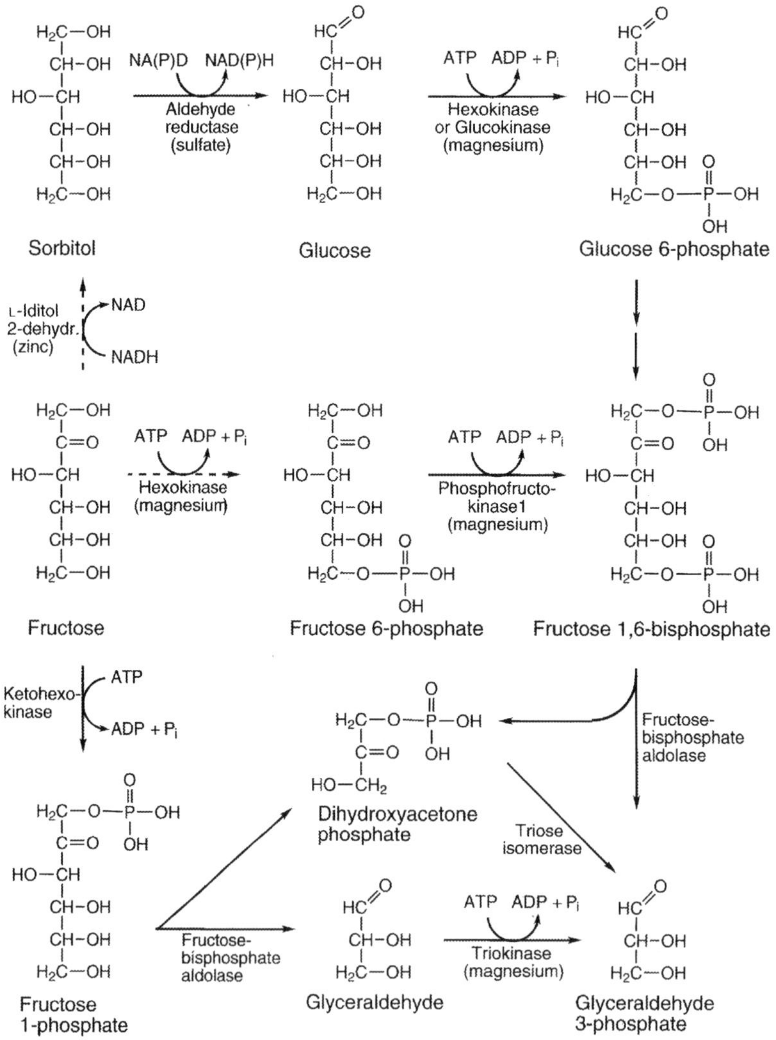

FIGURE 6.10

Fructose metabolism.

Another significant proportion of ingested fructose can be directly converted to glucose via sorbitol (Kawaguchi et al., 1996).

Catabolism via fructose 1-phosphate: Ketohexokinase (hepatic fructokinase, EC2.7.1.3) in the liver and in pancreatic islet cells phosphorylates Fru to fructose 1-phosphate. Fructose bisphosphate aldolase (aldolase, EC4.1.2.13) cleaves both fructose 1-phosphate and the glycolysis intermediate fructose 1,6-bisphosphate. There are three genetically distinct isoforms of this crucial enzyme: aldolase A, predominantly in muscle; aldolase B, in the liver; and aldolase C, in the brain. People with a lack of aldolase B cannot metabolize Fru properly (hereditary fructose intolerance). Glyceraldehyde is phosphorylated by triokinase (EC2.7.1.28). Triosephosphate isomerase (EC5.3.1.1) converts dihydroxyacetone phosphate into glyceraldehyde 3-phosphate, which can then continue along the glycolytic pathway or contributes to gluconeogenesis depending on prevailing conditions.

The sorbitol pathway: The zinc-containing L-iditol 2-dehydrogenase (sorbitol dehydrogenase, EC1.1.1.14) converts small amounts of Fru to sorbitol, which can then be oxidized to Glc by NADP-dependent aldehyde reductase (aldose reductase, EC1.1.1.21).

STORAGE

There is no significant specific accumulation of Fru that could be mobilized in times of need.

EXCRETION

Very little net loss of Fru occurs in healthy people, even when consumption is 100 g or more. Significant amounts of the plasma solute Fru are filtered in the renal glomerulus. Most of the Fru in renal ultrafiltrate is recovered from the proximal renal tubular lumen through the facilitative transporter GLUT5 (Sugawara-Yokoo et al., 1999) and returned into the bloodstream via GLUT2.

REGULATION

Absorption via GLUT2 at the luminal side of the enterocyte is rapidly and strongly upregulated in response to feeding and humoral factors (Helliwell et al., 2000). Stress and possibly hyperglycemia increase GLUT2 trafficking to the brush border membrane within minutes through activation of p38 MAP kinase signaling. Growth factors, insulin, and other factors also influence Fru absorption. Insulin enhances transcription of the aldolase B gene, and glucagon suppresses its transcription. The latter exerts its action by binding to a cAMP-responsive element in the promoter region (Takano et al., 2000).

The metabolism of Fru to pyruvate (glycolysis) is not subject to the regulatory factors that act on phosphofructokinase-1 and that play a great role in regulation of Glc metabolism. Small amounts of Fru have been suggested to improve control of Glc metabolism (Hawkins et al., 2002). On the other hand, high Fru levels can promote hexosamine synthesis and thereby slow insulin-dependent Glc utilization in muscle and adipose tissue (Wu et al., 2001).

FUNCTION

Energy fuel: The complete oxidation of Fru yields about 4 kcal/g and requires adequate supplies of thiamin, riboflavin, niacin, lipoate, ubiquinone, iron, and magnesium.

Unspecific precursor: All metabolizable sugars can provide carbons for numerous endogenously generated compounds such as amino acids (e.g., glutamate from the Krebs cycle intermediate α-ketoglutarate), cholesterol (from acetyl-coenzyme A), or the glycerol in triglycerides.

Hexosamines: Fru is a precursor for glucosamine 6-phosphate synthesis by glutamine: fructose-6-phosphate transaminase/isomerizing (GFAT, EC2.6.1.16). *N*-Acetyl glucosamine and other hexosamines are formed after the initial rate-limiting GFAT reaction. The addition of O-linked *N*-acetyl glucosamine to proteins can modify their signaling function and give them roles in nutrient sensing (Hanover, 2001). Fru- (and Glc)-derived hexosamines are critical constituents of glycans (chondroitins, keratans, dermatans, hyaluronan, heparans, and heparin) in the extracellular matrix of all tissues.

REFERENCES

Born, P., Zech, J., Stark, M., Classen, M., Lorenz, R., 1994. Zuckeraustauschstoffe: vergleichende untersuchung zur intestinalen resorption von fructose, sorbit und xylit. Med. Klin. 89, 575–578.

Colville, C.A., Seatter, M.J., Jess, T.J., Gould, G.W., Thomas, H.M., 1993. Kinetic analysis of the liver-type (GLUT2) and brain type (GLUT3) glucose transporters in Xenopus oocytes: substrate specificities and effects of transport inhibitors. Biochem. J. 290, 701–706.

Elliott, S.S., Keim, N.L., Stern, J.S., Teff, K., Havel, P.J., 2002. Fructose, weight gain, and the insulin resistance syndrome. Ant. J. Clin. Nutr. 76, 911–922.

Ganiats, T.G., Norcross, W.A., Halverson, A.L., Burford, P.A., Palinkas, L.A., 1994. Does Beano prevent "gas"? A double-blind crossover study of oral alpha-galactosidase to treat dietary oligosaccharide intolerance. J. Fam. Pract. 39, 441–445.

Hallfrisch, J., Ellwood, K., Michaelis IV, O.E., Reiser, S., Prather, E.S., 1986. Plasma fructose, uric acid, and inorganic phosphorus responses of hyperinsulinemic men fed fructose. J. Am. Coll. Nutr. 5, 61–68.

Hanover, J.A., 2001. Glycan-dependent signaling: *O*-linked *N*-acetylglucosamine. FASEB J. 15, 1865–1876.

Hawkins, M., Gabriely, I., Wozniak, R., Vilcu, C., Shamoon, H., Rossetti, L., 2002. Diabetes 51, 606–614.

Helliwell, P.A., Richardson, M., Affleck, J., Kellett, G.L., 2000. Regulation of GLUT5, GLUT2 and intestinal brush border fructose absorption by the extracellular signal-regulated kinase, p38 mitogen-activated kinase and phosphatidylinositol 3-kinase intracellular signaling pathways: implications for adaptation to diabetes. Biochem. J. 350, 163–169.

Kawaguchi, M., Fujii, T., Kamiya, Y., Ito, J., Okada, M., Sakuma, N., et al., 1996. Effects of fructose ingestion on sorbitol and fructose 3-phosphate contents oferythrocytes from healthy men. Acta Diabetol. 33, 100–102.

Ruxton, C.H., Garceau, F.J., Cottrell, R.C., 1999. Guidelines for sugar consumption in Europe: is a quantitative approach justified? Eur. J. Clin. Nutr. 53, 503–513.

Stone, H., Oliver, S.M., 1969. Measurement of the relative sweetness of selected sweeteners and sweetener mixtures. J. Food Sci. 34, 215–222.

Sugawara-Yokoo, M., Suzuki, T., Matsuzaki, T., Naruse, T., Takata, K., 1999. Presence of fructose transporter GLUT5 in the S3 proximal tubules in the rat kidney. Kidney Int. 56, 1022–1028.

Takano, Y., Iuchi, Y., Ito, J., Otsu, K., Kuzumaki, T., Ishikawa, K., 2000. Characterization of the responsive elements to hormones in the rat aldolase B gene. Arch. Biochem. Biophys. 377, 58–64.

Tazawa, S., Yamato, T., Fujikura, H., Hiratochi, M., Itoh, F., Tomae, M., et al., 2005. SLC5A9/SGLT4, a new Na^+-dependent glucose transporter, is an essential transporter for mannose, 1,5-anhydro-D-glucitol, and fructose. Life Sci. 76, 1039–1050.

Wu, G., Haynes, T.E., Yan, W., Meininger, C.J., 2001. Presence of glutamine:fructose-6-phosphate amidotransferase for glucosamine-6-phosphate synthesis in endothelial cells: effects of hyperglycemia and glutamine. Diabetologia 44, 196–202.

GALACTOSE

Galactose (D-α-galactose, cerebrose; molecular weight 180) is a six-carbon aldose (aldohexose) (Figure 6.11).

ABBREVIATIONS

Gal	D-α-galactose
GalNAc	Gal N-acetylglucosamine
Glc	D-glucose
GLUT1	glucose transporter 1 (SLC2A1)
SGLT1	sodium-glucose cotransporter 1 (SLC5A1)

NUTRITIONAL SUMMARY

Function: Galactose (Gal) is used as an energy fuel and for the synthesis of glycoproteins and glycolipids.

Food sources: Most Gal is consumed with dairy products.

Requirements: No dietary Gal is needed since the required Gal amounts are modest and can easily be produced endogenously from glucose.

Deficiency: There are no health effects associated with low intake.

Excessive intake: Lactose intolerance is the most common complaint related to higher than minimal intakes. Much higher Gal intake than likely to ever occur in humans induces lens deposits (cataracts) in animal models, and increase mortality most likely due to oxidative damage. High intake appears to be harmful for adults (Michaëlsson et al., 2014).

ENDOGENOUS SOURCES

Enough Gal for all functional requirements is endogenously produced from D-glucose (Glc). First, Glc is activated by conjugation with UDP (UTP-glucose-1-phosphate uridylyltransferase, EC2.7.7.9; there are two genetically distinct isoforms). UDP glucose-4'-epimerase (GALE, EC5.1.3.2) can then generate UDP-galactose (UDPGal) in a reversible reaction. UDPGal is the direct precursor for lactose synthesis and for galactosyl-transfer for glycoprotein and glycolipid synthesis (Figure 6.12).

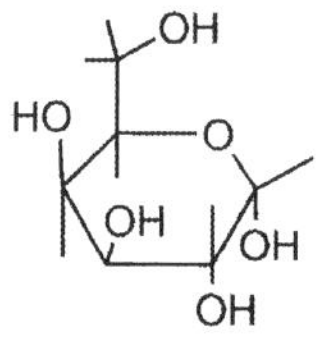

FIGURE 6.11

D-α-Galactose.

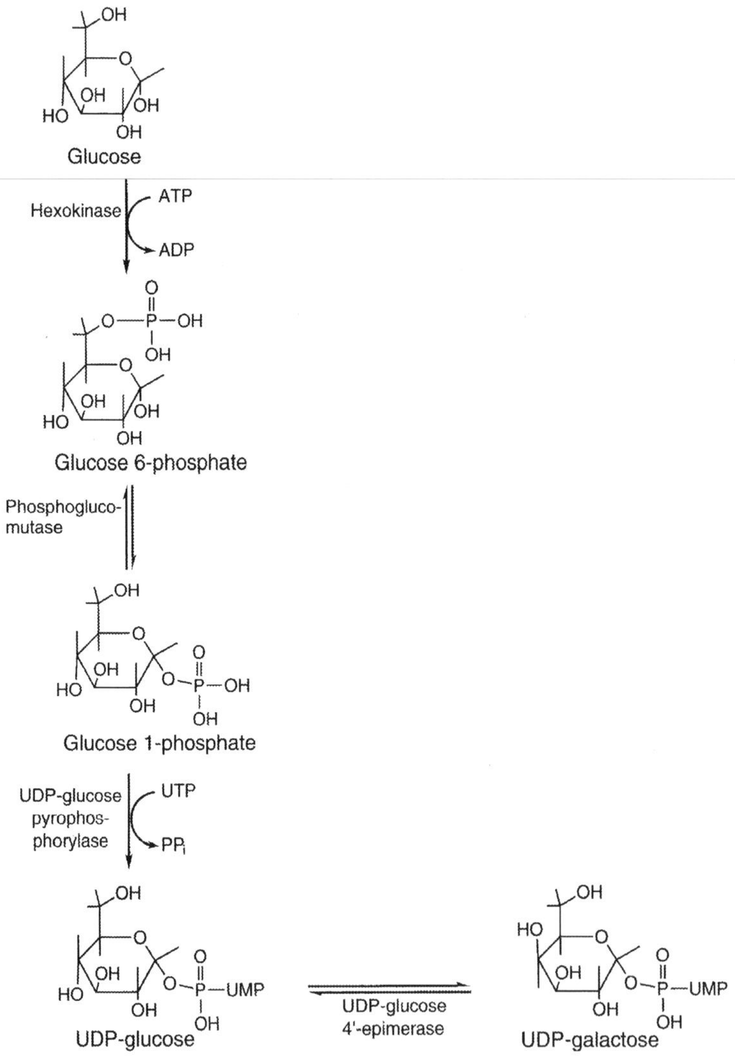

FIGURE 6.12

Gal is synthesized from glucose.

DIETARY SOURCES

Lactose, the α-D-galactopyranosyl-[1 > 4] D-glucose dimer, provides most of the carbohydrate in milk (including human milk), infant formula, and dairy products. Lactose provides about 40% of the energy for infants, and about 2% in a mixed diet of American adults (Perisse et al., 1969). Small amounts of Gal are also present in a wide variety of foods, including legumes and meats (Acosta and Gross, 1995).

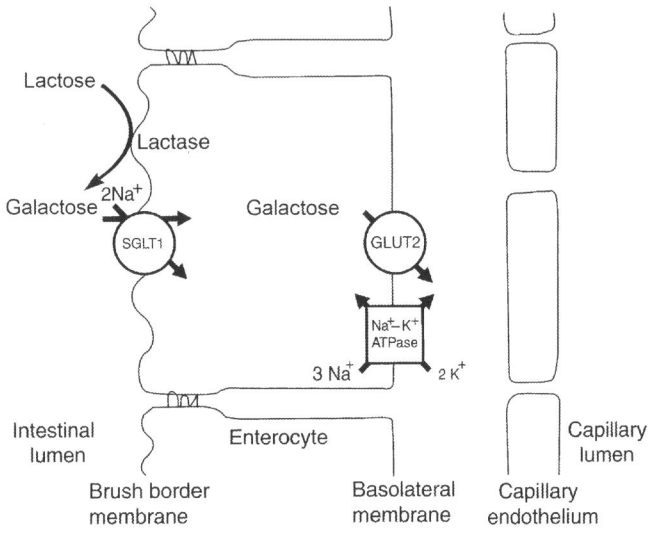

FIGURE 6.13

Intestinal absorption of Gal.

The Gal content of peas (boiled, ready to eat) can be as high as 35 mg/g (Peterbauer et al., 2002). Lactose is used as an extender (bulking agent) in some medications.

DIGESTION AND ABSORPTION

The hydrolysis of lactose by brush border lactase (EC3.2.1.108) generates Glc and Gal. Intestinal lactase expression persists to a considerable degree beyond childhood in most humans. Persistent expression is least prevalent in Asian populations and most common in Northern Europeans (Harvey et al., 1995). Two specific variants (−13910T and −22018A) are responsible for lactase persistence in European populations and predict intestinal enzyme activity with high specificity (Fang et al., 2012). Several other variants have the same effect in various pastoral populations in Africa and Asia.

Gal is taken up into enterocytes by active transport (Figure 6.13) via sodium-glucose cotransporter 1 (SGLT1, SLC5A1), and to a much lesser extent by SGLT2 (SLC5A2; Helliwell et al., 2000).

Some of the absorbed Gal is used to provide energy or precursors for the enterocyte's own needs or is lost into the intestinal lumen when the enterocyte is shed. Most Gal leaves the enterocyte via GLUT2 (SLC5A2) and diffuses into portal blood. The plant oligosaccharides raffinose (Gal α-[1 > 6] Glc α-[1 > β2] Fru), stachyose (Gal α-[1 > 6] Gal α-[1 > 6] Glc α-[1 > β2] Fru), and verbascose (Gal α-[>6] Gal α-[1 > 6] Gal α-[1 > 6] Glc α-[1 > β2] Fru) in beans, peas, and other plant foods are not well absorbed because the Gal in alpha-position blocks digestion. Alpha-galactosidase is active in the lysosomes of most cells but not in digestive secretions or the luminal side of the human intestine. It is not known to what extent Gal is released by unspecific digestive or bacterial enzyme action and how much of the released Gal is absorbed (particularly from the terminal ileum and colon). Ingestion of manufactured alpha-galactosidase (EC3.2.1.22, Beano) along with legumes promotes digestion and may decrease oligosaccharide utilization by gas-forming intestinal bacteria (Ganiats et al., 1994).

TRANSPORT AND CELLULAR UPTAKE

Blood circulation: Gal is transported in blood as a serum solute. The concentration in the blood of healthy people is under 0.22 mmol/l. Gal is taken up into cells via facilitative transporters, including GLUT1 (brain), GLUT2 (SLC5A2, liver, kidney), GLUT3 (SLC5A3, many tissues), and several related transporters.

BBB: GLUT1, which is present on both sides of the brain capillary epithelial cells, transports Gal as readily as glucose.

Materno-fetal transfer: GLUT1 mediates facilitative Gal transport across both sides of the syntrophoblast cell layer (Illsley, 2000). The net transfer of Gal to the fetus is unknown.

METABOLISM

Gal is mainly converted to glucose l-phosphate and then to glucose 6-phosphate in the liver (Figure 6.14). A minor alternate pathway exists but remains to be characterized (Berry et al., 2001). The initial critical step is phosphorylation by galactokinase (EC2.7.1.6). There are two genetically distinct isoforms of the enzyme with different tissue distribution. Galactitol accumulation in the lenses of individuals with defective galactokinase 1 can cause cataracts in childhood or early adulthood. The next step of Gal metabolism is the transfer of UDP by UDP-glucose-hexose-i-phosphate uridylyltransferase (EC2.7.7.12). UDP-glucose-4'-epimerase (EC5.1.3.2) epimerizes UDPGal to UDP-glucose. Since UDP-glucose provides the UDP again for the next Gal 1-phosphate molecule, this works like an autocatalytic mechanism with a net conversion of Gal 1-phosphate to glucose 1-phosphate. Magnesium-dependent phosphoglucomutase (EC5.4.2.2; two isoforms, PGM1 and PGM2) converts glucose 1-phosphate into the readily metabolizable intermediate glucose 6-phosphate. Gal can alternatively be reduced to galactitol by NADPH-dependent aldehyde reductase (aldose reductase, EC1.1.1.21), especially in the presence of Gal excess.

STORAGE

There is no significant specific accumulation of Gal that could be mobilized in times of need.

EXCRETION

Gal passes into renal primary filtrate owing to its small molecular size and complete water solubility. SGLTI actively transports Gal into the epithelium of proximal renal tubules from where it is exported into the blood via the high-capacity transporter GLUT2.

REGULATION

The regulation of tissue and whole-body Gal homeostasis is complex and tightly integrated into the regulation of carbohydrate metabolism.

FUNCTION

Lactose synthesis: Nursing infants depend on the high lactose content of milk as the main energy fuel. Women produce large amounts of lactose in their mammary glands after a normal pregnancy. Human milk contains 60–80 g/l of lactose and Gal-containing oligosaccharides. Gal and Glc are actively taken

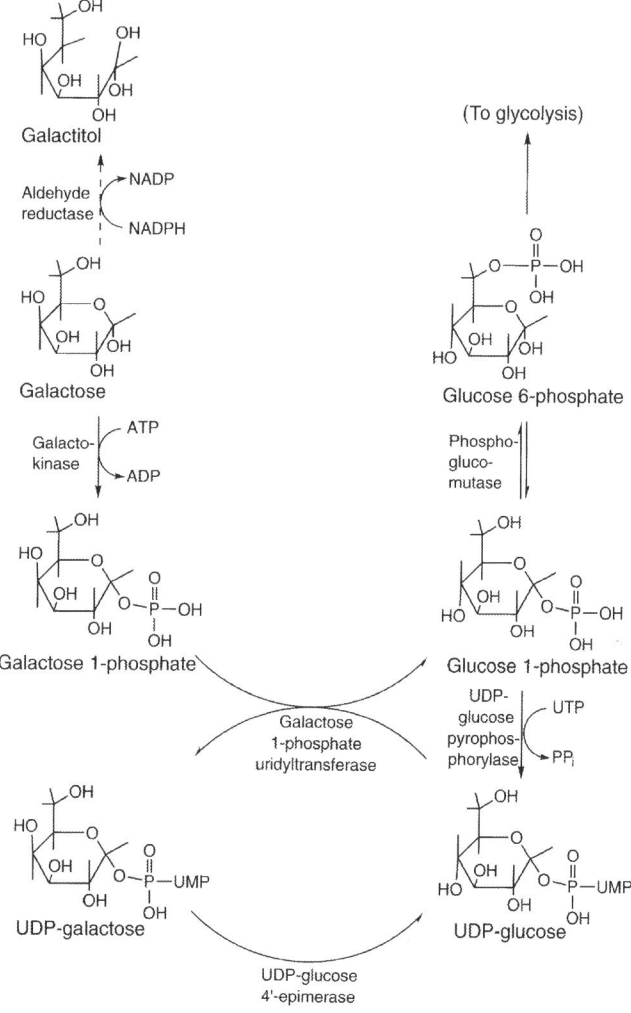

FIGURE 6.14

Metabolism of Gal.

up from maternal blood across the basolateral membrane into the mammary epithelial cell by SGLTI (Obermeier et al., 2000). Transport of Gal into the Golgi system may require a transporter, possibly GLUT1 (Nemeth et al., 2000). The presence of GLUT1 in human mammary glands is, however, in question (Obermeier et al., 2000).

UDP-galactose and glucose are linked in the Golgi complex by lactase synthase (EC2.4.1.22), a heterodimer consisting of the enzymatically active A-protein (the shortened version of beta-1,4-galactosyltransferase 1, which is transcribed from an alternative initiation site) and alpha-lactalbumin.

Energy fuel: The complete oxidation of Gal yields about 4 kcal/g and requires adequate supplies of thiamin, riboflavin, niacin, lipoate, ubiquinone, iron, and magnesium.

Unspecific precursor: Gal can provide carbons for numerous endogenously generated compounds such as amino acids (e.g., glutamate from the Krebs cycle intermediate alpha-ketoglutarate), cholesterol (from acetyl-coenzyme A), or the glycerol in triglycerides.

Glycoprotein synthesis: Gal and galactosamine are attached to numerous proteins. Examples of the presence of Gal in *O*-glycans are the Gal and Glc-α2-Gal side chains beta-linked to 5-hydroxylysine in collagens. Gal *N*-acetylglucosamine (GalNAc) is attached to serine and threonine residues in mucins. Glucosaminoglycans (chondroitin sulfate, dermatan sulfates, and keratan sulfates) contain Gal, GalNAc, or both.

Glycolipid synthesis: Gal is a structural component of both neutral and acidic cerebrosides and gangliosides. The brain and the myelin sheath of nerves contain particularly large amounts of Gal-linked glycolipids, typically in association with specific proteins. Neutral glycolipids are of the types Gal(β1–4) ceramide and Gal(β1–4) Glc(β1–1) ceramide. Acidic glycolipids are of the types Neu5Ac(α2–3) Gal(β1–4) Glc(β1–1) ceramide and Neu5Ac(α2-8)Neu5Ac(α2–3)Gal(β1–4)Glc(β1–1) ceramide (Neu5Ac = *N*-acetylneuraminic acid). The ceramides are sphingolipids with long-chain fatty acids.

REFERENCES

Acosta, P.B., Gross, K.C., 1995. Hidden sources of galactose in the environment. Eur. J. Ped. 154, S87–S92.

Berry, G.T., Leslie, N., Reynolds, R., Yager, C.T., Segal, S., 2001. Evidence for alternate galactose oxidation in a patient with deletion of the galactose-1-phosphate uridyltransferase gene. Mol. Genet. Metab. 72, 316–321.

Fang, L., Ahn, J.K., Wodziak, D., Sibley, E., 2012. The human lactase persistence-associated SNP-13910*T enables *in vivo* functional persistence of lactase promoter-reporter transgene expression. Hum. Genet. 131, 1153–1159.

Ganiats, T.G., Norcross, W.A., Halverson, A.L., Burford, P.A., Palinkas, L.A., 1994. Does Beano prevent "gas"? A double-blind crossover study of oral alpha-galactosidase to treat dietary oligosaccharide intolerance. J. Fam. Pract. 39, 441–445.

Harvey, C.B., Hollox, E.J., Poulter, M., Wang, Y., Rossi, M., Auricchio, S., et al., 1998. Lactase haplotype frequencies in Caucasians: association with the lactase persistence/non-persistence polymorphism. Ann. Hum. Genet. 62, 215–223.

Harvey, C.B., Pratt, W.S., Islam, I., Whitehouse, D.B., Swallow, D.M., 1995. DNA polymorphisms in the lactase gene: linkage disequilibrium across the 70-kb region. Eur. J. Hum. Genet. 3, 27–41.

Helliwell, P.A., Richardson, M., Affleck, J., Kellett, G.L., 2000. Regulation of GLUT5, GLUT2 and intestinal brush border fructose absorption by the extracellular signal-regulated kinase, p38 mitogen-activated kinase and phosphatidylinositol 3-kinase intracellular signaling pathways: implications for adaptation to diabetes. Biochem. J. 350, 163–169.

Illsley, N.P., 2000. Glucose transporters in the human placenta. Placenta 21, 14–22.

Michaëlsson, K., Wolk, A., Langenskiöld, S., Basu, S., Warensjö Lemming, E., Melhus, H., et al., 2014. Milk intake and risk of mortality and fractures in women and men: cohort studies. BMJ 349, g6015.

Nemeth, B.A., Tsang, S.W., Geske, R.S., Haney, P.M., 2000. Golgi targeting of the GLUT1 glucose transporter in lactating mouse mammary gland. Ped. Res. 47, 444–450.

Obermeier, S., Huselweh, B., Tinel, H., Kinne, R.H., Kunz, C., 2000. Expression of glucose transporters in lactating human mammary gland epithelial cells. Eur. J. Nutr. 39, 194–200.

Perisse, J., Sizaret, E., Francois, P., 1969. The effect of income on the structure of the diet. FAO Nutr. Newslett. 7, 1–9.

Peterbauer, T., Mucha, J., Mach, L., Richter, A., 2002. Chain elongation of raffinose in pea seeds. Isolation, characterization, and molecular cloning of multifunctional enzyme catalyzing the synthesis of stachyose and verbascose. J. Biol. Chem. 277, 194–200.

XYLITOL

Xylitol (xyloopentane-1,2,3,4,5-pentol; molecular weight 152) is a water-soluble sugar alcohol that sometimes is used as a glucose substitute for diabetics and in chewing gum (Figure 6.15).

ABBREVIATIONS

GAP	glyceraldehyde 3-phosphate
F6P	fructose 6-phosphate
X5P	xylulose 5-phosphate

NUTRITIONAL SUMMARY

Function: Xylitol is sometimes used as a sweetener in gum because it is less likely to promote caries than other sugars, or in dietetic foods for diabetics because its uptake into cells is not dependent on insulin.

Requirements: Dietary intake of xylitol is not necessary for optimal health.

Food sources: It is present in modest amounts in plums and a few other fruits.

Excessive intake: Intakes of 50 g/day or more may cause some malabsorption and gas production from bacterial fermentation of residual xylitol in the colon. Long-term health risks of high intakes have not been thoroughly evaluated.

ENDOGENOUS SOURCES

Daily endogenous production of xylitol is 1–4 g (Figure 6.16). Xylitol derives from the catabolism in the liver and kidney of D-glucuronate released from connective tissue and from glucuronate-containing proteoglycans. The enzyme that catalyzes the initial glucuronate reduction, glucuronate reductase (EC1.1.1.19), actually may be identical with aldehyde reductase (EC1.1.1.2). L-Gulonate can then be oxidized by L-gulonate 3-dehydrogenase (EC1.1.1.45). The third step is catalyzed by dehydro-L-gulonate decarboxylase (EC4.1.1.34), which can use either magnesium or manganese as a cofactor.

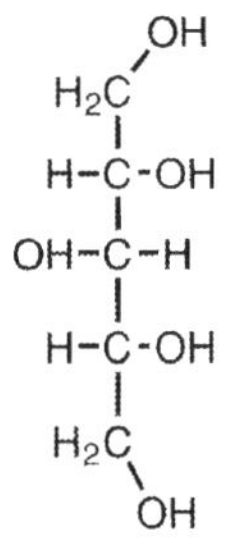

FIGURE 6.15

Xylitol.

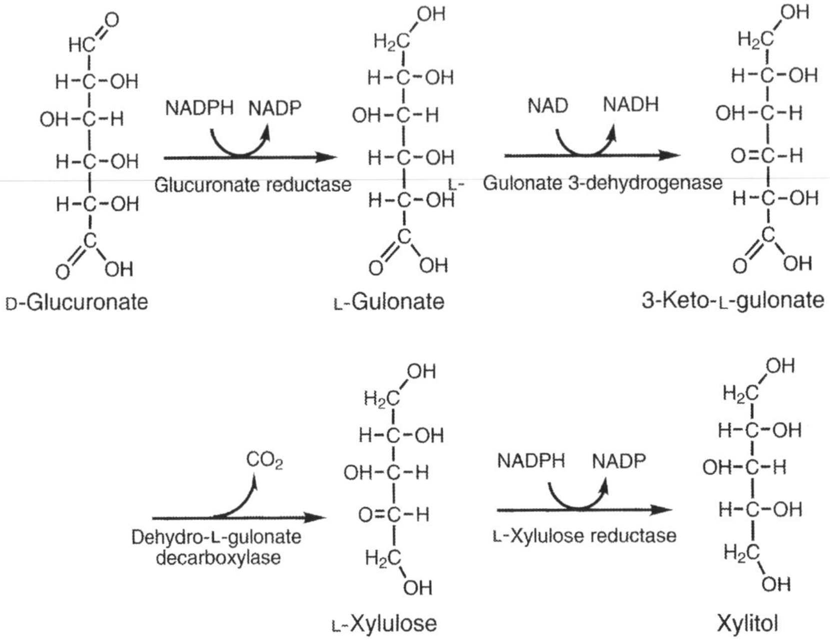

FIGURE 6.16

Endogenous xylitol synthesis.

Xylitol synthesis is then completed by L-xylulose reductase (EC1.1.1.10). Two distinct genes encode this enzyme. The major isoenzyme occurs in both cytosol and mitochondria, and the minor isoenzyme is limited to cytosol. A genetic defect of the major isoenzyme causes pentosuria. This benign condition disrupts the normal metabolism of glucuronate and is characterized by excretion of 1–4 g/day L-xylulose with urine.

DIETARY SOURCES

Small amounts of xylitol are naturally consumed with plums (9 mg/g), raspberries, spinach (1.1 mg/g), and carrots (0.9 mg/g); some sugarless chewing gums also contain xylitol. Dietary intakes may be significant, especially in diabetics consuming xylitol as a sugar replacer.

DIGESTION AND ABSORPTION

Xylitol is absorbed by passive diffusion, more slowly than glucose, presumably through one of the glucose transporters. It is much less likely to cause malabsorption symptoms than sorbitol or fructose (Born et al., 1994).

TRANSPORT AND CELLULAR UPTAKE

Blood circulation: After transport via portal blood in free form, xylitol is taken up into the liver and other tissues by an unknown mechanism. Xylitol is sometimes used as a sweetener for diabetics because most tissues can take it up without stimulation by insulin.

METABOLISM

Xylitol is an important intermediate of the pentitol pathway in the liver for the reutilization of glucuronide from connective tissue (proteoglycan) breakdown. The reactions of the pentose–phosphate pathway can partially convert xylitol into glucose 6-phosphate, which can then either provide glucose or reenter the pentose–phosphate pathway. The reactions require adequate availability of thiamin, niacin, magnesium, and manganese. Oxidation to D-xylulose by D-xylulose reductase (EC1.1.1.9, cofactor manganese) uses NAD, in contrast to the NADPH-dependent reduction of L-xylulose to xylitol (Figure 6.17).

The availability of NADPH from the early steps of the pentose–phosphate pathway can thus promote xylitol generation while its metabolism generates NADH. Phosphorylation to D-xylulose 5-phosphate (X5P) uses magnesium-dependent xylulokinase (EC2.7.1.17). Transketolase (EC2.2.1.1) with covalently bound thiamin-pyrophosphate catalyzes two rearrangement reactions. One of these converts two pentose phosphates (X5P and ribose 5-phosphate) into a set of compounds with 7 (D-sedoheptulose 7-phosphate) and 3 (glyceraldehyde 3-phosphate, GAP) carbons. The other one rearranges X5P plus erythrose 4-phosphate into glyceraldehyde 3-phosphate and fructose 6-phosphate (F6P). A third possibility is the rearrangement of two X5P molecules into two GAP molecules and one molecule of erythrulose (Bykova et al., 2001). The same reactions are catalyzed by transketolase 2 (Coy et al., 1996), with different isoforms in the brain and heart generated by alternative splicing. Variants of the *transketolase 2* gene may be implicated in the pathogenesis of Wernicke–Korsakoff syndrome. Transaldolase (EC2.2.1.2) complements the transketolase-catalyzed rearranging reactions by converting the compounds with 7 and 3 carbons into erythrose 4-phosphate (4 carbons) and F6P (6 carbons). Two additional steps, catalyzed by glucose 6-phosphate isomerase (EC5.3.1.9) and glucose 6-phosphatase (EC3.1.3.9), then can generate glucose. Alternatively, depending on feeding status, 6-phosphofructokinase (phosphofructokinase 1, EC2.7.1.11) can initiate utilization via glycolysis.

FUNCTION

Xylitol has both passive and active anti-caries properties (Levine, 1998; Makinen et al., 1998) and appears to be more effective in arrest of dental caries than sorbitol (Makinen et al., 1996).

Xylitol was found to inhibit the growth *of Streptococcus pneumoniae* and possibly thereby reduce the incidence of otitis media in children (Uhari et al., 1998). Oral intake may increase intestinal absorption of iron, copper, and calcium (Hamalainen and Makinen, 1985). Animal studies suggest that xylitol feeding may improve bone architecture and biomechanical stability of bone (Mattila et al., 2002).

Use of xylitol-containing solutions for parenteral nutrition support may lead to the development of oxalosis, with the risk of seizures and renal failure (Leidig et al., 2001); therefore, such use is banned in the United States.

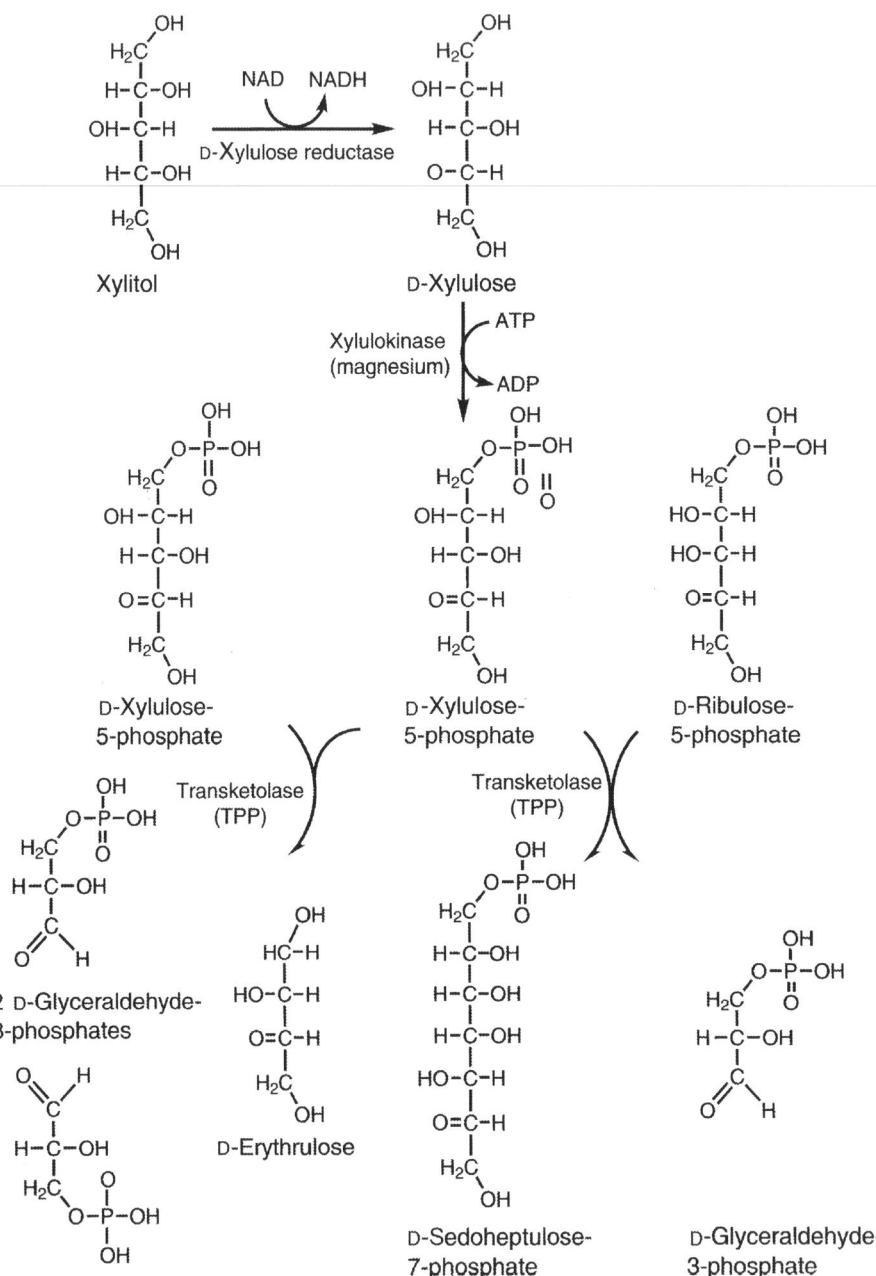

FIGURE 6.17

Xylitol metabolism.

REFERENCES

Born, P., Zech, J., Stark, M., Classen, M., Lorenz, R., 1994. Zuckeraustauschstoffe: Vergleichende Untersuchung zur intestinalen Resorption yon Fructose, Sorbit und Xylit. Med. Klin. 89, 575–578.

Bykova, I.A., Solovjeva, O.N., Meshalkina, L.E., Kovina, M.V., Kochetov, G.A., 2001. One-substrate transketolase-catalyzed reaction. Biochem. Biophys. Res. Commun. 280, 845–847.

Coy, J.E., Dubel, S., Kioschis, E., Thomas, K., Micklem, G., Delius, H., et al., 1996. Molecular cloning of tissue-specific transcripts ofa transketolase-related gene: implications for the evolution of new vertebrate genes. Genomics 32, 309–316.

Hamalainen, M.M., Makinen, K.K., 1985. Duodenal xanthine oxidase (EC1.2.3.2) and ferroxidase activities in the rat in relation to the increased iron absorption caused by peroral xylitol. Br. J. Nutr. 54, 493–498.

Leidig, E., Gerding, W., Arns, W., Ortmann, M., 2001. Renal oxalosis with renal failure after infusion of xylitol. Deutsche Med. Wochenschr. 126, 1357–1360.

Levine, R.S., 1998. Briefing paper: xylitol, caries and plaque. Br. Dent. J. 185, 520.

Makinen, K.K., Chiego Jr., D.J., Allen, E., Bennett, C., Isotupa, K.E., Tiekso, J., et al., 1998. Physical, chemical, and histologic changes in dentin caries lesions of primary teeth induced by regular use of polyol chewing gums. Acta Odontol. Scand. 56, 148–156.

Makinen, K.K., Makinen, P.L., Pape Jr., H.R., Peldyak, L., Hujoel, E., Isotupa, K.E., et al., 1996. Conclusion and review of the Michigan Xylitol Programme (1986–1995) for the prevention of dental caries. Int. Dent. J. 46, 22–34.

Mattila, P.T., Svanberg, M.J., Jamsa, T., Knuuttila, M.L., 2002. Improved bone biomechanical properties in xylitol-fed aged rats. Metab. Clin. Exp. 51, 92–96.

Uhari, M., Kontiokari, T., Niemela, M., 1998. A novel use of xylitol sugar in preventing acute otitis media. Pediatrics 102, 879–884.

PYRUVATE

Pyruvate (pyruvic acid, 2-oxopropanoic acid, alpha-ketopropionic acid, acetylformic acid, pyroracemic acid; molecular weight 88) is a keto-monocarboxylic acid.

ABBREVIATIONS	
CoA	coenzyme A
MCT1	proton/monocarboxylic acid cotransporter 1 (SLC16A1)
MCT2	proton/monocarboxylic acid cotransporter 2 (SLC16A7)
PLP	pyridoxal 5′-phosphate

NUTRITIONAL SUMMARY

Function: Pyruvate is the product of glucose, L-alanine, and L-serine breakdown.

Food sources: Insignificant amounts are present in foods from both animal and plant sources. In comparison, more than 100 g of pyruvate are generated daily from the breakdown of carbohydrates and protein.

Requirements: No dietary intake is needed. Pyruvate is commercially available as a single compound or in combination with other ingredients.

Deficiency: A lack of intake has no harmful consequences. The efficacy of oral pyruvate supplements to promote weight loss in conjunction with exercise and to improve exercise performance is uncertain. The intracoronary application for the salvage of ischemic myocardium has been described but requires further evaluation.

Excessive intake: The risks associated with use of supplemental pyruvate are not known.

ENDOGENOUS SOURCES

Carbohydrates: Several hundred grams of pyruvate are generated during the metabolism of glucose via glycolysis. As a rule of thumb, about 1 g of absorbed carbohydrate generates 1 g of pyruvate. As a result of the shuttling of anaerobic glucose metabolites from muscle to liver for complete utilization (Cori cycle), large amounts of pyruvate are generated from *S*-lactate by NADH-dependent L-lactate dehydrogenase (EC1.1.1.27).

Amino acids: Almost all ingested L-alanine is eventually broken down by alanine aminotransferase (EC2.6.1.2) to pyruvate. The L-alanine–pyruvate pair is critical for the transfer of protein-derived carbons for gluconeogensis from muscle to liver during fasting and severe illness (alanine–glucose cycle). A small proportion of L-serine is converted to pyruvate by the PLP-dependent enzymes serine dehydratase (EC4.2.1.13) and threonine dehydratase (EC4.2.1.16).

DIETARY SOURCES

While many foods contain some pyruvate from the metabolic processes occurring in the food source, the amounts tend to be very small. Dietary supplements with gram amounts of pyruvate are commercially available.

DIGESTION AND ABSORPTION

A proton/monocarboxylic acid cotransporter (MCT1, SLC16A1) is possibly present in the apical, and certainly in the basolateral membrane (Garcia et al., 1994; Orsenigo et al., 1999; Tamai et al., 1999) of the entire intestine, especially the jejunum (Tamai et al., 1995). In addition to lactate, acetate, acetoacetate, β-hydroxybutyrate, propionate, butyrate, and benzoic acid, this transporter mediates the uptake of pyruvate across both the brush border membrane and the basolateral membrane. Additional transporters or mechanisms of entry might exist.

TRANSPORT AND CELLULAR UPTAKE

Blood circulation: Pyruvate is present in the blood in free form. Tissues can take it up via several members of the proton/monocarboxylate cotransporter (MCT) family (Halestrap and Price, 1999). MCT1 is the predominant form responsible for pyruvate uptake from circulation in the liver, muscle, and brain. MCT2 (SLC16A7) is a high-affinity transporter with a preference for pyruvate in many tissues. Expression is especially high in testis and in some neoplastic cells (Lin et al., 1998). Once inside a cell, pyruvate can enter mitochondria via the mitochondrial tricarboxylate carrier, a six-transmembrane helix that is not related to the MCT family of genes (Halestrap and Price, 1999).

Materno-fetal transfer: Several members of the MCT family contribute to placental transport, but their individual locations still need to be clarified.

BBB: MCT1 is present at both sides of the brain endothelium. Ketosis increases MCT1 expression in these cells.

METABOLISM

Some of the energy content of carbohydrates can be utilized even in the absence of oxygen, especially in strenuously exercised muscle. In this case, pyruvate is reduced to L-lactate by L-lactate dehydrogenase (EC1.1.1.27) providing a renewed supply of oxidized NAD for continued glycolysis. The net yield of anaerobic glycolysis is 2 ATP for each glucose molecule metabolized to L-lactate. The shuttling of L-lactate from muscle to liver, eventual regeneration of pyruvate, glucose synthesis from pyruvate, and return transport of glucose to muscles are referred to as the *Cori cycle*. If enough oxygen is available, on the other hand, pyruvate is fully metabolized in mitochondria by oxidative decarboxylation to acetyl-CoA, oxidation in the citric acid cycle, and use of the resulting reductants (NADH and FADH2) for oxidative phosphorylation.

A smaller proportion of pyruvate is carboxylated to oxaloacetate in a biotin-dependent reaction. This latter reaction is called *anaplerotic* (Greek for "refilling"), because it replenishes the citric acid cycle intermediates and thus sustains their ability to metabolize acetyl-CoA. The metabolic fate of pyruvate is closely regulated. During glycolysis, most pyruvate is metabolized to acetyl-CoA. When the prevailing metabolic direction is toward gluconeogenesis, a large proportion of available pyruvate is converted to oxaloacetate, which is then used to resynthesize glucose.

Oxidative decarboxylation: Pyruvate dehydrogenase (EC3.1.3.43) in the mitochondrial matrix comprises multiple copies of three distinct moieties: E1, E2, and E3 (Figure 6.18). Thiamin pyrophosphate is covalently bound to E1. Each subunit E2 (dihydrolipoamide *S*-acetyltransferase, EC2.3.1.12) contains two lipoate molecules, which are covalently bound to lysines 99 and 226. These lipoamides serve as acceptors for the acetyl residues from pyruvate, transfer them to acetyl CoA, and reduce lipoamide to dihydrolipoamide in the process. Another component of the complex, dihydrolipoamide dehydrogenase (E3, EC1.8.1.4), transfers the hydrogen via FAD to NAD. A single gene encodes the dihydrolipoamide dehydrogenase of pyruvate dehydrogenase and the other two alpha-ketoacid dehydrogenases.

Carboxylation: The biotin-containing enzyme pyruvate carboxylase (EC6.4.1.1) generates oxaloacetate, a pivotal precursor for glucose synthesis in gluconeogenic tissues (liver and kidney). Pyruvate carboxylation is the only anaplerotic reaction that can replenish Krebs cycle intermediates without drawing on L-glutamate or other amino acids.

EXCRETION

Pyruvate is recovered from primary filtrate both in proximal renal tubules and collecting ducts. It has been suggested that a sodium-linked carrier (possibly MCT6) is responsible for uptake across the brush border membrane (Halestrap and Price, 1999). MCT1 mediates transport across basolateral membranes of proximal tubules. MCT2 performs this function in collecting ducts (Garcia et al., 1995).

REGULATION

Both synthesis from glycolytic precursors and breakdown to either acetyl-CoA or oxaloacetate are tightly regulated. Pyruvate dehydrogenase (EC3.1.3.43) is the enzyme that connects glycolysis to

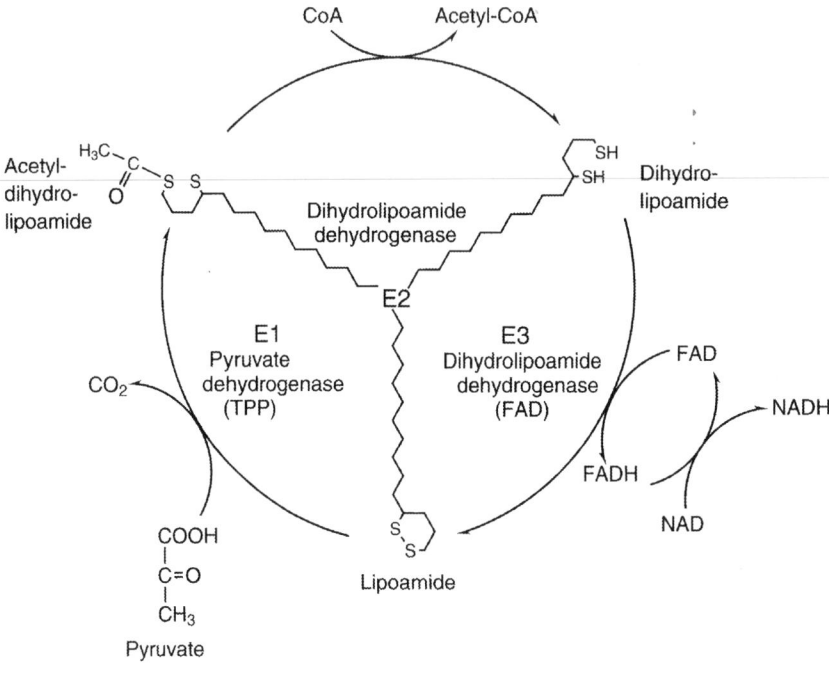

FIGURE 6.18

A multisubunit enzyme complex oxidizes pyruvate.

the citric acid cycle. This enzyme complex is inactivated by phosphorylation ([pyruvate dehydroge-nase (lipoamide)] kinase, EC2.7.1.99) of three serines in the E1 subunit and reactivated by removal of these phosphates by [pyruvate dehydrogenase (lipoamide)]–phosphatase (EC3.1.3.43). Inactivation is strongly subject to substrate and product feedback: ADP and pyruvate decrease the rate of inactiva-tion, while NADH and acetyl-CoA increase it. Activation, on the other hand, is mainly under hormonal control, mediated by calcium.

The activity of pyruvate carboxylase increases with rising acetyl-CoA concentration, which pre-vents further accumulation from pyruvate metabolism and makes more oxaloacetate available to form citrate condensation with acetyl-CoA.

Unlike glucose, pyruvate does not stimulate insulin secretion. This dissociation of insulin secretion from mitochondrial substrate oxidation has been called the *pyruvate paradox* (Ishihara et al., 1999).

FUNCTION

Fuel metabolism: As described previously, pyruvate is the linchpin between glucose metabolism and the citric acid cycle. Ingested pyruvate provides about 4 kcal/g. Its complete oxidation requires ade-quate supplies of thiamin, riboflavin, niacin, pantothenate, lipoate, ubiquinone, magnesium, and iron.

Amino acid synthesis: L-alanine aminotransferase (EC2.6.1.2) uses L-glutamate for transamination of the glycolysis metabolite pyruvate and production of L-alanine. Since the reaction operates near equilibrium, high availability of glucose (and consequently of pyruvate) increases L-alanine production. During fasting or severe illness, the alanine–glucose cycle uses pyruvate and the amino groups from protein catabolism to shuttle the gluconeogenesis precursor from extrahepatic tissues to the liver.

Enzyme cofactor: Pyruvate is an essential cofactor of several bacterial enzymes, but no human enzymes with this type of requirement are known.

Performance enhancement: A beneficial effect of pyruvate on myocardial contractility in patients with heart failure has been suggested (Hermann et al., 1999), which might be mediated by an increase of ionized calcium in the sarcoplasmic reticulum (Hermann et al., 2000). It has also been suggested that supplemental pyruvate (6 g/day) in combination with moderate exercise promotes weight loss (Kalman et al., 1999). This type of regimen did not increase short-term strength, however (Stone et al., 1999).

REFERENCES

Garcia, C.K., Brown, M.S., Pathak, R.K., Goldstein, J.L., 1995. cDNA cloning of MCT2, a second monocarboxylate transporter expressed in different cells than MCTI. J. Biol. Chem. 270, 1843–1849.

Garcia, C.K., Goldstein, J.L., Pathak, R.K., Anderson, R.G., Brown, M.S., 1994. Molecular characterization of a membrane transporter for lactate, pyruvate, and other monocarboxylates: implications for the Cori cycle. Cell 76, 865–873.

Halestrap, A.P., Price, N.T., 1999. The proton-linked monocarboxylate transporter (MCT) family: structure, function and regulation. Biochem. J. 343, 281–299.

Hermann, H.E., Pieske, B., Schwarzmüller, E., Keul, J., Just, H., Hasenfuss, G., 1999. Haemodynamic effects of intracoronary pyruvate in patients with congestive heart failure: an open study. Lancet 353, 1321–1323.

Hermann, H.E., Zeitz, O., Keweloh, B., Hasenfuss, G., Janssen, P.M., 2000. Pyruvate potentiates inotropic effects of isoproterenol and Ca^{2+} in rabbit cardiac muscle preparations. Am. J. Physiol. 279, H702–H708.

Ishihara, H., Wang, H., Drewes, L.R., Wollheim, C.B., 1999. Overexpression of monocarboxylate transporter and lactate dehydrogenase alters insulin secretory responses to pyruvate and lactate in beta cells. J. Clin. Invest. 104, 1621–1629.

Kalman, D., Colker, C.M., Wilets, I., Roufs, J.B., Antonio, J., 1999. The effects of pyruvate supplementation on body composition in overweight individuals. Nutrition 15, 337–340.

Lin, R.Y., Vera, J.C., Chaganti, R.S.K., Golde, D.W., 1998. Human monocarboxylate transporter 2 (MCT2) is a high affinity pyruvate transporter. J. Biol. Chem. 273, 28959–28965.

Orsenigo, M.N., Tosco, M., Bazzini, C., Laforenza, U., Faelli, A., 1999. A monocarboxylate transporter MCT1 is located at the basolateral pole of rat jejunum. Exp. Physiol. 84, 1033–1042.

Stone, M.H., Sanborn, K., Smith, L.L., O'Bryant, H.S., Hoke, T., Utter, A.C., et al., 1999. Effects of in-season (5 weeks) creatine and pyruvate supplementation on anaerobic performance and body composition in American football players. Int. J. Sport Nutr. 9, 146–165.

Tamai, I., Sai, Y., Ono, A., Kido, Y., Yabuuchi, H., Takanaga, H., et al., 1999. Immunohistochemical and functional characterization of pH-dependent intestinal absorption of weak organic acids by the monocarboxylic acid transporter MCT1. J. Pharm. Pharmacol. 51, 1113–1121.

Tamai, I., Takanaga, H., Maeda, H., Sai, Y., Ogihara, T., Higashida, H., et al., 1995. Participation of a proton/cotransporter, MCTI, in the intestinal transport of monocarboxylic acids. Biochem. Biophys. Res. Commun. 214, 482–489.

OXALIC ACID

Oxalic acid (oxalate, ethanedioic acid; molecular weight 90) is a dicarboxylic acid generated in the metabolism of most organisms (Figure 6.19).

ABBREVIATIONS

CLD	congenital diarrhea gene (SLC26A3)
DRA	down-regulated in adenoma gene (SLC26A3)
DTDST	diastrophic dysplasia sulfate transporter (SLC26A2)
PDS	Pendred syndrome gene (SLC26A4)

NUTRITIONAL SUMMARY

Function: Oxalate is a minor breakdown product of dehydroascorbate, ethanoloamine, serine, and glycine that has no known biological role.

Requirements: There is no dietary intake requirement.

Food sources: Spinach and rhubarb are especially rich in oxalate. Other fruits, vegetables, and herbs contain smaller amounts.

Deficiency: No adverse effects are known or expected from lack of intake.

Excessive intake: Dietary intakes increase oxalate concentration in urine and can thereby increase risk of renal stone formation (calcium oxalate calculi).

ENDOGENOUS SOURCES

Oxalate is generated mainly from the breakdown of dehydroascorbic acid and glyoxylate. Dehydroascorbic acid spontaneously decomposes to 2,3-ketogulonate and then to threonic acid and oxalate. This irreversible reaction contributes about 40% of total oxalate at moderate ascorbate intake levels.

Oxalate is also produced (Figure 6.20) from the oxidation of excess glyoxylate by (*S*)-2-hydroxy-acid oxidase (EC1.1.3.15) in liver peroxisomes (Seargeant et al., 1991). L-Serine, hydroxy-proline (Ichiyama et al., 2000), and ethanolamine contribute to glyoxylate production via glycoaldehyde and glycolate; glycolate itself is ingested with plant foods (Ichiyama et al., 2000). Another source of oxalate is glycine; a small percentage (0.1%) of the glycine pool ends up as oxalate. This might be due to a

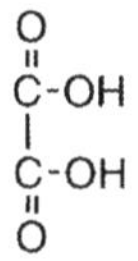

$$
\begin{array}{c}
O \\
\parallel \\
C\text{-OH} \\
\mid \\
C\text{-OH} \\
\parallel \\
O
\end{array}
$$

FIGURE 6.19

Oxalic acid.

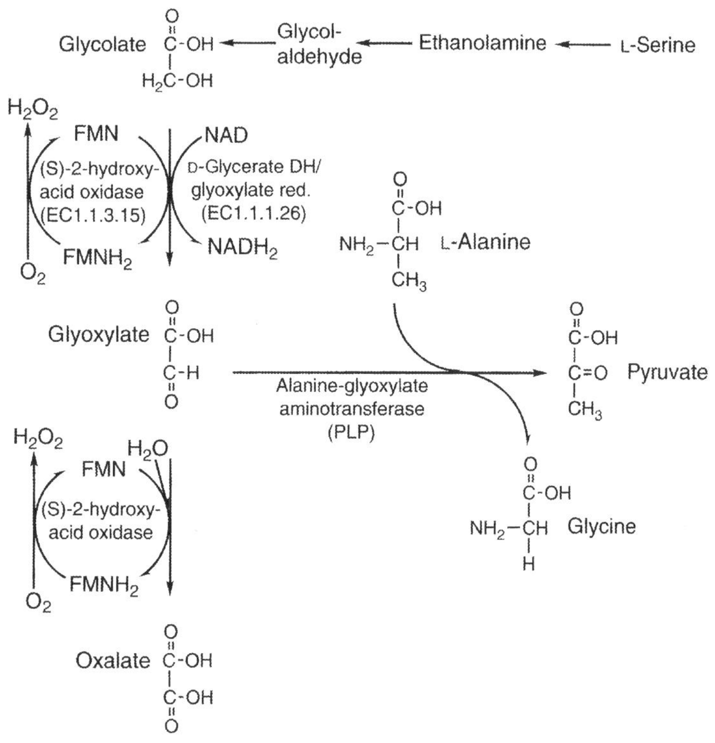

FIGURE 6.20

Endogenous sources of oxalate.

reversal of the alanine–glyoxylate pathway in the liver peroxisomes, whereby the pyridoxal-phosphate-dependent enzyme alanine–glyoxylate aminotransferase (EC2.6.1.44) would then convert pyruvate and glycine into glyoxylate plus alanine. A third potential glyoxylate precursor is ethylene glycol, though significant exposure other than from antifreeze ingestion is rare. Since the bifunctional enzyme D-glycerate dehydrogenase/glyoxylate reductase (EC1.1.1.26) can reduce glyoxylate to glycolate, decreased activity of this enzyme may contribute to glyoxylate accumulation.

A rise in urinary excretion of oxalate suggests that glucose ingestion promotes oxalate production, but the mechanism is not clear (Nguyen et al., 1998).

DIETARY SOURCES

Major sources of dietary oxalate are cassava (13 mg/g), spinach (10 mg/g), carrots and radishes (5 mg/g), rhubarb, bamboo shoots, parsley (17 mg/g), chives (15 mg/g), and chard (USDA, 1984). Human milk and formula contain about 8 mg/l, with a several-fold range of concentrations; this oxalate load can contribute to the increased kidney stone risk of premature infants (Hoppe et al., 1998).

DIGESTION AND ABSORPTION

Intestinal absorption is about 5%, both from the small intestine (Freel et al., 1998) and, possibly to a greater extent, from the cecum and colon. Increased calcium intake reduces urinary oxalate excretion and calculus formation (Takei et al., 1998), presumably because calcium oxalate is very insoluble and thus less bioavailable than other forms of oxalate.

Oxalate is taken up via a sodium-independent transporter (CLD/DRA, SLC26A3) whose main substrates are sulfate and chloride (Silberg et al., 1995). This or a parallel high-affinity transporter uses an oxalate-hydroxylate exchange mechanism for high-affinity transport (Tyagi et al., 2001). Defects in this transporter have been linked to congenital diarrhea (CLD) and colon adenoma (down-regulated in adenoma, or DRA). The transporter is highly expressed in the cecum and colon, and to a lesser extent in the small intestine. Additional transporters may similarly facilitate oxalate transport as a minor activity. Basolateral transport also appears to use a sulfate carrier.

TRANSPORT AND CELLULAR UPTAKE

Blood circulation: Oxalate is present in the blood in its free ionic form. Several anion transporters in various tissues are known to accept oxalate. One of these is the sulfate carrier related to CLD and familial risk of colon adenoma CLD/DRA (SLC26A3). Another is the diastrophic dysplasia sulfate transporter (DTDST, SLC26A2), and a third is the transporter related to Pendred syndrome (PDS, SLC26A4). Another member of the same transporter family, SLC26A6, is strongly expressed in the pancreas and kidneys (Lohi et al., 2000).

METABOLISM

Oxalate appears to be a dead-end product in humans that cannot be metabolized.

EXCRETION

Oxalate in primary filtrate is reabsorbed partially from the proximal renal tubule and papillary surface (Chandhoke and Fan, 2000); the transport across the basolateral membrane might be coupled indirectly with the reabsorption of chloride (Brandle et al., 1998). Normal oxalate excretion with urine is about 50 mg/day, depending on dietary intakes and metabolic and hormonal factors (Nguyen et al., 1998).

Note: High vitamin B6 intake from food (RR 0.66 highest vs. lowest intakes: >40 vs. <3 mg/day) has been found to decrease the risk of renal calculus formation (Curhan et al., 1999), possibly by promoting the removal of the oxalate precursor glyoxylate via the alanine–glyoxylate pathway described previously. Average B6 intake in the United States was 1.9 mg/day (USDA, 1986).

REFERENCES

Brandle, E., Bernt, U., Hautmann, R.E., 1998. *In situ* characterization of oxalate transport across the basolateral membrane of the proximal tubule. Pfl. Arch. Eur. J. Physiol. 435, 840–849.

Chandhoke, P.S., Fan, J., 2000. Transport of oxalate across the rabbit papillary surface epithelium. J. Urol. 164, 1724–1728.

Curhan, G.C., Willett, W.C., Speizer, F.E., Stampfer, M.J., 1999. Intake of vitamins B6 and C and the risk of kidney stones in women. J. Am. Soc. Nephrol. 10, 840–845.

Freel, R.W., Hatch, M., Vaziri, N.D., 1998. Conductive pathways for chloride and oxalate in rabbit ileal brush border membrane vesicles. Am. J. Physiol. 275, C748–C757.

Hoppe, B., Roth, B., Bauerfeld, C., Langman, C.B., 1998. Oxalate, citrate, and sulfate concentration in human milk compared with formula preparations: influence on urinary anion excretion. J. Ped. Gastroent. Nutr. 27, 383–386.

Ichiyama, A., Xue, H.H., Oda, T., Uchida, C., Sugiyama, T., Maeda-Nakai, E., et al., 2000. Oxalate synthesis in mammals: properties and subcellular distribution of serine:pyruvate/alanine:glyoxylate aminotransferase in the liver. Mol. Urol. 4, 333–340.

Lohi, H., Kujala, M., Kerkela, E., Saarialho-Kere, U., Kestila, M., Kere, J., 2000. Mapping of five new putative anion transporter genes in human and characterization of SLC26A6, a candidate gene for pancreatic anion exchanger. Genomics 70, 102–112.

Nguyen, U.N., Dumoulin, G., Henriet, M.T., Regnard, J., 1998. Aspartame ingestion increases urinary calcium, but not oxalate excretion, in healthy subjects. J. Clin. Endocrinol. Metab. 83, 165–168.

Seargeant, L.E., deGroot, G.W., Dilling, L.A., Mallory, C.J., Haworth, J.C., 1991. Primary oxaluria type 2 (L-glyceric aciduria): a rare cause of nephrolithiasis in children. J. Pediat. 118, 912–914.

Silberg, D.G., Wang, W., Moseley, R.H., Traber, P.G., 1995. The Down regulated in Adenoma (dra) gene encodes an intestine-specific membrane sulfate transport protein. J. Biol. Chem. 270, 11897–11902.

Takei, K., Ito, H., Masai, M., Kotake, T., 1998. Oral calcium supplement decreases urinary oxalate excretion in patients with enteric hyperoxaluria. Urologia Internationalis 61, 192–195.

Tyagi, S., Kavilaveettil, R.J., Alrefai, W.A., Alsafwah, S., Ramaswamy, K., Dudeja, P.K., 2001. Evidence for the existence of a distinct SO(4)(--)-OH(-) exchange mechanism in the human proximal colonic apical membrane vesicles and its possible role in chloride transport. Exp. Biol. Med. 226, 912–918.

USDA (United States Department of Agriculture). Vegetables and Vegetable products. Agriculture Handbook No. 8–11, 1984.

USDA (United States Department of Agriculture). Food Consumption Survey, Report 85-3, 1986.

ETHANOL

Ethanol (ethyl alcohol, alcohol; molecular weight 46) is readily water-soluble and has a limited capacity to dissolve slightly hydrophobic molecules (Figure 6.21).

ABBREVIATIONS	
ADH	alcohol dehydrogenase
ALDH	aldehyde dehydrogenase
CoA	coenzyme A
FAD	flavin adenine dinucleotide
FMN	flavin mononucleotide
MEOS	microsomal ethanol oxidizing system

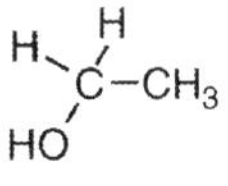

FIGURE 6.21

Ethanol.

NUTRITIONAL SUMMARY

Function: Ethanol is an energy-rich nutrient that provides about 7 kcal/g. Its conversion to acetyl-CoA requires riboflavin, niacin, pantothenate, zinc, heme iron, molybdenum, and magnesium; further oxidation of acetyl-CoA to water and carbon dioxide depends on thiamin, riboflavin, niacin, pantothenate, lipoate, ubiquinone, iron, and magnesium. The ethanol metabolite acetate can be used as a precursor for fatty acid and cholesterol synthesis.

Food sources: Alcoholic beverages typically contain 10–20 g ethanol per serving (glass of beer or wine, shot glass of liquor). Foods do not contain significant amounts of alcohol unless they are steeped in liquor. Ethanol is converted completely into acetyl-CoA, which is small compared to the several hundred grams that are generated daily during the breakdown of carbohydrates, fat, and protein.

Requirements: No dietary ethanol intake is necessary. The beneficial effect of moderate ethanol intake (one serving of an alcoholic beverage or less a day) on cardiovascular risk often is outweighed by adverse effects from violent death or injury and the potential for habitual or addicted excessive consumption.

Excessive intake: Even moderate intakes of ethanol (as little as one serving of an alcoholic beverage) slow response time, impair motor control, limit judgment, and thus increase the risk of accidents and perpetrating or suffering crimes. Acute poisoning with large amounts can induce coma and death. Chronic abuse can lead to dependency, increase cancer risk, induce bone loss, and cause damage to the liver, heart, pancreas, brain, and other organs.

ENDOGENOUS SOURCES

Pathological microorganisms in the small intestine (e.g., *Candida albicans*) can generate several grams of ethanol per day in some individuals. Normal intestinal flora can do the same when significant amounts of fermentable carbohydrates reach the terminal ileum and colon (Geertinger et al., 1982; Picot et al., 1997; Cope et al., 2000). A specific syndrome due to high gastrointestinal ethanol production (Mei-Tei-Sho), which may be related to high rice intake, has been described in Japan (Picot et al., 1997).

DIETARY SOURCES

Ethanol is consumed mostly with alcoholic beverages such as beer, wine, hard cider, liqueurs, and hard liquor. A typical serving of American beer (360 ml/12 fl oz) contains about 13 g of ethanol, and the same amount of light beer has about 11 g. About 10 g is consumed with a small glass of wine (100 ml/3.5 fl oz), and 16 g with a 40-ml (1.5 fl oz) jigger of hard liquor (whiskey, gin, vodka, etc.). Very small amounts may be taken up with fruits and vegetables, especially if they are fermented. Foods steeped in alcoholic beverages (e.g., fruitcake with rum) may contain unexpectedly large amounts depending on the recipe.

Ethanol is sensed by taste receptor TAS2R38 and tastes bitter to some (Mattes and DiMeglio, 2001) and not to others (Duffy et al., 2004). Acute ingestion decreases quinine bitterness but enhances its aftertaste. The ability to discriminate between different ethanol concentrations in alcoholic beverages is limited to mixtures under 40% by volume (Lachenmeier et al., 2014). That means that a beverage with 45% ethanol and another with 60% ethanol taste the same.

DIGESTION AND ABSORPTION

Ethanol is absorbed in the stomach and proximal small intestine. Since ethanol readily permeates lipid membranes, the main transfer mechanism appears to be diffusion. The lining of the mouth, esophagus, and stomach has significant capacity for the first steps of ethanol metabolism. Because of this, a small proportion of ingested ethanol is transferred into the blood as acetaldehyde and acetic acid.

TRANSPORT AND CELLULAR UPTAKE

Ethanol is readily soluble in plasma and not associated with macromolecules or cells. Blood concentrations are related to quantity and time since intake, but foods ingested with or prior to a dose, genetic disposition, and habituation greatly modify individual responses. Concentrations in excess of 4 g/l, which would be lethal for most people, have been measured in habituated individuals after consumption of large amounts. Ethanol rapidly crosses both the BBB and the placenta because it diffuses freely through cell membranes.

METABOLISM

Overview: Small doses of ingested ethanol are taken up and metabolized mainly in the stomach wall (Lim et al. 1993; Lieber, 2000). Larger doses that exceed the first-pass metabolic capacity of the gastrointestinal wall are metabolized in the liver. About three-quarters of a moderate dose is exported to peripheral tissues as acetate; less than 5% contribute to *de novo* lipogenesis in the liver (Siler et al., 1999).

Ethanol oxidation: After consumption of moderate amounts (<5–10 g/day), the zinc-containing cytosolic enzyme ADH (EC1.1.1.1) in the stomach and liver initiates ethanol metabolism in a NAD-dependent reaction (Figure 6.22). Several isoenzymes occur due to the formation of homodimers and heterodimers of alpha-, beta-, and gamma-chains. Four additional isoforms exist, formed as homodimers of chains from distinct genetic loci and with catalytic and other characteristics different from the first three dominant forms. Women have distinctly lower class III and IV ADH activities in the stomach than men; as a consequence, ethanol is metabolized less rapidly in the digestive tract lining and a higher percentage of an ingested large dose reaches the circulation (Seitz et al., 1993; Baraona et al., 2001). At higher consumption levels and in habituated individuals (those consuming in excess of 5–7 servings of alcoholic beverages a week or more), ethanol is increasingly oxidized by the microsomal ethanol oxidizing system (MEOS), which includes monooxygenases of the cytochrome P-450 family (EC1.14.14.1) with broad specificity. CYP2E1 and, to a lesser extent, CYP1A2 and CYP3A4 are the main ethanol-oxidizing enzymes of the MEOS (Salmela et al., 1998). A flavoprotein, NADPH ferrihemoprotein reductase (EC1.6.2.4), which contains both flavin mononucleotide (FMN) and FAD, is structurally associated with the cytochromes and uses the reducing equivalents to generate NADPH. Riboflavin deficiency in rats has been found to reduce the activity of this enzyme (Wang et al., 1985).

Acetaldehyde: The intermediate generated by ADH action can be metabolized to acetic acid mainly by two types of enzymes aldehyde dehydrogenase (AIDH, EC1.2.1.3/NAD-requiring, EC1.2.1.4/NADP-requiring, EC1.2.1.5/NAD- or NADP-requiring) and acetaldehyde oxidase (EC1.2.3.1), which generates hydrogen peroxide. Many distinct gene products conforming to three different enzyme types (differing in cofactor requirements) exert AIDH activity in both cytosol and mitochondria. ALDH2 (EC1.2.1.3) in the mitochondrial matrix of the liver appears to be of particular importance for the clearance of large ethanol-derived amounts of acetaldehyde since Asians and South American Indians with inactive variants (e.g., 504E > K) have greatly increased susceptibility to acute alcohol intoxication. Other forms of AIDH are more abundantly expressed in the tissues and organs of the upper digestive tract, such as the salivary glands, esophagus, and stomach. The cytoplasmic enzyme acetaldehyde

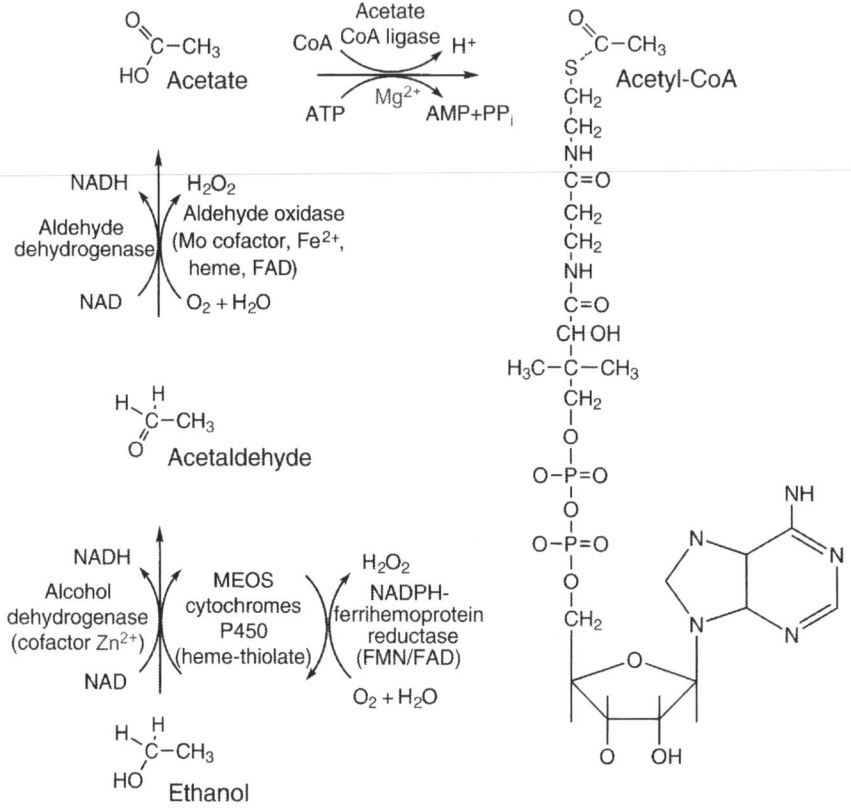

FIGURE 6.22

Ethanol oxidation and formation of acetyl-CoA.

oxidase (EC1.2.3.1) is abundantly expressed in the liver; in the kidney, heart, and brain, a less abundant variant is produced from a shortened transcript starting at an alternative polyadenylation site (Tomita et al., 1993). This enzyme requires FAD, the molybdenum cofactor, heme, and additional irons arranged in a 2Fe–2S cluster. When ethanol oxidation by the MEOS increases and acetaldehyde production greatly exceeds the capacity for its oxidation, tissue concentrations may reach levels that are directly toxic. Some individuals, especially of Asian descent, are susceptible to ethanol-induced flushing. This effect may be due to competition of acetaldehyde with the metabolites of histamine, methylimidazole acetaldehyde, and imidazole acetaldehyde (Zimatkin and Anichtchik, 1999). A single-base variant of ALDH2 increases susceptibility to alcohol-induced flushing (Crabb, 1990).

Acetate: While a significant proportion of unconjugated acetate is exported from liver to muscle and other tissues (Siler et al., 1999), large amounts are conjugated in mitochondria or cytosol by acetate-CoA ligase (thiokinase, EC6.2.1.1) and metabolized via the Krebs cycle. Alternatively, acetyl-CoA may enter one of several synthetic pathways for fatty acids, cholesterol, and other compounds. Condensation of two acetyl-CoA moieties (acetyl-CoA C-acetyltransferase, EC2.3.1.9), as well as addition of another acetyl-CoA (hydroxymethylglutaryl-CoA synthase, EC4.1.3.5), free up two CoA residues. The final CoA is released through the action of HMG-CoA lyase (EC4.1.3.4). Acetoacetate can be decarboxy-lated nonenzymatically to acetone or be reduced by 3-hydroxybutyrate dehydrogenase (EC1.1.1.30) to β-hydroxybutyrate. These three ketone bodies (acetone, acetoacetate, and β-hydroxybutyrate) are exported into the blood and utilized by the muscles, brain, and other extrahepatic tissues. The oxidative capacity for ketone bodies is exceeded when their concentration in arterial blood rises above 700 mg/l. Such high ketone body concentrations often occur with chronic consumption of large amounts of ethanol. In this case, noticeable amounts will appear in the breath and urine.

Free radicals: The metabolism of ethanol generates large amounts of oxygen free radicals. Hydrogen peroxide is produced by aldehyde oxidase–mediated oxidation of acetaldehyde in peroxisomes; in addition, acetaldehyde oxidase (EC1.2.3.1) can give rise to superoxide radicals from NADH generated by alcohol dehydrogenase (Mira et al., 1995). Carbon-centered radicals such as the hydroxyethyl radical can form adducts with cytochrome P450 2E1 and trigger specific autoimmune reactions associated with alcoholic liver cirrhosis (Clot et al., 1997).

Compound drugs: Users of cocaine and ethanol are exposed to a compound drug, cocaethylene, that arises from the *in vivo* transesterification of cocaine; cocaethylene is thought to be more neurotoxic than cocaine and readily passes across the placenta (Simone et al., 1997).

FUNCTION

Energy production: Ethanol has higher caloric density (6.9 kcal/g) than carbohydrates or proteins. There is increasing doubt about how much of the chemical energy content of ethanol can be utilized by humans (Siler et al., 1999).

Impact on intestinal function: Ethanol consumption increases intestinal motility and may cause diarrhea. Ingestion of significant amounts may interfere with the intestinal absorption of amino acids, folate (Feinman and Lieber, 1994), biotin (Said et al., 1990), and other nutrients.

Metabolic effects: Excessive ethanol oxidation increases the NADH/NAD ratio in the liver two-fold to threefold, which affects many other metabolic pathways, including oxidation of ethanol itself, gluconeogenesis, fatty acid oxidation, and lipoprotein secretion. At high concentrations, ethanol also competes with the metabolism of many xenobiotics, including medicinal drugs. Ethanol also increases greatly the proportion of 5-hydroxytryptamine (serotonin) that is converted to the dead-end product 5-hydroxytryptophol (5-HTOL) by alcohol dehydrogenase (EC1.1.1.1), enough to cause headaches, diarrhea, and fatigue in healthy subjects (Helander and Some, 2000).

Toxicity: Ethanol is acutely toxic for many tissues. Even relatively small doses (<0.1 g/kg body weight) increase reaction time, larger doses impair coordination, cloud judgment, and affect mood. Acute effects of large doses may also include loss of voluntary and involuntary motor control, vomiting, double vision, agitation, delusions, coma, and even death. Long-term overconsumption can cause fatty liver, hepatitis, liver cirrhosis, pancreatitis and cirrhosis of the pancreas, and various other damages.

Even at relatively modest intake levels, ethanol consumption appears to increase the risk of rectal and colon cancer (Seitz et al., 2001; Simanowski et al., 2001), and possibly also the risk of cancer at other sites. Ethanol consumption during pregnancy is often responsible for typical facial and other malformations and impaired mental development (fetal alcohol syndrome) in the children. Genetic differences in alcohol metabolism influence the vulnerability of the fetus (Streissguth and Dehaene, 1993;

McCarver, 2001). Even seemingly small amounts of alcoholic beverages (i.e., one serving) have the potential to cause irreversible harm.

Even modest amounts of ingested ethanol increase the permeability of the small and the large intestines (Elamin et al., 2014). The tight junction proteins ZO-1 and occluding were found to be redistributed in humans, and ZO-1 expression was down-regulated. These changes may be significant enough to disrupt the optimal barrier function of the intestinal wall and block the migration of pathogenic bacteria into adjacent tissues and the blood.

Habituation and alcoholism: Ethanol habituation induces several enzymes of the microsomal mixed-function oxidases, especially cytochrome P450 2E1. Induction of these enzymes greatly increases the capacity to metabolize ethanol (Lieber, 1999). Oxidation of ethanol without phosphorylation by the MEOS might explain the relatively inefficient utilization of its energy, especially in habituated drinkers (Feinman and Lieber, 1994).

As a consequence of MEOS induction, acetaminophen and other commonly used drugs are metabolized more rapidly to their toxic metabolites; similarly, the accelerated activation of dimethylnitrosamine can promote carcinogenesis. Catabolism of retinol is also accelerated. Finally, the induction of ethanol-metabolizing enzymes and various other liver proteins, including apolipoprotein A-I, affects synthesis and breakdown of lipids and lipoproteins (Luoma, 1988). Thus, increased omega-hydroxylation can change both type and rate of fatty acid metabolism (Laethem et al., 1993; Adas et al., 1998). Both beneficial and detrimental changes of lipoprotein profiles can result, depending on quantitative ethanol exposure, dietary habits, and individual disposition. About 10% of Americans suffer from alcoholism at some time in their lives (Garbutt et al., 1999). Drug treatment options with a reasonable level of evidence for effectiveness include opioid antagonists (naltrexone and nalmefene), acamprosate (Garbutt et al., 1999), and disulfiram. This drug (trade name Antabuse) for preventing relapse of alcoholism inhibits ethanol metabolism by irreversible carbamoylation and inhibition of mitochondrial aldehyde dehydrogenase (Shen et al., 2001).

REFERENCES

Adas, F., Betthou, F., Picart, D., Lozac'h, P., Beauge, F., Amet, Y., 1998. Involvement of cytochrome P450 E21 in the (omega l)-hydroxylation of oleic acid in human and rat liver microsomes. J. Lipid Res. 39, 1210–1219.

Baraona, E., Abittan, C.S., Dohmen, K., Moretti, M., Pozzato, G., Chayes, Z.W., et al., 2001. Gender differences in pharmacokinetics of alcohol. Alcohol Clin. Exp. Res. 25, 502–507.

Clot, P., Parola, M., Bellomo, G., Dianzani, U., Carini, R., Tabone, M., et al., 1997. Plasma membrane hydroxyethyl radical adducts cause antibody-dependent cytotoxicity in rat hepatocytes exposed to alcohol. Gastroenterology 113, 265–276.

Cope, K., Risby, T., Diehl, A.M., 2000. Increased gastrointestinal ethanol production in obese mice: implications for fatty liver disease pathogenesis. Gastroenterology 119, 1340–1347.

Crabb, D.W., 1990. Biological markers for increased risk of alcoholism and for quantitation of alcohol consumption. J. Clin. Invest. 85, 311–315.

Duffy, V.B., Davidson, A.C., Kidd, J.R., Kidd, K.K., Speed, W.C., Pakstis, A.J., et al., 2004. Bitter receptor gene (TAS2R38), 6-n-propylthiouracil (PROP) bitterness and alcohol intake. Alcohol Clin. Exp. Res. 28, 1629–1637.

Elamin, E., Masclee, A., Troost, F., Pieters, H.J., Keszthelyi, D., Aleksa, K., et al., 2014. Ethanol impairs intestinal barrier function in humans through mitogen activated protein kinase signaling: a combined *in vivo* and *in vitro* approach. PLoS One 9, e107421.

Feinman, L., Lieber, C.S., 1994. Nutrition and diet in alcoholism. In: Shils, M.E., Olson, J.A., Shike, M. (Eds.), Modern Nutrition in Health and Disease, eighth ed. Lea & Febiger, Philadelphia, PA, pp. 1081–1101.

Garbutt, J.C., West, S.L., Carey, T.S., Lohr, K.N., Crews, F.T., 1999. Pharmacological treatment of alcohol dependence: a review of the evidence. JAMA 281, 1318–1325.

Geertinger, E., Bodenhoff, J., Helweg-Larsen, K., Lund, A., 1982. Endogenous alcohol production by intestinal fermentation in sudden infant death. Z Rechtsmed. 89, 167–172.

Helander, A., Some, M., 2000. Dietary serotonin and alcohol combined may provoke adverse physiological symptoms due to 5-hydroxytryptophol. Life Sci. 67, 799–806.

Lachenmeier, D.W., Kanteres, F., Rehm, J., 2014. Alcoholic beverage strength discrimination by taste may have an upper threshold. Alcohol Clin. Exp. Res. 38, 2460–2467.

Laethem, R.M., Balaxy, M., Falck, J.R., Laethem, C.L., Koop, D.R., 1993. Formation of 19(S)-, 19(R)-, and 18(R)-hydroxyeicosatetraenoic acids by alcohol-inducible cytochrome P4502E1. J. Biol. Chem. 268, 12912–12918.

Lieber, C.S., 1999. Microsomal ethanol-oxidizing system (MEOS): the first 30 years (1968–1998)—a review. Alcohol. Clin. Exp. Res. 23, 991–1007.

Lieber, C.S., 2000. Alcohol: its metabolism and interaction with nutrients. Anmt. Rev. Nutr. 20, 395–430.

Lim Jr., R.T., Gentry, R.T., Ito, D., Yokoyama, H., Baraona, E., Lieber, C.S., 1993. First-pass metabolism of ethanol is predominantly gastric. Alcohol. Clin. Exp. Res. 17, 1337–1344.

Luoma, P.V., 1988. Microsomal enzyme induction, lipoproteins and atherosclerosis. Pharmacol. Toxicol. 62, 243–249.

Mattes, R.D., DiMeglio, D., 2001. Ethanol perception and ingestion. Physiol. Behav. 72, 217–229.

McCarver, D.G., 2001. ADH2 and CYP2EI genetic polymorphisms: risk factors for alcohol-related birth defects. Drug Metab. Disp. 29, 562–565.

Mira, L., Maia, L., Barreira, L., Manso, C.E., 1995. Evidence for free radical generation due to NADH oxidation by aldehyde oxidase during ethanol metabolism. Arch. Biochem. Biophys. 318, 53–58.

Picot, D., Lauvin, R., Hellegouarc'h, R., 1997. Intra-digestive fermentation in intestinal malabsorption syndromes: relations with elevated serum activity of gamma-glutamyltranspeptidase. Gastroenterol. Clin. Biol. 21, 562–566.

Said, H.M., Sharifan, A., Bagherzadeh, A., Mock, D., 1990. Chronic ethanol feeding and acute ethanol exposure *in vitro:* effect on intestinal transport of biotin. Am. J. Clin. Nutr. 52, 1083–1086.

Salmela, K.S., Kessova, I.G., Tsyrlov, I.B., Lieber, C.S., 1998. Respective roles of human cytochrome P-4502E1, 1A2, and 3A4 in the hepatic microsomal ethanol oxidizing system. Alcohol. Clin. Exp. Res. 22, 2125–2132.

Seitz, H.K., Egerer, G., Simanowski, U.A., Waldherr, R., Eckey, R., Agarwal, D.E., et al., 1993. Human gastric alcohol dehydrogenase activity: effect of age, sex, and alcoholism. Gut 34, 1433–1437.

Seitz, H.K., Matsuzaki, S., Yokoyama, A., Homann, N., Vakevainen, S., Wang, X.D., 2001. Alcohol and cancer. Alcohol. Clin. Exp. Res. 25, 137S–143S.

Shen, M.L., Johnson, K.L., Mays, D.C., Lipsky, J.J., Naylor, S., 2001. Determination of *in vivo* adducts of disulfiram with mitochondrial aldehyde dehydrogenase. Biochem. Pharmacol. 61, 537–545.

Siler, S.Q., Neese, R.A., Hellerstein, M.K., 1999. *De novo* lipogenesis, lipid kinetics, and whole-body lipid balances in humans after acute alcohol consumption. Am. J. Clin. Nutr. 70, 928–936.

Simanowski, U.A., Homann, N., Knuhl, M., Arce, L., Waldherr, R., Conradt, C., et al., 2001. Increased rectal cell proliferation following alcohol abuse. Gut 49, 418–422.

Simone, C., Byrne, B.M., Derewlany, L.O., Oskamp, M., Koren, G., 1997. The transfer of cocaethylene across the human term placental cotyledon perfused *in vitro*. Reprod. Toxicol. 11, 215–219.

Streissguth, A.P., Dehaene, P., 1993. Fetal alcohol syndrome in twins of alcoholic mothers: concordance of diagnosis and IQ. Am. J. Med. Genet. 47, 857–861.

Tomita, S., Tsujita, M., Ichikawa, Y., 1993. Retinal oxidase is identical to aldehyde oxidase. FEBS Lett. 336, 272–274.

Wang, T., Miller, K.W., Tu, Y.Y., Yang, C.S., 1985. Effects of riboflavin deficiency on metabolism of nitrosamines by rat liver microsomes. J. Natl. Cancer Inst. 74, 1291–1297.

Zimatkin, S.M., Anichtchik, O.V., 1999. Alcohol–histamine interactions. Alc Alcohol. 34, 141–147.

METHANOL

Methanol (methyl alcohol, wood alcohol; molecular weight 32) is readily water-soluble and can dissolve slightly hydrophobic molecules (Figure 6.23).

ABBREVIATIONS

ADH	alcohol dehydrogenase
AlDH	aldehyde dehydrogenase
CoA	coenzyme A
FAD	flavin adenine dinucleotide
FMN	flavin mononucleotide
MEOS	microsomal ethanol oxidizing system

NUTRITIONAL SUMMARY

Function: Methanol is an extremely toxic alcohol. Its conversion to acetyl-CoA requires riboflavin, niacin, pantothenate, zinc, (heme) iron, molybdenum, and magnesium; further oxidation of acetyl-CoA to water and carbon dioxide depends on thiamin, riboflavin, niacin, pantothenate, lipoate, ubiquinone, iron, and magnesium.

Food sources: Small amounts of methanol occur in poorly distilled spirits (such as moonshine). Rubbing alcohol, which may contain significant amounts of methanol, can also be a source, especially when ingested as alcohol abusers sometimes do. Inhalation or skin exposure can be important in some industrial settings.

Requirements: No beneficial effect of methanol intake is known.

Excessive intake: Methanol is harmful with intakes of just a few grams, including by inhalation or skin exposure. The main danger is related to the rapidly accumulating metabolite formic acid. At greater than the low concentration generated by normal metabolism, this organic acid causes direct toxicity to most other tissues and depletes active folate.

ENDOGENOUS SOURCES

Minute amounts of methanol are released with the cleavage of protein-linked carboxylic esters (e.g., methylated adrenocorticotropic hormone, growth hormone, luteinizing hormone, calmodulin, gamma-globulin, and histones) by protein-glutamate methylesterase (EC3.1.1.61) and with the detoxification of

OH
|

FIGURE 6.23

Methanol.

1-methyl-4-phenyl-1,2,3,6-tetrahydropyridine by methylphenyltetrahydropyridine N-monooxygenase (EC1.13.12.11, contains FAD).

$$1\text{-Methyl-4-phenyl-1,2,3,6-tetrahydropyridine} + O_2$$
$$\rightarrow 1\text{-methyl-4-phenyl-1,2,3,6-tetrahydropyridine N-oxide} + methanol$$

Microbial fermentation in the intestines of undigested carbohydrates and fiber from fruits and vegetables can generate biologically relevant amounts of methanol (Lee et al., 2012).

Histone deacetylation in the nucleus as part of normal genomic regulation generates significant amounts of formaldehyde, which have to be sequestered through conversion to formate and linking to tetrahydrofolate by the formate–tetrahydrofuran (THF) ligase component of methylene tetrahydrofolate dehydrogenase 1 (EC6.3.4.3).

DIETARY SOURCES

Methanol ingestion often comes with adulterated alcoholic beverages from illegal or home-brewed sources. Methanol tastes bitter, which prevents most people from ingesting significant quantities. Another route of exposure is through accidental or intentional inhalation of fumes. Carburetor cleaner and windshield washer fluids are typical sources of inhalatory poisoning that need emergency treatment (Givens et al., 2008).

Ingested ethylene glycol is converted metabolically to methanol and can cause significant toxicity (McMahon et al., 2009).

INTESTINAL ABSORPTION

Methanol is absorbed in the stomach and proximal small intestine. Since methanol readily permeates lipid membranes, the main transfer mechanism appears to be diffusion.

TRANSPORT AND CELLULAR UPTAKE

Blood circulation: Methanol is readily soluble in plasma and not associated with macromolecules or cells. Blood concentrations are related to quantity and time since intake, but foods ingested with or prior to a dose, genetic disposition, and habituation greatly modify individual responses.

BBB: Methanol rapidly crosses both the BBB because it diffuses freely through cell membranes.

Materno-fetal transfer: The methanol metabolite from maternal metabolism readily passes through the placenta and can expose the fetus to harmful amounts (Hutson et al., 2013).

METABOLISM

Most ingested methanol is oxidized via formaldehyde to formate, and this can be oxidized to CO_2 or used for purine synthesis (Figure 6.24). Complete oxidation of methanol depends on adequate supplies of niacin, folate, glutathione, zinc, and magnesium.

Methanol oxidation: After consumption of moderate amounts (<5–10 g/day) the zinc-containing cytosolic enzyme ADH (EC1.1.1.1) in the stomach and liver initiates methanol metabolism in a NAD-dependent reaction. Several isoenzymes occur due to the formation of homodimers and heterodimers of

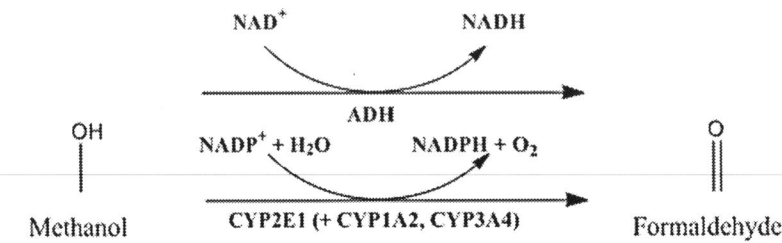

FIGURE 6.24

Metabolism of methanol.

alpha-, beta-, and gamma-chains. Four additional isoforms exist, formed as homodimers of chains from distinct genetic loci and with catalytic and other characteristics different from the first three dominant forms. Women have distinctly lower class III and IV ADH activities in the stomach than men, and as a consequence, methanol is metabolized less rapidly in the digestive tract lining and a higher percentage of an ingested large dose reaches circulation (Seitz et al., 1993; Baraona et al., 2001). At higher consumption levels and in habituated individuals (those consuming in excess of 5–7 servings of alcoholic beverages a week or more), ethanol is increasingly oxidized by the MEOS, which includes monooxygenases of the cytochrome P-450 family (EC1.14.14.1) with broad specificity. CYP2E1 and, to a lesser extent, CYP1A2 and CYP3A4 are the main methanol-oxidizing enzymes of the MEOS (Salmela et al., 1998). A flavoprotein, NADPH-ferrihemoprotein reductase (EC1.6.2.4), which contains both FMN and FAD, is structurally associated with the cytochromes and uses the reducing equivalents to generate NADPH. Riboflavin deficiency in rats has been found to reduce the activity of this enzyme (Wang et al., 1985).

Formaldehyde: The intermediate generated by ADH action can be metabolized to formic acid mainly by two types of enzymes: aldehyde dehydrogenase (ALDH, EC1.2.1.3/NAD-requiring, EC1.2.1.4/NADP-requiring, EC1.2.1.5/NAD or NADP-requiring) and acetaldehyde oxidase (EC1.2.3.1), which generates hydrogen peroxide.

Formaldehyde dismutase (EC1.2.99.4) is a class II alcohol dehydrogenase (Zn) in the liver with isoenzymes beta1, beta1 and gamma1, and gamma1 that convert formaldehyde into equal amounts of methanol and formic acid (Kapur et al., 2007).

$$2\,\text{Formaldehyde} + H_2O \rightarrow \text{methanol} + \text{formic acid}$$

Many distinct gene products conforming to three different enzyme types (differing in cofactor requirement) exert ALDH activity in both cytosol and mitochondria. ALDH2 (EC1.2.1.3) in the mitochondrial matrix of the liver appears to be of particular importance for the clearance of large ethanol-derived amounts of acetaldehyde since Orientals and South American Indians with an inactive variant (ALDH2 504E > K) have greatly increased susceptibility to acute alcohol intoxication. Other forms of ALDH are more abundantly expressed in the tissues and organs of the upper digestive tract, such as the salivary glands, esophagus, and stomach. The cytoplasmic enzyme acetaldehyde oxidase (EC1.2.3.1) is abundantly expressed in the liver; in the kidney, heart, and brain, a less abundant variant is produced from a shortened transcript starting at an alternative polyadenylation site (Tomita et al., 1993). This enzyme requires FAD, the molybdenum cofactor, heme, and additional irons arranged in a 2Fe–2S cluster.

When ethanol oxidation by the MEOS increases and acetaldehyde production greatly exceeds the capacity for its oxidation, tissue concentrations may reach levels that are directly toxic. Some individuals, especially of Asian descent, are susceptible to ethanol-induced flushing. This effect may be due to competition of acetaldehyde with the metabolites of histamine, methylimidazole acetaldehyde, and imidazole acetaldehyde (Zimatkin and Anichtchik, 1999). A single-base variant of ALDH2 increases susceptibility to alcohol-induced flushing (Crabb, 1990).

Formic acid: Adequate folate intake optimizes formic metabolism and provides some protection of the fetus in a mother exposed to methanol (Hutson et al., 2013).

A major objective of treatment is the controlled administration of ethanol, which slows formic acid generation from methanol by competition for the ADH enzyme (Coulter et al., 2011). The inhibitor fomepizole has also been used to slow the harmful conversion of methanol to formic acid, a toxic metabolite (Burns et al., 1997).

Hemodialysis with the use of appropriate filters can remove methanol from the blood (Kute et al., 2012).

FUNCTIONS AND EFFECTS

Toxicity: Methanol is acutely toxic and often fatal due to irreversible brain damage. However, methanol amounts that would be lethal for most people have been survived in habituated individuals.

The toxic effects come partially from direct actions and partially through the effects of its metabolites, formic acid and formaldehyde. Methanol and its metabolites damage brain tissue, the retina, the optic nerve, and other neural tissues. Rapid generation of formic acid often leads to severe acidosis and can be the cause of death after methanol intoxication (Kruse, 2012). Formic acid is a potent inhibitor of cytochrome oxidase in the mitochondrial respiratory chain (Eells et al., 2003).

Even relatively small doses of methanol (<0.1 g/kg body weight) increase reaction time, larger doses impair coordination, cloud judgment, and affect mood. Cell damage in the retina causes loss of vision at the center (central scotoma) and the periphery of the visual field, photophobia, diminished light adaptation, and even complete blindness (Eells et al., 2003). Fundoscopic examination may reveal hyperemia and disk edema, and with prolonged exposure, atrophy of the optic nerve as well.

REFERENCES

Burns, MJ, Graudins, A, Aaron, CK, McMartin, K, Brent, J., 1997. Treatment of methanol poisoning with intravenous 4-methylpyrazole. Ann. Emerg. Med. 30, 829–832.

Colter, C.V., Isbister, G.K., Duffull, S.B., 2011. The pharmacokinetics of methanol in the presence of ethanol: a case study. Clin. Pharmacokinet. 50, 245–251.

Eells, J.T., Henry, M.M., Summerfelt, P., Wong-Riley, M.T., Buchmann, E.V., Kane, M., et al., 2003. Therapeutic photobiomodulation for methanol-induced retinal toxicity. Proc. Nat. Acad. Sci. USA 100, 3439–3444.

Givens, M., Kalbfleisch, K., Bryson, S., 2008. Comparison of methanol exposure routes reported to Texas poison control centers. West. J. Emerg. Med. 9, 150–153.

Hutson, J.R., Lubetsky, A., Eichhorst, J., Hackmon, R., Koren, G., Kapur, B.M., 2013. Adverse placental effect of formic acid on hCG secretion is mitigated by folic acid. Alcohol Alcohol 48, 283–287.

Kapur, B.M., Vandenbroucke, A.C., Adamchik, Y., Lehotay, D.C., Carlen, P.L., 2007. Formic acid, a novel metabolite of chronic ethanol abuse, causes neurotoxicity, which is prevented by folic acid. Alcohol Clin. Exp. Res. 31, 2114–2120.

Kruse, J.A., 2012. Methanol and ethylene glycol intoxication. Crit. Care Clin. 28, 661–711.

Kute, V.B., Godara, S.M., Shah, P.R., Gumber, M.R., Goplani, K.R., Vanikar, A.V., et al., 2012. Hemodialysis for methyl alcohol poisoning: a single-center experience. Saudi J. Kidney Dis. Transpl. 23, 37–43.

Lee, H.J., Pahl, M.V., Vaziri, N.D., Blake, D.R., 2012. Effect of hemodialysis and diet on the exhaled breath methanol concentration in patients with ESRD. J. Ren. Nutr. 22, 357–364.

McMahon, D.M., Winstead, S., Weant, K.A., 2009. Toxic alcohol ingestions: focus on ethylene glycol and methanol. Adv. Emerg. Nurs. J. 31, 206–213.

NONNUTRIENTS AND BIOACTIVES

7

CHAPTER OUTLINE

Indigestible Carbohydrates 243
Flavonoids and Isoflavones 247
Garlic Bioactives 260

INDIGESTIBLE CARBOHYDRATES

The bulk of material that is not absorbed by the time ingested food reaches the terminal ileum is one or another form of indigestible or poorly digestible carbohydrates. These carbohydrates are not digested because humans do not have the biochemical tools to break the connecting bonds. Long chains of indigestible carbohydrates are conventionally called *dietary fiber*. The usual distinction between insoluble and soluble fiber is not just descriptive; it also predicts chemical and biological properties. A somewhat related group comprises the resistant starches. The term *resistant* refers to the fact that a significant percentage of these starch molecules escapes digestion because it is partially encapsulated (mostly as part of whole grains) or is not fully hydrated (incompletely cooked al dente pasta).

The seven types of enzymes in the mouth and gastrointestinal tract cleave only a limited number of sugar bonds (Table 7.1). They cannot break down sugar-containing polymers with other types of sugar links.

ABBREVIATIONS	
Fru	D-fructose
Glc	D-glucose
Gal	d-α-galactose
GalNAc	Gal *N*-acetylglucosamine
Man	D-mannose

NUTRITIONAL SUMMARY

Function: Dietary fiber provides bulk to the intestinal contents; binds bile acids, carcinogens, and other potentially harmful food constituents; and feeds as a prebiotic beneficial gut microbes.

Food sources: Whole grains, fruits, and vegetables are excellent sources. Oats provide significant amounts of beneficial soluble fiber.

Nutrient Metabolism.

243

Table 7.1 The Seven Types of Human Digestive Enzymes		
Glc–Glc α-[1>4] linkages	Starches	Alpha-amylase (EC3.2.1.1)
Glc–Glc α-[1>6] linkages	Starches	Alpha-amylase (EC3.2.1.1)
Glc–Glc terminal nonreducing 1,4-linkages	Starches	Glucoamylase (EC3.2.1.3)
Glc–Glc terminal nonreducing 1,4-linkages	Maltose	Maltase (EC3.2.1.20)
1,6-α-D-glucosidic bonds of branched oligomers	Isomaltose	Isomaltase (EC3.2.1.10)
Glc–Fru 1,6-α-D-glucosidic	Sucrose	Sucrase (EC3.2.1.48)
Glc–Glc α-glucosidase	Trehalose	Trehalase (EC3.2.1.28)
Gal–Glc β-glucosidase	Lactose, cellobiose	Lactase (EC3.2.1.108)

Requirements: Men should get 38 g of total fiber per day. Women need at least 25 g. More than a third of these amounts should consist of soluble fiber.

Deficiency: Low fiber intake increases the risk of constipation, cardiovascular disease, and cancer.

Excessive intake: There is no credible evidence that very high intake of dietary fiber with food causes harm. Consumption of large amounts of dietary fiber as supplements may cause constipation and fecal compaction, particularly when used without the requisite large amounts of liquids.

INDIGESTIBLE POLYSACCHARIDES

Most plant carbohydrates cannot be absorbed in the human digestive tract. Nonetheless, many of these nondigestible polysaccharides as part of plant foods, usually called *dietary fiber* in this context, provide important health benefits. Daily adequate intake of total dietary fiber is 38 g for men and 25 g for women (King et al., 2012). Actual intake by Americans still hovers around 16 g/day, and many people miss health benefits that would be available with higher intake.

INSOLUBLE FIBER

Cellulose is a relatively linear glucose polymer with α-[1>4] linkages. This is the main and poorly soluble dietary fiber in grains and most vegetables.

Hemicelluoses have backbones of α-[1>4] linked glucose molecules with various branches and side chains.

Most insoluble fiber is consumed with grains, particularly whole grains. The typical insoluble fiber intake of Americans is around 13 g/day (Peery et al., 2012).

Likely health benefits with adequate consumption of insoluble fiber include lower cardiovascular risk, lower risk of colorectal cancer, and lower prostate cancer risk (Deschasaux et al., 2014). Unsupported beliefs in specific health benefits of dietary fiber still persist and need to be corrected. For instance, a high-fiber diet usually increases stool volume and bowel movement frequency but is not likely to protect against diverticulosis (Peery et al., 2012).

SOLUBLE FIBER

Beta-glucans, particularly in oats, eggplants, mushrooms, yeast, and psyllium seeds (ispaghula, plantago, mucilage, and Metamucil), consist of glucose with mixed [1>3][1>4] linkage.

Galactomannans are the main polysaccharides in guar gum and locust bean gum.

Pectins are acidic polymers with frequent side chains. Galacturonic acids and a small percentage of α-L-rhamnose constitute the main chains, and galactose and arabinose form the side chains. Pectins from different sources vary in the extent of methoxylation of the galacturonate moieties. Apple, pear, and guava are particularly rich sources. Natural (but not commercial purified) pectin is largely insoluble in water.

Lignin is mainly consumed with whole seeds, such as flax, sesame, and poppy seeds.

Carageenan is a type of polysaccharides from algae. Despite the pronouncement by the International Agency for Research on Cancer (IARC) long ago that carrageenan breakdown products are carcinogenic in humans, carageenan is added to many commercial food products (Tobacman, 2001).

Xanthan gum is an industrially produced polysaccharide, originally from slime molds, that is added as a gelling agent to numerous commercial food products. Repeating pentasaccharide units consist of a β-1,4 glucan backbone dimer with a trisaccharide side chain of acetylated α-mannose, β-glucuronate, partially (30%) pyruvylated β-mannose. Sometimes the side chains do not contain the terminal mannose.

Alginate is a copolymer of α-L-guluronic acid and β-D-mannuronic acid. It is not only commercially produced from brown algae but also secreted by many bacteria and other organisms.

Most soluble fiber is consumed with oats, fruits, juices, and vegetables. Dietary supplements are a significant source for a small minority of consumers. The typical insoluble fiber intake of Americans is around 7 g/day (Peery et al., 2012).

Long-term benefits from high intake of soluble fiber from foods may include a reduced risk of breast cancer (Deschasaux et al., 2013) and lower cardiovascular risk. Generous dietary fiber intake is very likely to reduce the risk of a first stroke, but there is uncertainty about the amount and type of dietary fiber that is most effective for prevention (Threapleton et al., 2013). Beta-glucans, in particular, are quite effective for lowering atherogenic lipoprotein concentrations in the blood (Charlton et al., 2012). Soluble fiber has been found to be helpful for controlling energy intake and manage weight (Carter et al., 2012). It is important to remember that soluble fiber can be added to beverages, which improves acceptability for some.

INTESTINAL FERMENTATION

Specialized bacteria in the stomachs of ruminants can utilize the large quantities of cellulose ingested with grass in these animals, but humans do not have such an arrangement. However, various indigestible polysaccharides are effective prebiotics, which are utilized by the human intestinal flora. As a result, the bacteria generate significant quantities of short-chain fatty acids (acetate, propionate, and butyrate), which contribute to the nutriture of the distal intestinal epithelia; help with the regulation of electrolyte absorption, increasing the water content and viscosity of stool; and may have additional metabolic effects in the liver and other tissues (Puertollano et al., 2014). Other effects of dietary fiber with likely or potential significance for function and health include the bulking action of fermentable dietary fiber, a delay of carbohydrate absorption, and the sequestration of bile acids, cholesterol, and toxic food substances.

EXCESSIVE INTAKE

There are persistently voiced concerns with excessively high dietary fiber intake, particularly as concentrated dietary supplements, because of possible interference with mineral and trace element absorption.

Lower calcium, iron, copper, and zinc absorption has been reported. The extent to which indigestible polysaccharides by themselves and associated nonabsorbed compounds (such as phytates) decrease metal ion absorption is still a matter of debate.

INDIGESTIBLE OLIGOSACCHARIDES

Plant-based indigestible oligosaccharides

The mixed oligosaccharides raffinose (Gal α-[1>6] Glc α-[1>β2] Fru), stachyose (Gal α-[1>6] Gal α-(1>6] Glc α-[1>β2] Fru), and verbascose (Gal α-[1>6] Gal α-[1>6] Gal α-[1>6] Glc α-[1>β2] Fru) in beans, peas, and other plant-derived foods are not digestible in humans and may cause abdominal distention and flatulence due to bacterial metabolism. Melezitose (Glc α-[1>3] Fru α-Glc) and turanose (Glc α-[1>3] Fru), minor constituents of conifer-derived honey, are additional examples of Fru-containing oligosaccharides to which humans may be exposed. Use of oral α-galactosidase (EC3.2.1.22, trade name Beano) along with a meal can initiate the digestion of these oligosaccharides (Ganiats et al., 1994).

The indigestible carbohydrates (dietary fiber) inulin and various smaller oligofructose species are composed of β[2-1] fructose-fructosyl chains with a minor contribution of glucose residues (Flamm et al., 2001). Inulin from chicory roots consists of 10–20 fructose residues. Oligofructoses contain fewer than 10 fructose residues. Microflora, especially *Bifidobacteria*, of the distal ileum and colon thrive on these indigestible carbohydrates. The bacterial breakdown generates short-chain fatty acids that are utilized by local enterocytes (mainly butyrate) or transported to the liver (mainly acetate and propionate).

Milk indigestible oligosaccharides

The colostrum (first milk) from humans and other mammals contains diverse indigestible oligosaccharides with particular prebiotic characteristics (Shen et al., 2001). Typical individual constituents include 3′- and 6′-sialyllactose, sialyllacto-N-tetraoses a and b, and sialyllacto-N-neotetraose c. These oligosaccharides constitute less than 5% of the total milk carbohydrates and are not likely to be important energy sources due to their low bioavailability. It is known, however, that they enhance in breast-fed infants the intestinal colonization with beneficial *Bifidobacteria*. Another protective effect may rely on their ability to bind to specific surface components (e.g., S-fimbriae) of potentially pathogenic bacteria (Schwertmann et al., 1999). Similarly beneficial probiotic effects may accrue to adult consumers of milk and milk products. Legumes (beans, peas, and lentils) and other plant-derived foods contain oligosaccharides which are not well digested. These include raffinose (Gal α-[1>6] Glc α-[1>β2] Fru), stachyose (Gal α-[1>6] Gal α-[1>6] Glc α-[1>β2] Fru), and verbascose (Gal α-[l>6] Gal α-[l>6] Gal α-[l>6] Glc α-[l>β2] Fru). Since human digestive enzymes cannot cleave the galactose in alpha-position, hydrogen- and methane-producing bacteria in the distal digestive tract ferment these carbohydrates. The use of a dietary supplement with alpha galactosidase (EC3.2.1.22, Beano) whenever legumes or other gas-inducing plant foods are consumed relieves symptoms (Ganiats et al., 1994). The enzyme cleaves the galactose from the offending oligosaccharides, and the resulting galactose and sucrose then can be digested and absorbed normally. Synthetic low-energy sugar substitutes include the indigestible sucrose-linkage isomers isomaltulose (6-O-alpha-D-glucopyranosyl-D-fructose) and leucrose (5-O-alpha-D-glucopyranosyl-D-Fructose).

REFERENCES

Carter, B.E., Drewnowski, A., 2012. Beverages containing soluble fiber, caffeine, and green tea catechins suppress hunger and lead to less energy consumption at the next meal. Appetite 59, 755–761.

Charlton, K.E., Tapsell, L.C., Batterham, M.J., O'Shea, J., Thorne, R., Beck, E., et al., 2012. Effect of 6 weeks' consumption of β-glucan-rich oat products on cholesterol levels in mildly hypercholesterolaemic overweight adults. Br. J. Nutr. 107, 1037–1047.

Deschasaux, M., Pouchieu, C., His, M., Hercberg, S., Latino-Martel, P., Touvier, M., 2014. Dietary total and insoluble fiber intakes are inversely associated with prostate cancer risk. J. Nutr. 144, 504–510.

Deschasaux, M., Zelek, L., Pouchieu, C., His, M., Hercberg, S., Galan, P., et al., 2013. Prospective association between dietary fiber intake and breast cancer risk. PLoS One 8 (11), e79718.

Flamm, G., Olinsmann, W., Kritchevsky, D., Prosky, L., Roberfroid, M., 2001. Inulin and oligofructose as dietary fiber: a review of the evidence. Crit. Rev. Food Sci. Nutr. 41, 353–362.

Ganiats, T.G., Norcross, W.A., Halverson, A.L., Burford, P.A., Palinkas, L.A., 1994. Does Beano prevent gas? A double-blind crossover study of oral alpha-galactosidase to treat dietary oligosaccharide intolerance. J. Fam. Pract. 39, 441–445.

King, D.E., Mainous III, A.G., Lambourne, C.A., 2012. Trends in dietary fiber intake in the United States, 1999–2008. J. Acad. Nutr. Diet 112, 642–648.

Peery, A.F., Barrett, P.R., Park, D., Rogers, A.J., Galanko, J.A., Martin, C.F., et al., 2012. A high-fiber diet does not protect against asymptomatic diverticulosis. Gastroenterology 142, 266–272.e1.

Puertollano, E., Kolida, S., Yaqoob, P., 2014. Biological significance of short-chain fatty acid metabolism by the intestinal microbiome. Curr. Opin. Clin. Nutr. Metab. Care 17, 139–144.

Schwertmann, A., Schroten, H., Hacker, J., Kunz, C., 1999. S-fimbriae from *Escherichia coli* bind to soluble glycoproteins from human milk. J. Ped. Gastroent. Nutr. 28, 257–263.

Shen, Z., Warren, C.D., Newburg, D.S., 2001. Resolution of structural isomers of sialylated oligosaccharides by capillary electrophoresis. J. Chromatogr. A 921, 315–321.

Threapleton, D.E., Greenwood, D.C., Evans, C.E., Cleghorn, C.L., Nykjaer, C., Woodhead, C., et al., 2013. Dietary fiber intake and risk of first stroke: a systematic review and meta-analysis. Stroke 44, 1360–1368.

Tobacman, J.K., 2001. Review of harmful gastrointestinal effects of carrageenan in animal experiments. Env. Hlth. Persp. 109, 983–994.

FLAVONOIDS AND ISOFLAVONES

Flavonoids are polyphenolic compounds with structural similarities to the oxygen heterocycle flavan. Members of this very diverse and ubiquitous group of plant molecules (phytochemicals) include flavones, isoflavans, flavonols, flavanols, flavanones, anthocyanins, and proanthocyanidins. The structurally and functionally closely related isoflavones are also discussed in this section.

ABBREVIATIONS

MDR1	multidrug resistance gene 1, P-glycoprotein 1 (Pgp1, ABCB1)
MDR3	multidrug resistance gene 3, P-glycoprotein 3 (Pgp3, ABCB4)
MRP1	multidrug resistance protein 1 (ABCC1)
SGLT1	sodium-glucose cotransporter 1 (SLC5A1)

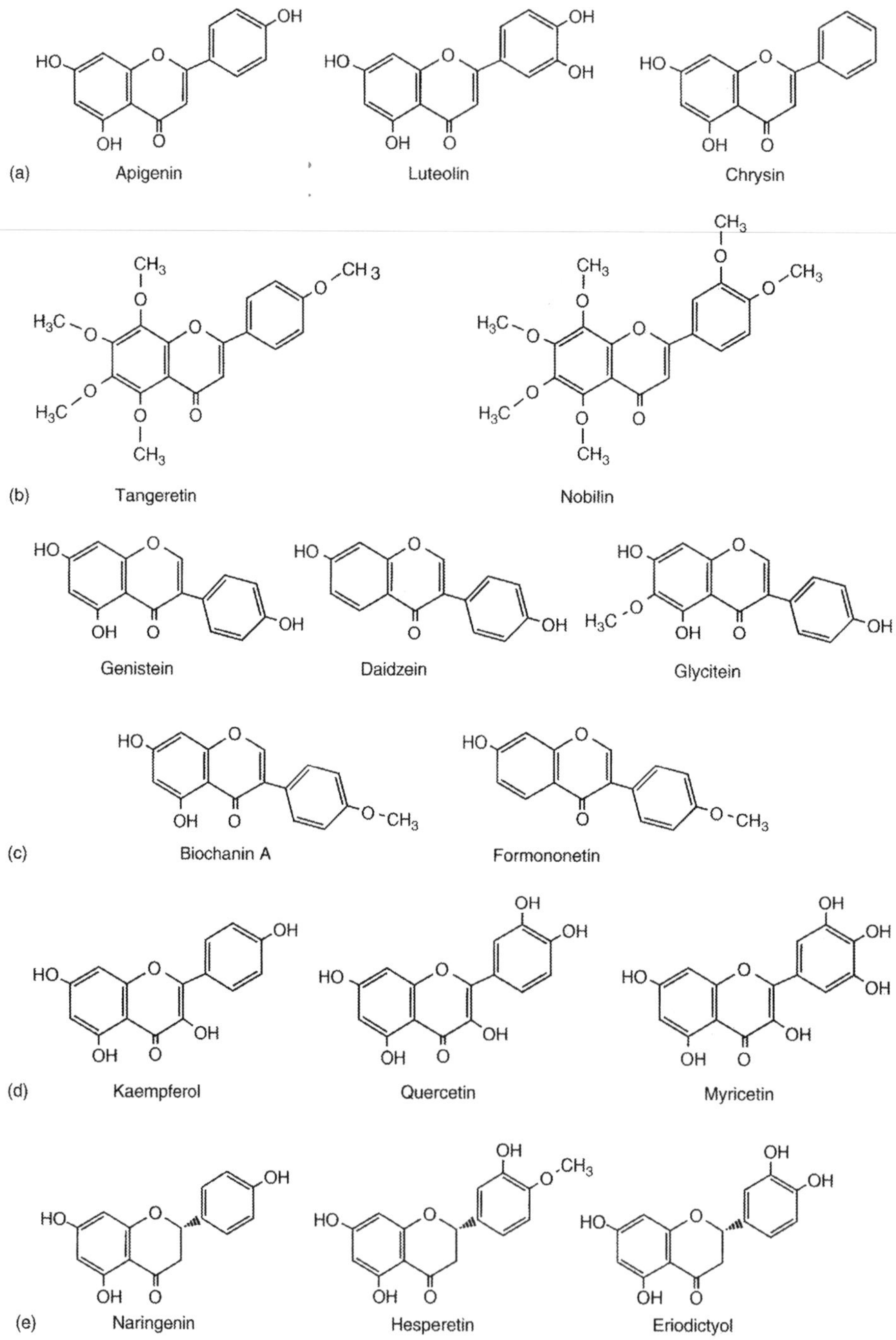

FIGURE 7.1

Structures of common types of dietary flavonoids. (A) flavones; (B) polymethoxylated flavones; (C) isoflavones; (D) flavonols; (E) flavanones; (F) catechins; (G) anthocyanidins.

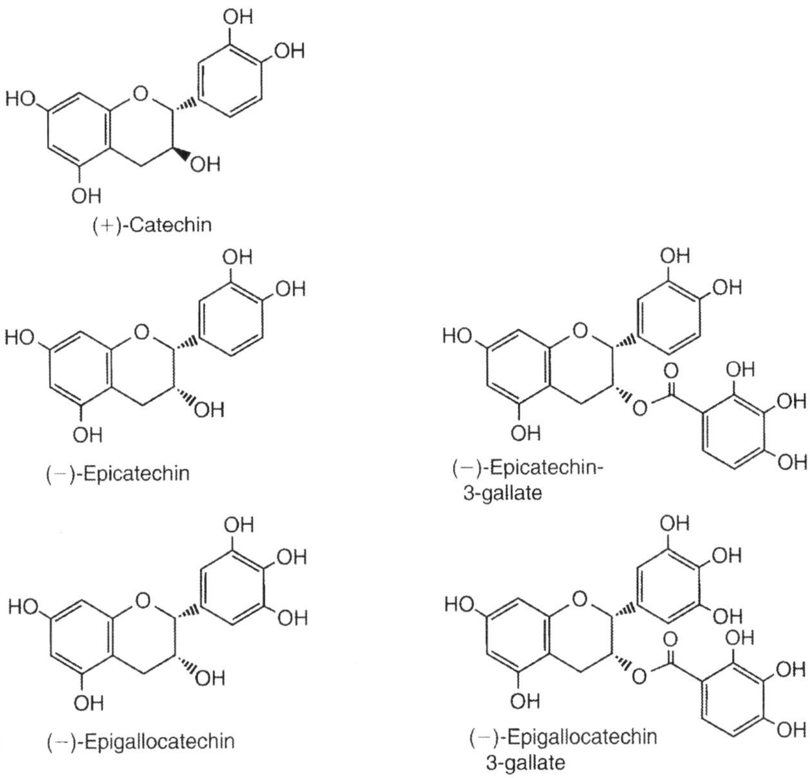

(+)-Catechin

(−)-Epicatechin

(−)-Epicatechin-
3-gallate

(f)　(−)-Epigallocatechin

(−)-Epigallocatechin
3-gallate

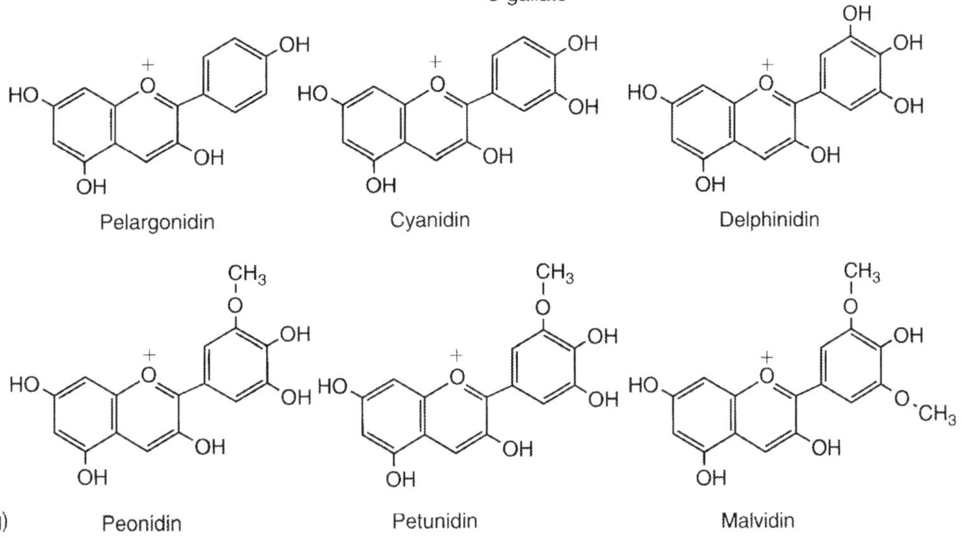

Pelargonidin

Cyanidin

Delphinidin

(g)　Peonidin

Petunidin

Malvidin

FIGURE 7.1

(Continued)

NUTRITIONAL SUMMARY

Function: The diversity of flavonoids is too great to assign them specific properties. It is particularly important to emphasize that many flavonoids are extremely toxic for humans. However, various flavonoids in edible foods are likely to promote health through antioxidant effects, actions on cellular signaling events, binding to sex-hormone receptors, and modulation of detoxifying enzymes. Flavonoid-rich fruits and vegetables (but not necessarily pharmaceutical products) are likely to decrease the risk of atherosclerosis, cancer, and osteoporosis.

Food sources: Many fruits and vegetables are rich in flavonoids. Well-researched examples include soybeans (with the isoflavones genistein, daidzein, and glycitein), onions (with the flavonols quercetin, kaempferol, and luteolin), grapefruit (with the flavanone naringenin), oranges (with the flavanone hesperitin and polymethoxylated flavones like tangeretin and nobiletin), chocolate (with anthocyanins), and tea (with catechins and quercetin).

Requirements: While there are no distinct requirements, a healthy diet should contain several servings of diverse fruits and vegetables. This implicitly ensures a moderate to high intake of various flavonoids.

Deficiency: No specific functional deterioration has been linked to low or absent intake of flavonoids. However, people with low intake miss out on likely beneficial effects.

Excessive intake: While the consumption of commonly consumed flavonoid-rich foods in typical quantities is likely to be beneficial, the safety of high-dose food extracts or synthetic products is uncertain.

STRUCTURES

All flavonoids (Figure 7.1) contain a benzopyran ring system (rings A and C) with an aromatic ring (ring B) linked to the oxygen heterocycle (Merken and Beecher, 2000). The classification of flavonoids is based on variations in the heterocyclic ring. The oxygen heterocyclic ring in flavones such as apigenin, luteolin (5,7,3′,4′-tetrahydroxyflavone), tricetin (5,7,3′,4′,5′-pentahydroxyflavone), and chrysin (5,7-dihydroxyflavone), contains two double bonds and no substitution. Methoxyl instead of hydroxyl groups may also be attached to the basic flavone structure, as in tangeretin (5,6,7,8,4′ pentamethoxyflavone), nobiletin (5,6,7,8,3′,4′-hexamethoxyflavone), and 3,5,6,7,8,3′,4′-heptamethoxyflavone. The B-ring in isoflavones (genistein, daidzein, glycitein, biochanin A, and formonetin) is linked to carbon 3 of the oxygen heterocycle. The flavonols, which include quercetin (3,5,7,3′,4′-pentahydroxyflavone), kaempferol (3,5,7,4′-tetrahydroxyflavone), and myricetin (3,5,7,3′,4′,5′-hexahydroxyflavone), have an additional hydroxy group at carbon 3. In flavanones, such as naringenin (5,7,4′-trihydroxyflavanone), hesperitin (5,7,3′-trihydroxy4′-methoxyflavanone), and eriodictyol (5,7,3′,4′ tetrahydroxyflavanone), the heterocyclic ring is saturated. The catechins and anthocyanidins lack the carbonyl function at carbon 4. An excellent discussion of flavonoid structures and analytical methods for flavonoid measurement in foods and other biological samples has been presented elsewhere (Merken and Beecher, 2000).

Flavonoids in plants are predominantly linked in beta-configuration to sugars (Figure 7.2) such as glucose, rhamnose (6-deoxy-L-mannose), rutinose (6-*O*-α-L-deoxymannosyl-D-glucose), neohesperidose (6-deoxy-α-L-mannopyranosyl-β-D-glucose), galactose, and xylose (Merken and Beecher, 2000). Genistein is ingested mainly as genistein-7-β-glucoside (genistin), naringenin as naringenin-7-β-rhamnoglucoside (naringin), naringenin-7-β-rutinoside (narirutin), or naringenin-7-β-glucoside (prunin), and quercetin in the forms of quercetin-3-β-rutinoside (rutin), quercetin-4′-β-glucoside, quercetin-3-β-glucoside

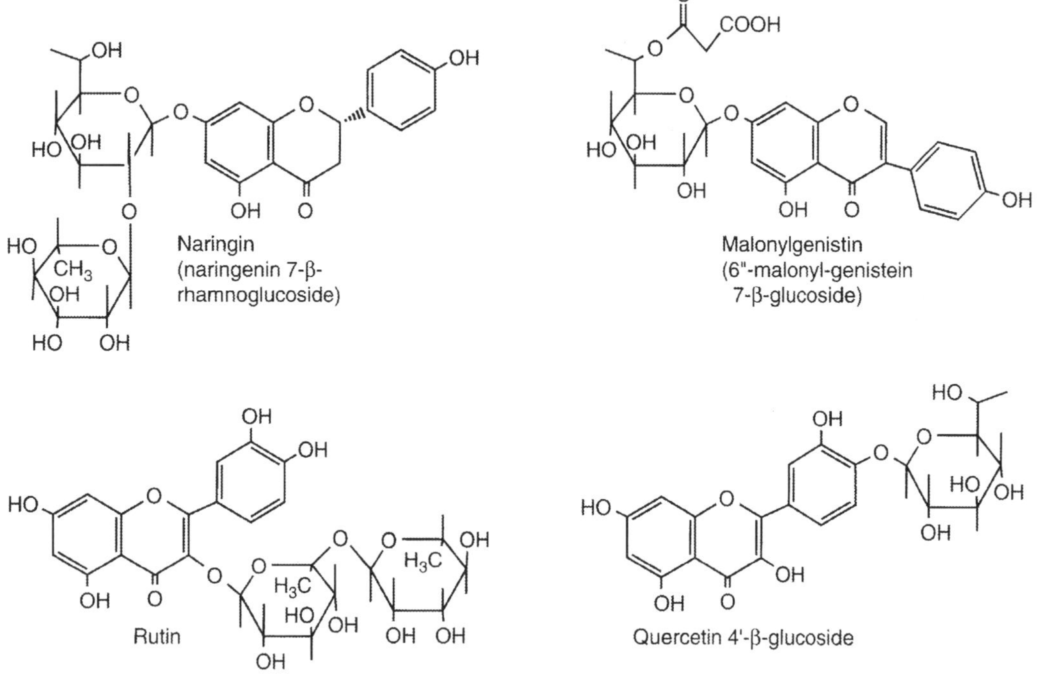

FIGURE 7.2

Most flavonoids in foods are linked to sugars and other molecules.

(isoquercitrin), quercetin-3-β-rhamnoside (quercitrin), or quercetin-3-galactoside (hyperoside). Flavonoid β-glycosides also may carry additional small molecular modifiers, such as malonyl and acetyl groups. Aglycones are flavonoid molecules without any attached sugars or other modifiers. Sugar-linked flavonoids are called *glycosides*. The term *glucoside* applies only to flavonoids linked to glucose.

DIETARY SOURCES

Isoflavones: The main flavonoid compounds in soy are genistein, daidzein, and glycitin. Coumestrol is found in lima beans, peas, alfalfa sprouts, and clover sprouts. Formononetin is in green beans and clover sprouts. Biochanin A is in garbanzo, kidney, pinto, and other kinds of beans, but not in soybeans (Franke et al., 1995). These isoflavones are present as aglycones, beta-glucosides, acetylglucosides, and malonylglucosides.

The amounts and proportions of these types depend on species, variety, growing conditions, and processing. Median daily isoflavone consumption of postmenopausal women in the United States is a fraction of a milligram (Framingham Heart Study; de Kleijn et al., 2001). Mean daily isoflavone intake of soy-consuming individuals, such as in some East Asian countries, often is well above 30 mg.

Flavones: Apigenin is consumed with parsley, chamomile tea, celery, and other fruits and vegetables. Many fruits and vegetables, including olives (Blekas et al., 2002), contain luteolin. The peels and expressed juice of oranges and other citrus fruit contain various polymethoxylated flavones (Takanaga et al., 2000), including tangeretin (5,6,7,8,4′-pentamethoxyflavone, 0.3–1.6 µg/ml), nobiletin (5,6,7,8,3′,4′-hexamethoxyflavone, 0.8–4.3 µg/ml), and 3,5,6,7,8,3′,4′-heptamethoxyflavone (0.2–1.5 µg/ml).

Flavonols: Yellow onions and berries are the main source of flavonols. The principal flavonol in onions is quercetin (about 0.4 mg/g), 45% of it as quercetin-4′-glucoside. Berries and the peels of apples, tomatoes (especially cherry tomatoes), and other fruits and vegetables contain quercetin-3-*O*-rutinoside (rutin). The quercetin content of olives can be as high as 0.6 mg/g (Blekas et al., 2002). Blackcurrants and other berries contain relatively large amounts of myricetin (Hakkinen et al., 1999). Kaempferol is not only found in gooseberries and strawberries, but also in many other fruits and vegetables. Average daily quercetin intake in the Netherlands is 16 mg/day (Hertog et al., 1993).

A much lower intake (3.3 mg) was found in Finland (Knekt et al., 2002). The Finnish survey also estimated mean daily intakes of kaempferol (0.6 mg) and myricetin (0.12 mg). Development of a flavonoid database has been undertaken to facilitate meaningful assessment of intake in the United States (Pillow et al., 1999).

Flavanones: The principal flavonoids in grapefruit are naringin (naringenin-7-rutinoside, 0.35 mg/ml) and naringenin-7-rhamnoglucoside (0.1–0.5 mg/ml). The amount of bitterness is an approximate indicator of the naringenin content of different grapefruit varieties. Oranges contain mainly hesperitin-7-rutinoside (0.2 mg/ml) and a smaller amount of naringin (0.04 mg/ml). Beer produces small amounts of hop-derived 8-prenylnaringenin (Rong et al., 2001), xanthohumol (3′-prenyl-6′-*O*-methylchalconaringenin), desmethylxanthohumol, and 3′-geranylchalconaringenin (Stevens et al., 2000). Tomato peels contain naringenin chalcone. Owing to widespread use of flavanone-containing foods and beverages, intake is significant. A survey in Finland (Knekt et al., 2002) assessed mean daily consumption of naringenin (5.1 mg) and hesperitin (15.1 mg).

Catechins: Green tea is a particularly rich source of (+)-catechin, (−)-epicatechin, (−)-epigallocatechin, and the 3-*O*-gallate derivatives of (−)-epicatechin and (−)-epigallocatechin. Black and other teas, red wine, apple cider, and beer are less concentrated sources.

Anthocyanidins: This class of colorful compounds, many of which are toxic, has many representatives in herbs and flowers. Many berries and other fruits and vegetables get their vibrant colors from anthocyanidins. The depth of the coloring tends to indicate the total flavonoid content of different fruit cultivars (Liu et al., 2002). Relatively common derivatives are cyanidin (in cocoa, chocolate, and fruits), pelargonidin (the red dye in strawberries), delphinidin (gives pomegranates their violet hue), malvidin (red to violet color), petunidin (violet), and peonidin (reddish-brown to purple). Anthocyanidins usually have one, two, or three sugars attached (and are then called *anthocyanins*) and may be acylated. The poor solubility and strong coloring of anthocyanins guarantee fairly permanent stains on fabrics (e.g., when blueberry juice spills on a white shirt).

ABSORPTION

Flavonoid absorption ranges from virtually nonexistent for some to extensive for others. There is good reason to believe that limited bioavailability, particularly of carbohydrate-linked flavonoids, affords some protection from the potent toxicity of many plant glycosides. However, some absorption does occur. Some glycosides can be cleaved by lactase (EC3.2.1.108) at the brush border of the proximal small intestine (Day et al., 2000; Kohlmeier et al., 2000; Setchell et al., 2002) and by bacterial β-glucosidases (EC3.2.1.21) in the terminal ileum and colon (Setchell et al., 1984). Hydrolysis improves bioavailability if the aglycone is better absorbed than the glycoside. The sodium-glucose cotransporter 1 (SGLT1, SLC5A1) is a port of entry for quercetin-4'-glucoside and its aglycone (Walgren et al., 2000), but the efficiency and selectivity of this pathway for other flavonoids remain to be investigated (Wolffram et al., 2002). The multidrug resistance protein 2 (MRP2, ABCC2) actively pumps conjugated flavonoids such as genistein-7-glucoside back into the intestinal lumen (Walle et al., 1999). P-glycoprotein (Pgp, MDR1, and ABCB1) and MDR3 (ABCB4) at the luminal membrane show similar activity, the former with particularly high affinity for flavones (Comte et al., 2001). In addition, MRP 1 (ABCC1) and the organic anion transporter OATP8 (SLC21A8) at the basolateral membrane import some flavonoid conjugates across the basolateral membrane from the pericellular space.

Intestinal uridine diphosphate (UDP)–glucuronosyltransferase UGT1A1 (EC2.4.1.17) conjugates most flavonoids. Sulfation by aryl sulfotransferase (phenol sulfotransferase, SULT, EC2.8.2.1) is not as extensive, but it is of equal importance since the sulfate conjugates may be the most active form of many flavonoids.

The mechanisms for export of flavonoids across the basolateral membrane toward pericapillary space remain uncertain but may involve active transport via MRP3 (ABCC3), MRP5 (ABCC5), and/or MRP6 (ABCC6). Effective fractional absorption therefore depends on the net transport rates across each side of the enterocyte, as well as on losses due to intestinal catabolism. Extensive enterohepatic circulation is likely to occur with isoflavones (Sfakianos et al., 1997) and presumably other flavonoids.

Isoflavones: Absorption is maximal within 4–6 h in healthy subjects and complete within 18 h. Urinary excretion is maximal within 6–8 h, and nearly complete within 12 h (Lu and Anderson, 1998; Fanti et al., 1999). Fractional intestinal absorption may be (Xu et al., 1994) as low as 9% for genistin (genistein-7-O-β-glucoside) and 21% for daidzin (daidzein 7-O-β-glucoside) or even less (Hutchins et al., 1995), but possibly higher with ingestion of aglycones. Others have found that glucoside and aglycone of genistein are similarly well absorbed, presumably after hydrolysis of the glucoside (King et al., 1996). Recent evidence strongly indicates that net absorption of uncleaved isoflavone glucosides is minimal (Setchell et al., 2002). Aglycones are converted to 7-O-β-glucuronides, and to a lesser extent to the sulfate conjugate. The mechanism of export is unknown.

Flavones: A very small but biologically significant fraction of ingested apigenin is absorbed, as indicated by the fact that 0.6% of an oral dose can be recovered from urine (Nielsen et al., 1999). The bioavailability of chrysin seems to be similarly low (Walle et al., 2001). Flavones bind to the reverse transporter P-glycoprotein (Pgp, MDR1, and ABCB1) with greater affinity than isoflavones, flavanones, and glycosylated derivatives (Comte et al., 2001). It should be noted that polymethoxylated flavones in orange juice are potent inhibitors of this transporter and therefore may increase net absorption of other flavonoids (Takanaga et al., 2000).

Flavonols: Quercetin rutinoside is very poorly absorbed. The bioavailability of the aglycon is slightly greater, and quercetin glucosides from onions are absorbed best (Hollman et al., 1995).

Flavanones: Cytosolic β-glucosidase (EC3.21.2.1) in enterocytes and hepatocytes can cleave naringenin-7-glucoside, but not naringenin-7-rhamnoglucoside (Day et al., 1998).

TRANSPORT AND CELLULAR UPTAKE

Blood circulation: Flavonoids are present in the blood mainly as glucuronides and to a much lesser extent as sulfates (Setchell et al., 2002). Concentrations depend entirely on recent intake and easily vary by several orders of magnitude.

Blood–brain barrier (BBB): The extent to which flavonoids can reach the brain is not well investigated.

Materno-fetal transfer: Both the aglycones of isoflavones and their conjugates pass across the placenta (Adlercreutz et al., 1999). The underlying mechanisms are not well understood.

METABOLISM

Many flavonoids are rapidly broken down either in the intestine before absorption is completed or in the liver and other metabolically active organs. Most of these reactions are not well characterized because no metabolic products have yet been identified. The picture is complicated by bacterial action, either prior to absorption or upon enterohepatic cycling. Methylation by catechol-*O*-methyltransferase (COMT, EC2.1.1.6) is common. Alternative splicing of this enzyme, whose main physiological substrates are catecholamines, produces a soluble cytosolic (S-COMT) and a membrane-bound (MB-COMT) form. Glucuronide and sulfate groups are both added and removed, while the flavonoids and their breakdown products move through different organs and target tissues. The fate and biological activity of particular compounds and intermediates are too diverse to predict without concrete experimental evidence.

Isoflavones: Daidzein is metabolized extensively, either to equol or to *O*-desmethylangolensin (Figure 7.3). Only some people (about one in three) generate and excrete equol with some regularity (Setchell et al., 1984; Hutchins et al., 1995). One pathway of daidzein metabolism proceeds via dihydrodaidzein and tetrahydrodaidzein to equol. This pathway has raised considerable interest since equol has higher estrogenic potency than its precursor, daidzein. A different pathway generates 2-dehydro-*O*-desmethylangolensin and then *O*-desmethylangolensin (Joannou et al., 1995). Bacterial enzymes are likely to play at least some part in these poorly understood conversions. The genetic basis for this difference is not well understood since the enzymes responsible for the various steps are still unknown. Surprisingly, equol was found to be highest in plasma from infants fed cow milk–based formula, but low in plasma from infants fed soy-based formula (Setchell et al., 1998). Genistein catabolism (Figure 7.4) occurs via dihydrogenistein to 6′-hydroxy-*O*-desmethylangolensin (Joannou et al., 1995). Another possible product is p-ethylphenol (Adlercreutz et al., 1995), but questions remain about this pathway. Unabsorbed genistein and daidzein are completely lost to degradation in the lower intestinal tract (Xu et al., 1994).

Just as with estrogen conjugates, sulfatases in tissues release the isoflavone aglycones from sulfate conjugates (Pasqualini et al., 1989). It is not clear whether isoflavones are released in tissues from their glucuronides (or even transported into cells) and thus exert activity.

Flavones: Metabolism to 4′-methoxyapigenin (acacetine) appears to be minimal, if it occurs at all (Nielsen et al., 1999).

Flavonols: Quercetin is methylated (Figure 7.5) extensively by catechol-*O*-methyltransferase (EC2.1.1.6; DuPont et al., 2002). This reaction is easily saturated at higher than minimal intake. The main metabolite is 3′-*O*-methylquercetin (isorhamnetin). Some individuals also generate the 4′-*O*-methylquercetin derivative (tamarixetin). Degradation of quercetin through unknown pathways in the kidney or other sites may affect 95% or more of the absorbed flavonoid (Hollman et al., 1995).

Flavanones: A significant fraction of absorbed naringenin is broken down to p-hydroxyphenylpropionic acid (p-HPPA), p-coumaric acid, and p-hydroxybenzoic acid in rats (Felgines et al., 2000). The extent of naringenin catabolism in humans and the underlying mechanisms remain to be investigated.

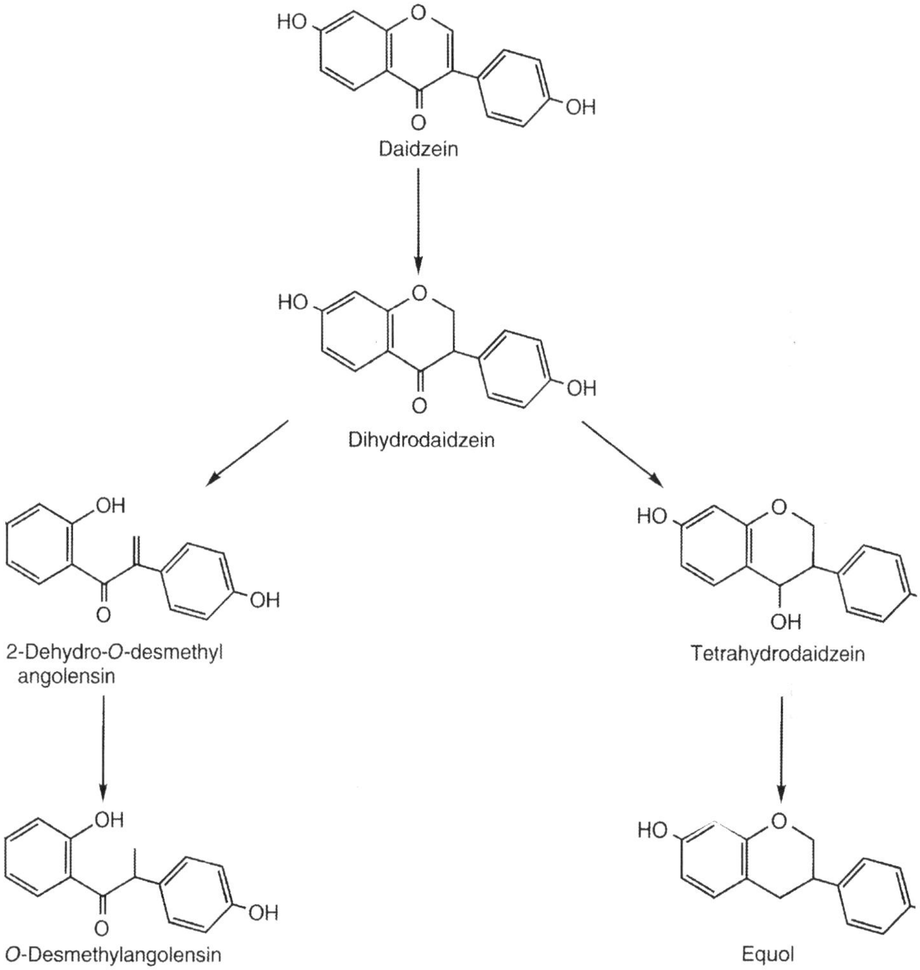

FIGURE 7.3

Only some people produce the estrogenic metabolite equol from daidzein.

Catechins: Catechins are metabolized and excreted mainly as glucuronides and the glucuronide/ sulfate conjugates of 3′-*O*-methylcatechin, and less as the analogous conjugates of catechin (Donovan et al., 1999).

STORAGE

Retention of flavonoids in specific tissues and later release does not account for significant amounts available to tissues.

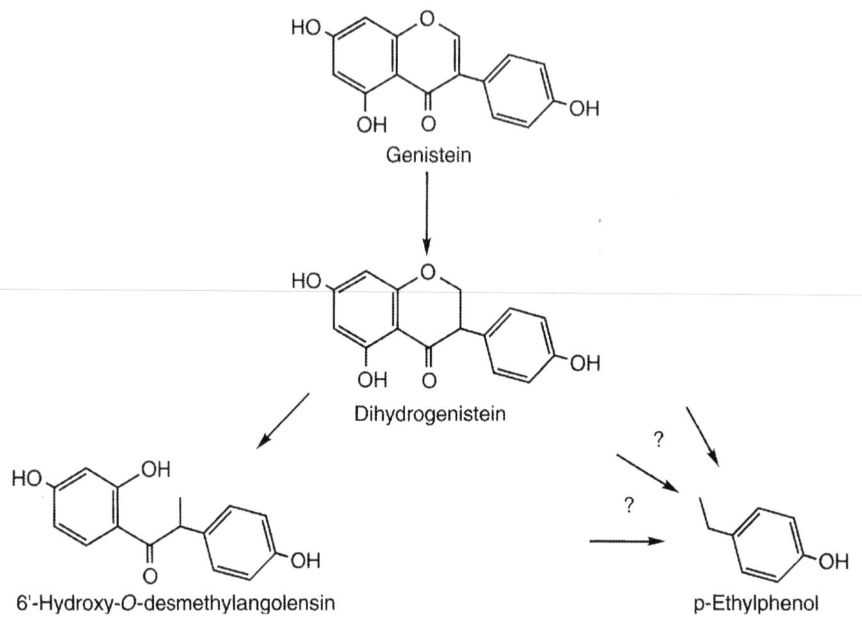

FIGURE 7.4

Catabolism irreversibly degrades genistein.

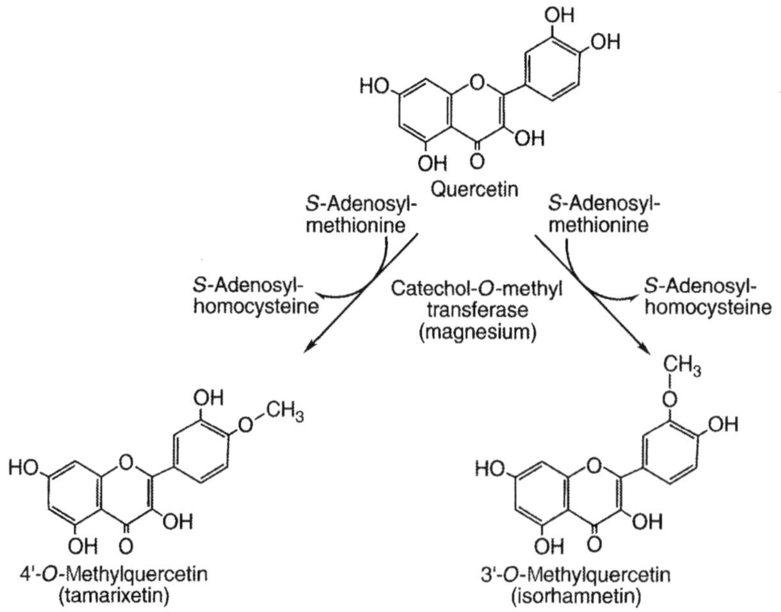

FIGURE 7.5

Quercetin is extensively methylated.

EXCRETION

The xenobiotic transporters at the basolateral membrane (MDR3/ABCB4) and the brush border membrane (MDR1/ABCB1, MRP1/ABCC1) transport many conjugates of flavonoids. This means that active tubular secretion occurs in addition to glomerular filtration. What is not yet known is the extent of net secretion into urine.

Isoflavones: Genistein and daidzein are excreted mostly with urine (Xu et al., 1994) as glucuronides and sulfate conjugates (Fanti et al., 1999). Clearance is very rapid once the compounds appear in the blood.

EFFECTS

The numerous biological actions of flavonoids reflect the diversity of these compounds and their origin. It should be remembered that many of the classical plant-derived medications are flavonoids (one of the best known of these is the heart medication digitalis). A renewed interest in such botanicals has identified many candidate compounds that might help to fight cancer, arthritis, and other common ailments. The many ways in which flavonoids in herbs or foods affect human cell biology (including the slowing of oxidation, proliferation, bacterial and viral infection, and inflammation) cannot be enumerated in detail here; only a few examples will be given.

Sex-hormone-like actions: Their structural similarity with estrogen and gestagen enables many flavonoids to bind to human sex-hormone receptors (Tang and Adams, 1980) and activate estrogen-regulated genes (Ratna and Simonelli, 2002). Typically, it is the diphenolic compounds that have such phytoestrogen activity. Observed responses may differ due to differences in approaches for the assessment (human studies, animal models, and *in vitro* experiments). Genistein and biochanin A tend to have relatively high estrogenic potency, only two orders of magnitude below that of estrogen, which is present at a much lower concentration *de novo* (Zand et al., 2000). The estrogenic potency of luteolin and naringenin is about three orders of magnitude below that of estrogen. Apigenin has considerable gestagen-like activity. Some isoflavones and other flavonoids do appear to be potent enough to reduce menopausal complaints, such as hot flashes (Chen et al., 2014).

Phase I enzymes: The potent inhibition of cytochrome P450 3A4 (CYP3A4) by naringenin slows the elimination of medications such as statins and contraceptive hormones (Hodek et al., 2002).

Phase II enzymes: The flavone chrysin and the flavonol quercetin potently induce UDP-glucuronosyltransferase (EC2.4.1.17) UGT1A1 (Galijatovic et al., 2001). On the other hand, quercetin inhibits phenol sulfotransferase A1 (SULT1A1, EC2.8.2.1) in the liver (De Santi et al., 2002).

Drug absorption: Since many flavonoids bind to transporters that pump many drugs back into the intestinal lumen, they have the potential to increase drug bioavailability. Moderate doses of orange juice and other citrus products contain enough tangeretin and other polymethoxylated phytochemicals to inhibit the reverse transport of digoxin and other drugs and increase their effectiveness, as well as the risk of side effects (Takanaga et al., 2000).

Antioxidant action: Many flavonoids are redox active and their consumption with a varied diet rich in fruits and vegetables contributes significantly to free radical defense and possibly reduction of disease (Kris-Etherton and Keen, 2002). Quercetin and other common dietary flavonoids have been shown to be very effective scavengers of peroxynitrite and other potent oxidants (Terao et al., 2001).

REFERENCES

Adlercreutz, H., Fotsis, T., Kurzer, M.S., Wahala, K., Makela, T., Hase, T., 1995. Isotope dilution gas chromato-graphic-mass spectrometric method for the determination of unconjugated lignans and isoflavonoids in human feces, with preliminary results in omnivorous and vegetarian women. Anal. Biochem. 225, 101–108.

Adlercreutz, H., Yamada, T., Wahala, K., Watanabe, S., 1999. Maternal and neonatal phytoestrogens in Japanese women during birth. Am. J. Obst. Gynecol. 180, 737–743.

Blekas, G., Vassilakis, C., Harizanis, C., Tsimidou, M., Boskou, D.G., 2002. Biophenols in table olives. J. Agric. Food Chem. 50, 3688–3692.

Chen, M.N., Lin, C.C., Liu, C., 2014. Efficacy of phytoestrogens for menopausal symptoms: a meta-analysis and systematic review. Climacteric, 1–21.

Comte, G., Daskiewicz, J.B., Bayet, C., Conseil, G., Viornery-Vanier, A., Dumontet, C., et al., 2001. C-Isoprenylation of flavonoids enhances binding affinity toward P-glycoprorein and modulation of cancer cell chemoresistance. J. Med. Chem. 44, 763–768.

Day, A.J., Canada, F.J., Diaz, J.C., Kroon, P.A., Mclauchlan, R., Faulds, C.B., et al., 2000. Dietary flavonoid and iso-flavone glycosides are hydrolysed by the lactase site of lactase phlorizin hydrolase. FEBS Lett. 468, 166–170.

Day, A.J., DuPont, M.S., Ridley, S., Rhodes, M., Rhodes, M.J., Morgan, M.R., et al., 1998. Deglycosylation of flavonoid and isoflavonoid glycosides by human small intestine and liver beta-glucosidase activity. FEBS Lett. 436, 71–75.

de Kleijn, M.J.J., van der Schouw, Y.T., Wilson, P.W.E., Grobbee, D.E., Jacques, P.E., 2001. Intake of dietary phy-toestrogens is low in postmenopausal women: the framingham heart study. J. Nutr. 131, 1826–1832.

De Santi, C., Pietrabissa, A., Mosca, E., Rane, A., Pacifici, G.M., 2002. Inhibition of phenol sulfotransferase (SULT1A1) by quercetin in human adult and foetal livers. Xenobiotica 32, 363–368.

Donovan, J.L., Bell, J.R., Kasim-Karakas, S., German, J.B., Walzem, R.L., Hansen, R.J., et al., 1999. Catechin is present as metabolites in human plasma after consumption of red wine. J. Nutr. 129, 1662–1668.

DuPont, M.S., Bennett, R.N., Mellon, F.A., Williamson, G., 2002. Polyphenols from alcoholic apple cider are absorbed, metabolized and excreted by humans. J. Nutr. 132, 172–175.

Fanti, P., Sawaya, B.E., Custer, L.J., Franke, A.A., 1999. Levels and metabolic clearance of the isoflavones genis-tein and daidzein in hemodialysis patients. J. Am. Soc. Nephrol. 10, 864–871.

Felgines, C., Texier, O., Morand, C., Manach, C., Scalbert, A., Regerat, E., et al., 2000. Bioavailability of the flavanone naringenin and its glycosides in rats. Am. J. Physiol. Gastroint. Liver Physiol. 279, G1148–G1154.

Franke, A.A., Custer, L.J., Cerna, C.M., Narala, K., 1995. Rapid HPLC analysis of dietary phytoestrogens from legumes and from human urine. Proc. Soc. Exp. Biol. Med. 208, 18–26.

Galijatovic, A., Otake, Y., Walle, U.K., Walle, T., 2001. Induction of UDP-glucuronosyltransferase UGT1A1 by the flavonoid chrysin in Caco-2 cells—potential role in carcinogen bioinactivation. Pharm. Res. 18, 374–379.

Hakkinen, S.H., Karenlampi, S.O., Heinonen, I.M., Mykkanen, H.M., Torronen, A.R., 1999. Content of the flavo-nols quercetin, myricetin, and kaempferol in 25 edible berries. J. Agr. Food Chem. 47, 2274–2279.

Hertog, M.G., Hollman, P.C., Katan, M.B., Kromhout, D., 1993. Intake of potentially anticarcinogenic flavonoids and their determinants in adults in the Netherlands. Nutr. Cancer 20, 21–29.

Hodek, E., Trefil, P., Stiborova, M., 2002. Flavonoids—potent and versatile biologically active compounds inter-acting with cytochromes P450. Chem. Biol. Interact. 139, 1–21.

Hollman, P.C.H., de Vries, J.H.M., van Leeuwen, S.D., Mengelers, M.J.B., Katan, M.B., 1995. Absorption of dietary quercetin glycosides and quercetin in healthy ileostomy volunteers. Am. J. Clin. Nutr. 62, 1276–1282.

Hutchins, A.M., Slavin, J.L., Lampe, J.W., 1995. Urinary isoflavonoid phytoestrogen and lignan excretion after consumption of fermented and unfermented soy products. J. Am. Diet Assoc. 95, 545–551.

Joannou, G.E., Kelly, G.E., Reeder, A.Y., Waring, M., Nelson, C., 1995. A urinary profile study of dietary phy-toestrogens. The identification and mode of metabolism of new isoflavonoids. J. Steroid. Biochem. Mol. Biol. 54, 167–184.

King, R.A., Broadbent, J.L., Head, R.J., 1996. Absorption and excretion of the soy isoflavone genistein in rats. J. Nutr. 126, 176–182.

Knekt, E., Kumpulainen, J., Järvinen, R., Rissanen, H., Heliövaara, M., Reunanen, A., et al., 2002. Flavonoid intake and risk of chronic diseases. Am. J. Clin. Nutr. 76, 560–568.

Kohlmeier, M., Muldrow, W., Switzer, B., 2000. Supplemental beta-galactosidase cleaves soy milk isoflavone beta-7-glucosides. J. Bone Miner Res. 15, S311.

Kris-Etherton, P.M., Keen, C.L., 2002. Evidence that the antioxidant flavonoids in tea and cocoa are beneficial for cardiovascular health. Curr. Opin. Lipidol. 13, 41–49.

Liu, M., Li, X.Q., Weber, C., Lee, C.Y., Brown, J., Liu, R.H., 2002. Antioxidant and antiproliferative activities of raspberries. J. Agric. Food Chem. 50, 2926–2930.

Lu, L.J.W., Anderson, K.E., 1998. Sex and long-term soy diets affect the metabolism and excretion of soy isoflavones in humans. Am. J. Clin. Nutr. 68, 1500S–1504S.

Merken, H.M., Beecher, G.R., 2000. Liquid chromatographic method for the separation and quantification of prominent flavonoid aglycones. J. Chromatogr. A 897, 177–184.

Nielsen, S.E., Young, J.E., Daneshvar, B., Lauridsen, S.T., Knuthsen, P., Sandstrom, B., et al., 1999. Effect of parsley (Petroselinum crispum) intake on urinary apigenin excretion, blood antioxidant enzymes and biomarkers for oxidative stress in human subjects. Br. J. Nutr. 81, 447–455.

Pasqualini, J.R., Gelly, C., Nguyen, B.L., Vella, C., 1989. Importance of estrogen sulfates in breast cancer. J. Steroid. Biochem. 34, 155–163.

Pillow, P.C., Duphorne, C.M., Chang, S., Contois, J.H., Strom, S.S., Spitz, M.R., et al., 1999. Development of a database for assessing dietary phytoestrogen intake. Nutr. Cancer 33, 3–19.

Ratna, W.N., Simonelli, J.A., 2002. The action of dietary phytochemicals quercetin, catechin, resveratrol and naringenin on estrogen-mediated gene expression. Life Sci. 70, 1577–1589.

Rong, H., Boterberg, T., Maubach, J., Stove, C., Depypere, H., Van Slambrouck, S., et al., 2001. 8-Prenylnaringenin, the phytoestrogen in hops and beer, upregulates the function of the E-cadherin/catenin complex in human mammary carcinoma cells. Eur. J. Cell Biol. 80, 580–585.

Setchell, K.D., Boriello, S.E., Hulme, P., Kirk, D.N., Axelson, M., 1984. Nonsteroidal estrogens of dietary origin: possible roles in hormone-dependent disease. Am. J. Clin. Nutr. 40, 569–578.

Setchell, K.D., Brown, N.M., Zimmer-Nechemias, L., Brashear, W.T., Wolfe, B.E., Kirschner, A.S., et al., 2002. Evidence for lack of absorption of soy isoflavone glycosides in humans, supporting the crucial role of intestinal metabolism for bioavailability. Am. J. Clin. Nutr. 76, 447–453.

Setchell, K.D.R., Zimmer-Nechemias, L., Cai, J., Heubi, J.E., 1998. Isoflavone content of intant formulas and the metabolic fate of these phytoestrogens in early life. Am. J. Clin. Nutr. 68, 1453S–1461S.

Sfakianos, J., Coward, L., Kirk, M., Barnes, S., 1997. Intestinal uptake and biliary excretion of the isoflavone genistein in rats. J. Nutr. 127, 1260–1268.

Stevens, J.F., Taylor, A.W., Nickerson, G.B., Ivancic, M., Henning, J., Haunold, A., et al., 2000. Prenylflavonoid variation in Humulus lupulus: distribution and taxonomic significance of xanthogalenol and 4′-O-methylxanthohumol. Phytochemistry 53, 759–775.

Takanaga, H., Ohnishi, A., Yamada, S., Matsuo, H., Morimoto, S., Shoyama, Y., et al., 2000. Polymethoxylated flavones in orange juice are inhibitors of P-glycoprotein but not cytochrome P450 3A4. J. Pharmacol. Exp. Ther. 293, 230–236.

Tang, B.Y., Adams, N.R., 1980. The effect of equol on estrogen receptors and on synthesis of DNA and protein in the immature rat uterus. J. Endocrinol. 85, 291–297.

Terao, J., Yamaguchi, S., Shirai, M., Miyoshi, M., Moon, J.H., Oshima, S., et al., 2001. Protection by quercetin and quercetin 3-O-beta-D-glucuronide of peroxynitrite-induced antioxidant consumption in human plasma low-density lipoprotein. Free Rad. Res. 35, 925–931.

Walgren, R.A., Lin, J.T., Kinne, R.K.H., Walle, T., 2000. Cellular uptake of dietary flavonoid quercetin 4′-β-glucoside by sodium-dependent glucose transporter SGLT1. J. Pharmacol. Exp. Ther. 294, 837–843.

Walle, T., Otake, Y., Brubaker, J.A., Walle, U.K., Halushka, P.V., 2001. Disposition and metabolism of the flavo-noid chrysin in normal volunteers. Br. J. Clin. Pharmacol. 51, 143–146.

Walle, U.K., French, K.L., Walgren, R.A., Walle, T., 1999. Transport of genistein-7-glucoside by human intestinal CACO-2 cells: potential role for MRP2. Res. Comm. Mol. Path. Pharmacol. 103, 45–56.

Wolffram, S., Blöck, M., Ader, P., 2002. Quercetin-3-glucoside is transported by the glucose carrier SGLT1 across the brush border membrane of rat small intestine. J. Nutr. 132, 630–635.

Xu, X., Wang, H.L., Murphy, P.A., Cook, L., Hendrich, S., 1994. Daidzein is a more bioavailable soymilk isofla-vone than is genistein in adult women. J. Nutr. 124, 825–832.

Zand, R.S., Jenkins, D.J., Diamandis, E.P., 2000. Steroid hormone activity of flavonoids and related compounds. Breast Cancer Res. Treat. 62, 35–49.

GARLIC BIOACTIVES

Food plants of the genus *Allium* (leek family) contain sulfur compounds with potent effects on human metabolism and health. *S*-allyl cysteine (SAC), *S*-ethyl cysteine (SEC), and *S*-propyl cysteine (SPC); gamma-glutamyl-*S*-allyl cysteine, gamma-glutamyl-*S*-methyl cysteine, gamma-glutamyl-*S*-propyl cysteine, (+)-*S*-(2-propenyl)-L-cysteine sulfoxide (alliin), *S*-allyl-N-acetyl cysteine, *S*-allylsulfonyl alanine, *S*-allyl mercaptocysteine, and *S*-methyl cysteine are water-soluble. Diallyl disulfide (DADS), diallyl trisulfide (DATS), dipropyl disulfide (DPDS), diallyl sulfide, dipropyl sulfide, and methyl allyl sulfide are fat soluble.

ABBREVIATIONS

DADS	diallyl disulfide
DADSO	diallyl thiosulfinate
DATS	diallyl trisulfide
DPDS	dipropyl disulfide
SAC	*S*-allyl cysteine
SEC	*S*-ethyl cysteine
SPC	*S*-propyl cysteine

NUTRITIONAL SUMMARY

Function: Several of the sulfur-containing compounds in garlic and related edible plants of the leek family are potent inducers of phase I enzymes and are likely to have beneficial health effects.

Food sources: Several varieties of garlic are edible vegetables. Leeks, spring onions, and other members of the leek family are also of interest.

Requirements: There are no dietary requirements.

Deficiency: No harm is directly related to low or absent intake of garlic or other members of the plant genus *Allium* (onions). However, the missed opportunity of health benefits should be considered.

Excessive intake: There is no credible evidence that very high intake of garlic causes harm. The unavoidable odor can be a nuisance to others and thus counts as a negative.

DIETARY SOURCES

Onions, garlic, chives, and leeks contain various odorless (+)-S-alk(en)yl-L-cysteine sulfoxides, including (+)-S-methyl-L-cysteine sulfoxide, (+)-S-(2-propenyl)-L-cysteine sulfoxide (alliin), and (+)-S-propyl-L-cysteine sulfoxide. The S-alk(en)ylcysteine sulfoxide content of garlic is between 0.5% and 1.3% flesh weight, mainly as alliin (Kubec et al., 1999).

When the plants are crushed, the plant enzyme alliinase (EC4.4.1.4) converts these precursors into volatile thiosulfinates, including diallyl sulfide, diallyl disulfide (DADS), diallyl trisulfide (DATS), and diallyl thiosulfinate (DADSO, allicin). The enzyme may be reactivated in garlic powder by moistening. The activity in powder is stabilized for several months by adding sodium chloride, sucrose, and pyridoxal-5′-phosphate (Figure 7.6).

METABOLISM

Nicotine adenine dinucleotide phosphate (NADPH)–dependent CYP2E1 in liver microsomes converts DADS into DADSO; other cytochrome P450 monooxygenases and flavin-containing monoxygenase contribute to this conversion to a lesser degree (Teyssier et al., 1999). It has been proposed that allyl mercaptan is formed after the ingestion of allicin or its decomposition products (Lawson, 1998). S-adenosylmethionine-dependent methylation of allyl mercaptan by an as-yet-unidentified methyltransferase then may give rise to allyl methyl sulfide.

EFFECTS

Garlic compounds have antithrombotic, antimicrobial, antifungal, antitumoral, hypoglycemic, hypolipemic, hypotensive, and brain-protective properties. Extensive knowledge exists on the activity of particular compounds. The following is just to give an indication of the wide spectrum of biological effects.

Malodor: The major garlic-derived odorous compound in breath and exuded through skin is allyl methyl sulfide (Rosen et al., 2001). Minute quantities of hydrogen sulfide may also be exhaled. The sulfur compounds most prominently responsible for the typical odor of crushed garlic are not exhaled to any significant extent. Diallyl sulfides (DAS, DADS, and DATS), presumably acting in the liver, increase acetone formation, which adds to breath odor (Amagase et al., 2001). The odor can trigger migraine headaches in some individuals exposed to the emanations from others (Roussos and Hirsch, 2014).

Antioxidant properties: Allicin quenches hydroxyl radicals generated by Fenton-type reactions (Rabinkov et al., 1998).

Lipoprotein metabolism: DADS inhibited cholesterol synthesis in whole-liver preparations (Liu and Yeh, 2000); at the most effective concentrations, strong cytotoxic effects were observed. DADS, DATS, SAC, SEC, and SPC, but not alliin, were found to be potent inhibitors of cholesterol synthesis in cultured cells (Liu and Yeh, 2000; Gupta and Porter, 2001). Crude garlic preparations inhibited HMG-CoA reductase (EC4.1.3.5), the key enzyme of cholesterol synthesis. Squalene monooxygenase (EC1.14.99.7) appears to be especially sensitive to inhibition by compounds that react with its sulfhydryl groups. Allicin is also known to inactivate rapidly thiol-containing enzymes (Rabinkov et al., 1998).

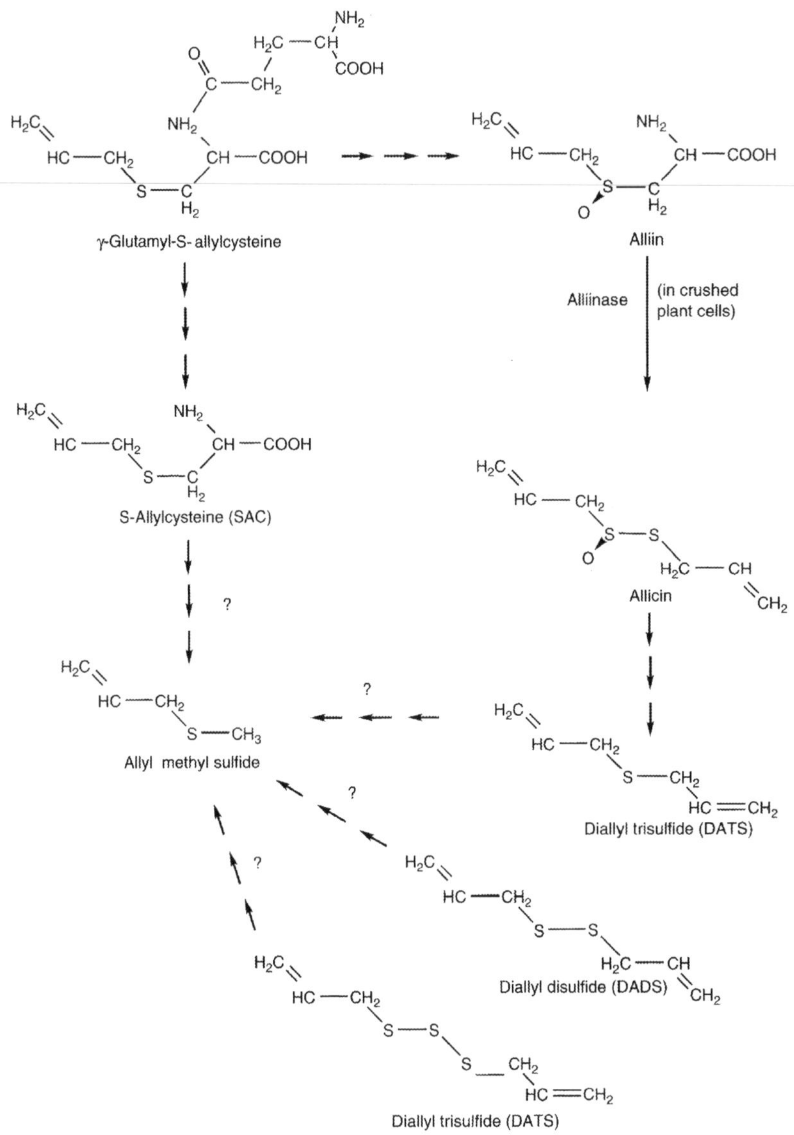

FIGURE 7.6

Bioactive compounds are created when the plant cells are crushed.

Energy metabolism: DADS, DATS, and alliin enhance thermogenesis in rodents by increasing uncoupling protein content of brown adipose tissue, as well as noradrenaline (norepinephrine) and adrenaline (epinephrine) secretion (Oi et al., 1999).

Brain function: Allicin has neurotrophic activity (Moriguchi et al., 1997).

REFERENCES

Amagase, H., Petesch, B.L., Matsuura, H., Kasuga, S., Itakura, Y., 2001. Intake of garlic and its bioactive components. J. Nutr. 131, 955S–962S.

Gupta, N., Porter, T.D., 2001. Garlic and garlic-derived compounds inhibit human squalene monooxygenase. J. Nutr. 131, 1662–1667.

Kubec, R., Svobodova, M., Velisek, J., 1999. Gas chromatographic determination of S-alk(en)ylcysteine sulfoxides. J. Chromatogr. A 862, 85–94.

Lawson, L.D., 1998. Garlic: a review of its medicinal effects and indicated active compounds. In: Lawson, L.D., Bauer, R. (Eds.), Phytomedicines of Europe American Chemical Society, Washington, DC, pp. 176–209.

Liu, L., Yeh, Y.Y., 2000. Inhibition of cholesterol biosynthesis by organosulfur compounds derived from garlic. Lipids 35, 197–203.

Moriguchi, T., Matsuura, H., Kodera, Y., Itakura, Y., Katsuki, H., Saito, H., et al., 1997. Neurotrophic activity of organosulfur compounds having a thioallyl group on cultured rat hippocampal neurons. Neurochem. Res. 22, 1449–1452.

Oi, Y., Kawada, T., Shishido, C., Wada, K., Kominato, Y., Nishimura, S., et al., 1999. Allylcontaining sulfides in garlic increase uncoupling protein content in brown adipose tissue, and noradrenaline and adrenaline secretion in rats. J. Nutr. 129, 336–342.

Rabinkov, A., Miron, T., Konstantinovski, L., Wilchek, M., Mirelman, D., Weiner, L., 1998. The mode of action of allicin: trapping of radicals and interaction with thiol containing proteins. Biochim. Biophys. Acta 1379, 233–244.

Rosen, R.T., Hiserodt, R.D., Fukuda, E.K., Ruiz, R.J., Zhou, Z., Lech, J., et al., 2001. Determination of allicin, S-allylcysteine and volatile metabolites of garlic in breath, plasma or simulated gastric fluids. J. Nutr. 131, 968S–971S.

Roussos, A.P., Hirsch, A.R., 2014. Alliaceous migraines. Headache 54, 378–382.

Teyssier, C., Guenot, L., Suschetet, M., Siess, M.H., 1999. Metabolism of diallyl disulfide by human liver microsomal cytochromes P-450 and flavin-containing monooxygenases. Drug Metab. Disp. 27, 835–841.

AMINO ACIDS AND NITROGEN COMPOUNDS

CHAPTER OUTLINE

Structure and Function of Amino Acids...266
Starvation...291
Glutamate...294
Glutamine...302
Glycine...309
Threonine...316
Serine...321
Alanine...329
Phenylalanine..336
Tyrosine...342
Tryptophan..349
Methionine...359
Cysteine...368
Lysine..376
Leucine...382
Valine...389
Isoleucine...396
Aspartate..401
Asparagine...408
Arginine...413
Proline...423
Histidine..431
Citrulline..440
Taurine...443
Creatine..450
Carnitine..454
Melatonin...461
Choline...468

STRUCTURE AND FUNCTION OF AMINO ACIDS

The amino acids relevant for human nutrition comprise dozens of compounds with a broad variation of characteristics and functions. Amino acids, in the form of proteins of connective tissues and muscles, hold the body together and give it strength to move around. Many other proteins catalyze reactions, transport compounds, and coordinate many events. Amino acids and their derivatives also serve virtually all categories of signaling, metabolism, and functional support.

ABBREVIATIONS

CoA	coenzyme A
CssC	L-cystine
GABA	gamma-aminobutyric acid
NMDA	*N*-methyl D-aspartate

COMPOSITION AND STRUCTURE

By definition, amino acids are always made up of an amino group, an acid, and a side chain. The amino group is most often adjacent to the primary carboxyl group (alpha-position), but may be in the beta-position instead (e.g., in beta-alanine). As in carnitine, a tertiary amine may serve the same function. Similarly, the acid group may be represented by a sulfoxy function, as in taurine. The side chain may be just a single hydrogen (in glycine), straight (alanine), or branched (valine, leucine, and isoleucine) aliphatic chains, contain aromatic rings (phenylalanine, tyrosine, and tryptophan), sulfur (methionine, cysteine, and taurine), selenium (selenocysteine), hydroxyl groups (serine, threonine, hydroxyproline, and hydroxylysine), second carboxyl groups (glutamate and aspartate) or third carboxyl groups (gamma-carboxyl glutamate and gammacarboxyl aspartate), amido groups (glutamine and asparagine), or additional nitrogen-containing groups (lysine, arginine, and histidine).

Since the alpha-carbon is chiral for all amino acids except glycine, two stereoisomeric conformations are possible. When the carboxylic group in structural diagrams is depicted at the top and the side chain at the bottom, the L form is the one that has the amino group pointing to the left, the D form has it pointing to the right. Human proteins contain only L-amino amino acids (and glycine). D-Amino acids are quantitatively less common in mammals than in fungi and bacteria, but a few (including D-aspartate and D-serine) are synthesized by humans (D'Aniello et al., 1993; Wolosker et al., 1999, 2000).

Some amino acids form only a few specific peptides, such as taurine as part of glutaurine, and beta-alanine as part of carnosine and anserine. Several other amino acids are not a regular part of human proteins but serve specific important functions. This is the case with carnitine (fatty acid transport), taurine (osmolyte, neuronal agent, and antioxidant), ornithine (urea cycle and polyamine synthesis), and citrulline (urea cycle) (Figures 8.1–8.3, Tables 8.1 and 8.2).

ENDOGENOUS SOURCES

Precursors from intermediary metabolism: Several amino acids are synthesized through the addition of an amino group to a common alpha-keto acid. Alpha-ketoglutarate and oxaloacetate are readily

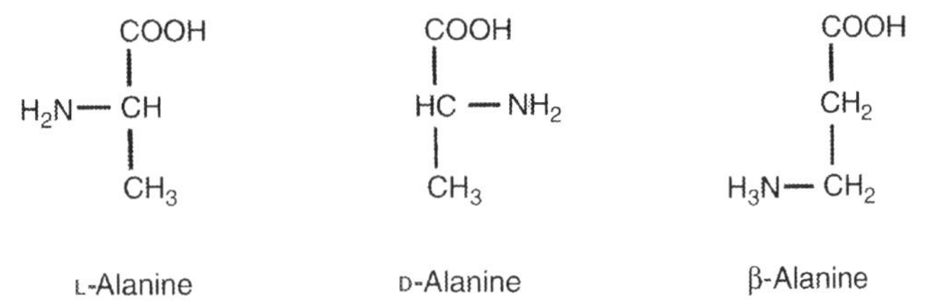

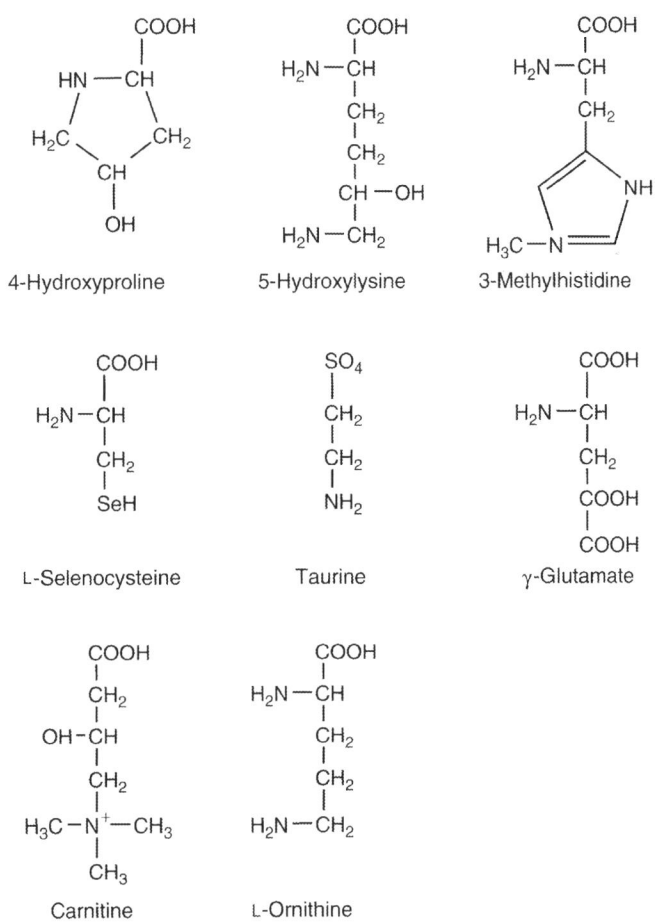

FIGURE 8.1

Different isomeric forms of alanine.

FIGURE 8.2

The most common amino acids in proteins.

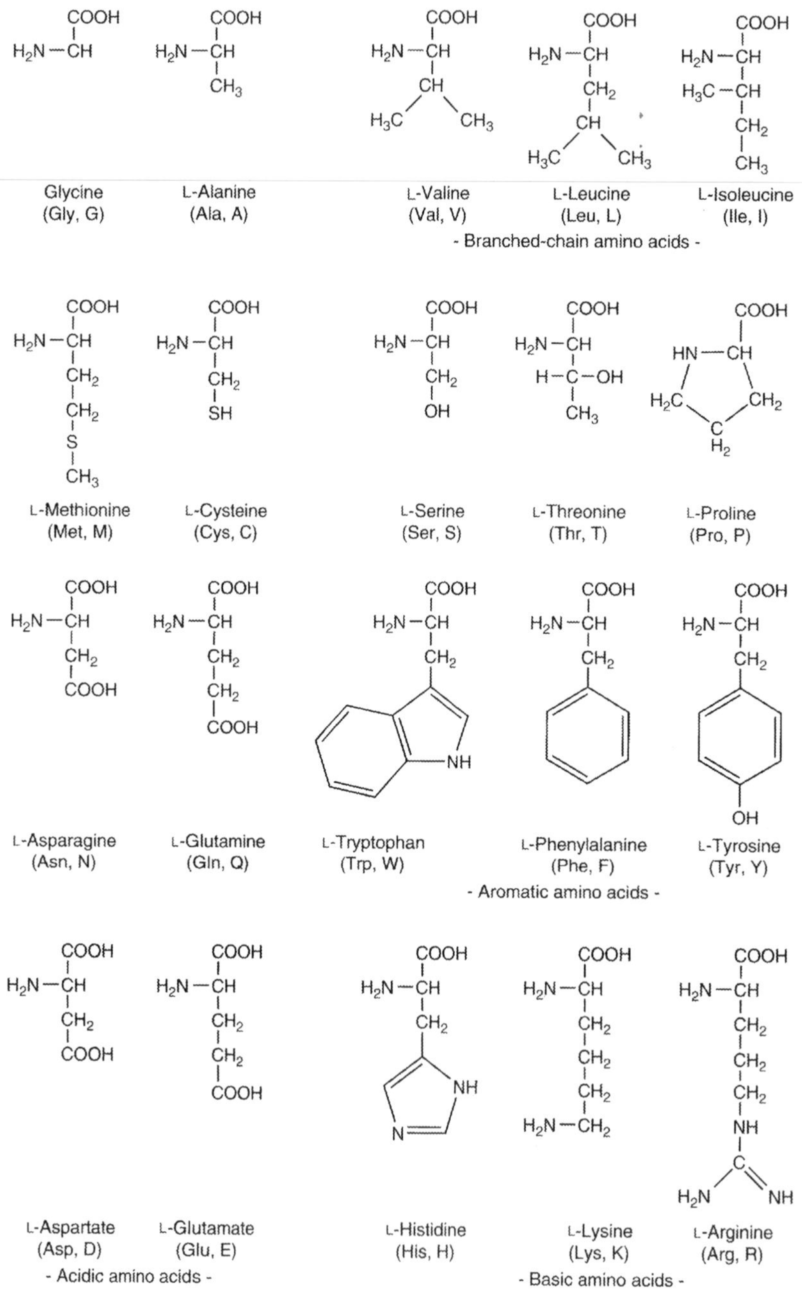

Glycine (Gly, G)

L-Alanine (Ala, A)

L-Valine (Val, V)

L-Leucine (Leu, L)

L-Isoleucine (Ile, I)

- Branched-chain amino acids -

L-Methionine (Met, M)

L-Cysteine (Cys, C)

L-Serine (Ser, S)

L-Threonine (Thr, T)

L-Proline (Pro, P)

L-Asparagine (Asn, N)

L-Glutamine (Gln, Q)

L-Tryptophan (Trp, W)

L-Phenylalanine (Phe, F)

L-Tyrosine (Tyr, Y)

- Aromatic amino acids -

L-Aspartate (Asp, D)

L-Glutamate (Glu, E)

L-Histidine (His, H)

L-Lysine (Lys, K)

L-Arginine (Arg, R)

- Acidic amino acids -

- Basic amino acids -

FIGURE 8.3

Less common forms of bases and acids in amino acids.

Table 8.1 Common Amino Acids

Amino Acid	Metabolic Fate		Chemical Structure	
Protein amino acids				
Glycine	Gly	G	One carbon	Aliphatic
L-Alanine	Ala	A	Glucogenic	Aliphatic
L-Proline	Pro	P	Glucogenic	Aliphatic imino
L-Valine	Val	V	Glucogenic	Branched-chain
L-Leucine	Leu	L	Ketogenic	Branched-chain
L-Isoleucine	Ile	I	Glucogenic + ketogenic	Branched-chain
L-Phenylalanine	Phe	F	Glucogenic + ketogenic	Aromatic
L-Tyrosine	Tyr	Y	Glucogenic + ketogenic	Aromatic
L-Tryptophan	Trp	W	Glucogenic + ketogenic	Aromatic
L-Methionine	Met	M	Glucogenic	Sulfurous
L-Cysteine	Cys	C	Glucogenic	Sulfurous
L-Serine	Ser	S	Glucogenic	Alcoholic
L-Threonine	Thr	T	One carbon or glucogenic	Alcoholic
L-Aspartate	Asp	D	Glucogenic	Acidic
L-Glutamate	Glu	E	Glucogenic	Acidic
L-Asparagine	Asn	N	Glucogenic	Neutral amide
L-Glutamine	Gln	Q	Glucogenic	Neutral amide
L-Histidine	His	H	Glucogenic	Basic
L-Lysine	Lys	K	Ketogenic	Basic
L-Arginine	Arg	R	Glucogenic	Basic
L-Hydroxyproline	Hyp		Glucogenic	Modified Pro
L-Hydroxylysine				Modified Lys
L-γ-carboxy glutamate	Gla			Modified Glu
3-Methyl-histidine				Modified His
Selenocysteine	Sec			Selenous
Nonprotein amino acids				
D-Serine				Alcoholic
D-Aspartate				Acidic
D-Leucine				Branched-chain
D-Methionine				Sulfurous
L-Carnitine				
Taurine			Glucogenic	
L-Ornithine			Glucogenic	
L-Citrulline			Glucogenic	
Beta-alanine				Aliphatic
Beta-leucine				Branched-chain

Table 8.2 Amino Acid Precursors

Protein amino acids	
Glycine	Serine, Threonine ≫ Choline, Betaine, Glyoxylate
L-Alanine	Pyruvate ≫ Tryptophan, Aspartate, Taurine, Thymine, Secys
L-Proline	Glutamate
L-Valine	None
L-Leucine	None
L-Isoleucine	None
L-Phenylalanine	None
L-Tyrosine	Phenylalanine
L-Tryptophan	None
L-Methionine	Extensive regeneration from homocysteine
L-Cysteine	Methionine
L-Serine	Glycine, 3-phosphoglycerate
L-Threonine	None
L-Aspartate	Oxaloacetate
L-Glutamate	Alpha-ketoglutarate
L-Asparagine	Aspartate
L-Glutamine	Glutamate
L-Histidine	None
L-Lysine	None
L-Arginine	Glutamine, Glutamate, Proline, L-Ornithine, L-Citrulline
L-Hydroxyproline	Proline
L-Hydroxylysine	Lysine
L-γ-carboxy glutamate	Glutamate
3-Methyl-histidine	Histidine
Selenocysteine	Serine
Nonprotein amino acids	
D-Serine	L-Serine
D-Aspartate	L-Aspartate
L-Carnitine	Lysine
Taurine	Cysteine
L-Ornithine	Arginine ≫ Glutamine, Glutamate, Proline
L-Citrulline	L-Ornithine
Beta-alanine	Uracil, carnosine, balenine
Beta-leucine	L-Leucine

available Krebs-cycle intermediates, and pyruvate is a glycolysis product. Glyoxylate is a much less abundant intermediary metabolite, which arises from the metabolism of hydroxyproline, ethanolamine, glycolate, ethylene glycol, and a few other minor sources. A slightly more complex case is serine, which is used to synthesize 3-phosphoglycerate in a three-step process. The main source of the amino groups is other amino acids. The exchange between glutamate and aspartate (aspartate aminotransferase,

EC2.6.1.1) and glutamate and alanine (alanine aminotransferase, EC2.6.1.2) is particularly important, as underscored by the abundance of the respective aminotransferases in the muscle, liver, and other tissues. The branched-chain amino acids (BCAAs) leucine, valine, and isoleucine are another abundant source of amino groups for transamination in muscle, especially in the postprandial phase. All of these aminotransferases require pyridoxal 5-phosphate (PHP) as a covalently bound cofactor. Ammonium ions can also contribute directly to amino acid synthesis, but the extent is limited. Glutamate dehydrogenase (EC1.4.1.3) catalyzes the nicotine adenine dinucleotide phosphate (NADPH)–dependent amidation of alpha-ketoglutarate in mitochondria. Alternatively, ammonia can also be linked to glutamate by glutamate dehydrogenase (EC1.4.1.3) in a distinct, NADPH-dependent reaction. Ammonia in cytosol, on the other hand, is used by glutamate-ammonia ligase (EC6.3.1.2) for glutamate amidation in an adenosine triphosphate (ATP)–energized reaction.

Amino acids as precursors: Tyrosine, cysteine, glutamine, asparagine, arginine, ornithine, citrulline, proline, taurine, and carnitine are derived from other amino acids, some of which are essential and must come from food. Tyrosine can be derived from the essential amino acid phenylalanine by hydroxylation. Cysteine is synthesized via the trans-sulfuration pathway from methionine. Glutamine and asparagine are the amidation products of their respective dicarboxylate amino acid precursors. Arginine, ornithine, citrulline, and proline start from glutamate; net synthesis occurs only in the intestinal wall. The precursor for routine synthesis is cysteine.

Posttranslational modification: The precursors of a few amino acids are incorporated into specific proteins and modifications take place only during or after protein synthesis. Many amino acid residues in proteins are modified, and they often acquire full functionality only after such changes. Common modifications include phosphorylation, acylation, and glycosylation. However, after breakdown of a protein, most of the modified amino acids can still be reutilized for protein synthesis. This is not true for a small group of amino acids that are permanently altered by modification of the protein. This group includes hydroxyproline, hydroxylysine, methylhistidine, gamma-carboxyglutamate, gamma-carboxyaspartate, and carnitine. The posttranslational hydroxylation of several procollagens and elastins generates hydroxyproline and hydroxylysine. Methylhistidine arises from the posttranslational methylation of actin and myosin in muscle and the synthesis of anserine and a few other specific peptides. A vitamin K–dependent reaction carboxylates the γ-carbon of specific glutamate residues in several coagulation factors and a few other proteins. Analogous reactions with aspartate residues may also occur. The selenocysteine residues, which are constituents of the reactive centers in a few enzymes, are produced by a complex series of reactions from the serine precursor. Serine-tRNA ligase (EC6.1.1.11) charges tRNA(Ser)[sec] with serine. The pyridoxal phosphate-dependent L-seryl-tRNA[sec] selenium transferase (EC2.9.1.1) then substitutes the hydroxyl group of the serine with selenophosphate. The tRNA is now ready to add its load of selenocysteine to an emerging peptide strand in an unusually complex process. Carnitine synthesis is similarly complicated. In this case, three methyl groups are added to specific lysines in myosin, actin, histones, and other proteins by histone–lysine N-methyltransferase (EC2.1.1.43). After the proteins have been broken down in the course of their normal turnover, carnitine synthesis requires another four steps for completion.

DIETARY SOURCES

Most amino acids are consumed with a wide variety of proteins and peptides in foods of animal and plant origin. These foods provide all 20 amino acids used for making the body's own proteins, peptides, and other compounds that require amino acids as precursors. Phenylalanine, tryptophan, methionine, histidine, lysine, valine, leucine, isoleucine, and threonine are called *essential* because they must be provided with food. Current recommendations are to consume daily at least 0.8 g of protein per kilogram body weight, and more during pregnancy (Food and Nutrition Board, Institute of Medicine, 2005). Children (Elango et al., 2011) and some adults (Kohlmeier, 2012) may have significantly higher requirements than suggested by current recommendations.

Several other amino acids, which are not incorporated in human proteins, have nutritional value nonetheless as nonessential precursors for the endogenous synthesis of important compounds, or just as a source of energy; examples include carnitine, taurine, and ornithine. The unusual amino acid L-theanine (5-*N*-ethylglutmine) constitutes half of the amino acids in tea (*Camellia sinensis*) leaves and appears to confer some beneficial effects (Unno et al., 2013). Finally, there are amino acids that cannot be utilized well or even are harmful, such as some D-amino acids. Particular peptides, of course, may have their own specific effects. Some are potent hormones (though most will not be taken up intact and active); others may be potent toxins (e.g., amanitin from death cap mushrooms).

Cooking of foods tends to improve the overall bioavailability of protein and confers desirable flavors and aromas. However, heating can modify amino acids in numerous ways and may lead to significant losses of cysteine, methionine, threonine, serine, and tryptophan (Dworschak, 1980). Heating also can promote the cross-linking of L-alanine with other amino acids in food proteins (generating lysinoalanine, ornithinoalanine, histidinoalanine, and phenylethylaminoalanine), the formation of dehydroalanine, methyldehydroalanine, beta-aminoalanine, and racemization to D-amino acids (Friedman, 1999). Another typical effect of heating is the cross-linking of amino acids and sugars. The attractive browning that develops during the cooking, frying, or baking of foods reflects these reactions. Initially, Schiff bases form between lysines, arginines, and sugars, which rearrange to more stable Amadori products (Biemel et al., 2001). With continued heating, these intermediates react further and form insoluble protein complexes with extensive lysine-arginine cross-links. These complexes are commonly referred to as *Maillard products* or *advanced glycation end products*. Presumably due to their low solubility and the extensive chemical modifications of their amino acid constituents, such Maillard products have extremely low bioavailability and food value (Erbersdobler and Faist, 2001).

Another problem related to heating foods to high temperatures, such as with broiling, is the formation of cancer-inducing compounds. Condensation and pyrolysis of creatine with aromatic amino acids (2-amino-1-methyl-6-phenylimidazo[4,5-b] pyridine, PhIP), glycine (2-amino-3,8-dimethylimidazo[4,5-f]quinoxaline, MeIQx), or other amino acids generate very potent carcinogens (Oguri et al., 1998; Schut and Snyderwine, 1999). These compounds readily attach to DNA (adduct formation) and cause mutations.

Daily protein intake in the United States declines with age, from about 68 g in young women to 60 g in women over the age of 60 (Smit et al., 1999). Similar trends are seen for men, though at overall higher intake levels.

DIGESTION

The chewing and mixing with saliva in the mouth (mastication) breaks solid food into small particles and initiate digestion. The importance of oral protein-cleaving enzymes appears to be minimal, however. Predigestion in the stomach initiates protein digestion. The acidity of the added hydrogen chloride denatures ingested proteins. Pepsin (EC3.4.23.1) and gastricsin (pepsinogen C, EC3.4.23.3) hydrolyze ingested proteins with broad specificity.

A second tier of protein-digesting enzymes comes from the pancreas. These include several forms of trypsin, alpha-chymotrypsin, chymotrypsin C, carboxypeptidase B, and two forms each of elastase II, endopeptidase E (EC3.4.21.70), and carboxypeptidase A (EC3.4.2.1). Like most pancreas enzymes, these proteases are not active when they reach the small intestine. Duodenal glands secrete duodenase (EC3.4.21. B3, Zamolodchikova et al., 2000). This serine protease activates the brush border protease enteropeptidase (enterokinase, EC3.4.21.9) by cleaving it. Enterokinase, in turn, cleaves and activates trypsin (EC3.4.21.4), and trypsin finally activates the other pancreas proteases. The activity of the protease at the top of this activating cascade is inhibited by alpha 1-proteinase inhibitor (Gladysheva et al., 2001). Additional proteases may have more narrow specificity. One such enzyme is tissue kallikrein (EC3.4.21.35), a kinin activator, which is expressed in acinar cells together with its inhibitor kallistatin (Wolf et al., 1998).

The third tier of enzymes works near or directly at the brush border membrane on a mixture that by now contains relatively small peptides. This means that individual amino acids, dipeptides, or tripeptides are released predominantly near the surface of the brush border membrane, from where they can be taken up into the epithelial cell layer. A few aminopeptidases, particularly membrane alanine aminopeptidase (aminopeptidase N, EC3.4.11.2), are relatively abundant. This enzyme releases N-terminal alanine and a broad spectrum of other amino acids from peptides, amides, or arylamides. There is a considerable diversity of enzymes commensurate with the different substrates. An example of an enzyme with a relatively narrow mission is folylpoly-γ-glutamate carboxypeptidase, which converts food folate into the absorbable free form. The many people who have a slightly less active form absorb food folate less well and are more likely to suffer detrimental consequences of low intake (Devlin et al., 2000) (Table 8.3).

ABSORPTION

Digested protein products are absorbed with very high efficiency (more than 80%) from the proximal small intestine. What helps is that some peptidases at the brush border are associated with an amino acid transporter of broad specificity (Fairweather et al., 2012). Some further absorption continues as necessary throughout the small and large intestine. A minute portion of larger peptides and proteins can migrate rapidly through enterocytes (Ziv and Bendayan, 2000). Transcytosis of relatively intact proteins can explain some phenomena relating to intestinal hormone function and food allergies but has no relevance for absorption efficiency.

Enterohepatic circulation of amino acids is significant (Chang and Lister, 1989). Two proton-driven cotransporters are available for the uptake of dipeptides and tripeptides from the intestinal lumen. The hydrogen ion/peptide cotransporter 1 (SLC15A1, PepT1) is most abundant and has broad specificity. The hydrogen ion/peptide cotransporter 2 (SLC15A2, PepT2) is much less abundant and has higher affinity.

Amino acids are taken up through mechanisms that are much more specific. This helps to reduce the uptake of some D-amino acids and other potentially toxic compounds. Absorption is mainly driven by the sodium gradient that is established by the sodium/potassium-exchanging ATPase (EC3.6.3.9) at the basolateral membrane. This process also generates a proton gradient and a negative membrane potential that favors the influx of cations. The most immediate use of the sodium gradient for amino acid absorption is the sodium-driven transport of amino acids from the intestinal lumen into the enterocyte. Transport system $B^{0}AT1$ (SLC6A19) is especially important for bulk uptake of neutral amino acids (Tümer et al., 2013). Low-activity variants of the *SLC6A19* gene cause the Hartnup syndrome with impaired intestinal absorption and increased renal loss of neutral amino acids (Bröer et al., 2011). The transcription factor SOX9 inhibits expression of the *SLC6A19* gene while the nascent enterocyte in the crypts is still immature. Transporter gene expression (like expression of gene coding for various other transporters) then increases in response to other transcription factors (including HNF1a and HNF4a) as

Table 8.3 A Multitude of Protein-Hydrolyzing Enzymes Act on Food throughout the Gastrointestinal Tract

	EC Code	Activation	Preferential Activity
Saliva			
N-acetylmuramoyl-L-alanine amidase	3.5.1.28		Cleaves N-acetylmuramidic acid from glycopeptides
Kallikrein 1	3.4.21.35		Leu-l-Xaa, Met-l-Xaa, Arg-l-Xaa
Stomach			
Pepsin A	3.4.23.1		Phe-l-Xaa, Leu-l-Xaa, broad specificity
Gastricsin (pepsinogen C)	3.4.23.3		Tyr-l-Xaa
Pancreas			
Trypsin I (cationic)	3.4.21.4	Enterokinase	Arg-l-Xaa or Lys-l-Xaa
Trypsin II (anionic)	3.4.21.4	Enterokinase	Arg-l-Xaa or Lys-l-Xaa
Trypsin III	3.4.21.4	Enterokinase	Arg-l-Xaa or Lys-l-Xaa
Alpha-chymotrypsin	3.4.21.1	Enterokinase	Tyr-l-Xaa, or Trp-l-Xaa, or Phe-l-Xaa, or Leu-l-Xaa
Chymotrypsin C	3.4.21.2		Leu-l-Xaa, or Tyr-l-Xaa, or Phe-l-Xaa, or Met-l-Xaa, or Trp-l-Xaa, or Gln-l-Xaa, or Asn-l-Xaa
Pancreatic elastase IIA	3.4.21.71	Trypsin	Leu-l-Xaa, Met-l-Xaa, Phe-l-Xaa
Pancreatic elastase IIB	3.4.21.71	Trypsin	Leu-l-Xaa, Met-l-Xaa, Phe-l-Xaa
Pancreatic endopeptidase E (EL3A)	3.4.21.70		Ala-l-Xaa (not elastin)
Pancreatic endopeptidase E (EL3B)	3.4.21.70		Ala-l-Xaa (not elastin)
Carboxypeptidase A1	3.4.2.1	Trypsin	Terminal Xaa except Asp, Glu, Arg, Lys or Pro
Carboxypeptidase A2	3.4.2.1	Trypsin	Terminal Xaa except Asp, Glu, Arg, Lys or Pro
Carboxypeptidase B	3.4.17.2	Trypsin	Terminal Lys or Arg
Tissue kallikrein	3.4.21.35	Trypsin	Leu-l-Xaa, Met-l-Xaa, Arg-l-Xaa
Small-intestinal brush border			
Gamma-glutamyl transpeptidase (γ-GT)	2.3.2.2		5-L-Glutamyl-l-Xaa
Duodenase	3.4.21.B3		Activates enteropeptidase
Enteropeptidase (Enterokinase)	3.4.21.9		Trypsinogen ITPKllVGG
Leucine aminopeptidase (LAP)	3.4.11.1	Zinc	N-terminal Leu
Membrane alanine aminopeptidase	3.4.11.2	Zinc	Ala-l-Xaa, others more slowly
X-Pro aminopeptidase	3.4.11.9	Manganese	Pro-l-Xaa
Glycylleucyl dipeptidase	3.4.13.18	Zinc	Hydrophobic dipeptides
Membrane dipeptidase	3.4.13.19	Zinc	Hydrophobic dipeptides
Dipeptidyl-peptidase IV	3.4.14.5		Xaa- Pro-l-Xcc
Angiotensin I-converting enzyme	3.4.15.1	Zinc	Carboxy terminal ProlXaa
Neprilysin	3.4.24.11	Zinc	Hydrophobic aa in specific proteins
Meprin A (PABA-peptide hydrolase)	3.4.24.18	Zinc/Trypsin	Carboxyl side chains of hydrophobic residues
Meprin B	3.4.24.63	Zinc/Trypsin	
Glutamate carboxypeptidase II	3.4.17.21	Zinc	Gamma-peptide bonds in Ac–Asp–Glu, Asp–Glu, Glu–Glu, pteroyl-gamma-peptide

The symbol | indicates the typical cleavage site in a model peptide.

the enterocyte matures and moves from the crypt to the villus tip (Tümer et al., 2013). A partner protein is needed to stabilize the transporter B^oAT1 in luminal plasma membranes. Angiotensin-converting enzyme 2 (ACE2, EC3.4.17.22) serves this function in the luminal membrane of enterocytes in the small intestine (Singer et al., 2012). ACE2 is notable because it counteracts the action of angiotensin-converting enzyme (ACE, EC3.4.15.1). Transmembrane protein 27 (Tmem27; collectrin) has an analogous function in the apical membranes of proximal tubules in the kidneys (Malakauskas et al., 2007).

Additional sodium cotransport systems are $B^{o,+}$ (Nakanishi et al., 2001a), ASC, beta, IMINO, a little characterized leucine transporter, and the anionic transporter EAAC1/XAG for glutamate and aspartate. Uptake of cationic amino acids (arginine, lysine, and ornithine) through y^+/CAT-1 (SLC7A1) and of carnitine through OCTN2 (SLC22A5) is driven by the voltage differential. The high intracellular concentration of imported amino acids then drives the uptake of other amino acids from the intestinal lumen through a neutral exchange mechanism. The same concentration difference can drive the export of amino acids across the basolateral membrane if the gradient during the absorptive phase is strong enough to overcome the steep sodium gradient. The direction of the substrate flux across the basolateral membrane reverses in the postabsorptive phase, especially during fasting, and the enterocytes receive amino acids for their own nutriture from the capillary bloodstream. Sodium-coupled transporters at the basolateral membrane include systems A (ATA2 and SLC38A2; Sugawara et al., 2000), and N (SN2 and SLC38A5; Nakanishi et al., 2001b). The taurine transporter (SLC6A6) is present only in the distal small intestine, where large amounts of taurine have to be recovered after the bacterial hydrolysis of bile acid conjugates. The equilibrative transporter TAT1 (SLC16A10) carries the aromatic amino acids tryptophan, tyrosine, and phenylalanine (Kim et al., 2001).

Both sides of the enterocytes have transporters that operate in exchange mode. These transporters are anchored to the plasma membrane by two distinct glycoprotein components: rBAT (SLC3A1) at the luminal side and 4F2 (SLC3A2) at the basolateral side. The heteroexchanger BAT1/$b^{o,+}$ (SLC7A9) in combination with rBAT (SLC3A1) at the luminal side transports a very broad range of amino acids, including BCAAs, cystine (CssC), and aromatic amino acids. Several heteroexchangers operate at the basolateral side, including y(+)LAT1, y(+)LAT2, LAT2, and asc-1 (Verrey et al., 1999; Fukasawa et al., 2000; Wagner et al., 2001). All of these are functional only as a complex with the much larger membrane-anchored 4F2 (SLC3A2) universal transporter component. Net transfer of individual amino acids across the basolateral membrane is driven by their own concentration gradients. However, a high concentration of one neutral amino acid can indirectly drive the transfer of another amino acid. The facilitated diffusion transporter TAT1 (SLC16A10) at the basolateral membrane colocalizes with the exchangers and recycles aromatic amino acids to help the efflux of other amino acids through the exchangers (Ramadan et al., 2007).

While amino acid absorption is most active in the proximal small intestine, uptake also occurs from the ileum (predominantly for taurine) and colon. It should be pointed out that the gut is a tissue with particularly rapid turnover and therefore in constant need of amino acids for protein synthesis, as an energy fuel, and for functionally important metabolites. These are supplied across the basolateral membrane (Table 8.4).

TRANSPORT AND CELLULAR UPTAKE

Blood circulation: The bulk of amino acids circulates with the blood as components of proteins, which is taken up by tissues according to their specific properties. The concentration of total free amino acids in plasma tends to be around 2.3 mmol/l. The concentrations of individual amino acids range from

Table 8.4 Amino Acid Transporters in the Small Intestine

Brush Border Membrane	In	Out	In
PAT1 (SLC36A1)	H$^+$		naa, zwitterionic amino/imino acids, heterocyclic aa
B^{0}AT1 (XT2s1, SLC6A19)	Na$^+$		Neutral amino acids
ASCT1 (SLC1A4)			
B^0/B/NBB/ASCT2 (SLC1A5)	Na$^+$		A, S, T, Q, N, D-Glu, D-Asn, D-Cys > M, G, L > V
B$^{0,+}$ (SLC6A14)		2 Na$^+$ + 1 Cl$^-$	H, C, R, taurine, beta-leucine, carnitine
ASC	Na$^+$		G, A, S, C, T, D-Set, D-Thr, D-Cys
X$\bar{C}$ (xCT (SLC7A11) + 4F2 (SLC3A2)		Glu	Cystine
Beta/TAUT/GAT- (SLC6A6)	Cl$^-$		β-leucine, taurine, GABA, aminobutyrate
IMINO (SIT1, SLC6A20)	Na$^+$		P,OH-P, taurine, β-leucine
Leucine transporter	Na$^+$		L
EAAT3 (XAG, SLC1A1)	3 Na$^+$	K$^+$	D,E
BAT1(b$^{0,+}$+rBAT (SLC7A9 + SLC3A1)	Neutral aa		CssC, M, L, R, K, F, ornithine, S, V, I, Y, W, H, A, N, Q, Hcys > D-Arg, D-Lys
y$^+$/CAT-1 (SLC7A1)			R, K, ornithine
OCTN2			Carnitine
Glyt1 (SLC6A9)			
Basolateral membrane			
A (ATA1, SLC38A1)	Na$^+$		A, S, M, C, P, N, Q > H, G
N/SN2 (SLC38A5)	Na$^+$		Q, N, G, A, S, H
Taurine transporter (SLC6A6)	Na$^+$Cl$^-$		Taurine, beta-leucine
System T (TAT1, SLC16A10)	Neutral aa$\leftrightarrow$		F, Y, W
y$^+$LAT1 + 4F2 (SLC7A7 + SLC3A2)	Neutral aa$\leftrightarrow$		K, R, H, Q, N
y$^+$LAT2 + 4F2 (SLC7A6 + SLC3A2)	Neutral aa$\leftrightarrow$		R, K, ornithine
LAT2 + 4F2 (SLC7A8 + SLC3A2)	Neutral aa$\leftrightarrow$		Y, F, W, T, N, I, C, S, L, V, Q > H, A, M, G
Asc-1 + 4F2 (SLC7A10 + SLC3A2)	Neutral aa$\leftrightarrow$		A, S, C, G, T, D-Ala, D-Ser, β-Ala > V, M, H, I, L, F

around 10 pmol/l (homocysteine, cysteine, and hydroxyproline) to over 100 µmol/l (glutamine, alanine, glycine, valine, lysine, leucine, threonine, and serine). Plasma concentrations of glutamine, glutamate, alanine, and other major amino acids increase significantly after meals (Tsai and Huang, 1999). Proline follows a distinctly different time pattern, with considerably delayed plasma concentration increases in response to a protein-rich meal.

The transporters mediating uptake of amino acids from circulation are of the same type described for the intestinal tract. Expression of specific forms in a particular tissue reflects the adaptation to the needs

Table 8.5 Typical Concentrations of Free Amino Acids in the Blood	
Amino Acid	**Plasma Concentration (µmol/l)**
L-Glutamine	655
L-Alanine	316
Glycine	248
L-Valine	220
L-Lysine	195
L-Proline	170
L-Threonine	128
L-Leucine	120
L-Arginine	115
L-Serine	114
L-Histidine	87
L-Ornithine	66
L-Isoleucine	63
L-Phenylalanine	53
L-Cystine	60
L-Tyrosine	60
Taurine	49
L-Asparagine	47
L-Tryptophan	46
L-Citrulline	34
L-Glutamate	32
L-Methionine	30
5-Hydroxy-L-proline	11
L-Cysteine	9
Homocysteine	9
L-Aspartate	2

Source: Data from Divino-Filho et al. (1997), Katrusiak et al. (2001) (cysteine), Hung et al. (2002) (cystine, homocysteine, and hydroxyproline), Midttun et al. (2013) (arginine and methionine).

of these cells. Differential expression of the three known members of the transport system A for small neutral amino acids may serve as an illustration. Liver expresses all three known forms. Stimulation with glucagon increases the expression of ATA2 (SLC38A2) and ATA1 (SLC38A1) but suppresses the expression of liver-specific ATA3 (SLC38A3; Hatanaka et al., 2001a). ATA3 transports the cationic amino acids arginine and lysine with high efficiency, whereas ATA2 and ATA1 do not (Hatanaka et al., 2001b). Fasting increases the flow of alanine from muscles into circulation. Glucagon simultaneously induces this shift in expression pattern, which increases the uptake of ATA2/ATA1-favored amino acids such as alanine without increasing the uptake of arginine or lysine. Thus, selective uptake into the liver allows the utilization of muscle-derived alanine for gluconeogenesis while sparing other amino acids for use by extrahepatic tissues. The glucagon effect on liver cells subsides after a meal, and the increasingly active ATA3 promptly expands the spectrum of amino acids utilized in the liver (Table 8.5).

Blood–brain barrier (BBB): The endothelial cells of brain capillaries form a highly effective seal that separates the luminal space from the brain (Pardridge, 1998). Passage of amino acids from circulation into the brain has to be mediated by specific transporters on both sides of the endothelial cell. Amino acids that are not accepted by one of the available transporters cannot cross. Amino acids are also transported out of the brain into the capillary lumen, which is essential to remove excess excitatory amino acids (Hosoya et al., 1999). A few specific proteins can cross the blood–brain barrier (BBB) to a limited extent by endocytosis (Tamai et al., 1997).

The heteroexchanger LAT1, which is expressed at both sides of the brain capillary cell epithelial cell (Duelli et al., 2000), provides the main route for large neutral amino acids across the BBB (Killian and Chikhale, 2001). LAT2 is also expressed in the brain (Wagner et al., 2001). Sodium-dependent system A also appears to mediate transfer of neutral amino acids from the blood into the brain (Kitazawa et al., 2001). Arginine and other cationic amino acids can cross via the 4F2-anchored exchange complex y^+ LAT2 (Bröer et al., 2000). A high-affinity transporter with system N properties mediates uptake of glutamate from the brain capillaries (Ennis et al., 1998). However, a sodium-dependent glutamate transporter at the abluminal side of the BBB facilitates efflux under most conditions and maintains a low brain concentration of this excitatory amino acid compared to circulation (Smith, 2000). It is presumably these mechanisms that mediate the ready efflux of excess L-aspartate, but not D-aspartate, from the brain into circulation (Hosoya et al., 1999).

Materno-fetal transfer: All essential amino acids, as well as large amounts of nonessential amino acids, pass from maternal circulation via the placenta to the fetus. The rates at which individual amino acids are transferred from mother to fetus are coordinated with the rate at which they are used for fetal growth (Paolini et al., 2001). Leucine, isoleucine, valine, serine, and glutamine constitute the bulk of net amino acid transfer from maternal circulation into the placenta (Cetin, 2001). Methionine and phenylalanine are also among the amino acids that are transferred with particular efficiency. Glycine, relatively little of which is taken from maternal blood, is a major product of placental amino acid metabolism, and much of it is transferred to the fetus. The major transporters at the maternal side of the syntrophoblast in the placenta are the sodium-dependent transporters A (SNAT1, SNAT2, and SNAT4; Desforges et al., 2010), B°, XAG (including EAAT1-4; Cramer et al., 2002), and beta (Norberg et al., 1998). LAT1 can take up most neutral and some cationic amino acids in exchange for other neutral amino acids. Several other transporters also contribute to amino acid uptake or release at the maternal side of the placental interface. System ASC is the main sodium cotransporter on the fetal side, ATA2, ATA1, or both play a quantitatively smaller role (Jansson, 2001). The sodium-independent systems LAT2 (Wagner et al., 2001), y^+ LAT1 (Kudo and Boyd, 2001), Asc-1 (Fukasawa et al., 2000), Asc-2 (Chairoungdua et al., 2001), TAT1 (Kim et al., 2001), and others all contribute to transporting from the placenta to the fetus (Table 8.6).

METABOLISM

One of the first steps in the breakdown of amino acids is the removal of the amino group. Typically, this involves a transamination reaction, often with alpha-ketoglutarate as the acceptor. All transaminases require PHP as a covalently bound cofactor. Ammonia may also be released directly, such as in the deamination reaction catalyzed by D-amino acid oxidase (EC1.4.3.3), which occurs especially in the kidneys (Hasegawa et al., 2011). The breakdown of a few amino acids involves nothing more than the reversal of the reactions responsible for their synthesis. Glutamate, for example, may shed its

Table 8.6 Amino Acid Transporters in Human Placenta

	In	Out	In
Maternal side			
ASC	Na^+		G, A, S, C, T
B°/B/NBB (SLC1A5)	Na^+		V, I, L, T, F, W > > A, S, C
Beta/TAUT/GAT-(SLC6A6)	Na^+		Beta, taur, GABA, aminobutyrate
X^-_{AG} (EAAT1, SLC1A3)	$3Na^+$	K^+	D, E
X^-_{AG} (EAAT2, SLC1A2)	$3Na^+$	K^+	D, E
X^-_{AG} (EAAT3, SLC1A1)	$3Na^+$	K^+	D, E
X^-_{AG} (EAAT4, SLC1A6)	$3Na^+$	K^+	D, E
LAT1 + 4F2 (SLC7A5 + SLC3A2)		naa	Y, F, W, T, N, I, C, S, L, V, Q > > H, A, M, G
y^+ (CAT-1, CAT-4, CAT-2B)			R, K, orn, choline, polyamines
Fetal side			
A (ATA2, SLC38A2)	Na^+		A, S, M, P, N > > G, Q, H
ASC	Na^+		G, A, S, C, T
System T (TAT1)			F, Y, W
LAT2 + 4F2 (SLC7A8 + SLC 3A2)		naa	Y, F, W, T, N, I, C, S, L, V, Q > > H, A, M, G
y^+ LAT1 + 4F2 (SLC7A7 + SLC3A2)		naa	K, R, H, Q, N
Asc-1 + heavy chain		naa	A, S, C, G, T, D-Ala, -Ala > > V, M, H, I, L, F
Asc-2 + heavy chain		naa	S, A, G, T > > C, V. I, L, F, T, D-Ala, D-Ser
naa = neutral amino acid			

amino group in any of numerous possible transamination reactions, and the resulting alpha-ketoglutarate can be utilized through the Krebs cycle. Some other amino acids require a much larger number of reactions and may depend on several vitamin cofactors. The complete oxidation of tryptophan, for example, takes more than 20 steps and requires adequate supplies of thiamin, riboflavin, vitamin B6, niacin, pantothenate, lipoate, ubiquinone, iron, and magnesium.

Most of the amino acids in proteins (alanine, valine, isoleucine, proline, phenylalanine, tyrosine, methionine, cysteine, serine, threonine, aspartate, glutamate, glutamine, aspartate, asparagine, histidine, and arginine) can be converted into glucose and are referred to as *glucogenic;* therefore, isoleucine, lysine, phenylalanine, tyrosine, and tryptophan are considered ketogenic because their catabolism generates ketone bodies or their precursors (acetoacetate, acetate, or acetyl-CoA).

Glycine: Most glycine is converted in the mitochondria into serine (using 5,10-methylene tetrahydrofolate) or undergoes deamination, decarboxylation, and one-carbon transfer to folate (generating 5,10-methylene tetrahydrofolate). Much smaller amounts are converted into glyoxylate. Glycine is used for purine nucleotide synthesis. Glycine breakdown requires riboflavin, vitamin B6, niacin, folate, lipoate, ubiquinone, iron, and magnesium.

L-*Alanine*: Transamination generates pyruvate, which can be metabolized to acetyl-CoA (which would make alanine ketogenic), or used for glucose synthesis. D-alanine can also be utilized after conversion to pyruvate (Ogawa and Fujioka, 1981) by glycine hydroxymethyltransferase (EC2.1.2.1). Complete alanine breakdown requires thiamin, riboflavin, vitamin B6, niacin, lipoate, ubiquinone, iron, and magnesium.

L-*Proline*: The bulk of proline is broken down via glutamate to alpha-ketoglutarate, which can then be used for glucose synthesis or further metabolism through the Krebs cycle. A very significant amount is used in the intestines as a precursor for the synthesis of citrulline, ornithine, and arginine. The oxidation of proline is remarkable for its generation of oxygen free radicals (Donald et al., 2001). Complete oxidation of proline depends on thiamin, riboflavin, vitamin B6, niacin, lipoate, ubiquinone, iron, and magnesium. Only the kidneys metabolize significant amounts of hydroxyproline.

The intermediary products are pyruvate and glyoxylate. The latter can be transaminated to glycine.

L-*Valine*: A total of 10 reactions is required to metabolize valine to succinyl-CoA, which can then be utilized further through the Krebs cycle and oxidative phosphorylation. Complete oxidation requires thiamin, riboflavin, niacin, vitamin B6, vitamin B12, pantothenate, biotin, lipoate, ubiquinone, iron, and magnesium.

L-*Leucine*: Most leucine is metabolized to acetoacetate and acetyl-CoA. This makes leucine the main ketogenic amino acid. About 5–10% is oxidized via beta-hydroxy beta-methylbutyrate (HMB) to acetyl-CoA. Even smaller amounts (mainly in the testis) are converted in an adenosylcobalamin-dependent reaction to beta-leucine. Normal leucine metabolism uses thiamin, riboflavin, vitamin B6, niacin, vitamin B12, biotin, pantothenate, lipoate, ubiquinone, iron, and magnesium.

L-*Isoleucine*: This BCAA is broken down to succinyl-CoA and acetyl-CoA in six steps. Utilization depends on adequate availability of thiamin, riboflavin, niacin, vitamin B6, vitamin B12, pantothenate, biotin, lipoate, ubiquinone, iron, and magnesium.

L-*Phenylalanine*/L-*tyrosine*: The tetrahydrobiopterin-dependent hydroxylation of phenylalanine generates tyrosine. Catabolism of tyrosine generates the glucogenic Krebs-cycle intermediate fumarate and the ketogenic metabolite acetoacetate in five reaction steps. Complete oxidation of phenylalanine requires biopterin, ascorbate, thiamin, riboflavin, niacin, vitamin B6, pantothenate, lipoate, ubiquinone, iron, and magnesium. Tyrosine is also a precursor of catecholamines and melanin.

L-*Tryptophan*: Almost all tryptophan is eventually metabolized through a long sequence of reactions to alanine and two acetyl-CoA molecules. Alternative tryptophan derivatives of biological importance include serotonin, melatonin, nicotinamide adenine dinucleotide (NAD), and nicotinamide adenine dinucleotide phosphate (NADP). Adequate supplies of thiamin, riboflavin, vitamin B6, niacin, pantothenate, lipoate, ubiquinone, iron, and magnesium are necessary for normal utilization.

L-*Methionine*/L-*cysteine*: Methionine is converted into cysteine via homocysteine. The intermediate homocysteine is extensively remethylated to methionine in a folate- and vitamin B12–requiring reaction. Most cysteine is metabolized to pyruvate through several alternative pathways. Smaller amounts are converted into alanine and taurine. These metabolic pathways require thiamin, riboflavin, niacin, vitamin B6, pantothenate, lipoate, ubiquinone, iron, and magnesium. Disposal of the toxic sulfite requires molybdenum.

L-*Serine*: Breakdown of serine can proceed with pyruvate as an intermediate, but most is used for the synthesis of glycine, cysteine, alanine, selenocysteine, and choline (via phosphatidylserine). These are then eventually catabolized. Complete oxidation requires thiamin, riboflavin, vitamin B6, niacin, lipoate, ubiquinone, iron, and magnesium.

L-*Threonine*: The mitochondria catabolize threonine to acetate and glycine. An alternative mitochondrial pathway can generate a spectrum of metabolites including acetol, lactaldehyde or D-lactate. Cytosolic metabolism leads to the glucogenic Krebs-cycle intermediate succinyl-CoA. Threonine breakdown uses thiamin, riboflavin, niacin, vitamin B6, folate, vitamin B12, pantothenate, biotin, lipoate, ubiquinone, zinc, iron, and magnesium.

L-*Asparagine*/L-*aspartate*: Asparagine can be deaminated to aspartate and then this amino acid is converted to the glucogenic Krebs-cycle intermediate oxaloacetate. Aspartate is a precursor of purine and pyrimidine nucleotide synthesis. Complete oxidation of either one depends on thiamin, riboflavin, niacin, vitamin B6, lipoate, ubiquinone, iron, and magnesium.

L-*Glutamine*/L-*glutamate*: Glutamine can be deaminated to glutamate, and then to the glucogenic Krebs-cycle intermediate alpha-ketoglutarate. Glutamine is a precursor of purine and pyrimidine nucleotide synthesis. Complete oxidation of either glutamine or glutamate depends on thiamin, riboflavin, niacin, vitamin B6, ubiquinone, iron, and magnesium.

L-*Histidine*: Most histidine is metabolized to glutamate. A minor alternative pathway generates imidazole pyruvate, imidazole acetate, and imidazole lactate, which are excreted. The methylhistidine from modified proteins and anserine cannot be utilized as an energy fuel. Histidine breakdown requires thiamin, riboflavin, vitamin B6, niacin, pantothenate, lipoate, ubiquinone, iron, and magnesium.

L-*Lysine*: This cationic amino acid is broken down into two molecules of acetyl-CoA in 10–13 steps, depending on the pathway. Complete oxidation depends on thiamin, riboflavin, vitamin B6, niacin, pantothenate, lipoate, ubiquinone, iron, and magnesium.

L-*Arginine*: Breakdown of arginine proceeds via glutamate to alpha-ketoglutarate, which is metabolized further through the Krebs cycle or used for glucose synthesis. Arginine is also a direct precursor of creatine and nitric oxide. Thiamin, riboflavin, vitamin B6, niacin, pantothenate, lipoate, ubiquinone, and magnesium are needed for the utilization of arginine.

Urea synthesis: Most of the nitrogen ingested with protein eventually ends up as urea in urine. Urea synthesis occurs mainly in the liver (Figure 8.4). Carbamoyl phosphate synthase I (EC6.3.4.16) in the mitochondrial matrix condenses ammonia and bicarbonate in an ATP-driven reaction. Mitochondrial ornithine carbamoyltransferase (OTC EC2.1.3.3) can then join L-ornithine and carbamoyl phosphate to form citrulline. The ornithine for this reaction comes from the cytosol, from where the mitochondrial ornithine transporter 1 (SLC25A15) shuttles it in exchange for citrulline across the inner mitochondrial membrane (Camacho et al., 1999). In the cytosol, citrulline is condensed with L-aspartate in an ATP-driven reaction (argininosuccinate synthase, EC6.3.4.5). Argininosuccinate lyase (EC4.3.2.1) then cleaves this intermediate into fumarate and L-arginine. The cycle is finally completed when the manganese-requiring enzyme arginase (EC3.5.3.1) cleaves L-arginine into urea and L-ornithine. While ornithine is mostly recycled during urea synthesis, it comes ultimately from *de novo* synthesis in the small intestine. Smaller amounts may be taken up from food sources.

Ornithine synthesis from L-glutamate occurs in the kidneys, intestine, brain, and other tissues. Glycine amidinotransferase (EC2.1.4.1) in the kidneys, which is the rate-limiting enzyme for creatine synthesis, produces L-ornithine and guanidinium acetate from L-arginine and glycine. An alternative synthesis pathway, mainly in the small intestine, starts with the phosphorylation by mitochondrial gamma-glutamate 5-kinase (EC2.7.2.11). In humans, the same protein also catalyzes the subsequent reduction to glutamate gamma semialdehyde (glutamate gamma-semialdehyde dehydrogenase, EC1.2.1.41). Ornithine-delta-aminotransferase (EC2.6.1.13) can then complete ornithine synthesis by transferring the amino group from L-glutamate or a number of other amino acids to glutamate gamma-semialdehyde.

A constant proportion of L-glutamate in liver mitochondria is acetylated by the amino acid *N*-acetyltransferase (EC2.3.1.1). Since *N*-acetyl glutamate activates carbamoyl phosphate synthase, the concentration of L-glutamate sets the pace of urea synthesis.

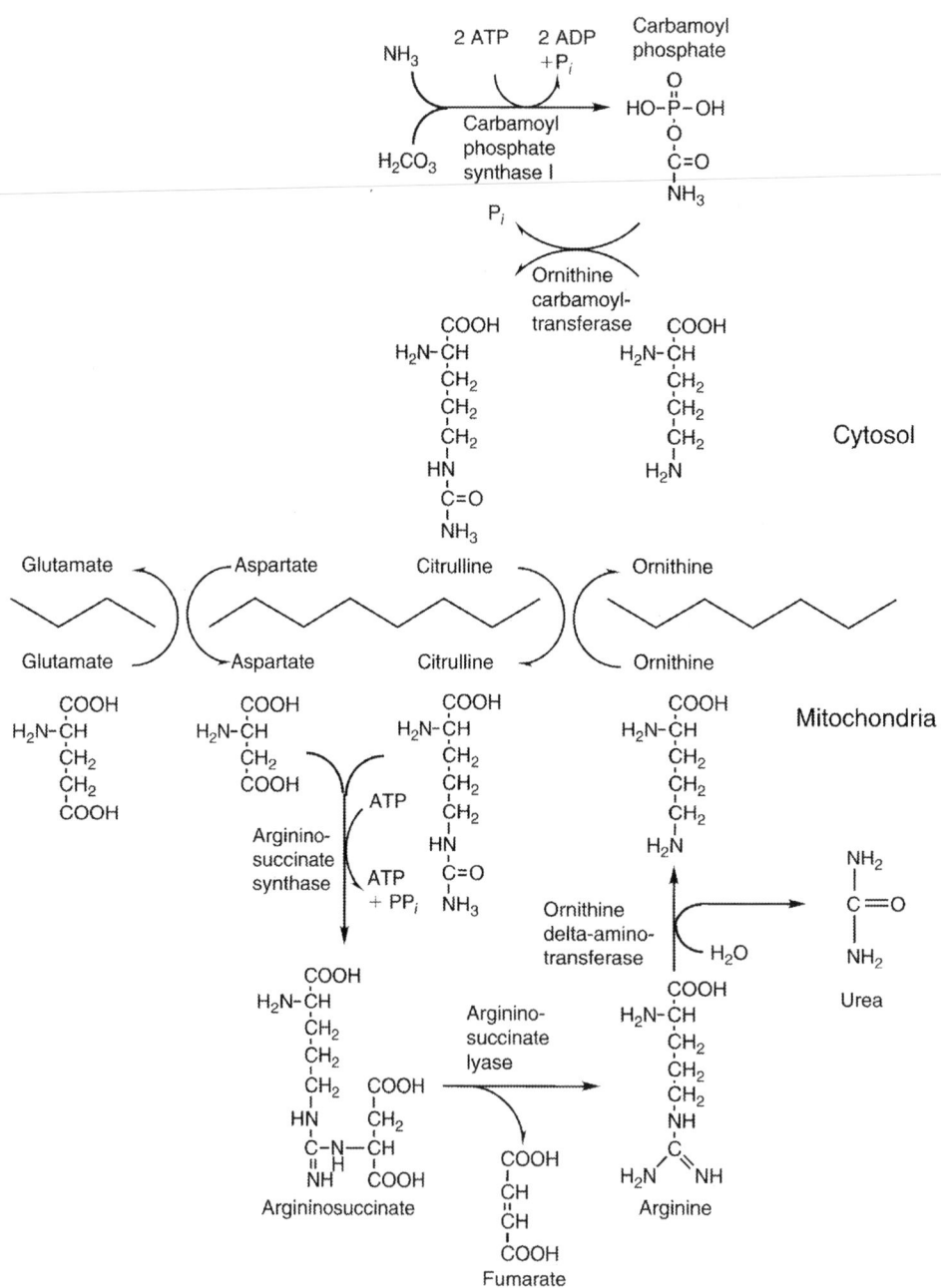

FIGURE 8.4

Urea synthesis combines two amino moieties for excretion.

STORAGE

A healthy young adult male contains about 14% protein, much of this in muscle tissue. Due to the constant rapid turnover of proteins in muscle and some other tissues, about 3–5 g/kg in healthy adults (Raguso et al., 2000) is broken down daily. Inadequate intake of a particular amino acid can be covered temporarily, though at the expense of muscle mass. Insufficient or imbalanced protein intake, immobilization, hormonal dysfunction, excessive cytokine action, and other factors are most commonly the underlying cause of low muscle mass (sarcopenia), especially in older people (Bales and Ritchie, 2002).

EXCRETION

Renal salvage: Only minimal amounts of amino acids are lost with urine intact. While several grams are filtered per day, almost all of this is recovered by a highly effective combination of sodium-driven and exchange transport systems. Proteins and peptides are cleaved by various brush border exoenzymes, including membrane Pro-X carboxypeptidase (EC3.4.17.16) and angiotensin 1–converting enzyme (ACE, EC3.4.15.1). Dipeptides and tripeptides can be taken up via sodium/peptide cotransporter 1 (PepT1, SLC15A1) in the S1 segment of the proximal tubule and sodium/peptide cotransporter 2 (PepT2, SLC15A2) in the S3 segment (Shen et al., 1999). Neutral amino acids enter epithelial cells mainly via the sodium-dependent neutral amino acid transporters B^0 (Bröer et al., 2011), ASC, and $B^{0,+}$. Glutamate and aspartate use the XAG transport system. The sodium-dependent transporters GAT-1 and GAT-3, which are better known for their role in neurotransmitter recovery in the brain, ferry gamma-aminobutyric acid (GABA), hypotaurine, and beta-alanine across the proximal tubular brush border membrane (Muth et al., 1998). Proline, hydroxyproline, taurine, and beta-alanine are taken up by the sodium-dependent imino transporter (Urdaneta et al., 1998), and betaine enters via the sodium- and chloride-dependent betaine transporter (SLC6A12). Taurine uptake via the taurine transporter (TAUT, SLC6A6) is sodium- and chloride-dependent (Chesney et al., 1990). High concentrations of osmolytes, such as taurine and betaine, protect epithelial cells against the high osmotic pressure in the medulla.

Specificity and capacity of the sodium-dependent transporters is expanded considerably by the rBAT (SLC3A1)-linked transporter BAT1 (SLC7A9). This transporter, which accounts for most of, if not all, the activity of system $b^{0,+}$, shuttles small and large neutral amino acids across the brush border membrane in exchange for other neutral amino acids. Carnitine enters the cell via the organic cation transporter OCTN2 in exchange for tetraethylammonium or other organic cations (Ohashi et al., 2001).

Amino acids are utilized to some extent in tubular epithelial cells for protein synthesis, energy production, and other metabolic pathways. The case of hydroxyproline is somewhat special, because the kidneys are the main sites of its metabolism, mainly to serine and glycine (Lowry et al., 1985). Hydroxyproline is derived from dietary collagen and from endogenous muscle, connective tissue, and bone turnover. It reaches the mitochondria of the tubular epithelial cells through a translocator that is distinct from that for proline (Atlante et al., 1994). Hydroxyproline is then oxidized by 4-oxoproline reductase (hydroxyproline oxidase, EC1.1.1.104) to 4-oxoproline (Kim et al., 1997). Then 4-hydroxy-2-oxoglutarate aldolase (EC4.1.3.16) generates pyruvate and glyoxylate. Glycine is produced when the pyridoxal-phosphate-dependent alanine-glyoxylate aminotransferase (EC2.6.1.44) uses alanine for the amination of glyoxylate.

The main sodium-dependent amino acid transporters of the basolateral membrane are system A (preferentially transporting alanine, serine, and glutamine) and ASCT1 (alanine, serine, cysteine, threonine, and glycine). It is important to note that net transfer of individual amino acids depends on their own concentration gradient. As on the luminal side, some transporters operate in exchange mode. Small neutral amino acids are the main counter molecules because their concentrations are the highest. Functional studies have characterized transport system Asc o be used for small neutral amino acids, but no corresponding gene or protein has been identified yet. Glycoprotein 4F2 anchors the amino acid exchangers typical for this side to the basolateral membrane (Verrey et al., 1999). The L-type transporter LAT2 (SLC7A8) accepts most neutral amino acids for transport in either direction.

Arginine and other cationic amino acids can pass through related heterodimers; one of these is 4F2 in combination with y$^+$LAT1 (SLCA7), another consists of 4F2 and y$^+$LAT2 (SLC7A6). These transporters can exchange a cationic amino acid for a neutral amino acid plus a sodium ion. GAT-2 mediates betaine, beta-alanine, and some taurine transport. The same compounds may also leave via the sodium chloride–dependent taurine transporter (SLC6A6).

Excretion with urine: Most nitrogen is excreted with urine as urea (0.16 g of nitrogen per gram of ingested protein, corresponding to 0.34 g of urea per gram of protein), creatinine (0.5 g/day), uric acid (0.2 g/day), and a few minor nitrogen-containing compounds. Elimination of these metabolic end products is important because they are toxic at higher than normal concentrations. Urea freely passes through the renal glomeruli, which means that a healthy adult male produces ultrafiltrate with more than 20 g of urea nitrogen. Since some reabsorption occurs, only about half of this amount is excreted with urine. A much smaller amount of protein-derived nitrogen is excreted as ammonia.

Fecal losses: Much smaller amounts of protein-derived nitrogen are lost from the intestinal tract. Losses are due to the shedding of intestinal epithelia (desquamation) and incomplete absorption of dietary proteins. More than one-tenth of ingested protein is not absorbed in healthy people, and the unabsorbed percentage may be much higher for some proteins with low digestibility or in people with digestive or absorptive disorders (Heymsfield et al., 1994).

REGULATION

Nitrogen balance: Healthy adults usually maintain constant lean body mass and neither accumulate protein nor lose protein mass. Since their combined nitrogen intake (mainly as protein) more or less equals their nitrogen losses, they are said to be in *nitrogen balance.*

Growing children and adolescents accumulate nitrogen and are therefore said to be in positive nitrogen balance. Starving, immobilized, and severely ill people, in contrast, break down tissue protein and lose more nitrogen than they take in; they are said to be in negative nitrogen balance.

Glucagon, catecholamines, cortisol, thyroid hormones, and cytokines promote the breakdown of tissue protein and its use for gluconeogensis. Excessive release of cytokines, such as tumor necrosis factor, interleukin 1 (IL-1), and interleukin 6 (IL-6), may be responsible for the accelerated protein catabolism in conditions such as tumor cachexia, but the details are not well understood (Tisdale, 1998). Several hormones promote protein synthesis (anabolic hormones), including insulin, insulin-like growth factor 1 (IGF-1), growth hormone, and testosterone. Muscle use and the abundance of free amino acids (especially BCAAs) are potent determinants of the rate of protein synthesis in muscle (Tipton et al., 2001). Leucine and its metabolites increase protein synthesis (Wilkinson et al., 2013) by activating a distinct signaling cascade (Anthony et al., 2001; Liu et al., 2001) that includes ribosomal protein S6 kinase (S6K1) and eukaryotic initiation factor 4E binding protein 1 (4E-BPI).

FUNCTION

Protein synthesis: Daily protein turnover may be as much as 300 g, which means that the same amount has to be resynthesized. The 20 basic amino acids are required for the synthesis of most of the more than 30,000 different proteins that constitute the human body. Deficiency of any single one affects all body functions and is ultimately not compatible with life.

Gluconeogenesis: Brain needs glucose as its main energy fuel. When carbohydrate sources and intermediary metabolites are depleted, amino acids are used for the synthesis of glucose (gluconeogenesis). Skeletal muscle is the major source due to its large mass, but proteins from all other tissues are also utilized. The alanine cycle mediates the transfer from muscle to the liver. The amino groups from muscle amino acids are preferentially transferred first to alpha-ketoglutarate and then from glutamate to pyruvate. Various minor pathways accomplish the same thing. The carbon skeletons of glucogenic amino acids are mostly oxidized locally. Alanine, on the other hand, is exported into the blood. The liver extracts alanine from the blood, incorporates the amino group into urea for excretion, and uses the pyruvate for glucose synthesis (Figure 8.5).

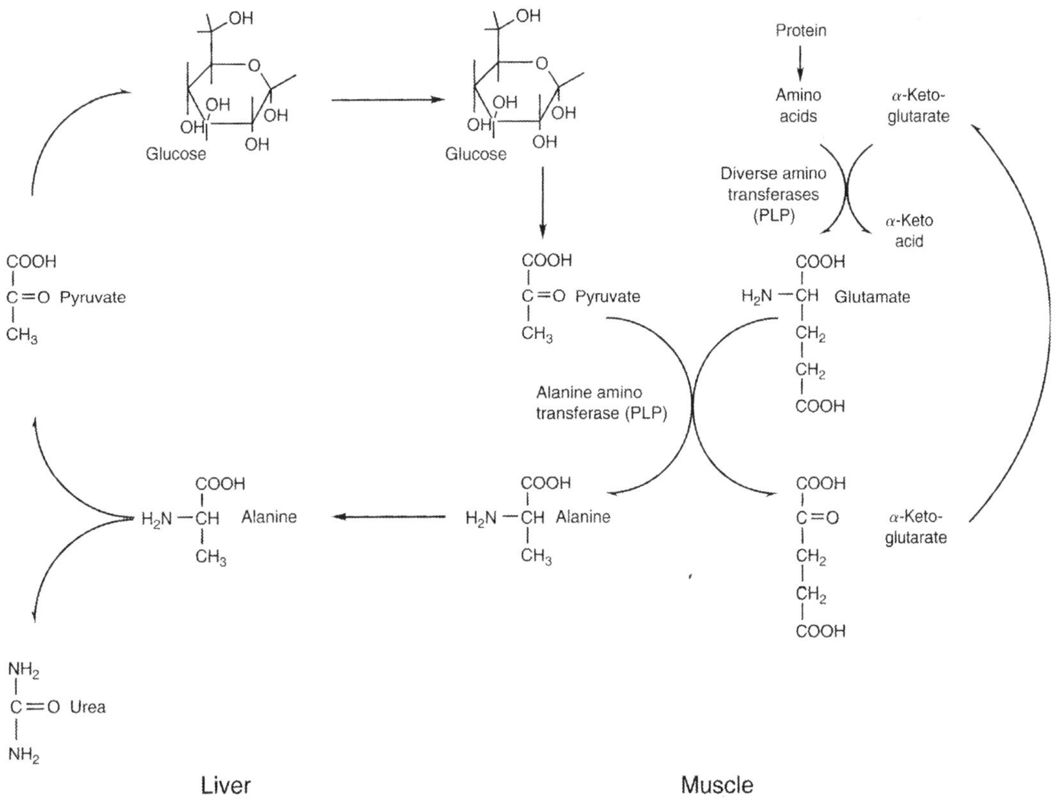

FIGURE 8.5

The alanine cycle allows the utilization of muscle proteins for gluconeogenesis in the liver.

Table 8.7 Energy Yield of Individual Amino Acids

Amino Acid	Typical Intake (mg/g)	Energy (kcal/g)
L-Alanine	53	3.4
Glycine	42	2.0
L-Leucine	92	6.0
L-Isoleucine	49	5.9
L-Valine	56	5.3
L-Lysine	70	4.9
L-Proline	73	5.0
5-Hydroxy-L-proline	-	5.0
L-Threonine	45	3.4
L-Serine	59	2.3
L-Phenylalanine	50	6.3
L-Tyrosine	41	5.4
L-Tryptophan	13	5.8
L-Methionine	26	5.3
L-Cysteine	17	4.3
L-Cystine	-	4.1
L-Histidine	30	3.4
L-Arginine	57	3.3
L-Glutamate	95	3.1
L-Glutamine	90	3.1
L-Aspartate	48	2.3
L-Asparagine	36	2.3

Source: Data from Lenders et al. (2009) and May and Hill (1990).

Energy fuel: Eventually, most amino acid molecules are fully oxidized to carbon dioxide, water, and urea. Only very minor amounts of a few amino acids are converted into compounds that are excreted in a more complex form. On average, the oxidation of the amino acids in proteins provides 4 kcal/g, but the yields differ between amino acids, as shown in Table 8.7.

Nonprotein mediator synthesis: Several hormones are derived from amino acids but are not peptides. This category includes catecholamines, serotonin, and melatonin. Virtually all organic compounds involved in neurotransmission or modulation of neuron excitation are either amino acids or amino acid metabolites. Amino acids with such functions include glutamate, aspartate, glycine, serine, and proline. Amino acid metabolites, which participate in neurotransmission, include GABA, N-methyl D-aspartate (NMDA), D-serine, nitric oxide, serotonin, melatonin, histamine, and agmatine. D-Serine in astrocytic glia cells has been identified as a "gliotransmitter," which can modify the responsiveness of nearby synapses (Martineau et al., 2013).

Nucleotide synthesis: Two of the four carbon atoms and one of the nitrogen atoms in purines come from glycine. Aspartate provides two of the five nitrogen atoms in adenosine nucleotides, one of the four nitrogens in guanosine nucleotides, and one of the nitrogens in pyrimidine nucleotides (uridine, thymine, and cytosine).

NUTRITIONAL ASSESSMENT

Protein deficiency (protein malnutrition) and suboptimal status may be detected by measuring the concentration of serum albumin or of particular intake-sensitive proteins, such as pre-albumin (Herselman et al., 2010). However, the predictive power of protein biomarkers is so fraught with error and confounding by disease that their utility is in doubt. Dietary protein intake is often estimated from the amount of excreted urinary nitrogen (Bingham, 2003).

REFERENCES

Anthony, J.C., Anthony, T.G., Kimball, S.R., Jefferson, L.S., 2001. Signaling pathways involved in translational control of protein synthesis in skeletal muscle by leucine. J. Nutr. 131, 856S–860S.

Atlante, A., Passarella, S., Quagliariello, E., 1994. Spectroscopic study of hydroxyproline transport in rat kidney mitochondria. Biochem. Biophys. Res. Commun. 202, 58–64.

Bales, C.W., Ritchie, C.S., 2002. Sarcopenia, weight loss, and nutritional frailty in the elderly. Annu. Rev. Nutr. 22, 309–323.

Biemel, K.M., Reihl, O., Conrad, J., Lederer, M.O., 2001. Formation pathways of lysine–arginine cross-links derived from hexoses and pentoses by Maillard processes. J. Biol. Chem. 276, 23405–23412.

Bingham, S.A., 2003. Urine nitrogen as a biomarker for the validation of dietary protein intake. J. Nutr. 133, 921S–924S.

Bröer, A., Juelich, T., Vanslambrouck, J.M., Tietze, N., Solomon, P.S., Holst, J., et al., 2011. Impaired nutrient signaling and body weight control in a Na^+ neutral amino acid cotransporter (Slc6a19)-deficient mouse. J. Biol. Chem. 286, 26638–26651.

Bröer, A., Wagner, C.A., Lang, F., Bröer, S., 2000. The heterodimeric amino acid transporter 4F2 hc/y^+ LAT2 mediates arginine efflux in exchange with glutamine. Biochem. J. 349, 787–795.

Camacho, J.A., Obie, C., Biery, B., Goodman, B.K., Hu, C.A., Almashanu, S., et al., 1999. Hyperornithinaemia–hyperammonaemia–homocitrullinuria syndrome is caused by mutations in a gene encoding a mitochondrial ornithine transporter. Nat. Genet. 22, 151–158.

Cetin, I., 2001. Amino acid interconversions in the fetal–placental unit: the animal model and human studies *in vivo*. Ped. Res. 49, 148–153.

Chairoungdua, A., Kanai, Y., Matsuo, H., Inatomi, J., Kim, D.K., Endou, H., 2001. Identification and characterization of a novel member of the heterodimeric amino acid transporter family presumed to be associated with an unknown heavy chain. J. Biol. Chem. 276, 49390–49399.

Chang, T.M., Lister, C., 1989. Plasma/intestinal concentration patterns suggestive of entero-portal recirculation of amino acids: effects of oral administration of asparaginase, glutaminase and tyrosinase immobilized by microencapsulation in artificial cells. Biomat. Art. Cells Art. Organs 16, 915–926.

Chesney, R.W., Zelikovic, I., Jones, D.P., Budreau, A., Jolly, K., 1990. The renal transport of taurine and the regulation of renal sodium-chloride-dependent transporter activity. Ped. Nephrol. 4, 399–407.

Cramer, S., Beveridge, M., Kilberg, M., Novak, D., 2002. Physiological importance of system A-mediated amino acid transport to rat fetal development. Am. J. Physiol. Cell Physiol. 282, C153–C160.

Devlin, A.M., Ling, E.H., Peerson, J.M., Fernando, S., Clarke, R., Smith, A.D., et al., 2000. Glutamate carboxypeptidase II: a polymorphism associated with lower levels of serum folate and hyperhomocysteinemia. Hum. Mol. Genet. 9, 2837–2844.

Divino-Filho, J.C., Barany, P., Stehle, P., Fürst, P., Bergstrom, J., 1997. Free amino-acid levels simultaneously collected in plasma, muscle, and erythrocytes of uraemic patients. Nephrol. Dial. Transplant. 12, 2339–2348.

Donald, S.P., Sun, X.Y., Hu, C.A., Yu, J., Mei, J.M., Valle, D., et al., 2001. Proline oxidase, encoded by p53-induced gene-6, catalyzes the generation of proline-dependent reactive oxygen species. Cancer Res. 61, 1810–1815.

Duelli, R., Enerson, B.E., Gerhart, D.Z., Drewes, L.R., 2000. Expression of large amino acid transporter LAT1 in rat brain endothelium. J. Cereb. Blood Flow Metab. 20, 1557–1562.

Dworschak, E., 1980. Nonenzyme browning and its effect on protein nutrition. Crit. Rev. Food Sci. Nutr. 13, 1–40.

Ennis, S.R., Kawai, N., Ren, X.D., Abdelkarim, G.E., Keep, R.E., 1998. Glutamine uptake at the blood–brain barrier is mediated by N-system transport. J. Neurochem. 71, 2565–2573.

Elango, R., Humayun, M.A., Ball, R.O., Pencharz, P.B., 2011. Protein requirement of healthy school-age children determined by the indicator amino acid oxidation method. Am. J. Clin. Nutr. 94, 1545–1552.

Erbersdobler, H.E., Faist, V., 2001. Metabolic transit of Amadori products. Nahrung 45, 177–181.

Fairweather, S.J., Bröer, A., O'Mara, M.L., Bröer, S., 2012. Intestinal peptidases form functional complexes with the neutral amino acid transporter B(0)AT1. Biochem. J. 446, 135–148.

Food and Nutrition Board, Institute of Medicine, 2005. Dietary Reference Intakes: Energy, Carbohydrates, Fiber, Fat, Fatty Acids, Cholesterol, Protein and Amino Acids. National Academy Press, Washington, DC.

Friedman, M., 1999. Chemistry, biochemistry, nutrition, and microbiology of lysinoalanine, lanthionine, and histidinoalanine in food and other proteins. J. Agric. Food Chem. 47, 1295–1319.

Fukasawa, Y., Segawa, H., Kim, J.Y., Chairoungdua, A., Kim, D.K., Matsuo, H., et al., 2000. Identification and characterization of a Na(+)-independent neutral amino acid transporter that associates with the 4F2 heavy chain and exhibits substrate selectivity for small neutral D- and L-amino acids. J. Biol. Chem. 275, 9690–9698.

Gladysheva, I.P., Popykina, N.A., Zamolodchikova, T.S., Larionova, N.I., 2001. Interaction between duodenase and alpha 1-proteinase inhibitor. Biochemistry (Russia) 66, 682–687.

Hasegawa, H., Matsukawa, T., Shinohara, Y., Konno, R., Hashimoto, T., 2004. Role of renal D-amino-acid oxidase in pharmacokinetics of D-leucine. Am. J. Physiol. Endocrinol. Metab. 287, E160–E165.

Hasegawa, H., Shinohara, Y., Akahane, K., Hashimoto, T., Ichida, K., 2011. Altered D-methionine kinetics in rats with renal impairment. Amino Acids 40, 1205–1211.

Hatanaka, T., Huang, W., Ling, R., Prasad, P.D., Sugawara, M., Leibach, F.H., et al., 2001a. Evidence for the transport of neutral as well as cationic amino acids by ATA3, a novel and liver-specific subtype of amino acid transport system A. Biochim. Biophys. Acta 1510, 10–17.

Hatanaka, T., Huang, W., Martindale, R.G., Ganapathy, V., 2001b. Differential influence of cAMP on the expression of three subtypes (ATA1, ATA2, ATA3) of the amino acid transport system A. FEBS Lett. 505, 317–320.

Herselman, M., Esau, N., Kruger, J.M., Labadarios, D., Moosa, M.R., 2010. Relationship between serum protein and mortality in adults on long-term hemodialysis: exhaustive review and meta-analysis. Nutrition 26, 10–32.

Heymsfield, S.B., Tighe, A., Wang, Z.M., 1994. Nutritional assessment by anthropometric and biochemical methods. In: Shils, M.E., Olson, J.A., Shike, M. (Eds.), Modern Nutrition in Health and Disease Lea & Febiger, Philadelphia, PA, pp. 812–841.

Hosoya, K., Sugawara, M., Asaba, H., Terasaki, T., 1999. Blood–brain barrier produces significant efflux of L-aspartic acid but not D-aspartic acid: in vivo evidence using the brain efflux index method. J. Neurochem. 73, 1206–1211.

Hung, C.J., Huang, P.C., Lu, S.C., Li, Y.H., Huang, H.B., Lin, B.E., et al., 2002. Plasma homocysteine levels in Taiwanese vegetarians are higher than those of omnivores. J. Nutr. 132, 152–158.

Jansson, T., 2001. Amino acid transporters in the human placenta. Pediatr. Res. 49, 141–147.

Katrusiak, A.E., Paterson, P.G., Kamencic, H., Shoker, A., Lyon, A.W., 2001. Pre-column derivatization high-performance liquid chromatographic method for determination of cysteine, cysteinyl-glycine, homocysteine and glutathione in plasma and cell extracts. J. Chromatogr. B 758, 207–212.

Killian, D.M., Chikhale, P.J., 2001. Predominant functional activity of the large, neutral amino acid transporter (LAT1) isoform at the cerebrovasculature. Neurosci. Lett. 306, 1–4.

Kim, D.K., Kanai, Y., Chairoungdua, A., Matsuo, H., Cha, S.H., Endou, H., 2001. Expression cloning of a Na+-independent aromatic amino acid transporter with structural similarity to H+/monocarboxylate transporters. J. Biol. Chem. 276, 17221–17228.

Kim, S.Z., Varvogli, L., Waisbren, S.E., Levy, H.L., 1997. Hydroxyprolinemia: comparison of a patient and her unaffected twin sister. J. Pediat. 130, 437–441.

Kitazawa, T., Hosya, K., Watanabe, M., Takashima, T., Ohtsuki, S., Takanaga, H., et al., 2001. Characterization of the amino acid transport of new immortalized choroid plexus epithelial cell lines: a novel *in vitro* system for investigating transport functions at the blood–cerebrospinal fluid barrier. Pharmaceut. Res. 18, 16–22.

Kohlmeier, M., 2012. Practical uses of nutrigenetics. In: Kohlmeier, M. (Ed.), Nutrigenetics. Applying the science of personal nutrition Academic Press, San Diego, CA, pp. 330–331.

Kudo, Y., Boyd, C.A., 2001. Characterisation of L-tryptophan transporters in human placenta: a comparison of brush border and basal membrane vesicles. J. Physiol. 531, 405–416.

Lenders, C.M., Liu, S., Wilmore, D.W., Sampson, L., Dougherty, L.W., Spiegelman, D., et al., 2009. Evaluation of a novel food composition database that includes glutamine and other amino acids derived from gene sequencing data. Eur. J. Clin. Nutr. 63, 1433–1439.

Liu, Z., Jahn, L.A., Long, W., Fryburg, D.A., Wei, L., Barrett, E.J., 2001. Branched chain amino acids activate messenger ribonucleic acid translation regulatory proteins in human skeletal muscle, and glucocorticoids blunt this action. J. Clin. Endocrinol. Metab. 86, 2136–2143.

Lowry, M., Hall, D.E., Brosnan, J.T., 1985. Hydroxyproline metabolism by the rat kidney: distribution of renal enzymes of hydroxyproline catabolism and renal conversion of hydroxyproline to glycine and serine. Metab. Clin. Exp. 34, 955–961.

Malakauskas, S.M., Quan, H., Fields, T.A., McCall, S.J., Yu, M.J., Kourany, W.M., et al., 2007. Aminoaciduria and altered renal expression of luminal amino acid transporters in mice lacking novel gene collectrin. Am. J. Physiol. Renal. Physiol. 292, F533–F544.

Martineau, M., Shi, T., Puyal, J., Knolhoff, A.M., Dulong, J., Gasnier, B., et al., 2013. Storage and uptake of D-serine into astrocytic synaptic-like vesicles specify gliotransmission. J. Neurosci. 33, 3413–3423.

Midttun, O., Kvalheim, G., Ueland, P.M., 2013. High-throughput, low-volume, multianalyte quantification of plasma metabolites related to one-carbon metabolism using HPLC-MS/MS. Anal. Bioanal. Chem. 405, 2009–2017.

Muth, T.R., Ahn, J., Caplan, M.J., 1998. Identification of sorting determinants in the C-terminal cytoplasmic tails of the gamma-aminobutyric acid transporters GAT-2 and GAT-3. J. Biol. Chem. 273, 25616–256127.

Nakanishi, T., Hatanaka, T., Huang, W., Prasad, P.D., Leibach, F.H., Ganapathy, M.E., et al., 2001a. Na^+- and Cl^--coupled active transport of carnitine by the amino acid transporter ATB(O, +) from mouse colon expressed in HRPE cells and Xenopus oocytes. J. Physiol. 532, 297–304.

Nakanishi, T., Sugawara, M., Huang, W., Martindale, R.G., Leibach, F.H., Ganapathy, M.E., et al., 2001b. Structure, function, and tissue expression pattern of human SN2, a subtype of the amino acid transport system N. Biochem. Biophys. Res. Commun. 281, 1343–1348.

Norberg, S., Powell, T.L., Jansson, T., 1998. Intrauterine growth restriction is associated with a reduced activity of placental taurine transporters. Ped. Res. 44, 233–238.

Ogawa, H., Fujioka, M., 1981. Purification and characterization of cytosolic and mitochondrial serine hydroxymethyltransferases from rat liver. J. Biochem. 90, 381–390.

Oguri, A., Suda, M., Totsuka, Y., Sugimura, T., Wakabayashi, K., 1998. Inhibitory effects of antioxidants on formation of heterocyclic amines. Mut. Res. 402, 237–245.

Ohashi, R., Tamai, I., Nezu Ji, J., Nikaido, H., Hashimoto, N., Oku, A., et al., 2001. Molecular and physiological evidence for multifunctionality of carnitine/organic cation transporter OCTN2. Mol. Pharmacol. 59, 358–366.

Paolini, C.L., Meschia, G., Fennessey, P.V., Pike, A.W., Teng, C., Battaglia, F.C., et al., 2001. An *in vivo* study of ovine placental transport of essential amino acids. Am. J. Physiol. Endocrinol. Metab. 280, E31–E39.

Pardridge, W.M., 1998. Blood–brain barrier carrier-mediated transport and brain metabolism of amino acids. Neurochem. Res. 23, 635–644.

Raguso, C.A., Regan, M.M., Young, V.R., 2000. Cysteine kinetics and oxidation at different intakes of methionine and cystine in young adults. Am. J. Clin. Nutr. 71, 491–499.

Ramadan, T., Camargo, S.M., Herzog, B., Bordin, M., Pos, K.M., Verrey, F., 2007. Recycling of aromatic amino acids via TAT1 allows efflux of neutral amino acids via LAT2-4F2hc exchanger. Pflügers Arch. 454, 507–516.

Ritchie, J.W., Taylor, P.M., 2001. Role of the System L permease LAT1 in amino acid and iodothyronine transport in placenta. Biochem. J. 356, 719–725.

Schut, H.A., Snyderwine, E.G., 1999. DNA adducts of heterocyclic amine food mutagens: implications for mutagenesis and carcinogenesis. Carcinogenesis 20, 353–368.

Shen, H., Smith, D.E., Yang, T., Huang, Y.G., Schnermann, J.B., Brosius III, F.C., 1999. Localization of PEPT1 and PEPT2 proton-coupled oligopeptide transporter mRNA and protein in rat kidney. Am. J. Physiol. 276, F658–F665.

Singer, D., Camargo, S.M., Ramadan, T., Schäfer, M., Mariotta, L., Herzog, B., et al., 2012. Defective intestinal amino acid absorption in Ace2 null mice. Am. J. Physiol. Gastrointest. Liver Physiol. 303, G686–G695.

Smit, E., Nieto, F.J., Crespo, C.J., Mitchell, P., 1999. Estimates of animal and plant protein intake in US adults: results from the Third National Health and Nutrition Examination Survey, 1988-1991. J. Am. Diet Assoc. 99, 813–820.

Smith, Q.R., 2000. Transport of glutamate and other amino acids at the blood–brain barrier. J. Nutr. 130, 1016S–1022S.

Sugawara, M., Nakanishi, T., Fei, Y.J., Huang, W., Ganapathy, M.E., Leibach, F.H., et al., 2000. Cloning of an amino acid transporter with functional characteristics and tissue expression pattern identical to that of system A. J. Biol. Chem. 275, 16473–16477.

Tamai, I., Sai, Y., Kobayashi, H., Kamata, M., Wakamiya, T., Tsuji, A., 1997. Structure-internalization relationship for adsorptive-mediated endocytosis of basic peptides at the blood brain barrier. J. Pharmacol. Exp. Ther. 280, 410–415.

Tipton, K.D., Rasmussen, B.B., Miller, S.L., Wolf, S.E., Owens-Stovall, S.K., Petrini, B.E., et al., 2001. Timing of amino acid-carbohydrate ingestion alters anabolic response of muscle to resistance exercise. Am. J. Physiol. Endocrinol. Metab. 281, E197–E206.

Tisdale, M.J., 1998. New cachexic factors. Curr. Opin. Clin. Nutr. Metab. Care 1, 253–256.

Tsai, P.J., Huang, P.C., 1999. Circadian variations in plasma and erythrocyte concentrations of glutamate, glutamine, and alanine in men on a diet without and with added monosodium glutamate. Metab. Clin. Exp. 48, 1455–1460.

Tümer, E., Bröer, A., Balkrishna, S., Jülich, T., Bröer, S., 2013. Enterocyte-specific regulation of the apical nutrient transporter SLC6A19 (B(0)AT1) by transcriptional and epigenetic networks. J. Biol. Chem. 288, 33813–33823.

Unno, K., Iguchi, K., Tanida, N., Fujitani, K., Takamori, N., Yamamoto, H., et al., 2013. Ingestion of theanine, an amino acid in tea, suppresses psychosocial stress in mice. Exp. Physiol. 98, 290–303.

Urdaneta, E., Barber, A., Wright, E.M., Lostao, M.P., 1998. Functional expression of the rabbit intestinal $Na^+/$ L-proline cotransporter (IMINO system) in *Xenopus laevis* oocytes. J. Physiol. Biochem. 54, 155–160.

Verrey, F., Jack, D.L., Paulsen, I.T., Saier Jr, M.H., Pfeiffer, R., 1999. New glycoprotein-associated amino acid transporters. J. Membrane Biol. 172, 181–192.

Wagner, C.A., Lang, F., Bröer, S., 2001. Function and structure of heterodimeric amino acid transporters. Am. J. Physiol. Cell Physiol. 281, C1077–C1093.

Wilkinson, D.J., Hossain, T., Hill, D.S., Phillips, B.E., Crossland, H., Williams, J., et al., 2013. Effects of leucine and its metabolite β-hydroxy-β-methylbutyrate on human skeletal muscle protein metabolism. J. Physiol. 591 (Pt 11), 2911–2923.

Wolf, W.C., Harley, R.A., Sluce, D., Chao, L., Chao, J., 1998. Cellular localization of kallistatin and tissue kallikrein in human pancreas and salivary glands. Histochem. Cell Biol. 110, 477–484.

Wolosker, H., D'Aniello, A., Snyder, S.H., 2000. D-aspartate disposition in neuronal and endocrine tissues: ontogeny, biosynthesis and release. Neuroscience 100, 183–189.

Wolosker, H., Sheth, K.N., Takahashi, M., Mothet, J.P., Brady Jr, R.O., Ferris, C.D., et al., 1999. Purification of serine racemase: biosynthesis of the neuromodulator D-serine. Proc. Natl. Acad. Sci. USA 96, 721–725.

Zamolodchikova, T.S., Sokolova, E.A., Lu, D., Sadler, J.E., 2000. Activation of recombinant proenteropeptidase by duodenase. FEBS Lett. 466, 295–299.

Ziv, E., Bendayan, M., 2000. Intestinal absorption of peptides through the enterocytes. Microsc. Res. Tech. 49, 346–352.

STARVATION

ABBREVIATION

PEM   protein-energy malnutrition

STARVATION IS A SERIOUS HEALTH THREAT

When nutrient intake does not meet requirements, for whatever reason, stores can fill the gap and allow the body to continue to function normally for a while. Eventually, however, the deficits start to affect physical performance and health. At this point, the body suffers from starvation. The word *malnutrition* refers to the state induced by starvation, and *hunger disease* expands this definition to include the typical health consequences of malnutrition. The primary concern is protein-energy malnutrition (PEM), but additional deficiencies (e.g., iron and vitamin A) usually amplify the health threats from starvation. The ability of people to escape the most serious health consequences of starvation for a while depends on the extent of the energy deficit, the range of nutrients that are lacking, and on individual predisposition.

A number of typical metabolic and physical adaptations occur in response to inadequate energy and amino acid supplies (Hoffer, 1994). Such changes favor the increased utilization of fat stores and the breakdown of body proteins. The body's metabolism also becomes more parsimonious and expends less energy (Minghelli et al., 1991). Much of that is due to a rapid decline in thyroxine and catecholamine activity. At the same time, the expression of various heat shock proteins increases as a basic protective response.

The ability to survive even prolonged starvation must have been optimized in the offspring of countless generations of famine survivors. Nonetheless, malnutrition harms the health of everyone affected, and the consequences are often severe to fatal. Even in affluent societies, malnutrition is not rare in very old people, patients with severe diseases such as cancer, those with eating disorders, and dieters seeking weight loss. Malnutrition makes survival of serious injury, infection and disease less likely, slows tissue repair, and delays recovery after illness. The effects of malnutrition may even carry over to the next generation, when it occurs during a pregnancy. The children of malnourished women are at increased risk for obesity and diabetes later in life (Birgisdottir et al., 2002).

PROTEIN TURNOVER

While considerable amounts of body fat may be lost during starvation, the main danger comes from the breakdown of protein. The obvious general muscle wasting is an ominous feature that points to the parallel loss of functional proteins in the heart, liver, kidneys, and other tissues.

Healthy males of average weight need daily about 150 g of essential amino acids and a similar quantity of nonessential amino acids for the synthesis of proteins (Munro, 1975). Much of these amounts can be derived from the breakdown of body proteins. When energy intake falls short of expenditures in a previously well nourished person and the 1-day supply of glycogen is used up, cellular free amino acids, regular protein turnover, and increased mobilization of a small pool of proteins with particularly rapid turnover ("labile protein") can cover glucose needs during the first few hours of fasting.

It is mainly the brain's heavy reliance on a steady supply of glucose that makes gluconeogenesis from amino acids so important.

Increasing secretion of glucagon in response to declining blood glucose levels promotes the mobilization of protein in skeletal muscle and other tissues. Most of the hydrolyzed protein reaches the blood as alanine and glutamine. The liver and kidneys can use these nonessential amino acids for gluconeogenesis and send the glucose out into circulation again for use by the brain, muscle, and other glucose-dependent tissues. The liver also secretes glutamine into the blood as a nitrogen and fuel carrier for other tissues (Watford et al. 2002). The term *alanine cycle* alludes to the use of alanine as a shuttle for nitrogen and carbon transfer between the muscle and liver. The increased breakdown of tissue protein and oxidation of the released amino acids are evident from the doubling of urea production during the first days of fasting (Giesecke et al., 1989). Hormones (adrenaline and cortisol) and cytokines (e.g., interleukin 6) that increase in circulation after traumatic stress and infection rapidly draw on intracellular free amino acids (Hammarqvist et al., 2001) and accelerate protein catabolism (Smeets et al., 1995). If starvation continues, the body's fat store needs can meet an increasing proportion of energy. Even the brain adapts increasingly to the use of the fatty acid metabolites acetoacetate and beta-hydroxybutyrate. It has been suggested, however, that tissue protein must be mobilized to replenish the Krebs-cycle intermediates alpha-ketoglutarate and oxaloacetate (Owen et al., 1998). The unusually large glucose needs of the human brain, which may account for as much as half of a child's total energy use, cause a constant drain on Krebs-cycle intermediates. Glycerol from mobilized triglycerides can meet the needs for glucose synthesis of other small-brained mammals, but in humans, this is not enough. Nonetheless, obese people lose tissue protein during starvation only half as fast as lean people (Elia et al., 1999). Availability of glycerol from fat may make the difference.

Sustained starvation changes the activity of key enzymes of protein and amino acid metabolism (Young and Marchini, 1990). Overall, there is a decline in the activity of enzymes involved in irreversible degradation and an increase in enzymes for amino acid recycling and utilization.

As a result of starvation the oxidation of amino acids from protein turnover drops to less than 10% (Tomkins et al., 1983). The activities of amino acid dehydrogenases in muscle and of the urea cycle in the liver and kidneys decline. In the absence of dietary intake, the ammonia uptake from portal blood greatly exceeds aspartate uptake. The consequence is that liver proteins have to be mobilized to provide aspartate for the detoxification of the ammonia as urea (Brosnan et al., 2001).

PROTEIN AVAILABILITY

Beyond the quantity of dietary protein, the combination of foods and their preparation matters, particularly for people on the verge of starvation. The protein in some foods is not fully available because it cannot be digested efficiently. Legumes contain various protease inhibitors that interfere with protein hydrolysis. These inhibitors not only limit absorption of ingested legume protein but also decrease absorption of other proteins consumed with the same meal and cause significant loss of endogenous proteins. Since protein secretion and loss of aged intestinal cells account for large amounts of protein that are normally recovered from the intestinal lumen, the consumption of protease inhibitors can lead to a net protein loss, the high protein content of inhibitor-containing foods (i.e., legumes) notwithstanding (Barth et al., 1993).

Bowman-Birk-type inhibitors in many beans potently diminish the activities of trypsin and chymotrypsin. A Kunitz-type trypsin inhibitor in soybeans and lima beans strongly inhibits duodenase, the brush border enzyme needed to initiate the activation of the proteases from pancreas (Zamolodchikova et al., 1995). Similar protease inhibitors are also present in some other plant-derived foods, including potatoes (Seppala et al., 2001).

Lectins are proteins that bind with high affinity to carbohydrate side chains of proteins (and other food compounds) and can thereby protect them from digestion. Kidney beans have particularly high lectin content, but most legumes contain lectins. Both protease inhibitors and lectins in foods become less active with cooking. Poor chewing and inadequate predigestion can also limit intestinal protein digestion. Cooking practices can also decrease protein availability. Moderate to high temperatures can convert alanine and other L-amino acids in food proteins to their D-isomers. Browning promotes the cross-linking of proteins (generating lysinoalanine, ornithinoalanine, histidinoalanine, and phenylethylaminoalanine), and the formation of derivatives, such as dehydroalanine, methyldehydroalanine, and beta-aminoalanine (Friedman, 1999). Oxidation can damage protein-bound methionine. All of these modified proteins tend to be less well digested and many of the chemically altered amino acids are not readily absorbed. Lysine, cysteine, and methionine losses from cooking can be particularly significant.

PROTEIN QUALITY

The amino acid composition of protein is of great importance if only small amounts of protein are consumed (or absorbed). The bottom line is that protein synthesis depends on adequate amounts of each constituent amino acid. If even a single amino acid is acids cannot be produced in the body, by definition, and each has to be supplied in the needed quantity with food. If one is in short supply, malnutrition will result over the long term. The limiting amino acid in protein from a particular food is indicated in the US Department of Agriculture (USDA) nutrient database. Various protein quality measures aim to capture the ability of food protein to supply all necessary amino acids. The classical assessment of the biological value of a protein uses animals to determine net nitrogen utilization.

What counts in the end, however, is the combination of food proteins consumed over the short term, not single food items. This means that foods that are low in one essential amino acid may complement other foods that have enough of this amino acid but lack another.

REFERENCES

Barth, C.A., Lunding, B., Schmitz, M., Hagemeister, H., 1993. Soybean trypsin inhibitor(s) reduce absorption of exogenous and increase loss of endogenous protein in miniature pigs. J. Nutr. 123, 2195–2200.

Birgisdottir, B.E., Gunnarsdottir, I., Thorsdottir, I., Gudnason, V., Benediktsson, R., 2002. Size at birth and glucose intolerance in a relatively genetically homogeneous, high-birth weight population. Am. J. Clin. Nutr. 76, 399–403.

Brosnan, J.T., Brosnan, M.E., Yudkoff, M., Nissim, I., Daikhin, Y., Lazarow, A., et al., 2001. Alanine metabolism in the perfused rat liver. Studies with (15)N. J. Biol. Chem. 276, 31876–31882.

Elia, M., Stubbs, R.J., Henry, C.J., 1999. Differences in fat, carbohydrate, and protein metabolism between lean and obese subjects undergoing total starvation. Obes. Res. 7, 597–604.

Friedman, M., 1999. Chemistry, biochemistry, nutrition, and microbiology of lysinoalanine, lanthionine, and histidinoalanine in food and other proteins. J. Agric. Food Chem. 47, 1295–1319.

Giesecke, K., Magnusson, I., Ahlberg, M., Hagenfeldt, L., Wahren, J., 1989. Protein and amino acid metabolism during early starvation as reflected by excretion of urea and methylhistidines. Metab. Clin. Exp. 38, 1196–1200.

Hammarqvist, E., Ejesson, B., Wernerman, J., 2001. Stress hormones initiate prolonged changes in the muscle amino acid pattern. Clin. Physiol. 21, 44–50.

Hoffer, L.J., 1994. Starvation. In: Shils, M.E., Olson, J.A., Shike, M. (Eds.), Modern Nutrition in Health and Disease, eighth ed. Lea & Febiger, Philadelphia, PA, pp. 927–949.

Minghelli, G., Schutz, Y., Whitehead, R., Jequier, E., 1991. Seasonal changes in 24-h and basal energy expenditures in rural Gambian men as measured in a respiration chamber. Am. J. Clin. Nutr. 53, 14–20.

Munro, H.N., 1975. Regulation of protein metabolism in relation to adequacy of intake. Infusionsther. Klin. Ern. 2, 112–117.

Owen, O.E., Smalley, K.J., D'Alessio, D.A., Mozzoli, M.A., Dawson, E.K., 1998. Protein, fat, and carbohydrate requirements during starvation: anaplerosis and cataplerosis. Am. J. Clin. Nutr. 68, 12–34.

Seppala, U., Majamaa, H., Turjanmaa, K., Helin, J., Reunala, T., Kalkkinen, N., et al., 2001. Identification of four novel potato (*Solarium tuberosum*) allergens belonging to the family of soybean trypsin inhibitors. Allergy 56, 619–626.

Smeets, H.J., Kievit, J., Harinck, H.I., Frolich, M., Hermans, J., 1995. Differential effects of counter-regulatory stress hormones on serum albumin concentrations and protein catabolism in healthy volunteers. Nutrition 11, 423–427.

Tomkins, A.M., Garlick, P.J., Schofield, W.N., Waterlow, J.C., 1983. The combined effects of infection and malnutrition on protein metabolism in children. Clin. Sci. 65, 313–324.

Watford, M., Chellaraj, V., Ismat, A., Brown, P., Raman, P., 2002. Hepatic glutamine metabolism. Nutrition 18, 301–303.

Young, V.R., Marchini, J.S., 1990. Mechanisms and nutritional significance of metabolic responses to altered intakes of protein and amino acids, with reference to nutritional adaptation in humans. Am. J. Clin. Nutr. 51, 270–289.

Zamolodchikova, T.S., Vorotyntseva, T.I., Antonov, V.K., 1995. Duodenase, a new serine protease of unusual specificity from bovine duodenal mucosa. Purification and properties. Eur. J. Biochem. 227, 866–872.

GLUTAMATE

The acidic amino acid L-glutamate (L-glutamic acid, 2-aminopentanedioic acid, MSG, one-letter code E; molecular weight 147) contains 9.5% nitrogen (Figure 8.6).

ABBREVIATIONS	
GABA	gamma-aminobutyric acid
Glu	L-glutamate
MSG	monosodium glutamate
RDA	recommended dietary allowance

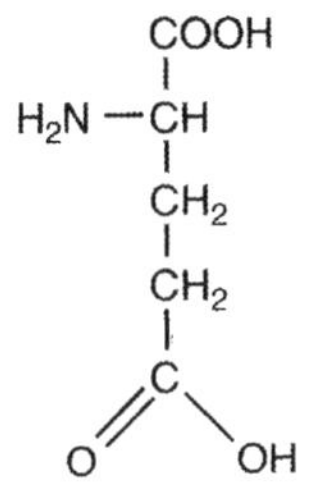

FIGURE 8.6

L-Glutamate.

NUTRITIONAL SUMMARY

Function: The nonessential amino acid L-glutamate (Glu) is used for the synthesis of L-glutamine, L-proline, L-arginine, proteins, and functional folate. Glu is essential for brain function. It also contributes, through L-glutamine, to the synthesis of purines, NAD, flavin adenine dinucleotide (FAD), and many other essential compounds. Glu plays an important role in the regulation of energy and nitrogen metabolism and is a major energy fuel; its complete oxidation requires thiamin, riboflavin, niacin, vitamin B6, pantothenate, lipoate, ubiquinone, iron, and magnesium. Glu in foods is specifically recognized by taste receptors that convey a meaty flavor referred to as *umami*.

Food sources: Dietary proteins from different sources all contain Glu. Dietary supplements containing crystalline Glu, monosodium glutamate (MSG), or other salts are commercially available.

Requirements: Since it can be synthesized from alpha-ketoglutarate, dietary intake of Glu is not necessary so long as enough total protein is available.

Deficiency: Prolonged lack of total protein causes growth failure, loss of muscle mass, and organ damage.

Excessive intake: Very high intake of protein and mixed amino acids (i.e., more than three times the RDA, which is 2.4 g/kg) is thought to increase the risk of renal glomerular sclerosis and accelerate osteoporosis. Controlled studies have not detected any specific health risks or discomfort in conjunction with the consumption of 10 g of MSG.

ENDOGENOUS SOURCES

Most Glu in the blood comes from muscle (50%), kidneys (15%), and liver (5–10%), largely synthesized *de novo* by transamination or amination of alpha-ketoglutarate (Gerich et al., 2000). Glutamate dehydrogenase (EC1.4.1.3) catalyzes the NADPH-dependent amidation of alpha-ketoglutarate. This reaction binds potentially toxic free ammonia as amino acid. Almost any amino acid can donate its amino group to the abundant Krebs-cycle intermediate alpha-ketoglutarate for the synthesis of Glu. Numerous PLP-dependent enzymes catalyze these transfers, usually with some specificity. Examples include aspartate aminotransferase (EC2.6.1.1) and alanine aminotransferase (EC2.6.1.2). A somewhat unique case is the synthesis from the urea-cycle intermediate L-ornithine by ornithine-oxo-acid aminotransferase (OAT, EC2.6.1.13) because this reaction releases both Glu and L-glutamate 5-semialdehyde; the latter can then be converted into a second Glu molecule by glutamate-5-semialdehyde dehydrogenase (EC1.2.1.41). Small amounts of Glu are generated from the breakdown of L-histidine and L-proline and by deamidation of L-glutamine by glutaminase (EC3.5.1.2) at the cytosolic face of the inner mitochondrial membrane.

DIETARY SOURCES

Particularly Glu-rich proteins are consumed with wheat (324 mg/g protein) and rye (279 mg/g). The proportions are intermediate in oats (220 mg/g), rice (195 mg/g), dairy products (209 mg/g), and soy products (200 mg/g). Lower relative contents are found in beef, pork, chicken, fish, eggs, and beans (all around 150 mg/g). Daily dietary Glu intake has been estimated to be about 28 g in a man weighing 70 kg (Garattini, 2000).

Foods that naturally contain a relatively large concentration of free glutamate include some forms of culinary seaweed and blue cheeses (Daniels et al., 1995). The rapid consumption of as much as 10 g of MSG, commonly used as a flavor enhancer, is not likely to pose any health risk or cause discomfort ("Chinese restaurant syndrome") as previously suspected (Walker and Lupien, 2000).

Prolonged heating, particularly at alkaline pH, promotes the generation of protein-bound D-glutamate (de Vrese et al., 2000). Some D-glutamate may be ingested with marine bivalves and plants (Man and Bada, 1987).

Taste perception: In addition to the four basic flavors recognized by European tradition, a fifth has been characterized that is specifically linked to the detection of Glu by a metabotrophic glutamate receptor on taste buds of the tongue (Kurihara and Kashiwayanagi, 2000). The technical term for the Glu flavor is *umami* (savory).

DIGESTION AND ABSORPTION

Healthy individuals absorb amino acids and proteins in the proximal small intestine nearly completely (Figure 8.7). Food proteins are hydrolyzed by an array of gastric, pancreatic, and enteral enzymes, which generate Glu as part of oligopeptides and in free form. The former can be taken up through the hydrogen ion/peptide cotransporters 1 (SLC15A1, PepT1) and 2 (SLC15A2, PepT2).

Enterocytes take up free Glu from the intestinal lumen via the high-affinity sodium cotransporter EAAT3 (SLC1A1), corresponding to system $X\overline{AG}$ (SLC1A1). Each glutamate anion is transported with three sodium ions in exchange for one potassium ion. The sodium-dependent system ASC provides an

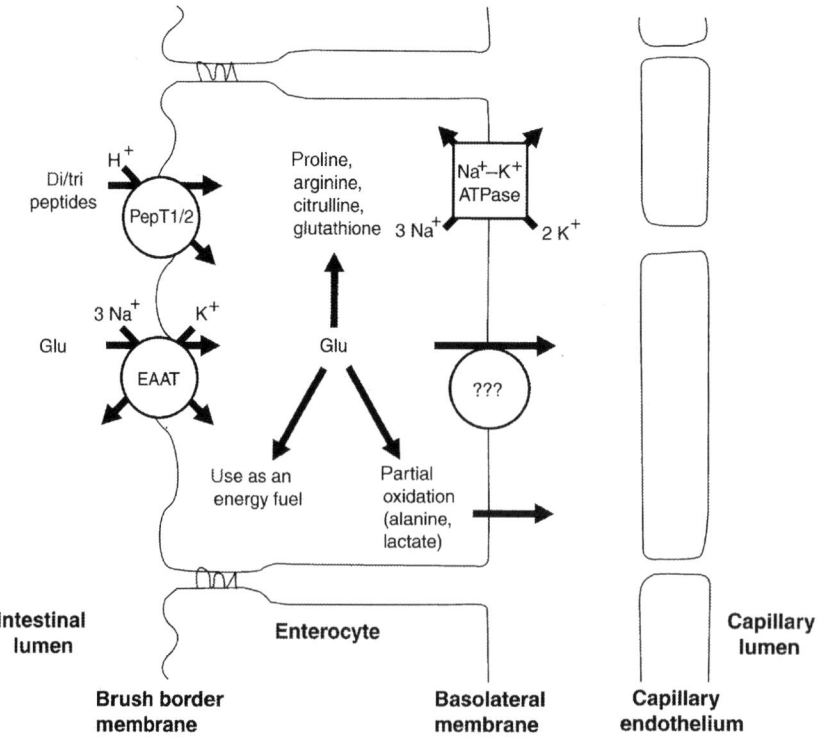

FIGURE 8.7

Intestinal absorption of L-glutamate.

additional minor route (Mordrelle et al., 1997), as may various other transporters with low efficiency. Very little of the ingested Glu reaches portal blood or systemic circulation unchanged (Young and Ajami, 2000). About a third is oxidized completely in enterocytes, and nearly as much is exported after partial oxidation to L-alanine or lactate (Reeds et al., 2000); smaller amounts are incorporated into mucosal proteins or converted to L-proline (6%), glutathione (6%), L-arginine (4%), and citrulline.

The neutral amino acid exchanger complex LAT2-4F2hc (SLC7A8-SLC3A2) works in conjunction with the aromatic amino acid diffusion transporter TAT1 to move Glu from the enterocyte into the blood (Ramadan et al., 2007).

TRANSPORT AND CELLULAR UPTAKE

Blood circulation: Normal plasma concentrations of free Glu are around 14 μmol/l (Hammarqvist et al., 2001). Glu is largely taken up by tissues via several members of the sodium-dependent system X~, including EAATI (GLAST1, SLC1A3), EAAT2 (GLT1, SLC1A2), EAAT3 (EAAC1, SLC1A1), EAAT4 (SLC1A6), and EAAT5 (SLC1A7).

Glu can be transported into mitochondria by glutamate-aspartate translocase in exchange for L-aspartate, and by the glutamate-hydroxide translocase in exchange for a hydroxyl ion. The former exchange is especially important in the liver to provide mitochondrial L-aspartate for the cytosolic steps of the urea cycle. Intracellular Glu can be taken out of some cells, particularly macrophages and other immune cells, in exchange for cystine via the glycoprotein 4F2-linked transporter xCT, SLC7A11 (Verrey et al., 1999), which is expressed in the liver and a few other tissues.

BBB: Transport into the brain uses mainly a sodium-independent transporter, which is not the sodium independent $X\overline{C}$ system (Benrabh and Lefauconnier, 1996). System $X\overline{C}$ exchanges L-cystine for Glu, thereby providing cystine to neurons for glutathione (Kim et al., 2001).

Materno-fetal transfer: Glu passes both apical and basal membranes via the sodium-dependent system $X\overline{AG}$ transporters EAAC1, GLT1, and GLAST1 (Novak et al., 2001); the exact location of each of these transporters remains unclear. The relative contribution of individual transporters appears to vary with developmental stages, but has not been assessed in humans yet. It should also be pointed out that relatively large amounts of Glu are produced by the fetal liver from maternal L-glutamine and taken up by the placenta as an important energy fuel. Thus, there is little Glu transfer from maternal to fetal circulation (Garattini, 2000); rather, the net flux of Glu is from the fetus to the placenta.

METABOLISM

The metabolism of Glu (Yang and Brunengraber, 2000) (Figure 8.8) starts either with the transfer of its amino group to an α-keto acid by one of the many aminotransferases or the oxidative elimination of ammonia by glutamate dehydrogenase (EC1.4.1.3). The resulting alpha-ketoglutarate can then be utilized via the Krebs cycle. The successive steps are catalyzed by the alpha-ketoglutarate dehydrogenase complex (EC1.2.4.2, contains thiamin pyrophosphate and lipoate), succinate CoA ligase, GDP-forming (succinyl-CoA synthase, EC6.2.1.4) or ADP-forming (EC6.2.1.5), the ubiquinone-linked succinate dehydrogenase complex (EC1.3.5.1, contains FAD and iron), fumarate hydratase (fumarase, EC4.2.1.2), malate dehydrogenase (EC1.1.1.37), citrate (Si)-synthase (EC4.1.3.7), aconitate hydratase (aconitase, EC4.2.1.3, contains iron), and isocitrate dehydrogenase, both NAD-using (EC1.1.1.41) and NADP-using (EC1.1.1.42). D-Glutamate can be metabolized in peroxisomes by D-aspartate oxidase (EC1.4.3.1) and other D-amino oxidases in the liver and kidneys. The metabolite D-5-oxoproline is excreted with urine (Sekura et al., 1976).

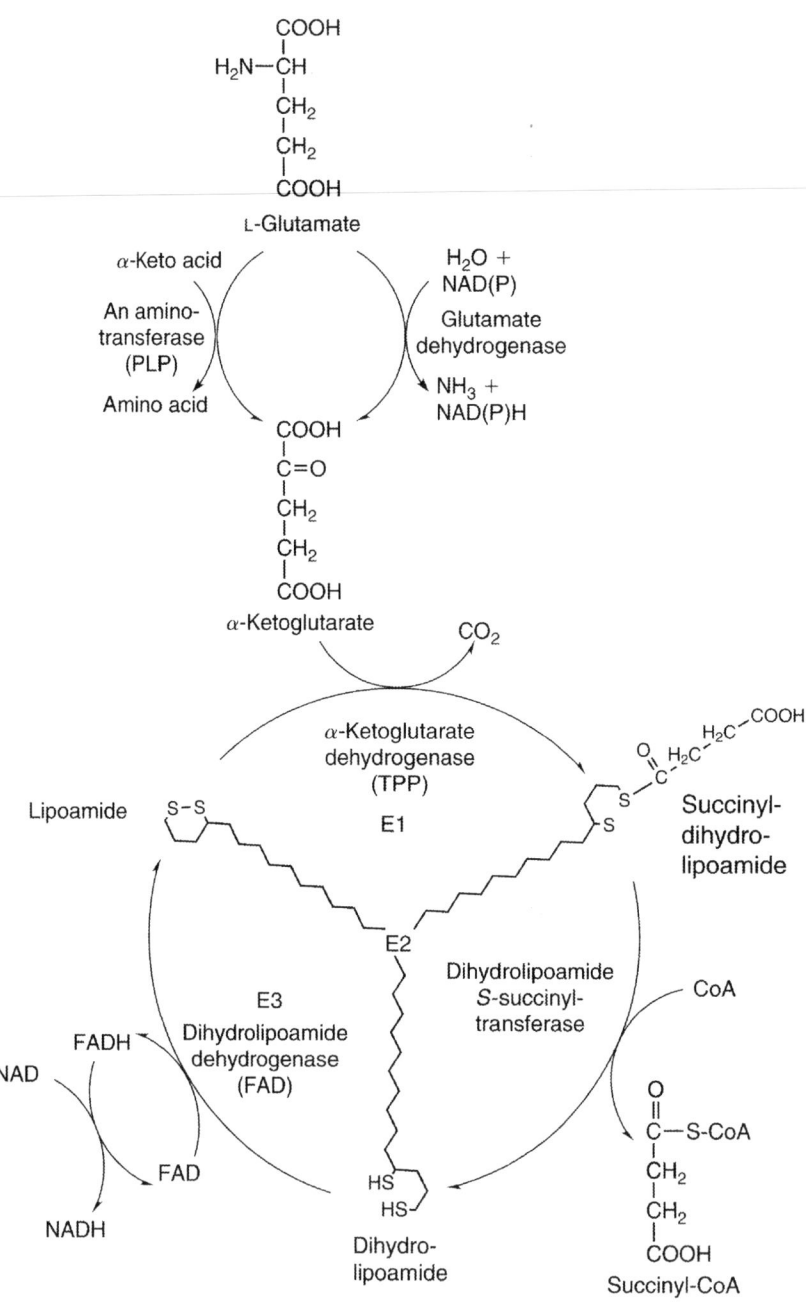

FIGURE 8.8

The oxidation of the glutamate metabolite alpha-ketoglutarate requires thiamin pyrophosphate, lipoamide, FAD, and NAD.

STORAGE

Glu content in muscle in women (17.2%) is higher than in men (13.1%; Kuhn et al., 1999). While turnover of protein in skeletal muscles and other tissues constantly releases some free Glu, starvation, trauma, and infection accelerate this protein catabolism.

EXCRETION

Very little Glu is lost with feces or urine because reabsorption is highly effective in healthy people. PepT1 and pepT2 recover dipeptides and tripeptides. Uptake from the tubular lumen proceeds via the sodium-linked $\overline{XAG}$ transport system (EAAC1, SLC1A1) and the sodium-independent rBAT (SLC3A1)-linked transporter BAT1 (SLC7A9). The sodium-independent aspartate/glutamate transporter 1 (AGT1, SLC7A13), associated with an as-yet-unidentified membrane-anchoring glycoprotein, moves Glu across the basolateral membrane (Matsuo et al., 2002). Sodium- and potassium-dependent transport has also been reported (Sacktor et al., 1981), but the identity of the responsible transporter(s) remains unknown.

Most nitrogen from metabolized Glu is excreted into urine as urea, a much smaller amount as free ammonium ions (from deamination by glutamate dehydrogenase and from oxidative deamination).

REGULATION

Accelerated protein catabolism and the ensuing rise in free amino acid concentration increase Glu concentration. Through mass action, more Glu is converted by amino acid N-acetyltransferase (EC2.3.1.1) to N-acetylglutamate. Because N-acetylglutamate is a strong activator of carbamoyl phosphate synthase 1 (EC6.3.4.16), the use of Glu-derived ammonia for urea increases.

Glu and L-glutamine compete for glutaminase (EC3.5.1.2); the rate of Glu production from L-glutamine thus increases with decreasing intracellular Glu concentration (Welbourne and Nissim, 2001). During the first few minutes of intense exercise, Glu is the most important source of anapleurotic Krebs-cycle intermediates (Gibala et al., 1997). Endurance exercise promotes glutamate dehydrogenase and glutamine formation (Graham et al., 1997).

FUNCTION

Energy fuel: Complete oxidation of Glu yields 3.09 kcal/g (May and Hill, 1990) and depends on adequate supplies of thiamin, riboflavin, niacin, vitamin B6, pantothenate, lipoate, ubiquinone, iron, and magnesium. Since glucose can be synthesized from the Glu metabolite alpha-ketoglutarate, Glu is a glucogenic amino acid.

Protein synthesis: Glutamate-tRNA ligase (EC6.1.1.17) links Glu with its correspondent tRNA in an ATP/magnesium-dependent reaction. Posttranslational gamma-carboxylation of specific glutamyl-residues in a few coagulation factors and other proteins is vitamin-dependent and usually confers calcium and phospholipid-binding ability. Another specific posttranslational modification is the formation of intermolecular epsilon (gamma-glutamyl)lysine cross-links by various isoforms of protein-glutamine gamma-glutamyltransferase (transglutaminase, EC2.3.2.13, calcium-dependent).

Glutamylation: The attachment of multiple glutamyl residues is essential for the function and metabolism of specific proteins, such as the tubulin of the cytoskeleton. The bulk of folate in foods also contains five or more glutamyl residues, which must be removed by digestion before the vitamin can be absorbed. Glutamyl residues are attached again in target tissues as required for full folate function.

Amino acid synthesis: Glu is the donor or recipient of amino groups in numerous transamination reactions and is thus the major currency for amino acid metabolism. It is a direct precursor of L-glutamine and thus relevant for the synthesis of purine nucleotides, including NAD and FAD. Proline synthesis from Glu proceeds in four steps, which are catalyzed by glutamate 5-kinase (EC2.7.2.11), glutamate gamma-semialdehyde dehydrogenase (EC1.2.41), and delta 1-pyrroline 5-carboxylate reductase (EC1.5.1.1). Glu in the small intestine, but not in other tissues, is the precursor of citrulline, which is used in the liver and kidneys for the synthesis of both urea and arginine.

Metabolic effects: Due to its central metabolic position, Glu affects the regulation of macronutrient disposition in numerous ways. It is a regulator of glycogen synthesis, gluconeogenesis, and lipolysis (Stumvoll et al., 1999).

Krebs cycle anapleurosis: Another important function is to replenish the supply of Krebs-cycle intermediates. Some leakage and breakdown of these intermediates occur inevitably, and without constant addition of new intermediates, the Krebs cycle would slow down and even cease to work. Dietary Glu/glutamine provides the bulk of intermediates (as alpha-ketoglutarate); smaller amounts come from glucose (via biotin-dependent pyruvate carboxylation), aspartate/asparagine breakdown, the biotin- and vitamin B12-dependent metabolism of propionyl-CoA (from odd chain fatty acids, valine, threonine, and methionine), fumarate from phenylalanine/tyrosine catabolism, and dietary intake of Krebs-cycle intermediates.

Glutathione: Glu is a constituent of the tripeptide gamma-glutamyl-cysteinyl-glycine (glutathione), which is an important intracellular and extracellular antioxidant. The exoenzyme γ-glutamyltransferase (EC2.3.2.2) initiates glutathione degradation by transferring the glutamyl residue to other peptides.

Neurotransmission: Glu is the main excitatory neurotransmitter in the brain and appears to be especially important for memory and learning. Glu acts by binding to ligand-gated ion channels and to G-protein linked (metabotropic) receptors. Overall, there are three families of Glu receptors with numerous subtypes, which serve a broad range of signaling functions in addition to neurotransmission. Another Glu-derived neurotransmitter in the brain is gamma-aminobutyric acid (GABA). It is synthesized by two genetically distinct isoforms of the PEP-dependent glutamate decarboxylase (EC4.1.1.15), DEC1, and DEC2. GABA metabolism depends on 4-aminobutyrate aminotransferase (GABA-transaminase, EC2.6.1.19) and succinate semialdehyde dehydrogenase (ALDH5A1, EC1.2.1.24). GABA and histidine are the constituents of the dipeptide homocarnosin, which is likely to be important for GABA storage in the brain.

Health effects: Generous Glu intake from plant-based foods may have a modest blood-pressure lowering effect (Stamler et al., 2009).

REFERENCES

Benrabh, H., Lefauconnier, J.M., 1996. Glutamate is transported across the rat blood–brain barrier by a sodium-independent system. Neurosci. Lett. 210, 9–12.

Daniels, D.H., Joe Jr, F.L., Diachenko, G.W., 1995. Determination of free glutamic acid in a variety of foods by high performance liquid chromatography. Food Add. Cont. 12, 21–29.

de Vrese, M., Frik, R., Roos, N., Hagemeister, H., 2000. Protein-bound ᴅ-amino acids, and to a lesser extent lysinoalanine, decrease true ileal protein digestibility in minipigs as determined with (15)N-labeling. J. Nutr. 130, 2026–2031.

Garattini, S., 2000. Glutamic acid, twenty years later. J. Nutr. 130, 901S–909S.

Gerich, J.E., Meyer, C., Stumvoll, M.W., 2000. Hormonal control of renal and systemic glutamine metabolism. J. Nutr. 130, 995S–1001S.

Gibala, M.J., MacLean, D.A., Graham, T.E., Saltin, B., 1997. Anaplerotic processes in human skeletal muscle during brief dynamic exercise. J. Physiol. 502, 703–713.

Graham, T.E., Turcotte, L.E., Kiens, B., Richter, E.A., 1997. Effect of endurance training on ammonia and amino acid metabolism in humans. Med. Sci. Sports Exerc. 29, 646–663.

Hammarqvist, F., Ejesson, B., Wernerman, J., 2001. Stress hormones initiate prolonged changes in the muscle amino acid pattern. Clin. Physiol. 21, 44–50.

Kim, J.Y., Kanai, Y., Chairoungdua, A., Cha, S.H., Matsuo, H., Kim, D.K., et al., 2001. Human cystine/glutamate transporter: cDNA cloning and upregulation by oxidative stress in glioma cells. Biochim. Biophys. Acta 1512, 335–344.

Kuhn, K.S., Schuhmann, K., Stehle, E., Darmaun, D., Furst, P., 1999. Determination of glutamine in muscle protein facilitates accurate assessment of proteolysis and *de novo* synthesis-derived endogenous glutamine production. Am. J. Clin. Nutr. 70, 484–489.

Kurihara, K., Kashiwayanagi, M., 2000. Physiological studies on umami taste. J. Nutr. 130, 931S–934S.

Man, E.H., Bada, J.L., 1987. Dietary ᴅ-amino acids. Annu. Rev. Nutr. 7, 209–225.

Matsuo, H., Kanai, Y., Kim, J.Y., Chairoungdua, A., Kim do, K., Inatomi, J., et al., 2002. Identification of a novel Na^+-independent acidic amino acid transporter with structural similarity to the member of a heterodimeric amino acid transporter family associated with unknown heavy chains. J. Biol. Chem. 277, 21017–21026.

May, M.E., Hill, J.O., 1990. Energy content of diets of variable amino acid composition. Am. J. Clin. Nutr. 52, 770–776.

Mordrelle, A., Huneau, J.E., Tome, D., 1997. Sodium-dependent and independent transport of ʟ-glutamate in the rat intestinal crypt-like cell line IEC-17. Biochem. Biophys. Res. Commun. 233, 244–247.

Novak, D., Quiggle, E., Artime, C., Beveridge, M., 2001. Regulation of glutamate transport and transport proteins in a placental cell line. Am. J. Physiol. Cell Physiol. 281, C1014–C1022.

Ramadan, T., Camargo, S.M., Herzog, B., Bordin, M., Pos, K.M., Verrey, F., 2007. Recycling of aromatic amino acids via TAT1 allows efflux of neutral amino acids via LAT2-4F2hc exchanger. Pflügers Arch. 454, 507–516.

Reeds, P.J., Burrin, D.G., Stoll, B., Jahoor, E., 2000. Intestinal glutamate metabolism. J. Nutr. 130, 978S–982S.

Sacktor, B., Rosenbloom, I.L., Liang, C.T., Cheng, L., 1981. Sodium gradient- and sodium plus potassium gradient-dependent ʟ-glutamate uptake in renal basolateral membrane vesicles. J. Memb. Biol. 60, 63–71.

Sekura, R., vander Werf, E., Meister, A., 1976. Mechanism and significance of the mammalian pathway for elimination of ᴅ-glutamate, inhibition of glutathione synthesis by ᴅ-glutamate. Biochem. Biophys. Res. Commun. 71, 11–18.

Stamler, J., Brown, I.J., Daviglus, M.L., Chan, Q., Kesteloot, H., Ueshima, H., INTERMAP Research Group, 2009. Glutamic acid, the main dietary amino acid, and blood pressure: the INTERMAP Study (International Collaborative Study of Macronutrients, Micronutrients and Blood Pressure). Circulation 120, 221–228.

Stumvoll, M., Perriello, G., Meyer, C., Gerich, J., 1999. Role of glutamine in human carbohydrate metabolism in kidney and other tissues. Kidney Int. 55, 778–792.

Verrey, E., Jack, D.L., Paulsen, I.T., Saier Jr., M.H., Pfeiffer, R., 1999. New glycoprotein-associated amino acid transporters. J. Membrane Biol. 172, 181–192.

Walker, R., Lupien, J.R., 2000. The safety evaluation of monosodium glutamate. J. Nutr. 130, 1049S–1052S.

Welbourne, T., Nissim, I., 2001. Regulation of mitochondrial glutamine/glutamate metabolism by glutamate transport: studies with [15]N. Am. J. Physiol. Cell Physiol. 280, C1151–C1159.

Yang, D., Brunengraber, H., 2000. Glutamate, a window on liver intermediary metabolism. J. Nutr. 130, 991S–994S.

Young, V.R., Ajami, A.M., 2000. Glutamate: an amino acid of particular distinction. J. Nutr. 130, 892S–900S.

GLUTAMINE

The neutral, aliphatic amino acid L-glutamine (glutamic acid 5-amide, 2-aminoglutaramic acid, one-letter code Q; molecular weight 146) contains 15.7% nitrogen. Glutamine is unstable in aqueous solutions, particularly upon heating (sterilization), because it forms a cyclical deamination product (Figure 8.9).

ABBREVIATIONS

Gln	L-glutamine
PepT1	hydrogen ion/peptide cotransporter (SLC15A1)
PLP	pyridoxal-5′-phosphate
RDA	recommended dietary allowance

NUTRITIONAL SUMMARY

Function: The nonessential amino acid L-glutamine (Gln) is needed for the synthesis of proteins, neurotransmitters, nucleotides, glycoproteins, and glycans. It is also a valuable precursor for glucose and other important metabolites. Because of the easy interchangeability with L-glutamate, it essentially serves all the functions of this amino acid too. Complete oxidation of Gln as an energy fuel requires niacin, thiamin, riboflavin, pyridoxine, pantothenate, lipoate, ubiquinone, iron, and magnesium.

Food sources: Proteins from plant and animal sources contain Gln and provide precursors for additional endogenous synthesis.

Requirements: Adequate amounts are consumed when total protein intake meets recommendations, since dietary proteins from all typical food sources contain a significant percentage of Gln and the body can use other amino acids for additional synthesis.

Deficiency: Prolonged lack of total protein intake as a direct source of Gln or for Gln synthesis causes growth failure, loss of muscle mass, and organ damage.

Excessive intake: Very high intake of protein and mixed amino acids (i.e., more than three times the RDA, which is 2.4 g/kg) is thought to increase the risk of renal glomerular sclerosis and accelerate osteoporosis. The consequences of very high intake of Gln have not been adequately evaluated.

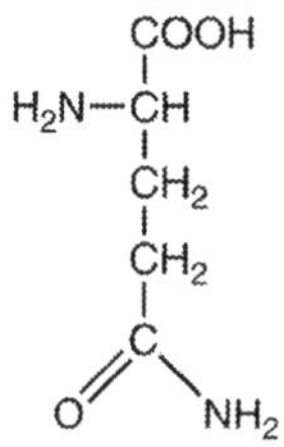

FIGURE 8.9

L-Glutamine.

ENDOGENOUS SOURCES

Muscles, the liver, the small intestine, and other tissues produce significant amounts of Gln from glutamate (Labow et al., 2001). The only enzyme capable of catalyzing the ATP-driven single-step reaction is glutamate ammonia ligase (glutamine synthase, EC6.3.1.2) (Figure 8.10).

DIETARY SOURCES

All food proteins contain Gln. Therefore, Gln intake depends more on the amount than on the type of protein consumed. Synthetic peptides, such as L-alanyl-L-glutamine and glycyl-L-glutamine, are commercially available for clinical nutritional support.

DIGESTION AND ABSORPTION

Denaturation and hydrolysis of Gln-containing proteins begin with mastication of food in the mouth and continues in the stomach under the influence of hydrochloric acid, pepsin A (EC3.4.23.1), and gastricin (EC3.4.23.3). Several pancreatic and brush border enzymes continue protein hydrolysis, though none specifically cleaves peptide bonds adjacent to Gln.

Gln is taken up from the small intestinal lumen mainly via the sodium–amino acid cotransport systems $B^{\circ}AT1$ (SLC6A19; Ducroc et al., 2010) and B° (ASCT2, SLC1A5; Bröer et al., 2011). The sodium-independent transport system $b^{\circ,+}$ comprises a light subunit BAT1 (SLC7A9) and a heavy subunit rBAT (SLC3A1), exchanges Gln for another neutral amino acid. Gln as a component of dipeptides or tripeptides can also be taken up via the hydrogen ion/peptide cotransporters 1 (SLC15A1, PepT1) and 2 (SLC15A2, PepT2) (Figure 8.11).

Export across the basolateral membrane uses the sodium–amino acid cotransport systems ATA2 (SLC38A2, Sugawara et al., 2000) and the sodium-independent transporter heterodimers LAT2 + 4F2 (SLC7A8 + SLC3A2), y + LAT1 + 4F2 (SLC7A7 + SLC3A2), y + LAT2 + 4F2 (SLC7A6 + SLC3A2).

Starvation increases expression of the transport systems A and L, whereas ASC and nonmediated uptake are not affected (Muniz et al., 1993).

TRANSPORT AND CELLULAR UPTAKE

Blood circulation: Gln has a higher plasma concentration (typically 300–600 µmol/l) than any other amino acid (Hammarqvist et al., 2001). Uptake from the blood into tissues relies largely on

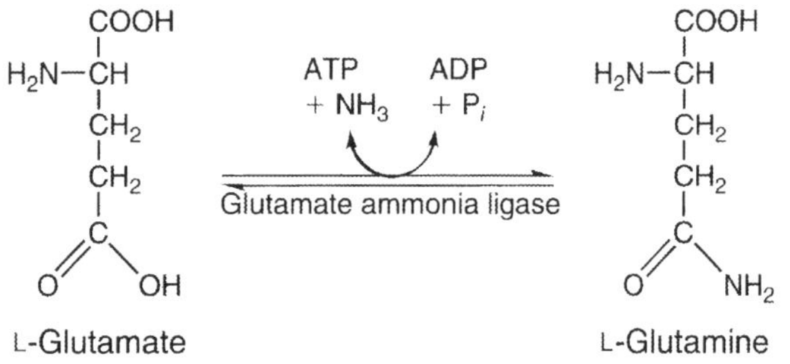

FIGURE 8.10

Endogenous synthesis of L-glutamine.

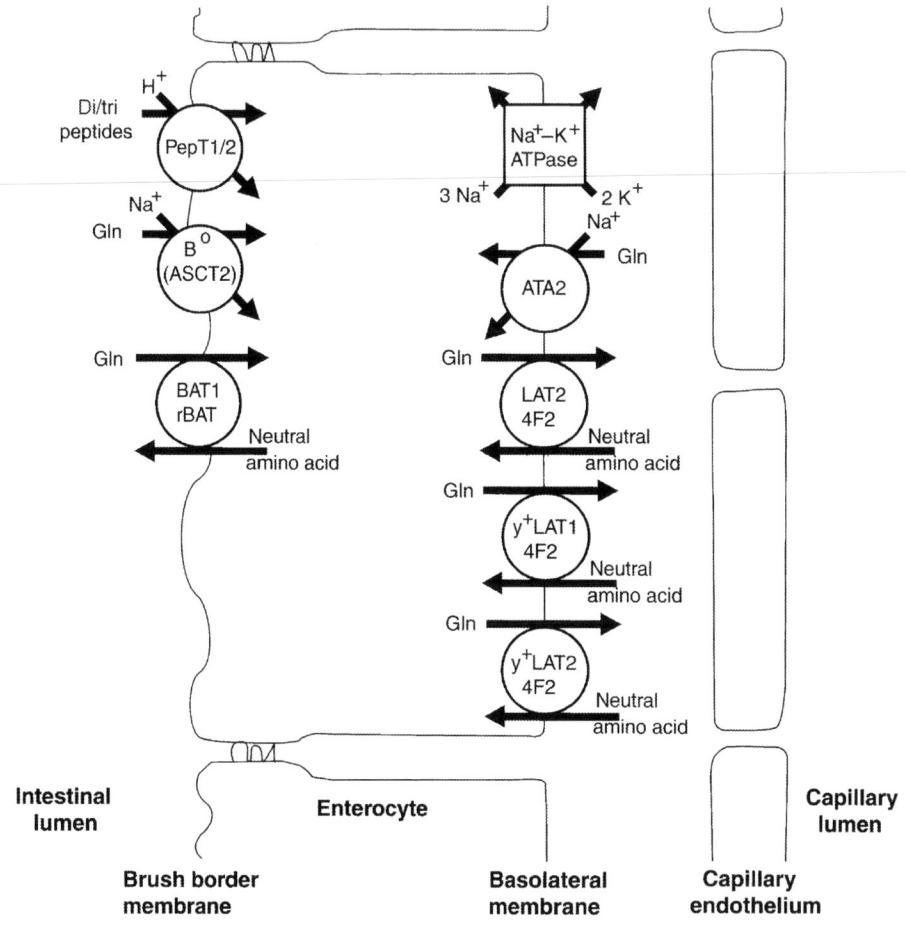

FIGURE 8.11

Intestinal absorption of L-glutamine.

the sodium-dependent transport systems N (SN1 and SN2, in the liver, muscle, and brain), ASC/B° (ASCT2, SLC1A5; particularly in the lung, muscle, pancreas, and neuronal glia), B°,+ (cotransports with one chloride and two sodium ions, in the lung, mammary gland, and other tissues), and A. Expression patterns of the system A transporters vary in characteristic fashion between different cell types. ATA2 is present in most cells (Sugawara et al., 2000), while ATA3 is restricted to the liver (Hatanaka et al., 2001).

Gln- and L-asparagine-specific uniporters, citrin (SLC25A13), and aralar 1 (SLC25A12) mediate the transfer from cytosol to mitochondria (Indiveri et al., 1998). Citrin, expressed mainly in the liver, kidneys, and other tissues, otherwise transports citrate.

Gln constitutes about 60% of the free amino acids in muscle cells (Hammarqvist et al., 2001).

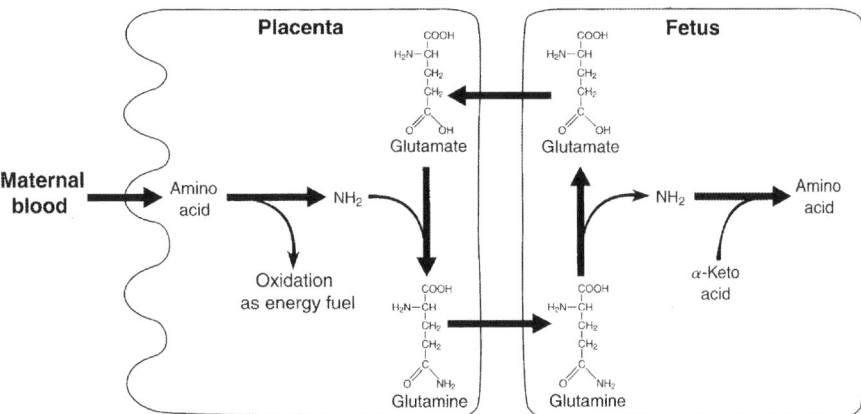

FIGURE 8.12

The glutamine–glutamate shuttle moves surplus placental nitrogen to the fetus.

BBB: Transfer of Gln in and out of the brain is tightly controlled since its immediate metabolite, L-glutamate, is a potent neurotransmitter that is toxic in excess. LAT1 (SLC7A5) and ATA1 (Varoqui et al., 2000) are expressed in brain capillary endothelial cells and contribute to Gln transport, but their locations, relative importance, and the role of other transporters are not completely understood. A model for the transfer of Gln from circulating blood into the brain (Bode, 2001) envisions uptake into astrocytes via system B° (ASCT2) and export into the intercellular space through the system N transporter (SN1). Neurons take up Gln via transporters ATA1 and ATA2.

Materno-fetal transfer: The sodium–amino acid cotransport system A (possibly also ASC and N), and the sodium-independent exchanger LAT1 mediate Gln uptake from maternal blood across the brush border membrane of the syncytiotrophoblast (Figure 8.12). Transfer across the basolateral membrane proceeds via the sodium-independent transporters LAT1 and LAT2 (SLC7A8) and the sodium–amino acid cotransport system A. The system A transporter ATA1 is expressed at a high level in the placenta (Varoqui et al., 2000), but its exact location is not yet known. The placenta takes up significant amounts of E-glutamate from the fetus, amidates some of it, and returns the resulting Gln back to the fetal liver. This placental glutamine–glutamate shuttle captures amino groups that are released during amino acid catabolism in the placenta and conserves them for use by the fetus.

METABOLISM

The γ-amido group of Gln can be released directly as ammonia or used for the synthesis of urea, amino acids, nucleotides, amino sugars, and glycoproteins. These reactions generate L-glutamate that can be deaminated to alpha-ketoglutarate and then metabolized through the Krebs cycle.

Glutaminase (EC3.5.1.2) in mitochondria is not only the key enzyme for Gln breakdown, but it is also important because it generates glutamate (needed especially in the brain as a neurotransmitter) and releases ammonia (which contributes in the kidneys to pH regulation). Two distinct genes encode proteins with glutaminase activity. Alternative splicing of one of these, the K-glutaminase gene, gives

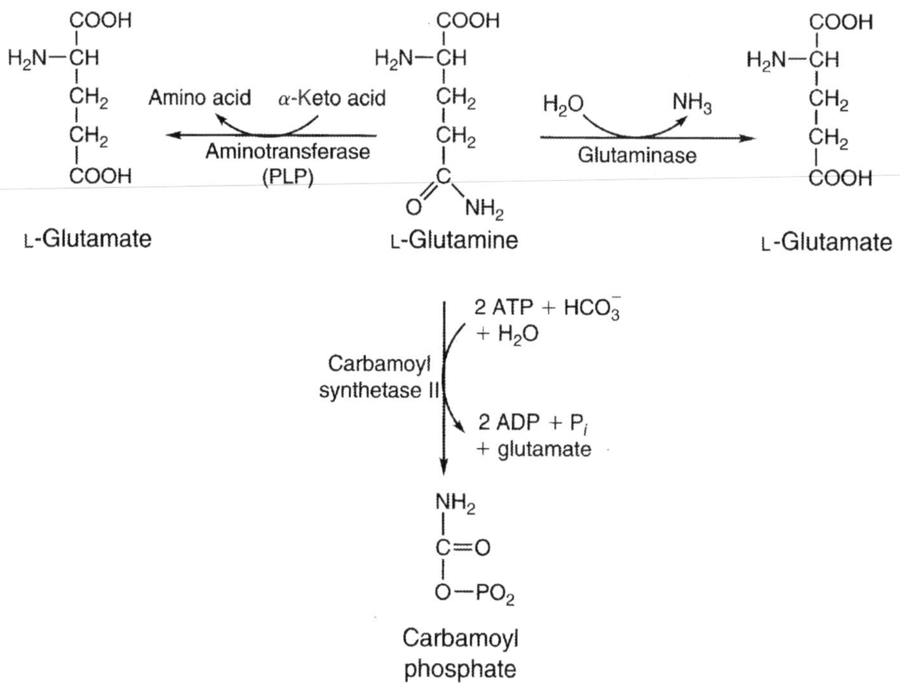

FIGURE 8.13

L-Glutamine metabolism.

rise to at least three isoforms with tissue-specific expression patterns. The ammonia can be bound in carbamoyl phosphate (for urea synthesis in the liver and kidneys) if it is not excreted directly or reincorporated into glutamate.

Enzymes that transfer the γ-amido group of Gln to ketoacids include glutamine pyruvate aminotransferase (EC2.6.1.15). Additional aminotransferases act on Gln with low activity. All aminotransferases require pyridoxal 5'-phosphate (PLP) as a prosthetic group. Carbamoyl synthetase II (EC6.3.5.5) is a multifunctional cytosolic liver enzyme that uses the γ-amido group of Gln for the synthesis of carbamoyl phosphate in an ATP-dependent reaction (Figure 8.13).

STORAGE

Gln is the most abundant free amino acid in most cell types, with concentrations of several millimoles per liter. Muscle proteins contain 4–5% Gln (Kuhn et al., 1999), of which about 0.01% is turned over per hour. A larger percentage can be mobilized in response to starvation, trauma, or infection.

EXCRETION

Losses via feces are negligible so long as gastrointestinal function is normal. The kidneys filter about 3.3 g/day, most of which is reabsorbed under most conditions. The proximal renal tubules take up free

Gln mainly through the sodium–amino acid cotransport systems B°AT1 (SLC6A19; Ducroc et al., 2010) and B° (ASCT2, SLC1A5; Bröer et al., 2011), dipeptides, and tripeptides via pepT1 and pepT2. Gln, if it is not metabolized in the renal epithelial cells, is then exported across the basolateral membrane via the sodium-dependent transporters ATA2 and ASCT1.

Most nitrogen from metabolized Gln is excreted into urine as urea, and to a much lesser extent as free ammonium ions.

REGULATION

Fasting, starvation, a lack of glucose, and high protein intake promote Gln deamination. Glucagon is a major mediator of this effect. A cyclic adenosine monophosphate (cAMP)–responsive element increases expression of glutaminase (EC3.5.1.2) in the liver, kidneys, and other tissues (Labow et al., 2001).

Stress and infection increase Gln synthesis, acting to a large extent through glucocorticoid hormones. Several glucocorticoid response elements in the upstream transcription initiation site and the first intron of the glutamate ammonia ligase (EC6.3.1.2) gene promote its expression (Labow et al., 2001), particularly in the lung. Additional posttranslational mechanisms appear to contribute to the tuning of enzyme activity. Gln concentration decreases synthesis by accelerating glutamate ammonia ligase degradation (Labow et al., 2001).

FUNCTION

Protein synthesis: Gln is a constituent of most proteins and peptides synthesized in the body. Glutamine-tRNA ligase (EC6.1.1.18) links Gln with its correspondent tRNA in an ATP/magnesium-dependent reaction.

Energy fuel: The alpha-ketoglutarate released upon transamination of Gln can be completely oxidized as an energy fuel yielding 3.089 kcal/g (May and Hill, 1990). The necessary reactions depend on niacin, thiamin, riboflavin, pyridoxine, pantothenate, lipoate, ubiquinone, iron, and magnesium. Gln is a major fuel for lymphocytes, macrophages, and other immune cells (Newsholme, 2001), which may explain a potentially beneficial role in times of heightened immune activity (trauma and infection).

Neurotransmitter cycling: Gln is the critical precursor of L-glutamate and GABA in the brain (Figure 8.14). After their release from neurons, these neurotransmitters are taken up by astrocytes, converted into Gln, and transferred to neurons for renewed use (Behar and Rothman, 2001). Here, 4-aminobutyrate aminotransferase (GABA transaminase, EC2.6.1.19) uses the amino group of GABA for the synthesis of glutamate; and succinate semialdehyde dehydrogenase (EC1.2.1.24) shunts the carbon skeleton into the Krebs cycle. Continued metabolism through the Krebs cycle and addition of acetyl-CoA from glucose utilization then generates alpha-ketoglutarate for the abovementioned glutamate synthesis. An astroglia-specific glutamate ammonia ligase (glutamine synthase, EC6.3.1.2) amidates glutamate. The sodium-dependent transporters N and ASCT2 export Gln and neurons can take it up via transporters ATA1 and ATA2 (Behar and Rothman, 2001). Phosphate-activated glutaminase (EC3.5.1.2) in neuronal mitochondria then produces glutamate as needed.

Nucleotide synthesis: Gln provides two of the purine nucleotide nitrogens and one of the pyrimidine nucleotide nitrogens. The first step of purine synthesis is the transfer of the γ-amino group from Gln to 5-phospho-a-D-ribosyl l-pyrophosphate (PRPP), catalyzed by amidophosphoribosyltransferase

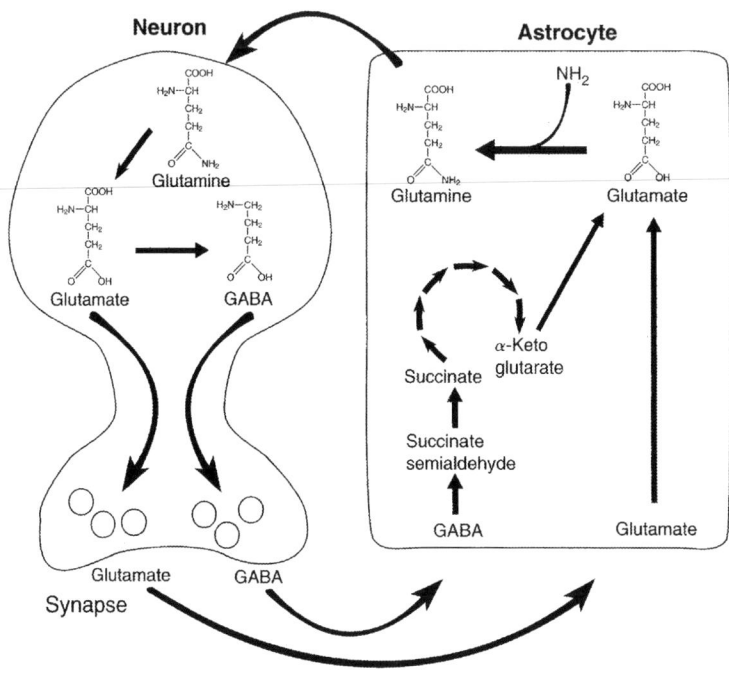

FIGURE 8.14

GABA–glutamate–glutamine cycle refurbishes neurotransmitters in astroglia cells for renewed use by neurons.

(EC2.4.2.14). Another amino group is transferred in the fourth step by phosphoribosylformylglycina-midine synthase (FGAM synthase, EC6.3.5.3). Pyrimidine nucleotide synthesis starts with the condensation reaction of Gln and bicarbonate to carbamoyl phosphate, catalyzed by glutamine-hydrolyzing carbamoyl-phosphate synthase (EC6.3.5.5).

Hexosamine synthesis: The synthesis of amino sugars in glycoproteins (e.g., gap junction proteins) and glycans (glucosaminoglycans in mucus, chondroitins, keratans, dermatans, hyaluronan, heparans, and heparin in an extracellular matrix) depends on an adequate supply of Gln. Glutamine:fructose-6-phosphate transaminase/isomerizing (GFAT, EC2.6.1.16) transfers the γ-amide from Gln to fructose-6-phosphate to produce glucosamine-6-phosphate. This key intermediate is the precursor of *N*-acetyl glucosamine, *N*-acetyl galactosamine, and other important hexosamines.

pH regulation: An excess of free acids in the blood (metabolic acidosis) increases the activity of glutaminase (EC3.5.1.2) and glutamate dehydrogenase NAD(P) 1 (EC1.4.1.3) in renal distal and proximal tubular epithelia and causes the rapid release of ammonia into urine. The ammonia cations allow the increased secretion of anions and thereby facilitate an increase in blood pH. Since Gln transport across the basolateral tubular membrane reverses in response to low intracellular concentration, additional Gln for ammonia production can be drawn from venous blood.

REFERENCES

Behar, K.L., Rothman, D.L., 2001. *In vivo* nuclear magnetic resonance studies of glutamate-γ-aminobutyric acid-glutamine cycling in rodent and human cortex: the central role of glutamine. J. Nutr. 131, 2498S–2504S.

Ducroc, R., Sakar, Y., Fanjul, C., Barber, A., Bado, A., Lostao, M.P., 2010. Luminal leptin inhibits L-glutamine transport in rat small intestine: involvement of ASCT2 and B0AT1. Am. J. Physiol. Gastrointest. Liver Physiol. 299, G179–G185.

Hatanaka, T., Huang, W., Ling, R., Prasad, P.D., Sugawara, M., Leibach, F.H., et al., 2001. Evidence for the transport of neutral as well as cationic amino acids by ATA3, a novel and liver-specific subtype of amino acid transport system A. Biochim. Biophys. Acta 1510, 10–17.

Indiveri, C., Abbruzzo, G., Stipani, I., Palmieri, E., 1998. Identitication and purification of the reconstitutively active glutamine carrier from rat kidney mitochondria. Biochem. J. 333, 285–290.

Kuhn, K.S., Schuhmann, K., Stehle, E., Darmaun, D., Fürst, P., 1999. Determination of glutamine in muscle protein facilitates accurate assessment of proteolysis and *de novo* synthesis derived endogenous glutamine production. Ant. J. Clin. Nutr. 70, 484–489.

Labow, B.I., Souba, W.W., Abcouwer, S.E., 2001. Mechanisms governing the expression of the enzymes of glutamine metabolism—glutaminase and glutamine synthetase. J. Nutr. 131, 2467S–2474S.

May, M.E., Hill, J.O., 1990. Energy content of diets of variable anaino acid composition. Am. J. Clin. Nutr. 52, 770–776.

Muniz, R., Burguillo, L., del Castillo, J.R., 1993. Effect of starvation on neutral amino acid transport in isolated small-intestinal cells from guinea pigs. Pfl. Arch. Eur. J. Physiol. 423, 59–66.

Newsholme, P., 2001. Why is L-glutamine metabolism important to cells of the immune system in health, postinjury, surgery or infection? J. Nutr., 2515S–2522S.

Sugawara, M., Nakanishi, T., Fei, Y.L., Huang, W., Ganapathy, M.E., Leibach, F.H., et al., 2000. Cloning of an amino acid transporter with functional characteristics and tissue expression pattern identical to that of system A. J. Biol. Chem. 275, 16473–16477.

Varoqui, H., Zhu, H., Yao, D., Ming, H., Erickson, J.D., 2000. Cloning and functional identification of a neuronal glutamine transporter. J. Biol. Chem. 275, 4049–4054.

GLYCINE

Glycine (aminoacetic acid, aminoethanoic acid, glycocoll, Gyn-Hydralin, Glycosthène, one-letter code G; molecular weight 75) is a nonessential small neutral amino acid containing 18.7% nitrogen (Figure 8.15).

ABBREVIATIONS

CoA	coenzyme A
Gly	glycine
GSH	glutathione (reduced)
RDA	recommended dietary allowance

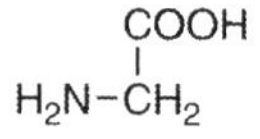

$$COOH$$
$$|$$
$$H_2N-CH_2$$

FIGURE 8.15

Glycine.

NUTRITIONAL SUMMARY

Function: The nonessential amino acid glycine (Gly) is an inhibitory neurotransmitter. It is needed for the synthesis of peptides and proteins, creatine, glutathione, porphyrins, and purines, and for the conjugation of bile acids and xenobiotics. Gly breakdown requires thiamin, riboflavin, niacin, vitamin B6, folate, vitamin B12, pantothenate, lipoate, ubiquinone, iron, and magnesium.

Food sources: Adequate amounts are consumed when total protein intake meets recommendations. Dietary supplements containing crystalline Gly are commercially available.

Requirements: With adequate total protein intake, enough Gly is available directly and from the conversion of serine and threonine.

Deficiency: Prolonged lack of Gly in protein deficiency causes growth failure, loss of muscle mass, and organ damage.

Excessive intake: Very high intake of protein, amino acids, or both (i.e., more than three times the RDA, equivalent to 2.4 g/kg) is thought to increase the risk of renal glomerular sclerosis and accelerate osteoporosis. A small percentage is converted to oxalate, which may increase the risk of kidney stones.

ENDOGENOUS SOURCES

Daily endogenous production of glycine is around 125 mg/kg body weight with plentiful total protein intake, and slightly higher with marginally low protein intake (Gibson et al., 2002). Glycine is generated from serine (by glycine hydroxymethyltransferase, EC2.1.2.1), threonine, choline, or betaine, and transamination of glyoxylate.

The threonine cleavage complex generates a small amount of glycine in a two-step sequence. L-Threonine dehydrogenase (EC1.1.103) oxidizes threonine to 2-amino 3-ketobutyrate. The PLP-dependent 2-amino-3-ketobutyrate coenzyme A (CoA) ligase (EC2.3.1.29) then links the acetyl moiety to CoA and releases glycine. Choline and its metabolites can be metabolized to Gly via betaine aldehyde, betaine, dimethyl glycine, and sarcosine.

PLP-dependent transamination of glyoxylate by glycine aminotransferase (EC2.6.1.4), alanine-glyoxylate aminotransferase (EC2.6.1.44), aromatic amino acid–glyoxylate aminotransferase (EC2.6.1.60), or kynurenineglyoxylate aminotransferase (EC2.6.1.63) provides additional small quantities of Gly (Figure 8.16).

DIETARY SOURCES

All food proteins contain Gly. Proteins with a relatively high percentage of Gly are in beef (55 mg/g protein) and pork (57 mg/g). Milk is at the other end of the spectrum, with only 21 mg/g. The typical intake of American adults is 38 mg/kg body weight or more (Gibson et al., 2002).

DIGESTION AND ABSORPTION

Protein in food is denatured by gastric acid and the action of gastric, pancreatic, and enteric enzymes, many of them cleaving peptide bonds between specific amino acids and Gly. Proteins as well as free Gly are nearly completely taken up from the small intestine; small residual amounts may also be absorbed from parts of the large intestine. Gly as a component of dipeptides or tripeptides is taken up via hydrogen ion/peptide cotransporter 1 (SLC15A1, PepT1) and, to a lesser extent, 2 (SLC15A2, PepT2).

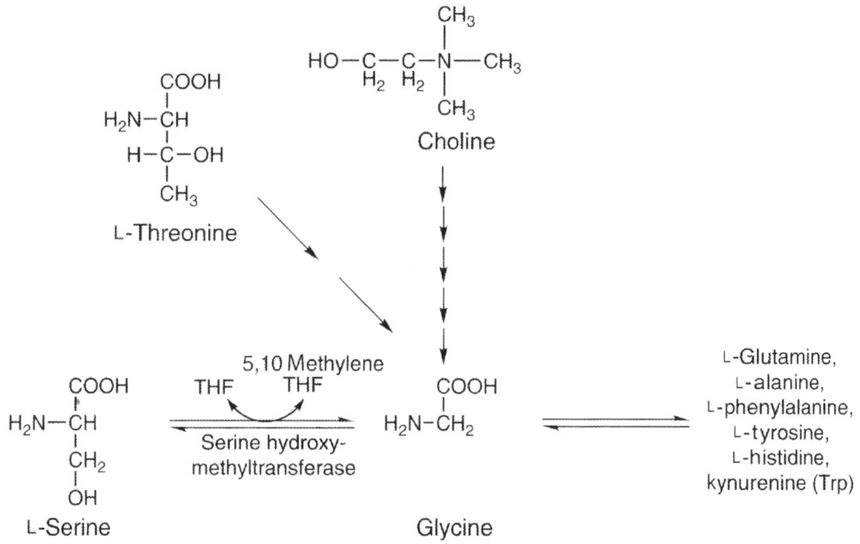

FIGURE 8.16

Endogenous glycine synthesis.

The main conduit for Gly uptake from the intestinal lumen is the sodium–amino acid cotransport system B° (Avissar et al., 2001). The sodium-independent rBAT (SLC3A1) glycoprotein-anchored transporter BAT1/b$^{o,+}$ (SLC7A9) uses serine in most situations as a counter molecule in exchange for the transport of other neutral amino acids and usually effects net serine transport into the lumen.

The sodium- and chloride-dependent high-affinity transporter GLYT1 at the basolateral membrane is a major route for concentration-dependent Gly exchange with the pericapillary space (Christie et al., 2001). The sodium-dependent transport systems A (ATA2), ASC (ASCT1), and N also accept Gly. The sodium-independent complex 4F2/LAT2 (SLC3A2/SLC7A8) mediates transport in either direction in exchange for another neutral amino acid (Figure 8.17).

TRANSPORT AND CELLULAR UPTAKE

Blood circulation: Plasma concentrations of Gly are typically around 248 μmol/l. Uptake from the blood into tissues occurs via transport systems A, ASC, N, and others (Barker et al., 1999). Expression patterns vary greatly between different cell types.

BBB: LAT1 (SLC7A5) is expressed in brain capillary endothelial cells and is certain to contribute to Gly transport, but its relative importance and the role of other transporters are not completely understood.

Materno-fetal transfer: While the sodium–amino acid cotransport systems A, ASC, and N are available for uptake from maternal blood (Jansson, 2001), there is no significant net extraction of Gly from maternal blood (Cetin, 2001). However, substantial amounts of Gly are produced in the placenta from serine, and this is transported across the basolateral membrane by the sodium–amino acid cotransport system A (Anand et al., 1996). Transport system A and two additional sodium-dependent systems appear to contribute to Gly transfer into milk (Rehan et al., 2000).

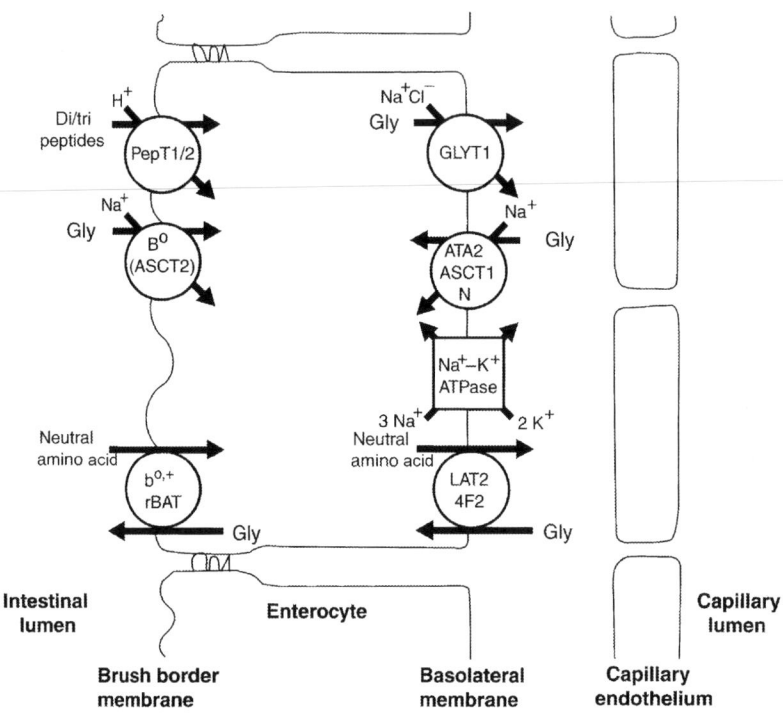

FIGURE 8.17

Intestinal absorption of glycine.

METABOLISM

Conversion to serine: The glycine-cleavage system (glycine hydroxymethyltransferase, EC2.1.2.1) converts Gly into serine by one-carbon transfer from 5,10-methylenetetrahydrofolate (Figure 8.18). Under some circumstances, this reaction runs in the reverse direction.

The glycine cleavage system: Glycine is decarboxylated in mitochondria by a large PHP-dependent glycine dehydrogenase (EC1.4.4.2) complex composed of multiple subunits (namely, P, T, L, and H); the H subunit contains lipoamide. In a fashion similar to the three lipoate-dependent alpha-keto acid dehydrogenases, the lipoamide arm acts as an acceptor for a methylene group from glycine, transfers it to folate, and the group is reduced in the process. The T subunit then transfers the hydrogen via FAD to NAD.

Minor pathways: Alanine-glyoxylate aminotransferase (EC2.6.1.44) in liver peroxisomes normally generates Gly by transferring the amino group from alanine to glyoxylate. Since this reaction is reversible, a small percentage (0.1%) of Gly is converted to glyoxylate and then to oxalate. Glyoxylate is also the product of the reaction of oxidative Gly deamination by the FAD-containing D-amino acid oxidase (EC1.4.3.3) in peroxisomes, and of several PLP-dependent transamination reactions (glycine aminotransferase, EC2.6.1.4; alanine-glyoxylate aminotransferase, EC2.6.1.44; aromatic amino acid-glyoxylate aminotransferase, EC2.6.1.60; and kynurenine-glyoxylate aminotransferase, EC2.6.1.63).

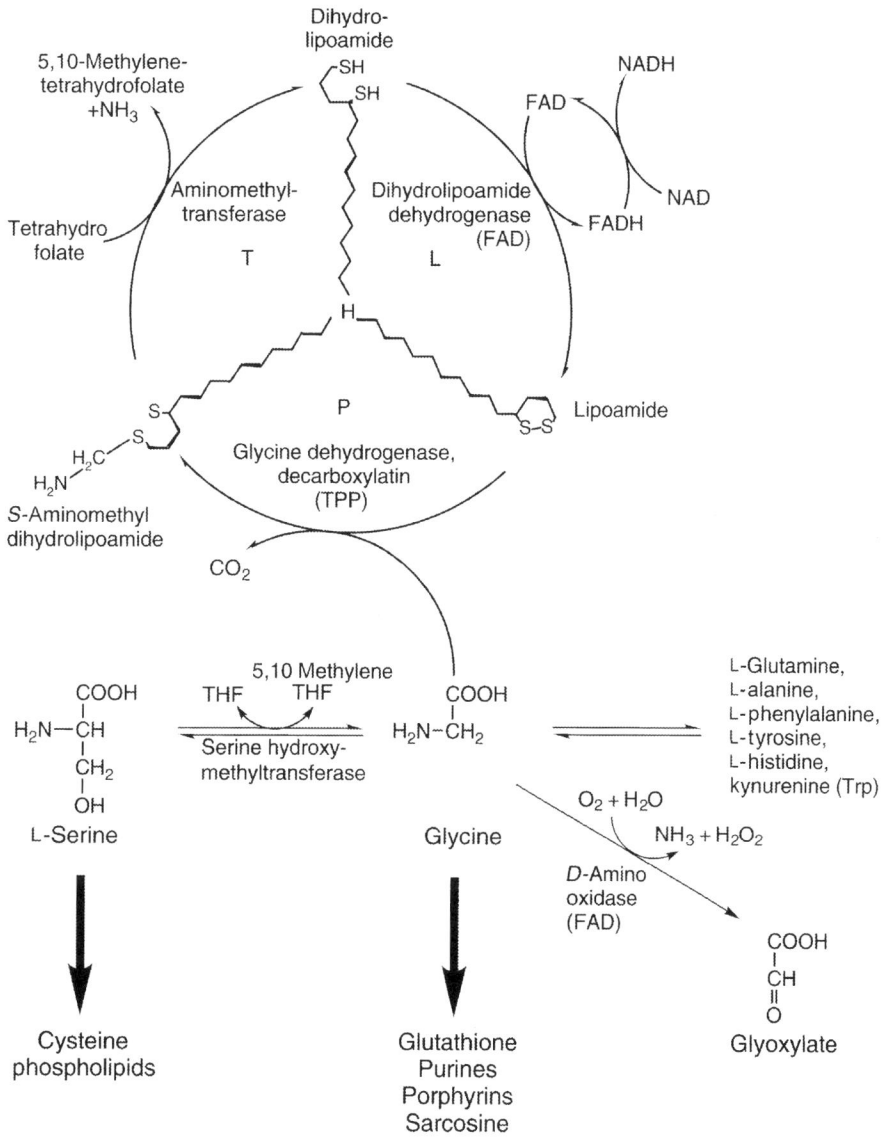

FIGURE 8.18

Glycine metabolism.

STORAGE

On average, the proteins in the human body contain more than 5%. Gly is released during normal protein turnover.

EXCRETION

Filtered free Gly is taken up into proximal renal tubules mainly by the sodium–amino acid cotransport system B° (Avissar et al., 2001), dipeptides, and tripeptides via PepT1 and PepT2. Gly is then exported across the basolateral membrane via the sodium-dependent systems N (SN2), A (ATA2), and ASC (ASCT1). As a result of very efficient reabsorption, the loss of Gly into urine is minimal in healthy people. Losses via feces are negligible while gastrointestinal function is normal.

Nitrogen from metabolized Gly is excreted into urine as urea or as uric acid. Nearly a quarter of the nitrogen in uric acid comes from Gly.

REGULATION

Gly homeostasis is maintained through the modulation of overall protein synthesis and breakdown and through changes in the activities of the glycine-cleavage system and glycine hydroxymethyltransferase (EC2.1.2.1). Glucagon promotes Gly catabolism in the liver by stimulating the glycine-cleavage system (Jois et al., 1989).

FUNCTION

Energy fuel: Daily Gly oxidation is about 90 mg/kg body weight (Gibson et al., 2002). The glycine-cleavage system (glycine dehydrogenase, EC1.4.4.2) generates one NADH and one 5,10-methylene–tetrahydrofuran (THF) per metabolized Gly molecule. Transfer of the one-carbon unit from 5,10-methylene-THF to another Gly molecule generates serine, which can then be utilized via pyruvate. The energy yield from complete oxidation is 2.011 kcal/g (May and Hill, 1990), requiring adequate supplies of thiamin, riboflavin, niacin, vitamin B6, folate, vitamin B12, pantothenate, lipoate, ubiquinone, iron, and magnesium.

Protein synthesis: Glycine-tRNA ligase (EC6.1.1.14) loads Gly to its specific tRNA for protein synthesis.

Methionine metabolism: One-carbon transfer from *S*-adenosyl methionine (SAM) to glycine by glycine *N*-methyltransferase (EC2.1.1.20) plays a crucial role in methionine metabolism that is not completely understood yet. Lack of enzyme causes hypermethioninemia and mild liver enlargement.

Glutathione synthesis: Gly is one of the three amino acids that make up the vital antioxidant peptide GSH. The second step of GSH synthesis, catalyzed by glutathione synthase (EC6.3.2.3), links Gly to gamma-L-glutamyl-L-cysteine.

Porphyrine synthesis: Condensation of succinyl-CoA and Gly by 5-aminolevulinic acid synthase (EC2.3.1.37) provides the initial precursor for porphyrins.

Purine synthesis: Two of the four carbons and one of the nitrogen atoms in purines come from Gly (Figure 8.19). The second step of *de novo* purine synthesis links Gly to phosphoribosylamine in an ATP-dependent reaction catalyzed by phosphoribosylamine-glycine ligase (GAR synthetase, EC6.3.4.13).

Creatine synthesis: About 5 mg Gly/kg body weight is used for daily creatine synthesis. Glycine amidinotransferase (EC2.1.4.1) in the kidneys condenses Gly and arginine (Wyss and Kaddurah-Daouk, 2000).

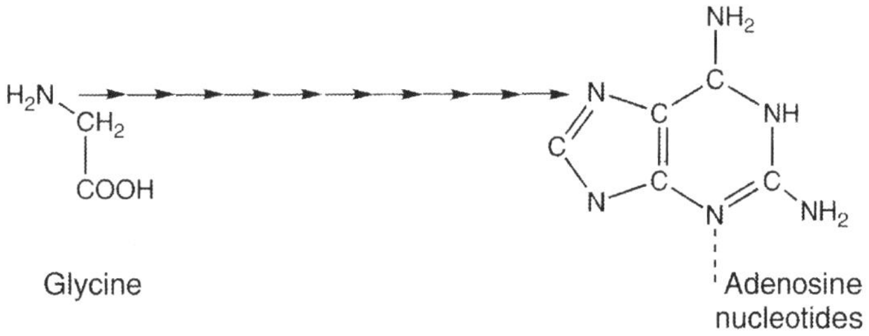

FIGURE 8.19

Glycine is a precursor of purine nucleotides.

The resulting guanidinoacetic acid is then methylated in a SAM-dependent reaction by guanidinoacetate *N*-methyltransferase (EC2.1.1.2) in the liver.

Bile acid conjugation: Prior to secretion into the bile duct a large proportion of the bile acids is conjugated by glycine *N*-choloyltransferase (bile acid-CoA:amino acid *N*-acyltransferase, BAT, EC2.3.1.65) in liver microsomes. While most of the bile acids are recovered from the ileum as free acids, the extent of Gly recovery is not clear.

Detoxification: Many nutrient metabolites and xenobiotics (e.g., salicylates) are conjugated to Gly. For example, conjugation of Gly to benzoyl-CoA by glycine *N*-acyltransferase (EC2.3.1.13) and by glycine *N*-benzoyltransferase (EC2.3.1.71) generates hippurate. Patients with genetic urea cycle disorders sometimes respond well to benzoate intake, which can enhance nitrogen elimination as hippurate.

Brain function: Gly is a major inhibitory neurotransmitter in the brain. The NMDA receptor in the brain activates its ion channel when both Gly and glutamate bind to it.

BIOMARKER OF DIETARY AVAILABILITY

Urinary 5-hydroxyproline excretion is related to Gly availability; this measure is also influenced by sulfur amino acid (SAA) availability (Metges et al., 2000).

REFERENCES

Anand, R.J., Kanwar, U., Sanyal, S.N., 1996. Transport of glycine in the brush border and basal cell membrane vesicles of the human term placenta. Biochem. Mol. Biol. Int. 38, 21–30.

Avissar, N.E., Ryan, C.K., Ganapathy, V., Sax, H.C., 2001. Na$^+$-dependent neutral amino acid transporter ATB$^\circ$ is a rabbit epithelial cell brush border protein. Am. J. Physiol. Cell Physiol. 281, C963–C971.

Barker, G.A., Wilkins, R.J., Golding, S., Ellory, J.C., 1999. Neutral amino acid transport in bovine articular chondrocytes. J. Physiol. 514, 795–808.

Cetin, I., 2001. Amino acid interconversions in the fetal-placental unit: the animal model and human studies *in vivo*. Ped. Res. 49, 148–153.

Christie, G.R., Ford, D., Howard, A., Clark, M.A., Hirst, B.H., 2001. Glycine supply to human enterocytes mediated by high-affinity basolateral GLYT 1. Gastroenterology 120, 439–448.

Gibson, N.R., Jahoor, F., Ware, L., Jackson, A.A., 2002. Endogenous glycine and tyrosine production is maintained in adults consuming a marginal-protein diet. Am. J. Clin. Nutr. 75, 511–518.

Jansson, T., 2001. Amino acid transporters in the human placenta. Pediatr. Res. 49, 141–147.

Jois, M., Hall, B., Fewer, K., Brosnan, J.T., 1989. Regulation of hepatic glycine catabolism by glucagon. J. Biol. Chem. 264, 3347–3351.

May, M.E., Hill, J.O., 1990. Energy content of diets of variable amino acid composition. Am. J. Clin. Nutr. 52, 770–776.

Metges, C.C., Yu, Y.M., Cai, W., Lu, X.M., Wong, S., Regan, M.M., et al., 2000. Oxoproline kinetics and oxoproline urinary excretion during glycine- or sulfur amino acid-free diets in humans. Am. J. Physiol. Endocrinol. Metab. 278, E868–E876.

Rehan, G., Kansal, V.K., Sharma, R., 2000. Mechanism ofglycine transport in mouse mammary tissue. J. Dairy Res. 67, 475–483.

Wyss, M., Kaddurah-Daouk, R., 2000. Creatine and creatinine metabolism. Physiol. Rev. 80, 1107–1213.

THREONINE

The neutral essential amino acid alcohol L-threonine (2-amino-3-hydroxybutyric acid, alpha-amino-beta-hydroxybutyric acid, 2-amino-3-hydroxybutanoic acid, one-letter code T; molecular weight 119) contains 11.8% nitrogen (Figure 8.20).

ABBREVIATIONS

CoA	coenzyme A
PepT1	hydrogen ion/peptide cotransporter 1 (SLC15A1)
RDA	recommended dietary allowance
Thr	L-threonine

NUTRITIONAL SUMMARY

Function: The essential amino acid L-threonine (Thr) is needed for the synthesis of proteins, is a precursor of glycine, and serves as an energy fuel depending on adequate supplies of thiamin, riboflavin, niacin, vitamin B6, vitamin B I2, biotin, pantothenate, lipoate, ubiquinone, iron, and magnesium.

Food sources: Adequate amounts are consumed when total protein intake meets recommendations. Dietary proteins from different sources all contain Thr, but animal and bean proteins have slightly higher content (4–5% of total protein) than grain proteins (3–4% of total protein).

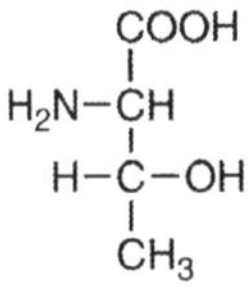

$$
\begin{array}{c}
COOH \\
|\\
H_2N-CH \\
|\\
H-C-OH \\
|\\
CH_3
\end{array}
$$

FIGURE 8.20

L-Threonine.

Requirements: The Thr requirement of adults is thought to be more than 500 mg/day.

Deficiency: Prolonged lack of Thr, as with all essential amino acids and a lack of protein, causes growth failure, loss of muscle mass, and organ damage.

Excessive intake: Very high intake of protein and mixed amino acids (i.e., more than three times the RDA, equivalent to 2.4 g/kg) is thought to increase the risk of renal glomerular sclerosis and accelerate osteoporosis.

ENDOGENOUS SOURCES

Thr is an essential nutrient in humans since no metabolic pathways allowing net Thr synthesis have been detected.

DIETARY SOURCES

The Thr content of most food proteins is about 41–45 mg/g, and slightly more in eggs (48 mg/g). Wheat (30 mg/g), oats (34 mg/g), rice (36 mg/g), corn (38 mg/g), and other grains have slightly less content. At adequate daily protein intake (0.8 g/kg body weight), Thr requirements (above 500 mg/day) are certain to be exceeded several-fold. Heating foods to high temperatures causes slight losses of Thr (Dworschak, 1980).

DIGESTION AND ABSORPTION

As food passes through the digestive tract, gastric and pancreatic enzymes continue protein breakdown, many of them cleaving peptide bonds *between* specific amino acids and Thr. Thr is taken up from the small intestinal lumen by the sodium–amino acid cotransport systems B° and ASC (Munck and Munck, 1999; Avissar et al., 2001). Thr as a component of dipeptides or tripeptides can also be taken up via hydrogen ion/peptide cotransporter 1 (SLC15A1, PepT1) and, to a lesser extent, 2 (SLC15A2, PepT2).

Export across the basolateral membrane uses the sodium–amino acid cotransport systems A (ATA2) and ASC (ASCT1), and possibly the incompletely characterized sodium independent system asc. So long as the amino acid concentration inside the cell is higher than in the basolateral space, transport can occur against the considerable sodium gradient. When the intracellular amino acid concentration falls below blood levels during fasting, transport turns toward the enterocyte and supplies needed amino acids.

Both the brush border membrane and the basolateral membrane contain transporters that exchange amino acids. Neutral amino acids that are enriched in the cell by sodium cotransporters are exchanged for a wide spectrum of other amino acids. This means that net Thr transport via the heteroexchangers is in the direction of intestinal lumen. The heteroexchanger in the brush border membrane is the rBAT (SLC3A1) glycoprotein-anchored transporter BAT1/$b^{0,+}$ (SLC7A9). On the basolateral side are the transporters LAT2, y^+ LAT1, and y^+ LAT2, all of them anchored to the membrane by glycoprotein 4F2 (SLC3A2) (Figure 8.21).

TRANSPORT AND CELLULAR UPTAKE

Blood circulation: Plasma concentration of Thr (typically around 128 µmol/l) increases significantly after meals and is lowest during the early morning hours (Tsai and Huang, 1999). Uptake from the blood

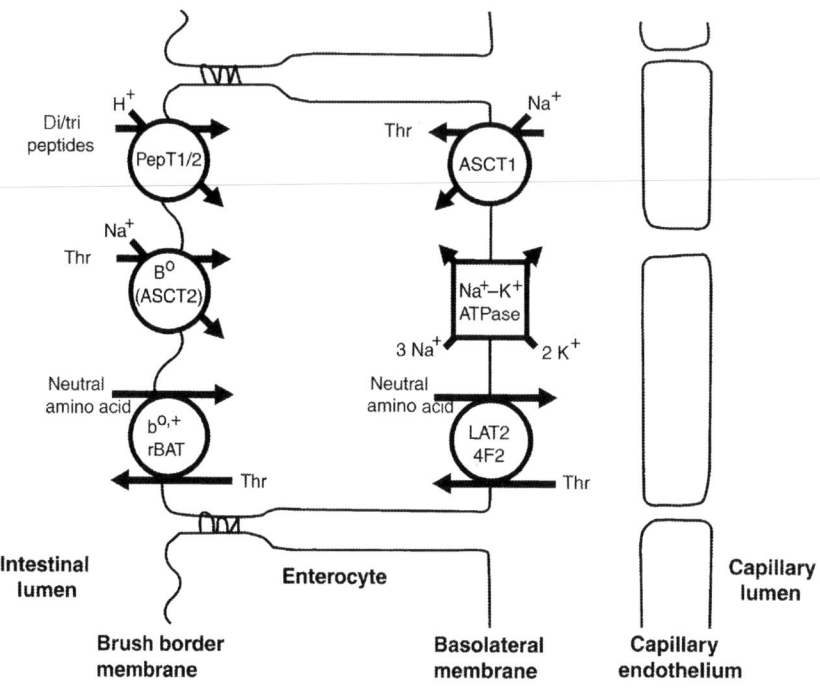

FIGURE 8.21

Intestinal absorption of L-threonine.

into tissues occurs via an array of transporters, many of them identical or similar to those described for intestinal absorption. Expression patterns vary greatly between different cell types.

Materno-fetal transfer: Uptake of Thr from maternal blood across the brush border membrane of the syncytiotrophoblast is mediated by the sodium–amino acid cotransport system A (possibly also ASC and N), and the sodium-independent system L. Transfer across the basolateral membrane proceeds via the sodium–amino acid cotransport system A and the sodium-independent transporters EAT1 and LAT2.

BBB: LAT1 is expressed in brain capillary endothelial cells and is certain to contribute to Ala transport, but its relative importance and the role of other transporters are not completely understood.

METABOLISM

Cytosol: The main flux of Thr catabolism occurs in cytosol with the generation of CO_2 and propionyl-CoA (Darling et al., 1997). This pathway (Figure 8.22) starts with the pyridoxal-phosphate-dependent deamination of Thr by threonine dehydratase (EC4.2.1.16), followed by the decarboxylation of alpha-ketobutyrate to propionyl CoA; both pyruvate dehydrogenase (EC3.1.3.43) and branched-chain alpha-ketoacid dehydrogenase (EC1.2.4.4) can catalyze the second reaction, and both enzymes require thiamin pyrophosphate, lipoic acid, and CoA. Cleavage into glycine and acetaldehyde by the two

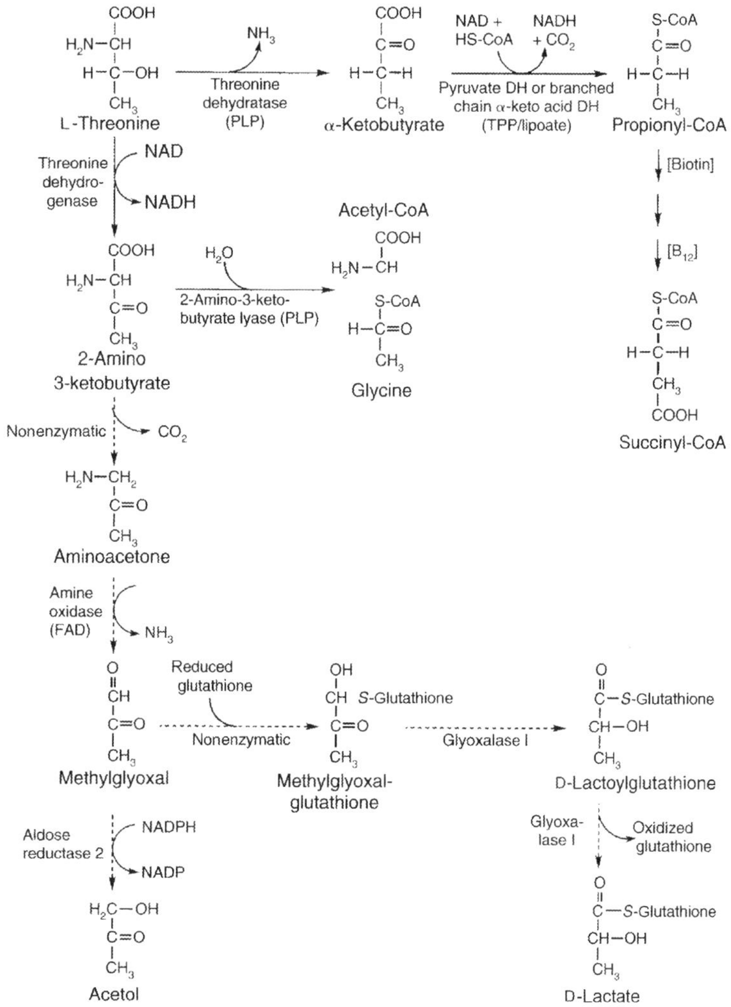

FIGURE 8.22

Metabolism of L-threonine.

PLP-containing cytosolic enzymes threonine aldolase (EC4.1.2.5) and glycine hydroxymethyltrans-ferase (EC2.1.2.1) appears to contribute only minimally to threonine breakdown (Ogawa et al., 2000).

Mitochondria: The threonine cleavage complex generates acetyl-CoA and glycine in a two-step sequence. L-threonine dehydrogenase (EC1.1.103) oxidizes Thr to 2-amino 3-ketobutyrate. The PLP-dependent 2-amino-3 ketobutyrate CpA ligase (EC2.3.1.29) links the acetyl moiety to CoA and cleaves off glycine. Both acetyl-CoA and glycine then can be utilized via their respective usual pathways. Also, 2-amino-3-ketobutyrate can spontaneously decarboxylate to aminoacetone, which is deaminated by FAD-dependent semicarbazide-sensitive amine oxidase (EC1.4.3.4). With increasing availability of Thr or with low availability of free CoA, a greater proportion enters this alternative pathway. The cytotoxic metabolite methylglyoxal can be converted to either acetol or lactaldehyde by aldehyde reductase (EC1.1.1.21), the relative proportions depending on reduced glutathione concentration (Vander Jagt et al., 2003). Alternatively, methylglyoxal can spontaneously form a hemithioacetal complex with glutathione, which is a preferred substrate for lactoylglutathione lyase (glyoxalase I, EC4.4.1.5). This zinc enzyme generates lactoylglutathione, which in turn is converted by another zinc enzyme, hydroxyacylglutathione hydrolase (glyoxalase II, EC3.1.2.6) to the dead-end metabolite D-lactate. Preference for this glyoxalase pathway increases with higher glutathione concentrations (Vander Jagt et al., 2003).

STORAGE

Assuming an average Thr content of human proteins of 48 mg/g (Smith, 1980), a 70-kg man may have a mobilizable reserve of about 288 g. The Thr content of hemoglobin is 59 mg/g. At a hypothetical breakdown rate of 20 ml red blood cells or 6 g hemoglobin per day, this would correspond to a daily release of 354 mg of Thr.

EXCRETION

Filtered free Thr is taken up into proximal renal tubules by sodium–amino acid cotransport systems B° and ASC (Avissar et al., 2001), dipeptides, and tripeptides via PepT1 and PepT2. As in the small intestine, the various heteroexchangers tend to use Thr as a counter molecule for the transport of other amino acids. Nonetheless, recovery of Thr tends to be nearly complete in healthy people. Losses via feces also are negligible when gastrointestinal function is normal. Most nitrogen from metabolized Thr is excreted into urine as urea. Aminoacetone and D-lactate are minor metabolic products whose urinary excretion increases disproportionately with high Thr intake and when availability of free CoA is limited.

REGULATION

Starvation increases expression of intestinal heteroexchangers, whereas ASC and nonmediated uptake are not affected (Muniz et al., 1993).

FUNCTION

Energy fuel: Eventually, most ingested Thr is utilized for energy production, yielding 3.4 kcal/g (May and Hill, 1990). Complete oxidation depends on adequate supplies of thiamin, riboflavin, niacin, vitamin B6, vitamin B12, biotin, pantothenate, lipoate, ubiquinone, iron, and magnesium.

Protein synthesis: Practically all human proteins and peptides contain Thr. Threonine tRNA ligase (EC6.1.1.3) loads Thr onto a specific tRNA in an ATP/magnesium-dependent reaction, which is then used for protein synthesis.

Glycine synthesis: A large proportion of Thr is metabolized to glycine and acetyl-CoA by the threonine cleavage complex, as detailed previously.

REFERENCES

Avissar, N.E., Ryan, C.K., Ganapathy, V., Sax, H.C., 2001. Na$^+$-dependent neutral amino acid transporter ATB$^\circ$ is a rabbit epithelial cell brush border protein. Am. J. Physiol. Cell Physiol. 281, C963–C971.

Darling, E., Rafii, M., Grunow, J., Ball, R.O., Brookes, S., Pencharz, P.B., 1997. Threonine dehydrogenase is not the major pathway ofthreonine catabolism in adult humans. FASEB J. 11, A149.

Dworschak, E., 1980. Nonenzyme browning and its effect on protein nutrition. Crit. Rev. Food Sci. Nutr. 13, 1–40.

May, M.E., Hill, J.O., 1990. Energy content of diets of variable amino acid composition. Am. J. Clin. Nutr. 52, 770–776.

Munck, B.G., Munck, L.K., 1999. Effects of pH changes on systems ASC and B in rabbit ileum. Am. J. Physiol. 276, G173–G184.

Muniz, R., Burguillo, L., del Castillo, J.R., 1993. Effect of starvation on neutral amino acid transport in isolated small-intestinal cells from guinea pigs. Pfl. Arch. Eur. J. Physiol. 423, 59–66.

Ogawa, H., Gomi, T., Fujioka, M., 2000. Serine hydroxymethyltransferase and threonine aldolase: are they identical? Int. J. Biochem. Cell Biol. 32, 289–301.

Smith, R.H., 1980. Comparative amino acid requirements. Proc. Nutr. Soc. 39, 71–78.

Tsai, P.J., Huang, P.C., 1999. Circadian variations in plasma and erythrocyte concentrations of glutamate, glutamine, and alanine in men on a diet without and with added monosodium glutamate. Metab. Clin. Exp. 48, 1455–1460.

Vander Jagt, D.L., Hunsaker, L.A., 2003. Methylglyoxal metabolism and diabetic complications: roles of aldose reductase, glyoxalase-I, betaine aldehyde dehydrogenase and 2-oxoaldehyde dehydrogenase. Chem. Biol. Interact. 143–144, 341–351.

SERINE

The neutral amino acid alcohol L-serine (2-amino-3-hydroxypropionic acid, 2-amino-3-hydroxypropanoic acid, alpha-amino-beta-hydroxypropionic acid, betahydroxyalanine, one-letter code S; molecular weight 105) contains 13.3% nitrogen (Figure 8.23).

ABBREVIATIONS

LAT2	L-type amino acid transporter 2 (SLC7A8)
PepT1	hydrogen ion/peptide cotransporter 1 (SLC15A1)
PepT2	hydrogen ion/peptide cotransporter 2 (SLC15A2)
PLP	pyridoxal 5-phosphate
RDA	recommended dietary allowance
Ser	L-serine

NUTRITIONAL SUMMARY

Function: The nonessential amino acid L-serine (Ser) is needed for the synthesis of proteins, selenocysteine, and 3-dehydro-D-sphinganine, and is a potential precursor of glycine, L-cysteine, and L-alanine. Its use as an energy fuel depends on adequate availability of thiamin, riboflavin, niacin, vitamin B6, folate, pantothenate, lipoate, ubiquinone, iron, and magnesium.

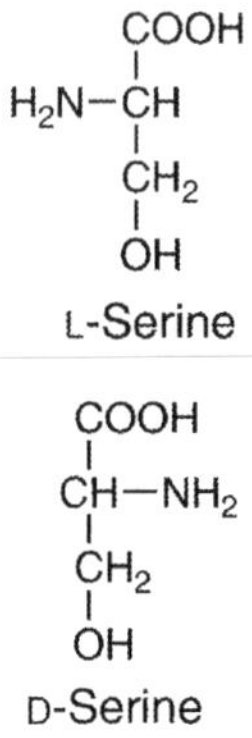

COOH
|
H₂N—CH
|
CH₂
|
OH

L-Serine

COOH
|
CH—NH₂
|
CH₂
|
OH

D-Serine

FIGURE 8.23

Serine enantiomers.

Food sources: Adequate amounts are consumed when total protein intake meets recommendations. Exposure to high heat causes losses due to oxidation.

Requirements: There are no specific requirements so long as total protein intake is adequate.

Deficiency: Since lack of Ser occurs only in severe protein deficiency, the symptoms are those of severe starvation.

Excessive intake: Very high intake of protein and mixed amino acids (i.e., more than three times the RDA, which is 2.4 g/kg) is thought to increase the risk of renal glomerular sclerosis and accelerate osteoporosis. The consequences of very high intake of Ser have not been adequately evaluated.

ENDOGENOUS SOURCES

Ser is released with the breakdown of proteins and is newly synthesized from the glycolysis intermediate 3-phosphoglycerate, from glycine, and indirectly from hydroxyproline in the kidneys (Lowry et al., 1985) (Figure 8.24).

Synthesis from glycolysis intermediates: Especially when protein intake is low, Ser synthesis can originate from 3-phosphoglycerate. Phosphoglycerate dehydrogenase (EC1.1.1.95), phosphoserine aminotransferase (EC2.6.1.52, PLP-dependent), and phosphoserine phosphatase (EC3.1.3.3, requires magnesium) catalyze this sequence of reactions.

Synthesis from glycine: Glycine hydroxymethyltransferase (EC2.1.2.1) mediates the one-carbon transfer from 5,10-methylenetetrahydrofolate to glycine. Under most circumstances, however, this reaction runs in the reverse direction.

Synthesis of D-serine: A specific enzyme in the brain, serine racemase (EC5.1.1.18), converts L-serine into its enantiomer D-serine (Wolosker et al., 1999).

DIETARY SOURCES

Most dietary proteins contain about 4–5% Set. Moderately enriched sources are eggs (7.4%), beans (5.5%), milk (5.4%), and rice (5.3%); chicken meat contains a slightly smaller percentage (3.4%). Heat

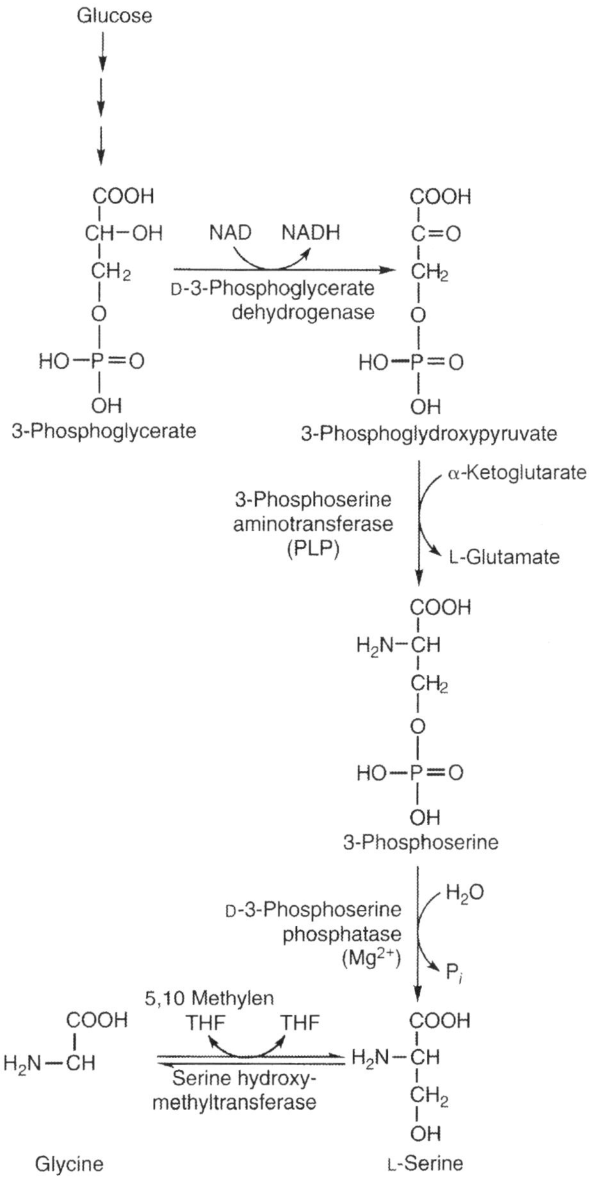

FIGURE 8.24

Endogenous synthesis of L-serine.

treatment of foods decreases the amount of bioavailable Ser (Dworschak, 1980). Many foods also contain some phospho-L-serine; minor amounts of the enantiomer D-serine also may be present.

DIGESTION AND ABSORPTION

Various enzymes from the stomach, pancreas, and intestinal wall break down Ser-containing proteins (see the overview discussion of amino acids earlier in this chapter), none of them with particular preference for Ser residues (Figure 8.25). Ser-containing dipeptides and tripeptides can also be taken up via the hydrogen ion/peptide cotransporter 1 (SLC15A1, PepT1) and, to a much lesser extent, via the hydrogen ion/peptide cotransporter 2 (SLC15A2, PepT2). Measurements in the small intestine of rabbit indicate that about half of Ser uptake across the brush border membrane is mediated by sodium–amino acid cotransport system B°, and the other half by system ASC (Munck and Munck, 1999; Avissar et al., 2001). The sodium-independent rBAT (SLC3A1) glycoprotein-anchored transporter BAT1/b°,+ (SLC7A9) uses Ser in most situations as a counter molecule in exchange for the transport of other neutral amino acids and usually effects net Ser transport into the lumen. Some of the Ser taken up is used for the enterocyte's own protein synthesis; significant amounts are converted to glycine.

Export of Ser across the basolateral membrane uses mainly the sodium–amino acid cotransport systems A (ATA2) and ASC (ASCT1). LAT2 (SLC7A8), a 4F2-glycoprotein (SLC3A2)–anchored member of the system L family, can transport Ser in either direction across the basolateral membrane

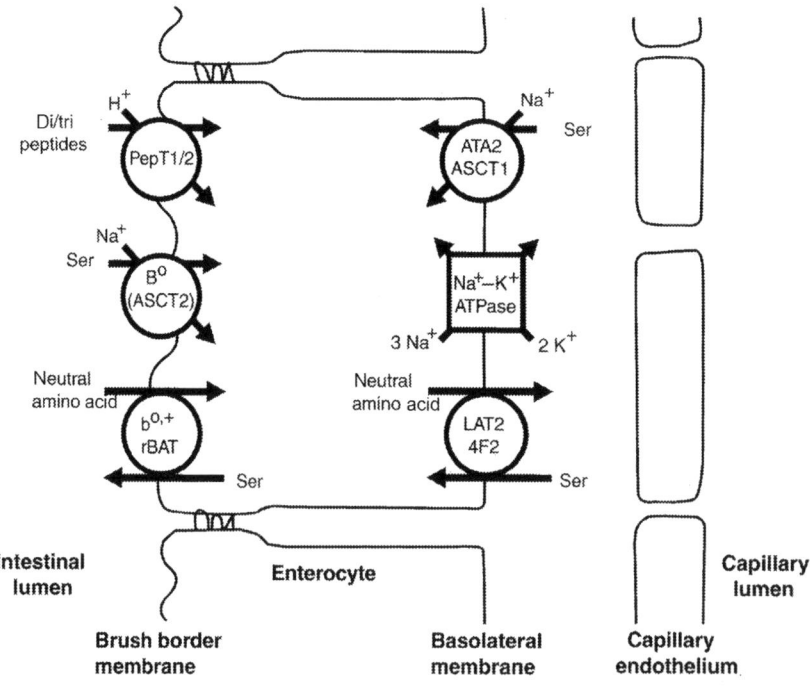

FIGURE 8.25

Intestinal absorption of serine.

in exchange for another neutral amino acid. Depending on the difference between intracellular and interstitial Ser concentrations, LAT2 and the sodium-independent transporter asc can also achieve net export. Starvation increases expression of the transport systems A and L, whereas ASC-mediated uptake will not be affected (Muniz et al., 1993).

TRANSPORT AND CELLULAR UPTAKE

Blood circulation: The plasma concentration of Ser (typically around 120 µmol/l) increases significantly after meals and is lowest during the early morning hours (Tsai and Huang, 1999). Uptake from the blood into tissues occurs via an array of transporters, many of them identical or similar to those described for intestinal absorption. Expression patterns vary greatly between different cell types. EAAT1 and EAAT2 (Vandenberg et al., 1998) can transport 1-serine-*O*-sulfate.

Materno-fetal transfer: Uptake of Ala from maternal blood across the brush border membrane of the syncytiotrophoblast is mediated by the sodium–amino acid cotransport system A (possibly also ASC and N), and the sodium-independent system L. Transfer across the basolateral membrane proceeds via the sodium–amino acid cotransport system A and the sodium-independent transporters LAT1 and LAT2.

BBB: LAT1 is expressed in brain capillary endothelial cells and certain to contribute to Ala transport, but questions remain about its relative importance and the contribution of other transporters. D-Alanine is transported in the brain by Asc-1, a sodium-independent transporter for neutral amino acids that is associated with the glycoprotein 4F2 (Fukasawa et al., 2000).

METABOLISM

Most ingested Ser is converted into glycine or used directly for protein synthesis: smaller amounts serve as precursors for the synthesis of L-cysteine, selenocysteine, or phospholipids. The remainder is converted to glycolysis intermediates.

Major catabolic pathway: Ser can be transaminated by one of several PLP-dependent aminotransferases to 3-hydroxypyruvate (Figure 8.26). One of these, serine-pyruvate aminotransferase (EC2.6.1.51), has high activity in liver peroxisomes; this enzyme transfers the amino group from Ser to pyruvate and generates L-alanine. Peroxisomal transamination via serine-glyoxylate aminotransferase (EC2.6.1.45) is not just important for serine breakdown, but at the same time removes the oxalate precursor glyoxylate by using it as an acceptor. Glyoxylate is a product of hydroxyproline (Ichiyama et al., 2000) and ethanolamine catabolism. The oxidation of glyoxylate by (S)-2-hydroxy-acid oxidase (EC1.1.3.15) in liver peroxisomes generates the dead-end product oxalate (Seargeant et al., 1991). Excessive oxalate production increases the risk of kidney stone formation. Ser transamination may also be catalyzed by phenylalanine (histidine) aminotransferase (EC2.6.1.58), but it is unlikely that this reaction contributes significantly to Ser metabolism under normal circumstances.

Several enzymes have the capacity to reduce 3-hydroxypyruvate to D-glycerate. One is a flavin mononucleotide (FMN)–containing isoform of (S)-2-hydroxy-acid oxidase (EC1.1.3.15), HAOX1; this peroxisomal reaction generates hydrogen peroxide. Alternatively, D-glycerate dehydrogenase (EC1.1.1.29) can accomplish the reaction with NAD as cosubstrate. A possible contribution by hydroxypyruvate reductase (EC1.1.1.81) is uncertain. Conversion to pyruvate can then be completed by the successive actions of D-glycerate kinase (EC2.7.1.31), phosphopyruvate hydratase (EC4.2.1.11), and pyruvate kinase (EC2.7.1.40).

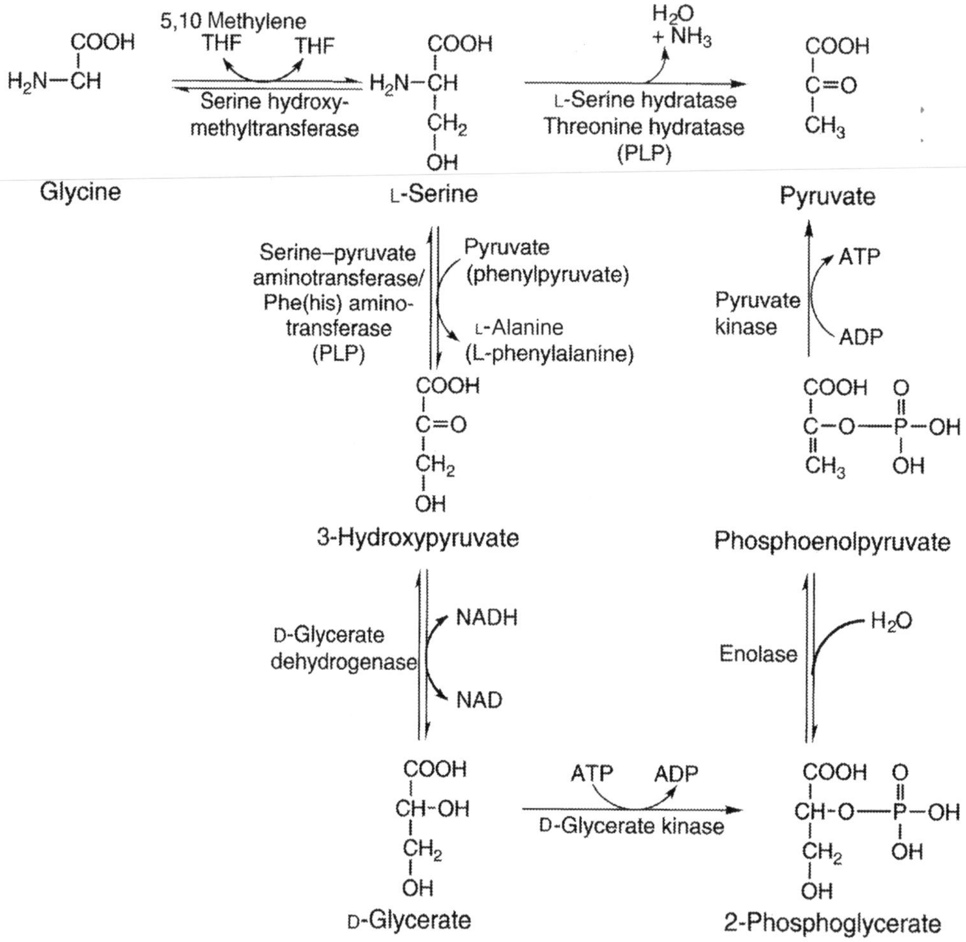

FIGURE 8.26

Serine metabolism.

Minor catabolic pathway: Some Ser can be converted directly to pyruvate by the PLP-dependent enzymes serine dehydratase (EC4.2.1.13) and threonine dehydratase (EC4.2.1.16). Unlike the much more complex transaminating pathway, direct deamination does consume NADH. Both dehydratases are present in the liver and other tissues, though they may account for only a small portion of total Ser breakdown in humans. High protein intake increases direct deamination and thereby gluconeogenesis; high carbohydrate intake decreases activity.

L-*Serine-sulfate:* Serine-sulfate ammonia-lyase (EC4.3.1.10) cleaves L-serine-sulfate into pyruvate, sulfate, and ammonia; serine racemase (EC5.1.1.18) does the same with high specific activity (Panizzutti et al., 2001).

STORAGE

Body proteins contain on average 4–5% Ser, which is released during protein turnover. The amounts that might be available from the breakdown of phospholipids are small.

EXCRETION

Very little Ser is lost via urine by healthy people. Filtered dipeptides and tripeptides are taken up from the lumen of proximal renal tubules via PepT1 and PepT2. The sodium–amino acid cotransport system $B°$ is the main conduit for recovery of free Ser (Avissar et al., 2001). Export across the basolateral membrane uses the sodium-dependent systems A (ATA2) and ASC (ASCTI). As with intestinal absorption, Ser serves mainly as a counter molecule for the transport of other neutral amino acids via the heteroexchangers (BAT1/rBAT in the brush border membrane; LAT2/4F2 and others in the basolateral membrane).

Virtually all nitrogen from Ser metabolism is excreted as urea. Losses of Ser via feces are negligible so long as gastrointestinal function is normal.

REGULATION

High protein intake increases, and high carbohydrate intake decreases Ser conversion to pyruvate, thereby providing substrate for gluconeogenesis.

FUNCTION

Energy fuel: Most Ser is oxidized eventually as an energy fuel providing 2.53 kcal/g (May and Hill, 1990). Its utilization depends on adequate availability of thiamin, riboflavin, niacin, vitamin B6, folate, pantothenate, lipoate, ubiquinone, iron, and magnesium.

Protein synthesis: While all human proteins contain Ser, their synthesis does not depend on dietary intake so long as total protein consumption provides amino group for endogenous Ser synthesis. Serine-tRNA ligase (EC6.1.1.11) loads Ser onto a specific tRNA for the synthesis of proteins and peptides.

Glycine synthesis: Ser is the major endogenous precursor of glycine. The one-step reaction catalyzed by glycine hydroxymethyltransferase (EC2.1.2.1) depends on an adequate supply of tetrahydrofolate and requires PLP as a cofactor; the reaction is reversible and thus also can generate serine as needed.

Cysteine synthesis: Ser is needed for the endogenous synthesis of L-cysteine from L-methionine. Ser condenses with homocysteine to cystathionine (cystathionine betasynthase, EC4.2.1.22), which then is cleaved by cystathionine gamma-lyase (EC4.4.1.1) to cysteine and alpha-ketobutyrate. Both of these reactions are PLP-dependent. The clearance of the potentially toxic (free-radical-forming) homocysteine may be as important as the generation of cysteine.

Selenocysteine synthesis: Ser is the required precursor for the synthesis of selenoproteins (Hubert et al., 1996). First, tRNA(Ser) sec is charged by serine-tRNA ligase (EC6.1.1.11) with serine. Then the pyridoxal phosphate-dependent L-seryl-tRNA sec selenium transferase (EC2.9.1.1) uses selenophosphate to link selenium to the hydroxyl group of serine. The selenocysteine on the tRNA(Ser) sec is then ready for incorporation into specific proteins.

Phospholipid synthesis: Ser can be a precursor for the synthesis of several types of phospholipids, including phosphatidylethanolamine, phosphatidylcholine, phosphatidylserine, sphingolipids, ceramides, and myelin. Phosphatidylethanolamine is produced via O-*sn*-phosphatidyl-L-serine from Ser and cytidyl diphosphate (CDP)–diacylglycerol (CDP-diacylglycerol-serine O-phosphatidyltransferase, EC2.7.8.8). Phosphatidylethanolamine then can be methylated to phosphatidylcholine in three successive, but as yet poorly characterized, reactions. However, the capacity of this pathway appears to be limited. Phosphatidylserine plays important roles in cell signaling, apoptosis, and blood coagulation, and as a major (5–10%) constituent of the cell membrane. This type of phospholipid is produced when the calcium-dependent phosphatidylserine synthase (EC2.7.8.8) exchanges Ser for the polar groups in phosphatidylethanolamine and phosphatidylcholine. Sphingosine-derived phospholipids, including myelin, ceramides, and sphingolipids in the brain, nerves, and other tissues, also depend on Ser as a precursor. The first step of sphingosin synthesis from Ser and palmitoyl-CoA is catalyzed by manganese- and PLP-dependent serine C-palmitoyltransferase (EC2.3.1.50) located at the endoplasmic reticulum.

Posttranslational protein modification: Specific Ser residues are the target of enzymes that phosphorylate tropomyosin (tropomyosin kinase, EC2.7.1.132), histones (protamine kinase, EC2.7.1.70), tau protein ([tau protein] kinase, EC2.7.1.135), and the cytoplasmic domain of the LDL-receptor ([low-density lipoprotein receptor] kinase, EC2.7.1.131), thereby providing important mechanisms for the regulation of protein function.

Connective tissue: Glycosaminoglycans are large polymeric constituents of the extracellular matrix and connective tissue in bone, cartilage, skin, blood vessels, the cornea, and many other tissues. Several types (chondroitin sulfate, dermatan sulfate, heparin sulfate, and heparin) consist of repeating linear disaccharide chains linked via Ser to the protein component. A variety of other cytosolic and nuclear proteins are O-glycosylated, mainly with *N*-acetylglucosamine or *N*-acetylgalactosamine.

Neuronal function: D-Serine, which is produced from Ser by serine racemase (EC5.1.1.18, PLP-dependent), specifically modulates NMDA receptor transmission in the brain. When the neurotransmitter L-glutamate stimulates non-NMDA receptors of astrocytes, they release free D-serine and provide the essential coactivation of glutamate/*N*-methyl-D-aspartate receptors (Wolosker et al., 1999). D-Serine has been reported to improve symptoms in schizophrenic patients (Javitt, 1999). The forebrains of mice were found to contain as much as 40 μg/g of wet tissue (Morikawa et al., 2001).

REFERENCES

Avissar, N.E., Ryan, C.K., Ganapathy, V., Sax, H.C., 2001. Na$^+$-dependent neutral amino acid transporter ATB° is a rabbit epithelial cell brush border protein. Am. J. Physiol. Cell Physiol. 281, C963–C971.

Dworschak, E., 1980. Nonenzyme browning and its effect on protein nutrition. Crit. Rev. Food Sci. Nutr. 13, 1–40.

Fukasawa, Y., Segawa, H., Kim, J.Y., Chairoungdua, A., Kim, D.K., Matsuo, H., et al., 2000. Identification and characterization ofa Na$^+$-independent neutral amino acid transporter that associates with the 4F2 heavy chain and exhibits substrate selectivity for small neutral D- and L-amino acids. J. Biol. Chem. 275, 9690–9698.

Hubert, N., Walczak, R., Sturchler, C., Myslinski, E., Schuster, C., Westhof, E., et al., 1996. RNAs mediating cotranslational insertion of selenocysteine in eukaryotic selenoproreins. Biochimie 78, 590–596.

Ichiyama, A., Xue, H.H., Oda, T., Uchida, C., Sugiyama, T., Maeda-Nakai, E., et al., 2000. Oxalate synthesis in mammals: properties and subcellular distribution of serine:pyruvate/alanine:glyoxylate aminotransferase in the liver. Mol. Urol. 4, 333–340.

Javitt, D.C., 1999. Treatment of negative and cognitive symptoms. Curr. Psychiat. Rep. 1, 25–30.

Lowry, M., Hall, D.E., Brosnan, J.T., 1985. Hydroxyproline metabolism by the rat kidney: distribution of renal enzymes of hydroxyproline catabolism and renal conversion of hydroxyproline to glycine and serine. Metab. Clin. Exp. 34, 955–961.

May, M.E., Hill, J.O., 1990. Energy content of diets of variable amino acid composition. Am. J. Clin. Nutr. 52, 770–776.

Morikawa, A., Hamase, K., Inoue, T., Konno, R., Niwa, A., Zaitsu, K., 2001. Determination of free D-aspartic acid, D-serine and D-alanine in the brain of mutant mice lacking D-amino acid oxidase activity. J. Chrom. Biomed. Sci. Appl. 757, 119–125.

Munck, B.G., Munck, L.K., 1999. Effects of pH changes on systems ASC and B in rabbit ileum. Am. J. Physiol. 276, G173–G184.

Muniz, R., Burguillo, L., del Castillo, J.R., 1993. Effect of starvation on neutral amino acid transport in isolated small-intestinal cells from guinea pigs. Pfl. Arch. Eur. J. Physiol. 423, 59–66.

Panizzutti, R., De Miranda, J., Ribeiro, C.S., Engelender, S., Wolosker, H., 2001. A new strategy to decrease N-methyl-D-aspartate (N M DA) receptor coactivation: inhibition of D-serine synthesis by converting serine racemase into an eliminase. Proc. Natl. Acad. Sci. USA 98, 5294–5299.

Seargeant, L.E., deGroot, G.W., Dilling, L.A., Mallory, C.J., Haworth, J.C., 1991. Primary oxaluria type 2 (L-glyceric aciduria): a rare cause of nephrolithiasis in children. J. Pediatr. 118, 912–914.

Tsai, P.J., Huang, P.C., 1999. Circadian variations in plasma and erythrocyte concentrations of glutamate, glutamine, and alanine in men on a diet without and with added monosodium glutamate. Metab. Clin. Exp. 48, 1455–1460.

Vandenberg, R.J., Mitrovic, A.D., Johnston, G.A., 1998. Serine-O-sulphate transport by the human glutamate transporter, EAAT2. Br. J. Pharmacol. 123, 1593–1600.

Wolosker, H., Sheth, K.N., Takahashi, M., Mothet, J.P., Brady Jr, R.O., Ferris, C.D., et al., 1999. Purification of serine racemase: biosynthesis of the neuromodulator D-serine. Proc. Natl. Acad. Sci. USA 96, 721–725.

ALANINE

The neutral, aliphatic amino acid L-alanine (L-alpha-alanine, L-alpha-aminopropionic acid, (S)-2-aminopropionic acid, one-letter code A; molecular weight 89) contains 15.7% nitrogen (Figure 8.27).

ABBREVIATIONS

Ala	L-alanine
PLP	pyridoxal 5'-phosphate
RDA	recommended dietary allowance

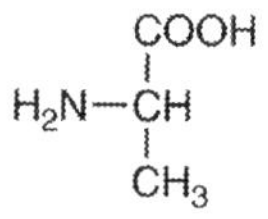

$$
\begin{array}{c}
COOH \\
| \\
H_2N-CH \\
| \\
CH_3
\end{array}
$$

FIGURE 8.27

L-Alanine.

NUTRITIONAL SUMMARY

Function: The nonessential amino acid L-alanine (Ala) is needed for the synthesis of proteins. It is also used as an energy fuel; its complete oxidation requires thiamin, riboflavin, niacin, vitamin B6, pantothenate, lipoate, ubiquinone, iron, and magnesium.

Food sources: Adequate amounts are consumed when total protein intake meets recommendations since dietary proteins from different sources all contain Ala, and the body can produce additional amounts from other amino acids.

Requirements: Since adequate amounts can be produced endogenously, no Ala has to be consumed, so long as total protein intake is adequate.

Deficiency: Prolonged lack of Ala due to low total protein intake causes growth failure, loss of muscle mass, and organ damage.

Excessive intake: Very high intake of protein and mixed amino acids (i.e., more than three times the RDA, which is 2.4 g/kg) is thought to increase the risk of renal glomerular sclerosis and accelerate osteoporosis. The consequences of very high intake of Ala have not been adequately evaluated.

ENDOGENOUS SOURCES

Large amounts of Ala are produced in muscles, the liver, small intestine, and some other tissues. L-Alanine aminotransferase (EC2.6.1.2) uses L-glutamate to transaminate the glycolysis metabolite pyruvate and produce Ala. Since the reaction operates near equilibrium, high availability of glucose (and, consequently, of pyruvate) can increase Ala production. Much smaller amounts of Ala arise from the metabolism of L-tryptophan (kynureninase, EC3.7.1.3), thymine (R-3-amino-2-methylpropionate-pyruvate aminotransferase, EC2.6.1.40), L-aspartate and 3-sulfino-L-alanine (both by aspartate 4-decarboxylase, EC4.1.1.12), and selenocysteine (L-selenocysteine selenide-lyase, EC4.4.1.16). All of these enzymes use PLP as a prosthetic group (Figure 8.28).

DIETARY SOURCES

Most food proteins contain about 4–6% Ala. Accordingly, Ala intake depends more on the amount than on the type of protein consumed. Small amounts of the D-isomer are present in many plant-derived foods, but typical intake has not been characterized. Cooking promotes the cross-linking of Ala with other amino acids in food proteins (generating lysinoalanine, ornithinoalanine, histidinoalanine, and phenylethylaminoalanine), the formation of dehydroalanine, methyldehydroalanine, beta-aminoalanine, and racemization to D-alanine (Friedman, 1999).

DIGESTION AND ABSORPTION

Hydrolysis of Ala-containing proteins begins with mastication of foods in the mouth. However, most of this activity, such as salivary *N*-acetylmuramyl-L-alanine amidase (EC3.5.1.28) and kallikrein 2, appears to be antibacterial in nature. A large spectrum of gastric and pancreatic enzymes continues protein breakdown, many of them cleaving peptide bonds between specific amino acids and Ala. Pancreatic endopeptidase E (EC3.4.21.70) specifically cleaves at Ala residues within proteins; the pancreatic carboxypeptidases A1 and A2 (EC3.4.2.1) cleave carboxyterminal amino acid Ala residues. Alanine aminopeptidase (AAP, EC3.4.11.2) at the small intestinal brush border membrane releases N-terminal Ala

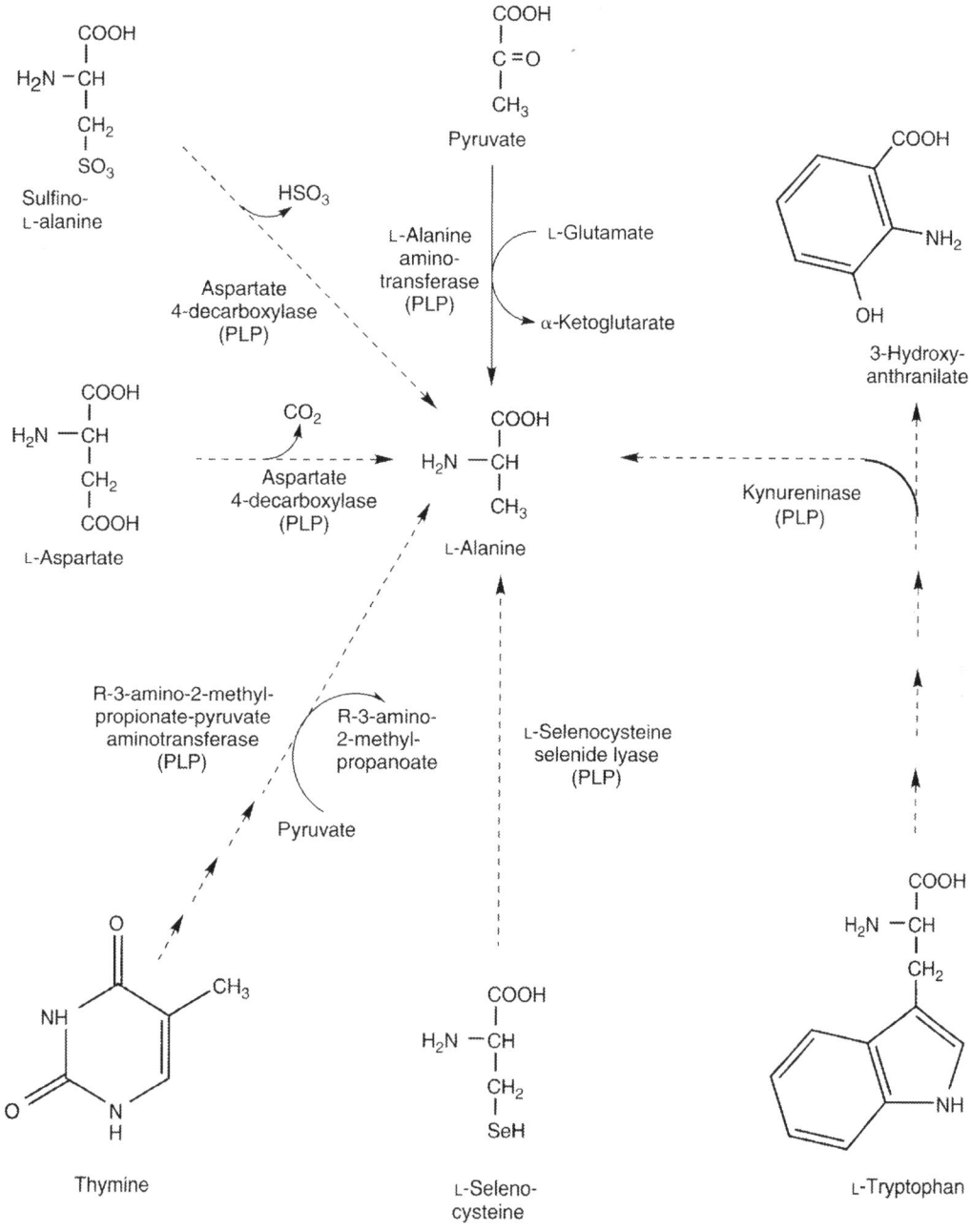

FIGURE 8.28

Endogenous sources of L-alanine.

from peptides, amides, or arylamides preferentially, but not exclusively. The presence of lysinoalanine and other modified alanine residues decreases the digestibility of food proteins (Friedman, 1999).

Ala is taken up from the small intestinal lumen mainly via the sodium–amino acid cotransport system B° (Avissar et al., 2001) (Figure 8.29). The transporter ASC works predominantly by exchanging Ala for another small neutral amino acid, as does the sodium-independent transport system $b^{0,+}$ a transporter comprised of a light subunit BAT1 (SLC7A9) and a heavy subunit rBAT (SLC3A1). Passive nonmediated uptake has also been suggested. Ala as a component of dipeptides or tripeptides can also be taken up via the hydrogen ion/peptide cotransporter (SLC15A1, PepT1).

Export across the basolateral membrane uses the sodium–amino acid cotransport systems ASCT1 (SLC1A4) and ATA2 (Sugawara et al., 2000), the sodium-independent transporter LAT2/4F2, and possibly a sodium-independent transporter with the properties of asc.

Starvation increases expression of the transport systems A and L (LAT2/4F2), whereas ASC and nonmediated uptake are not affected (Muniz et al., 1993).

TRANSPORT AND CELLULAR UPTAKE

Blood circulation: Plasma concentration of Ala (typically between 270 and 500 µmol/l) increases significantly after meals and is lowest during the early morning hours (Tsai and Huang, 1999). Uptake from the blood into tissues relies largely on the sodium-dependent transport systems A and ASC. Expression

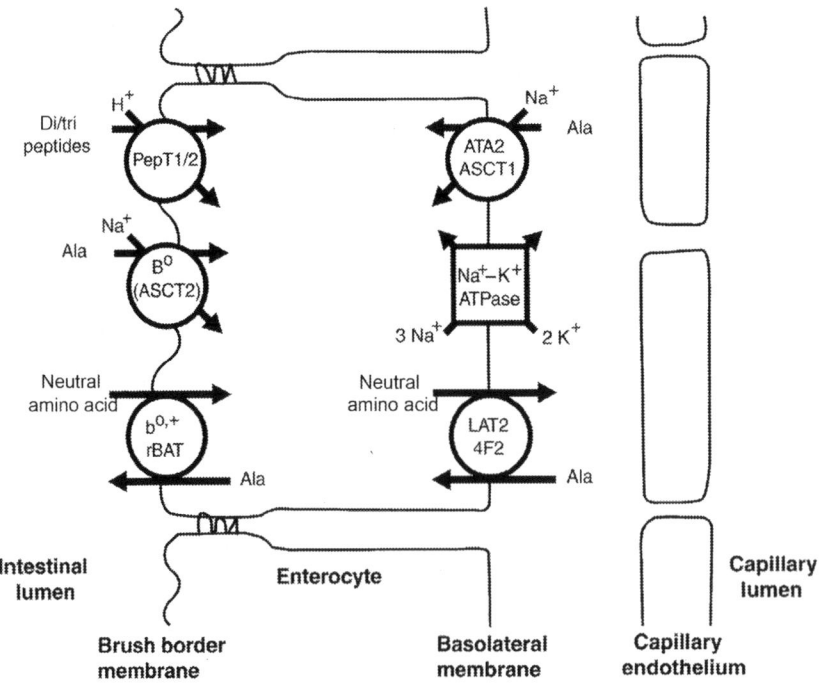

FIGURE 8.29

Intestinal absorption of ʟ-alanine.

patterns of individual transporters vary greatly between different cell types. ATA2 is present in most cells (Sugawara et al., 2000), while ATA3 is restricted to the liver (Hatanaka et al., 2001).

BBB: LAT1 (SLC7A5) and ATA1 (Varoqui et al., 2000) are expressed in brain capillary endothelial cells and contribute to Ala transport, but their locations, relative importance, and the role of other transporters are not completely understood. D-Alanine is transported into the brain by Asc-1 (SLC7A10), a recently described sodium-independent transporter for neutral amino acids, which is associated with the glycoprotein 4F2 (SLC3A2) (Fukasawa et al., 2000).

Materno-fetal transfer: The sodium–amino acid cotransport system A (possibly also ASC and N), and the sodium-independent exchanger LAT1 mediate Ala uptake from maternal blood across the brush border membrane of the syncytiotrophoblast. Transfer across the basolateral membrane proceeds via the sodium-independent transporters LAT1 and LAT2 (SLC7A8) and the sodium–amino acid cotransport system A (ATA1). This system A transporter is expressed at high levels in the placenta (Varoqui et al., 2000), but its exact location is not yet known.

METABOLISM

The amino group of Ala can be used to generate glutamate (alanine aminotransferase, EC2.6.1.2, in cytosol and mitochondria), glutamine (glutamine-pyruvate aminotransferase, EC2.6.1.15), glycine (alanine-glyoxylate aminotransferase, EC2.6.1.44), L-phenylalanine (phenylalanine/histidine aminotransferase, EC2.6.1.58), L-serine (serine-pyruvate aminotransferase, EC2.6.1.51), and aminomalonate (alanineoxomalonate aminotransferase, EC2.6.1.47). Additional enzymes may transfer the amino group to keto-acids with low activity. All aminotransferases require PLP tightly bound to a specific lysine residue at the catalytic center (Figure 8.30).

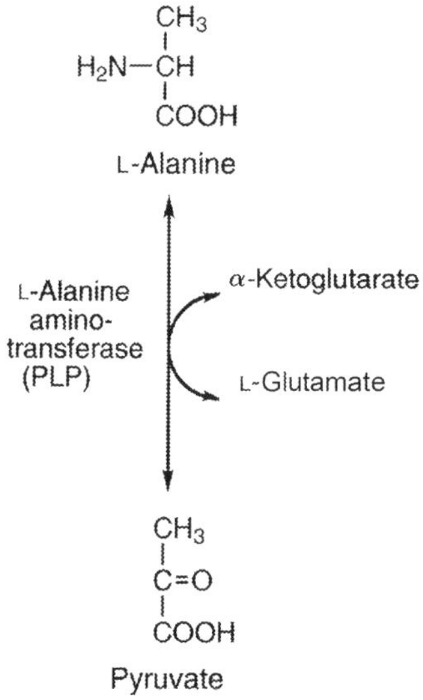

FIGURE 8.30

L-Alanine transamination.

As a result of the transamination reaction pyruvate is generated, which can be utilized via the pyruvate dehydrogenase complex, the citric acid cycle, and oxidative phosphorylation. D-Alanine probably can be converted to pyruvate by glycine hydroxymethyltransferase (EC2.1.2.1); the amino group is transferred in this reaction to the enzyme-bound PLP, generating pyridoxamine phosphate in the process (Ogawa and Fujioka, 1981).

STORAGE

Most Ala is bound in proteins, particularly in muscle, which contains about 6%. As protein is turned over, Ala becomes available. Ala mobilized from muscle protein can be exported to the liver for urea synthesis and gluconeogenesis.

EXCRETION

Filtered free Ala is taken up into proximal renal tubules mainly by the sodium–amino acid cotransport system B° (Avissar et al., 2001), dipeptides, and tripeptides via pepT1 and pepT2. Ala is then exported across the basolateral membrane via the sodium-dependent transporters ATA2 and ASCT1. As a result of very efficient reabsorption, the loss of Ala into urine is minimal in healthy people. Losses via feces are negligible while gastrointestinal function is normal. Most nitrogen from metabolized Ala is excreted into urine as urea.

REGULATION

Early starvation increases Ala synthesis. When increased amounts of amino acids are released from muscle, especially in response to starvation, severe infection, inflammation, and injury, the amino group is transported (preferentially as Ala) from muscles to the liver (glucose–alanine cycle; see the discussion later in this chapter). Cytokines such as TNF-alpha can acutely alter amino acid transport, thereby depleting muscle Ala concentration (Tayek, 1996).

FUNCTION

Protein synthesis: Ala is a constituent of practically all proteins and peptides synthesized in the body. Alanine-tRNA ligase (EC6.1.1.7) loads Ala onto a specific tRNA in an ATP/magnesium-dependent reaction.

Energy fuel: The pyruvate released by transamination of Ala may be completely oxidized as an energy fuel yielding 3.425 kcal/g (May and Hill, 1990). The necessary reactions depend on thiamin, riboflavin, niacin, vitamin B6, pantothenate, lipoate, ubiquinone, iron, and magnesium.

Glucose–alanine cycle: Most of the free Ala from muscles is exported into circulation. In the liver, the amino group is used for urea synthesis and the residual pyruvate is used for gluconeogenesis via conversion to oxaloacetate (pyruvate carboxylase, EC6.4.1.1) and phosphoenolpyruvate (phosphoenolpyruvate carboxylase, EC4.1.1.31). The reactions involved in glucose synthesis from pyruvate depend on niacin, biotin, and magnesium (see the section entitled "Pyruvate" in Chapter 6). The cycle is completed when glucose returns to muscles, is metabolized to pyruvate, and then is transaminated to Ala. The glucose–alanine cycle facilitates the utilization of muscle amino acids in the liver during fasting, starvation, and traumatic stress. Ala accounts for more than a quarter of total amino acids taken up by the liver from circulation.

Heme synthesis: Aminolevulinate aminotransferase (EC2.6.1.43, PLP-dependent) synthesizes the heme precursor delta-aminolevulinate by transferring the amino group from Ala to 4,5-dioxovalerate. It has been suggested that this housekeeping enzyme ensures a minimal level of 5-aminolevulinic acid, in addition to the more variable and regulated amounts produced by aminolevulinate synthase.

Other nitrogen compounds: The amino group of Ala can be used for the synthesis of other amino acids or any of numerous other nitrogen compounds.

Glyoxylate metabolism: The transfer of the amino group from Ala to glyoxylate by alanine-glyoxylate aminotransferase (EC2.6.1.44, PLP-dependent) in liver peroxisomes is critical for the conversion of this metabolite to glycine. The alternative fate of glyoxalate is nonenzymatic conversion into the dead-end metabolite oxalate.

REFERENCES

Avissar, N.E., Ryan, C.K., Ganapathy, V., Sax, H.C., 2001. Na^+-dependent neutral amino acid transporter ATB° is a rabbit epithelial cell brush border protein. Am. J. Physiol. Cell Physiol. 281, C963–C971.

Friedman, M., 1999. Chemistry, biochemistry, nutrition, and microbiology of lysinoalanine, lanthionine, and histidinoalanine in food and other proteins. J. Agric. Food Chem. 47, 1295–1319.

Fukasawa, Y., Segawa, H., Kiln, J.Y., Chairoungdua, A., Kim, D.K., Matsuo, H., et al., 2000. Identitication and characterization of a Na(+)-independent neutral amino acid transporter that associates with the 4F2 heavy chain and exhibits substrate selectivity for small neutral D- and L-amino acids. J. Biol. Chem. 275, 9690–9698.

Hatanaka, T., Huang, W., Ling, R., Prasad, P.D., Sugawara, M., Leibach, F.H., et al., 2001. Evidence for the transport of neutral as well as cationic amino acids by ATA3, a novel and liver-specific subtype of amino acid transport system A. Biochim. Biophys. Acta 1510, 10–17.

May, M.E., Hill, J.O., 1990. Energy content of diets of variable amino acid composition. Am. J. Clin. Nutr. 52, 770–776.

Muniz, R., Burguillo, L., del Castillo, J.R., 1993. Effect of starvation on neutral amino acid transport in isolated small-intestinal cells from guinea pigs. Pfl. Arch. Eur. J. Physiol. 423, 59–66.

Ogawa, H., Fujioka, M., 1981. Purification and characterization of cytosolic and mitochondrial serine hydroxymethyltransferases from rat liver. J. Biochem. 90, 381–390.

Sugawara, M., Nakanishi, T., Fei, Y.J., Huang, W., Ganapathy, M.E., Leibach, F.H., et al., 2000. Cloning of an amino acid transporter with functional characteristics and tissue expression pattern identical to that of system A. J. Biol. Chem. 275, 16473–16477.

Tayek, J.A., 1996. Effects of tumor necrosis factor alpha on skeletal muscle amino acid metabolism studied *in vivo*. J. Am. Coll. Nutr. 15, 164–168.

Tsai, P.J., Huang, P.C., 1999. Circadian variations in plasma and erythrocyte concentrations of glutamate, glutamine, and alanine in men on a diet without and with added monosodium glutamate. Metab. Clin. Exp. 48, 1455–1460.

Varoqui, H., Zhu, H., Yao, D., Ming, H., Erickson, J.D., 2000. Cloning and functional identification of a neuronal glutamine transporter. J. Biol. Chem. 275, 4049–4054.

PHENYLALANINE

L-Phenylalanine (beta-phenylalanine, alpha-aminohydrocinnamic acid, alpha-amino beta-phenylpropionic acid; one-letter code F; molecular weight 165) is an essential amino acid with a nitrogen content of 8.5% (Figure 8.31).

ABBREVIATIONS

BH4	5,6,7,8-tetrahydrobiopterin
LAT2	L-type amino acid transporter 2
Phe	L-phenylalanine
PKU	phenylketonuria
PLP	pyridoxal 5-phosphate
TAT1	T-type amino acid transporter 1
Tyr	L-tyrosine

NUTRITIONAL SUMMARY

Function: The essential amino acid L-phenylalanine (Phe) is needed for the synthesis of proteins, catecholamines, and melanin; it is also an important precursor of the amino acid L-tyrosine (Tyr). Phe is used as an energy fuel; its complete oxidation requires biopterin, ascorbate, thiamin, riboflavin, niacin, vitamin B6, pantothenate, lipoate, ubiquinone, magnesium, and iron.

Food sources: Adequate amounts are consumed when total protein intake meets recommendations.

Requirements: The combined daily adult requirement for Phe and Tyr is 39 mg/kg (Young and Borgonha, 2000).

Deficiency: Prolonged lack of Phe, as with all essential amino acids or a lack of protein, causes growth failure, loss of muscle mass, and organ damage.

Excessive intake: Adults with phenylketonuria (PKU) should not exceed the Phe requirement, particularly not during pregnancy. There have been anecdotal reports of illness similar to eosinophilia-myalgia syndrome (EMS) and of seizures following the use of manufactured Phe.

DIETARY SOURCES

The protein in legumes, such as soybeans (5.5%), eggs (5.3%), and rice (5.3%), has slightly higher content than cow milk (4.8%), wheat (4.9%), and corn (4.9%). Pork (4.0%), beef (3.9%), and chicken (3.9%) contain slightly smaller percentages. In the end, however, total protein content is the main

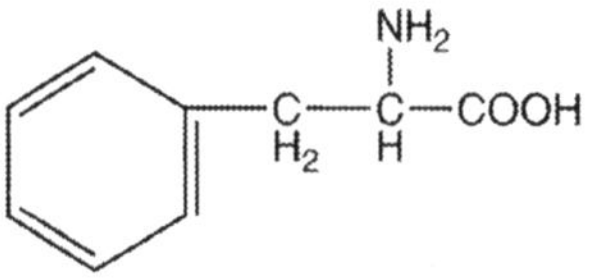

FIGURE 8.31

L-Phenylalanine.

determinant of Phe intake. A typical US diet provides 40 mg/kg body weight or more (Gibson et al., 2002). A commonly used synthetic source of Phe is aspartame (N-L-alpha-aspartyl-L-phenylalanine L-methyl ester), which provides about 40 mg of Phe per serving of sweetener.

DIGESTION AND ABSORPTION

Chewing of food in the mouth, denaturation by hydrochloric acid and various proteases (including pepsin in the stomach), and hydrolysis by pancreatic and enteric enzymes present a mixture of free amino acids and small peptides that are suitable for absorption (Figure 8.32). Dipeptides and tripeptides are taken up via the hydrogen ion/peptide cotransporter 1 (PepT1, SLC15A1) and, to a lesser extent, 2 (PepT2, SLC15A2). Transport of free Phe is mediated by the sodium–amino acid cotransport system $B°$ (Avissar et al., 2001). Exchange for other neutral amino acids via the sodium-independent transporter complex BAT1/$b^{o,+}$ (SLC7A9-SLC3A1) extends the capacity of the sodium-dependent transporters (Verrey et al., 1999; Mizoguchi et al., 2001). Nearly a third of the Phe taken up appears to be used by enterocytes themselves, mostly for protein synthesis, and 80% of it is after conversion to tyrosine (Matthews et al., 1993). The T-type amino acid transporter 1 (TAT1, SLC16A10) facilitates diffusion across the basolateral membrane of enterocytes in the jejunum, ileum, and colon (Ramadan et al., 2007). Additional transport capacity is provided by the 4F2-glycoprotein-anchored exchanger LAT2 (Rossier et al, 1999; Rajan et al., 2000).

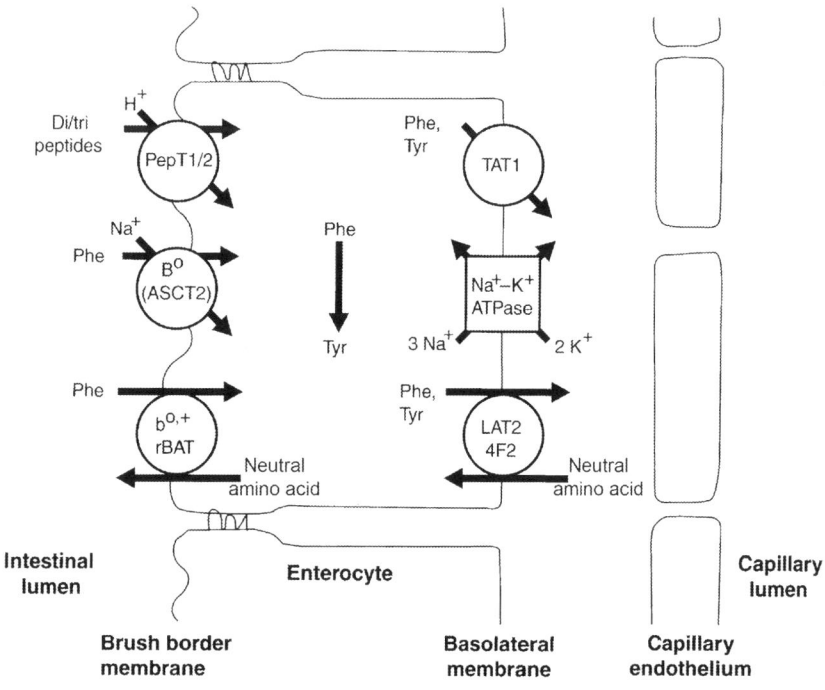

FIGURE 8.32

Intestinal absorption of L-phenylalanine.

TRANSPORT AND CELLULAR UPTAKE

Blood circulation: Normal plasma concentrations of Phe are around 50 µmol/l (Artuch et al., 1999). Concentrations are not significantly influenced by Phe intake (Fox et al., 2000). The amino acid enters cells from the blood via several transporters, including system T (TAT1) and LAT1, whose expression patterns vary considerably between specific tissues. LAT1 also transports D-phenylalanine (Yanagida et al., 2001).

BBB: The sodium-independent transporter TAT1 and the glycoprotein anchored exchangers LATI and LAT2 in brain capillary endothelial cells are involved in Phe transport, but their relative importance and location and the role of other transporters are not completely understood. Phe competes with other large neutral amino acids (valine, leucine, isoleucine, methionine, tyrosine, tryptophan, and histidine) and its own metabolite L-DOPA for the transport mechanisms from the blood into the brain (Pietz et al., 1999).

Materno-fetal transfer: The exchanger LAT1 appears to be the major route for Phe traveling from maternal blood into the syncytiotrophoblast (Ritchie and Taylor, 2001). Transfer across the basolateral membrane may proceed predominantly via LAT1 and LAT2 (Ritchie and Taylor, 2001); a contribution of TAT1, which is strongly expressed in the placenta (Kim et al., 2001), has been disputed (Ritchie and Taylor, 2001).

METABOLISM

Most Phe is eventually metabolized via L-tyrosine (Tyr) (Figure 8.33) giving rise to the glucogenic Krebs-cycle intermediate fumarate and the ketogenic metabolite acetoacetate (Figure 8.34). The six

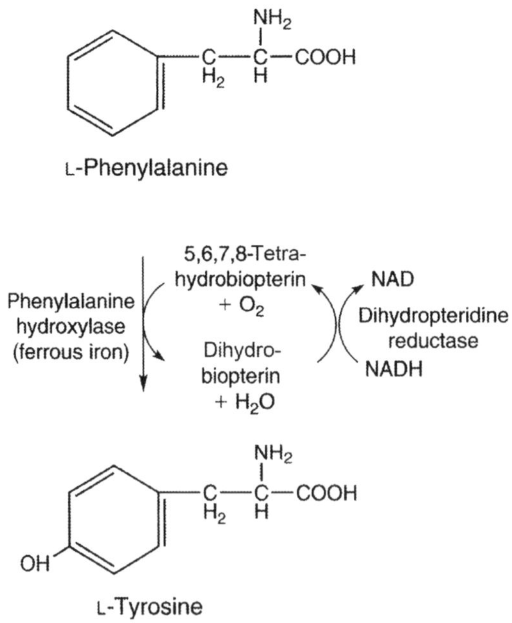

FIGURE 8.33

Conversion to L-tyrosine is the first step of L-phenylalanine catabolism.

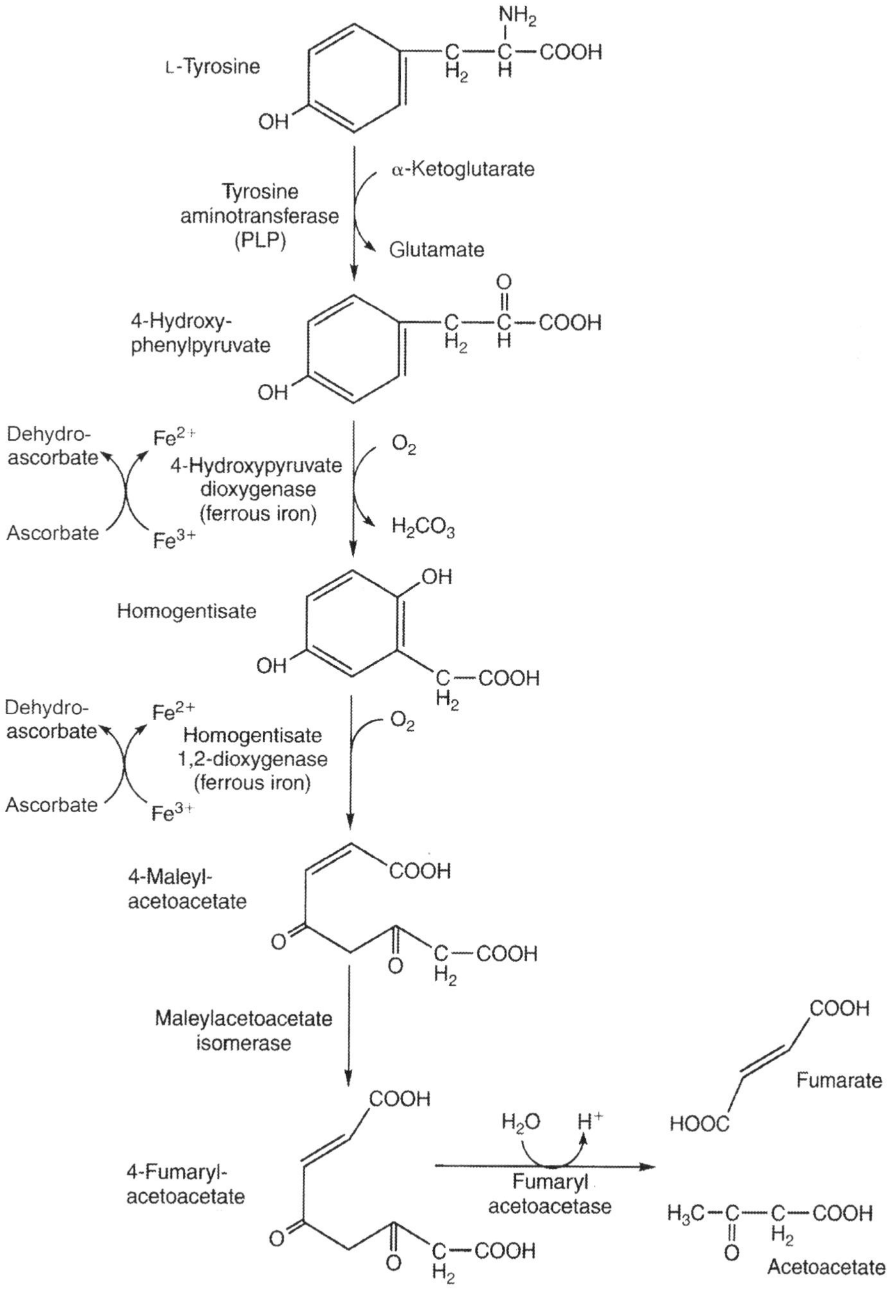

FIGURE 8.34

L-Tyrosine is broken down into fumerate and acetoacetate.

reaction steps require adequate supplies of biopterin, pyridoxin, ascorbate, and iron. Direct transamination to the toxic intermediate phenylpyruvate is not significant under most normal conditions. Small amounts of Phe-derived Tyr are used for the synthesis of catecholamines, melanins, and ubiquinone.

Conversion to L-tyrosine: The ferroenzyme phenylalanine hydroxylase (EC1.14.16.1) uses 5,6,7,8-tetrahydrobiopterin (BH4) to convert Phe to Tyr. The reaction oxidizes BH4 to 4a-hydroxytetrahydrobiopterin (4a-carbinolamine). BH4 is regenerated by dihydropteridine reductase (EC1.6.99.7) and NADPH or NADH. Tetrahydrobiopterin is also an essential cofactor for the hydroxylation of tyrosine (tyrosine 3-monooxygenase, EC1.14.16.2) and tryptophan (tryptophan hydroxylase, EC1.14.16.4). Loss-of-function mutations of phenylalanine hydroxylase or (in about 3% of cases) of enzymes involved in biopterin metabolism are the cause of PKU. The toxic effects of usually minor phenolic metabolites cause severe mental retardation unless effective treatment is initiated during pregnancy and after birth.

Tyrosine catabolism: Both mitochondrial and cytosolic forms of tyrosine aminotransferase (EC2.6.1.5, PLP-dependent) move the amino group from tyrosine to alpha-ketoglutarate. Mitochondrial and cytosolic aspartate aminotransferase (AST, EC2.6.1.1) can also catalyze the same reaction. The mitochondrial AST, which contributes to tyrosine metabolism especially in the small intestinal mucosa, can be generated by partial hydrolysis of aromatic amino acid transferase (EC2.6.1.57). The next step, oxidative decarboxylation to homogentisate, is facilitated by 4-hydroxyphenylpyruvate dioxygenase (EC1.13.11.27). Ascorbate reverts the enzyme-bound iron to the ferrous state whenever it gets oxidized. Vitamin C deficiency impairs Tyr catabolism and causes tyrosinemia. Homogentisate 1,2-dioxygenase (EC1.13.11.5) is another ferroenzyme whose iron must be maintained by reduced ascorbate. Maleylacetoacetate isomerase (EC5.2.1.2) catalyzes the next step. This enzyme is bifunctional since it also has glutathione *S*-transferase (EC2.5.1.18) activity (GST zeta 1). Fumarylacetoacetase (EC3.7.1.2) completes the conversion of Phe to acetoacetate and fumarate.

Fumarate can be utilized directly through the Krebs cycle. Acetoacetate can be activated to acetyl-CoA through the successive actions of 3-oxoacid CoA-transferase (succinyl-CoA transferase, EC2.8.3.5) and acetyl-CoA C-acetyltransferase (thiolase, EC2.3.1.9). Depending on the redox potential, it may also be reduced to beta-hydroxybutyrate by NADH-dependent 3-hydroxybutyrate dehydrogenase (EC1.1.1.30).

Direct transamination: Phenylalanine (histidine) aminotransferase (EC2.6.1.58), glutamine-phenylpyruvate aminotransferase (EC2.6.1.64), aromatic-amino acid-glyoxylate aminotransferase (EC2.6.1.60), and kynurenine-glyoxylate aminotransferase (EC2.6.1.63)—all of which are PLP-dependent—accept Phe as a substrate. The resulting phenylacetate is activated in the liver by phenylacetate-CoA ligase (EC6.2.1.21) to phenylacetyl-CoA; conjugation with L-glutamine by glutamine *N*-phenylacetyltransferase (EC2.3.1.14) forms the dead-end product alpha-*N*-phenylacetylglutamine (Yang et al., 1996).

STORAGE

Proteins in skeletal muscle and other tissues contain 5.3% Phe (Smith, 1980). During normal protein turnover, Phe is released and becomes available for renewed protein synthesis or other metabolic pathways. Daily release of Phe from protein catabolism in healthy people is about 12 mg/kg body weight. Severe trauma, systemic infection, and immobilization increase protein catabolism.

EXCRETION

At normal plasma concentrations, several grams of Phe appear in renal ultrafiltrate in a day. Almost all of this is taken up by proximal tubular endothelial cells via the sodium–amino acid cotransport system B° (Avissar et al., 2001) and the sodium-independent transporter complex BAT1/$b^{0,+}$ (Verrey et al., 1999; Mizoguchi et al., 2001). Export across the basolateral membrane is mediated by the 4F2- glycoprotein-anchored exchanger LAT2 (Rossier et al., 1999; Rajan et al., 2000). Excretion of complex breakdown products, such as phenylacetylglutamine (Yang et al., 1996), is usually minimal. Typical excretion as the catecholamine metabolites vanillylmandelic acid and homovanillic acid is about 3–6 mg/day.

REGULATION

A decrease of protein intake near or below adequacy slows the conversion of Phe to Tyr (Gibson et al., 2002). The activity of phenylalanine hydroxylase (EC1.14.16.1) is allosterically regulated by Phe.

FUNCTION

Energy fuel: Eventually most Phe is broken down and oxidized to carbon dioxide, water, and urea with an energy yield of about 6.3 kcal/g. Its complete oxidation requires adequate availability of biopterin, ascorbate, thiamin, riboflavine, niacin, vitamin B6, pantothenate, lipoate, ubiquinone, magnesium, and iron.

Protein synthesis: Typically about phenylalanine-tRNA ligase (EC6.1.1.20) is responsible for loading Phe on its tRNA. Daily Phe incorporation into body proteins uses less than 10 mg/kg body weight in healthy people (Gibson et al., 2002).

Catecholamine synthesis: Specialized tissues, including the brain, some extracerebral neurons, and the adrenal gland, produce catecholamines. Tyrosine 3-monooxygenase (EC1.14.16.2) facilitates the initial step of 3,4-dihydroxyphenylalanine (DOPA) synthesis from Tyr. The iron-enzyme requires tetrahydropterine as a cosubstrate.

Melanin synthesis: Monophenol monooxygenase (EC1.14.18.1) uses dihydroxyphenylalanine (DOPA) as a cosubstrate during the oxidation of tyrosine to dopaquinone, a precursor of the skin and eye pigment melanin.

Ubiquinone synthesis: Tyr provides the ring moiety, 4-hydroxybenzoate, for ubiquinone synthesis (Artuch et al., 1999). The reactions responsible for the conversion of L-tyrosine to 4-hydroxybenzoate have not been characterized yet. The 12 mg of ubiquinone produced daily (Elmberger et al., 1987) utilize <2 mg Tyr.

Thyroid hormone synthesis: Tyr is needed for thyroid hormone production in the thyroid and, to a much lesser extent, in the brain and possibly some other nonthyroid organs. Thyroxin and triiodothyronin are the major active thyroid hormones.

REFERENCES

Artuch, R., Vilaseca, M.A., Moreno, J., Lambruschini, N., Cambra, F.J., Campistol, J., 1999. Decreased serum ubiquinone-10 concentrations in phenylketonuria. Am. J. Clin. Nutr. 70, 892–895.

Avissar, N.E., Ryan, C.K., Ganapathy, V., Sax, H.C., 2001. Na^+-dependent neutral amino acid transporter ATB° is a rabbit epithelial cell brush border protein. Am. J. Physiol. Cell Physiol. 281, C963–C971.

Elmberger, P.G., Kalen, A., Appelkvist, E.L., Dallner, G., 1987. *In vitro* and *in vivo* synthesis of dolichol and other main mevalonate products in various organs of the rat. Eur. J. Biochem. 168, 1–11.

Fox, C., Marquis, J., Kipp, D.E., 2000. Nutritional factors affecting serum phenylalanine concentration during pregnancy for identical twin mothers with phenylketonuria. Acta Paediatr. 89, 947–950.

Gibson, N.R., Jahoor, F., Ware, L., Jackson, A.A., 2002. Endogenous glycine and tyrosine production is maintained in adults consuming a marginal-protein diet. Am. J. Clin. Nutr. 75, 511–518.

Matthews, D.E., Marano, M.A., Campbell, R.G., 1993. Splanchnic bed utilization of leucine and phenylalanine in humans. Am. J. Physiol. 264, E109–E118.

Mizoguchi, K., Cha, S.H., Chairoungdua, A., Kim, D.K., Shigeta, Y., Matsuo, H., et al., 2001. Human cystinuria-related transporter: localization and functional characterization. Kidney Int. 59, 1821–1833.

Pietz, J., Kreis, R., Rupp, A., Mayatepek, E., Rating, D., Boesch, C., et al., 1999. Large neutral amino acids block phenylalanine transport into brain tissue in patients with phenylketonuria. J. Clin. Invest. 103, 1169–1178.

Rajan, D.P., Kekuda, R., Huang, W., Devoe, L.D., Leibach, F.H., Prasad, P.D., et al., 2000. Cloning and functional characterization of a Na(+)-independent, broad-specific neutral amino acid transporter from mammalian intestine. Biochim. Biophys. Acta 1463, 6–14.

Ramadan, T., Camargo, S.M., Herzog, B., Bordin, M., Pos, K.M., Verrey, F., 2007. Recycling of aromatic amino acids via TAT1 allows efflux of neutral amino acids via LAT2-4F2hc exchanger. Pflügers Arch. 454, 507–516.

Ritchie, J.W., Taylor, P.M., 2001. Role of the System L permease LAT1 in amino acid and iodothyronine transport in placenta. Biochem. J. 356, 719–725.

Rossier, G., Meier, C., Bauch, C., Summa, V., Sordat, B., Verrey, F., et al., 1999. LAT2, a new basolateral 4F2hc/CD98-associated amino acid transporter of kidney and intestine. J. Biol. Chem. 274, 34948–34954.

Smith, R.H., 1980. Comparative amino acid requirements. Proc. Nutr. Soc. 39, 71–78.

Verrey, F., Jack, D.L., Paulsen, I.T., Saier Jr., M.H., Pfeiffer, R., 1999. New glycoprotein-associated amino acid transporters. J. Membrane Biol. 172, 181–192.

Yanagida, O., Kanai, Y., Chairoungdua, A., Kim, D.K., Segawa, H., Nii, T., et al., 2001. Human L-type amino acid transporter 1 (LAT1): characterization of function and expression in tumor cell lines. Biochim. Biophys. Acta 1514, 291–302.

Yang, D., Previs, S.F., Fernandez, C.A., Dugelay, S., Soloviev, M.V., Hazey, J.W., et al., 1996. Non-invasive probing of liver citric acid cycle intermediates with phenylacetylglutamine. Am. J. Physiol. 270, E882–E889.

Young, V.R., Borgonha, S., 2000. Nitrogen and amino acid requirements: the Massachusetts Institute of Technology amino acid requirement pattern. J. Nutr. 130, 1841S–1849S.

TYROSINE

L-Tyrosine (beta-(p-hydroxyphenyl)alanine, alpha-amino-p-hydroxyhydrocinnamic acid, one-letter code Y; molecular weight 181) is an essential amino acid with a nitrogen content of 7.7% (Figure 8.35).

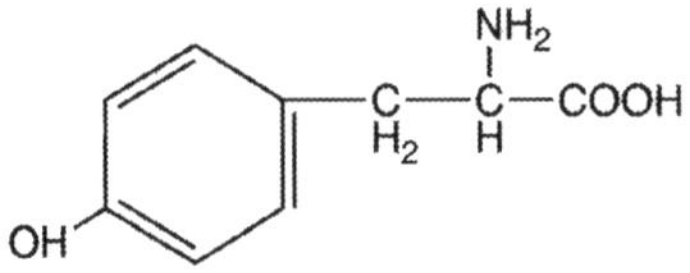

FIGURE 8.35

L-Tyrosine.

ABBREVIATIONS

BH4	5,6,7,8-tetrahydrobiopterin
CoA	coenzyme A
LAT2	L-type amino acid transporter 2
Phe	L-phenylalanine
PKU	phenylketonuria
PLP	pyridoxal 5′-phosphate
RDA	recommended dietary allowance
TAT1	T-type amino acid transporter 1
Tyr	L-tyrosine

NUTRITIONAL SUMMARY

Function: The essential amino acid L-tyrosine (Tyr) is needed for the synthesis of proteins, catecholamines, and thyroid hormones. Most Tyr is used eventually as an energy fuel; its complete oxidation requires ascorbate, thiamin, riboflavin, niacin, vitamin B6, pantothenate, lipoate, ubiquinone, iron, and magnesium.

Food sources: Adequate amounts are consumed when total protein intake meets recommendations.

Requirements: Tyr can be derived from L-phenylalanine (Phe); the combined daily adult requirement for Phe and Tyr is 39 mg/kg body weight (Young and Borgonha, 2000). Individuals with PKU have an increased Tyr requirement due to the impaired conversion from Phe. In controlled clinical experiments, supplemental Tyr did not increase burst exercise performance, boost growth hormone output, or lower increased blood pressure.

Deficiency: Prolonged lack of Tyr, as with all essential amino acids or a lack of protein, causes growth failure, loss of muscle mass, and organ damage.

Excessive intake: Very high intake of protein and mixed amino acids (i.e., more than three times the RDA, which is 2.4 g/kg) is thought to increase the risk of renal glomerular sclerosis and accelerate osteoporosis.

ENDOGENOUS SOURCES

The ferroenzyme phenylalanine hydroxylase (EC1.14.16.1) uses 5,6,7,8-tetrahydrobiopterin (BH4) for the conversion of Phe to Tyr. The reaction oxidizes BH4 to 4a hydroxytetrahydrobiopterin (4a-carbinolamine). BH4 is regenerated by dihydropteridine reductase (EC1.6.99.7) and NADPH or NADH. Tetrahydrobiopterin is also an essential cofactor for the hydroxylation of tyrosine itself (tyrosine 3-monooxygenase, EC1.14.16.2) and tryptophan (tryptophan hydroxylase, EC1.14.16.4). Loss-of-function mutations of this enzyme or (in about 3% of cases) of enzymes involved in biopterin metabolism cause PKU. The toxic effects of accumulating phenylalanine metabolites cause severe mental retardation unless effective treatment is initiated during pregnancy and after birth. Inadequate Tyr synthesis in PKU patients increases their reliance on dietary intake (Figure 8.36).

DIETARY SOURCES

All protein-containing foods provide Tyr. Relatively Tyr-rich protein sources include soy (55 mg/g), cow milk (48 mg/g), and eggs (41 mg/g). The Tyr content of flesh foods (beef, pork, chicken, and fish)

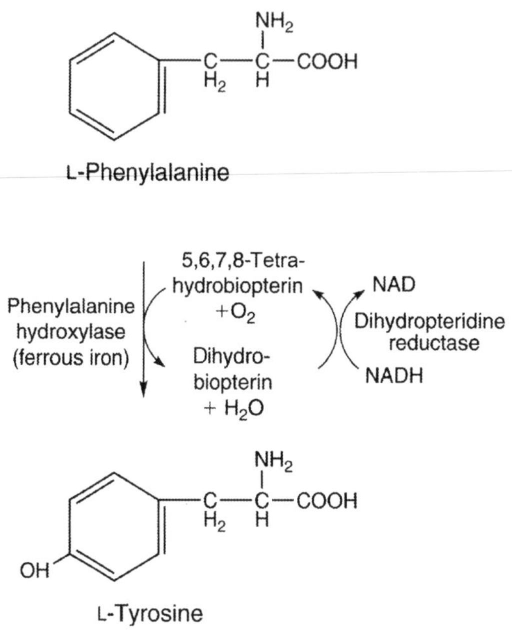

FIGURE 8.36

L-Tyrosine can be synthesized from L-phenylalanine.

is around 34 mg/g. The Tyr content of grain protein ranges widely. At the high end is corn (maize, 41 mg/g); at the low end is rye (20 mg/g). Wheat, oats, and rice, as well as legumes, contain around 30 mg/g. The typical Tyr intake of American adults is about 38 mg/kg body weight (Gibson et al., 2002).

Tyrosine in foods may be degraded by spoilage or processing. Bacterial EC4.1.1.25 tyrosine decarboxylase generates tyramine. Some fermented foods, especially cheeses, contain significant amounts of tyramine. Catechol oxidase (EC1.10.3.1) in mushrooms, apples, potatoes, and other fresh plant foods causes newly cut surfaces to darken due to the oxidation of Tyr to monohydric and O-dihydric phenols.

DIGESTION AND ABSORPTION

As food passes through the digestive tract, gastric and pancreatic enzymes continue protein breakdown, many of them cleaving peptide bonds between specific amino acids and Tyr (Figure 8.37). The sodium-independent transport system $b^{o,+}$ ferries Tyr across the intestinal brush border membrane in exchange for another neutral amino acid. This system consists of a light-chain BAT1 (SLC7A9) and a heavy-chain rBAT (SLC3A1) glycoprotein that anchors the complex to the membrane (Verrey et al., 1999; Mizoguchi et al., 2001). The T-type amino acid transporter 1 (TAT1, SLC16A10) facilitates diffusion across the basolateral membrane of enterocytes in the jejunum, ileum, and colon (Ramadan et al., 2007). Additional transport capacity is provided by the 4F2-glycoprotein-anchored exchanger LAT2 (Rossier et al., 1999; Rajan et al., 2000).

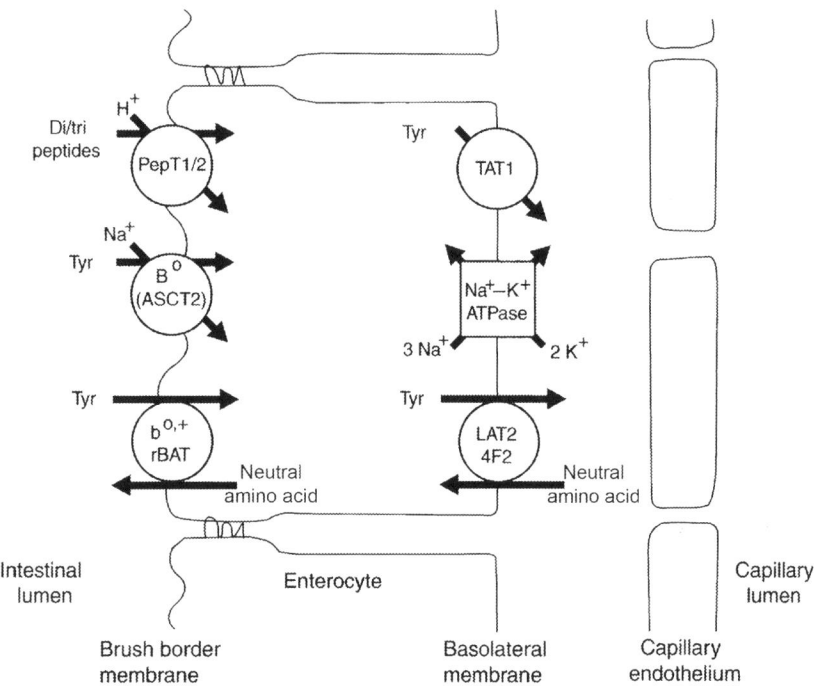

FIGURE 8.37

Intestinal absorption of L-tyrosine.

TRANSPORT AND CELLULAR UPTAKE

Blood circulation: Normal plasma concentrations of Tyr are around 66 μmol/l (Artuch et al., 1999). The amino acid enters cells from the blood via several transporters, including system T (TAT1) and LAT1, whose expression patterns vary considerably between specific tissues. LAT1 also transports D-phenylalanine (Yanagida et al., 2001).

BBB: The sodium-independent transporter TAT1 and the glycoprotein-anchored exchangers LAT1 and LAT2 in the endothelial cells of brain capillary are involved in Tyr transport, but their relative importance, location, and the role of other transporters are not completely understood. Tyr competes with other large neutral amino acids (valine, leucine, isoleucine, methionine, phenylalanine, tryptophan, and histidine) for the transport mechanisms from the blood into the brain (Pietz et al., 1999).

Materno-fetal transfer: The exchanger LAT1 appears to be the major route for Tyr traveling from maternal blood into the syncytiotrophoblast (Ritchie and Taylor, 2001). Transfer across the basolateral membrane may proceed predominantly via LAT1 and LAT2 (Ritchie and Taylor, 2001); a contribution of TAT1, which is strongly expressed in the placenta (Kim et al., 2001), has been disputed (Ritchie and Taylor, 2001).

METABOLISM

Both mitochondrial and cytosolic forms of tyrosine aminotransferase (PLP-dependent, EC2.6.1.5) move the amino group from Tyr to alpha-ketoglutarate (Figure 8.38). Mitochondrial and cytosolic aspartate aminotransferase (AST, EC2.6.1.1) can also catalyze the same reaction.

The mitochondrial AST, which contributes to Tyr metabolism (especially in the mall intestinal mucosa), can be generated by partial hydrolysis of aromatic amino acid transferase (EC2.6.1.57). The next step, oxidative decarboxylation to homogentisate, is facilitated by 4-hydroxyphenylpyruvate dioxygenase (EC1.13.11.27). Ascorbate reverts the enzyme-bound iron to the ferrous state whenever the iron gets oxidized. Vitamin C deficiency impairs Tyr catabolism and causes tyrosinemia. Homogentisate 1,2-dioxygenase (EC1.13.11.5) is another ferroenzyme whose iron must be maintained by reduced ascorbate. Maleylacetoacetate isomerase (EC5.2.1.2) catalyzes the next step. This enzyme is bifunctional since it also has glutathione S-transferase (EC2.5.1.18) activity (GST zeta 1). Fumarylacetoacetase (EC3.7.1.2) completes the conversion of Tyr to acetoacetate and fumarate.

Fumarate can be utilized directly through the Krebs cycle. Acetoacetate can be activated to acetyl-CoA through the successive actions of 3-oxoacid CoA-transferase (succinyl-CoA transferase, EC2.8.3.5) and acetyl-CoA C acetyltransferase (thiolase, EC2.3.1.9). Depending on the redox potential, it may also be reduced to beta-hydroxybutyrate by NADH-dependent 3-hydroxybutyrate dehydrogenase (EC1.1.1.30).

STORAGE

On average, the body contains 33 mg of Tyr per gram of total protein (Smith, 1980). Amino acids are released as functional proteins are broken down during normal turnover. Accelerated protein catabolism occurs in response to starvation, severe trauma, or infection.

EXCRETION

Due to very effective renal reabsorption of filtered Tyr, very little is lost in urine. Uptake from the proximal tubular lumen uses mainly the sodium-dependent system B° (Avissar et al., 2001) and is augmented by the action of the sodium-independent transporter complexes $b^{o,+}$ and BAT1-rBAT (Verrey et al., 1999). Impaired reabsorption in Hartnup disease, due to defects in the transporter B°AT1 (XT2s1, SLC6A19), causes excessive renal losses (Seow et al., 2004).

Normal daily excretion of breakdown products of Tyr-derived catecholamine, such as vanillylmandelic acid and homovanillic acid, adds up to several milligrams.

REGULATION

The rate of Tyr uptake into various tissues (including the brain) appears to be governed largely by availability (as reflected by blood concentration) from dietary intake and endogenous synthesis from phenylalanine without significant intervening regulatory events (Garabal et al., 1988). Oral contraceptives decrease Tyr availability during the luteal phase by promoting transamination (Moiler et al., 1995).

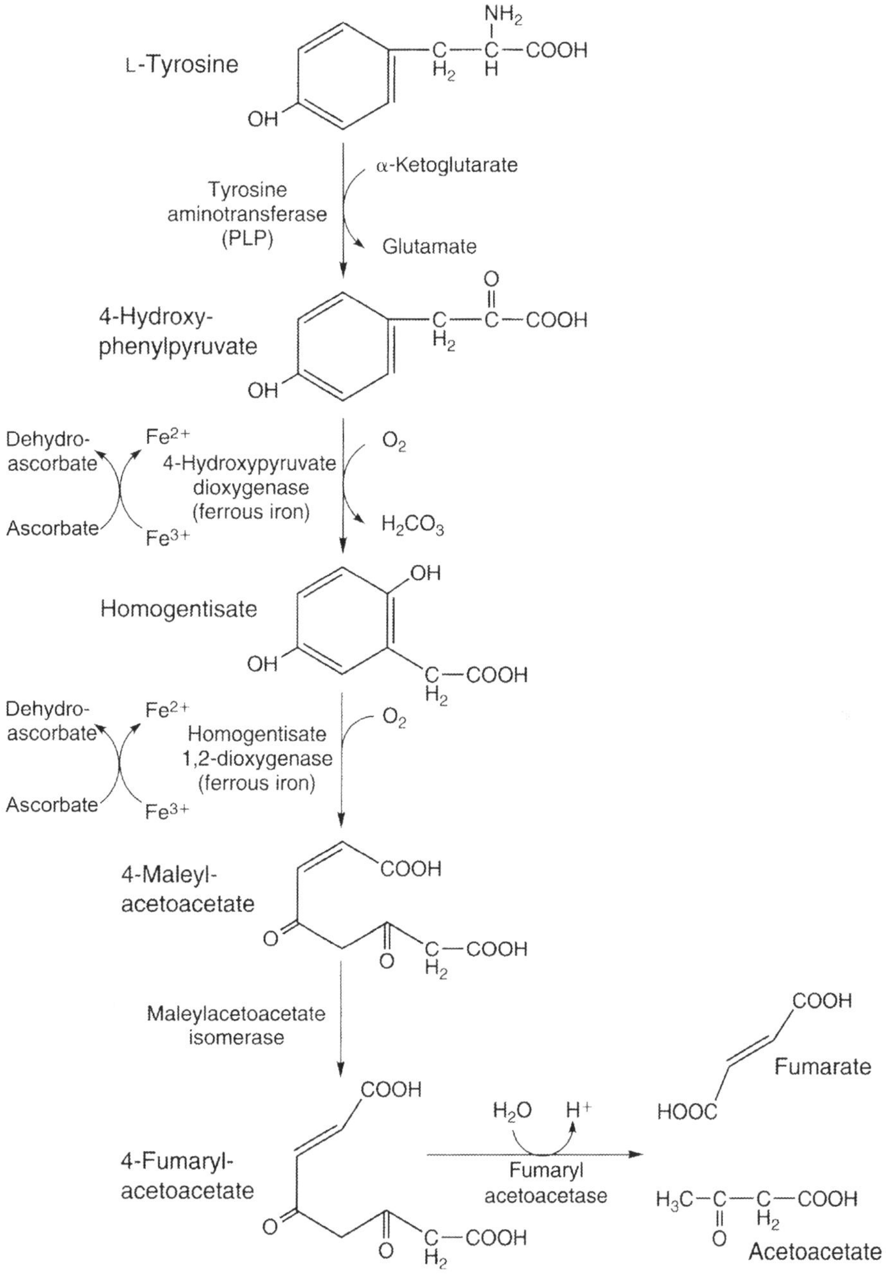

FIGURE 8.38

Metabolism of L-tyrosine.

FUNCTION

Energy fuel: Eventually, most Tyr is broken down and oxidized to carbon dioxide, water, and urea with an energy yield of 5.4 kcal/g (May and Hill, 1990). Its complete oxidation requires adequate availability of ascorbate, thiamin, riboflavin, niacin, vitamin B6, pantothenate, lipoate, ubiquinone, iron, and magnesium.

Protein synthesis: Tyrosine-tRNA ligase (EC6.1.1.1) loads Tyr on its tRNA for peptide synthesis. The posttranslational phosphorylation of specific Tyr residues in various proteins is an important mechanism for the regulation of numerous physiological functions.

Catecholamine synthesis: Specialized tissues (including the brain, some extracerebral neurons, and the adrenal gland) produce catecholamines. Tyrosine 3-monooxygenase (EC1.14.16.2) facilitates the initial step of 3,4-DOPA synthesis from Tyr. The iron-enzyme requires tetrahydropterine as a cosubstrate.

Melanin synthesis: Monophenol monooxygenase (EC1.14.18.1) uses DOPA as a cosubstrate during the oxidation of Tyr to dopaquinone, a precursor of the skin and eye pigment melanin.

Ubiquinone synthesis: Tyr provides the ring moiety, 4-hydroxybenzoate, for ubiquinone synthesis (Artuch et al., 1999). The reactions responsible for the conversion of L-tyrosine to 4-hydroxybenzoate have not been characterized yet. The daily production of 12 mg (Elmberger et al., 1987) utilizes less than 2 mg of Tyr.

Thyroid hormone synthesis: Tyr is needed for thyroid hormone production in the thyroid and, to a much lesser extent, in the brain and possibly some other nonthyroid organs. Thyroxin and triiodothyronin are the major active thyroid hormones.

REFERENCES

Artuch, R., Vilaseca, M.A., Moreno, J., Lambruschini, N., Cambra, F.J., Campistol, J., 1999. Decreased serum ubiquinone-10 concentrations in phenylketonuria. Am. J. Clin. Nutr. 70, 892–895.

Avissar, N.E., Ryan, C.K., Ganapathy, V., Sax, H.C., 2001. Na^+-dependent neutral amino acid transporter ATB^0 is a rabbit epithelial cell brush border protein. Am. J. Physiol. Cell Physiol. 281, C963–C971.

Elmberger, P.G., Kalen, A., Appelkvist, E.L., Dallner, G., 1987. *In vitro* and *in vivo* synthesis of dolichol and other main mevalonate products in various organs of the rat. Eur. J. Biochem. 168, 1–11.

Garabal, M.V., Arevalo, R.M., Diaz-Palarea, M.D., Castro, R., Rodriguez, M., 1988. Tyrosine availability and brain noradrenaline synthesis in the fetus: control by maternal tyrosine ingestion. Brain Res. 457, 330–337.

Gibson, N.R., Jahoor, F., Ware, L., Jackson, A.A., 2002. Endogenous glycine and tyrosine production is maintained in adults consuming a marginal-protein diet. Am. J. Clin. Nutr. 75, 511–518.

Kim, D.K., Kanai, Y., Chairoungdua, A., Matsuo, H., Cha, S.H., Endou, H., 2001. Expression cloning of a Na^+-independent aromatic amino acid transporter with structural similarity to H^+/monocarboxylate transporters. J. Biol. Chem. 276, 17221–17228.

May, M.E., Hill, J.O., 1990. Energy content of diets of variable amino acid compostion. Am. J. Clin. Nutr. 52, 770–776.

Mizoguchi, K., Cha, S.H., Chairoungdua, A., Kim, D.K., Shigeta, Y., Matsuo, H., et al., 2001. Human cystinuria-related transporter: localization and functional characterization. Kidney Int. 59, 1821–1833.

Moiler, S.E., Maach-Moller, B., Olesen, M., Madsen, B., Madsen, E., Fjalland, B., 1995. Tyrosine metabolism in users of oral contraceptives. Life Sci. 56, 687–695.

Pietz, J., Kreis, R., Rupp, A., Mayatepek, E., Rating, D., Boesch, C., et al., 1999. Large neutral amino acids block phenylalanine transport into brain tissue in patients with phenylketonuria. J. Clin. Invest. 103, 1169–1178.

Rajan, D.E., Kekuda, R., Huang, W., Devoe, L.D., Leibach, F.H., Prasad, P.D., et al., 2000. Cloning and functional characterization of a Na(+)-independent, broad-specific neutral amino acid transporter from mammalian intestine. Biochim. Biophys. Acta 1463, 6–14.

Ramadan, T., Camargo, S.M., Herzog, B., Bordin, M., Pos, K.M., Verrey, F., 2007. Recycling of aromatic amino acids via TAT1 allows efflux of neutral amino acids via LAT2-4F2hc exchanger. Pflügers Arch. 454, 507–516.

Ritchie, J.W., Taylor, P.M., 2001. Role of the System L permease LAT1 in amino acid and iodothyronine transport in placenta. Biochem. J. 356, 719–725.

Rossier, G., Meier, C., Bauch, C., Summa, V., Sordat, B., Verrey, E., et al., 1999. LAT2, a new basolateral 4F2hc/CD98-associated amino acid transporter of kidney and intestine. J. Biol. Chem. 274, 34948–34954.

Seow, H.F., Bröer, S., Bröer, A., Bailey, C.G., Potter, S.J., Cavanaugh, J.A., et al., 2004. Hartnup disorder is caused by mutations in the gene encoding the neutral amino acid transporter SLC6A19. Nat. Genet. 36, 1003–1007.

Smith, R.H., 1980. Comparative amino acid requirements. Proc. Nutr. Soc. 39, 71–78.

Verrey, E., Jack, D.L., Paulsen, I.T., Saier Jr., M.H., Pfeiffer, R., 1999. New glycoprotein-associated amino acid transporters. J. Membrane. Biol. 172, 181–192.

Yanagida, O., Kanai, Y., Chairoungdua, A., Kim, D.K., Segawa, H., Nil, T., et al., 2001. Human L-type amino acid transporter 1 (LAT1): characterization of function and expression in tumor cell lines. Biochim. Biophys. Acta 1514, 291–302.

Young, V.R., Borgonha, S., 2000. Nitrogen and amino acid requirements: the Massachusetts Institute of Technology amino acid requirement pattern. J. Nutr. 130, 1841S–1849S.

TRYPTOPHAN

L-Tryptophan (1-alpha-aminoindole-3-propionic acid, 1-alpha-amino-3-indolepropionic acid, 2-amino-3-indolylpropanoic acid, 1-beta-3-indolylalanine, one-letter code W; molecular weight 205) is an essential amino acid (Figure 8.39).

ABBREVIATIONS

CoA	coenzyme A
EMS	eosinophilia-myalgia syndrome
ETF	electron-transferring flavoprotein
5-HT	5-hydroxytryptamine
5-HTOL	5-hydroxytryptophol
KAT	kynurenine-oxoglutarate aminotransferase
PLP	pyridoxal-5′-phosphate
RDA	recommended dietary allowance
Trp	L-tryptophan

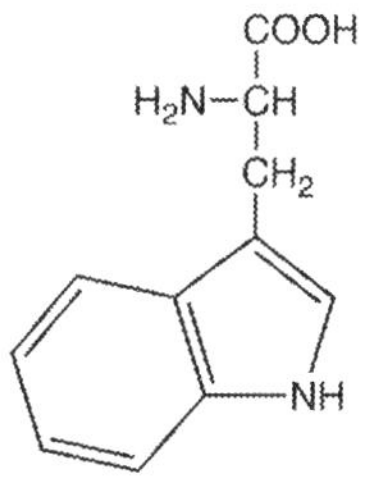

FIGURE 8.39

L-Tryptophan.

NUTRITIONAL SUMMARY

Function: The essential amino acid L-tryptophan (Trp) is needed for the synthesis of proteins, serotonin, melatonin, and niacin. Utilization as an energy fuel depends on adequate availability of thiamin, riboflavin, vitamin B6, niacin, pantothenate, lipoate, ubiquinone, magnesium, and iron.

Food sources: Adequate amounts are consumed when total protein intake meets recommendations since dietary proteins from different sources all contain Trp. The protein in milk and dairy products contains slightly more than most other food proteins; corn protein contains less. Exposure to high heat as in grilling and frying can reduce the Trp content of food. Manufactured Trp is little used now, because in the past, contaminants have caused severe and irreversible harm (i.e., EOS).

The same contaminants, as well as related ones, have been found in manufactured 5-hydroxytryptophan and melatonin.

Requirements: Adults are thought to require 6 mg/kg per clay (Young and Borgonha, 2000).

Deficiency: Prolonged lack of Trp, as with all essential amino acids or a lack of protein, causes growth failure, loss of muscle mass, and organ damage.

Excessive intake: Very high intake of protein and mixed amino acids (i.e., more than three times the RDA, equivalent to 2.4 g/kg) is thought to increase the risk of renal glomerular sclerosis and accelerate osteoporosis. The consequences of very high intake of Trp, other than the known risks associated with toxic contaminants, have not been adequately evaluated.

DIETARY AND OTHER SOURCES

Foods contain Trp, mostly bound in proteins, 5-hydroxytryptophan, 5-hydroxytryptamine (5-HT), and tryptamine; synthetic supplements sometimes contain tryptophan ethyl ester, whose uptake is not limited by Hartnup disease (Anonymous, 1990). Foods containing significant amounts of 5-HT include bananas (0.02 mg/g), pineapples (0.01 mg/g), and black walnuts (0.3 mg/g; Bruce. 1960; Feldman and Lee, 1985; Helander and Some, 2000).

Milk protein is relatively richer in Trp (14.1 mg/g protein) than the protein in eggs (12.2 mg/g), white bread (11.7 mg/g), and beef (11.2 mg/g); corn contains much less (7 mg/g). High heat can induce the condensation of Trp with aldohexoses and thereby generate various Amadori rearrangement products, including N(1)-(beta-D-hexopyranosyl)-1-tryptophan, 2-(beta-D-hexopyranosyl)-1-tryptophan, and 1-(1,2,3,4,5-pentahydroxypent-1-yl)-1,2,3,4-tetrahydro-beta-carboline-3-carboxylic acid derivatives (Gutsche et al., 1999); thus, grilling, frying, and possibly other types of cooking can reduce the Trp content of food (Dworschak, 1980). The comutagenic Trp derivative norharman can form at relatively low temperatures in heat-processed food (Ziegenhagen et al., 1999). Another important mechanism of Trp loss during heating (e.g., in ultrahigh-temperature-treated milk) is oxidation, particularly in the presence of iron or copper (Birlouez-Aragon et al., 1997).

Note: EMS with symptoms that include fatigue, pain, depression, sleep disturbance, and verbal memory impairment can be caused by Trp contaminated with one or more of at least six compounds, including 2-(3'-indolylmethyl)-indole (Müller et al., 1999), 3a-hydroxy-1,2,3,3a,8,Sa-hexahydropyrrolo-[2–3b]-indole-2-carboxylic acid, 2-(2-hydroxy-indoline)-tryptophan (Naylor et al., 1999), and a family of 234 Da compounds (Klarskov et al., 1999). Some of these EMS-related contaminants are also present in commercial preparations of 5-hydroxytryptophan (Klarskov et al., 1999) and melatonin (Laforce et al., 1999). EMS may have developed in patients primarily as an allergic reaction toward a more immunogenic L-tryptophan preparation (Barth et al., 1999).

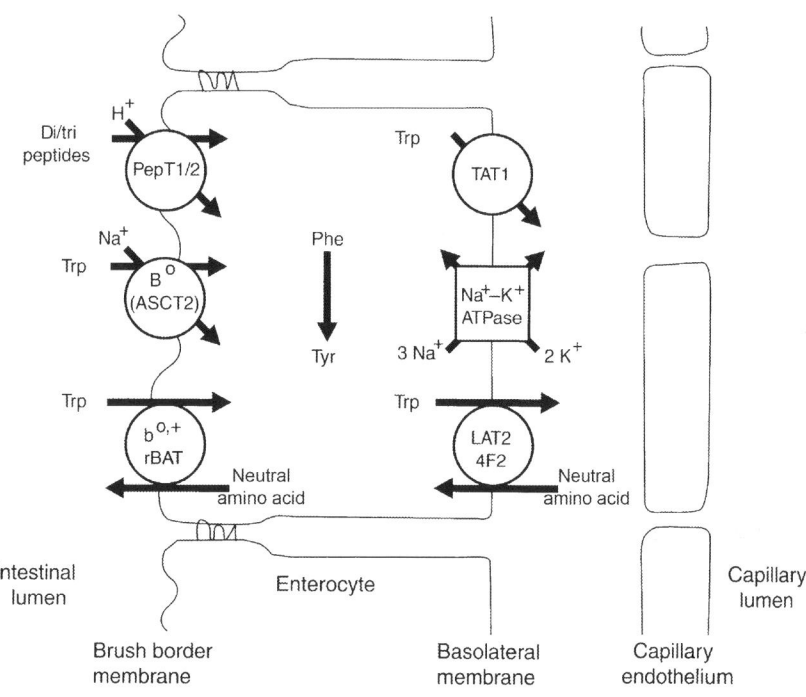

FIGURE 8.40

Intestinal absorption of L-tryptophan.

DIGESTION AND ABSORPTION

Mastication of foods in the mouth, denaturation by hydrochloric acid, and unspecific protein hydrolysis by pepsin in the stomach initiate the breakdown of Trp-containing proteins (Figure 8.40). Pancreatic proteases and aminopeptidases on the intestinal brush border membrane can then generate free Trp, as well as dipeptides and tripeptides, which are suitable for absorption. Dipeptides and tripeptides are taken up via the proton-peptide cotransporter (SLC15A1, PepT1). The sodium–amino acid cotransport system B° is the main conduit for the intestinal uptake of free Trp (Avissar et al., 2001); the molecular defect responsible for impaired uptake (and increased renal losses due to diminished recovery) in Hartnup disease is due to defective B°AT1 (SLC6A19), which is one of the B° transporters. Exchange for other neutral amino acids via the sodium-independent transporter complex BAT1-rBAT (SLC7A9-SLC3A1) augments this pathway (Verrey et al., 1999; Mizoguchi et al., 2001). The T-type amino acid transporter 1 (TAT1, SLC16A10) facilitates ion-independent diffusion across the basolateral membrane of enterocytes in the jejunum, ileum, and colon (Ramadan et al., 2007). Additional transport capacity is provided by the 4F2-glycoprotein-anchored exchanger LAT2 (Rossier et al., 1999; Rajan et al., 2000).

TRANSPORT AND CELLULAR UPTAKE

Blood circulation: The plasma concentration of Trp (typically around 50 μmol/l) decreases in response to low dietary intake (Kaye et al., 2000). Uptake from the blood into tissues uses various transporters, including system T (TAT1), LAT1, and LAT2, whose expression patterns vary considerably between specific tissues.

BBB: The sodium-independent transporter TAT1 and the glycoprotein anchored complex LAT1 are expressed in brain capillary endothelial cells and certainly contribute to Trp transport, but their relative importance, location, and the role of other transporters are not completely understood. Trp competes with the BCAAs (valine, leucine, and isoleucine) and other large neutral amino acids (methionine, tyrosine, tryptophan, and histidine) for transport into the brain. This may mean that increased blood concentrations of phenylalanine (especially in patients with PKU, an inborn error of metabolism with defective phenylalanine utilization) or BCAAs (due to a high-carbohydrate diet) limit Trp availability in the brain.

Materno-feta/transfer: The exchanger LAT1 appears to be the major route for Trp traveling from maternal blood into the syncytiotrophoblast (Ritchie and Taylor, 2001). Transfer across the basolateral membrane may proceed predominantly via LAT1 and LAT2 (Ritchie and Taylor, 2001); a contribution by TAT1, which is strongly expressed in the placenta (Kim et al., 2001), has been disputed (Ritchie and Taylor, 2001).

METABOLISM

Trp can be converted into very different components depending on location and regulation. While most Trp is eventually broken down into carbon dioxide, water, and urea, intermediary or alternative products of biological importance include alanine, acetyl-CoA, serotonin, melatonin, NAD, and NADP. It should also be noted that the minor metabolite kynurenic acid is a potent antagonist of excitatory N-methyl-D-aspartate receptors of brain neurons, while quinolinate is a potent agonist of these same receptors (Luthman, 2000). The precise balance of the alternative Trp degradation reactions in the brain is likely to be of great physiological importance and may be amenable to pharmacological modulation. In specific instances, genetic variants affecting Trp metabolism may promote the development of epileptic seizures and other pathologies.

Catabolism: Complete oxidation of Trp to L-alanine and two acetyl-CoA molecules proceeds in 11 steps, followed by one cycle of beta-oxidation (of crotonyl CoA), and depends on thiamin, riboflavin, vitamin B6, niacin, pantothenate, lipoate, ubiquinone, magnesium, and iron (Figure 8.41).

The heme-enzyme tryptophan 2,3-dioxygenase (EC1.13.11.11) opens the pyrrole ring of Trp, but also of D-tryptophan, 5-hydroxytryptophan, tryptamine, and 5-HT (serotonin). Indoleamine-pyrrole 2,3-dioxygenase (EC1.13.11.42), which does the same, has slightly different substrate specificity and tissue expression pattern. Alternatively, the PLP-dependent enzyme tryptophan aminotransferase (EC2.6.1.27) can deaminate Trp; the fate of the resulting indole 3-pyruvate is uncertain, but it may also undergo ring opening and rejoin the main catabolic pathway. Arylformamidase (EC3.5.1.9) generates kynurenine by hydrolyzing N-formylkynurenine.

Most kynurenine is then oxidized by the flavoenzyme kynurenine 3-monooxygenase (EC1.14.13.9). Two quantitatively minor alternative reactions also occur. One is the irreversible transamination, which is dependent on PLP, to kynurenic acid (4-(2-aminophenyl)-2,4-dioxobutanoate). Kynurenine-glyoxylate aminotransferase (EC2.6.1.63), aromatic-amino-acid-glyoxylate aminotransferase (EC2.6.1.60), or kynurenine-oxoglutarate aminotransferase (KAT, EC2.6.1.7) can catalyze this reaction. Two genetically distinct forms of the latter enzyme, KATI and KATII, occur in many tissues, including the liver, kidneys, and brain. Kynurenic acid is a potent inhibitor of all three known ionotropic excitatory amino acid receptors, including the NMDA receptor.

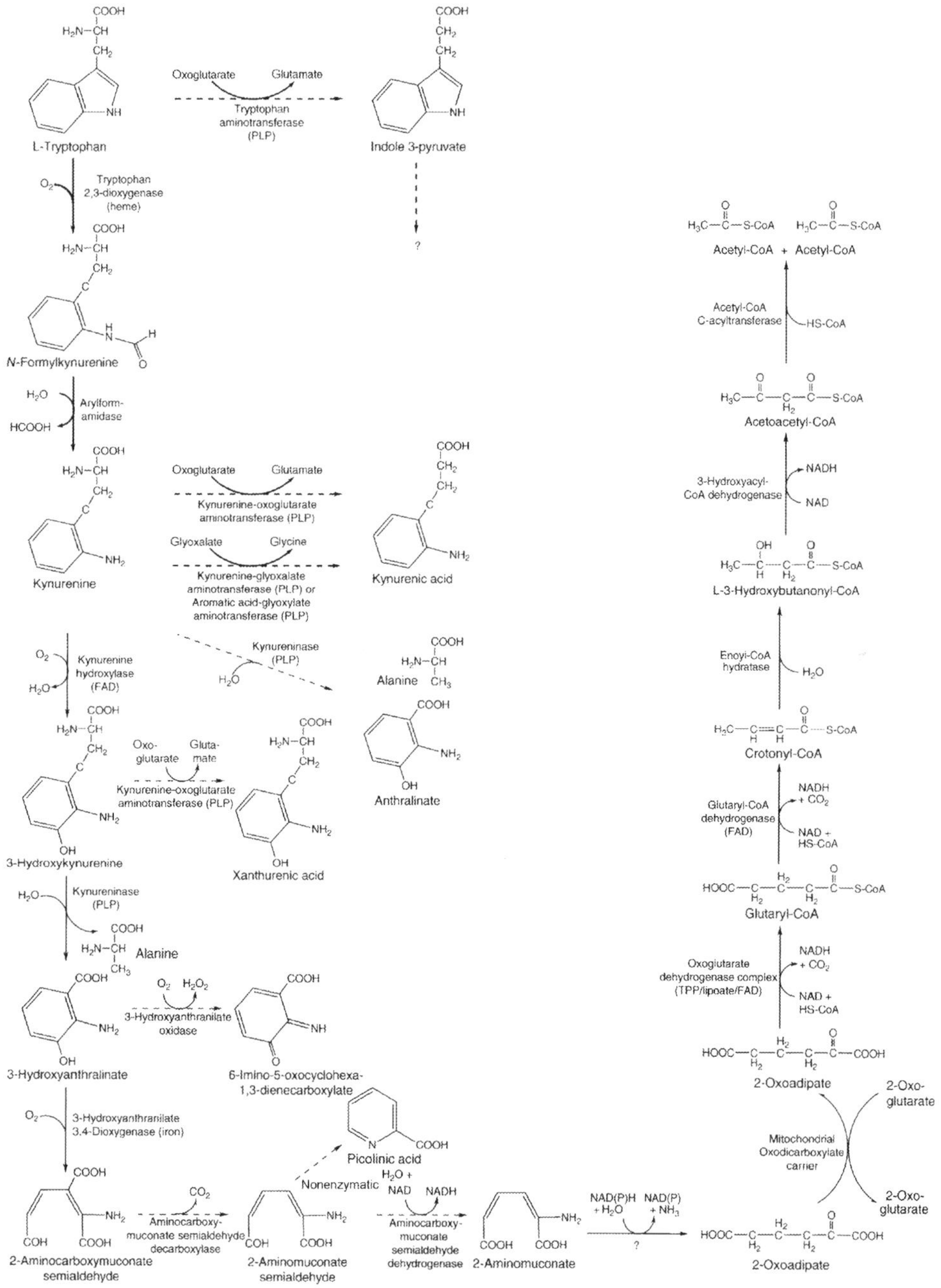

FIGURE 8.41

Pathways of L-tryptophan breakdown.

Another minor reaction is the premature cleavage by PLP-dependent kynureninase (EC3.7.1.3), which releases anthralinate and L-alanine. Both kynurenic acid and anthralinate are dead-end products that can be excreted after glucuronidation. The preferred reaction of kynureninase, however, is the hydrolysis of 3-hydroxy-kynurenine to 3-hydroxyanthralinate and L-alanine. The alternatively possible transamination of 3-hydroxy-kynurenine by kynurenine-oxoglutarate aminotransferase (EC2.6.1.7) to xanthurenic acid is insignificant so long as kynureninase activity is normal.

Oxidative cleavage of 3-hydroxy-anthralinate by the iron-enzyme 3-hydroxyanthranilate 3,4-dioxygenase (EC1.13.11.6) generates 2-amino-3-carboxymuconate semialdehyde. This is the branch point for nicotinate synthesis, since a small percentage of it spontaneously rearranges to the NAD precursor quinolinic acid. The bulk of 2-amino-3-carboxymuconate semialdehyde is decarboxylated by aminocarboxymuconate-semialdehyde decarboxylase (EC4.1.1.45). Some of the product, 2-amino-muconate semialdehyde (2-amino-3-(3-oxoprop-2-enyl)-but-2-enedioate), may rearrange nonenzymically to picolinate. Most of it, however, is oxidized to 2-aminomuconate by aminomuconate-semialdehyde dehydrogenase (EC1.2.1.32) and then reduced again to 2-oxoadipate by an as-yet-unknown enzyme.

In exchange for alpha-ketoglutarate, the oxodicarboxylate carrier (Fiermonte et al., 2001) transports 2-oxoadipate, an intermediate of lysine catabolism, across the inner mitochondrial membrane. Continued catabolism to acetyl-CoA takes place at the mitochondrial matrix. An enzyme complex whose identity is not entirely clear catalyzes the oxidative phosphorylation and conjugation to CoA. The oxoglutarate dehydrogenase (EC1.2.4.2) complex is able to facilitate the reaction (Bunik and Pavlova, 1997), but the existence of a closely related 2-oxoadipate dehydrogenase has not been ruled out. The alpha-ketoglutarate dehydrogenase complex contains thiamin pyrophosphate (bound to the E1 subunits), and lipoic acid (bound to the E2 subunits); a third type of subunit, dihydrolipoamide dehydrogenase (E3, EC1.8.1.4), then uses covalently bound FAD to reduce NAD. The resulting glutaryl-CoA is oxidized and decarboxylated by glutaryl-CoA dehydrogenase (EC1.3.99.7), which contains covalently bound FAD. Crotonyl-CoA completes beta-oxidation to 2 moles of acetyl-CoA. The three successive steps are catalyzed by mitochondrial enoyl-CoA hydratase (EC4.2.1.17), mitochondrial short-chain 3-hydroxyacyl-CoA dehydrogenase (EC1.1.1.35), and mitochondrial acetyl-CoA C-acyltransferase (3-ketoacyl-CoA thiolase, EC2.3.1.16).

Most 5-HT and some tryptamine are normally metabolized by monoamine oxidase A (MAOA, EC1.4.3.4) and one or more of the aldehyde dehydrogenases (ALDH, EC1.2.1.3/NAD-requiring, EC1.2.1.4/NADP-requiring, and EC1.2.1.5/NAD or NADP-requiring); 5-HT is converted into 5-hydroxyindole-3-acetate and tryptamine into indole-3-acetate. Both 5-HT and tryptamine can also be shunted into the main Trp catabolic pathway by tryptophan 2,3-dioxygenase (EC1.13.11.11, see previous discussion).

Some 5-HT can also be converted by alcohol dehydrogenase (EC1.1.1.1) into the dead-end product 5-hydroxytryptophol (5-HTOL), which has biological effects similar to 5-HT. Ethanol acutely increases the proportion converted into 5-HTOL. A single dose of ethanol (one liter of beer in a 70 kg man) ingested with several bananas was enough to cause headaches, diarrhea, and fatigue in healthy subjects (Helander and Some, 2000).

STORAGE

Body protein contains about 14 mg Trp/g (Smith, 1980). Hemoglobin is a moderately Trp-rich protein (3.8%).

EXCRETION

Due to very effective renal reabsorption of filtered Trp very little is lost with urine. Uptake from the proximal tubular lumen uses mainly the sodium-dependent system B° (Avissar et al., 2001), and is augmented by the action of the sodium-independent transporter complexes $b^{o,+}$ and BAT1-rBAT (Verrey et al., 1999). Impaired reabsorption in Hartnup disease, due to defective system $B^\circ AT1$ (SLC1A4) causes excessive renal losses.

Minor Trp metabolites in urine include 5-hydroxyindoleacetic acid, from serotonin breakdown, and 2-(alpha-mannopyranosyl)-1-tryptophan, from a pathway involving mannosylation of Trp (Gutsche et al., 1999).

REGULATION

High tissue concentration of Trp stabilizes tryptophan 2,3-dioxygenase (EC1.13.11.11), thereby promoting Trp breakdown. A carbohydrate-rich, protein-poor diet aids in the retention of Trp in the brain (Kaye et al., 2000). Enhanced tryptophan catabolism through selective upregulation of indoleaminepyrrole 2,3-dioxygenase (EC1.13.11.42) expression in trophoblasts and macrophages appears to suppress T-cell activity and contribute in a critical way to the immune tolerance of genetically different fetal tissues during pregnancy (Munn et al., 1998).

FUNCTION

Protein synthesis: Trp is a constituent of practically all proteins and peptides synthesized in the body. Tryptophan-tRNA ligase (EC6.1.1.2) loads Trp onto a specific tRNA in an ATP/magnesium-dependent reaction.

Energy fuel: Eventually, most Trp is completely oxidized providing about 5.8 kcal/g. The proportion that is converted to dead-end products such as kynurenic acid or used for the synthesis of products with lower energy yield is insignificant.

NAD synthesis: A small proportion of catabolized Trp (as well as tryptamine, 5-hydroxytryptophan, serotonin, and melatonin) gives rise to nicotinamide-containing compounds. Impaired absorption in Hartnup syndrome is associated with pellagra-like skin changes that are typical for niacin deficiency. The branch point is at the conversion of 2-amino-3-carboxymuconate semialdehyde. While most of this intermediate is enzymically decarboxylated, a small proportion undergoes spontaneous dehydration to quinolinic acid (pyridine-2,3-dicarboxylate). Nicotinate-nucleotide pyrophosphorylase (EC2.4.2.19) decarboxylates quinolinate and attaches a ribose phosphate moiety. Nicotinate-nucleotide adenylyltransferase (EC2.7.7.18) can then add adenosine phosphate. Amination by either of two NAD synthases (EC6.3.5.1 and EC6.3.1.5) then completes NAD synthesis. It is usually assumed that about one-sixtieth of the metabolized Trp goes to NAD synthesis (Horwitt et al., 1981).

Serotonin synthesis: The two-step synthesis of serotonin from Trp takes place in pinealocytes (in the pineal gland), raphe neurons of the brain, beta cells of the islets of Langerhans, enterochromaffin cells of the pancreas and small intestine, mast cells, and mononuclear leukocytes and requires biopterin and vitamin B6 (Figure 8.42).

First, tryptophan 5-monooxygenase (EC1.14.16.4) hydroxylates Trp in a tetrahydropteridine-dependent reaction. The enzyme in neurons of the cerebral cortex is activated by protein kinase-dependent protein 14-3-3. The PLP-dependent aromatic-L-amino-acid decarboxylase (EC4.1.1.28) then completes the synthesis of serotonin. Tryptophan 5-monooxygenase can also act directly on tryptamine (which may be ingested with cheeses or other fermented or spoiled foods) to synthesize serotonin. Serotonin is an important neurotransmitter in the brain (serotonergic system) and a hormone-like substance in other tissues. Several synaptic receptors initiate neuron depolarization upon binding serotonin. Modulation of serotonin reuptake is an important drug target for the treatment of mood disorders, as well as appetite control.

Various observations have suggested links between Trp intake and serotonin-mediated actions in the brain, at least in particularly susceptible individuals. Thus, men with aggressive histories may be more

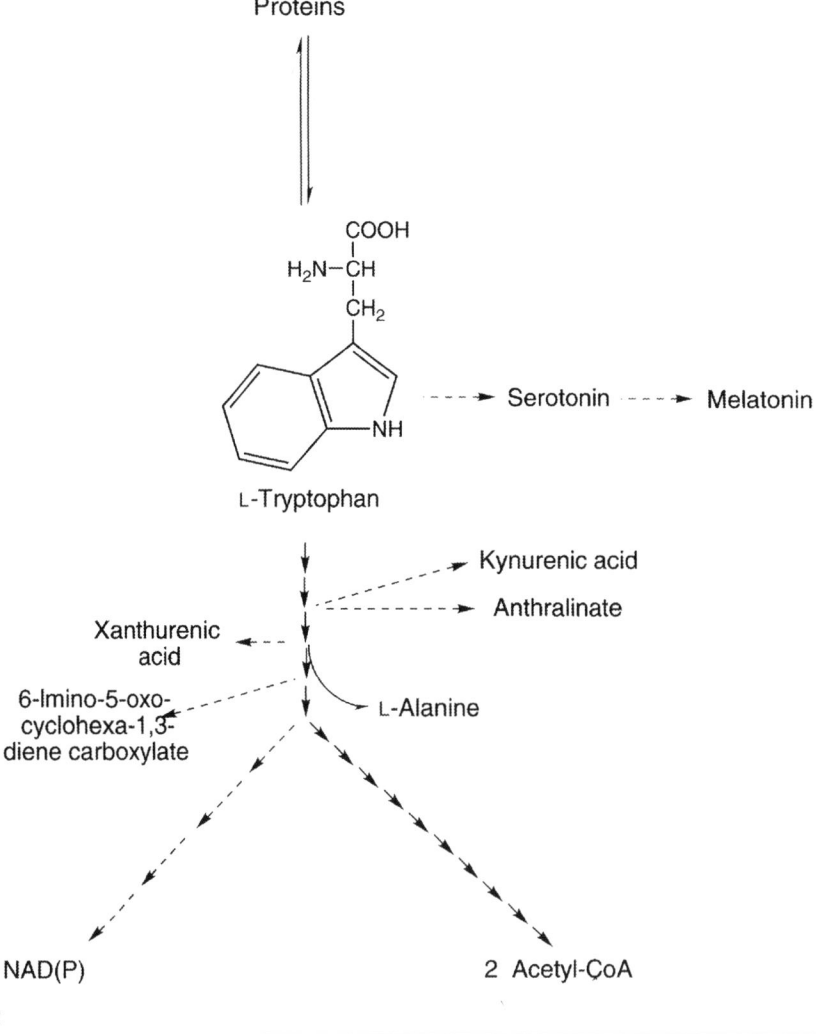

FIGURE 8.42

Uses of L-tryptophan.

prone to aggressive behavior during Trp depletion (Bjork et al., 2000). During Trp depletion, women with bulimia nervosa were more likely to experience depression, mood lability, sadness, and desire to binge than other women (Kaye et al., 2000).

Melatonin synthesis: Serotonin serves as the precursor for melatonin (*N*-acetyl-5-methoxytryptamine) synthesis in the pineal gland, other brain regions, the retina, and the small intestine (Zagajewski et al., 2012). Synthesis proceeds in two steps that depend on adequate supplies of methionine, folate, and vitamin B12. Serotonin enters pinealocytes through the Na^+- and Cl^--dependent transporter SLC6A4, whose activity appears to increase at night (Lima and Schmeer, 1994). Serotonin is then acetylated by aralkylamine *N*-acetyltransferase (EC2.3.1.87) and methylated by acetylserotonin *O*-methyltransferase (EC2.1.1.4).

Melatonin participates in the regulation of circadian (Hebert et al., 1999) and seasonal rhythms (Pevet, 2000), influences growth hormone and thyroid hormone status (Meeking et al., 1999), increases pigmentation (Iyengar, 2000), promotes immune function (Guerrero et al., 2000), and may contribute to free radical scavenging (Karbownik et al., 2000). Intake as a dietary supplement may help to promote sleep-cycle adjustment and reducing the impact of time-zone shifts (jet lag, Wyatt et al. 2006).

Photoprotection: Human lens epithelial cells produce 3-hydroxykynurenine glycoside. This pigment absorbs shortwave light and thereby appears to help protect the eye from ultraviolet (UV)–induced photodamage (Wood and Truscott, 1993).

REFERENCES

Anonymous, 1990. Use of ethyl esters of tryptophan to bypass the absorption defect in Hartnup disease. Nutr. Rev. 48, 22–24.

Avissar, N.E., Ryan, C.K., Ganapathy, V., Sax, H.C., 2001. Na^+-dependent neutral amino acid transporter ATB^0 is a rabbit epithelial cell brush border protein. Am. J. Physiol. Cell Physiol. 281, C963–C971.

Barth, H., Berg, P.A., Klein, R., 1999. Is there any relationship between eosinophilia myalgia syndrome (EMS) and fibromyalgia syndrome (FMS)? An analysis of clinical and immunological data. Adv. Exp. Med. Biol. 467, 487–496.

Birlouez-Aragon, I., Moreaux, V., Nicolas, M., Ducauze, C.J., 1997. Effect of iron and lactose supplementation of milk on the Maillard reaction and tryptophan content. Food Addit. Contam. 14, 381–388.

Bjork, J.M., Dougherty, D.M., Moeller, F.G., Swann, A.C., 2000. Differential behavioral effects of plasma tryptophan depletion and loading in aggressive and nonaggressive men. Neuropsychopharnlacology 22, 357–369.

Bruce, D.W., 1960. Serotonin in pineapple. Nature 188, 147.

Bunik, V.I., Pavlova, O.G., 1997. Inactivation of alpha-ketoglutarate dehydrogenase during its enzymatic reaction. Biochemistry (Moscow) 62, 973–982.

Dworschak, E., 1980. Nonenzyme browning and its effect on protein nutrition. Crit. Rev. Food Sci. Nutr. 13, 1–40.

Feldman, JM, Lee, EM., 1985. Serotonin content of foods: effect on urinary excretion of 5-hydroxyindoleacetic acid. Am. J. Clin. Nutr. 42, 639–643.

Fiermonte, G., Dolce, V., Palmieri, L., Ventura, M., Runswick, M.J., Palmieri, E., et al., 2001. Identification of the human mitochondrial oxodicarboxylate carrier. J. Biol. Chem. 276, 8225–8230.

Guerrero, J.M., Pozo, D., Garcia-Maurino, S., Carrillo, A., Osuna, C., Molinero, E., et al., 2000. Nuclear receptors are involved in the enhanced IL-6 production by melatonin in U937 cells. Biol. Signals Receptotw 9, 197–202.

Gutsche, B., Diem, S., Herderich, M., 1999. Electrospray ionization-tandem mass spectrometry for the analysis oftryptophan derivatives in food. Adv. Exp. Med. Biol. 467, 757–767.

Hebert, M., Martin, S.K., Eastman, C.I., 1999. Nocturnal melatonin secretion is not suppressed by light exposure behind the knee in humans. Netuvsci. Lett. 274, 127–130.

Helander, A., Some, M., 2000. Dietary serotonin and alcohol combined may provoke adverse physiological symptoms due to 5-hydroxytryptophol. Life Sci. 67, 799–806.

Horwitt, M.K., Harper, A.E., Henderson, L.M., 1981. Niacin-tryptophan relationships for evaluating niacin equivalents. Am. J. Clin. Nutr. 34, 423–427.

Iyengar, B., 2000. Melatonin and melanocyte functions. Biol. Signals Receptors 9, 260–266.

Karbownik, M., Tan, D.X., Reiter, R.J., 2000. Melatonin reduces the oxidation of nuclear DNA and membrane lipids induced by the carcinogen delta-aminolevulinic acid. Int. J. Cancer 88, 7–11.

Kaye, W.H., Gendall, K.A., Fernstrom, M.H., Femstrom, J.D., McConaha, C.W., Weltzin, T.E., 2000. Effects of acute tryptophan depletion on mood in bulimia nervosa. Biol. Psychiatr. 47, 151–157.

Klarskov, K., Johnson, K.L., Benson, L.M., Gleich, G.L., Naylor, S., 1999. Eosinophilia-myalgia syndrome case-associated contaminants in commercially available 5-hydroxytryptophan. Adv. Exp. Med. Biol. 467, 461–468.

Laforce, R., Rigozzi, K., Paganetti, M., Mossi, W., Guainazzi, E., Calderari, G., 1999. Aspects of melatonin manufacturing and requirements for a reliable active component. Biol. Signals Receptors 8, 143–146.

Lima, L., Schmeer, C., 1994. Characterization of serotonin transporter in goldfish retina by the binding of [3H] paroxetine and the uptake of [3H]serotonin: modulation by light. J. Neurochem. 62, 528–535.

Luthman, J., 2000. The kynurenine pathway of tryptophan degradation as a target for neuroprotective therapies. Amino Acids 19, 273–274.

Meeking, D.R., Wallace, J.D., Cuneo, R.C., Forsling, M., Russell-Jones, D.L., 1999. Exercise-induced GH secretion is enhanced by the oral ingestion of melatonin in healthy adult male subjects. Eur. J. Endocrinol. 141, 22–26.

Mizoguchi, K., Cha, S.H., Chairoungdua, A., Kim, D.K., Shigeta, Y., Matsuo, H., et al., 2001. Human cystinuria-related transporter: localization and functional characterization. Kidney Int. 59, 1821–1833.

Müller, B., Pacholski, C., Simat, T., Steinhart, H., 1999. Synthesis and formation of an EMS correlated contaminant in biotechnologically manufactured L-tryptophan. Adv. Exp. Med. Biol. 467, 481–486.

Munn, D.H., Zhou, M., Attwood, J.T., Bondarev, I., Conway, S.L., Marshall, B., et al., 1998. Prevention of allogeneic fetal rejection by tryptophan catabolism. Science 281, 1191–1193.

Naylor, S., Williamson, B.L., Johnson, K.L., Gleich, G.J., 1999. Structural characterization of case-associated contaminants peak C and FF in L-tryptophan implicated in eosinophiliamyalgia syndrome. Adv. Exp. Med. Biol. 467, 453–460.

Pevet, P., 2000. Melatonin and biological rhythms. Biol. Signals Receptors 9, 203–912.

Rajan, D.R., Kekuda, R., Huang, W., Devoe, L.D., Leibach, F.H., Prasad, P.D., et al., 2000. Cloning and functional characterization of a Na(+)-independent, broad-specific neutral amino acid transporter from mammalian intestine. Biochim. Biophys. Acta 1463, 6–14.

Ramadan, T., Camargo, S.M., Herzog, B., Bordin, M., Pos, K.M., Verrey, F., 2007. Recycling of aromatic amino acids via TAT1 allows efflux of neutral amino acids via LAT2-4F2hc exchanger. Pflügers Arch. 454, 507–516.

Ritchie, J.W., Taylor, P.M., 2001. Role of the System L permease LAT1 in amino acid and iodothyronine transport in placenta. Biochem. J. 356, 719–725.

Rossier, G., Meier, C., Bauch, C., Summa, V., Sordat, B., Verrey, F., et al., 1999. LAT2, a new basolateral 4F2hc/CD98-associated amino acid transporter of kidney and intestine. J. Biol. Chem. 274, 34948–34954.

Smith, R.H., 1980. Comparative amino acid requirements. Proc. Nutr. Soc. 39, 71–78.

Verrey, F., Jack, D.L., Paulsen, I.T., Saier Jr., M.H., Pfeiffer, R., 1999. New glycoprotein-associated amino acid transporters. J. Membrane Biol. 172, 181–192.

Wood, A.M., Truscott, R.J., 1993. UV filters in human lenses: tryptophan catabolism. Exp. Eye Res. 56, 317–325.

Wyatt, J.K., Dijk, D.J., Ritz-de Cecco, A., Ronda, J.M., Czeisler, C.A., 2006. Sleep-facilitating effect of exogenous melatonin in healthy young men and women is circadian-phase dependent. Sleep 29, 609–618.

Young, V.R., Borgonha, S., 2000. Nitrogen and amino acid requirements: the Massachusetts Institute of Technology amino acid requirement pattern. J. Nutr. 130, 1841S–1849S.

Zagajewski, J., Drozdowicz, D., Brzozowska, I., Hubalewska-Mazgaj, M., Stelmaszynska, T., Laidler, P.M., et al., 2012. Conversion L-tryptophan to melatonin in the gastrointestinal tract: the new high performance liquid chromatography method enabling simultaneous determination of six metabolites of L-tryptophan by native fluorescence and UV–Vis detection. J. Physiol. Pharmacol 63, 613–621.

Ziegenhagen, R., Boczek, P., Viell, B., 1999. Formation of the comutagenic beta-carboline norhannan in a simple tryptophan-containing model system at low temperature (40–80°C). Adv. Exp. Med. Biol. 467, 693–696.

METHIONINE

The polar neutral amino acid L-methionine (2-amino-4-(methylthio)butyric acid, alpha-amino-gamma-methylmercaptobutyric acid, 2-amino-4-methylthiobutanoic acid, gamma-methylthio-alpha-aminobutyric acid, one-letter code M; molecular weight 149) contains 9.4% nitrogen and 21.5% sulfur (one sulfhydryl group) (Figure 8.43).

ABBREVIATIONS

CssC	L-cystine
Cys	L-cysteine
Met	L-methionine
MTA	5′-methylthioadenosine
PLP	pyridoxal-5′-phosphate
RDA	recommended dietary allowance
SAM	S-adenosylmethionine

NUTRITIONAL SUMMARY

Function: The essential amino acid L-methionine (Met) is needed for the synthesis of proteins and is a precursor for L-cysteine (Cys). It has a special role as the precursor of SAM. SAM provides methyl groups for the synthesis of adrenaline (epinephrine), creatine, melatonin, phosphatidylcholine, carnitine, and numerous other essential compounds and is the precursor of essential polyamines. Met is also an energy fuel; its complete oxidation requires thiamin, riboflavin, niacin, vitamin B6, vitamin B12,

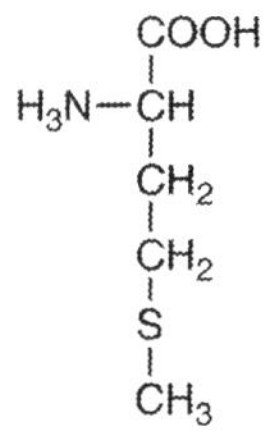

$$
\begin{array}{l}
\text{COOH} \\
| \\
\text{H}_3\text{N—CH} \\
| \\
\text{CH}_2 \\
| \\
\text{CH}_2 \\
| \\
\text{S} \\
| \\
\text{CH}_3
\end{array}
$$

FIGURE 8.43

L-Methionine.

pantothenate, biotin, lipoate, ubiquinone, iron, and magnesium; disposal of the sulfur in Met requires molybdenum.

Food sources: Plant-derived foods contain much less Met than meats and other animal-derived foods. Nonetheless, all foods are likely to provide adequate amounts so long as total protein intake meets recommendations. The body cannot produce additional amounts of Met from other amino acids.

Requirements: Combined daily Met and Cys intake of healthy adults should be at least 13 mg/kg.

Deficiency: Prolonged lack of Met, as with all essential amino acids or a lack of protein, causes growth failure, loss of muscle mass, and organ damage.

Excessive intake: Very high intake of protein and mixed amino acids (i.e., more than three times the RDA, which is 2.4 g/kg) is thought to increase the risk of renal glomerular sclerosis and accelerate osteoporosis. Consumption of Met greatly in excess of recommendations or high doses of SAM supplements may increase potentially damaging homocysteine concentrations and promote bone mineral loss. Long-term health risks from the consumption of such supplements have not been adequately evaluated.

DIETARY SOURCES

The Met content of proteins varies considerably depending on the food source. Foods with a particularly high percentage include eggs (31 mg/g protein), cod (30 mg/g), and chicken (28 mg/g). Intermediate content is in beef (26 mg/g), pork (26 mg/g), milk (25 mg/g), and rice (24 mg/g). Grains and other plant-derived protein sources tend to contain a lower percentage. Examples are corn (21 mg/g), wheat and oats (18 mg/g), rye and beans (15 mg/g), and cauliflower (14 mg/g). Cooking foods at high temperatures (browning) can decrease Met bioavailability due to oxidation (Dworschak, 1980). Since Met cannot be synthesized in the body, adequate amounts have to be supplied.

Met and Cys are closely linked metabolically, and therefore, recommendations are often given for the sum of both SAAs. Healthy adults should get at least 13 mg/kg per day in combination (Raguso et al., 2000). Adequate Cys intake minimizes Met requirements (Di Buono et al., 2001), but high Cys intake has no additional sparing effect beyond that in healthy people (Raguso et al., 2000).

Average daily Met intake in elderly nonvegetarian Americans was estimated at 1450 mg, and that of ovo-lacto-vegetarians at 770 mg (Sachan et al., 1997).

DIGESTION AND ABSORPTION

Proteases and aminopeptidases from the stomach, pancreas, and small-intestinal wall digest Met-containing proteins. Several small intestinal brush border membrane enzymes act on small peptides. Met-containing dipeptides and tripeptides are taken up via the hydrogen ion/peptide cotransporter 1 (SLC15A1) and, to a lesser extent, the hydrogen ion/peptide cotransporter 2 (SLC15A2). Free Met is taken up from the proximal small intestine by the sodium-dependent transporters B° (ASCT2, SLC1A5) and ASC, but the evidence for the latter is controversial (Chen et al., 1994; Munck et al., 2000). In addition, the glycoprotein-anchored transport system b$^{o,+}$ (consisting of BAT1/SLC7A9 and rBAT/SLC3A1) can take up Met in exchange for a neutral amino acid (Wagner et al., 2001). Sodium-coupled amino acid transport system A (ATA1, SLC38A1) moves Met across the basolateral membrane toward the pericapillary space when the intracellular concentration is high after a meal. Between meals, the flux tends to reverse, and Met from the blood is supplied for the enterocyte's own great need. The heterodimeric transporter LAT2 + 4F2 (SLC7A8 + SLC3A2) augments Met transport in either direction through exchange for another neutral amino acid (Figure 8.44).

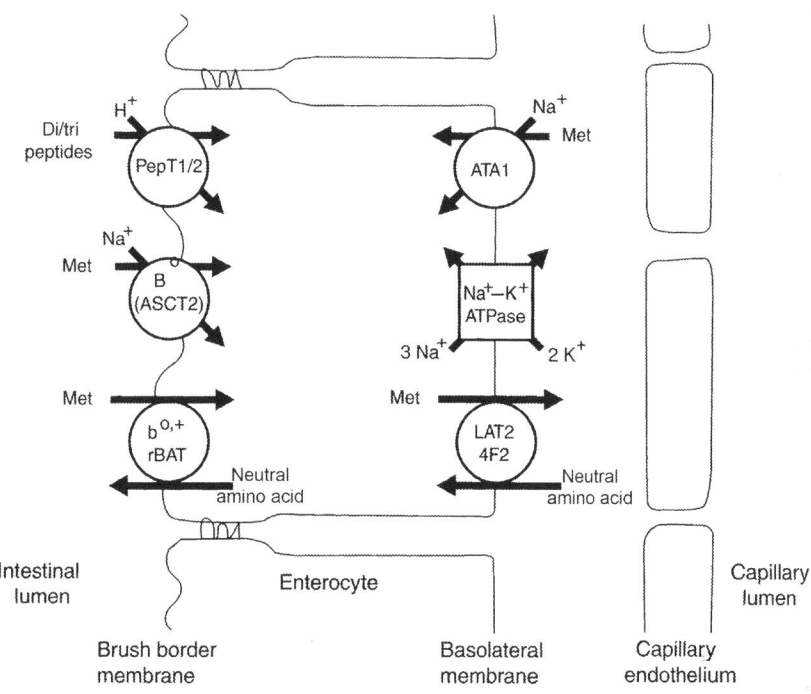

FIGURE 8.44

Intestinal absorption of ʟ-methionine.

TRANSPORT AND CELLULAR UPTAKE

Blood circulation: Plasma concentration of free Met is about 25 μmol/l. ATA2 (Hatanaka et al., 2000) mediates the sodium-driven uptake of Met into skeletal and heart muscle and other tissues. ATA3 (Hatanaka et al., 2001) is important for uptake into the liver.

  BBB: The heteroexchanger LAT1, and to a lesser extent LAT2, at both sides of the brain capillary cell epithelial cell (Duelli et al., 2000) move Met across the BBB (Killian and Chikhale, 2001). The sodium-dependent system A is a minor contributor to Met transport (Kitazawa et al., 2001).

  Materno-fetal transfer: Transport of Met from maternal circulation into the syntrophoblast uses mainly the sodium-dependent ATA2 (SLC38A2; Cramer et al., 2002). Members of system A then mediate Met movement along its concentration gradient to the fetal side (Jansson, 2001).

METABOLISM

About half of all Met metabolism takes place in the liver (Mato et al., 2002). The main pathway for Met breakdown is transsulfuration to Cys (Figure 8.45). Keep in mind that the first part of this pathway is partially cyclical because half or more of the generated homocysteine reverts to Met. Decarboxylation of the central metabolite SAM by SAM decarboxylase (EC4.1.1.50, contains pyruvate) and use for the synthesis of spermidine and spermine accounts for less than 10% of total turnover (Mato et al., 2002).

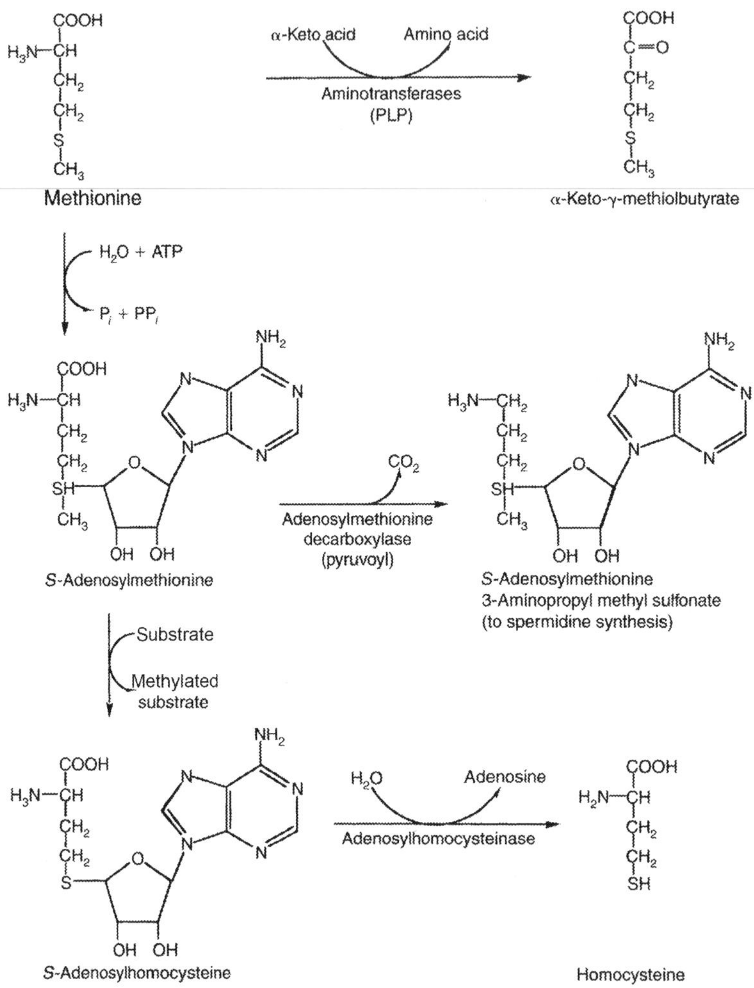

FIGURE 8.45

Metabolic fates of L-methionine.

Transamination of Met to α-keto-γ-methiolbutyrate is quantitatively of little significance (Raguso et al., 2000). Met can act as a donor in reactions catalyzed by cytoplasmic and mitochondrial glutamine-pyruvate aminotransferase (EC2.6.1.15) and, to a minor extent, phenylalanine (histidine) aminotransferase (EC2.6.1.58).

The SAM cycle: Met is activated by ATP:L-methionine-S-adenosyltransferase (methionine adenosyltransferase, MAT, EC2.5.1.6). Isoforms I and III are expressed in the liver, and isoform II is present in the kidneys, brain, and other tissues. Hydrolysis of pyrophosphate from the adenosylation reaction by ubiquitous pyrophosphatase (EC3.6.1.1) keeps the equilibrium in favor of SAM synthesis. Many methyltransferases, of which glycine *N*-methyltransferase (EC2.1.1.20) is the most abundant in the

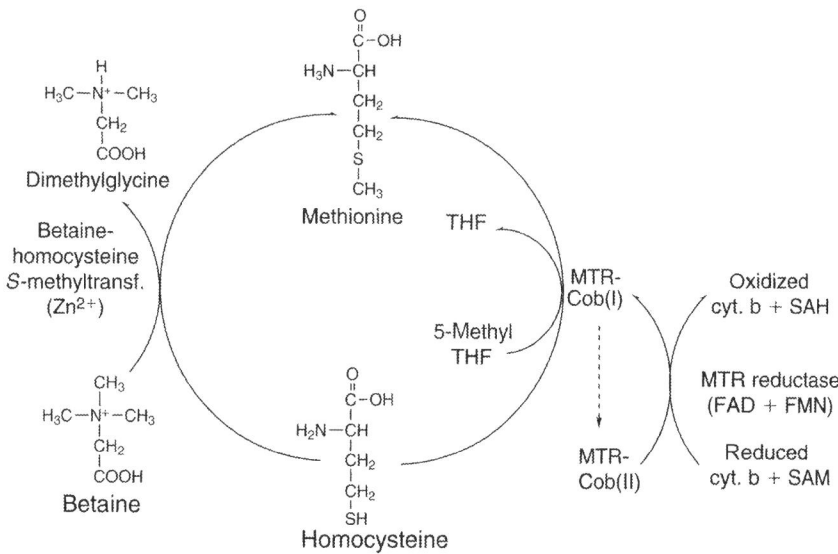

FIGURE 8.46

Homocysteine remethylation.

liver, use SAM as a methyl group donor. Adenosylhomocysteinase (EC3.3.1.1) hydrolyzes the resulting adenosylhomocysteine. Here, 5-methyl-tetrahydrofolate-homocysteine methyltransferase (MTR, methionine synthase, EC2.1.1.13) provides the main pathway for the regeneration of Met (Figure 8.46). Cob(I)alamin in the active center of the enzyme receives the methyl group from 5-methyl-tetrahydrofolate (5-methyl-THF) and moves it to homocysteine. Oxidation of the enzyme-bound cob(I)alamin without methyl group transfer, which occurs in a small percentage of reaction cycles, and possibly with high oxidative stress (McCaddon et al., 2002), inactivates MTR. In this case, the enzyme has to be remethylated by methionine synthase reductase (EC2.1.1.135), which uses both FAD and FMN as cofactors (Leclerc et al., 1998). SAM serves as the methyl-group donor, and cytochrome b5 provides the reducing potential. Cytochrome b5 is regenerated by NADPH-dependent cytochrome P450 reductase (EC1.6.2.4), another enzyme that has both FAD and FMN as prosthetic groups. Inhalation of nitrous oxide irreversibly inactivates MTR (Home et al., 1989; Riedel et al., 1999). Low MTR activity causes the accumulation of 5-methyl-THF because there are no other major uses for this folate metabolite.

The liver and kidneys have an additional route for Met remethylation that uses the zinc-enzyme betaine-homocysteine S-methyltransferase (EC2.1.1.5). The choline metabolite betaine can regenerate Met directly without requiring folate or vitamin B12 like the main remethylation pathway. However, the metabolism of the dimethylglycine arising from this reaction generates 3 mol of 5-methyl-THF, and this causes an even greater sequestration (trapping) and potential functional deficiency of folate.

Transsulfuration: The successive actions of cystathionine beta-synthase (EC4.2.1.22) and cystathionine-gamma-lyase (EC4.4.1.1) convert homocysteine to Cys and alpha-ketobutyrate (Figure 8.47). Both enzymes contain pyridoxal phosphate as a prosthetic group. The main fate of Cys is metabolism to pyruvate via cysteinesulfinate, CssC, mercaptopyruvate, or directly. Much smaller amounts of Cys are

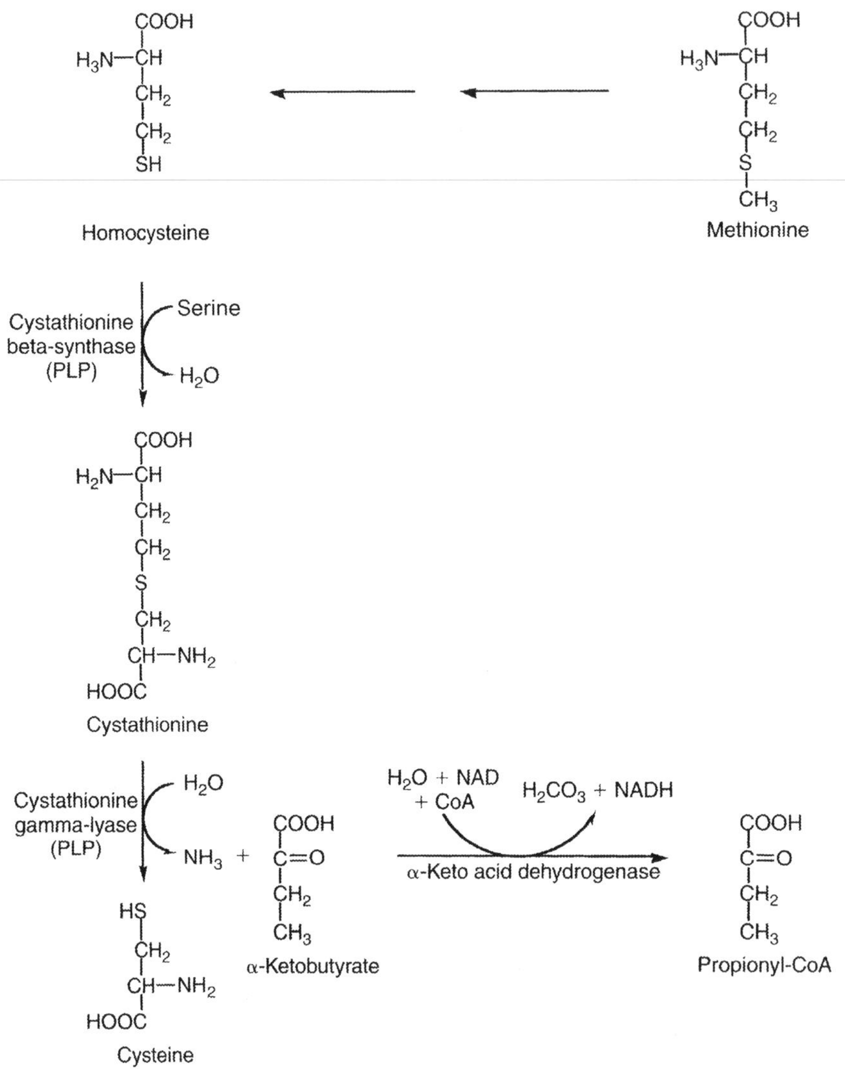

FIGURE 8.47

The transsulfuration pathway.

converted to taurine, L-alanine, and cysteamine. Detoxification of the sulfite produced by Cys oxidation depends on the molybdenum enzyme sulfite oxidase (EC1.8.3.1).

Repair of oxidative damage: The sulfur in Met is highly susceptible to oxidation. Therefore, exposure of proteins to molecular oxygen or other reactive oxygen species, particularly in the mitochondrial matrix, can result in their inactivation. A thioredoxin-using enzyme, protein-methionine-*S*-oxide reductase (EC1.8.4.6), can repair such oxidative damage in proteins, but it does not work on free Met.

STORAGE

Body proteins on average contain about 120 mg/g (Raguso et al., 2000). Small amounts of Met (2.1 mg/kg body weight) are released with normal turnover of tissue proteins. Very low intake may promote breakdown of protein from muscle and other tissues.

EXCRETION

Virtually no Met is lost via feces or urine in healthy people because of the high efficiency of the absorption systems. More than 600 mg of Met is filtered by the kidneys each day. Met is quantitatively reabsorbed from the proximal tubular lumen by amino acid transporter B° and returned across the basolateral membrane into the blood by amino acid transport system A. Transport is augmented by the heterodimeric transport system $b^{0,+}$ (consisting of BAT1 and rBAT) at the luminal side, and LAT2/4F2 at the basolateral side (Wagner et al., 2001).

REGULATION

The main factors affecting Met homeostasis are the utilization of SAM as a methyl group donor, the remethylation of homocysteine, and the rate of conversion to cysteine (Mato et al., 2002). Glycine N-methyltransferase is the major SAM-utilizing enzyme in the liver that helps to metabolize excess Met. An inborn lack of the enzyme is associated with elevated Met concentration in the blood.

FUNCTION

Protein synthesis: Met is a constituent of practically all proteins and peptides synthesized in the body. Methionine-tRNA ligase (EC6.1.1.10) loads Met onto a specific tRNA in an ATP/magnesium-dependent reaction. About 2.1 mg/kg body weight are incorporated into body proteins per day (Raguso et al., 2000).

Energy fuel: Most (90% or more) Met is eventually metabolized to carbon dioxide, water, and urea. The complete oxidation of Met (via cysteine and propionyl-CoA) requires adequate availability of thiamin, riboflavin, niacin, vitamin B6, vitamin B12, pantothenate, biotin, lipoate, ubiquinone, iron, and magnesium; disposal of the sulfur in Met requires molybdenum. Met provides 5.3 kcal/g (May and Hill, 1990).

Methyl-group transfer: ATP:L-methionine-S-adenosyltransferase (EC2.5.1.6) generates the essential methyl group donor SAM. Genetically distinct isoforms of this enzyme exist, which are expressed in a tissue-specific manner.

A large number of enzyme-catalyzed reactions depend on adequate supplies of SAM. Important examples are the synthesis of phosphatidylcholine, carnitine, catecholamines, and melatonine. SAM-dependent creatine synthesis constitutes a very significant drain on methyl group donors, drawing about 70% of the available pool (Wyss and Kaddurah-Daouk, 2000). Methylation silences the expression of specific segments of DNA and regulates gene function. Inadequate SAM availability may impair proper DNA methylation, thereby increasing the risk of some cancers and other diseases.

However, adequacy of the nutrients enabling methionine remethylation (folate, vitamin B12) is likely to be more important than Met availability. Indeed, increased risk of cancer of the colon (Giovannucci et al., 1993), stomach (La Vecchia et al., 1997), or other sites may be associated with high Met intake.

Homocysteine: Use of SAM as a methyl donor generates homocysteine. When remethylation does not keep pace with production, homocysteine will spill over into extracellular fluid and blood circulation. Homocysteine is a potent oxidant since it circulates in the blood almost exclusively as homocystine, homocysteine-cysteine mixed disulfide, and protein-bound disulfides. Less than 2% is present in the thiol form (Lentz, 2002). Elevated blood homocysteine concentrations and increased homocysteine excretion are associated with increased cardiovascular risk (atherosclerosis, thrombosis, and myocardial infarction) and accelerated cognitive decline.

Cysteine synthesis: The transsulfuration pathway contributes a significant percentage to the body's Cys input. When cysteine intake is low, a greater percentage of the available Met is metabolized via the

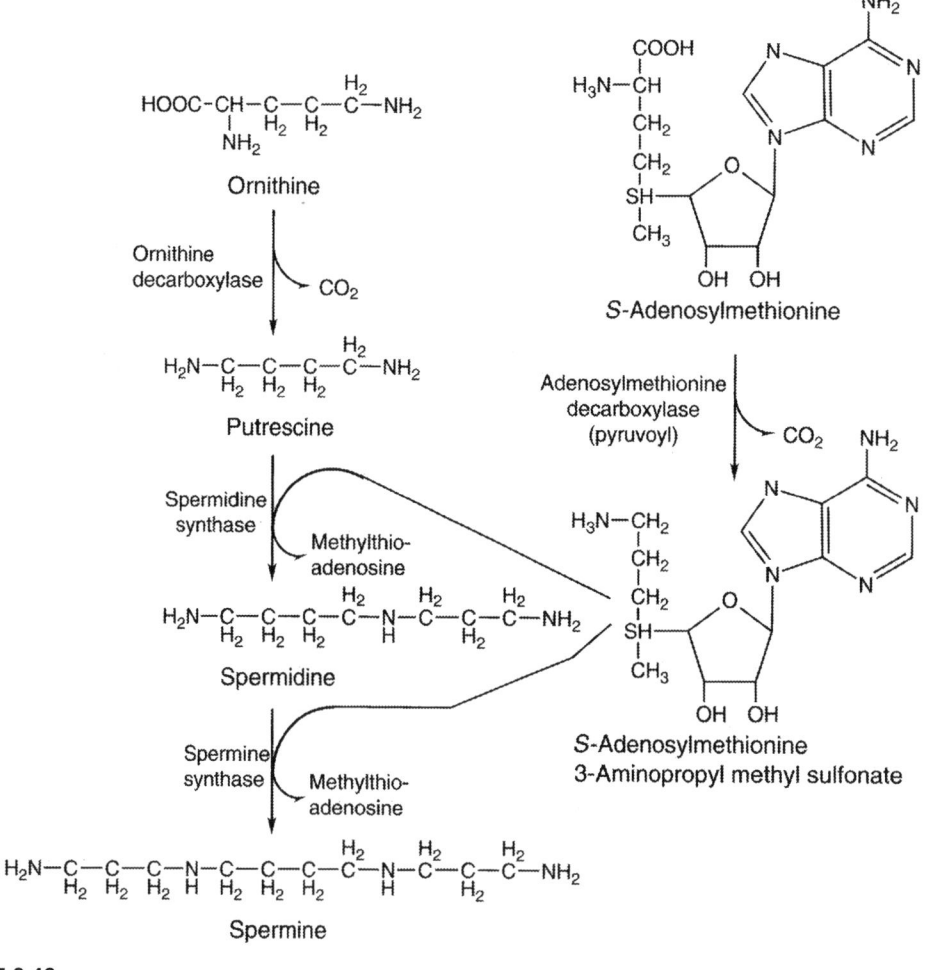

FIGURE 8.48

SAM is a precursor for polyamine synthesis.

transsulfuration pathway. Adequate Cys intake minimizes this draw on Met supplies (Di Buono et al., 2001). Cys is also an important precursor for the synthesis of glutathione, CoA, taurine, sulfate (for phosphoadenosyl phosphosulfate synthesis), and reactive sulfur compounds.

Polyamine synthesis: Decarboxylation of SAM by adenosylmethionine decarboxylase (EC4.1.1.50) generates methyl-*S*-adenosylthiopropylamine (Figure 8.48). Spermidine synthase (EC2.5.1.16) generates spermidine by transferring the aminopropyl moiety of this SAM metabolite to putrescine (decarboxylation product of ornithine). Spermine synthase (EC2.5.1.22) is a different enzyme that can add another aminopropyl group to spermine. The polycations spermidine and spermine are essential growth factors for all cells, but it is not yet known how they act. The residual metabolite 5'-MTA itself has distinct biological properties, such as promotion of apoptosis in abnormal (i.e., transformed) cells (Ansorena et al., 2002). Both 5'-methyl-thioadenosine phosphorylase (EC2.4.2.28) and adenosylhomocysteinase (EC3.3.1.1) can salvage the nucleoside moiety of MTA (Smolenski et al., 1992).

Acidity: A large portion of titratable acid in urine comes from the production of sulfuric acid from Met and Cys. High intake of sulfur-containing amino acids drains amino groups (used for neutralization in the kidneys) and may accelerate bone mineral loss (Marsh et al., 1988).

REFERENCES

Ansorena, E., Garcia-Trevijano, E.R., Martinez-Chantar, M.L., Huang, Z.Z., Chen, L., Mato, J.M., et al., 2002. *S*-adenosylmethionine and methylthioadenosine are antiapoptotic in cultured rat hepatocytes but proapoptotic in human hepatoma cells. Hepatology 35, 274–280.

Chen, J., Zhu, Y., Hu, M., 1994. Mechanisms and kinetics of uptake and efflux of L-methionine in an intestinal epithelial model (Caco-2). J. Nutr. 124, 1907–1916.

Cramer, S., Beveridge, M., Kilberg, M., Novak, D., 2002. Physiological importance of system A-mediated amino acid transport to rat fetal development. Am. J. Physiol. Cell Physiol. 282, C153–C160.

Di Buono, M., Wykes, L.J., Ball, R.O., Pencharz, P.B., 2001. Dietary cysteine reduces the methionine requirement in men. Ant. J. Clin. Nutr. 74, 761–766.

Duelli, R., Enerson, B.E., Gerhart, D.Z., Drewes, L.R., 2000. Expression of large amino acid transporter LAT1 in rat brain endothelium. J. Cereb. Blood Flow Metab. 20, 1557–1562.

Dworschak, E., 1980. Nonenzyme browning and its effect on protein nutrition. Crit. Rev. Food Sci. Nutr. 13, 1–40.

Giovannucci, E., Stampfer, M.L., Colditz, G.A., Rimm, E.B., Trichopoulos, D., Rosner, B.A., et al., 1993. Folate, methionine, and alcohol intake and risk of colorectal adenoma. J. Natl. Cancer Inst. 85, 875–884.

Hatanaka, T., Huang, W., Ling, R.H., Prasad, P.D., Sugawara, M., Leibach, F.H., et al., 2001. Evidence for the transport of neutral as well as cationic amino acids by ATA3, a novel and liver-specific subtype of human ATA2, a subtype of amino acid transport system A. Biochim. Biophys. Acta 1510, 10–17.

Hatanaka, T., Huang, W., Wang, H., Sugawara, M., Prasad, P.D., Leibach, F.H., et al., 2000. Primary structure, functional characteristics and tissue expression pattern of human ATA2, a subtype of amino acid transport system A. Biochim. Biophys. Acta 1467, 1–6.

Home, D.W., Patterson, D., Cook, R.J., 1989. Effect of nitrous oxide inactivation of vitamin B12-dependent methionine synthetase on the subcellular distribution of folate coenzymes in rat liver. Arch. Biochem. Biophys. 270, 729–733.

Jansson, T., 2001. Amino acid transporters in the human placenta. Pediatr. Res. 49, 141–147.

Killian, D.M., Chikhale, P.J., 2001. Predominant functional activity of the large, neutral amino acid transporter (EAT1) isoform at the cerebrovasculature. Neurosci. Lett. 306, 1–4.

Kitazawa, T., Hosya, K., Watanabe, M., Takashima, T., Ohtsuki, S., Takanaga, H., et al., 2001. Characterization of the amino acid transport of new immortalized choroid plexus epithelial cell lines: a novel *in vivo* system for investigating transport functions at the blood–cerebrospinal fluid barrier. Pharmaceut. Res. 18, 16–22.

La Vecchia, C., Negri, E., Franceschi, S., Decarli, A., 1997. Case–control study on influence of methionine, nitrite, and salt on gastric carcinogenesis in northern Italy. Nutr. Cancer 27, 65–68.

Leclerc, D., Wilson, A., Dumas, R., Gafuik, C., Song, D., Watkins, D., et al., 1998. Cloning and mapping of a cDNA for methionine synthase reductase, a flavoprotein defective in patients with homocystinuria. Proc. Natl. Acad. Sci. USA 95, 3059–3064.

Lentz, S.R., 2002. Does homocysteine promote atherosclerosis? Arterioscl. Thromb. Vasc. Biol. 21, 1385–1386.

Marsh, A.G., Sanchez, T.V., Michelsen, O., Chaffee, F.L., Fagal, S.M., 1988. Vegetarian lifestyle and bone mineral density. Am. J. Clin. Nutr. 48, 837–841.

Mato, J.M., Corrales, F.J., Lu, S.C., Avila, M.A., 2002. S-Adenosylmethionine: a control switch that regulates liver function. FASEB J. 16, 15–26.

May, M.E., Hill, J.O., 1990. Energy content of diets of variable amino acid composition. Am. J. Clin. Nutr. 52, 770–776.

McCaddon, A., Regland, B., Hudson, P., Davies, G., 2002. Functional vitamin B(12) deficiency and Alzheimer disease. Neurology 58, 1395–1399.

Munck, L.K., Grondahl, M.L., Thorboll, J.E., Skadhauge, E., Munck, B.G., 2000. Transport of neutral, cationic and anionic amino acids by systems B, b(o,+), X(AG), and ASC in swine small intestine. Comp. Biochem. Physiol. A Mol. Integ. Physiol. 126, 527–537.

Raguso, C.A., Regan, M.M., Young, V.R., 2000. Cysteine kinetics and oxidation at different intakes of methionine and cystine in young adults. Am. J. Clin. Nutr. 71, 491–499.

Riedel, B., Fiskerstrand, T., Refsum, H., Ueland, P.M., 1999. Co-ordinate variations in methylmalonyl-CoA mutase and methionine synthase, and the cobalamin cofactors in human glioma cells during nitrous oxide exposure and the subsequent recovery phase. Biochem. J. 341, 133–138.

Sachan, D.S., Daily III, J.W., Munroe, S.G., Beauchene, R.E., 1997. Vegetarian elderly women may risk compromised carnitine status. Veg. Nutr. 1, 64–69.

Smolenski, R.T., Fabianowska-Majewska, K., Montero, C., Duley, J.A., Fairbanks, L.D., Marlewski, M., et al., 1992. A novel route of ATP synthesis. Biochem. Pharmacol. 43, 2053–2057.

Wagner, C.A., Lang, E., Bröer, S., 2001. Function and structure of heterodimeric amino acid transporters. Am. J. Physiol. Cell Physiol. 281, C1077–C1093.

Wyss, M., Kaddurah-Daouk, R., 2000. Creatine and creatinine metabolism. Physiol. Rev., 1107–1213.

CYSTEINE

L-Cysteine (beta-mercaptoalanine, 2-amino-3-mercaptopropanoic acid, 2-amino-3- mercaptopropionic acid, alpha-amino-beta-thiolpropionic acid, one-letter code C; molecular weight 121) contains 11.6% nitrogen and 26.5% sulfur (Figure 8.49).

ABBREVIATIONS

CssC	L-cystine (oxidized L-cysteine)
Cys	L-cysteine
Met	L-methionine
RDA	recommended dietary allowance

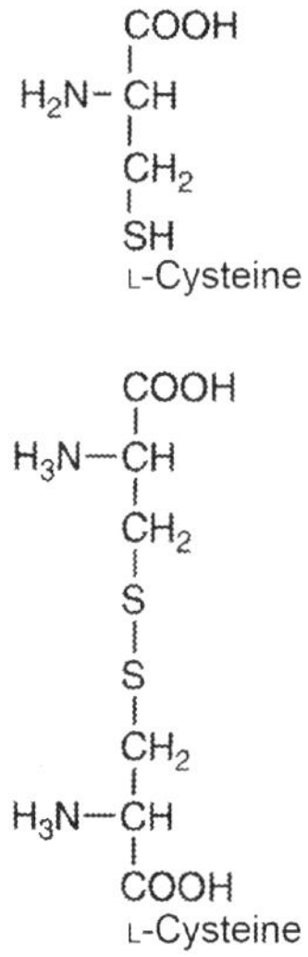

FIGURE 8.49

L-Cysteine and L-cystine.

NUTRITIONAL SUMMARY

Function: L-Cysteine (Cys) is a component of proteins and provides the sulfur in iron sulfur proteins, pantothenate, tRNA, proteoglycans, and many other compounds. It is a precursor of glutathione (anti-oxidation) and taurine (conjugation, membranes, and vision). Cys is also an energy fuel; its complete oxidation requires thiamin, riboflavin, niacin, vitamin B6, pantothenate, lipoate, ubiquinone, iron, zinc, and magnesium; disposal of the sulfur in Cys requires molybdenum.

Food sources: Plant-derived foods contain much less L-methionine than meats and other animal-derived foods. Nonetheless, all foods are likely to provide adequate amounts so long as total protein intake meets recommendations.

Requirements: The combined requirement for L-methionine (Met) and Cys is 845 mg/d (FAO/WHO/UNU, 1985). Very young infants and severely ill adults need some dietary Cys because they may not be able to generate enough Cys from Met.

Deficiency: Prolonged lack of both Cys and Met causes growth failure, loss of muscle mass, and organ damage.

Excessive intake: Very high intake of protein and mixed amino acids (i.e., more than three times the RDA, which is 2.4 g/kg) is thought to increase the risk of renal glomerular sclerosis and accelerate osteoporosis. Consumption of Cys greatly in excess of recommendations generates large amounts of sulfite, which is not well tolerated by sensitive individuals. Long-term health risks from the consumption of high-dose supplements have not been adequately evaluated.

ENDOGENOUS SOURCES

About half of all available Cys is metabolically derived from the Met metabolite L-homocysteine. The PLP-dependent enzyme cystathionine beta-synthase (EC4.2.1.22) condenses homocysteine and L-serine to cystathionine, which is then cleaved by another PLP-dependent enzyme, [namely, cystathionine-gamma-lyase (EC4.4.1.1)], to release Cys, ammonia, and 2-oxobutanoate (transsulfuration pathway).

DIETARY SOURCES

Foods contain Cys and cystine (CssC) as part of their proteins, which makes total protein consumption the main determinant of Cys intake. The proteins in eggs, human milk, and grains (e.g., wheat, rye, and rice) contain about 20 mg/g; corn contains less than half as much. Meats and milk also contain only 9–13 mg/g. Heat treatment of foods slightly decreases the amount of Cys that can be absorbed and utilized (Dworschak, 1980). The currently recommended dietary intake for sulfur-containing amino acids (Cys, CssC, and methionine) is 13 mg/kg per day (Raguso et al., 2000), but actual requirements may be considerably higher (Di Buono et al., 2001). Adequate Cys intake can reduce Met requirements by about 8 mg/kg (Di Buono et al., 2001).

DIGESTION AND ABSORPTION

Healthy individuals absorb amino acids and proteins in the proximal small intestine nearly completely. Food proteins are hydrolyzed by an array of gastric, pancreatic, and enteral enzymes, which generate Cys and cystine as part of oligopeptides and in free form. The former can be taken up through the hydrogen ion/peptide cotransporter (PepT1, SLC15A1). The main port of entry for free Cys is the sodium-driven transporter B° (Avissar et al., 2001). Smaller amounts enter with the sodium cotransporter ASC. One of the transporters comprising the ASC family, ASCT2 (SLC1A5), is expressed in the colon, suggesting a capacity for Cys uptake beyond the small intestine (Utsunomiya-Tate et al., 1996).

Both Cys and CssC are shuttled across the brush border membrane in either direction in exchange for neutral amino acids by the amino acid transporter BAT1/b$^{o,+}$ (SLC7A9); this transporter is anchored to the membrane and complemented by the glycoprotein subunit rBAT (SLC3A1) (Chairoungdua et al., 1999).

System X$\overline{C}$ is a sodium-independent transporter that consists of 4F2 and xCT (SLC7A11) and facilitates the exchange of cystine for glutamate (Burdo et al., 2006). Once the cysteine has crossed the luminal membrane, it can be reduced to cysteine as needed.

Significant amounts of Cys, both from the intestinal lumen and from blood circulation, are used for the enterocyte's own needs. The remainder is transported across the basolateral membrane mainly by the sodium/amino acid cotransporters A and ASC and the sodium-independent transporter asc. The sodium-independent transporter LAT2 (SLC7A8) can exchange Cys and CssC for other neutral amino acids in either direction. Like the heteroexchanger in the brush border membrane, LAT2 is anchored to the basolateral membrane by a complementary glycoprotein, 4F2 (SLC3A2) in this case (Rajan et al., 2000) (Figure 8.50).

TRANSPORT AND CELLULAR UPTAKE

Blood circulation: Only a small proportion circulates as the reduced free amino acid (Cys, about 12 μmol/l), more is in the oxidized form (CssC, about 70 μmol/l), and most (about 210 μmol/l) is

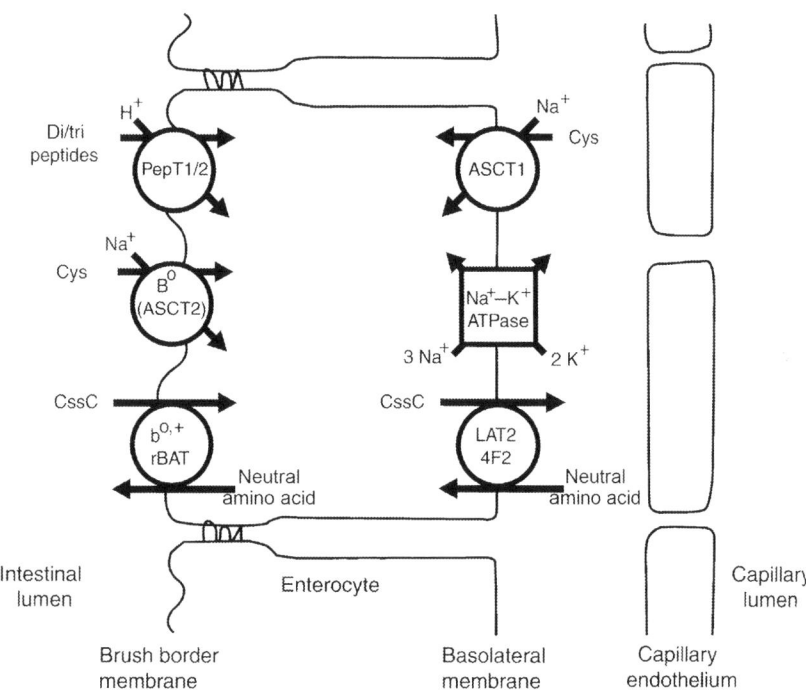

FIGURE 8.50

Intestinal absorption of L-cysteine and L-cystine.

bound in proteins and peptides (Wlodek et al., 2001). An important Cys-containing transport peptide is glutathione (gamma-glutamyl-cysteinylglycine), originating mainly from the liver. The exoenzyme gamma-glutamyl transpeptidase (γ-GT, EC2.3.2.2) can cleave glutathione in extracellular fluid. Various dipeptidases can then cleave the resulting cysteinyl-glycine again and release Cys.

The kidneys have particularly high γ-GT activity and utilize a large proportion of circulating glutathione. Free Cys can be taken up by cells via several transporters. Sodium-dependent transport occurs with systems ASC (specifically ASCT1/SLC1A4) and A. ATA2 is ubiquitously expressed (Sugawara et al., 2000), ATA3 is restricted to the liver (Hatanaka et al., 2001). The glycoprotein-linked heteroexchangers include LAT1 (SLC7A5), LAT2 (SLC7A8), and system $\overline{XC}$ (consisting of transporter xCT, SLC7A11, and the 4F2 cell surface antigen heavy-chain, SLC3A2). System $\overline{XC}$ is the major conduit for CssC uptake into macrophages (Sato et al., 2001) and glial cells (Kim et al., 2001), which have a great need for synthesis of the antioxidant tripeptide glutathione (Sato et al., 2001).

Hydrolysis of proteins with disulfide bridges releases CssC; this is also the predominant form in circulation. Glutathione-cystine transhydrogenase (EC1.8.4.4) uses the reducing equivalents of glutathione in the reduction of CssC to Cys. Cystine reductase (EC1.6.4.1) accomplishes the same using NADH. Much less is known about intracellular transport. Cystinosin is a transmembrane protein involved in the export of CssC from lysosomes.

Note: Cystine is much less soluble in aqueous media than Cys and tends to precipitate at concentrations above 1 mmol/l.

Materno-fetal transfer: The sodium-dependent system A is the main conduit for Cys uptake from the maternal side across the syntrophoblast brush border membrane (Jansson, 2001), the transporter ATA1 (SLC38A1) is expressed at a high level in the placenta (Varoqui et al., 2000). The sodium-dependent system ASC appears to be located mainly at the basal membrane. The presence of transporter $B^{o,+}$ also has been suggested. These transporters establish the gradient that drives Cys toward the fetal side. As in the small intestine, the glycoprotein-anchored transporters LAT1 and LAT2 mediate the transfer of both Cys and cystine in both directions in exchange for neutral amino acids.

BBB: CssC is the main species that crosses the BBB, mainly via high-affinity uptake by system x-C into BBB epithelial cells. NADH-dependent cystine reductase (EC1.6.4.1) can then convert the oxidized form to Cys prior to export. A cystine-glutamate antiporter mediates CssC uptake into glial cells in exchange for glutamate.

METABOLISM

The main pathways of Cys metabolism to pyruvate proceed via cysteinesulfinate, CssC, mercaptopyruvate, or directly (Figure 8.51). Much smaller amounts of Cys are converted to taurine, L-alanine, and cysteamine.

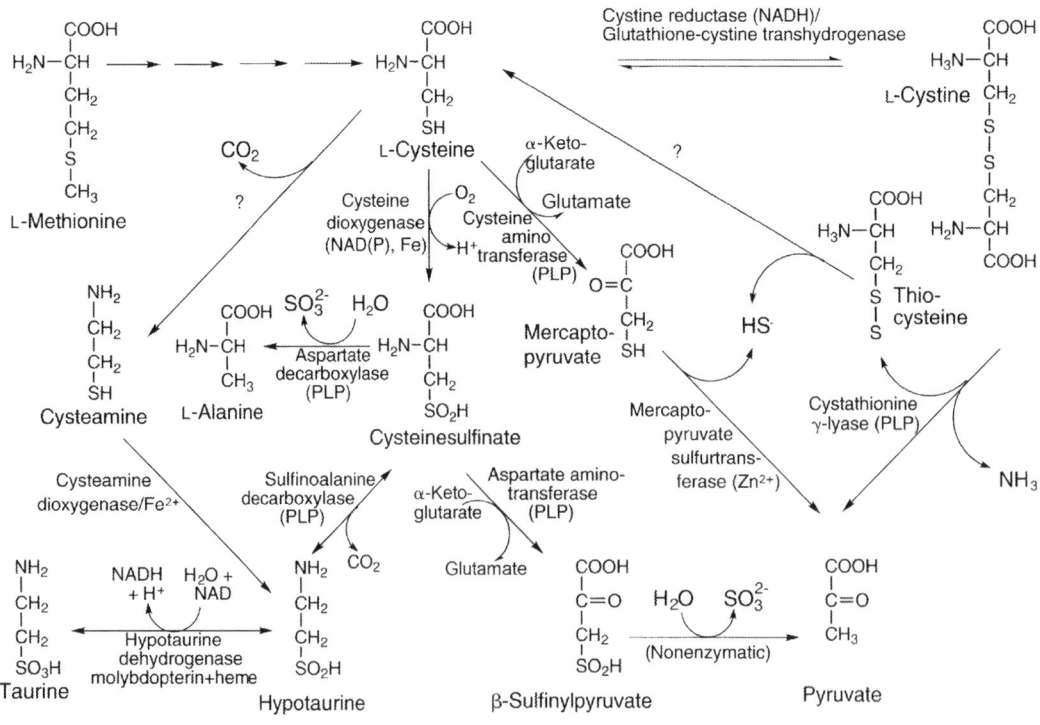

FIGURE 8.51

Metabolism of L-cysteine and L-cystine.

Cysteinesulfinate: Cysteine dioxygenase (EC1.13.11.20) with iron and FAD at its catalytic center oxidizes cysteine to cysteinesulfinate, mainly in the liver. The second step can be catalyzed by the PLP-dependent aspartate aminotransferase (EC2.6.1.1) in cytosol. The enzyme is generated by partial hydrolysis of aromatic amino acid transferase (EC2.6.1.57). This cytosolic enzyme is not identical with the more abundant mitochondrial aspartate aminotransferase (EC2.6.1.1). The product of this reaction, beta-sulfinylpyruvate, decomposes spontaneously into sulfite and pyruvate. The potentially toxic sulfite can be converted to sulfate by sulfite oxidase (EC1.8.3.1) in the mitochondrial intermembrane space. Sulfite oxidase contains both molybdenum cofactor and heme as prosthetic groups, and uses cytochrome c as an electron acceptor. Aspartate 4-decarboxylase (EC4.1.1.12) is present in mammalian tissues (Rathod and Fellman, 1985). The desulfinase activity of this PLP-dependent enzyme facilitates the conversion of cysteinesulfinate to sulfite and alanine.

Cystine: Cys can be converted to cystine in most tissues by glutathione-cystine transhydrogenase (EC1.8.4.4) or cystine reductase (EC1.6.4.1) depending on the prevailing redox state. Cleavage of thiocysteine by PLP-dependent cystathionine gammalyase (EC4.4.1.1) generates thiocysteine, ammonia, and pyruvate. Nonenzymic hydrolysis of thiocysteine releases Cys and persulfide.

Mercaptopyruvate: Cysteine aminotransferase (EC2.6.1.3) catalyzes the PLP dependent transfer of the Cys amino group to oxoglutarate. The resulting 3-mercaptopyruvate can then donate its persulfide group to a protein (to insert iron-sulfur clusters), thiol compounds (for thiosulfate biosynthesis), or another accepter (e.g., cyanide to generate thiocyanate); this reaction is facilitated by the zinc enzyme mercaptopyruvate sulfurtransferase (EC2.8.1.2).

Single-step conversion: Cystathionine gamma-lyase (EC4.4.1.1) can produce pyruvate by direct elimination of ammonia and persulfide from Cys. The relative contribution of this reaction to Cys catabolism is unclear.

Taurine synthesis: Oxidation by cysteine dioxygenase (EC1.13.11.20) followed by decarboxylation (PLP-dependent sulfinoalanine decarboxylase, EC4.1.1.29), and oxidation again (molybdenum-cofactor-dependent hypotaurine dehydrogenase, EC1.8.1.3) convert Cys into taurine (see the section entitled "Taurine," later in this chapter). Alternatively, the iron-containing enzyme cysteamine dioxygenase (EC1.13.11.19) can synthesize hypotaurine from cysteamine.

STORAGE

Most of the body's Cys is contained in functional peptides and proteins from where it is released, as proteins are hydrolyzed during normal tissue turnover. Glutathione is an important example of such Cys-containing peptides with dual storage and antioxidant functions. The tripeptide is produced in two steps. The first, rate-limiting step links L-glutamate to Cys in an ATP-driven reaction catalyzed by gamma-glutamylcysteine synthetase (EC6.3.2.2). Glutathione synthase (EC6.3.2.3) then adds a glycine residue. Glutathione can be broken down again by the successive actions of gamma-glutamyl transpeptidase (EC2.3.2.2) and any of a number of alternative dipeptidases (e.g., EC3.13.18, EC3.4.13.6, or EC3.4.11.2). Additional amounts of cystine are membrane-bound in lysosomes. The export from this compartment is an ATPase-dependent process that involves the transmembrane protein cystinosin (Touchman et al., 2000).

EXCRETION

Cys from renal ultrafiltrate is recovered in the proximal tubule via the sodium-dependent transporters ASC and B° and the exchanger BAT1/rBAT. It is of note that only the latter also mediates uptake of

CssC. Individuals with defective BAT1/rBAT have cystinuria, nephrolithiasis, and other complications because they can recover Cys, but not its oxidized form. Dipeptides and tripeptides are recovered via the proton/peptide cotransporter 2 (PepT2, SLC15A2). Export across the basolateral membrane can use the sodium-dependent transporter ASCT1 (only Cys) and the exchanger LAT2/4F2. Most (80%) of the sulfur from Cys (and Met) is excreted as sulfate, and smaller amounts as taurine (Bella and Stipanuk, 1995). The amino groups are excreted mainly after incorporation into the urea.

REGULATION

The complexity of the systems that maintain Cys homeostasis is becoming increasingly apparent. Regulation affects tissue uptake, metabolism (synthesis from Met and breakdown of Cys), and secretion from cells (as by astrocytes in the brain). The phosphorylation of amino acid exchangers BAT1/$b^{o,+}$, LAT1, and LAT2 is under the control of cell-signaling pathways; both cAMP and phytoestrogens influence the activity of these transporters (Mizoguchi et al., 2001).

FUNCTION

Energy fuel: Eventually, most of the ingested Cys is broken down and its carbon skeleton converted into pyruvate (see the previous discussion). The complete oxidation of Cys yields 4.31 kcal/g. Oxidation of CssC provides 4.09 kcal/g (May and Hill, 1990). Utilization of either amino acid requires thiamin, riboflavin, niacin, vitamin B6, pantothenate, lipoate, ubiquinone, iron, zinc, and magnesium; disposal of the sulfur in Cys requires molybdenum.

Protein synthesis: Cys is a constituent of most proteins and peptides. Its ability to form disulfide bridges is of crucial importance for protein folding and other specific functions.

tRNA sulfuration: The thiolation of some tRNA species appears to be important for maintaining proper acylation and ensuring high translational fidelity (Agris et al., 1983). A complex composed of tRNA-methyltransferase and aminoacyl-tRNA synthetase activities appears to contain cysteine-tRNA sulfurtransferase (EC2.8.1.4) activity, which uses Cys to generate 4-thiouridine and other thionucleotides in tRNAs.

Pantothenate sulfuration: Cys is needed for the synthesis of CoA. Phosphopantothenate-cysteine ligase (EC6.3.2.5) catalyzes the second synthesis step, which bonds Cys to phosphopantothenate.

Glutathione metabolism: Glutathione plays a critical role in the regulation of the sulfhydryl redox state (disulfide bridges in proteins and peptides), detoxification of endogenous steroids and xenobiotics, and control of oxygen free radicals. In the oxygen-rich environment of red blood cells, particularly large amounts of hydrogen peroxide and related oxygen free radical compounds are produced. Lipid hydroperoxide residues are particularly dangerous, because the free radical group can propagate throughout complex structures such as membranes. The selenium-dependent enzyme glutathione peroxidase (EC1.11.1.9) uses reduced glutathione to quench oxygen free radicals and terminate free-radical generating chain reactions. The flavoprotein glutathione reductase (EC1.6.4.2) can then use NADPH to restore the active form of glutathione. Glutathione plays an additional important role in the detoxification of numerous endogenous and xenobiotic compounds (van Bladeren, 2000). Several glutathione *S*-transferases (EC2.5.1.18) facilitate the conjugation of glutathione to numerous hydrophobic and electrophilic compounds (van Bladeren, 2000).

Taurine: About half of bile acids secreted into bile is conjugated with taurine. In addition, taurine is an important osmolyte and antioxidant, and it may be involved in the regulation of hormone activity and brain function.

Cyanide detoxification: Cassava and other food plants can supply considerable amounts of cyanide or cyanide-generating precursors. An important detoxification pathway converts cyanide into less toxic thiocyanate, which is excreted through the kidneys. PLP-dependent transfer of the amino group from cysteine to L-glutamate (cysteine aminotransferase, EC2.6.1.3) generates mercaptopyruvate, which then reacts with cyanide (mercaptopyruvate sulfurtransferase, EC2.6.1.3) to generate thiocyanate and pyruvate. Inefficiency of this pathway is thought to cause persistent spastic weakness of the legs and degeneration of corticospinal pathways in people with high dietary cyanide load (Spencer, 1999).

Supply of reduced sulfur: Cleavage of beta-mercaptopyruvate, a Cys metabolite, by 3-mercaptopyruvate sulfurtransferase (EC2.8.1.2) or the action of a PLP-dependent cysteine desulfurase (EC2.8.17) can provide persulfide for the synthesis of iron-sulfur clusters in proteins. An analogous transfer provides the dithiolene group for the molybdenum cofactor (Leimkühler and Rajagopalan, 2001). Similarly, persulfides can be generated from the cystine metabolite thiocysteine, possibly by nonenzymic decomposition.

Supply of sulfate: Most of the excreted sulfate is derived from Cys and from L-methionine via Cys. A bifunctional protein comprising the activities of sulfate adenylyltransferase (EC2.7.7.4) and adenylylsulfate kinase (EC2.7.1.25) can activate sulfate in humans to 3'-phosphoadenosine 5'-phosphosulfate (PAPS), which is the essential cosubstrate for the synthesis of sulfa-esters of steroids, catecholamines, and many other compounds and xenobiotics.

REFERENCES

Agris, P.F., Playl, T., Goldman, L., Horton, E., Woolverton, D., Setzer, D., et al., 1983. Processing of tRNA is accomplished by a high-molecular-weight enzyme complex. Recent Results Cancer Res. 84, 237–254.

Avissar, N.E., Ryan, C.K., Ganapathy, V., Sax, H.C., 2001. Na^+-dependent neutral amino acid transporter ATBo is a rabbit epithelial cell brush border protein. Am. J. Physiol. Cell Physiol. 281, C963–C971.

Bella, D.L., Stipanuk, M.H., 1995. Effects of protein, methionine, or chloride on acid–base balance and on cysteine catabolism. Am. J. Physiol. 269, E910–E917.

Burdo, J., Dargusch, R., Schubert, D., 2006. Distribution of the cystine/glutamate antiporter system xc- in the brain, kidney, and duodenum. J. Histochem. Cytochem. 54, 549–557.

Chairoungdua, A., Segawa, H., Kim, J.Y., Miyamoto, K., Haga, H., Fukui, Y., et al., 1999. Identification of an amino acid transporter associated with the cystinuria-related type II membrane glycoprotein. J. Biol. Chem. 274, 28845–28848.

Di Buono, M., Wykes, L.J., Bali, R.O., Pencharz, P.B., 2001. Dietary cysteine reduces the methionine requirement in men. Am. J. Clin. Nutr. 74, 761–766.

Dworschak, E., 1980. Nonenzyme browning and its effect on protein nutrition. Crit. Rev. Food Sci. Nutr. 13, 1–40.

FAO/WHO/UNU, 1985. Energy and protein requirements. World Health Organ. Tech. Rep. Ser. 724, 1–206. (p. 204).

Hatanaka, T., Huang, W., Ling, R., Prasad, P.D., Sugawara, M., Leibach, F.H., et al., 2001. Evidence for the transport of neutral as well as cationic amino acids by ATA3, a novel and liver-specific subtype of amino acid transport system A. Biochim. Biophys. Acta 1510, 10–17.

Jansson, T., 2001. Amino acid transporters in the human placenta. Pediatr. Res. 49, 141–147.

Kim, J.Y., Kanai, Y., Chairoungdua, A., Cha, S.H., Matsuo, H., Kim, D.K., et al., 2001. Human cystine/glutamate transporter: cDNA cloning and upregulation by oxidative stress in glioma cells. Biochim. Biophys. Acta 1512, 335–344.

Leimkühler, S., Rajagopalan, K.V., 2001. A sulfurtransferase is required in the transfer of cysteine sulfur in the *in vitro* synthesis of molybdopterin from precursor Z in *Escherichia coli*. J. Biol. Chem. 276, 22024–22031.

May, M.E., Hill, J.O., 1990. Energy content of diets of variable amino acid composition. Ant. J. Clin. Nutr. 52, 770–776.

Mizoguchi, K., Cha, S.H., Chairoungdua, A., Kim, D.K., Shigeta, Y., Matsuo, H., et al., 2001. Human cystinuria-related transporter: localization and functional characterization. Kidney Int. 59, 8121–8133.

Raguso, C.A., Regan, M.M., Young, V.R., 2000. Cysteine kinetics and oxidation at different intakes of methionine and cystine in young adults. Am. J. Clin. Nutr. 71, 491–499.

Rajan, D.P., Kekuda, R., Huang, W., Devoe, L.D., Leibach, F.H., Prasad, P.D., et al., 2000. Cloning and functional characterization of a Na(+)-independent, broad-specific neutral amino acid transporter from mammalian intestine. Biochim. Biophys. Acta 1463, 6–14.

Rathod, P.K., Fellman, J.H., 1985. Identification of mammalian aspartate-4-decarboxylase. Arch. Biochem. Biophys. 238, 435–446.

Sato, H., Kuriyama-Matsumara, K., Hashimotro, T., Sasaki, H., Wang, H., Ishii, T., et al., 2001. Effect of oxygen on induction of the cystine transporter by bacterial lipopolysaccharide in mouse peritoneal macrophages. J. Biol. Chem. 276, 10407–10412.

Spencer, P.S., 1999. Food toxins, ampa receptors, and motor neuron diseases. Drug Metab. Rev. 31, 561–587.

Sugawara, M., Nakanishi, T., Fei, Y.J., Huang, W., Ganapathy, M.E., Leibach, F.H., et al., 2000. Cloning of an amino acid transporter with functional characteristics and tissue expression pattern identical to that of system A. J. Biol. Chem. 275, 16473–16477.

Touchman, J.W., Anikster, Y., Dietrich, N.L., Maduro, V.V., McDowell, G., Shotelersuk, V., et al., 2000. The genomic region encompassing the nephropathic cystinosis gene (CTNS): complete sequencing of a 200-kb segment and discovery of a novel gene within the common cystinosis-causing deletion. Genome Res. 10, 165–173.

Utsunomiya-Tate, N., Endou, H., Kanai, Y., 1996. Cloning and functional characterization of a system ASC-like Na$^+$-dependent neutral amino acid transporter. J. Biol. Chem. 271, 14883–14890.

van Bladeren, P.J., 2000. Glutathione conjugation as a bioactivation reaction. Chem. Biol. Interact. 129, 61–76.

Varoqui, H., Zhu, H., Yao, D., Ming, H., Erickson, J.D., 2000. Cloning and functional identification of a neuronal glutamine transporter. J. Biol. Chem. 275, 4049–4054.

Wlodek, P.J., Iciek, M.B., Milkowski, A., Smolenski, O.B., 2001. Various forms of plasma cysteine and its metabolites in patients undergoing hemodialysis. Clin. Chim. Acta 304, 9–18.

LYSINE

L-Lysine (2,6-diaminohexanoic acid, alpha, epsilon-diaminocaproic acid, one-letter code K; molecular weight 146) is an essential amino acid with 19.2% nitrogen (Figure 8.52).

ABBREVIATION

Lys   L-lysine

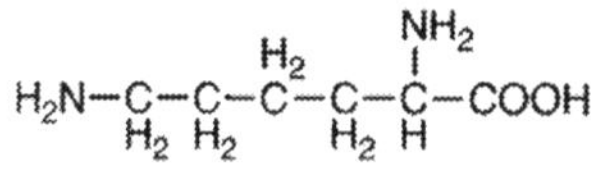

FIGURE 8.52

L-Lysine is an essential amino acid.

NUTRITIONAL SUMMARY

Function: The essential amino acid L-lysine (Lys) is needed for the synthesis of peptides and proteins and as a precursor of carnitine. Use as an energy fuel requires adequate supplies of thiamin, riboflavin, niacin, vitamin B6, pantothenate, lipoate, ubiquinone, iron, and magnesium.

Food sources: Adequate amounts are consumed when total protein intake meets recommendations. Dietary supplements containing crystalline Lys are commercially available.

Requirements: Estimates of daily Lys requirements vary, but they appear to be near 30 mg/kg in healthy adults (El-Khoury et al., 2000).

Deficiency: Prolonged lack of Lys, as with all essential amino acids or a lack of protein, causes growth failure, loss of muscle mass, and organ damage.

Excessive intake: A very high intake of protein and mixed amino acids (i.e., more than three times the RDA, which is 2.4 g/kg) is thought to increase the risk of renal glomerular sclerosis and accelerate osteoporosis. There have been anecdotal reports of symptoms alike to EMS and other severe illnesses following the use of some synthetic Lys preparations.

DIETARY SOURCES

All protein-containing foods provide some Lys. In addition, significant amounts of hydroxylysine are consumed with meats. Foods relatively enriched in Lys (as a percentage of total protein) include pork (9.0%), beef (8.3%), chicken (8.0%), cow's milk (7.9%), eggs (7.2%), and soy (6.5%). Grains contain a much smaller percentage, as seen with rice (3.6%), wheat (2.7%), and corn (2.8%), making Lys a limiting amino acid and reducing the protein quality of these grains when consumed without significant amounts of Lys-rich foods. Amaranth, in contrast, has more substantial Lys content (5.5%).

The typical Lys intake of American adults is 4.5 g/day; vegetarians get considerably less (Sachan et al., 1997). Heat treatment (e.g., baking, broiling, grilling, and frying) can chemically alter Lys and induce the formation of so-called Maillard products, conjugates of Lys, and sugars (Dworschak, 1980).

DIGESTION AND ABSORPTION

Various enzymes from the stomach, pancreas, and intestine hydrolyze food proteins (Figure 8.53). The hydrogen ion/peptide cotransporter 1 (SLC15A1, PepT1) and, to a much lesser extent, the hydrogen ion/peptide cotransporter 2 (SLC15A2, PepT2) mediate the uptake of dipeptides and tripeptides with broad specificity.

Free Lys can enter intestinal cells via the y^+, sodium-independent cationic amino acid transporters 1 (CAT-1, SLC7A1) and 2 (CAT-2, SLC7A2). The sodium-dependent system $B^{o,+}$ seems to constitute only a minor transport route, possibly in the distal small intestine. Uptake can also proceed through the rBAT (SLC3A1)–anchored amino acid transporter BAT1/$b^{o,+}$ (SLC7A9) at the brush border membrane, which can shuttle Lys into the enterocyte in exchange for a neutral amino acid plus a sodium ion (Chairoungdua et al., 1999).

Export of Lys toward the pericapillary space uses two transporters that are anchored to the basolateral membrane by glycoprotein 4F2 (SLC3A2): both y^+LAT1 (SLC7A7) and y^+ (SLC7A6) move Lys into the basolateral space in exchange for a neutral amino acid and a sodium ion.

Note: The conjugates Lys forms with other amino acids or sugars during the heating of foods in a browning (Maillard) reaction are not cleaved or absorbed in the small intestine. Examples of such heat-generated compounds include iysinoalanine, fructoselysine, and N epsilon-carboxymethyllysine.

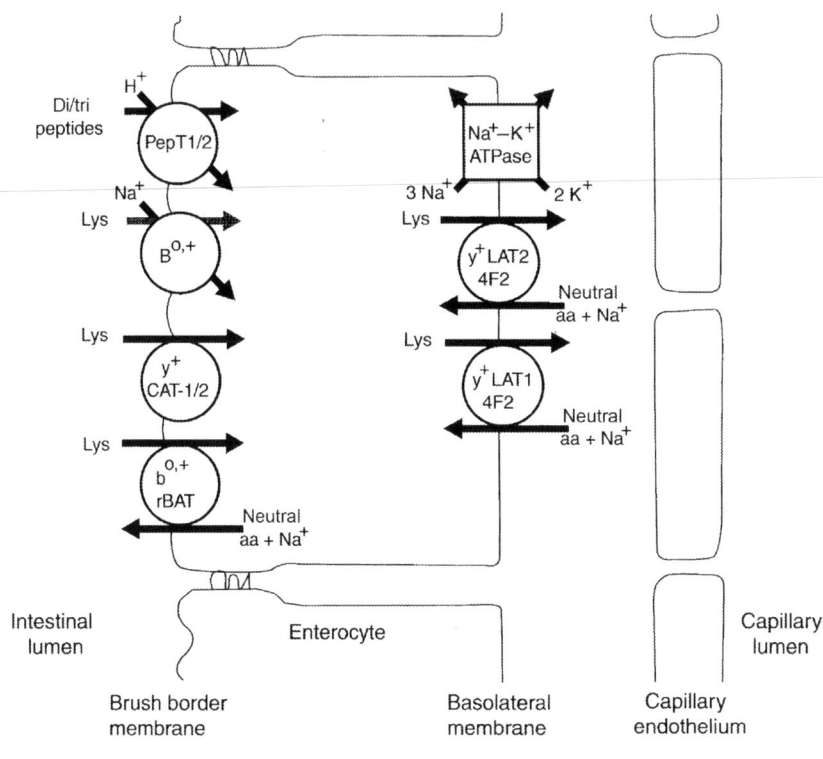

FIGURE 8.53

Intestinal absorption of L-lysine.

TRANSPORT AND CELLULAR UPTAKE

Blood circulation: While most Lys in blood is part of proteins, the concentration of free Lys in plasma is around 195 μmol/l. Uptake into tissues proceeds mainly via various members of system y^+. The glycoprotein-anchored heteroexchangers y^+LAT1 and y^+ contribute to a lesser extent to Lys uptake in some tissues. At least one of these transporters is also active in red blood cells.

The mitochondrial ornithine transporter 1 (ORNT1, ornithine/citrulline carrier; SLC25A15) can move Lys from cytosol into mitochondria (Indiveri et al., 1999).

Materno-fetal transfer: Since Lys is an essential amino acid, the fetus is fully dependent on transfer across the placenta. Several members of system y^+ (CAT-1, CAT-4, and C AT-2B) mediate uptake from maternal circulation into the syntrophoblast layer. Export toward fetal circulation uses mainly the membrane-anchored heterodimer composed of y^+LAT1 (SLC7A7) and glycoprotein 4F2 (SLC3A2). Lys is exchanged by the y^+LAT1 transporter for a neutral amino acid and a sodium ion.

BBB: While there is no doubt that circulating blood has to supply the essential nutrient Lys to the brain, knowledge about the mechanism for transfer across the BBB is limited. The 4F2-anchored exchange complex y$^+$LAT2 is known to contribute significantly (Bröer et al., 2000). Adequacy of Lys intake influences flux into the brain (Tews et al., 1988).

METABOLISM

Lys catabolism transfers the two amino groups to alpha-ketoglutarate and generates two molecules of acetyl-CoA (Figure 8.54). The main pathway proceeds via saccharopine to alpha-ketoadipate in liver cytosol, and then continues in mitochondria to acetyl-CoA. An alternative peroxisomal pathway via L-pipecolate is most important in the brain and also contributes to some extent to Lys breakdown in other extrahepatic tissues.

The first two steps of Lys breakdown via the main pathway use the bifunctional protein semialdehyde synthase that combines the activities of lysine ketoglutarate reductase (EC1.5.1.8) and saccharopine dehydrogenase (EC1.5.1.9). In the end, these activities move the epsilon amino group from Lys to alpha-ketoglutarate.

Oxidation by aminoadipate-semialdehyde dehydrogenase (EC1.2.1.31, magnesium-dependent) and PLP-dependent transamination by 2-aminoadipate aminotransferase (EC2.6.1.39) generate alpha-ketoadipate. The mitochondrial oxodicarboxylate carrier (ODC, SLC25A21) moves this intermediate from cytosol into mitochondria (Fiermonte et al., 2001), where oxidative decarboxylation by oxoglutarate dehydrogenase (EC1.2.4.2) continues its metabolism. This multisubunit enzyme contains thiamin pyrophosphate and lipoamide as covalently bound cofactors. Glutaryl-CoA dehydrogenase (EC1.3.99.7) then catalyzes both FAD-dependent oxidation and decarboxylation to crotonyl-CoA. Beta-oxidation of crotonyl-CoA finally releases two acetyl-CoA molecules. Alpha-deamination by L-lysine oxidase (EC1.4.3.14) starts the pipecolate pathway of Lys breakdown. Two as-yet-uncharacterized steps then generate pipecolate. This Lys intermediate is transported into brain neurons via the high-affinity proline transporter (PROT) and may influence excitation (Galli et al., 1999). Oxidation by the FAD-containing L-pipecolate oxidase (EC1.5.3.7; IJlst et al., 2000) and nonenzymic hydration generate alpha-aminoadipate semialdehyde and thus rejoin the main pathway.

STORAGE

Body proteins contain about 82 mg/g (Smith, 1980). Normal turnover of body proteins will release significant amounts of Lys. In the case of isolated Lys deficiency, protein mobilization may increase through incompletely understood mechanisms.

EXCRETION

Losses of intact Lys are minimal due to efficient recovery both from the intestines and kidneys. In contrast, the dead-end metabolite methyllysine is not well reabsorbed, and most of it is excreted with urine.

Nearly 5 g of Lys passes across the renal glomeruli, and most of it is reabsorbed from the proximal tubular lumen. Uptake proceeds via the sodium-dependent system B°, the sodium-independent system y$^+$ (CAT-1, SLC7A1) and 2 (CAT-2, SLC7A2), and the transporter heterodimer, consisting of BAT1/b$^{o,+}$ (SLC7A9) and rBAT (SLC3A1). The glycoprotein 4F2-linked transporters y$^+$LAT1 (SLC7A7) and y$^+$LAT2 (SLC7A6) on the basolateral side mediate export toward the capillaries (Bröer et al., 2000; Bode, 2001).

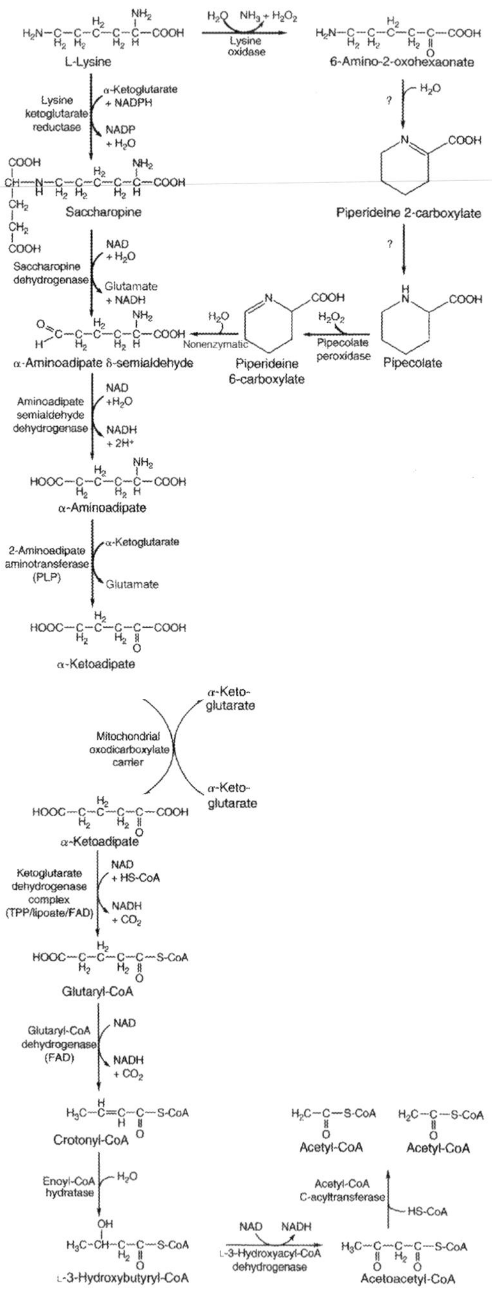

FIGURE 8.54

Alternate pathways degrade L-lysine via saccharopine or pipecolate.

REGULATION

Lys homeostasis is maintained in part by changes in the rate of protein catabolism and Lys metabolic utilization, as well as oxidation as an energy fuel, but details of the responsible mechanisms remain to be elucidated. Lys catabolism increases greatly in response to intake (El-Khoury et al., 1998).

FUNCTION

Energy fuel: Eventually, most of the ingested Lys is completely metabolized. The daily rate of Lys oxidation in healthy adults with low to moderate intake is about 27 mg/kg body weight (El-Khoury et al., 2000). Lys is a particularly important energy fuel for muscles. Complete oxidation provides 4.92 kcal/g (May and Hill, 1990) and depends on adequate supplies of thiamin, riboflavin, niacin, vitamin B6, pantothenate, lipoate, ubiquinone, iron, and magnesium.

Protein and peptide synthesis: Lys is a regular component of most proteins and many peptides. Posttranslational reactions can modify the amino group that is not engaged by the peptide bond in proteins. An important example is the conjugation of specific Lys residues in a few proteins to biotin by biotin-[propionyl-CoA-carboxylase (ATP-hydrolyzing)] ligase (EC6.3.4.10). Another type of posttranslational modification involving Lys residues consists of hydroxylation and the subsequent formation of cross-links in collagens. First, procollagen-lysine 5-dioxygenase (EC1.14.11.4) attached to the rough endoplasmic reticulum uses alpha-ketoglutarate and oxygen to hydroxylate Lys residues adjacent to glycines in procollagen. Ascorbate keeps iron in this ferroenzyme in a reduced state. Several genetically distinct isoforms exist. As procollagen extrudes into the extracellular space and forms the typical triple helix arrangements, the copper-enzyme lysyl oxidase (protein-lysine 6-oxidase, EC1.4.3.13) links strands through the formation of bonds between Lys and hydroxylysine residues. Another example is the methylation of specific Lys residues in histones, which is critical to maintaining certain chromatin segments in the inactive state (Peters et al., 2002).

Carnitine synthesis: Lys is the critical precursor for endogenous carnitine synthesis in the liver, kidneys, and some other tissues. Daily production is about 0.2 mg/kg and depends on the adequate availability of niacin, vitamin B6, folate, ascorbate, SAM, and iron. Specific lysine residues of myosin, actin, histones, and a few other proteins are trimethylated. Hydrolysis of these proteins during normal tissue turnover releases trimethyllysine, and this is hydroxylated and modified further to finally yield carnitine.

Polyamine synthesis: The PLP-containing ornithine decarboxylase (EC4.1.1.17) converts Lys into cadaverine, which plays an important role in intracellular and intercellular signaling. One example of its functions is the ability of cadaverine to prevent the escape of *Shigella flexneri* from phagolysosomes by blocking transepithelial signaling to polymorphonuclear cells (Fernandez et al., 2001).

REFERENCES

Bode, B.P., 2001. Recent molecular advances in mammalian glutamine transport. J. Nutr. 131, 2475S–2485S.

Bröer, A., Wagner, C.A., Lang, F., Bröer, S., 2000. The heterodimeric amino acid transporter 4F2hc/y⁺LAT2 mediates arginine efflux in exchange with glutamine. Biochem. J. 349, 787–795.

Chairoungdua, A., Segawa, H., Kim, J.Y., Miyamoto, K., Haga, H., Fukui, Y., et al., 1999. Identification of an amino acid transporter associated with the cystinuria-related type II membrane glycoprotein. J. Biol. Chem. 274, 28845–28848.

Dworschak, E., 1980. Nonenzyme browning and its effect on protein nutrition. Crit. Rev. Food Sci. Nutr 13, 1–40.

EI-Khoury, A.E., Basile, A., Beaumier, L., Wang, S.Y., Al-Amiri, H.A., Selvaraj, A., et al., 1998. Twenty-four-hour intravenous and oral tracer studies with L-[1-^{13}C]-2-aminoadipic acid and L-[1-^{13}C] lysine as tracers at generous nitrogen and lysine intakes in healthy adults. Am. J. Clin. Nutr. 68, 827–839.

EI-Khoury, A.E., Pereira, P.C., Borgonha, S., Basile-Filho, A., Beaumier, L., Wang, S.Y., et al., 2000. Twenty-four-hour oral tracer studies with L-[1-^{13}C] lysine at a low (15 mg. kg (-1). d (-1)) and intermediate (29 mg. kg (-1)-d(-1)) lysine intake in healthy adults. Am. J. Clin. Nutr. 72, 122–130.

Fernandez, I.M., Silva, M., Schuch, R., Walker, W.A., Siber, A.M., Maurelli, A.T., et al., 2001. Cadaverine prevents the escape *of Shigella.flexneri* from the phagolysosome: a connection between bacterial dissemination and neutrophil transepithelial signaling. J. Inf. Dis. 184, 743–753.

Fiermonte, G., Dolce, V., Palmieri, L., Ventura, M., Runswick, M.J., Palmieri, E., et al., 2001. Identification of the human mitochondrial oxodicarboxylate carrier. Bacterial expression, reconstitution, functional characterization, tissue distribution, and chromosomal location. J. Biol. Chem. 276, 8225–8230.

Galli, A., Jayanthi, L.D., Ramsey, I.S., Miller, J.W., Fremeau Jr., R.T., DeFelice, L.J., 1999. L-proline and L-pipecolate induce enkephalin-sensitive currents in human embryonic kidney 293 cells transfected with the high-affinity mammalian brain L-proline transporter. J. Neurosci. 19 (15), 6290–6297.

IJlst, L., de Kromme, I., Oostheim, W., Wanders, R.J., 2000. Molecular cloning and expression of human L-pipecolate oxidase. Biochem. Biophys. Res. Commun. 270, 1101–1105.

Indiveri, C., Tonazzi, A., Stipani, I., Palmieri, E., 1999. The purified and reconstituted ornithine/citrulline carrier from rat liver mitochondria catalyses a second transport mode: ornithine$^+$/H$^+$ exchange. Biochem. J. 341, 705–711.

May, M.E., Hill, J.O., 1990. Energy content of diets of variable amino acid composition. Am. J. Clin. Nutr. 52, 770–776.

Peters, A.H., Mermoud, J.E., O'Carroll, D., Pagani, M., Schweizer, D., Brockdorff, N., et al., 2002. Histone H3 lysine 9 methylation is an epigenetic imprint of facultative heterochromatin. Nat. Genet. 30, 77–80.

Sachan, D.S., Daily III, J.W., Munroe, S.G., Beauchenne, R.E., 1997. Vegetarian elderly women may risk compromised carnitine status. Veg. Nutr. 1, 64–69.

Smith, R.H., 1980. Comparative amino acid requirements. Proc. Nutr. Soc. 39, 71–78.

Tews, J.K., Greenwood, J., Pratt, O.E., Harper, A.E., 1988. Dietary amino acid analogues and transport of lysine or valine across the blood–brain barrier in rats. J. Nutr. 118, 756–763.

LEUCINE

The hydrophobic neutral BCAA L-leucine (2-amino-4-methylvaleric acid, alpha-aminoisocaproic acid, 2-amino-4-methylpentanoic acid, one-letter code L; molecular weight 131) contains 10.7% nitrogen (Figure 8.55).

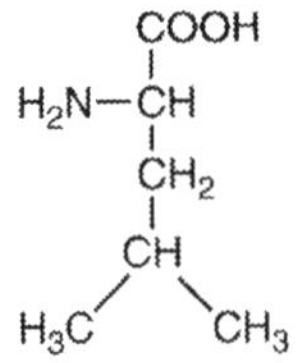

FIGURE 8.55

L-Leucine.

NUTRITIONAL SUMMARY

Function: The essential amino acid L-leucine (Leu) is needed for the synthesis of proteins. It is also an important energy fuel, especially in muscles; its breakdown requires thiamin, riboflavin, pyridoxine, niacin, cobalamin, biotin, pantothenate, lipoate, ubiquinone, magnesium, and iron.

Food sources: Adequate amounts are consumed when total protein intake meets recommendations since dietary proteins from different sources all contain Leu. Corn and dairy protein contains slightly more Leu than protein from most other foods.

Requirements: Adults are thought to require more than 40 mg/kg per day (Kurpad et al., 2002).

Deficiency: Prolonged lack of Leu, as with all essential amino acids or a lack of protein, causes growth failure, loss of muscle mass, and organ damage. Additional Leu might increase lean body mass in people with muscle wasting, and possibly also concurrent with exercise.

Excessive retake: Very high intake of protein and mixed amino acids (i.e., more than three times the RDA, which is 2.4 g/kg) is thought to increase the risk of renal glomerular sclerosis and accelerate osteoporosis. The consequences of very high intake of Leu have not been adequately evaluated.

DIETARY SOURCES

Leu is an essential amino acid that cannot be synthesized by humans from other amino acids or keto acids and therefore must be derived from food. All food proteins contain Leu, usually between 7% and 9%. Slightly richer sources are the proteins in milk and dairy products (9.8%), corn (12.3%), and sorghum (13.2%). Mature human milk protein contains 10.4% Leu.

DIGESTION AND ABSORPTION

Proteins are hydrolyzed by various enzymes; most of them are derived from the stomach wall and pancreas. The hydrogen ion/peptide cotransporters 1 (PepT1, SLC15A1) and, to a lesser extent, 2 (PepT2, SLC15A2) mediate uptake of dipeptides and tripeptides, including those containing Leu (Figure 8.56). Free Leu can cross the small intestinal brush border membrane via the sodium-driven transporter B° (SLC1A5, Avissar et al., 2001).

Like most neutral amino acids, Leu can be exchanged for other neutral amino acids across the brush border membrane in either direction by the rBAT (SLC3A1)-anchored amino acid transporter BAT1/ $b^{\circ,+}$ (SLC7A9) (Chairoungdua et al., 1999). The 4F2 (SLC3A2) glycoprotein-anchored transporter LAT2 (SLC7A8) can exchange Leu across the basolateral membrane in both directions in exchange for other neutral amino acids.

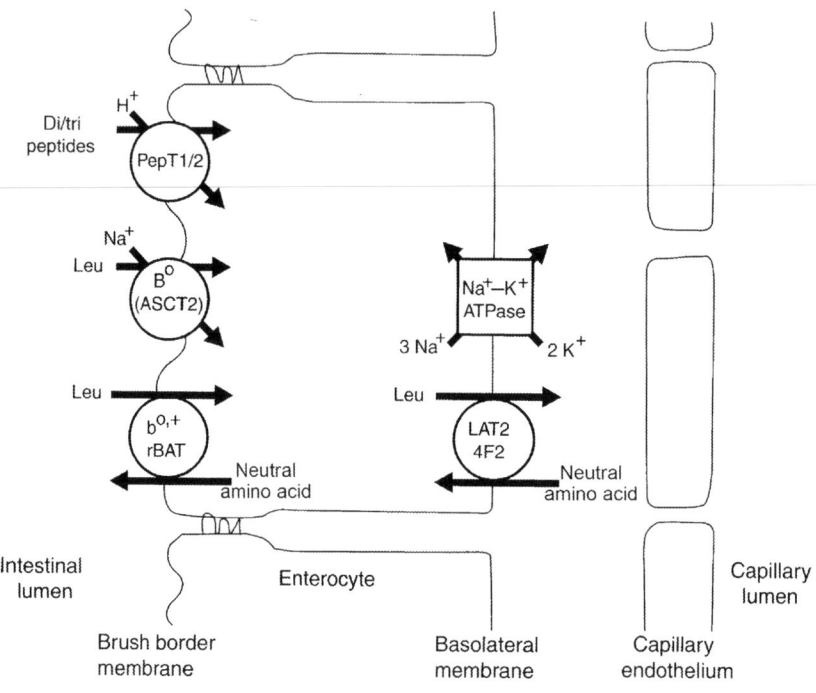

FIGURE 8.56

Intestinal absorption of L-leucine.

Leu can also be absorbed from the colon lumen, but the extent of this is unknown (Utsunomiya-Tate et al., 1996). Uptake proceeds via ASCT2 (SLC1A4) and possibly other transporters. Little is known about Leu transport across the basolateral membrane of colonic enterocytes.

The intestines sequester about 20% of ingested Leu after uptake. Only a small portion of this is oxidized; most of it is incorporated into newly synthesized proteins or converted to alpha-ketoisocaproate and released (Matthews et al., 1993).

TRANSPORT AND CELLULAR UPTAKE

Blood circulation: Leu is a ubiquitous component of proteins in the blood and can be taken up by cells via mechanisms specific to the respective proteins. Free Leu (plasma concentrations are around 120 μmol/l) enters cells mainly via system L, including the specifically identified heteroexchanger LAT1 (SLC7A5). The BCAA transaminase (EC2.6.1.42), which is located at the inner mitochondrial membrane, may function as an importer of BCAAs.

Materno-fetal transfer: Leu uptake across the microvillous membrane of the syncytiotrophoblast is mediated by LAT1 (Ritchie and Taylor, 2001) and transport across the basal membrane proceeds via LAT2 (SLC7A8), both in exchange for other neutral amino acids (Jansson, 2001). The driving force for LAT1/LAT2-mediated transport is the concentration gradient of small neutral amino acids (glycine,

L-alanine, and L-cysteine) established by the sodium-dependent transport systems A and ASC. Only some of the Leu taken up by the placenta from maternal circulation reaches the fetus, and a significant proportion is metabolized (Scislowski et al., 1983).

BBB: System L mediates Leu transport across both sides of the neuroendothelial cell layer. The molecular identity and location of the responsible transporter(s) remain unclear.

METABOLISM

Most Leu is broken down via the main catabolic pathway into acetoacetate and acetyl-CoA in a sequence of six enzyme-catalyzed steps that are initiated by transfer of the amino group to alpha-ketoglutarate (Figure 8.57). Complete oxidation of Leu via this pathway depends on adequate availability of thiamin, riboflavin, pyridoxine, niacin, pantothenate, biotin, ubiquinone, and lipoic acid. A much smaller amount (5–10%) is oxidized via HMB. Another alternative metabolic sequence, the beta-keto pathway, appears to be significant only in testes. This pathway is remarkable because its initial reaction is catalyzed by one of only three enzymes that require cobalamin as a prosthetic group.

Main catabolic pathway: Breakdown of Leu in mitochondria can begin with the transfer of its amino group to alpha-ketoglutarate by BCAA transaminase (EC2.6.1.42). The activity of an alternative enzyme, L-leucine-aminotransferase (LAT, EC2.6.1.6) is significant only in a few specialized tissues, such as the testes (Sertoli cells) and pancreas. The resulting alpha-ketoisocaproic acid then undergoes oxidative decarboxylation; this reaction is irreversible. The enzyme responsible for this activity, branched-chain alpha-keto acid dehydrogenase (EC1.2.4.4), is a large complex in the mitochondrial matrix consisting of multiple copies of three distinct subunits. The subunit E1 catalyzes the decarboxylation reaction with reduced CoA as a cosubstrate. E1 itself is a heterodimer of an alpha chain with thiamin pyrophosphate as a prosthetic group, and a beta chain. Subunit E2 anchors the lipoic acid residue, which serves as an acceptor for the decarboxylated substrate, transfers it to acetyl-CoA, and reduces lipoamide to dihydrolipoamide in the process. The lipoamide dehydrogenase component, subunit E3 (EC1.8.1.4), transfers the hydrogen from dihydrolipoamide via its FAD group to NAD. The glycine cleavage system (EC1.4.4.2) and the dehydrogenases for pyruvate (EC1.2.4.1) and alpha-ketoglutarate (EC2.3.1.61) use the same enzyme subunit. The enzyme complex is inactivated by phosphorylation ([3-methyl-2-oxobutanoate dehydrogenase (lipoamide)] kinase, EC2.7.1.115), and reactivated by dephosphorylation ([3-methyl-2-oxobutanoate dehydrogenase (lipoamide)]-phosphatase, EC3.1.3.52). Branched-chain alpha-keto acid dehydrogenase is defective in maple syrup urine disease, which affects the breakdown of Leu as well as the other BCAAs, L-isoleucine and L-valine.

The next step, conversion of isovaleryl-CoA into 3-methylcrotonyl-CoA, is catalyzed by isovaleryl-CoA dehydrogenase (EC1.3.99.10). This enzyme is not identical to the analogous enzyme (2-methylacyl-CoA dehydrogenase, EC1.3.99.12) for the acyl-CoA products of L-valine and L-isoleucine metabolism. Isovaleryl-CoA dehydrogenase is a flavoprotein closely associated with electron-transferring flavoprotein (ETF) and another flavoprotein, ETF dehydrogenase (EC1.5.5.1). FAD is used in each of these three components of the mitochondrial electron transfer system to form a cascade that eventually reduces ubiquinone to ubiquinol; this can then be utilized directly for ATP synthesis through oxidative phosphorylation. Excessive amounts of isovaleryl-CoA, which accumulate to neurotoxic levels in patients with deficiency of isovaleryl-CoA dehydrogenase, can be conjugated by glycine-N-acylase (EC2.3.1.13), and excreted via urine as isovalerylglycine.

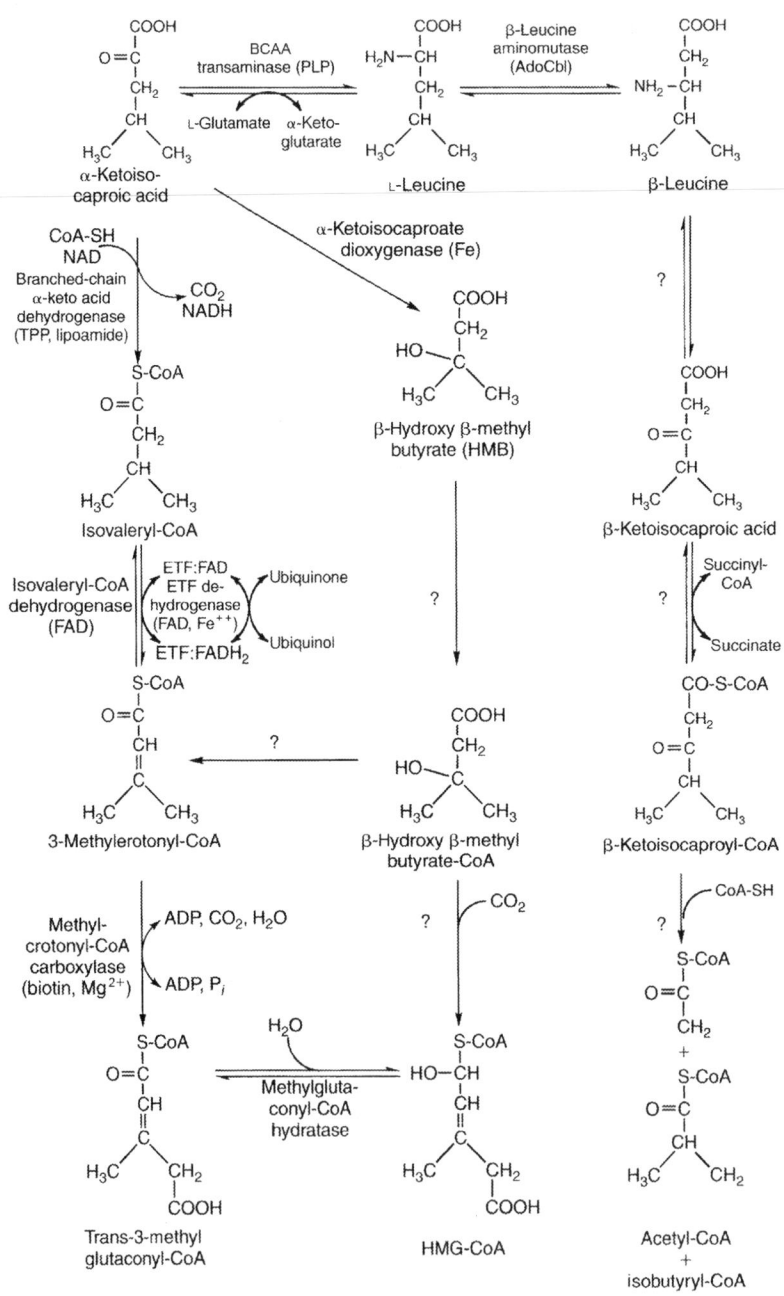

FIGURE 8.57

Metabolism of L-leucine.

A biotin-containing enzyme, methylcrotonyl-CoA carboxylase (EC6.4.1.4), catalyzes the next, irreversible conversion to trans-3-methyl glutaconyl-CoA, followed by hydration (methylglutaconyl-CoA hydratase, EC4.2.1.18) to hydroxymethylglutaryl-CoA (HMG-CoA). HMG-CoA reductase (EC1.1.1.34) catalyzes the reaction that commits HMG-CoA to the cholesterol synthesis pathway. Indeed, Leu has been found to provide a significant proportion of the carbon in *de novo* synthesized cholesterol (Bloch et al., 1954). Alternatively, HMG-CoA can be cleaved into acetyl-CoA and acetoacetate by HMG-CoA lyase (EC4.1.3.4).

Beta-hydroxy beta-methylbutyrate: Conversion to HMB accounts for 5–10% of Leu catabolism (Nissen and Abumrad, 1997). Alpha-ketoisocaproate dioxygenase, the cytosolic iron-containing enzyme responsible for oxidative decarboxylation, may actually be 4-hydroxyphenylpyruvate dioxygenase (EC1.13.11.27, catalyzes the second step of tyrosine catabolism) judging from its very similar properties (Nissen and Abumrad, 1997). The subsequent steps are known to alternatively rejoin the main pathway at the level of 3-methylcrotonyl-CoA or HMG-CoA. The enzymes necessary for these reactions are not yet well characterized.

Beta-keto pathway: An alternative pathway of Leu degradation proceeds via L-beta-leucine aminomutase (EC5.4.3.7) and generates beta-hydroxy-beta-methylbutyrate; this enzyme requires adenosylcobalamin. Significant L-beta-leucine aminomutase activity is present in the testes, where about a third of Leu metabolism proceeds via the beta-keto pathway. In all other investigated tissues leucine flux through the beta-keto pathway accounts for less than 5% of Leu metabolism (Poston, 1984). The enzymes responsible for this pathway are not yet well characterized.

STORAGE

Body protein contains 5–10% Leu (Mero, 1999). Protein can be mobilized rapidly, especially from muscle, and used as an energy fuel. In the fasting state, most of the amino groups are transferred to pyruvate; the resulting L-alanine is exported to the liver for urea synthesis and gluconeogenesis. Glucose is returned to the muscle to complete this alanine-glucose cycle. In the fed state, L-glutamine is the predominant transamination product.

EXCRETION

The organization of Leu reabsorption from the proximal tubule lumen closely resembles small intestinal absorption. The bulk action of the sodium-driven brush border transporter B° (Avissar et al., 2001) is supplemented by the rBAT (SLC3A1)-anchored heteroexchanger BAT1/b$^{o,+}$ (SLC7A9; Chairoungdua et al., 1999). Transport across the basolateral membrane depends mainly on the 4F2 (SLC3A2) glyco-protein-anchored transporter LAT2 (SLC7A8).

REGULATION

Branched-chain alpha-keto acid dehydrogenase (EC2.6.1.42) is the key regulatory enzyme for Leu (and L-valine/L-isoleucine) metabolism, which helps preserve this essential amino acid for protein synthesis during starvation. The enzyme is inactivated by a specific, calmodulin-dependent kinase ([3-methyl-2-oxobutanoate dehydrogenase (lipoamide)] kinase, EC2.7.1.115) and activated by a specific phosphatase ([3-methyl-2-oxobutanoate dehydrogenase (lipoamide)]-phosphatase, EC3.1.3.52). Branched-chain alpha-keto acid dehydrogenase is also inhibited by the products of the reaction that it catalyzes (namely, isovaleryl-CoA, isobutyryl-CoA, and 2-methylbutyryl-CoA).

An intake imbalance with inadequacy of one of the BCAAs (valine, leucine, or isoleucine) causes the activation of orbitofrontal cortex, frontal gyrus, and thalamus and leads to the development of conditioned taste aversion (Gietzen and Magrum, 2001). Expression of several genes in response to inadequacy changes, but the mechanisms responsible for learning to balance BCAA intake remain elusive.

FUNCTION

Energy fuel: Eventually, most Leu is broken down, providing about 6.0 kcal/g. About 60% of ingested Leu is oxidized within a few hours (Toth et al., 2001). The main site of Leu oxidation is skeletal muscle. The liver, intestines, and other organs are responsible for much smaller percentages of Leu utilization as an energy fuel.

Protein synthesis: Virtually all proteins contain Leu as part of their specific sequence. A lack of Leu, as with all essential amino acids, limits protein synthesis. High concentration of free Leu in muscle, as after a protein-rich meal, stimulates protein synthesis (Mero, 1999). Leu and the other BCAAs have anabolic effects on skeletal muscle (Anthony et al., 2001). The protein kinase mammalian target of rapamycin (mTOR) is a key target: high Leu availability increases mTOR activity, which leads to the hyperphosphorylation of ribosomal protein S6 kinase (S6K1) and eukaryotic initiation factor 4E binding protein 1 (4E-BP1). Signaling downstream of 4E-BP1 initiates the translation of capped messenger RNA (mRNA); events following S6K1 activation eventually increase the synthesis of proteins needed for translation. Both Leu and its minor (5–10%) metabolite HMB are sometimes used to boost muscle mass in people with muscle wasting due to AIDS or other diseases (Kreider, 1999; Mero, 1999; Nissen et al., 2000).

Cell proliferation: It has been suggested that the Leu metabolite alpha-ketoisocaproate stimulates the phosphorylation of PHAS-1 (phosphorylated heat- and acid-stable protein regulated by insulin), a recently discovered regulator of translation initiation during cell mitogenesis (Xu et al., 1998). This effect appears to be particularly important for the promotion of beta-cell proliferation and augments the growth-stimulating effects of insulin and insulinlike growth factor 1 (IGF-1).

Ketogenesis: A significant proportion (40% of an ingested dose) is converted into acetyl-CoA and thereby contributes to the synthesis of ketones, steroids, fatty acids, and other compounds (Toth et al., 2001). Adipocytes in particular have a high capacity to generate ketone bodies from Leu.

Neurotransmitter metabolism: It has been suggested that Leu and the other BCAAs promote nitrogen transfer between astrocytes and neurons in the brain. This BCAA shuttle might be critical for the synthesis of the neurotransmitter L-glutamate (Hutson et al., 2001).

Toxicity: High Leu concentrations in the blood are potentially harmful. The mechanisms responsible for such toxicity are not yet well understood.

REFERENCES

Anthony, J.C., Anthony, T.G., Kimball, S.R., Jefferson, L.S., 2001. Signaling pathways involved in translational control of protein synthesis in skeletal muscle by leucine. J. Nutr. 131, 856S–860S.

Avissar, N.E., Ryan, C.K., Ganapathy, V., Sax, H.C., 2001. Na+-dependent neutral amino acid transporter ATB° is a rabbit epithelial cell brush border protein. Am. J. Physiol. Cell Physiol. 281, C963–C971.

Bloch, K., Clark, L.C., Haray, I., 1954. Utilization of branched chain acids in cholesterol synthesis. J. Biol. Chem. 211, 687–699.

Chairoungdua, A., Segawa, H., Kim, J.Y., Miyamoto, K., Haga, H., Fukui, Y., et al., 1999. Identification of an amino acid transporter associated with the cystinuria-related type II membrane glycoprotein. J. Biol. Chem. 274, 28845–28848.

Gietzen, D.W., Magrum, L.J., 2001. Molecular mechanisms in the brain involved in the anorexia of branched-chain amino acid deficiency. J. Nutr. 851S–855S.

Hutson, S.M., Lieth, E., LaNoue, K.F., 2001. Function of leucine in excitatory neurotransmitter metabolism in the central nervous system. J. Nutr. 131, 846S–850S.

Jansson, T., 2001. Amino acid transporters in the human placenta. Pediatr. Res. 49, 141–147.

Kreider, R.B., 1999. Dietary supplements and the promotion of muscle growth with resistance exercise. Sports Med. 27, 97–110.

Kurpad, A.V., Regas, M.M., Raj, T., Masuthy, K., Gnanou, J., Young, V.R., 2002. Intravenously infused ^{13}C-leucine is retained in fasting healthy adult men. J. Nutr. 132, 1906–1908.

Matthews, D.E., Marano, M.A., Campbell, R.G., 1993. Splanchnic bed utilization of leucine and phenylalanine in humans. Am. J. Physiol. 264, E109–E118.

Mero, A., 1999. Leucine supplementation and intensive training. Sports Med. 27, 347–358.

Nissen, S., Sharp, R.L., Panton, L., Vukovich, M., Trappe, S., Fuller Jr., J.C., 2000. β-hydroxy-β-methylbutyrate (HMB) supplementation in humans is safe and may decrease cardiovascular factors. J. Nutr. 130, 1937–1945.

Nissen, S.L., Abumrad, N.N., 1997. Nutritional role of the leucine metabolite β-hydroxy-β-methylbutyrate (HMB). J. Nutr. Biochem. 8, 300–311.

Poston, J.M., 1984. The relative carbon flux through the alpha- and the beta-keto pathways of leucine metabolism. J. Biol. Chem. 259, 2059–2061.

Ritchie, J.W., Taylor, P.M., 2001. Role of the System L permease LAT1 in amino acid and iodothyronine transport in placenta. Biochem. J. 356, 719–725.

Scislowski, P.W., Zolnierowicz, S., Swierczynski, J., Elewski, L., 1983. Leucine catabolism in human term placenta. Biochem. Med. 30, 141–145.

Toth, M.J., MacCoss, M.J., Poehlman, E.T., Matthews, D.E., 2001. Recovery of (13)CO(2) from infused [1-(13)C]leucine and [1,2-(13)C(2)]leucine in healthy humans. Am. J. Physiol. Endocrinol. Metab. 281, E233–E241.

Utsunomiya-Tate, N., Endou, H., Kanai, Y., 1996. Cloning and functional characterization of a system ASC-like Na$^+$-dependent neutral amino acid transporter. J. Biol. Chem. 271, 14883–14890.

Xu, G., Kwon, G., Marshall, C.A., Lin, T.A., Lawrence Jr., J.C., McDaniel, M.L., 1998. Branched-chain amino acids are essential in the regulation of PHAS-1 and p70 S6 kinase by pancreatic beta-cells. A possible role in protein translation and mitogenic signaling. J. Biol. Chem. 273, 28178–28184.

VALINE

The hydrophobic neutral amino acid L-valine (2-aminoisovaleric acid, 2-amino-3-methylbutylbutyric acid, alpha-aminoisovaleric acid, 2-amino-3-methylbutanoic acid, one-letter code V; molecular weight 117) contains 12.0% nitrogen (Figure 8.58).

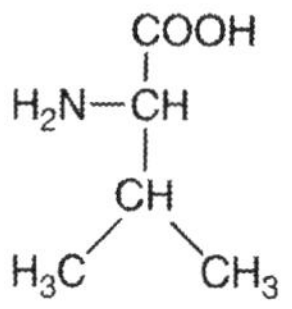

FIGURE 8.58

L-Valine.

ABBREVIATIONS

CoA	coenzyme A
ETF	electron-transferring flavoprotein
LAT1	L-type amino acid transporter 1 (SLC7A5)
LAT2	L-type amino acid transporter 2 (SLC7A8)
RDA	recommended dietary allowance
Val	L-valine

NUTRITIONAL SUMMARY

Function: The essential amino acid L-valine (Val) is needed for the synthesis of proteins. It is also used as an energy fuel; its complete oxidation requires thiamin, riboflavin, niacin, vitamin B6, vitamin B12, pantothenate, biotin, lipoate, ubiquinone, magnesium, and iron.

Food sources: The proteins in foods from animal and plant-sources typically contain about 40–60 mg/g Val. Dietary supplements with crystalline Val, often in combination with other amino acids, are commercially available.

Requirements: Adults are thought to require at least 20 mg/kg per day (Young and Borgonha, 2000).

Deficiency: Prolonged lack of Val, as with all essential amino acids or a lack of protein, causes growth failure, loss of muscle mass, and organ damage.

Excessive intake: Very high intake of protein and mixed amino acids (i.e., more than three times the RDA, which is 2.4 g/kg) is thought to increase the risk of renal glomerular sclerosis and accelerate osteoporosis.

DIETARY SOURCES

Dietary proteins from different sources all contain Val. Protein from milk, eggs, and rice has slightly higher content (more than 60 mg/g of total protein) than protein from meats, legumes, and corn (about 50 mg/g), or wheat (44 mg/g).

DIGESTION AND ABSORPTION

Proteins are hydrolyzed by various enzymes from gastric, pancreatic, and intestinal secretions. The hydrogen ion/peptide cotransporters (SLC15A1, PepT1), and to a lesser degree, PepT2 (SLC15A2) mediate uptake of dipeptides and tripeptides, including those containing Val. Free Val crosses the small intestinal brush border membrane via the sodium-driven transporter B° (Avissar et al., 2001). Like many other neutral amino acids, Val can be exchanged for other neutral amino acids across the brush border membrane in either direction by the rBAT (SLC3A1)–anchored amino acid transporter BAT1/b°,+ (SLC7A9) (Chairoungdua et al., 1999).

The 4F2 (SLC3A2) glycoprotein-anchored transporter LAT2 (SLC7A8) can move Val across the basolateral membrane in both directions in exchange for other neutral amino acids. The intestines sequester about 20% of ingested BCAAs after uptake. Only a small portion of this is oxidized, and most of it is incorporated into newly synthesized proteins or converted into alpha-keto acids and released (Matthews et al., 1993) (Figure 8.59).

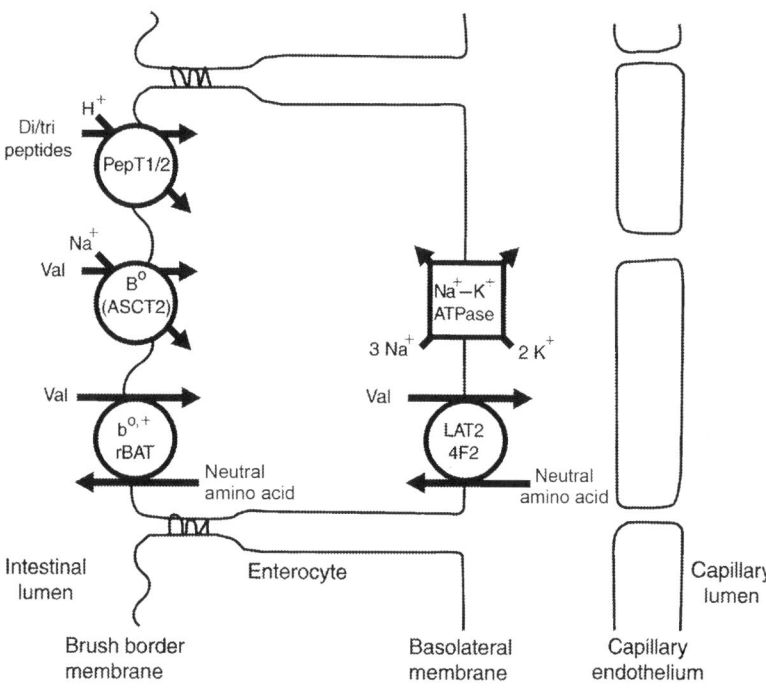

FIGURE 8.59

Intestinal absorption of L-valine.

TRANSPORT AND CELLULAR UPTAKE

Blood circulation: Cells can take up Val bound in the various blood proteins via mechanisms specific to the respective proteins. Free Val enters cells mainly via the sodium-driven transporter B° and transporters of the sodium-independent system L, including the glycoprotein 4F2 (SLC3A2)–linked LAT1 (SLC7A5) (Verrey et al., 1999). The BCAA transaminase (EC2.6.1.42), which is located at the inner mitochondrial membrane, may function as a transporter of BCAAs.

BBB: One or more members of the L-type amino acid transporter family mediate the exchange of Val for other neutral amino acids across the membranes on both sides of the brain. Their molecular identity is not yet established.

Materno-fetal transfer: The sodium-driven transporter B° in the brush border membrane mediates Val uptake into the syncytiotrophoblast; its transport capacity is expanded by the heteroexchanger LAT1/4F2 (Jansson, 2001).

A significant percentage of the Val taken up from maternal circulation is metabolized in the placenta (Cetin, 2001). The remainder is taken across the basolateral membrane by LAT2/4F2 in exchange for another neutral amino acid.

METABOLISM

The main pathway of Val breakdown to propionyl-CoA is initiated by transamination, followed by another six reactions (Figure 8.60). Propionyl-CoA can then be converted to the Krebs-cycle intermediate succinyl-CoA via three additional steps. The entire metabolic sequence requires vitamin B6, thiamin, riboflavin, niacin, pantothenic acid, lipoate, biotin, and vitamin B12. Val is glucogenic since it is broken down to

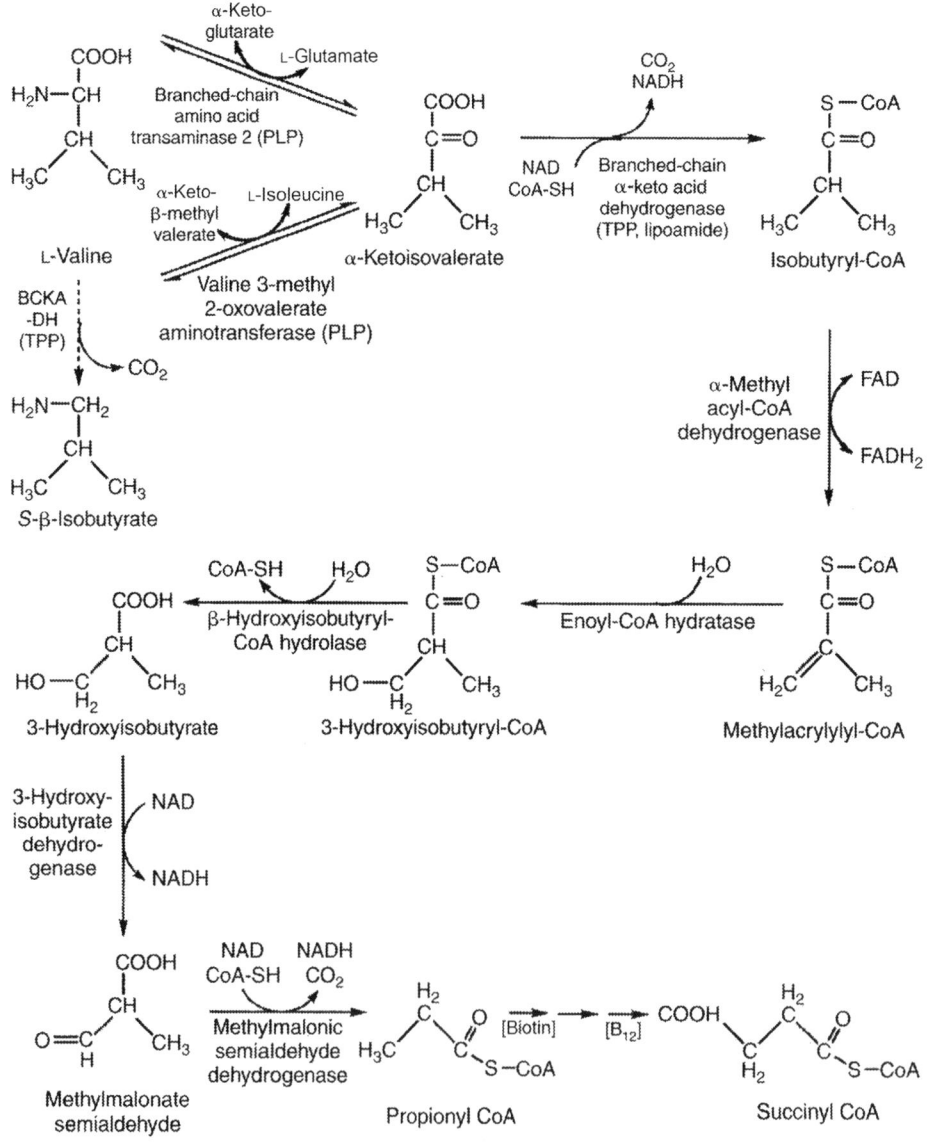

FIGURE 8.60

Metabolism of L-valine.

succinyl-CoA. The BCAA aminotransferase 2 (EC2.6.1.42) moves the amino group of Val to alpha-ketoglutarate. Alternatively, valine-3-methyl-2-oxovalerate aminotransferase (EC2.6.1.32) can use the amino group of Val to reconstitute the L-isoleucine metabolite (S)-3-methyl-2-oxopentanoate (Dancis et al., 1967). While the relative importance of this pathway remains to be elucidated, it has been found that Val concentrations in the blood are elevated in people with half the normal activity of this particular aminotransferase.

The next step is catalyzed by branched-chain alpha-keto acid dehydrogenase (EC1.2.4.4), which is a large complex in the mitochondrial matrix consisting of multiple copies of three distinct subunits. The subunit E1 catalyzes the decarboxylation reaction with reduced CoA as a cosubstrate. E1 itself is a heterodimer of an alpha chain, with thiamin pyrophosphate (TPP) as a prosthetic group, and a beta chain. Subunit E2 anchors the lipoic acid residue, which serves as an acceptor for the decarboxylated substrate, transfers it to acetyl-CoA, and reduces lipoamide to dihydrolipoamide in the process. The lipoamide dehydrogenase component, subunit E3 (EC1.8.1.4), transfers the hydrogen from dihydrolipoamide via its FAD group to NAD. The glycine cleavage system (EC1.4.4.2) and the dehydrogenases for pyruvate (EC1.2.4.1) and alphaketoglutarate (EC2.3.1.61) use the same enzyme subunit. The enzyme complex is inactivated by phosphorylation ([3-methyl-2-oxobutanoate dehydrogenase (lipoamide)] kinase, EC2.7.1.15), and reactivated by dephosphorylation ([3-methyl-2-oxobutanoate dehydrogenase (lipoamide)]-phosphatase, EC3.1.3.52). Branched-chain alpha-keto acid dehydrogenase is defective in maple syrup urine disease, which affects the breakdown of all three BCAAs, including L-leucine and L-isoleucine.

A small amount of Val appears to be decarboxylated directly to S-beta-aminoisobutyrate (BAIB) by branched-chain alpha-keto acid dehydrogenase (EC1.2.4.4). Most of the BAIB in the blood is the S-stereoisomer and thus derived from Val. R-BAIB, in contrast, which is the dominant stereoisomer excreted with urine, is derived from thymine breakdown. Methylmalonic semialdehyde dehydrogenase (EC1.2.1.27) can metabolize both forms to propionyl CoA.

Acyl-CoA dehydrogenase-8 (medium-chain acyl-CoA dehydrogenase, EC1.3.8.7) oxidizes isobutyryl-CoA to methylacrylyl-CoA (Telford et al., 1999; Andresen et al., 2000), and reduces FAD in the closely associated mitochondrial ETF. Its deficiency causes the accumulation and urinary excretion of isobutyrate derivatives (Roe et al., 1998). The extent to which other acyl-CoA dehydrogenases, particularly short-chain acyl-CoA dehydrogenase (EC1.3.99.2), also contribute to isobutyryl-CoA metabolism in various tissues remains to be seen.

A separate iron-sulfur flavoenzyme associated with the acyl-CoA dehydrogenases, ETF dehydrogenase (EC1.5.5.1), uses FADH2 to generate the oxidative phosphorylation fuel ubiquinol. Subsequent reactions are catalyzed by 3-hydroxybutyryl-CoA dehydratase (crotonase, EC4.2.1.55), 3-hydroxyisobutyryl-CoA hydrolase (EC3.1.2.4), 3-hydroxybutyrate dehydrogenase (EC1.1.1.31), and methylmalonic semialdehyde dehydrogenase (EC1.2.1.27). The product of the last reaction is propionyl-CoA. Conversion of propionyl-CoA in three additional steps to the Krebs-cycle intermediate succinyl-CoA requires biotin and cobalamin.

While there is little transamination of Val in the liver due to the low activity (compared to muscle) of the BCAA aminotransferase (EC2.6.1.42), the potentially toxic product methacrylyl-CoA is rapidly metabolized in the liver due to the relatively high activity of oxoglutarate dehydrogenase (Taniguchi et al., 1996).

The concentration of the Val transamination product, alpha-ketoisovaleric acid, does not increase after a meal. Its concentration in muscle is only twice that of its hydroxylated metabolite, alpha-hydroxyisovaleric acid (Hoffer et al., 1993). This pattern differs from those resulting with the other two BCAAs: the concentrations of their transamination products increase after a meal and are a 100-fold greater than the respective hydroxylated metabolites.

STORAGE

Most Val is bound in proteins, much of which is in muscle, which contains around 60 mg/g protein. A 70-kg healthy adult might thus have reserves of about 360 g mobilizable Val.

EXCRETION

Recovery of Val from glomerular filtrate in the proximal renal tubule closely resembles absorption from the small intestinal lumen. The sodium-driven transporter B° (Avissar et al., 2001) carries the bulk into the cell with some help from the heteroexchanger BAT1/b$^{o,+}$ (Chairoungdua et al., 1999). Another heteroexchanger, LAT2 (SLC7A8) linked to the membrane glycoprotein 4F2 (SLC3A2), completes the process by transporting Val across the basolateral membrane in exchange for other neutral amino acids. Dipeptides and tripeptides are taken up via PepT1 (SLC15A1) and PepT2 (SLC15A2), and cleaved in the tubular cell. Since most Val is reabsorbed, healthy people lose very little via urine. Losses via feces are also negligible, so long as gastrointestinal function remains normal.

Most nitrogen from metabolized Val is excreted eventually via urine as urea.

REGULATION

Branched-chain alpha-keto acid dehydrogenase (EC2.6.1.42) is the key regulatory enzyme for Val metabolism, which helps preserve this essential amino acid for protein synthesis during starvation. The enzyme is inactivated by a specific, calmodulin-dependent kinase ([3-methyl-2-oxobutanoate dehydrogenase (lipoamide)] kinase, EC2.7.1.115) and activated by a specific phosphatase ([3-methyl-2-oxobutanoate dehydrogenase (lipoamide)]-phosphatase, EC3.1.3.52). Branched-chain alpha-keto acid dehydrogenase is also inhibited by the products of the reaction it catalyzes (namely, isovaleryl-CoA, isobutyryl-CoA, and 2-methylbutyryl-CoA).

An intake imbalance with inadequacy of one of the BCAAs (Val, leucine, or isoleucine) causes the activation of orbitofrontal cortex, frontal gyrus, and thalamus and leads to the development of conditioned taste aversion (Gietzen and Magrum, 2001). Expression of several genes in response to inadequacy changes, but the mechanisms responsible for learning to balance BCAA intake remain elusive.

FUNCTION

Energy fuel: Eventually, most Val is broken down, providing about 5.3 kcal/g. A large percentage of ingested Val is oxidized soon after absorption. The main site of Val breakdown is skeletal muscle. The liver, intestines, and other organs are responsible for much smaller percentages of Val utilization as an energy fuel.

Protein synthesis: Virtually all proteins contain Val as part of their specific sequence. Valine-tRNA ligase (EC6.1.1.9) is responsible for loading Val onto its specific tRNA. A lack of Val, as with all essential amino acids, slows protein synthesis. Like other BCAAs, Val has anabolic effects on skeletal muscle (Anthony et al., 2001). The effect of L-leucine is better understood, however.

Neurotransmitter metabolism: It has been suggested that Val and the other BCAAs promote nitrogen transfer between astrocytes and neurons in the brain. This BCAA shuttle might be critical for the synthesis of the neurotransmitter L-glutamate (Hutson et al., 2001).

REFERENCES

Andresen, B.S., Christensen, E., Corydon, T.J., Bross, P., Pilgaard, B., Wanders, R.J., et al., 2000. Isolated 2-methylbutyrylglycinuria caused by short/branched-chain acyl-CoA dehydrogenase deficiency: identification of a new enzyme defect, resolution of its molecular basis, and evidence for distinct acyl-CoA dehydrogenases in isoleucine and valine metabolism. Am. J. Hum. Genet. 67, 1095–1103.

Anthony, J.C., Anthony, T.G., Kimball, S.R., Jefferson, L.S., 2001. Signaling pathways involved in translational control of protein synthesis in skeletal muscle by leucine. J. Nutr. 131, 856S–860S.

Avissar, N.E., Ryan, C.K., Ganapathy, V., Sax, H.C., 2001. Na^+-dependent neutral amino acid transporter ATB^0 is a rabbit epithelial cell brush border protein. Am. J. Physiol. Cell Physiol. 281, C963–C971.

Cetin, I., 2001. Amino acid interconversions in the fetal-placental unit: the animal model and human studies *in vivo*. Ped. Res. 49, 148–153.

Chairoungdua, A., Segawa, H., Kim, J.Y., Miyamoto, K., Haga, H., Fukui, Y., et al., 1999. Identification of an amino acid transporter associated with the cystinuria-related type II membrane glycoprotein. J. Biol. Chem. 274, 28845–28848.

Dancis, J., Hutzler, J., Tada, K., Wada, Y., Morikawa, T., Arakawa, Y., 1967. Hypervalinemia: a defect in valine transamination. Pediatrics 39, 813–817.

Gietzen, D.W., Magrum, L.J., 2001. Molecular mechanisms in the brain involved in the anorexia of branched-chain amino acid deficiency. J. Nutr., 851S–855S.

Hoffer, L.J., Taveroff, A., Robitaille, L., Mamer, O.A., Reimer, M.E., 1993. Alpha-keto and alpha-hydroxy branched-chain acid interrelationships in normal humans. J. Nutr. 123, 1513–1521.

Hutson, S.M., Lieth, E., LaNoue, K.E., 2001. Function of leucine in excitatory neurotransmitter metabolism in the central nervous system. J. Nutr. 131, 846S–850S.

Jansson, T., 2001. Amino acid transporters in the human placenta. Pediatr. Res. 49, 141–147.

Matthews, D.E., Marano, M.A., Campbell, R.G., 1993. Splanchnic bed utilization of leucine and phenylalanine in humans. Am. J. Physiol. 264, E109–E118.

Roe, C.R., Cederbaum, S.D., Roe, D.S., Mardach, R., Galindo, A., Sweetman, L., 1998. Isolated isobutyryl-CoA dehydrogenase deficiency: an unrecognized defect in human valine metabolism. Mol. Genet. Metab. 65, 264–271.

Taniguchi, K., Nonami, T., Nakao, A., Harada, A., Kurokawa, T., Sugiyama, S., et al., 1996. The valine catabolic pathway in human liver: effect of cirrhosis on enzyme activities. Hepatology 24, 1395–1398.

Telford, E.A., Moynihan, L.M., Markham, A.E., Lench, N.J., 1999. Isolation and characterization of a cDNA encoding the precursor for a novel member of the acyl-CoA dehydrogenase gene family. Biochim. Biophys. Acta 1446, 371–376.

Verrey, F., Jack, D.E., Paulsen, I.T., Saier Jr., M.H., Pfeiffer, R., 1999. New glycoprotein-associated amino acid transporters. J. Membrane Biol. 172, 181–192.

Young, V.R., Borgonha, S., 2000. Nitrogen and amino acid requirements: the Massachusetts Institute of Technology amino acid requirement pattern. J. Nutr. 130, 1841S–1849S.

ISOLEUCINE

The hydrophobic neutral amino acid L-isoleucine (2-amino-3-methylvaleric acid, alpha-amino-beta-methylvaleric acid, 2-amino-3-methylpentanoic acid, one-letter code I; molecular weight 131) contains 10.7% nitrogen (Figure 8.61).

ABBREVIATIONS

BCAA	branched-chain amino acid
CoA	coenzyme A
ETF	electron-transferring flavoprotein
FAD	flavin adenine dinucleotide
Ile	L-isoleucine
LAT1	L-type amino acid transporter 1 (SLC7A5)
LAT2	L-type amino acid transporter 2 (SLC7A8)
RDA	recommended dietary allowance
TPP	thiamin pyrophosphate

NUTRITIONAL SUMMARY

Function: The essential amino acid L-isoleucine (Ile) is needed for the synthesis of proteins. It is also an important energy fuel, especially in skeletal muscle; its breakdown requires thiamin, riboflavin, niacin, vitamin B6, vitamin B12, pantothenate, biotin, lipoate, ubiquinone, magnesium, and iron.

Food sources: Dietary proteins from different sources all contain Ile. Protein from dairy has a slightly higher content (6% of total protein) than protein from meats and soy (around 5% of total protein) or grains (3–4%). Dietary supplements containing crystalline Ile, often in combination with other amino acids, are commercially available.

Requirements: Adults are thought to require at least 23 mg/kg per day (Young and Borgonha, 2000).

Deficiency: Prolonged lack of Ile, as with all essential amino acids or a lack of protein, causes growth failure, loss of muscle mass, and organ damage.

Excessive intake: Very high intake of protein and mixed amino acids (i.e., more than three times the RDA, which is 2.4 g/kg) is thought to increase the risk of renal glomerular sclerosis and accelerate osteoporosis.

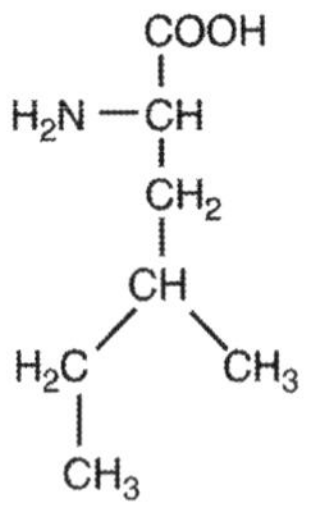

FIGURE 8.61

L-Isoleucine.

DIETARY SOURCES

All food proteins contain Ile, though the relative content varies slightly between different types of foods. Relatively rich sources are proteins from milk (60 mg/g), eggs (55 mg/g), and soy (52 mg/g). Intermediate content is found in meats, such as chicken (49 mg/g), pork (47 mg/g), and beef (45 mg/g), as well as fish (46 mg/g). Lower content is typical for grains such as corn (36 mg/g) and wheat (39 mg/g), and slightly more in rice (43 mg/g). Overall, the amount of protein consumed has greater impact on Ile intake than the source of the protein.

DIGESTION AND ABSORPTION

Proteins are hydrolyzed by various enzymes; most of them are derived from the stomach wall and pancreas. The hydrogen ion/peptide cotransporters 1 (PepT1, SLC15A1) and, to a lesser extent, 2 (PepT2, SLC 15A2) mediate uptake of dipeptides and tripeptides, including those containing Leu. Free Leu can cross the small intestinal brush border membrane via the sodium-driven transporter B° (identical with system B/NBB, Avissar et al., 2001), sodium-dependent system L, and a specific sodium/leucine cotransporter, which is only partially characterized to date (Figure 8.62).

Like many other neutral amino acids, Ile can be exchanged for other neutral amino acids across the brush border membrane in either direction by the rBAT (SLC3A1)–anchored amino acid transporter BAT1/$b^{0,+}$ (SLC7A9, Chairoungdua et al., 1999). The 4F2 (SLC3A2) glycoprotein-anchored

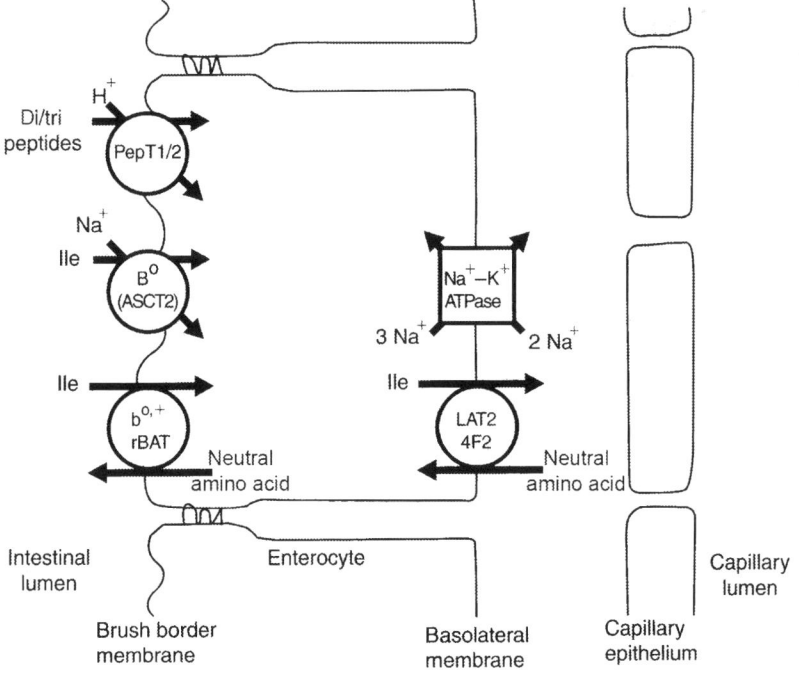

FIGURE 8.62

Intestinal absorption of L-isoleucine.

transporter LAT2 (SLC7A8) can exchange Ile across the basolateral membrane in both directions in exchange for other neutral amino acids.

TRANSPORT AND CELLULAR UPTAKE

Blood circulation: Ile is a ubiquitous component of proteins in the blood and can be taken up by cells via mechanisms specific to the respective proteins. Free Ile (plasma concentrations are around 65 µmol/l) enters cells mainly via system L, including the specifically identified heteroexchanger LAT1 (SLC7A5). The BCAA transaminase (EC2.6.1.42), which is located at the inner mitochondrial membrane, may function as an importer of BCAAs.

Materno-fetal transfer: Ile uptake across the microvillous membrane of the syncytiotrophoblast is mediated by LAT1 (Ritchie and Taylor, 2001) and transport across the basal membrane proceeds via LAT2, both in exchange for other neutral amino acids (Jansson, 2001). The driving force for LAT1/LAT2-mediated transport is the concentration gradient of small neutral amino acids (glycine, L-alanine, L-cysteine) established by the sodium-dependent transport systems A and ASC.

BBB: System L mediates Ile transport across both sides of the neuroendothelial cell layer. The molecular identity and location of the responsible transporter(s) remain unclear.

METABOLISM

Ile breakdown to propionyl-CoA and acetyl-CoA proceeds in six steps that depend on adequate supplies of riboflavin, pyridoxine, niacin, and pantothenate (Figure 8.63). Complete oxidation of propionyl-CoA and acetyl-CoA then requires thiamin, cobalamin, biotin, and ubiquinone.

The PLP-dependent BCAA aminotransferase 2 (EC2.6.1.42) starts Ile metabolism by moving its amino group to alphaketoglutarate. Another minor PLP-dependent enzyme, valine-3-methyl-2-oxovalerate aminotransferase (EC2.6. 1.32), can transfer the amino group to alpha-ketovalerate, thus reconstituting L-valine (Dancis et al., 1967). The reverse reaction, which would reconstitute Ile, might actually be more important. Branched-chain alpha-keto acid dehydrogenase (EC1.2.4.4) in the mitochondrial matrix acts on all three BCAAs. This large enzyme complex consists of multiple copies of three distinct subunits. The subunit El catalyzes the decarboxylation reaction with reduced CoA as a cosubstrate. E1 itself is a heterodimer of an alpha chain, with TPP as a prosthetic group, and a beta chain. Subunit E2 anchors the lipoic acid residue, which serves as an acceptor for the decarboxylated substrate, transfers it to acetyl-CoA, and reduces lipoamide to dihydrolipoamide in the process. The lipoamide dehydrogenase component, subunit E3 (EC1.8.1.4), transfers the hydrogen from dihydrolipoamide via its FAD group to NAD. The glycine cleavage system (EC1.4.4.2) and the dehydrogenases for pyruvate (EC1.2.4.1) and alpha-ketoglutarate (EC2.3.1.61) use the same enzyme subunit. The enzyme complex is inactivated by phosphorylation ([3-methyl-2-oxobutanoate dehydrogenase (lipoamide)] kinase, EC2.7.1.115), and reactivated by dephosphorylation ([3-methyl-2-oxobutanoate dehydrogenase (lipoamide)]-phosphatase, EC3.1.3.52). Branched-chain alpha-keto acid dehydrogenase is defective in maple syrup urine disease, which affects the breakdown of all three BCAAs.

The acyl-CoA dehydrogenase responsible for oxidation of alpha-methylbutyryl-CoA to tiglyl-CoA (Andresen et al., 2000) is short-branched-chain acyl-CoA dehydrogenase (EC1.3.99.2, FAD-containing). Short-chain acyl-CoA dehydrogenase (EC1.3.99.2), and possibly additional acyl-CoA dehydrogenases, also appears to oxidize alphamethylbutyryl-CoA, though with less specific activity.

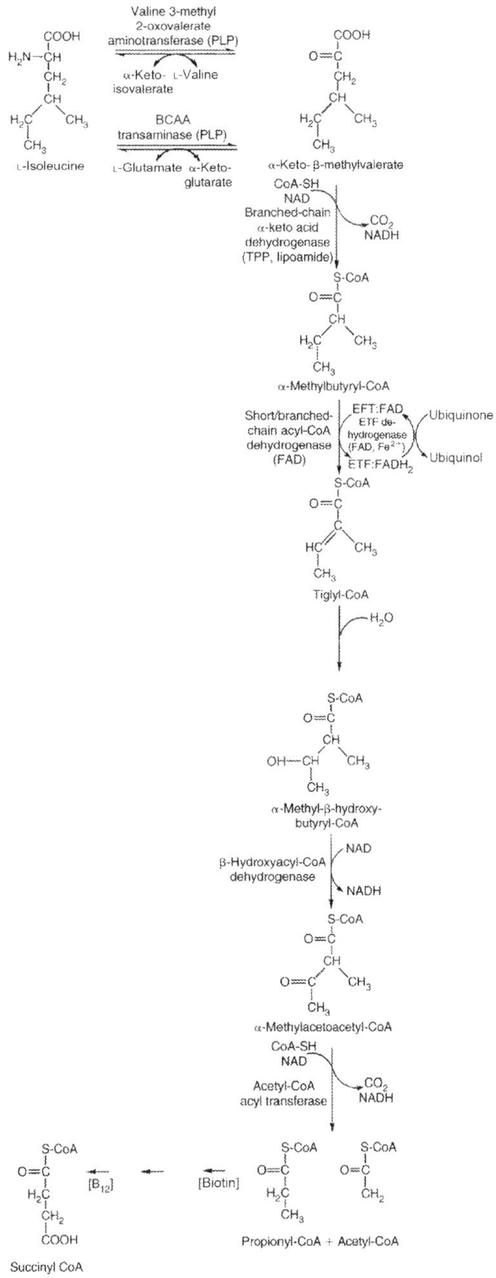

FIGURE 8.63

Metabolism of L-isoleucine.

An iron-sulfur flavoenzyme associated with the acyl-CoA dehydrogenases, ETF dehydrogenase (EC1.5.5.1, FAD-containing), uses the reduced FAD-moiety of ETF to generate the oxidative phosphorylation fuel ubiquinol. The next three steps are catalyzed by enzymes otherwise used for short-chain fatty acid beta-oxidation, 3-hydroxybutyryl-CoA dehydratase (crotonase, EC4.2.1.55), beta-hydroxyacyl-CoA dehydrogenase (EC1.1.1.35), and acetyl-CoA acyltransferase (EC2.3.1.16). The intermediates generated by the last step, propionyl-CoA and acetyl-CoA, can then be metabolized further via typical reactions.

STORAGE

Body protein contains 48 mg/g (Smith, 1980). Proteins of skeletal muscle are especially Ile-rich (Mero, 1999). Protein breakdown during fasting, starvation, or severe illness provides much-needed amino acid precursors for the synthesis of acute-phase proteins.

EXCRETION

Healthy individuals excrete very little Ile. Whatever is shed or excreted into the intestines is reabsorbed quite efficiently via mechanisms described previously. Most of the free Ile filtered by the renal glomerulum is taken up from the proximal tubule via the sodium-driven brushborder transporter B° (Avissar et al., 2001). The rBAT (SLC3A1)–anchored heteroexchanger BAT1/$b^{o,+}$ (SLC7A9, Chairoungdua et al., 1999) can augment the action of the bulk transporter. Transport across the basolateral membrane depends mainly on the 4F2 (SLC3A2) glycoprotein-anchored transporter LAT2 (SLC7A8).

REGULATION

Branched-chain alpha-keto acid dehydrogenase (EC2.6.1.42) is the key regulatory enzyme for Ile (and equally for L-valine and L-leucine) metabolism, which helps preserve this essential amino acid for protein synthesis during starvation. The enzyme is inactivated by a specific, calmodulin-dependent kinase ([3-methyl-2-oxobutanoate dehydrogenase (lipoamide)] kinase, EC2.7.1.115) and activated by a specific phosphatase ([3-methyl-2-oxobutanoate dehydrogenase (lipoamide)]-phosphatase, EC3.1.3.52).

Branched-chain alpha-keto acid dehydrogenase is also inhibited by the products of the reaction it catalyzes (namely, isovaleryl-CoA, isobutyryl-CoA, and 2-methylbutyryl-CoA). An intake imbalance with inadequacy of one of the BCAAs (valine, leucine, or isoleucine) causes activation of the orbitofrontal cortex, frontal gyrus, and thalamus and leads to the development of conditioned taste aversion (Gietzen and Magrum, 2001). The expression of several genes in response to inadequacy has changed, but the mechanisms responsible for learning to balance BCAA intake remain elusive.

FUNCTION

Energy fuel: Eventually, most Ile is broken down, providing about 5.9 kcal/g. As for L-leucine and L-valine, the main site of Ile oxidation is skeletal muscle. The liver, intestines, and other organs are responsible for much smaller percentages of Ile utilization as an energy fuel.

Protein synthesis: Most human proteins contain Ile as part of their specific sequence. Therefore, a lack of this essential amino acid would limit their synthesis.

REFERENCES

Andresen, B.S., Christensen, E., Corydon, T.J., Bross, P., Pilgaard, B., Wanders, R.J., et al., 2000. Isolated 2-methylbutyrylglycinuria caused by short/branched-chain acyl-CoA dehydrogenase deficiency: identification of a new enzyme defect, resolution of its molecular basis, and evidence for distinct acyl-CoA dehydrogenases in isoleucine and valine metabolism. Am. J. Human Genet. 67, 1095–1103.

Avissar, N.E., Ryan, C.K., Ganapathy, V., Sax, H.C., 2001. Na$^+$-dependent neutral amino acid transporter ATB^0 is a rabbit epithelial cell brush border protein. Am. J. Physiol. Cell Physiol. 281, C963–C971.

Chairoungdua, A., Segawa, H., Kim, J.Y., Miyamoto, K., Haga, H., Fukui, Y., et al., 1999. Identification of an amino acid transporter associated with the cystinuria-related type II membrane glycoprotein. J. Biol. Chem. 274, 28845–28848.

Dancis, J., Hutzler, J., Tada, K., Wada, Y., Morikawa, T., Arakawa, T., 1967. Hypervalinemia: a defect in valine transamination. Pediatrics 39, 813–817.

Gietzen, D.W., Magrum, L.J., 2001. Molecular mechanisms in the brain involved in the anorexia of branched-chain amino acid deficiency. J. Nutr., 851S–855S.

Harris, R.A., Hawes, J.W., Popov, K.M., Zhao, Y., Shimomura, Y., Sato, J., et al., 1997. Studies on the regulation of the mitochondrial alpha-ketoacid dehydrogenase complexes and their kinases. Adv. Enzyme Reg. 37, 271–293.

Jansson, T., 2001. Amino acid transporters in the human placenta. Pediatr. Res. 49, 141–147.

Mero, A., 1999. Leucine supplementation and intensive training. Sports Med. 27, 347–358.

Ritchie, J.W., Taylor, P.M., 2001. Role of the System L permease LAT1 in amino acid and iodothyronine transport in placenta. Biochem. J. 356, 719–725.

Smith, R.H., 1980. Comparative amino acid requirements. Proc. Nutr. Soc. 39, 71–78.

Young, V.R., Borgonha, S., 2000. Nitrogen and amino acid requirements: the Massachusetts Institute of Technology amino acid requirement pattern. J. Nutr. 130, 1841S–1849S.

ASPARTATE

The acidic amino acid L-aspartate (aspartic acid, asparagic acid, aminosuccinic acid, 2-aminobutanedioic acid, one-letter code D; molecular weight 133) contains 10.5% nitrogen (Figure 8.64).

ABBREVIATIONS

Asp   L-aspartate
PLP   pyridoxal-5′-phosphate

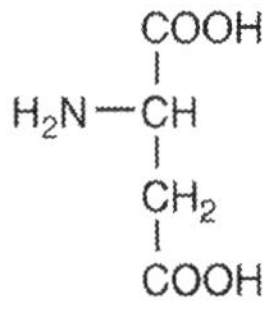

FIGURE 8.64

L-Aspartate.

NUTRITIONAL SUMMARY

Function: The nonessential amino acid L-aspartate (Asp) is used for the synthesis of L-asparagine, proteins, pyrimidine and purine nucleotides, and the neurotransmitter *N*-methyl-D-aspartate. Its complete oxidation requires thiamin, riboflavin, niacin, vitamin B6, pantothenate, lipoate, ubiquinone, iron, and magnesium.

Food sources: Dietary proteins from different sources all contain Asp.

Requirements: Since it can be synthesized from oxaloacetate, dietary intake of Asp is not necessary, so long as enough total protein as a source of the amino group is available.

Deficiency: Prolonged lack of total protein causes growth failure, loss of muscle mass, and organ damage.

Excessive intake: Very high intake of protein and mixed amino acids (i.e., more than three times the RDA, which is 2.4 g/kg) is thought to increase the risk of renal glomerular sclerosis and accelerate osteoporosis. Information on the specific risks from high intake of Asp alone is lacking.

ENDOGENOUS SOURCES

The Krebs-cycle intermediate oxaloacetate is the abundantly available precursor for Asp synthesis. Aspartate aminotransferase (EC2.6.1.1), which contains PLP like all known aminotransferases, uses the amino group from L-glutamate for Asp synthesis. A few other enzymes, such as alanine-oxomalonate aminotransferase (EC2.6.1.47), catalyze similar but quantitatively minor reactions. Another minor source of Asp is the deamination of L-asparagine.

DIETARY SOURCES

Since all food proteins contain a significant percentage of Asp, total protein intake is a major determinant of Asp intake. Relatively Asp-rich proteins are consumed with soybeans (124 mg/g total protein) and other legumes. Proteins with intermediate Asp content are consumed with fish (100 mg/g), eggs (100 mg/g), rice (94 mg/g), and most meats (around 90 mg/g). Wheat (50 mg/g) is at the low end of this distribution.

DIGESTION AND ABSORPTION

Hydrochloric acid and various proteases from the stomach, pancreas, and small intestine break down food proteins. The resulting mixture of free amino acids and small peptides is absorbed nearly completely from the duodenum and jejunum (Figure 8.65).

The hydrogen ion/peptide cotransporters PepT1 (SLC15A1) and PepT2 (SLC15A2) take up Asp-containing dipeptides and tripeptides. Free Asp can enter intestinal cells via the XAG transport system (including EAAT3/SLC1A1). Three sodium ions are cotransported with each aspartate ion in exchange for one potassium ion. The molecular identity of the transporter(s) that mediate efflux mechanism is not yet established.

TRANSPORT AND CELLULAR UPTAKE

Blood circulation: Most Asp in the blood is bound in proteins, and only minimal amounts (2 μmol/l or less) circulate as free amino acid (Hammarqvist et al., 2001). Several high-affinity glutamate transporters also accept Asp, including the excitatory amino acid transporters 1 (EAAT1, GLAST1, SLC1A3), 3 (EAAT3, SLC1A1), and 4 (EAAT4, SLC1A6).

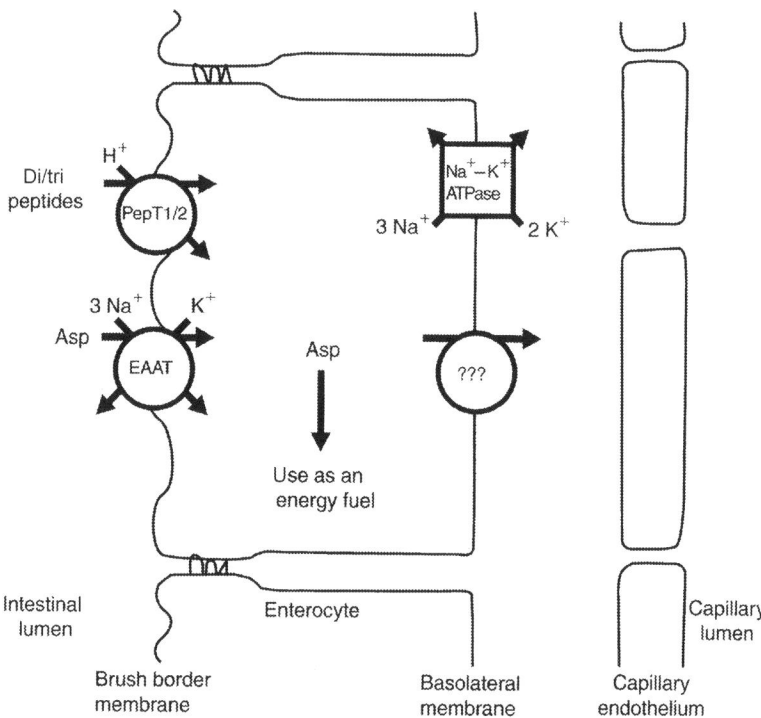

FIGURE 8.65

Intestinal absorption of ʟ-aspartate.

Mitochondrial translocation: The malate–aspartate shuttle moves the carbon skeleton and amino group for Asp synthesis from cytosol into mitochondria, assembles Asp there, and transports it back out into cytosol (Figure 8.66). Malate from the cytosolic Krebs cycle crosses into mitochondria via dicarbo-xylate translocase (SLC25A11) and is oxidized to oxaloacetate by malate dehydrogenase (EC1.1.1.37). The glutamate for oxaloacetate transamination is brought into mitochondria by glutamate-aspartate translocase in exchange for Asp. This exchange is especially important in the liver to provide Asp for the cytosolic steps of the urea cycle. In all other tissues including the liver, the shuttle moves reducing equivalents from cytosol into mitochondria to fuel oxidative phosphorylation.

METABOLISM

The two major catabolic fates of Asp are transamination and utilization of the resulting oxaloacetate, and utilization in the urea cycle. A small amount of Asp is directly converted to alanine by aspartate 4-decarboxylase (EC4.1.1.12, PLP-dependent) in the liver and kidneys (Rathod and Fellman, 1985).

Transamination: The PLP-dependent aspartate aminotransferase (EC2.6.1.1) moves the amino group from Asp to alpha-ketoglutarate. Both cytosolic and mitochondrial isoforms are abundant. Quantitatively minor transamination reactions are also catalyzed by a few other enzymes. Oxaloacetate then can be metabolized further through the standard Krebs-cycle reactions.

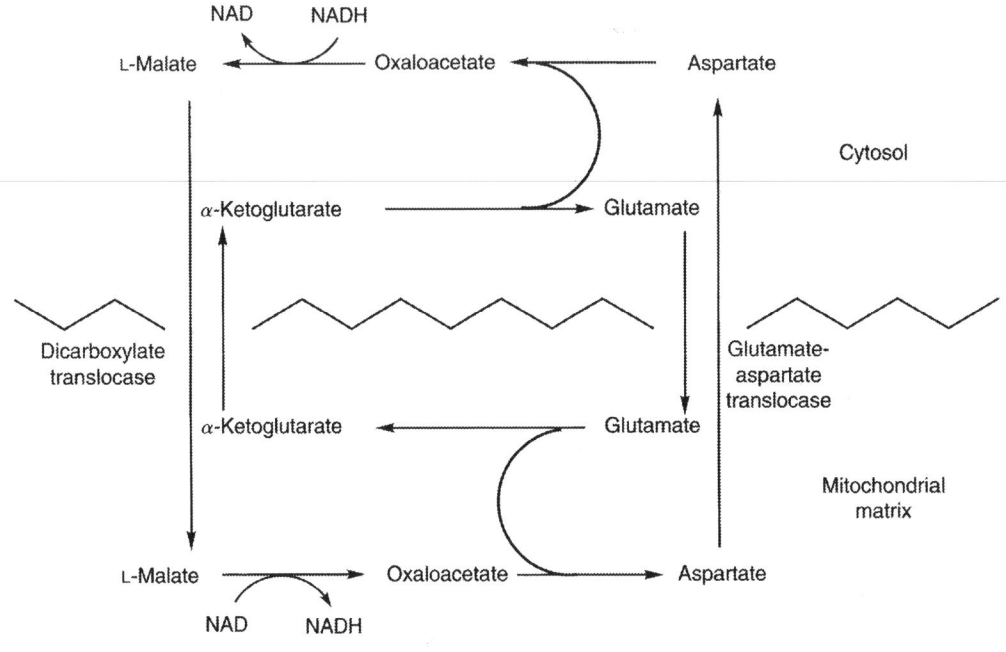

FIGURE 8.66

Malate–aspartate shuttle moves reducing equivalents and the precursors for aspartate synthesis into mitochondria.

Urea cycle: Asp functions as a critical substrate for urea synthesis in the mitochondria of the liver and kidneys. Argininosuccinate synthase (EC6.3.4.5) catalyzes the initial synthesis of argininosuccinate from Asp and citrulline. Argininosuccinate lyase (EC4.3.2.1) releases L-arginine and fumarate, and arginase (EC3.5.3.1) finally releases urea and ornithine (Figure 8.67).

Protein modification: Amino-acid N-acetyltransferase (EC2.3.1.1) and aspartate N-acetyltransferase (EC2.3.1.17) catalyze the acylation of aspartyl residues in some proteins, especially in the brain. Aspartoacylase (EC3.5.1.15) deacetylates these aspartyl residues again.

Specific aspartyl residues in some vitamin K–dependent coagulation factors are hydroxylated by the iron-containing peptide-aspartate beta-dioxygenase (EC1.14.11.16). Aspartyl residues in myelin basic protein (MBP), a long-lived brain protein, gradually racemize to their D form (Shapira et al., 1988). Protein-L-isoaspartate (D-aspartate) O-methyltransferase (EC2.1.1.77) facilitates the SAM-dependent methylation of chemically altered aspartyl residues in proteins. This modification is a critical step in the removal of damaged proteins.

STORAGE

Proteins in muscles and other organs contain large amounts of Asp that can be released in response to low protein intake.

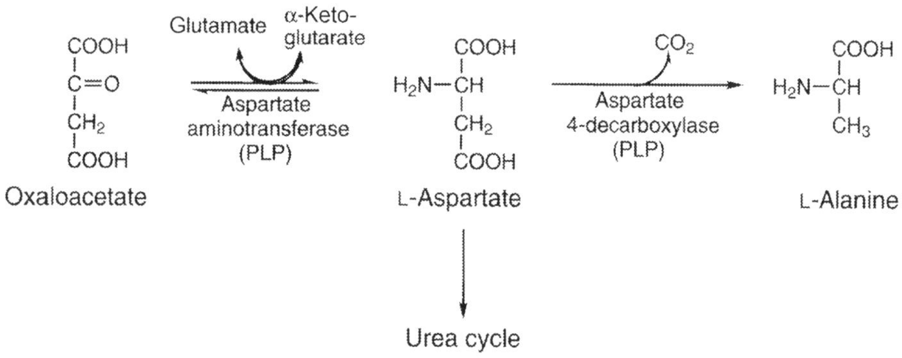

FIGURE 8.67

L-Aspartate catabolism by transamination, decarboxylation, or via urea synthesis.

EXCRETION

Asp losses with urine and feces are minimal in healthy people. Due to the low concentration in plasma, very little is filtered through the kidneys. Uptake from the tubular lumen proceeds via the sodium-linked XAG transport system (SLC1A1) and the sodium-independent rBAT (SLC3A1)–linked transporter BAT1 (SLC7A9). The sodium-independent aspartate/glutamate transporter 1 (AGT1, SLC7A13), associated with an as-yet-unidentified membrane-anchoring glycoprotein, moves Asp across the basolateral membrane (Matsuo et al., 2002). Sodium- and potassium-dependent transport has also been reported (Sacktor et al., 1981), but the identity of the responsible transporter(s) remains unknown.

REGULATION

Urea cycle: If Asp is abundantly available, the amino group is used to a significant extent for carbamoylphosphate synthesis (via glutamate and ammonia). If Asp supplies are on the low side, on the other hand, additional amounts are generated by transaminating oxaloacetate. Balancing Asp synthesis and breakdown thus maintain a steady stream of both carbamoylphosphate and Asp for urea synthesis.

FUNCTION

Protein synthesis: Asp is a constituent of most proteins and peptides synthesized in the body. Aspartate-tRNA ligase (EC6.1.1.12) links Asp with its correspondent tRNA in an ATP/magnesium-dependent reaction. Asp residues in proteins can be posttranslationally carboxylated, paralleling the vitamin K–dependent gamma-carboxylation of L-glutamate residues. Information on this type of protein modification is still very sparse.

Energy fuel: Eventually, most ingested Asp is used as an energy fuel (2.3 kcal/g). Asp transamination and subsequent oxidation requires adequate supplies of thiamin, riboflavin, niacin, vitamin B6, pantothenate, lipoate, ubiquinone, iron, and magnesium. Asp is a gluconeogenic amino acid since its metabolite oxaloacetate can be converted to glucose.

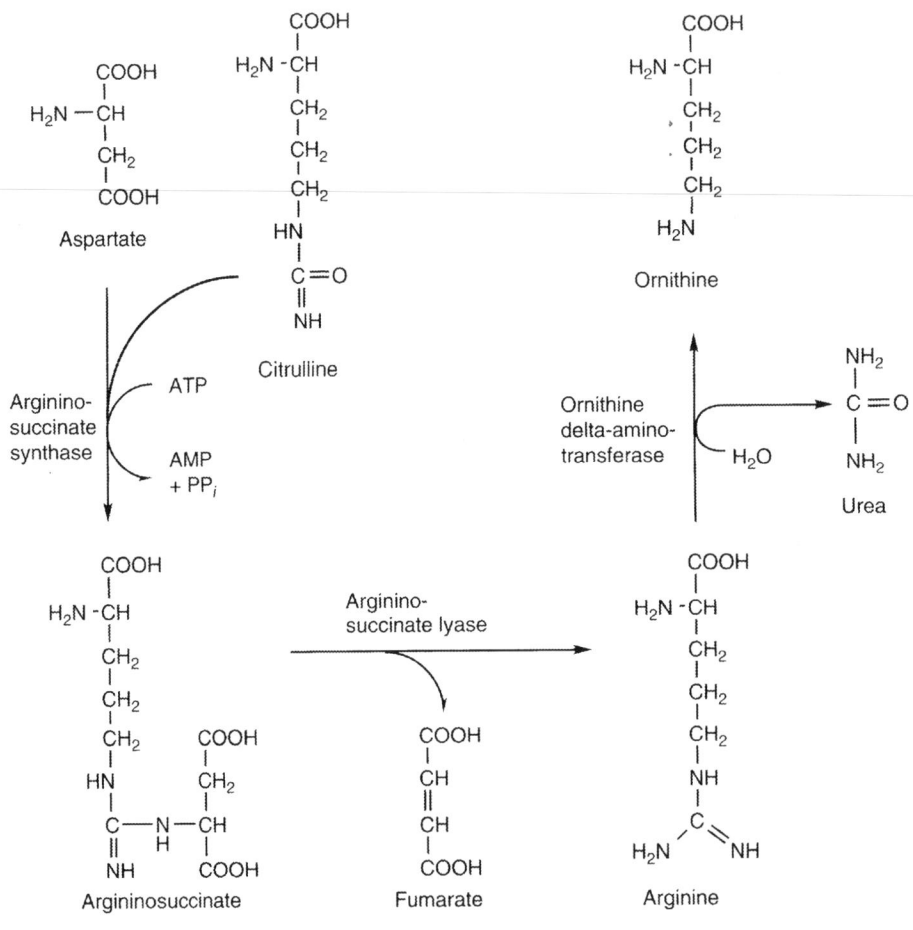

FIGURE 8.68

L-Aspartate is an important precursor for urea synthesis.

Asparagine synthesis: Glutamine-hydrolyzing asparagine synthase (EC6.3.5.4) and aspartate-ammonia ligase (EC6.3.1.1) catalyze the ATP-driven transamination of Asp to L-asparagine.

Arginine synthesis: Mitochondrial argininosuccinate synthase (EC6.3.4.5) condenses Asp and citrulline to argininosuccinate. The cleavage of this intermediate by argininosuccinate lyase (EC4.3.2.1) releases L-arginine and fumarate (Figure 8.68). While urea is the main product of this pathway, some arginine is not immediately hydrolyzed and becomes available to cover about one-fifth of the body's arginine needs (Wu and Morris, 1998).

Nucleotide synthesis: Asp provides two of the five nitrogen atoms in adenosine nucleotides, one of the four nitrogens in guanosine nucleotides, and one of the nitrogens in pyrimidine nucleotides (uridine, thymine, and cytosine) (Figure 8.69).

Vitamin metabolism: Asp can donate an amino group donor in the conversion of pyridoxal to pyridoxamine by pyridoxamine-oxaloacetate aminotransferase (EC2.6.1.31).

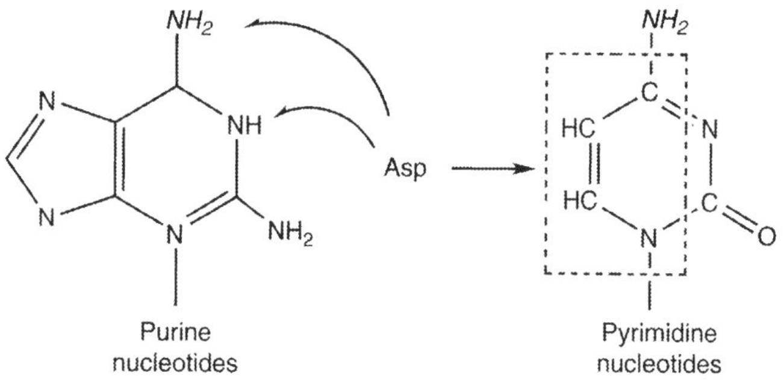

FIGURE 8.69

Nucleotide synthesis depends on L-aspartate.

D-*aspartate*: Incompletely characterized enzymes (probably PLP-dependent) in the cerebral cortex and other brain structures, adrenals, testis, and possibly other tissues racemize L-aspartate into D-aspartate (Wolosker et al., 2000). Since concentrations are highest during embryonic development and infancy, some of the specific functions may relate specifically to that period in life. A subtype of glutamate receptors in the brain responds to N-methyl-D-aspartate, which is presumably produced by local methylation of D-aspartate. Peroxisomal D-aspartate oxidase (EC1.4.3.1), which can contain either FAD or 6-hydroxyflavin adenine dinucleotide, catabolizes both N-methyl-D-aspartate and D-aspartate.

N-acetylaspartate: The amount of N-acetyl aspartate in the brain is regulated, and the genetic impairment of its deacetylation (Canavan disease) causes severe spongy degeneration of cortical white matter. Low N-acetyl aspartate concentration in the brain, on the other hand, may be associated with certain chronic pain syndromes (Grachev et al., 2002). Nonetheless, the function of N-acetyl aspartate remains unclear.

REFERENCES

Grachev, I.D., Thomas, P.S., Ramachandran, T.S., 2002. Decreased levels of N-acetylaspartate in dorsolateral prefrontal cortex in a case of intractable severe sympathetically mediated chronic pain (complex regional pain syndrome, type I). Brain Cogn. 49 (102), 113.

Hammarqvist, E., Ejesson, B., Wernerman, J., 2001. Stress hormones initiate prolonged changes in the muscle amino acid pattern. Clin. Physiol. 21, 44–50.

Matsuo, H., Kanai, Y., Kim, J.Y., Chairoungdua, A., Kim do, K., Inatomi, J., et al., 2002. Identification of a novel Na+-independent acidic amino acid transporter with structural similarity to the member of a heterodimeric amino acid transporter family associated with unknown heavy chains. J. Biol. Chem. 277, 21017–21026.

Rathod, P.K., Fellman, J.H., 1985. Identification of mammalian aspartate-4-decarboxylase. Arch. Biochem. Biophys. 238, 435–446.

Sacktor, B., Rosenbloom, I.L., Liang, C.T., Cheng, L., 1981. Sodium gradient- and sodium plus potassium gradient-dependent L-glutamate uptake in renal basolateral membrane vesicles. J. Membr. Biol. 60, 63–71.

Shapira, R., Wilkinson, K.D., Shapira, G., 1988. Racemization of individual aspartate residues in human myelin basic protein. J. Neurochem. 50, 649–654.

Wolosker, H., D'Aniello, A., Snyder, S.H., 2000. D-aspartate disposition in neuronal and endocrine tissues: ontogeny, biosynthesis and release. Neuroscience 100, 183–189.

Wu, G., Morris, S.M., 1998. Arginine metabolism: nitric oxide and beyond. Biochem. J. 336, 1–17.

ASPARAGINE

The neutral amino acid L-asparagine (L-beta-asparagine, alpha-aminosuccinamic acid, aspartic acid beta-amide, atheine, asparamide~ agedoite, one-letter code N; molecular weight 132) contains 21.2% nitrogen (Figure 8.70).

ABBREVIATIONS

Asn	L-asparagine
PepT1	hydrogen ion/peptide cotransporter (SLC15A1)
PepT2	hydrogen ion/peptide cotransporter (SLC15A2)
PLP	pyridoxal-5'-phosphate

NUTRITIONAL SUMMARY

Function: The nonessential amino acid L-asparagine (Asn) is used for the synthesis of L-aspartate and of proteins. Its complete oxidation requires thiamin, riboflavin, niacin, vitamin B6, pantothenate, lipoate, ubiquinone, iron, and magnesium.

Food sources: Dietary proteins from different sources all contain Asn.

Requirements: Since it can be synthesized from L-aspartate, and L-aspartate from oxaloacetate, dietary intake of Asn is not necessary, so long as enough total protein as a source of the amino group is available.

Deficiency: Prolonged lack of total protein causes growth failure, loss of muscle mass, and organ damage. Asn depletion through the use of exogenous asparaginase is employed for the antitumor treatment of some leukemias.

Excessive intake: Very high intake of protein and mixed amino acids (i.e., more than three times the RDA, which is 2.4 g/kg) is thought to increase the risk of renal glomerular sclerosis and accelerate osteoporosis. Information on the specific risks from high intake of Asn alone is lacking.

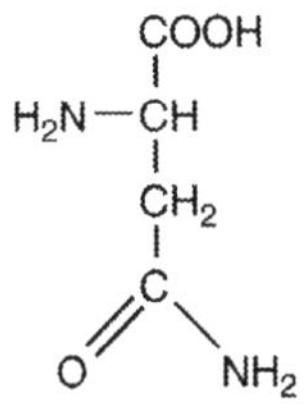

FIGURE 8.70

L-Asparagine.

ENDOGENOUS SOURCES

Asparagine synthase (EC6.3.5.4) uses L-glutamine as the amino group donor to generate Asn from aspartate. An alternative pathway is catalyzed by aspartate-ammonia ligase (EC6.3.1.1). Amination of the Krebs-cycle intermediate oxaloacetate by aspartate aminotransferase (EC2.6.1.1) can always generate ample amounts of the precursor L-aspartate, so long as total protein and vitamin B6 supplies are adequate.

DIETARY SOURCES

All food proteins contain Asn, but quantitative information is not readily available from publications or standard food composition tables. Grain proteins are relatively rich in Asn. Prolonged heating, particularly at alkaline pH, promotes the generation of protein-bound D-asparagine (de Vrese et al., 2000). Deamidation with the release of ammonia can also occur to some extent during heating or hydrolysis.

DIGESTION AND ABSORPTION

Denaturation and hydrolysis of proteins begins with mastication of foods in the mouth and continues in the stomach under the influence of hydrochloric acid, pepsin A (EC3.4.23.1), and gastricin (EC3.4.23.3). Several pancreatic and brush border enzymes continue protein hydrolysis, although none specifically cleaves peptide bonds adjacent to Asn. Asn as part of dipeptides and tripeptides is taken up via the hydrogen ion/peptide cotransporters PepT1 (SLC15A1) and PepT2 (SLC15A2). The sodium–amino acid cotransport system B° (ASCT2, SLC1A5; Avissar et al., 2001; Bode, 2001) at the small intestinal brush border membrane accepts Ash with high affinity. The sodium-independent transport system $b^{o,+}$ comprised of a light subunit BAT1 (SLC7A9) and a heavy subunit rBAT (SLC3A1), exchanges Asn for most other neutral amino acids. Asn can move across the basolateral membrane via the sodium amino acid cotransport systems ATA2 (Sugawara et al., 2000) and the sodium-independent transporter heterodimers LAT2 + 4F2 (SLC7A8 + SLC3A2), y^+LAT1 + 4F2 (SLC7A7 + SLC3A2), y^+LAT2 + 4F2 (SLC7A6 + SLC3A2). The directionality of the transport depends on the prevailing intracellular concentrations (Figure 8.71).

TRANSPORT AND CELLULAR UPTAKE

Blood circulation: Asn concentration in plasma tends to be around 30–60 µmol/l (Hammarqvist et al., 2001). Uptake from the blood into tissues relies largely on the sodium-dependent transport systems N (SN1 and SN2; in the liver, muscle, and brain), ASC/B° (SLC1A5, ASCT2; particularly in the lung, muscle, pancreas, and neuronal glia), $B^{o,+}$ (SLC6A14; cotransports with one chloride and two sodium ions, in the lung, mammary gland, and other tissues), and A. Expression patterns of the system A transporters vary in characteristic fashion between different cell types. ATA2 (SLC38A2) is present in most cells (Sugawara et al., 2000), while ATA3 (SLC38A4) is restricted to the liver (Hatanaka et al., 2001).

Gln- and L-asparagine-specific uniporters, citrin (SLC25A13) and aralar 1 (SLC25A12) mediate the transfer from cytosol into mitochondria (Indiveri et al., 1998). Citrin, expressed mainly in the liver, kidneys, and other tissues, otherwise transports citrate.

BBB: Transfer of Asn into and out of the brain is tightly controlled, since L-glutamine, which is carried by the same transporters, generates the potent neurotransmitter L-glutamate. LAT1 (SLC7A5)

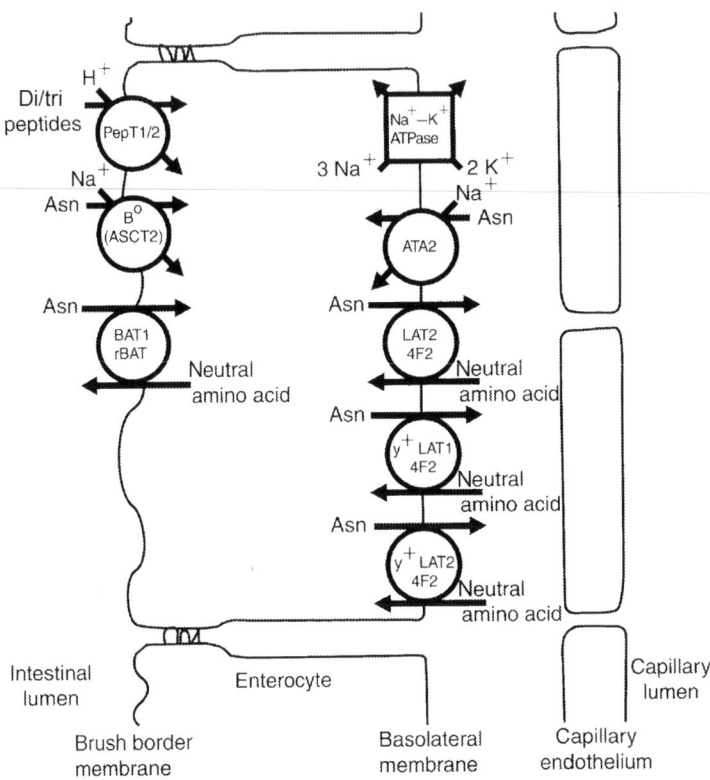

FIGURE 8.71

Intestinal absorption of L-asparagine.

and ATA1 (SLC38A1) (Varoqui et al., 2000) are expressed in brain capillary endothelial cells and contribute to Asn transport, but their locations, relative importance, and the role of other transporters are not completely understood. A model for the transfer of Asn from circulating blood into the brain (Bode, 2001) envisions uptake into astrocytes via system B° (ASCT2) and export into the intercellular space through SN 1. Neurons take up Ash via transporters ATA1 and ATA2.

Materno-fetal transfer: The sodium–amino acid cotransport system A (possibly also ASC and N), and the sodium-independent exchanger LAT I mediate Asn uptake from maternal blood across the brush border membrane of the syncytiotrophoblast. Transfer across the basolateral membrane proceeds via the sodium-independent transporters LAT1 and LAT2 (SLC7A8) and the sodium–amino acid cotransport system A. The system A transporter ATAI is expressed at a high level in the placenta (Varoqui et al., 2000), but its exact location is not yet known.

METABOLISM

Removal of the γ-amido group by asparaginase (EC3.5.1.1, requires PLP like all transaminases) in liver and kidney cytosol is usually the first step of Asn breakdown (Figure 8.72). The PLP-dependent

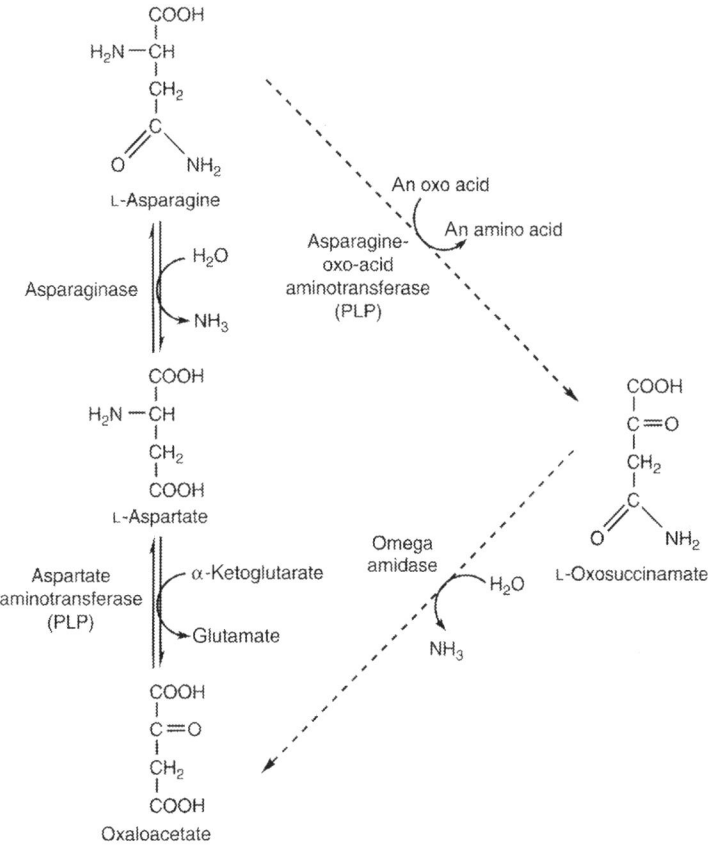

FIGURE 8.72

L-Asparagine catabolism.

aspartate aminotransferase (EC2.6.1.1) removes the α-amino group and produces the Krebs-cycle intermediate oxaloacetate. The urea cycle in liver and kidney cytosol provides an alternative pathway for aspartate metabolism. Argininosuccinate synthase (EC6.3.4.5) condenses aspartate and citrulline, and the successive actions of argininosuccinate lyase (EC4.3.2.1) and arginase (EC3.5.3.1) then release urea and ornithine. Other transaminases, including asparagine-oxo-acid aminotransferase (EC2.6.1.14), are of lesser quantitative significance. The L-oxosuccanamate resulting from the transfer of the α-amino group can be converted to oxaloacetate by omega-amidase (EC3.5.1.1).

STORAGE

Proteins in muscles and other organs contain some Asn that can be released in response to low protein intake.

EXCRETION

Losses via feces are negligible so long as gastrointestinal function is normal. The kidneys filter a gram or more per day, most of which is reabsorbed in healthy people. The proximal renal tubules take up free Asn mainly through the sodium–amino acid cotransport system B° (ASCT2) (Avissar et al., 2001; Bode, 2001), dipeptides and tripeptides via pepT1 and pepT2. Asn, if it is not metabolized in the renal epithelial cells, is then exported across the basolateral membrane via the sodium-dependent transporters ATA2 and ASCT1. Most nitrogen from metabolized Asn is excreted into urine as urea, a much smaller amount as free ammonium ions.

REGULATION

Starvation promotes Asn transport by increasing expression of the transport systems A and L, whereas ASC and nonmediated uptake are not affected (Muniz et al., 1993). Depletion of amino acids, through activation of the amino acid response pathway, increases asparagine synthase (EC6.3.5.4) activity (Leung-Pineda and Kilberg, 2002). Glucose depletion does the same through activation of endoplasmic reticulum (ER) stress response pathway. Both pathways act by binding to nutrient-sensing response elements 1 and 2 (NSRE-1 and NSRE-2), thereby increasing asparagine synthase transcription.

FUNCTION

Protein synthesis: Asn is a constituent of most proteins and peptides synthesized in the body. Asparagine-tRNA ligase (EC6.1.1.22) links Asn with its correspondent tRNA in an ATP/magnesium-dependent reaction. Asn residues can serve as anchors for glucosyl-mannosyl-glucosamine polysaccharide side-chains in glycoproteins (attached via dolichyl diphosphooligosaccharides) and be the site of other types of glycosylation.

Energy fuel: Eventually, most ingested Asn is used as an energy fuel yielding 2.3 kcal/g (May and Hill, 1990). The release of both amino groups generates oxaloacetate, which can be utilized via the Krebs cycle. Complete Asn oxidation requires adequate supplies of thiamin, riboflavin, niacin, vitamin B6, pantothenate, lipoate, ubiquinone, iron, and magnesium.

Chemotherapy: Asn depletion of rapidly growing tumor cells by treatment with bacterial or synthetic asparaginase (EC3.5.1.1) is effective for some leukemias, especially in children with acute lymphoblastic leukemia (Ramakers-van Woerden et al., 2000).

REFERENCES

Avissar, N.E., Ryan, C.K., Ganapathy, V., Sax, H.C., 2001. Na$^+$-dependent neutral amino acid transporter ATB° is a rabbit epithelial cell brush border protein. Am. J. Physiol. Cell Physiol. 281, C963–C971.

Bode, B.P., 2001. Recent molecular advances in mammalian glutamine transport. J. Nutr. 131, 2475S–2485S.

de Vrese, M., Frik, R., Roos, N., Hagemeister, H., 2000. Protein-bound D-amino acids, and to a lesser extent lysinoalanine, decrease true ileal protein digestibility in minipigs as determined with (15)N-labeling. J. Nutr. 130, 2026–2031.

Hammarqvist, E., Ejesson, B., Wernerman, J., 2001. Stress hormones initiate prolonged changes in the muscle amino acid pattern. Clin. Physiol. 21, 44–50.

Hatanaka, T., Huang, W., Ling, R., Prasad, P.D., Sugawara, M., Leibach, F.H., et al., 2001. Evidence for the transport of neutral as well as cationic amino acids by ATA3, a novel and liver-specific subtype of amino acid transport system A. Biochim. Biophys. Acta 1510, 10–17.

Indiveri, C., Abbruzzo, G., Stipani, I., Palmieri, E., 1998. Identification and purification of the reconstitutively active glutamine carrier from rat kidney mitochondria. Biochem. J. 333, 285–290.

Leung-Pineda, V., Kilberg, M.S., 2002. Role of Sp1 and Sp3 in the nutrient-regulated expression of the human asparagine synthetase gene. J. Biol. Chem. 277, 16585–16591.

May, M.E., Hill, J.O., 1990. Energy content of diets of variable amino acid composition. Am. J. Clin. Nutr. 52, 770–776.

Muniz, R., Burguillo, L., del Castillo, J.R., 1993. Effect of starvation on neutral amino acid transport in isolated small-intestinal cells from guinea pigs. Pfl. Arch. Eur. J. Physiol. 423, 59–66.

Ramakers-van Woerden, N.L., Pieters, R., Loonen, A.H., Hubeek, I., van Drunen, E., Beverloo, H.B., et al., 2000. *TEL/AML1* gene fusion is related to *in vitro* drug sensitivity for L-asparaginase in childhood acute lymphoblastic leukemia. Blood 96, 1094–1099.

Sugawara, M., Nakanishi, T., Fei, Y.J., Huang, W., Ganapathy, M.E., Leibach, F.H., et al., 2000. Cloning of an amino acid transporter with functional characteristics and tissue expression pattern identical to that of system A. J. Biol. Chem. 275, 16473–16477.

Varoqui, H., Zhu, H., Yao, D., Ming, H., Erickson, J.D., 2000. Cloning and functional identification of a neuronal glutamine transporter. J. Biol. Chem. 275, 4049–4054.

ARGININE

The conditionally essential amino acid arginine (2-amino-5-guanidinovaleric acid, one-letter code R; molecular weight 174) contains two amino groups, one imido group, and another nitrogen atom, and thus contains 32.1% nitrogen (Figure 8.73).

ABBREVIATIONS

Arg	L-arginine
BH4	tetrahydrobiopterin
Glu	L-glutamate
OAT	ornithine-oxo-acid aminotransferase (EC2.6.1.13)
OTC	ornithine carbamoyltransferase (EC2.1.3.3)

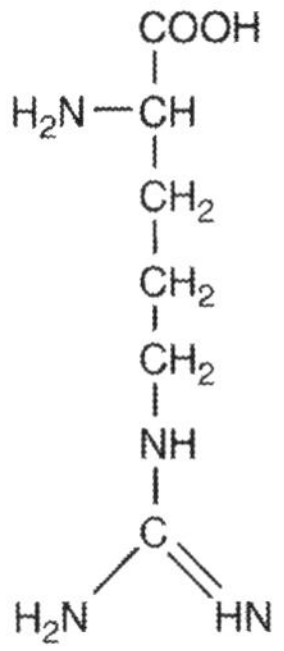

FIGURE 8.73

L-Arginine.

NUTRITIONAL SUMMARY

Function: The conditionally essential amino acid L-arginine (Arg) is used for high-energy phosphate storage in muscle (phosphoarginine and creatine), protein synthesis, and nitric oxide production. It is also used as an energy fuel; its complete oxidation requires thiamin, riboflavin, niacin, vitamin B6, pantothenate, lipoate, ubiquinone, iron, and magnesium.

Food sources: Adequate amounts are consumed with protein from most sources. Dietary supplements with manufactured Arg are commercially available.

Requirements: Due to the possibility of endogenous synthesis from L-glutamate (Glu), dietary intake is not usually needed except in newborn infants. Increased dietary intake may be beneficial for wound healing, tissue repair, and immune function when needs are increased due to severe illness, infection, or injury.

Deficiency: Prolonged lack of protein causes growth failure, loss of muscle mass, and organ damage.

Excessive intake: Very high intake of protein and mixed amino acids (i.e., more than three times the RDA, which is 2.4 g/kg) is thought to increase the risk of renal glomerular sclerosis and accelerate osteoporosis. The risk from high intake of manufactured supplements is not completely understood but may include liver and kidney failure, mental disturbances, and other severe dysfunction.

ENDOGENOUS SOURCES

Endogenous synthesis in the small intestine from L-glutamine, L-glutamate (Glu), and L-proline contributes about one-fifth (Wu and Morris, 1998) of the Arg utilized in the body for protein synthesis, energy generation, and synthesis of functional compounds, including nitric oxide, creatine (>1 g/day), and phosphoarginine (Wu et al., 1997). Endogenous synthesis requires adequate intake of niacin and vitamin B6 (Figure 8.74). It takes place in two separate stages. The first stage takes place in the small and large intestine and converts the precursors (glutamine, Glu, and proline) into ornithine and citrulline. Arg synthesis is completed mainly in the kidneys and, to a lesser extent, in the liver. The first two steps of citrulline synthesis in the small intestine from Glu (and indirectly from glutamine) use the bifunctional pyrroline 5-carboxylate synthase (no EC number assigned). This enzyme comprises phosphorylation (glutamate 5-kinase, EC2.7.2.11) and reduction (glutamate 5-semialdehyde dehydrogenase, EC1.2.1.41) activities (Wu and Morris, 1998). Citrulline synthesis is completed by ornithine-oxo-acid aminotransferase (ornithine transaminase or ornithine aminotransferase, OAT, EC2.6.1.13, PLP-dependent) and ornithine carbamoyltransferase (OTC, EC2.1.3.3). The ornithine/citrulline carrier (SLC25AI5) can then shuttle citrulline into the cytosol in exchange for ornithine.

After transport via the bloodstream and uptake through the transport system y$^+$ into proximal tubular epithelium of the kidneys, citrulline is converted by another two enzymatic reactions into Arg. Argininosuccinate synthase (EC6.3.4.5) adds aspartate in an ATP-energized reaction. Argininosuccinate lyase (EC4.3.2.1) finally releases Arg and fumarate.

DIETARY SOURCES

Proteins have a particularly high Arg content in rice (83 mg/g), oats (71 mg/g), and soybeans (78 mg/g). The protein in meats, fish, eggs, and most beans contains around 60 mg/g. Protein from cow milk and wheat contains less than 40 mg/g. The variation of the relative Arg content of food proteins notwithstanding, total protein intake remains the major determinant of Arg intake. Daily intake of Arg may be 56 mg/kg (Beaumier et al., 1995) or more (Wu and Morris, 1998).

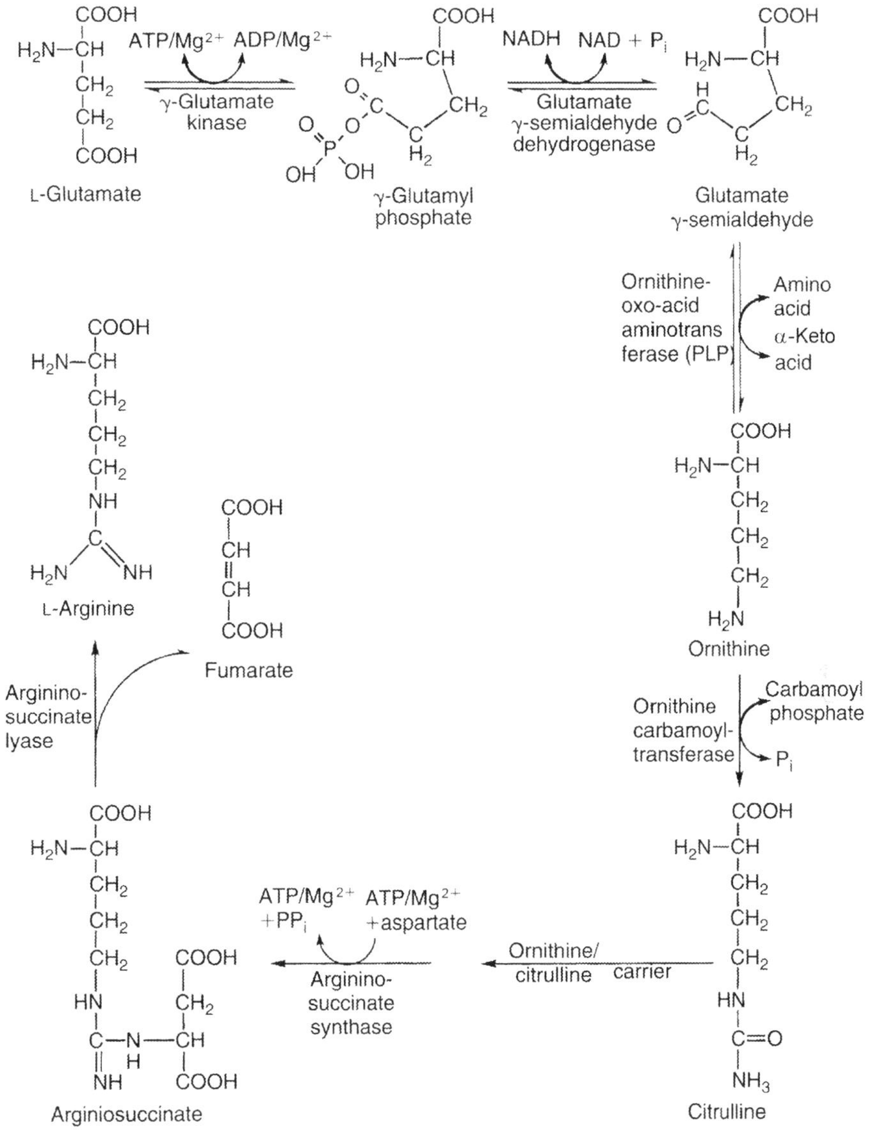

FIGURE 8.74

Glutamate is an important precursor for intestinal arginine synthesis.

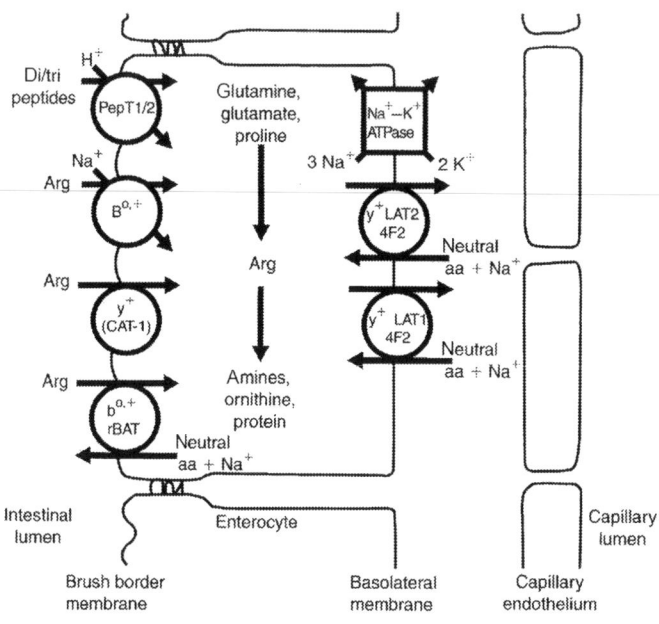

FIGURE 8.75

Intestinal absorption of arginine.

DIGESTION AND ABSORPTION

Arg-containing proteins are digested, as all other proteins, by an array of secreted enzymes from the stomach, pancreas, and small intestine, and by aminopeptidases at the brush border membrane (Figure 8.75). The combined action of these enzymes releases small peptides and free amino acids. Dipeptides and tripeptides are taken up through the hydrogen ion/peptide cotransporters 1 (PepT1, SLC15A1) and 2 (PepT2, SLC15A2).

Sodium-dependent transport systems are known to facilitate Arg uptake into enterocytes of the distal small intestine, including at least one high-capacity, low-affinity system and a low-capacity, high-affinity system (Iannoli et al., 1998). One of these is the sodium-dependent transport system B°. Even more important is the sodium-independent uptake via system y$^+$ and BAT1/b$^{o,+}$ (SLC7A9, linked to rBAT) in exchange for a neutral amino acid plus a sodium ion.

The transporters y$^+$LAT1 (SLC7A7) and y$^+$LAT2 (SLC7A6) have characteristics of amino acid system y$^+$ and work only in conjunction with the same membrane glycoprotein anchor 4F2 (SLC3A2). Both mediate Arg export across the basolateral membrane in equimolar exchange for glutamine or another neutral amino acid (Bröer et al., 2000; Bode, 2001). While these systems are not driven by the sodium gradient, they transport a sodium ion together with glutamine in compensation for the charge in Arg. Enterocytes use about 40% of the dietary Arg for their own energy needs and protein synthesis. Only about 60% of the absorbed Arg reach the portal bloodstream. On the other hand, the enterocytes are a major site of *de novo* Arg synthesis from glutamine, glutamate, and proline (Wu and Morris, 1998).

TRANSPORT AND CELLULAR UPTAKE

Blood circulation: Plasma concentration of Arg, typically around 94 µmol/l (Teerlink et al., 2002), is lowest during the early morning hours and increases significantly after meals (Tsai and Huang, 1999). Uptake from the blood into tissues occurs mainly via sodium-independent transporters that have the characteristics of system y^+ (Wu and Morris, 1998). CAT-1, CAT-2A, and CAT-2B are representatives of that family of transporters. The CAT-2 isoforms A and B are especially expressed in muscle and macrophages (Kakuda et al., 1998). The Arg-derived amine agmatine enters cells via the polyamine transporter (Satriano et al., 2001).

Materno-fetal transfer: Several members of the y^+ amino acid transporter family (CAT-1, CAT-4, and CAT-2B) are present at the maternal side of the syntrophoblast, but Arg uptake from maternal circulation appears to be made up of only a small proportion of total amino acid transfer (Cetin, 2001). Transfer across the basal membrane into fetal blood uses the heterodimeric complex, consisting of y^+ LATI and 4F2 (SLC7A7 + SLC3A2).

BBB: System y^+ is the main conduit of transport for Arg into the brain. CAT-3 is a brain-specific form in rats. Additional transporters may contribute to transport in either direction.

METABOLISM

Arg breakdown in the liver proceeds mainly via ornithine and glutamate to alpha-ketoglutarate and releases four nitrogens (two with urea and another two in transamination reactions). Several of the necessary steps are the same used for Arg synthesis operating in the reverse direction (Figure 8.76). Complete oxidation requires adequate supplies of thiamin, riboflavin, vitamin B6, niacin, pantothenate, lipoate, ubiquinone, iron, and magnesium.

The first step of Arg catabolism (in cytosol) uses the final enzyme of urea synthesis, the manganese-dependent arginase (EC3.5.3.1). Transport of the resulting ornithine into mitochondria by the ornithine/citrulline carrier (SLC25A15) is the rate-limiting step of Arg catabolism. This transporter usually exchanges an inwardly carried ornithine molecule for an outwardly transported citrulline molecule. Since a proton instead of citrulline can also serve as the counter ion (Indiveri et al., 1999), removal of ornithine from the urea cycle sequence does not impede the functioning of the ornithine/citrulline carrier. The delta-amino group of ornithine can then be moved by mitochondrial OAT (ornithine transaminase, EC2.6.1.13, PLP-dependent) to alpha-ketoglutarate, pyruvate, or glyoxylate. Glutamate-5-semialdehyde dehydrogenase (EC1.2.1.41) produces glutamate, which can then be transaminated by more than a dozen PLP-dependent aminotransferases, including aspartate aminotransferase (EC2.6.1.1) and alanine aminotransferase (EC2.6.1.2), to the Krebs-cycle intermediate alpha-ketoglutarate.

Protein arginine *N*-methyltransferases use SAM to methylate a small portion of arginyl residues in specific proteins. This activity is important for mRNA splicing, RNA transport, transcription control, signal transduction, and maturation of protein such as the MBP. Type I enzymes dimethylate asymmetrically, type II enzymes methylate proteins such as the MBP symmetrically. Protein–arginine

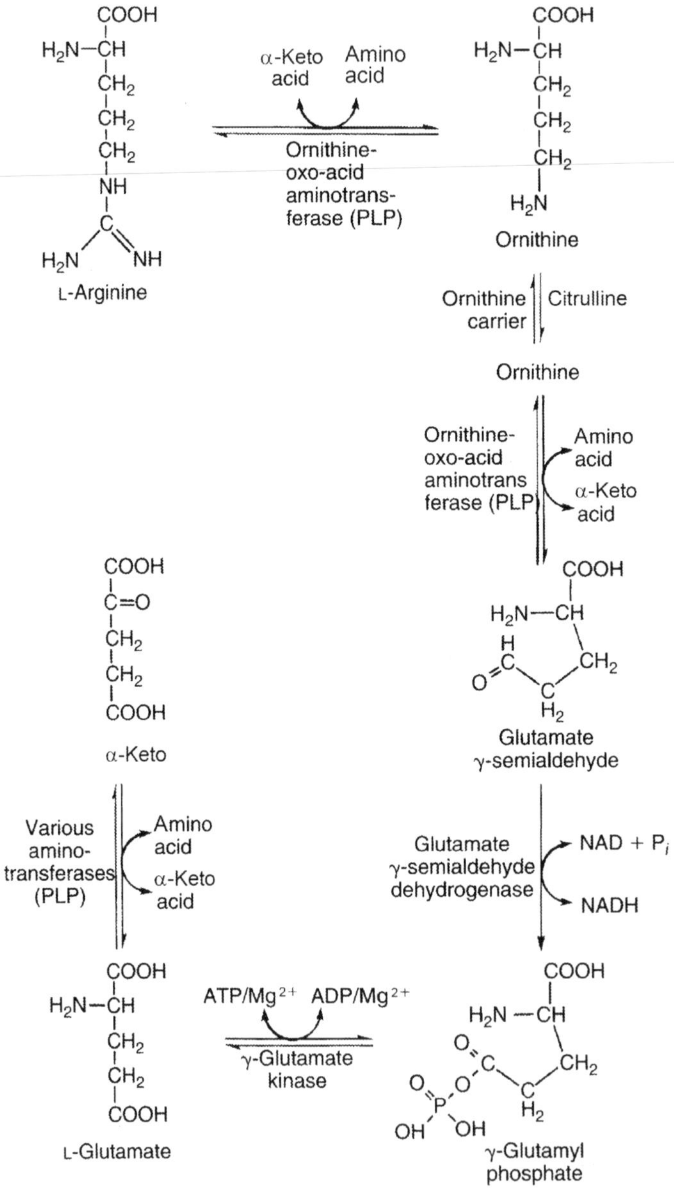

FIGURE 8.76

Arginine catabolism in the liver depends on the transport of ornithine from cytosol into mitochondria.

N-methyltransferases 1, 3, 4, and 6 are type I enzymes, whereas isoform 5 is a type II enzyme (Frankel et al., 2002). Breakdown of symmetrically methylated proteins releases symmetric N(G),N'(G)-dimethylarginine, and breakdown of asymmetrically dimethylated proteins generates asymmetric N(G),N(G)-dimethylarginine (ADMA). Dimethylargininase (dimethylarginine dimethylaminohydrolase, EC3.5.3.18, contains zinc), which is present in two isoforms in most tissues, converts these metabolites to L-citrulline by cleaving off methylamine or dimethylamine, respectively.

STORAGE

The Arg content of skeletal muscles at rest is about 0.35 mmol/kg (Hammarqvist et al., 2001). Increased metabolic demand, through the action of adrenaline, cortisol, glucagon, and other mediators, can trigger the release of 40% of these reserves within hours. Significant mobilizable Arg stores are also present in the liver, kidneys, and other organs, but their quantitative contribution is less well understood.

EXCRETION

At its typical plasma concentration, more than 6 g of Arg is filtered daily in a 70-kg man. Most of this is recovered from the proximal tubular lumen through the sodium-dependent system $B^{0,+}$ (SLC6A14) and the sodium-independent systems y^+ and BAT1/$b^{0,+}$ (SLC7A9, linked to rBAT). The mechanisms are the same as those mediating small intestinal Arg uptake.

On the basolateral side, again the transporters y^+LAT1 (SLC7A7) and y^+LAT2 (SLC7A6) mediate export toward the capillaries (Bröer et al., 2000; Bode, 2001). The inner medullary collecting ducts in the kidneys take up Arg via CAT1 (system y^+) but do not express CAT2A, CAT2B, or CAT3 (Wu et al., 2000).

REGULATION

Information on regulatory events in Arg homeostasis is still very incomplete. The control of dietary intake as well as differential distribution to the liver and other tissues contribute to a steady supply for vital functions and the avoidance of excess. Selection of foods may be influenced by their Arg content as suggested by observations in rats (Yamamoto and Muramatsu, 1987). Subjects with citrullinemia, an inborn error of arginine synthesis, have been reported to crave beans, peas, and peanuts, which are high in arginine (Walser, 1983); this observation could be another indication of some kind of feedback control of Arg intake.

Arg potentiates glucose-induced insulin secretion (Thams and Capito, 1999), which in turn shifts Arg away from use for gluconeogenesis and toward use for protein synthesis. The rate of endogenous Arg synthesis appears to be little affected by intake levels (Castillo et al., 1994). Inflammatory cytokines and endotoxin acting on system y^+ transporters enhance uptake of Arg into particular tissues. This system is expressed only at a low level in hepatocytes but strongly induced by cytokines (Kakuda et al., 1998). The control of CAT-2 expression indicates the level of complexity at work. Not only does

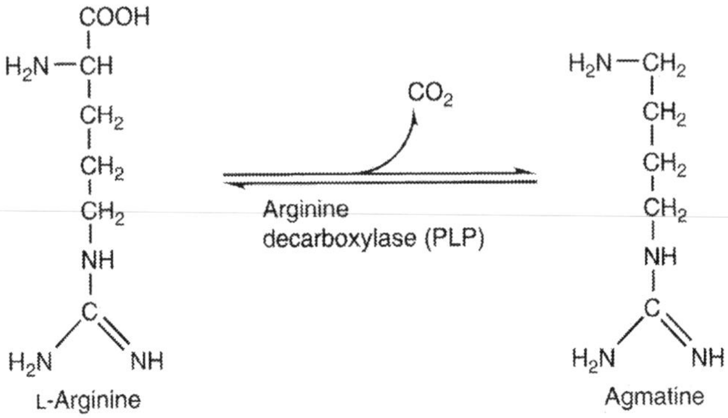

FIGURE 8.77

Decarboxylation of arginine generates a potent amine.

this gene encode for two distinct isoforms (CAT-2 and CAT-2A) with 10-fold difference in substrate-binding affinity, but it also has four separate fully functional promoters that allow differential response to a particular stimulant (Kakuda et al., 1998).

FUNCTION

Energy fuel: Most catabolic pathways of Arg lead to complete oxidation with an energy yield of 3.3 kcal/g (May and Hill, 1990). The necessary reactions depend on thiamin, riboflavin, niacin, vitamin B6, pantothenate, lipoate, ubiquinone, iron, and magnesium.

Protein synthesis: Significant amounts of Arg are needed for protein synthesis. Arginine-tRNA ligase (EC6.1.1.19) loads Arg onto its specific tRNA. Hair protein has a relatively high Arg content and is adversely affected by Arg deficiency.

Agmatin: The amine (Figure 8.77) produced through the mitochondrial decarboxylation of Arg (arginine decarboxylase, EC4.1.1.19, PLP-dependent) interacts with neurotransmitter receptors in the brain, is an acceptor of ADP-ribose (immunemodulator), tempers the proliferative effects of polyamines (through its effects on ornithine decarboxylase and the putrescine transporter), causes vasodilatation and increases the renal glomerular filtration rate (GFR), and slows nitric oxide production (Blantz et al., 2000).

Nitric oxide synthesis: All isoforms (NOS1, NOS2, and NOS3) of nitric oxide synthase (EC1.14.13.39) produce nitric oxide from Arg (Figure 8.78). They require 1 mole of BH4 and 1 mole of heme per dimer as cofactors (Rafferty et al., 1999). Diminished BH4 synthesis and the resulting decrease in nitric oxide production has been suggested to contribute importantly to impaired angiogenesis (Marinos et al., 2001), epithelial dysfunction, and insufficient vasodilation (Gruhn et al., 2001). Nitric oxide synthesis is inhibited by ADMA at moderately elevated concentrations, but not by symmetric N(G),N'(G)-dimethylarginine (Masuda et al., 2002). Plasma concentrations of both compounds are below 0.5 μmol/l in normal subjects (Teerlink et al., 2002) but distinctly elevated in people with diabetes, renal failure, and other diseases.

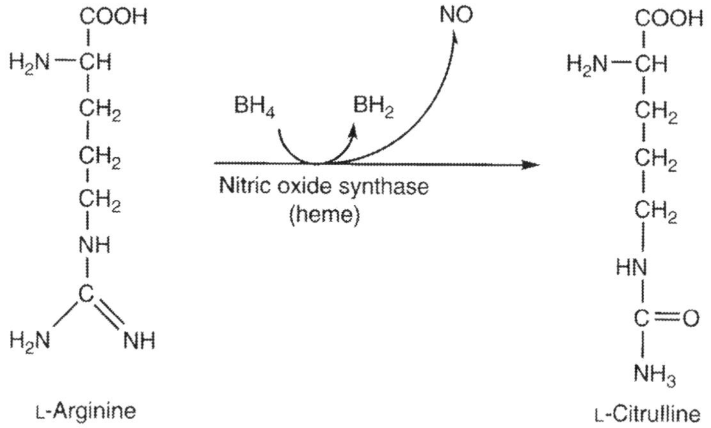

FIGURE 8.78

Nitric oxide synthesis from arginine depends on BH_4 and heme as cofactors.

High-energy phosphates: Daily production of the high-energy storage compound creatine is about 15 mg/kg, using Arg and glycine as precursors. The synthesis takes place in two stages, starting in the kidneys, and coming to completion in the liver. Another functionally related energy-storage compound for muscle is phosphoarginine.

Hormone stimulation: Increased Arg intake elicits the release of prolactine, insulin, glucagon, growth hormone, and enhances the number and responsiveness of circulating lymphocytes to mitogens. Arginine deiminase may inhibit the proliferation of human leukemia cells (by inducing cell growth arrest in the G1- and/or S-phase and apoptosis) more potently than analogous asparaginase treatment (Gong et al., 2000).

REFERENCES

Beaumier, L., Castillo, L., Ajami, A.M., Young, V.R., 1995. Urea cycle intermediate kinetics and nitrate excretion at normal and "therapeutic" intakes of arginine in humans. Am. J. Physiol. 269, E884–E896.

Blantz, R.C., Satriano, J., Gabbai, E., Kelly, C., 2000. Biological effects of arginine metabolites. Acta. Physiol. Scand. 168, 21–25.

Bode, B.E., 2001. Recent molecular advances in mammalian glutamine transport. J. Nutr. 131, 2475S–2485S.

Bröer, A., Wagner, C.A., Lang, E., Bröer, S., 2000. The heterodimeric amino acid transporter 4F2hc/y⁺ LAT2 mediates arginine efflux in exchange with glutamine. Biochem. J. 349, 787–795.

Castillo, L., Sanchez, M., Chapman, T.E., Ajami, A., Burke, J.F., Young, V.R., 1994. The plasma flux and oxidation rate of ornithine adaptively decline with restricted arginine intake. Proc. Natl. Acad. Sci. USA 91, 6393–6397.

Cetin, I., 2001. Amino acid interconversions in the fetal-placental unit: the animal model and human studies *in vivo*. Pediatr. Res. 49, 148–153.

Frankel, A., Yadav, N., Lee, J., Branscombe, T.L., Clarke, S., Bedford, M.T., 2002. The novel human protein arginine *N*-methyltransferase PRMT6 is a nuclear enzyme displaying unique substrate specificity. J. Biol. Chem. 277, 3537–3543.

Gong, H., Zolzer, F., yon Recklinghausen, G., Havers, W., Schweigerer, L., 2000. Arginine deiminase inhibits proliferation of human leukemia cells more potently than asparaginase by inducing cell cycle arrest and apoptosis. Leukemia 14, 826–829.

Gruhn, N., Aldershvile, J., Boesgaard, S., 2001. Tetrahydrobiopterin improves endothelium-dependent vasodilation in nitroglycerin-tolerant rats. Eur. J. Pharmacol. 416, 245–249.

Hammarqvist, F., Ejesson, B., Wernerman, J., 2001. Stress hormones initiate prolonged changes in the muscle amino acid pattern. Clin. Physiol. 21, 44–50.

Iannoli, P., Miller, J.H., Sax, H.C., 1998. Epidermal growth factor and human growth hormone induce two sodium-dependent arginine transport systems after massive enterectomy. J. Parent. Ent. Nutr. 22, 326–330.

Indiveri, C., Tonazzi, A., Stipani, I., Palmieri, F., 1999. The purified and reconstituted ornithine/citrulline carrier from rat liver mitochondria catalyses a second transport mode:ornithine$^+$/H$^+$ exchange. Biochem. J. 341, 705–711.

Kakuda, D.K., Finley, K.D., Maruyama, M., MacLeod, C.L., 1998. Stress differentially induces cationic amino acid transporter gene expression. Biochim. Biophys. Acta 1414, 75–84.

Marinos, R.S., Zhang, W., Wu, G., Kelly, K.A., Meininger, C.J., 2001. Tetrahydrobiopterin levels regulate endothelial cell proliferation. Am. J. Physiol. Heart Circ. Physiol. 281, H482–H489.

Masuda, H., Tsujii, T., Okuno, T., Kihara, K., Goto, M., Azuma, H., 2002. Accumulated endogenous NOS inhibitors, decreased NOS activity, and impaired cavernosal relaxation with ischemia. Am. J. Physiol. Reg. Integ. Comp. Physiol. 282, R1730–R1738.

May, M.E., Hill, J.O., 1990. Energy content of diets of variable amino acid composition. Am. J. Clin. Nutr. 52, 770–776.

Rafferty, S.P., Boyington, J.C., Kulansky, R., Sun, P.D., Malech, H.L., 1999. Stoichiometric arginine binding in the oxygenase domain of inducible nitric oxide synthase requires a single molecule of tetrahydrobiopterin per dimer. Biochem. Biophys. Res. Comm. 257, 344–347.

Satriano, J., Isome, M., Casero Jr., R.A., Thomson, S.C., Blantz, R.C., 2001. Polyamine transport system mediates agmatine transport in mammalian cells. Am. J. Physiol. Cell Physiol. 281, C329–C334.

Teerlink, T., Nijveldt, R.J., de Jong, S., van Leeuwen, P.A., 2002. Determination of arginine, asymmetric dimethylarginine, and symmetric dimethylarginine in human plasma and other biological samples by high-performance liquid chromatography. Anal. Biochem. 303, 131–137.

Thams, E., Capito, K., 1999. L-arginine stimulation of glucose-induced insulin secretion through membrane depolarization and independent of nitric oxide. Eur. J. Endocrinol. 140, 87–93.

Tsai, P.J., Huang, P.C., 1999. Circadian variations in plasma and erythrocyte concentrations of glutamate, glutamine, and alanine in men on a diet without and with added monosodium glutamate. Metab. Clin. Exp. 48, 1455–1460.

Walser, M., 1983. Urea cycle disorders and other hereditary hyperammonemic syndromes. In: Stanbury, J.B., Wyngaarden, J.B., Fredrickson, D.S., Goldstein, J.L., Brown, M.S. (Eds.), The Metabolic Basis of Inherited Disease, fifth ed. McGraw-Hill, New York, NY, pp. 402–438.

Wu, E., Cholewa, B., Mattson, D.L., 2000. Characterization of L-arginine transporters in rat renal inner medullary collecting duct. Am. J. Physiol. Reg. Integr. Comp. Physiol. 278, R1506–R1512.

Wu, G., Davis, P.K., Flynn, N.E., Knabe, D.A., Davidson, J.T., 1997. Endogenous synthesis of arginine plays an important role in maintaining arginine homeostasis in postweaning growing pigs. J. Nutr. 127, 2342–2349.

Wu, G., Morris, S.M., 1998. Arginine metabolism: nitric oxide and beyond. Biochem. J. 336, 1–17.

Yamamoto, Y., Muramatsu, K., 1987. Self-selection of histidine and arginine intake and the requirements for these amino acids in growing rats. J. Nutr. Sci. Vitaminol. 33, 245–253.

PROLINE

The cyclic neutral amino acid L-proline (2-pyrrolidinecarboxylic acid; one-letter code P; molecular weight 115) contains 12.2% nitrogen. Hydroxyproline is a proline metabolite with distinct properties and metabolic fate (Figure 8.79).

ABBREVIATIONS

Hypro	4-hydroxyproline
P5C	delta-1-pyrroline-5-carboxylate
PLP	pyridoxal-5'-phosphate
Pro	L-proline
RDA	recommended dietary allowance
HIF-1	hypoxia-inducible factor 1

NUTRITIONAL SUMMARY

Function: The nonessential amino acid L-proline (Pro) is used as an energy fuel, as a precursor of L-glutamate, and for the synthesis of proteins. Its complete oxidation requires thiamin, riboflavin, niacin, vitamin B6, pantothenate, lipoate, ubiquinone, iron, and magnesium.

Food sources: Dietary proteins from different sources all contain Pro. Collagen is particularly rich in Pro. Dietary supplements containing crystalline Pro are commercially available.

Requirements: Since it can be synthesized from L-glutamate, L-glutamine, or L-arginine, dietary intake of Pro may not be necessary, so long as enough total protein is available.

Defidency: Prolonged lack of total protein causes growth failure, loss of muscle mass, and organ damage.

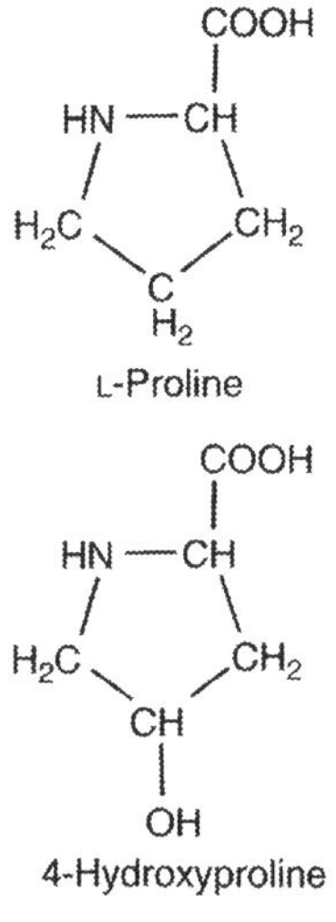

FIGURE 8.79

L-Proline and 4-hydroxyproline.

Excessive intake: Very high intake of protein and mixed amino acids (i.e., more than the RDA, equivalent to 2.4 g/kg) is thought to increase the risk of renal glomerular sclerosis and accelerate osteoporosis. Information on the specific risks from high intake of Pro alone is lacking.

ENDOGENOUS SOURCES

De novo synthesis of Pro begins with the phosphorylation of L-glutamate by glutamate 5-kinase (EC2.7.2.11). The same bifunctional protein also catalyzes the next step, an NADPH-dependent reduction (glutamate gamma-semialdehyde dehydrogenase, EC1.2.1.41). The resulting glutamate-gamma-semialdehyde is in spontaneous equilibrium with delta L-pyrroline 5-carboxylate, which traverses the inner mitochondrial membrane by an as-yet-unclear mechanism. It can then be converted into Pro by NADPH-dependent delta L-pyrroline 5-carboxylate reductase (EC1.5.1.1) in the cytosol. Glutamate gamma-semialdehyde is produced only in the mitochondrial matrix of the small intestine. Pro synthesis may be completed there or proceed after transfer of the intermediate into liver cells (competing there with the synthesis of the key urea cycle metabolite L-ornithine). Here, 4-hydroxyproline (Hypro) arises from the posttranslational modification of collagen precursor proteins by prolyl 4-hydroxylase (EC1.14.11.2, iron- and ascorbate-dependent) (Figure 8.80).

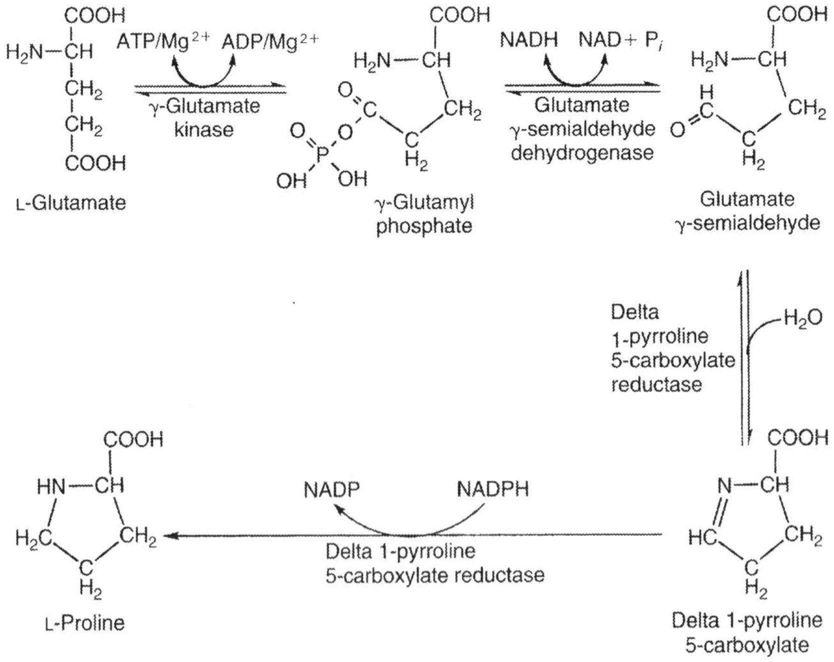

FIGURE 8.80

Endogenous synthesis of L-proline.

DIETARY SOURCES

Reliable information on the Pro content of individual foods is very limited. There is no indication that some commonly consumed foods contain a much higher percentage of Pro than others do. Collagen in bones and connectives tissues contains about 13% Hypro. About 10% of the protein in plant cell walls consists of glycoproteins with high Hypro content.

Note: Casein and soy protein hydrolysates in some dietary supplements contain high levels of the diketopiperazine cyclo(His-Pro), which can be absorbed from the intestine (Prasad, 1998). Cyclo(His-Pro), which is normally produced during the endogenous hydrolysis of thyroid-releasing hormone (TRH), affects motor functions, influences body temperature, inhibits prolactin secretion (Prasad, 1998), and might stimulate growth hormone output (Kagabu et al., 1998).

DIGESTION AND ABSORPTION

Protein hydrolysis by various gastric, pancreatic, and enterai enzymes generates a mixture of free amino acids and oligopeptides. Pro-containing dipeptides and tripeptides are efficiently taken up throughout the small intestine via the hydrogen ion/peptide cotransporters 1 (PepT1; SLC15A1), and to a much lesser extent, 2 (PepT2; SLC15A2). Free Pro and Hypro enter the small intestinal cell via the sodium-dependent transporter IMINO (Urdaneta et al., 1998); this transporter may also be chloride-dependent. The rBAT (SLC3A1)–anchored amino acid transporter BAT1/$b^{0,+}$ (SLC7A9) facilitates entry by exchange with another neutral amino acid (Chairoungdua et al., 1999); this heteroexchange can occur in either direction, depending on the prevalent concentration gradient (Figure 8.81).

TRANSPORT AND CELLULAR UPTAKE

Blood circulation: Pro is a part of most proteins in the blood and can be taken up by cells via mechanisms specific to each of them. Free Pro (about 170 µmol/l in the blood) enters tissues mainly via system L, including the specifically identified heteroexchanger LAT1. Typical Hypro concentrations in the blood are around 10 µmol/l in meat consumers and about half as much in vegetarians (Hung et al., 2002).

Materno-feta/transfer: Pro transfer across the microvillous membrane of the syncytiotrophoblast is mediated by LAT1 (Ritchie and Taylor, 2001), transport across the basal membrane proceeds via LAT2, both in exchange for other neutral amino acids (Kudo and Boyd, 2001). The driving force for LAT1/LAT2-mediated transport is the concentration gradient of small neutral amino acids (glycine, L-alanine, and cysteine) established by the sodium-dependent transport systems A and ASC.

BBB: System L mediates Pro transport across the neuroendothelial cell layer. Within the brain itself the sodium chloride-dependent PROT mediates uptake into neurons.

METABOLISM

Proline: Pro breakdown (Figure 8.82) in the mitochondrial matrix of muscle, the liver, kidneys, and other tissues begins with its oxidation by proline dehydrogenase (proline oxidase, EC1.5.99.8); this enzyme contains covalently bound ADP as a functional group and uses a cytochrome c as an electron acceptor. Because some leakage of reactive oxygen species occurs, the oxidation of Pro generates free radicals (Donald et al., 2001).

Sarcosin oxidase (EC1.5.3.1, contains FAD) in peroxisomes also converts Pro into delta-1-pyrroline-5-carboxylate (P5C) but uses oxygen and generates hydrogen peroxide (Reuber et al., 1997).

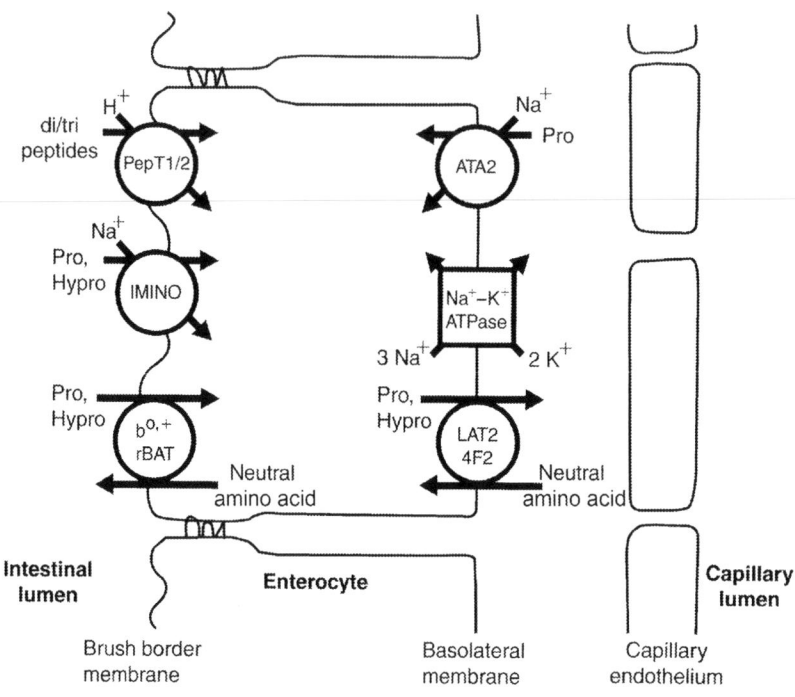

FIGURE 8.81

Intestinal absorption of L-proline.

Here, 1-Pyrroline 5-carboxylic acid dehydrogenase (EC1.5.1.12, aldehyde dehydrogenase 4) catalyzes the second step of Pro conversion to glutamate. Alternatively, OAT (EC2.6.1.13, PLP-dependent) can convert the intermediate glutamate semialdehyde into ornithine. Indeed, Pro metabolism in the small intestine is the major source of citrulline, ornithine, and arginine in the body (Wu, 1997; Dillon et al., 1999).

Hydroxyproline: The catabolism of Hypro differs from that of Pro. Hypro is derived from dietary collagen and endogenously from the turnover of muscle, connective tissue, and bone (Figure 8.83). Most Hypro is catabolized in the epithelial cells of renal proximal tubules, and smaller amounts in the liver. The amino acid reaches the mitochondria through a translocator that is distinct from that for Pro (Atlante et al., 1994). The initial step is catalyzed by 4-oxoproline reductase (hydroxyproline oxidase, EC1.1.1.104) in the kidneys (Kim et al., 1997). The enzyme responsible for Hypro breakdown in the liver is proline dehydrogenase (EC1.5.99.8). The next step, irreversible conversion of 3-hydroxy-1-pyrroline- 5-carboxylate into 4-hydroxyglutamate is catalyzed by I-pyrroline 5-carboxylic acid dehydrogenase (aldehyde dehydrogenase 4, EC1.5.1.12), the same enzyme that acts on the Pro breakdown product L-pyrroline-5-carboxylate. Nonenzymic conversion into 4-hydroxy glutamate semialdehyde occurs in the liver. Aspartate aminotransferase (PLP-dependent, EC2.6.1.1) moves the amino group to alpha-ketoglutarate. Then, 4-Hydroxy-2-oxoglutarate aldolase (EC4.1.3.16) catalyzes the final step of Hypro catabolism, hydrolysis of 4-hydroxy-2-oxoglutarate to pyruvate plus glyoxylate. Some of the glyoxylate generated by Hypro catabolism can be converted into oxalate, which may increase the risk of kidney stone formation (Ichiyama et al., 2000).

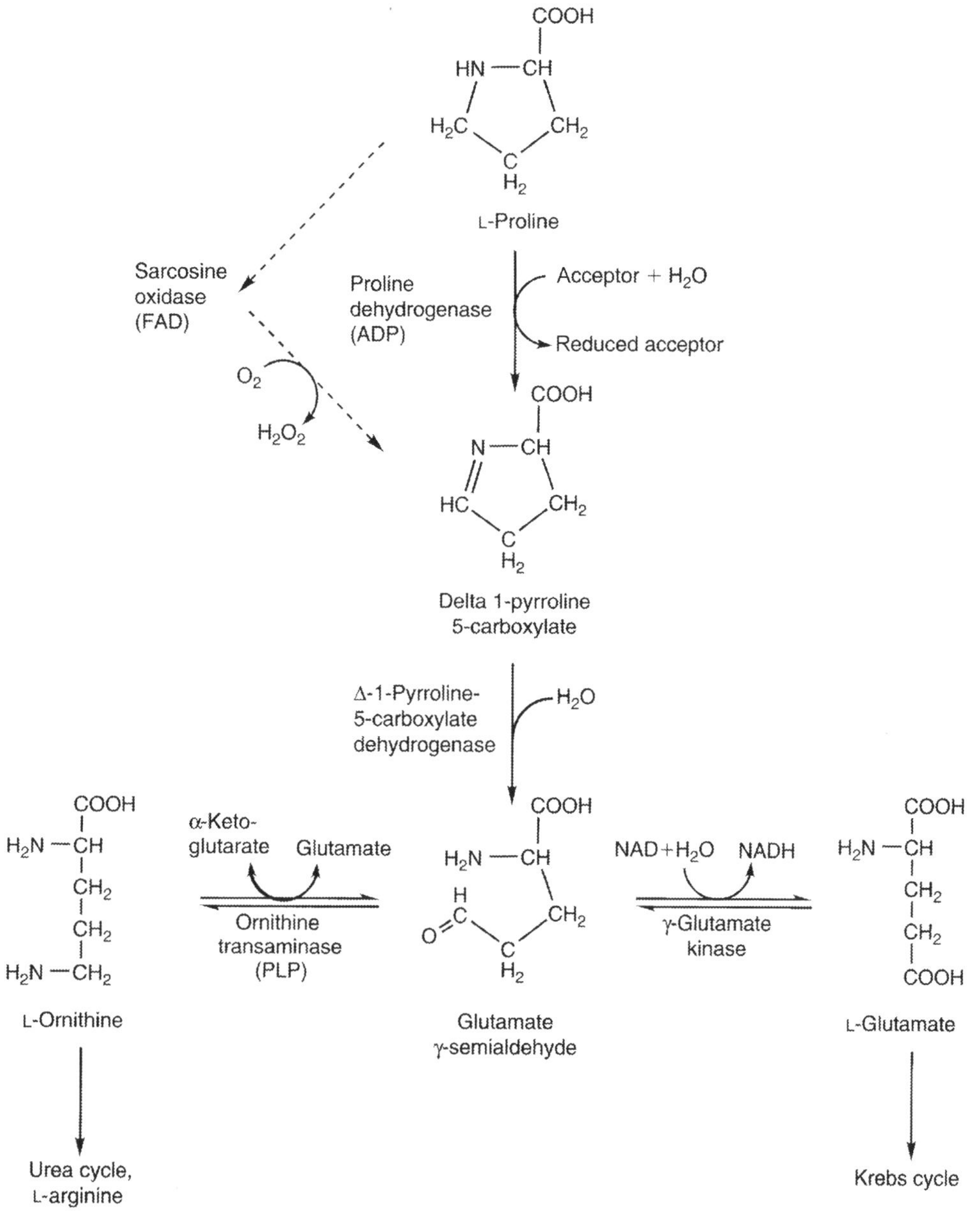

FIGURE 8.82

Breakdown of L-proline.

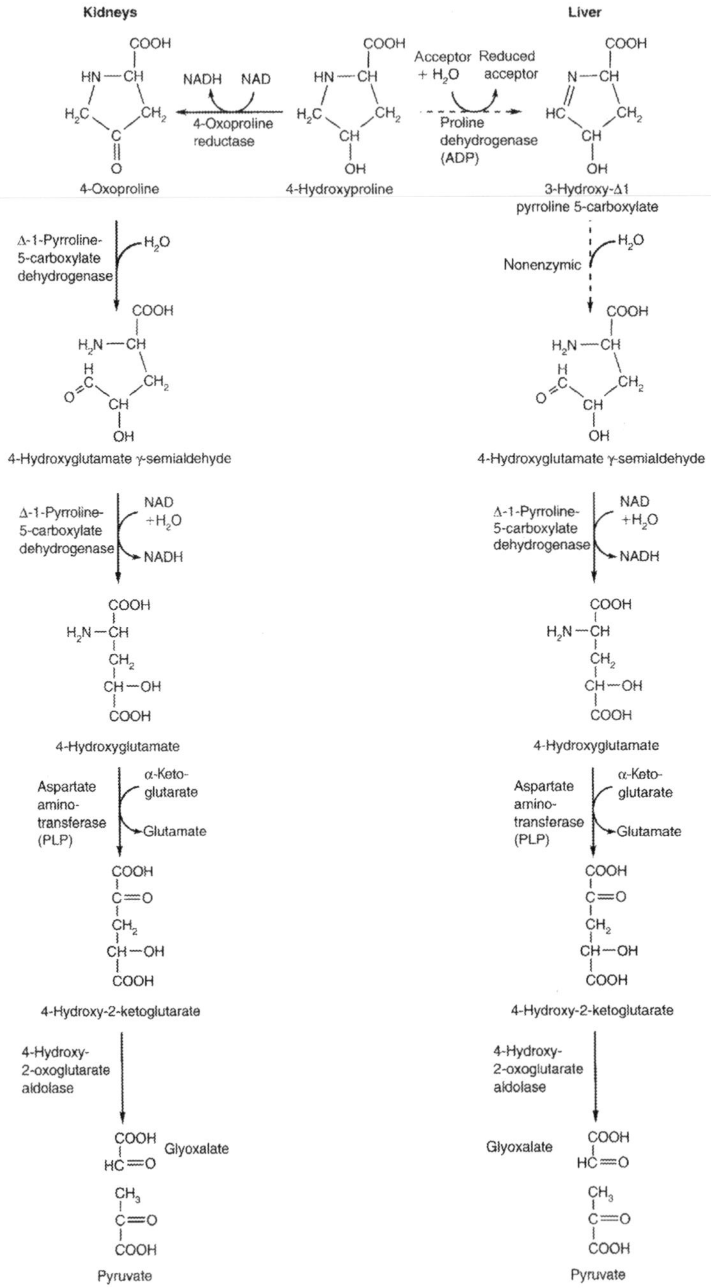

FIGURE 8.83

Breakdown of 4-hydroxyproline.

STORAGE

The body's proteins contain about 4.2% Pro, which is released as proteins are broken down during normal turnover. Accelerated protein catabolism occurs in response to starvation, severe trauma, or infection.

EXCRETION

The kidneys filter more than 3 g of Pro daily. Recovery from the proximal tubular lumen can use the rBAT-linked heteroexchanger BAT1/b$^{o,+}$ (SLC7A9). The presence of the IMINO transporter in the renal brush border membrane is uncertain. The 4F2-1inked heteroexchanger LAT2 mediates export across the basolateral membrane.

REGULATION

Lactate inhibits the synthesis of arginine, ornithine, and citrulline from Pro in the small intestine (Dillon et al., 1999). Conditions that increase the concentration of lactate in the blood can interfere with the incorporation of ammonia into urea and with the generation of nitric oxide (from arginine).

FUNCTION

Energy fuel: Most of the ingested Pro and Hypro is eventually broken down to carbon dioxide, water, and urea, yielding 5.0 kcal/g (May and Hill, 1990). Its complete oxidation depends on adequate supplies of thiamin, riboflavin, niacin, vitamin B6, pantothenate, lipoate, ubiquinone, iron, and magnesium.

Protein synthesis: Proline-tRNA ligase (EC6.1.1.15) loads Pro onto its specific tRNA for incorporation into proteins. Prolyl 4-hydroxylase (iron- and ascorbate-dependent, EC1.14.11.2) hydroxylates prolyl residues in preprocollagens. The active enzyme is a heterotetramer whose two beta subunits consist of the active enzyme disulfide isomerase (EC5.3.4.1), which is also independently active.

Urea cycle: Pro catabolism in small intestine provides the precursor glutamate semialdehyde for the synthesis of ornithine (Dillon et al., 1999).

Arginine synthesis: The bulk of endogenously synthesized arginine is generated in the small intestine from Pro (Dillon et al., 1999).

Brain function: The fact that a specific PROT exists in the brain has suggested to some an important role there, possibly as a synaptic regulatory molecule. One function in some brain regions could be to provide glutamate, gamma-aminobutyric acid (GABA), and aspartate (Gogos et al., 1999).

Redox shuttle: The reduction of P5C in cytosol to Pro oxidizes NADPH. Pro can then be transferred into mitochondria. The oxidation of Pro to P5C feeds reducing equivalents into the electron transport chain for the synthesis of ATP. Transport of P5C back into the cytosol completes this redox shuttle. Proline dehydrogenase (proline oxidase, EC1.5.99.8) is inducible by cytotoxic agents, such as adriamycin, and may play a role in the regulation of p53-dependent apoptosis (Maxwell and Davis, 2000; Donald et al., 2001).

Oxygen sensing: A heme-containing oxygen sensor (hypoxia-inducible factor 1, HIF-1) stimulates erythropoietin production. The oxygen- and iron-dependent HIF-proline dioxygenase (EC1.14.11.29) activates HIF-1 by hydroxylating Pro564 (Zhu and Bunn, 2001).

REFERENCES

Atlante, A., Passarella, S., Quagliariello, E., 1994. Spectroscopic study of hydroxyproline transport in rat kidney mitochondria. Biochem. Biophys. Res. Commun. 202, 58–64.

Chairoungdua, A., Segawa, H., Kim, J.Y., Miyamoto, K., Haga, H., Fukui, Y., et al., 1999. Identification of an amino acid transporter associated with the cystinuria-related type II membrane glycoprotein. J. Biol. Chem. 274, 28845–28848.

Dillon, E.L., Knabe, D.A., Wu, G., 1999. Lactate inhibits citrulline and arginine synthesis from proline in pig enterocytes. Am. J. Physiol. 276, G1079–G1086.

Donald, S.P., Sun, X.Y., Hu, C.A., Yu, J., Mei, J.M., Valle, D., et al., 2001. Proline oxidase, encoded by p53-induced gene-6, catalyzes the generation of proline-dependent reactive oxygen species. Cancer Res. 61, 1810–1815.

Gogos, J.A., Santha, M., Takacs, Z., Beck, K.D., Luine, V., Lucas, L.R., et al., 1999. The gene encoding proline dehydrogenase modulates sensorimotor gating in mice. Nat. Genet. 21, 434–439.

Hung, C.J., Huang, P.C., Lu, S.C., Li, Y.H., Huang, H.B., Lin, B.F., et al., 2002. Plasma homocysteine levels in Taiwanese vegetarians are higher than those of omnivores. J. Nutr. 132, 152–158.

Ichiyama, A., Xue, H.H., Oda, T., Uchida, C., Sugiyama, T., Maeda-Nakai, E., et al., 2000. Oxalate synthesis in mammals: properties and subcellular distribution of serine:pyruvate/alanine:glyoxylate aminotransferase in the liver. Mol. Urol. 4, 333–340.

Kagabu, Y., Mishiba, T., Okino, T., Yanagisawa, T., 1998. Effects of thyrotropin-releasing hormone and its metabolites, Cyclo(His-Pro) and TRH-OH, on growth hormone and prolactin synthesis in primary cultured pituitary cells of the common carp, Cyprinus carpio. Gen. Comp. Endocrinol. 111, 395–403.

Kim, S.Z., Varvogli, L., Waisbren, S.E., Levy, H.L., 1997. Hydroxyprolinemia: comparison of a patient and her unaffected twin sister. J. Pediatr. 130, 437–441.

Kudo, Y., Boyd, C.A., 2001. Characterization of L-tryptophan transporters in human placenta: a comparison of brush border and basal membrane vesicles. J. Physiol. 531, 405–416.

Maxwell, S.A., Davis, G.E., 2000. Differential gene expression in p53-mediated apoptosis-resistant vs. apoptosis-sensitive tumor cell lines. Proc. Natl. Acad. Sci. USA 97 (24), 13009–13014.

May, M.E., Hill, J.O., 1990. Energy content of diets of variable amino acid composition. Am. J. Clin. Nutr. 52, 770–776.

Prasad, C., 1998. Limited proteolysis and physiological regulation: an example from thyrotropin-releasing hormone metabolism. Thyroid 8, 969–975.

Prasad, C., Hilton, C.W., Svec, E., Onaivi, E.S., Vo, P., 1991. Could dietary proteins serve as cyclo(His-Pro) precursors? Neuropeptides 19, 17–21.

Reuber, B.E., Karl, C., Reimann, S.A., Mihalik, S.J., Dodt, G., 1997. Cloning and functional expression of a mammalian gene for a peroxisomal sarcosine oxidase. J. Biol. Chem. 272, 6766–6776.

Ritchie, J.W., Taylor, P.M., 2001. Role of the System L permease LAT1 in amino acid and iodothyronine transport in placenta. Biochem. J. 356, 719–725.

Urdaneta, E., Barber, A., Wright, E.M., Lostao, M.R., 1998. Functional expression of the rabbit intestinal Na+/L-proline cotransporter (IMINO system) in Xenopus laevis oocytes. J. Physiol. Biochem. 54, 155–160.

Wu, G., 1997. Synthesis of citrulline and arginine from proline in enterocytes of postnatal pigs. Am. J. Physiol. 272, G1382–G1390.

Zhu, H., Bunn, H.E., 2001. Signal transduction. How do cells sense oxygen? Science 292, 449–451.

HISTIDINE

L-Histidine   ((*S*)-alpha-amino-1H-imidazole-4-propanoic   acid,   L-2-amino-3-(1H-imidazol-4-yl) propionic acid, alpha-amino-4-imidazolepropionic acid, alpha-amino-5-imidazolepropionic acid, levo-histidine, glyoxaline-5-alanine, one-letter code H; molecular weight 155) is a basic amino acid with 27% nitrogen (Figure 8.84).

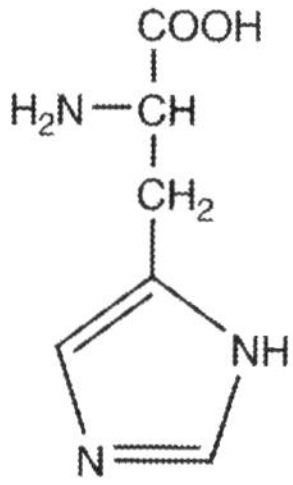

FIGURE 8.84

L-Histidine.

ABBREVIATIONS

GABA	gamma-aminobutyric acid
His	L-histidine
LAT2	L-type amino acid transporter 2 (SLC7A8)
PLP	pyridoxal-5′-phosphate
RDA	recommended dietary allowance

NUTRITIONAL SUMMARY

Function: The essential amino acid L-histidine (His) is used as a precursor of the mediator histamine, and for the synthesis of proteins. Several dipeptides involved in signaling and cell protection contain His. His is also an energy fuel; its complete oxidation requires thiamin, riboflavin, niacin, vitamin B6, folate, pantothenate, lipoate, ubiquinone, iron, and magnesium.

Food sources: Dietary proteins from different sources all contain His. Pork and beef protein has a slightly higher content (3–4% of total protein) than milk, chicken, egg, or grain proteins (2–3% of total protein). Dietary supplements containing crystalline His are commercially available.

Requirements: Adults are thought to require about 520–780 mg/day (FAO/WHO/UNU, 1985). Whether individuals with rheumatoid arthritis and other chronic inflammatory conditions benefit from additional intake is uncertain.

Deficiency: Prolonged lack of His, as with all essential amino acids or a lack of protein, causes growth failure, loss of muscle mass, and organ damage.

Excessive intake: Very high intake of protein and mixed amino acids (i.e., more than three times the RDA, which is 2.4 g/kg) is thought to increase the risk of renal glomerular sclerosis and accelerate osteoporosis. Dermatomyositis has been reported in a user of crystalline His. The health consequences of accumulation of the metabolite formiminoglutamic in folate-deficient consumers of His supplements are not known.

DIETARY SOURCES

His is an essential amino acid that cannot be produced in the body and must be supplied from external sources. All dietary proteins contain His, though the quantities vary between types of food. Pork (38 mg/g), beef (34 mg/g), and other meats contain slightly more His than legumes (around 27 mg/g) and grains (around 23 mg/g); corn (maize) has a slightly higher content (about 30 mg/g) than other grains. The histidyl dipeptides carnosine and homocarnosine represent a significant proportion of the His in meats. Anserine (beta-alanyl-1-methyl-L-histidine) is not biologically equivalent to His because it contains methylhistidine. Bacterial synthesis in the ileum and colon can be a significant source (Metges, 2000).

Note: Casein and soy protein hydrolysates in some dietary supplements contain significant amounts of the diketopiperazine cyclo(His-Pro) (Figure 8.85), which can be absorbed from the intestine (Mizumo et al., 1997; Song et al., 2005). Cyclo(His-Pro), which is normally produced during the specific degradation of TRH (pyroglutamylhistidylprolinamide) by pyroglutamyl-peptidase II (EC3.4.19.6), affects motor functions, influences body temperature, inhibits prolactin secretion (Prasad, 1998), and might stimulate growth hormone output (Kagabu et al., 1998). Normal daily excretion with urine, which is likely to reflect endogenous production, is about 0.5 mg (Prasad et al., 1990).

Fish histamine (Scombroid) poisoning: When fish of the *Scombroid* family (includes tuna, mackerel, sardines, mahimahi, and bluefish) is stored too long, particularly without proper refrigeration, naturally occurring bacteria on the fish can convert significant amounts of His to histamine, which can cause significant abdominal and systemic discomfort and illness within less than an hour of consumption. Symptoms vary and often include headache, dizziness, nausea, diarrhea, itching, flushing, and skin rashes (Al Bulushi et al., 2009). The dark meat of the fish tends to have the highest His concentration (Ruiz-Capillas et al., 2004). Antihistamines are effective for symptom relief.

DIGESTION AND ABSORPTION

Healthy individuals absorb amino acids and proteins in the proximal small intestine nearly completely (Figure 8.86). Food proteins are hydrolyzed by an array of gastric, pancreatic, and enteral enzymes, which generate His as part of oligopeptides, and in free form. The former can be taken up through the hydrogen ion/peptide cotransporter (PepT1; SLC15A1). His can be taken up by the sodium-driven transport system B°. The capacity of this transporter is expanded by the action of the rBAT (SLC3A1)–glycoprotein-anchored transporter BAT1/b°,+ (SLC7A9), which can exchange His for other neutral amino acids in either direction across the brush border membrane (Chairoungdua et al., 1999).

The heteroexchanger y+LAT1 (SLC7A7) and, with lower affinity, LAT2 (SLC7A8) in combination with their glycoprotein membrane anchor 4F2 (SLC3A2) can shuttle His across the basolateral membrane in exchange for other neutral amino acids in combination with a sodium ion.

TRANSPORT AND CELLULAR UPTAKE

Blood circulation: Most of the His in the blood is part of proteins and peptides, some of which are taken up into tissues via their specific mechanisms. The small amounts of His circulating in free form (typically around 87 µmol/l) can enter cells through different transporters, depending on the particular tissue. The sodium-independent transporter y+ L, AT2 (SLC7A6) can exchange His for another neutral amino acid plus one sodium ion in many tissues. Another glycoprotein-linked heteroexchanger, LAT1 (SLC7A5), has

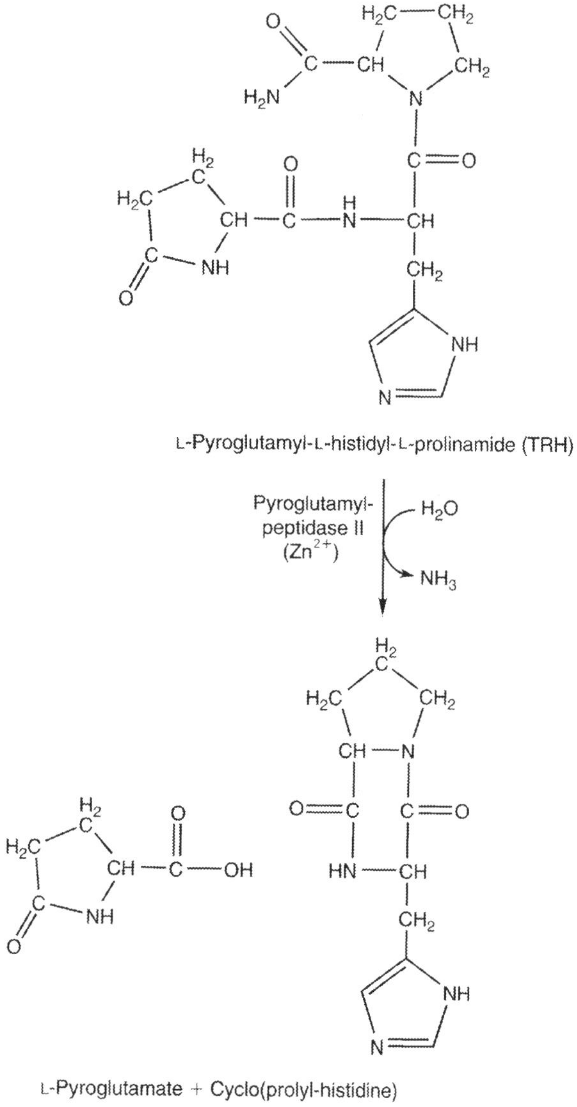

L-Pyroglutamyl-L-histidyl-L-prolinamide (TRH)

Pyroglutamyl-
peptidase II
(Zn^{2+})

H_2O

NH_3

L-Pyroglutamate + Cyclo(prolyl-histidine)

FIGURE 8.85

Some protein hydrolysates contain bioactive cyclo (His–Pro).

lower affinity. ATA2, a representative of sodium-dependent system A in most tissues, can also accept His for transport, though with lower affinity (Sugawara et al., 2000). The same applies to ATA3 in the liver.

Materno-fetal transfer: Net His transfer from the mother to the fetus appears to occur. The sodium-independent transporter y^+ and system L-related transporters, which exchange His for other neutral

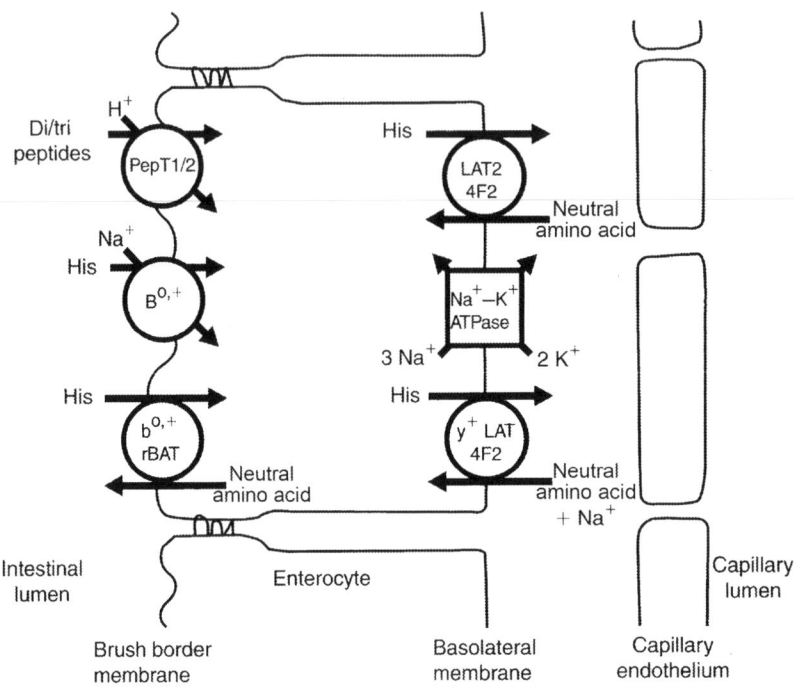

FIGURE 8.86

Intestinal absorption of L-histidine.

amino acids, are expressed in the microvillous and basal membranes (Jansson, 2001). One of these, the heteroexchanger y⁺LAT1 (SLC7A7, associated with 4F2), mediates transfer across the basal membrane (Kudo and Boyd, 2001). There is still some uncertainty about the exact location and contribution of the various transporters.

BBB: Transport of His across the BBB involves one or more of the system L, transporters (Reichel et al., 2000). ATA1 in the brain also accepts His. The exact location of these and other transporters remains uncertain.

METABOLISM

Main pathway: His breakdown is usually initiated by histidine ammonia-lyase (histidase, EC4.3.1.3). The next step is catalyzed by urocanate hydratase (urocanase, EC4.2.1.49). This enzyme is somewhat unusual because it contains a tightly bound NAD molecule, which serves as a prosthetic group, not as a redox cosubstrate, imidazolonepropionase (EC3.5.2.7) and tetrahydrofolate-dependent glutamate formiminotransferase (EC2.1.2.5) then complete the conversion to L-glutamate (Figure 8.87).

Alternative pathways: His can also be transaminated by PLP requiring amino transferases, including phenylalanine(histidine) aminotransferase (EC2.6.1.58) and aromatic-amino-acid-glyoxylate aminotransferase (EC2.6.1.60). The resulting imidazole pyruvate and its metabolites imidazole acetate

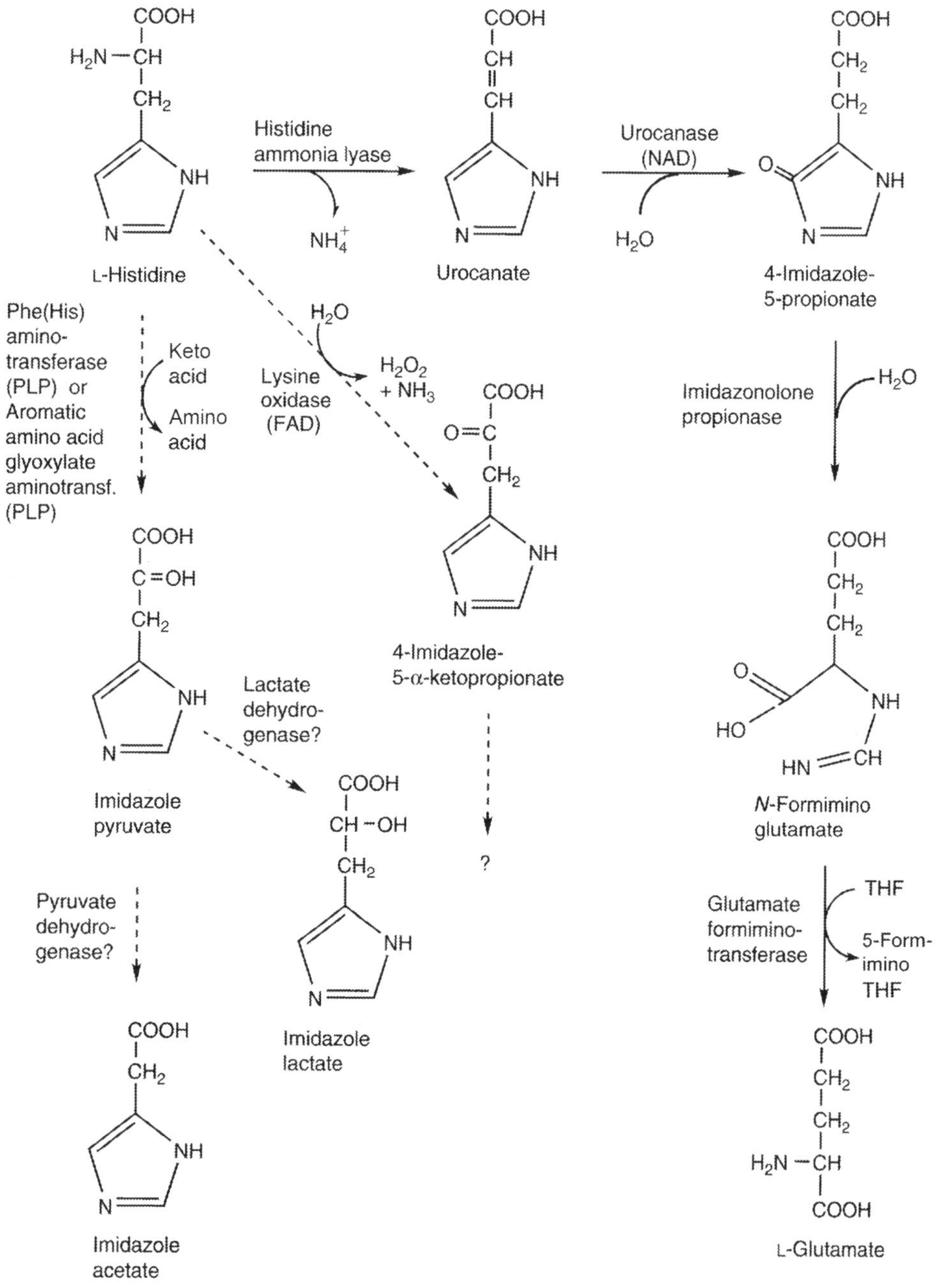

FIGURE 8.87

Metabolic fates of L-histidine.

and imidazole lactate cannot be utilized and excreted with urine. A small amount of His undergoes oxidative deamination by the peroxisomal flavoenzyme L-lysine oxidase (EC1.4.3.14); the further fate of the resulting metabolite is unclear. Despite the significant activities of these enzymes, particularly of phenylalanine(histidine) aminotransferase in the liver, almost all His is metabolized via the main histidase pathway.

Protein histidyl methylation: Specific histidyl residues in actin, myosin, and other proteins are methylated by protein-histidine *N*-methyltransferase (EC2.1.1.85) which uses *S*-adenosyl-L-methionine as a methyl group donor. The daily rate of His 3-methylation in humans has been estimated to be about 3.1 μmol/kg (Rathmacher and Nissen, 1998). It is of note that the methylhistidine from modified proteins and anserine cannot be incorporated into proteins nor can it be metabolized and utilized as an energy fuel.

STORAGE

On average, the His content of human proteins is 24 mg/g (Smith, 1980), presumably corresponding in a 70-kg man to a mobilizable reserve of about 144 g. Hemoglobin is a particularly His-rich protein (84 mg/g) whose normal turnover alone provides more than 0.5 g/day of His. Carnosine (beta-alanyl-L-histidine) in muscle and homocarnosine (gamma-aminobutyryl-L-histidine) in the brain are His-rich peptides that can provide significant amounts of His in times of need (Figure 8.88). Carnosine is synthesized by carnosine synthase (EC6.3.2.11) from His and beta-alanine. This enzyme in the brain, muscles, and other tissues uses NAD as a prosthetic group. Beta-alanine feeding, but not His feeding, was found to increase carnosine stores in horses (Dunnett and Harris, 1999); this has not yet been investigated in humans. His stored with carnosine and other dipeptides can be released in the blood, brain, and other tissues by the zinc-enzyme carnosinase (EC3.4.13.3). Another enzyme, beta-ala-His dipeptidase (EC3.4.13.20), appears to exist only in humans and other primates. The enzyme is activated best by cadmium, and only half as well by manganese; the physiological significance of this unusual activation pattern is unclear.

Hemoglobin is another His-rich protein (91.3 mg/g) whose degradation can provide significant amounts of His when dietary sources are inadequate. At a breakdown rate of 20 ml red blood cells or 6 g hemoglobin per day this corresponds to a daily release of 548 mg His.

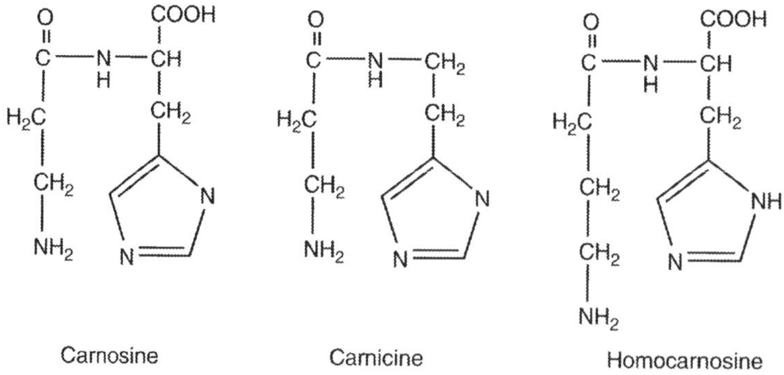

Carnosine   Carnicine   Homocarnosine

FIGURE 8.88

L-Histidine-rich peptides can cover the needs of the human body for some time.

EXCRETION

Free His is filtered in the kidneys and reabsorbed in the proximal tubule more or less via the same transporters that mediate absorption from the small intestinal lumen.

Transporters $B^{o,+}$ (SLC6A14) and $b^{o,+}$ (SLC7A9) are located on the luminal side, transporters y^{+}LAT1 (SLC7A7) and LAT2 (SLC7A8) on the basolateral membrane side. Little of the filtered methylhistidine will be reabsorbed and most is excreted with urine.

REGULATION

Little is known about how, if at all, the body keeps its His content constant or ensures adequate availability for critical functions.

FUNCTION

Energy fuel: Eventually a large proportion of ingested His is oxidized with an energy yield of 3.4 kcal/g (May and Hill, 1990). Complete oxidation depends on adequate supplies of thiamin, riboflavin, niacin, vitamin B6, folate, pantothenate, lipoate, ubiquinone, iron, and magnesium. The fraction that is converted to methylhistidine does not contribute to energy production.

Protein synthesis: Histidine-tRNA ligase (EC6.1.1.21) loads His onto tRNA His that can then proceed to participate in protein synthesis.

Histidine-containing dipeptides: Carnosine (20 mmol/l in the heart) is likely to be important as a His storage molecule, but it also contributes to the proton buffering (Dunnett and Harris, 1999), free radical scavenging (Choi et al., 1999), and aldehyde scavenging (Swearengin et al., 1999) capacity of muscle. Dietary carnosine (in the millimole range) appears to have vasodilatory effects on aortic muscle (Ririe et al., 2000).

Homocarnosin (gamma-aminobutyryl-L-histidine) is synthesized from the neurotransmitter GABA and His by carnosine synthase (EC6.3.2.11) in the brain. A role of this dipeptide in the synthesis of GABA has been suggested. Another important role of carnosine appears to be the scavenging of peroxyl and other free radicals *in vitro*. Its potency is enhanced when it is chelated with copper (Decker et al., 2000). It has been suggested that due to its high concentration in muscle, carnosine is the main protector against polyunsaturated and saturated aldehydes. Carnosine may also affect immune function by potentiating the respiratory burst of neutrophils and protecting them against free-radical–induced apoptosis (Tan and Candlish, 1998). Carnosine can be decarboxylated through an as-yet-unknown mechanism to carnicine. Another His-derivative, ergothioneine (2-mercaptohistidine betaine), is present in erythrocytes, ocular lens, the liver, and a few other tissues in millimolar concentration (Misiti et al., 2001) (Figure 8.89). Since no biosynthetic pathways have been identified in humans or other mammals, it has been suggested that all ergothioneine in humans must be derived from food. However, the consistently high concentrations of this compound in the blood and tissues of diverse populations appear to argue against this assumption. Ergothioneine acts as an antioxidant and facilitates the decomposition of S-nitrosoglutathione in the blood, the liver, and kidneys.

Balenine (beta-alanyl-L-3-methylhistidine, ophidine) is generated by proteolysis of methylated proteins. Swine and other mammals produce much larger quantities than humans do. Antioxidant and other specific properties have been attributed to this compound, but the significance in humans is uncertain.

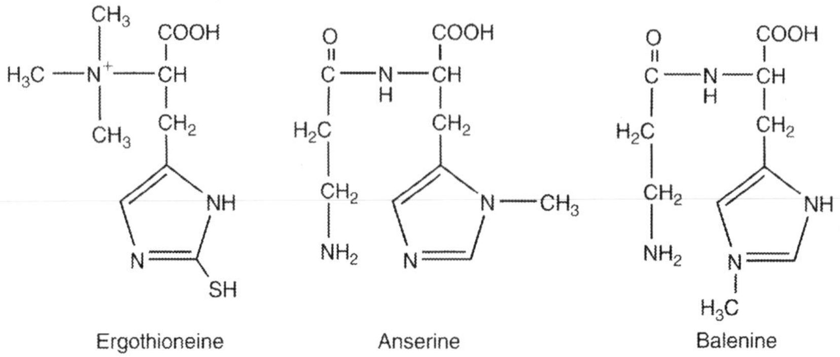

FIGURE 8.89

L-Histidine-containing dipeptides have diverse biological functions.

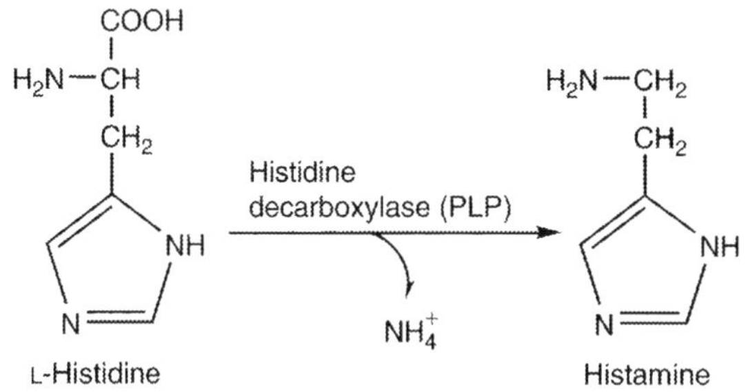

FIGURE 8.90

Histamine is a potent neurotransmitter and paracrine agent.

Additional histidyl dipeptides, including acetylcarnosine, anserine, and acetylanserine, are present in various tissues; some may be primarily or exclusively of dietary origin. Such carnosine-related compounds could be important for the regulation of intracellular oxygen free radical concentrations (Boldyrev and Abe, 1999), but questions about their precise role remain (Mitsuyama and May, 1999).

Receptor ligands: Many tissues contain specific PLP-dependent histidine decarboxylases (EC4.1.1.22), which can release histamine (Figure 8.90), a potent neurotransmitter and paracrine agent. H1-histamine receptors (in mast cells and many peripheral tissues) and H2-histamine receptors (parietal cells of the stomach) respond to binding of agonists by increasing histamine release, which in turn affects tissue functions such as hydrochloric acid release in the stomach. Carnosine complexed with zinc also appears to specifically interact with both H1 and H2 receptors. This complex has been found to provoke saphenous vein contractions with a potency comparable to that of histamine (Miller and O'Dowd, 2000).

Effect on trace metal bioavailability: Dietary His has been reported to improve intestinal zinc absorption (Lönnerdal, 2000).

BIOMARKERS OF DIETARY INTAKE

Urinary 3-methylhistidine excretion correlates with His intake and may be used to assess adequacy (Prescott et al., 1988).

REFERENCES

Al Bulushi, I., Poole, S., Deeth, H.C., Dykes, G.A., 2009. Biogenic amines in fish: roles in intoxication, spoilage, and nitrosamine formation—a review. Crit. Rev. Food Sci. Nutr. 49, 369–377.

Boldyrev, A., Abe, H., 1999. Metabolic transformation of neuropeptide carnosine modifies its biological activity. Cell Mol. Neurobiol. 19, 163–175.

Chairoungdua, A., Segawa, H., Kim, J.Y., Miyamoto, K., Haga, H., Fukui, Y., et al., 1999. Identification of an amino acid transporter associated with the cystinuria-related type II membrane glycoprotein. J. Biol. Chem. 274, 28845–28848.

Choi, S.Y., Kwon, H.Y., Kwon, O.B., Kang, J.H., 1999. Hydrogen peroxide-mediated Cu, Zn-superoxide dismutase fragmentation: protection by carnosine, homocarnosine and anserine. Biochim. Biophys. Acta 1472, 651–657.

Decker, E.A., Livisay, S.A., Zhou, S., 2000. A re-evaluation of the antioxidant activity of purified carnosine. Biochemistry (Moscow) 65, 766–770.

Dunnett, M., Harris, R.C., 1999. Influence of oral beta-alanine and L-histidine supplementation on the carnosine content of the gluteus medius. Equine Vet. J. 30, 499–504.

FAO/WHO/UNU, 1985. Energy and protein requirements. World Health Organ. Tech. Rep. Ser. 724, 204.

Jansson, T., 2001. Amino acid transporters in the human placenta. Pediatr. Res. 49, 141–147.

Kagabu, Y., Mishiba, T., Okino, T., Yanagisawa, T., 1998. Effects of thyrotropin-releasing hormone and its metabolites, Cyclo(His-Pro) and TRH-OH, on growth hormone and prolactin synthesis in primary cultured pituitary cells of the common carp, Cyprinus carpio. Gen. Comp. Endocrinol. 111, 395–403.

Kudo, Y., Boyd, C.A., 2001. Characterisation of L-tryptophan transporters in human placenta: a comparison of brush border and basal membrane vesicles. J. Physiol. 531, 405–416.

Lönnerdal, B., 2000. Dietary factors influencing zinc absorption. J. Nutr. 130, 1378S–1383S.

May, M.E., Hill, J.O., 1990. Energy content of diets of variable amino acid composition. Am. J. Clin. Nutr. 52, 770–776.

Miller, D.J., O'Dowd, A., 2000. Vascular smooth muscle actions of carnosine as its zinc complex are mediated by histamine H(1) and H(2) receptors. Biochemistry (Moscow) 65, 798–806.

Misiti, E., Castagnola, M., Zuppi, C., Giardina, B., Messana, I., 2001. Role of ergothioneine on S-nitrosoglutathione catabolism. Biochem. J. 356, 799–804.

Mitsuyama, H., May, J.M., 1999. Uptake and antioxidant effects of ergothioneine in human erythrocytes. Clin. Sci. 97, 407–411.

Mizumo, H., Svec, E., Prasad, C., Hilton, C., 1997. Cyclo(His-Pro) augments the insulin response to oral glucose in rats. Life Sci. 60, 369–374.

Prasad, C., 1998. Limited proteolysis and physiological regulation: an example from thyrotropinreleasing hormone metabolism. Thyroid 8, 969–975.

Prasad, C., Ragan Jr., F.A., Hilton, C.W., 1990. Isolation of cyclo(His-Pro)-like immunoreactivity from human urine and demonstration of its immunologic, pharmacologic, and physico-chemical identity with the synthetic peptide. Biochem. Int. 21, 425–434.

Prescott, S.L., Jenner, D.A., Beilin, L.J., Margetts, B.M., Vandongen, R.A., 1988. Randomized controlled trial of the effect on blood pressure of dietary non-meat protein versus meat protein in normotensive omnivores. Clin Sci. 74, 665–672.

Rathmacher, J.A., Nissen, S.L., 1998. Development and application of a compartmental model of 3-methylhistidine metabolism in humans and domestic animals. Adv. Exp. Med. Biol. 445, 303–324.

Reichel, A., Begley, D.J., Abbott, N.J., 2000. Carrier-mediated delivery of metabotrophic glutamate receptor iigands to the central nervous system: structural tolerance and potential of the L-system amino acid transporter at the blood–brain barrier. J. Cereb. Blood Flow Metab. 20, 168–174.

Ririe, D.G., Roberts, P.R., Shouse, M.N., Zaloga, G.P., 2000. Vasodilatory actions of the dietary peptide carnosine. Nutr 16, 168–172.

Ruiz-Capillas, C., Moral, A., 2004. Free amino acids and biogenic amines in red and white muscle of tuna stored in controlled atmospheres. Amino Acids 26, 125–132.

Song, MK, Rosenthal, MJ, Song, AM, Yang, H, Ao, Y, Yamaguchi, DT., 2005. Raw vegetable food containing high cyclo (his-pro) improved insulin sensitivity and body weight control. Metabolism 54, 1480–1489.

Smith, R.H., 1980. Comparative amino acid requirements. Proc. Nutr. Soc. 39, 71–78.

Sugawara, M., Nakanishi, T., Fei, Y.J., Huang, W., Ganapathy, M.E., Leibach, F.H., et al., 2000. Cloning of an amino acid transporter with functional characteristics and tissue expression pattern identical to that of system A. J. Biol. Chem. 275, 16473–16477.

Swearengin, T.A., Fitzgerald, C., Seidler, N.W., 1999. Carnosine prevents glyceraldehyde 3-phosphate-mediated inhibition of aspartate aminotransferase. Arch. Toxicol. 73, 307–309.

Tan, K.M., Candlish, J.K., 1998. Carnosine and anserine as modulators of neutrophil function. Clin. Lab. Haematol. 20, 239–344.

CITRULLINE

Citrulline (2-amino-5-(carbamoylamino)pentanoic acid; molecular weight 175.19) is a nonessential zwitterionic alpha-amino acid (Figure 8.91).

NUTRITIONAL SUMMARY

Function: Citrulline is used as an energy fuel; its complete oxidation requires niacin, thiamin, riboflavine, vitamin B6, pantothenate, lipoate, and ubiquinone.

Food sources: Citrulline is not a constituent of proteins or peptides. Most plant- and animal-derived foods contain only the small amounts that reflect its role as a metabolite in the urea cycle and other pathways. Modest amounts may be consumed with watermelons (about 3 mg/g fresh weight).

Requirements: There is no nutritional requirement for citrulline.

Deficiency: Low citrulline intake is not known to cause specific symptoms or conditions.

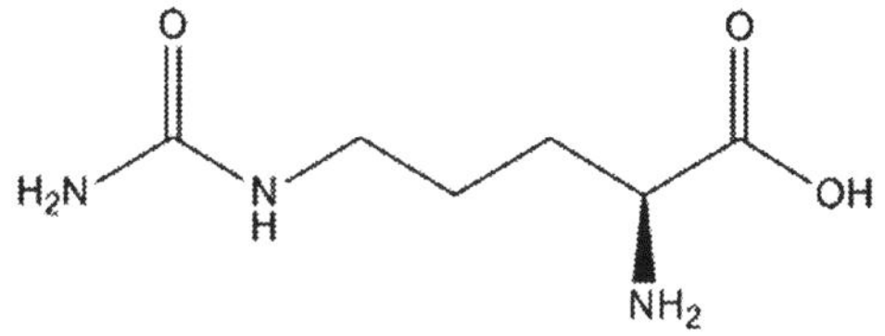

FIGURE 8.91

Citrulline.

Excessive intake: Consumption of citrulline from dietary supplements carries all the potential risks of such products, such as the presence of chemical contaminants or by-products. Specific risks related to large amounts are poorly understood.

ENDOGENOUS PRODUCTION

The small intestine and liver are the major sites of citrulline synthesis. Glutamate, glutamine, proline, histidine, and arginine can serve directly or indirectly as precursors. Transamination of alpha-ketoglutarate generates additional glutamate from other amino acids, especially in small intestinal enterocytes after a protein-rich meal. Adequate supplies of niacin, magnesium, and manganese are needed for maximal endogenous production.

Urea cycle: L-Ornithine and carbamoyl phosphate are condensed by ornithine carbamoyltransferase (OTC, EC2.1.3.3). The carbamoyl phosphate comes from a reaction that combines ammonia and bicarbonate, catalyzed by carbamoyl-phosphate synthase (ammonia) (EC6.3.4.16). This enzyme requires the presence of its allosteric activator *N*-acetylglutamate.

Intestinal synthesis: Another pathway for citrulline synthesis starts in the small intestine from Glu (and indirectly from glutamine) and uses the bifunctional pyrroline 5-carboxylate synthase (no EC number assigned). This enzyme comprises phosphorylation (glutamate 5-kinase, EC2.7.2.11) and reduction (glutamate 5-semialdehyde dehydrogenase, EC1.2.1.41) activities (Wu and Morris, 1998). Citrulline synthesis is completed by OAT aminotransferase (ornithine transaminase or ornithine aminotransferase, OAT, EC2.6.1.13, PLP-dependent) and OTC (EC2.1.3.3).

From arginine: Much smaller amounts are generated by nitric oxide synthase (EC1.14.13.39). This reaction generates equimolar amounts of citrulline from L-arginine. In a somewhat similar reaction, peptidylarginine deiminases (EC3.5.3.15) in the inner root sheath and medulla of hair follicles convert arginine residues of nascent trichohyalin to citrullines, thereby strengthening this hair protein (Steinert et al., 2003).

DIETARY SOURCES

Watermelons are the only commonly consumed citrulline-rich food. Typical citrulline content in fresh fruit is about 3 mg/g (Davis et al., 2010–2011). This means that a typical wedge (286 g) contains almost 1 g of citrulline. Some other fruits also contain citrulline, but in much smaller amounts.

ABSORPTION

In rats, citrulline is absorbed best in the middle to lower ileum, possibly via the amino acid–sodium cotransporter ASC (Vadgama and Evered, 1992).

TRANSPORT AND CELLULAR UPTAKE

Blood circulation: Serum concentration of citrulline in healthy people is around 40 μmol/l. A concentration under 30 μmol/l may be indicative of villous atrophy (Crenn et al., 2003).

The mitochondrial ornithine transporter 1 (ORC1, encoded by SLC25A15) exchanges ornithine from cytosol for citrulline from mitochondria, where it has been produced from carbamoyl phosphate and orthinine. This ornithine–citrulline exchange effectively moves the ammonia-derived nitrogen from mitochondria into the cytosol for completion of urea synthesis.

METABOLISM

Citrulline is mostly used as a vehicle for urea synthesis in the small intestine and liver. In other tissues, such as endothelial cells of the arteries, the main significance of citrulline is as a precursor of L-arginine for nitric oxide synthesis.

Urea synthesis: The ATP-powered addition of aspartic acid to citrulline by argininosuccinate synthase (EC6.3.4.5) in the cytosol generates argininosuccinic acid. Argininosuccinate lyase (EC4.3.2.1) then cleaves this intermediate into fumarate and L-arginine. The cycle is completed when the manganese-requiring enzyme arginase (EC3.5.3.1) cleaves L-arginine into urea and L-ornithine. While ornithine is mostly recycled during urea synthesis, it comes ultimately from *de novo* synthesis in the small intestine. Smaller amounts may be taken up from food sources.

L-Arginine synthesis: Argininosuccinate synthase (EC6.3.4.5) adds aspartate to citrulline in an ATP-dependent reaction. Argininosuccinate lyase (EC4.3.2.1) then releases arginine and fumarate.

FUNCTION

Urea synthesis: Citrulline is as an essential intermediate of urea synthesis in the liver and kidneys, arising from the combination of L-ornithine and carbamoyl phosphate. Dietary citrulline intake is not relevant for the normal functioning of the urea cycle, and there is no indication that urea cycle efficacy is improved by intake.

Arginine synthesis: Citrulline is a precursor of L-arginine, which has vasodilating activity and appears to counter erectile dysfunction with modest effectiveness (Cormio et al., 2011).

Hair protein modification: Peptidylarginine deiminase (EC3.5.3.15) in the inner root sheath and medulla of hair follicles converts arginine residues of nascent trichohyalin to citrullines, thereby strenghening this hair protein (Steinert et al., 2003).

Reverse cholesterol transport: Experiments with macrophages suggest that citrulline increases expression of the reverse-cholesterol transporters ABCA1 and ABCG1 (Uto-Kondo et al., 2014).

REFERENCES

Cormio, L., De Siati, M., Lorusso, F., Selvaggio, O., Mirabella, L., Sanguedolce, F., et al., 2011. Oral L-citrulline supplementation improves erection hardness in men with mild erectile dysfunction. Urology 77, 119–122.

Crenn, P., Vahedi, K., Lavergne-Slove, A., Cynober, L., Matuchansky, C., Messing, B., 2003. Plasma citrulline: a marker of enterocyte mass in villous atrophy-associated small bowel disease. Gastroenterology 124, 1210–1219.

Davis, A.R., Fish, W.W., Levi, A., King, S., Wehner, T., 2010-2011. Perkins-Veazie. L-Citrulline levels in watermelon cultivars from three locations. Cucurbit Genet. Coop. Rep. 33–34, 36–39.

Steinert, P.M., Parry, D.A., Marekov, L.N., 2003. Trichohyalin mechanically strengthens the hair follicle: multiple cross-bridging roles in the inner root shealth. J. Biol. Chem. 278, 41409–41419.

Uto-Kondo, H., Ayaori, M., Nakaya, K., Takiguchi, S., Yakushiji, E., Ogura, M., et al., 2014. Citrulline increases cholesterol efflux from macrophages *in vitro* and *ex vivo* via ATP-binding cassette transporters. J. Clin. Biochem. Nutr. 5, 32–39.

Vadgama, J.V., Evered, D.F., 1992. Characteristics of L-citrulline transport across rat small intestine *in vitro*. Ped. Res. 32, 472–478.

Wu, G., Morris, S.M., 1998. Arginine metabolism: nitric oxide and beyond. Biochem. J. 336, 1–17.

TAURINE

The ampholytic amino acid taurine (2-aminoethane sulfonic acid; molecular weight 125) is a conditionally essential nutrient (Figure 8.92).

ABBREVIATION

TAUT   taurine transporter (SLC6A6)

NUTRITIONAL SUMMARY

Function: Critical for kidney function and protection of cells against dehydration, eye and brain function, hormonal regulation, nutrient absorption (bile) and protection against oxygen free radicals (hypotaurine).

Requirements: In most people, endogenous synthesis is adequate (dependent on cystein, niacin, vitamin B6, iron, and molybdenum). Prematurely born infants and possibly even healthy infants may depend partially on dietary intake; a newborn's requirement from endogenous and dietary sources combined is at least 6 mg/day. Adults may need as little as 30–40 mg/day from endogenous synthesis and diet combined.

Food sources: Clams, fish, and meat are the best sources. Human milk is a good source for infants. Taurine is absent from a lacto-ovo-vegetarian or vegan diet.

Deficiency: A lack of taurine in infants may cause irreversible degeneration of the retina, limit brain maturation, and slow growth and weight gain.

Excessive intake: The risks from very high taurine intake (>1–2 g/day) are not known.

ENDOGENOUS SOURCES

Taurine is produced endogenously, mainly in liver cytosol; its synthesis requires cysteine, niacin, iron, pyridoxine, and molybdenum (Figure 8.93). In the main metabolic pathway, cysteine is oxidized to 3-sulfinoalanine. The cysteine dioxygenase (EC1.13.11.20) responsible for this reaction uses iron and NAD or NADP as prosthetic groups. Sulfinoalanine decarboxylase (EC4.1.1.29) with pyridoxal phosphate as a prosthetic group generates hypotaurine, which is finally converted into taurine by hypotaurine

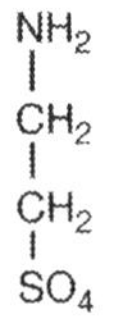

$$\begin{array}{c} NH_2 \\ | \\ CH_2 \\ | \\ CH_2 \\ | \\ SO_4 \end{array}$$

FIGURE 8.92

Taurine.

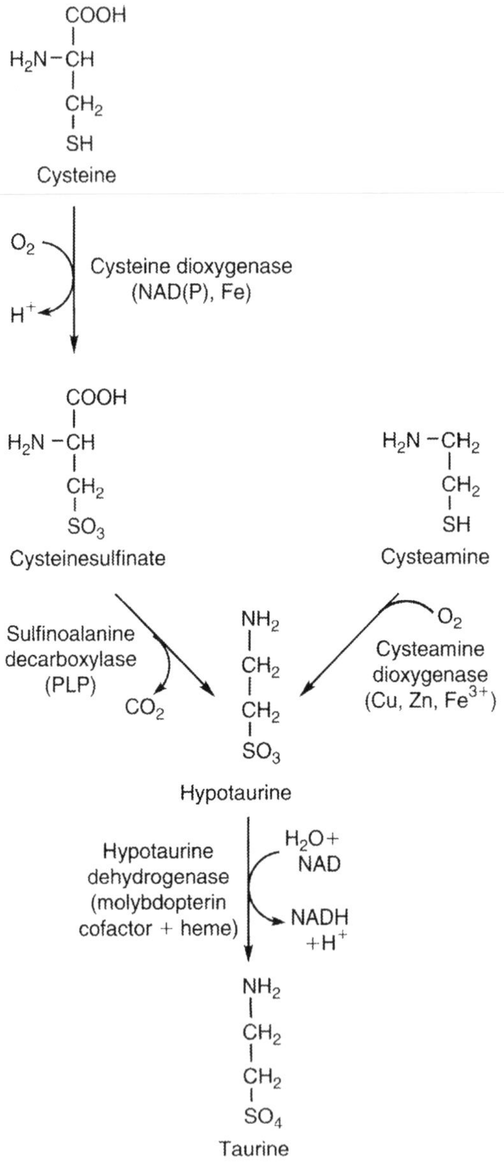

COOH
|
H$_2$N−CH
|
CH$_2$
|
SH
Cysteine

O$_2$ ⟍
    Cysteine dioxygenase
    (NAD(P), Fe)
H$^+$ ⟍

COOH
|
H$_2$N−CH
|
CH$_2$
|
SO$_3$
Cysteinesulfinate

H$_2$N−CH$_2$
|
CH$_2$
|
SH
Cysteamine

Sulfinoalanine
decarboxylase
(PLP)
  CO$_2$

NH$_2$
|
CH$_2$
|
CH$_2$
|
SO$_3$
Hypotaurine

O$_2$
Cysteamine
dioxygenase
(Cu, Zn, Fe^{3+})

Hypotaurine
dehydrogenase
(molybdopterin
cofactor + heme)

H$_2$O+
NAD

NADH
+H$^+$

NH$_2$
|
CH$_2$
|
CH$_2$
|
SO$_4$
Taurine

FIGURE 8.93

Endogeneous taurine synthesis.

dehydrogenase (EC1.8.1.3) in an NAD-dependent reaction. Hypotaurine dehydrogenase contains both home and molybdenum cofactor (molybdopterin, a pterine with molybdenum coordinated to it).

Another pathway in the cytosol of most tissues uses cysteamine for hypotaurine and taurine synthesis; this reaction is catalyzed by cysteamine dioxygenase (EC1.13.11.19), a metalloprotein that contains one atom each of copper, zinc, and ferric iron.

DIETARY SOURCES

Most of the body's taurine comes from meat (200–400 mg/kg), fish (300–700 mg/kg), clams (1500–2400 mg/kg), milk (6 mg/l), and a few other foods of animal origin; it is not a component of plant foods. Typical intake in Chinese omnivores was 30–80 mg/day (Zhao et al., 1998). Human milk provides 40 mg/l, and colostrum is an even richer source.

DIGESTION AND ABSORPTION

Significant amounts of taurine (more than 2000 mg/day) are secreted with bile as conjugates with bile acids; these conjugates are cleaved by bacterial chenodeoxycholoyltaurine hydrolase (EC3.5.1.74). Free taurine is taken up from the small intestinal (ileal) lumen by a sodium chloride–dependent cotransporter. The GABA transporter 1 is expressed in the small intestine (Jin et al., 2001) and might be solely or in part responsible for this transport activity.

Taurine is transported out of the enterocytes by the NaCl-dependent taurine transporter (TAUT; SLC6A6).

TRANSPORT AND CELLULAR UPTAKE

Blood circulation: Uptake of taurine into cells is mediated by a sodium/taurine cotransporter that is osmotically induced and upregulated by protein kinase C (PKC) (Stevens et al., 1999). Hypotaurine, but not taurine, is transported by the NaCl-dependent GAT-2 transporter in liver cells (Liu et al., 1999).

Materno-fetal transfer: The same taurine transporter has also been found in the placenta, brain, and kidneys. It is not clear how taurine is exported from the syntrophoblast into the intercapillary space on the fetal side of the placental membrane.

BBB: GAT2/BGT-1 contributes to the transport of taurine from blood circulation into the brain (Takanaga et al., 2001). The NaCl-dependent taurine transporter TAUT is also involved. The entire sequence of events contributing to crossing the BBB remains to be resolved. Some taurine is also transported in the reverse direction (Kang, 2000).

METABOLISM

Taurine breakdown uses the reactions of synthesis in reverse, mainly in the liver. Hypotaurine dehydrogenase (EC1.8.1.3) contains molybdenum cofactor (contains molybdopterin, Figure 8.94) and heme and reduces taurine again to hypotaurine, which can be converted to cysteine sulfinate (sulfinoalanine) by sulfinoalanine decarboxylase (EC4.1.1.29, PLP-dependent). At this point, taurine breakdown joins the pathway for cysteine catabolism (Figure 8.95). The abundant cytosolic aspartate aminotransferase

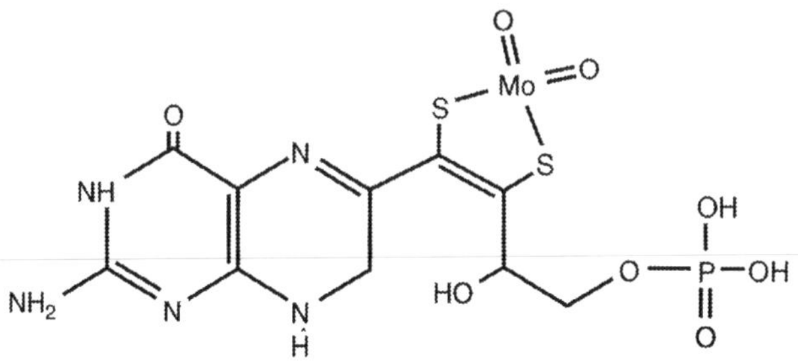

FIGURE 8.94

Molybdenum cofactor.

(PLP-dependent, EC2.6.1.1) generates beta-sulfinylpyruvate, which decomposes spontaneously into sulfite and pyruvate. Alternatively, aspartate 4-decarboxylase (EC4.1.1.12) can cleave cysteinesulfinate into sulfite and alanine (Rathod and Fellman, 1985). Finally, the molybdenum-containing sulfite oxidase (EC1.8.3.1) oxidizes sulfite oxidation to sulfate with cytochrome c as the electron acceptor. Taurine metabolism is thus intimately dependent on molybdenum since two key reactions use molybdenum cofactor.

STORAGE

Muscle cells contain most of the body's taurine; smaller amounts are in the liver and other tissues, as well as bile acid-taurine conjugates (1%) in the liver and intestinal lumen. Glutaurine (γ-L-glutamyltaurine) may be an intracellular storage form.

EXCRETION

Normally, about 2–3% of excreted bile acids are lost via feces, and the attached taurine with them. Daily taurine losses may be estimated at 60 mg, assuming total bile acid losses of 500 mg, about half of which will be conjugated with taurine.

The taurine transporter (TAUT, SLC6A6) cotransports taurine, along with two sodium ions and one chloride ion, into endothelial cells of the proximal renal tubule (Chesney et al., 1990; Bitoun et al., 2001). TAUT also mediates transport across the basolateral membrane.

Hypotaurine can be transported in kidney cells by the NaCl-dependent GAT-2 transporter (Liu et al., 1999), but the quantitative significance of this pathway is unknown.

REGULATION

With high taurine intake, the capacity of the taurine transport system is exceeded and most taurine is excreted unchanged via urine (Wang and Zhao, 1998). Regulation of TAUT expression in straight proximal tubule cells, and thereby modulation of taurine recovery (Matsell et al., 1997), also seems to be an important mechanism for the maintenance of a stable body pool. The expression of TAUT is down-regulated when protein kinase C-mediated signaling is activated (Hart et al., 1999).

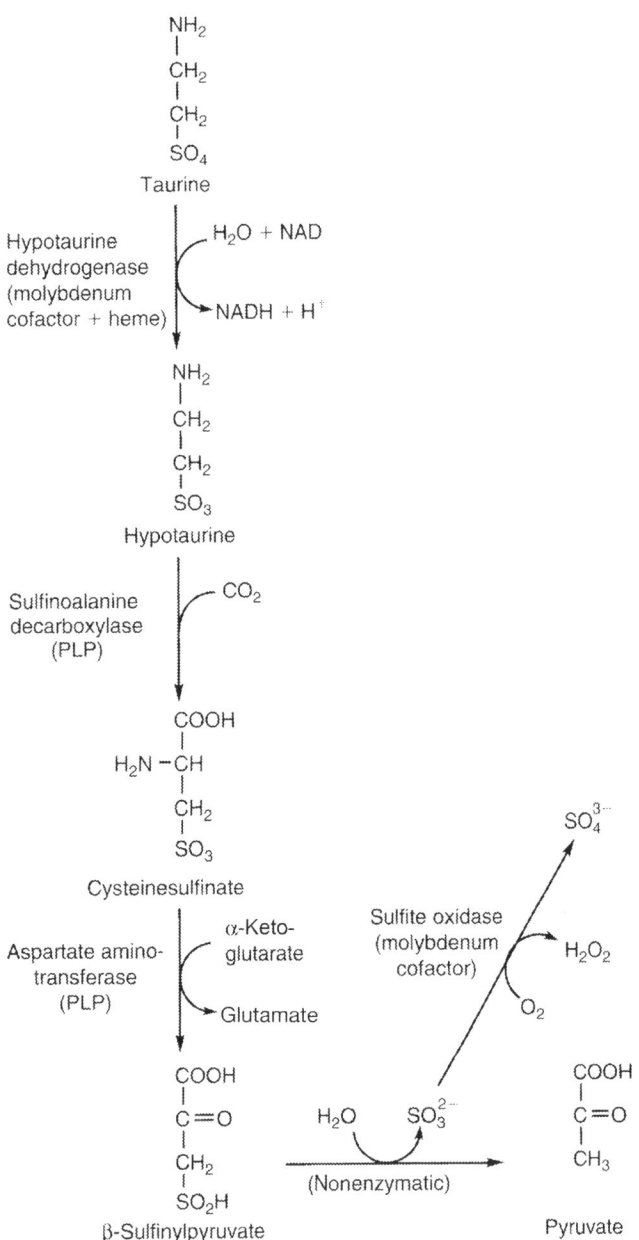

FIGURE 8.95

Taurine metabolism.

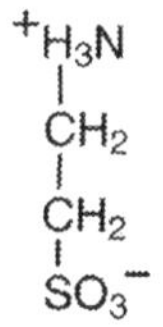

$$
\begin{array}{c}
^+\mathrm{H_3N} \\
| \\
\mathrm{CH_2} \\
| \\
\mathrm{CH_2} \\
| \\
\mathrm{SO_3^-}
\end{array}
$$

FIGURE 8.96

Taurine is an ampholyte.

FUNCTION

Osmolar buffering: Taurine is the most abundant intracellular free amino acid and a potent ampholyte (Figure 8.96). This allows it to stabilize intracellular pH and protect against water loss to extracellular environment with high osmotic pressure. Such protection is of particular importance in the kidneys where the osmolarity is more than twofold higher toward the tip of the loop of Henle than in other regions. Without the ability to withstand such high osmotic pressure, the countercurrent amplification of sodium chloride and water transport from primary filtrate back into circulation would not be possible.

The great importance of taurine for kidney function is underscored by the fact that the highest expression of a key enzyme of taurine synthesis, sulfinoalanine decarboxylase (EC4.1.1.29), occurs in proximal straight tubules of the kidneys (Reymond et al., 2000). Taurine may also increase expression of the osmolarity sensor protein ENVZ (Moenkemann et al., 1999), which modulates water and electrolyte transport by altering aquaporin expression.

In response to low osmolarity, taurine can exit brain cells such as astrocytes through specific channels formed by phospholemman. Such fluxes of taurine, which contribute to regulatory volume decrease of the cells, are modulated by protein kinase A (Moran et al., 2001).

Bile acid conjugation: CoA-conjugated bile acids can be conjugated to taurine through the action of glycine *N*-choloyltransferase (EC2.3.1.65). The enzyme links nearly 50% of bile acids with taurine.

Glutaurine: The dipeptide from taurine and glutamate is called *glutaurine*. It has been reported to participate in neuroexcitation (Wu et al., 1992) and to have antiamnesic potential when given orally (Balazs et al., 1988). Glutaurine may also be involved in the regulation of thyroid and parathyroid activity (Baskin et al., 1987).

Antioxidation: Hypotaurine acts as an antioxidant (Shi et al., 1997; Devamanoharan et al., 1997), particularly by converting the oxidant hypochiorous acid into taurochloramine. Macrophages generate hypochlorous acid during antibacterial action (free radical burst) and phagocytosis.

REFERENCES

Balazs, M., Telegdy, G., 1988. Effects of glutaurine treatment on electroshock-induced amnesia. Antiamnesic action of glutaurine. Neuropeptides 12, 55–58.

Baskin, S., Bartuska, D., Thampi, N., McBride, M., Finnigan, J., 1987. The effect of glutaurine on thyroid hormones in the rat. Neuropeptides 9, 45–50.

Bitoun, M., Levillain, O., Tappaz, M., 2001. Gone expression of the taurine transporter and taurine biosynthetic enzymes in rat kidney after antidiuresis. Pfl. Arch. Eur. J. Physiol. 442, 87–95.

Chesney, R.W., Zelikovic, I., Jones, D.E., Budreau, A., Jolly, K., 1990. The renal transport of taurine and the regulation of renal sodium-chloride-dependent transporter activity. Pediatr. Nephrol. 4, 399–407.

Devamanoharan, P.S., Ali, A.H., Varma, S.D., 1997. Prevention of lens protein glycation by taurine. Mol. Cell Biochem. 177, 245–250.

Han, X., Budreau, A.M., Chesney, R.W., 1999. Ser-322 is a critical site for PKC regulation of the MDCK cell taurine transporter (pNCT). J. Am. Soc. Nephrol. 10, 1874–1879.

Jin, X.P., Huang, E., Yang, N., Lu, B.F., Fei, J., Guo, L.H., 2001. GABA transporter 1 transcriptional starting site exhibiting tissue specific difference. Cell Res. 11, 161–163.

Kang, Y.S., 2000. Taurine transport mechanism through the blood–brain barrier in spontaneously hypertensive rats. Adv. Exp. Med. Biol. 483, 321–324.

Liu, M., Russell, R.L., Beigelman, L., Handschumacher, R.E., Pizzorno, G., 1999. beta-alanine and alpha-fluoro-beta-alanine concentrative transport in rat hepatocytes is mediated by GABA transporter GAT-2. Am. J. Physiol. 276, G206–G210.

Matsell, D.G., Bennett, T., Han, X.B., Budreau, A.M., Chesney, R.W., 1997. Regulation of the taurine transporter gene in the $3 segment of the proximal tubule. Kidney Int. 52, 748–754.

Moenkemann, H., Labudova, O., Yeghiazarian, K., Rink, H., Hoeger, H., Lubec, G., 1999. Evidence that taurine modulates osmoregulation by modification of osmolarity sensor protein ENVZ-expression. Amino Acids 17, 347–355.

Moran, J., Morales-Mulia, M., Pasantes-Morales, H., 2001. Reduction of phospholemman expression decreases osmosensitive taurine efflux in astrocytes. Biochim. Biophys. Acta 1538, 313–320.

Rathod, P.K., Fellman, J.H., 1985. Identification of mammalian aspartate-4-decarboxylase. Arch. Biochem. Biophys. 238, 435–446.

Reymond, I., Bitoun, M., Levillain, O., Tappaz, M., 2000. Regional expression and histological localization of cysteine sulfinate decarboxylase mRNA in the rat kidney. J. Histochem. Cytochem. 48, 1461–1468.

Shi, X., Flynn, D.C., Porter, D.W., Leonard, S.S., Vallyathan, V., Castranova, V., 1997. Efficacy of taurine based compounds as hydroxyl radical scavengers in silica induced peroxidation. Ann. Clin. Lab. Sci. 27, 365–374.

Stevens, M.J., Hosaka, Y., Masterson, J.A., Jones, S.M., Thomas, T.P., Larkin, D.D., 1999. Downregulation of the human taurine transporter by glucose in cultured retinal pigment epithelial cells. Am. J. Physiol. 277, E760–E771.

Takanaga, H., Ohtsuki, S., Hosoya, K., Terasaki, T., 2001. GAT2/BGT-1 as a system responsible for the transport of gamma-aminobutyric acid at the mouse blood–brain barrier. J. Cereb. Blood Flow Metab. 21, 1232–1239.

Wang, X.B., Zhao, X.H., 1998. The effect of dietary sulfur-containing amino acids on calcium excretion. Adv. Exp. Med. Biol. 442, 495–499.

Wu, J.Y., Tang, X.W., Tsai, W.H., 1992. Taurine receptor: kinetic analysis and pharmacological studies. Adv. Exp. Med. Biol. 315, 263–268.

Zhao, X., Jia, J., Lin, Y., 1998. Taurine content in Chinese food and daily intake of Chinese men. Adv. Exp. Med. Biol. 442, 501–505.

CREATINE

Creatine (*N*-(aminoiminomethyl)-*N*-methylglycine; molecular weight 131) is an amino compound that plays an important role in energy metabolism in muscles and the brain (Figure 8.97).

ABBREVIATION

SAM   *S*-adenosylmethionine

NUTRITIONAL SUMMARY

Function: The phosphorylated form creatine phosphate provides an instantly available and vital source of energy in muscles and the brain.

Requirements: Dietary intake is not necessary; precursors are available from a balanced diet. High intake increases muscle and brain stores slightly and may provide a modest, short-term increase in exercise performance.

Food sources: Present in significant amounts only in muscle proteins. Meat contains 300–500 mg per serving (100 g). Milk, eggs, and plant-derived foods do not contain creatine; lacto-ovo vegetarians and vegans have zero intake.

Deficiency: Creatine is produced in adequate amounts by the body, so long as protein and vitamin (folate, vitamin B12) intake is adequate.

Excessive intake: While short-term studies have not observed harmful effects on the liver, kidneys, or other functions even at much higher-than-normal dietary intake levels (i.e., 20 g/day for several days, or 10 g/day for 8 weeks), there have been anecdotal reports of deaths, seizures, arrhythmia, ventricular fibrillation, muscle cramping, and other health problems. Long-term health risks of high intake thus remain a concern.

ENDOGENOUS SOURCES

Daily endogenous production of creatine is about 15 mg/kg requiring arginine, glycine, SAM, folate, and vitamin B12. Creatine synthesis constitutes a very significant drain on methyl group donors, drawing about 70% of the available pool (Wyss and Kaddurah-Daouk, 2000). Synthesis begins in the kidneys with the rate-limiting condensation of arginine plus glycine to guanidinoacetic acid plus ornithine by glycine amidinotransferase (EC2.1.4.1). The second step is a SAM-requiring reaction, which takes place

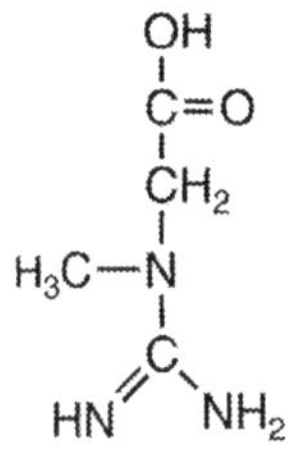

FIGURE 8.97

Creatine.

in the liver and is catalyzed by guanidinoacetate *N*-methyltransferase (EC2.1.1.2). Daily arginine intake typically is about 4 g, and a similar amount is synthesized from glutamate in the intestinal mucosa.

DIETARY SOURCES

The creatine (free plus phosphorylated) content of meat is about 3–5 mg/g, which provides about 1000 mg/day to meat eaters; vegetarians have very low intake since neither plants, eggs, nor dairy products contain significant amounts.

Note: Heating of meats generates a wide variety of creatine-derived carcinogens (Schut and Snyderwine, 1999). One of these is 2-amino-1-methyl-6-phenylimidazo[4,5-b]pyridine (PhIP), generated by the pyrolysis of creatine in the presence of phenylalanine or tyrosine; another is 2-amino-3,8-dimethylimidazo[4,5-f]quinoxaline (MeIQx) derived from creatine and glycine (Oguri et al., 1998) (Figure 8.98).

DIGESTION AND ABSORPTION

Creatine phosphate from the diet is readily cleaved by alkaline phosphatase (EC3.1.3.1) in the proximal intestinal lumen. While it is likely that creatine, but not phosphocreatine, is absorbed from the small intestine, the mechanism of uptake from the lumen into enterocytes and export into portal blood remains unclear. The cationic amino acid transporters 1 and/or 2 (SLC7A1 and SLC7A2), which have a wide specificity, are likely candidates as the conduit for uptake.

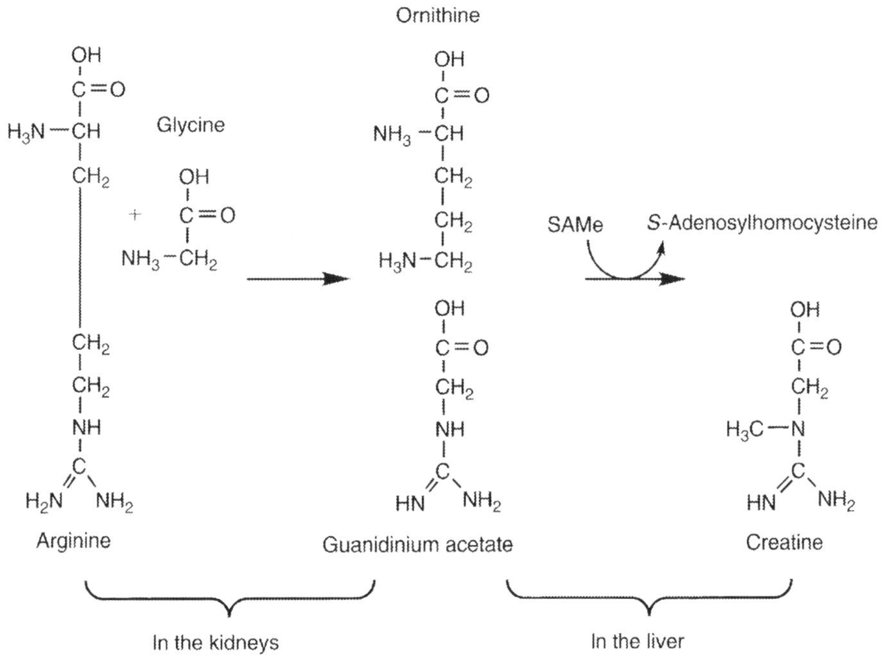

FIGURE 8.98

Endogenous creatine synthesis.

TRANSPORT AND CELLULAR UPTAKE

Blood circulation: Creatine is taken up into muscle and brain cells via the sodium-dependent choline transporter (SLC7A1; Schloss et al., 1994). The existence of distinct sodium-dependent creatine trans-porters in these and several other tissues has been demonstrated (Wyss and Kaddurah-Daouk, 2000). Rapid phosphorylation by creatine kinase (EC2.7.3.2) effectively traps the highly polar creatine phos-phate inside the cell and helps to sustain the several-fold concentration gradient versus the interstitium.

METABOLISM

In muscle and many other tissues, creatine is phosphorylated by creatine kinase (EC2.7.3.2) during periods of abundant ATP availability (Figure 8.99). Creatine phosphate can be dephosphorylated by phosphoamidase (EC3.9.1.1). This enzyme might actually be serine/threonine specific protein phos-phatase (EC3.1.3.16) and/or glucose-6-phosphatase (EC3.1.3.9). A normally minor metabolic pathway is the conversion into methylamine and then into formaldehyde. This route becomes more significant when several grams of creatine are consumed with supplements (Yu and Deng, 2000).

Both creatine and creatine phosphate decompose nonenzymically to creatinine (Figure 8.100). About 1.1% of the creatine and about 2.6% of the phosphocreatine in the body is thus converted to

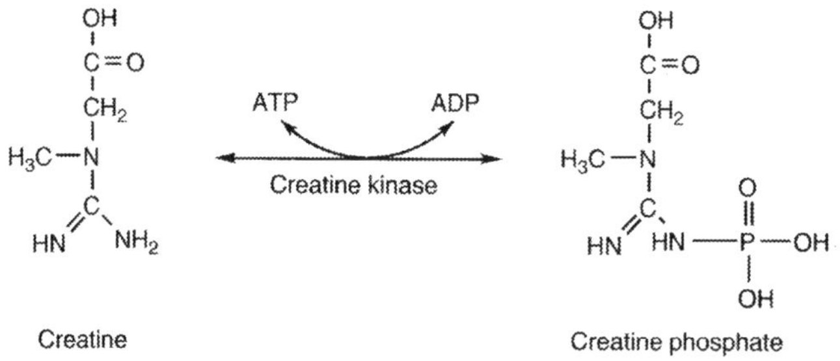

FIGURE 8.99

Phosphorylation by creatine kinase.

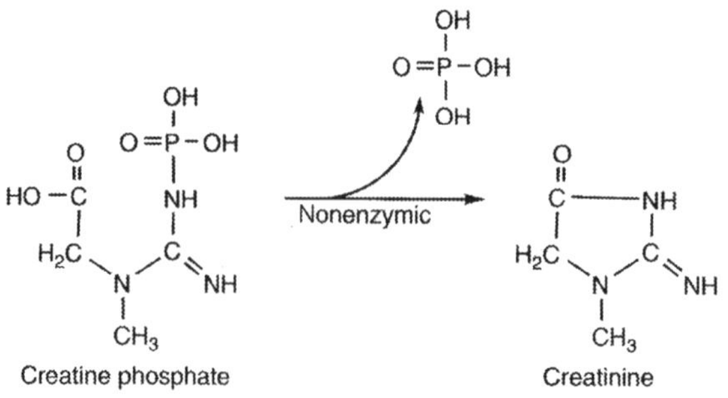

FIGURE 8.100

Creatinine arises from creatine phosphate and creatine.

creatinine (Wyss and Kaddurah-Daouk, 2000). The rate of creatinine generation correlates well with muscle mass, since this is where most of the more easily decomposing phosphocreatine resides.

There has been the suggestion that some creatinine may be metabolically recovered (Boroujerdi and Mattocks, 1983), but supportive evidence is limited.

STORAGE

While much of the body's creatine is in muscle (0.3–0.5% of muscle weight; Crim and Munro, 1994), several other tissues also contain high concentrations (Wyss and Kaddurah-Daouk, 2000). Very high intake (20 g/day) slightly increases creatine concentration in skeletal muscle (+15%; Smith et al., 1999) and in the brain (+9%; Dechent et al., 1999).

EXCRETION

A high-affinity, sodium-dependent creatine transporter with a 2:1 sodium:creatine transport ratio recovers creatine (but not creatinine) very efficiently from the proximal tubular lumen in the kidneys (Wyss and Kaddurah-Daouk, 2000).

More than a gram of creatine-derived creatinine is filtered daily by the kidneys. Little of this is reabsorbed, and there is no significant secretion into the tubular lumen so long as renal function is normal. Since the production rate of creatinine is reasonably constant, it is common clinical practice to use the creatinine concentration in plasma as an approximate indicator of the GFR. Greatly increased plasma creatinine concentration invariably indicates diminished filtration function and renal failure.

Daily creatinine excretion with urine is about 15 mg/kg body weight or about 1.7% of creatine body stores (Forbes and Bruining, 1976).

REGULATION

High creatine intake does not greatly increase plasma creatinine concentration, and it increases only the excretion of creatine, not of creatinine (Poortmans and Francaux, 1999).

Inhibition of expression and activity of glycine amidinotransferase (EC2.1.4.1) by its downstream product creatine appear to be the major regulatory events that control endogenous creatine synthesis (Wyss and Kaddurah-Daouk, 2000). Thyroxin and growth hormone stimulate this rate-limiting step. Low energy intake and vitamin E deficiency slow it down. The regulation of creatine synthesis in the brain, pancreas, testis, and other tissues is largely autonomous and independent of bulk production rates in the kidneys and liver.

FUNCTION

Creatine phosphate is the main high-energy, phosphate-storage molecule of muscle. In rested muscle, creatine phosphate is the predominant form (Demant and Rhodes, 1999); its maximal concentration is five times higher than that of ATP. During times of acute energy need, the creatine kinase (EC2.7.3.2) uses creatine phosphate for the ultrarapid phosphorylation of ADP to ATP. Spermatozoa and photoreceptor cells of the eyes also appear to depend critically on creatine phosphate.

Creatine phosphate may be equally important as a stabilizing energy source in the brain. It has been suggested that high-energy phosphates help to maintain membrane potentials, participate in neurotransmitter release, contribute to calcium homeostasis, and play roles in neuronal migration, survival, and apoptosis (Wyss and Kaddurah-Daouk, 2000). Creatine is a required cofactor of adenosylate kinase (EC2.7.4.3) and other enzymes.

REFERENCES

Boroujerdi, M., Mattocks, A.M., 1983. Metabolism of creatinine *in vivo*. Clin. Chem. 29, 1363–1366.

Crim, M.C., Munro, H.N., 1994. Proteins and amino acids. In: Shils, M.E., Olson, J.A., Shike, M. (Eds.), Modern Nutrition in Health and Disease Lea & Febiger, Philadelphia, PA, pp. 3–35.

Dechent, P., Pouwels, P.J., Wilken, B., Hanefeld, E., Frahm, J., 1999. Increase of total creatine in human brain after oral supplementation of creatine-monohydrate. Am. J. Physiol. 277, R698–R704.

Demant, T.W., Rhodes, E.C., 1999. Effects of creatine supplementation on exercise performance. Sports Med. 28, 49–60.

Forbes, G.B., Bruining, O.J., 1976. Urinary creatinine excretion and lean body mass. Am. J. Clin. Nutr. 29, 1359–1366.

Oguri, A., Suda, M., Totsuka, Y., Sugimura, T., Wakabayashi, K., 1998. Inhibitory effects of antioxidants on formation of heterocyclic amines. Mut. Res. 402, 237–245.

Poortmans, J.R., Francaux, M., 1999. Long-term oral creatine supplementation does not impair renal function in healthy athletes. Med. Sci. Sports Ex. 31, 1108–1110.

Schloss, P., Mayser, W., Betz, H., 1994. The putative rat choline transporter chotl transports creatine and is highly expressed in neural and muscle-rich tissues. Biochem. Biophys. Res. Commun. 198, 637–645.

Schut, H.A., Snyderwine, E.G., 1999. DNA adducts of heterocyclic amine food mutagens: implications for mutagenesis and carcinogenesis. Carcinogenesis 20, 353–368.

Smith, S.A., Montain, S.J., Matott, R.E., Zientara, G.P., Jolesz, F.A., Fielding, R.A., 1999. Effects of creatine supplementation on the energy cost of muscle contraction: a 31P-MRS study. J. Appl. Physiol. 87, 116–123.

Wyss, M., Kaddurah-Daouk, R., 2000. Creatine and creatinine metabolism. Physiol. Rev. 1107–1213.

Yu, P.H., Deng, Y., 2000. Potential cytotoxic effect of chronic administration of creatine, a nutrition supplement to augment athletic performance. Med. Hypotheses 54, 726–778.

CARNITINE

L-Carnitine (3-carboxy-2-hydroxy-*N,N,N*-trimethyl-l-propanaminium hydroxide, gamma-amino-beta-hydroxybutyric acid trimethylbetaine, 3-hydroxy-4-(trimethylammonio) butanoate; obsolete name vitamin BT; molecular weight 161) is a water-soluble amine (Figure 8.101).

ABBREVIATIONS

CACT	carnitine-acylcarnitine translocase
CoA	coenzyme A
SAM	*S*-adenosylmethionine

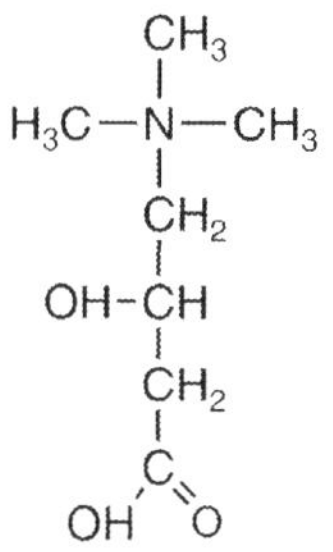

$$
\begin{array}{c}
\text{CH}_3 \\
| \\
\text{H}_3\text{C}-\text{N}-\text{CH}_3 \\
| \\
\text{CH}_2 \\
| \\
\text{OH}-\text{CH} \\
| \\
\text{CH}_2 \\
| \\
\text{C} \\
\text{OH} \quad \text{O}
\end{array}
$$

FIGURE 8.101

L-Carnitine.

NUTRITIONAL SUMMARY

Function: Shuttles fatty acids across mitochondrial membranes and, to a lesser extent, across the peroxisomal membrane. Conjugation to carnitine is important for the metabolism of lysine, BCAAs (namely, valine, leucine, and isoleucine), and acetyl-CoA. Adequate carnitine supplies ensure membrane stability, ketogenesis, urea production, and conjugation of toxins.

Food sources: Meats and fish provide most of the carnitine in food; the only other food sources (but much less important) are milk, peanut butter, asparagus, and whole wheat bread.

Requirements: Healthy people do not need dietary intake. Supplemental intake may be useful for patients with certain heart diseases and rare genetic disorders.

Deficiency: Endogenous synthesis may be impaired in severe malnutrition, liver disease, or rare genetic disorders. Renal losses may be abnormally high due to hemodialysis or genetic disorders. Only in such instances are symptoms of deficiency, including myopathy, fatty liver infiltrates, and central nervous system abnormalities, likely to occur. Choline deficiency causes a decrease of carnitine stores.

Excessive intake: In the short term, no harmful effects of several grams of dietary carnitine have been observed. The health consequences of long-term, high-dose supplementation are not known. Excessive intake with supplements (>3 g/day) may cause diarrhea and cause an offensive, fishy odor.

ENDOGENOUS SOURCES

Daily endogenous production has been estimated to be 0.2 mg/kg or less and depends on the adequate availability of lysine, SAM, ascorbate, iron, alpha-ketoglutarate, tetrahydrofolate, vitamin B6, and niacin.

L-Carnitine is produced in the liver, kidneys, and some other tissues, but not in skeletal and heart muscle. Synthesis proceeds in six steps (Figure 8.102). First, the lysine residues of myosin, actin, histones, or other proteins are methylated threefold by histone-lysine *N*-methyltransferase (EC2.1.1.43). In this reaction, SAM provides the methyl groups; the resulting *S*-adenosylhomocysteine is deadenylated to homocysteine, which can then be regenerated in a methylfolate and vitamin B12–dependent reaction to methionine. As the proteins containing trimethylated lysine residues eventually are broken down by intracellular proteases, trimethyllysine is released and then hydroxylated by trimethyllysine dioxygenase (EC1.14.11.8). This ferroenzyme uses alpha-ketoglutarate and oxygen; its divalent iron is prone to oxidation itself and has to be maintained in the reduced state by ascorbate.

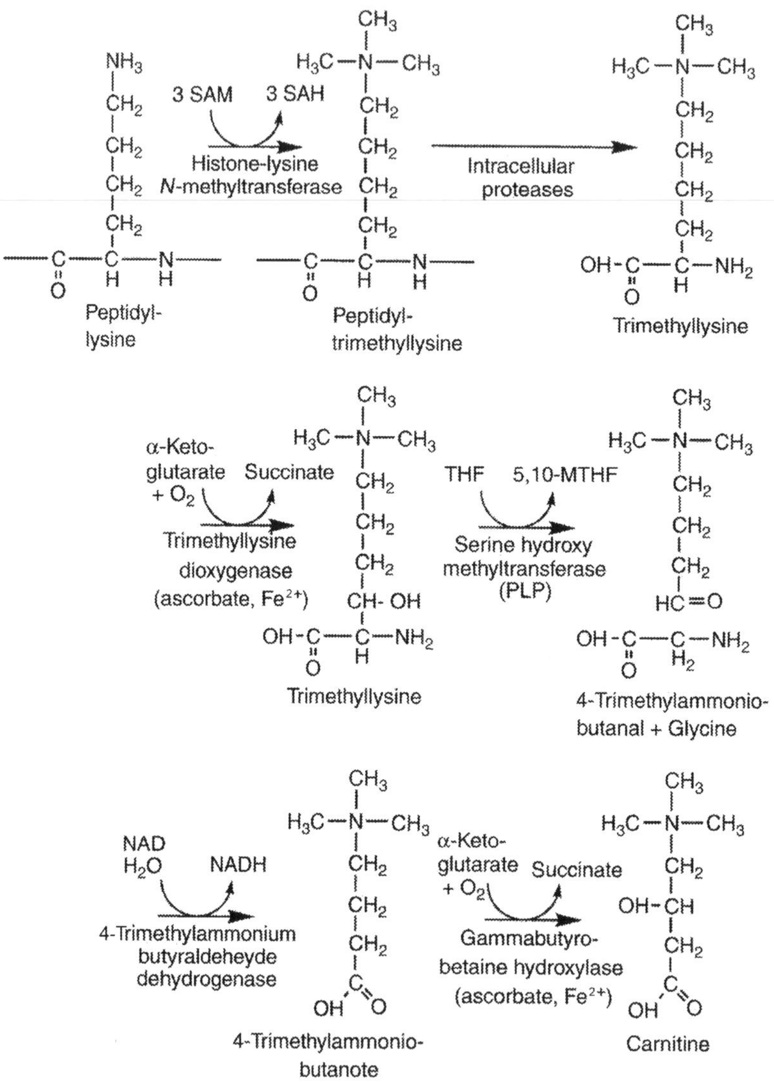

FIGURE 8.102

Endogenous synthesis of L-carnitine.

Here, 3-hydroxy trimethyllysine is split into glycine and 4-trimethyl ammoniobutanal by the PLP-dependent serine hydroxymethyl transferase (EC2.1.2.1). In the penultimate step, the product is oxidized by NAD-dependent 4-trimethyl ammoniobutyraldehyde dehydrogenase (EC1.2.1.47), and, finally, gamma-butyrobetaine hydroxylase (EC1.14.11.1) hydroxylates 4-trimethylammoniobutanoate to generate carnitine. This ultimate step of carnitine synthesis, just like the third reaction in the sequence using alpha-ketoglutarate and oxygen, depends on ferrous iron coordinated to the enzyme and on ascorbate to maintain this iron in the reduced state.

DIETARY SOURCES

Foods may contain carnitine in its free form and as acylcarnitine. The richest sources are red meat (50–120 mg/kg) and whole milk and cheese (3 mg/l); peanut butter, asparagus (0.2 mg/kg), and whole wheat bread contain relatively more than other plant foods.

The carnitine precursor trimethyllysine occurs in a few dietary proteins (i.e., histones, myosin, cytochrome c, actin, and calmodulin); but it is not well utilized, possibly due to its low bioavailability and high renal loss (Broquist, 1994), and high dietary intake may actually deplete carnitine stores (Melegh et al., 1999).

Typical daily carnitine consumption in the United States ranges between 100 and 300 mg (Broquist, 1994). Intake is lowest in people consuming little meat and dairy, the main dietary sources (Sachan et al., 1997). In this case, most carnitine has to be provided by endogenous synthesis, which may be inadequate by itself in very young infants (especially premature infants), in elderly people, and in severely malnourished individuals.

DIGESTION AND ABSORPTION

About 65–75% of ingested carnitine is absorbed from the small intestine. Acylcarnitine in the small intestinal lumen is cleaved by pancreatic carboxylester lipase (EC3.1.1.13). Free carnitine (both L- and D-forms) can be taken up via the sodium cotransporter OCTN2 (SLC22A5); carnitine and at least some of its esters, including propionylcarnitine and acetylcarnitine, are taken up through the amino acid transporter $B^{o,+}$ (SLC6A14) (Nakanishi et al., 2001). Since uptake through the latter transporter is driven by sodium and chloride ions at a 2:1 ratio, it has greater concentrative power than OCTN2, which may enable it to extract carnitine at low concentrations from bacterial sources in the colon. Passive diffusion becomes important when very large doses are ingested. Uptake is twofold greater in jejunum than ileum. D-Carnitine is absorbed but not further metabolized.

It is not clear how carnitine is exported from enterocytes into portal blood.

Intestinal bacteria convert some of the ingested carnitine into trimethylamine (TMA), which can then be metabolized further by flavin-containing monooxygenase 3 (FMO3, EC1.14.13.148) to trimethylamine oxide (Koeth et al., 2013).

TRANSPORT AND CELLULAR UPTAKE

Blood circulation: Both free and acylated carnitine is present in blood circulation. Carnitine enters liver cells with the sodium-dependent, plasmalemmal carnitine transporter OCTN2 (SLC22A5). This transporter is also present in the kidneys, skeletal muscle, heart, and placenta. Choline deficiency has been found to decrease tissue carnitine concentrations; this effect may be related to impaired function of the carnitine carrier (Zeisel, 1994).

BBB: Only free carnitine, not acylcarnitine, is transported across the BBB. The organic cation/carnitine transporter 2 (OCTN2/SLC22A5) appears to contribute significantly to this transport activity (Mroczkowska et al., 1997; Okura et al., 2014).

METABOLISM

An unknown proportion of carnitine in tissues is broken down by carnitine decarboxylase (EC4.1.1.42) to methylcholine. This decarboxylase requires ATP and magnesium as cofactors. The unphysiological isomer D-carnitine is converted to trimethylaminoacetone (Bremer, 1983).

Gut bacteria convert some of the ingested carnitine into TMA (Zhu et al., 2014). It is of note that this conversion occurs in people who eat meat, but only to a small extent (if at all) in vegans.

STORAGE

Most (95%) of the body's carnitine is stored as acylcarnitine in muscles, small amounts in the liver; these stores have a turnover rate of 8 days (Broquist, 1994). For an as-yet-poorly understood reason choline deficiency appears to decrease carnitine stores. It has been suggested that an involvement of choline in carrier-mediated carnitine export may be the reason (Zeisel, 1994).

Nonspecific carboxylesterases, acylcarnitine hydrolase (EC3.1.1.28), monoacylglycerol lipase (EC3.1.1.23), and palmitoyl-CoA hydrolase (EC3.1.2.2) in rough and smooth endoplasmic reticulum can release carnitine again from storage.

EXCRETION

Free carnitine and acylcarnitine are filtered freely in the kidneys. Reabsorption of carnitine in proximal and distal tubules from the primary filtrate uses the luminal transporters OCTN2 (SLC22A5) and $B^{o,+}$ (SLC6A14) (Nakanishi et al., 2001). Urine contains both free carnitine and acylcarnitine.

REGULATION

Recovery of carnitine from ultrafiltrate in the kidneys is nearly complete at normal concentrations, but less at higher concentrations. Homeostatic control, therefore, may be achieved through renal clearance (Broquist, 1994).

FUNCTION

Fatty acid transport into mitochondria: Medium- and long-chain fatty acids by themselves cannot enter mitochondria for beta-oxidation. Their translocation into the matrix depends on a shuttle system with carnitine acyltransferases on both sides of the inner mitochondrial membrane and an acylcarnitine translocase anchored to it (Figure 8.103).

Palmitoyl-CoA:L-carnitine *O*-palmitoyltransferase I is associated with the outer mitochondrial membrane; it links a medium- or long-chain fatty acid from fatty acyl-CoA to free carnitine in the intermembrane space. The acylcarnitine is then transported by carnitine–acylcarnitine translocase (CACT) across the inner mitochondrial membrane in exchange for a free carnitine. Finally, in the mitochondrial matrix, the fatty acid is transferred by palmitoyl-CoA:L-carnitine *O*-palmitoyltransferase II (EC2.3.1.21) to CoA and carnitine is released and ready for shuttling back into the intermembrane space. Short-chain fatty acids (2–6 carbons) are linked to carnitine by corresponding carnitine *O*-acetyltransferases (EC2.3.1.7) on the outer and inner mitochondrial membranes; carnitine *O*-octanoyltransferase (EC2.3.1.137) can deal with a wide spectrum of fatty acids.

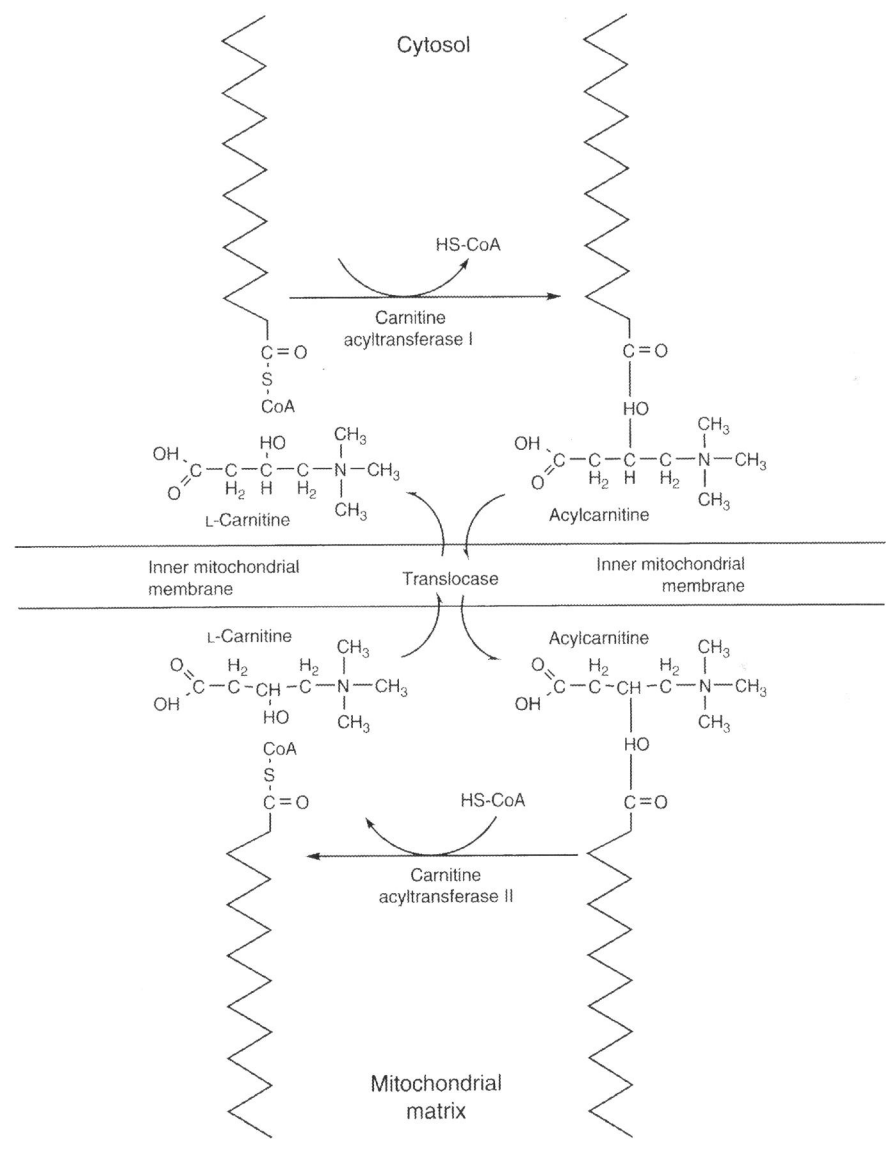

FIGURE 8.103

Carnitine ferries fatty acids across the inner mitochondrial membrane.

Fatty acid transport into peroxisomes: While carnitine is not necessary for the translocation of long-chain fatty acids into peroxisomes, carnitine acetyltransferase (specific for acyl groups with 2–6 carbons) and carnitine medium-chain acyltransferase facilitate the oxidation of acetyl-CoA and shortened fatty acids generated in the peroxisomes.

Amino acid and organic acid metabolism: Acetyl-CoA newly generated from pyruvate by pyruvate dehydrogenase is readily conjugated to carnitine by acetyl-CoA C-acetyltransferase (EC2.3.1.9) and exported to other tissues if carnitine concentrations are high (Lysiak et al., 1986).

Likewise, the alpha-keto acids from catabolism of lysine and the BCAAs valine, leucine, and isoleucine in mitochondria can be conjugated to carnitine. These and various other organic acids are exported as short-chain acylcarnitines into circulation (Ji et al., 1987; Bremer and Hokland, 1987; Bhuiyan et al., 1995). This is particularly important for short-chain fatty acids, which result from partial beta-oxidation. Their transconjugation from CoA to carnitine frees CoA for further use in beta-oxidation and tricarboxylic acid (Krebs) cycle reactions. This may be particularly important in the heart or skeletal muscle when short-chain beta-oxidation is less efficient than long-chain beta-oxidation, or if faster turnover in the Krebs cycle is needed during short-term exertion.

Carnitine may also play a role, which makes it essential for ketogenesis in the liver (Arenas et al., 1998).

Gene regulation: Carnitine deficiency appears to be associated with a reduced expression of urea cycle enzymes due to unknown mechanisms; large amounts of supplemental carnitine promote urea formation (Chapa et al., 1998). Another example is the modulation by carnitine of the extent to which expression of malic enzyme (EC1.1.1.38) and of fatty acid synthase (EC2.3.1.85) responds to triiodothyronine.

Conjugation of xenobiotics: Decreased carnitine availability can induce fatty liver following exposure to aflatoxin or carbon tetrachloride. The reason appears to be a role in conjugation and elimination of these toxins. Consequently, fewer adducts with DNA, RNA, and protein are formed (Sachan and Yatim, 1992).

Membrane stability: Carnitine appears to promote the replacement of peroxidized fatty acids in membrane phospholipids altered by oxygen free radical attack (Arenas et al., 1998); the exact mechanism remains to be elucidated.

Acylcarnitine also interacts specifically with the apical membranes of renal and intestinal epithelium, thereby increasing the intracellular calcium concentration.

Other effects: Carnitine esters of drugs can be used to enhance their absorption from the intestine and to improve their delivery into mitochondria.

Large doses of carnitine (2–5 g/day) are used by many athletes with the expectation to boost their energy, and by hyperlipidemic patients to lower their blood lipid levels.

REFERENCES

Arenas, J., Rubio, J.C., Martin, M.A., Campos, Y., 1998. Biological roles of L-carnitine in perinatal metabolism. Early Hum. Devel. 53, S43–S50.

Bhuiyan, A.K., 1995. Seccombe 13, Bartlett K. Production of acyl-carnitines from the metabolism of [U-^{14}C]3-methyl-2-oxopentanoate by rat liver and skeletal muscle mitochondria. Clin. Invest. Med. 18, 144–151.

Bremer, J., 1983. Carnitine—metabolism and functions. Physiol. Rev. 63, 1420–1480.

Bremer, J., Hokland, B., 1987. Role of carnitine-dependent metabolic pathways in heart disease without primary ischemia. Z Kardiol. 76, 9–13.

Broquist, H.P., 1994. Carnitine. In: Shils, M.E., Olson, J.A., Shike, M. (Eds.), Modern Nutrition in Health and Disease Lea and Febiger, Philadelphia, PA, pp. 459–465.

Chapa, A.M., Fernandez, J.M., White, T.W., Bunting, L.D., Gentry, L.R., Ward, T.L., et al., 1998. Influence of intravenous L-carnitine administration in sheep preceding an oral urea drench. J. Anim. Sci. 76, 2930–2937.

Ji, L.L., Miller, R.H., Nagle, F.J., Lardy, H.A., Stratman, F.W., 1987. Amino acid metabolism during exercise in trained rats: the potential role of carnitine in the metabolic fate of branchedchain amino acids. Metab. Clin. Exp. 36, 748–752.

Koeth, R.A., Wang, Z., Levison, B.S., Buffa, J.A., Org, E., Sheehy, B.T., et al., 2013. Intestinal microbiota metabolism of L-carnitine, a nutrient in red meat, promotes atherosclerosis. Nat. Med. 19, 576–585.

Lysiak, W., Toth, P.E., Suelter, C.H., Bieber, L.L., 1986. Quantitation of the efflux of acylcarnitines from rat heart, brain, and liver mitochondria. J. Biol. Chem. 261, 13698–13703.

Melegh, B., Toth, G., Adamovich, K., Szekely, G., Gage, D.A., Bieber, L.L., 1999. Labeled trimethyllysine load depletes unlabeled carnitine in premature infants without evidence of incorporation. Biol. Neonate 76, 19–25.

Mroczkowska, J.E., Galla, H.J., Nalecz, M.J., Nalecz, K.A., 1997. Evidence for an asymmetrical uptake of L-carnitine in the blood–brain barrier *in vitro*. Biochem. Biophys. Res. Comm. 241, 127–131.

Nakanishi, T., Hatanaka, T., Huang, W., Prasad, P.D., Leibach, F.H., Ganapathy, M.E., et al., 2001. Na^+ and Cl^--coupled active transport of carnitine by the amino acid transporter ATB(0,+) from mouse colon expressed in HRPE cells and Xenopus oocytes. J. Physiol. 532, 297–304.

Okura, T, Kato, S, Deguchi, Y., 2014. Functional expression of organic cation/carnitine transporter 2 (OCTN2/SLC22A5) in human brain capillary endothelial cell line hCMEC/D3, a human blood–brain barrier model. Drug Metab. Pharmacokinet. 29, 69–74.

Sachan, D.S., Daily III, J.W., Munroe, S.G., Beauchene, R.E., 1997. Vegetarian elderly women may risk compromised carnitine status. Veg. Nutr. 1, 64–69.

Sachan, D.S., Yatim, A.M., 1992. Suppression of aflatoxin B1-induced lipid abnormalities and macromolecule-adduct formation by L-carnitine. J. Env. Pathol. Toxicol. Oncol. 11, 205–210.

Zeisel, S.H., 1994. Choline. In: Shils, M.E., Olson, J.A., Shike, M. (Eds.), Modern Nutrition in Health and Disease Lea & Febiger, Philadelphia, PA, pp. 449–458.

Zhu, Y., Jameson, E., Crosatti, M., Schäfer, H., Rajakumar, K., Bugg, T.D., et al., 2014. Carnitine metabolism to trimethylamine by an unusual Rieske-type oxygenase from human microbiota. Proc. Natl. Acad. Sci. USA 111, 4268–4273.

MELATONIN

Melatonin (*N*-[2-(5-methoxy-lH-indol-3-yl)ethyl]acetamide, *N*-acetyl-5-methoxytryptamine; molecular weight 232) contains 12% nitrogen (Figure 8.104).

ABBREVIATIONS

5-HT	5-hydroxytryptamine (serotonin)
5-HTOL	5-hydroxytryptophol
EMS	eosinophilia-myalgia syndrome
PAPS	3′-phosphoadenosine 5′-phosphosulfate
6-SMT	6-sulfatoxymelatonin

NUTRITIONAL SUMMARY

Function: Melatonin, a hormone produced by the brain, retina, and pineal gland, participates in the coordination of sleep patterns, thermoregulation, and possibly reproductive cycles to daylight.

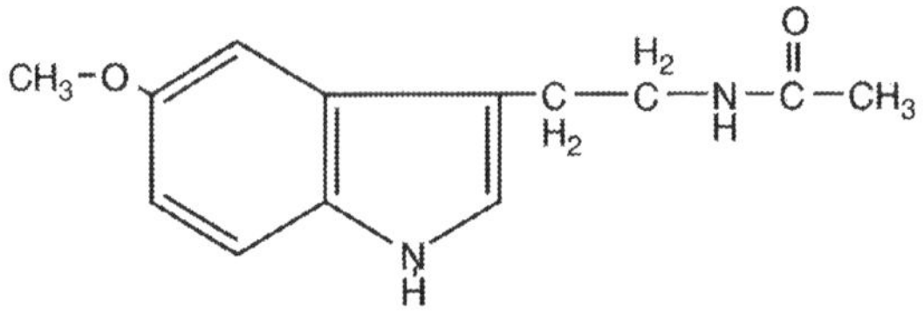

FIGURE 8.104

Melatonin.

Sources: The body's production depends on adequate availability of L-tryptophan, niacin, vitamin B6, endogenously generated SAM (methionine, folate, and vitamin B12), biopterin, pantotherate, ascorbate, and iron.

Requirements: Endogenous synthesis is usually adequate.

Deficiency: Inadequacies of nutrients needed for endogenous synthesis may sometimes limit production, possibly contributing to insomnia and sexual dysfunction.

Excessive intake: Many commercial melatonin preparations have been found to contain potentially toxic contaminants similar to those thought to have caused EMS in some users of tryptophan supplements. Other consequences of supplemental intake of melatonin have not been adequately evaluated.

ENDOGENOUS SOURCES

Melatonin is synthesized in the brain, retina, and pineal gland. This endogenous synthesis proceeds in steps catalyzed by four enzymes and requires L-tryptophan, biopterin, niacin, vitamin B6, methionine, folate, vitamin B12, pantotherate, ascorbate, and iron. Here, 5-hydroxytryptamine (5-HT, serotonin) is an intermediate of melatonin synthesis, produced by the first two steps (Figure 8.105).

The synthetic sequence is initiated by tryptophan 5-monooxygenase (EC1.14.16.4, requires iron and ascorbate) in pinealocytes and raphe neurons, as well as tissues outside the brain, including beta-cells of the islets of Langerhans, enterochromaffin cells of the pancreas and small intestine, mast cells, and mononuclear leukocytes. The hydroxylation of L-tryptophan requires tetrahydropteridine as a cofactor and is inhibited by high concentrations of L-phenylalanine. A calcium ion-triggered protein kinase phosphorylates and thereby activates tryptophan 5-monooxygenase; melatonin, the ultimate end product, inhibits it.

In the next step, aromatic-L-amino-acid decarboxylase (EC4.1.1.28) generates 5-HT. This PLP-dependent enzyme is also present in basal ganglia, sympathetic nerves, and adrenal medulla, where it acts mainly on L-tryptophan and DOPA. When aromatic-L-amino-acid decarboxylase generates tryptamine by acting on L-tryptophan, 5-HT synthesis can be completed subsequently by tryptophan 5-monooxygenase. Significant quantities of tryptamine may be ingested with certain types of cheese or with foods exposed to bacterial action.

5-HT can then be acetylated by aralkylamine *N*-acetyltransferase (EC2.3.1.87) and finally converted into *N*-acetyl-5-methoxytryptamine (melatonin) by acetylserotonin *O*-methyltransferase (EC2.1.1.4).

DIETARY SOURCES

Normally, very little melatonin is obtained from food. Commercial preparations are now widely available as dietary supplements. Foods containing significant amounts of 5-HT include bananas (0.02 mg/g), pineapples (0.01 mg/g), and walnuts (Bruce, 1960; Helander and Some, 2000).

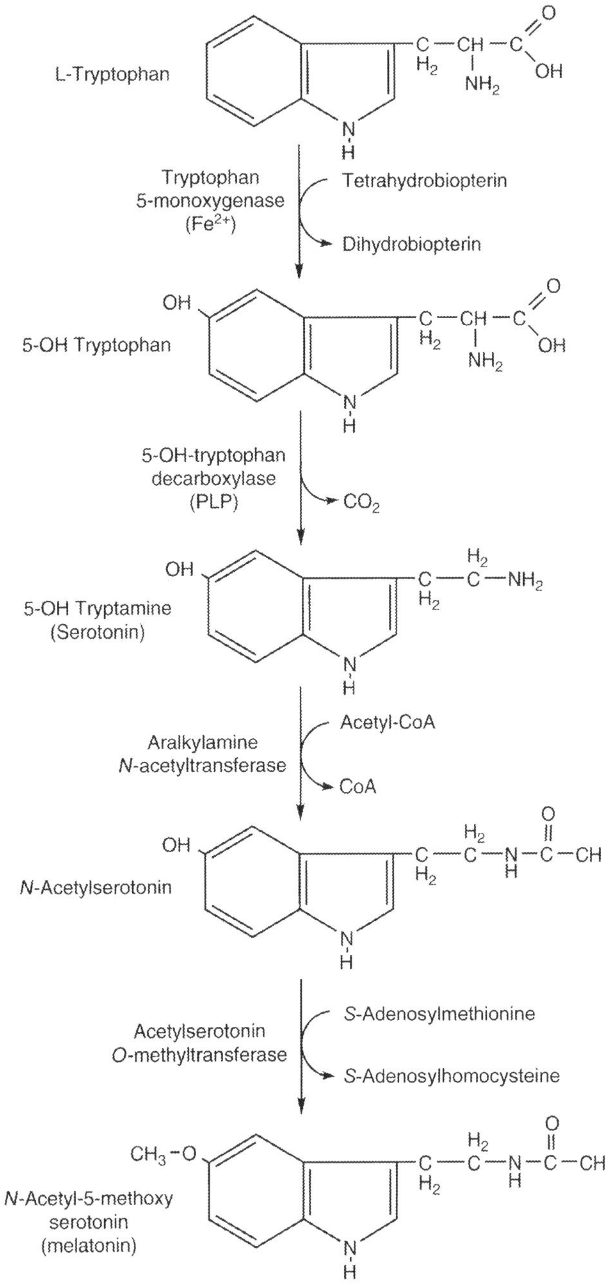

FIGURE 8.105

Melatonin synthesis in the pineal gland.

Note: EMS with symptoms that include fatigue, pain, depression, sleep disturbance, and verbal memory impairment may be caused by contaminants in commercial preparations of 5-hydroxytryptophan (Klarskov et al., 1999) and melatonin (Laforce et al., 1999).

DIGESTION AND ABSORPTION

Melatonin absorption proceeds rapidly from the proximal small intestine via an incompletely understood process (Aldhous et al., 1985). Fractional absorption has been found to vary between individuals by more than an order of magnitude (Waldhauser et al., 1984).

TRANSPORT AND CELLULAR UPTAKE

Blood circulation: Only very small amounts of melatonin, normally around 7 ng/l (Ludemann et al., 2001), are circulating with blood. The melatonin precursor 5-HT is taken up into pinealocytes by a specific sodium- and chloride-dependent transporter (SLC6A4). The importance of transporter-mediated uptake to provide the precursor for melatonin synthesis is suggested by the frequent occurrence of a less active form in patients with seasonal-affective disorder (Partonen and Lonnqvist, 1998).

METABOLISM

The main melatonin metabolite is 6-hydroxymelatonin; smaller amounts are broken down into the serotonin metabolites 5-OH-indole 3-acetate and N1-acetyl-N2-formyl-5-methoxykynuramine (Figure 8.106). Melatonin is hydroxylated at position 6 by CYP1A2 (EC1.14.14.1) and other cytochromes in the liver (Facciola et al., 2001). Conjugation by 3'-phosphoadenosine 5'-phosphosulfate (PAPS)–dependent sulfotransferase ST1A3 (EC2.8.2.3) then generates 6-sulfatoxymelatonin (6-SMT), the main melatonin metabolite (Honma et al., 2001).

Ring opening of melatonin and generation of N1-acetyl-N2-formyl-5-methoxykynuramine occurs either through enzyme action or by reaction with oxygen free radicals. The hemoproteins indoleamine-pyrrole 2,3-dioxygenase (EC1.13.11.42) and peroxidase (myeloperoxidase, EC1.11.1.7) generate N1-acetyl-N2-formyl-5-methoxykynuramine (Tan et al., 2001). The same intermediate appears to be formed nonenzymically when melatonin reacts with various oxygen free radicals. Melatonin can be metabolized to 5-OH-indole 3-acetate in the liver, presumably via serotonin. *N*-acetylserotonin may be generated (Beedham et al., 1987) by a reversal of the reaction catalyzed by the melatonin-synthesis enzyme acetylserotonin *O*-methyltransferase (EC2.1.1.4). Aryl-acylamidase (EC3.5.1.13) deacetylates melatonin (Rogawski et al., 1979).

Most 5-HT is metabolized by amine oxidase (monoamine oxidase A, MAO-A, EC1.4.3.4, FAD-containing) and one or more of the aldehyde dehydrogenases (ALDH, EC1.2.1.3/NAD-requiring, EC1.2.1.4/NADP-requiring, EC1.2.1.5/NAD or NADP-requiring) to 5-hydroxyindole-3-acetate. Smaller amounts of 5-HT enter the main tryptophan catabolic pathway through the action of tryptophan 2,3-dioxygenase (EC1.13.11.11). Alcohol dehydrogenase (EC1.1.1.1) converts some 5-HT into the dead-end product 5-HTOL. Ethanol acutely increases conversion of 5-HT into 5-HTOL (Helander and Some, 2000).

STORAGE

The dense core granules of pinealocytes contain melatonin; storage and release are mediated by vesicular monoamine transporters.

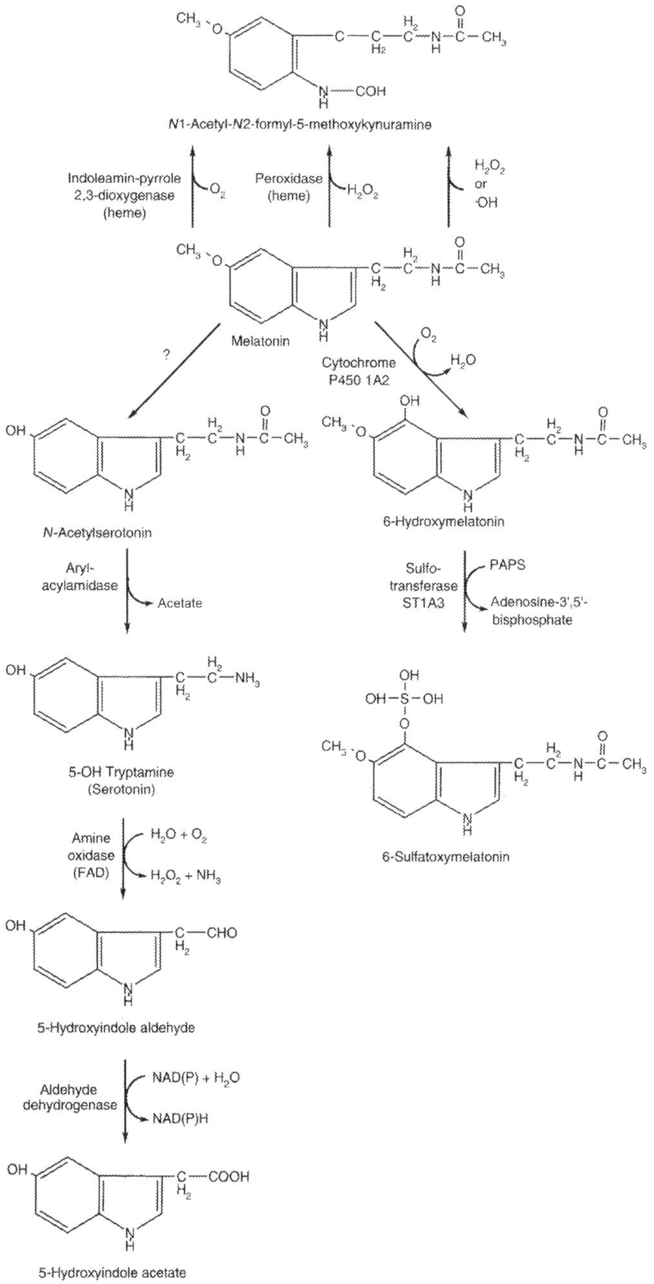

FIGURE 8.106

Melatonin catabolism.

EXCRETION

The main products of melatonin metabolism in urine are 6-SMT and 5-hydroxyindole 3-acetate. 6-SMT excretion is about 0.4 μg/kg body weight, slightly higher in winter than in summer (Davis et al., 2001). Supplemental intake of 5 mg of melatonin was found to double the amount of 6-SMT in urine (Pierard et al., 2001).

REGULATION

The activity of the 5-HT transporter that ferries the melatonin precursor into pinealocytes is tightly controlled. In the absence of light exposure, 5-HT-uptake and melatonin synthesis increase (Lima and Schmeer, 1994).

FUNCTION

Receptor binding: Melatonin binds to and activates specific G-protein-coupled membrane receptors. Nearly 20 genes have been identified that code for putative melatonin receptors, including mt1, MT2, and Mellc. Expression of mt1 is particularly high in neurons of the suprachiasmatic nucleus and the pars tuberalis of the pituitary.

Circadian rhythm: During the dark periods of the daily cycle, the suprachiasmatic nucleus stimulates the expression of the arylalkylamine N gene, thereby increasing the production and secretion of melatonin at night. Ocular light exposure, but not extraocular illumination, shifts circadian rhythms and suppresses nocturnal melatonin production (Lockley et al., 1998; Hebert et al., 1999). The efficacy of oral melatonin supplements for the relief of jet lag remains in doubt (Spitzer et al., 1999).

Seasonal rhythm: The modulation of melatonin rhythm by light exposure appears to entrain seasonal temporal organization (Pevet, 2000). The investigation of seasonally recurring events influenced by melatonin cycles has focused mainly on estrus of horses and other mammals, and on the behavior of songbirds. The relevance for human performance or health remains unknown.

Influences on mood: Melatonin may have anxiolytic properties. Preoperative use of 0.05 mg of melatonin per kilogram of body weight has reduced anxiety and relaxation in one study without impairing cognitive and psychomotor skills (Naguib and Samarkandi, 2000). Abnormal attenuation of the nocturnal rise in plasma melatonin may be related to night-eating syndrome (NES), which consists of morning anorexia, evening hyperphagia, and insomnia (Birketvedt et al., 1999).

Skin pigmentation: Uptake of melatonin increases pigment production of melanocytes and stimulates their cell division (Iyengar, 2000).

Free radical scavenging: Protective effects against oxidative DNA damage (and thereby antitumorigenic properties) have been observed in some animal studies (Karbownik et al., 2000). It has been suggested that at submicromolar concentrations, melatonin is 50–70 times more effective as an antioxidant compared to ascorbate or alphatocopherol. This high potency might be related to its ability to scavenge hydroxyl radicals directly and thereby terminate Fenton-type reactions (Qi et al., 2000). The metabolite N1-acetyl-N2-formyl-5-methoxykynuramine may be an even more potent antioxidant than melatonin itself (Tan et al., 2001).

Immune function: Circulating melatonin may be an important element of neuroimmune communication that periodically stimulates immune response. Production of cytokines (e.g., IL-6) in lymphocytic and monocytic cells is stimulated through the binding of melatonin to nuclear receptors (Guerrero et al., 2000). The pineal gland may exert oncostatic actions on distant tissues, and melatonin has been suggested to aid in the suppression of tumor proliferation (Cos and Sanchez-Barcelo, 2000).

Hormonal function: There have been reports that growth hormone secretion is influenced by melatonin at a hypothalamic level (Meeking et al., 1999). Melatonin reduces the sensitivity of the thyroid to thyrotropin (thyroid-stimulating hormone, TSH), inhibits thyroid cell proliferation, and lowers thyroxine secretion.

Fertility: Sperm cell maturation is promoted through a stimulating effect of melatonin on epididymal cells.

REFERENCES

Aldhous, M., Franey, C., Wright, J., Arendt, J., 1985. Plasma concentrations of melatonin in man following oral absorption of different preparations. Br. J. Clin. Pharmacol. 19, 517–521.

Beedham, C., Smith, J.A., Steele, D.L., Wright, P.A., 1987. Chlorpromazine inhibition of melatonin metabolism by normal and induced rat liver microsomes. Eur. J. Drug Metab. Pharmacokin. 12, 299–302.

Birketvedt, G.S., Florholmen, J., Sundsfjord, J., Osterud, B., Dinges, D., Bilker, W., et al., 1999. Behavioral and neuroendocrine characteristics of the night-eating syndrome. JAMA 282, 657–663.

Bruce, D.W., 1960. Serotonin in pineapple. Nature 188, 147.

Cos, S., Sanchez-Barcelo, E.J., 2000. Melatonin, experimental basis for a possible application in breast cancer prevention and treatment. Histol. Histopathol. 15, 637–647.

Davis, S., Kaune, W.T., Mirick, D.K., Chen, C., Stevens, R.G., 2001. Residential magnetic fields, light at-night, and nocturnal urinary 6-sulfatoxymelatonin concentration in women. Am. J. Epidemiol. 154, 591–600.

Facciola, G., Hidestrand, M., von Bahr, C., Tybring, G., 2001. Cytochrome P450 isoforms involved in melatonin metabolism in human liver microsomes. Eur. J. Clin. Pharmacol. 56, 881–888.

Guerrero, J.M., Pozo, D., Garcia-Maurino, S., Carrillo, A., Osuna, C., Molinero, E., et al., 2000. Nuclear receptors are involved in the enhanced IL-6 production by melatonin in U937 cells. Biol. Signals Receptors 9, 197–202.

Hebert, M., Martin, S.K., Eastman, C.I., 1999. Nocturnal melatonin secretion is not suppressed by light exposure behind the knee in humans. Neurosci. Lett. 274, 127–130.

Helander, A., Some, M., 2000. Dietary serotonin and alcohol combined may provoke adverse physiological symptoms due to 5-hydroxytryptophol. Life Sci. 67, 799–806.

Honma, W., Kamiyama, Y., Yoshinari, K., Sasano, H., Shimada, M., Nagata, K., et al., 2001. Enzymatic characterization and interspecies difference of phenol sulfotransferases, ST1A forms. Drug Metab. Disp. 29, 274–281.

Iyengar, B., 2000. Melatonin and melanocyte functions. Biol. Signals Receptors 9, 260–266.

Karbownik, M., Tan, D.X., Reiter, R.J., 2000. Melatonin reduces the oxidation of nuclear DNA and membrane lipids induced by the carcinogen delta-aminolevulinic acid. Int. J. Cancer 88, 7–11.

Klarskov, K., Johnson, K.L., Benson, L.M., Gleich, G.J., Naylor, S., 1999. Eosinophilia-myalgia syndrome case-associated contaminants in commercially available 5-hydroxytryptophan. Adv. Exp. Med. Biol. 467, 461–468.

Laforce, R., Rigozzi, K., Paganetti, M., Mossi, W., Guainazzi, P., Calderari, G., 1999. Aspects of melatonin manufacturing and requirements for a reliable active component. Biol. Signals Receptors 8, 143–146.

Lima, L., Schmeer, C., 1994. Characterization of serotonin transporter in goldfish retina by the binding of [3H] paroxetine and the uptake of [^{3}H]serotonin: modulation by light. J. Neurochem. 62, 528–535.

Lockley, S.W., Skene, D.J., Thapan, K., English, J., Ribeiro, D., Haimov, I., et al., 1998. Extraocular light exposure does not suppress plasma melatonin in humans. J. Clin. Endocrinol. Metab. 83, 3369–3372.

Ludemann, E., Zwernemann, S., Lerchl, A., 2001. Clearance of melatonin and 6-sulfatoxymelatonin by hemodialysis in patients with end-stage renal disease. J. Pineal. Res. 31, 222–227.

Meeking, D.R., Wallace, J.D., Cuneo, R.C., Forsling, M., Russell-Jones, D.L., 1999. Exercise-induced GH secretion is enhanced by the oral ingestion of melatonin in healthy adult male subjects. Eur. J. Endocrinol. 141, 22–26.

Naguib, M., Samarkandi, A.H., 2000. The comparative dose–response effects of melatonin and midazolam for premedication of adult patients: a double-blinded, placebo-controlled study. Anesth. Analg. 91, 473–479.

Partonen, T., Lonnqvist, J., 1998. Seasonal affective disorder. Lancet 352, 1369–1374.

Pevet, P., 2000. Melatonin and biological rhythms. Biol. Signals Receptors 9, 203–212.

Pierard, C., Beaumont, M., Enslen, M., Chauffard, E., Tan, D.X., Reiter, R.J., et al., 2001. Resynchronization of hormonal rhythms after an eastbound flight in humans: effects of slow-release caffeine and melatonin. Eur. J. Appl. Physiol. 85, 144–150.

Qi, W., Reiter, R.J., Tan, D.X., Garcia, J.J., Manchester, L.C., Karbownik, M., et al., 2000. Chromium(III)-induced 8-hydroxydeoxyguanosine in DNA and its reduction by antioxidants: comparative effects of melatonin, ascorbate, and vitamin E. Env. Health Persp. 108, 399–402.

Rogawski, M.A., Roth, R.H., Aghajanian, G.K., 1979. Melatonin: deacetylation to 5-methoxytryptamine by liver but not brain aryl acylamidase. J. Neurochem. 32, 1219–1226.

Spitzer, R.L., Terman, M., Williams, J.B., Terman, J.S., Malt, U.E., Singer, E., et al., 1999. Jet lag: clinical features, validation of a new syndrome-specific scale, and lack of response to melatonin in a randomized, double-blind trial. Am. J. Psychiatr. 156, 1392–1396.

Tan, D.X., Manchester, L.C., Burkhardt, S., Sainz, R.M., Mayo, J.C., Kohen, R., et al., 2001. N1-acetyl-N2-formyl-5-methoxykynuramine, a biogenic amine and melatonin metabolite, functions as a potent antioxidant. FASEB J. 15, 2294–2296.

Waldhauser, E., Waldhauser, M., Lieberman, H.R., Deng, M.H., Lynch, H.L., Wurtman, R.J., 1984. Bioavailability of oral melatonin in humans. Neuroendocrinology 39, 307–313.

CHOLINE

Choline (2-hydroxyethyl[trimethyl]azanium hydrochloride, 2-hydroxy-*N,N,N*-trimethylethanaminium, [beta-hydroxyethyl]trimethylammonium, bilineurine; molecular weight 104) is a weak base (Figure 8.107).

ABBREVIATIONS

BGT1	sodium chloride–dependent betaine transporter (SLC6A12)
CAT-2	cationic amino acid transporter 2 (y$^+$, SLC7A2)
OCT1	organic cation transporter 1 (SLC22A1)
OCT2	organic cation transporter 2 (SLC22A2)
PEMT	phosphatidylethanolamine-*N*-methyltransferase

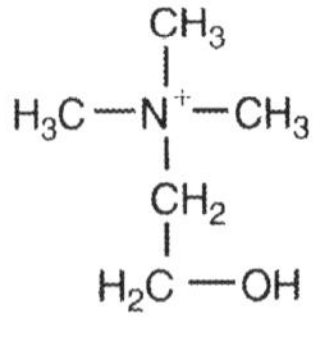

FIGURE 8.107

Choline.

NUTRITIONAL SUMMARY

Function: Choline is essential for the synthesis of phospholipids, betaine, and the neurotransmitter acetylcholine, as well as to regenerate methionine from homocysteine.

Requirements: Adequate daily intake is 550 mg/day for men and 425 mg/day for women. Pregnancy and lactation increase needs slightly.

Sources: Egg yolk, organ meats, peanuts, and legumes are good sources. Many other foods provide smaller but significant quantities. Lecithin supplements provide choline.

Deficiency: Inadequate intake causes liver damage and possibly increases the risk of cancer and atherosclerosis. Young infants may suffer growth retardation and suboptimal cognitive development.

Excessive intake: Consumption of more than 3500 mg/day can cause sweating, salivation, and diarrhea; induce a fishy body odor; and cause mild liver damage.

ENDOGENOUS SOURCES

The liver, and to a lesser extent the brain and mammary gland, produce choline from phosphatidylethanolamine (Figure 8.108). However, the capacity of this pathway appears to be limited. *De novo* choline synthesis requires adequate supplies of serine, methionine, niacin, folate, vitamin B12, pantothenate, and magnesium. The immediate choline precursor phosphatidylethanolamine is produced from serine and CDP-diacylglycerol via synthesis of *O*-sn-phosphatidyl-L-serine (CDP diacylglycerol-serine *O*-phosphatidyltransferase, EC2.7.8.8) and decarboxylation of the resulting phospholipid by pyruvate-containing phosphatidylserine decarboxylase (EC4.1.1.65; Dowhan, 1997). Phosphatidylethanolamine can be methylated to phosphatidylcholine in three successive reactions catalyzed by magnesium-dependent phosphatidylethanolamine-*N*-methyltransferase (PEMT, EC2.1.1.17). PEMT1 in the endoplasmic reticulum is responsible for most of this activity. The genetically distinct isoform PEMT2 is located on mitochondrial membranes.

Choline can be released from phosphatidylcholine through successive cleavage by phospholipase A2 (EC3.1.1.4), lysophospholipase (EC3.1.1.5), and glycerylphosphocholine phosphodiesterase (EC3.1.4.2). Carnitine decarboxylase (EC4.1.1.42; Habibulla and Newburgh, 1969) may also generate some choline (Figure 8.109). This enzyme, which contains a prosthetic complex of ATP and magnesium, generates 2-methylcholine. Cholinesterase (EC3.1.1.8, magnesium-dependent), which cleaves a wide spectrum of organoesters, is likely to demethylate 2-methylcholine. Note that choline deficiency reduces carnitine concentrations in tissues and that choline administration immediately relieves carnitine depression. Carnitine may serve as an acute supply for choline in tissues.

DIETARY SOURCES

Foods contain free choline, choline-containing phosphatidyl choline (lecithin), sphingomyelin, and minimal amounts of acetylcholine. Good choline sources are egg yolk (5.6 mg/g), liver (5.3 mg/g), kidneys, and legumes (e.g., peanuts, with 1 mg/g). Significant amounts of free choline are consumed in liver, oatmeal, soybeans, kale, and cabbages. Smaller quantities are present in many other foods. Milk provides only 40 mg/l. Average daily intake of adults in the United States has been estimated at 600–1000 mg (Zeisel and Da Costa, 2009).

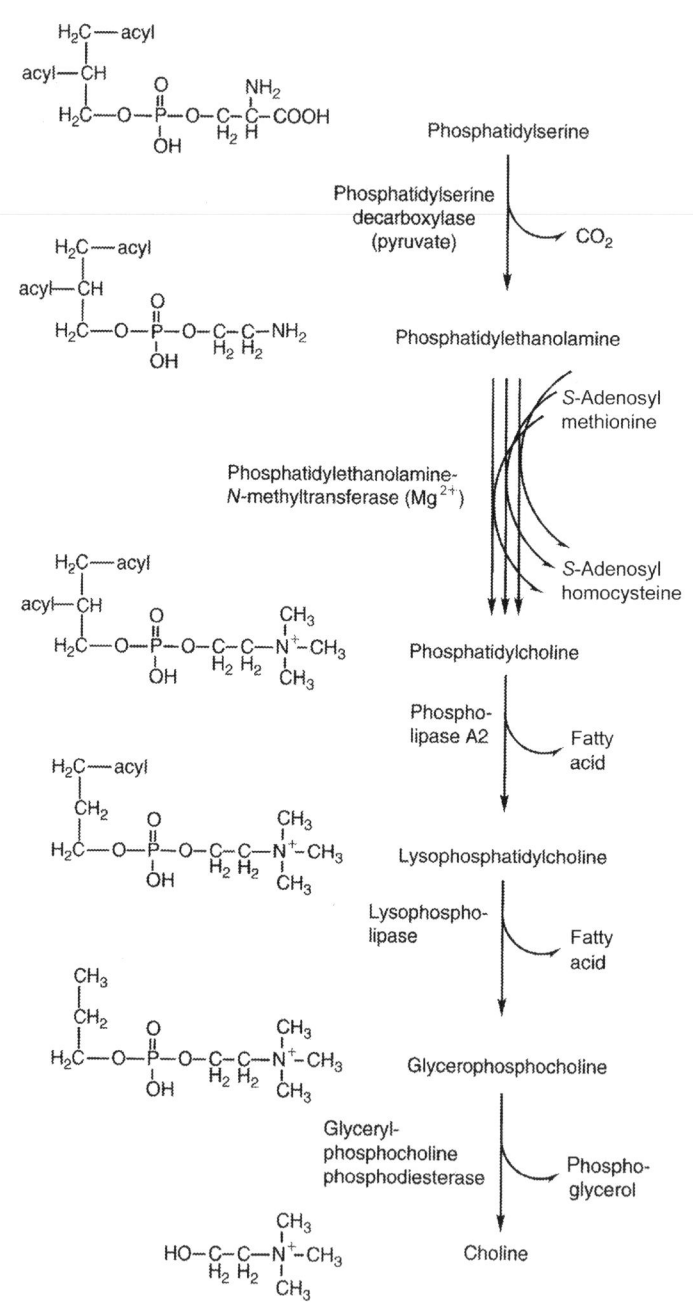

FIGURE 8.108

Choline is synthesized from phosphotidylethanolamine.

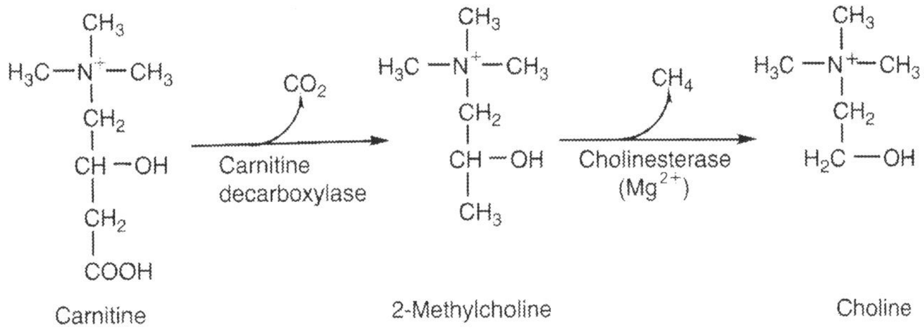

FIGURE 8.109

Decarboxylation and demethylation of carnitine generates choline.

DIGESTION AND ABSORPTION

Choline-containing phospholipids are cleaved by pancreatic phospholipase A2 (EC3.1.1.4), and the resulting lysophosphatidylcholine (which becomes part of mixed micelles) is taken up into the small intestinal enterocyte.

Free choline is taken up both by mediated transport and diffusion all along the small intestine (Zeisel, 1994). The cationic amino acid transporter 2 (CAT-2, y^+, SLC7A2) is expressed in the brush border of the small intestine and specifically mediates choline uptake into enterocytes. A large proportion of absorbed choline is incorporated into phospholipids and secreted with chylomicrons into lymph.

Lysophospholipase (EC3.1.1.5) cleaves lysophosphatidylcholine and then glycerylphosphocholine phosphodiesterase (EC3.1.4.2) finally releases choline. Export of free choline across the basolateral membranes is not well understood yet. Betaine uptake from the intestinal lumen also needs further investigation. In chicks, it was found to involve sodium-dependent and sodium-independent components (Kettunen et al., 2001).

Note: Intestinal bacteria degrade a significant proportion of ingested choline and choline phospholipids to betaine and to trimethylamine (TMA). Both betaine and TMA are absorbed. Flavin-containing monooxygenase (FMO, EC1.14.13.8) N-oxidizes TMA. The main isoform responsible for metabolism in the liver is FMO3 (Treacy et al., 1998). FMO2 contributes to a lesser degree. Nearly 4% of subjects with suspected body malodor were found to have severely impaired TMA N-oxidation (fish odor syndrome). Their parents had a less pronounced (but still noticeable) decrease in FMO activity (Ayesh et al., 1993).

TRANSPORT AND CELLULAR UPTAKE

Blood circulation: Plasma contains about 15 μmol/l free choline, and 50–300 mg/l phosphatidylcholine in lipoproteins. The red blood cell membranes consist mostly of choline-containing phospholipids. Free choline from the blood enters cells through specific transport systems that are as-yet-insufficiently characterized. Uptake into the liver may be mediated in part by the organic cation transporter OCT1

(SLC22A1). Neurons contain a high-affinity sodium chloride-dependent choline transporter that provides these cells with the essential precursor for production of the neurotransmitter acetylcholine (Kobayashi et al., 2002). Choline transporter-like proteins 1 (CDw92 is a C-terminal variant), 2, and 4 are expressed at the plasma membrane of neurons, endothelial cells, and leukocytes and play incompletely understood roles in choline transport.

The transporter SLC44A1 moves large amounts of choline into mitochondria, where a much higher choline concentration is maintained than in cytosol (Traiffort et al., 2013).

Choline-containing phospholipids are taken up with lipoproteins through receptor-mediated endocytosis. Significant amounts of phospholipid-rich membranes are cleared by phagocytosis in the spleen and reticuloendothelial system. The choline exchanger couples the influx of choline chloride into erythrocytes to the efflux of magnesium or other cations (Ebel et al., 2002). The concentration gradient drives choline via a specific choline transporter from cytosol into mitochondria of the liver and kidneys (Porter et al., 1992), where most oxidation takes place.

BBB: Choline is transported into the brain by both high- and low-affinity systems (Lockman et al., 2001). Choline from neuronal release (acetylcholine) can leak into the extracellular space, and from there into cerebrospinal fluid. The organic cation transporter 2 (SLC7A2) helps to maintain cerebral choline balance by moving excess choline across the choroid plexus into the blood (Sweet et al., 2001). The sodium chloride–dependent betaine transporter 1 (BGT1, SLC6A12) carries betaine from the blood into the brain (Takanaga et al., 2001).

Materno-fetal transfer: Specific transporters on maternal and fetal sides of the syncytiotrophoblast mediate transfer of choline across the placenta. The placenta contains members of the y^+ amino acid transport system, including the cationic amino acid transporters 1 (SLC7A1) and 4 (SLC7A4) at the maternal side (Ayuk et al., 2000), which contribute to choline transfer. A y^+ transporter in conjunction with its 4F2 (CD98, SLC3A2) glycoprotein anchor is likely to move choline across the basal membrane into fetal circulation. Cross-placental transport of choline is inhibited by many common medical drugs, including propranolol, quinine, imipramine, verapamil, flurazepam, amiloride, and ritodrin (Grassl, 1994).

METABOLISM

The breakdown of choline occurs mainly in the liver and kidneys and depends on adequate supplies of riboflavin, niacin, and folate (Figure 8.110). Choline in the liver can be transported from cytosol into mitochondria and oxidized to betaine aldehyde (by choline dehydrogenase, EC1.1.99.1), and then to betaine (by betaine aldehyde dehydrogenase, EC1.2.1.8). Betaine provides one methyl group for the remethylation of homocysteine (by betaine homocysteine methyltransferase, EC2.1.1.5, contains zinc). The oxidation of the choline metabolite dimethylglycine by the flavoenzyme dimethylglycine dehydrogenase (EC1.5.99.2, FAD) releases formaldehyde. The next step of choline metabolism, catalyzed by FMN-containing sarcosine dehydrogenase (EC1.5.99.1), also releases formaldehyde. An alternative pathway in the kidneys uses H_2O_2-producing sarcosine oxidase (Reuber et al., 1997). Glycine may be used for one of its various metabolic functions or oxidized by the glycine cleavage system.

Phospholipase C (EC3.1.4.3) is a specialized zinc-containing enzyme in seminal plasma that generates phosphocholine.

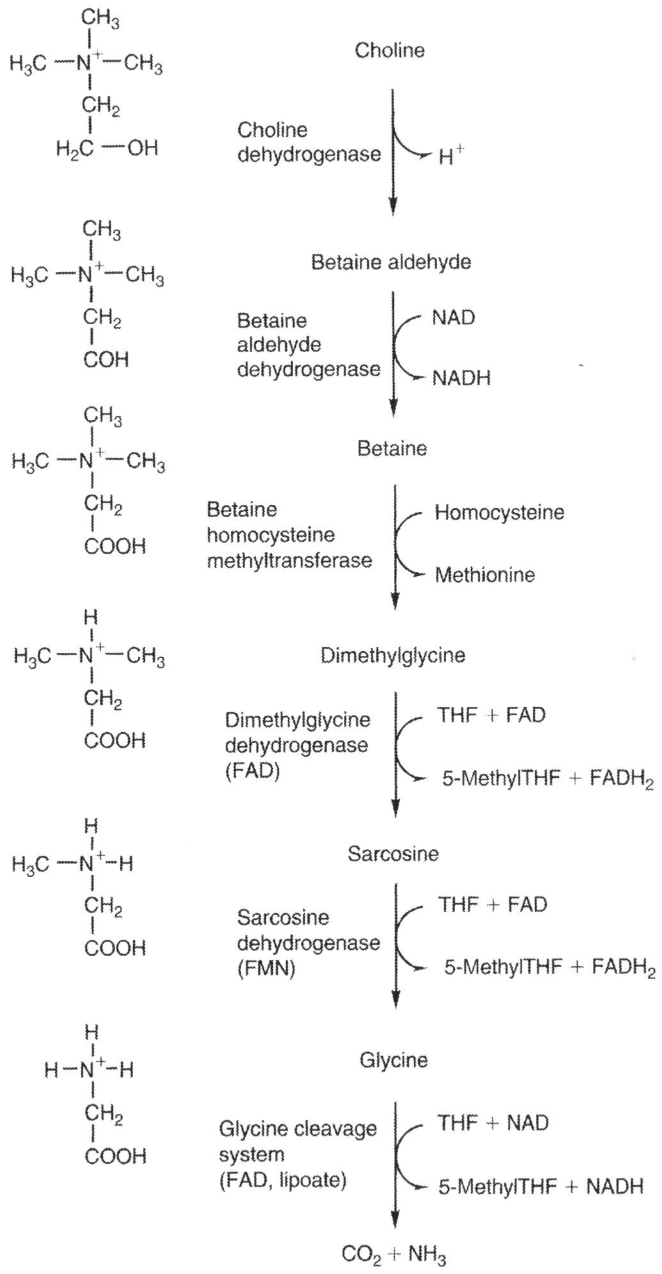

FIGURE 8.110

Choline breakdown uses oxidation and successive demethylation.

STORAGE

Choline is a component of the majority of phospholipids in membranes and other functional structures, and thus it is present in all tissues in significant amounts. The successive action of phospholipase A2 (EC3.1.1.4), lysophospholipase (EC3.1.1.5), and glycerylphosphocholine phosphodiesterase (EC3.1.4.2) releases choline from phosphatidylcholine.

EXCRETION

MDR2 (ABCB4) transports phospholipids actively into bile. Due to very effective intestinal absorption, however, healthy people lose little choline via feces. About 270 mg of choline is filtered daily in the kidneys, and most is reabsorbed through incompletely understood mechanisms. The organic cation transporters OCT1 (SLC22A1) at the basolateral membrane of the proximal tubules and OCT2 (SLC22A2) in the distal tubules are likely to be involved in choline transport (Arndt et al., 2001), but their significance may be as conduits for uptake from the pericapillary space.

The choline metabolite betaine is taken up from the proximal tubule and the descending limb of Henle's loop via a brush border membrane transporter with high affinity to L-proline. Export across the basolateral membrane (Pummer et al., 2000) uses the sodium- and chloride-dependent betaine transporter (BGT1, SLC6A12).

REGULATION

The rate of choline breakdown by oxidation depends mainly on the activity of the mitochondrial transporter (Kaplan et al., 1998). The mechanisms controlling this activity are not yet known.

FUNCTION

Methylgroup transfer: Choline is a major source of one-carbon groups. Breakdown of its metabolite betaine is directly coupled with the remethylation of methionine, and each of the three catabolic steps after that generates the one-carbon donor methylene tetrahydrofolate. This gives choline a central role in the homeostasis of hormone synthesis, DNA methylation, and other critical metabolic events.

Neurotransmission: A small amount of the intracellular choline, probably from phosphatidylcholine, is used for the synthesis of acetylcholine in neurons of the brain and the parasympathetic nervous system (Figure 8.111). Choline O-acetyltransferase (choline acetyltransferase, EC2.3.1.6) is responsible for the synthesis. Acetylcholine is cleaved again by acetylcholine esterase (YT blood group antigen, EC3.1.1.7).

Complex lipids: Choline is an important constituent of structural lipids such as phosphatidylcholine (membranes, digestive micelles, lipoproteins, and intracellular signaling), sphingomyelin (enhances neural conductivity), and platelet-activating factor (hormone-like action). The synthesis of phosphatidyl choline (Figure 8.112) starts with choline phosphorylation by choline kinase (EC2.7.1.32). CTP phosphocholine cytidyltransferase (EC2.7.7.15) and then generates the activated form and magnesium-dependent diacylglycerol cholinephosphotransferase (CDP-choline:diacyl glycerol choline phosphotransferase, EC2.7.8.2) adds the 12-diacylglycerol moiety.

Countercurrent hypertonia: The choline metabolite betaine is an important osmolyte in epithelial cells from the renal inner medulla that helps to concentrate urine. Expression of the betaine transporter

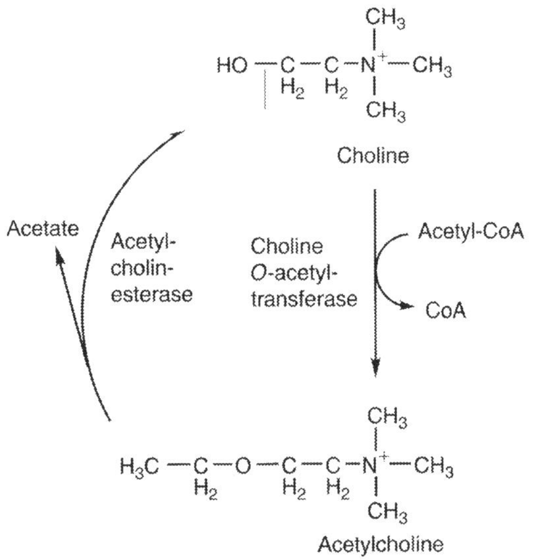

FIGURE 8.111

Acetylcholine synthesis and breakdown.

(SLC6A12) in the renal medulla is induced by hypertonicity. A similar osmoprotective function may be important for the intestine (Kidd et al., 1997).

Enzyme activation: The activity of some enzymes is increased by phosphatidylcholine. A pertinent example is the allosteric activation of 3-hydroxybutyrate dehydrogenase (EC1.1.1.30), an enzyme of ketone body metabolism.

Cancer: Choline deficiency increases cancer risk in various rodent models. An important mechanism may be inadequate DNA methylation due to a diminished pool of one-carbon groups. Many cancers are characterized by increased choline kinase activity, which effectively traps choline in affected cells (Roivainen et al., 2000). This phenomenon can be employed for radiological identification of some cancer cells by positron emission tomography (PET) after administration of ^{11}C-labeled choline.

Cell cycle regulation: Choline deficiency increases the rate of apoptotic cell death in many tissues (Shin et al., 1997).

NUTRITIONAL ASSESSMENT

Choline deficiency and suboptimal status are reflected with modest specificity by the concentration of choline in plasma (Wallace et al., 2012). If any information is to be gained from such measurements, blood samples must be immediately cooled and separated and then stored at −80°C within 45 min. Otherwise, hydrolysis of phospholipids, of which there are much larger amounts in the blood than choline, will irreversibly foul up the results.

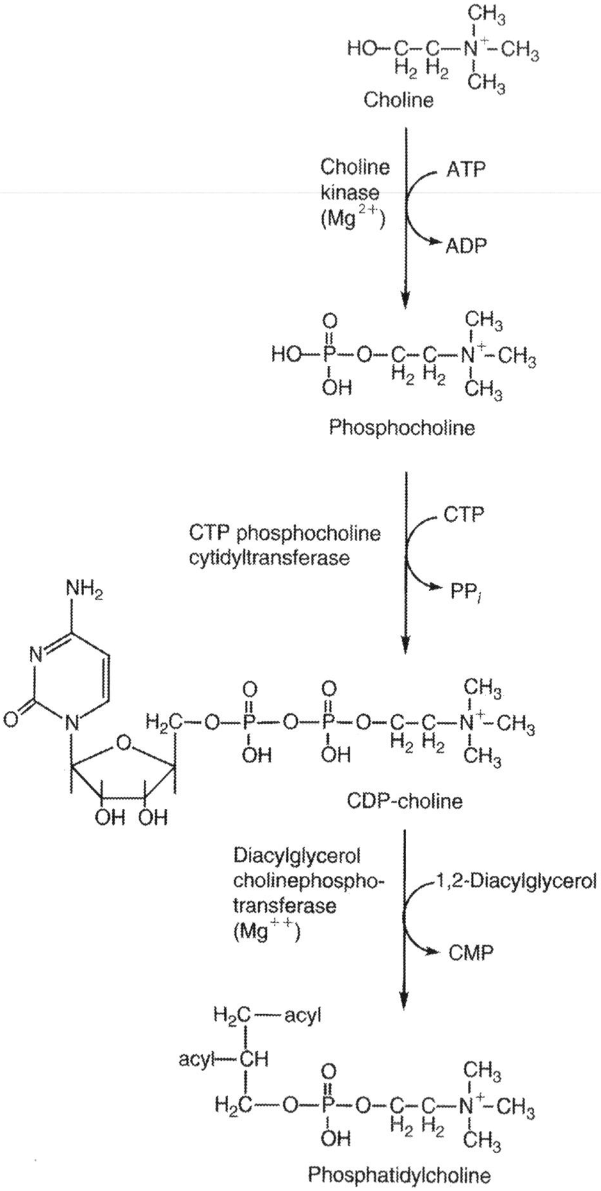

FIGURE 8.112

Phosphatidylcholine synthesis from choline.

REFERENCES

Arndt, P., Volk, C., Gorboulev, V., Budiman, T., Popp, C., Ulzheimer-Teuber, I., et al., 2001. Interaction of cations, anions, and weak base quinine with rat renal cation transporter rOCT2 compared with rOCT1. Am. J. Physiol. Ren. Fluid Electrolyte Physiol. 281, F454–F468.

Ayesh, R., Mitchell, S.C., Zhang, A., Smith, R.L., 1993. The fish odour syndrome: biochemical, familial, and clinical aspects. Br. Med. J. 307, 655–657.

Ayuk, P.T.Y., Sibley, C.P., Donnai, P., D'Souza, S., Glazier, J.D., 2000. Development and polarization of cationic amino acid transporters and regulators in the human placenta. Am. J. Physiol. Cell Physiol. 278, C1162–C1171.

Dowhan, W., 1997. Phosphatidylserine decarboxylases:pyruvoyl-dependent enzymes from bacteria to mammals. Methods Enzymol. 280, 81–88.

Ebel, H., Hollstein, M., Gunther, T., 2002. Role of the choline exchanger in Na(+)-independent Mg(2+) efflux from rat erythrocytes. Biochim. Biophys. Acta 1559, 135–144.

Grassl, S.M., 1994. Choline transport in human placental brush border membrane vesicles. Biochim. Biophys. Acta 1194, 203–213.

Habibulla, M., Newburgh, R.W., 1969. Carnitine decarboxylase and phosphokinase in *Phormia regina*. J. Insect. Physiol. 15, 2245–2253.

Kaplan, C.E., Porter, R.K., Brand, M.D., 1998. The choline transporter is the major site of control of choline oxidation in isolated rat liver mitochondria. FEBS Lett. 321, 24–26.

Kettunen, H., Peuranen, S., Tiihonen, K., Saarinen, M., 2001. Intestinal uptake ofbetaine *in vitro* and the distribution of methyl groups from betaine, choline, and methionine in the body of broiler chicks. Comp. Biochem. Physiol. A 128, 269–278.

Kidd, M.T., Ferket, P.R., Garlich, J.D., 1997. Nutritional and osmoregulatory functions of betaine. Poult. Sci. 53, 125–139.

Kobayashi, Y., Okuda, T., Fujioka, Y., Matsumura, G., Nishimura, Y., Haga, T., 2002. Distribution of the high-affinity choline transporter in the human and macaque monkey spinal cord. Neurosci. Lett. 317, 25–28.

Lockman, P.R., Roder, K.E., Allen, D.D., 2001. Inhibition of the rat blood–brain barrier choline transporter by manganese chloride. J. Neurochem. 79, 588–594.

Porter, R.K., Scott, J.M., Brand, M.D., 1992. Choline transport into rat liver mitochondria. Characterization and kinetics of a specific transporter. J. Biol. Chem. 267, 14637–14646.

Pummer, S., Dantzler, W.H., Lien, Y.H., Moeckel, G.W., Volker, K., Silbernagl, S., 2000. Reabsorption of betaine in Henle's loops of rat kidney *in vivo*. Am. J. Physiol. Ren. Fluid Electrolyte Physiol. 278, F434–F439.

Reuber, B.E., Karl, C., Reimann, S.A., Mihalik, S.J., Dodt, G., 1997. Cloning and functional expression of a mammalian gene for a peroxisomal sarcosine oxidase. J. Biol. Chem. 272, 6766–6776.

Roivainen, A., Forsback, S., Grönroos, T., Lehikoiunen, E., Käihkönen, M., Sutinene, E., et al., 2000. Blood metabolism of [methyl-^{11}C]choline; implications for *in vivo* imaging with positron emission tomography. Eur. J. Nucl. Med. 27, 25–32.

Shin, O.H., Mar, M.H., Albright, C.D., Citarella, M.T., da Costa, K.A., Zeisel, S.H., 1997. Methyl-group donors cannot prevent apoptotic death of rat hepatocytes induced by choline deficiency. J. Cell Biochem. 64, 196–208.

Sweet, D.H., Miller, D.S., Pritchard, J.B., 2001. Ventricular choline transport: a role for organic cation transporter 2 expressed in choroid plexus. J. Biol. Chem. 276, 41611–41619.

Takanaga, H., Ohtsuki, S., Hosoya, K., Terasaki, T., 2001. *GAT2/BGT-I* as a system responsible for the transport of gamma-aminobutyric acid at the mouse blood–brain barrier. J. Cereb. Blood Flow Metab. 21, 1232–1239.

Traiffort, E., O'Regan, S., Ruat, M., 2013. The choline transporter-like family SLC44: properties and roles in human diseases. Mol. Asp. Med. 34, 646–654.

Treacy, E.P., Akerman, B.R., Chow, L.M.L., Youil, R., Bibeau, C., Lin, J., et al., 1998. Mutations of the flavin-containing monooxygenase gene (FMO3) cause trimethylaminuria, a defect in detoxication. Hum. Mol. Genet. 7, 839–845.

Wallace, J.M., McCormack, J.M., McNulty, H., Walsh, P.M., Robson, P.J., Bonham, M.P., et al., 2012. Choline supplementation and measures of choline and betaine status: a randomised, controlled trial in postmenopausal women. Br. J. Nutr. 108, 1264–1271.

Zeisel, S.H., da Costa, K.A., 2009. Choline: an essential nutrient for public health. Nutr. Rev. 67, 615–623.

FAT-SOLUBLE VITAMINS AND NONNUTRIENTS

CHAPTER OUTLINE

Free Radicals and Antioxidants 479
Vitamin A 486
Vitamin D 501
Vitamin E..514
Vitamin K 526
Cholesterol...539
Lipoic Acid 553
Ubiquinone...560

FREE RADICALS AND ANTIOXIDANTS

ABBREVIATION

ROS   reactive oxygen species

COMPOSITION AND STRUCTURE

Various normal reactions and functions generate reactive oxygen species (ROS) and other compounds that are characterized by their high potential for causing oxidative damage to the body's DNA, proteins, membranes, and other components. Several of these compounds are called *free radicals* because they contain an unpaired electron. Free radicals have a strong propensity to donate their unpaired electron to another compound or to abstract an electron from elsewhere to complement their own unpaired one.

Their high and unspecific reactivity gives them the power to modify most biological macromolecules and disrupt their structure. These relentless attacks are thought to be a major cause of progressive functional decline with aging (e.g., macular degeneration) and major chronic diseases of adulthood, including cardiovascular disease, cancer, and rheumatoid arthritis. Nonetheless, some ROS are of vital importance for signaling and immune defense, and their elimination would probably be harmful. Several enzyme-catalyzed reactions detoxify ROS, and various redox-active metabolites provide additional protection. Adequate availability of diet-derived cofactors, such as vitamins C and E, selenium, zinc, and manganese, maintains the body's natural antioxidant protection.

Food contains a wide range of additional antioxidants that may have beneficial properties. The potential for inadvertent suppression of vital ROS functions, their conversion into free radical metabolites, or activity unrelated to their free-radical fighting properties make the intake of large amounts of exogenous antioxidants a double-edged sword. There is unequivocal evidence that at least some antioxidants, such as beta-carotene and vitamin E, cause harm when taken as high-dosed supplements for a long time (Albanes et al., 1995). The same compounds protect against atherosclerosis, cancer, and other diseases when consumed in modest quantities from a mixed diet rich in fruits and vegetables.

TYPES OF OXYGEN FREE RADICALS

ROS are to a large extent the nonstoichiometric by-products of oxidative phosphorylation. When an electron in the respiratory chain moves to oxygen instead of the next acceptor in line, superoxide anion forms (Cadenas and Davies, 2000). A healthy 70-kg man may be expected to generate about 190–380 mmol/day, based on the assumption that 1–2% of the oxygen consumption produces superoxide anion. An additional mechanism for the production of ROS during oxidative phosphorylation occurs with the transfer of less than the required four electrons to oxygen. The transfer of only one electron generates superoxide anion, two added electrons yield hydrogen peroxide, and three electrons give rise to hydroxyl radical ($\cdot$OH). Partial disruption of oxidative phosphorylation by alcohol (Bailey and Cunningham, 2002), methamphetamine (Virmani et al., 2002), other compounds, illness or genetic variants increases the production of ROS by-products. Another major source of ROS is the breakdown of purine nucleotides (adenosine and guanosine) in peroxisomes. The final conversion to uric acid by xanthine oxidase [EC1.1.3.22, contains flavin adenine dinucleotide (FAD), iron, and molybdenum] produces hydrogen peroxide. The typical daily production of about 400–800 mg of uric acid generates 2–5 mmol H_2O_2.

Many other peroxisomal reactions also generate hydrogen peroxide. Several nicotinamide adenine dinucleotide/nicotine adenine dinucleotide phosphate (NADH/NADPH) oxidases, lipoxygenases, cyclooxygenase, and P450 monooxygenases in other cellular compartments also contribute to ROS production. Ozone (O_3) is an inhaled ROS that readily reaches any tissue along with normal oxygen. At moderate levels around 50 ppb, about 1 μmol is inhaled per day. This amount adds little to total oxidant burden, but local effects at the point of first contact; that is, respiratory mucosa and lung alveoli, are much more significant.

Sunlight and other types of radiation are potent inducers of ROS production. Ultraviolet (UV) light is particularly damaging when acting on the skin with little protection from pigments or sunscreen or on the retina of the eye.

One of the most harmful ROS, the hydroxyl radical, is mainly generated during metal-catalyzed secondary reactions (Fenton reactions). Superoxide anion transfers its excess electron to a metal ion, usually iron or copper (reaction equation 1). The reduced metal ion can then abstract an electron again from hydrogen peroxide and cleave it into a hydroxyl ion and a hydroxyl radical (reaction equation 2). It is this reaction that makes unbound iron and copper so toxic even at micromolar concentrations, particularly in the presence of hydrogen peroxide.

$$O_2^- + Fe^{3+} \rightarrow O_2 + Fe^{2+} \qquad \text{(reaction equation 1)}$$

$$H_2O_2 + Fe^{2+} \rightarrow \;^{\cdot}OH + OH^- + Fe^{3+} \qquad \text{(reaction equation 2)}$$

Table 9.1 Free Radicals, Compounds with an Unpaired Electron, Are Common	
Ozone	O_3
Superoxide anion	O_2^-
Singlet oxygen	$O^{\cdot}$
Hydroxyl radical	$^{\cdot}OH$
Nitric oxide	$NO^{\cdot}$
Peroxinitrite	$ONOO^{\cdot}$
Lipid peroxyl	$LOO^{\cdot}$
Tryptophan radical	$^{\cdot}Trp$
Tyrosine	$TyrO^{\cdot}$
Ascorbyl radical anion	$Asc^{\cdot-}$
α-Tocopheroxyl radical	$\alpha\text{-}TO^{\cdot}$

The signaling compound nitric oxide (NO*), which is produced from arginine, can combine with a superoxide anion and form the highly reactive peroxynitrite (reaction equation 3).

$$O_2 + NO \rightarrow ONOO^- \qquad \text{(reaction equation 3)}$$

Protonation of peroxynitrite forms the unstable intermediate HONOO, which rapidly decomposes with the release of a hydroxyl radical. The reaction of primary ROS with additional susceptible targets can convert them into radicals. Important examples include tryptophan (·Trp), tyrosine (TyrO·), and bilirubin radicals. Polyunsaturated fatty acids are particularly susceptible targets. Their oxidation initiates a rapidly cascading chain reaction because each radical generates two new oxidized fatty acid radicals. The metabolites of oxidized fatty acid, including 4-hydroxy-2,3-trans-nonenal (HNE), crotonaldehyde, and malondialdehyde, are highly reactive compounds that can cross-link proteins and engage in other harmful reactions. The major antioxidants also come out of each encounter with ROS as free radicals that have to be detoxified by auxiliary reactions as described later in this chapter. This is the case with vitamin E (tocopheroxyl radical), ascorbate (ascorbyl radical anion), and flavonoids.

Some commonly consumed foods, including coffee (Ruiz-Laguna and Pueyo, 1999), fried foods (Wilson et al., 2002), and even wine (Rossetto et al., 2001), can be a source of exogenous ROS and other free radical species (Table 9.1).

PHYSIOLOGICAL FUNCTIONS

The notion that certain ROS play important roles in normal body function is gaining momentum. ROS provide signals that can trigger mitogen-activated protein kinases (Klotz et al., 2002), modulate the adhesion of neutrophils to target sites and initiate their activation (Guo and Ward, 2002), and stimulate hormonal responses (Hsieh et al., 2002). The production of ROS is clearly regulated in some instances. A key step of programmed cell death (apoptosis) is the inactivation of mitochondrial cytochrome c for the enhanced production of ROS (Moncada and Erusalimsky, 2002). The inactivation can be spontaneous (indicating defective cell function) or the result of a regulatory event (to initiate the removal of a targeted cell). The ensuing high level of ROS then contributes to the fragmentation of the cell's DNA and its ultimate demise. ROS also enable immune cells to destroy and remove pathogens. Before macrophages and other immune cells engulf bacteria, they can disrupt them with a directed stream of corrosive reactants. This oxidative blast uses ROS in addition to hypochlorous acid and other chemicals (Vazquez-Torres et al., 2000). An NADPH oxidase (phagocyte oxidase, EC1.6.3.1) uses FAD and cytochrome b_{558} to transfer a single electron to oxygen and generate superoxide anion (Seguchi and Kobayashi, 2002). The nitric oxide produced by nitric oxide synthase (EC1.14.13.39, contains heme and uses tetrahydrobiopterin as a cofactor) can then be combined by the lysosomal hemeenzyme myeloperoxidase (EC1.11.1.7) with superoxide to generate peroxynitrite (Eiserich et al., 2002). Superoxide also drives the production of hydrogen peroxide by superoxide dismutase (EC1.15.1.1).

ROS-INDUCED DAMAGE

The main characteristic of ROS is their high reactivity with low specificity. They oxidize and cross-link proteins, fragment DNA and alter its bases, and disrupt membranes by oxidizing their fatty acids.

DNA and RNA: The molecular structures of more than 70 ROS-induced DNA modifications have been identified (Pouget et al., 2002). Among the most common lesions are 8-oxo-7,8-dihydro-2′-deoxyguanosine (8-oxo-dGuo), 5-formyl-2′-deoxyuridine (5-FordUrd), 5-(hydroxymethyl)-2′-deoxyuridine (5-HmdUrd), and 5,6-dihydroxy-5,6-dihydrothymidine (dThdGly).

Polyunsaturated fatty acids: The changes to cholesteryllinoleate of low-density lipoprotein (LDL) exposed to monocytes should illustrate oxidative damage occurring with exposure to ROS. In this scenario, the oxidative blast of activated monocytes has released superoxide and hydrogen peroxide and a Fenton reaction in the presence of free ionic iron has generated hydroxyl radicals. Lipid oxidation starts with the abstraction of a proton from carbon 11 between the two double bonds, but without its electron. This converts the hydroxyl radical into water but leaves the cholesteryllinoleate with a supernumerary electron at carbon 11. Shift of the 12,13 double bond and movement of the extra electron to carbon 13 form an unstable conjugated diene. The shift can also occur in the other direction (Folcik and Cathcart, 1994). The peroxyllinoleate generated by the addition of two oxygen molecules to carbon 13 can react with an adjacent polyunsaturated linoleate and thus spread the damage. The resulting 13-hydroperoxylinoleate, however, can react with Fe^{2+} and generate a 13-hydroxy radical. It is this second free-radical-generating reaction that causes the amplification of the initial ROS attack (Figures 9.1 and 9.2).

MECHANISMS OF ANTIOXIDANT ACTION

Metal ion chelation: The formation of ROS is effectively reduced by maintaining iron and copper in tightly bound form that cannot participate in Fenton-type reactions. The metal-chelating capacity of some food-derived compounds, such as flavonoids, may provide additional protection. The relevance of this effect at the cellular level remains unclear, however.

Enzyme-catalyzed reactions: The body has an elaborate system to protect against ROS. These systems tend to be most active at the sites of greatest ROS release. Catalase (EC 1.11.1.6), which contains both heme and manganese, dissipates hydrogen peroxide in peroxisomes to oxygen and water. This enzyme with its high capacity and low affinity is best suited to detoxify overflow quantities and sudden bursts of hydrogen peroxide. Other enzymes with peroxidase activity have lower capacity, but their

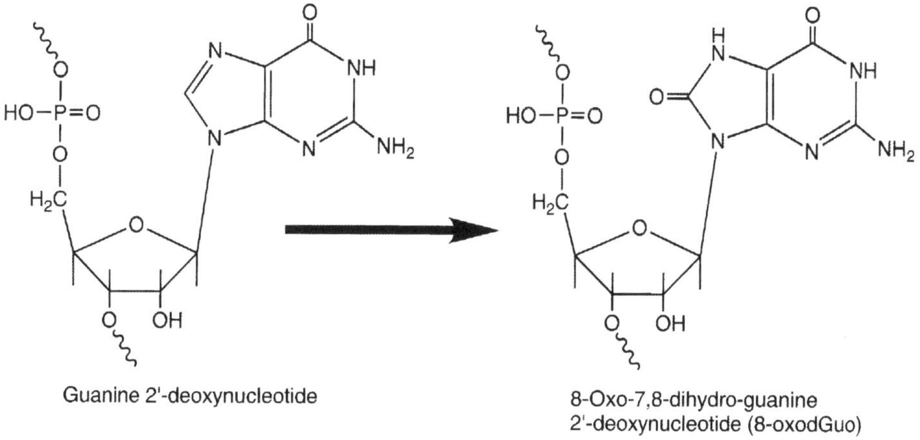

Guanine 2'-deoxynucleotide

8-Oxo-7,8-dihydro-guanine
2'-deoxynucleotide (8-oxodGuo)

FIGURE 9.1

Polynucleotides are susceptible to oxidative damage.

high substrate affinity keeps hydrogen peroxide concentrations very low. This group of high affinity peroxidases includes the peroxiredoxins (Prx), which are closely related heme enzymes. Different superoxide dismutase (EC 1.15.1.l) isoenzymes in cytosol and the extracellular space convert superoxide radicals to hydrogen peroxide (reaction equation 4). All isoenzymes contain copper and a second transition metal. The isoenzyme in mitochondria contains manganese, the ones in cytosol and extracellular fluids contain zinc or iron.

$$2O_2^{\cdot -} + 2H^+ \rightarrow O_2 + H_2O_2 \qquad \text{(reaction equation 4)}$$

Another high-capacity free radical scavenger in extracellular fluid is the copper enzyme ferroxidase (ceruloplasmin, iron (II):oxygen oxidoreductase, EC1.16.3.l). Thioredoxin reductase (EC1.6.4.5) is a ubiquitous NADPH-dependent selenoenzyme in cytosol that reduces both dehydroascorbate and the semidehydroascorbate radical to ascorbate (May et al., 1998). A different protective strategy seeks to remedy the damage. Four different selenium-containing glutathione peroxidases (EC1.11.1.9) with distinct tissue distributions and activity profiles use glutathione (GSH) for the reduction of peroxides of free fatty acids and other lipids. Another example is the activity of arylesterase (paraoxonase 1, PAN1, EC3.1.1.2) in high-density lipoprotein (HDL). This enzyme cleaves the fatty aldehydes from damaged phospholipids and releases them from the lipoprotein (Lp) particle for further metabolic treatment in the liver and other tissues (Ahmed et al., 2001).

Antioxidants: The body uses both fat-soluble and water-soluble compounds to reach all cellular compartments (Tables 9.2 and 9.3).

The essential nutrient ascorbate is a particularly versatile antioxidant since it can quench radicals that have one or two excess electrons. The systems for the regeneration of the oxidized forms include NADH-dependent monodehydroascorbate reductase (EC1.6.5.4), thioredoxin reductase (EC1.6.4.5), and an NADH-dependent dehydroascorbate-reducing transporter in erythrocytes (May et al., 1998). Thioredoxin is a small peptide with two redox-active cysteines that potently quenches singlet oxygen and hydroxyl radicals. The oxidation of its cysteines reduces oxidants or oxidized compounds.

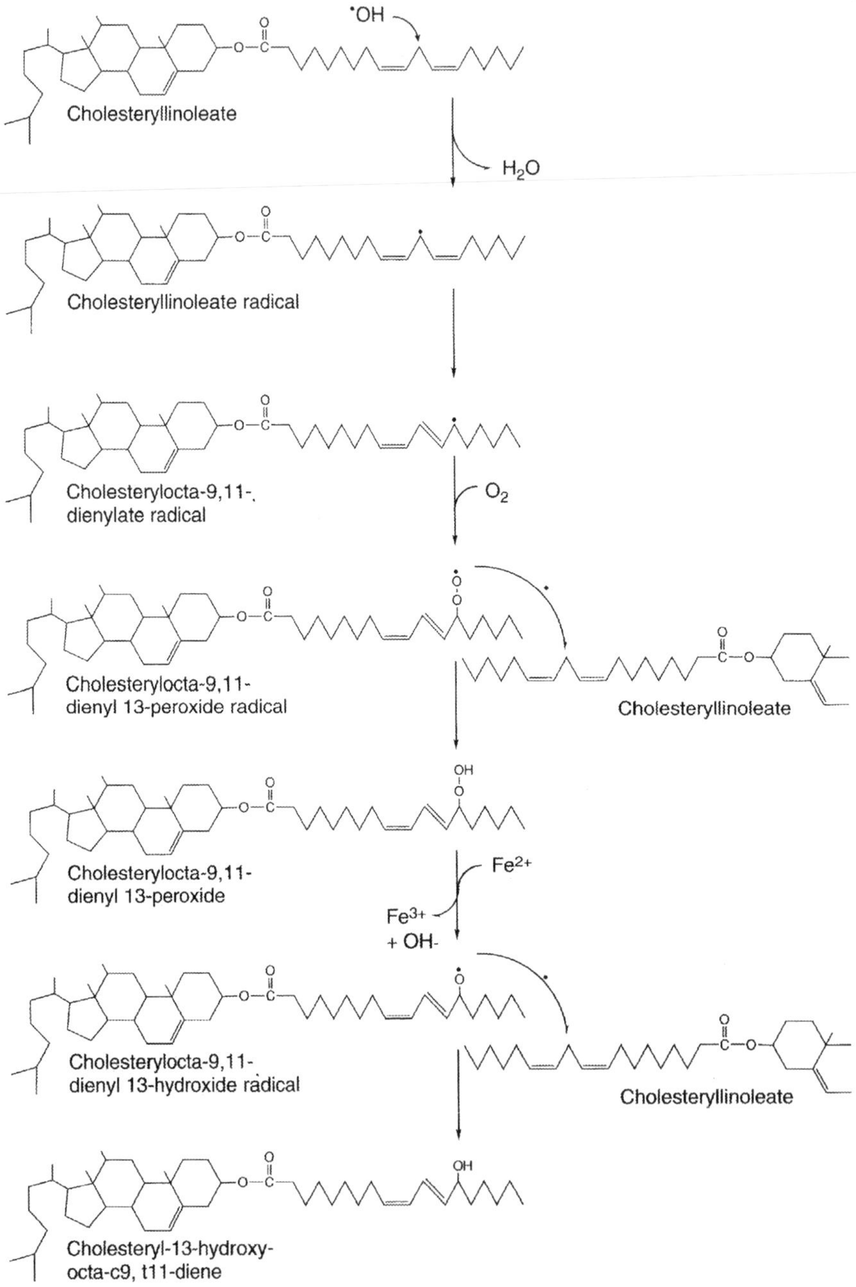

FIGURE 9.2

Oxidative damage to polyunsaturated lipids tends to spread.

Table 9.2 Antioxidant Enzymes

Catalase	EC1.1 1.1.6, heme
Superoxide dismutase	EC1.1 S.1.1, iron, manganese, zinc
Peroxidase	EC1.11.1.7, heme
Glutathione peroxidases	EC1.11.1.9, selenium
Thioredoxin reductase	EC1.6.4.S, selenium
Arylesterase	EC3.1.1.2

Table 9.3 Antioxidant Compounds

Ascorbatic acid	Vitamin E
Thioredoxin	Ubiquinone
Lipoate	Carotenoids
Tetrahydrobiopterin	Conjugated linoleic acid
Uric acid	Protein disulfides
Polyphenols/flavonoids	Melatonin

It also detoxifies hydrogen peroxide in conjunction with a group of enzymes called *peroxiredoxins*. Thioredoxin reductase (EC1.6.4.5) uses NADH to rapidly regenerate the oxidized thioredoxin. Lipoate, tetrahydrobiopterin, uric acid, phenols, flavonoids and isoflavones, additional protein disulfides, and possibly melatonin add to the mix of water-soluble antioxidants.

Vitamin E is particularly important for antioxidant protection in Lps, membranes, and other lipophilic environments. Since the interaction of ROS with vitamin E generates the tocopheroxyl radical, the net effectiveness depends on adequate availability of ascorbate and other co-antioxidants for regeneration (Terentis et al., 2002).

Ubiquinone and tetrahydrobiopterin have considerable antioxidant potential unrelated to their function as enzyme cofactors. In addition to these endogenous metabolites, a wide range of food-derived compounds is known to provide additional protection. Hundreds of carotenoids from fruits and vegetables increase the resistance of tissues to the harmful effects of ROS.

REFERENCES

Ahmed, Z., Ravandi, A., Maguire, G.F., Emili, A., Draganov, D., La Du, B.N., et al., 2001. Apolipoprotein A-I promotes the formation of phosphatidylcholine core aldehydes that are hydrolyzed by paraoxonase (PON-1) during high intensity lipoprotein oxidation with a peroxynitrite donor. J. Biol. Chem. 276, 24473–24481.

Albanes, D., Heinonen, O.P., Huttunen, J.K., Taylor, P.R., Virtamo, J., Edwards, B.K., et al., 1995. Effects of alpha-tocopherol and beta-carotene supplements on cancer incidence in the Alpha-Tocopherol Beta-Carotene Cancer Prevention Study. Am. J. Clin. Nutr. 62, 1427S–1430S.

Bailey, S.M., Cunningham, C.C., 2002. Contribution of mitochondria to oxidative stress associated with alcoholic liver disease. Free Rad. Biol. Med. 32, 11–16.

Cadenas, E., Davies, K.J., 2000. Mitochondrial free radical generation, oxidative stress, and aging. Free Rad. Biol. Med. 29, 222–230.

Eiserich, J.E., Baldus, S., Brennan, M.L., Ma, W., Zhang, C., Tousson, A., et al., 2002. Myeloperoxidase, a leuko-cyte-derived vascular NO oxidase. Science 296, 2391–2394.

Folcik, V.A., Cathcart, M.K., 1994. Predominance of esterified hydroperoxy-linoleic acid in human monocyte-oxidized LDL. J. Lipid Res. 35, 1570–1582.

Guo, R.E., Ward, P.A., 2002. Mediators and regulation of neutrophil accumulation in inflammatory responses in lung: insights from the IgG immune complex system. Free Rad. Biol. Med. 33, 303–310.

Hsieh, T.J., Zhang, S.L., Filep, J.G., Tang, S.S., Ingelfinger, J.R., Chan, J.S., 2002. High glucose stimulates angiotensinogen gene expression via reactive oxygen species generation in rat kidney proximal tubular cells. Endocrinologica 143, 2975–2985.

Klotz, L.O., Schroeder, P., Sies, H., 2002. Peroxynitrite signaling: receptor tyrosine kinases and activation of stress-responsive pathways. Free Rad. Biol. Med. 33, 737–743.

May, J.M., Cobb, C.E., Mendiratta, S., Hill, K.E., Burk, R.E., 1998. Reduction of the ascorbyl free radical to ascorbate by thioredoxin reductase. J. Biol. Chem. 273, 23039–23045.

Moncada, S., Erusalimsky, J.D., 2002. Does nitric oxide modulate mitochondrial energy generation and apoptosis? Nat. Rev. Mol. Cell Biol. 3, 214–220.

Pouget, J.P., Frelon, S., Ravanat, J.L., Testard, I., Odin, F., Cadet, J., 2002. Formation of modified DNA bases in cells exposed either to gamma radiation or to high-LET particles. Rad. Res. 157, 589–595.

Rossetto, M., Vianello, F., Rigo, A., Vrhovsek, U., Mattivi, F., Scarpa, M., 2001. Stable free radicals as ubiquitous components of red wines. Free Rad. Res. 35, 933–999.

Ruiz-Laguna, J., Pueyo, C., 1999. Hydrogen peroxide and coffee induce G:C→T:A transversions in the lac1 gene of catalase-defective Escherichia coli. Mutagenesis 14, 95–102.

Seguchi, H., Kobayashi, T., 2002. Study of NADPH oxidase-activated sites in human neutrophils. J. Electron. Microsc. 51, 87–91.

Terentis, A.C., Thomas, S.R., Burr, J.A., Liebler, D.C., Stocker, R., 2002. Vitamin E oxidation in human athero-sclerotic lesions. Circ. Res. 90, 333–339.

Vazquez-Torres, A., Jones-Carson, J., Mastroeni, P., Ischiropoulos, H., Fang, F.C., 2000. Antimicrobial actions of the NADPH phagocyte oxidase and inducible nitric oxide synthase in experimental salmonellosis. I. Effects on microbial killing by activated peritoneal macrophages in vitro. J. Exp. Med. 192, 227–236.

Virmani, A., Gaetani, E., Imam, S., Binienda, Z., Ali, S., 2002. The protective role of L-carnitine against neuro-toxicity evoked by drug of abuse, methamphetamine, could be related to mitochondrial dysfunction. Ann. NY Acad. Sci. 965, 225–232.

Wilson, R., Lyall, K., Smyth, L., Fernie, C.E., Riemersma, R.A., 2002. Dietary hydroxy fatty acids are absorbed in humans: implications for the measurement of "oxidative stress" in vivo. Free Rad. Biol. Med. 32, 162–168.

VITAMIN A

The term *vitamin A* usually refers to retinol (vitamin A, 1,3,7-dimethyl-9(2,6,6-trimethyl-1-cyclohexen-1-yl)-2,4,6,8-nonatetraen-1-ol, obsolete name ophthalmin; molecular weight 286), its esters, and the metabolically equivalent retinal (retinaldehyde, retinene, vitamin A aldehyde, axerophthal; molecu-lar weight 284). Dehydroretinol (3,4-dehydrporetinol, vitamin A2) is a retinol metabolite with much lower biological activity than retinol. Beta-carotene, alpha-carotene, beta-cryptoxanthin, and a few other carotenoids, which can be metabolically converted into retinal and retinol, are referred to as *provitamin A*. The term *vitamin A* should not be used for retinoic acid (vitamin A acid, 3,7-dimethyl-9-(2,6,6-trimethyl-1-cyclohexen-1-yl)-2,4,6,8-nonatetraenoic acid, tretinoin; molecular weight 300), a metabolite of retinol with important biological functions because it cannot be converted back into its precursor (Figure 9.3).

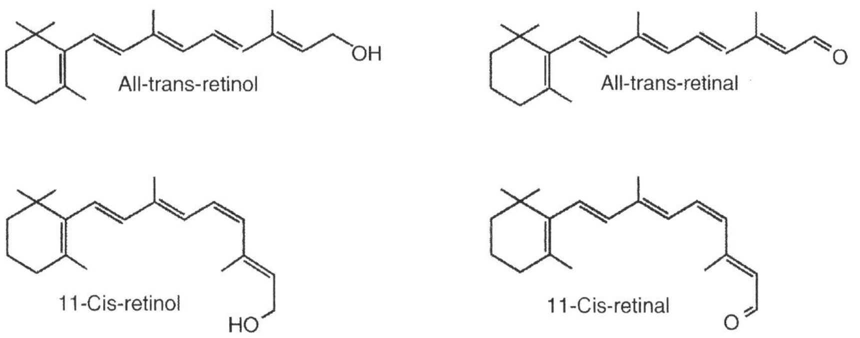

FIGURE 9.3

Several structurally related compounds have vitamin A activity.

ABBREVIATIONS

ADH	alcohol dehydrogenase (EC1.1.1.1)
CRBP2	cellular retinol-binding protein 2, RBP2
LDH	lactate dehydrogenase isoform (EC1.1.1.27)
RAE	retinol activity equivalent
RALDH1	retinal dehydrogenase 1 (EC1.2.1.36)
RALDH2	retinal dehydrogenase 2 (EC1.2.1.36)
RBP1	retinol-binding protein 1 (cellular RBP)
RBP2	retinol-binding protein 2 (cellular RBP of enterocytes)
RBP3	retinol-binding protein 3 (interstitial RBP)
RBP4	retinol-binding protein 4 (in the blood)
vitA	vitamin A (all biologically active forms)

NUTRITIONAL SUMMARY

Function: Vitamin A (vitA) is essential for vision, immune function, and regulation of cell growth.

Food sources: Only animal foods contain retinol. Particularly rich sources are the liver, flesh foods, eggs, and fortified milk. Good sources of provitamin A carotenoids are carrots, spinach, broccoli, yellow melons, mangos, and many other dark-green or orange-yellow fruits and vegetables. The activation of vitA requires riboflavin, niacin, zinc, and iron.

Requirements: To account for metabolic differences between vitA and its carotenoid precursors, dietary amounts are expressed as equivalents of 1 µg of retinol (retinol activity equivalent = RAE). 12 µg of beta-carotene and 24 µg of alpha-carotene or beta-cryptoxanthin corresponds to 1 RAE. Adults should get between 700 (women) and 900 (men) RAE. Women's needs are higher during lactation.

Deficiency: A lack of vitA (for which about one-third of the world population is at risk) initially causes reversible night blindness, later increasingly severe and irreversible loss of vision due to changes of the eye structure (xerophthalmia with drying of the conjunctiva and increasing opacity of the cornea).

Hyperkeratosis and other skin lesions are further typical effects of inadequate intake. Another concern with even mild deficiency is impaired immune function, especially in children.

Excessive intake: Retinol intake above 1000 µg/day increases bone fracture risk in older people. Moderately high intake of retinol (3000 µg/day), but not of provitamin A carotenoids, during early pregnancy increases the risk of birth defects. Daily ingestion of 15,000 µg retinol initially may lead to itching, scaling of skin, malaise, and loss of appetite. Cerebrospinal pressure may increase causing nausea, vomiting, headaches, and eventually seizures, coma, respiratory failure, and death.

ENDOGENOUS SYNTHESIS

Beta-carotene 15,15′-dioxygenase (EC1.14.99.36) in the cytosol of mature jejuna enterocytes, the liver, kidneys, and testes splits carotenoids in the middle (Duszka et al., 1996; Barua and Olson, 2000). This enzyme, which is actually a monooxygenase, requires bile acids and iron for its three-step activity (epoxidation at the 15,15′-double bond, hydration to the diol, and oxidative cleavage). The provitamin A carotenoids yield two (beta-carotene) or one (alpha-carotene, gamma-carotene, beta-cryptoxanthin, and beta-zeacarotene) retinal molecules (Rock, 1997). Some other carotenoids are also cleaved (e.g., lycopene, lutein, and zeaxanthin) but do not generate a fragment that can be converted into retinol (Figures 9.4 and 9.5).

Eccentric (asymmetric) cleavage at the 14′,13′ double (Dmitrovskii et al., 1997) or the 9′,10′ double bond (Kiefer et al., 2001) by 9-cis-beta-carotene 9′,10′-cleaving dioxygenase (BCDO2, EC1.13.11.68) appears to play a lesser role. Excentric cleavage products include 8′-β-apocarotenal (the largest metabolite), beta-ionone (compound contributing to rose fragrance), 10′-β-apocarotenal, 12′-β apocarotenal, and 14′-flapocarotenal. Many questions about the detailed metabolic pathways and actions of carotenoid cleavage products remain (Shete and Quadro, 2013). Vitamin C and vitamin E appear to promote symmetric at the expense of eccentric cleavage of beta-carotene (Liu et al., 2004).

Oxygen free radicals, such as fatty acid hydroperoxides, can also react with any of the conjugated polyene double bonds and initiate cleavage (Yeum et al., 1995). The antioxidant alpha-tocopherol can prevent such random cleavage and keep retinal yield near the theoretical optimum (Yeum et al., 2000). The side chain of longer fragments can be shortened by beta-carotene 15,15′-dioxygenase to generate retinal (Paik et al., 2001).

Retinal resulting from carotenoid cleavage is rapidly metabolized, mainly to retinol. A specific lactate dehydrogenase isoform (LDH-C, EC1.1.1.27) in testes is closely associated with beta-carotene 15,15′-dioxygenase, and the reduction of newly generated retinal may be its main physiological function (Paik et al., 2001). In most other tissues, class III ADH (EC1.1.1.1) may be more important (Molotkov et al., 2002). Retinoic acid is exported through an unknown pathway into portal blood and taken up by the liver.

DIETARY INTAKE

Foods contain retinol, retinylesters, provitamin A carotenoids, and also compounds with a restricted spectrum of vitamin A activity, such as retinoic acid. Several isomeric forms of both retinol derivatives and carotenoids are possible. The most common form of retinol in natural foods is the all-trans isomer. Much smaller amounts of 9-cis-retinol and other vitA metabolites are ingested with some foods. Synthetic retinol may also contain isomers in addition to the biologically active all-trans-retinol.

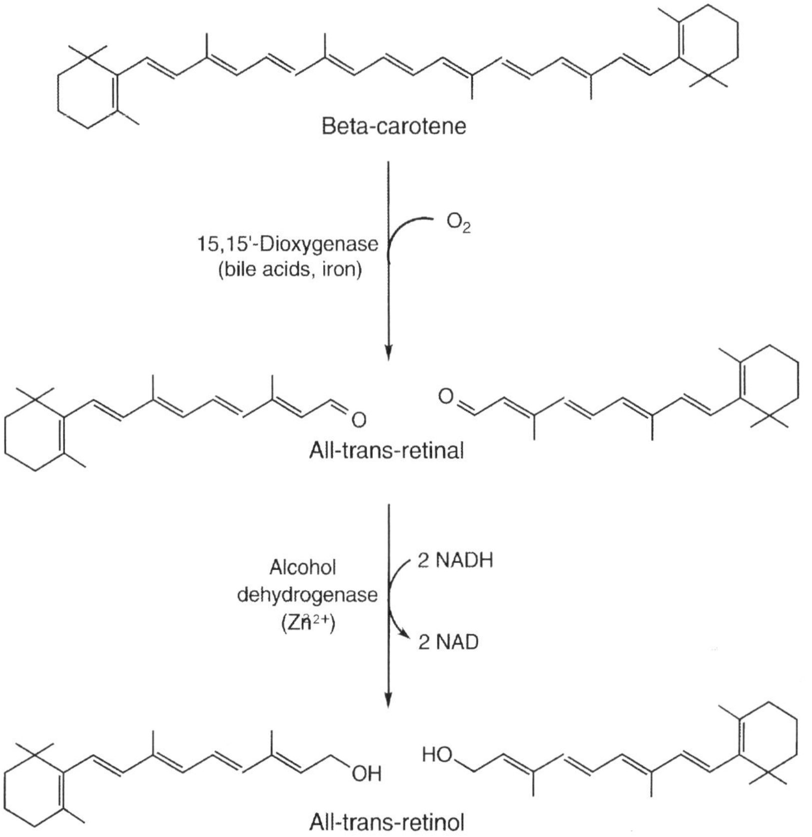

FIGURE 9.4

Beta-carotene and a few other carotenoids can be converted to retinol.

Compounds with vitA activity are very sensitive to oxidation, isomerization, and polymerization.

To facilitate assessment, intake of different forms of vitA is usually expressed as retinol activity equivalents (RAE) with 1 μg of all-trans-retinol as the basis (Food and Nutrition Board, Institute of Medicine, 2002). The conversion factors for nonretinol compounds take bioavailability and metabolic yield into account. Thus, 12 μg of ingested beta-carotene corresponds to 1 RAE, as do 24 μg of alpha-carotene and 24 μg of beta-cryptoxanthin. International units (IUs) for retinol are divided by 3.33 to convert into RAE, while those for 13-carotene are divided by 20.

Retinol and directly related compounds are only present in animal-derived foods. The highest concentration is in organ meats, such as beef liver (106 μg of RAE/g). Another group of retinol-rich foods are hard and cream cheeses such as Swiss cheese (2.5 μg/g), cheddar (2.8 μg/g), and regular cream cheese (3.8 μg/g). Eggs contain about 1.9 μg/g, and milk 0.6 μg/g.

Good plant-derived sources of vitamin A precursor carotenoids are green leafy vegetables, including spinach (8.2 μg of RAE/g), kale (7.4 μg/g), chard (3.1 μg/g), and broccoli (1.4 μg/g). Other

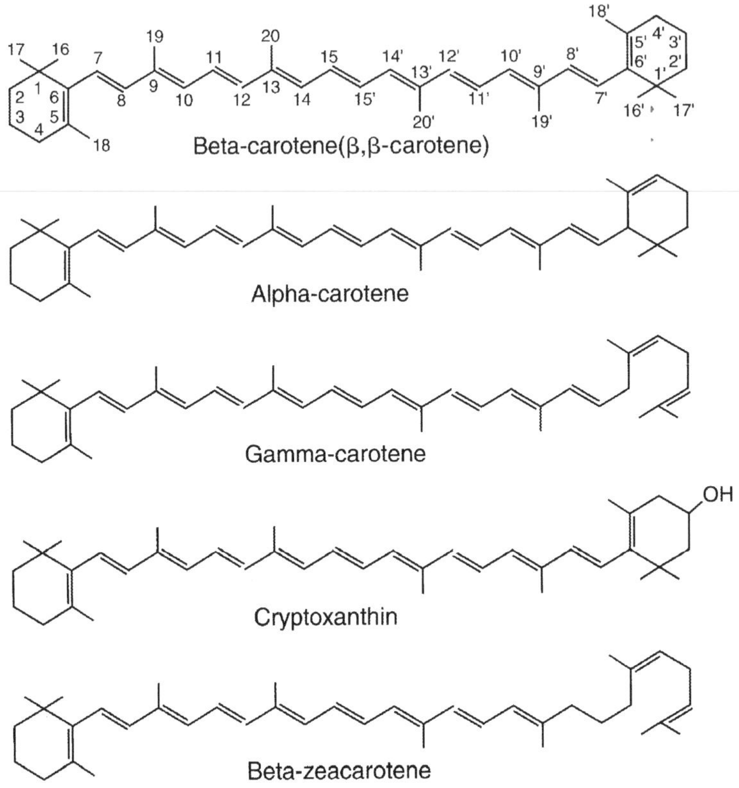

FIGURE 9.5

Carotenoids that can be converted to retinol.

good sources are orange or red vegetables and fruits, such as sweet potatoes (16.4 µg/g) and carrots (24.6 µg/g). Median daily intake of vitA from all sources were estimated to be around 900 µg for men and around 700 µg for women in the United States (Food and Nutrition Board, Institute of Medicine, 2002; Appendix C). More than half of their total vitA intake comes from preformed retinol.

Europeans tend to have higher intake. In a collection of representative surveys, men consumed 1060 µg of preformed retinol and a 7 µg of beta-carotene; women consumed 708 and 2651 µg, respectively (Jenab et al., 2009).

DIGESTION AND INTESTINAL ABSORPTION

Both retinol derivatives and carotenoids are absorbed from the proximal small intestine in a process that requires the formation of mixed micelles and concurrent fat absorption (Figure 9.6). About 70–90% of ingested retinol is absorbed (Sivakumar and Reddy, 1972), but only 3% or less of carotenoids (Edwards et al., 2002).

Retinol: Retinylesters are cleaved by lipase (EC3.1.1.3) and sterol esterase (EC3.1.1.13) from pancreas and at least one other brush border esterase. Retinol is absorbed efficiently only when it

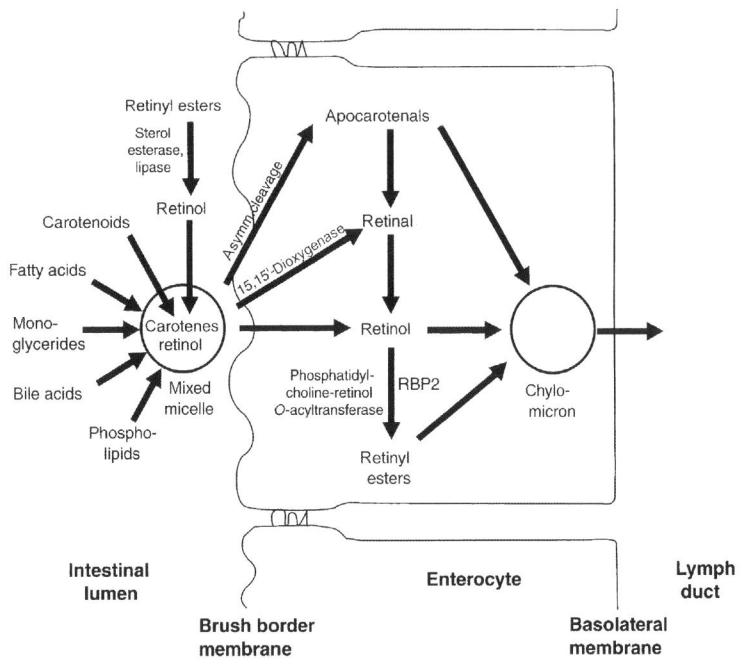

FIGURE 9.6

Intestinal absorption of vitamin A.

is incorporated into mixed micelles along with lipase-digested fat (fatty acids and monoglycerides), bile acids, and phospholipids (Harrison, 2012). It has been suggested that a specific transporter may be involved in retinol uptake, but its identity remains elusive. Retinol-binding protein 2 (RBP2, and cellular retinol-binding protein 2, CRBP2) is needed for retinol metabolism and trafficking across the enterocyte. For example, phosphatidylcholine-retinol *O*-acyltransferase (EC2.3.1.135) esterifies only retinol that is bound to RBP2. Retinol and retinylesters exit enterocytes as integral components of chylomicrons (Chylos). The transfer of retinylesters into Chylos is facilitated by CRBP2 and probably also cellular retinol-binding protein 1 (CRBP1), both of which are members of the lipocalin family (Harrison, 2012). Retinoic acid is exported through the lipid-transporter ABCA1 and possibly addition transporters (Harrison, 2012) into portal blood and taken up by the liver. A small amount of newly absorbed retinol is oxidized to retinoic acid by class III ADH (EC1.1.1.1; Molotkov et al., 2002). The mechanism of retinoic acid export from enterocytes is insufficiently understood.

Carotenoids: Digestion by the various proteases and peptidases releases carotenoids if they are attached to proteins. Since carotenoids in plants are embedded in the thylakoid matrix within cells with fairly digestion-resistant walls, cooking or extensive mechanical grinding (chewing) is usually necessary (Rock et al., 1998). The highest reported beta-carotene bioavailability from carrots, in a finely homogenized preparation, is still less than 3%, however (Edwards et al., 2002). Carotenoids have to be incorporated into mixed micelles before they can be absorbed (Yeum and Russell, 2002). Carotenoid uptake into enterocytes uses the mechanisms for fatty acid import, including the fatty acid transporter CD36, the scavenger receptor B1 (SRB1), and the Niemann–Pick-C-protein-like 1 (NPC1L1). It does not appear to be very selective.

Carotenoids are incorporated by unknown means into Chylos and exported into lymph. Carotenoids take about 8 h after a challenge meal to appear in the blood.

TRANSPORT AND CELLULAR UPTAKE

Blood circulation: The concentration of retinol (about one-tenth as retinyl ester) in plasma is homeostatically maintained around 2 μmol/l in men and closer to 1.7 μmol/l in women (Food and Nutrition Board, Institute of Medicine, 2002: Appendix G; Olmedilla et al., 1994). On its first pass from portal blood through the liver, retinol is taken up into hepatocytes through an unknown mechanism. Esterification by phosphatidylcholine-retinol *O*-acyltransferase (lecithin:retinol acyltransferase, LRAT, EC2.3.1.135) is critical for this cellular uptake (Amengual et al., 2012). It can then be resecreted with retinol-binding protein 4 (RBP4). RBPs are lipocalins, which means that the protein engulfs a retinol molecule and shields it from the hydrophilic environment in circulation or inside cells. After its release from the liver, RBP4 combines in the blood with transthyretin (Episkopou et al., 1993). Surface receptors on many peripheral cells bind RBP4 and mediate the uptake of the bound retinol. The RBP4-binding receptor STRA6 does so in the eye, but not in other tissues (Berry et al., 2013). Some epithelial cells, including those in epididymis, thyroid, parathyroid, and endometrium, express the facultive RBP receptor megalin. This member of the LDL receptor family delivers the RBP-bound retinol to the cell by endocytosis. Production and secretion of RBP4 by the liver increase during vitA deficiency.

About 0.19 μmol/l beta-carotene, 0.22 μmol/l beta-cryptoxanthin, and 0.04 μmol/l alpha-carotene circulate with plasma (Ruiz Rejon et al., 2002). Chylos carry much of the recently ingested carotenoids. They rapidly lose triglycerides, but not carotenoids, and the depleted Chylos remnants are taken up into hepatocytes and extrahepatic cells via diverse Lp receptors. LDLs and HDLs also carry some carotenoids and bring them into cells when they are taken up through their typical receptor-mediated endocytotic pathways.

Blood–brain barrier (BBB): STRA6-mediated uptake of the RBP4-retinol complex is important for the transfer of retinol into the brain (Amengual et al., 2014). The pathways for transfer of carotenoids from the blood into the brain are not well understood.

Materno-fetal transfer: Megalin, which is expressed in syntrophoblasts, binds and internalizes RBP (Christensen and Birn, 2001). How much this pathway contributes to vitA transfer across the placenta is not known. Another putative RBP receptor has been identified (Johansson et al., 1999). To some extent, retinol can also be esterified and stored in villous mesenchymal fibroblasts of the placenta (Sapin et al., 2000).

METABOLISM

Retinol and its provitamin A precursors are converted into active metabolites in tissue-specific patterns. Oxidation of retinoic acid is the only known inactivating catabolic pathway. Among the metabolites with functional importance are retinol itself, all-trans-retinal and 11-cis-retinal (vision), all-trans-retinoic acid, 9-cis-retinoic acid, 13-cis-retinoic acid, and 14-hydroxy-retro-retinol. It may be safely assumed that not all enzymes with significant vitA metabolic activity are known yet. The picture is additionally complicated by the extensive overlapping activity spectra of the various enzymes involved.

All-trans-retinoic acid synthesis: Conversion of retinol via retinal to retinoic acid requires two successive oxidation reactions (Figure 9.7). Among the widely expressed dehydrogenases suitable for

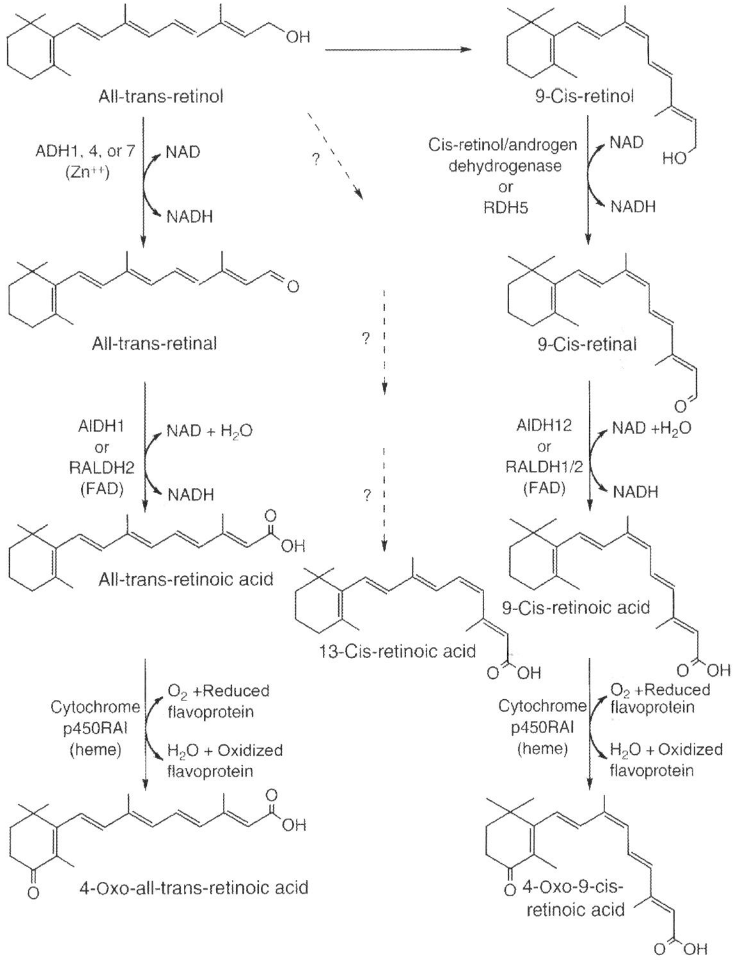

FIGURE 9.7

Retinol is the precursor of several retinoic acid isomers.

the first oxidation are alcohol dehydrogenases 1 (ADH1, EC1.1.1.1, contains zinc), 4 (ADH4), and 7 (ADH7, in the cytoplasma of epithelial cells of the stomach). Aldehyde dehydrogenase 1 (ALDH1) and retinal dehydrogenase 2 (RALDH2, EC1.2.1.36, contains FAD) usually complete retinoic acid synthesis (Duester, 2000). All-trans-retinoic acid is quantitatively the predominant isomer.

9-cis-retinoic acid: The liver, kidneys, small intestine, and other tissues produce the 9-cis isomer of retinoic acid that is functionally and metabolically distinct from the all-trans isomer (Zhuang et al., 2002). A specific enzyme responsible for conversion of all-trans-retinol to 9-cis-retinol is not known. It is possible that a slow isomerization reaction depends on prevalent nucleophilic metabolites such as dithiols, dihydroflavin, or dihydrofolate (Fan et al., 2003). The isoenzymes 1 and 3 of cis-retinol/androgen dehydrogenase (CRAD1 and CRAD3, EC1.1.1.105) and retinol dehydrogenase (RDH5,

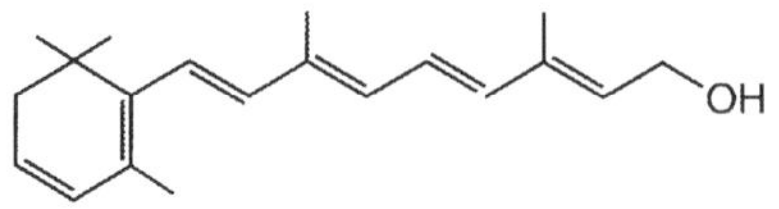

FIGURE 9.8

Potent retinol metabolite 3,4 didehydroretinoic acid is produced in skin.

EC1.1.1.315) oxidize 9-cis-retinol to 9 cis-retinal. Enzymes that accept this isomer as a substrate for further oxidation to 9-cis-retinoic acid include RALDH1, RALDH2, and ALDH12. A retinol dehydrogenase (cRDH, EC1.1.1.105) specifically oxidizes 9-cis-retinol to 9-cis-retinal, particularly early in fetal development (Gamble et al., 1999). Beta-carotene may also be converted directly into retinoic acid.

Retinoic acid breakdown: Cytochrome p450RAI-1 (CYP26A1) in the liver and many other tissues and p450RAI-2 (CYP26B1) in the brain oxidizes retinoic acid to 4-hydroxy retinoic acid, 4-oxo-retinoic acid, and 18-hydroxy retinoic acid, and thereby contributes to retinoic acid clearance (White et al., 2000). Microsomal NADPH-ferrihemoprotein reductase [NADPH-cytochrome P450 oxidoreductase, EC1.6.2.4, contains flavin mononucleotide (FMN) and FAD] is the flavoprotein that provides the electrons for this reaction. Both p450RAI-1 and p450RAI-2 prefer all-trans-retinoic acid, but also catabolize 9-cis-retinoic acid and 13-cis-retinoic acid, although with lower activity. The ethanol-inducible cytochrome CYP2E1 also inactivates retinoic acid (Liu et al., 2001).

3,4-didehydroretinol: This metabolite (vitamin A2) is produced in keratinocytes and other skin cells (Figure 9.8) through a poorly understood mechanism (Andersson et al., 1994). The conversion appears to be irreversible and controlled by all-trans-retinoic acid (Randolph and Simon, 1996). Further oxidation by one or several of the previously described retinal oxidizing enzymes can then generate the 3,4-didehydroretinoic acid, which is a potent metabolite (Sani et al., 1997). It may be lack of this derivative that causes the typical scaly skin lesions and the production of abnormally large keratins during vitA deficiency (Olson, 1994).

Hydroxylated retinoids: Most body tissues can add a hydroxyl group to retinol and generate metabolites with distinct biological and metabolic properties (Figure 9.9). Conversion of retinol to 14-hydroxy-4,14-retro-retinol (14-HRR), 13,14-dihydroxyretinol, anhydroretinol, and 4-hydro-5-hydroxy-anhydroretinol, but not to retinoic acid analogs, has been demonstrated in the liver (Mao et al., 2000). The enzymes and other specific proteins involved in most of these conversions remain to be identified, however. In 14-hydroxy-4, 14-HRR, and anhydroretinol, the double bonds of the retinoid side-chain are shifted toward the ring system. A phosphoadenosine 5′-phospho-sulfate (PAPS)–requiring retinol dehydratase (EC4.2.1.-) generates anhydroretinol from retinol in insects and fish but no corresponding human enzyme has yet been identified (Pakhomova et al., 2001).

STORAGE

Most (80%) of the body's vitA reserves (around 450 mg) of healthy vitA-replete adults are stored in the liver (500 µg/g wet tissue; Leo and Lieber, 1999). This amount can cover requirements for several months. The vitA in the liver resides mainly as retinyl palmitate in lipid globules of hepatocytes (10–20%) and of stellate (Ito) cells (80–90%). Two enzymes esterify free retinol—namely, retinol

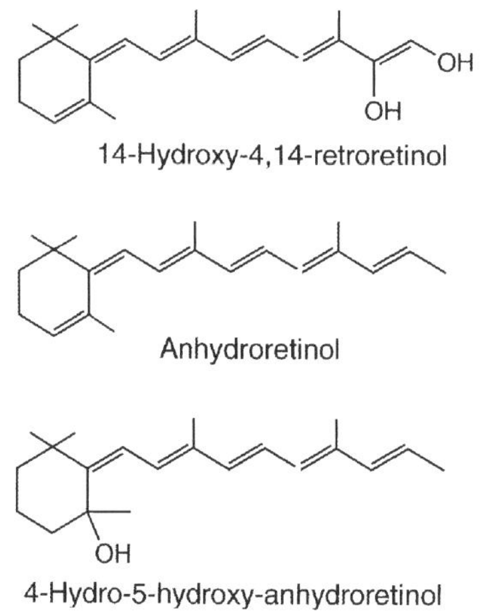

14-Hydroxy-4,14-retroretinol

Anhydroretinol

4-Hydro-5-hydroxy-anhydroretinol

FIGURE 9.9

Hydroxylated retinol metabolites.

O-fatty-acyltransferase (acyl CoA:retinol acyltransferase, ARAT, EC2.3.1.76) and phosphatidylcholine-retinol O-acyltransferase (lecithin:retinol acyltransferase, LRAT, EC2.3.1.135). The stored esters are released through the action of all-trans-retinyl-palmitate hydrolase (EC3.1.1.64) or 11-cis-retinyl-palmitate hydrolase (EC3.1.1.63). High alcohol intake can mobilize vitA and deplete stores (Leo and Lieber, 1999).

EXCRETION

Retinoic acid and other metabolites can be conjugated to glucuronide and excreted with bile. The relatively high efficiency of intestinal absorption for most retinoids minimizes losses, however, and maintains extensive enterohepatic cycling. Several milligrams of retinol are filtered in the kidneys, despite the considerable size of the RBP4-transthyretin complex to which retinol is attached. Megalin, a particularly large member of the LDL receptor family, specifically binds RBP4 and mediates its uptake by endocytosis. Recent evidence suggests that RBP4 and its retinol load can proceed through the epithelial cell and return intact into circulation (Marino et al., 2001).

REGULATION

The concentration of retinol in the blood is loosely maintained by moving excess into the liver and other tissues, and by release of retinol in the liver during times of need. The concentration of retinol-binding protein 4 (RBP4) influences uptake of retinol by the liver, adipocytes, and other tissues, including by modulating uptake of retinol-carrying Lps into liver cells (Yamaaki et al., 2013). Liver production of RBP4 increases during times of vitA deficiency, and this facilitates export of stored retinol from the liver.

The availability of retinoic acid in particular tissues is largely controlled by induction of producing (ALDH1 and RALDH2) and catabolic (cytochrome P450RAI) activities (White et al., 2000).

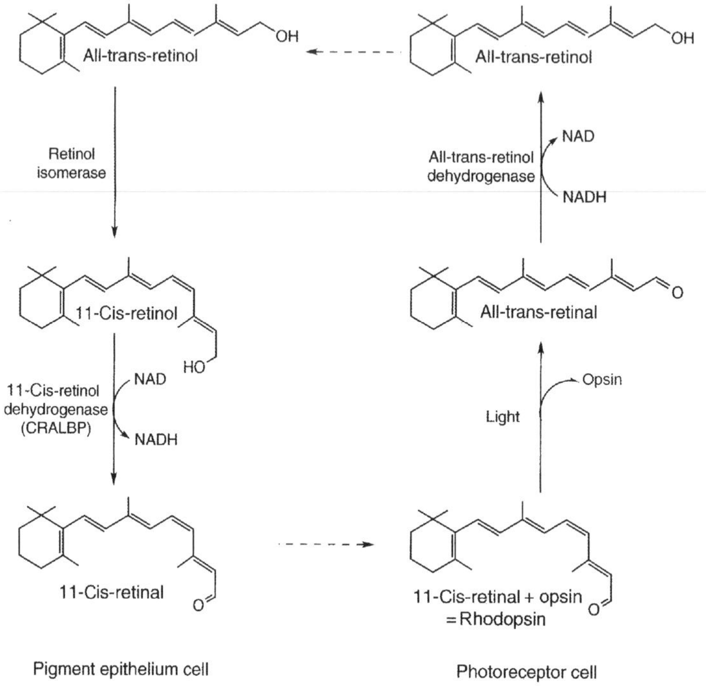

FIGURE 9.10

Vitamin A metabolism in the retina.

FUNCTION

Vision: The vitA metabolite 11-cis-retinal is a critical component of the light-detecting complex in the photoreceptor cells of the eye. Interaction with a single photon triggers the conversion of the 11-cis to the all-trans isoform, which leads to the release of retinal from the associated pigment protein (opsin) and the start of a signaling cascade (Figure 9.10).

Rapid reduction of the free all-trans-retinal by retinol dehydrogenase (all-trans-retinol dehydrogenase, EC1.1.1.105, NAD-dependent) prevents the reversion into the vision-active form (Saari et al., 1998). Four distinct 1 1-cis-retinal-containing rhodopsins (vision pigments) with specific properties enable humans to detect light at particular wavelengths. Rhodopsin in combination with 11-cis-retinal absorbs light with a maximum at 495 nm in retinal rods. This pigment gives dual-mode vision (black-and-white). Three pigments in the retinal cones, each consisting of a specific cone pigment and 1-cis retinal, provide color vision. Blue cone pigment absorbs best at 440 nm and gives blue vision. Green cone pigment detects green due to its absorption maximum at 535 nm, and red cone pigment detects red and yellow with a maximum at 560 nm.

The retinal pigment epithelium forms a layer between the photoreceptors and the capillary blood supply. The active metabolite, 11-cis-retinal, is produced from retinol by the successive action of retinol isomerase (EC5.2.1.7) and 11 cis-retinol dehydrogenase (RDH5, EC1.1.1.315; can use NAD as well as NADP). It has been suggested that cellular retinaldehyde binding protein (CRALBP) acts as

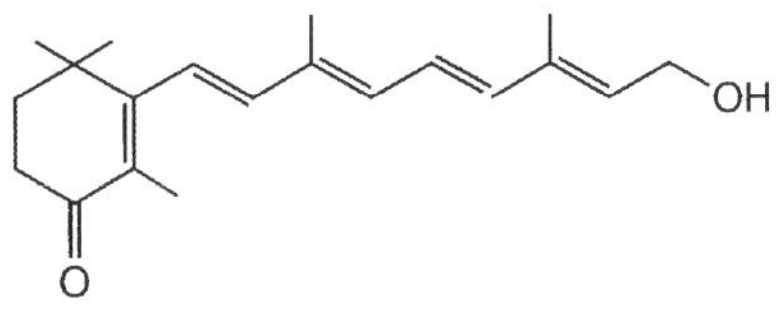

FIGURE 9.11

4-Oxoretinol.

an acceptor for the 11-cis-retinal intermediate (Saari et al., 2001). Inhibition of RDH5 might contribute to the transient blindness sometimes seen with excessive licorice consumption (Dobbins and Saul, 2000). All-trans-retinal can also be activated by retinal isomerase (EC5.2.1.3). The pigment epithelial cells also store limited amounts of vitA as retinylesters. For this purpose, phosphatidyl-choline-retinol O-acyltransferase (EC2.3.1.135) moves a fatty acid from phospholipid to the RBP1-bound retinol. The stored esters are released through the actions of all-trans-retinyl-palmitate hydrolase (EC3.1.1.64), 11-cis-retinylpalmitate hydrolase (EC3.1.1.63), or all-trans-retinylester isomerohydro-lase (EC3.1.1.64). RBP3 (interphotoreceptor retinoid-binding protein, IRBP) is a specific lipocalin in the interstitial space between pigment epithelium and photoreceptors.

Its importance may extend more to the survival of photoreceptor cells than to their supply of active 11-cis-retinal (Palczewski et al., 1999).

Nuclear actions: VitA is of vital importance for normal endodermal differentiation, morphogenesis, regulation of embryonic and childhood development, and sustaining balanced cell proliferation, differentiation, and apoptosis in adulthood. Many of the underlying events involve the binding of vitA-containing receptors to specific binding elements in nuclear DNA and control of the expression of associated genes. Two groups of retinoid receptors have been identified. The first includes the retinoic acid receptors (RARs) alpha, beta, and gamma. Their ligands are all-trans-retinoic acid, 9-cis-retinoic acid, 4-oxo-retinoic acid, S-4-oxo-9-cis-13,14-dihydro-retinoic acid, and a few other candidate retinoids. The second group comprises the retinoid X receptors alpha, beta, and gamma, with 9-cis-retinoic acid as the main activating ligand. The retinoid X receptor (RXR) group is particularly interesting since association with them enables the actions of numerous other nuclear receptors. The list of RXR-dependent nuclear binding proteins includes RAR, thyroid receptors (TRs), vitamin D receptor (VDR), peroxisome proliferation activating receptors (PPARs), pregnane X receptor (steroid and xenobiotic receptor, SXR/PXR), liver X receptors (LXRs), farnesoid X–activated receptors (FXRs), and benzoate X receptors (BXRs). The diversity of functions becomes evident just from the designations of these receptors. A compilation of published research data indicates more than 100 genes that are known or likely targets of retinoic acid–mediated action alone (Balmer and Blomhoff, 2002).

Cell cycle and apoptosis: In contrast to the generally growth-promoting properties of the major forms of vitA, anhydroretinol triggers nonclassical programmed cell death (Korichneva and Hammerling, 1999), possibly by inducing oxygen free radical production (Chen et al., 1999). Here, 4-oxoretinol (Figure 9.11) is able to induce growth arrest and promote differentiation of promyelocytes (Faria et al., 1998). Apocarotenoic acids, the metabolites of excentric beta-carotene cleavage products, inhibit tumor cell growth through mechanisms that are distinct from all-trans-retinoic acid (Tibaduiza et al., 2002).

Cell signaling: Retinol and some metabolites interact directly with phosphokinase C (PKC), a central element of the intracellular signaling cascade (Imam et al., 2001). They may do this through binding to specific sites and directing the functionally important oxidation of particular cysteines.

NUTRITIONAL ASSESSMENT

VitA deficiency and suboptimal status may be detected with the measurement of vitA metabolites in the blood or by measuring RBP.

Blood measurements: The concentrations of retinol and of RBP in plasma reflect current status reasonably. Measurements also work with dried blood spots (Fallah and Peighambardoust, 2012; Baingana et al., 2008).

More recent assays use stable isotope dilution methods with combined liquid chromatography-mass spectrometry to measure both active retinol and the pro-vitamin carotenoid precursors (Oxley et al., 2014).

REFERENCES

Amengual, J., Golczak, M., Palczewski, K., von Lintig, J., 2012. Lecithin:retinol acyltransferase is critical for cellular uptake of vitamin A from serum retinol-binding protein. J. Biol. Chem. 287, 24216–24227.

Amengual, J., Zhang, N., Kemerer, M., Maeda, T., Palczewski, K., Von Lintig, J., 2014. STRA6 is critical for cellular vitamin A uptake and homeostasis. Hum. Mol. Genet. 23, 5402–5417.

Andersson, E., Bjorklind, C., Torma, H., Vahlquist, A., 1994. The metabolism of vitamin A to 3,4-didehydroretinol can be demonstrated in human keratinocytes, melanoma cells and HeLa cells, and is correlated to cellular retinoid-binding protein expression. Biochim. Biophys. Acta 1224, 349–354.

Baingana, R.K., Matovu, D.K., Garrett, D., 2008. Application of retinol-binding protein enzyme immunoassay to dried blood spots to assess vitamin A deficiency in a population-based survey: the Uganda Demographic and Health Survey 2006. Food Nutr. Bull. 29, 297–305.

Balmer, J.E., Blomhoff, R., 2002. Gene expression regulation by retinoic acid. J. Lipid Res. 43, 1773–1808.

Barua, A.B., Olson, J.A., 2000. β-carotene is converted primarily to retinoids in rats *in vivo*. J. Nutr. 130, 1996–2001.

Berry, D.C., Jacobs, H., Marwarha, G., Gely-Pernot, A., O'Byrne, S.M., DeSantis, D., et al., 2013. The STRA6 receptor is essential for retinol-binding protein-induced insulin resistance but not for maintaining vitamin A homeostasis in tissues other than the eye. J. Biol. Chem. 288, 24528–24539.

Chen, Y., Buck, J., Derguini, F., 1999. Anhydroretinol induces oxidative stress and cell death. Cancer Res. 59, 3985–3990.

Christensen, E.I., Birn, H., 2001. Megalin and cubilin: synergistic endocytic receptors in renal proximal tubule. Am. J. Physiol. Renal. Fluid Electrolyte Physiol. 280, F562–F573.

Dmitrovskii, A.A., Gessler, N.N., Gomboeva, S.B., Ershov, Y.V., Bykhovsky, V.Y., 1997. Enzymatic oxidation of beta-apo-8′-carotenol to beta-apo-14′-carotenal by an enzyme different from beta-carotene-15,15′-dioxygenase. Biochemistry (Russia) 62, 787–792.

Dobbins, K.R., Saul, R.E., 2000. Transient visual loss after licorice ingestion. J. Neuroophthalmol. 20, 38–41.

Duester, G., 2000. Families of retinoid dehydrogenases regulating vitamin A function: production of visual pigment and retinoic acid. Eur. J. Biochem. 267, 4315–4324.

Duszka, C., Grolier, P., Azim, E.M., Alexandre-Gouabau, M.C., Borel, P., Azais-Braesco, V., 1996. Rat intestinal beta-carotene dioxygenase activity is located primarily in the cytosol of mature jejunal enterocytes. J. Nutr. 126, 2550–2556.

Edwards, A.J., Nguyen, C.H., You, C.S., Swanson, J.E., Emenhiser, C., Parker, R.S., 2002. Alpha- and beta-carotene from a commercial puree are more bioavailable to humans than from boiled mashed carrots, as determined using an extrinsic stable isotope reference method. J. Nutr. 132, 159–167.

Episkopou, V., Maeda, S., Nishiguchi, S., Shimada, K., Gaitanaris, G.A., Gottesman, M.E., et al., 1993. Disruption of the transthyretin gene results in mice with depressed levels of plasma retinol and thyroid hormone. Proc. Natl. Acad. Sci. 90, 2375–2379.

Fallah, E., Peighambardoust, S.H., 2012. Validation of the use of dried blood spot (DBS) method to assess vitamin A status. Health Promot. Perspect. 2, 180–189.

Fan, J., Rohrer, B., Moiseyev, G., Ma, J.X., Crouch, R.K., 2003. Isorhodopsin rather than rhodopsin mediates rod function in RPE65 knock-out mice. Proc. Natl. Acad. Sci. USA 100, 13662–13667.

Faria, T.N., Rivi, R., Derguini, F., Pandolfi, P.P., Gudas, L.J., 1998. 4-Oxoretinol, a metabolite of retinol in the human promyelocytic leukemia cell line NB4, induces cell growth arrest and granulocytic differentiation. Cancer Res. 58, 2007–2013.

Food and Nutrition Board, Institute of Medicine, 2002. Dietary Reference Intakes for Vitamin A, Vitamin K, Arsenic, Boron, Chromium, Copper, Iodine, Iron, Manganese, Molybdenum, Nickel, Silicon, Vanadium, and Zinc. National Academy Press, Washington, DC.

Gamble, M.V., Shang, E., Zott, R.P., Mertz, J.R., Wolgemuth, D.J., Blaner, W.S., 1999. Biochemical properties, tissue expression, and gene structure of a short chain dehydrogenase/reductase able to catalyze cis-retinol oxidation. J. Lipid Res. 40, 2279–2292.

Harrison, E.H., 2012. Mechanisms involved in the intestinal absorption of dietary vitamin A and provitamin A carotenoids. Biochim. Biophys. Acta 1821, 70–77.

Imam, A., Hoyos, B., Swenson, C., Levi, E., Chua, R., Viriya, E., et al., 2001. Retinoids as ligands and coactivators of protein kinase C alpha. FASEB J. 15, 28–30.

Jenab, M., Salvini, S., van Gils, C.H., et al., 2009. Dietary intakes of retinol, beta-carotene, vitamin D and vitamin E in the European Prospective Investigation into Cancer and Nutrition cohort. Eur. J. Clin. Nutr. 63 (Suppl. 4), S150–S178.

Johansson, S., Gustafson, A.L., Donovan, M., Eriksson, U., Dencker, L., 1999. Retinoid binding proteinsexpression patterns in the human placenta. Placenta 20, 459–465.

Kiefer, C., Hessel, S., Lampert, J.M., Vogt, K., Lederer, M.O., Breithaupt, D.E., et al., 2001. Identification and characterization of a mammalian enzyme catalyzing the asymmetric oxidative cleavage of provitamin A. J. Biol. Chem. 276, 14110–14116.

Korichneva, I., Hammerling, U., 1999. F-actin as a functional target for retro-retinoids: a potential role in anhydro-retinol-triggered cell death. J. Cell. Sci. 112, 2521–2528.

Leo, M.A., Lieber, C.S., 1999. Alcohol, vitamin A, and beta-carotene: adverse interactions, including hepatotoxicity and carcinogenicity. Am. J. Clin. Nutr. 69, 1071–1085.

Liu, C., Russell, R.M., Seitz, H.K., Wang, X.D., 2001. Ethanol enhances retinoic acid metabolism into polar metabolites in rat liver via induction of cytochrome P4502EI. Gastroenterology 120, 179–189.

Liu, C., Russell, R.M., Wang, X.D., 2004. Alpha-tocopherol and ascorbic acid decrease the production of beta-apo-carotenals and increase the formation of retinoids from beta-carotene in the lung tissues of cigarette smoke-exposed ferrets in vitro. J. Nutr. 134, 426–430.

Mao, G.E., Collins, M.E., Derguini, E., 2000. Teratogenicity, tissue distribution, and metabolism of the retro-retinoids, 14-hydroxy-4,14-retro-retinol and anhydroretinol, in the C57BL/6J mouse. Toxicol. Appl. Pharmacol. 163, 38–49.

Marino, M., Andrews, D., Brown, D., McCluskey, R.T., 2001. Transcytosis of retinol-binding protein across renal proximal tubule cells after megalin (gp330)-mediated endocytosis. J. Am. Soc. Nephrol. 12, 637–648.

Molotkov, A., Fan, X., Deltour, L., Foglio, M.H., Martras, S., Farres, J., et al., 2002. Stimulation of retinoic acid production and growth by ubiquitously expressed alcohol dehydrogenase Adh3. Proc. Natl. Acad. Sci. USA 99, 5337–5342.

Olmedilla, B., Granado, E., Blanco, I., Rojas-Hidalgo, E., 1994. Seasonal and sex-related variations in six serum carotenoids, retinol, and alpha-tocopherol. Am. J. Clin. Nutr. 60, 106–110.

Olson, J.A., 1994. Vitamin A, retinoids, and carotenoids. In: Shils, M.E., Olson, J.A., Shike, M. (Eds.), Modern Nutrition in Health and Disease Lea & Febiger, Philadelphia, PA, pp. 287–307.

Oxley, A., Berry, P., Taylor, G.A., Cowell, J., Hall, M.J., Hesketh, J., et al., 2014. An LC/MS/MS method for stable isotope dilution studies of β-carotene bioavailability, bioconversion, and vitamin A status in humans. J. Lipid Res. 55, 319–328.

Paik, J., During, A., Harrison, E.H., Mendelsohn, C.L., Lai, K., Blaner, W.S., 2001. Expression and characterization of a murine enzyme able to cleave beta-carotene. The formation of retinoids. J. Biol. Chem. 276, 32160–32168.

Pakhomova, S., Kobayashi, M., Buck, J., Newcomer, M.E., 2001. A helical lid converts a sulfotransferase to a dehydratase. Nat. Struct. Biol. 8, 447–451.

Palczewski, K., Van Hooser, J.P., Garwin, G.G., Chen, J., Liou, G.I., Saari, J.C., 1999. Kinetics of visual pigment regeneration in excised mouse eyes and in mice with a targeted disruption of the gene encoding interphotoreceptor retinoid-binding protein or arrestin. Biochemistry 38, 12012–12019.

Randolph, R.K., Simon, M., 1996. All-trans-retinoic acid regulates retinol and 3,4-didehydroretinol metabolism in cultured human epidermal keratinocytes. J. Invest. Dermatol. 106, 168–175.

Rock, C.L., 1997. Carotenoids: biology and treatment. Pharmacol. Ther. 75, 185–197.

Rock, C.L., Lovalvo, J.L., Emenhiser, C., Ruffin, M.T., Flatt, S.W., Schwartz, S.J., 1998. Bioavailability of beta-carotene is lower in raw than in processed carrots and spinach in women. J. Nutr. 128, 913–916.

Ruiz Rejon, F., Martin-Pena, G., Granado, F., Ruiz-Galiana, J., Blanco, I., Olmedilla, B., 2002. Plasma status of retinol, alpha- and gamma-tocopherols, and main carotenoids to first myocardial infarction: case control and follow-up study. Nutrition 18, 26–31.

Saari, J.C., Garwin, G.G., Van Hooser, J.E., Palczewski, K., 1998. Reduction of ail-trans-retinal limits regeneration of visual pigment in mice. Vision Res. 38, 1325–1333.

Saari, J.C., Nawrot, M., Kennedy, B.N., Garwin, G.G., Hurley, J.B., Huang, J., et al., 2001. Visual cycle impairment in cellular retinaldehyde binding protein (CRALBP) knockout mice results in delayed dark adaptation. Neuron 29, 739–748.

Sani, B.E., Venepally, P.R., Levin, A.A., 1997. Didehydroretinoic acid: retinoid receptor-mediated transcriptional activation and binding properties. Biochem. Pharmacol. 53, 1049–1053.

Sapin, V., Chaib, S., Blanchon, L., Alexandre-Gouabau, M.C., Lemery, D., Charbonne, E., et al., 2000. Esterification of vitamin A by the human placenta involves villous mesenchymal fibroblasts. Pediatr. Res. 48, 565–572.

Shete, V., Quadro, L., 2013. Mammalian metabolism of β-carotene: gaps in knowledge. Nutrients 5, 4849–4868.

Sivakumar, B., Reddy, V., 1972. Absorption of labelled vitamin A in children during infection. Br. J. Nutr. 27, 299–304.

Tibaduiza, E.C., Fleet, J.C., Russell, R.M., Krinsky, Nl, 2002. Excentric cleavage products of beta-carotene inhibit estrogen receptor positive and negative breast tumor cell growth in vitro and inhibit activator protein-l-mediated transcriptional activation. J. Nutr. 132, 1368–1375.

White, J.A., Ramshaw, H., Taimi, M., Stangle, W., Zhang, A., Everingham, S., et al., 2000. Identification of the human cytochrome P450, P450RAI-2, which is predominantly expressed in the adult cerebellum and is responsible for all-trans-retinoic acid metabolism. Proc. Natl. Acad. Sci. USA 97, 6403–6408.

Yamaaki, N., Yagi, K., Kobayashi, J., Nohara, A., Ito, N., Asano, A., et al., 2013. Impact of serum retinol-binding protein 4 levels on regulation of remnant-like particles triglyceride in type 2 diabetes mellitus. J. Diabetes Res. 2013, 143515.

Yeum, K.J., dos Anjos Ferreira, A.L., Smith, D., Krinsky, N.I., Russell, R.M., 2000. The effect of alpha-tocopherol on the oxidative cleavage of beta-carotene. Free Rad. Biol. Med. 29, 105–114.

Yeum, K.J., Lee-Kim, Y.C., Yoon, S., Lee, K.Y., Park, I.S., Lee, K.S., et al., 1995. Similar metabolites formed from beta-carotene by human gastric mucosal homogenates, lipoxygenase, or linoleic acid hydroperoxide. Arch. Biochem. Biophys. 321, 167–174.

Yeum, K.J., Russell, R.M., 2002. Carotenoid bioavailability and bioconversion. Annu. Rev. Nutr. 22, 483–504.
Zhuang, R., Lin, M., Napoli, J.L., 2002. cis-Retinol/androgen dehydrogenase, isozyme 3 (CRAD3): a short-chain dehydrogenase active in a reconstituted path of 9-cis-retinoic acid biosynthesis in intact cells. Biochemistry 41, 3477–3483.

VITAMIN D

The most common form of vitamin D (vitD) in foods is vitamin D_3 (9,10-seco(5Z,7E)-cholesta-5,7,10(19)-trien-3-ol, cholecalciferol, colecalciferol, oleovitamin D3; molecular weight 384). The less common form vitamin D_2 (ergocalciferol) is somewhat less effective (Figure 9.12).

ABBREVIATIONS

D2	vitamin D_2
D3	vitamin D_3
DBP	vitamin D–binding protein
25-D	25-hydroxyvitamin D
1,25-D	1α,25-dihydroxyvitamin D
24,25-D	24R,25-dihydroxyvitamin D
PTH	parathyroid hormone
UV-B	ultraviolet B light (290–315 nm)

NUTRITIONAL SUMMARY

Function: Promotes intestinal absorption of calcium and its retention in the body. Through its role in gene regulation the active form, 1,25-dihydroxyvitamin D (1,25-D), it influences the growth of bone and connective tissues and may protect against some forms of cancer.

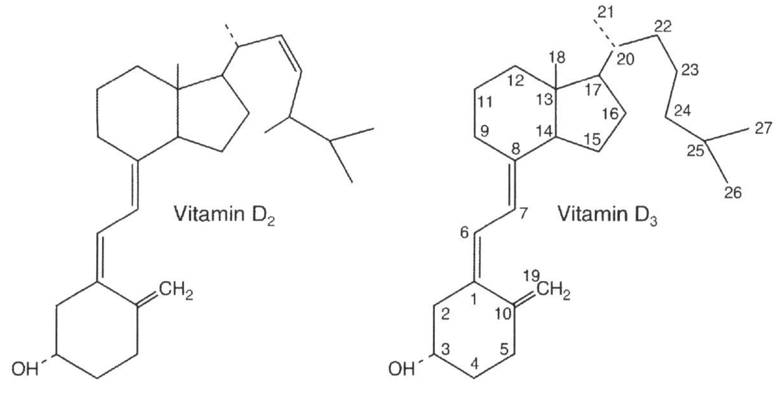

FIGURE 9.12

Dietary compounds with vitD activity.

Requirements: Adults should get at least 600 IU/day (15 µg/day), 800 IU/day (20 µg/day) with advanced age (Institute of Medicine (US) Committee to Review Dietary Reference Intakes for Vitamin D and Calcium, 2011).

Sources: Fatty fish and fortified milk are good dietary sources; eggs, fortified cereals, and fortified margarines contribute smaller amounts. A young person gets a full day's supplies from 10–15 min of face and arm exposure to summer sun (UV-B, 290–315 nm), while an older person needs several times greater exposure.

Deficiency: A severe lack during childhood causes rickets, characterized by bone deformities in lower limbs (bowlegs) and chest. Deficiency at a later age causes loss of bone minerals (osteoporosis), in the most severe cases of both minerals and connective tissue (osteomalacia). Tetany and severe bone pain are characteristic signs. People with suboptimal vitD status tend to have elevated parathyroid hormone levels and absorb dietary calcium less well. Indicative symptoms for mild vitD deficiency include fatigue, muscle ache, and diffuse bone pain (Nykjaer et al., 2001).

Excessive intake: Prolonged consumption of several hundred micrograms per day may cause hypercalcemia and soft tissue calcification. Continued exposure to doses of thousands of micrograms daily may cause coma and death in extreme cases.

ENDOGENOUS SOURCES

Exposure of skin to UV light with wavelengths between 290 and 315 nm (UV-B) converts some of the cholesterol precursor 7-dehydrocholesterol to previtamin D3, which rearranges spontaneously to vitamin D_3 (Holick et al., 1989). Suberythemal irradiation of skin with UV-B (0.5 J/cm; a dose that does not cause sunburn) was found to convert about one-third of endogenous 7-dehydrocholesterol (2.3 µg/cm^2) into previtamin D_3, and another third into the precursor lumisterol and the inactive metabolite tachysterol (Obi-Tabot et al., 2000). UV-B inactivates some of the newly generated vitD and its unstable precursors (Figure 9.13). VitD synthesis rapidly becomes maximal upon continued exposure when light-induced production and destruction of vitD reach equilibrium. It has been estimated that exposure of the entire body to summer sun for less than 20 min is sufficient to generate vitD in skin equivalent to an oral dose of 250 µg or more (Vieth, 1999).

Skin pigmentation decreases the effective light dose and greatly decreases vitD production with less than maximal sun exposure (Kreiter et al., 2000). The diminished vitD production in older people (75% decline by age 70) has been attributed in part to a lower concentration of unesterified 7 dehydrocholesterol in skin (Holick, 1999).

DIETARY SOURCES

Most natural vitD is consumed in the form of vitamin D_3 (D3, cholecalciferol) from animal-derived foods. The content of foods is expressed in micrograms or IUs (1 µg = 40 IU). The only natural foods that contain the structurally related vitamin D_2 (D2, ergocalciferol) are mushrooms. It is important to recognize that most D2 and D3 that are used for fortification or in dietary supplements come from the fortification of nonfood precursors.

Ocean fish is the main dietary source of D3. Particularly rich sources are the fatty types of fish, such as salmon (0.1–0.3 µg/g), sardines (0.4 µg/g), and mackerel (0.1 µg/g). Lean ocean fish, such as cod (0.01 µg/g), and freshwater fish, contain only a little vitD.

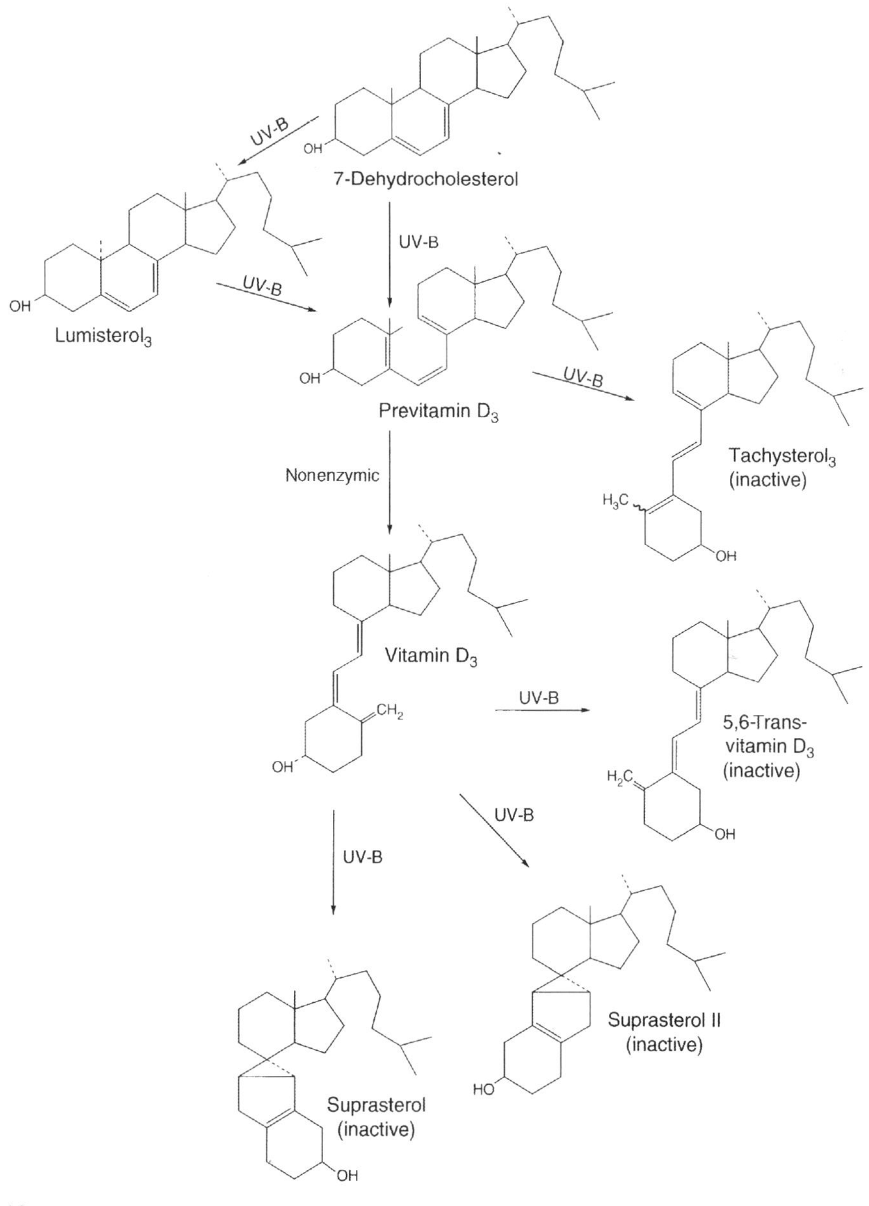

FIGURE 9.13

Light-induced synthesis of vitamin D₃.

Most milk in the United States is fortified at a level of 10 μg/l (400 IU/l). Considerable variation of the actual vitD content in milk has been observed in the past, however (Holick et al., 1992). Other dairy products, such as yogurt or cheese, are not usually fortified. D2, which is the compound originally used for fortification, has been replaced with D3 in the United States and many other countries. D2 can be produced relatively simply by UV light irradiation of lanosterol. D2 is biologically less active than D3 (Trang et al., 1998). The typical daily vitD intake in North American women may be as low as 2.5 μg (Krall et al., 1989), which usually is insufficient to prevent suboptimal vitD status in regions with low sunlight exposure (Vieth et al., 2001a,b). In Denmark, where milk does not contain added vitD, median daily intake around 3 μg in men, and around 2 μg in women were recorded (Osler and Heitmann, 1998). Even lower median intake (1.2 μg/day) was reported for Australians (Pasco et al., 2001). Older people in particular, who need much more vitD (at least 10 μg for ages 51–70, and 15 μg for people over 70) than younger people, commonly do not get enough (Kohlmeier et al., 1997).

Europeans also tend to have very low vitD intake. In a collection of representative surveys, men consumed 5.5 μg on average, and women consumed 3.6 μg (Jenab et al., 2009).

DIGESTION AND ABSORPTION

VitD is highly fat-soluble and becomes part of mixed micelles (consisting mainly of bile acids, phospholipids, fatty acids, and monoglycerides) during fat digestion. Nearly all of the ingested vitD is absorbed. The vitamin enters the small intestinal cell, along with fatty acids and other lipids, in an incompletely understood process. Chylos then carry vitD into lymph vessels and eventually into blood circulation. Little, if any, vitD is released while the Chylos circulate and rapidly lose most of their triglyceride load. The liver takes up about half of the triglyceride-depleted Chylos through a receptor-mediated process that involves apolipoprotein E and the LDL receptor (LDLR). Bone marrow and bone take up about 20%, and other extrahepatic tissues clear the remainder. VitD reaching the liver can be secreted again into circulation as a complex with DBP (group-specific component, Gc) (Figure 9.14).

TRANSPORT AND CELLULAR UPTAKE

Blood circulation: VitD and all its normal metabolites in the blood are bound to DBP. Almost all of the vitD in circulation is 25-hydroxyvitamin D (25-D); much smaller amounts are 1,25-dihydroxy–vitD (1,25-D). Typical 25-D concentrations in plasma of young adults living under sun-rich conditions are well in excess of 100 nmol/l (Vieth, 1999). Nonetheless, the lower limit of reference ranges is commonly set to 50 nmol/l or lower. Average 25-D concentrations of people living at latitudes of 50° or higher may be as low as 40 nmol/l during the winter months (Trang et al., 1998). Typical 1,25-D concentrations in vitD-replete people tend to be around 100 pmol/l. 25-D concentrations are low in vitD-deficient people, intermediate with adequate status, and increase further with excessive dietary intake (but not with very intense UV light exposure). 1,25-D concentrations also are low in vitD deficiency but do not increase further after adequate vitD intake is exceeded.

Thus, 25-D concentration in plasma is a good marker to reflect both inadequate and excessive vitD supplies.

It has been the traditional view that because of their high fat solubility vitamin metabolites can cross plasma membranes by simple diffusion. A more directed entry pathway may pertain in some tissues, however, such as endocytotic uptake mediated by cubilin and/or megalin (LDL receptor-related protein 2, LRP2).

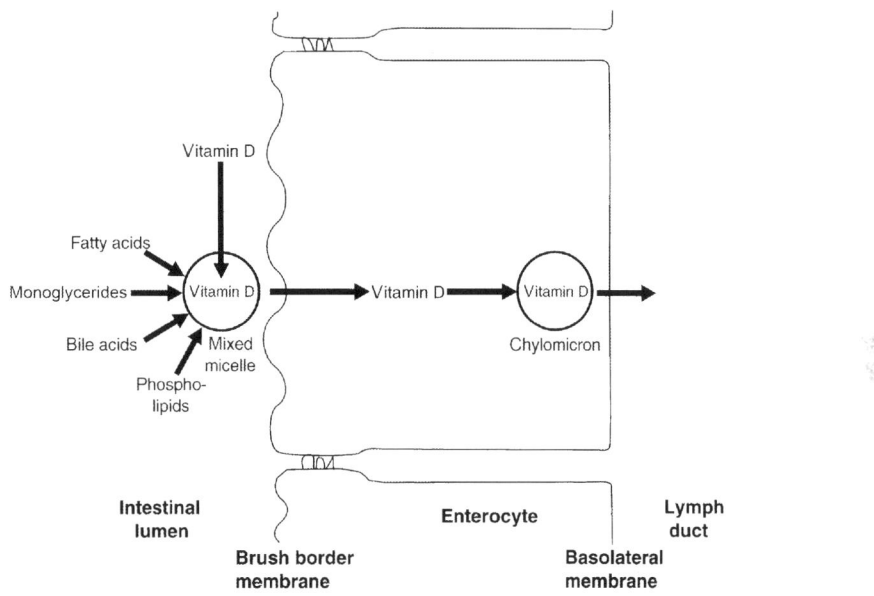

FIGURE 9.14

Intestinal absorption of vitD.

BBB: Transport of vitD metabolites from the blood into the brain is very limited (Pardridge et al., 1985). The underlying mechanisms are not well understood.

Materno-fetal transfer: 25-D is the main metabolite supplied by the mother to the fetus through incompletely understood mechanisms (Salle et al., 2000). Fetal concentrations are lower than on the maternal side. VitD is not only transferred to the fetus but also has important functions in the placenta itself. It is no surprise, therefore, that placenta expresses the vitD-activating enzyme 25-hydroxy-vitamin D(3)-1α-hydroxylase (Zehnder et al., 2001).

METABOLISM

VitD is metabolized extensively in the liver, intestines, and kidneys. More recent evidence shows that keratinocytes in skin are fully autonomous in respect to vitD metabolism and are capable of all major activating and inactivating reactions (Schuessler et al., 2001). Similarly, osteoblasts, parathyroid cells, myelocytes, and other cell types have relevant metabolic activity. Three best-recognized enzyme reactions produce two biologically active metabolites: 1α,25-dihydroxyvitamin D and 24R,25-dihydroxyvitamin D (Norman, 2006).

Microsomal vitamin D_3 25-hydroxylase (CYP2D25, identical with sterol-27 hydroxylase, CYP27A) in the liver and intestines catalyzes the essential first step in the bioactivation of the prohormone vitD (Theodoropoulos et al., 2001) (Figure 9.15). This reaction is so effective upon the first pass of newly absorbed vitD through the small intestine and liver that the blood contains very little unmetabolized vitD.

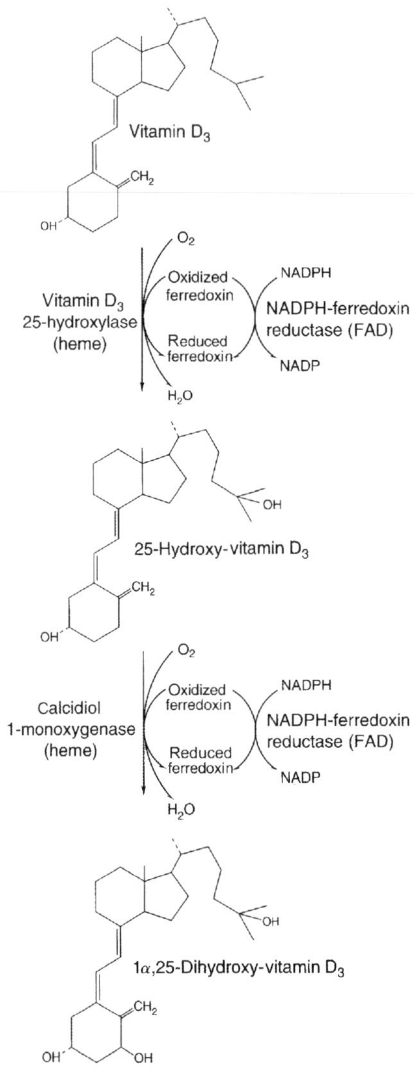

FIGURE 9.15

VitD activation depends on renal filtration and reabsorption.

Activation to $1\alpha,25$-dihydroxy-vitD (1,25-D) is completed in the mitochondria of proximal tubular epithelial cells in the kidneys by calcidiol 1-monooxygenase (25-hydroxyvitamin D–1α hydroxylase, CYP27B1, P450C1-α, EC1.14.13.13), which is a mitochondrial cytochrome P450 oxidase. Ferredoxin, which provides the reducing equivalents for all cytochrome P450 systems, is regenerated by ferredoxin-NADP reductase (EC 1.18.1,2, FAD-containing).

Before this hydroxylation can take place, however, the 25-D precursor has to reach the tubular cell. The main, and possibly exclusive, route is glomerular filtration of the 25-D/DBP complex and endocytosis mediated jointly by cubilin and megalin (Nykjaer et al., 2001). Decreased filtration rate (as with advancing age or in renal failure) or defective cubilin or megalin diminishes the production of 1,25-D. Some 25-D hydroxylation also occurs in extrarenal cells, including skin and white blood cells (Hewison et al., 2000; Schuessler et al., 2001). The glomerular filtration rate of a healthy man can be expected to fall from about 140 ml/min at age 20 to about 90 ml/min by age 70 (Lindeman, 1999). This age-typical decline in renal function raises the threshold plasma 25-D concentration that sustains adequate 1,25-D production by more than half. A several-fold increase in dietary intake or endogenous production is necessary to make up for this difference (Vieth, 1999). Current recommendations recognize that people over 70 years of age need three times more vitD than young adults (Institute of Medicine, 2011).

Another mitochondrial cytochrome P450 oxidase, 25-hydroxy-vitD-24Rhydroxylase (CYP24), converts 1,25-D into $1\alpha,24R,25$-trihydroxyvitamin D (24,25-D) (Figure 9.16). Alternatively, hydroxylation at carbon 23 may occur. Hydroxylations of the side chain, possibly with the involvement of CYP24 (Inouye and Sakaki, 2001), generate the water-soluble metabolite calcitroic acid (1α-hydroxy-24,25,26,27-tetranor-23-carboxyl–vitamin D).

The bulk of 25-D is catabolized through the C24 oxidation pathway with 24R,25-dihydroxyvitamin D_3 as an important intermediate metabolite (Henry, 2001), which may have its own specific biological activities (Figure 9.17). Numerous additional hydroxylated and otherwise modified metabolites are present in the blood and tissues. While it has been held that calcitroic acid and other metabolites are inactive, more recent investigations seem to indicate that they retain some typical vitD activity (Harant et al., 2000). Irreversible 3-epoxidation initiates a distinct metabolic pathway (Figure 9.18).

STORAGE

VitD is known to be stored extensively in the liver. These stores sustain normal vitD-dependent functions during the winter at high latitudes even in the absence of significant dietary intake. However, quantitative data on amounts stored, alternative storage sites, or the precise mechanisms for deposition and release are not available.

Smaller amounts of vitD are stored in extrahepatic tissues. The cartilage oligomeric matrix protein may provide a local storage mechanism that supports rapid delivery to nearby target structures (Guo et al., 1998).

EXCRETION

Calcitroic acid, the 3- and 24-glucuronides of 24,25-D, and additional vitD metabolites are excreted with bile. Since intestinal absorption of vitD is very efficient, losses of active vitD via this route are likely to be minor. Quantitative information in this regard is limited, however. Here, 25-D in plasma is bound to DBP (group-specific component, Gc), a single peptide chain with molecular weight of 52,000 (Witke et al., 1993). A small percentage of this complex is filtered in the renal glomeruli. DBP binds with high affinity to cubilin at the brush border membrane of the proximal renal tubule, as described previously. Megalin assists with the endocytosis and intracellular trafficking of cubilin and all its captured ligands (which include RBP and transferrin, among others). Due to the high efficiency of the process, very little of the filtered vitD escapes via urine. Calcitroic acid is a major catabolite of both vitamin D_2 and D_3 in urine (Zimmerman et al., 2001).

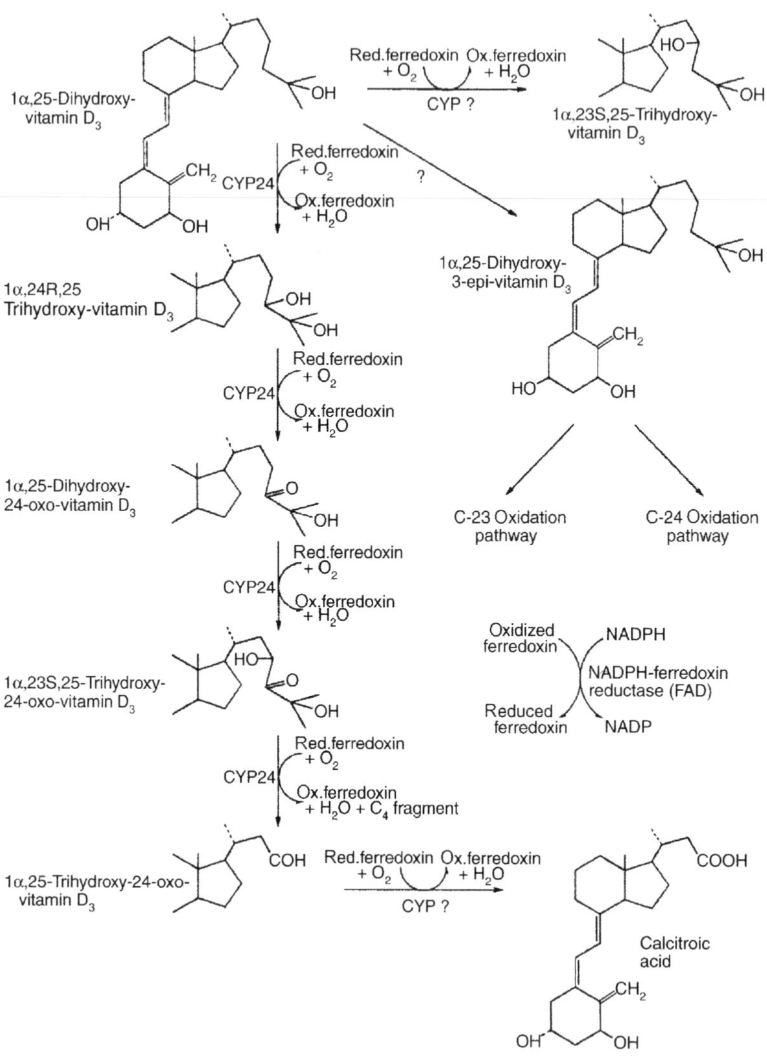

FIGURE 9.16

Catabolic pathways for 1-α,24R,25-trihydroxyvitamin D_3.

REGULATION

Feedback inhibition strongly limits 1,25-D production (Norman, 2006). Binding of 1,25-D to the VDR at the cytosolic side of the epithelial cells of proximal renal tubules causes the relocation of the complex to the nucleus where it decreases expression of 25-hydroxyvitamin D_3 1α-hydroxylase and increases expression of 25-hydroxyvitamin D_3 24-hydroxylase (Wang et al., 2015).

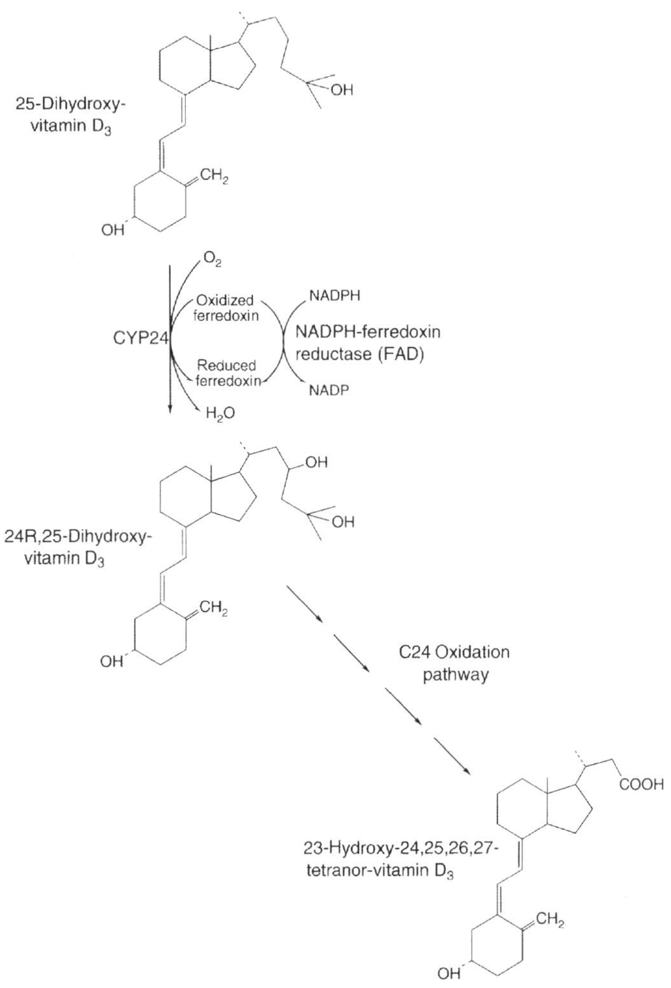

FIGURE 9.17

Catabolism of 25-hydroxyvitamin D.

The main hormonal activator of renal production of 1,25-D is PTH, which increases expression of 1α-hydroxylase (Theodoropoulos et al., 2001). Calcitonin, estrogen, prolactin, insulin, growth hormone, and glucocorticoids also activate this key enzyme. Conversely, 1,25-D decreases PTH secretion both directly by acting on a VDRE in the promoter of the PTH gene and indirectly through its effects on calcium and phosphate concentrations in the blood and extracellular fluid (Sela-Brown et al., 1999). PTH also promotes the conversion of 25-D to 24,25-D (Armbrecht et al., 1998) (Figure 9.19).

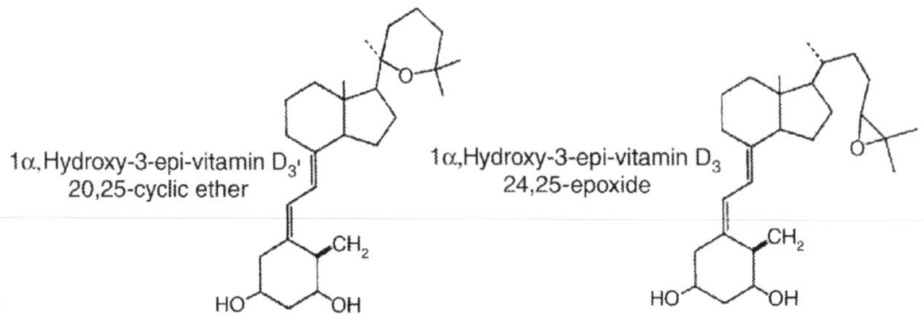

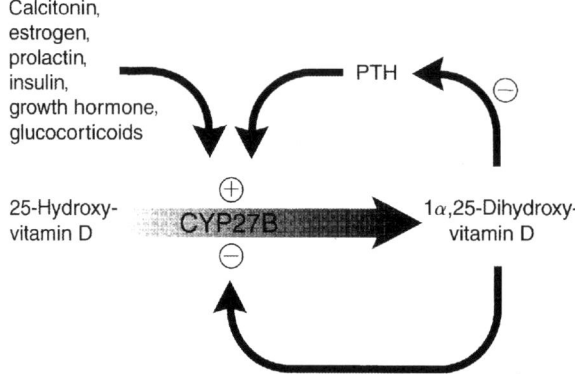

FIGURE 9.18

Metabolites derived from vitamin D_3–epi-intermediates.

FIGURE 9.19

Regulation of $1\alpha,25$-dihydroxyvitamin D synthesis.

FUNCTION

Nuclear effects: A complex consisting of 1,25-D, the nuclear VDR, and the RXR binds to specific vitamin D response elements (VDREs) in the nucleus and modifies the rate of expression of the associated genes. RXR must contain the vitamin A metabolite 9-cis-retinoic acid in order to form the active receptor complex. VDREs increase the expression of many genes, including osteocalcin, osteopontin, alkaline phosphatase, calbindin-9K, and calcium-transporting adenosine triphosphatease (ATPase). Down-regulated genes include those for many collagens, as well as for cell cycle regulators such as c-myc, c-fos, c-sis, and ubiquitin-conjugating enzyme 2 variant 2 (UBE2V2).

Intestinal calcium absorption: The overall consequence of improved vitD status in small intestinal cells is faster calcium influx from the lumen, more efficient transcellular calcium transfer, and accelerated calcium pumping toward the bloodstream. These effects all add up to considerably increased fractional calcium absorption. Phosphate absorption also increases slightly. Daily oral doses of 1,25-D between 0.5 and 3 µg are sufficient to increase intestinal calcium absorption. The effect of 25-D on intestinal calcium absorption is much smaller. It has been estimated that as much as one-eighth of the vitD-related absorption-enhancing action is due to 25-D (Heaney et al., 1997).

Fractional calcium absorption from small intestine increases much more rapidly (within seconds, sometimes referred to as *transcaltachia*) than could be explained by effects on protein expression. It has become apparent that these rapid changes are mediated by the classic 1,25-D receptor associated with caveolae in the plasma membrane (Norman, 2006). This configuration explains the rapid response within minutes—much faster than VDR interactions with nuclear DNA, which take hours.

Effects on bone: 1,25-D acts directly on hematopoietic stem cells and induces the generation of osteoclast cells. These migrate into bone, start breaking down bone matrix and minerals, and release calcium and phosphate. The various effects of 1,25-D on the osteoblast nucleus have been alluded to earlier in this chapter and are mainly mediated by the VDR/RXR complex. The increased expression of alkaline phosphatase, osteocalcin, and osteopontin supports the control of minerals released by osteoclast action and orderly redepositing of any excesses.

There is persuasive evidence that 24,25-D is needed for proper bone mineralization in addition to 1,25-D (Norman, 2006; van Leeuwen et al., 2001). Some of the actions of 24,25-D may be mediated through a plasma membrane receptor that is different from the nuclear VDR. The significance of 24,25-D remains in dispute, however, since information on this metabolite is much more limited than on 1,25-D.

Cell differentiation: An important effect of vitD is on growth and maturation of a wide range of cell types, including those in brown adipose tissue (Ricciardi et al., 2014). For example, 1,25-D regulates the differentiation of keratinocytes (Bikle et al., 2001; Lu et al., 2005). Here, 1,25-D also influences quiescence, modulates growth factors, promotes apoptosis of cancer cells, and slows metastasis in model systems. Suboptimal vitD status is likely to increase the risk of cancer of prostate (Polek and Weigel, 2002), breast (Colston and Hansen, 2002), and other sites. There are ongoing efforts to develop vitD analogs that retain their antiproliferative potency while minimizing hypercalcemic effects.

Other cellular effects: The plasma membrane receptor described previously (Norman, 2006) may mediate rapid effects of 1,25-D on various cells; 1,25-D stimulates, among others, the rapid calcium influx into osteoblasts and other cells through voltage-gated calcium channels, the release of calcium from stores in muscle cells, and the activation of mitogen-activated kinase. VitD metabolites act on VDR to limit mitochondrial respiration and this hormonal regulation is commonly defective in cancer cells (Ricciardi et al., 2014).

NUTRITIONAL ASSESSMENT

VitD deficiency and suboptimal status may be detected with the measurement of vitD metabolites in the blood.

Blood measurements: The concentration of active vitD forms in plasma reflects current status reasonably. Previous assays used antibodies for binding vitD, sometimes after an initial separation step with HPLC (Fraser and Milan, 2013). 1,25-dihydroxy vitamin D is usually measured with a separate immunoassay because the concentrations are several orders of magnitude lower (Clive et al., 2002).

More recently, combined chromatographic and mass spectrometric methods have been introduced. Measurement should resolve 25-hydroxyvitamin D_2 and 25-hydroxyvitamin D_3 (Bruce et al., 2013), 1,25-dihydroxy vitamin D, and 3-epi-25-hydroxyvitamin D_3 (van den Ouweland et al., 2014) in light of their very distinctive analytical and biological properties. These assays often also measure additional metabolites of importance.

REFERENCES

Armbrecht, H.J., Hodam, T.L., Boltz, M.A., Partridge, N.C., Brown, A.J., Kumar, V.B., 1998. Induction of the vitamin D 24-hydroxylase (CYP24) by 1,25-dihydroxyvitamin D3 is regulated by parathyroid hormone in UMR106 osteoblastic cells. Endocrinology 139, 3375–3381.

Bikle, D.D., Ng, D., Tu, C.L., Oda, Y., Xie, Z., 2001. Calcium- and vitamin D-regulated keratinocyte differentiation. Mol. Cell. Endocrinol. 177, 161–171.

Bruce, S.J., Rochat, B., Béguin, A., Pesse, B., Guessous, I., Boulat, O., et al., 2013. Analysis and quantification of vitamin D metabolites in serum by ultra-performance liquid chromatography coupled to tandem mass spectrometry and high-resolution mass spectrometry—a method comparison and validation. Rapid Commun. Mass Spectrom. 27, 200–206.

Clive, D.R., Sudhaker, D., Giacherio, D., Gupta, M., Schreiber, M.J., Sackrison, J.L., et al., 2002. Analytical and clinical validation of a radioimmunoassay for the measurement of 1,25 dihydroxy vitamin D. Clin. Biochem. 35, 517–521.

Colston, K.W., Hansen, C.M., 2002. Mechanisms implicated in the growth regulatory effects of vitamin D in breast cancer. Endocrin. Rel. Cancer 9, 45–59.

Fraser, W.D., Milan, A.M., 2013. Vitamin D assays: past and present debates, difficulties, and developments. Calcif. Tissue Int. 92, 118–127.

Guo, Y., Bozic, D., Malashkevich, V.N., Kammerer, R.A., Schulthess, T., Engel, J., 1998. All-trans retinol, vitamin D and other hydrophobic compounds bind in the axial pore of the five-stranded coiled-coil domain of cartilage oligomeric matrix protein. EMBO J. 17, 5265–5272.

Harant, H., Spinner, D., Reddy, G.S., Lindley, I.J., 2000. Natural metabolites of 1alpha,25-dihydroxyvitamin D(3) retain biologic activity mediated through the vitamin D receptor. J. Cell. Biochem. 78, 112–120.

Heaney, R.P., Barger-Lux, M.J., Dowell, M.S., Chen, T.C., Holick, M.E., 1997. Calcium absorptive effects of vitamin D and its major metabolites. J. Clin. Endocrinol. Metab. 82, 4111–4116.

Henry, H.L., 2001. The 25(OH)D3/1a,25(OH):D3-24R-hydroxylase: a catabolic or biosynthetic enzyme? Steroids 66, 391–398.

Hewison, M., Zehnder, D., Bland, R., Stewart, P.M., 2000. 1-alpha-Hydroxylase and the action of vitamin D. J. Mol. Endocrinol. 25, 141–148.

Holick, M.E., 1999. Vitamin D: photobiology, metabolism, mechanism of action, and clinical applications. In: Favus, M.J. (Ed.), Primer on the Metabolic Bone Diseases and Disorders of Mineral Metabolism, fourth ed. Lippincott Williams & Wilkins, Philadelphia, PA, pp. 92–98.

Holick, M.E., Matsuoka, L.Y., Wortsman, J., 1989. Age, vitamin D, and solar ultraviolet radiation. Lancet 4, 1104–1105.

Holick, M.E., Shao, Q., Liu, W.W., Chen, T.C., 1992. The vitamin D content of fortified milk and infant formula. N. Engl. J. Med. 326, 1178–1181.

Inouye, K., Sakaki, T., 2001. Enzymatic studies on the key enzymes of vitamin D metabolism; 1alpha-hydroxylase (CYP27BI) and 24-hydroxylase (CYP24). Biotechnol. Ann. Rev. 7, 179–194.

Institute of Medicine (US) Committee to Review Dietary Reference Intakes for Vitamin D and Calcium, 2011. In: Ross, A.C., Taylor, C.L., Yaktine, A.L., Del Valle, H.B. (Eds.), Dietary Reference Intakes for Calcium and Vitamin D National Academic Press (US), Washington, DC.

Jenab, M., Salvini, S., van Gils, C.H., et al., 2009. Dietary intakes of retinol, beta-carotene, vitamin D and vitamin E in the European Prospective Investigation into Cancer and Nutrition cohort. Eur. J. Clin. Nutr. 63 (Suppl. 4), S150–S178.

Kohlmeier, M., Garris, S., Anderson, J.J.B., 1997. Vitamin K: a vegetarian promoter of bone health. Veg. Nutr. 1, 53–57.

Krall, E.A., Sahyoun, N., Tannenbaum, S., Dallal, G.E., Dawson-Hughes, B., 1989. Effect of vitamin D intake on seasonal variations in parathyroid hormone secretion in postmenopausal women. N. Engl. J. Med. 321, 1777–1783.

Kreiter, S.R., Schwartz, R.E., Kirkman Jr, H.N., Charlton, P.A., Calikoglu, A.S., Davenport, M.L., 2000. Nutritional rickets in African American breast-fed infants. J. Pediatr. 137, 153–157.

Lindeman, R.D., 1999. The aging renal system. In: Chernoff, R. (Ed.), Geriatric Nutrition Aspen Publishers, Maryland, pp. 275–287.

Lu, J., Goldstein, K.M., Chen, P., Huang, S., Gelbert, L.M., Nagpal, S., 2005. Transcriptional profiling of kerati-nocytes reveals a vitamin D-regulated epidermal differentiation network. J. Invest. Dermatol. 124, 778–785.

Norman, A.W., 2006. Minireview: vitamin D receptor: new assignments for an already busy receptor. Endocrinology 147, 5542–5548.

Norman, P.E., Powell, J.T., 2014. Vitamin D and cardiovascular disease. Circ. Res. 114, 379–393.

Nykjaer, A., Fyfe, J.C., Kozyraki, R., Leheste, J.R., Jacobsen, C., Nielsen, M.S., et al., 2001. Cubilin dysfunction causes abnormal metabolism of the steroid hormone 25(OH) vitamin D(3). Proc. Natl. Acad. Sci. USA 98, 13895–13900.

Obi-Tabot, E.T., Tian, X.Q., Chen, T.C., Holick, M.E., 2000. A human skin equivalent model that mimics the pho-toproduction of vitamin D3 in human skin. In Vitro Cell. Dev. Biol. Anim. 36, 201–204.

Osler, M., Heitmann, B.L., 1998. Food patterns, flour fortification, and intakes of calcium and vitamin D: a longi-tudinal study of Danish adults. J. Epidemiol. Commun. Health 52, 161–165.

Pardridge, W.M., Sakiyama, R., Coty, W.A., 1985. Restricted transport of vitamin D and A derivatives through the rat blood–brain barrier. J. Neurochem. 44, 1138–1141.

Pasco, J.A., Henry, M.J., Nicholson, G.C., Sanders, K.M., Kotowicz, M.A., 2001. Vitamin D status of women in the Geelong Osteoporosis Study: association with diet and casual exposure to sunlight. Med. J. Aust. 175, 401–405.

Polek, T.C., Weigel, N.L., 2002. Vitamin D and prostate cancer. J. Androl. 23, 9–17.

Ricciardi, CJ, Bae, J, Esposito, D, Komarnytsky, S, Hu, P, Chen, J, et al., 2014. 1,25-Dihydroxyvitamin D3/vitamin D receptor suppresses brown adipocyte differentiation and mitochondrial respiration. Eur. J. Nutr. Oct 9. [Epub ahead of print].

Salle, B.L., Delvin, E.E., Lapillonne, A., Bishop, N.J., Glorieux, F.H., 2000. Perinatal metabolism of vitamin D. Am. J. Clin. Nutr. 71, 1317S–1324S.

Schuessler, M., Astecker, N., Herzig, G., Vorisek, G., Schuster, I., 2001. Skin is an autonomous organ in synthesis, two-step activation and degradation of vitamin D(3): CYP27 in epidermis completes the set of essential vita-min D(3)-hydroxylases. Steroids 66, 399–408.

Sela-Brown, A., Naveh-Many, T., Silver, J., 1999. Transcriptional and post-transcriptional regulation of PTH gene expression by vitamin D, calcium and phosphate. Min. Electrolyte Metab. 25, 342–344.

Theodoropoulos, C., Demers, C., Mirshahi, A., Gascon-Barre, M., 2001. 1,25-Dihydroxyvitamin D(3) down-regulates the rat intestinal vitamin D(3)-25-hydroxylase CYP27A. Am. J. Physiol. Endocrinol. Metab. 281, E315–E325.

Trang, H.M., Cole, D.E., Rubin, L.A., Pierratos, A., Siu, S., Vieth, R., 1998. Evidence that vitamin D3 increases serum 25-hydroxyvitamin D more efficiently than does vitamin D2. Am. J. Clin. Nutr. 68, 854–858.

van den Ouweland, J.M., Beijers, A.M., van Daal, H., 2014. Overestimation of 25-hydroxyvitamin D3 by increased ionisation efficiency of 3-epi-25-hydroxyvitamin D3 in LC–MS/MS methods not separating both metabolites as determined by an LC–MS/MS method for separate quantification of 25-hydroxyvitamin D3, 3-epi-25-hydroxyvitamin D3 and 25-hydroxyvitamin D2 in human serum. J. Chromatogr. B Analyt. Technol. Biomed. Life Sci. 967, 195–202.

van Leeuwen, J.R., van den Bemd, G.J., van Driel, M., Buurman, C.J., Pols, H.A., 2001. 24,25-Dihydroxyvitamin D(3) and bone metabolism. Steroids 66, 375–380.

Vieth, R., 1999. Vitamin D supplementation, 25-hydroxyvitamin D concentrations, and safety. Am. J. Clin. Nutr. 69, 842–856.

Vieth, R., Chan, P.C.R., MacFarlane, G.D., 2001a. Efficacy and safety of vitamin D3 intake exceeding the lowest observed adverse effect level. Am. J. Clin. Nutr. 73, 288–294.

Vieth, R., Cole, D.E., Hawker, G.A., Trang, H.M., Rubin, L.A., 2001b. Wintertime vitamin D insufficiency is common in young Canadian women, and their vitamin D intake does not prevent it. Eur. J. Clin. Nutr. 55, 1091–1097.

Wang, Y., Zhu, J., DeLuca, H.F., 2015. The vitamin D receptor in the proximal renal tubule is a key regulator of serum 1α,25-dihydroxyvitamin D3. Am. J. Physiol. Endocrinol. Metab. 308, E201–5.

Witke, F.W., Gibbs, P.E.M., Zielinski, R., Yang, E., Bowman, B.H., Dugaiczyk, A., 1993. Complete structure of the human Gc gene: differences and similarities between members of the albumin gene family. Genomics 16, 751–754.

Zehnder, D., Bland, R., Williams, M.C., McNinch, R.W., Howie, A.J., Stewart, P.M., et al., 2001. Extrarenal expression of 25-hydroxyvitamin d(3)-1 alpha-hydroxylase. J. Clin. Endocrinol. Metab. 86, 888–894.

Zimmerman, D.R., Reinhardt, T.A., Kremer, R., Beitz, D.C., Reddy, G.S., Horst, R.L., 2001. Calcitroic acid is a major catabolic metabolite in the metabolism of 1alpha-dihydroxyvitamin D(2). Arch. Biochem. Biophys. 392, 14–22.

VITAMIN E

Vitamin E (RRR-alpha-tocopherol, 3,4-dihydro-2,5,7,8-tetramethyl-2-(4,8,12-trimethyltridecyl)-2H-1-benzopyran-6-ol, 2,5,7,8-tetramethyl-2-(4′,8′, 12′-trimethyltrdecyl) 6-chromanol, 5,7,8-trimethyl-tocol; obsolete name antisterility vitamin; molecular weight 430); there are seven stereoisomers of alpha-tocopherol with lower activity.

The related compounds beta-, gamma-, and delta-tocopherol, and tocotrienols, as well as others, have more limited activity (Figure 9.20).

ABBREVIATIONS

HDL	high-density lipoprotein
LDL	low-density lipoprotein
TAP	tocopherol-associated protein
TBP	tocopherol-binding protein
TTP	alpha-tocopherol transfer protein
vitE	vitamin E (all forms)
VLDL	very-low-density lipoprotein

NUTRITIONAL SUMMARY

Function: Vitamin E (vitE) is a fat-soluble antioxidant that inactivates oxygen free radicals in membranes, Lps, and other lipid-rich compartments. A specific role in maintaining fertility may also exist.

Requirements: Adults should get at least 15 mg/day. Pregnancy, breast-feeding, and high intake of polyunsaturated fatty acids slightly increase the need for vitE. Adequate amounts of ascorbate are essential for reactivation.

Sources: Wheat germ and sunflower oils, as well as nuts, are good sources. Other plant oils, fruits, and vegetables provide smaller amounts.

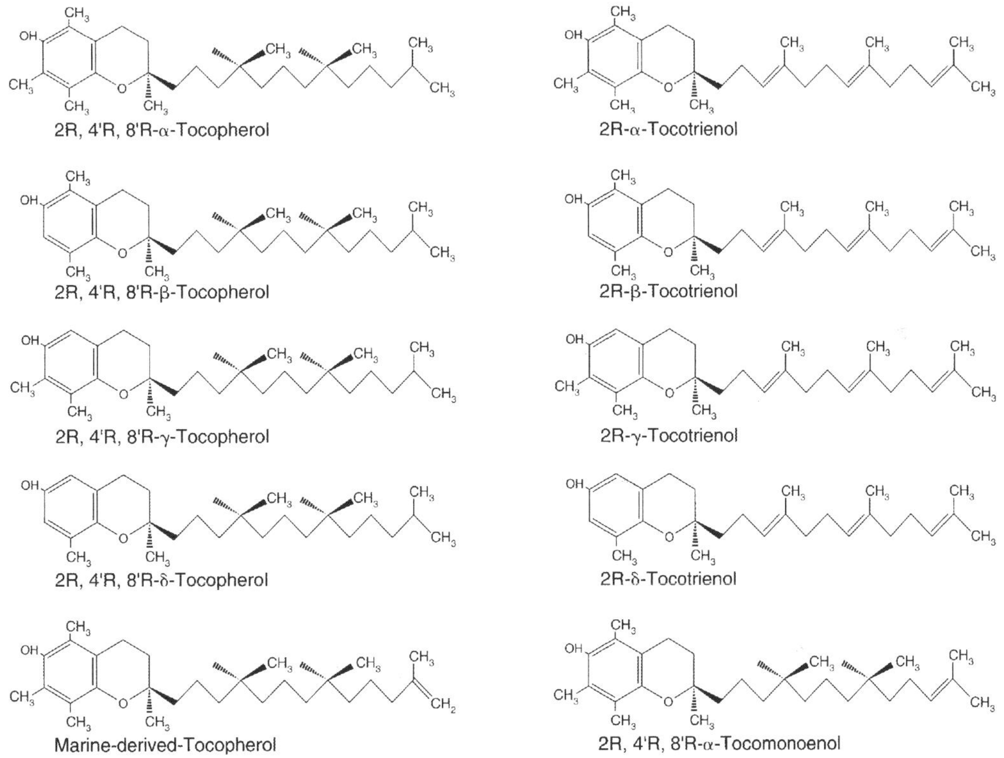

FIGURE 9.20

Naturally occurring forms of vitE.

Deficiency: There is limited evidence that inadequate intake promotes the progression of atherosclerosis, Parkinson's and Alzheimer's diseases, cancer, and cataracts, and impairs immune function.

Excessive intake: High doses (more than 1000 mg of any supplement form) can interfere with blood clotting and thus increase risk of hemorrhagic stroke. This adds to anticlotting effects of coumadins and salicylates.

DIETARY SOURCES

The most confusing aspect of vitE nutrition may come from the great diversity of compounds with some kind of related activity. So far, 12 compounds with characteristic vitE activity have been identified: alpha-tocopherol, beta-tocopherol, gamma-tocopherol, delta-tocopherol, alpha-tocotrienol, beta-tocotrienol, gammatocotrienol, delta-tocotrienol, the tocotrienols d-P(21)-T3 and d-P(25)-T3 (Qureshi et al., 2001), alpha-tocomonoenol, and marine-derived tocopherol (Yamamoto et al., 2001).

The food constituent most closely identified with vitE properties is RRR-alpha-tocopherol with methyl groups in the side chain at positions 5, 7, and 8. The side chain of RRR-beta-tocopherol is methylated at positions 5 and 8, and in RRR-gamma-tocopherol at positions 7 and 8. RRR-delta-tocopherol

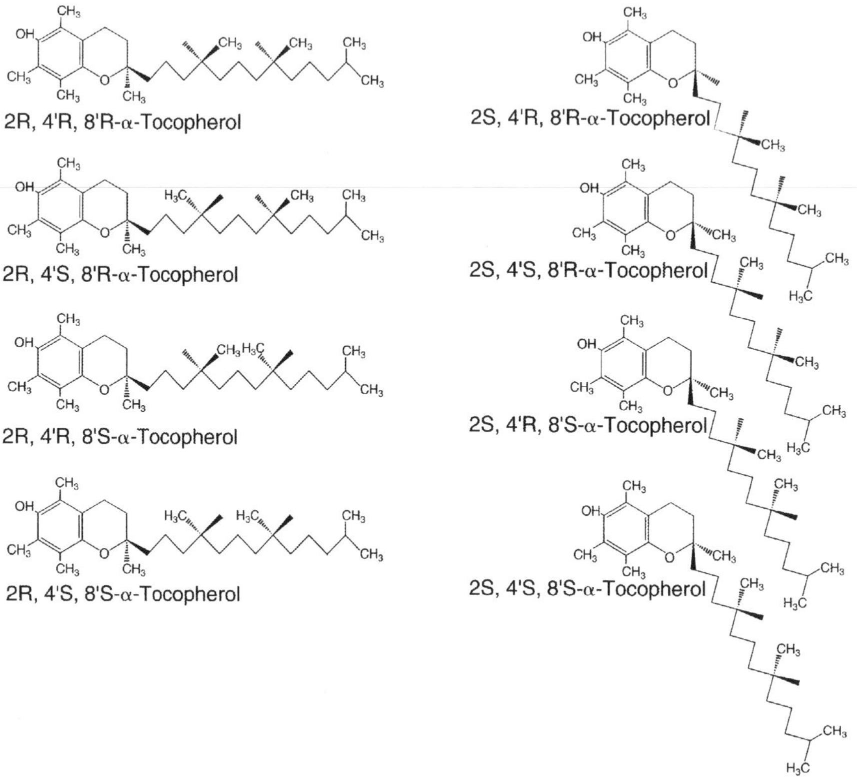

FIGURE 9.21

Synthetic all-rac alpha-tocopherol contains about equal amounts of eight isomers.

has only one methyl group in the side chain at position 8. The members of the analogous series of tocotrienols contain three double bonds in the side chain. All four members of the tocopherol series and the four members of the tocotrienol series are naturally present, although in varying amounts, in a wide range of foods.

Synthetic production of vitE usually yields about equal amounts of the eight possible isomers: RRR, RSS, RRS, RSR, SRR, SRS, SSR, and SSS (Figure 9.21). The first three in this list of isomers are often called the *2R isomers* because they are R-isomeric at position 2. The metabolic fate of the various isoforms differs and needs to be determined in every case.

The biological potency of vitE doses is often expressed as USP (US Pharmacopeia) vitE units or as IUs. One such unit corresponds to 1.0 mg of racemic (synthetic, all-rac) alpha-tocopheryl acetate (this is the original reference standard), or 1.1 mg of all-rac alpha-tocopherol, 1.36 mg RRR-alpha-tocopheryl acetate, 1.49 mg RRR-alpha-tocopherol, 0.89 mg all-rac-alpha-tocopheryl succinate, or 1.21 mg RRR-alpha-tocopheryl succinate. In the following discussion, vitE content will be expressed as an alpha-tocopherol equivalent (ATE), which is the amount of RRR alpha-tocopherol that is expected to have the same potency as all vitE forms in a food combined.

There are only a few good sources that provide one serving with at least 2.5 mg (one-sixth of the recommended intake). Since other tocopherols may differ in their action profiles from RRR-alpha-tocopherol, the exact composition (which is often not reliably known) may be as important as the ATE figure. Vegetable oils with high to moderate content include wheat germ oil (1.9 mg ATE/g, more than half in the form of alpha-tocopherol) and sunflower oil (0.5 mg/g, most in the form of alpha-tocopherol). Most other commonly consumed oils have a much lower content, such as corn and soybean oil (0.2 mg/g, most as gamma-tocopherol), canola oil (0.2 mg/g, most as alpha-tocopherol), or olive oil (0.1 mg/g, most as alpha-tocopherol). Sunflower seeds (0.5 mg/g) are also a good source.

Walnuts (0.03 mg ATE/g) contain nearly equal amounts of alpha-tocopherol, gamma-tocopherol, and delta-tocopherol.

American men have a daily vitE intake of about 8 mg; women get close to 6 mg (Phillips et al., 2000). American food consumption data indicate that only about 10% of men and virtually none of the women reach the recommended intake level (15 mg/day) with food alone (Food and Nutrition Board, Institute of Medicine, 2000, Appendix D).

Europeans tend to have higher intake levels. In a collection of representative surveys, men in Europe consumed 14.2 mg on average, and women 10.6 mg (Jenab et al., 2009).

DIGESTION AND ABSORPTION

Most forms of vitE are absorbed with similarly high efficiency from the proximal small intestine. VitE esters are cleaved by esterases from pancreas.

Uptake into enterocytes depends on the prior incorporation of free vitE into mixed micelles (Borel et al., 2001). This means that a small amount of fat has to be absorbed along with vitE. A very-low-fat meal or poor fat digestion effectively minimizes vitE absorption. The transfer of vitE from micelles to enterocyte depends on most of the actors involved in fatty acid uptake, including CD36 (Goncalves et al., 2014), scavenger receptor B1, Niemann–Pick C1-like protein 1, microsomal triglyceride transfer protein, apolipoprotein E (apoE), and apolipoprotein B (apoB). VitE is then extruded with Chylos into intestinal lymph and eventually reaches blood circulation.

TRANSPORT AND CELLULAR UPTAKE

Blood circulation: About 27 µmol/l alpha-tocopherol, 1.7 µmol/l gamma-tocopherol (Ruiz Rejon et al., 2002), and much smaller quantities of other vitE species are present in plasma. Virtually all vitE in plasma is associated with LDL (about 6 molecules of alpha-tocopherol and 0.5 gamma-tocopherol molecules per particle in vitE-replete people; Esterbauer et al., 1992) and HDL.

Liver cells take up chylomicron remnants with recently absorbed vitE via receptor-mediated endocytosis. The highly lipophilic vitE then redistributes rapidly to the plasma membrane and intracellular membranes. VitE, like all other plasma membrane constituents, is internalized into endocytic compartments (sorting endosomes) several times per hour (Hao and Maxfield, 2000). It seems that at this point, alpha-tocopherol transfer protein (TTP) plays a critical role by preferentially redirecting RRR-alpha-tocopherol back toward newly forming plasma and other membranes (Blatt et al., 2001) (Figure 9.22). Two additional proteins are likely to be involved in directing the intracellular recycling of vitE: tocopherol-associated protein (TAP) and tocopherol-binding protein (TBP). Due to their much lower affinity to TTP, a smaller percentage of 2R-alpha-tocopherols and RRR-beta-tocopherol is recycled to

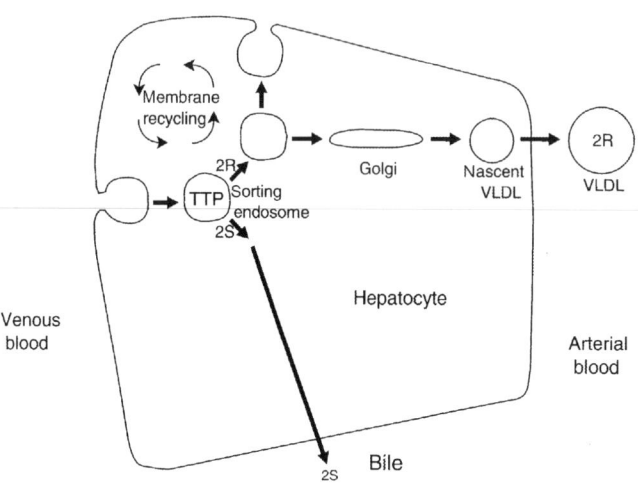

FIGURE 9.22

Alpha-tocopherol transfer protein preferentially recovers RRR-alpha-tocopherol from endosomes.

the membranes, as is an even smaller percentage of other forms of vitE. The portion left behind tends to be eliminated into bile. Newly secreted very low-density lipoproteins (VLDLs) preferentially carry RRR-alpha-tocopherol, with a lower preference for 2R-alpha-tocopherol and beta-tocopherol, and very little of the other forms. VitE may come mostly from plasma and endosomal membrane, hence the preference for the RRR-alpha-form (Blatt et al., 2001). A typical nascent VLDL contains about six molecules of vitE (Mertens and Holvoet, 2001), but only every other particle contains a gamma-tocopherol molecule (Esterbauer et al., 1992). Delivery of vitE to some tissues (e.g., lung, ovaries, and testes) involves the scavenger receptor class B, type 1 (SR-BI; Mardones et al., 2002).

Some vitE from extrahepatic tissues appears to be mobilized by the adenosine triphosphate (ATP)–dependent transporter ABCA1 (Oram et al., 2001), shuttled by phospholipid transfer protein (PLTP) to HDL (Huuskonen et al., 2001) and returned via this vehicle of reverse cholesterol transport to the liver (Mardones et al., 2002).

BBB: Very little vitE reaches the brain, which keeps the concentration in cerebrospinal fluid to about one-hundredth of the concentration in plasma (Pappert et al., 1996). The mechanisms whereby even those low concentrations are maintained remain uncertain. HDL is five times more effective than LDL in delivering tocopherol to the BBB (Goti et al., 2001). The transfer of vitE from HDL into the capillary epithelial cells uses the SR-BI (Mardones et al., 2002). TTP in glial cells helps to direct vitE into compartments that ultimately supply it to Purkinje cells and other neurons (Blatt et al., 2001).

Materno-fetal transfer: Understanding of the exact mechanism of vitE transport across the syntrophoblast is still incomplete. HDLs (Christiansen-Weber et al., 2000) and other Lps are known to deliver lipophilic compounds through receptor-mediated endocytosis. TTP plays an important role, possibly through a sequence of events as in the liver (redirecting of vitE in sorting endosomes).

METABOLISM

Side-chain breakdown: A combination of microsomal to-oxidation, peroxisomal beta-oxidation and less well characterized reactions generate a series of hydroxy-chroman metabolites (Parker et al., 2000; Birringer et al., 2001) (Figure 9.23). TTP in the liver shields RRR-alpha-tocopherol to a large extent from this type of breakdown. Cytochrome p450 3A4 and/or CYP4F2 initiates breakdown by oxidizing the methyl group at the terminal end of the side chain (w-oxidation). After another oxidation step, the molecule has a carboxyl group at the end of the side chain. Successive rounds of beta-oxidation can then shorten the side chain. Since the methyl groups of the side chain are in configurations relative to the carboxyl group, they do not block the initial step of beta-oxidation, catalyzed by long-chain acyl-CoA dehydrogenase (EC1.3.99.13). The final metabolite is the 2(2'-carboxyethyl)-6-hydroxychroman (CEHC) of the original vitE compound. Since the ring system is not affected, distinctive CEHCs result from side-chain metabolism. This means that all isomers of alpha-tocopherol and alphatocotrienol result in alpha-CEHC, the beta-tocopherols and beta-tocotrienols generate beta-CEHC, and so on (Lodge et al., 2001).

Ring modification: Reaction with two-electron oxidants, such as hypochlorous acid (from the myeloperoxidase-generated oxidative burst of leukocytes) and peroxynitrite (a reactive nitrogen species formed from the reaction of superoxide with nitric oxide) generates tocopheryl quinone in one step, without the chance of regeneration (Terentis et al., 2002) (Figure 9.24).

Reaction of vitE with one-electron oxidants, such as superoxide anions, produces a tocopheroxyl radical, which may be converted into tocopheryl quinone by another reaction with an oxidant. Reaction of a lipid peroxyl radical with the tocopheroxyl radical converts this to 5,6-epoxy-tocopherol or 2,3-epoxy-tocopherol. NAD(P)H:quinone oxidoreductase 1 (NQO1, EC1.6.5.2) reverts some of the tocopheroxyl radicals to tocopherol (Ross et al., 2000). The use of dietary supplements with large amounts of vitE (>400 IU) impairs the reactivation of vitamin K epoxide, and through this mechanism, it can trigger a hemorrhagic event, especially in patients treated with warfarin-type anticoagulants (Pastori et al., 2013).

More important may be the reduction of tocopheryl quinone to tocopheryl hydroquinone by this enzyme because the reduced metabolite has antioxidant activity. It should be noted that quercetin, the most abundant flavonoid in food, induces NAD(P)H:quinone oxidoreductase 1 (Valerio et al., 2001).

EXCRETION

Intact vitE and some of its apolar metabolites are excreted with bile. Various polar metabolites appear in urine. The scavenger receptor class B, type I (SR-BI) plays an important role in the transfer of vitE from HDL through the liver cells into bile (Mardones et al., 2002). This pathway is in some ways analogous to the reverse cholesterol transport from peripheral tissues into bile. The final step of translocation into the bile canaliculi uses the multidrug resistance P-glycoprotein 2 (MDR2, ABCB1; Mustacich et al., 1998). Biliary excretion of unchanged alpha-tocopherol increases when excess amounts of vitE are present.

The four different 2(2'-carboxyethyl)-6-hydroxychromans (CEHCs) from alpha-, beta-, gamma-, and delta-tocopherols and tocotrienols are excreted with urine (Lodge et al., 2001). It may be of note that all isomers of alpha-tocopherol generate the same alpha-CEHC and provide no indication whether RRR-alpha-tocopherol or all-rac-alpha-tocopherol has been ingested. The CEHC catabolites are mostly

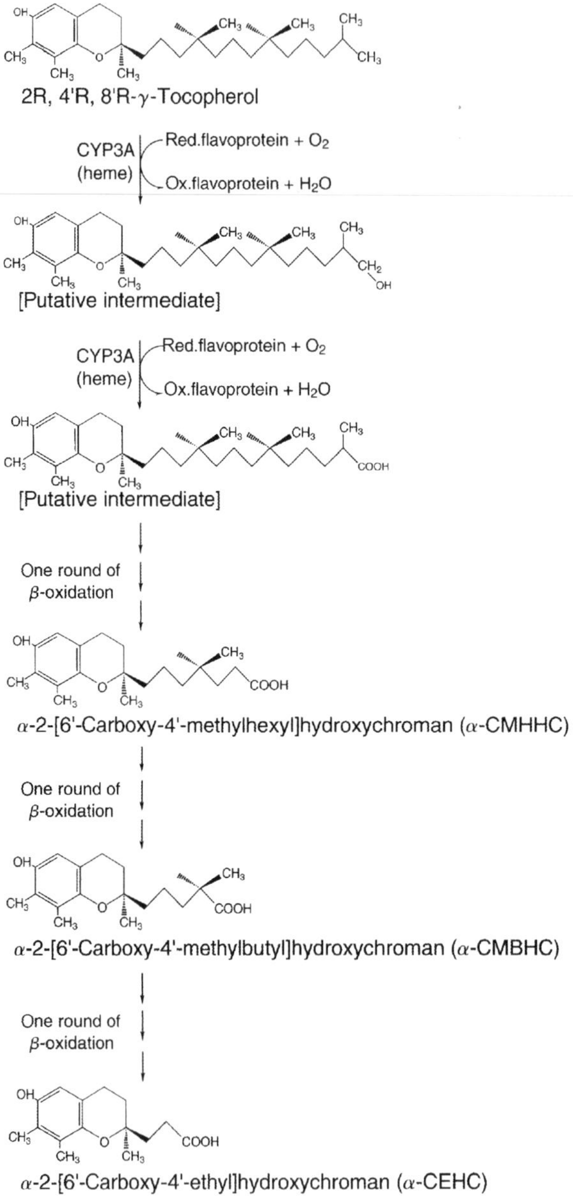

FIGURE 9.23

Metabolism of vitE side chains generates polar compounds.

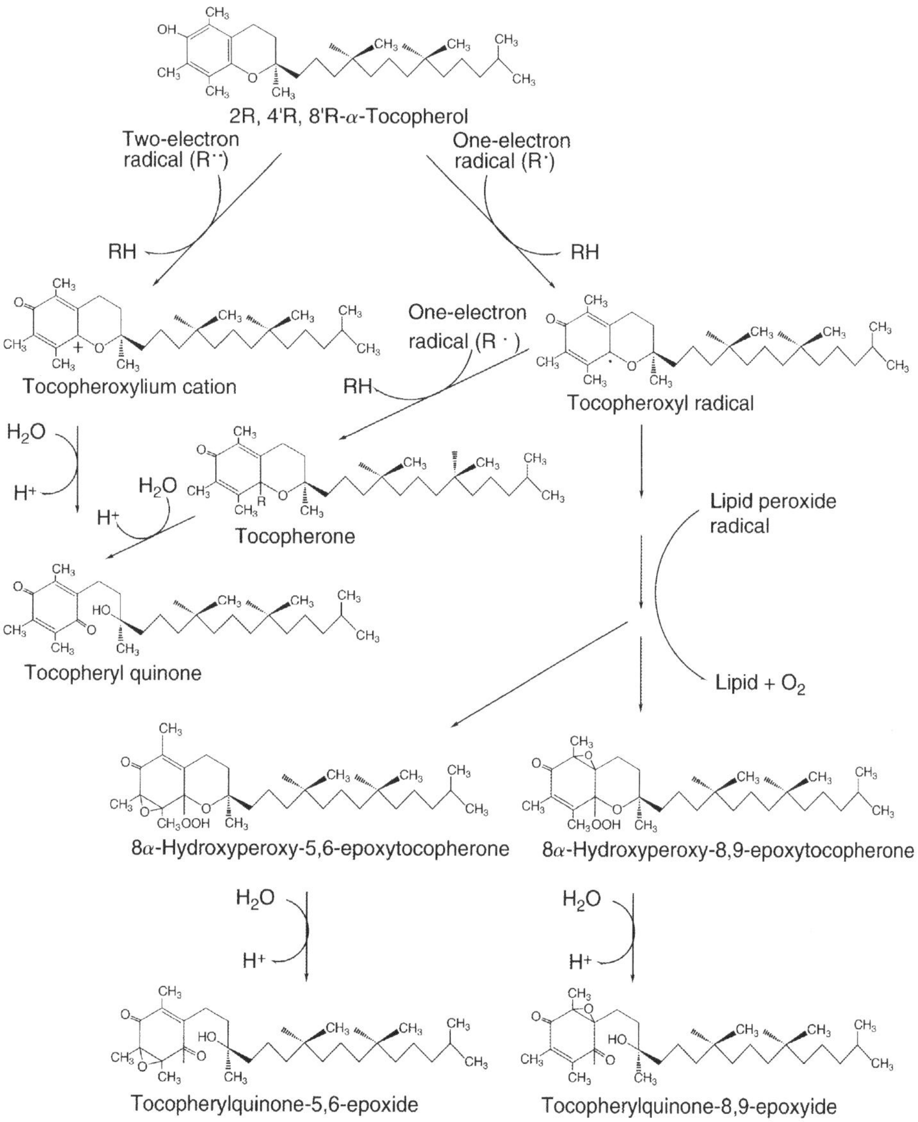

FIGURE 9.24

Reaction with free radicals initiates metabolism of the vitE ring.

excreted from the liver into bile. The exact mechanism is still uncertain. The SLC22A5 (OCTN2) organic cation transporter may be responsible since its expression increases in response to high vitE intake (Mustacich et al., 2009).

REGULATION

The limited capacity of TTP may be the most important protection against vitE excess. A modulating effect of alpha-tocopherol on the expression of TTP is likely (Azzi et al., 2001). However, experience clearly shows that high consumption levels can at least partially overwhelm such limitations, presumably by using unspecific pathways for transport.

FUNCTION

Antioxidant protection: Molecules with vitE biological activity can abstract free electrons from an oxygen free radical, thereby rendering it much less reactive (Figure 9.25). It is presumed that lack of this protective role underlies the progressive cerebellar ataxia, nerve damage, retinal atrophy, and other symptoms related to severe vitE deficiency, such as seen with defective TTP or abetalipoproteinemia. Increased breath exhalation of ethane and pentane due to increased lipid peroxidation reflects the inadequate antioxidant action even with more moderate depletion. Adequate vitE is particularly critical when other antioxidant systems fail. For example, after experimental ablation of the selenium-dependent

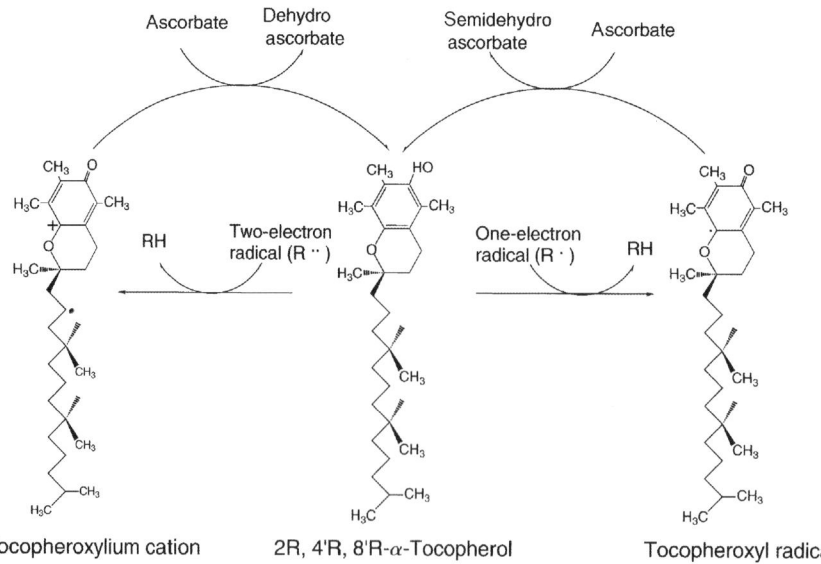

FIGURE 9.25

VitE quenches free radicals.

glutathione peroxidase 4 (GPX4, a repair enzyme for oxidized phospholipids), high vitE becomes critical for the survival of endothelial cells (Wortmann et al., 2013).

There is a downside to quenching one-electron oxidants, however, since this type of reaction turns the vitE molecule itself into a free radical. Recent observations on atherosclerotic lesions indicate, however, that the tocopheroxyl radical promotes lipid oxidation to a greater extent than providing protection against it (Terentis et al., 2002). Therefore, the presence of co-antioxidants, such as ascorbate, becomes critical to quench and reactivate the tocopheroxyl radical before it can attack a lipid double bond. VitE may lower HDL cholesterol (Mondul et al., 2011) and promote the activity of cholesteryl ester transfer protein (CETP) (Kuller, 2001).

Another concern is about the suppression of vitally important signaling events that use free radical reactions (Dröge, 2002). High vitE concentration may interfere with important functions such as apoptosis (programmed cell death), cell adherence, and immune response (Dröge, 2002).

Fertility: Inadequate vitE status interferes with normal oocyte production and placenta function in rodent animal models. This role was one of the earliest postulated functions that led to the discovery of vitE. However, the relevance of vitE status for human reproduction is not known.

Nonantioxidant functions: It has become increasingly clear that the needs for at least some vitE species extend beyond interactions with free radicals (Azzi et al., 2001; Ricciarelli et al., 2001; Lirangi et al., 2012). It was thus demonstrated that alpha-tocopherol, but not other vitE species, induces smooth muscle cell growth arrest by inhibiting phosphokinase C. Alpha-tocopherol also inhibits cell adhesion, platelet aggregation, and the production of oxygen free radicals by neutrophils and monocytes. Expression of several genes is inhibited by alpha-tocopherol, including liver collagen $\alpha 1$, α-tropomyosin, and collagenase MMP 1, ICAM-1, VCAM-1. Other effects are activation of PP2 A, diacylglycerol kinase, and inhibition of 5-lipoxygenase, scavenger receptor SR-A and scavenger receptor CD36.

Pathogen mutations: An otherwise benign strain of coxsackie virus B3 was shown to cause myocarditis in vitE-deficient mice (Beck et al., 2005). This sheds a new light on the role of nutrient deficiency in host–pathogen interactions and could pose a significant health risk on its own.

Hemorrhage: High levels of tocopherol quinone, which can arise from excessive vitE intake, competitively inhibit regeneration of vitamin K epoxide by NAD(P)H:quinine oxidoreductase 1 (EC1.6.99.2). A daily dose of 50 mg of all-rac-alpha-tocopherol in the largest long-term study conducted to date increased mortality from hemorrhagic stroke (ATBC Cancer Prevention Study Group, 1994). People with low vitamin K intake or on warfarin-type anticoagulation (Pastori et al., 2013) are at particularly high risk. Inhibiting effects on platelet aggregation (Calzada et al., 1997) may contribute slightly to a bleeding tendency but may be beneficial for the prevention of coronary thrombosis.

NUTRITIONAL ASSESSMENT

VitE deficiency and suboptimal status may be detected with the measurement of vitE metabolites in the blood. Dietary intake can be estimated from the amount of excreted urinary vitE metabolites.

Blood measurements: The concentration of active vitE forms in plasma reflects the current status reasonably. The measurement should resolve the major forms of vitE, particularly distinguishing between α-tocopherol and γ-tocopherol in light of their distinctive biological properties. A significant problem comes from increased vitE concentrations in people with elevated plasma triglyceride concentrations. The debate about using triglyceride concentrations for correction of measured vitE concentrations is still unresolved (Heseker et al., 1993).

Urine measurements: The amount of vitE metabolites excreted with a urine sample collected over a 24-h period or adjusted to creatinine concentration reflects correlates with vitE intake (Sharma et al., 2010).

The α-tocopherol-derived metabolites generated by opening of the chroman ring include α-tocopheronolactone (α-TL), α-tocopheryl quinone and α-tocopheronic acid. Side-chain shortening by ω-oxidation generates α-carboxy-ethyl-hydroxychroman (a-CEHC) and α-carboxymethyl-butylhydroxychroman.

REFERENCES

ATBC Cancer Prevention Study Group, 1994. The effect of vitamin E and beta carotene on the incidence of lung cancer and other cancers in male smokers. N. Engl. J. Med. 330, 1029–1035.

Azzi, A., Breyer, I., Feher, M., Ricciarelli, R., Stocker, A., Zimmer, S., et al., 2001. Nonantioxidant functions of alpha-tocopherol in smooth muscle cells. J. Nutr. 131, 378S–381S.

Beck, M.A., Shi, Q., Morris, V.C., Levander, O.A., 2005. Benign coxsackievirus damages heart muscle in iron-loaded vitamin E-deficient mice. Free Radic. Biol. Med. 38, 112–116.

Birringer, M., Drogan, D., Brigelius-Flohé, R., 2001. Tocopherols are metabolized in HepG2 cells by side chain omega-oxidation and consecutive beta-oxidation. Free Rad. Biol. Med. 31, 226–232.

Blatt, D.H., Leonard, S.W., Traber, M.G., 2001. Vitamin E kinetics and the function of tocopherol regulatory proteins. Nutrition 17, 799–805.

Borel, P., Pasquier, B., Armand, M., Tyssandier, V., Grolier, P., Alexandre-Gouabau, M.C., et al., 2001. Processing of vitamin A and E in the human gastrointestinal tract. Am. J. Physiol. Gastroint. Liver Physiol. 280, G95–G103.

Calzada, C., Bruckdorfer, K.R., Rice-Evans, C.A., 1997. The influence of antioxidant nutrients on platelet function in healthy volunteers. Atherosclerosis 128, 97–105.

Christiansen-Weber, T.A., Voland, J.R., Wu, Y., Ngo, K., Roland, B.L., Nguyen, S., et al., 2000. Functional loss of ABCAI in mice causes severe placental malformation, aberrant lipid distribution, and kidney glomerulonephritis as well as high-density lipoprotein cholesterol deficiency. Am. J. Pathol. 157, 1017–1029.

Dröge, W., 2002. Free radicals in the physiological control of cell function. Phys. Rev. 82, 47–95.

Esterbauer, H., Gebicki, J., Puhl, H., Jurgens, G., 1992. The role of lipid peroxidation and antioxidants in oxidative modification of LDL. Free Rad. Biol. Med. 13, 341–390.

Food and Nutrition Board, Institute of Medicine, 2000. Dietary reference intakes for vitamin C, vitamin E, Selenium, and Carotenoids. National Academy Press, Washington, DC.

Goncalves, A., Roi, S., Nowicki, M., Niot, I., Reboul, E., 2014. Cluster-determinant 36 impacts on vitamin E postprandial response. Mol. Nutr. Food Res. 58, 2297–2306.

Goti, D., Hrzenjak, A., Levak-Frank, S., Frank, S., van der Westerhuysen, D.R., Malle, E., et al., 2001. Scavenger receptor class B, type I is expressed in porcine brain capillary endothelial cells and contributes to selective uptake of HDL-associated vitamin E. J. Neurochem. 76, 498–508.

Hao, M., Maxfield, F.R., 2000. Characterization of rapid membrane internalization and recycling. J. Biol. Chem. 275, 15279–15287.

Heseker, H., Kohlmeier, M., Schneider, R., 1993. Lipid adjustment of alpha-tocopherol concentrations in plasma. Z. Ernahrungswiss. 32 (3), 219–228.

Huuskonen, J., Olkkonen, V.M., Jauhiainen, M., Ehnholm, C., 2001. The impact of phospholipid transfer protein (PLTP) on HDL metabolism. Atherosclerosis 155, 269–281.

Jenab, M., Salvini, S., van Gils, C.H., et al., 2009. Dietary intakes of retinol, beta-carotene, vitamin D and vitamin E in the European Prospective Investigation into Cancer and Nutrition cohort. Eur. J. Clin. Nutr. 63 (Suppl. 4), S150–S178.

Kuller, L.H., 2001. A time to stop prescribing antioxidant vitamins to prevent and treat heart disease? Arteriosclerosis 21, 1253.

Lirangi, M., Meydani, M., Zingg, J.M., Azzi, A., 2012. α-Tocopheryl-phosphate regulation of gene expression in preadipocytes and adipocytes. Biofactors 38, 450–457.

Lodge, J.K., Ridlington, J., Leonard, S., Vaule, H., Traber, M.G., 2001. Alpha- and gamma-tocotrienols are metabolized to carboxyethyl-hydroxychroman derivatives and excreted in human urine. Lipids 36, 43–48.

Mardones, E., Strobel, E., Miranda, S., Leighton, E., Quinones, V., Amigo, L., et al., 2002. Alpha-tocopherol metabolism is abnormal in scavenger receptor class B type I (SR-BI)-deficient mice. J. Nutr. 132, 443–449.

Mertens, A., Holvoet, P., 2001. Oxidized LDL and HDL: antagonists in atherothrombosis. FASEB J. 15, 2073–2084.

Mondul, A.M., Weinstein, S.J., Virtamo, J., Albanes, D., 2011. Serum total and HDL cholesterol and risk of prostate cancer. Cancer Causes Control 22, 1545–1552.

Mustacich, D.J., Gohil, K., Bruno, R.S., Yan, M., Leonard, S.W., Ho, E., et al., 2009. Alpha-tocopherol modulates genes involved in hepatic xenobiotic pathways in mice. J. Nutr. Biochem. 20, 469–476.

Mustacich, D.J., Shields, J., Horton, R.A., Brown, M.K., Reed, D.J., 1998. Biliary secretion of alpha-tocopherol and the role of the mdr2 P-glycoprotein in rats and mice. Arch. Biochem. Biophys. 350, 183–192.

Oram, J.E., Vaughan, A.M., Stocker, R., 2001. ATP-binding cassette transporter A I mediates cellular secretion of alpha-tocopherol. J. Biol. Chem. 276, 39898–39902.

Pappert, E.J., Tangney, C.C., Goetz, C.G., Ling, Z.D., Lipton, J.W., Stebbins, G.T., et al., 1996. Alpha-tocopherol in the ventricular cerebrospinal fluid of Parkinson's disease patients: dose-response study and correlations with plasma levels. Neurologica. 47, 1037–1042.

Parker, R.S., Sontag, T.J., Swanson, J.E., 2000. Cytochrome P4503A-dependent metabolism of tocopherols and inhibition by sesamin. Biochem. Biophys. Res. Commun. 277, 531–534.

Pastori, D., Carnevale, R., Cangemi, R., Saliola, M., Nocella, C., Bartimoccia, S., et al., 2013. Vitamin E serum levels and bleeding risk in patients receiving oral anticoagulant therapy: a retrospective cohort study. J. Am. Heart Assoc. 2 (6), e000364.

Phillips, E.L., Arnett, D.K., Himes, J.H., McGovern, P.G., Blackburn, H., Luepker, R.V., 2000. Differences and trends in antioxidant dietary intake in smokers and non-smokers, 1980–1992: the Minnesota Heart Survey. Ann. Epidemiol. 10, 417–423.

Qureshi, A.A., Peterson, D.M., Hasler-Rapacz, J.O., Rapacz, J., 2001. Novel tocotrienols of rice bran suppress cholesterogenesis in hereditary hypercholesterolemic swine. J. Nutr. 131, 223–230.

Ricciarelli, R., Zingg, J.M., Azzi, A., 2001. Vitamin E: protective role of a Janus molecule. FASEB J. 15, 2314–2325.

Ross, D., Kepa, J.K., Winski, S.L., Beall, H.D., Anwar, A., Siegel, D., 2000. NAD(P)H:quinone oxidoreductase 1 (NQO1): chemoprotection, bioactivation, gene regulation and genetic polymorphisms. Chemico-Biol. Interact. 129, 77–97.

Ruiz Rejon, E., Martin-Pena, G., Granado, E., Ruiz-Galiana, J., Blanco, I., Olmedilla, B., 2002. Plasma status of retinol, alpha- and gamma-tocopherols, and main carotenoids to first myocardial infarction: case control and follow-up study. Nutrition 18, 26–31.

Sharma, G., Muller, D., O'Riordan, S., Bryan, S., Hindmarsh, P., Dattani, M., et al., 2010. A novel method for the direct measurement of urinary conjugated metabolites of alpha-tocopherol and its use in diabetes. Mol. Nutr. Food Res. 54, 599–600.

Terentis, A.C., Thomas, S.R., Burr, J.A., Liebler, D.C., Stocker, R., 2002. Vitamin E oxidation in human atherosclerotic lesions. Circ. Res. 90, 333–339.

Valerio Jr, L.G., Kepa, J.K., Pickwell, G.V., Quattrochi, L.C., 2001. Induction of human NAD(P)H:quinine oxidoreductase (NQO1) gene expression by the flavonol quercetin. Toxicol. Lett. 119, 49–57.

Wortmann, M., Schneider, M., Pircher, J., Hellfritsch, J., Aichler, M., Vegi, N., et al., 2013. Combined deficiency in glutathione peroxidase 4 and vitamin E causes multiorgan thrombus formation and early death in mice. Circ. Res. 113, 408–417.

Yamamoto, Y., Fujisawa, A., Hara, A., Dunlap, W.C., 2001. An unusual vitamin E constituent (alpha-tocomonoenol) provides enhanced antioxidant protection in marine organisms adapted to cold-water environments. Proc. Natl. Acad. Sci. USA 98, 13144–13148.

VITAMIN K

Vitamin K designates a group of compounds characterized by a 2-methylnaphthalene-1,4-dione ring system with antihemorrhagic properties. The natural form in plants is phylloquinone (2-methyl-3-[3,7, 11, 15-tetramethyl-2 hexadecenyl]-1,4-naphthalenedione, phytonadione, 3-phytylmenadione; molecular weight 450). Bacteria produce menaquinones whose side chain consists of 7–13 isoprenyl units. The mixture of produced menaquinones is highly specific for individual species. Additional natural and synthetic compounds are covered under the definitions given previously, and are shown in Figure 9.26.

ABBREVIATIONS

IL-6	interleukin-6
K1	phylloquinone
MK4	menaquinone-4 (menatetrenone)
vitK	vitamin K (all quinones with vitamin K activity; phylloquinone, menaquinones)

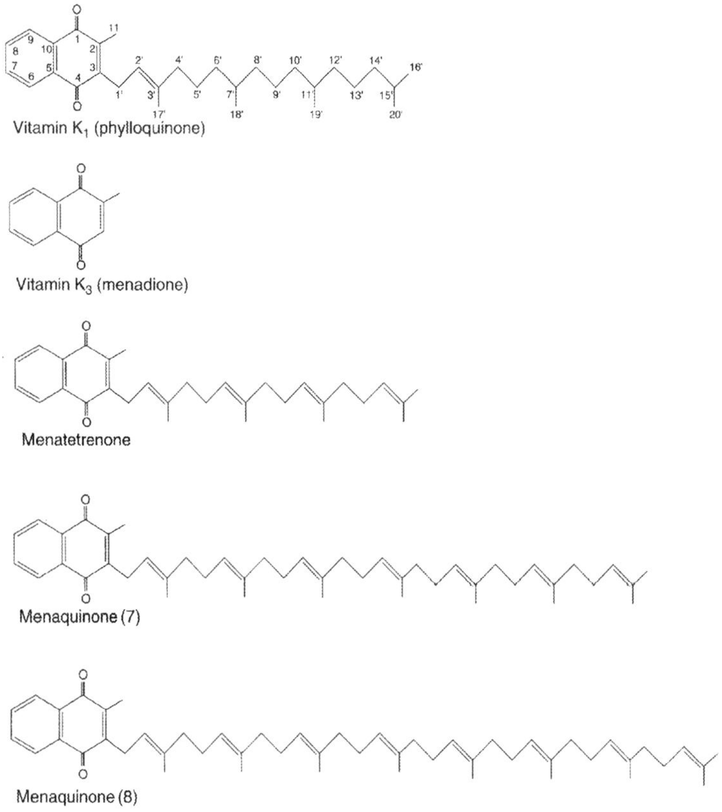

FIGURE 9.26

Common forms of vitamin K.

NUTRITIONAL SUMMARY

Function: Vitamin K (vitK) activates proteins needed for blood coagulation, counteracting the calcification of arteries and other soft tissues, promoting the mineralization of bone, and the regulation of cell division and differentiation.

Requirements: Newborn infants should get at least one supplemental dose right after birth to prevent cerebral hemorrhage; additional doses would be given during the first few months of life. Adequate daily intake (Dietary Reference Intake in the United States) for women is 90 µg; men need 120 µg/day.

Sources: Intestinal bacteria provide enough under optimal circumstances to prevent bleeding, but not enough to optimize other functions. Many antibiotics suppress this important baseline contribution. Cooked spinach, kale, and other green leafy vegetables provide much more than enough to achieve adequate intake with a single serving. Specific fermented foods including *natto* (soy fermented with *Bacterium subtilis natto*) and some cheeses contain particularly active menaquinones. The small amounts in other sources make it difficult to meet requirements without eating greens or eating *natto*. Human milk contains very little.

Deficiency: The greatest risk is for newborn breast-fed infants, especially those with disorders (e.g., biliary atresia) interfering with absorption. Brain hemorrhage in very young children is a rare but extremely serious affliction due to low vitK status. Oral antibiotics can cause bleeding in some adults who have persistently minimal intake. Suboptimal intake is likely to increase the risk of artery calcification, osteoporosis, and possibly accelerated cognitive decline.

Excessive intake: No unfavorable health effects have been observed following consumption of 45 mg/day for a year. Patients taking anti-vitK (coumadin type) anticoagulants have to maintain a constant intake, which is likely to be easier at a high than at a low level.

ENDOGENOUS SOURCES

Several normal intestinal bacteria produce considerable quantities of menaquinones, some of which is likely to be absorbed (Conly and Stein, 1992). How much is actually available from the distal intestines remains uncertain (Lipsky, 1994).

DIETARY SOURCES

Dark-green vegetables contain large amounts of phylloquinone (K1). VitK-rich foods are chard (8.3 µg/g), kale (7 µg/g), green asparagus (4 µg/g), spinach (3.8 µg/g), broccoli (2.7 µg/g), and Brussels sprouts (1.5 µg/g). Soya oil (1.5 µg/g), canola oil (1.1 µg/g), and virgin olive oil (0.8 µg/g) also can contribute significantly. Natto, a fermented bean curd commonly consumed in some regions of Japan (Kaneki et al., 2001), is a very rich source of menaquinones (9 µg/g). Some cheeses (for instance, Roquefort, with about 0.3 µg/g) provide small but significant amounts of menaquinone. Most other oils and foods do not provide enough for adequate intake in the absence of vitK-rich foods (Booth et al., 2001; Shearer et al., 1996). Hydrogenation also saturates K1 in oils and appears to render it largely inactive (Booth et al., 2001) (Figure 9.27). Phylloquinone and the menaquinones are very sensitive to light and alkali, which induce decomposition to inactive chromenol compounds.

Typical daily intake levels in the United States and in the United Kingdom (Thane et al., 2002) are commonly below what are considered adequate intake levels, but they may be much higher in some Asian countries due to their high intake of soybeans and soy products.

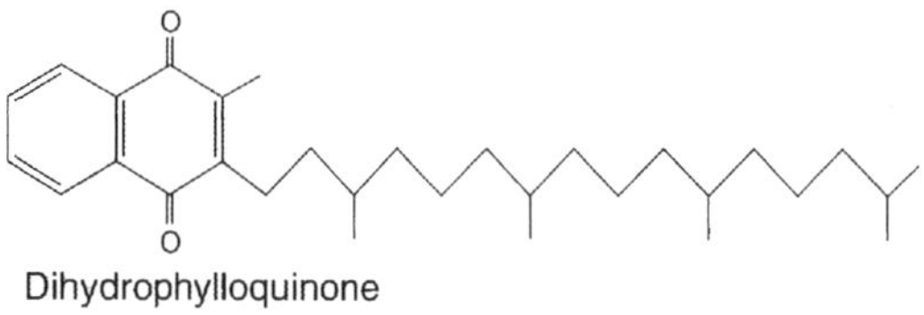

Dihydrophylloquinone

FIGURE 9.27

FIGURE 9.27

Industrial hydrogenation of fats generates dihydrovitamin K.

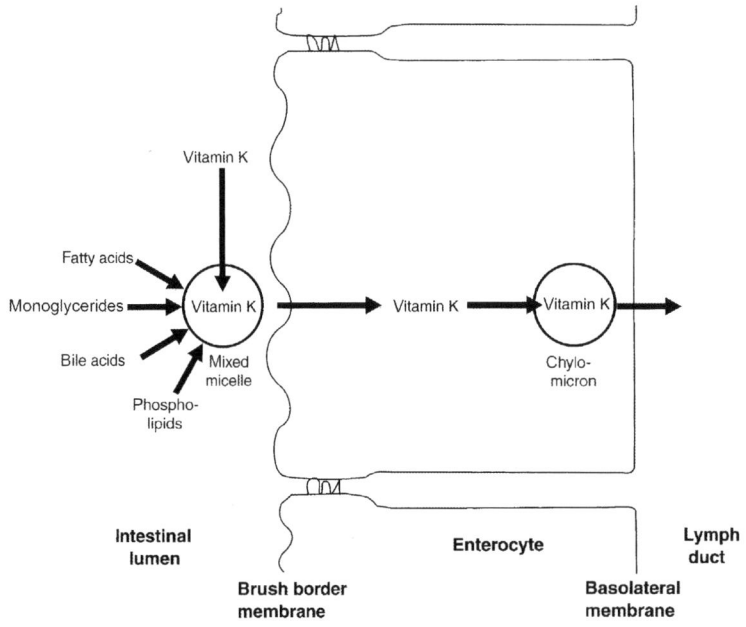

FIGURE 9.28

Intestinal absorption of vitK.

DIGESTION AND ABSORPTION

The absorption of vitK from the small intestine is tightly coupled to fat absorption (Figure 9.28). Most of a micellar K1 emulsion is absorbed (Gijsbers et al., 1996). The K1 in leafy vegetables and lettuces, on the other hand, is firmly embedded in the chloroplast cotyledons and not easily released. Less than 20% is absorbed from cooked spinach served with fat, and almost nothing from uncooked spinach (Gijsbers et al., 1996; Garber et al., 1999; Schurgers and Vermeer, 2000). Menaquinones and K1, all of which are highly lipophilic, become part of mixed micelles formed during fat digestion in the small intestine. Micelles and the vitK inside them are taken up by small intestinal enterocytes through a process involving the fatty acid transporter CD36, SRB1, and NPC1L1. During the formation of Chylos, vitK is incorporated into these triglyceride-rich Lp particles. Chylos with vitK are secreted into intestinal lymph ducts and reach blood circulation from there. There may also be some interference from other fat-soluble micronutrients. High intake of lutein and vitamin E in combination were found to reduce bioavailability of K1 and menaquinone-4 in rats (Mitchell et al., 2001).

TRANSPORT AND CELLULAR UPTAKE

Blood circulation: VitK concentrations typically are around 0.1–2 µg/l and increase considerably after a vitK-rich meal. VitK is transported in the blood primarily with Chylos, and to a much smaller extent with other Lps (Kohlmeier et al., 1995). Upon exposure to Lp lipase, Chylos release most of their lipid load but retain vitK. This enters cells in the liver, bone, and other tissues during receptor-mediated uptake of chylomicron remnants. Uptake requires the binding of apolipoprotein E at the surface of chylomicron remnants to specific receptors including LDLR and the LDL receptor-related protein 1 (LRP1, apolipoprotein E receptor). Common variant forms of apolipoprotein E (E2, E3, E4) mediate uptake with differing efficacy and are associated with different residual vitK plasma concentrations after an overnight fast. Individuals with the E4 variant tend to have lower concentrations than those who have only the E3 isoform (Kohlmeier et al., 1995). People with the E2 isoform tend to have the highest concentrations. There is no indication of significant transport of intact vitK between cells or organs.

It has been suggested that the multidrug transporter ABCG2 (BCRP, MXR) mediates the transfer of vitK from mammary cells into milk (Vlaming et al., 2009).

BBB: The transport of vitK into the brain is not well understood.

Materno-fetal transfer: The fetal blood concentration (around 0.01 µg/l without maternal supplementation) is held to much lower levels than in the maternal blood. While some transfers undoubtedly take place, possibly with a delay and slow release from placenta to the fetus (Iioka et al., 1991), the precise mechanisms are not known. Uptake of vitK-containing Lps from maternal blood is certain to involve one or more of the many Lp receptors at the microvillous side of the syntrophoblast.

METABOLISM

VitK cycle: The best-characterized function of vitK is the gamma-carboxylation of specific glutamyl residues in a handful of proteins. VitK-dependent carboxylase (EC4.1.1.90), in combination with phylloquinone monooxygenase (EC1.14.99.20), uses the large redox potential of hydroquinone to drive the reaction. For each carboxylated glutamyl residue, 1 mol of vitK 2,3-epoxide is generated. A two-step system of reductases regenerates vitK hydroquinone for the next round of carboxylation (Figure 9.29). Warfarin-sensitive vitamin-K–epoxide reductase (EC1.1.4.1, encoded by the *VKORC1* gene), which is strongly expressed in the liver, uses thioredoxin or another reduced thiol for reduction to the vitK form. A distinct form of vitamin-K–epoxide reductase that is distinctly less warfarin sensitive (EC1.1.4.2, encoded by the VKORC1L1 gene) has the highest activity in osteoblasts, the lung, and the testis, which makes these cells and tissues insensitive to the effects of warfarin (Hammed et al., 2013). The importance of this enzyme is highlighted by the fact that the protein encoded by VKORC1L1 has twice the catalytic efficiency of the one encoded by VKORC1 (Hammed et al., 2013).

The oxidized thioredoxin, in turn, is restored to its reduced form again by NADPH-dependent thioredoxin reductase (EC1.6.4.5). The second step is catalyzed alternatively by the vitamin K–epoxide reductases or by NAD(P)H:quinone oxidoreductases 1 (NQO1, EC1.6.5.2) and 2 (NQO2,

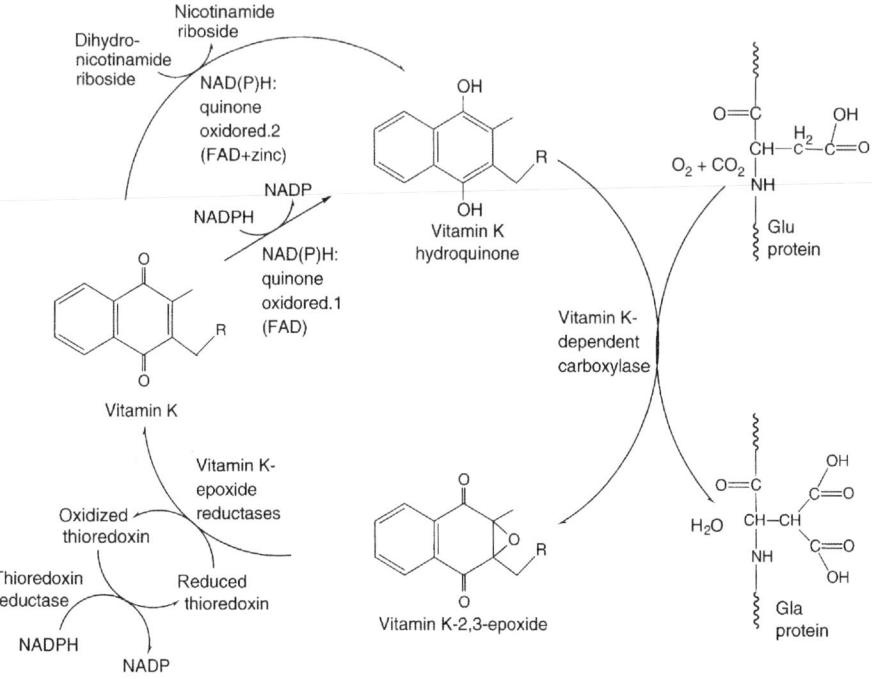

FIGURE 9.29

The vitK cycle.

EC1.10.99.2). Both NQO1 and NQO2 contain FAD, and both also convert dietary vitK into the active hydroquinone form. NQO1 is inhibited by warfarin, whereas NQO2 (a zinc metalloenzyme) is not (Zhu and Li, 2012). It should also be noted that NQO1 is an acute phase protein, while NQO2 is not. Quercetin, which is the most abundant flavonoid in food, induces NQO1 (Valerio et al., 2001). Another difference between these two enzymes is that NQO2 uses dihydronicotinamide riboside (NRH) instead of NADPH as the electron donor. Both alpha-tocopherol quinone, the oxidized form of vitamin E, and ubiqunone (coenzyme Q) compete with vitK epoxide for NQO1 activity. High vitamin E intake with dietary supplements can cause bleeding due to impaired reactivation of vitK epoxide, especially in patients treated with warfarin-type anticoagulants (Pastori et al., 2013).

Side-chain resynthesis: Menaquinone-4 (menatetrenone, MK4) is the main vitamer of vitK in the brain and is present in most other tissues. Normally very little MK4 is ingested and intestinal bacteria are not necessary for its appearance (Ronden et al., 1998). This is because in humans 5–25% of pre-nylated quinones (phylloquinone, menaquinones) are converted shortly after ingestion into menaquinone (Figure 9.30), probably in the small intestine (Thijssen et al., 2006). Experiments in rats have demonstrated that phylloquinone conversion to MK4 proceeds via side-chain removal and resynthesis by UbiA prenyltransferase containing 1 (UBIAD1, EC2.5.1.74), particularly in the brain (Al-Rajabi et al., 2012).

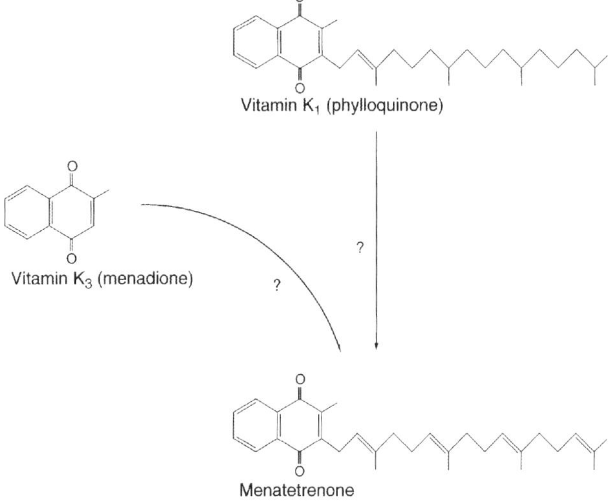

FIGURE 9.30

Menaquinone-4 is an important vitK metabolite.

Catabolism: The side chains of K1 and prenylated menaquinones are shortened by omega-oxidation to the acidic catabolites 2-methyl-3-(5′-carboxy-3′-methyl-2′pentenyl)-l,4-naphtoquinone and 2-methyl-3-(3′-carboxy-3′-methylpropyl)-1,4-naphthoquinone (Shearer et al., 1972) (Figure 9.31).

STORAGE

The liver contains the largest amount (typically 112 nmol) of vitK in the body (Thijssen and Drittij-Reinders, 1996), mostly as menaquinones (64 nmol/kg), and much less as phylloquinone (10.6 nmol/kg). Bone also contains significant amounts (about 53 nmol/kg), as does the fat tissue (about 22 nmol/kg) in bone (Hodges et al., 1993). Brain contains almost exclusively menaquinone-4 (Ferland, 2012). However, since no mechanism seems to provide for the mobilization and transport of stored vitK, each cell or tissue appears to be on its own in times of need.

EXCRETION

About 60% of vitK losses are into bile, the remainder goes into urine (Shearer et al., 1972). The bile contains conjugated inactive metabolites, but little, if any, intact vitK. Since most circulating vitK is contained in Lps, very little gets into glomerular ultrafiltrate in the kidney, and those small amounts are mostly recovered from the proximal tubulus. The main catabolites in urine are glucuronides of 2-methyl-3-(5′-carboxy-3′-methyl-2′-pentenyl)-1,4-naphtoquinone   and   2-methyl-3-(3′-carboxy-3′-methylpropyl)-1,4-naphthoquinone (Shearer et al., 1972; Harrington et al., 2007).

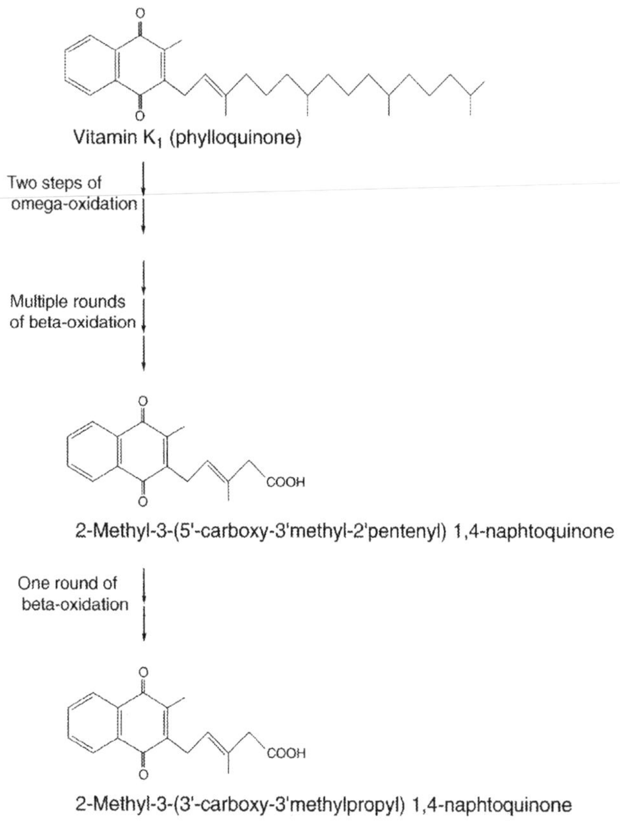

Vitamin K₁ (phylloquinone)

Two steps of omega-oxidation

Multiple rounds of beta-oxidation

2-Methyl-3-(5'-carboxy-3'methyl-2'pentenyl) 1,4-naphtoquinone

One round of beta-oxidation

2-Methyl-3-(3'-carboxy-3'methylpropyl) 1,4-naphtoquinone

FIGURE 9.31

The vitK side chain is metabolized by omega-oxidation.

REGULATION

It is doubtful whether there is any homeostatic regulation of tissue concentration or stores of vitK in the body.

FUNCTION

Blood coagulation: The first property assigned to vitK was the ability of compounds to prevent bleeding. Indeed, *K*, derived from the German word for coagulation (*Koagulation*), became the vitamin's moniker. Four clotting factors (II, VII, IX, and X) are known to require vitK-dependent modifications. In addition, and of equal importance, there are three vitK-dependent proteins (proteins C, S, and Z) that counteract and regulate the action of the clotting factors. Thrombinogen (coagulation factor II) and the blood coagulation factors VII, IX, and X have to be posttranslationally modified to be active. Membrane-bound microsomal gamma-carboxylase uses the very high activation energy of vitK hydroquinone to attach an additional carboxyl group in gamma-position to a few specific glutamyl residues in the target proteins.

Regulation of tissue mineralization: The vitK-dependent matrix Gla protein (MGP) and other calcium-complexing proteins suppress the calcification of soft tissues, particularly of the arterial wall (Theuwissen et al., 2012). Genetic deficiency of MGP in animal models causes fatal blood vessel calcification in a very short time. In humans, however, the rare inherited loss of MGP activity results in characteristic facial features (Keutel syndrome), recurring otitis and hearing loss, sinusitis, pulmonary artery stenosis, and cartilage calcification (Weaver et al., 2014). The promotion of aortal calcification by low vitK intake in adults is a much slower process (Jie et al., 1996). Insufficient MGP carboxylation, often due to anti-vitK treatment, is a powerful driver of ectopic tissue calcification (Schurgers et al., 2013).

Gla-rich protein (GRP, also called the *upper zone of growth plate* and *cartilage matrix-associated protein*) is another vitK-dependent calcium-binding protein that helps to regulate tissue mineralization, particularly in cartilage and the arterial walls (Cancela et al., 2012). It has the highest Gla content of all known Gla proteins. Cartilage from patients with osteoarthritis has been found to be less carboxylated than material from healthy controls (Rafael et al., 2014). Undercarboxylated GRP may also be related to ectopic calcification in cancer of the breast and other tissues (Viegas et al., 2014). Insufficient GRP carboxylation may contribute to embryopathy with typical facial malformations in women treated with warfarin during pregnancy.

Bone mineralization: Adequate vitK intake is critical for bone mineral retention and prevents fractures, and high intake often is beneficial in this respect (Vermeer and Theuwissen, 2011). The role of the vitK-dependent proteins osteocalcin, MGP, and protein S in bone remains unclear. The attenuation of PTH-stimulated IL-6 production by vitK (which is not abolished by warfarin) or other mechanisms may be important (Kohlmeier et al., 1998). Indeed, the vitK catabolite 2-methyl, 3-(2'-methyl)-hexanoic acid-l,4-naphthoquinone, which does not support gamma-carboxylation, is a more potent suppressor of stimulated IL-6 secretion than vitK (Reddi et al., 1995).

Glucose homeostasis: It is now understood that undercarboxylated osteocalcin from osteoblasts acts as a hormone that links the endocrine pancreas to the metabolic and nutritional state of bone (Lacombe et al., 2013). Osteocalcin promotes acute insulin release both by directly stimulating beta-cells and by acting through glucagon-like peptide-1 (GLP-1) from the small intestine. Osteocalcin also sensitizes cells in the liver and other tissues. In the long run, osteocalcin promotes beta-cell proliferation by acting on their *Ccnd2* and *Cdk4* genes. It is important to emphasize again that it is the incompletely carboxylated form of osteocalcin released during osteoclastic bone turnover that has these homeostatic effects. The relevance of vitK status on these functions is still evolving (Willems et al., 2014).

Fertility: Undercarboxylated osteocalcin binds to the G-protein-coupled receptor Gprc6a on Leydig cells and through this mechanism promotes testosterone production (Oury et al., 2013). Insufficient stimulation in men with low concentration of undercarboxylated osteocalcin impairs sperm cell maturation. It is unclear what happens when vitK supplies are particularly ample or insufficient.

Signaling: At least four of the vitK-dependent proteins (thrombin, factor Xa, protein S, and gas6) bind to specific cell surface receptors and elicit typical responses. Growth arrest-specific protein 6 (gas6) binds to the tyrosine kinase receptors Axl (Ufo/Ark), Dtk (Sky/Rse/Tyro3/Brt/Tif), and Mer (Eyk). Protein S also binds to Dtk, but it may not be the native ligand. Binding to Axl affects proliferation and differentiation through its effects on cell cycle progression (Goruppi et al., 1999), and possibly through protection against apoptosis (Bellosta et al., 1997; Lee et al., 2002). The gas6-induced effects are at least partially mediated through different signaling pathways that involve phosphatidylinositol

3-hydroxy kinase (PI3K), the oncogen src, p38 MAPK (Goruppi et al., 1999), and Akt (Lee et al., 2002). Some tissues, such as the renal mesangium (Yanagita et al., 1999), are critically dependent on adequate vitK supplies for normal cell growth in adulthood. Gas6 is also essential for the initiation of phagocytosis of photoreceptor outer segments in the human retina (Hall et al., 2001), which degenerates without this activity. The brain is similarly dependent on gas6 activity, with suggested roles in oligodendride function and axon myelination (Ferland, 2012). Protein S has a comparable function in the brain. Other tissues may have critical vitK needs during fetal or childhood development. Anti-vitK medication during early pregnancy is known to cause typical craniofacial malformations (Howe and Webster, 1994).

The vitK-dependent protein periostin plays a critical role in embryonic heart valve development and the remodeling of extracellular matrix (Coutu et al., 2008). It suppresses calcification of mesenchymal cells. Low activity due to genetic variants or impaired vitK status may be a primary cause for the development of atherosclerosis (Hixson et al., 2011) and aortic valve degeneration (Hakuno et al., 2010) later in life. Elevated periostin concentration in the blood appears to predict cardiovascular events in patients with recent heart attacks (Ling et al., 2014).

Transforming growth factor beta induced (TGFBI, keratoepithelin), a paralogue of periostin, is a Gla protein (Coutu et al., 2008) present in many tissues. It appears to promote microtubulin stability. Down-regulation, hypermethylation, or other genetic disruption in cancer tissues may be important for cell adhesion (Ahmed et al., 2007) and sensitivity to chemotherapy (Wang et al., 2012).

Proline-rich Gla protein 2 (PRGP2) binds to the transcriptional coactivator Yes-associated protein and may be part of a signaling pathway that modulates apoptosis (Kulman et al., 2007).

Thrombin binds to specific G-protein-coupled receptors and induces the synthesis of several humoral messengers, including endothelin, vasopressin, nerve growth factor, and platelet-derived growth factor.

Gla-proteins with unknown function: Proline-rich Gla protein 1 (PRGP1) is another orphan vitK-dependent protein. The transmembrane Gla proteins 3 and 4 (TMG3 and TMG4) also still await insight into what they do (Kulman et al., 2001). Periostin-like protein is another protein that is now known to be vitK-dependent because it contains Gla in its amino acid sequence. Its function remains unclear (Table 9.4).

Sulfur metabolism: The activities of galactocerebroside sulfotransferase (EC2.8.2.11) and arylsulfatase (EC3.1.6.1) in the brain are vitK-dependent (Sundaram and Lev, 1992), though the precise nature of this requirement is not known. VitK deficiency (warfarin treatment) induces typical changes in the profiles of complex lipids in the brain (Ferland, 2012). Arylsulfatase E (ARSE, EC3.1.6.1) in bone and

Table 9.4 VitK-Dependent Proteins

Blood Coagulation	Mineralization	Cell Signaling	Other Function
Prothrombin	Osteocalcin	Gas6	VitK-dependent carboxylase
Factor VII	MGP	Periostin	Arylsulfatase E
Factor IX	GRP	PRGP2	PRGP1
Factor X		TGFBI	TMG3
Protein C			TMG4
Protein S			Periostin-like
Protein Z			

cartilage is another enzyme with direct or indirect vitK dependence (Franco et al., 1995; Daniele et al., 1998). Genetic inactivation (Nino et al., 2008) or inhibition by maternal anti-vitK medication during early pregnancy is related to typical craniofacial defects (Herman and Siegel, 2010; Matos-Miranda et al., 2013).

Prostaglandin metabolism: VitK inhibits the activity of prostaglandin H synthase (COX-2, EC1.14.99.1) in bone and thereby decreases the production of prostaglandin E2 (Koshihara et al., 1993). This may explain the inhibiting effect of vitK on stimulated interleukin-6 (IL-6) production in fibroblasts (Reddi et al., 1995) and bone cells (Kohlmeier et al., 1998), since prostaglandin E2 is a potent activator of IL-6 synthesis. The reported antinocive effect of vitK (Onodera et al., 2000) may be due to a similar mechanism.

Mitochondria: VitK is oxidized very actively in mitochondria (Inyangetor and Thierry-Palmer, 1988). Since all currently known posttranslational protein carboxylation reactions take place at the endoplasmic reticulum, the target of the oxidative activity in mitochondria remains unclear.

NUTRITIONAL ASSESSMENT

VitK deficiency and suboptimal status may be detected with the measurement of vitK metabolites in the blood. Dietary intake can be estimated from the amount of excreted urinary vitK metabolites.

Blood measurements: The concentration of active vitK species (phylloquinone and the menaquinones with diverse side-chain lengths) in plasma reflects current status reasonably. Dihydrophylloquinone is a common low-activity form that should be distinguished from the more active natural forms. The significance of increased vitK concentrations in people with elevated plasma triglyceride concentrations is not clear. The concentrations of vitK metabolites are influenced by common genetic variants, such as the apoE isoforms (Kohlmeier et al., 1996).

Urine measurements: The amount of vitK metabolites excreted with a urine sample collected over a 24-h period or related to creatinine concentration indicates current consumption levels (Harrington et al., 2007). The key metabolites are 2-methyl-3-(5′-carboxy-3′-methyl-2′-pentenyl)-1,4-naphthoquinone and 2-methyl-3-(3′-3′-carboxymethylpropyl)-1,4-naphthoquinone.

REFERENCES

Ahmed, A.A., Mills, A.D., Ibrahim, A.E., Temple, J., Blenkiron, C., Vias, M., et al., 2007. The extracellular matrix protein TGFBI induces microtubule stabilization and sensitizes ovarian cancers to paclitaxel. Cancer Cell 12, 514–527.

Al Rajabi, A., Booth, S.L., Peterson, J.W., Choi, S.W., Suttie, J.W., Shea, M.K., et al., 2012. Deuterium-labeled phylloquinone has tissue-specific conversion to menaquinone-4 among Fischer 344 male rats. J. Nutr. 142, 841–845.

Bellosta, P., Zhang, Q., Goff, S.E., Basilico, C., 1997. Signaling through the ARK tyrosine kinase receptor protects from apoptosis in the absence of growth stimulation. Oncogene 15, 2387–2397.

Booth, S.L., Lichtenstein, A.H., O'Brien-Morse, M., McKeown, N.M., Wood, R.J., Saltzman, E., et al., 2001. Effects of a hydrogenated form of vitamin K on bone formation and resorption. Am. J. Clin. Nutr. 74, 783–790.

Cancela, M.L., Conceição, N., Laizé, V., 2012. Gla-rich protein, a new player in tissue calcification? Adv. Nutr. 3, 174–181.

Conly, J.M., Stein, K., 1992. The production of menaquinones (vitamin K2) by intestinal bacteria and their role in maintaining coagulation homeostasis. Progr. Food Nutr. Sci. 16, 307–343.

Coutu, D.L., Wu, J.H., Monette, A., Rivard, G.E., Blostein, M.D., Galipeau, J., 2008. Periostin, a member of a novel family of vitamin K-dependent proteins, is expressed by mesenchymal stromal cells. J. Biol. Chem. 283, 17991–18001.

Daniele, A., Parenti, G., d'Addio, M., Andria, G., Ballabio, A., Meroni, G., 1998. Biochemical characterization of arylsulfatase E and functional analysis of mutations found in patients with X-linked chondrodysplasia punctata. Am. J. Hum. Genet. 62, 562–572.

Ferland, G., 2012. Vitamin K and the nervous system: an overview of its actions. Adv. Nutr. 3, 204–212.

Franco, B., Meroni, G., Parenti, G., Levilliers, J., Bernard, L., Gebbia, M., et al., 1995. A cluster of sulfatase genes on Xp22.3: Mutations in chondrodysplasia punctata (CDPX) and implications for warfarin embryopathy. Cell 81, 15–25.

Garber, A.K., Binkley, N.C., Krueger, D.C., Suttie, J.W., 1999. Comparison of phylloquinone bioavailability from food sources or a supplement in human subjects. J. Nutr. 129, 1201–1203.

Gijsbers, B.L., Jie, K.S., Vermeer, C., 1996. Effect of food composition on vitamin K absorption in human volunteers. Br. J. Nutr. 76, 223–229.

Goruppi, S., Ruaro, E., Varnum, B., Schneider, C., 1999. Gas6-mediated survival in NIH3T3 cells activates stress signalling cascade and is independent of Ras. Oncogene 18, 4224–4236.

Hakuno, D., Kimura, N., Yoshioka, M., Mukai, M., Kimura, T., Okada, Y., et al., 2010. Periostin advances atherosclerotic and rheumatic cardiac valve degeneration by inducing angiogenesis and MMP production in humans and rodents. J. Clin. Invest. 120, 2292–2306.

Hall, M.O., Prieto, A.L., Obin, M.S., Abrams, T.A., Burgess, B.L., Heeb, M.J., et al., 2001. Outer segment phagocytosis by cultured retinal pigment epithelial cells requires Gas6. Exp. Eye Res. 73, 509–520.

Hammed, A., Matagrin, B., Spohn, G., Prouillac, C., Benoit, E., Lattard, V., 2013. VKORC1L1, an enzyme rescuing the vitamin K 2,3-epoxide reductase activity in some extrahepatic tissues during anticoagulation therapy. J. Biol. Chem. 288, 28733–28742.

Harrington, D.J., Booth, S.L., Card, D.J., Shearer, M.J., 2007. Excretion of the urinary 5C- and 7C-aglycone metabolites of vitamin K by young adults responds to changes in dietary phylloquinone and dihydrophylloquinone intakes. J. Nutr. 137, 1763–1768.

Herman, T.E., Siegel, M.J., 2010. Warfarin-induced brachytelephalangic chondrodysplasia punctata. J. Perinatol. 30, 437–438.

Hixson, J.E., Shimmin, L.C., Montasser, M.E., Kim, D.K., Zhong, Y., Ibarguen, H., et al., 2011. Common variants in the periostin gene influence development of atherosclerosis in young persons. Arterioscler. Thromb. Vasc. Biol. 31, 1661–1667.

Hodges, S.J., Bejui, J., Leclercq, M., Delmas, P.D., 1993. Detection and measurement of vitamins K1 and K2 in human cortical and trabecular bone. J. Bone Min. Res. 8, 1005–1008.

Howe, A.M., Webster, W.S., 1994. Vitamin K-its essential role in craniofacial development. A review of the literature regarding vitamin K and craniofacial development. Austral. Dent. J. 39, 88–92.

Iioka, H., Moriyama, I.S., Morimoto, K., Akada, S., Hisanaga, H., Ishihara, Y., et al., 1991. Pharmacokinetics of vitamin K in mothers and children in the perinatal period: transplacental transport of vitamin K2 (MK-4). Asia Oceania J. Obstet. Gynaecol. 17, 97–100.

Inyangetor, P.T., Thierry-Palmer, M., 1988. Synthesis of vitamin K1 2,3-epoxide in rat liver mitochondria. Arch. Biochem. Biophys. 262, 389–396.

Jie, K.G., Bots, M.L., Vermeer, C., Witteman, J.C., Grobbee, D.E., 1996. Vitamin K status and bone mass in women with and without aortic atherosclerosis: a population-based study. Calc. Tissue Int. 59, 352–356.

Kaneki, M., Hedges, S.J., Hosoi, T., Fujiwara, S., Lyons, A., Crean, S.J., et al., 2001. Japanese fermented soybean food as the major determinant of the large geographic difference in circulating levels of vitamin K2: possible implications for hip-fracture risk. Nutrition 17, 315–321.

Kohlmeier, M., Chen, X.W., Anderson, J.J.B., 1998. Vitamin K reduces the IL-6 stimulating effect of PTH in murine osteoblast-like cells. Bone 23, S564.

Kohlmeier, M., Salomon, A., Saupe, J., Shearer, M.J., 1996. Transport of vitamin K to bone in humans. J. Nutr. 126, 1192S–1196S.

Kohlmeier, M., Saupe, J., Drossel, H.J., Shearer, M.J., 1995. Variation of phylloquinone (vitamin K1) concentrations in hemodialysis patients. Thromb. Haemostas. 74, 1252–1254.

Koshihara, Y., Hoshi, K., Shiraki, M., 1993. Vitamin K2 (menatetrenone) inhibits prostaglandin synthesis in cultured human osteoblast-like periosteal cells by inhibiting prostaglandin H synthase activity. Biochem. Pharmacol. 46, 1355–1362.

Kulman, J.D., Harris, J.E., Xie, L., Davie, E.W., 2001. Identification of two novel transmembrane gamma-carboxyglutamic acid proteins expressed broadly in fetal and adult tissues. Proc. Natl. Acad. Sci. USA 98, 1370–1375.

Kulman, J.D., Harris, J.E., Xie, L., Davie, E.W., 2007. Proline-rich Gla protein 2 is a cell-surface vitamin K-dependent protein that binds to the transcriptional coactivator Yes-associated protein. Proc. Natl. Acad. Sci. USA 104, 8767–8772.

Lacombe, J., Karsenty, G., Ferron, M., 2013. *In vivo* analysis of the contribution of bone resorption to the control of glucose metabolism in mice. Mol. Metab. 2, 498–504.

Lee, W.E., Wen, Y., Varnum, B., Hung, M.C., 2002. Akt is required for Axl-Gas6 signaling to protect cells from E1A-mediated apoptosis. Oncogene 21, 329–336.

Ling, L., Cheng, Y., Ding, L., Yang, X., 2014. Association of serum periostin with cardiac function and short-term prognosis in acute myocardial infarction patients. PLoS One 9, e88755.

Lipsky, J.J., 1994. Nutritional sources of vitamin K. Mayo Clin. Proc. 69, 462–466.

Matos-Miranda, C., Nimmo, G., Williams, B., Tysoe, C., Owens, M., Bale, S., et al., 2013. A prospective study of brachytelephalangic chondrodysplasia punctata: identification of arylsulfatase E mutations, functional analysis of novel missense alleles, and determination of potential phenocopies. Genet. Med. 58, 2297–2306.

Mitchell, G.V., Cook, K.K., Jenkins, M.Y., Grundel, E., 2001. Supplementation of rats with a lutein mixture preserved with vitamin E reduces tissue phylloquinone and menaquinone-4. Int. J. Vit. Nutr. Res. 71, 30–35.

Nino, M., Matos-Miranda, C., Maeda, M., Chen, L., Allanson, J., Armour, C., et al., 2008. Clinical and molecular analysis of arylsulfatase E in patients with brachytelephalangic chondrodysplasia punctata. Am. J. Med. Genet. A 146A, 997–1008.

Onodera, K., Shinoda, H., Zushida, K., Taki, K., Kamei, J., 2000. Antinociceptive effect induced by intraperitoneal administration of vitamin K2 (menatetrenone) in ICR mice. Life Sci. 68, 91–97.

Oury, F., Ferron, M., Huizhen, W., Confavreux, C., Xu, L., Lacombe, J., et al., 2013. Osteocalcin regulates murine and human fertility through a pancreas-bone-testis axis. J. Clin. Invest. 123, 2421–2433.

Pastori, D., Carnevale, R., Cangemi, R., Saliola, M., Nocella, C., Bartimoccia, S., et al., 2013. Vitamin E serum levels and bleeding risk in patients receiving oral anticoagulant therapy: a retrospective cohort study. J. Am. Heart Assoc. 2 (6), e000364.

Rafael, M.S., Cavaco, S., Viegas, C.S., Santos, S., Ramos, A., Willems, B.A., et al., 2014. Insights into the association of Gla-rich protein and osteoarthritis, novel splice variants and γ-carboxylation status. Mol. Nutr. Food Res. 58, 1636–1646.

Reddi, K., Henderson, B., Meghji, S., Wilson, M., Poole, S., Hopper, C., et al., 1995. Interleukin 6 production by lipopolysaccharide-stimulated human fibroblasts is potently inhibited by naphthoquinone (vitamin K) compounds. Cytokine 7, 287–290.

Ronden, J.E., Drittij-Reijnders, M.J., Vermeer, C., Thijssen, H.H., 1998. Intestinal flora is not an intermediate in the phylloquinone-menaquinone-4 conversion in the rat. Biochim. Biophys. Acta 1379, 69–75.

Schurgers, L.J., Uitto, J., Reutelingsperger, C.P., 2013. Vitamin K-dependent carboxylation of matrix Gla-protein: a crucial switch to control ectopic mineralization. Trends Mol. Med. 19, 217–226.

Schurgers, L.J., Vermeer, C., 2000. Determination of phylloquinone and menaquinones in food. Effect of food matrix on circulating vitamin K concentrations. Haemostasis 30, 298–307.

Shearer, M.J., Bach, A., Kohlmeier, M., 1996. Chemistry, nutritional sources, tissue distribution and metabolism of vitamin K with special reference to bone health. J. Nutr. 126, 1181S–1186S.

Shearer, M.J., Mallinson, C.N., Webster, G.R., Barkhan, P., 1972. Clearance from plasma and excretion in urine, faeces and bile of an intravenous dose of tritiated vitamin K1 in man. Br. J. Haematol. 22, 579–588.

Shiraki, M., Shiraki, Y., Aoki, C., Miura, M., 2000. Vitamin K2 (menatetrenone) effectively prevents fractures and sustains lumbar bone mineral density in osteoporosis. J. Bone Min. Res. 15, 515–521.

Sundaram, K.S., Lev, M., 1992. Purification and activation of brain sulfotransferase. J. Biol. Chem. 267, 24041–24044.

Thane, C.W., Paul, A.A., Bates, C.J., Bolton-Smith, C., Prentice, A., Shearer, M.J., 2002. Intake and sources of phylloquinone (vitamin K1): variation with socio-demographic and lifestyle factors in a national sample of British elderly people. Br. J. Nutr. 87, 605–613.

Theuwissen, E., Smit, E., Vermeer, C., 2012. The role of vitamin K in soft-tissue calcification. Adv. Nutr. 3, 166–173.

Thijssen, H.H., Vervoort, L.M., Schurgers, L.J., Shearer, M.J., 2006. Menadione is a metabolite of oral vitamin K. Br. J. Nutr. 95, 260–266.

Thijssen, H.H.W., Drittij-Reinders, M.J., 1996. Vitamin K status in human tissues: tissue specific accumulation of phylloquinone and menaquinone-4. Br. J. Nutr. 75, 121–127.

Valerio Jr, L.G., Kepa, J.K., Pickwell, G.V., Quattrochi, L.C., 2001. Induction of human NAD(P)H:quinine oxido-reductase (NQO1) gene expression by the flavonol quercetin. Toxicol. Lett. 119, 49–57.

Viegas, C.S., Herfs, M., Rafael, M.S., Enriquez, J.L., Teixeira, A., Luís, I.M., et al., 2014. Gla-rich protein is a potential new vitamin K target in cancer: evidences for a direct GRP-mineral interaction. Biomed. Res. Int. 2014, 340216.

Vermeer, C., Theuwissen, E., 2011. Vitamin K, osteoporosis and degenerative diseases of ageing. Menopause Int. 17, 19–23.

Vlaming, M.L., Lagas, J.S., Schinkel, A.H., 2009. Physiological and pharmacological roles of ABCG2 (BCRP): recent findings in Abcg2 knockout mice. Adv. Drug. Deliv. Rev. 61, 14–25.

Wang, N., Zhang, H., Yao, Q., Wang, Y., Dai, S., Yang, X., 2012. TGFBI promoter hypermethylation correlating with paclitaxel chemoresistance in ovarian cancer. J. Exp. Clin. Cancer Res. 31, 6.

Weaver, K.N., El Hallek, M., Hopkin, R.J., Sund, K.L., Henrickson, M., Del Gaudio, D., et al., 2014. Keutel syndrome: report of two novel MGP mutations and discussion of clinical overlap with arylsulfatase E deficiency and relapsing polychondritis. Am. J. Med. Genet. A 164A, 1062–1068.

Willems, B.A., Vermeer, C., Reutelingsperger, C.P., Schurgers, L.J., 2014. The realm of vitamin K dependent proteins: Shifting from coagulation toward calcification. Mol. Nutr. Food Res. 58, 1620–1635.

Yanagita, M., Ishii, K., Ozaki, H., Arai, H., Nakano, T., Ohashi, K., et al., 1999. Mechanism of inhibitory effect of warfarin on mesangial cell proliferation. J. Am. Soc. Nephrol. 10, 2503–2509.

Zhu, H., Li, Y., 2012. NAD(P)H: quinone oxidoreductase 1 and its potential protective role in cardiovascular diseases and related conditions. Cardiovasc. Toxicol. 12, 39–45.

CHOLESTEROL

Cholesterol (molecular weight 387) is the principal neutral sterol in mammals.

Note: The mechanisms involved in the synthesis, intestinal absorption, transport, and maintenance of cellular and whole-body homeostasis of cholesterol are likely to involve significantly more than 100 distinct genes and are by far the most complex for any nutrient (Figure 9.32).

ABBREVIATIONS

ABCA1	ATP-binding cassette A1
ABCG5	ATP-binding cassette G5
ABCG8	ATP-binding cassette G8
apoB	apolipoprotein B
apoB48	apolipoprotein B48
apoB100	apolipoprotein B100
apoE	apolipoprotein E
Chol	cholesterol
Chylos	chylomicrons
CYP	cytochrome p450
HDL	high-density lipoprotein
LDL	low-density lipoprotein
LDLR	LDL receptor
Lp	lipoprotein
LRP	LDL receptor-related protein
SR-B1	scavenger receptor class B type 1
SREBP	sterol regulatory element-binding protein
VLDL	very-low-density lipoprotein

NUTRITIONAL SUMMARY

Function: Cholesterol (Chol) is an essential component of membranes and serves as a precursor for the synthesis of bile acids and steroid hormones. Intermediates of Chol synthesis are also used for protein modification and for the synthesis of ubiquinone, dolichol, and vitamin D.

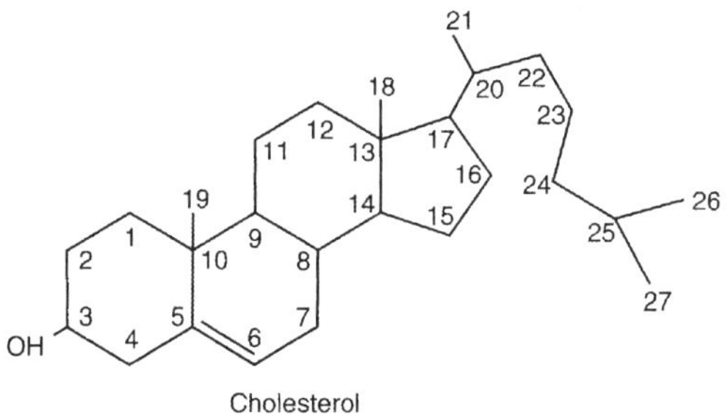

Cholesterol

FIGURE 9.32

Cholesterol.

Food sources: Most foods of animal origin contain Chol; the largest amounts come from fatty foods and from the liver. Plant-derived foods contain no significant amounts of Chol, regardless of their fat content.

Requirements: More than adequate amounts can be produced by the body, and there is no evidence that intake at any level improves health.

Deficiency: No adverse effects of low or absent intake are known.

Excessive intake: High Chol intake increases blood Chol concentrations, and possibly cardiovascular risk, in some people but not others.

ENDOGENOUS SOURCES

Daily Chol synthesis in healthy people is about 15 mg/kg body weight (Rajaratnam et al., 2001). Chol intake decreases endogenous synthesis moderately (Jones et al., 1996), but this effect is seen more clearly in some people (known as *responders*) than in others. High fat intake (Linazasoro et al., 1958) and many other factors increase Chol synthesis. Consumption of the Chol precursor squalene at levels comparable to those in normal foods also promotes Chol synthesis (Relas et al., 2000).

The main sites of production are the liver and intestines; smaller amounts come from many other tissues. Chol synthesis proceeds in cytosol with acetyl-CoA as a precursor and requires adequate availability of riboflavin, niacin, pantothenate, magnesium, and iron (Figure 9.33). The rate-limiting step of Chol synthesis is the production of mevalonate, which is then converted to the 5-carbon isoprenoid isopentenyl pyrophosphate.

Addition of six isoprenoid units generates squalene, which is cyclized and then converted in two more steps to Chol. The first reactions serve the synthesis of both Chol and several important isoprenoid compounds. A parallel system for the synthesis of isoprenoids, which operates in peroxisomes, uses some genetically distinct isoenzymes and is likely to be regulated differently. Farnesyl pyrophosphate provides a lipophilic membrane anchor for specific proteins and is the precursor for the synthesis of ubiquinone (oxidative phosphorylation) and dolichol (synthesis of N-linked glycoproteins). Isopentenyl phosphate might also be important as a precursor for the side-chain resynthesis of vitamin K. The acetate precursor for Chol synthesis is shuttled from mitochondria (where it is generated by the catabolism of fatty acids and some amino acids) to the cytosol as citrate by the citrate transport system. Acetyl-CoA is then released from citrate in cytosol by ATP-citrate (pro-S-)-lyase (EC4.1.3.8), a magnesium-dependent enzyme. The activity of this critical branch-point enzyme is regulated in part through phosphorylation by cyclic adenosine monophosphate (cAMP)–dependent protein kinase. Hydroxymethylglutaryl-CoA synthase (HMG-CoA synthase, EC4.1.3.5) condenses acetyl-CoA and acetoacetyl-CoA. Acetyl-CoA C-acetyltransferase (thiolase, EC2.3.1.9) produces the latter by linking two acetyl-CoA molecules. The cytoplasmic forms of both enzymes for sterol synthesis are genetically distinct from the mitochondrial forms for ketone body synthesis.

The next step, NADPH-dependent reduction of hydroxymethylglutaryl-CoA to mevalonate by hydroxymethylglutaryl-CoA reductase (HMG CoA reductase, EC1.1.1.34), is a tightly controlled reaction that sets the pace of Chol production. Phosphorylation by hydroxymethylglutaryl-CoA reductase (NADPH) kinase (EC2.7.1.109) inactivates HMG CoA reductase; dephosphorylation by hydroxymethylglutaryl-CoA reductase (NADPH)-phosphatase (EC3.1.3.47) reactivates the phosphorylated enzyme. Mevalonate is then diphosphorylated in two ATP-consuming steps (catalyzed by mevalonate kinase, EC2.7.1.36; and phosphomevalonate kinase, EC2.7.4.2) and converted to isopentenyl phosphate by diphosphomevalonate decarboxylase (EC4.1.1.33). The magnesium-dependent enzyme isopentenyl diphosphate delta-isomerase (EC5.3.3.2) produces dimethylallyl phosphate, the second five-carbon precursor used in isoprenoid and Chol synthesis.

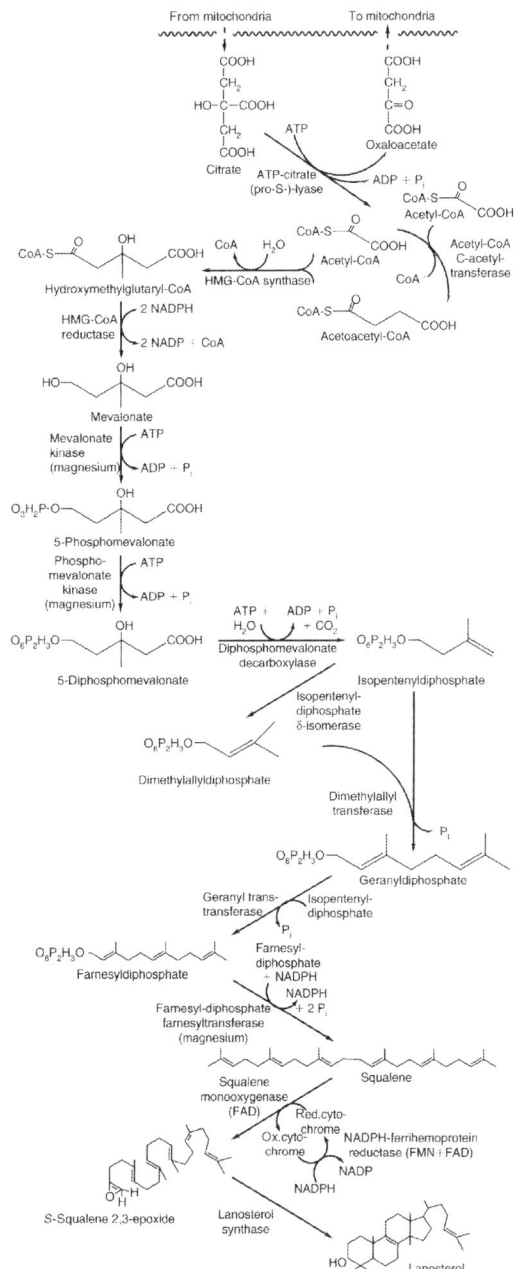

FIGURE 9.33

Endogenous Chol synthesis starts from acetyl-CoA.

A protein expressed in the liver and many other tissues first generates geranyl diphosphate by linking isopentenyl phosphate and dimethylallyl phosphate (dimethylallyltransferase, EC2.5.1.1) and then produces trans, trans-farnesyl diphosphate by adding another isopentenyl diphosphate moiety (geranyltranstransferase, EC2.5.1.10). A genetically distinct protein, which is strongly expressed in the testis and at lower levels in most other tissues, has the additional capacity to synthesize by adding yet another isopentenyl diphosphate to trans, trans-farnesyl diphosphate (farnesyltranstransferase, EC2.5.1.29). This 20-carbon intermediate is not only a key Chol precursor but also provides the starting point for dolichol synthesis and has important functional roles as a membrane anchor for farnesylated proteins.

Two farnesyl pyrophosphates are joined by farnesyl-diphosphate farnesyltransferase (squalene synthase, EC2.5.1.21, cofactor magnesium) and then reduced by the same NADPH-dependent enzyme. The linear 30-carbon precursor squalene is oxidized in an NADPH-consuming reaction by the FAD-enzyme squalene monooxygenase (squalene epoxidase, EC1.14.99.7) and cyclized to the first sterol intermediate lanosterol by lanosterol synthase (EC5.4.99.7).

To complete Chol synthesis, the methyl groups at positions 4-alpha, 4-beta, and 14 have to be removed, the double bond at position saturated, and the double bond at position 8 moved to position 5. These reactions are catalyzed by lanosterol demethylase (uses NADPH for the decarboxylation reaction, EC1.14.13.70), lathosterol oxidase (EC1.3.3.2), and 7-dehydrocholesterol reductase (EC1.3.1.21). Several pathways are possible, depending on the relative activities of the involved enzymes, but the preferred sequence may be lanosterol > 14-alpha desmethyllanosterol > zymosterol > 5-alpha-cholest-8-en 3-beta-ol > lathosterol > 7 dehydrocholesterol > cholesterol. Desmosterol (delta-3,24-cholestadien-3-beta-ol) is one of several alternative intermediates. 7-dehydrocholesterol is important since it is the precursor for UV-B-light-induced synthesis of vitamin D in the skin (Obi-Tabot et al., 2000).

DIETARY SOURCES

Membranes and fat deposits of animal-derived foods contain Chol, while plant-derived foods contain plant sterols instead (Figure 9.34). A large portion of Chol in animals is linked to fatty acids, predominantly the long-chain monounsaturated and polyunsaturated ones. The single largest source of Chol in the American diet is eggs (4.3 mg/g), which provide about 215 mg per serving (50 g). Other Chol-rich sources include organ meats, such as the liver (3.9 mg/g), and animal fats, such as tallow (1.1 mg/g), lard (1.0 mg/g) and butter (2.2 mg/g), and all meats and fish. The fat-free part of meats still contains between 0.7 and 0.9 mg/g. Typical daily intake in the United States is around 300–500 mg, but much lower in vegans and other people with minimal intake of animal-derived products.

Significant amounts of squalene and other Chol precursors are also present in some foods. Plant-derived foods contain a wide range of neutral sterols (phytosterols) that resemble Chol structurally but cannot substitute for most of its functions and have different metabolic fates. Several hundred milligrams of such phytosterols are consumed daily with typical mixed diets.

DIGESTION AND ABSORPTION

Dietary Chol is absorbed best from mixed micelles that contain bile acid, phospholipids, and monoglycerides (Figure 9.35). Bile acids and phospholipids come from bile. Gastric and pancreatic lipases (EC3.1.1.3) act on dietary triglycerides and release the monoglycerides. Bile-salt-activated lipase from pancreas and mammary gland has carboxylester lipase (EC3.1.1.13) activity and cleaves Chol esters

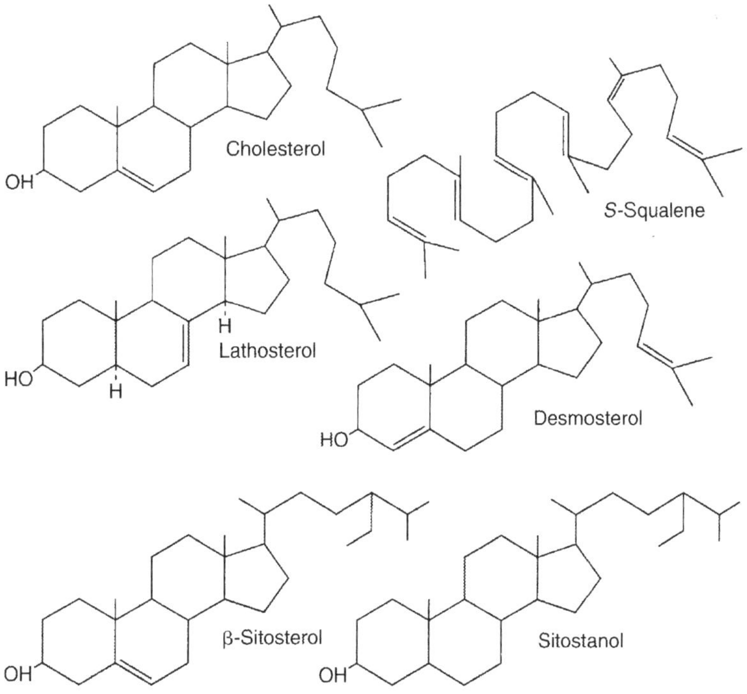

FIGURE 9.34

Sterols in foods.

in concert with pancreatic lipase (EC3.1.1.3) and colipase. Most people absorb 40–45% of a moderate Chol dose ingested with triglycerides (Rajaratnam et al., 2001). Absorption occurs mainly in the proximal small intestine, is selective (high compared to closely related plant sterols), and responds to whole-body Chol status (Lu et al., 2001).

The mechanism whereby Chol moves from the mixed micelles (containing Chol, phospholipids, and bile acids) across the enterocyte brush border membrane is still incompletely understood (Hui et al., 2008). Both the SR-BI and the fatty acid transporter CD36 (SCARB3, thrombospondin receptor) help to move Chol from the intestinal lumen into enterocytes. SR-BI is located mainly in the duodenum and proximal part of the jejunum, while CD36 is in the distal parts of the jejunum and the proximal ileum. The interaction takes place at Chol and sphingomyelin-enriched domains of the plasma membrane called caveolae, which facilitate lipid exchange between micelles and enterocytes (Graf et al., 1999). The caveolae contain caveolin-1, a protein with as-yet-incompletely understood function. NPC1L1 works with both SR-BI and CD36, helping to traffic absorbed Chol from the apical membrane to the cells interior (Hui et al., 2008). Absorption efficiency is modulated by a heterocomplex containing ABCG5 and ABCG8 that actively pumps Chol (and plant sterols) back into the intestinal lumen. In some as-yet-unexplained fashion, this complex assists in Chol absorption (Nguyen et al., 2012).

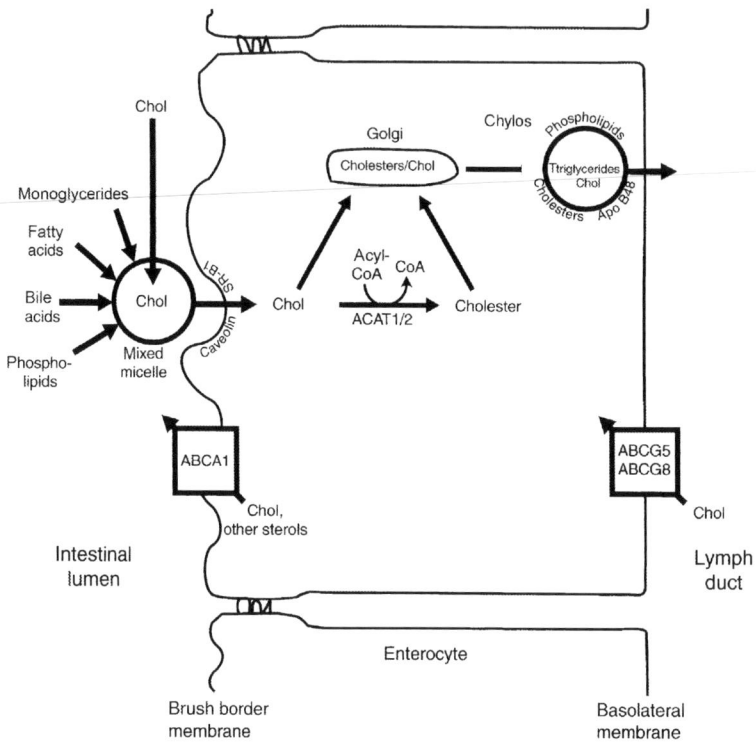

FIGURE 9.35

Intestinal absorption of Chol.

Once in the enterocyte, most Chol is esterified by at least two genetically distinct forms of sterol *O*-acyltransferase (EC2.3.1.26), ACAT, and ACAT2 (Buhman et al., 2000). Both forms specifically esterify neutral sterols with omega-9 fatty acids, such as oleic acid.

Chol is exported from enterocytes into adjacent lymph ducts with Chylos, which are very-triglyceride-rich Lps. Each chylomicron contains a single copy of apolipoprotein B48 (apoB48). A highly specific cytidine deaminase (C > U editing enzyme APOBEC-1, EC3.5.4.-) is part of a multiprotein complex that modifies cytidine 6666 of the apoB mRNA to uridine and thereby introduces a stop codon into the sequence. Due to this modification, the intestinal protein transcripts are shortened to about 48% of the full-length version (hence B48). In humans, this enzyme is expressed only in the small intestine (Chen et al., 2001), which means that all intestinal apoB is of the apoB48 variety. Rodents and many other mammals, in contrast, produce in their enterocytes both apoB48 and the full-length version, apoB100.

The ATP-binding cassette transporter A1 (ABCA1) transfers some intracellular-free Chol back into the intestinal lumen and another portion with apolipoprotein AI (apoAI) into the blood (Abumrad and Davidson, 2012). Only some of the Chol exported by ABCA1 comes from the diet, since most newly absorbed Chol is rapidly esterified.

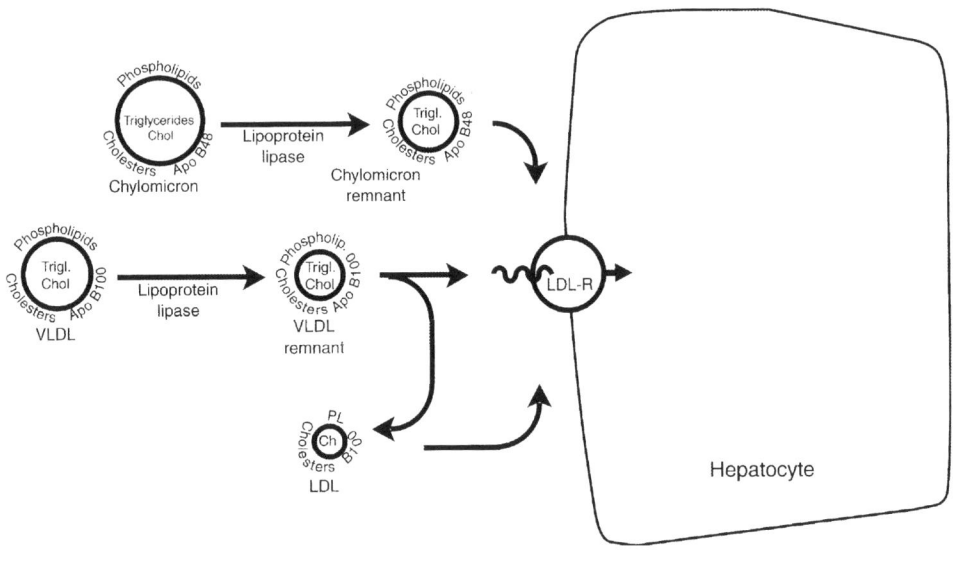

FIGURE 9.36

Modification and uptake of Chol-carrying Lps.

TRANSPORT AND CELLULAR UPTAKE

Blood circulation: All Chol in the blood is contained in Lps, and all Lps contain Chol. Lps are classi-fied based on their buoyant density and additional pathogenetic characteristics. The major Lps in the blood are LDLs, HDLs, VLDLs, Lp(a), Chylos, VLDL remnants, and Chylo remnants. Chylos, which carry triglycerides from small intestinal absorption, release much of their triglycerides as they pass through small arterioles and arterial capillaries, due to triglyceride hydrolysis by Lp lipase (EC3.1.1.3). The resulting Chylo remnants are taken up mainly into the liver and bone marrow (Cooper, 1997). VLDLs, which transport excess triglycerides from the liver, release triglycerides in the same way. The liver and peripheral tissues take up some of the resulting VLDL remnants, while others are con-verted into LDL. All LDLs in the blood are derived from VLDLs. While circulating with HDLs in the blood, Chol can be esterified by phosphatidylcholine-sterol *O*-acyltransferase (LCAT, EC2.3.1.43). Since ω6-polyunsaturated fatty acids (mostly linoleic acid) are preferentially used, about two-thirds of the Chol esters in the blood contain linoleic acid. In addition, lipids can be transferred from one Lp to another; CETPs and PLTPs facilitate such exchanges.

Numerous membrane-linked receptors recognize one or several of the circulating Lps, usually lead-ing to the endocytotic uptake of the bound Lp. The LDL receptor (LDLR) is present at the surface of hepatocytes and other cells and mediates the uptake of LDL and remnants (Figure 9.36). Some VLDLs can be taken up, mainly into heart and skeletal muscle and adipose tissue, via the VLDL receptor (VLDLR), another member of the LDLR family (Zenimaru et al., 2008). The LDL receptor-related protein 1 (LRP1) is also involved in Lp uptake and is especially important for the hepatic uptake of Chylo remnants (Herz and Strickland, 2001). Other members of the LDLR family (VLDLR, APOER2,

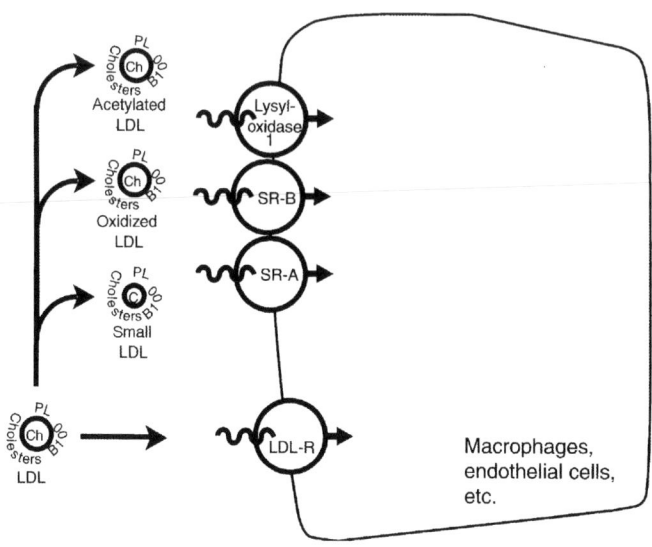

FIGURE 9.37

Lp modification changes Chol flux into tissues.

LRP1, LRP2, LRP6, and others) have much more diverse roles (Go and Mani, 2012). The receptor-associated protein (RAP) acts as a chaperone for such receptors and modulates their activity (Willnow et al., 1995). LDLR clusters in coated pits, which are destined for endocytosis (Figure 9.37).

As these coated pits are folded into endosomes, moved through cells, and transformed into lysosomes, the Chol esters inside are hydrolyzed and free Chol is delivered by the NPC1 protein (defective in Niemann–Pick disease type C) to the endoplasmic reticulum (Skov et al., 2011). Most Chol in the ESR is esterified again by sterol O-acyltransferase (ACAT, EC 2.3.1.26). An alternative and less explored pathway for the uptake of VLDL-derived particles in smooth muscle cells may be mediated directly by heparan sulfate proteoglycans on the cell surface of smooth muscle and other cells with low expression of LDLR family members (Weaver et al., 1997). Apolipoprotein C-I (ApoC-I) is a small peptide constituent of Chylo and VLDL remnants. ApoC-I inhibits the binding of Lps to the LDLR, LRP1, and the VLDL receptor, slows lipid exchange mediated by the CETP, and decreases cellular fatty acid uptake (Shachter, 2001).

As Lps circulate in the blood, they age due to exposure to enzymatic and chemical modification. A relatively common type of Lp modification is the unspecific oxidation of the protein or lipid constituents upon reaction with oxygen free radicals. Modification of LDL diminishes binding to LDLR but opens up new epitopes for binding to scavenger receptors, including scavenger receptors of classes A and B and lysyl-oxidase-1 (Krieger, 2001; Linton and Fazio, 2001; Mertens and Holvoet, 2001). Chol entering cells via scavenger receptors is sequestered in lysosomes and is less available for reverse transport. Since the liver contains mainly LDLR and few scavenger receptors, hepatic uptake and excretion of modified LDL are blocked and redirected to macrophages and other extrahepatic tissues.

Reverse Chol transport: Some tissues, especially endothelial cells and macrophages, can export excess Chol (Fielding and Fielding, 2001). Since Chol transport is directed predominantly from the

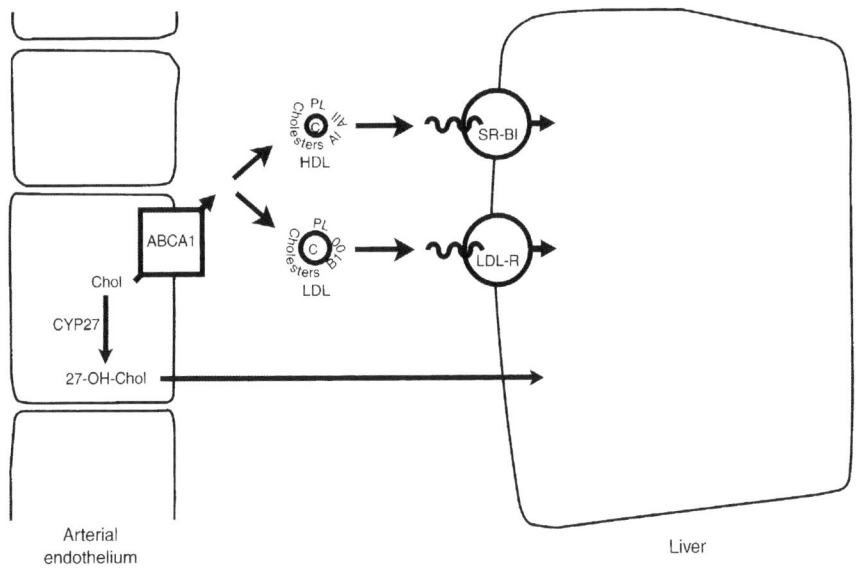

FIGURE 9.38

Reverse transport is crucial for the removal of excess tissue Chol.

blood into tissues, this is commonly called *reverse Chol transport* (Figure 9.38). This type of transfer involves the active transport of Chol by several ATP-binding cassette (ABC) transporters, including ABCA1 (Liao et al., 2002), ABCG1 (Kennedy et al., 2001), and ABCG4 (Engel et al., 2001) to LDL and HDL or precursors of HDL (Fielding and Fielding, 2001). The precise mechanisms are still under intense investigation. LDL and ligands of the RXR and LXR increase ABCA1 expression (Liao et al., 2002). Scavenger receptor-B1 (SRB1) appears to be instrumental for the binding of HDLs (Silver and Tall, 2001) and the internalization of their Chol load in the liver (Thuahnai et al., 2001). A polymorphic variant of this receptor may confer survival benefits in populations exposed to parasite infections. Another mode of reverse transport uses the conversion of Chol into 27-hydroxycholesterol by the mitochondrial sterol-27 hydroxylase (CYP27, EC1.14.13.15), followed by export of this oxysterol into the blood, its uptake into the liver, and metabolism to bile acids (Russell, 2000). An analogous mechanism with the initial synthesis of 24S hydroxycholesterol in the brain is described later in this chapter.

Materno-fetal transfer: Transfer of Chol from mother to fetus is minimal and not enough to supply the needs of the fetus (DeBarber et al., 2011). Maternal Chol may be more important as a precursor for steroid hormone synthesis in the placenta than for use by the fetus. Without the adequate delivery of maternal Chol by HDL, the placenta does not develop normally and fetal death is possible (Christiansen-Weber et al., 2000). Endocytotic uptake of maternal HDL is mediated by the intrinsic factor receptor (cubilin), a member of the LDL receptor-protein family (Kozyraki et al., 1999). Significant amounts of Chol may also enter the syntrophoblast without the mediation of receptors (Wyne and Woollett, 1998). Less certain are the individual contributions of the numerous other Lp receptors at the maternal face of the placenta, which include the LDL, acetylated LDL, apoE, and VLDL receptors, as well as the SR-BI and megalin. Chol efflux across the basal membrane into fetal circulation is mediated by the ATP-driven transporter ABCA1 (Christiansen-Weber et al., 2000).

BBB: Excess Chol (about 3 mg/day) can move from the brain into the blood as 24S-hydroxycholesterol, partially mediated by ABCA1 (Saint-Pol et al., 2012). Very little Chol moves from the blood into the brain (DeBarber et al., 2011).

METABOLISM

Oxysterols: The hydroxylation of Chol by specific enzymes generates oxysterols, particularly 24(S)-hydroxycholesterol, 25-hydroxycholesterol, and 27-hydroxycholesterol (Russell, 2000). These oxysterols are minor precursors for bile acid synthesis, which help some tissues to export excess Chol as described previously (Figure 9.39). Even more important, these oxysterols serve as ligands for a

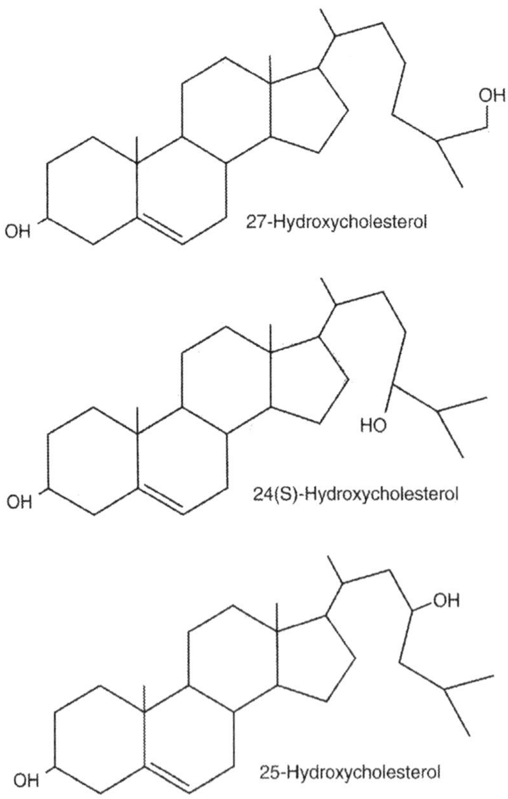

FIGURE 9.39

Oxysterols are both bile acid precursors and important signaling compounds.

group of nuclear receptors (including LXRα and LXRβ) that regulate Chol synthesis, intestinal absorption, and movement into and out of tissues. Sterol 27-hydroxylase (CYP27, EC1.14.13.15) is a mitochondrial cytochrome P450 enzyme that generates 27-hydroxycholesterol in the liver and other tissues; lack of this activity is the cause of the rare genetic disease cerebrotendinous xanthomatosis characterized by accumulation of Chol and cholestanol in the brain and tendons (Björkhem, 2013). Cholesterol 24-hydroxylase (CYP46, EC1.14.13.98) generates 24(S)-hydroxycholesterol in the brain, but not in other tissues to any significant extent. Cholesterol 25-hydroxylase (EC 1.14.99.38) is expressed at very low levels in many tissues.

Bile acids: About half of the newly synthesized and ingested Chol is metabolized, much of this directed toward bile acid synthesis in the liver (Figure 9.40). Smaller amounts of bile acids and oxysterols are also produced in extrahepatic tissues, not least as a means of eliminating excess Chol (Russell, 2000; Björkhem, 2013).

A membrane-bound heme enzyme of the endoplasmic reticulum, cytochrome P450 7A1 (cholesterol 7-alpha-monooxygenase, EC1.14.13.17), catalyzes the initial rate-limiting step of the classic (neutral) pathway for bile acid synthesis from Chol in the liver. The NADPH-dependent reaction generates 7-alpha-cholesterol, which then can be converted by subsequent reactions into cholic acid or chenodesoxycholic acid. The alternative acidic pathway starts from the oxysterol 27-hydroxycholesterol in the liver and other tissues via 27-hydroxycholesterol 7-alpha-monooxygenase (EC1.14.13.60).

EXCRETION

Daily secretion of Chol in bile amounts to about 18 mg/kg body weight; about 50–60% of this biliary Chol is usually lost via feces (Rajaratnam et al., 2001). In addition, large amounts of Chol-derived bile

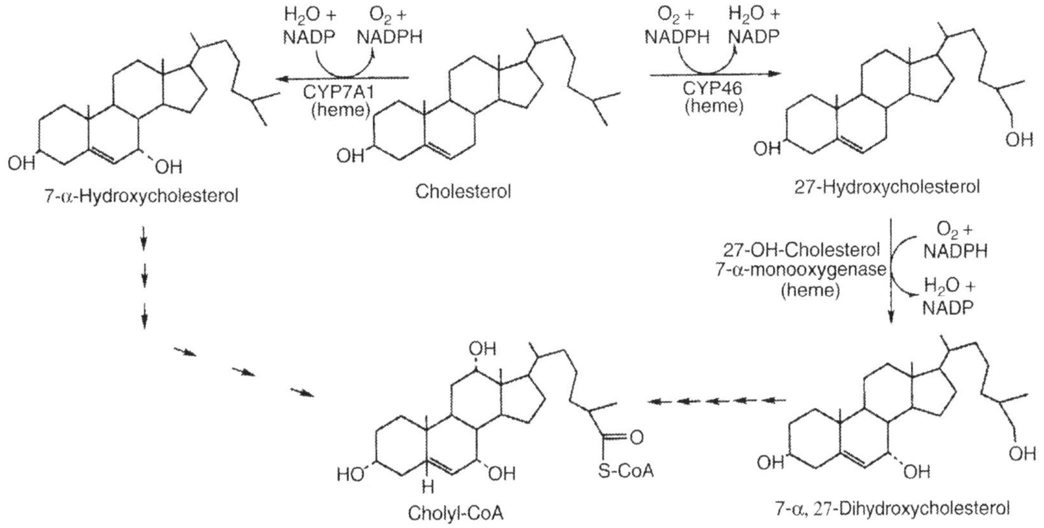

FIGURE 9.40

The main metabolic fate of Chol is conversion into bile acids.

acids are secreted with bile. Due to the very efficient recovery from the ileum via the sodium/bile acid cotransporter (SLC10A2), which works in conjunction with the intestinal bile-acid-binding protein, the loss of bile acids amounts to less than 2% (about 6 mg/kg body weight) of the amount secreted with bile. Taken together, daily excretion of Chol and Chol-derived products is typically around 17 mg/kg. Much smaller amounts are lost with skin and the secretions of seborrheic glands.

REGULATION

Chol homeostasis depends mainly on the rate of endogenous synthesis, the effectiveness of intestinal absorption of Chol from diet and bile, and the rate of reverse Chol transport from tissues back to the liver for excretion.

It appears that import of sterols from circulation via this latter pathway, particularly from apoE-containing Lps, provides the enterocyte with the necessary signal to indicate Chol availability in the system.

The SR-BI senses the Chol content of cell membranes of enterocytes and other endothelial cells (Saddar et al., 2013). SR-BI also functions as an HDL receptor and mediates Chol uptake from intestinal micelles. Its signals to intracellular compartments influence trafficking of apoB, a key event of Chol transport. SR-BI also modulates HDL-dependent activation of endothelial NO synthase (EC1.14.13.39) and induction of angiogenesis. Whole-body Chol status is known to influence fractional Chol absorption from the small intestine (Lu et al., 2001). While the active sterol transporter complex ABCG5/ABCG8 is known to play a critical role in this regulation, there is still considerable uncertainty about the specifics.

The key enzyme of Chol synthesis is HMG CoA reductase (EC1.1.1.34). Inactivation by phosphorylation (hydroxymethylglutaryl-CoA reductase [NADPH] kinase, EC2.7.1.109) reactivation by dephosphorylation (hydroxymethylglutaryl-CoA reductase [NADPH]-phosphatase, EC3.1.3.47) modulate HMG CoA reductase activity. Considerable uncertainty still exists about the precise manner in which kinase and phosphatase activities are controlled by various regulators, including various receptors (e.g., LXR and RXR). In addition, every individual step of the Chol synthetic pathway is activated by the sterol regulatory element-binding proteins (SREBPs) la and 2 (Sakakura et al., 2001).

Uptake into tissues: Lp transport into cells is differentially regulated in tissues and influenced by an exceedingly complex web of numerous receptors and modulators affecting their cell-specific expression. Only two examples will be mentioned here. Expression of the LDLR, the main conduit for Chol from circulation into the liver, is down-regulated by high Chol intake (Jones et al., 1996). Phosphorylation of the LDLR by low-density Lp receptor kinase (EC2.7.1.131) is another tool for the rapid fine-tuning of Chol uptake to the individual needs of target cells (Kishimoto et al., 1987).

Reverse Chol transport: Heterodimers comprising RXRs in combination with the bile acid receptor and similar ones containing oxysterol receptors increase expression of the reverse Chol transporter ABCA1 in response to oxysterols and bile acids (Repa et al., 2000). The same oxysterols and bile acids also drive the expression of SR-BI, another critical actor of reverse Chol transport. There is increasing evidence that the gut microbiome influences reverse Chol transport by modifying bile acids and other metabolites with signaling properties (Lee-Rueckert et al., 2013). However, considerable gaps still persist in the understanding of all the different mechanisms that control reverse Chol transport.

FUNCTION

Bile acid synthesis: Chol is the precursor for bile acid synthesis as detailed previously. Acids in bile contribute to the formation of mixed micelles, a prerequisite for the digestion and absorption of all fat-soluble nutrients, including triglycerides, sterols, and vitamins A, D, E, and K.

Steroid hormone synthesis: Chol is the direct precursor of corticosteroid, mineralocorticoid, and steroid sex hormones. Cytochrome P450 11A1 (Chol side-chain cleavage enzyme, P450scc, EC1.14.15.6), a mitochondrial heme enzyme, catalyzes the initial and rate-limiting step of steroid hormone synthesis. This oxidative cleavage reaction uses reduced adrenal ferredoxin and generates pregnenolone. Hydroxylation at the 3-beta position generates progesterone, 17-alpha-hydroxylation and side-chain cleavage gives rise to the important precursor dehydroepiandrosterone (DHEA). The tissue-specific actions of various enzymes on these precursors produce androgens, gestagens, estrogens, and other gonadal steroid hormones. Similar reactions in the adrenal cortex convert pregnenolone to glucocorticosteroids (cortisol and corticosterone) and mineralocorticosteroids (aldosterone). The synthesis rate of all steroid hormones is tightly regulated and not influenced by variations in Chol supply.

Vitamin D synthesis: Exposure of skin to UV light with wavelengths between 290 and 315 nm can convert the Chol precursor 7-dehydrocholesterol to previtamin D_3, which rearranges spontaneously to form vitamin D_3 (Holick et al., 1989). The diminished vitamin D production in older people has been attributed in part to a lower concentration of unesterified 7-dehydrocholesterol in their skin (Holick, 1999).

Protein lipidation: The covalent attachment of Chol to its C-terminus modifies the signaling properties of sonic hedgehog (Shh) protein (Resh, 2012). Disrupted Chol attachment interferes with embryonic patterning of limbs and organs.

REFERENCES

Abumrad, N.A., Davidson, N.O., 2012. Role of the gut in lipid homeostasis. Physiol. Rev. 92, 1061–1085.

Björkhem, I., 2013. Five decades with oxysterols. Biochimie 95, 448–454.

Buhman, K.K., Accad, M., Novak, S., Choi, R.S., Wang, J.S., Hamilton, R.L., et al., 2000. Resistance to diet-induced hypercholesterolemia and gallstone formation in ACAT2-deficient mice. Nat. Med. 6, 1341–1347.

Chen, Z., Eggerman, T.L., Patterson, A.P., 2001. Phosphorylation is a regulatory mechanism in apolipoprotein B mRNA editing. Biochem. J. 357, 661–672.

Christiansen-Weber, T.A., Voland, J.R., Wu, Y., Ngo, K., Roland, B.L., Nguyen, S., et al., 2000. Functional loss of ABCA1 in mice causes severe placental malformation, aberrant lipid distribution, and kidney glomerulonephritis as well as high-density lipoprotein cholesterol deficiency. Am. J. Pathol. 157, 1017–1029.

Cooper, A.D., 1997. Hepatic uptake of chylomicron remnants. J. Lipid Res. 38, 2173–2192.

DeBarber, A.E., Eroglu, Y., Merkens, L.S., Pappu, A.S., Steiner, R.D., 2011. Smith-Lemli-Opitz syndrome. Expert Rev. Mol. Med. 13, e24.

Engel, T., Lorkowski, S., Lueken, A., Rust, S., Schluter, B., Berger, G., et al., 2001. The human ABCG4 gene is regulated by oxysterols and retinoids in monocyte-derived macrophages. Biochem. Biophys. Res. Commun. 288, 483–488.

Fielding, C.J., Fielding, P.E., 2001. Cellular cholesterol efflux. Biochim. Biophys. Acta 1533, 175–189.

Go, G.W., Mani, A., 2012. Low-density lipoprotein receptor (LDLR) family orchestrates cholesterol homeostasis. Yale. J. Biol. Med. 85, 19–28.

Graf, G.A., Matveev, S.V., Smart, E.J., 1999. Class B scavenger receptors, caveolae and cholesterol homeostasis. Trends. Cardiovasc. Med. 9, 221–225.

Herz, J., Strickland, D.K., 2001. LRP: a multifunctional scavenger and signaling receptor. J. Clin. Invest. 108, 779–784.

Holick, M.E., 1999. Vitamin D: Photobiology, metabolism, mechanism of action, and clinical applications. In: Favus, M.J. (Ed.), Primer on the Metabolic Bone Diseases and Disorders of Mineral Metabolism, fourth ed. Lippincott Williams & Wilkins, Philadelphia, PA, pp. 928.

Holick, M.E., Matsuoka, L.Y., Wortsman, J., 1989. Age, vitamin D, and solar ultraviolet radiation. Lancet 4, 1104–1105.

Hui, D.Y., Labonté, E.D., Howles, P.N., 2008. Development and physiological regulation of intestinal lipid absorption. III. Intestinal transporters and cholesterol absorption. Am. J. Physiol. Gastrointest. Liver Physiol. 294, G839–G843.

Jones, P.J., Pappu, A.S., Hatcher, L., Li, Z.C., Illingworth, D.R., Connor, W.E., 1996. Dietary cholesterol feeding suppresses human cholesterol synthesis measured by deuterium incorporation and urinary mevalonic acid levels. Arterioscl. Thromb. Vasc. Biol. 16, 1222–1228.

Kennedy, M.A., Venkateswaran, A., Tarr, P.T., Xenarios, I., Kudoh, J., Shimizu, N., et al., 2001. Characterization of the human ABCG1 gene: liver X receptor activates an internal promoter that produces a novel transcript encoding an alternative form of the protein. J. Biol. Chem. 276, 39438–39447.

Kishimoto, A., Goldstein, J.L., Brown, M.S., 1987. Purification of catalytic subunit of low density lipoprotein receptor kinase and identification of heat-stable activator protein. J. Biol. Chem. 262, 9367–9373.

Kozyraki, R., Fyfe, J., Kristiansen, M., Gerdes, C., Jacobsen, C., Cui, S., et al., 1999. The intrinsic factor-vitamin B I2 receptor, cubilin, is a high-affinity apolipoprotein A-I receptor facilitating endocytosis of high-density lipoprotein. Nature Med. 5, 656–661.

Krieger, M., 2001. Scavenger receptor class B type I is a multiligand HDL receptor that influences diverse physiologic systems. J. Clin. Invest. 108, 793–797.

Lee-Rueckert, M., Blanco-Vaca, F., Kovanen, P.T., Escola-Gil, J.C., 2013. The role of the gut in reverse cholesterol transport—focus on the enterocyte. Prog. Lipid Res. 52, 317–328.

Liao, H., Langmann, T., Schmitz, G., Zhu, Y., 2002. Native LDL upregulation of ATP-binding cassette transporter-1 in human vascular endothelial cells. Arterioscl. Thromb. Vasc. Biol. 22, 127–132.

Linazasoro, J.M., Hill, R., Chevallier, E., Chaikoff, I.L., 1958. Regulation of cholesterol synthesis in the liver: the influence of dietary fat. J. Exp. Med. 107, 813–820.

Linton, M.E., Fazio, S., 2001. Class A scavenger receptors, macrophages, and atherosclerosis. Curr. Opin. Lipid 12, 489–495.

Lu, K., Lee, M.H., Patel, S.B., 2001. Dietary cholesterol absorption; more than just bile. Trends Endocrinol. Metab. 12, 314–320.

Mertens, A., Holvoet, P., 2001. Oxidized LDL and HDL: antagonists in atherothrombosis. FASEB J. 15, 2073–2084.

Nguyen, T.M., Sawyer, J.K., Kelley, K.L., Davis, M.A., Kent, C.R., Rudel, L.L., 2012. ACAT2 and ABCG5/G8 are both required for efficient cholesterol absorption in mice: evidence from thoracic lymph duct cannulation. J. Lipid Res. 53, 1598–1609.

Obi-Tabot, E.T., Tian, X.Q., Chen, T.C., Holick, M.E., 2000. A human skin equivalent model that mimics the photoproduction of vitamin D3 in human skin. In Vitro Cell Dev. Biol. Anim. 36, 201–204.

Rajaratnam, R.A., Gylling, H., Miettinen, T.A., 2001. Cholesterol absorption, synthesis, and fecal output in postmenopausal women with and without coronary artery disease. Arterioscl. Thromb. Vasc. Biol. 21, 1650–1655.

Relas, H., Gylling, H., Miettinen, T.A., 2000. Dietary squalene increases cholesterol synthesis measured with serum non-cholesterol sterols after a single oral dose in humans. Atherosclerosis 152, 377–383.

Repa, J.J., Turley, S.D., Lobaccaro, J.A., Medina, J., Li, L., Lustig, K., et al., 2000. Regulation of absorption and ABC1-mediated efflux of cholesterol by RXR heterodimers. Science 289, 1524–1529.

Resh, M.D., 2012. Targeting protein lipidation in disease. Trends Mol. Med. 18, 206–214.

Russell, D.W., 2000. Oxysterol biosynthetic enzymes. Biochim. Biophys. Acta 1529, 126–135.

Saddar, S., Carriere, V., Lee, W.R., Tanigaki, K., Yuhanna, I.S., Parathath, S., et al., 2013. Scavenger receptor class B type I is a plasma membrane cholesterol sensor. Circ. Res. 112, 140–151.

Saint-Pol, J., Vandenhaute, E., Boucau, M.C., Candela, P., Dehouck, L., Cecchelli, R., et al., 2012. Brain pericytes ABCA1 expression mediates cholesterol efflux but not cellular amyloid-β peptide accumulation. J. Alzheimers Dis. 30, 489–503.

Sakakura, Y., Shimano, H., Sone, H., Takahashi, A., Inoue, N., Toyoshima, H., et al., 2001. Sterol regulatory element-binding proteins induce an entire pathway of cholesterol synthesis. Biochem. Biophys. Res. Commun. 286, 176–183.

Shachter, N.S., 2001. Apolipoproteins C-I and C-III as important modulators of lipoprotein metabolism. Curr. Opin. Lipid 12, 297–304.

Silver, D.L., Tall, A.R., 2001. The cellular biology of scavenger receptor class B type I. Curr. Opin. Lipid 12, 497–504.

Skov, M., Tønnesen, C.K., Hansen, G.H., Danielsen, E.M., 2011. Dietary cholesterol induces trafficking of intestinal Niemann–Pick Type C1 Like 1 from the brush border to endosomes. Am. J. Physiol. Gastrointest. Liver Physiol. 300, G33–G40.

Thuahnai, S.T., Lund-Katz, S., Williams, D.L., Phillips, M.C., 2001. Scavenger receptor class B, type I-mediated uptake of various lipids into cells. Influence of the nature of the donor particle interaction with the receptor. J. Biol. Chem. 276, 43801–43808.

Weaver, A.M., Lysiak, J.J., Gonias, S.L., 1997. LDL receptor family-dependent and -independent pathways for the internalization and digestion of lipoprotein lipase-associated beta-VLDL by rat vascular smooth muscle cells. J. Lipid Res. 38, 1841–1850.

Willnow, T.E., Armstrong, S.A., Hammer, R.E., Herz, J., 1995. Functional expression of low density lipoprotein receptor-related protein is controlled by receptor-associated protein *in vivo*. Proc. Natl. Acad. Sci. USA 92, 4537–4541.

Wyne, K.L., Woollett, L.A., 1998. Transport of maternal LDL and LDL to the fetal membranes and placenta of the golden Syrian hamster is mediated by receptor-dependent and receptor-independent processes. J. Lipid Res. 39, 518–530.

Zenimaru, Y., Takahashi, S., Takahashi, M., Yamada, K., Iwasaki, T., Hattori, H., et al., 2008. Glucose deprivation accelerates VLDL receptor-mediated TG-rich lipoprotein uptake by AMPK activation in skeletal muscle cells. Biochem. Biophys. Res. Commun. 368, 716–722.

LIPOIC ACID

Lipoic acid (lipoate, 6,8-dithiooctanoic amide, thioctic acid, alpha-lipoic acid; obsolete names include protogen, thiocytin, factor 11 or 11A, and pyruvic oxidation factor; molecular weight 208) serves as an enzyme cofactor and an antioxidant. The oxidized form is water-soluble and the reduced form is fat-soluble (Figure 9.41).

ABBREVIATIONS

CoA	coenzyme A
LA	lipoic acid
TPP	thiamin pyrophosphate

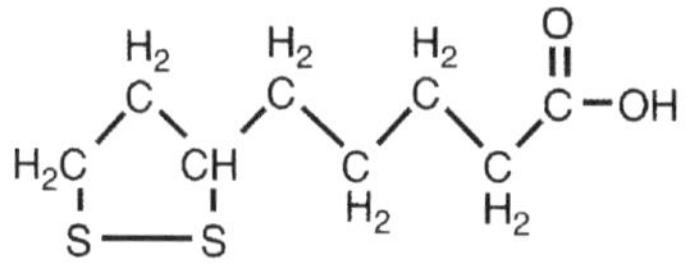

FIGURE 9.41

R-(+)-Lipoate.

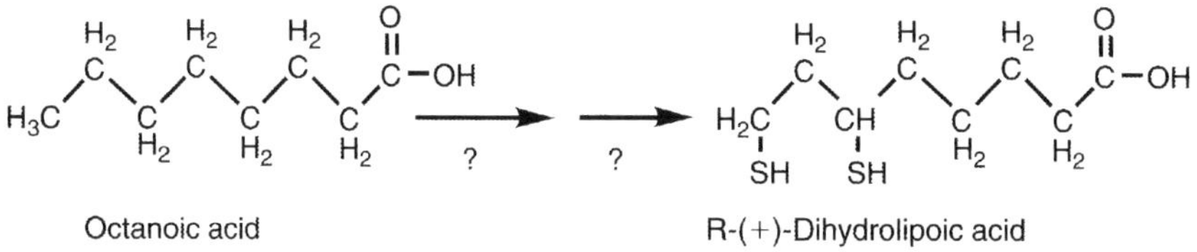

Octanoic acid   R-(+)-Dihydrolipoic acid

FIGURE 9.42

Endogenous R-(+)-lipoate synthesis.

NUTRITIONAL SUMMARY

Function: Lipoic acid (LA) is a cofactor of several enzymes essential for fuel metabolism. It is a potent antioxidant and protects the liver and other organs against some toxins.

Sources: Endogenous synthesis and absorption of LA from bacterial production in the colon account for most of the body's lipoate. Yeast, liver, and kidney are the best dietary sources. Many other foods provide much smaller amounts.

Requirements: Most people get enough LA from endogenous synthesis and bacterial production alone. Dietary intake may provide some additional benefits, especially with regard to antioxidant action.

Deficiency: No symptomatic deficiency has been observed in humans.

Excessive intake: The elimination of some potentially harmful substances, including mercury compounds and benzoic acid, may be compromised by very high intake.

ENDOGENOUS SOURCES

Most lipoate is synthesized locally in mitochondria, which may explain the presence of related enzymes and proteins (including acyl carrier protein) in this compartment (Wada et al., 1997). The principal steps are synthesis of capric acid, an 8-carbon fatty acid, and the sulfhydrylation of the carbon at the acyl end. While some progress has been made in the elucidation of the responsible enzymes in unicellular organisms, little is known about the corresponding human enzymes. The end product of endogenous synthesis is the R-(+) enantiomer (see Figure 9.42).

INTESTINAL FLORA

A significant amount of LA may be derived from normal intestinal bacteria, but the contribution from this source remains unclear.

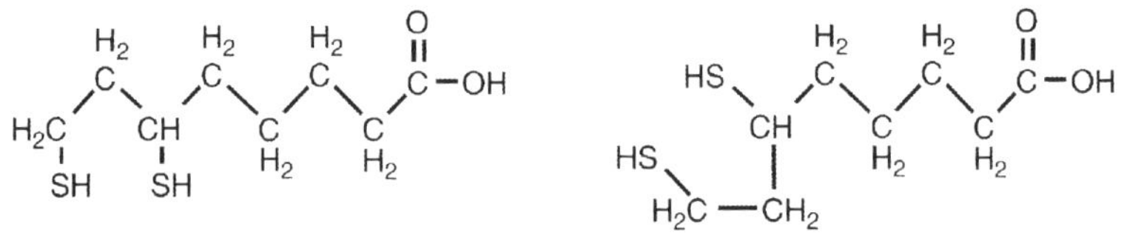

FIGURE 9.43

Synthetic versions of lipoate contain both R-(+) and S-(−) enantiomers.

DIETARY SOURCES

Most foods contain some LA. The highest amounts are in yeast, liver, heart, and kidney, with much less in muscle. Good plant-based sources include spinach, broccoli, Brussels sprouts, tomatoes, and peas (Kataoka, 1998). Average or typical dietary intake has not been estimated. Dietary supplements containing synthetic products contain equal amounts of the R-(+)- and S-(−)-enantiomers (Figure 9.43).

DIGESTION AND ABSORPTION

Much of the LA in foods and intestinal bacteria is covalently bound to lysine in enzymes. Lipoamidase (lipoyl-X hydrolase, EC 3.5.1.-), which is present in milk, can release free lipoate from its protein-bound lipoamide form (Yoshikawa et al., 1996). It is not clear whether there are enteric forms of this enzyme that could act on ingested food.

A sodium-dependent carrier for LA (SLC5A6), which also transports pantothenate and biotin, is located in the apical membrane of enterocyte of both the small and large intestine (Prasad et al., 1998). Two sodium ions enter the cell with each LA molecule. The mechanism whereby lipoate is transferred into portal blood may involve a stereospecific transport system that prefers the R-(+)-enantiomer (Breithaupt-Grogler et al., 1999).

Some of the LA is reduced in the enterocytes to dihydrolipoic acid (Takaishi et al., 2007). More than 80% of an ingested modest dose of LA is usually absorbed (Takaishi et al., 2007).

TRANSPORT AND CELLULAR UPTAKE

Blood circulation: Transport with the blood occurs as free LA (Breithaupt-Grogler et al., 1999). The liver takes up nearly a third of newly absorbed lipoate during the first pass. Uptake appears to be mediated by an as-yet-unspecified carrier (Peinado et al., 1989), possibly identical with the sodium-multivitamin transporter. Uptake of LA into other tissues might use the same transporter.

Materno-fetal transfer: The sodium-multivitamin transporter (SLC5A6) is expressed in the placenta (Wang et al., 1999) and participates in the transfer of LA to the fetus.

METABOLISM

Lipoyltransferase (EC2.3.1.181) uses lipoyl-AMP generated by lipoate ligase (EC2.7.7.63) to covalently bond lipoate to a lysine residue of nascent proteins (Fujiwara et al., 1999). Another pathway in mitochondria that does not require the ATP-dependent activation of LA appears to be through a lipoate

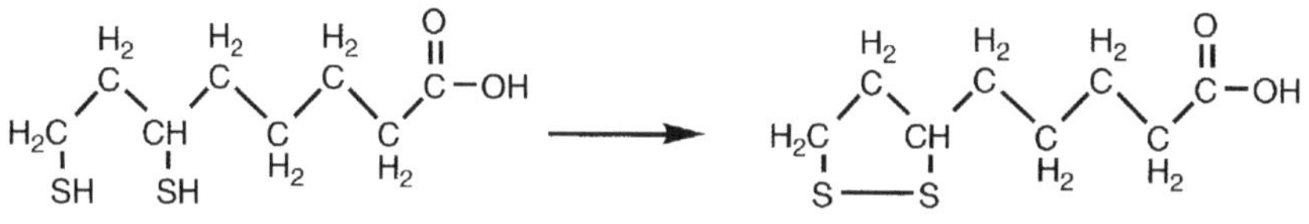

R-(+)-Dihydrolipoic acid   R-(+)-Lipoic acid (oxidized)

FIGURE 9.44

LA is a potent redox reagent.

transferase (EC2.3.1.181) using the lipoyl-acyl carrier protein (Jordan and Cronan, 1997). Protein-bound lipoamide can be recycled in tissues through the action of lipoamidase (SIRT4, EC3.5.1.-); lipoamide-releasing activity has been detected in the gray matter of the brain, the liver, serum, and milk (Yoshikawa et al., 1996; Mathias et al., 2014).

LA is a potent antioxidant (Figure 9.44). Oxidized lipoate can be reduced to dihydrolipoate through two distinct mechanisms: the mitochondrial NADH-dependent dihydrolipoamide dehydrogenase, which exhibits a marked preference for R(+)-lipoate; and NADPH-dependent glutathione reductase, which shows slightly greater activity toward the S(−)-lipoate stereoisomer.

Mitochondrial NADH-dependent dihydrolipoamide dehydrogenase is most active in heart and skeletal muscle. Both enzymes have similar activity in erythrocytes (Haramaki et al., 1997).

STORAGE

The mechanisms for storage and release of lipoate, tissue distribution of stores, and total amounts stored are uncertain.

EXCRETION

A small percentage of hepatic LA is excreted in bile (Peinado et al., 1989), and more than 80% is recovered by absorption from both the small and large intestine (Takaishi et al., 2007). Renal excretion of the water soluble oxidized form may occur, but the extent has not been determined.

REGULATION

The processes maintaining adequate LA supplies are incompletely understood.

FUNCTION

Pyruvate dehydrogenase (EC 1.2.4.1): Pyruvate is decarboxylated by a large enzyme complex composed of multiple copies of three distinct subunits. The reaction catalyzed by subunit E1 and the overall stability of the complex is dependent on thiamin pyrophosphate (TPP). Each subunit E2 (dihydrolipoamide *S*-acetyltransferase, EC2.3.1.12) contains two LA molecules, which are covalently bound to lysines 99 and 226. These lipoamides serve as acceptors for the acetyl residues from pyruvate, transfer them to acetyl-CoA, and reduce lipoamide to dihydrolipoamide in the process (Figure 9.45). Only the R-(+)-alpha-lipoic acid isomer is effective *in vivo*, not the S isomer (Frölich et al., 2004).

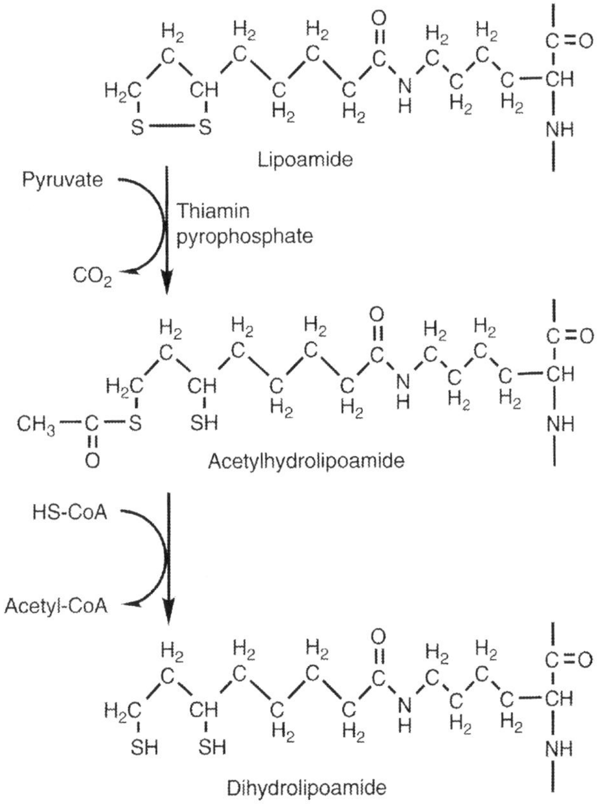

FIGURE 9.45

Lipoamide is a covalently bound prosthetic group of pyruvate dehydrogenase and four other enzymes.

Another component of the complex, dihydrolipoamide dehydrogenase (E3, EC1.8.1.4), transfers the hydrogens to NAD via FAD. The same gene encodes the dihydrolipoamide dehydrogenase of pyruvate dehydrogenase and the other two alpha-ketoacid dehydrogenases. The enzyme complex is inactivated by phosphorylation ([pyruvate dehydrogenase (lipoamide)] kinase, EC2.7.1.99) of three serines in the E1 subunit and reactivated by removal of these phosphates by [pyruvate dehydrogenase (lipoamide)]-phosphatase (EC3.1.3.43).

2-oxoglutarate dehydrogenase (EC2.3.1.61): The Krebs-cycle intermediate 2-oxoglutarate is metabolized to succinyl-CoA by a large TPP-dependent multienzyme complex containing 24 copies of the lipoamide-containing subunit E2 (dihydrolipoamide succinyltransferase) with octahedral symmetry; these subunits contain a single lipoamide attached to lysine 110.

Branched-chain alpha-keto acid dehydrogenase (EC1.2.4.4): The alpha-ketoacids 3-methyl-2-oxobutanoate, 4-methyl-2-oxopentanoate, and (*S*)-3-methyl-2-oxopentanoate, generated by deamination of the branched-chain amino acids valine, leucine, and isoleucine, are decarboxylated by another very large TPP-dependent enzyme complex containing multiple lipoamide-containing subunits E2 (3-methyl-2-oxobutanoate dehydrogenase (lipoamide), EC1.2.4.4). The enzyme complex is inactivated by phosphorylation ([3-methyl-2-oxobutanoate dehydrogenase (lipoamide)] kinase, EC2.7.1.115), and reactivated by dephosphorylation ([3-methyl-2-oxobutanoate dehydrogenase (lipoamide)]-phosphatase, EC3.1.3.52).

Glycine dehydrogenase (EC1.4.4.2): Glycine is decarboxylated in mitochondria by a large pyridoxal phosphate-dependent enzyme complex composed of multiple subunits E, T, L, and H; the H subunit contains lipoamide. In a similar fashion to the three LA-dependent alpha-ketoacid dehydrogenases, the lipoamide arm acts as an acceptor for a methylene group from glycine, transfers it to folate, and is reduced in the process. The T subunit then transfers the hydrogen to NAD via FAD.

Antioxidation: Dehydrolipoate reduces ubiquinone and semiubiquinone to ubiquinol, thereby enhancing the antioxidant potential of ubiquinol and preventing the potent oxidant-free-radical action of the semiquinone (Kozlov et al., 1999). LA may also support the reactivation of oxidized forms of vitamins C and E (Gorąca et al., 2011). Since the mitochondrial oxidation of pyruvate, alpha-ketoglutarate, branched-chain alpha-keto acids, and glycine continuously regenerates oxidized LA, there is a constant supply of antioxidant dehydrolipoate. The high iron and copper-binding potential of lipoate also reduces the risk of oxygen free radical–producing Fenton reactions. LA can function as an oxygen free radical scavenger and decrease LDL oxidation and the production of F2-isoprostanes (Marangon et al., 1999). The mitigation of neuroleptic action (haloperidol) may be due to the protection of enzymes (mitochondrial complex I) from oxidation (Balijepalli et al., 1999). Another important mechanism, whereby LA protects against toxic effects of cisplatin and other compounds, may be the maintenance of reduced glutathione concentrations and inhibiting lipid peroxidation. As with other antioxidants, LA may become a pro-oxidant under some conditions (Mottley and Mason, 2001), and the potential risks of large supplemental doses remain to be evaluated.

Glucose metabolism: LA increased insulin sensitivity (Jacob et al., 1999) and cellular glucose uptake. Improvement of glucose transport may be the mechanism underlying the prevention of polyneuropathy by lipoate administration in an animal model (Kishi et al., 1999). A protective effect against diabetic embryopathy (neural tube defect) and vascular placenta damage has been suggested (Wiznitzer et al., 1999).

Liver protection: LA has been used with some success in the mitigation of the effects of poisoning with death cap mushrooms (*Amanita phalloides*), carbontetrachloride (CCl_4), acetaminophen (in rats; Sudheesh et al., 2013), and toxic metals (Bustamante et al., 1998), largely through the scavenging of free radicals and protein stabilization. It may also protect hepatocytes through the activation of uroporphyrinogen decarboxylase (EC4.1.1.7; Vilas et al., 1999).

Acetylcholine: Dihydrolipoic acid is a powerful activator of choline *O*-acetyltransferase (EC2.3.1.6) and may have an important regulatory effect on the synthesis of acetylcholine.

Other effects: Depletion of coenzyme A (CoA) may impair glycine conjugation of benzoic acid by LA, possibly compromising the tubular secretion of benzoylglycine and causing acute renal failure in an animal model of benzoic acid exposure (Gregus et al., 1996).

REFERENCES

Balijepalli, S., Boyd, M.R., Ravindranath, V., 1999. Inhibition of mitochondrial complex I by haloperidol: the role of thiol oxidation. Neuropharmacol 38, 567–577.

Breithaupt-Grogler, K., Niebch, G., Schneider, E., Erb, K., Hermann, R., Blume, H.H., et al., 1999. Dose-proportionality of oral thioctic acid—coincidence of assessments via pooled plasma and individual data. Eur. J. Pharm. Sci. 8, 57–65.

Bustamante, J., Lodge, J.K., Marcocci, L., Tritschler, H.J., Packer, L., Rihn, B.H., 1998. Alpha-lipoic acid in liver metabolism and disease. Free Radic. Biol. Med. 24, 1023–1039.

Frölich, L., Götz, M.E., Weinmüller, M., Youdim, M.B., Barth, N., Dirr, A., et al., 2004. (r)-, but not (s)-alpha lipoic acid stimulates deficient brain pyruvate dehydrogenase complex in vascular dementia, but not in Alzheimer dementia. J. Neural Transm. 111, 295–310.

Fujiwara, K., Suzuki, M., Okumachi, Y., Okamura-lkeda, K., Fujiwara, T., Takahashi, E., et al., 1999. Molecular cloning, structural characterization and chromosomal localization of human lipoyltransferase gene. Eur. J. Biochem. 260, 761–767.

Goraca, A., Huk-Kolega, H., Piechota, A., Kleniewska, P., Ciejka, E., Skibska, B., 2011. Lipoic acid—Biological activity and therapeutic potential. Pharmacol. Rep. 63, 849–858.

Gregus, Z., Fekete, T., Halaszi, E., Klaassen, C.D., 1996. Lipoic acid impairs glycine conjugation of benzoic acid and renal excretion of benzoylglycine. Drug Metab. Disp. 24, 682–688.

Haramaki, N., Han, D., Handelman, G.J., Tritschler, H.J., Packer, L., 1997. Cytosolic and mitochondrial systems for NADH- and NADPH-dependent reduction of alpha-lipoic acid. Free Rad. Biol. Med. 22, 535–542.

Jacob, S., Ruus, P., Hermann, R., Tritschler, H.J., Maerker, E., Renn, W., et al., 1999. Oral administration of RAC-alpha-lipoic acid modulates insulin sensitivity in patients with type-2 diabetes mellitus: a placebo-controlled pilot trial. Free Rad. Biol. Med. 27, 309–314.

Jordan, S.W., Cronan Jr., J.E., 1997. A new metabolic link. The acyl carrier protein of lipid synthesis donates lipoic acid to the pyruvate dehydrogenase complex in *Escherichia coli* and mitochondria. J. Biol. Chem. 272, 17903–17906.

Kataoka, H., 1998. Chromatographic analysis of lipoic acid and related compounds. J. Chromatogr. B 717, 247–262.

Kishi, Y., Schmelzer, J.D., Yao, J.K., Zollman, P.J., Nickander, K.K., Tritschler, H.J., et al., 1999. Alpha-lipoic acid: effect on glucose uptake, sorbitol pathway, and energy metabolism in experimental diabetic neuropathy. Diabetes 48, 2045–2051.

Kozlov, A.M., Gille, L., Staniek, K., Nohl, H., 1999. Dihydrolipoic acid maintains ubiquinone in the antioxidant active form by two-electron reduction of ubiquinone and one-electron reduction of ubisemiquinone. Arch. Biochem. Biophys. 363, 148–154.

Marangon, K., Devaraj, S., Tirosh, O., Packer, L., Jialal, I., 1999. Comparison of the effect of alpha-lipoic acid and alpha-tocopherol supplementation on measures of oxidative stress. Free Rad. Biol. Med. 27, 1114–1121.

Mathias, R.A., Greco, T.M., Oberstein, A., Budayeva, H.G., Chakrabarti, R., Rowland, E.A., et al., 2014. Sirtuin 4 is a lipoamidase regulating pyruvate dehydrogenase complex activity. Cell 159, 1615–1625.

Mottley, C., Mason, R.P., 2001. Sulfur-centered radical formation from the antioxidant dihydrolipoic acid. J. Biol. Chem. 276, 42677–42683.

Peinado, J., Sies, H., Akerboom, T.P., 1989. Hepatic lipoate uptake. Arch. Biochem. Biophys. 273, 389–395.

Prasad, P.D., Wang, H., Kekuda, R., Fujita, T., Fei, Y.J., Devoe, L.D., et al., 1998. Cloning and functional expression of a cDNA encoding a mammalian sodium-dependent vitamin transporter mediating the uptake of pantothenate, biotin, and lipoate. J. Biol. Chem. 273, 7501–7506.

Sudheesh, N.P., Ajith, T.A., Janardhanan, K.K., 2013. Hepatoprotective effects of DL-α-lipoic acid and α-Tocopherol through amelioration of the mitochondrial oxidative stress in acetaminophen challenged rats. Toxicol. Mech. Methods 23, 368–376.

Takaishi, N., Yoshida, K., Satsu, H., Shimizu, M., 2007. Transepithelial transport of alpha-lipoic acid across human intestinal Caco-2 cell monolayers. J. Agric. Food Chem. 55, 5253–5259.

Vilas, G.L., Aldonatti, C., San Martin de Viale, L.C., Rios de Molina, M.C., 1999. Effect of alpha lipoic acid amide on hexachlorobenzene porphyria. Biochem. Mol. Biol. Int. 47, 815–823.

Wada, H., Shintani, D., Ohlrogge, J., 1997. Why do mitochondria synthesize fatty acids? Evidence for involvement in lipoic acid production. Proc. Natl. Acad. Sci. USA 94, 1591–1596.

Wang, H., Huang, W., Fei, Y.L., Xia, H., Yang-Feng, T.L., Leibach, F.H., et al., 1999. Human placental Na$^+$-dependent multivitamin transporter. Cloning, functional expression, gene structure, and chromosomal localization. J. Biol. Chem. 274, 14875–14883.

Wiznitzer, A., Ayalon, N., Hershkovitz, R., Khamaisi, M., Reece, E.A., Trischler, H., et al., 1999. Lipoic acid prevention of neural tube defects in offspring of rats with streptozocin-induced diabetes. Am. J. Obstet. Gynecol. 180, 188–193.

Yoshikawa, K., Hayakawa, K., Katsumata, N., Tanaka, T., Kimura, T., Yamauchi, K., 1996. High-performance liquid chromatographic determination of lipoamidase (lipoyl-X hydrolase) activity with a novel substrate, lipoyl-6-aminoquinoline. J. Chrom. Biomed. App. 679, 41–47.

UBIQUINONE

The term *ubiquinone* (coenzyme Q, mitoquinone, SA, Q-275, 272-substance) comprises fat-soluble benzoquinones that differ in the length of the isoprenyl side chain. The functional compound in humans (ubiquinone-10, coenzyme Q10, Q-199, ubidecarenone, 2,3-dimethoxy-5-methyl-6-decaprenyl-1,4-benzoquinone; molecular weight 862) has 10 repeating isoprenyl units in the side chain (Figure 9.46).

ABBREVIATIONS

CoA	coenzyme A
Q10	ubiquinone-10
SAM	S-adenosyl methionine

NUTRITIONAL SUMMARY

Function: Ubiquinone-10 (Q10) is needed for the electron transport of oxidative phosphorylation, is a cofactor of pyrimidine nucleotide synthesis, aids nitric oxide recycling, and acts as an intracellular antioxidant.

Food sources: Most Q10 is consumed in meat and poultry, as well as some germ oils.

Requirements: Normally adequate amounts are synthesized endogenously, requiring sufficient availability of tyrosine and unrestricted isoprenyl synthesis.

Deficiency: Genetically low synthesis can cause muscle weakness, fatigability, mental impairment, and seizures (Ogasahara et al., 1989; Boitier et al., 1998). In some circumstances, such as heart failure

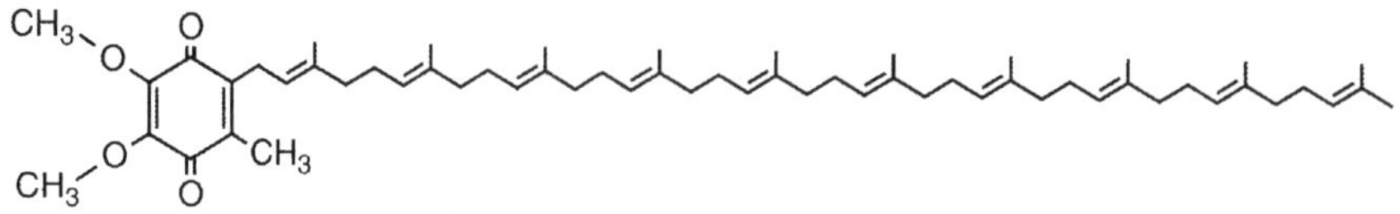

FIGURE 9.46

Ubiquinone-10.

and during treatment with HMG-CoA reductase inhibitors (statins), endogenous production may not be adequate, and additional dietary intake of up to 30 mg/day may be beneficial.

Excessive intake: Supplementation with 100 mg Q10/day has been used for up to 6 years without adverse effects.

ENDOGENOUS SOURCES

Most tissues synthesize Q10 from farnesyl diphosphate and tyrosine (Nagata et al., 1990) via a multistep process that requires vitamin B6, S-adenosyl methionine (methionine, folate, B12), iron, and magnesium. Daily production is about 12 mg (Elmberger et al., 1987). Decreased availability of tyrosine in phenylketonuria (PKU) lowers Q10 concentrations (Artuch et al., 1999). The relevance of additional dietary deficiencies and metabolic factors remains unclear. Synthesis of the ring system appears to take place in mitochondria. The side chain is produced in the Golgi system (Appelkvist et al., 1994). Since peroxisomal inducers promote endogenous synthesis, at least some steps may also occur in peroxisomes (Turunen et al., 2000).

Much of the information about *de novo* synthesis of ubiquinone comes from experiments with model systems such as the roundworm *Caenorhabditis elegans* (Gomez et al., 2012) and still needs to be confirmed in humans. The ring moiety, 4-hydroxybenzoate, is derived mainly from L-tyrosine (Artuch et al., 1999). The reactions responsible for the conversion of L-tyrosine to 4-hydroxybenzoate have not been characterized yet. Synthesis of the side-chain moiety of coenzyme Q uses farnesyl diphosphate, which is extended in several as-yet-unresolved steps to octaprenyl diphosphate (solanyl diphosphate). Further extension of the chain to the nonaprenyl, and presumably then to the decaprenyl, is catalyzed by transoctaprenyltranstransferase (EC2.5.1.11). The side chain can then be joined to the ring by 4-hydroxybenzoate nonaprenyltransferase (EC2.5.1.39) or a similar magnesium-dependent enzyme. Only a few of the following steps have been elucidated. Here, 4,5-dihydroxybenzoate is 5-methylated by an enzyme identical or similar to the yeast protein Coq3 (hexaprenyl dihydroxybenzoate methyltransferase, EC2.1.1.114) in an S-adenosyl methionine (SAM)–requiring reaction that might be a rate-limiting step. Synthesis is completed by the ferroenzyme 3-demethylubiquinol 3-O-methyltransferase (EC2.1.1.64), which again requires SAM. CLK-1 (homologue of yeast Coq7p/Cat5p) is another mitochondrial inner membrane protein directly involved in ubiquinone biosynthesis (Vajo et al., 1999), apparently by hydroxylation of the hydroquinone ring. O-methyltransferase catalyzes the SAM-dependent methylation of 3,4-dihydroxy-5-polyprenyl benzoic acid to 3-methoxy-4-hydroxy-5-polyprenyl benzoic acid and of 2-polyprenyl-3-methyl-5-hydroxy-6-methoxy-1,4 benzoquinol (demethyl ubiquinol) to ubiquinol (Jonassen and Clarke, 2000) (Figure 9.47).

DIETARY SOURCES

Additional amounts of Q10 are derived from food, typically about 3–5 mg/day (Weber et al., 1997). Good sources are meat, poultry, cereals, and corn oil (Dupont et al., 1990; Weber et al., 1997).

DIGESTION AND ABSORPTION

Q10 is not only absorbed from the intestine, particularly the duodenum and slightly less from the rest of the small intestine, but also from the colon (Miles, 2007). Both active and passive transport

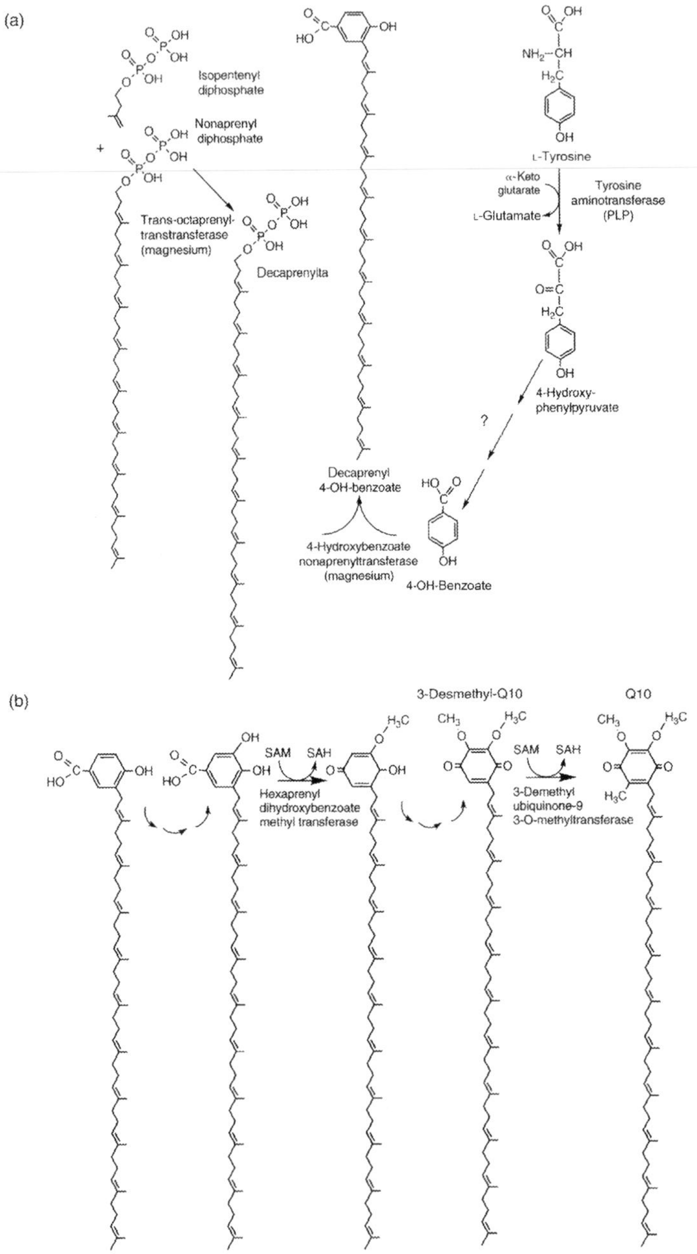

FIGURE 9.47

(a) Endogenous ubiquinone-10, part 1; (b) endogenous ubiquinone-10, part 2.

mechanisms appear to contribute, but the responsible molecular structure remains to be elucidated. So far as Q10 is absorbed and incorporated into Chylos, the fatty acid transporter CD36, SRB1, and NPC1L1 are likely to be involved.

TRANSPORT AND CELLULAR UPTAKE

Blood circulation: Normal Q10 concentrations in serum are around 0.6 μmol/l, slightly higher in early childhood (Artuch et al., 1999). Circulating Q10 is taken up rapidly from the blood into the liver, then resecreted with very-low-density Lp (Yuzuriha et al., 1983). The mechanisms whereby Q10 reaches other tissues are not well understood, however (Turunen et al., 1999). Both the oxidized and reduced forms of Q10 are apolar and readily permeate mitochondrial and other membranes.

CATABOLISM AND EXCRETION

The steps involved in Q10 breakdown and excretion are not well characterized yet.

REGULATION

Little is known about the mechanisms that maintain a constant supply of Q10.

FUNCTION

Electron transport: Complex I of the oxidative phosphorylation system is the electron-transferring-flavoprotein (ETF) dehydrogenase (EC1.5.5.1), which catalyzes the electron transfer from primary flavoprotein dehydrogenases in the mitochondrial matrix to Q10 in the inner membrane.

Q10 also participates in electron transport of oxidative phosphorylation from succinate (which is converted to fumarate) to oxygen at the matrix face of the inner mitochondrial membrane by complex II (succinate dehydrogenase/ubiquinone, EC1.3.5.1). As a result of this oxidation, protons are pumped into the intermembrane space for the eventual capturing of this energy by ATP synthase during the flow of protons back across the intermembrane into the matrix.

At the cytosolic face, the reduced form, ubisemiquinone (QH2), is oxidized by complex III to its semiquinone by transferring an electron via an Fe-S cluster to cytochrome c1, and then to Q10 by transferring another electron via b566 to b560. At the matrix face, the transfer of two electrons from b560 to Q10 reduces this again via the semiquinone to QH2. This ubiquinone cycle is possible because Q10 and QH2 are uncharged and diffuse freely from one face of the inner mitochondrial membrane to the other.

The importance of Q10 for ion transport and ATP production, especially during rapid growth, is underscored by the finding that the viability of embryos depends on adequate Q10 availability (Stojkovic et al., 1999).

Q10 is also a constituent of a lysosomal electron transport chain. The redox potentials carried through this electron transport chain drive the transport of protons across the lysosomal membrane and help to build up the acid environment of lysosomes (Gille and Nohl, 2000).

Redox reactions: Ubiquinone is an electron acceptor for various mitochondrial enzymes, such as dihydroorotate dehydrogenase (EC1.3.99.11) for uridine synthesis.

Another redox reaction involving Q10 is the removal of nitrite, the end product of intracellular nitric oxide degradation. A mitochondrial nitrite reductase (EC1.7.2.1) uses ubisemiquinone associated with the bcl complex to convert the potentially toxic nitrite back to nitric oxide, thus providing an alternative source for this signaling compound, which is independent of arginine (Kozlov et al., 1999b). Ubiquinol also acts as a general intracellular antioxidant (Ernster and Dallner, 1995). In these various reactions, the reduced form ubiquinole is oxidized to ubiquinone or to ubisemiquinone, which has pro-oxidant properties itself (Nohl et al., 1999). The oxidized forms of Q10 can then be reactivated to ubiquinole by dihydrolipoic acid (Kozlov et al., 1999a).

Supplement use: Contrary to common expectations, dietary supplements did not improve aerobic power in healthy people (Bonetti et al., 2000), nor was ejection fraction, peak oxygen consumption, or exercise duration increased in patients with congestive heart failure receiving standard medical therapy (Khatta et al., 2000). Large amounts of supplemental Q10 may impair vitamin K reactivation because of completion for NAD(P)H:quinone oxidoreductase 1 (NQO1, EC1.6.5.2) and cause episodic bleeding (Shalansky et al., 2007).

NUTRITIONAL ASSESSMENT

Q10 deficiency may be detected with the measurement of Q10 concentration in the blood. This is most helpful for the evaluation of patients with clearly abnormal concentrations due to suspected genetic defects (Emmanuele et al., 2012), but less so to capture variations in normal populations. As with most other fat-soluble nutrients, individuals with elevated triglyceride concentration tend to have higher Q10 concentrations.

REFERENCES

Appelkvist, E.L., Aberg, F., Guan, Z., Parmryd, I., Dallner, G., 1994. Regulation of coenzyme Q biosynthesis. Mol. Asp. Med. 15, S37–S46.

Artuch, R., Vilaseca, M.A., Moreno, J., Lambruschini, N., Cambra, F.J., Campistol, J., 1999. Decreased serum ubiquinone-10 concentrations in phenylketonuria. Am. J. Clin. Nutr. 70, 892–895.

Bonetti, A., Solito, E., Carmosino, G., Bargossi, A.M., Fiorella, P.L., 2000. Effect of ubidecarenone oral treatment on aerobic power in middle-aged trained subjects. J. Sports Med. Phys. Fitness 40, 51–57.

Boitier, E., Degoul, F., Desguerre, I., Charpentier, C., Francois, D., Ponsot, G., et al., 1998. A case of mitochondrial encephalomyopathy associated with a muscle coenzyme Q10 deficiency. J. Neurol. Sci. 156, 41–46.

Dupont, J., White, P.J., Carpenter, M.E., Schaefer, E.J., Meydani, S.N., Elson, C.E., et al., 1990. Food uses and health effects of corn oil. J. Am. Coll. Nutr. 9, 438–470.

Elmberger, P.G., Kalen, A., Appelkvist, E.L., Dallner, G., 1987. *In vitro* and *in vivo* synthesis of dolichol and other main mevalonate products in various organs of the rat. Eur. J. Biochem. 168, 1–11.

Emmanuele, V., López, L.C., Berardo, A., Naini, A., Tadesse, S., Wen, B., et al., 2012. Heterogeneity of coenzyme Q10 deficiency: patient study and literature review. Arch. Neurol. 69, 978–983.

Ernster, L., Dallner, G., 1995. Biochemical, physiological and medical aspects of ubiquinone function. Biochim. Biophys. Acta 1271, 195–204.

Gille, L., Nohl, H., 2000. The existence of a lysosomal redox chain and the role of ubiquinone. Arch. Biochem. Biophys. 375, 347–354.

Gomez, F., Saiki, R., Chin, R., Srinivasan, C., Clarke, C.F., 2012. Restoring *de novo* coenzyme Q biosynthesis in *Caenorhabditis elegans* coq-3 mutants yields profound rescue compared to exogenous coenzyme Q supplementation. Gene 506, 106–116.

Jonassen, T., Clarke, C.E., 2000. Isolation and functional expression of human COQ3, a gene encoding a methyltransferase required for ubiquinone biosynthesis. J. Biol. Chem. 275, 12381–12387.

Khatta, M., Alexander, B.S., Krichten, C.M., Fisher, M.L., Freudenberger, R., Robinson, S.W., et al., 2000. The effect of coenzyme Q10 in patients with congestive heart failure. Ann. Int. Med. 132, 636–640.

Kozlov, A.V., Gille, L., Staniek, K., Nohl, H., 1999a. Dihydrolipoic acid maintains ubiquinone in the antioxidant active form by two-electron reduction of ubiquinone and one-electron reduction of ubisemiquinone. Arch. Biochem. Biophys. 363, 148–154.

Kozlov, A.V., Staniek, K., Nohl, H., 1999b. Nitrite reductase activity is a novel function of mammalian mitochondria. FEBS Lett. 454, 127–130.

Miles, M.V., 2007. The uptake and distribution of coenzyme Q10. Mitochondrion (Suppl. 7), S72–S77.

Nagata, Y., Hidaka, Y., Ishida, E., Kamei, T., 1990. Effects of simvastatin (MK-733) on branched pathway of mevalonate. Japan J. Pharmacol. 54, 315–324.

Nohl, H., Gille, L., Kozlov, A.V., 1999. Critical aspects of the antioxidant function of coenzyme Q in biomembranes. Biofactors 9, 155–161.

Ogasahara, S., Engel, A.G., Frens, D., Mack, D., 1989. Muscle coenzyme Q deficiency in familial mitochondrial encephalomyopathy. Proc. Natl. Acad. Sci. USA 86, 2379–2382.

Shalansky, S., Lynd, L., Richardson, K., Ingaszewski, A., Kerr, C., 2007. Risk of warfarin-related bleeding events and supratherapeutic international normalized ratios associated with complementary and alternative medicine: a longitudinal analysis. Pharmacotherapy 27, 1237–1247.

Stojkovic, M., Westesen, K., Zakhartchenko, V., Stojkovic, E., Boxhammer, K., Wolf, E., 1999. Coenzyme Q(10) in submicron-sized dispersion improves development, hatching, cell proliferation, and adenosine triphosphate content of *in vitro*-produced bovine embryos. Biol. Reprod. 61, 541–547.

Turunen, M., Appelkvist, E.L., Sindelar, E., Dallner, G., 1999. Blood concentration of coenzyme Q(10) increases in rats when esterified forms are administered. J. Nutr. 129, 2113–2118.

Turunen, M., Peters, J.M., Gonzalez, F.J., Schedin, S., Dallner, G., 2000. Influence of peroxisome proliferator-activated receptor alpha on ubiquinone biosynthesis. J. Mol. Biol. 297, 607–614.

Vajo, Z., King, L.M., Jonassen, T., Wilkin, D.J., Ho, N., Munnich, A., et al., 1999. Conservation of the Caenorhabditis elegans timing gene clk-1 from yeast to human: a gene required for ubiquinone biosynthesis with potential implications for aging. Mamm. Genome 10, 1000–1004.

Weber, C., Bysted, A., Holmer, G., 1997. Coenzyme Q10 in the diet—daily intake and relative bioavailability. Mol. Asp. Med. 18, S251–S254.

Yuzuriha, T., Takada, M., Katayama, K., 1983. Transport of [^{14}C]coenzyme Q 10 from the liver to other tissues after intravenous administration to guinea pigs. Biochim. Biophys. Acta 759, 286–291.

WATER-SOLUBLE VITAMINS AND NONNUTRIENTS

10

CHAPTER OUTLINE

Methylation 567
Vitamin C 570
Thiamin 580
Riboflavin 589
Niacin 599
Vitamin B6 610
Folate 620
Vitamin B12 632
Biotin 642
Pantothenate 647
Queuine 654
Biopterin 657
Inositol 663

METHYLATION

Metabolism is an integrated system of reactions that cannot be viewed from the aspect of just one component, such as a particular nutrient. Silencing of gene expression by DNA methylation, which depends on an adequate supply of *S*-adenosylmethione (SAM), provides an illustration of the interdependence of methionine, folate, riboflavin, vitamin B6, vitamin B12, choline, and other critical nutrients within a network of transporters, binding proteins, and enzymes. When intake level of one nutrient is at the low end, the outcome often depends on adequacy of the other nutrients, as well as on individual genetic disposition. Lack of one nutrient can also affect the use of another nutrient and has indirect consequences in seemingly distant metabolic pathways. Thus, weaknesses in one-carbon metabolism have broad consequences. Such weaknesses can cause epigenetic disorders (even before conception), birth defects during early pregnancy, slowed brain growth and cognitive development, anemia, cancer, diabetes, atherosclerosis, thrombosis, and other diseases. Obviously, optimal nutrition is a complex goal that requires consideration of nutrient intake and health outcomes in a broad context.

ABBREVIATIONS	
Hcys	homocysteine
Met	L-methionine
MTHFR	methylenetetrahydrofolate reductase (EC1.5.1.20)
MTR	5-methyltetrahydrofolate-homocysteine S-methyltransferase (EC2.1.1.1 3)
SAM	S-adenosylmethionine

SAM-DEPENDENT METHYLATION

De novo creatine synthesis uses about 70% of the SAM available for methylation (Wyss and Kaddurah-Daouk, 2000). SAM is also the prerequisite precursor for the production of the polyamines spermine and spermidine (Figure 10.1). Much smaller amounts go to the synthesis of carnitine (0.3 mmol/day), choline, estrogen, and other compounds. As the example of DNA methylation emphasizes, the quantity of the product does not indicate the importance of the reaction.

Specific DNA target sequences are methylated by several DNA (cytosine-5)-methyltransferase (EC2.1.1.37, zinc-binding) isoenzymes. The presence of a five-methyl group on cytosine usually blocks transcription directly or through attached binding proteins. Current evidence indicates that silencing by selective methylation regulates developmentally appropriate expression of genes and suppresses parasitic insertions. Hypomethylation of DNA due to suboptimal nutrient intake and metabolic disposition has been linked to an increased risk of cancer (Ehrlich, 2002). Cancer of the colon, cervix, breast, stomach, esophagus, and other sites occurs with increased frequency in people with low habitual intake

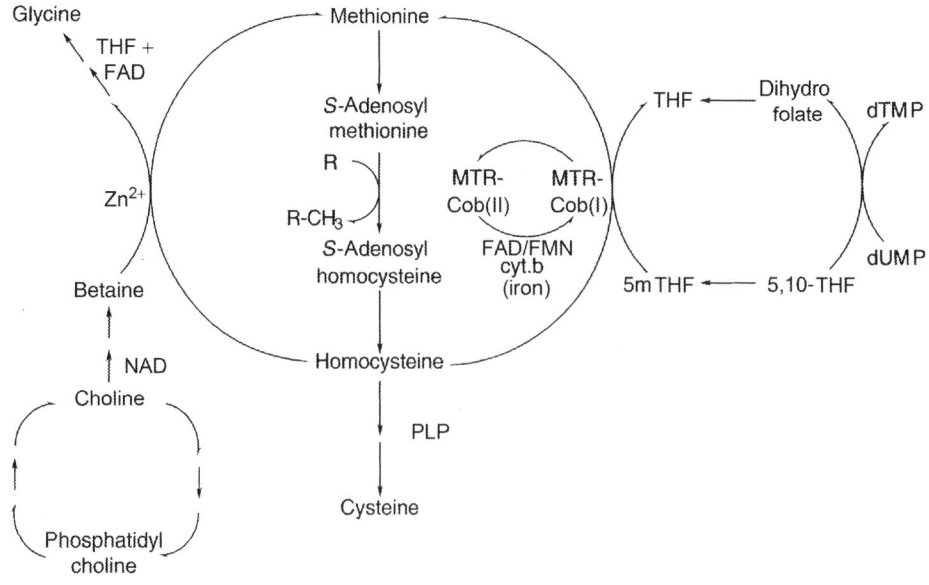

FIGURE 10.1

DNA methylation depends on a web of nutrients and metabolic events.

of folate. It is likely that additional nutrients involved in one-carbon metabolism influence cancer risk, but this issue needs more investigation (Ames, 2001). Hypomethylation may also contribute to the deterioration of brain function with aging and age-typical afflictions (Selhub, 2002).

Methylation of parental DNA imprints some genomic regions. Faulty imprinting is linked to severe developmental defects in Prader-Willi and Angelman syndromes (Court et al., 2011) and possibly much more common adulthood conditions, such as obesity and diabetes (Dyer and Rosenfeld, 2011).

SOURCES OF METHYL GROUPS

While methyl groups abound in nutrients and their metabolites, only a few of them can be used for SAM-dependent methylation. The major sources are methionine (Met), choline, and 5-methyltetrahydrofolate (5-mTHF). Met intake depends to a significant degree on dietary habits. Compared with meat eaters, ovo-lacto-vegetarians have much lower (770 vs. 1450 mg/day) average intake (Sachan et al., 1997).

The second step of choline breakdown, catalyzed by the zinc-enzyme betaine homocysteine methyltransferase (EC2.1.1.5), remethylates homocysteine (Hcys) to Met, thus providing a methyl group for SAM synthesis. Since the *de novo* synthesis of choline requires the SAM-mediated transfer of three methyl groups, only dietary choline can contribute to net SAM synthesis. The median daily choline intake has been estimated at about 300 mg in men and 270 mg in women (Bidulescu et al., 2009).

A much more significant pathway of Hcys remethylation is catalyzed by 5-methyltetrahydrofolate-homocysteine S-methyltransferase (MTR, EC2.1.1.13). Very little of the necessary cosubstrate 5-mTHF is generated by direct methylation of tetrahydrofolate (THF), such as in the betaine homocysteine methyltransferase-catalyzed reaction; rather, it comes from the reduction of 5,10-methylene-THF by methylenetetrahydrofolate reductase (MTHFR, EC1.5.1.20). Large quantities of this metabolite arise from the metabolism of serine (40 mmol/day), glycine (150 mmol/day), histidine (17 mmol/day), and formate. Both a low supply of folate and reduced MTHFR activity diminish the availability of the 5-mTHF cosubstrate for optimal Met recycling.

The active cob(I)alamine form of MTR becomes slowly oxidized to the inactive cob(II)alamine form. The flavin adenine dinucleotide (FAD)– and flavin mononucleotide (FMN)–dependent methionine synthase reductase (EC2.1.1.135) can reactivate the cob(I)alamine-containing form again by reductive methylation (Wolthers and Scrutton, 2009). This reaction uses cytochrome b5, which in turn is regenerated by NADPH-dependent cytochrome P450 reductase (EC1.6.2.4), another enzyme with both FAD and FMN as prosthetic groups (Chen and Banerjee, 1998). Inhalation of nitrous oxide irreversibly inactivates MTR (Horne et al., 1989; Riedel et al., 1999).

REFERENCES

Ames, B.N., 2001. DNA damage from micronutrient deficiencies is likely to be a major cause of cancer. Mutat. Res. 475, 7–20.

Bidulescu, A., Chambless, L.E., Siega-Riz, A.M., Zeisel, S.H., Heiss, G., 2009. Repeatability and measurement error in the assessment of choline and betaine dietary intake: the Atherosclerosis Risk in Communities (ARIC) study. Nutr. J. 8, 14.

Chen, Z., Banerjee, R., 1998. Purification of soluble cytochrome b5 as a component of the reductive activation of porcine methionine synthase. J. Biol. Chem. 273, 26248–26255.

Court, F., Martin-Trujillo, A., Romanelli, V., Garin, I., Iglesias-Platas, I., Salafsky, I., et al., 2011. Metabolic imprinting by prenatal, perinatal, and postnatal overnutrition: a review. Semin. Reprod. Med. 29, 266–276.

Dyer, J.S., Rosenfeld, C.R., 2011. Metabolic imprinting by prenatal, perinatal, and postnatal overnutrition: a review. Semin. Reprod. Med. 29, 266–276.

Ehrlich, M., 2002. DNA methylation, cancer, the immunodeficiency, centromeric region instability, facial anomalies syndrome and chromosomal rearrangements. J. Nutr. 132, 2424S–2429S.

Horne, D.W., Patterson, D., Cook, R.J., 1989. Effect of nitrous oxide inactivation of vitamin B12-dependent methionine synthetase on the subcellular distribution of folate coenzymes in rat liver. Arch. Biochem. Biophys. 270, 729–733.

Lapunzina, P., Monk, D., 2013. Genome-wide allelic methylation analysis reveals disease-specific susceptibility to multiple methylation defects in imprinting syndromes. Hum. Mutat. 34, 595–602.

Riedel, B., Fiskerstrand, T., Refsum, H., Ueland, P.M., 1999. Co-ordinate variations in methylmalonyl-CoA mutase and methionine synthase, and the cobalamine cofactors in human glioma cells during nitrous oxide exposure and the subsequent recovery phase. Biochem. J. 341, 133–138.

Sachan, D.S., Daily III, J.W., Munroe, S.G., Beauchene, R.E., 1997. Vegetarian elderly women may risk compromised carnitine status. Veg. Nutr. 1, 64–69.

Selhub, J., 2002. Folate, vitamin B12 and vitamin B6 and one carbon metabolism. J. Nutr. Health Aging. 6, 39–42.

Wolthers, K.R., Scrutton, N.S., 2009. Cobalamin uptake and reactivation occurs through specific protein interactions in the methionine synthase-methionine synthase reductase complex. FEBS. J. 276, 1942–1951.

Wyss, M., Kaddurah-Daouk, R., 2000. Creatine and creatinine metabolism. Physiol. Rev. 107–213.

VITAMIN C

The hexuronic lactone L–ascorbic acid is a water-soluble essential nutrient (ascorbate, L-xyloascorbic acid, antiscorbutic factor, antiscorbutic vitamin, L-3-ketothreohexuronic acid, 3-oxo-L-gulofuranolactone[enol form] molecular weight 176). Vitamin C occurs in three different redox states: fully reduced as ascorbate (ASC), partially oxidized as semidehydroascorbate (SDA), and fully oxidized as dehydro-L-ascorbic acid (DHA). Several derivatives are also biologically active (Figure 10.2).

ABBREVIATIONS

ASC	L-ascorbic acid (specifically to indicate the reduced form)
A2S	ascorbate-2-sulfate
DHA	dehydro-L-ascorbic acid
GLUT1	glucose transporter 1
PAPS	3′-phosphoadenosine 5′-phosphosulfate
SDA	semidehydroascorbate
SGLT1	sodium-glucose cotransporter
SVCT1	sodium-ascorbate transporter 1 (SLC23A2)
SVCT2	sodium-ascorbate transporter 2 (SLC23A1)

NUTRITIONAL SUMMARY

Function: Vitamin C is essential for gums, arteries, other soft tissues, and bone (collagen synthesis), for brain and nerve function (neurotransmitter and hormone synthesis), for nutrient metabolism (especially iron, protein, and fat), and for antioxidant defense (directly and by reactivating vitamin E) against free radicals (free radicals increase the risk of cancer and cardiovascular disease).

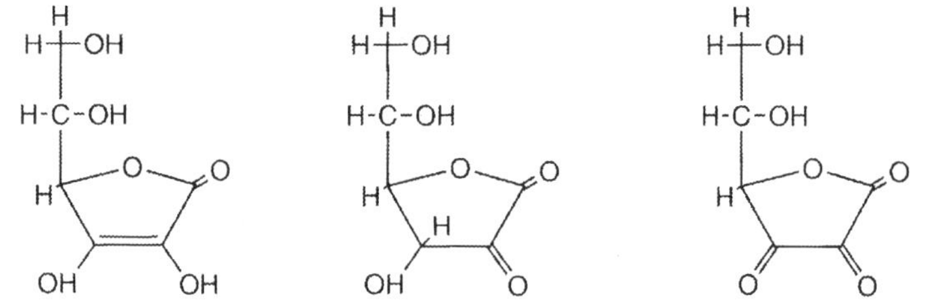

FIGURE 10.2

Vitamin C has three different redox states.

Food sources: Many fruits and vegetables provide at least 20% of the recommended daily intake per serving; citrus fruits, berries, and tomatoes are especially rich sources. Prolonged storage, extensive processing, and overcooking all diminish vitamin C content of foods considerably.

Requirements: Adult women should get at least 75 mg/day, and men at least 90 mg/day (Food and Nutrition Board, Institute of Medicine, 2000). Being over 50, smoking, strenuous exercise, heat, infections, and injuries each may increase needs.

Deficiency: Scurvy (symptoms include painful swelling and bleeding into gums, joints, and extremities, poor wound healing, fatigue, and confusion) has become rare in most countries; 10 mg/day helps prevent it. Lower than optimal intake may diminish immune function and wound healing and increase the risk of heart disease and cancer, especially in susceptible individuals. If intake is low, stores last only a few weeks.

Excessive intake: Daily doses of 2000 mg or more may irritate stomach and bowels, cause kidney stones, and interfere with copper status.

ENDOGENOUS SOURCES

Unlike most other vertebrates, humans (just like all primates and guinea pigs) lack the enzyme for the final step of ascorbate synthesis (L-gulonolactone oxidase, EC1.1.3.8) from hexose precursors due to multiple deletions and point mutations in the responsible gene. Thus, compounds with vitamin C activity have to be obtained from food.

DIETARY SOURCES

Foods contain a mixture of compounds with vitamin C activity: ASC (the reduced form), SDA (the partially oxidized form), DHA (the oxidized form), and small amounts of ascorbate-2-sulfate (A2S). Acylated ASC, ascorbyl phosphate, and other derivatives are sometimes used as food additives. (*Note*: Food tables usually report only the ASC content of food, not total bioactive vitamin C.) ASC and SDA are oxidized rapidly when exposed to air without losing their biological activity for humans. All forms of vitamin C decompose during prolonged heating or storage of foods.

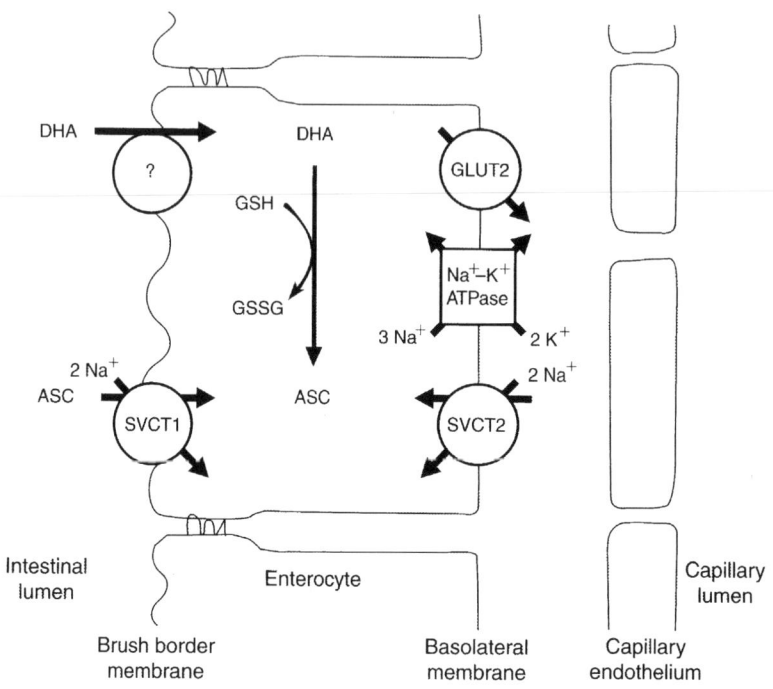

FIGURE 10.3

Intestinal uptake of vitamin C.

Typical intake of vitamin C from food in the United States is 77 mg/day in women and 109 mg/day in men. The best food sources are fruits and vegetables, especially citrus, berries, peppers, tomatoes, broccoli, and greens. On average, a serving of fruits or vegetables provides about 30 mg of vitamin C (Lykkesfeldt et al., 2000). Consumption of nonfood sources is common, most often as multivitamin supplements, which usually provide 50–60 mg/day, or as megadose preparations that may contain 500 mg or more.

DIGESTION AND ABSORPTION

Small amounts of ingested ascorbate are absorbed nearly completely (Figure 10.3), but fractional absorption rapidly drops to less than 20% when daily intake is above 200 mg (Blanchard et al., 1997). ASC and SDA may be oxidized in the lumen of the intestine, partially through the activity of ceruloplasmin (EC1.16.3.1). On the other hand, DHA may be reduced by the duodenal cytochrome b561 (DcytB) to ASC (Corpe et al., 2013). DHA is taken up slightly more efficiently in the proximal than in the distal small intestine (Malo and Wilson, 2000) by facilitated transport with GLUT2 and GLUT8 (Corpe et al., 2013). ASC uptake, in contrast, is more efficient in the distal than in the proximal small intestine. The sodium-dependent ascorbate transporter 1 (SVCT1, SLC23A2) mediates its uptake with high efficiency and specificity (only the L-form of ASC is transferred). Two sodium ions are needed for the transport of one ASC; one of these sodium ions may be actually coupled to the simultaneous transport of glucose via the sodium-glucose cotransporter (SGLT1, Malo and Wilson, 2000). Absorption is efficient, with a maximum of around 3 h after ingestion (Piotrovskij et al., 1993). Ascorbate-2-sulfate (A2S) is not effectively hydrolyzed in the intestine and not absorbed. Ascorbate-dependent glutathione reductase (EC1.8.5.1), and reduced glutathione (nonenzymically) in the enterocyte reduce DHA to ASC; the oxidized glutathione, in turn, is regenerated by a system of thioredoxin and nicotine adenine dinucleotide phosphate (NADPH)-dependent thioredoxin reductase (EC1.6.4.5). The high intracellular concentration generates enough of a gradient to drive ASC via the sodium-ascorbate transporter 2 (SVCT2, SLC23A1) across the basolateral membrane (Rose and Wilson, 1997; Liang et al., 2001). The colon might also have some absorptive capacity since both sodium-ascorbate transporters are expressed there.

TRANSPORT AND CELLULAR UPTAKE

Blood circulation: ASC, the main form in the blood (95%), is transferred from the blood to some tissues (chromaffin cells, osteoblasts, and fibroblasts) predominantly by the two sodium-dependent ascorbate transporters (SVCTI and SVCT2) in an energy-dependent, concentrative process. SVCT1 is expressed in most tissues, while the expression of SVCT2 is not expressed in skeletal muscle and the lung. On the basis of recent observations in erythrocytes, the existence of an nicotinamide adenine dinucleotide (NADH)-dependent, reducing DHA transmembrane transporter has been proposed (Himmelreich et al., 1998). If the same transporter system were present in enterocytes, this would provide another explanation for the near-absence of DHA in the blood.

Blood–brain barrier (BBB): ASC in interstitial fluids is oxidized (by ceruloplasmin, iron compounds, or other agents) to DHA, which is then transferred via glucose transporters (GLUT1, GLUT3, and GLUT4) both into and out of cells (Liang et al., 2001). Reduction of DHA to ASC blocks reverse transport since GLUTs are impervious to ASC and SDA; the sodium-ascorbate cotransporter 2 (SLC23A1, distinct from the ubiquitous SVCT1) completes the concentrative transport into the brain. This mechanism may explain the 10-fold higher concentration of ASC in the brain compared to blood.

Materno-fetal transfer: Similar to the arrangement in the small intestines, DHA transports into the syntrophoblast using glucose transporters, while ASC is taken up via SVCT1. DHA can be reduced inside the syntrophoblast layer. ASC is then exported to the fetal side via SVCT2.

METABOLISM

DHA is reduced to ASC in cytosol by ascorbate-dependent glutathione reductase (EC1.8.5.1), NADH-dependent monodehydroascorbic acid reductase (EC1.6.5.4), the NADPH-dependent selenoenzyme thioredoxin reductase (EC1.6.4.5), or nonenzymatically by reduced glutathione (since this is a near-equilibrium reaction, ASC can also reduce oxidized glutathione) (Figure 10.4). SDA, which is the unstable radical intermediate generated by some reactions, is reduced by NADH-dependent monodehydroascorbate reductase (EC1.6.5.4) and by thioredoxin reductase (EC1.6.4.5) (May et al., 1998); another important mechanism may be the NADH-dependent, DHA-reducing transporter found in erythrocytes. Additional DHA- and SDA-reducing enzymes, including NADPH-dependent DHA-reductase, have been described but are not fully characterized yet. L-Ascorbate-cytochrome-b5 reductase (EC1.10.2.1) catalyzes the shuttling of reducing equivalents with cytochrome b561 across the phospholipid bilayer enveloping microsomes, neurovesicles, and chromaffin granules, thereby coupling the oxidation of ASC to SDA inside to the reduction of SDA to ASC outside (Figure 10.5).

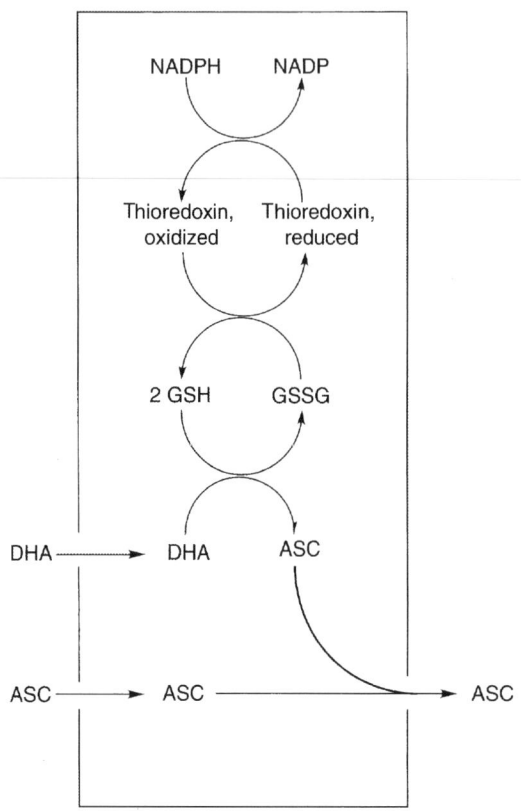

FIGURE 10.4

Intracellular metabolism of vitamin C.

A small amount of available ASC is converted by alcohol sulfotransferase (EC2.8.2.2) into A2S using 3′-phosphoadenosine 5′-phosphosulfate (PAPS) as the cosubstrate (Figure 10.6); this enzyme is otherwise important for the conjugation of a wide range of steroids (e.g., dehydroepiandrosterone), bile acids, and drugs in the adrenals and liver. ASC is released again when A2S is hydrolyzed by arylsulfatase A (EC3.1.6.8) or arylsulfatase B (EC3.1.6.12).

DHA in the blood is broken down rapidly and irreversibly by nonenzymic hydrolysis to 2,3-ketogulonate, some of which is hydrolyzed further to oxalate and L-threonate (Figure 10.7). The extent of this degradation is proportional to DHA concentration. Increased amounts are lost by tobacco smokers due to accelerated ASC oxidation upon exposure to free oxygen radicals (Lykkesfeldt et al., 2000).

STORAGE

The highest total ascorbate tissue concentrations are in leukocytes, adrenal glands, the pituitary, and the brain. Replete body stores are in the range of 1–2 g depending on habitual intake. Stores below 300 mg are thought to cause the development of severe deficiency symptoms (i.e., scurvy). The highest

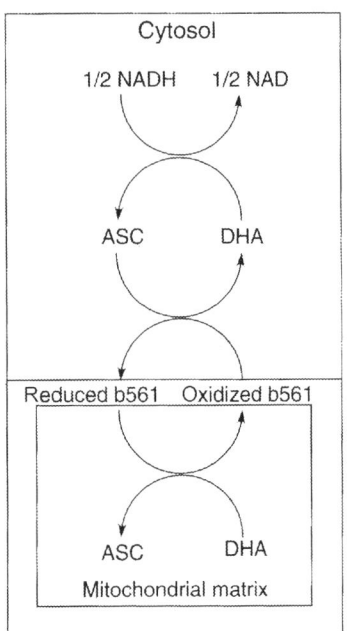

FIGURE 10.5

Vitamin C redox equivalents are shuttled across the mitochondrial membrane.

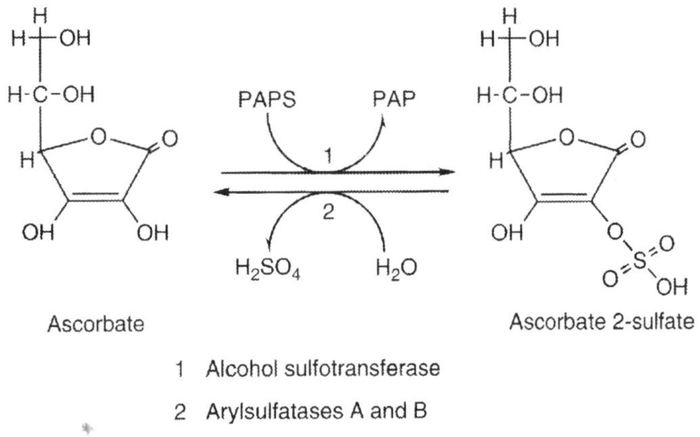

Ascorbate

Ascorbate 2-sulfate

1 Alcohol sulfotransferase

2 Arylsulfatases A and B

FIGURE 10.6

Ascorbate 2-sulfate metabolism.

concentrations of bioactive metabolites are in the adrenal and pituitary glands, corpus luteum, brain, liver, pancreas, and spleen. About 3% of the total body pool turns over per day. The biological half-life of stored ascorbate increases with depletion; it ranges from 8 to 40 days (Kallner et al., 1979). A2S may have special importance for the storage of ASC.

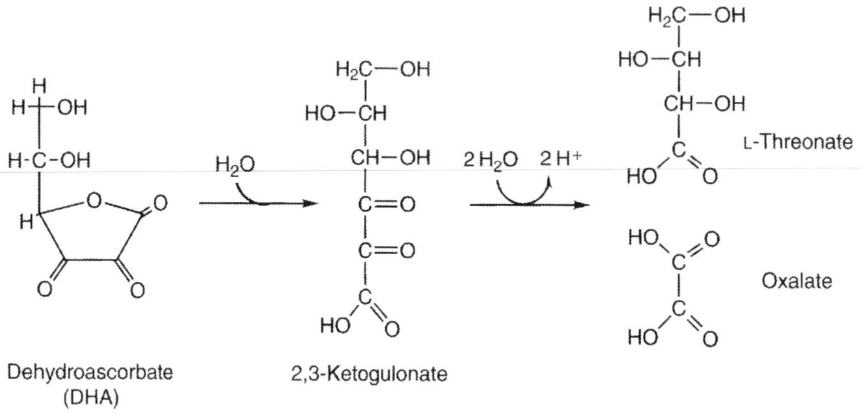

FIGURE 10.7

Nonenzymic degradation of vitamin C.

REGULATION

Information is lacking on mechanisms that maintain the body's stores of the various ascorbate metabolites. Follicle-stimulating hormone (FSH) and insulin-like growth factor 1 (IGF-1) induce the transport into follicular granuiosa cells of ASC via ascorbate transporter and of DHA via GLUT1. Luteinizing hormone (LH) and PGF2cz stimulate the energy- and sodium-dependent efflux of ASC from luteal cells. The role of hormonal regulation of ASC and DHA transport in other tissues is unclear.

EXCRETION

ASC, SDA, DHA, 2,3-ketogulonate, and other catabolic products pass completely into primary glomerular filtrate owing to their small size. ASC is reabsorbed from renal tubules by the (electrogenic) sodium-dependent ascorbate transporter. Nearly all ingested vitamin C is eventually excreted with urine as ASC, 2,3-ketogulonate, or oxalate. Usually losses via other body fluids or feces are not significant.

FUNCTION

Free radical scavenging: Free radicals are highly reactive chemicals with at least one unpaired electron. They are generated in the course of normal biological functions, particularly from oxidation reactions in the mitochondria and from the activity of white blood cells. ASC in its various redox states comprises a particularly versatile free radical scavenging system in the aqueous phase, since either one or two electrons can be accepted. ASC is particularly effective in quenching the hydroxyl radical. The resulting oxidized forms (ascorbyl radical, SDA) are reduced again to ASC by the enzyme systems described previously. ASC is essential for the nonenzymic reactivation of tocopheryl quinone, the product of the reaction of alpha-tocopherol with an oxygen free radical. ASC also protects folate and other compounds that are highly susceptible to oxidation. While ASC is an antioxidant under most circumstances, it is important to point to the pro-oxidant potential of ASC when promoting iron-catalyzed reactions.

ASC can reduce iron that has been oxidized in a Fenton-type reaction:

$$(LOOH + Fe^{2+} \rightarrow LO \cdot + OH^- + Fe^{3+}; \quad or \quad H_2O_2 + Fe^{2+} \rightarrow \cdot OH + OH + Fe^{3+})$$

This iron-recycling reaction (Fe^{3+}+ASC→Fe^{2+}+SDA) can perpetuate the generation of oxygen free radicals.

ASC also protects nonenzymically against the oxidation of cholesterol in tissues by reactive oxygen species (Menéndez-Carreño et al., 2008).

Protein modification: ASC maintains the reduced form of iron in two types of metalloenzymes that posttranslationally hydroxylate lysine and proline residues in collagen (EC1.14.11.2 procollagen-proline, 2-oxoglutarate-dioxygenase and EC1.14,11.4 procollagen-lysine, 5-dioxygenase: the latter activity is exerted by two genetically distinct isoenzymes). Each reduction of Fe(III) to Fe(II) oxidizes one ASC to SDA. Another ascorbate-dependent copper enzyme (EC1.14.17.3 peptidyl-glycine alpha-amidating monooxygenase) activates numerous proteohormones, including thyroid-releasing hormone (TRH), CRE GnRH, neuropeptide Y (NPY), endorphins, gastrin, pancreatic polypeptide, atrial natriuretic factor, and arginine vasopressin, in neurosecretory vesicles; each modifier reaction oxidizes one ASC molecule to DHA.

Amino acid metabolism: ASC participates in the metabolism of phenylalanine and tyrosine and the synthesis of carnitine. ASC is one of several alternative reductants for the synthesis of homogentisinate, an intermediary metabolite of phenylalanine and tyrosine catabolism, by 4-hydroxyphenylpyruvate dioxygenase (EC1.13.11.27). For the following hydroxylation of homogentisate, ASC is again necessary to maintain the ferrous state of enzyme-bound iron in homogentisate 1,2-dioxygenase (EC1.13.11.5). In a similar way, ASC keeps the copper of dopamine-beta-monooxygenase (EC1.14.17.1) in its reduced state, thereby ensuring the synthesis of noradrenaline (norepinephrine) from dopamine. At least two steps of carnitine synthesis from lysine are mediated by ASC-dependent ferroenzymes. Trimethyllysine is hydroxylated by trimethyllysine dioxygenase (EC1.14.11.8), and the ultimate precursor, trimethyl-ammoniobutanoate, is converted into carnitine by gamma-butyrobetaine hydroxylase (EC1.14.11.1).

Steroids and lipids: ASC is important for the last three steps of aldosterone synthesis, 11-beta- and 18-hydroxylation (EC1.14.15.4 steroid 11-beta-monooxygenase and EC1.14.15.5 corticosterone 18-monooxygenase) and oxidation of 11-deoxycorticosterone, in the adrenal cortex. All three reactions are mediated by cytochrome P-450scc, the activity of which decreases with decreasing availability of ASC. The same is true for another cytochrome, CYP11B2, which mediates cholesterol 7-alpha-hydroxylation in liver microsomes, a key step of bile acid synthesis. The mechanism and stoichiometry of ASC in these reactions are not well characterized. Cytochromes generate profuse amounts of oxygen free radicals, and it has been suggested that ASC is essential to protect the enzyme protein and the lipids in the surrounding microsomal membrane.

In a reaction catalyzed by L-ascorbate cytochrome-b5 reductase (EC1.10.2.1) ASC generates the reducing equivalents (ferrocytochrome b5) for the desaturation of fatty acids in microsomes. ASC can provide, as an alternative to NADPH, reducing equivalents for the hydroxylation of *N*-acetylneuraminic acid (EC1.14.99.18, CMP-*N*-acetylneuraminate monooxygenase), a critical step in the synthesis of GM3(NeuGc) gangliosides in many tissues.

DNA modification: ASC is a cofactor of the ferrozyme Tet methylcytosine dioxygenase (EC1.14.11. n2), which hydroxylates 5-methylcytosine in DNA (Minor et al., 2013). The importance appears to be particularly great in embryonic stem cells, where this ASC-dependent reaction has been found to demethylate nearly 10% of all genes (Chung et al., 2010).

Iron metabolism: ASC enhances absorption of nonheme iron from the small intestine. Generally, it has been assumed that this is due to a reduction of ferric iron by ASC, but this mechanism has been disputed. ASC is also necessary for the mobilization of ferritin iron deposits. It is able to penetrate via molecular pores of the ferritin-Fe(III) complex, where it reduces iron; the resulting ejection of Fe(II) from the crystal lattice makes iron available for binding to transferrin and transport away from storage sites. For each mobilized iron one, ASC is oxidized to DHA.

Sulfate transfer: A2S appears to be a reservoir of sulfate groups that become available when A2S is hydrolyzed by arylsulfatase A (EC3.1.6.8) or arylsulfatase B (EC3.1.6.12). This may explain how A2S promotes the sulfation of cholesterol and of glucosaminoglycans such as chondroitin and dermatan sulfate. Since the activities of both arylsulfatase A and arylsulfatase B are inhibited by ASC, replete vitamin C status not only slows the release of ASC from A2S, but also the hydrolytic release of noradrenaline and a wide range of other sulfoconjugated compounds.

Other functions: Dietary ASC in the stomach suppresses the reaction of ingested nitrites with food proteins to nitrosamines, thereby decreasing carcinogen exposure (Helser et al., 1992). ASC also has been reported to directly inhibit the growth of some tumor cell lines *in vitro*. ASC can reduce various elements, such as chromium, and thereby affect their bioavailability and biological activity. It has been proposed that ASC reduces selenite to elemental Se, which then can form links with selenocysteine in proteins. The biological significance of such an effect is not known. ASC has also been suggested to act as an aldose reductase, thereby diminishing the risk of free radical production from sorbitol in diabetics (Crabbe and Goode, 1998). ASC is also a reducing agent for enzymes involved in prostaglandin synthesis (Horrobin, 1996), the mixed-function oxidases, and the cytochrome p450 electron transport system (Tsao, 1997) contributing to the metabolism of xenobiotics.

NUTRITIONAL ASSESSMENT

Vitamin C deficiency and suboptimal status may be detected by measuring the concentration of ascorbic acid in the blood (Wannamethee et al., 2013). Dietary intake can be estimated with modest accuracy from the amount of vitamin C in urine (Tsuji et al., 2010).

Hyperpolarized ^{13}C magnetic resonance (MR) spectroscopy is an experimental method for quantitating the concentration of dehydroascorbic acid and evaluating the vitamin C reactivation capacity *in vivo* (Keshari et al., 2014).

REFERENCES

Blanchard, J., Tozer, T.N., Rowland, M., 1997. Pharmacokinetic perspectives on megadoses of ascorbic acid. Am. J. Clin. Nutr. 66, 1165–1171.

Chung, T.L., Brena, R.M., Kolle, G., Grimmond, S.M., Berman, B.P., Laird, P.W., et al., 2010. Vitamin C promotes widespread yet specific DNA demethylation of the epigenome in human embryonic stem cells. Stem Cells 28, 1848–1855.

Corpe, C.P., Eck, P., Wang, J., Al-Hasani, H., Levine, M., 2013. Intestinal Dehydroascorbic Acid (DHA) transport mediated by the facilitative sugar transporters, GLUT2 and GLUT8. J. Biol. Chem. 288, 9092–9101.

Crabbe, M.J., Goode, D., 1998. Aldose reductase: a window to the treatment of diabetic complications? Progr. Ret. Eve. Res. 17, 313–383.

Food and Nutrition Board, Institute of Medicine, 2000. Dietary Reference Intakes for Vitamin C, Vitamin E, Selenium, and Carotenoids. National Academy Press, Washington, DC.

Helser, M.A., Hotchkiss, J.H., Roe, D.A., 1992. Influence of fruit and vegetable juices on the endogenous formation of *N*-nitrosoproline and *N*-nitrosothiazolidine-4-carboxylic acid in humans on controlled diets. Carcinogenesis 13, 2277–2280.

Himmelreich, U., Drew, K.N., Serianni, A.S., Kuchel, P.W., 1998. ^{13}C NMR studies of vitamin C transport and its redox cycling in human erythrocytes. Biochemistry 37, 7578–7588.

Horrobin, D.E., 1996. Ascorbic acid and prostaglandin synthesis. Subcell. Biochem. 25, 109–115.

Kallner, A., Hartmann, D., Hornig, D., 1979. Steady-state turnover and body pool of ascorbic acid in man. Am. J. Clin. Nutr. 32, 530–539.

Keshari, K.R., Wilson, D.M., Sai, V., Bok, R., Jen, K.Y., Larson, P., et al., 2015. Non-invasive *in vivo* imaging of diabetes-induced renal oxidative stress and response to therapy using hyperpolarized ^{13}C dehydroascorbate magnetic resonance. Diabetes 64, 344–352.

Liang, W.J., Johnson, D., Jarvis, S.M., 2001. Vitamin C transport systems of mammalian cells. Mol. Membr. Biol. 18, 87–95.

Lykkesfeldt, J., Christen, S., Wallock, L.M., Chang, H.H., Jacob, R.A., Ames, B.N., 2000. Ascorbate is depleted by smoking and repleted by moderate supplementation: a study in male smokers and nonsmokers with matched dietary antioxidant intakes. Am. J. Clin. Nutr. 71, 530–536.

Malo, C., Wilson, J.X., 2000. Glucose modulates vitamin C transport in adult human small intestinal brush border membrane vesicles. J. Nutr. 130, 63–69.

May, J.M., Cobb, C.E., Mendiratta, S., Hill, K.E., Burk, R.E., 1998. Reduction of the ascorbyl free radical to ascorbate by thioredoxin reductase. J. Biol. Chem. 273, 23039–23045.

Menéndez-Carreño, M., Ansorena, D., Milagro, F.I., Campión, J., Martínez, J.A., Astiasarán, I., 2008. Inhibition of serum cholesterol oxidation by dietary vitamin C and selenium intake in high fat fed rats. Lipids 43, 383–390.

Minor, E.A., Court, B.L., Young, J.I., Wang, G., 2013. Ascorbate induces ten-eleven translocation (Tet) methyl-cytosine dioxygenase-mediated generation of 5-hydroxymethylcytosine. J. Biol. Chem. 288, 13669–13674.

Piotrovskij, V.K., Kallay, Z., Gajdos, M., Gerykova, M., Trnovec, T., 1993. The use of a nonlinear absorption model in the study of ascorbic acid bioavailability in man. Biopharm. Drug Disp. 14, 429–442.

Rose, R.C., Wilson, J.X., 1997. Ascorbate membrane transport properties. In: Packer, L., Fuchs, J. (Eds.), Vitamin C in Health and Disease Marcel Dekker, New York, NY, pp. 143–162.

Tsao, C.S., 1997. An overview of ascorbic acid chemistry and biochemistry. In: Packer, L., Fuchs, J. (Eds.), Vitamin C in Health and Disease Marcel Dekker, New York, NY, pp. 25–58.

Tsuji, T., Fukuwatari, T., Sasaki, S., Shibata, K., 2010. Twenty-four-hour urinary water-soluble vitamin levels correlate with their intakes in free-living Japanese university students. Eur. J. Clin. Nutr. 64, 800–807.

Wannamethee, S.G., Bruckdorfer, K.R., Shaper, A.G., Papacosta, O., Lennon, L., Whincup, P.H., 2013. Plasma vitamin C, but not vitamin E, is associated with reduced risk of heart failure in older men. Circ. Heart Fail. 6, 647–654.

THIAMIN

Thiamin (thiamine, vitamin B 1, 3-[4-amino-2-methyl-5-pyrimidinylmethyl]-5-(2-hydroxyethyl)-4-methylthiazolium, aneurine, molecular weight of the hydrochloride 337) is a water-soluble vitamin whose heterocyclic rings contain both nitrogen and sulfur (Figure 10.8).

ABBREVIATIONS	
RFC1	reduced folate carrier (SLC19A1)
TCP	thiamin carrier protein
ThTr2	thiamin transporter 2 (SLC19A3)
TMP	thiamin monophosphate
TPP	thiamin pyrophosphate
TTP	thiamin triphosphate
AK	adenylate kinase

NUTRITIONAL SUMMARY

Function: Thiamin pyrophosphate (TPP) is an essential cofactor of five enzymes involved in carbohydrate, amino acid, intermediary (tricarboxylic acid cycle), and phytol metabolism; thiamin triphosphate (TTP) is important for brain function.

Food sources: Good sources are yeast, legumes, enriched grain products, and pork.

Requirements: Men should get 1.2 mg/day and women 1.1 mg/day; women's needs are slightly higher during pregnancy and lactation. Use of diuretics increases dietary needs.

Deficiency: Early signs may include anorexia, weight loss, muscle weakness, apathy, confusion, and irritability. Later consequences include edema (wet beriberi), muscle wasting (dry beriberi), and psychosis (Wernicke–Korsakoff syndrome). Onset of cardiac failure in young infants can be very sudden and rapidly lethal. Alcohol abuse is often associated with deficiency, possibly because alcohol interferes with thiamin uptake and metabolism (Latt and Dore, 2014). A rare genetic disturbance of transporter-mediated cellular uptake results in megaloblastic anemia, diabetes mellitus, and sensorineural deafness (Labay et al., 1999).

Excessive intake: No adverse effects from high oral doses (as much as 10 mg/day) have been reported. Anaphylactic reactions following parenteral administration have been observed.

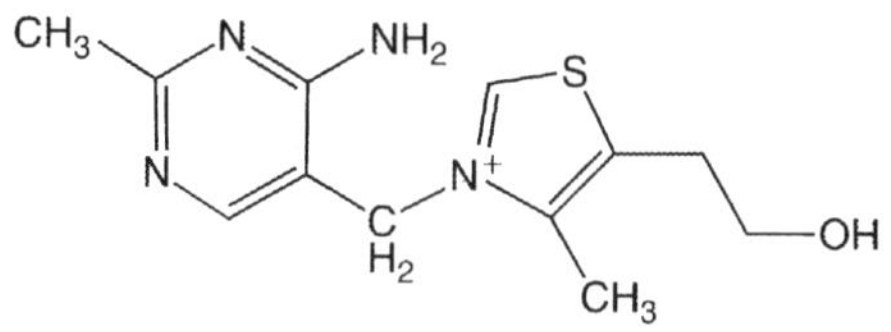

FIGURE 10.8

Thiamin.

DIETARY AND OTHER SOURCES

Foods may contain TPP, TTP, thiamin monophosphate (TMP), thiamin, thiamin sulfide, thiamin disulfide, thiamin tetrahydrofurfuryl disulfide (sometimes added to wine), thiamin hydrochloride; thiamin is a monovalent cation at physiological pH. The median intake of adult men in the United States is about 1.9 mg/day (Food and Nutrition Board, Institute of Medicine, 1998).

Good sources include yeast, legumes, enriched or whole-grain products, bran, and pork. By law, bread in the United States is enriched with 1.8 mg/lb, and flour with 2.9 mg/lb.

Significant amounts of thiamin are lost when food is cooked in water and the vitamin leaches out. Thiamin also is unstable when exposed to high temperatures, irradiation, alkaline medium (bicarbonate treatment of peas and beans), or metabisulphite (preservatives in dried fruits and wine); oxidation generates thiochrome and other oxidation products. Antithiamin factors are present in betel nuts (tannins), tea (caffeic acid, tannic acid, and chlorogenic acid), berries (flavonoids), Brussels sprouts (isothiocyanates), and red beets (anthocyanines), mostly by forming nonabsorbable thiamindisulfide. Ascorbate and other organic acids may protect thiamin.

Thiaminase (EC3.5.99.2) is active in various microorganisms, including *Saccharomyces cerevisiae* and *Staphylococcus aureus*. Thiamin in foods contaminated with thiaminase is cleaved into the inactive breakdown products 4-amino-5-hydroxymethyl-2-methylpyrimidine and 5-(2-hydroxyethyl)-4-methylthiazole.

Thiamin pyridinylase (EC2.5.1.2) from viscera of some fish and shellfish diminishes the vitamin content of exposed foods by convening thiamin to heteropyrithiamin. Large amounts of thiamin are produced in the colon by normal enteric bacteria and secreted in free form. Enterocytes of the colon express specific thiamin transport capacity and thus can take up thiamin synthesized by enteric bacteria (Said et al., 2001). The quantitative contribution from this source is not known, however.

DIGESTION AND ABSORPTION

Dietary thiamin is absorbed throughout the small intestine (Figure 10.9), maximally in the duodenum. Evidence is mounting that some thiamin from bacterial production can be absorbed in the colon (Said et al., 2001). Absorption of microgram amounts from the small intestine may be nearly complete; it increases with thiamin deficiency (Laforenza et al., 1997), and decreases with folate deficiency. Aging (Rindi and Laforenza, 2000) and alcohol intake (Breen et al., 1985) decrease fractional absorption; concomitant alcohol ingestion may decrease absorption from a single thiamin dose by nearly 30%. At doses around 5 mg/day, absorption becomes very ineffective (Hayes and Hegsted, 1973). Only about 5% of a 50-mg dose is absorbed (Tallaksen et al., 1993).

The main forms of thiamin in unfortified foods, TPP, TTP, and TMP are hydrolyzed by nucleotide pyrophosphatase (EC3.6.1.9) from pancreatic exocrine secretions (Beaudoin et al., 1983) and brush border alkaline phosphatase (EC3.1.3.1, zinc- and magnesium-dependent). Unphosphorylated thiamin can then be taken up by the thiamin transporter 2 (THTR2, SLC19A3). The accessory protein transmembrane 4 superfamily 4 (TM4SF4) participates in the uptake process and improves its efficiency (Subramanian et al., 2014).

The reduced folate carrier (RFC1, SLC19A1) does not usually play a role, in spite of the fact that it can transport phosphorylated thiamin (Zhao and Goldman, 2013). Once in the enterocyte, the free thiamin is phosphorylated to TPP by thiamin pyrophosphokinase (EC2.7.6.2) and thereby prevented from returning to the intestinal lumen.

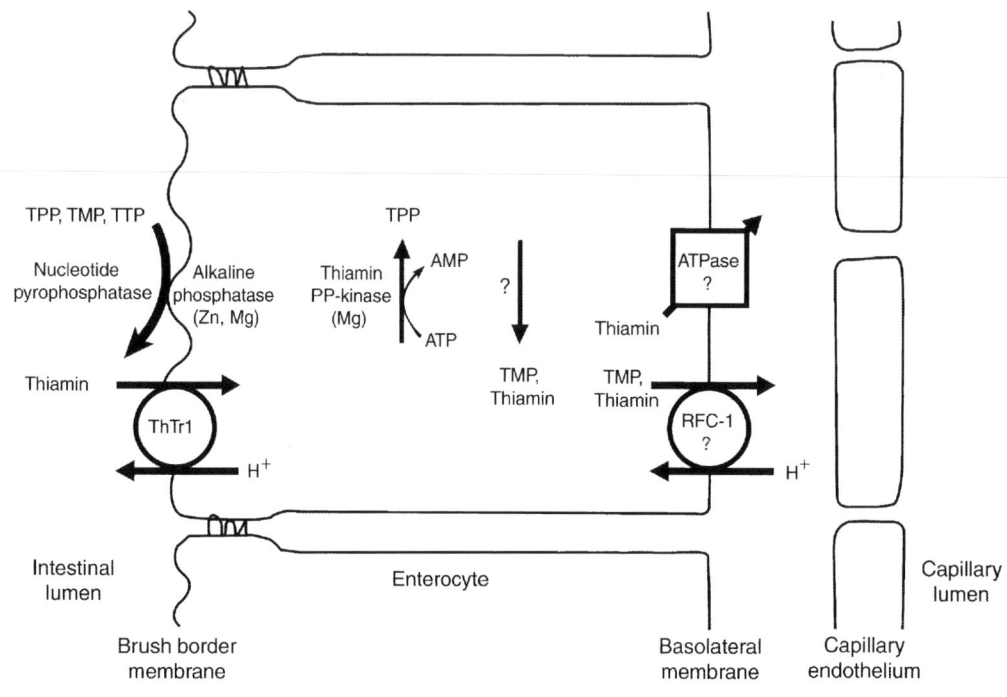

FIGURE 10.9

Intestinal absorption of thiamin. The question marks indicate that the molecular nature responsible for the activity is still uncertain.

Transport across the basolateral membrane into portal blood uses the thiamin transporter 1 (ThTr1, SLC19A2). The phosphorylated thiamin must be cleaved prior to export by an incompletely characterized thiamin pyrophosphatase (TPPase) or thiamin monophosphatase (TMPase).

High alcohol (ethanol) consumption interferes with both intestinal absorption and transfer to tissues by down-regulating the thiamin transporters (Subramanya et al., 2011).

TRANSPORT AND CELLULAR UPTAKE

Blood circulation: Transport in blood plasma occurs mostly as thiamin bound to albumin (about 10–20 nmol/l); the concentrations of TMP are slightly lower (Gangolf et al., 2010a,,b). Concentrations in whole blood cells are an order of magnitude higher (around 200 nmol/l). The predominant species in red blood cells is TPP; TMP and thiamin contribute much less (Tallaksen et al., 1997). Thiamin can be taken up into hepatocytes, erythrocytes, and other cells by active transport (Labay et al., 1999) and from there into mitochondria through the high-affinity thiamin transporter 1 (SLC19A2). The closely related thiamin transporter (ThTr2, SLC19A3) in the liver, heart, and other tissues also mediates selective high-affinity uptake of free thiamin (Zhao and Goldman, 2013). The reduced folate carrier 1 (SLC19A1) provides an alternative minor access route for TMP. Limited TPP efflux via the reduced folate carrier 1 appears to help regulating cellular thiamin homeostasis (Zhao and Goldman, 2013).

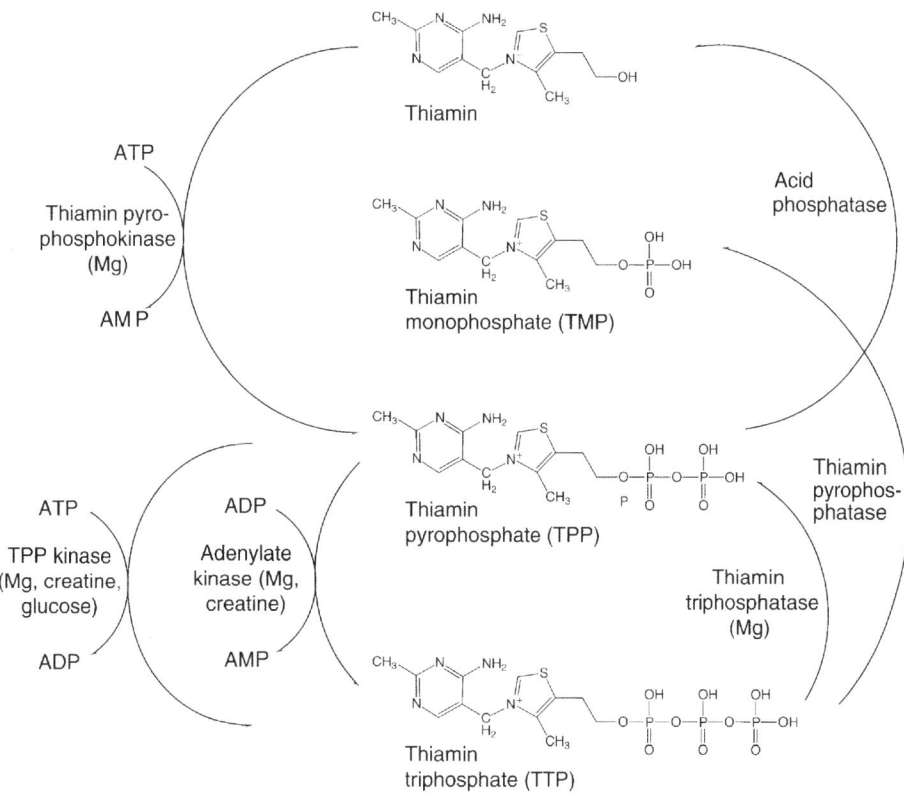

FIGURE 10.10

Thiamin metabolism.

Yet another transporter, SLC35F3, also appears to contribute to thiamin transfer into red blood cells and other tissues from circulation (Zhang et al., 2014). *BBB*: Free thiamin and, to a lesser extent, TMP are transported across the BBB (Patrini et al., 1988). Thiamin may share a carrier-mediated transport system with choline that mediates uptake (Kang et al., 1990). The thiamin transporter 1 (SLC19A2) is expressed in the brain, but its exact location and role remain to be elucidated.

Materno-fetal transfer: Transport across the placental membrane is known (Dutta et al., 1999) to involve both the thiamin transporter 1 (SLC19A2) and ThTr2 (SLC19A3; Zhao and Goldman, 2013), but it is not yet completely characterized. In the process, thiamin is concentrated in the placenta, relative to both the maternal and the fetal side. Since transport in the materno-fetal direction is slower than in the fetomaternal direction, the net transfer is from mother to fetus (Schenker et al., 1990).

METABOLISM

Thiamin can be phosphorylated into TPP in most tissues (Zhao and Goldman, 2013) by thiamin pyrophosphokinase (EC2.7.6.2). This enzyme converts free thiamin and TMP to TPP and TTP (Figure 10.10). In the brain and other tissues, a significant proportion of TPP is phosphorylated again to TTP by thiamin-diphosphate kinase (EC2.7.4.15) in an adenosine triphosphate (ATP)–dependent reaction. Adenosylate kinase (EC2.7.4.3) facilitates (Gangolf et al., 2010a) transphosphorylation (TPP + ADP→AMP + TTP). Both of these enzymes are unusual in that they require creatine as cofactors (Shikata et al., 1986, 1989). Thiamin-diphosphate kinase may also require glucose as an activating factor (Nishino et al., 1983). An alternative pathway in brain mitochondria appears to transfer phosphate from oxidative phosphorylation directly to TPP (Gangolf et al., 2010b).

A broad array of enzymes in various tissues dephosphorylates the thiamin phosphates. Magnesium-dependent thiamin triphosphatase (EC3.6.1.28), which generates TPP from TTP, is present in many tissues, both as cytosolic and membrane-bound form. TPP in mitochondria can be hydrolyzed by a heterodimeric isoenzyme of acid phosphatase (EC3.1.3.2). Thiamin pyrophosphatase, which converts TTP to TMP, could be a modified form of type B nucleoside diphosphatase (EC3.6.1.6) in the Golgi apparatus. Additional less specific phosphatases also act on thiamin phosphates.

Thiamin can be metabolized to thiamin acetate, thiamin sulfide, pyrimidine carboxylic acid, thiazole acetate, 2-methyl-4-amino-5-formylaminomethylpyrimidine, thiochrome, and other compounds (Pearson and Darby, 1967; White et al., 1970), but the exact nature or location of the involved metabolic processes is not well understood.

It has been suggested that some of these compounds may be absorbed after intestinal bacteria have acted on thiamin or its derivates, but the exact nature and location of the involved reactions still need to be clarified.

STORAGE

A healthy adult has thiamin stores of about 30 mg, half of them in muscle, less in the liver and kidneys. The biological half-life of thiamin is 9–18 days (Ariaey-Nejad et al., 1970). About 80% of total body thiamin is TPP (mostly bound to pyruvate dehydrogenase and alpha-ketoglutarate dehydrogenase in mitochondria), 10% is TTP; smaller amounts are present as free thiamin and TMP (McCormick, 2000).

EXCRETION

With high thiamin intake, most of the excess is rapidly excreted via urine (Davis et al., 1984), and very little with bile. A mean creatinine/thiamin renal clearance ratio of 2.4 indicates that the thiamin excess is actively excreted. At low to moderate concentrations, on the other hand, recovery of intact thiamin from primary filtrate in the proximal tubule is very effective due to mechanisms very similar to those responsible for intestinal absorption. A thiamin/H^+ antiporter with a 1:1 stoichiometric ratio in the brush border recovers filtered thiamin (Gastaldi et al., 2000), and an ATP-driven thiamin carrier completes transport across the basolateral membrane. All diuretics appear to increase thiamin losses via urine (Suter and Vetter, 2000).

Most thiamin losses with urine are in the form of pyrimidine carboxylic acid, thiazole acetic acid, or thiamin acetic acid, and a considerable number of additional minor metabolites (White et al., 1970).

REGULATION

Thiamin homeostasis is maintained both at the level of intestinal absorption and of renal tubular recovery, both of which are tightly limited. Expression of the thiamin transporter gene is induced by the p53 tumor suppressor (Lo et al., 2001) and regulated via intracellular calcium/calcmodulin signaling (Said et al., 2001). Export of both TMP and TPP from extraintestinal tissues by the reduced folate carrier (SLC19A1) is likely to limit concentrations, thereby contributing to the maintenance of homeostasis (Zhao and Goldman, 2013).

FUNCTION

Only five enzymes have been identified so far that have a strict thiamin requirement. All of them use TPP coordinated with magnesium as a cofactor. Additional actions of thiamin, especially as TTP in the brain, appear to be similarly essential, but they have not yet been completely characterized.

Transketolases: Transketolase (EC2.2.1.1) is needed for glucose metabolism via the pentose-phosphate pathway, the only pathway that generates significant amounts of NADPH. Two distinct genes are now known to encode proteins with transketolase activity. Alternative splicing of the more recently discovered one, transketolase 2 (Coy et al., 1996), gives rise to different isoforms in the brain and heart. Decreased activity of this enzyme may contribute to the Wernicke–Korsakoff syndrome observed in alcohol abusers (see the "Brain function" section that follows shortly). The transketolase 2 gene locus is immediately adjacent to the protein-coding regions of the retina color pigment genes on the X chromosome (Hanna et al., 1997), which might suggest a particular importance for vision.

Pyruvate dehydrogenase: This key enzyme (EC3.1.3.43) of glucose metabolism is embedded in the mitochondrial matrix and contains multiple copies of three distinct moieties: E1, E2, and E3. TPP is associated with E1.

Alpha-ketoglutarate dehydrogenase: This enzyme (EC1.2.4.2) of the tricarboxylic acid cycle consists of three distinct moieties: E1, E2, and E3; TPP is associated with E1. The enzyme probably also participates in the breakdown of tryptophan, lysine, and hydroxylysine.

Branched-chain alpha-keto acid dehydrogenase: This enzyme with the systematic name 3-methyl-2-oxobutanoate dehydrogenase (EC1.2.4.4) comprises three distinct subunits, E1 (with TPP bound to His292), E2, and E3. The enzyme is needed for the catabolism of the branched-chain amino acids valine, isoleucine, and leucine. Branched-chain alpha-keto acid dehydrogenase and pyruvate dehydrogenase also cleave alpha-ketobutyrate (from L-threonine and homocysteine metabolism) into CO_2 and propionyl-CoA (Paxton et al., 1986: Pettit and Reed, 1988).

Phytanic acid metabolism: Alpha-oxidation of 3-methyl fatty acids such as phytanic acid (Foulon et al., 1999) involves as the third step a reaction catalyzed by the TPP-dependent enzyme 2-hydroxy-phytanoyl-CoA lyase (EC4.1.2.n2).

Brain function: TTP appears to be important for brain function, possibly by participating in the function of maxi-Cl-channels (chloride channels of large unitary conductance). Deficiency due to genetic causes during early fetal development or infancy may cause progressive degeneration of the cerebral cortex (Laurence and Cavanagh, 1968). Leigh syndrome is characterized by the degeneration and focal necrosis of gray matter, as well as capillary proliferation in the brain stem. Reduced production of TTP, possibly through inhibition of thiamin-diphosphate kinase (EC2.7.4.15), has been suggested as a causative factor, pointing to the critical importance of this thiamin metabolite (Ortigoza-Escobar et al., 2014). Severe confusion and agitation characterizes Wernicke–Korsakoff syndrome (WKS). It is seen most often in chronic alcohol abusers and usually responds well to thiamin administration. Working memory of alcohol abusers in a detoxification program appeared to improve in a dose-dependent manner with intramuscular injection of thiamin (Ambrose et al., 2001). However, there are considerable doubts whether alcohol actually plays a causative role (Scalzo et al., 2015). Systematic review of applicable trial reports found insufficient evidence to settle on dose, frequency, route, or duration of thiamin use for prevention or treatment of WKS (Day et al., 2013).

Cardiovascular function: The original clinical descriptions of thiamin deficiency included the wet beriberi type, characterized by severe edemas and heart failure. In regions with significant food insecurity, the use as a staple food of polished rice, which lacks thiamin-containing germ and bran, this presentation was and still is particularly insidious and has a high fatality rate. In the more affluent regions of the world, heart failure due to thiamin deficiency is most often related to heavy alcohol use. Another often underappreciated cause of thiamin deficiency is increased loss via urine in patients treated with furosemide or related diuretics, which may present in a fulminant and particularly deadly form (Misumida et al., 2014). There is also evidence that inadequate thiamin availability to tissues is associated with increased blood pressure (Zhang et al., 2014). Ultimately, there are still significant evidence gaps for defining the best protocol for thiamin supplementation in the management of heart failure (Ahmed et al., 2015).

Mitochondria: A facilitating role of thiamin for mitochondrial function has been suggested (Sato et al., 2000).

NUTRITIONAL ASSESSMENT

Thiamin deficiency and suboptimal status may be detected by enzyme activation assays or blood concentration of thiamin metabolites. Dietary intake is often estimated from the amount of excreted urinary thiamin metabolites.

The enzyme activation assays measure transketolase activity in untreated and TPP-treated red blood cells (Michalak et al., 2013). An activation coefficient (the ratio of activity in supplemented red blood cells divided by the activity in native red blood cells) in excess of 1.15 indicates less-than-optimal TPP saturation of the enzyme. Others measure directly the TPP concentration in red blood cells (Hanninen et al., 2006). A recognized weakness of using red blood cells is that deficiency-induced increases of transporter activity in red blood cell membranes masks true deficiency states (Thornalley et al., 2007).

The sum of active thiamin metabolites (thiamin, TMP, and TPP) in plasma is thought to reflect thiamin status. Such measurements indicated, for example, that individuals with type 1 or type 2 diabetes have greatly diminished plasma concentrations (Thornalley et al., 2007).

The urinary excretion of thiamin or of the thiamin-derived metabolites (pyrimidine carboxylic acid, thiazole acetic acid, thiamin acetic acid, and minor metabolites) can be used to estimate dietary intake (Tasevska et al., 2008).

REFERENCES

Ahmed, M., Azizi-Namini, P., Yan, A.T., Keith, M., 2015. Thiamin deficiency and heart failure: the current knowledge and gaps in literature. Heart Fail Rev. 20, 1–11.

Ambrose, M.L., Bowden, S.C., Whelan, G., 2001. Thiamin treatment and working memory function of alcohol-dependent people: preliminary findings. Alc. Clin. Exp. Res. 25, 112–116.

Ariaey-Nejad, M.R., Balaghi, M., Baker, E.M., Sauberlich, H.E., 1970. Thiamin metabolism in man. Am. J. Clin. Nutr. 23, 764–778.

Beaudoin, A.R., Grondin, G., Lord, A., Roberge, M., St-Jean, P., 1983. The origin of the zymogen granule membrane of the pancreatic acinar cell as examined by ultrastructural cytochemistry of acid phosphatase, thiamine pyrophosphatase, and ATP-diphosphohydrolase activities. Eur. J. Cell Biol. 29, 218–225.

Breen, K.J., Buttigieg, R., Iossifidis, S., Lourensz, C., Wood, B., 1985. Jejunal uptake of thiamin hydrochloride in man: influence of alcoholism and alcohol. Am. J. Clin. Nutr. 42, 121–126.

Coy, J.F., Dubel, S., Kioschis, P., Thomas, K., Micklem, G., Delius, H., et al., 1996. Molecular cloning of tissue-specific transcripts of a transketolase-related gene: implications for the evolution of new vertebrate genes. Genomics 32, 309–316.

Davis, R.E., Icke, G.C., Thorn, L., Riley, W.J., 1984. Intestinal absorption of thiamin in man compared with folate and pyridoxal and its subsequent urinary excretion. J. Nutr. Sci. Vitaminol 30, 475–482.

Day, E, Bentham, PW, Callaghan, R, Kuruvilla, T, George, S., 2013. Thiamine for prevention and treatment of Wernicke-Korsakoff Syndrome in people who abuse alcohol. Cochrane Database Syst. Rev. 7, CD004033.

Dudeja, P.K., Tyagi, S., Kavilaveettil, R.J., Gill, R., Said, H.M., 2001. Mechanism of thiamine uptake by human jejunal brush border membrane vesicles. Am. J. Physiol. Cell Physiol. 281, C786–C792.

Dutta, B., Huang, W., Molero, M., Kekuda, R., Leibach, F.H., Devoe, L.D., et al., 1999. Cloning of the human thiamine transporter, a member of the folate transporter family. J. Biol. Chem. 274, 31925–31929.

Food and Nutrition Board, Institute of Medicine, 1998. Dietary Reference Intakes for Thiamin, Riboflavin, Niacin, Vitamin B6, Folate, Vitamin B12, Pantothenic Acid, Biotin, and Choline. National Academy Press, Washington, DC, pp. 58–86; 480–481.

Foulon, V., Antonenkov, V.D., Croes, K., Waelkens, E., Mannaerts, G.E., Van Veldhoven, P.E., et al., 1999. Purification, molecular cloning, and expression of 2-hydroxyphytanoyl-CoA lyase, a peroxisomal thiamine pyrophosphate-dependent enzyme that catalyzes the carbon–carbon bond cleavage during alpha-oxidation of 3-methyl-branched fatty acids. Proc. Natl. Acad. Sci. USA 96, 10039–10044.

Gangolf, M., Czerniecki, J., Radermecker, M., Detry, O., Nisolle, M., Jouan, C., et al., 2010a. Thiamine status in humans and content of phosphorylated thiamine derivatives in biopsies and cultured cells. PLoS One 5 (10), e13616.

Gangolf, M., Wins, P., Thiry, M., El Moualij, B., Bettendorff, L., 2010b. Thiamine triphosphate synthesis in rat brain occurs in mitochondria and is coupled to the respiratory chain. J. Biol. Chem. 285, 583–594.

Gastaldi, G., Cova, E., Verri, A., Laforenza, U., Faelli, A., Rindi, G., 2000. Transport of thiamin in rat renal brush border membrane vesicles. Kidney Int. 57, 2043–2054.

Gregory III, J.F., 1997. Bioavailability of thiamin. Eur. J. Clin. Nutr. 51, S34–S37.

Hanna, M.C., Platts, J.T., Kirkness, E.E., 1997. Identification of a gene within the tandem array of red and green color pigment genes. Genomics 43, 384–386.

Hanninen, S.A., Darling, P.B., Sole, M.J., Barr, A., Keith, M.E., 2006. The prevalence of thiamin deficiency in hospitalized patients with congestive heart failure. J. Am. Coll. Cardiol. 47, 354–361.

Hayes, K.C., Hegsted, D.M., 1973. Toxicity of the vitamins Toxicants Occurring Naturally in Foods. Food and Nutrition Board, National Research Council, National Academy Press, Washington, DC, pp. 235–253.

Kang, Y.S., Terasaki, T., Ohnishi, T., Tsuji, A., 1990. In vivo and in vitro evidence for a common carrier mediated transport of choline and basic drugs through the blood–brain barrier. J. Pharmacobio-Dyn. 13, 353–360.

Labay, V., Raz, T., Baron, D., Mandel, H., Williams, H., Barrett, T., et al., 1999. Mutations in SLC19A2 cause thiamine-responsive megaloblastic anemia associated with diabetes mellitus and deafness. Nat. Genet. 22, 300–304.

Laforenza, U., Patrinin, C., Alvisi, C., Faelli, A., Licandro, A., Rindi, G., 1997. Thiamine uptake in human intestinal biopsy specimens, including observations from a patient with acute thiamine deficiency. Am. J. Clin. Nutr. 66, 320–326.

Latt, N., Dore, G., 2014. Thiamine in the treatment of Wernicke encephalopathy in patients with alcohol use disorders. Intern. Med. J. 44, 911–915.

Laurence, K.M., Cavanagh, J.B., 1968. Progressive degeneration of the cerebral cortex in infancy. Brain 91, 261–280.

Lo, P.K., Chen, J.Y., Tang, P.P., Lin, J., Lin, C.H., Su, L.T., et al., 2001. Identification of a mouse thiamine transporter gene as a direct transcriptional target for p53. J. Biol. Chem. 276, 37186–37193.

McCormick, D.B., 2000. Niacin, riboflavin, and thiamin. In: Stipanuk, M.H. (Ed.), Biochemical and Physiological Aspects of Human Nutrition W.B. Saunders, Philadelphia, PA, pp. 458–482.

Michalak, S., Michałowska-Wender, G., Adamcewicz, G., Wender, M.B., 2013. Erythrocyte transketolase activity in patients with diabetic and alcoholic neuropathies. Folia Neuropathol. 51, 222–226.

Misumida, N., Umeda, H., Iwase, M., 2014. Shoshin beriberi induced by long-term administration of diuretics: a case report. Case Rep. Cardiol. 2014, 878915.

Nishino, K., Itokawa, Y., Nishino, N., Piros, K., Cooper, J.R., 1983. Enzyme system involved in the synthesis of thiamin triphosphate. I. Purification and characterization of protein-bound thiamin diphosphate: ATP phosphoryltransferase. J. Biol. Chem. 258, 11871–11878.

Ortigoza-Escobar, J.D., Serrano, M., Molero, M., Oyarzabal, A., Rebollo, M., Muchart, J., et al., 2014. Thiamine transporter-2 deficiency: outcome and treatment monitoring. Orphanet J. Rare Dis. 9, 92.

Patrini, C., Reggiani, C., Laforenza, U., Rindi, G., 1988. Blood–brain transport of thiamine monophosphate in the rat: a kinetic study *in vivo*. J. Neurochem. 50, 90–93.

Paxton, R., Scislowski, P.W.D., Davis, E.J., Harris, R.A., 1986. Role of branched-chain 2-oxo acid dehydrogenase and pyruvate dehydrogenase in 2-oxobutyrate metabolism. Biochem. J. 234, 295–303.

Pearson, W.N., Darby Jr., W.J., 1967. Catabolism of ^{14}C-labeled thiamine by the rat as influenced by dietary intake and body thiamine stores. J. Nutr. 93, 491–498.

Pettit, F.H., Reed, L.J., 1988. Branched-chain alpha-keto acid dehydrogenase complex from bovine kidney Methods Enzymol. 166, 309–312.

Quadri, G, McQuillin, A, Guerrini, I, Thomson, AD, Cherian, R, Saini, J, et al., 2014. Evidence for genetic susceptibility to the alcohol dependence syndrome from the thiamine transporter 2 gene solute carrier SLC19A3. Psychiatr. Genet. 24, 122–123.

Rindi, G., Laforenza, U., 2000. Thiamine intestinal transport and related issues: recent aspects. Proc. Soc. Exp. Biol. Med. 224, 246–255.

Said, H.M., Ortiz, A., Subramanian, V.S., Neufeld, E.J., Moyer, M.E., Dudeja, P.K., 2001. Mechanism of thiamine uptake by human colonocytes: studies with cultured colonic epithelial cell line NCM460. Am. J. Physiol. Gastrointest. Liver Physiol. 281, G144–G150.

Sato, Y., Nakagawa, M., Higuchi, I., Osame, M., Naito, E., Oizumi, K., 2000. Mitochondrial myopathy and familial thiamine deficiency. Muscle Nerve 23, 1069–1075.

Scalzo, SJ, Bowden, SC, Ambrose, ML, Whelan, G, Cook, MJ., 2015. Wernicke-Korsakoff syndrome not related to alcohol use: a systematic review. J. Neurol. Neurosurg. Psychiatry pii: jnnp-2014 309598.

Schenker, S., Johnson, R.E., Hoyumpa, A.M., Henderson, G.I., 1990. Thiamine-transfer by human placenta: normal transport and effects of ethanol. J. Lab. Clin. Med. 116, 106–115.

Shikata, H., Egi, Y., Koyama, S., Yamada, K., Kawasaki, T., 1989. Properties of the thiamin triphosphate- synthesizing activity catalyzed by adenylate kinase (isoenzyme I). Biochem. Int. 18, 943–949.

Shikata, H., Koyama, S., Egi, Y., Yamada, K., Kawasaki, T., 1986. Identification of creatine as a cofactor of thiamin-diphosphate kinase. FEBS Lett. 201, 101–104.

Subramanian, V.S., Nabokina, S.M., Said, H.M., 2014. Association of TM4SF4 with the human thiamine transporter-2 in intestinal epithelial cells. Dig. Dis. Sci. 59, 583–590.

Subramanya, S.B., Subramanian, V.S., Sekar, V.T., Said, H.M., 2011. Thiamin uptake by pancreatic acinar cells: effect of chronic alcohol feeding/exposure. Am. J. Physiol. Gastrointest. Liver Physiol. 301, G896–G904.

Suter, P.M., Vetter, W., 2000. Diuretics and vitamin B1: are diuretics a risk factor for thiamin malnutrition? Nutr. Rev. 58, 319–323.

Tallaksen, C.M., Bohmer, T., Karlsen, J., Bell, H., 1997. Determination of thiamin and its phosphate esters in human blood, plasma, and urine. Methods Enzymol. 279, 67–74.

Tallaksen, C.M., Sande, A., Bohmer, T., Bell, H., Karlsen, J., 1993. Kinetics of thiamin and thiamin phosphate esters in human blood, plasma and urine after 50 mg intravenously or orally. Eur. J. Clin. Pharmacol. 44, 73–78.

Tasevska, N., Runswick, S.A., McTaggart, A., Bingham, S.A., 2008. Twenty-four-hour urinary thiamine as a biomarker for the assessment of thiamine intake. Eur. J. Clin. Nutr. 62, 1139–1147.

Thornalley, P.J., Babaei-Jadidi, R., Al Ali, H., Rabbani, N., Antonysunil, A., Larkin, J., et al., 2007. High prevalence of low plasma thiamine concentration in diabetes linked to a marker of vascular disease. Diabetologia 50, 2164210.

White III, W.W., Amos Jr., W.H., Neal, R.A., 1970. Isolation and identification of the pyrimidine moiety of thiamin in rat urine using gas chromatography-mass spectrometry. J. Nutr. 100, 1053–1056.

Zhang, K., Huentelman, M.J., Rao, F., Sun, E.I., Corneveaux, J.J., Schork, A.J., International Consortium for Blood Pressure Genome-Wide Association Studies, 2014. Genetic implication of a novel thiamine transporter in human hypertension. J. Am. Coll. Cardiol. 63, 1542–1555.

Zhao, R., Goldman, I.D., 2013. Folate and thiamine transporters mediated by facilitative carriers (SLC19A1-3 and SLC46A1) and folate receptors. Mol. Aspects Med. 34, 373–385.

RIBOFLAVIN

Riboflavin (7,8-dimethyl-10-(D-ribo-2,3,4,5-tetrahydroxypentyl)isoalloxazine, vitamin B2; obsolete names vitamin G, lyochrome, lactoflavin, hepatoflavin, ooflavin, uroflavin; molecular weight 376) is a water-soluble, heat-stable, and light- and alkali-sensitive vitamin (Figure 10.11).

ABBREVIATIONS	
AMP	adenosine monophosphate
ETF	electron-transfer flavoprotein
FAD	flavin adenine dinucleotide
FMN	flavin mononucleotide
RCP	riboflavin carrier protein
RDA	recommended dietary allowance

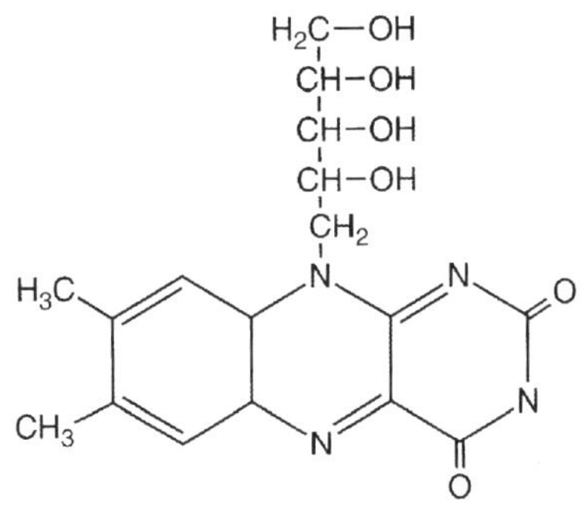

FIGURE 10.11

Riboflavin.

NUTRITIONAL SUMMARY

Function: Riboflavin is the precursor of FMN and FAD, which serve both as prosthetic groups and as cofactors for a wide range of enzymes important for beta-oxidation and oxidative phosphorylation, antioxidant defense, vitamin metabolism (folate, niacin, vitamins A, C, B6, and B12), amino acid utilization, hormone synthesis, and many other functions.

Food sources: Milk and dairy products, meat, poultry and fish, cereals and bread, and green vegetables each can provide at least one-sixth of recommended intake per serving.

Requirements: The current RDA for women is 1.1 mg/day, and for men, it is 1.3 mg/day (Food and Nutrition Board, Institute of Medicine, 1998). Pregnancy, lactation, and increased energy intake and expenditure all increase requirements. Since only 1–2 weeks' requirement of riboflavin is stored, regular adequate intake is important.

Deficiency: Prolonged low intake causes cracking and swelling of the lips (cheilosis), cracking and inflammation of the angles of the mouth (angular stomatitis), dark-red tongue (glossitis), skin changes at other sites (seborrheic dermatitis), and normocytic anemia. Deficiency during infancy and childhood impairs growth.

Excessive intake: There is little danger even when intake exceeds the RDA many times. Since excess riboflavin is lost rapidly, there is no additional benefit over recommended intake levels and stores will not increase.

DIETARY SOURCES

Foods contain free and beta-glucosylated riboflavin (Gregory, 1998) as well as FMN and FAD, some of the latter covalently bound to proteins. Riboflavin is heat-stable, but light- and alkali-sensitive; the inactive form lumiflavin (7,8,10-trimethylisoalloxazine) (Figure 10.12) is a product of photodecomposition (Chastain and McCormick, 1987). Best sources of riboflavin are milk and dairy products, meat, poultry and fish, green vegetables, cereals, and bread. Grain products in the United States must be fortified with 4 mg/kg.

Median intake in the United States is about 2 mg/day, much of this from fortified foods and dietary supplements.

DIGESTION AND ABSORPTION

Riboflavin is present in foods mostly (80–90%) as FAD and FMN cofactors of proteins. Hydrochloric acid from the stomach readily releases the flavins that are only loosely bound to their proteins. A small percentage of food flavin is bound to a histidyl-nitrogen or cysteinyl-sulfur, and proteolysis results in

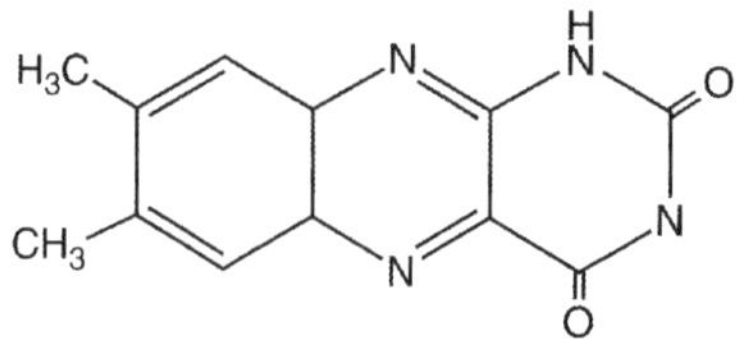

FIGURE 10.12

Lumiflavin is a light-inactivated product of riboflavin.

the release of amino acid–linked 8-alpha-FAD, which is biologically inactive. FMN is dephosphorylated to riboflavin by alkaline phosphatase (EC3.1.3.1) in the small intestine. FAD is broken up by nucleotide pyrophosphatase (EC3.6.1.9) at the brush border of villous tip cells into AMP and FMN, from which riboflavin can then be released (Byrd et al., 1985).

Some of the riboflavin in plant-derived foods is present as beta-glucoside, which has to be cleaved by a beta-glucosidase (possibly lactase) prior to absorption. Fractional intestinal absorption of riboflavin and related compounds is high over a large range of intakes (75% of a 20-mg dose) and declines with intakes beyond that (Zempleni et al., 1996).

Riboflavin is absorbed mainly from the jejunum, and only to a much lesser degree from the large intestine (Said et al., 2000). Uptake proceeds through the proton-cotransporter RFVT3 (SLC52A3). Another two transporters, RFVT1 (SLC52A1) and RFVT2 (SLC52A2), mediate efflux across the basolateral membrane through an as-yet-unexplored mechanism (Yonezawa and Inui, 2013).

At higher concentrations, passive diffusion into the enterocyte becomes increasingly relevant. Retention in the enterocyte does not entail metabolic modification of free riboflavin (Said and Ma, 1994). The maximal amount that can be absorbed from a single dose appears to be about 27 mg (Zempleni et al., 1996).

The breast cancer resistance protein (BCRP, ABCG2) is an intestinal efflux transporter that moves free riboflavin back into the intestinal lumen (Tan et al., 2013). Metabolic conversion prevents the loss of newly absorbed riboflavin. Phosphorylation of free riboflavin by riboflavin kinase (flavokinase, EC2.7.1.26, zinc) to FMN is critical for retaining riboflavin in the enterocyte (Gastaldi et al., 2000). FMN can then be convened to FAD by ATP:FMN adenylyltransferase (FAD synthetase, EC2.7.7.2). About 60% of absorbed riboflavin is exported as FMN or FAD (Gastaldi et al., 2000). It is not clear whether riboflavin and its metabolites leave the enterocyte by simple diffusion or by another process.

TRANSPORT AND CELLULAR UPTAKE

Blood circulation: FAD, riboflavin, and FMN are (in descending concentration order) the main forms in plasma; in severe malnutrition, FAD concentration may be lower than FMN concentration (Capochichi et al., 2000). Neither the plasma concentration of flavo-coenzymes (around 79 nmol/l) nor of riboflavin concentration in plasma (around 13 nmol/l) varies much in response to different levels of intake (Zempleni et al., 1996). Most riboflavin metabolites in plasma are associated with immunoglobulins, albumin, and other proteins (Innis et al., 1985). Riboflavin is taken up into the liver by an energy-dependent process (Said et al., 1998). Uptake into the liver and other tissues first requires the hydrolysis of FAD and FMN to riboflavin (Lee and Ford, 1988). Some inactive metabolites, such as 2′-hydroxyethylflavin, impede cellular uptake of riboflavin by competition (Aw et al., 1983).

The riboflavin carrier protein (RCP) mediates intracellular riboflavin transport (Schneider, 1996). RCP is expressed in the placenta, Sertoli and Leydig cells of the testes, and spermatozoa. The chicken vitellogenin receptor imports very-low-density lipoprotein (VLDL), riboflavin-binding protein, and aipha-2-macroglobulin into growing oocytes. The similarity of vitellogenin receptor and VLDL receptor raises the question whether the latter may contribute to the uptake of riboflavin-binding protein in humans (Schneider, 1996).

The multidrug transporter ABCG2 (BCRP, MXR) appears to move riboflavin from the mammary glands into milk (Vlaming et al., 2009).

Materno-fetal transfer: Riboflavin-containing nucleotides have to be cleaved prior to uptake into syntrophoblasts via an as-yet-unknown carrier. RCP is critical for riboflavin transfer across the placental membrane and is inducible by estrogen. Antibodies against RCP reduce fertility (Adiga et al., 1997). RFVT1 (SLC52A1) is highly expressed in the placenta and clearly constitutes one part of the materno-fetal transport mechanism (Yonezawa and Inui, 2013). The complete details remain to be investigated.

BBB: Riboflavin rapidly crosses from blood circulation into the brain and is converted into FMN and FAD (Spector, 1980). Reverse transport is also readily possible. A highly specific, sodium-dependent carrier mediates this transfer (Patel et al., 2012).

METABOLISM

Activation: Flavin cofactor synthesis occurs in the liver and most other tissues (Figure 10.13). In the initial rate-limiting step, the zinc-dependent riboflavin kinase (EC2.7.1.26) phosphorylates riboflavin. From

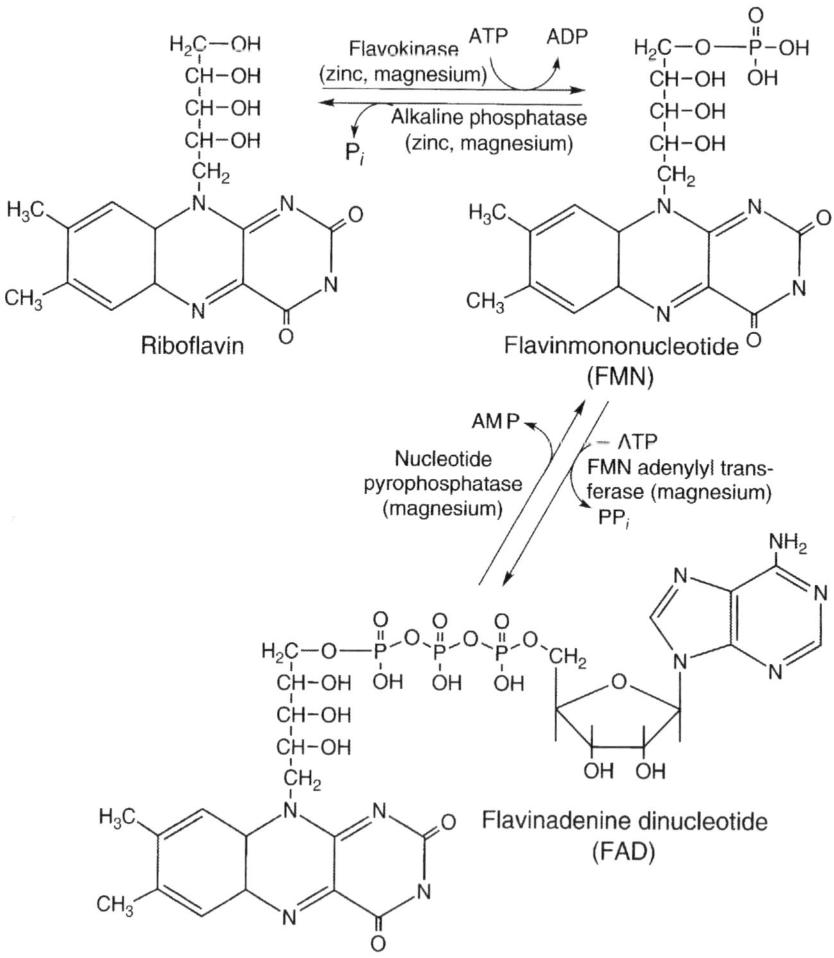

FIGURE 10.13

Riboflavin metabolism.

FMN, the more commonly used cofactor FAD is produced by FMN adenylyltransferase (EC2.7.7.2). This magnesium-dependent reaction links the phosphate group of the AMP moiety to the phosphate group of FMN. There is no information about the mechanisms whereby FMN or FAD becomes covalently bound to specific histidyl or cysteinyl residues of just a few of the numerous flavoproteins.

Catabolism: Intracellular FAD can be cleaved into FMN and AMP by FAD pyrophosphatase (EC3.6.1.1.8), and FMN can be dephosphorylated by alkaline phosphatase. Some of the free intracellular riboflavin is metabolized to 7-alpha-hydromethyl riboflavin and 8-alpha-hydromethyl riboflavin by microsomal oxidation. The mechanism for production of the antivitamin 10-(2'-hydroxyethyl) flavin (Figure 10.14), which is present in human (and bovine) milk, is not known (Aw et al., 1983).

In the few instances where flavin moiety is covalently bound, the intracellular degradation of the protein may generate 8-alpha-*S*-cysteinyl riboflavin. This metabolite is converted to 8-alpha-sulfonylriboflavin upon release from FAD-containing amine oxidase (monoamine oxidase, EC1.4.3.4; Chastain and McCormick, 1987). Histidine-linked flavins are metabolized to histidyl riboflavine or related catabolites.

STORAGE

There is general agreement that riboflavin is not stored in specific compartments and ingested vitamin is turned over and excreted with a half-life of a few days. Several metabolites, including some of the protein-bound FMN and FAD, are reutilized upon protein turnover and provide a riboflavin pool that can cover needs during times of low intake. Unfortunately, information is lacking on the precise size of this pool in different states of repletion and how much biologically active vitamin is released per day in the absence of dietary intake.

EXCRETION

Less than 1% of absorbed riboflavin is eliminated with bile. About half can be recovered as riboflavin or known riboflavin metabolites via urine (Zempleni et al., 1996). The remainder leaves the body by as-yet-unknown pathways (Zempleni et al., 1996). The portion of riboflavin and its various circulating metabolites that is not bound to albumin or other proteins readily passes into primary filtrate in the renal glomerulus and depends for reuptake in the proximal renal tubule on sodium-independent, carrier-mediated transport (Kumar et al., 1998). In addition, riboflavin is secreted into the tubular lumen.

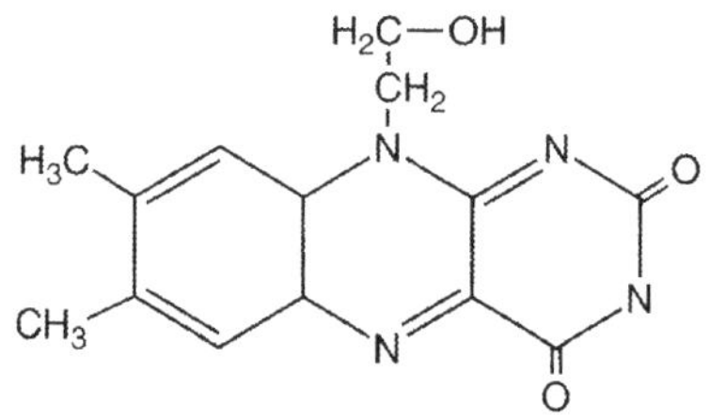

FIGURE 10.14

The riboflavin catabolite 10-(2'-hydroxyethyl) flavin acts as an antivitamin.

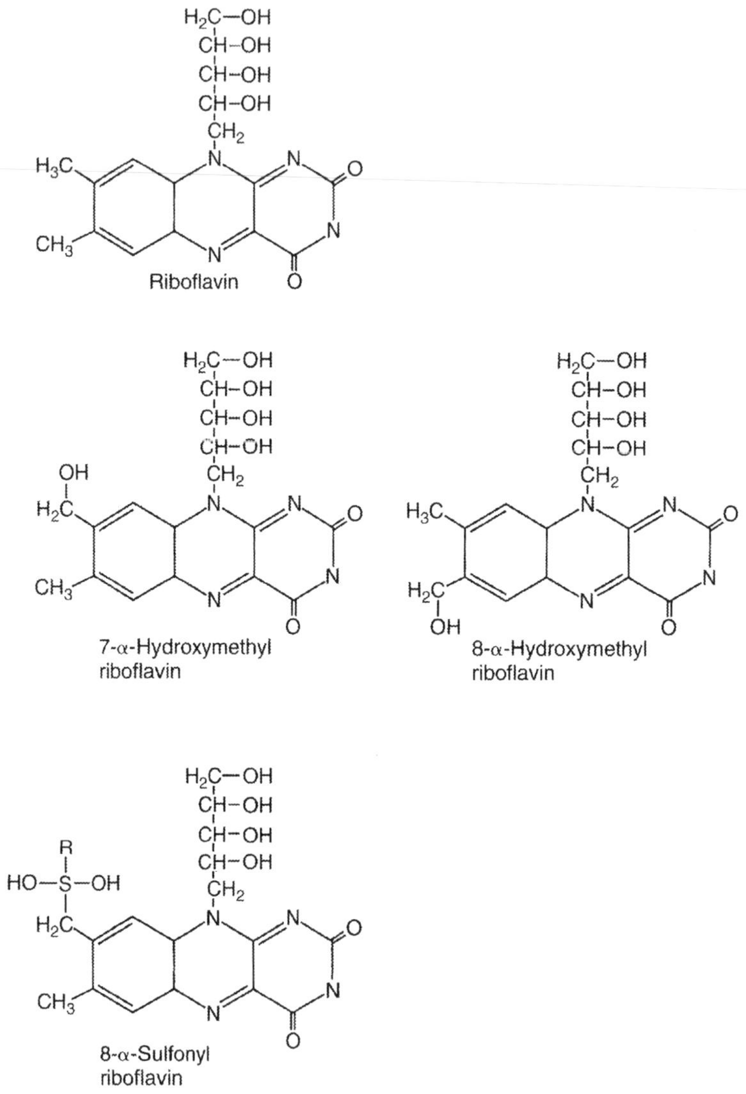

FIGURE 10.15

Major urinary metabolites related to riboflavin ingestion.

With high intake, the secreted amount far exceeds the reabsorbed amount. Less than a third of ingested riboflavin at normal intake levels is excreted with urine (Zempleni et al., 1996). Urine normally contains mainly riboflavin (60–70%), and much smaller amounts of 7-hydromethylriboflavin (10–15%), 8-hydromethylriboflavin (4–7%), and 8-alpha-sulfonylriboflavin (Chastain and McCormick, 1987; McCormick, 1994) (Figure 10.15).

REGULATION

Several hormones, including triiodothyronine, adrenocorticotrophic hormone (ACTH), and aldosterone tightly control FMN synthesis by enhancing riboflavin kinase activity. FMN synthesis declines in severe malnutrition, apparently due to diminished triiodothyronine production (Capo-chichi et al., 2000).

FUNCTION

FAD in flavoproteins: Most human flavoproteins contain one or more loosely bound FAD moieties. In a few specific instances, the 8-alpha methyl group of FAD is covalently linked to a peptidyl residue. Enzymes with a histidyl-linked FAD include succinate dehydrogenase (EC1.3.5.1), several acyl-CoA dehydrogenases, and polyamine oxidase (EC1.5.3.11). In both monoamine oxidase A and B (EC1.4.3.4), the 8-alpha methyl group of FAD is linked to an *S*-cysteinyl residue.

FMN in flavoproteins: FMN is used by very few human enzymes. Where it occurs, it is usually loosely bound to the enzyme. FMN-containing proteins include the 51-kD subunit of NADH reductase (respiratory chain complex 1, EC1.6.5.3), pyridoxamine phosphate oxidase [EC1.4.3.5L (*S*)-2-hydroxy-acid oxidase (EC1.1.3.15)], NADPH ferrihemoprotein reductase (EC1.6.2.4, together with FAD), and possibly the NADH-dependent (EC1.6.1.3) and NADPH-dependent (EC1.6.1.5) aquacobalamin reductases. In sarcosine dehydrogenase (EC1.5.99.1), FMN is covalently linked through its 8-alpha methyl group to a histidyl residue of the protein.

Oxidative phosphorylation: The complex I of mitochondrial respiratory electron transport (NADH dehydrogenase, EC1.6.99.3) contains a 42-kD subunit with FAD as a prosthetic group, and a 51-kD subunit (flavoprotein I) with FMN. Complex 11 (succinate ubiquinone dehydrogenase, EC1.3.5.1) contains one covalently bound FAD.

FAD-containing NAD(P) transhydrogenases use the reducing equivalents for ATP-linked proton pumping across the inner mitochondrial membrane. The mitochondrial component of the glycerol phosphate shuttle, the FAD-enzyme glycerol 3-P dehydrogenase (EC1.1.99.5), works together with a cytoplasmic glycerol 3-P dehydrogenase (which does not contain a flavin) to transfer reducing equivalents from cytoplasmic glycolysis into mitochondria.

Glutathione-linked reactions: Numerous flavoproteins help to maintain the intracellular redox potential and protect sulfur compounds against oxidation. An important example is the ubiquitous cytoplasmic glutathione reductase (EC1.6.4.2), which uses FAD and NADPH to reduce oxidized glutathione. The percentage increase of *in vitro* enzyme activity in red blood cells upon addition of FAD is commonly used as an indicator of riboflavin status; greater activity indicates incomplete saturation of FAD-binding sites of the enzyme and thus poorer riboflavin status. The roles of the FAD enzymes glutathione oxidase (EC1.8.3.3) and CoA-glutathione reductase (EC1.6.4.6) need further exploration.

Redox reactions: NADPH-ferrihemoprotein reductase (EC1.6.2.4) is a FAD-containing enzyme that reduces heme-thiolate-dependent monooxygenases such as the unspecific monooxygenase (EC1.14.14.1), which is part of the microsomal hydroxylating system. It seems ironic that this latter enzyme in turn inactivates some of the free riboflavin by generating the metabolites 7-hydroxymethyl riboflavin and 8-hydroxymethyl riboflavin. Another microsomal flavoenzyme involved in redox reaction is NADPH-cytochrome c2 reductase (EC1.6.2.5).

Intermediary metabolism: D-2-hydroxy-acid dehydrogenase (EC1.1.99.6) metabolizes hydroxy acids, including (R)-lactate.

Fatty acid beta-oxidation: Three distinct mitochondrial fatty acyl dehydrogenases oxidize acyl-CoA of varying chain length. As a long-chain fatty acyl-CoA successively gets shortened during cycles of beta-oxidation, the appropriate enzyme can take over, starting with long-chain acyl-CoA dehydrogenase (EC1.3.99.13), to acyl-CoA dehydrogenase (EC1.3.99.3), and finally to butyryl-CoA dehydrogenase (EC1.3.99.2).

These enzymes possess a covalently N(5)-linked FAD and use the FAD-containing electron-transfer flavoprotein (ETF) as an electron acceptor. ETF is then reduced by ETF:ubiquinone oxidoreductase (EC1.5.5.1), generating ubiquinol for use in the respiratory chain. Peroxisomal beta-oxidation, in contrast, uses only a single, FAD-dependent acyl-CoA oxidase (EC1.3.3.6) for chain lengths between 18 and 8 and does not use ETF as an acceptor.

Lipid metabolism: The FAD-containing sphinganine oxidase (with no EC number assigned) is needed for the synthesis of sphingosin for a wide range of phospholipids and other complex lipids. FAD is also participating in cholesterol synthesis as the prosthetic group of squalene monooxygenase (EC1.14.99.7), which initiates the cyclization of squalene.

Vitamin metabolism: The metabolism of several vitamins involves flavoproteins. Thioredoxin reductase (EC1.6.4.5) regenerates reduced glutathione, which is used for dehydroascorbate reduction. Pyridoxamine-phosphate oxidase (EC1.4.3.5) interconverts the B6 vitamers pyridoxine, pyridoxamine, and pyridoxal, as well as their phosphates.

Kynurenine 3-monooxygenase (EC1.14.13.9) is a key enzyme in the formation of nicotinate from tryptophan. A lack of riboflavin is known to diminish vitamin B6 sufficiency. The FAD-dependent methylene tetrahydrofolate reductase (EC1.5.1.20) is needed for folate metabolite recycling; with a reduction of its activity, higher folate intake is needed to avoid deficiency. Vitamin B12 requires three flavoenzymes for its metabolism: cob(ll)alamin reductase (EC1.6.99.9), aquacobalamin reductase/NADPH (EC1.6.99.11), and aquacobalamin reductase/NADH (EC1.6.99.8). Retinal dehydrogenase (EC1.2.1.36) is the enzyme that generates retinoic acid from retinal. NADPH dehydrogenase (EC1.6.99.6) and two forms of NAD(P)H dehydrogenase (EC1.6.99.2) reactivate vitamin K (dicoumarol inhibitable) and also provide important antioxidant protection.

Heme metabolism: Protoporphyrinogen oxidase (EC1.3.3.4) at the inner mitochondrial membrane contains one FAD moiety per homodimer. Protoporphyrinogen-IX is oxidized to protoporphyrin-IX, into which iron can then be inserted by another (not flavin-dependent) enzyme. NADPH dehydrogenase (EC1.6.99.1) reduces biliverdin to bilirubin in the liver and also may protect against oxidative damage.

Nucleotide metabolism: In the third-to-last step of pyrimidine synthesis, the FAD containing dihydroorotate oxidase (EC1.3.3.1) generates orotate. Xanthine dehydrogenase (EC1.1.3.22), which contains FAD, molybdopterin, and iron-sulfur clusters, catalyzes the final step of purine catabolism to uric acid.

Amino acid and amine metabolism: Dihydrolipoamide dehydrogenase (EC1.8.1.4) uses FAD to transfer reducing equivalents to NAD. Flavoprotein is part of enzyme complexes that participate in the catabolism of glycine, glutamate, valine, leucine, and isoleucine. This category includes glutaryl-CoA dehydrogenase (EC1.3.99.7), 2-methylacyl-CoA dehydrogenase (EC1.3.99.12, branched-chain amino acid catabolism), *N*-methyl-L-amino-acid oxidase (EC1.5.3.2), kynurenine 3-monooxygenase (EC1.14.13.9), methionine synthase reductase (EC2.1.1.135, contains FAD, FMN, and cobalamin), *N*-methyl-L-amino acid oxidase (EC1.5.3.2), L-amino acid oxidase (EC1.4.3.2), (*S*)-2-hydroxy-acid oxidase (EC1.1.3.15, peroxisomal, also acts as an L-amino acid oxidase), D-aspartate oxidase (EC1.4.3.1), and D-amino acid oxidase (EC1.4.3.3).

Two FAD enzymes that participate in choline catabolism are dimethylglycine dehydrogenase (EC1.5.99.2) and sarcosine dehydrogenase (EC1.5.99.1). Polyamine oxidase (EC1.5.3.11) is one of two key enzymes in polyamine catabolism.

Hormones and cell signaling: The monoamine oxidases A and B (EC1.4.3.4), which are needed for the catabolism of adrenaline, noradrenaline, and serotonin, contain FAD. Nitric oxide, which acts on blood vessels and many other tissues, is generated by several forms of nitric oxide synthase (EC1.14.13.39, contains FAD, FMN, heme, and biopterin). Synthesis of steroid hormones depends on ketosteroid monoxygenase (EC1.14.13.54). Ferredoxin-NADP reductase (adrenodoxin reductase, EC1.18.1.2) mediates the initial electron transfer for all mitochondrial p450 systems, including those responsible for steroid 11-beta hydroxylation in the adrenal cortex and 24-hydroxylation (inactivation) of vitamin D.

Detoxification: Several of the flavoenzymes mentioned previously play a role in the breakdown and removal of potentially toxic xenobiotics. Methylphenyltetrahydropyridine *N*-monooxygenase (EC1.13.12.11) and albendazole monooxygenase (EC1.14.13.32, albendazole is a benzimidazole anthelmintic drug) are further microsomal enzymes that help with the elimination of complex xenobiotics.

Compliance monitoring: A larger than normally consumed dose (e.g., 28 mg) of riboflavin added to foods or liquids helps to determine whether study subjects have consumed the full prescribed amount. Such large amounts of riboflavin are almost completely excreted via urine and then can be easily measured with a fluorometric assay (Switzer et al., 1997; Ramanujam et al., 2011).

NUTRITIONAL ASSESSMENT

Riboflavin deficiency and suboptimal status may be detected by enzyme activation assays or with the measurement of riboflavin metabolites in the blood. Dietary intake can be estimated from the amount of excreted urinary riboflavin metabolites.

Blood measurement: The enzyme activation assays measure glutathione reductase (GR, EC1.6.4.2) activity in untreated and FAD-treated red blood cells (Dror et al., 1994). An activation coefficient (ratio of activity in supplemented red blood cells divided by activity in native red blood cells) in excess of 1.15 indicates lower than optimal FAD saturation of the enzyme.

Urine measurement: The amount of riboflavin and its metabolites excreted with a urine sample collected over a 24-h period reflects recent daily riboflavin intake (Roughead and McCormick, 1991).

REFERENCES

Adiga, P.R., Subramanian, S., Rao, J., Kumar, M., 1997. Prospects of riboflavin carrier protein (RCP) as an anti-fertility vaccine in male and female mammals. Hum. Reprod. Update 3, 325–334.

Aw, T.Y., Jones, D.R., McCormick, D.B., 1983. Uptake of riboflavin by isolated liver cells. J. Nutr. 113, 1249–1254.

Byrd, J.C., Fearney, F.J., Kim, Y.S., 1985. Rat intestinal nucleotide-sugar pyrophosphatase. Localization, partial purification, and substrate specificity. J. Biol. Chem. 260, 7474–7480.

Capo-chichi, C.D., Feillet, E., Guéant, J.L., Amouzou, K.S., Sanni, A., Lefebvre, E., et al., 2000. Concentrations of riboflavin and related organic acids in children with protein-energy malnutrition. Am. J. Clin. Nutr. 71, 978–986.

Chastain, J.L., McCormick, D.B., 1987. Flavin catabolites: identification and quantitation in human urine. Am. J. Clin. Nutr. 46, 830–834.

Dror, Y., Stern, F., Komarnitsky, M., 1994. Optimal and stable conditions for the determination of erythrocyte glutathione reductase activation coefficient to evaluate riboflavin status. Int. J. Vitam. Nutr. Res. 64, 257–262.

Food and Nutrition Board, Institute of Medicine, 1998. Dietary Reference Intakes for Thiamin, Riboflavin, Niacin, Vitamin B6, Folate, Vitamin B12, Pantothenic Acid, Biotin, and Choline. National Academy Press, Washington, DC, pp. 87–122.

Gastaldi, G., Ferrari, G., Verri, A., Casirola, D., Orsenigo, M.N., Laforenza, U., 2000. Riboflavin phosphorylation is the crucial event in riboflavin transport by isolated rat enterocytes. J. Nutr. 130, 2556–2561.

Gregory III, J.F., 1998. Nutritional properties and significance of vitamin glycosides. Annu. Rev. Nutr. 18, 277–296.

Innis, W.S., McCormick, D.B., Merrill Jr., A.H., 1985. Variation in riboflavin binding by human plasma: identification of immunoglobulins as the major proteins responsible. Biochem. Med. 34, 151–165.

Kumar, C.K., Yanagawa, N., Ortiz, A., Said, H.M., 1998. Mechanism and regulation of riboflavin uptake by human renal proximal tubule epithelial cell line HK-2. Am. J. Physiol. 274, F104–F110.

Lee, R.S., Ford, H.C., 1988. 5′-Nucleotidase of human placental trophoblastic microvilli possesses cobalt-stimulated FAD pyrophosphatase activity. J. Biol. Chem. 263, 14878–14883.

McCormick, D.B., 1994. Riboflavin. In: Shils, M.E., Olson, J.A., Shike, M. (Eds.), Modern Nutrition in Health and Disease Lea & Febiger, Philadelphia, PA, pp. 366–375.

Patel, M., Vadlapatla, R.K., Pal, D., Mitra, A.K., 2012. Molecular and functional characterization of riboflavin specific transport system in rat brain capillary endothelial cells. Brain Res. 1468, 1–10.

Ramanujam, V.M., Anderson, K.E., Grady, J.J., Nayeem, F., Lu, L.J., 2011. Riboflavin as an oral tracer for monitoring compliance in clinical research. Open Biomark J. 2011, 1–7.

Roughead, Z.K., McCormick, D.B., 1991. Urinary riboflavin and its metabolites: effects of riboflavin supplementation in healthy residents of rural Georgia (USA). Eur. J. Clin. Nutr. 45, 299–307.

Said, H.M., Ma, T.Y., 1994. Mechanism of riboflavine uptake by Caco-2 human intestinal epithelial cells. Am. J. Physiol. 266, G15–G21.

Said, H.M., Ortiz, A., Ma, T.Y., McCloud, E., 1998. Riboflavin uptake by the human-derived liver cells Hep G2: mechanism and regulation. J. Cell. Physiol. 176, 588–594.

Said, H.M., Ortiz, A., Moyer, M.P., Yanagawa, N., 2000. Riboflavin uptake by human-derived colonic epithelial NCM460 cells. Am. J. Physiol. 278, C270–C276.

Schneider, W.J., 1996. Vitellogenin receptors: oocyte-specific members of the low-density lipoprotein receptor supergene family. Int. Rev. Cytol. 166, 103–137.

Spector, R., 1980. Riboflavin homeostasis in the central nervous system. J. Neurochem. 35, 202–209.

Switzer, B.R., Stark, A.H., Atwood, J.R., Ritenbaugh, C., Travis, R.G., Wu, H.M., 1997. Development of a urinary riboflavin adherence marker for a wheat bran fiber community intervention trial. Cancer Epidemiol. Biomarkers Prev. 6, 439–442.

Tan, K.W., Li, Y., Paxton, J.W., Birch, N.P., Scheepens, A., 2013. Identification of novel dietary phytochemicals inhibiting the efflux transporter breast cancer resistance protein (BCRP/ABCG2). Food Chem. 138, 2267–2274.

Vlaming, M.L., Lagas, J.S., Schinkel, A.H., 2009. Physiological and pharmacological roles of ABCG2 (BCRP): recent findings in Abcg2 knockout mice. Adv. Drug Deliv. Rev. 61, 14–25.

Yonezawa, A., Inui, K., 2013. Novel riboflavin transporter family RFVT/SLC52: identification, nomenclature, functional characterization and genetic diseases of RFVT/SLC52. Mol. Aspects Med. 34, 693–701.

Zempleni, J., Galloway, J.R., McCormick, D.B., 1996. Pharmacokinetics of orally and intravenously administered riboflavin in healthy humans. Am. J. Clin. Nutr. 63, 54–66.

NIACIN

The term *niacin* comprises the two main water-soluble forms nicotinamide (3-pyridinecarboxamide, nicotinic acid amide, and niacinamide; obsolete names vitamin B3 and vitamin PP; molecular weight 122) and nicotinic acid (nicotinate, 3-pyridinecarboxylic acid, pyridine-beta-carboxylic acid, and niacin; obsolete names include P.P. factor and antipellagra vitamin; molecular weight 123) (Figure 10.16).

ABBREVIATIONS

ADPR	ADP ribose
cADPR	cyclic ADP ribose
MNA	*N*-methylnicotinamide
NMN	nicotinamide mononucleotide
NAD	nicotinamide adenine dinucleotide
NADP	nicotinamide adenine dinucleotide phosphate
NE	niacin equivalent
P-cADPR	2′-phospho-cyclic ADP-ribose
SAM	*S*-adenosylmethionine
SAH	*S*-adenosylhomocysteine
Trp	L-tryptophan

NUTRITIONAL SUMMARY

Function: Niacin, the precursor of NAD and NADP, is essential for normal fuel metabolism, as well as for the synthesis or removal of numerous compounds.

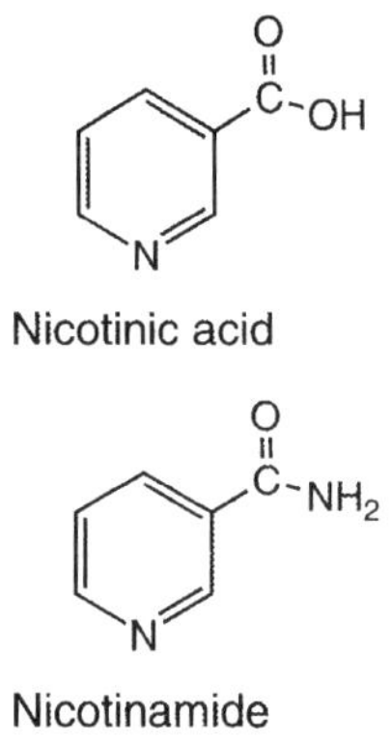

Nicotinic acid

Nicotinamide

FIGURE 10.16

Niacin.

Food sources: Good sources include enriched grain products, yeast, seeds, legumes, dairy, meat, poultry, fish, green leafy vegetables, and coffee.

Requirements: Dietary intake are expressed as milligram niacin equivalents (NEs), which correspond to 1 mg of pure niacin or 60 mg of tryptophan (the alternate precursor of NAD and NADP). Men should get 16 mg and women 14 mg of NE per day. Dietary requirements are slightly higher for women during pregnancy and lactation.

Deficiency: Late symptoms of severe deficiency (pellagra) include fatigue, headache, apathy, depression, memory loss, dementia, pigmented skin rash after sun exposure, bright red tongue, vomiting, diarrhea, and constipation.

Excessive intake: Flushing (burning and itching of the face, arms, and chest) and stomach irritation are the main side effects of moderately high supplemental niacin intake (more than 35 mg/day). Liver damage that may culminate in irreversible liver failure is a risk associated with long-term use of very high doses ($\geq$3000 mg/day) as a cholesterol-lowering drug. Such high doses must never be used without close monitoring of liver function. High intake of nicotinic acid may interfere with the effects of sulfinpyrazone (Anturane).

ENDOGENOUS SOURCES

L-tryptophan (Trp), obtained both from food and from tissue protein breakdown, is an important alternative for preformed niacin. A small proportion of the available tryptophan is metabolized. A small proportion of catabolized Trp (as well as tryptamine, 5-hydroxytryptophan, serotonin, and melatonin) gives rise to nicotinamide-containing compounds (Magni et al., 1999). Impaired absorption in Hartnup syndrome is associated with pellagra-like skin changes that are typical for niacin deficiency. The branch point is at the conversion of 2-amino-3-carboxymuconate semialdehyde. While most of this intermediate is enzymatically decarboxylated, a small proportion undergoes spontaneous dehydration to quinolinic acid (pyridine 2,3-dicarboxylate). The ratio of NAD derived from tryptophan to total metabolized tryptophan is usually assumed to be about 1:60 (Horwitt et al., 1981). Nicotinate-nucleotide pyrophosphorylase (EC2.4.2.19) decarboxylates quinolinate and attaches a ribose phosphate moiety. Magnesium-dependent nicotinatenucleotide adenylyltransferase (EC2.7.7.18; Schweiger et al., 2001) adds adenosine phosphate and amination by either of two NAD synthases (EC6.3.5.1 and EC6.3.1.5), and then completes the synthesis of NAD (nicotinamide adenine dinucleotide). Breakdown of NAD can generate both nicotinamide and nicotinate (Figures 10.17 and 10.18).

DIETARY SOURCES

Niacin-rich foods include liver, meat, and fish, mainly as NAD, NADP, nicotinate, and nicotinamide. The outer (aleuron) layer of grains (but not the endospermium) is relatively niacin-rich, though much of this may be in the form of complexes with carbohydrates (niacytins) or peptides (niacinogens). These niacin-containing macromolecules have low bioavailability (Carter and Carpenter, 1982). Niacin can be released from niacytin complexes, such as maize (corn) and millet, by treatment with alkali (Van den Berg, 1997). Pretreatment with lime has been a common practice in some traditional cultures. Coffee beverages can be another relevant source (1–2 mg of NE per cup). Coffee beans contain significant amounts (10–40 mg/100 g) of trigonelline (*N*-methylnicotinic acid) (Figure 10.19), which is partially converted into nicotinate by roasting (Adrian and Frangne, 1991; Casal et al., 2000). Flour sold in the United States has to be fortified with 52.9 mg of niacin per kilogram (DHHS-FDA, 1996); this translates into an added 33 mg/kg in bread.

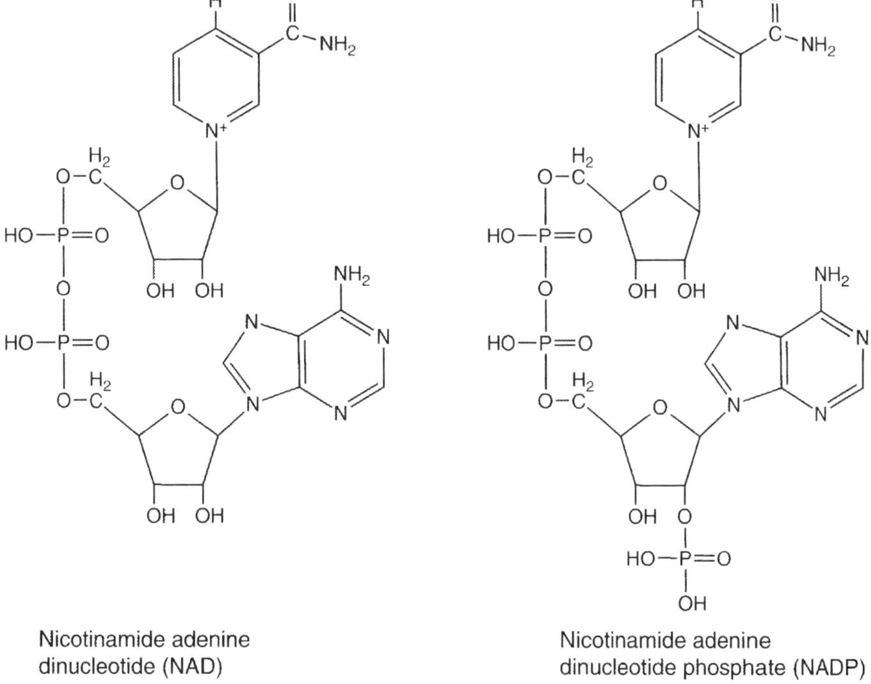

FIGURE 10.17

Niacin-derived dinucleotides.

Median daily intake of preformed niacin in the United States, not taking into account endogenous synthesis from tryptophan, in women was reported to be around 21 mg, and in men, around 25 mg (Food and Nutrition Board, Institute of Medicine, 1998: Appendices G and H), clearly above estimated minimal needs. Reported intake in Canada was even higher (Food and Nutrition Board, Institute of Medicine, 1998: Appendix I). Nicotinic acid and various derivatives at dosages of several grams per day are used for the medical treatment of hyperlipidemias.

Since both preformed niacin and tryptophan can contribute to niacin adequacy, intake is often assessed in NEs. Calculations of NEs are based on the assumption that 60 mg of tryptophan contribute as much as 1 mg of nicotinamide or nicotinic acid.

DIGESTION AND ABSORPTION

Nucleotide pyrophosphatase (EC3.6.1.9) from pancreatic secretions, intestinal secretions, or both cleave NAD/NADP into 5′-AMP and nicotinamide ribonucleotide. An intestinal brush border enzyme, whose identity remains obscure, can then release nicotinamide, though with low activity (Gross and Henderson, 1983).

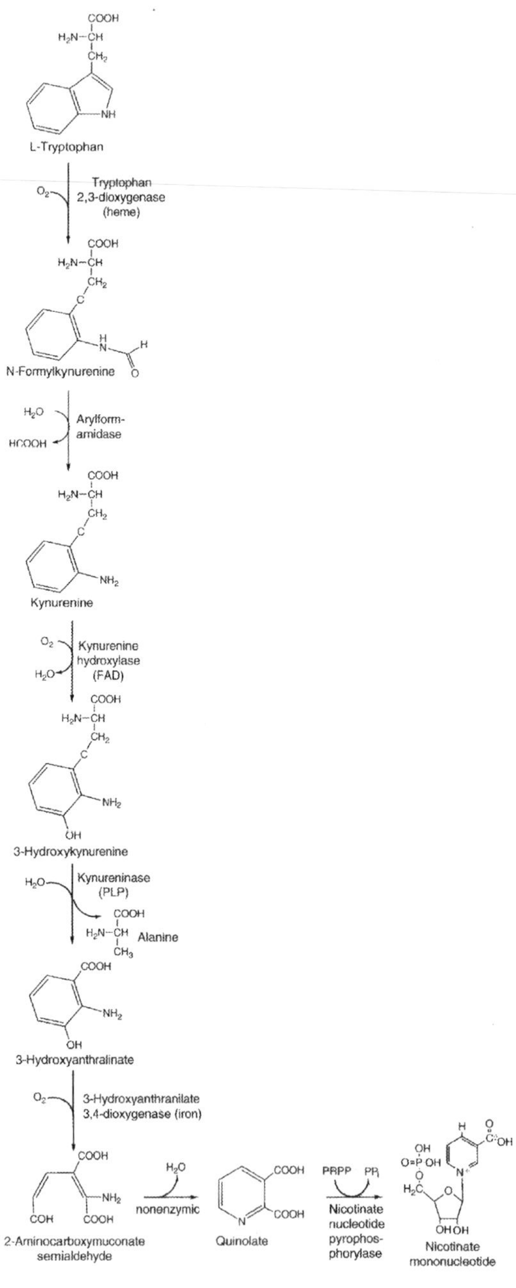

FIGURE 10.18

Endogenous synthesis of endogenous NAD precursors.

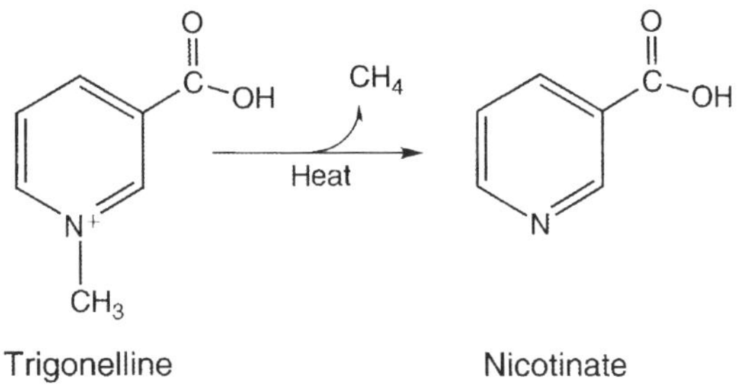

FIGURE 10.19

Roasting of trigonelline releases nicotinate.

Nicotinamide and nicotinate are absorbed from stomach and proximal small intestine (Bechgaard and Jespersen, 1977). Uptake of these compounds proceeds by sodium-dependent, facilitated diffusion at low concentrations, and by passive diffusion at higher concentrations. A proton cotransporter and a distinct anion antiporter for nicotinic acid appear to operate in the small intestine (Takanaga et al., 1996). Bilitranslocase in gastric mucosa has been shown to mediate the uptake of nicotinate, but not of nicotinamide (Passamonti et al., 2000). Since this carrier does not seem to be expressed in the intestines, the transport activity in the stomach may be of little significance.

TRANSPORT AND CELLULAR UPTAKE

Blood circulation: The main vitamers transported in the blood are nicotinate and nicotinamide. An as-yet-incompletely-characterized anion transporter facilitates the diffusion of NAD into red blood cells, and possibly other tissues as well. NAD also exchanges between adjacent cells. Connexin 43 (Cx43), a component of intercellular gap junctions, functions as an NAD transporter that facilitates diffusion between cells (Bruzzone et al., 2001).

BBB: Both nicotinamide and nicotinate are transported rapidly into the brain, as demonstrated by positron emission tomography (PET) scans (Hankes et al., 1991). Nicotinamide transfer may be more efficient than nicotinate transfer, but the underlying mechanisms are not well understood.

Materno-fetal transfer: Transport of niacin derivatives across the placenta is a necessity since the fetus depends on this vitamin, just as the mother does. Nicotinate, however, does not appear to cross easily (Baker et al., 1981), and other forms may be more important for transfer to the fetus.

METABOLISM

Nucleotide synthesis: NAD and NADP are readily synthesized from nicotinate or nicotinamide, or from L-tryptophan-derived nicotinate mononucleotide. Synthesis takes place in the liver, red blood cells, and other tissues.

Nicotinate mononucleotide may be considered the starting point of the predominant pathway (Preiss–Handler pathway) in the liver and red blood cells. This intermediate is produced largely from nicotinate by nicotinate phosphoribosyltransferase (EC2.4.2.11, magnesium-dependent). Nicotinate, in turn, may have been generated from nicotinamide through the action of nicotinamidase (EC3.5.1.19, magnesium-dependent). Only a much smaller percentage comes from L-tryptophan catabolism (Figure 10.20).

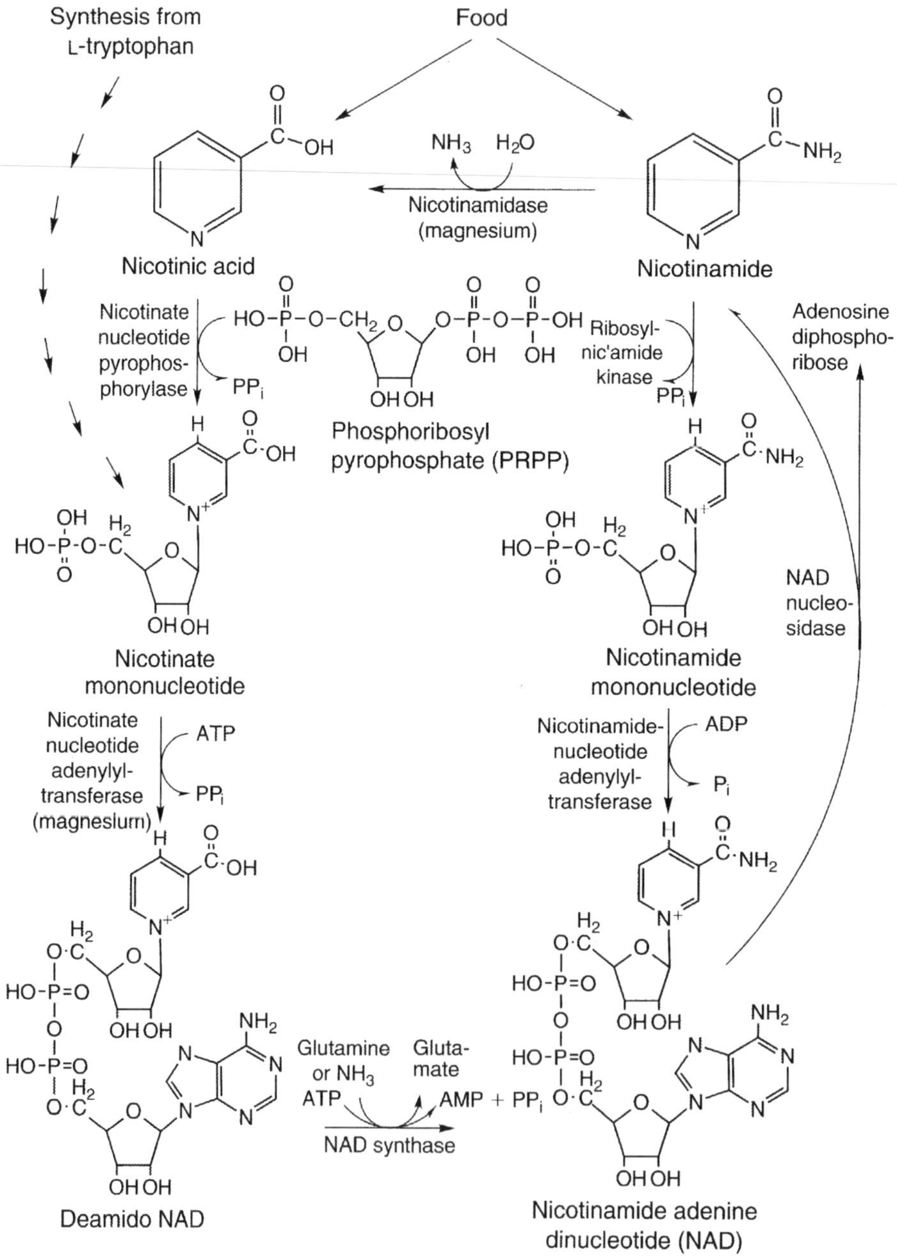

FIGURE 10.20

Metabolism of niacin compounds.

Addition of adenosine phosphate by magnesium-dependent nicotinate-nucleotide adenylyltransferase (EC2.7.7.18; Schweiger et al., 2001) generates deamido-NAD. The glutamine-hydrolyzing NAD synthase (EC6.3.5.1) can use either the amino group from glutamine or ammonium ions. Lead exposure inhibits the activity of this enzyme in red blood cells (Morita et al., 1997). Even modestly elevated blood lead levels ($<600\,\mu g/l$) diminish NAD synthase activity by two-thirds compared to low lead levels ($<200\,\mu g/l$). NADP is produced through additional phosphorylation of NAD by NAD kinase (EC2.7.1.23). Equally important is NAD synthesis directly from nicotinamide in just two steps (Dietrich pathway), which occurs in red and white blood cells, as well as in other tissues.

Nicotinate phosphoribosyltransferase (EC2.4.2.11) first links phosphoribosyl pyrophosphate to nicotinamide. The next and final step can then be catalyzed by nicotinamide-nucleotide adenylyltransferase (EC2.7.7.1).

ADP-ribose synthesis: Endocytotic vesicles have the ability to generate cyclic ADP ribose (cADPR). First, a distinct dinucleotide transport system imports NAD from cytosol (Zocchi et al., 1999). NAD hydrolysis by NAD nucleosidase (EC3.2.2.5) then generates ADP-ribose (ADPR).

Breakdown: Nicotinamide can be converted to *N*-methylnicotinamide (MNA) by nicotinamide N-methyltransferase (EC2.1.1.1). The use of SAM as the methyl group donor generates *S*-adenosyl-L-homocysteine (SAH) and links the reaction to adequate availability of folate and vitamin B12. This reaction metabolically sequesters excessive amounts of nicotinamide, especially in the liver (Figure 10.21). Aldehyde oxidase (EC1.2.3.1, contains molybdopterin and heme) can metabolize MNA further to N^1-methyl-2-pyridone-5-carboxamide or N^1-methyl-4-pyridone-3-carboxamide. Nicotinuric acid is a major metabolite of nicotinate that arises from its conjugation to glycine (catalyzed by glycine N-benzoyltransferase, EC2.3.1.71, glycine N-acyltransferase, EC2.3.1.13, or both), especially when a pharmacological dose of nicotinate is ingested (Neuvonen et al., 1991).

Niacin salvage pathways: NAD, NADP, and related metabolites are recycled extensively. Extracellular NAD can be broken down by NAD nucleosidase (NAD-glycohydrolase, EC3.2.2.5), which releases nicotinamide and ADPR, or by nucleotide pyrophosphatase (EC3.6.1.9), which releases nicotinamide mononucleotide (NMN) and AMP. Both enzymes are active on the external face of cell membranes (Aleo et al., 2001). NAD and NADP are readily resynthesized from these products by the pathways described earlier. Whether humans have additional salvage reactions, which are known to be important in bacteria (Magni et al., 1999), remain to be seen. Salvage reactions operating in these organisms include NAD and NADH pyrophosphatases (EC3.6.1.22, generate NMN), nicotinamide deamidase, and NMN deamidase.

STORAGE

The liver, kidney, heart, and skeletal muscle contain relatively high concentrations of NAD, which can become available in times of low niacin intake. Stores can cover the needs for only a few weeks, however. Nicotinamide, stored mainly in the liver as NAD, can be released by NAD nucleosidase (NAD-glycohydrolase, EC3.2.2.5) in response to needs.

EXCRETION

Filtered nicotinate is recovered from the renal proximal tubular lumen by a sodium cotransporter (Boumendil-Podevin and Podevin, 1981). The main niacin excretory products in urine are

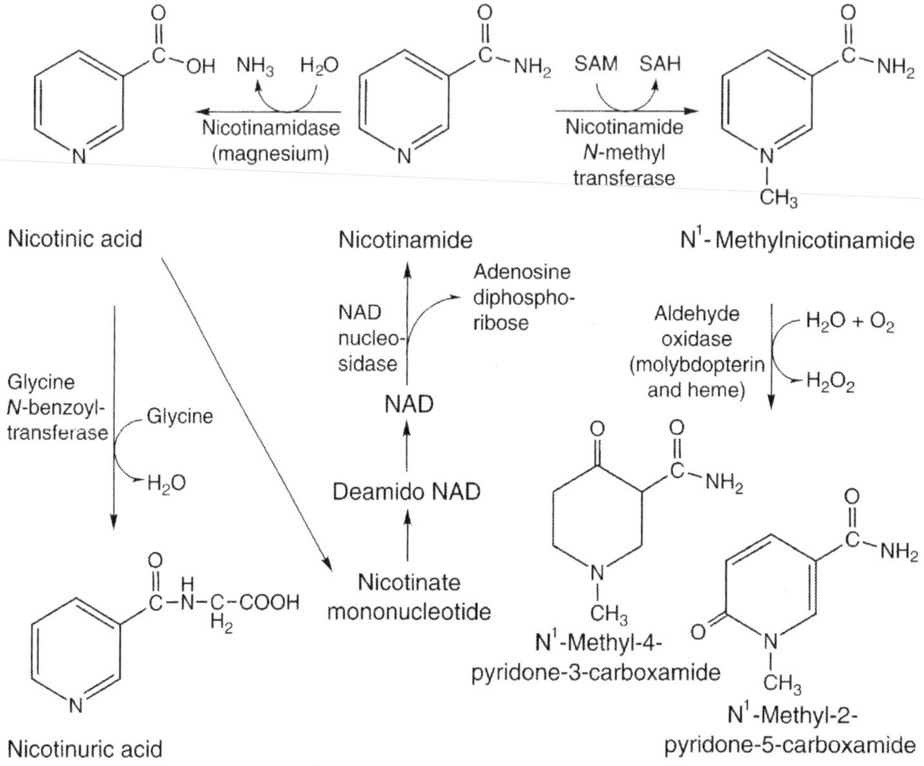

FIGURE 10.21

Catabolism of niacin compounds.

N-methyinicotinamide, N^1-methyl-2-pyridone-5-carboxamide, N^1-methyl-4-pyridone-5-carboxamide, nicotinuric acid (glycinyl nicotinate), nicotinic acid N-oxide, and hydroxyl nicotinate (Mrochek et al., 1976). N-methylnicotinamide is secreted into the tubular lumen by cationic ion transporters and is commonly used as a model compound for the functional analysis of such transporters.

REGULATION

Even extremely high intake of nicotinamide or nicotinate will not result in greatly increased NAD production, despite the capacity for promoting NAD synthesis via the tryptophan/quinolate pathway (McCreanor and Bender, 1986). Knowledge is incomplete, however, how such regulation is achieved. Likely targets for niacin homeostasis regulation are the rates of *de novo* synthesis from L-tryptophan, intestinal absorption efficiency, catabolism, or excretion. The NadR regulator in bacteria is a multifunctional protein that has nicotinamide-nucleotide adenylyltransferase (EC2.7.7.1) activity, participates in the transport of NMN, and represses the transcription of several enzymes for synthesis and salvage of NAD (Penfound and Foster, 1999; Raffaelli et al., 1999). It will have to be seen whether such multifunctionality also applies to a potential (as-yet-unidentified) mammalian homolog. A decrease in the activity of quinolinate phosphoribosyl transferase with high tryptophan intake has been reported (Satyanarayana and Rao, 1980), but the suppression of synthesis from tryptophan appears less stringent than suppression of vitamin-derived synthesis (McCreanor and Bender, 1986). It has been postulated that sex hormones may decrease NAD production from Trp (Shibata and Toda, 1997).

FUNCTION

Enzyme cofactor NAD: More than 200 well-characterized, NAD-dependent dehydrogenases use NAD as an electron acceptor or NADH as a hydrogen donor for intracellular respiration and oxidoreductive reactions. Typical examples are alcohol dehydrogenase (ADH, EC1.1.1.1) in cytosol and malate dehydrogenase (EC1.1.1.37, the Krebs cycle enzyme) in mitochondria. Reduced equivalents generated within the mitochondria can then be transferred to complex I of the oxidative phosphorylation system (ETF dehydrogenase, EC1.5.5.1) for the generation of ATP.

Enzyme cofactor NADP: The functions of NADP are mostly analogous to those of NAD. Many enzymes actually are equally active with NADP and NAD. NADP has an additional specific role as a hydrogen donor for the reductive synthesis of fatty acids and steroids.

Enzyme cosubstrate NMN: Some reactions also use NMN. An example is the flavoprotein NAD(P)H dehydrogenase 2 (EC1.6.99.2), which reactivates vitamin K and modifies xenobiotic quinones.

NAD as a prosthetic group: Some enzymes, such as urocanate hydratase (EC4.2.1.49) and myo-inositol-1-phosphate synthase (EC5.5.1.4), use covalently bound NAD as an electrophilic prosthetic group, not as a redox cosubstrate.

ADP ribosylation of proteins: NAD can be hydrolyzed by NAD nucleosidase (EC3.2.2.5) to generate ADPR, the substrate for covalent modification of arginine residues in proteins. NAD(P)-arginine ADP-ribosyltransferase (EC2.4.2.31) attaches a single ADPR to arginine-residues of specific proteins in an NAD-dependent reaction. An important example is the activation of adenylate cyclase (EC4.6.1.1). Free arginine, guanidine, and related compounds also can be linked to ADPR by this enzyme. Several distinct genes exist with distinct tissue expression patterns.

Niacin receptors: The G-protein-coupled receptor 109A (GPR109A) binds and responds to nicotinic acid or niacin (Hanson et al., 2010). The flush induced by pharmacological doses (i.e., several grams) of nicotinic acid appears to be mediated by this receptor. The receptor is also expressed in immune cells in the spleen and bone marrow. Activation of GPR109A by niacin has been reported to affect sleep characteristics (latency and efficiency) in healthy people (Wakade et al., 2014).

DNA repair: Poly(ADPR) polymerase (EC2.4.2.30) attaches several ADPR molecules to histones, topoisomerases, and other proteins. The ADPR precursor NAD is thus critical for the repair of single-strand DNA breaks, maintenance of chromosome stability, and regulation of teiomer length (Jacobson and Jacobson, 1999).

DNA deacetylation: NAD-dependent histone acetylases are encoded by silent mating type information regulation 2 homolog genes (sirtuins, EC3.5.1.-). These enzymes cleave acetyl side chains from lysines in modified histones and thereby usually terminate DNA expression. Sirtuins are critically important regulators of the epigenetic structure and minute-to-minute function of DNA. For each removed acetyl molecule, one molecule of NAD (and ultimately niacin) is lost irreversibly. High sirtuin 1 (SIRT1) expression increases insulin sensitivity in many tissues and has been linked to longevity. Among patients with type 2 diabetes, low niacin intake appears to shorten life span by many years in those with susceptible SIRT1 forms (Zillikens et al., 2009).

Calcium signaling: Metabolism of NAD and NADP can generate cADPR and 2′-phospho-cyclic ADP-ribose (P-cADPR), respectively. Both compounds function as calcium-mobilizing second messengers in neutrophils (Partida-Sanchez et al., 2001), muscle, and other tissues (Higashida et al., 2001).

Glucose metabolism: A synergistic promotion of glucose uptake by nicotinic acid and chromium has been suggested (Urberg and Zemel, 1987), but not enough is known yet, about the nature of the presumed glucose tolerance factor.

Pharmacological uses: Niacin and various related derivatives in daily doses of 1 g to several grams are used for the treatment of hyperlipidemias. Typical effects include decreases of low-density lipoprotein (LDL) cholesterol and Lp(a) concentrations, and an increase of high-density lipoprotein (HDL) cholesterol concentrations (Kamanna and Kashyap, 2000; Kwiterovich, 2000). Large doses ingested as dietary supplements or medical drugs usually induce a typical skin flush, possibly mediated through an increase of prostaglandin D2 release (Stern et al., 1991). A similar reaction can be observed when these compounds are applied topically.

NUTRITIONAL ASSESSMENT

Niacin deficiency and suboptimal status may be detected with the measurement of niacin metabolites in the blood. Dietary intake can be estimated from the amount of excreted urinary niacin metabolites.

Blood measurement: Low concentrations of N^1-methylnicotinamide (MNA) and N^1-methyl-2-pyridone-5-carboxamide (2-pyr) indicate suboptimal niacin supplies (Jacob et al., 1989). Here, 2-pyr is a particularly sensitive indicator of deficiency and concentrations cease to be detectable with severe niacin restriction. Neither of these metabolites is a useful indicator for suboptimal status. The ratio of MNA and 2-pyr may be a slightly better indicator.

Urine measurement: Low excretion of MNA with urine indicates inadequate niacin intake.

REFERENCES

Adrian, J., Frangne, R., 1991. Synthesis and availability of niacin in roasted coffee. Adv. Exp. Med. Biol. 289, 49–59.

Aleo, M.E., Giudici, M.L., Sestini, S., Danesi, E., Pompucci, G., Preti, A., 2001. Metabolic fate of extracellular NAD in human skin fibroblasts. J. Cell. Biochem. 80, 360–366.

Baker, H., Frank, O., Deangelis, B., Feingold, S., Kaminetzky, H.A., 1981. Role of placenta in maternal-fetal vitamin transfer in humans. Am. J. Obstet. Gynecol. 141, 792–796.

Bechgaard, H., Jespersen, S., 1977. GI absorption of niacin in humans. J. Pharmaceut. Sci. 66, 871–872.

Boumendil-Podevin, E.E., Podevin, R.A., 1981. Nicotinic acid transport by brush border membrane vesicles from rabbit kidney. Am. J. Physiol. 240, F185–F191.

Bruzzone, S., Guida, L., Zocchi, E., Franco, L., De Flora, A., 2001. Connexin 43 hemi channels mediate Ca^{2+}-regulated transmembrane NAD^+ fluxes in intact cells. FASEB J. 15, 10–12.

Carter, E.G.A., Carpenter, K.J., 1982. The bioavailability for humans of bound niacin from wheat bran. Am. J. Clin. Nutr. 36, 855–861.

Casal, S., Oliveira, M.B., Alves, M.R., Ferreira, M.A., 2000. Discriminate analysis of roasted coffee varieties for trigonelline, nicotinic acid. and caffeine content. J. Agric. Food. Chem. 48, 3420–3424.

DHHS-FDA, 1996. Federal Register (March 5), vol. 61, No. 44, pp. 8781–8797.

Food and Nutrition Board, Institute of Medicine, 1998. Dietary Reference Intakes for Thiamin, Riboflavin, Niacin, Vitamin B6, Folate, Vitamin B12, Pantothenic Acid, Biotin, and Choline. National Academy Press, Washington, DC, pp. 123–149.

Gross, C.J., Henderson, L.M., 1983. Digestion and absorption of NAD by the small intestine of the rat. J. Nutr. 113, 412–420.

Hankes, L.V., Coenen, H.H., Rota, E., Langen, K.J., Herzog, H., Wutz, W., et al., 1991. Effect of Huntington's and Alzheimer's diseases on the transport of nicotinic acid or nicotinamide across the human blood–brain barrier. Adv. Exp. Med. Biol. 294, 675–678.

Hanson, J., Gille, A., Zwykiel, S., Lukasova, M., Clausen, B.E., Ahmed, K., et al., 2010. Nicotinic acid- and monomethyl fumarate-induced flushing involves GPR109A expressed by keratinocytes and COX-2-dependent prostanoid formation in mice. J. Clin. Invest. 120, 2910–2919.

Higashida, H., Hashii, M., Yokoyama, S., Hoshi, N., Chen, X.L., Egorova, A., et al., 2001. Cyclic ADP-ribose as a second messenger revisited from a new aspect of signal transduction from receptors to ADP ribosyl cyclase. Pharmacol. Ther. 90, 283–296.

Horwitt, M.K., Harper, A.E., Henderson, L.M., 1981. Niacin-tryptophan relationships for evaluating niacin equivalents. Am. J. Clin. Nutr. 34, 423–427.

Jacob, R.A., Swendseid, M.E., McKee, R.W., Fu, C.S., Clemens, R.A., 1989. Biochemical markers for assessment of niacin status in young men: urinary and blood levels of niacin metabolites. J. Nutr. 119, 591–598.

Jacobson, M.K., Jacobson, E.L., 1999. Discovering new ADP-ribose polymer cycles: protecting the genome and more. Trends Biochem. Sci. 24, 415–417.

Kamanna, V.S., Kashyap, M.E., 2000. Mechanism of action of niacin on lipoprotein metabolism. Curr. Atheroscl. Rep. 2, 36–46.

Kwiterovich Jr., P.O., 2000. The metabolic pathways of high-density lipoprotein, low-density lipoprotein, and triglycerides: a current review. Am. J. Cardiol. 86, 5L–10L.

Magni, G., Amici, A., Emanuelli, M., Raffaelli, N., Ruggieri, S., 1999. Enzymology of NAD$^+$ synthesis. Adv. Enzymol. Rel. Areas Mol. Biol. 73, 135–182.

McCreanor, G.M., Bender, D.A., 1986. The metabolism of high intakes of tryptophan, nicotinamide and nicotinic acid in the rat. Br. J. Nutr. 56, 577–586.

Morita, Y., Sakai, T., Araki, S., Araki, T., Masuyama, Y., 1997. Nicotinamide adenine dinucleotide synthetase activity in erythrocytes as a tool for the biological monitoring of lead exposure. Int. Arch. Occ. Env. Health 70, 195–198.

Mrochek, J.E., Jolley, R.E., Young, D.S., Turner, W.J., 1976. Metabolic response of humans to ingestion of nicotinic acid and nicotinamide. Clin. Chem. 22, 1821–1827.

Neuvonen, P.J., Roivas, L., Laine, K., Sundholm, O., 1991. The bioavailability of sustained release nicotinic acid formulations. Br. J. Clin. Pharmacol. 32, 473–476.

Partida-Sanchez, S., Cockayne, D.A., Monard, S., Jacobson, E.L., Oppenheimer, N., Garvy, B., et al., 2001. Cyclic ADP ribose production by CD38 regulates intracellular calcium release, extracellular calcium influx and chemotaxis in neutrophils and is required for bacterial clearance *in vivo*. Nat. Med. 7, 1209–1216.

Passamonti, S., Battiston, L., Sottocasa, G.L., 2000. Gastric uptake of nicotinic acid by bilitranslocase. FEBS Lett. 482, 167–168.

Penfound, T., Foster, J.W., 1999. NAD-dependent DNA-binding activity of the bifunctional NadR regulator of *Salmonella typhimurium*. J. Bacteriol. 181, 648–655.

Raffaelli, N., Lorenzi, T., Mariani, P.L., Emanuelli, M., Amici, A., Ruggieri, S., et al., 1999. The *Escherichia coli* NadR regulator is endowed with nicotinamide mononucleotide adenylyltransferase activity. J. Bacteriol. 181, 5509–5511.

Satyanarayana, U., Rao, B.S., 1980. Dietary tryptophan level and the enzymes of tryptophan NAD pathway. Br. J. Nutr. 43, 107–113.

Schweiger, M., Hennig, K., Lerner, E., Niere, M., Hirsch-Kauffmann, M., Specht, T., et al., 2001. Characterization of recombinant human nicotinamide mononucleotide adenylyl transferase (NMNAT), a nuclear enzyme essential for NAD synthesis. FEBS Lett. 492, 95–100.

Shibata, K., Toda, S., 1997. Effects of sex hormones on the metabolism of tryptophan to niacin and to serotonin in male rats. Biosci. Biotechnol. Biochem. 61, 1200–1202.

Stern, R.H., Spence, J.D., Freeman, D.J., Parbtani, A., 1991. Tolerance to nicotinic acid flushing. Clin. Pharmacol. Ther. 50, 66–70.

Takanaga, H., Maeda, H., Yabuuchi, H., Tamai, I., Higashida, H., Tsuji, A., 1996. Nicotinic acid transport mediated by pH-dependent anion antiporter and proton cotransporter in rabbit intestinal brush border membrane. J. Pharm. Pharmacol. 48, 1073–1077.

Urberg, M., Zemel, M.B., 1987. Evidence for synergism between chromium and nicotinic acid in the control of glucose tolerance in elderly humans. Metab. Clin. Exp. 36, 896–899.

van den Berg, H., 1997. Bioavailability of niacin. Eur. J. Clin. Nutr. 511, S64–S65.

Wakade, C., Chong, R., Bradley, E., Thomas, B., Morgan, J., 2014. Upregulation of GPR109A in Parkinson's Disease. PLoS One 9 (10), e109818.

Zillikens, M.C., van Meurs, J.B., Sijbrands, E.J., Rivadeneira, F., Dehghan, A., van Leeuwen, J.P., et al., 2009. SIRT1 genetic variation and mortality in type 2 diabetes: interaction with smoking and dietary niacin. Free Radic. Biol. Med. 46, 836–841.

Zocchi, E., Usai, C., Guida, L., Franco, L., Bruzzone, S., Passalacqua, M., et al., 1999. Ligand-induced internalization of CD38 results in intracellular Ca^{2+} mobilization: role of NAD^+ transport across cell membranes. FASEB J. 13, 273–283.

VITAMIN B6

The term *vitamin B6* comprises several compounds that can be metabolically converted into the biologically active pyridoxal-5′-phosphate (PLP; molecular weight 247). B6 vitamers in foods and human tissues include PLP, pyridoxal (3-hydroxymethyl)-2-methyl-4-pyridinecarboxaldehyde, molecular weight 167), pyridoxine, pyridoxine phosphate, pyridoxamine, pyridoxamine 5′-phosphate (PMP), and 5′-*O*-(beta-D-glycopyranosyl)pyridoxine.

ABBREVIATIONS	
B6	vitamin B6
GABA	gamma-aminobutyric acid
PLP	pyridoxal-5′-phosphate
PMP	pyridoxamine 5′-phosphate
PNP	pyridoxine 5′-phosphate

NUTRITIONAL SUMMARY

Function: Vitamin B6 (B6) participates in more than 100 transaminations, decarboxylations, and other types of reactions, including the initial step of porphyrin synthesis, glycogen mobilization, amino acid transsulfuration, and neurotransmitter synthesis.

Food sources: Good sources include fortified cereals, organ meats, muscle foods, potatoes, and fruits other than citrus.

Requirements: Current intake recommendations are 1.3 mg/day for young women and men, and slightly higher for older adults. Higher needs are also likely during pregnancy (1.9 mg/day) and lactation (2.0 mg/day).

Deficiency: Low intake may cause microcytic anemia, brain current (electroencephalogram) abnormalities and epileptic seizures, depression, confusion, seborrheic dermatitis, and possibly also platelet and clotting dysfunction.

Excessive intake: Consumption of very high doses of pyridoxine (more than 100 mg/day in adults) may cause peripheral sensory neuropathy, and possibly dermatological lesions. Increased intake from supplements may interact with the action of some drugs, including levodopa (Dopar, Larodopa, Sinemet, and Atamet), phenobarbital (Luminal, Solfoton), and phenytoin (Dilantin).

DIETARY SOURCES

Animal-derived foods contain mainly PLP and PMP (Figure 10.22). Plant-derived foods contain mainly pyridoxine, pyridoxine 5′-phosphate (PNP), 4′-*O*-(β-D-glycopyranosyl) pyridoxine, 5′-*O*-(β-D-glycopyranosyl) pyridoxine, pyridoxine-5′-β-D-cellobioside, pyridoxine-4′-oligosaccharides, pyridoxine-5′-oligosaccharides, pyridoxine-5′-(β-D-glucosyl-malonyl ester), pyridoxine-5′-(β-D-glucosyl-6-hydroxymethylglutarylester), and pyridoxine-5′-β-D-cellobiosyl-indoleacetyl ester, among others (Gregory, 1998) (Figure 10.23).

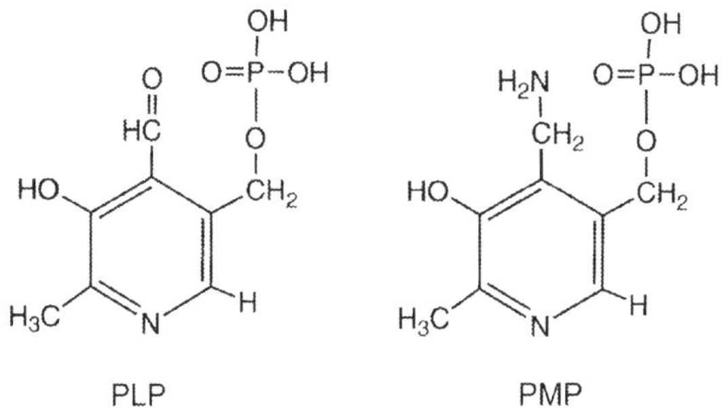

FIGURE 10.22

Vitamin B6 in animal-derived foods.

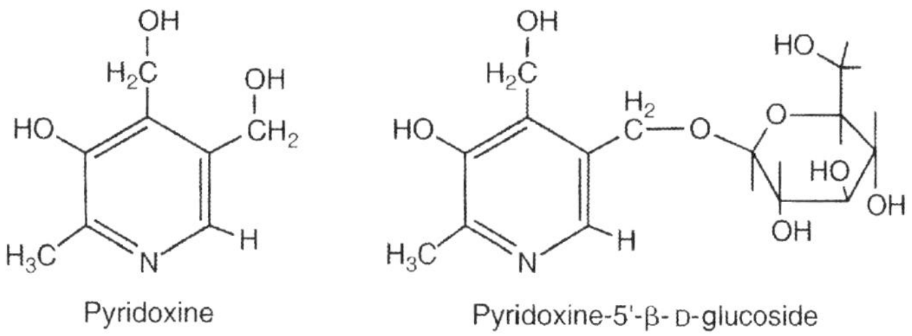

FIGURE 10.23

Vitamin B6 in plant-derived foods.

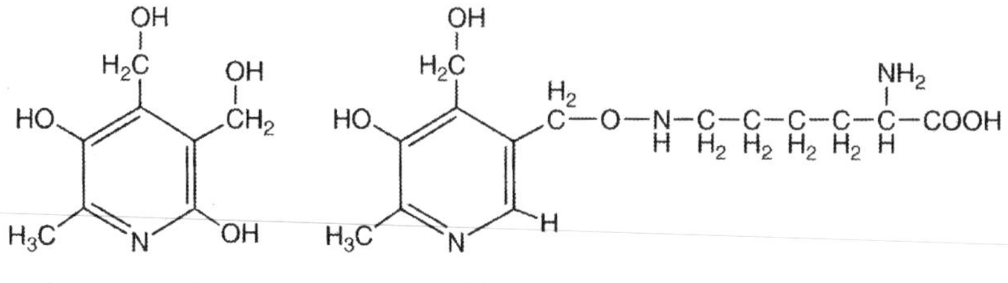

FIGURE 10.24

Inactive derivatives of vitamin B6.

Good B6 sources include bananas (6 mg/kg), potatoes (3.0 mg/kg), and other tubers, watermelon (1.4 mg/kg), and fortified cereals, as well as liver (9 mg/kg) and other organ meats, beef (3.9 mg/kg), pork (4.0 mg/g), poultry (4.7 mg/kg), and fish (2.8 mg/kg).

Median daily intake in the United States has been estimated to be 1.5 mg in women and 2 mg in men (Food and Nutrition Board, Institute of Medicine, 1998), but much lower in elderly women (1.0 mg) and men (1.2 mg).

Cooking may lead to losses due to leaching into discarded water. Heating or prolonged storage can promote the reaction of PLP or pyridoxal with unspecific protein lysyl residues and the formation of ε-pyridoxyllysine with low bioavailability. Ascorbate and other components in foods or in a meal can promote at modestly elevated temperatures (<50°C) the conversion of pyridoxine to the inactive form 6-hydroxypyridoxine (Tadera et al., 1986) (Figure 10.24).

DIGESTION AND ABSORPTION

More than 75% of B6 in a mixed meal is absorbed in the small intestine, if digestive and absorptive functions are normal (Figure 10.25). PLP, PMP, and PNP are dephosphorylated in the intestinal lumen by alkaline phosphatase (EC3.1.3.1) and other phosphatases. Dietary 5'-0-(β-D-glycopyranosyl) pyridoxine may be cleaved by an incompletely characterized mucosal glucosidase (Gregory, 1998). The nonphosphorylated and nonglycosylated forms can enter duodenal and jejunal enterocytes by nonsaturable diffusion through unidentified carriers. B6 is trapped metabolically by phosphorylation in the cytosol (pyridoxal kinase, EC2.7.1.35). In addition, some of the pyridoxine glycoside can enter mucosal cells intact, possibly via the sodium-glucose cotransporter 1 (SLC5A1), and be cleaved in mucosal cytosol of the jejunum by pyridoxine-beta-D-glucoside hydrolase (Trumbo et al., 1990; McMahon et al., 1997). Most B6 metabolites in the enterocyte are converted to pyridoxal and appear to cross the basolateral membrane without a phosphate group attached (Albersen et al., 2013). The mechanism of this transfer is still uncertain.

The bioavailability of dietary B6 glucoside (about 50%) is low compared to other B6 vitamers (Gregory, 1998). Incomplete hydrolysis of the glucoside and inhibition of interconversion reactions may be responsible; ε-pyridoxyllysine is a B6 form with low bioavailability that may be generated by heating or prolonged storage under reducing conditions. The normal Schiff base linkage with specific sites of B6-containing proteins readily dissolves in the inactive enzyme, especially upon hydrolysis of

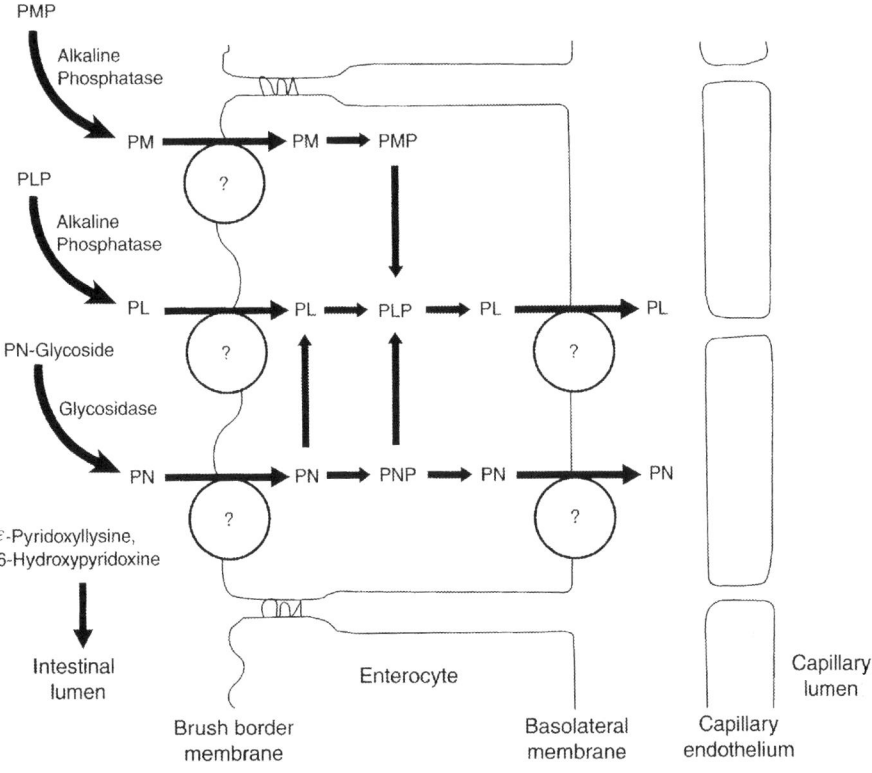

FIGURE 10.25

Intestinal absorption of vitamin B6.

the protein. The covalent bond in ε-pyridoxyllysine, on the other hand, is less likely to dissociate in the intestinal lumen and thus lost to absorption. Bioavailability of B6 from foods can also be decreased by high fiber content of a meal due to slowed dephosphorylation.

TRANSPORT AND CELLULAR UPTAKE

Blood circulation: Transport of B6 in the blood occurs as pyridoxine and pyridoxal, and as PLP bound to albumin or other proteins. Hemoglobin becomes a significant B6 binder when pharmacological doses are ingested. Protein-bound B6 in circulation has to be released from its carrier, and phosphate has to be hydrolyzed by phosphatases before tissues can take it up. The free pyridoxine is taken up into the liver and other tissues by a nonconcentrative process, followed by metabolic trapping (phosphorylation by pyridoxal kinase, EC2.7.1.35) as PLP or PMP (Mehansho et al., 1980).

Since many of the B6-dependent enzymes are in the mitochondria, a mechanism must exist for the transfer of the coenzyme or of precursors across the mitochondrial membrane. The mitochondrial transporter SLC25A38 may serve that function.

Materno-fetal transfer: Transport across the placental membrane of pyridoxal is much greater than transport of PLP. It has been suggested that free pyridoxal is taken up by passive transport, binding, and phosphorylation in the placenta, and released to the fetal side as PLP (Schenker et al., 1992).

BBB: It is obvious that adequate B6 supplies are of vital importance for brain function since the vitamin is needed for the synthesis and catabolism of many neurotransmitters, including glutamate, gamma-aminobutyric acid (GABA), serotonin, catecholamines, histamine, tryptamine, taurine, D-serine, *N*-methyl-D-aspartate, glycine, and proline. This does not even take into account numerous other B6 functions. The brain also has a very active B6 metabolism with demonstrated activities of pyridoxamine phosphate oxidase (EC1.4.3.5; Bahn et al., 2002) and pyridoxal kinase (EC2.7.1.35; Lee et al., 2000). Nonetheless, detailed knowledge on the mechanisms that govern transfer of B6 into the brain is lacking.

METABOLISM

The ubiquitous cytosolic enzyme pyridoxal kinase (EC2.7.1.35) phosphorylates pyridoxine, pyridoxal, and pyridoxamine (Figure 10.26). This activity is important for two reasons: First, since the transporters accept only the free forms, the phosphorylated forms become metabolically trapped inside the cell. Second, the reaction produces PLP, the cofactor of most B6-dependent enzymes. The flavoprotein pyridoxamine phosphate oxidase (pyridoxine-5′-phosphate oxidase, EC1.4.3.5, FMN-containing) can then convert PNP and PMP into PLR. This enzyme also oxidizes pyridoxine to pyridoxal.

Alternatively (Merrill et al., 1984), PEP and PMP can be interconverted in the liver and kidney by transamination (pyridoxamine-oxaloacetate aminotransferase, EC2.6.1.31). Several intracellular phosphatases cleave the phosphorylated B6 vitamers unless they are not bound as cofactors to proteins. In the liver, but not in muscle or erythrocytes, pyridoxal is irreversibly oxidized to the inactive metabolite 4-pyridoxic acid by FAD-dependent pyridoxamine-phosphate oxidase (EC1.4.3.5), aldehyde dehydrogenase (ALDH, EC1.2.1.3), or aldehyde oxidase (EC1.2.3.1, contains FAD, molybdenum cofactor, heme, and additional irons arranged in a 2Fe–2S cluster).

STORAGE

Most of the body's B6 content, about 110 mg in young men and 60 mg in young women (Johansson et al., 1966), is stored in muscle (90% of total body content), mainly as PLP bound to phosphorylase. B6 deficiency by itself will not promote the release of the vitamin from phosphorylase (Black et al., 1977), but regular or accelerated (due to an energy deficit) enzyme turnover will make it available for use by other enzymes and/or tissues.

Pyridoxine in people with full stores has an estimated half-life of 25 days (Shane, 1978).

EXCRETION

Proximal renal tubular cells take up pyridoxine and pyridoxamine by a facilitated, sodium-dependent process (Zhang and McCormick, 1991). Most of the body's B6 is lost eventually with urine, about half of it as 4-pyridoxic acid. Significant amounts of pyridoxine are excreted with high intake from supplements.

REGULATION

An overflow mechanism protects tissues against excessive intracellular B6 concentrations, which would increasingly promote the unspecific binding to lysyl residues. Phosphatases hydrolyze nonprotein-bound

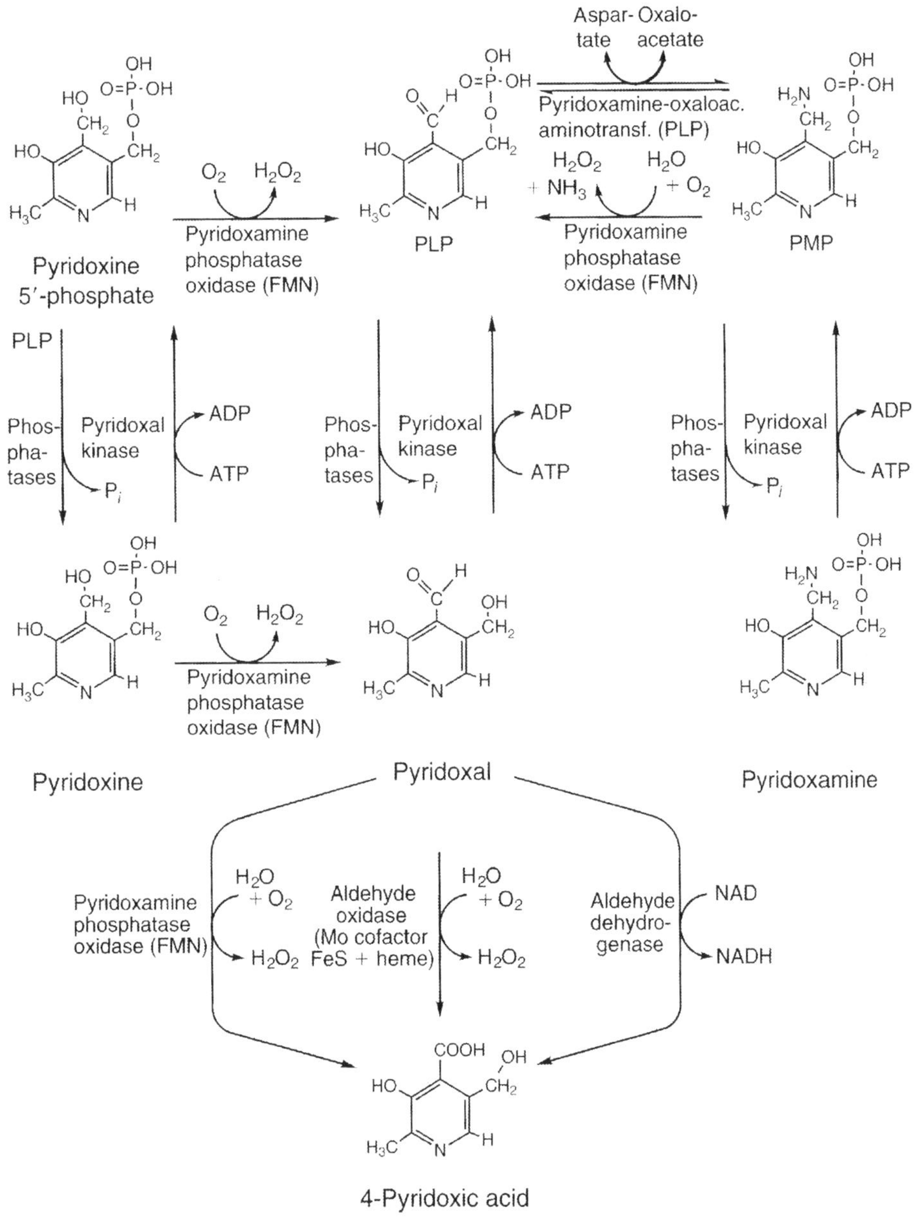

FIGURE 10.26

Vitamin B6 metabolism.

PLP and PLP preferentially. With increasing intracellular concentration, therefore, a greater percentage of the B6 is converted into the nonphosphorylated form and can leave the cell.

FUNCTION

Typically, PLP forms a Schiff base with the ε-amino group of a lysine residue, which is then stabilized further by other neighboring residues (Figure 10.27). B6-dependent enzymes facilitate transaminations, decarboxylations, and other reactions.

Carbohydrate metabolism: There are at least three distinct glycogen phosphorylases (EC2.4.1.1) with tissue-specific expression (liver, muscle, and brain types) that require lysine-bound PLP as a cofactor for the mobilization of glucose from glycogen. Glucosamine-fructose-6-phosphate aminotransferase (EC2.6.1.16, PLP-dependent) is the rate-limiting enzyme for the synthesis of hexosamine and thus important for the N- and O-glycosylation of proteins.

Heme synthesis: Delta-aminolevulinate synthase (EC2.3.1.37) catalyzes the initial step of porphyrin synthesis—namely, the condensation of succinyl-CoA and glycine to delta-aminolevulinic acid. Since the oxygen-transport system and numerous enzymes use porphyrin-derived cofactors or prosthetic groups (including heme, myoglobin, and cytochromes), this reaction is crucial for all tissues.

Neurotransmitter synthesis: PLP is a known or presumed component of several enzymes involved in the synthesis and catabolism of compounds that affect brain function, including glutamate, GABA, serotonin, catecholamines, histamine, tryptamine, taurine, D-serine, N-methyl-D-aspartate, glycine, and proline. Tryptamine, dopamine, and serotonin are produced by DOPA decarboxylase (EC4.1.1.28), histamine by histidine decarboxylase (EC4.1.1.22), and L-DOPA by tyrosine phenol-lyase (EC4.1.99.2). GABA is synthesized by glutamate decarboxylase (EC4.1. 1.15) and inactivated by 4-aminobutyrate

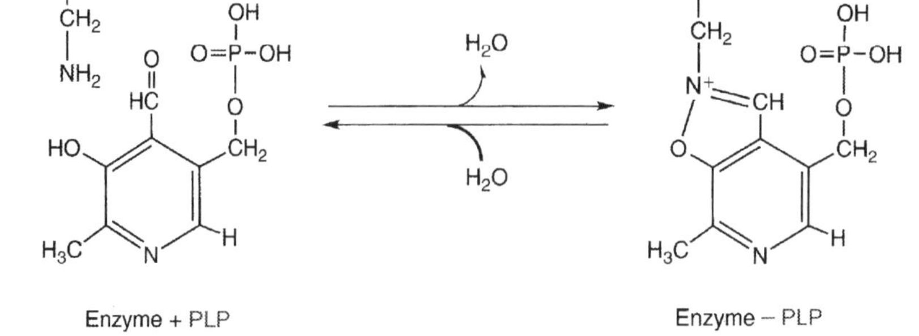

Enzyme + PLP   Enzyme − PLP

FIGURE 10.27

PLP forms a Schiff base with a specific lysyl ε-amino group in enzymes.

aminotransferase (EC2.6.1.19). Sulfinoalanine decarboxylase (EC4.1.1.29) catalyzes the initial step of taurine synthesis. D-Aspartate for *N*-methyl-D-aspartate synthesis is generated by aspartate racemase (EC5.1.1.13) and D-serine is produced by serine racemase (EC5.1.1.18). These compounds are endogenous ligands of the *N*-methyl-D-aspartate receptor (Wolosker et al., 1999).

Urea cycle: A PLP-containing enzyme of the mitochondrial matrix (ornithine oxo-acid aminotransferase; EC2.6.1.13) is needed for the second step of urea synthesis.

Amino acid synthesis and catabolism: The vast majority of all known B6-dependent enzymes moves amino groups to or from amino acids and their precursors. The only amino acid that does not rely exclusively on at least one such reaction is proline. Additional PLP-dependent enzymes acting on amino acids include decarboxylases (such as ornithine decarboxylase, EC4.1.1.17), a racemase (serine racemase, EC5.1.1.18), aldolases (threonine aldolase; EC4.1.2.5), and dehydratases (e.g., threonine dehydratase, EC4.2.1.16).

Glycine: Enzymes contributing to glycine synthesis include 2-amino-3-ketobutyrate coenzyme A (CoA) ligase (EC2.3.1.29), alanine-glyoxylate aminotransferase (EC2.6.1.44), glycine hydroxymethyltransferase (EC2.1.2.1), and, to a much lesser extent, kynurenineglyoxylate aminotransferase (EC2.6.1.63), and aromatic amino acid–glyoxylate aminotransferase (EC2.6.1.60). A catabolic enzyme is the P protein of the glycine cleavage complex (EC2.1.2.10).

L-*Alanine*: Alanine aminotransferase (ALT, EC2.6.1.2) is one of the major enzymes in muscle and the liver.

Valine/leucine/isoleucine: The breakdown of these three bulky amino acids is initiated by branched-chain amino acid aminotransferase (EC2.6.1.42).

Proline: The only step involving a PLP-dependent enzyme is ornithine-oxo-acid aminotransferase (EC2.6.1.13, PLP-dependent), which participates in the synthesis of arginine from proline in the gut.

Phenylalanine/tyrosine: Catabolic enzymes include phenylalanine(histidine) aminotransferase (EC2.6.1.58), glutamine-phenylpyruvate aminotransferase (EC2.6.1.64), kynurenine-glyoxylate aminotransferase (EC2.6.1.63), aromatic amino acid glyoxylate aminotransferase (EC2.6.1.60), dihydroxyphenylalanine aminotransferase (EC2.6.1.49), tyrosine aminotransferase (EC2.6.1.5), and aromatic amino acid glyoxylate aminotransferase (EC2.6.1.60).

Tryptophan: Several PLP-dependent enzymes participate in its metabolism, including tryptophan aminotransferase (EC2.6.1.27), kynureninase (EC3.7.1.3), kynurenine-oxoglutarate aminotransferase (EC2.6.1.7), kynurenine glyoxylate aminotransferase (EC2.6.1.63), and aromatic amino acid-glyoxylate aminotransferase (EC2.6.1.60).

Methionine/cysteine: The transsulfuration pathway uses the PLP-dependent enzymes cystathionine-beta-synthase (EC4.2.1.22) and cystathionine-gamma-lyase (EC4.4.1.1). The deamination of cysteine can then be catalyzed by cysteine aminotransferase (EC2.6.1.3). An alternative transamination reaction for L-methionine can be catalyzed by glutamine-pyruvate aminotransferase (EC2.6.1.15).

Serine: Conversion to and from glycine depends on glycine hydroxymethyltransferase (EC2.1.2.1). Serine dehydratase (EC4.2.1.13) and threonine dehydratase (EC4.2.1.16), both PLP-dependent enzymes, converted some serine directly to pyruvate. Serine racemase (with no EC number assigned) generates D-serine, an important neuromodulator in the brain.

Threonine: The main cytosolic pathway of threonine degradation starts with PLP-containing threonine dehydratase (EC4.2.1.16). Cleavage into glycine and acetaldehyde by the two PLP-containing cytosolic enzymes threonine aldolase (EC4.1.2.5) and glycine hydroxymethyltransferase (EC2.1.2.1) is a minor catabolic pathway in cytosol. The second step of threonine catabolism in mitochondria uses PLP-dependent 2-amino-3-ketobutyrate CoA ligase (EC2.3.1.29).

Glutamate/glutamine/aspartate/asparagine: Numerous PLP-dependent enzymes move amino groups to and from these pivotal amino acids, including aspartate aminotransferase (AST, EC2.6.1.1), alanine aminotransferase (ALT, EC2.6.1.2), and 4-butyrate aminotransferase (EC2.6.1.19). Another important PLP-dependent enzyme is the GABA-synthesizing glutamate decarboxylase (EC4.1.1.15).

Histidine: Aromatic-amino-acid-glyoxylate aminotransferase (EC2.6.1.60) catalyzes transamination of L-histidine.

Lysine: Its breakdown uses 2-aminoadipate aminotransferase (EC2.6.1.39). Ornithine decarboxylase (EC4.1.1.17) catalyzes lysine conversion into cadaverine.

Arginine: Both endogenous synthesis and catabolism rely on PLP-containing enzymes, most important ornithine-oxo-acid aminotransferase (EC2.6.1.13).

Carnitine: The third step of carnitine synthesis is catalyzed by PLP-dependent serine hydroxymethyltransferase (EC2.1.2.1).

Taurine: The initial step of taurine synthesis depends on sulfinoalanine decarboxylase (EC4.1.1.29).

Lipid metabolism: B6 is also important for the synthesis and metabolism of various lipids. Sphingosine, produced by PLP-containing serine C-palmitoyltransferase (EC2.3.1.50), is part of myelin, ceramides, and sphingolipids in the brain, nerves, and other tissues. For still unclear reasons, B6 in some form is also needed for full activity of 6-desaturase (EC1.14.99.25), which extends the chain length of omega-3 delta fatty acids. Reduced activity impairs the production of EPA and DHA from dietary precursors, such as alpha-linolenic acid (Tsuge and Hotta, 2000). Similarly, B6 deficiency greatly diminishes peroxisomal acyl-CoA oxidase (EC1.3.3.6) activity, which catalyzes the initial reaction of fatty acid beta-oxidation in peroxisomes (Tsuge and Hotta, 2000).

Selenium metabolism: Selenocysteine lyase (EC4.4.1.16, PLP-dependent) catalyzes the decomposition of L-selenocysteine to L-alanine and elemental selenium, thereby providing selenium to selenophosphate synthetase in selenoprotein biosynthesis. L-Seryl-tRNA^{Sec} selenium transferase (EC2.9.1.1, PLP-dependent) uses selenophosphate to substitute the hydroxyl oxygen of serine with selenium.

Hormone metabolism: Diiodotyrosine aminotransferase (EC2.6.1.24), thyroid hormone aminotransferase (EC2.6.1.26), or both catabolize and thereby inactivate thyroid hormones.

Vitamin metabolism: Kynureninase (EC3.7.1.3, PLP-dependent) is the key enzyme for niacin synthesis from tryptophan.

Xenobiotic metabolism: Cysteine conjugate beta-lyase (EC4.4.1.13) cleaves cysteine conjugates in the kidneys, releasing ammonia and pyruvate.

Metabolic regulation: The activity of several enzymes is inhibited when PLP firmly binds to the protein. Understanding the significance of such actions is very preliminary. Examples include the effects on succinic semialdehyde dehydrogenase in the brain (Choi et al., 2001) and on tyrosine phosphatase in many tissues (Zhou and Van Etten, 1999). The binding to steroid hormone receptors may be similar in some ways, but the significance is poorly understood.

NUTRITIONAL ASSESSMENT

B6 deficiency and suboptimal status may be detected by enzyme activation assays or with the measurement of B6 metabolites in the blood. Dietary intake can be estimated from the amount of excreted urinary B6 metabolites.

Blood measurement: The enzyme activation assays usually measure alanine aminotransferase (ALT, EC2.6.1.2) activity in untreated and PLP-treated red blood cells. An activation coefficient (ratio of activity in supplemented red blood cells divided by activity in native red blood cells) in excess of 1.15 indicates lower than optimal PLP saturation of the enzyme.

Urine measurement: The amount of B6 metabolites, particularly 4-pyridoxic acid , excreted with a urine sample collected over a 24-h period reflects recent B6 intake (Tsuji et al., 2010).

REFERENCES

Albersen, M., Bosma, M., Knoers, N.V., de Ruiter, B.H., Diekman, E.F., de Ruijter, J., et al., 2013. The intestine plays a substantial role in human vitamin B6 metabolism: a Caco-2 cell model. PLoS One 8, e54113.

Bahn, J.H., Kwon, O.S., Joo, H.M., Ho Jang, S., Park, J., Hwang, I.K., et al., 2002. Immunohistochemical studies of brain pyridoxine-5′-phosphate oxidase. Brain Res. 925, 159–168.

Black, A.L., Guirard, B.M., Snell, E.E., 1977. Increased muscle phosphorylase in rats fed high levels of vitamin B6. J. Nutr. 107, 1962–1968.

Choi, S.Y., Bahn, J.H., Lee, B.R., Jeon, S.G., Jang, J.S., Kim, C.K., et al., 2001. Brain succinic semialdehyde dehydrogenase: identification of reactive lysyl residues labeled with pyridoxal-5′-phosphate. J. Neurochem. 76, 919–925.

Food and Nutrition Board, Institute of Medicine, 1998. Dietary Reference Intakes for Thiamin, Riboflavin, Niacin, Vitamin B6, Folate, Vitamin B12, Pantothenic Acid, Biotin, and Choline. National Academic Press, Washington, DC, pp. 150–195.

Gregory III, J.F., 1998. Nutritional properties and significance of vitamin glycosides. Annu. Rev. Nutr. 18, 277–296.

Johansson, S., Lindstedt, S., Register, U., Wadstrom, L., 1966. Studies on the labeled pyridoxine in man. Am. J. Clin. Nutr. 18, 185–196.

Lee, H.S., Moon, B.J., Choi, S.Y., Kwon, O.S., 2000. Human pyridoxal kinase: overexpression and properties of the recombinant enzyme. Mol. Cells 10, 452–459.

McMahon, L.G., Nakano, H., Levy, M.D., Gregory III, J.F., 1997. Cytosolic pyridoxine-beta-D-glucoside hydrolase from porcine jejunal mucosa. Purification, properties, and comparison with broad specificity beta-glucosidase. J. Biol. Chem. 272, 32025–32033.

Mehansho, H., Buss, D.D., Hamm, M.W., Henderson, L.M., 1980. Transport and metabolism of pyridoxine in rat liver. Biochim. Biophys. Acta 631, 112–123.

Merrill Jr., A.H., Henderson, J.M., Wang, E., McDonald, B.W., Millikan, W.J., 1984. Metabolism of vitamin B-6 by human liver. J. Nutr. 114, 1664–1674.

Schenker, S., Johnson, R.E., Mahuren, J.D., Henderson, G.I., Coburn, S.P., 1992. Human placental vitamin B6 (pyridoxal) transport: normal characteristics and effects of ethanol. Am. J. Physiol. 262, R966–R974.

Shane, B., 1978. Vitamin B6 and blood Human Vitamin B6 Requirements: Proceedings of a Workshop. National Academy of Sciences, Washington, DC, pp. 111–128.

Tadera, K., Arima, M., Yoshino, S., Yagi, F., Kobayashi, A., 1986. Conversion of pyridoxine into 6-hydroxypyridoxine by food components, especially ascorbic acid. J. Nutr. Sci. Vitaminol. 32, 267–277.

Trumbo, P.R., Banks, M.A., Gregory III, J.F., 1990. Hydrolysis ofpyridoxine-5′-beta-D-glucoside by a broad-specificity beta-glucosidase from mammalian tissues. Proc. Soc. Exp. Biol. Med. 195, 240–246.

Tsuge, H., Hotta, N.T., 2000. Effects of vitamin B-6 on $(n-3)$ polyunsaturated fatty acid metabolism. J. Nutr. 130, 333S–334S.

Tsuji, T., Fukuwatari, T., Sasaki, S., Shibata, K., 2010. Twenty-four-hour urinary water-soluble vitamin levels correlate with their intakes in free-living Japanese university students. Eur. J. Clin. Nutr. 64, 800–807.

Wolosker, H., Sheth, K.N., Takahashi, M., Mothet, J.P., Brady Jr., R.O., Ferris, C.D., et al., 1999. Purification of serine racemase: biosynthesis of the neuromodulator D-serine. Proc. Natl. Acad. Sci. USA 96, 721–725.

Zhang, Z.M., McCormick, D.B., 1991. Uptake of N-(4′-pyridoxyl)amines and release of amines by renal cells: a model for transporter-enhanced delivery of bioactive compounds. Proc. Natl. Acad. Sci. USA 88, 10407–10410.

Zhou, M., Van Etten, R.L., 1999. Structural basis of the tight binding of pyridoxai 5′-phosphate to a low molecular weight protein tyrosine phosphatase. Biochemistry 38, 2636–2646.

FOLATE

Folate (folic acid, *N*-[4-[[(2-amino-1,4-dihydro-4-oxo-6-pteridinyl)methyl]amino]benzoyl]- L-glutamic acid, pteroylglutamic acid, PGA, obsolete names vitamin B8, vitamin B9, vitamin Bc, and vitamin M; molecular weight 441) is one of several biological active vitamers; compounds with comparable activity include the physiological metabolites 7,8-dihydrofolate, 5,6,7,8-THF, 5,10-methylene-THF, methyl-THF, 5-formimino THF, 10-formyl-THF, and 5-formyl-THF (folinic acid, citrovorum factor, and leucovorin) (Figure 10.28).

ABBREVIATIONS

apBGlu	p-acetamidobenzoylglutamate
DHF	dihydrofolate
FIGLU	formiminoglutarnate
FOLR	folate receptor
MRP2	multidrug resistance protein 2 (ABCC2)
MTHFD	methylenetetrahydrofolate dehydrogenase (EC1.5.1.5)
pABA	para-aminobenzoic acid
RFC1	reduced folate carrier gene (SLC19A1)
THF	5,6,7,8-tetrahydrofolate

NUTRITIONAL SUMMARY

Function: Folate mediates the transfer of one carbon in numerous reactions, including the synthesis of purine nucleotides, amino acids, carnitine, creatine, lipids, hormones, and also serves as a cofactor for proteins involved in the control of circadian rhythm.

Requirements: Adults should get at least 400 μg/day. Women who may become pregnant should use a supplement with this amount. Needs during pregnancy and breastfeeding are slightly higher.

Sources: Fortified cereals, dark-green vegetables, legumes, oranges, and liver are very good sources. In the United States, bread, pasta, flour, and other grain products are fortified.

Deficiency: Inadequate intake predisposes to anemia and abnormal white blood cells, thrombosis, cancer, and to birth of children with severe defects. People with certain common genetic dispositions are most sensitive to low intake.

Excessive intake: Folate from dark-green vegetables, citrus fruits, and other plant-derived foods will not cause harm at intake levels of several times the recommended amount, but consumption of more than 1000 μg folic acid from dietary supplements and fortified foods may be unhealthy for some susceptible individuals and should be avoided.

ENDOGENOUS SOURCES

The normal microflora of the distal ileum and throughout the colon produces significant amounts of folate. While it is certain that some of this bacterial folate contributes to folate status in humans, the amounts are not known (Dudeja et al., 2001).

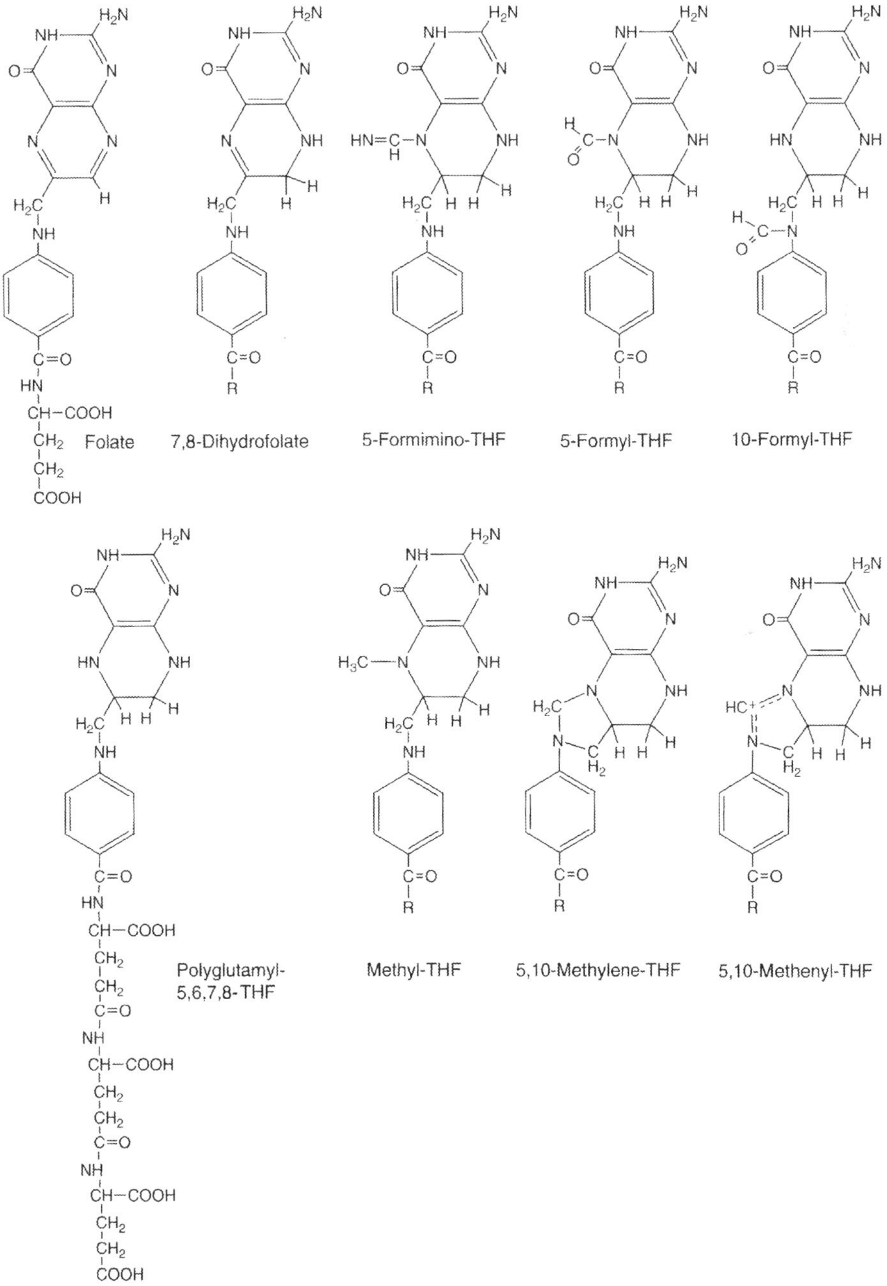

FIGURE 10.28

Important folate vitamers.

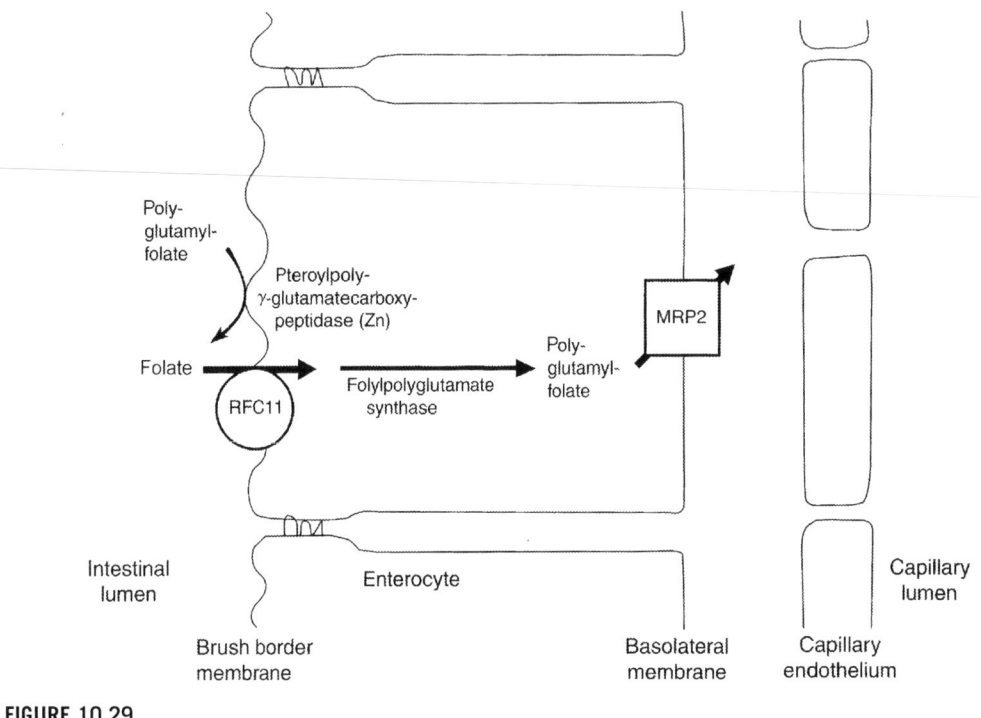

FIGURE 10.29

Intestinal absorption of folate.

DIETARY SOURCES

Foods contain folylpolyglutamate and free folate (mainly from dietary supplements and fortified foods). These compounds are easily degraded during cooking and in the acid environment of the stomach. Ascorbate and reducing thiol protect folate during processing and after ingestion (Seyoum and Selhub, 1998). Oxidation of the C9–N10 bond splits the folate molecule and causes its irreversible loss. Here, 5-methyltetrahydrofolate is more stable than other reduced folate metabolites (Suh et al., 2001).

Bioavailability of folate from different food sources ranges (Seyoum and Selhub, 1998) from very high for fortified foods (cereals and bread) and supplements (over 85%), egg yolk (72%) and liver (56%) to intermediate for orange juice (21%) to marginal in vegetables (3–6%). Very little folate is available from yeast (<1%). Bioavailability of folate, 5-methyl-THF, and 5-formyl-THF appears to be similar in rats (Bhandari and Gregory, 1992).

DIGESTION AND ABSORPTION

Folate absorption is most active in the jejunum (Figure 10.29). In healthy people, about 85% of a moderate dose of pure folic acid is absorbed; absorption of folate from nonfortified foods is only about 50% (Food and Nutrition Board, Institute of Medicine, 1998). Absorption also occurs from other segments of the small intestine and from the colon, but the amounts available from those sites are not known.

The transporters responsible for absorption in the distal intestine appear to be the same as in jejunum (Dudeja et al., 2001; Wang et al., 2001).

Before folylpolyglutamate, the form present in unfortified foods, can be taken up, it has to be cleaved by pteroylpoly-gamma-glutamate carboxypeptidase (EC3.4.17.21), a membrane-bound, zinc-dependent exopeptidase of the jejunal brush border. A relatively common variant (H475Y) of this enzyme has considerably reduced activity which limits bioavailability of food folate (Devlin et al., 2000). The proton-coupled folate transporter (PCFT, SLC46A1) is the main conduit for folate absorption from the small intestine (Le Blanc et al., 2012). It should be noted that the same transporter also mediates the uptake of heme. Monoglutamyl folate (and analogs such as methotrexate) can also be taken up from small intestine, as well as from the colon via the high-affinity folate receptor 1 (reduced folate carrier gene, RFC1, SLC19A1). The transport is a facilitative process that is optimal at low or neutral pH, depending on as-yet-unidentified tissue-specific modulators (Wani et al., 2012).

Free folate is rapidly returned into the intestinal lumen by the breast cancer resistance protein (BCRP, ABCG2) (Tan et al., 2013). Metabolic conversion and export into the bloodstream prevent the loss of newly absorbed folate.

Export of folate across the basolateral membrane may use the ATP-driven ABC transporter multidrug resistance protein 2 (MRP2; ABCC2) and related proteins (Wang et al., 2001).

TRANSPORT AND CELLULAR UPTAKE

Blood circulation: The main form of folate in circulation is 5-methyltetrahydrofolate. Typical concentrations of total folate in red blood cells are around 400 nmol/l, while concentrations in plasma are around 16 nmol/l (Fohr et al., 2002). Folate intake and common genetic variants influence blood concentrations.

Here, 5-methyltetrahydrofolate and other reduced folate metabolites in circulation reach tissues through the bidirectional anion transporter RFC1 (SLC19A1). Additional folate metabolites are taken up by folate receptors (FOLRs), which comprise a family of glycosylphosphatidylinositol (GPI)–anchored membrane proteins. The high-affinity folate receptor 1 (FOLR1, FOLR type alpha) is present in the placenta, thymus, lung, brain, and liver. A soluble form of this receptor circulates in the blood, but its function is not well understood. Folate receptor 3 (FOLR3) is expressed mainly in hematogenic tissues. Folate receptor 2 (FOLR2, FOLR type beta) is expressed during fetal development and in the placenta. FOLR2 has a lower affinity for methyl-THF than FRI (Maziarz et al., 1999). Multidrug resistance proteins, such as MDR1 (Gifford et al., 1998), are thought to export folate from cells with a strong preference for the monoglutamyl forms.

Trapping as folylpolyglutamate is an important mechanism for the maintenance of folate supplies in tissues. Folylpolyglutamate synthase (EC6.3.2.17) successively adds glutamyl residues to folate metabolites and thereby blocks their transporter-mediated efflux. Translation beginning from alternative initiation sites generates distinct isoforms of the enzyme. Transcription at the upstream initiation site generates the form with a mitochondrial leader peptide, which is absent in the cytosolic form. Methyltetrahydrofolate is a relatively poor substrate for folylpolyglutamate synthase and therefore is not well retained in tissues. Gamma-glutamyl hydrolase (EC3.4.19.9) releases unsubstituted folate. The enzyme resides in lysosomes and contains four zinc ions. Exocarboxypeptidase II (EC3.4.17.21) in the prostate and possibly other tissues cleaves off specific gamma-linked glutamates. This enzyme is identical with prostate-specific membrane (PSM) antigen, whose serum concentration is often elevated in patients with hormone-refractory and metastatic prostate cancer (Tiffany et al., 1999). Folate reaches mitochondria through an as-yet-unidentified carrier (Suh et al., 2001). Only folylmonoglutamate can enter. In the reverse direction, folates with longer side chains appear to be transported.

BBB: RFC1 carries folates across the apical membrane of BBB epithelial cells and choroid plexus epithelial cells (Wang et al., 2001). FOLRs facilitate transport across the basolateral membrane in these cells.

Materno-fetal transfer: Here, 5-methyltetrahydrofolate is concentrated several-fold into intervillous blood space by binding to FOLRs on the maternally facing chorionic surface. This provides a concentration gradient that drives folates across the placental barrier (Henderson et al., 1995). Both FOLR1 and FOLR2 are expressed in the placenta.

METABOLISM

Polyglutamyl side chains: At least five folate metabolites have direct functional importance, including THF, dihydrofolate, 5,10-methylene-THF 5-methyl-THF, 5-formyl-THF, 10-formyl-dihydrofolate, and 10-formyl-THF. Additional forms are crucial intermediates of folate metabolism. All of these vitamers require polyglutamyl side chains of specific length to be fully active. The removal and addition of glutamyl groups as described previously is an integral part of folate metabolism that controls both retention and function.

Folate activation: Dihydrofolate reductase (EC1.5.1.3) catalyzes the NADP-dependent reduction of folate to 7,8-dihydrofolate and then to 5,6,7,8-THF. This folate-activating enzyme is inhibited by the anticancer agent methotrexate (Figure 10.30).

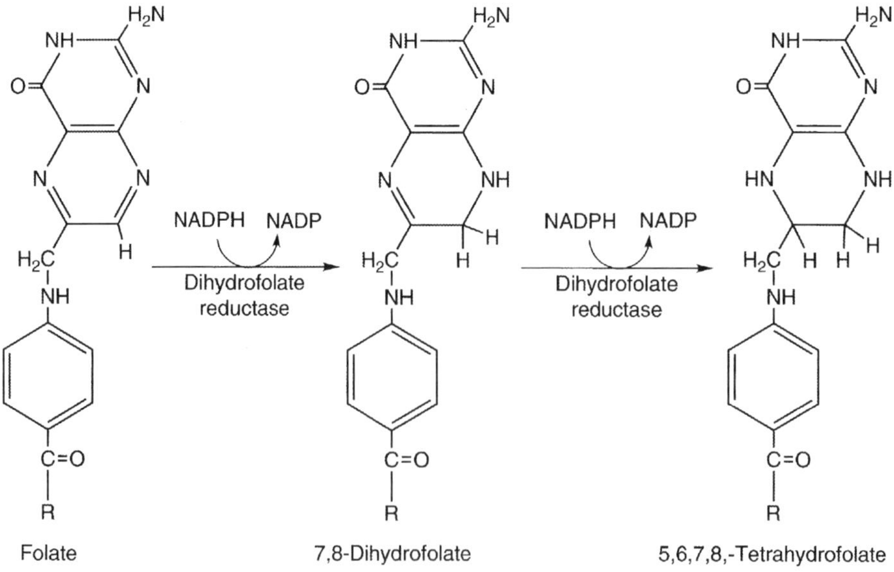

FIGURE 10.30

Dihydrofolate reductase activates folate.

Folate reactivation: The reactions of folate metabolites are numerous and complex. Reactions that use one active form often generate another functionally important form. On the other hand, such reactions are essential to unload one-carbon groups and regenerate free folate for other reactions. The picture is further complicated by the fact that some of these activities are bundled into multifunctional proteins.

5,10-methylene-THF: The conversion of serine to glycine by glycine hydroxymethyltransferase (EC2.1.2.1) generates large quantities of 5,10-methylene-THF. Catabolism of glycine through the mitochondrial glycine cleavage system (aminomethyltransferase, EC2.1.2.10) also eliminates a carbon with 5,10-methylene-THF. NADPH-dependent reduction of 5,10-methenyl-THF by methylenetetrahydrofolate dehydrogenase (MTHFD1, EC1.5.1.5), an activity of the C1-THF synthase, also generates 5,10-methylene-THF. The trifunctional C1-THF synthase protein combines in humans the activities of MTHFD1, formate-THF ligase (EC6.3.4.3), and methenyl-THF cyclohydrolase (EC3.5.4.9). In addition, 5,10-methylene-THF can be generated from 10-formyl-THF by the methenyl-THF cyclohydrolase (EC3.5.4.9) activity of C1-THF synthase, and it is the cofactor for deoxythymidine monophosphate (dTMP) synthesis by thymidylate synthase (EC2.1.1.45). This reaction releases dihydrofolate, which can be reduced to THF again. Alternatively, both the NADPH-dependent methylenetetrahydrofolate reductase (EC1.5.1.20) and the FADH-dependent 5,10-methylenetetrahydrofolate reductase (EC1.7.99.5) can convert 5,10-methylene-THF into 5-methyl-THF, which then must be disposed of in turn. Both enzymes contain FAD as a prosthetic group (Figure 10.31).

5,10-methenyl-THF: The PLP-dependent glutamate formimidoyltransferase (EC2.1.2.5) is a key enzyme of histidine metabolism, which moves a formidoyl group to THF. The resulting 5-formimino-THF is converted to 5,10-methenyl-THF by formiminotetrahydrofolate cyclodeaminase (EC4.3.1.4), the second activity of the bifunctional protein formiminotransferase-cyclodeaminase.

5-methyl-THF: The methylcobalamin-containing 5-methyltetrahydrofolate-homocysteine *S*-methyltransferase (EC2.1.1.13) and betaine:homocysteine methyltransferase (EC2.1.1.5) are the only enzymes that utilize 5-methyl-THF in significant quantities. A reduction of their activities (particularly due to vitamin B12 deficiency) decreases the availability of THF for other reactions.

10-formyl-THF: Phosphoribosylglycinamide formyltransferase (EC2.1.2.2) and phosphoribosyl aminoimidazole carboxamide formyltransferase (AICAR transformylase, EC2.1.2.3) use 10-formyl-THF in purine synthesis. Both reactions release THF. Alternatively, formyltetrahydrofolate dehydrogenase (EC1.5.1.6) can regenerate THF from 10-formyl-THF by splitting off carbon dioxide in an NADPH-generating reaction.

5-formyl-THF: This is a minor folate metabolite, which may come from food or from the reaction of *N*-formyl-L-glutamate with THF, catalyzed by glutamate formimidoyltransferase (EC2.1.2.5, PLP). The ATP-driven enzyme 5-formyltetrahydrofolate cyclo-ligase (EC6.3.3.2, requires PLP) converts 5-formyl-THF to 5,10-methenyl-THF (Baggott et al., 2001).

10-formyl dihydrofolate: This metabolite arises from the reaction of formate with dihydrofolate, catalyzed by formate-dihydrofolate ligase (EC6.3.4.17) as detailed next. Here, 10-formyl dihydrofolate can serve as a cofactor for the penultimate step of inosine-5′-monophosphate (IMP; purine) synthesis catalyzed by phosphoribosylaminoimidazole carboxamide formyltransferase (AICAR transformylase, EC2.1.2.3). Spontaneous oxidation is also possible, which generates the metabolically inert metabolite 10-formyl folate (Baggott et al., 2001).

Folate catabolism: Dihydrofolate and 10-formyl-THF are particularly sensitive to oxidative degradation with the release of para-aminobenzoic acid (pABA). It has been proposed that ferritin specifically facilitates the oxidative cleavage of 10-formyl-THF and that the degradation of this metabolite is

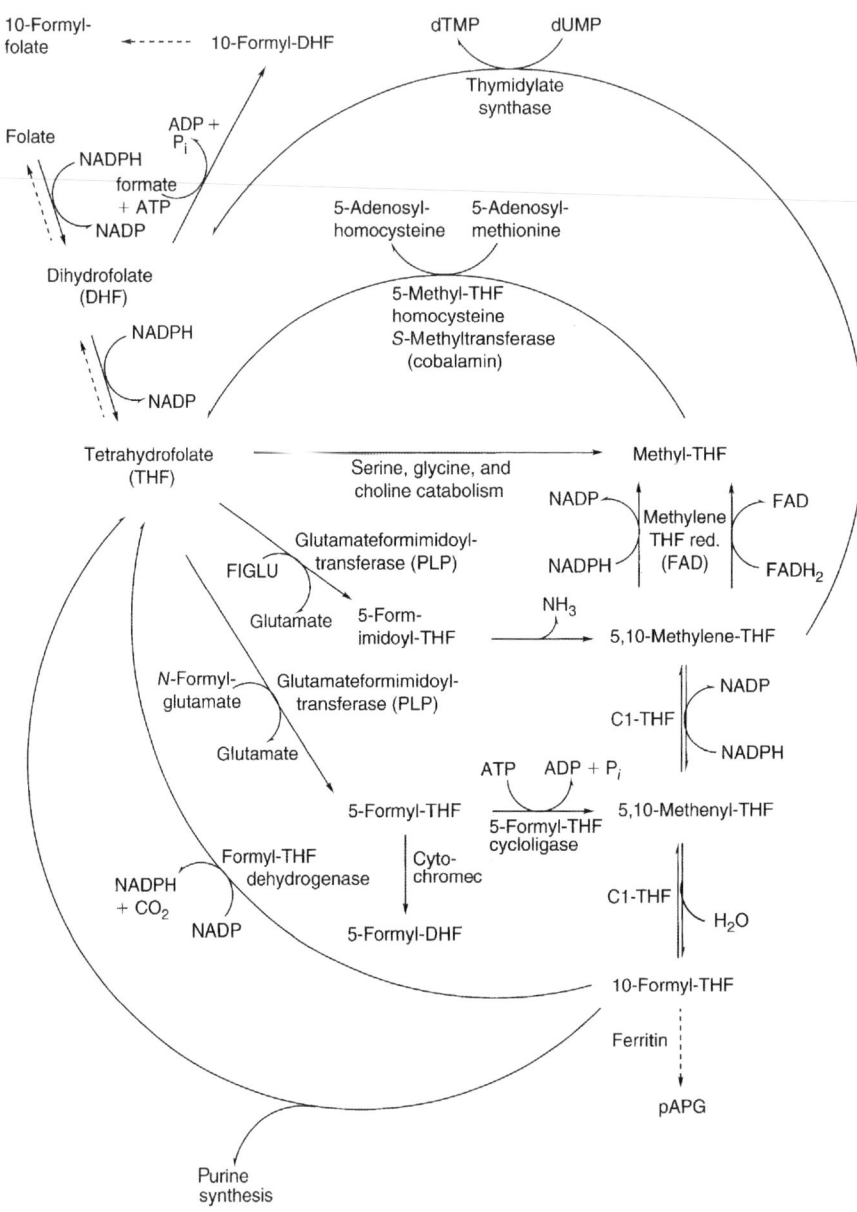

FIGURE 10.31

Folate interconversions.

an important modulator of intracellular folate concentration (Suh et al., 2001). Acetylation of pAPG by arylamine *N*-acetyltransferases (EC2.3.1.5) facilitates the removal of inactive folate catabolites from cells.

STORAGE

The liver contains relatively high concentrations of folate (up to 15 μg/g), mainly as polyglutamylfolate (Hoppner and Lampi, 1980). Other tissues contain smaller amounts.

Short-term (40–80 min) exposure to nitrous oxide anesthesia reduces folate concentrations in serum, but not in red cells (Deleu et al., 2000).

EXCRETION

Losses occur mostly through metabolic inactivation and via urine. Significant amounts of folate (100 μg/day) are secreted into bile and largely recovered again by intestinal absorption (hepatobiliary cycling; Herbert, 1968). Some folate is lost with the feces, however (Caudill et al., 1998). One hypothesis suggests that photolysis in skin accounts for a significant proportion of total folate losses (Jablonski, 1999). Prolonged whole-body exposure to ultraviolet (UV) light with sunbathing or in tanning beds may increase the risk of neural tube defects (NTDs) in the case of an incipient pregnancy. The main functional metabolite in ultrafiltrate of the kidneys, 5-methyltetrahydrofolate, is recovered from the proximal tubular lumen via an FOLR that acts in conjunction with megalin (Birn et al., 2005). The reduced folate carrier 1 (SLC19A1) then completes transport across the basolateral membrane by exchange for organic phosphate (Sikka and McMartin, 1998; Wang et al., 2001). This arrangement differs completely from the one operating in small intestine. The organic anion carriers (OATK1 and OATK2) at the brush border membrane may provide additional absorptive capacity.

Total folate losses have been estimated to be about 60 μg/day (Gailani et al., 1970; National Research Council, 1989). An important metabolite in urine is *N*-acetyl-pABA (McPartlin et al., 1993), but it appears to represent only about 5–7% of ingested supplements (Kownacki-Brown et al., 1993). This assessment may be obscured, however, by prolonged storage of breakdown products and delayed release for excretion (Suh et al., 2001).

REGULATION

Absorption, renal salvage, and uptake into tissue via RFC1 are regulated, but the responsible mechanisms and modulators remain to be elucidated (Wani et al., 2012).

FUNCTION

Thymidylate synthesis: Thymidylate synthetase (EC2.1.1.45) methylates 2′-deoxyuridine 5′-monophosphate (dUMP) to dTMP using 5,10-methylene-THF as one-carbon donor.

Purine synthesis: Both the third step of purine synthesis, catalyzed by glycinamide ribonucleotide (GAR) transformylase (EC2.1.2.2), and the penultimate step, catalyzed by the bifunctional purine biosynthesis protein (phosphoribosyl aminoimidazole carboxamide formyltransferase, EC2.1.2.3), utilize 10-formyl-THF as one-carbon donor.

Amino acid metabolism: Homocysteine is remethylated to methionine by 5-methyltetrahydrofolate-homocysteine *S*-methyltransferase (EC2.1.1.13), a cytoplasmic enzyme with covalently bound methyl-cobalamin. Homocysteine remethylation converts the cosubstrate 5-methyl-THF back into THF and makes this form available again for other reactions. The methionine metabolite *S*-adenosyl-methionine is the major methyl-group donor for DNA methylation and for the synthesis of numerous essential compounds, including catecholamines, carnitine, choline, melatonin, and creatine.

Glycine hydroxymethyltransferase (EC2.1.2.1), cytosolic and mitochondrial, methylates glycine to serine or catalyzes the reverse reaction. The final step of the conversion of L-histidine to L-glutamate is catalyzed by the PLP-dependent enzyme glutamate formiminotransferase (EC2.1.2.5). The metabolite generated in this reaction and excreted with urine in response to a histidine load, formiminoglutamate (FIGLU), has been used in the past as a marker of folate deficiency.

Formate utilization: The metabolism of choline generates formaldehyde in the last two oxidative steps. Formaldehyde can react nonenzymatically with THF and ATP to generate 10-formyl-THF (Figure 10.31) or it is converted into formate in an NAD-requiring reaction catalyzed by glutathione-dependent formaldehyde dehydrogenase (EC1.2.1.l, identical with alcohol dehydrogenase class III chi chain, contains zinc). An important source of formate is the nonoxidative release from 10-formyl-THF by formyltetrahydrofolate dehydrogenase (EC1.5.1.6). The main activity of this dehydrogenase is the oxidative release of the formyl group from 10-formyl-THF in an NADPH-generating reaction. Formyltetrahydrofolate dehydrogenase contains pentaglutamyl-THF as a tightly bound noncatalytic cofactor. The amounts of formate generated from methanol metabolism are usually small but can be significant with high dietary or industrial exposure (Bouchard et al., 2001). Formate moves into cytosol where formate-THF ligase (EC6.3.4.3, an activity of the trifunctional protein C1-THF synthase, MTHFD1) can link it to THF in an ATP-dependent reaction. An alternative formate-removing enzyme is formate-dihydrofolate ligase (EC6.3.4.17). This magnesium-dependent cytosolic enzyme links formate to dihydrofolate in an ATP-dependent reaction. 10-formyl dihydrofolate can then be used by phosphoribosylaminoimidazole carboxamide formyltransferase (AICAR transformylase, EC2.1.2.3) to generate 5-formarnido-l-(5-phospho-D-ribosyl)imidazole-4-carboxarnide (formyl-AICAR), the precursor for purine synthesis. Some 10-formyl-dihydrofolate may oxidize nonenzymatically to the dead-end product 10-formyl-folate (Baggott et al., 2001).

Formate is also a product released in the nucleus as a result of DNA demethylation and needs to be detoxified locally by conversion to 10-formyl-THF.

Circadian rhythm: Cryptochrome 1 and cryptochrome 2 are proteins in mitochondria that appear to act as photoreceptors that help maintain circadian period length and rhythmicity; they use FAD and folate as cofactors (Sancar, 2000).

Fetal development: Inadequate folate supplies during the first weeks of pregnancy increase the risk of NTDs, cleft palate (Martinelli et al., 2001), congenital heart disease (Kapusta et al., 1999), and other organ malformations. The exact causal mechanisms are still not well understood. Less active forms of several gene products involved in folate metabolism are known or suspected to increase NTD risk, including methylene THF reductase (EC1.5.1.20; Rosenberg et al., 2002), pteroylpoly-gamma-glutamate carboxypeptidase (EC3.4.17.21; Devlin et al., 2000), methylenetetrahydrofolate-dehydrogenase (MTHFD, C1-THF synthase, Hal et al., 1998), and RFC1 (De Marco et al., 2001). The risk of placental abruption and pregnancy failure may also be related in part to inadequate folate availability (Eskes, 2001).

Folate synthesis in microorganisms: Sulfa drugs inhibit conjugation of para-aminobenzoic acid (pABA) to pterin in bacteria. Since humans cannot synthesize folate through this reaction, they are not affected by its inhibition (Figure 10.32).

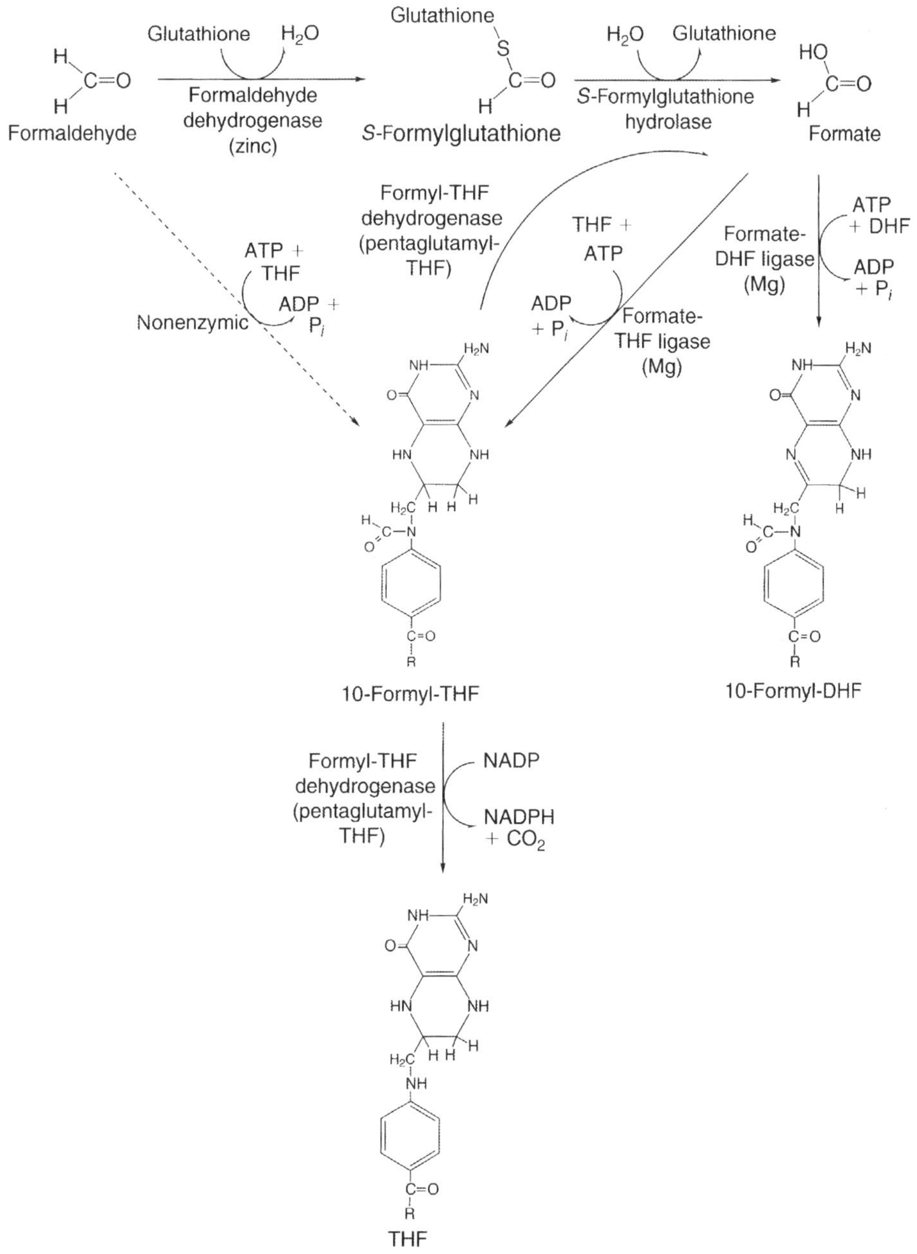

FIGURE 10.32

Formate detoxification requires THF both as a cosubstrate and as a tightly bound enzyme cofactor.

NUTRITIONAL ASSESSMENT

Folate deficiency and suboptimal status may be detected by microbial assays or with the measurement of active folate metabolites in plasma or red blood cells. Dietary intake can be estimated from the amount of excreted urinary folate metabolites.

Blood measurement: Measurements of blood plasma reflect intake during the previous several days. Folate in red blood cells reflects intake during the previous 3–4 weeks. Most commonly, a competitive binding assay with a high-affinity folate binder is used for folate measurements.

Measurement of all active folate metabolites in red blood cells or plasma is also possible (Fazili et al., 2013). Assays usually remove enzymatically the polyglutamate side chains from the metabolites to simplify chromatographic analyses.

Microbial assays use auxotroph *Lactobacillus casei* cultures that grow only with added folate in the medium. A set of calibrators with known folate concentrations and the samples for measurement are added to wells with culture medium and inoculated with the bacteria. After incubating for a day or two, the growth in each well is optically measured and the concentration in the samples calculated in comparison to the calibration curve. The assay is more cumbersome than automated binder assays, but it also is more accurate because it captures most bioactive metabolites.

Alternatively, homocysteine concentration can be measured. Concentration of this metabolite increases with inadequate folate status because it has to be metabolized by 5-methyltetrahydrofolate-homocysteine *S*-methyltransferase (MTR, EC2.1.1.13). This biomarker is not very specific because it is also increased by inadequate vitamin B12 status.

REFERENCES

Baggott, J.E., Robinson, C.B., Johnston, K.E., 2001. Bioactivity of [6R]-5-formyltetrahydrofolate, an unusual isomer, in humans and *Enterococcus hirae*, and cytochrome c oxidation of 10-formytetrahydrofolate to 10 formyldihydrofolate. Biochem. J. 354, 115–122.

Bhandari, S.D., Gregory III, J.F., 1992. Folic acid, 5-methyl-tetrahydrofolate and 5-forrnyl-tetrahydrofolate exhibit equivalent intestinal absorption, metabolism and *in vivo* kinetics in rats. J. Nutr. 122, 1847–1854.

Birn, H., Zhai, X., Holm, J., Hansen, S.I., Jacobsen, C., Christensen, E.I., et al., 2005. Megalin binds and mediates cellular internalization of folate binding protein. FEBS J. 272, 4423–4430.

Bouchard, M., Brunet, R.C., Droz, P.O., Carrier, G., 2001. A biologically based dynamic model for predicting the disposition of methanol and its metabolites in animals and humans. Toxicol. Sci. 64, 169–184.

Caudill, M.A., Gregory, J.E., Hutson, A.D., Bailey, L.B., 1998. Folate catabolism in pregnant and nonpregnant women with controlled folate intakes. J. Nutr. 128, 204–208.

Deleu, D., Louon, A., Sivagnanam, S., Sundaram, K., Okereke, P., Gravell, D., et al., 2000. Long-term effects of nitrous oxide anesthesia on laboratory and clinical parameters in elderly Omani patients: a randomized double-blind study. J. Clin. Pharmacol. Themp. 25, 271–277.

De Marco, R., Calevo, M.G., Moroni, A., Arata, L., Merello, E., Cama, A., et al., 2001. Polymorphisms in genes involved in folate metabolism as risk factors for NTDs. Eur. J. Ped. Surg. 11, S14–S17.

Devlin, A.M., Ling, E.H., Peerson, J.M., Fernando, S., Clarke, R., Smith, A.D., et al., 2000. Glutamate carboxypeptidase II: a polymorphism associated with lower levels of serum folate and hyperhomocysteinemia. Hum. Mol. Genet. 9, 2837–2844.

Dudeja, P.K., Kode, A., Alnoudu, M., Tyagi, S., Torania, S., Subramanian, V.S., et al., 2001. Mechanism of folate transport across the human colonic basolateral membrane. Am. J. Physiol. Gastrointest. Liver Physiol. 281, G54–G60.

Eskes, T.K., 2001. Clotting disorders and placental abruption: homocysteine—a new risk factor. Eur. J. Obstet. Gynecol. Reprod. Biol. 95, 206–212.

Fazili, Z., Whitehead Jr., R.D., Paladugula, N., Pfeiffer, C.M., 2013. A high-throughput LC–MS/MS method suitable for population biomonitoring measures five serum folate vitamers and one oxidation product. Anal. Bioanal. Chem. 405, 4549–4560.

Fohr, I.E., Prinz-Langenohl, R., Bronstrup, A., Bohlmann, A.M., Nau, H., Berthold, H.K., et al., 2002. 5,10-Methylenetetrahydrofolate reductase genotype determines the plasma homocysteine-lowering effect of supplementation with 5-methyltetrahydrofolate or folic acid in healthy young women. Am. J. Clin. Nutr. 75, 275–282.

Food and Nutrition Board, Institute of Medicine, 1998. Dietary Reference Intakes for Thiamin, Riboflavin, Niacin, Vitamin B6, Folate, Vitamin B12, Pantothenic Acid, Biotin, and Choline. National Academy Press, Washington, DC, pp. 196–305.

Gailani, S.D., Carey, R.W., Holland, J.E., O'Malley, J.A., 1970. Studies of folate deficiency in patients with neoplastic diseases. Cancer Res. 30, 327–333.

Gifford, A.J., Kavallaris, M., Madafiglio, J., Matherly, L.H., Stewart, B.W., Haber, M., et al., 1998. P-glycoprotein-mediated methotrexate resistance in CCRF-CEM sublines deficient in methotrexate accumulation due to a point mutation in the reduced folate carrier gene. Int. J. Cancer 78, 176–181.

Henderson, G.I., Perez, T., Schenker, S., Mackins, J., Antony, A.C., 1995. Maternal-to-fetal transfer of 5-methyltetrahydrofolate by the perfused human placental cotyledon: evidence for a concentrative role by placental folate receptors in fetal folate delivery. J. Lab. Clin. Med. 126, 184–203.

Herbert, V., 1968. Nutritional requirements for vitamin B12 and folic acid. Am. J. Clin. Nutr. 21, 743–752.

Hol, F.A., van der Put, N.M., Geurds, M.E., Heil, S.G., Trijbels, F.J., Hamel, B.C., et al., 1998. Molecular genetic analysis of the gene encoding the trifunctional enzyme MTHFD (methylenetetrahydrofolate-dehydrogenase, methenyltetrahydrofolate cyclohydrolase, formyltetrahydrofolate synthetase) in patients with neural tube defects. Clin. Genet. 53, 119–125.

Hoppner, K., Lampi, B., 1980. Folate levels in human liver from autopsies in Canada. Am. J. Clin. Nutr. 33, 862–864.

Jablonski, N.G., 1999. A possible link between neural tube defects and ultraviolet light exposure. Med. Hypotheses 52, 581–582.

Kapusta, L., Haagmans, M.L., Steegers, E.A., Cuypers, M.H., Blom, H.L., Eskes, T.K., 1999. Congenital heart defects and maternal derangement of homocysteine metabolism. J. Pediatr. 135, 773–774.

Kownacki-Brown, P.A., Wang, C., Bailey, L.B., Toth, J.P., Gregory III, J.F., 1993. Urinary excretion of deuterium-labeled folate and the metabolite p-aminobenzoylglutamate in humans. J. Nutr. 123, 1101–1108.

Le Blanc, S., Garrick, M.D., Arredondo, M., 2012. Heme carrier protein 1 transports heme and is involved in heme-Fe metabolism. Am. J. Physiol. Cell Physiol. 302, C1780–C1785.

Martinelli, M., Scapoli, L., Pezzetti, E., Carinci, E., Carinci, P., Stabellini, G., et al., 2001. C677T variant form at the MTHFR gene and CL/P: a risk factor for mothers? Am. J. Med. Genet. 98, 357–360.

Maziarz, K.M., Monaco, H.L., Shen, F., Ratnam, M., 1999. Complete mapping of divergent amino acids responsible for differential ligand binding of folate receptors alpha and beta. J. Biol. Chem. 274, 11086–11091.

McPartlin, J., Halligan, A., Scott, J.M., Darling, M., Weir, D.G., 1993. Accelerated folate breakdown in pregnancy. Lancet 341, 148–149.

National Research Council, 1989. Folate Recommended Dietary Allowances. National Academy Press, Washington, DC, pp. 150–158.

Rosenberg, N., Murata, M., Ikeda, Y., Opare-Sem, O., Zivelin, A., Geffen, E., et al., 2002. The frequent 5,10-methylenetetrahydrofolate reductase C677T polymorphism is associated with a common haplotype in whites, Japanese, and Africans. Am. J. Hum. Genet. 70, 758–762.

Sancar, A., 2000. Cryptochrome: the second photoactive pigment in the eye and its role in circadian photoreception. Ann. Rev. Biochem. 69, 31–67.

Seyoum, E., Selhub, J., 1998. Properties of food folates determined by stability and susceptibility to intestinal pteroylpolyglutamate hydrolase action. J. Nutr. 128, 1956–1960.

Sikka, P.K., McMartin, K.E., 1998. Determination of folate transport pathways in cultured rat proximal tubule cells. Chem. Biol. Interact. 114, 15–31.

Suh, J.R., Herbig, A.K., Stover, P.J., 2001. New perspectives on folate catabolism. Annu. Rev. Nutr. 21, 255–282.

Tan, K.W., Li, Y., Paxton, J.W., Birch, N.P., Scheepens, A., 2013. Identification of novel dietary phytochemicals inhibiting the efflux transporter breast cancer resistance protein (BCRP/ABCG2). Food Chem. 138, 2267–2274.

Tiffany, C.W., Lapidus, R.G., Metion, A., Calvin, D.C., Slusher, B.S., 1999. Characterization of the enzymatic activity of PSM: comparison with brain NAALADase. Prostate 39, 28–35.

Wang, Y., Zhao, R., Russell, R.G., Goldman, I.D., 2001. Localization of the murine reduced folate cartier as assessed by immunohistochemical analysis. Biochim. Biophys. Acta 1513, 49–54.

Wani, N.A., Thakur, S., Kaur, J., 2012. Mechanism of intestinal folate transport during folate deficiency in rodent model. Indian J. Med. Res. 136, 758–765.

VITAMIN B12

The water-soluble compounds cyanocobalamin, aquacobalamin (Figure 10.33), 5-deoxyadenosylco balamin, and methylcobalamin are able to satisfy human requirements for vitamin B12 (Figure 10.34).

ABBREVIATIONS

B12	vitamin B12
LRP2	LDL receptor-related protein 2
MTR	5-methyltetrahydrofolate-homocysteine *S*-methyltransferase
RAP	receptor-associated protein
SAM	*S*-adenosyl methionine

NUTRITIONAL SUMMARY

Function: Vitamin B12 (B12) is an essential cofactor for only three enzymes, but they have critical impact on the metabolism of amino acids, fatty acids, phospholipids, hormones, and numerous other compounds.

Food sources: Rich sources include clams, crabs, liver, beef, and lamb. Pork, milk, dairy foods, and eggs contain somewhat less.

Requirements: Current intake recommendations are 2.4 μg/day for adults, and slightly more for women during pregnancy (2.6 μg/day) and lactation (2.8 μg/day).

Deficiency: Low intake may cause decreased production and abnormal constitution of blood cells (red, white, and thrombocytes), irreversible neurological damage with tingling and numbness of the lower limbs as well as loss of vibratory and position sense, progressive memory loss, and dementia. Infertility and recurrent fetal loss recently have been attributed to B12 deficiency in some cases (Bennett, 2001).

Excessive intake: There is no indication that high intake causes harm.

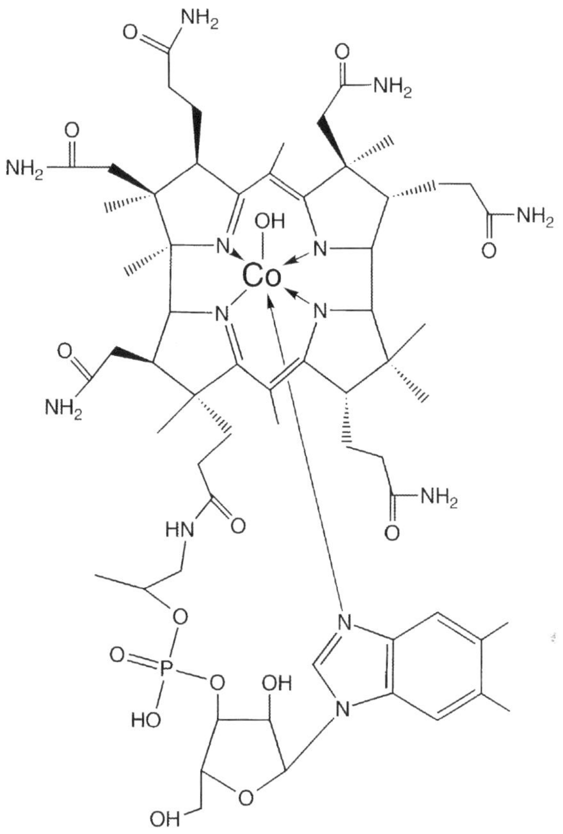

FIGURE 10.33

Aquacobalamin (hydroxycobalamin).

Dietary and other sources

Only bacteria are thought to produce B12. The synthesis is more complex than that of most other small molecules in nature and involves 30 or more steps (Raux et al., 2000). Foods contain mainly 5-deoxy-adenosylcobalamin and methylcobalamin; the latter is relatively heat resistant; photolysis can convert these forms into aquacobalamin or (in the presence of cyanide) cyanocobalamin. A high concentration of ascorbic acid in foods can degrade B12 in foods.

B12-rich foods are liver, meat and fish, eggs, and milk. Plant-derived foods do not contain significant amounts. Certain algae and unicellular organisms contain significant amounts of biologically active B12 (Miyamoto et al., 2001). Typical daily intake of nonvegan adults in the United States is between 4 and 5 pg/day (Food and Nutrition Board, Institute of Medicine, 1998). People who avoid animal-derived foods are most likely to have very low B12 intake. In particular, this includes anyone who avoids clams, crab, liver, beef, or lamb or eats fewer than three servings a day of pork, milk, dairy, egg, or sausage.

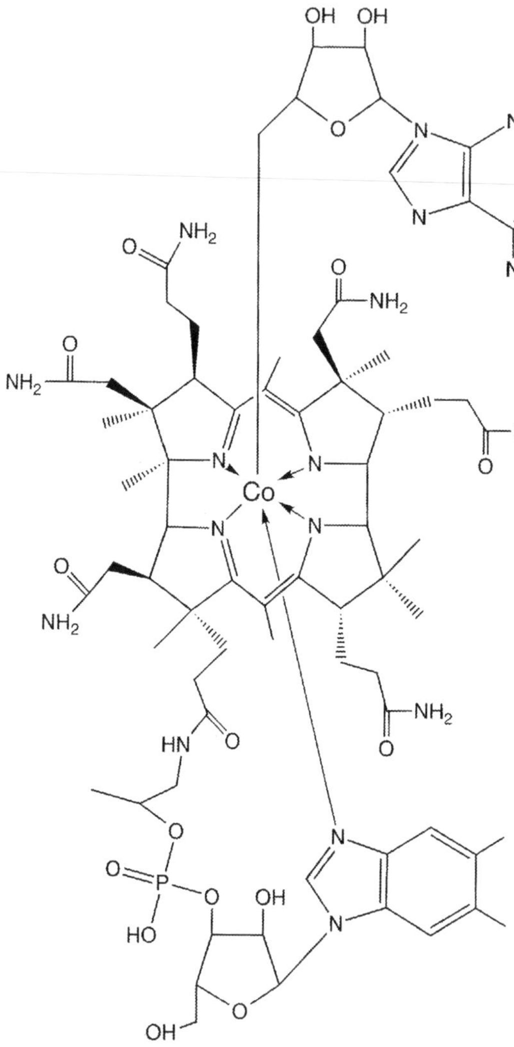

FIGURE 10.34

5'-Deoxyadenosylcobalamin.

DIGESTION AND ABSORPTION

Healthy people absorb about 50% of ingested food B12 in the distal small intestine. The absorption process is more complex than that of most other nutrients (Figure 10.35). Intestinal B12 absorption involves proteolytic release from food proteins, binding to a series of specific carrier proteins, eventually to intrinsic factor, receptor-mediated endocytosis, transport out of lysosomal vesicles, reassociation with another carrier protein, and transport out of the enterocyte into circulation. A much smaller

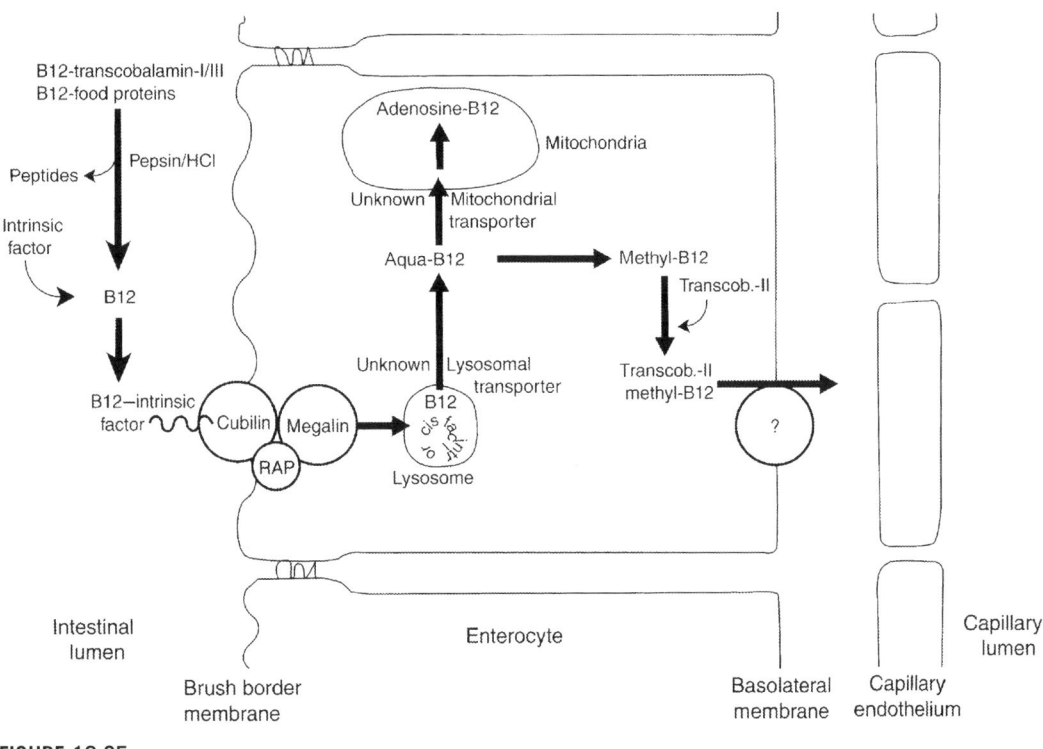

FIGURE 10.35

Intestinal absorption of vitamin B12.

percentage of ingested B12 (1% or less) can be absorbed even in the absence of intrinsic factor (Andres et al., 2001). The mechanism responsible for such intrinsic factor–independent uptake is not fully understood.

Digestion and endogenous B12-binders: Free B12 binds to transcobalamins I and III (R-binders, haptocorrins) from saliva and gastric secretions (Russell-Jones and Alpers, 1999). These B12-binding proteins have an amino acid sequence similar to that of intrinsic factor but differ in carbohydrate content. Their physiological significance remains unclear since a lack does not appear to affect B12 status noticeably.

Digestion by pepsin in the acid environment of the stomach releases B12 from dietary proteins or endogenous B12-binding proteins. The alkaline milieu of the small intestine favors binding of B12 to intrinsic factor, which is secreted by parietal cells of the stomach together with hydrochloric acid. Normally, about 2–4 µg B12 can be loaded onto the amount of intrinsic factor secreted per meal. Histamine H2-receptor antagonists do not significantly affect intrinsic factor secretion (Kittang et al., 1985). However, suppression of acid (and pepsin) output by antacids limits release of B12 from food proteins and haptocorrins and thereby impairs absorption of food B12, but not of free B12 from supplements (Force and Nahata, 1992; Carmel, 1997). The Schilling test, which uses free B12 to assess absorption, will not detect a failure to release B12 from protein binders. Transcobalamins I and III are also the main ligands of B12 secreted with bile. Trypsin cleaves these binders and thereby releases B12.

Uptake into ileal enterocytes: The complex of intrinsic factor and B12 binds with high affinity to a receptor on the apical surface of enterocytes in the distal ileum. This intrinsic factor receptor (cubilin) is a giant glycoprotein (460 kDa) that belongs to the LDL-receptor protein (LRP) family and is concentrated in coated pits of the apical membrane (Christensen and Birn, 2001). Binding of the intrinsic factor/B12 complex to cubilin is calcium-dependent (Kozyraki et al., 1998). Cubilin is also a high-affinity receptor for the HDL constituent apolipoprotein AI (apoAI; Kozyraki et al., 1999).

Transcellular transport: Trafficking of cubilin into lysosomes is facilitated when it is associated with an even larger member of the same giant receptor family, megalin (LDL receptor-related protein 2, LRP2). Both cubilin and megalin bind to the receptor-associated protein (RAP), a smaller endoplasmic reticulum protein that appears to function as a chaperone. Megalin mediates the merging of its endocytotic vesicle with lysosome. Intrinsic factor is degraded by lysosomal proteases, and the B12 component is transported into cytosol by a specific, though as-yet-uncharacterized, lysosomal transporter. Megalin (and possibly cubilin) is recycled to the apical membrane via dense apical tubules (Christensen and Birn, 2001).

Most of the B12 taken up by the enterocyte is converted to methylcobalamine or adenosylcobalamin in mitochondria, as described next. Intestinal lysosomes contain aquacobalamin as the main form (Toyoshima and Grasbeck, 1987). When and how this form is generated remains to be determined. Eventually, B12 (mainly methylcobalamin) attaches to transcobalamin-II and crosses the basolateral membrane by an unknown mechanism.

Transcobalamin-II-mediated transport: Megalin is the specific receptor for transcobalamin-II/B12 complexes. This receptor is located predominantly on the basolateral side; much smaller amounts are present at the apical side of the ileal enterocyte (Bose et al., 1997). This giant transporter has three distinct functions. The first function, facilitating of cubilin endocytosis and trafficking to lysosomes, has been mentioned previously. A second proposed function might be apical to basolateral transcytosis of the B12/transcobalamin-II complex. The third function is the endocytotic uptake of B12/transcobalamin-II complexes across the basolateral membrane and steering them toward lysosomes. This pathway provides B12 for the enterocyte's own uses.

TRANSPORT AND CELLULAR UPTAKE

Blood circulation: Most of the B12 in the blood is methyl cobalamin; the same is true for other fluids, including milk. Here, 5′-deoxyadenosylcobalamin and aquacobalamin account for much smaller percentages: minor amounts of other, inactive, corrins also may be present. B12 in the blood is bound either to transcobalamin or to haptocorrin (von Castel-Roberts et al., 2007). Most of the physiological forms of B12, as well as inactive corrinoids, are complexed to haptocorrin. A much smaller amount, usually less than 20% is bound to transcobalamins. Only B12 bound to transcobalamins can be taken up by tissues. Transcobalamin-II carries a much greater percentage than transcobalamin-I or transcobalamin-III. One very common polymorphic variant of transcobalamin-II (259Arg) is less effective in delivering B12 to tissues than the reference form 259Pro (Namour et al., 2001).

Tissues can take up the transcobalamin-II/B12 complex from circulating blood via its specific receptor (cubilin), which is ubiquitously expressed. The mechanism of cubilin-mediated endocytosis is the same as in enterocytes. The transcobalamins I or III, in contrast, are cleared only in the liver after binding to the asialoglycoprotein receptor. It has been suggested that this constitutes a protective mechanism for the selective excretion of potentially toxic corrins, since the transcobalamins I or III preferentially bind nonphysiological B12 analogs.

BBB: Cerebral microvessels contain megalin, suggesting its role in mediating the transfer of the B12/transcobalamin-II complex into cerebrospinal fluid (Zlokovic et al., 1996).

Materno-fetal transfer: The B12/transcobalamin-II complex is taken up by the syntrophoblast via megalin; the complex is the predominant form of maternally derived B12 in cytosol (Perez-D'Gregorio and Miller, 1998). The mode of B12 exit into fetal circulation is less clear.

METABOLISM

B12 enters the cell fully oxidized to cob(III)alamin. Before newly absorbed B12 can attach to an enzyme, it must be reduced to cob(I)alamin. An as-yet-uncharacterized enzyme is thought to reduce cob(III)alamin to cob(II)alamin. Methionine synthase reductase (EC2.1.1.135, contains FMN and FAD) might be the enzyme that reduces cob(II)alamin in a separate reaction to the metabolically active form, cob(I)alamin. The most common inborn error of B12 metabolism, cbl C, appears to interfere with this step. A thiolatocobalamin, possibly glutathionyl-cobalamin, may be an intermediate of cobalamin prosthetic group synthesis (Pezacka et al., 1990).

The other biologically active form of B12 is 5′-deoxyadenosyl-cobalamin. Prior to its synthesis in mitochondria, it has to cross the mitochondrial membrane by an unknown mechanism. Depending on the form that is transported, it also has to be reduced to cob(I)alamin in two NADH-dependent steps. A series of flavoproteins in conjunction with FAD and FMN was shown to mediate these reductions in a bacterial model system (Fonseca and Escalante-Semerena, 2001). The corresponding human enzymes have not yet been identified. Aquacob(I)alamin adenosyltransferase (EC2.5.1.17, probably magnesium-dependent) transfers the adenosyl group from ATP to B12, releasing phosphate and pyrophosphate (Fonseca and Escalante-Semerena, 2001). Since this enzyme has been characterized only in microorganisms so far, considerable uncertainties about its characteristics remain. Here, 5′-Deoxyadenosylcobalamin appears to be preferentially retained in mitochondria, but the exact mechanism is not well understood, nor is it known how B12 is released again. Just as the import of B12 into mitochondria, the mechanism of its export remains uncertain (Figure 10.36).

STORAGE

The main storage form of B12 is 5′-deoxyadenosylcobalamin in mitochondria. The liver contains about 2–5 mg B12 in people with adequate B12 status. About 0.1% of the body's B12 stores are turned over per day. There is a lack of information on the processes that govern mobilization of stored 5′-deoxyadenosylcobalamin.

Exposure to nitrous oxide (during anesthesia or as substance abuse) can deplete B12 stores in people with mild deficiency (Hathout and El-Saden, 2011; Cheng et al., 2013). However, not all studies found a significant effect of nitrous oxide anesthesia on B12 stores (Deleu et al., 2000).

EXCRETION

Bile: Significant amounts of cobalamin and related compounds are secreted in bile. Since cobalamin is absorbed from the distal small intestine, extensive cycling between intestinal absorption and hepatic secretion takes place. The enterohepatic cycling effectively removes noncobalamin

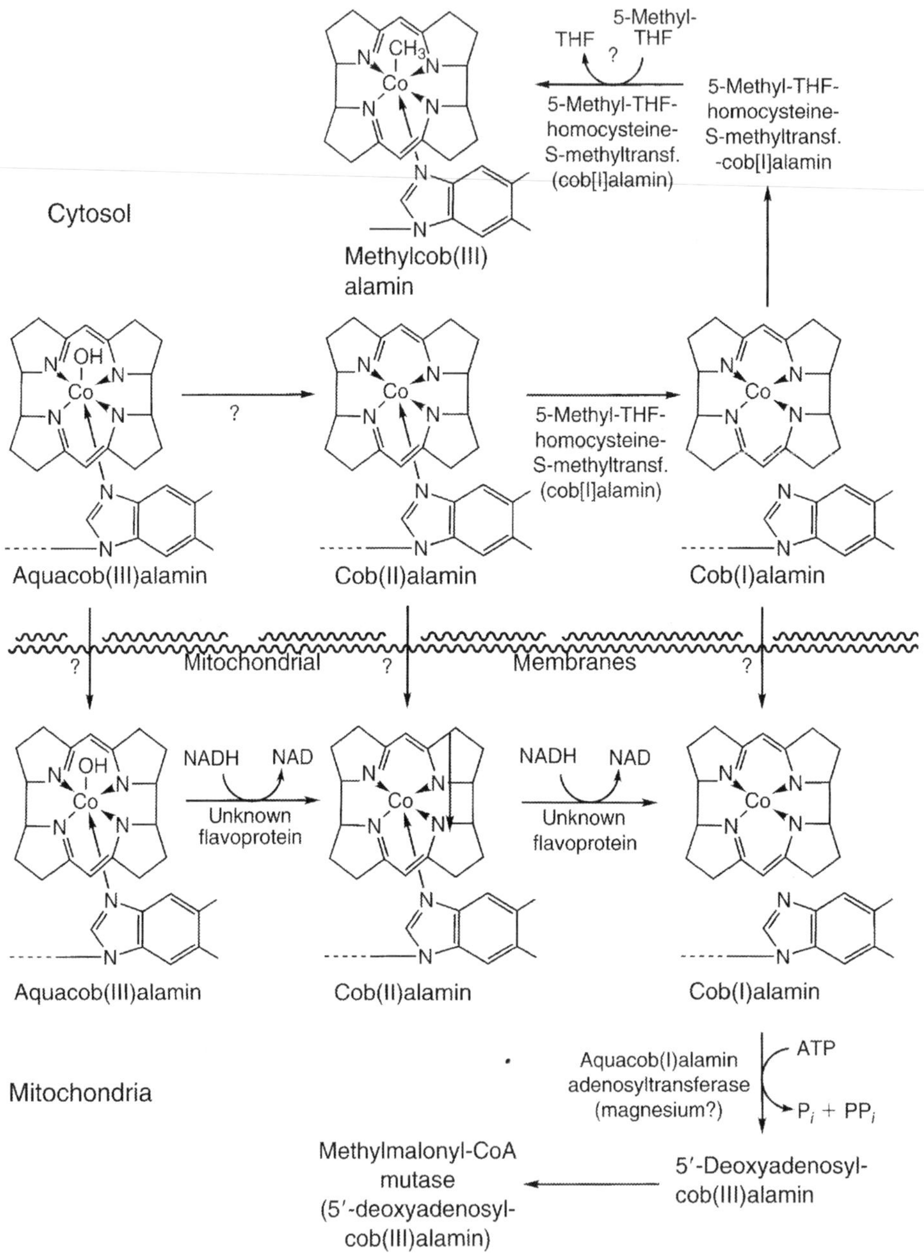

FIGURE 10.36

Metabolic activation of vitamin B12. The question marks indicate that the molecular nature responsible for the activity is still uncertain.

corrinoids from the body because they are not reabsorbed well, while cobalamin is recovered with very high efficiency.

Kidneys: The transcobalamin-II-B12 complex, which is slightly over 50 kD in size, is filtered by the renal glomeruli. In the proximal tubules, the complex is then taken up by the amply expressed endocytic receptor megalin (Nielsen et al., 2001). An alternative mechanism for B12 recovery uses intrinsic factor, which is produced locally in the kidney and secreted by the proximal tubuli. Any free B12 in the primary filtrate can thus be complexed and taken up via cubilin, the intrinsic factor/B12 receptor. Megalin and cubilin then cooperatively target the complexes toward lysosomes. Thus, the mechanism of cellular uptake and transport across the basolateral membrane is the same as in the distal ileum. Under most normal circumstances, the capacity for reabsorption is not exceeded by the amount entering the proximal tubules, and very little, if any, B12 is lost into urine.

REGULATION

Growth hormone increases secretion of salivary haptocorrin and of gastric intrinsic factor (Lobie and Waters, 1997).

Reabsorption of B12 from proximal renal tubules plays an important role in the regulation of B12 homeostasis, but the involved mechanisms are not yet fully understood (Nielsen et al., 2001).

FUNCTION

Homocysteine remethylation: SAM is the main methyl-group donor for methylations and other one-carbon transfer reactions. The product of this reaction, *S*-adenosylhomocysteine, can be remethylated to the SAM precursor methionine by 5-methyltetrahydrofolate-homocysteine *S*-methyltransferase (MTR, EC2.1.1.13). The reaction is initiated by the transfer of a methyl group from 5-methyltetrahydrofolate to the prosthetic group of the enzyme, cob(I)alamin. The methyl group linked to cob(III)alamin is then transferred to *S*-adenosylhomocysteine.

The cob(I)alamin prosthetic group of MTR is easily oxidized, and a small percentage is converted nonenzymically into the nonreactive cob(II)alamin form. A FAD- and FMN-dependent enzyme, methionine synthase reductase (EC2.1.1.135), can revert the oxidized form to its active cob(I)alamin form by reductive methylation (Wolthers and Scrutton, 2009). SAM serves as the methyl-group donor and cytochrome b5 provides the reducing potential. Cytochrome b5 is regenerated by NADPH-dependent cytochrome P450 reductase (EC1.6.2.4), another enzyme with both FAD and FMN as prosthetic groups (Chen and Banerjee, 1998). Inhalation of nitrous oxide irreversibly inactivates MTR (Horne et al., 1989; Riedel et al., 1999).

It has been suggested that infertility and recurrent fetal loss is due to poor B12 status in some cases. The underlying cause might be hypercoagulation related to elevated homocystein levels (Bennett, 2001).

Propionyl-CoA metabolism: As a result of the breakdown of L-methionine, L-valine, L-threonine, the cholesterol side-chain, and of fatty acids with an odd number of carbons, propionyl-CoA is generated. The final step of propionyl-CoA conversion to succinyl-CoA is facilitated by methylmalonyl-CoA mutase (EC5.4.99.2). This enzyme contains 5'-deoxyadenosylcobalamin as a prosthetic group.

Leucine metabolism: It has been reported that a small percentage of L-leucine is metabolized via an alternative beta-keto acid pathway distinct from the prevalent of alpha-keto pathway (Poston, 1984;

Ward et al., 1988). Another investigator did not detect such activity (Aberhart, 1988). The first step is catalyzed by L-beta-leucine aminomutase (EC5.4.3.7), which requires 5′-deoxyadenosylcobalamin. Subsequent reactions then convert beta-leucine to beta-ketoisocaproic acid, ligate this intermediate with CoA, and finally generate the valine metabolite isobutyryl-CoA.

Since all of these reactions are reversible, the importance of this pathway may be as much L-leucine synthesis as catabolism. Significant activity is present in testes, where about a third of L-leucine metabolism proceeds via this pathway. In all other investigated tissues, the beta-keto pathway accounts for less than 5% of L-leucine metabolism.

Cyanide antidote: Aquacobalamin readily binds free cyanide and thus is an effective and relatively safe antidote against cyanide poisoning (Sauer and Keim, 2001).

NUTRITIONAL ASSESSMENT

B12 deficiency and suboptimal status may be detected with the direct measurement of B12 in plasma or by detecting increased amounts of unmetabolized substrates in the blood or urine.

Blood measurement: Most commonly, a competitive binding assay is used that exploits the high affinity of active B12 metabolites to intrinsic factor. The advantage is that intrinsic factor does not bind structurally similar non-B12 corrinoids. Predictive power can be increased further by also measuring B12 bound to transcobalamin II (holo-transcobalamin) because only this form is available to tissues (von Castel-Roberts et al., 2007). Alternatively, methylmalonate concentration can be measured, which increases with inadequate B12 status because it cannot be metabolized effectively without adequate B12 cofactor for methylmalonyl-CoA mutase (EC5.4.99.2). The previously used measurement of homocysteine concentration, which depends on B12-dependent 5-methyltetrahydrofolate-homocysteine *S*-methyltransferase (MTR; EC2.1.1.13) is less specific because it will also be increased by inadequate folate status.

Urine measurement: The amount of methylmalonate excreted with a urine sample collected over a 24-h period reflects B12 status. Alternatively, the ratio of methylmalonate to creatinine can be used.

REFERENCES

Aberhart, D.J., 1988. Separation by high-performance liquid chromatography of alpha- and beta-amino acids: application to assays of lysine 2,3-aminomutase and leucine 2,3-aminomutase. Anal. Biochem. 169, 350–355.

Andres, E., Kurtz, J.E., Perrin, A.E., Maloisel, F., Demangeat, C., Goichot, B., et al., 2001. Oral cobalamin therapy for the treatment of patients with food-cobalamin malabsorption. Am. J. Med. 111, 126–129.

Bennett, M., 2001. Vitamin B12 deficiency, infertility and recurrent fetal loss. J. Reprod. Med. 46, 209–212.

Bose, S., Seetharam, S., Dahms, N.M., Seetharam, B., 1997. Bipolar functional expression of transcobalamin II receptor in human intestinal epithelial Caco-2 cells. J. Biol. Chem. 272, 3538–3543.

Carmel, R., 1997. Cobalamin, the stomach, and aging. Am. J. Clin. Nutr. 66, 750–759.

Chen, Z., Banerjee, R., 1998. Purification of soluble cytochrome b5 as a component of the reducrive activation of porcine methionine synthase. J. Biol. Chem. 273, 26248–26255.

Cheng, H.M., Park, J.H., Hernstadt, D., 2013. Subacute combined degeneration of the spinal cord following recreational nitrous oxide use. BMJ Case Rep. 2013.

Christensen, E.I., Birn, H., 2001. Megalin and cubilin: synergistic endocytic receptors in renal proximal tubule. Am. J. Physiol. Renal. Physiol. 280, F562–F573.

Deleu, D., Louon, A., Sivagnanam, S., Sundaram, K., Okereke, P., Gravell, D., et al., 2000. Long-term effects of nitrous oxide anesthesia on laboratory and clinical parameters in elderly Omani patients: a randomized double-blind study. J. Clin. Pharm. Ther. 25, 271–277.

Fonseca, M.V., Escalante-Semerena, J.C., 2001. An *in vitro* reducing system for the enzymic conversion of cobalamine to adenosylcobalamin. J. Biol. Chem. 276, 32101–32108.

Food and Nutrition Board, Institute of Medicine, 1998. Dietary Reference Intakes for Thiamin, Riboflavin, Niacin, Vitamin B6, Folate, Vitamin B12, Pantothenic Acid, Biotin, and Choline. National Academy Press, Washington, DC, pp. 306–356.

Force, R.W., Nahata, M.C., 1992. Effect of histamine H2-receptor antagonists on vitamin B12 absorption. Ann. Pharmacother. 26, 1283–1286.

Hathout, L., El-Saden, S., 2011. Nitrous oxide-induced B_{12} deficiency myelopathy: perspectives on the clinical biochemistry of vitamin B_{12}. J. Neurol. Sci. 301, 1–8.

Horne, D.W., Patterson, D., Cook, R.J., 1989. Effect of nitrous oxide inactivation of vitamin B12-dependent methionine synthetase on the subcellular distribution of folate coenzymes in rat liver. Arch. Biochem. Biophys. 270, 729–733.

Kittang, E., Aadland, E., Schjonsby, H., 1985. Effect of omeprazole on the secretion of intrinsic factor, gastric acid and pepsin in man. Gut 26, 594–598.

Kozyraki, R., Fyfe, J., Kristiansen, M., Gerdes, C., Jacobsen, C., Cui, S., et al., 1999. The intrinsic factor-vitamin B12 receptor, cubilin, is a high-affinity apolipoprotein A-I receptor facilitating endocytosis of high-density lipoprotein. Nat. Med. 5, 656–661.

Kozyraki, R., Kristiansen, M., Silahtaroglu, A., Hansen, C., Jacobsen, C., Tommerup, N., et al., 1998. The human intrinsic factor-vitamin B12 receptor, cubilin: molecular characterization and chromosomal mapping of the gene to 10p within the autosomal recessive megaloblastic anemia (MGA1) region. Blood 91, 3593–3600.

Lobie, P.E., Waters, M.J., 1997. Growth hormone (GH) regulation of submandibular gland structure and function in the GH-deficient rat: upregulation of haptocorrin. J. Endocrinol. 154, 459–466.

Miyamoto, E., Watanabe, E., Ebara, S., Takenaka, S., Takenaka, H., Yamaguchi, Y., et al., 2001. Characterization of a vitamin B12 compound from unicellular coccolithophorid alga *(Pleurochrysis carterae)*. J. Agric. Food Chem. 49, 3486–3489.

Namour, E., Olivier, J., Abdelmouttaleb, I., Adjalla, C., Debard, R., Salvat, C., et al., 2001. Transcobalamin codon 259 polymorphism in HT-29 and Caco-2 cells and in Caucasians: relation to transcobalamin and homocysteine concentration in blood. Blood 97, 1092–1098.

Nielsen, R., Sorensen, B.S., Birn, H., Christensen, E.I., Nexo, E., 2001. Transcellular transport of vitamin B(12) in LLC-PKI renal proximal tubule cells. J. Am. Soc. Nephrol. 12, 1099–1106.

Perez-D'Gregorio, R.E., Miller, R.K., 1998. Transport and endogenous release of vitamin B12 in the dually perfused human placenta. J. Pediatr. 132, S35–S42.

Pezacka, E., Green, R., Jacobsen, D.W., 1990. Glutathionylcobalamin as an intermediate in the formation of cobalamin coenzymes. Biochem. Biophys. Res. Commun. 169, 443–450.

Poston, J.M., 1984. The relative carbon flux through the alpha- and the beta-keto pathways of leucine metabolism. J. Biol. Chem. 259, 2059–2061.

Raux, E., Schubert, H.L., Warren, M.J., 2000. Biosynthesis of cobalamin (vitamin B12): a bacterial conundrum. Cell. Mol. Life Sci. 57, 1880–1893.

Riedel, B., Fiskerstrand, T., Refsum, H., Ueland, P.M., 1999. Co-ordinate variations in methylmalonyl-CoA mutase and methionine synthase, and the cobalamin cofactors in human glioma cells during nitrous oxide exposure and the subsequent recovery phase. Biochem. J. 341, 133–138.

Russell-Jones, G.J., Alpers, D.H., 1999. Vitamin B12 transporters. Pharmaceut. Biotechnol. 12, 493–520.

Sauer, S.W., Keim, M.E., 2001. Hydroxocobalamin: improved public health readiness for cyanide disasters. Ann. Emerg. Med. 37, 635–641.

Toyoshima, M., Grasbeck, R., 1987. Cobalamin derivatives in subcellular fractions of porcine ileal enterocytes. Scand. J. Clin. Lab. Invest. 47, 277–284.

von Castel-Roberts, K.M., Morkbak, A.L., Nexo, E., Edgemon, C.A., Maneval, D.R., Shuster, J.J., et al., 2007. Holo-transcobalamin is an indicator of vitamin B-12 absorption in healthy adults with adequate vitamin B-12 status. Am. J. Clin. Nutr. 85, 1057–1061.

Ward, N.E., Jones, J., Maurice, D.V., 1988. Essential role of adenosylcobalamin in leucine synthesis from beta-leucine in the domestic chicken. J. Nutr. 118, 159–164.

Wolthers, K.R., Scrutton, N.S., 2009. Cobalamin uptake and reactivation occurs through specific protein interactions in the methionine synthase-methionine synthase reductase complex. FEBS J. 276, 1942–1951.

Zlokovic, B.V., Martel, C.L., Matsubara, E., McComb, J.G., Zheng, G., McCluskey, R.T., et al., 1996. Glycoprotein 330/megalin: probable role in receptor-mediated transport of apolipoprotein J alone and in a complex with Alzheimer disease amyloid beta at the blood–brain and blood–cerebrospinal fluid barriers. Proc. Natl. Acad. Sci. USA 93, 4229–4234.

BIOTIN

Biotin (hexahydro-2-oxo-1H-thieno[3,4-d]imidazole-4-pentanoic acid; obsolete names vitamin H, coenzyme R, vitamin B7; molecular weight 244) is a water-soluble vitamin (Figure 10.37).

ABBREVIATIONS	
CoA	coenzyme A
SMVT	sodium-dependent multivitamin transporter

NUTRITIONAL SUMMARY

Function: Biotin is essential for lipid metabolism, amino acid breakdown, replenishing of tricarboxylic acid cycle intermediates, and some nuclear functions.

Requirements: Adequate daily intake for infants is 0.7 μg/kg. Children need 8–20 μg/day, depending on their age. Adults should get 30 μg/day. Lactation increases this amount to 35 μg/day for women. Hemodialysis increases requirements.

Sources: Intestinal bacteria probably provide enough to satisfy requirements under most circumstances. Soybeans, liver, cauliflower, mushrooms, bean sprouts, and eggs each provide at least 20% of adult needs per serving.

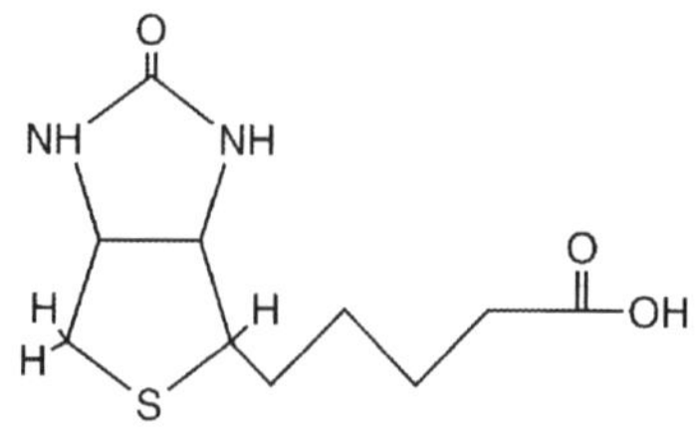

FIGURE 10.37

Biotin.

Deficiency: The consequences of diminished biotin status include impaired glucose tolerance, mental dysfunction (Bregola et al., 1996), myalgia, hyperesthesia and paresthesia, anorexia, nausea, dry eyes, maculo-squamous, seborrheic dermatitis (similar to the changes seen with essential fatty acid deficiency), angular cheilitis, hair loss (which may be complete in patients with short-bowel syndrome), impaired immune response, and possibly teratogenicity. Biotin deficiency may be aggravated by pantothenic acid deficiency.

Excessive intake: No toxicity has been observed following consumption of 20 mg/day.

DIETARY AND OTHER SOURCES

Foods contain both free and bound biotin (epsilon-*N*-biotinyl-L-lysine); metabolites such as bisnorbiotin and biotin sulfoxide have no biotin activity. Good sources include soybeans (0.6 µg/g), liver (1.0 µg/g), cauliflower and mushrooms (0.17 µg/g), egg yolk (8.5 µg/yolk), legumes (especially sprouts), grains, and nuts. Milk contains only 3 5 µg/l; most vegetables, fruits, and meats are poor sources. Intakes may be between 28 and 100 µg/day (Dakshinamurti, 1994).

A wide variety of intestinal flora probably produces much more biotin than is normally consumed with food, but only a small fraction of this appears to contribute to the body's supplies.

DIGESTION AND ABSORPTION

Biotin-containing food proteins are broken down by the usual digestive proteases; biotin is then released by biotinidase (EC3.5.1.12) from the resulting lysylbiotin (biocytin) or lysylbiotin-containing peptides. A high-affinity, low-capacity sodium-dependent multivitamin transporter (SMVT, SLC5A6) in the small intestine mediates apical uptake of biotin (and also of pantothenate and lipoate) into enterocytes (Ghosal et al., 2013). The fact that SMVT is also expressed in the colon (Said, 1999) may explain why deficiency of this critically important vitamin is so rare. Biotin transport across the basolateral membrane is mediated by a sodium-independent, electrogenic mechanism (Said et al., 1988).

Note: Heat-labile avidin in egg white tightly binds biotin and prevents its uptake from the intestinal lumen.

TRANSPORT AND CELLULAR UPTAKE

Blood circulation: Biotin in the blood is transported, at least in part, via biotinidase (EC3.5.1.12), which acts as a biotinyl-transferase in this case (Hymes and Wolf, 1999). One or more anion acid carriers, sharing characteristics or being identical with the intestinal transporter, mediate uptake by liver cells. Inside the hepatocyte (and probably all other cells), some biotin is transported with biotinidase into the nucleus and transferred to histones.

BBB: The SMVT (SLC5A6) mediates biotin transfer across the BBB (Prasad et al., 1998).

Materno-fetal transfer: The SMVT participates in biotin transport across the brush border membrane of the syntrophoblast of the placenta, but knowledge on the involvement of additional transporters is still evolving (Prasad et al., 1998).

METABOLISM

Biotin can be linked covalently to four carboxylases (acetyl-CoA carboxylase, EC6.4.1.2; propionyl-CoA-carboxylase, EC6.4.1.3; pyruvate carboxylase, EC6.4.1.1; 3-methylcrotonyl-CoA carboxylase,

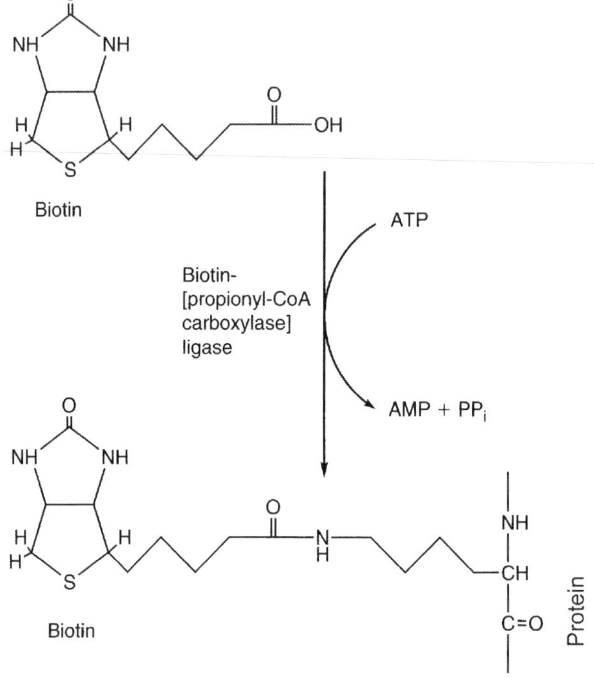

FIGURE 10.38

Biotin becomes linked to carboxylases.

EC6.4.1.4) through the action of biotin-[propionyl-CoA-carboxylase (ATP-hydrolyzing)] ligase (EC6.3.4.10). The reaction is driven by the hydrolysis of ATP to AMP (Figure 10.38).

Protein-bound biotin is recycled very efficiently. As the biotin-containing enzymes are broken down in due course by intracellular proteases, lysylbiotin or peptides containing lysylbiotin are generated. Biotinidase (EC3.5.1.12) then releases free biotin, which thus becomes available again at the site where it is needed for incorporation into proteins.

Beta-oxidation successively shortens the side chain to bisnorbiotin and tetranorbiotin. A small amount of bisnorbiotin methylketone arises from the nonenzymic decarboxylation of beta-keto-biotin, an intermediate product of biotin beta-oxidation.

In the ring moiety (thiophane) of biotin, the thioether sulfur can be oxidized to biotin-1-sulfoxide, biotin-d-sulfoxide, or biotin sulfone; the mechanism of this oxidation is not known. The combination of side-chain catabolism and sulfur oxidation generates a wide spectrum of catabolites, none of which has significant biotin activity (Figure 10.39); some of them (such as tetranorbiotin-1-sulfoxide) have greatly decreased affinity to avidin, which can make them difficult to detect with binding assays (Zempleni et al., 1997).

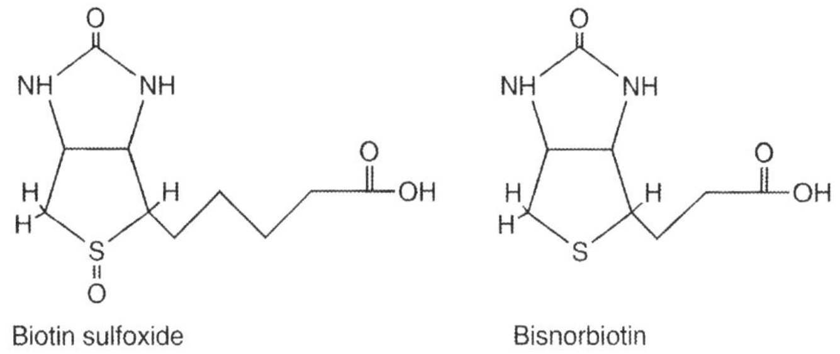

FIGURE 10.39

Inactive biotin metabolites.

STORAGE

Mitochondrial biotinyl-acetyl-CoA carboxylase (EC6.4.1.2) appears to comprise the majority of biotin reserves in a wide range of tissues (1.2 mg in the liver alone). Stores may be depleted within 5–7 weeks of intake cessation.

EXCRETION

The high-affinity SMVT (Prasad et al., 1998) recovers filtered biotin from the tubular lumen. Little is known about biotin transport across the cell and into the pericapillary space. Excess vitamin is excreted via urine as biotin, bisnorbiotin, biotin sulfoxide, bisnorbiotin methyl ketone, and biotin sulfone (3:2:1:0.4:0.1); additional intermediates of biotin catabolism are excreted in much smaller amounts (Zempleni et al., 1997).

REGULATION

Modulation of SMVT activity by protein kinase C and Ca^{2+}/calmodulin signaling adapts intestinal absorption to intake levels (Said, 1999). Similar mechanisms appear to be involved in the regulation of cellular uptake.

FUNCTION

Biotinylated carboxylases: Biotin is a prosthetic group of four carboxylases. In active enzymes, it is covalently linked to the epsilon amino group of a specific lysyl residue near the active center.

Acetyl-CoA carboxylase (EC6.4.1.2) is present in both cytosol and mitochondria, promotes elongation of fatty acids in adipose tissue, the placenta, kidney, and pancreas, but not in the brain. Alternative splicing in nonlipogenic tissues (heart and skeletal muscle, liver) produces a distinct isoform of the enzyme; it acts in conjunction with fatty acid beta-oxidation through control of carnitine palmitoyl transferase I activity. Pyruvate carboxylase (EC6.4.1.1) provides oxaloacetate (Figure 10.40), the precursor for glucose synthesis in gluconeogenic tissues (liver, kidneys). This is the only anaplerotic (refilling) reaction that can replenish TCA cycle intermediates without drawing on glutamate or other amino acids. Pyruvate carboxylase activity increases with rising acetyl-CoA concentration.

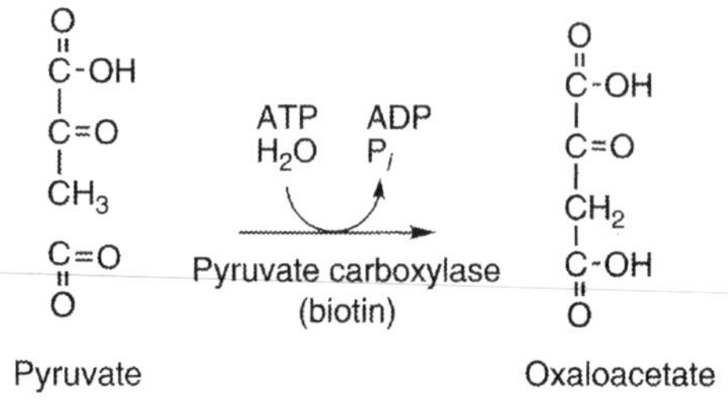

FIGURE 10.40

Biotin-dependent pyruvate carboxylase.

Propionyl-CoA-carboxylase (EC6.4.1.3) in mitochondria catalyzes one of the final steps of valine, leucine, isoleucine, methionine, threonine, and odd-chain fatty acids catabolism. The mitochondrial enzyme 3 methylcrotonyl-CoA carboxylase (EC6.4.1.4) participates in the catabolism of leucine.

Protein expression: Biotin has been found to enhance the availability for translation of messenger RNA (mRNA) for hepatic glucokinase (Romero-Navarro et al., 1999), ornithine transcarbamylase mRNA (Maeda et al., 1996), and the asialoglycoprotein receptor (Collins et al., 1988). Low biotin availability decreases the expression of biotin-[propionyl-CoA-carboxylase] (EC6.3.4.10) and limits pyruvate and propionyl CoA-carboxylase masses (Rodríguez-Meléndez et al., 2001).

One of the mechanisms that mediates biotin-dependent repression of specific genes is the interaction with DNA (cytosine-5)-methyltransferase 1 (DNMT1, EC2.1.1.37), methylated cytosine-binding protein 2 (MeCP2), and euchromatic histone-lysine *N*-methyltransferase 1 (EHMT1, EC2.1.1.43) (Xue and Zempleni, 2013).

Alternative mechanisms through which biotin might influence transcription might involve guanosine 3′,5′-cyclic monophosphate as a second messenger or the biotinidase-catalyzed biotinylation of the nuclear histones H1, H2A, H3, and H4 (Zempleni et al., 2012). Since only a tiny fraction of these histones is biotinylated (<0.1%), the functional significance of this type of protein modification remains unclear.

NUTRITIONAL ASSESSMENT

Biotin deficiency can be recognized by measuring the abundance of biotinylated carboxylases in lymphocytes. The concentrations of biotinylated 3-methylcrotonyl-CoA carboxylase and propionyl-CoA carboxylase greatly increase within a few weeks in response to a biotin-deficient diet protocol (Eng et al., 2013).

Another approach for developing a deficiency biomarker has been to measure in urine the substrates of biotin-dependent reactions, such as 3-hydroxyisovaleric acid. This compound was indeed found to be abundant in some individuals with severe biotin deficiency. In the end, however, it was not sufficiently specific to be useful (Eng et al., 2013).

REFERENCES

Bregola, G., Muzzolini, A., Mazzari, S., Leon, A., Skaper, S.D., Beani, L., et al., 1996. Biotin deficiency facilitates kindling hyperexcitability in rats. Neuroreport 7, 1745–1748.

Collins, J.C., Paietta, E., Green, R., Morell, A.G., Stockert, R.J., 1988. Biotin-dependent expression of the asialo-glycoprotein receptor in HepG2. J. Biol. Chem. 263, 11280–11283.

Dakshinamurti, K., 1994. Biotin. In: Shils, M.E., Olson, J.A., Shike, M. (Eds.), Modern Nutrition in Health and Disease Lea & Febiger, Philadelphia, PA, pp. 426–431.

Eng, W.K., Giraud, D., Schlegel, V.L., Wang, D., Lee, B.H., Zempleni, J., 2013. Identification and assessment of markers of biotin status in healthy adults. Br. J. Nutr. 110, 321–329.

Food and Nutrition Board, Institute of Medicine, 1998. Dietary Reference Intakes for Thiamin, Riboflavin, Niacin, Vitamin B6, Folate, Vitamin B12, Pantothenic Acid, Biotin, and Choline. National Academy Press, Washington, DC, pp. 374–389.

Ghosal, A., Lambrecht, N., Subramanya, S.B., Kapadia, R., Said, H.M., 2013. Conditional knockout of the Slc5a6 gene in mouse intestine impairs biotin absorption. Am. J. Physiol. Gastrointest. Liver Physiol. 304, G64–G71.

Hymes, J., Wolf, B., 1999. Human biotinidase isn't just for recycling biotin. J. Nutr. 129, 485S–489S.

Maeda, Y., Kawata, S., Inui, Y., Fukuda, K., Tgura, T., Matsuzawa, Y., 1996. Biotin deficiency decreases ornithine transcarbamylase activity and mRNA in rat liver. J. Nutr. 126, 61–66.

Prasad, P.D., Wang, H., Kekuda, R., Fujita, T., Fei, Y.J., Devoe, L.D., et al., 1998. Cloning and functional expression of a cDNA encoding a mammalian sodium-dependent vitamin transporter mediating the uptake of pantothenate, biotin, and lipoate. J. Biol. Chem. 273, 7501–7506.

Rodriguez-Meléndez, R., Cano, S., Mendez, S.T., Velazquez, A., 2001. Biotin regulates the genetic expression of holocarboxylase synthetase and mitochondriai carboxylases in rats. J. Nutr. 131, 1909–1913.

Romero-Navarro, G., Cabrera-Valladares, G., German, M.S., Matschinsky, F.M., Velazquez, A., Wang, J., et al., 1999. Biotin regulation of pancreatic glucokinase and insulin in primary cultured rat islets and in biotin-deficient rats. Endocrinology 140, 4595–4600.

Said, H.M., 1999. Cellular uptake of biotin: mechanisms and regulation. J. Nutr. 129, 490S–493S.

Said, H.M., Redha, R., Nylander, W., 1988. Biotin transport in basolateral membrane vehicles of human intestine. Gastroenterology 94, 1157–1163.

Xue, J., Zempleni, J., 2013. Epigenetic synergies between biotin and folate in the regulation of pro-inflammatory cytokines and repeats. Scand. J. Immunol. 78, 419–425.

Zempleni, J., McCormick, D.B., Mock, D.M., 1997. Identification of biotin sulfone, bisnorbiotin methylketone, and tetranorbiotin-sulfoxide in human urine. Am. J. Clin. Nutr. 65, 508–511.

Zempleni, J., Teixeira, D.C., Kuroishi, T., Cordonier, E.L., Baier, S., 2012. Biotin requirements for DNA damage prevention. Mutat. Res. 733, 58–60.

PANTOTHENATE

Pantothenate (pantothenic acid, (R)-N-(2,4-dihydroxy-3,3-dimethyl-L-oxobutyl)-beta-alanine, D(+)-N-(2,4-dihydroxy-3,3-dimethylbutyryl)-beta-alanine, dihydroxybeta, beta-dimethylbutyryl-beta-alanine, vitamin B5, chick antidermatitis factor; molecular weight 219) is a water-soluble vitamin (Figure 10.41).

ABBREVIATIONS	
CoA	coenzyme A
PANK	pantotheine kinase

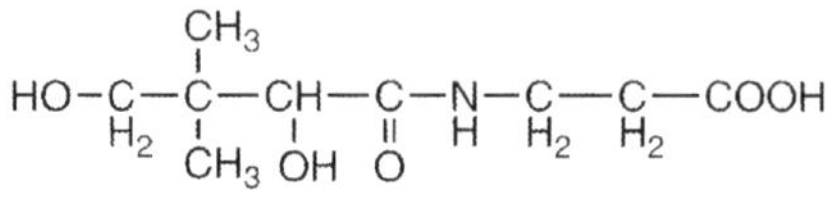

$$HO-\underset{H_2}{\underset{|}{C}}-\underset{\underset{CH_3}{|}}{\overset{\overset{CH_3}{|}}{C}}-CH-\underset{\underset{OH}{|}}{C}-\underset{\overset{||}{O}}{C}-N-\underset{H_2}{C}-\underset{H_2}{C}-COOH$$

FIGURE 10.41

Pantothenate.

NUTRITIONAL SUMMARY

Function: Pantothenate is essential for fuel metabolism (especially of fat and alcohol), brain function (neurotransmitter and hormone synthesis), growth, and regeneration (synthesis of membranes, heme, hormones, and functional proteins).

Food sources: All food groups provide at least a modest amount of pantothenate. Good sources (at least 20% of recommended intake per serving) include sweet potato, lentils, split peas, yogurt, avocado, chicken, and liver. Refining of grains and overcooking of foods reduce their pantothenate content.

Requirements: Adults should get about 5 mg/day (Food and Nutrition Board, Institute of Medicine, 1998). Smoking, strenuous exercise, heat, infections, and injuries each may increase needs by 50% or more.

Deficiency: The widespread occurrence of pantothenate in all food groups makes inadequate intake by healthy people very unlikely. The interference by a few prescription drugs (antimetabolites), used in rare specific instances, can interfere with absorption and cause problems. The few people with confirmed deficiency (due to medication) had relatively mild and reversible symptoms, which included tingling sensations of the toes and feet, fatigue, vomiting, sleeplessness, and increased susceptibility to infection.

Excessive intake: No harmful effects of doses many times above recommended levels have been observed.

DIETARY AND OTHER SOURCES

Foods contain coenzyme A (CoA) (Figure 10.42), 4-phosphopantothenate, and pantothenate. Panthenol may be consumed with dietary supplements. Good sources include liver, yeast, egg yolk, fresh vegetables, and bee royal jelly. About half of the pantothenate in meat, but not in vegetables, may be lost during cooking. Average pantothenate consumption of Americans is about 10 mg/2500 kcal. The amounts available from production by intestinal flora in the terminal ileum and colon are unknown, but are likely to be significant. This may explain why pantothenate deficiency symptoms are almost never observed, despite the critical importance of this vitamin for a broad spectrum of metabolic functions.

DIGESTION AND ABSORPTION

Small doses of crystalline pantothenate are absorbed almost completely (Shibata et al., 1983). Dietary CoA and similar compounds can be absorbed only when digestive enzymes, including alkaline phosphatase (EC3.1.3. 1) and pantetheine hydrolase (EC3.5.1.-), release their pantothenate moieties.

The sodium-dependent multivitamin carrier (SLC5A6) is responsible for uptake of pantothenate into enterocytes of the small and large intestine. Pantothenate uptake via SLC5A6 is driven by two

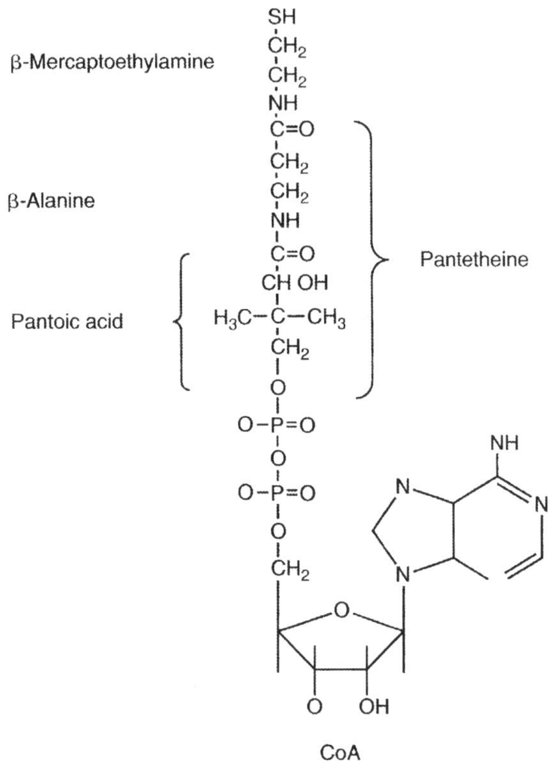

FIGURE 10.42

Coenzyme A (CoA).

sodium ions, analogous to the mechanism of the structurally related sodium-glucose cotransporter 1 (Prasad et al., 2000). Several molecular variants of this carrier have been found in the small and large intestines (Chatterjee et al., 1999). The significance of this diversity is not yet understood. The sodium-dependent multivitamin carrier also mediates uptake of biotin and lipoate. The mechanism of pantothenate transport across the basolateral membrane and onward into portal blood has not been characterized as yet.

TRANSPORT AND CELLULAR UPTAKE

Blood circulation: The main transport form in the blood is pantothenate; red blood cells contain pantothenate, pantetheine, and 4'-phosphopantothenate, but little CoA (Annous and Song, 1995). Red cell membranes are normally impermeable for pantothenate, but infection with *plasmodium falciparum* (malaria parasite) induces the opening of new permeation pathways to supply the parasite (Saliba et al., 1998). CoA does not cross cell membranes; uptake into most tissues proceeds as pantothenate followed by intracellular synthesis of CoA (see the discussion that follows). The concentration of CoA in mitochondria is much higher than in cytosol, much of this as fatty acyl-CoA; mitochondria are capable of transporting CoA from and into the cytosol (Tahiliani, 1991) using L-carnitine in the process.

Materno-fetal transfer: Increased maternal pantothenate intake raises pantothenate concentration in fetal blood (Baker et al., 1981). Transport across the placenta involves a distinct variant of the sodium-dependent multivitamin carrier (SLC5A6), but mechanistic details remain to be elucidated.

BBB: Transport into the brain does not appear to be sodium-dependent and may thus not involve SLC5A6.

METABOLISM

Synthesis: CoA synthesis uses pantothenate, cysteine, one adenylate, three phosphates, and the energy of six high-energy phosphates from ATP (Figure 10.43). Significant amounts of pantothenate are generated from pantetheine through the action of pantetheine hydrolase (EC3.5.1.-), which is expressed in many tissues. Panthenol and panthenal may also be converted to a limited extent into pantothenate by alcohol dehydrogenase (EC1.1.1.1) and aldehyde dehydrogenase (EC1.2.1.3).

The initial phosphorylation of pantothenate by pantothenate kinase (PANK, EC2.7.1.33) is the rate-limiting step of CoA synthesis. In addition to the enzyme present in most tissues, a brain-specific form (PANK2) has been identified (Zhou et al., 2001). Phosphopantothenate can then be linked to cysteine by phosphopantothenate-cysteine ligase (EC6.3.2.5) and decarboxylated by pantothenoylcysteine decarboxylase (EC4.1.1.36). An alternative pathway catalyzed by pantothenoylcysteine decarboxylase (EC4.1.1.30) and pantetheine kinase (EC2.7.1.34) exists in the liver, and possibly other tissues as well. Whether this is a salvage pathway for inappropriately dephosphorylated pantothenoylcysteine or has other significance remains uncertain.

CoA synthesis is completed either in cytosol or in mitochondria by a bifunctional CoA synthase complex that comprises both pantetheine phosphate adenylyltransferase (EC2.7.7.3) and dephospho-CoA kinase (EC2.7.1.24) activities. Different genes encode the cytosolic and mitochondrial forms of the CoA synthase complex.

Significant transport of both pantotheine 4′-phosphate and CoA (Tahiliani, 1991) into mitochondria occurs. Much more CoA is inside mitochondria than in cytosol (75–95% depending on tissue), both due to the direction of the normal electrochemical gradient and metabolic trapping as acyl-CoA.

Breakdown: CoA is hydrolyzed in multiple steps by as-yet-incompletely-characterized phosphatases and pyrophosphatases. The final step is the hydrolysis of pantetheine to pantothenate and cysteamine by pantetheine hydrolase (EC3.5.1.-). This enzyme circulates with blood, is present in mucosal membranes, and is anchored to microsomal membranes.

STORAGE

Limited amounts of pantothenate are stored, especially in red blood cells and adipose tissue. Pantothenate is mobilized again by the hydrolysis of pantetheine to pantothenate.

EXCRETION

Most pantothenate losses occur in the kidneys. Pantothenate is the main form lost via urine. Significant amounts of CoA and other pantothenate-related compounds are filtered in the kidneys and have to be

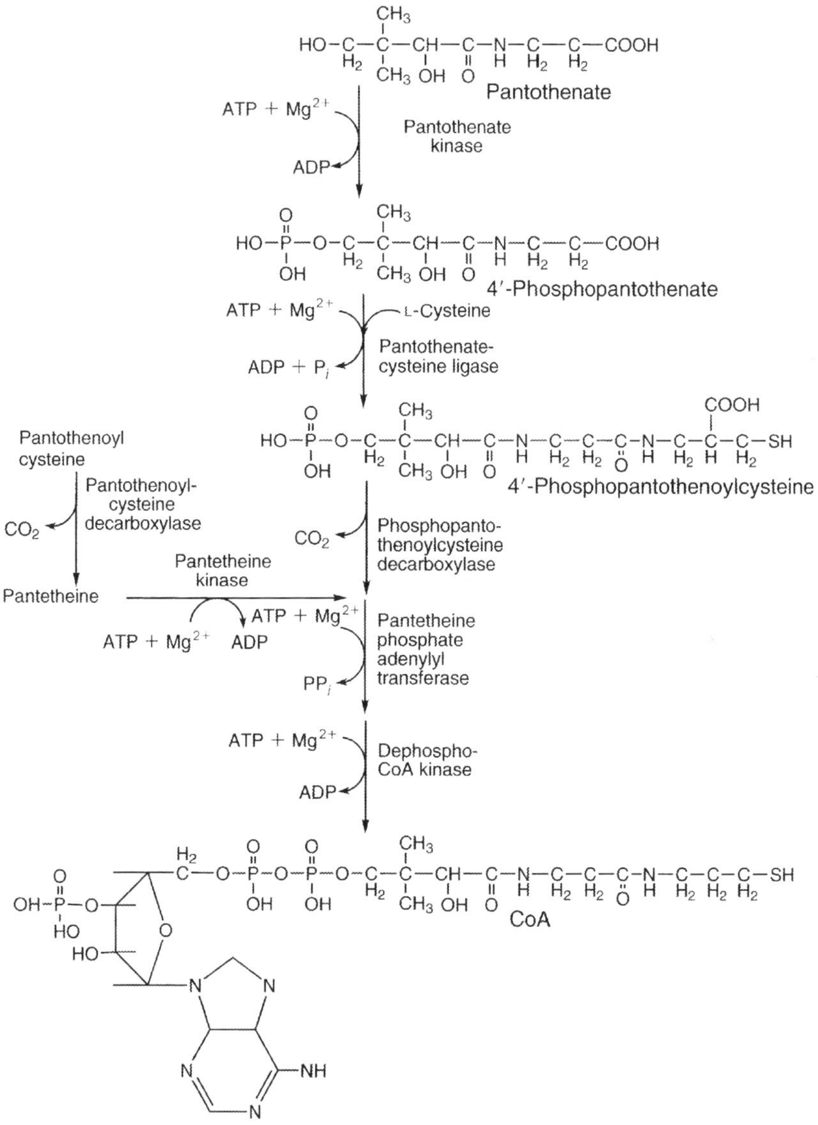

FIGURE 10.43

Pantothenate metabolism.

recovered from the tubular lumen. Nucleotide pyrophosphatase (EC3.6.1.9) at the brush border of proximal tubular epithelium releases pantothenate from CoA (Byrd et al., 1985). Another brush border enzyme that generates transportable pantothenate is pantetheine hydrolase (EC3.5.1.-) in the proximal tubular epithelium. Transport of pantothenate across the tubular brush border membrane is driven by the inward sodium gradient (Barbarat and Podevin, 1986), presumably via the sodium-dependent multivitamin carrier (SLC5A6). Little is known about how pantothenate crosses the basolateral membrane and returns into circulation. Transport also occurs in the opposite direction, and pantothenate excess can be secreted into tubules (Karnitz et al., 1984).

REGULATION

Protein kinase C and calcium/calmodulin regulate the activity of the sodium-dependent multivitamin carrier (SLC5A6).

FUNCTION

Numerous metabolic activities depend on adequate availability of pantothenate, only a few of which can be mentioned here. Most pantothenate-dependent reactions use CoA as the near-universal donor and acceptor of acetyl and acyl groups. CoA is thus indispensable for the metabolism of carbohydrates, fatty acids, ethanol, and amino acids (with the exception of a small proportion of glycine). The consequences of a lack of CoA synthesis in the brain have become more apparent with the identification of a defective brain-specific form of pantotheine kinase 2 (PANK2) as the cause of Hallervorden–Spatz syndrome (Zhou et al., 2001). This severe neurodegenerative disorder with onset in childhood is characterized by increasing iron accumulation in the basal ganglia of the brain and progressive extrapyramidal dysfunction. It is not known, however, which pantothenate-requiring processes are involved. A few reactions use 4′-phosphopantetheine, most of them are related to lipid synthesis. In all these instances, the substrates form a thiolester with CoA or 4′-phosphopantetheine, which can then be metabolized further. An example of an enzyme with 4′-phosphopantetheine as a prosthetic group is the guanosine triphosphate (GTP)–dependent acyl-CoA synthetase (acid-CoA ligase/GDP-forming, EC6.2.1.10). This enzyme is distinct from other acyl-CoA synthetases that are energized by ATP and do not contain phosphopantetheine. Effects or functions of other pantothenate-related compounds or precursors, such as pantethine, are less certain.

Intermediary metabolism: The production of acetyl-CoA from pyruvate and succinyl-CoA from alpha-ketoglutarate constantly consumes large amounts of CoA. The reactivation of the ketone bodies acetoacetate and hydroxybutyrate also draws on the CoA pool of a cell. The same is true for other metabolites that feed into the Krebs cycle, such as acetate and ethanol.

Lipid metabolism: Fatty acid synthesis starts from acetyl-CoA, cholesterol synthesis from hydroxymethyl glutaryl-CoA. Fatty acids also have to be linked to CoA before they can be metabolized via beta-oxidation. Fatty acids with an odd number of carbons, methionine, valine, threonine, and the side-chain of cholesterol can be metabolized only after ligation to CoA. Bile acids, which are essential for fat absorption, undergo CoA-dependent conjugation to taurine or glycine before they are secreted in bile.

Protein acylation: During or after translation, many proteins must be modified to become fully functional, and many of these modifications use CoA as a cofactor. Important types of modification are the acetylation of protein N-termini, agylation of proteins with acetate (e.g., lysine of alpha-tubuline), myristate, or palmitate.

Xenobiotic detoxification: Conjugation of aromatic, heterocyclic, or other complex compounds to glycine, taurine, or glucuronate typically requires an initial activation step. The metabolism of benzoate in the liver is a typical example. The free acid is first joined to CoA by either of two mitochondrial xenobiotic/medium-chain fatty acid:CoA ligases (XM-ligases, Vessey et al., 1999). Glycine *N*-acyltransferase (EC2.3.1.13) or glycine *N*-benzoyltransferase (EC2.3.1.71) can then use glycine to transform the activated benzoate into hippurate and release CoA again (Gregus et al., 1999).

Acyl carrier protein: Holo-[acyl-carrier protein] synthase (EC2.7.8.7) uses CoA to attach a 4'-phosphopantetheine residue to the acyl carrier protein subunit of fatty acid synthase (EC2.3.1.85). The 4'-phosphopantetheine anchors the nascent fatty acid and pivots it to the active centers of other components of the fatty acid synthase complex. A distinct form of the mitochondrial acyl carrier protein (NDUFAB1) constitutes the 9.6-kD subunit of NADH-ubiquinone oxidoreductase, a component of the respiratory complex I (Triepels et al., 1999). The precise function of the acyl carrier protein in complex I is not known, but a role in fatty acid and polypeptide synthesis has been suggested.

Other forms: Pantethine may enhance the anti-aggregation activity of the chaperone alpha-crystallin (Clark and Huang, 1996), but the physiological significance of such an action needs further clarification.

REFERENCES

Annous, K.F., Song, W.O., 1995. Pantothenic acid uptake and metabolism by red blood cells of rats. J. Nutr. 125, 2586–2593.

Baker, H., Frank, O., Deangelis, B., Feingold, S., Kaminetzky, H.A., 1981. Role of placenta in maternalfetal vitamin transfer in humans. Am. J. Obstet. Gynecol. 141, 792–796.

Barbarat, B., Podevin, R.A., 1986. Pantothenate-sodium cotransport in renal brush border membranes. J. Biol. Chem. 261, 14455–14460.

Byrd, J.C., Fearney, F.J., Kim, Y.S., 1985. Rat intestinal nucleotide-sugar pyrophosphatase. Localization, partial purification, and substrate specificity. J. Biol. Chem. 260, 7474–7480.

Chatterjee, N.S., Kumar, C.K., Ortiz, A., Rubin, S.A., Said, H.M., 1999. Molecular mechanism of the intestinal biotin transport process. Am. J. Physiol. 277, C605–C613.

Clark, J.L., Huang, Q.L., 1996. Modulation of the chaperone-like activity of bovine alpha-crystallin. Proc. Natl. Acad. Sci. USA 93, 15185–15189.

Food and Nutrition Board, Institute of Medicine, 1998. Dietary Reference Intakes for Thiamin, Riboflavin, Niacin, Vitamin B6, Folate, Vitamin B12, Pantothenic Acid, Biotin, and Choline. National Academy Press, Washington, DC, pp. 357–373.

Gregus, Z., Halaszi, E., Klaassen, C.D., 1999. Effect of chlorophenoxyacetic acid herbicides on glycine conjugation of benzoic acid. Xenobiotica 29, 547–559.

Karnitz, L.M., Gross, C.L., Henderson, L.M., 1984. Transport and metabolism of pantothenic acid by rat kidney. Biochim. Biophys. Acta 769, 486–492.

Prasad, P.D., Wang, H., Kekuda, R., Fujita, T., Fei, Y.J., Devoe, L.D., et al., 1998. Cloning and functional expression of a cDNA encoding a mammalian sodium-dependent vitamin transporter mediating the uptake of pantothenate, biotin, and lipoate. J. Biol. Chem. 273, 7501–7506.

Prasad, P.D., Srinivas, S.R., Wang, H., Leibach, F.H., Devoe, L.D., Ganapathy, V., 2000. Electrogenic nature of rat sodium-dependent multivitamin transport. Biochem. Biophys. Res. Commun. 270, 836–840.

Saliba, K.J., Horner, H.A., Kirk, K., 1998. Transport and metabolism of the essential vitamin pantothenic acid in human erythrocytes infected with the malaria parasite plasmodium falciparum. J. Biol. Chem. 273, 10190–10195.

Shibata, K., Gross, C.J., Henderson, L.M., 1983. Hydrolysis and absorption of pantothenate and its coenzymes in the rat small intestine. J. Nutr. 113, 2107–2115.

Tahiliani, A.G., 1991. Evidence for net uptake and efflux of mitochondrial coenzyme A. Biochim. Biophys. Acta 1067, 29–37.

Tahiliani, A.G., Beinlich, C.J., 1991. Pantothenic acid in health and disease. Vitam. Horm. 46, 165–228.

Triepels, R., Smeitink, J., Loeffen, J., Smeets, R., Buskens, C., Trijbels, E., et al., 1999. The human nuclear-encoded acyl carrier subunit (NDUFAB1) of the mitochondrial complex I in human pathology. J. Inherit. Metab. Dis. 22, 163–173.

Vessey, D.A., Kelley, M., Warren, R.S., 1999. Characterization of the CoA ligases of human liver mitochondria catalyzing the activation of short- and medium-chain fatty acids and xenobiotic carboxylic acids. Biochim. Biophys. Acta 1428, 455–462.

Zhou, B., Westaway, S.K., Levinson, B., Johnson, M.A., Gitschier, J., Hayflick, S.J., 2001. A novel pantothenate kinase gene (PANK2) is defective in Hallervorden–Spatz syndrome. Nat. Genet. 28, 345–349.

QUEUINE

Queuine [7-(((4,5-cis-dihydroxy-2-cyclopenten-1-yl)amino)methyl)-7-deazaguanosine] is a presumably essential nucleoside base resembling guanine. The corresponding nucleoside is queuosine (Figure 10.44).

ABBREVIATION	
TGT	RNA-guanine transglycosylase (queuine tRNA-ribosyltransferase)

NUTRITIONAL SUMMARY

Function: The queuosine nucleoside base influences the growth and differentiation of colonic enterocytes.

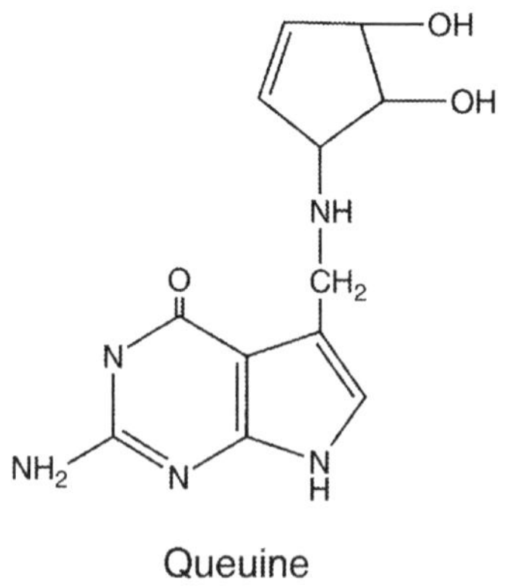

Queuine

FIGURE 10.44

Queuine.

Requirements: Most or all queuine comes from normal intestinal flora. It is likely to be essential, since a specific enzyme exists for its incorporation into certain tRNAs; however, the health consequences of reduced production due to a disturbance or elimination of intestinal bacteria (antibiotics, low-fiber diet) are not known.

Food sources: Only fermented foods are likely to contain significant amounts of queuine due to bacterial action. No information is available on the queuine content of fermented foods.

Deficiency: There is no information on the health consequences of low queuine availability in humans.

Excessive intake: The risks from dietary intake at any level are unknown.

DIETARY AND OTHER SOURCES

It is not known how much queuine is consumed with foods. Eubacteria of the terminal ileum and colon (specifically *Escherichia coli*) synthesize queuine.

ABSORPTION AND TRANSPORT

Some of the queuine produced by intestinal flora is taken up from the colonic lumen into enterocytes. It is not known whether there is significant transport out of the intestinal wall.

METABOLISM

Queuine-containing DNA in the intestinal wall can be salvaged by a specific protein with queuine-related ribonucleoside hydrolase activity (Zallot et al., 2014).

FUNCTION

Queuosine (Q) is incorporated into specific tRNAs through the specific exchange against guanine by queuine tRNA-ribosyltransferase (RNA-guanine transglycosylase, TGT, guanine insertion enzyme, EC2.4.2.29) (Slany and Muller, 1995; Deshpande et al., 1996; Deshpande and Katze, 2001). Q usually occurs at position 34, the first position of the anticodons GUT, GUC, GUA, and GUG (collectively referred to as *GUN anticodons*) of both in nuclear and mitochondrial tRNAs, specifying the amino acids asparagine, aspartate, histidine, and tyrosine (Figure 10.45).

Queuine deficiency interferes with efficient conversion of phenylalanine to tyrosine due to a lack of tetrahydrobiopterin (Rakovich et al., 2011). Tyrosine-deficient animals that lack a queuine source (because they are germ free) die within a few weeks. Interaction of queuine deficiency with a lack of other nutrients or metabolic inefficiencies is possible since tetrahydrobiopterin is a cofactor for further enzymes, particularly tyrosine hydroxylase (EC1.14.16.2), tryptophan hydroxylase (EC1.14.16.4), and nitric oxide synthase (EC1.14.13.39).

Q modification may have the potential to influence cellular growth and differentiation by codon bias–based regulation of protein synthesis for discrete mRNA transcripts (Morris et al., 1999).

Since queuine is essential for intestinal pathogens, such as *Shigella flexneri*, targeted inhibition of its synthesis might be promising (Gradler et al., 2001).

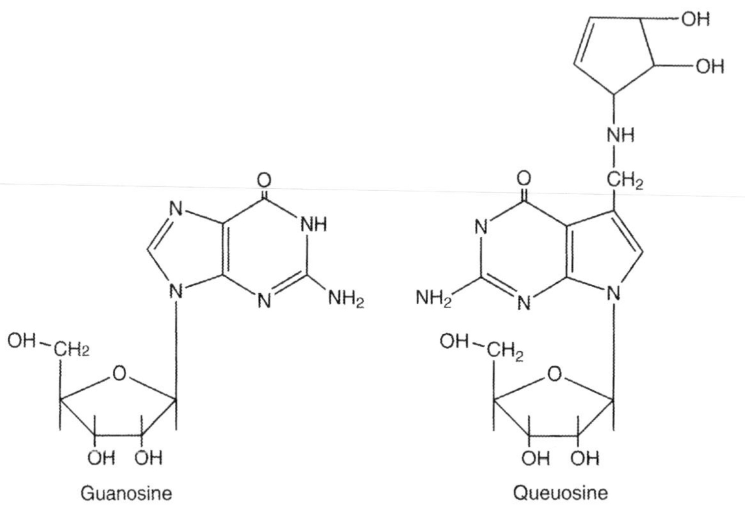

FIGURE 10.45

The bacterial nucleotide queuosine replaces guanosine in specific tRNAs.

REFERENCES

Deshpande, K.L., Katze, J.R., 2001. Characterization ofcDNA encoding the human tRNA-guanine transglycosylase (TGT) catalytic subunit. Gene 265, 205–212.

Deshpande, K.L., Seubert, P.H., Tillman, D.M., Farkas, W.R., Katze, J.R., 1996. Cloning and characterization of cDNA encoding the rabbit tRNA-guanine transglycosylase 60-kilodalton subunit. Arch. Biochem. Biophys. 326, 1–7.

Gradler, U., Gerber, H.D., Goodenough-Lashua, D.M., Garcia, G.A., Ficner, R., Reuter, K., et al., 2001. A new target for shigellosis: rational design and crystallographic studies of inhibitors of tRNA-guanine transglycosylase. J. Mol. Biol. 306, 455–467.

Morris, R.C., Brown, K.G., Elliott, M.S., 1999. The effect of queuosine on tRNA structure and function. J. Biomol. Struct. Dyn. 16, 757–774.

Rakovich, T., Boland, C., Bernstein, I., Chikwana, V.M., Iwata-Reuyl, D., Kelly, V.P., 2011. Queuosine deficiency in eukaryotes compromises tyrosine production through increased tetrahydrobiopterin oxidation. J. Biol. Chem. 286, 19354–19363.

Slany, R.K., Muller, S.O., 1995. tRNA-guanine transglycosylase from bovine liver. Purification of the enzyme to homogeneity and biochemical characterization. Eur. J. Biochem. 230, 221–228.

Zallot, R, Brochier-Armanet, C, Gaston, KW, Forouhar, F, Limbach, PA, Hunt, JF, et al., 2014. Plant, animal, and fungal micronutrient queuosine is salvaged by members of the DUF2419 protein family. ACS Chem. Biol. 9, 1812–1825.

BIOPTERIN

Biopterin (S-(R*,S*)-2-amino-6-(1,2-dihydoxypropyl)-4(1H)-pteridinone, 6,7-dihydropteridine, molecular weight 237) is a moderately water-soluble heterocyclic compound. The biologically active form is tetrahydrobiopterin (BH4) (Figure 10.46).

ABBREVIATIONS	
BH4	5,6,7,8-tetrahydrobiopterin
GTP	guanosine triphosphate
PTP	6-pyruvoyltetrahydropterin

NUTRITIONAL SUMMARY

Function: Biopterin is needed for the metabolism of phenylalanine, tyrosine, and tryptophan, and for the synthesis of hormones, neurotransmitters, and skin pigments (catecholamines, melanin, serotonin, and melanotonin), and cell signaling (nitric oxide). Roles in promoting angiogenesis, neuronal survival, and cellular immunity, and for protection against free radicals also are likely.

Requirements: The body can produce adequate amounts of biopterin in the absence of dietary intake; synthesis requires GTP, niacin, and magnesium.

Food sources: While many animal foods contain various bioactive forms of biopterin, the amounts have not been well investigated.

Deficiency: Genetic defects of biopterin synthesis or activation cause severe phenylketonuria (PKU) with neurological damage and weak muscle tonus. Deficiency has also been suggested to contribute to Alzheimer's disease, Parkinson's disease, autism, and depression. Trimethoprim inhibits the reactivation of oxidized biopterin; whether this causes deficiency symptoms is not known.

Excessive intake: The risk from using supplemental biopterin at any level is not well documented.

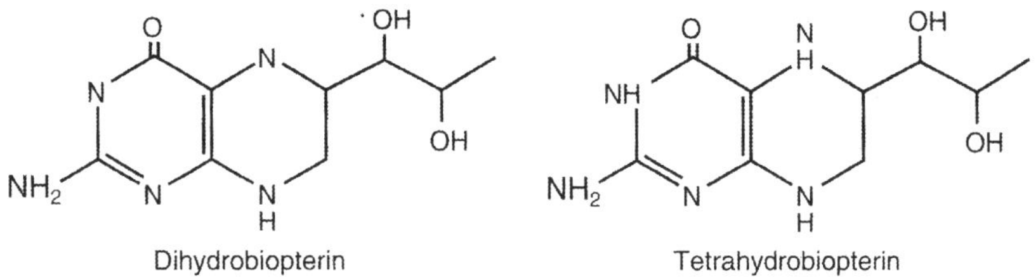

FIGURE 10.46

Dihydrobiopterin and tetrahydrobiopterin.

ENDOGENOUS SOURCES

Biopterin is synthesized endogenously in most tissues, starting with a molecular rearrangement of GTP (Bonafé et al., 2001). The reactions use NADPH, NADH, and magnesium as cofactors (Figure 10.47). GTP cyclohydrolase 1 (EC3.5.4.16) generates a neopterin intermediate, which is converted by the NADPH- and magnesium-requiring 6-pyruvoyltetrahydropterin synthase (EC4.6.1.10) into 6-pyruvoyltetrahydropterin (PTP). Both keto groups of the side chain then have to be reduced to arrive at the final product 5,6,7,8-tetrahydrobiopterin (BH4). Sepiapterin reductase (EC1.1.1.153) can accomplish this in three distinct steps, with 2′-oxo-tetrahydropterin and 1′-oxo-tetrahydropterin as intermediates. This pathway accounts for about half of the BH4 synthesis *in vivo*. The remainder is generated through alternative pathways that rely on aldehyde reductase (EC1.1.1.21) and carbonyl reductase (EC1.1.1.184), some in conjunction with sepiapterin reductase.

Lack of the nucleoside base queuine reduces tetrahydrobiopterine concentrations and is associated with an accumulation of dihydrobiopterine (Rakovich et al., 2011). Both of these metabolic changes reduce the activity of tetrahydrobiopterine-dependent enzymes.

The NADPH-requiring enzyme pterin-4a-carbinolamine dehydratase (EC4.2.1.96) appears to facilitate BH4 synthesis by accelerating the formation of quinoid dihydrobiopterin and preventing the formation of inactive 7-pterins. However, the mechanism of this reaction is still unclear.

DIETARY SOURCES

Forms present in food include tetrahydrobiopterin, dihydrobiopterin, and neopterin. The quantities available from specific foods or typical intake levels are not yet well characterized.

DIGESTION AND ABSORPTION

Some ingested tetrahydrobiopterin is absorbed, as demonstrated by the rapid lowering of phenylalanine levels in a patient with defective biopterin synthesis (Snyderman et al., 1987). The mechanisms of uptake and export are uncertain.

TRANSPORT AND CELLULAR UPTAKE

The plasma concentrations of total biopterin and BH4 in plasma were found to be highest at 9 A.M. and lowest shortly past midnight (Hashimoto et al., 1993). BH4 from the blood is rapidly taken up into the liver and kidneys and transferred across the placenta to the fetus, whereas transfer into the brain and other tissues appears to be very limited (Hoshiga et al., 1993). The extent of transport in blood circulation and mechanism of uptake into cells has not been well characterized yet.

METABOLISM

BH4-utilizing reactions generate several products, including 4a-hydroxytetrahydrobiopterin (4a-carbinolamine), quinonoid dihydrobiopterin, and 4a-cyclic-tetrahydrobiopterin. These reaction products are inactive as cofactors until they are converted again to BH4 (Figure 10.48). The reactivation of dihydrobiopterin is accomplished by an NAD(P)H-dependent enzyme, dihydropteridine reductase (EC1.6.99.7). This enzyme is inhibited by the antibiotic trimethoprim. Here, 4a-hydroxytetrahydrobiopterin and

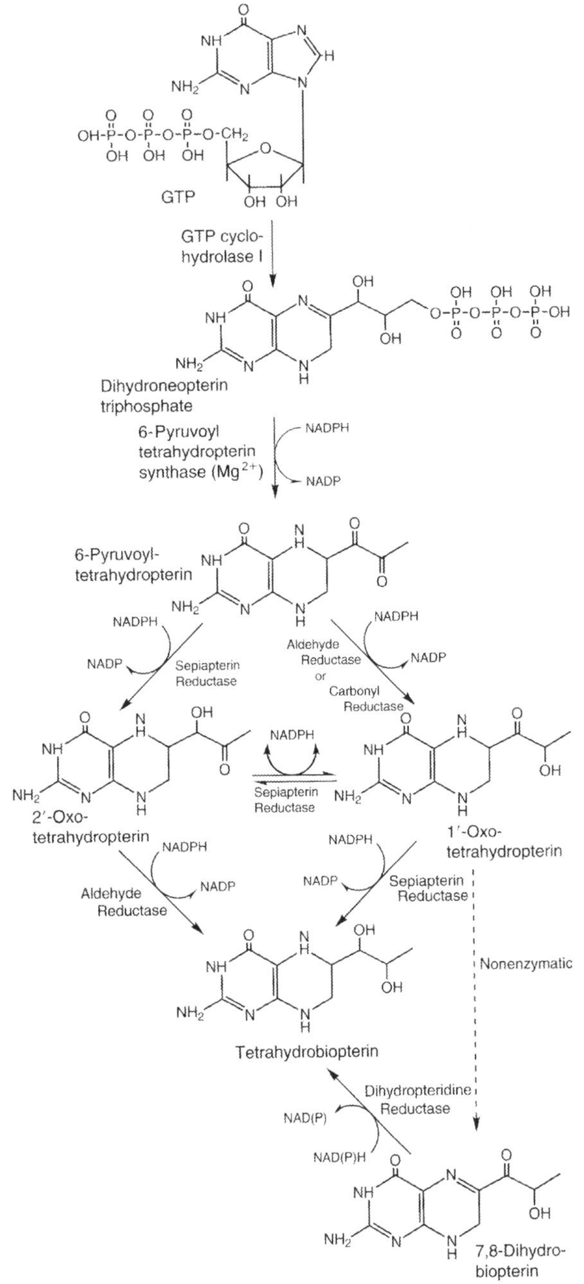

FIGURE 10.47

Tetrahydrobiopterin synthesis.

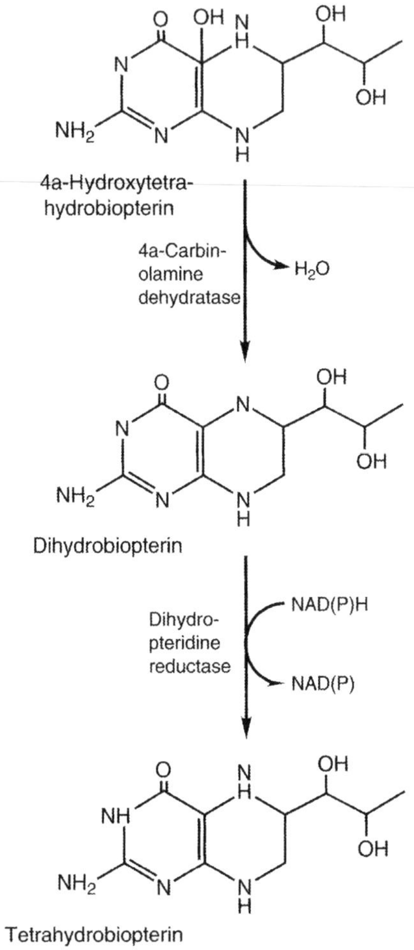

FIGURE 10.48

Tetrahydrobiopterin reactivation.

4a-cyclic-tetrahydrobiopterin can be reactivated to BH4 by pterin-4a-carbinolamine dehydratase (EC4.2.1.96) in conjunction with dihydropteridine reductase. It is of note that pterin-4a-carbinolamine dehydratase (EC4.2.1.96) participates in gene regulation as a cofactor for hepatocyte nuclear factor 1 homeobox transcription factors. It has been suggested that it might play a role in the development of colon carcinoma and melanoma. There is evidence that the peroxisomal enzyme xanthine dehydrogenase (EC1.1.3.22) participates in the conversion of pterin to isoxanthopterin (Blau et al., 1996). This enzyme contains molybdenum cofactor, iron-sulfur clusters, and FAD.

STORAGE

Neither the body content of biopterin and its precursors nor specific storage mechanisms are known.

EXCRETION

The oxidized form of biopterin is excreted much more rapidly than BH4 (Hoshiga et al., 1993). Urinary metabolites derived from biopterin synthesis and metabolism include D-threo-biopterin and L-threo-biopterin (Fukushima and Shiota, 1972), neopterin, isoxanthopterin, 7-biopterin (primapterin), and N2-(3-aminopropyl)biopterin (oncopterin). Low activity of pterin-4a-carbinolamine dehydratase (EC4.2.1.96) increases the excretion as 7-biopterin (primapterin). A methotrexate-derived compound coelutes on high-performance liquid chromatography (HPLC) with oncopterin and might be misinterpreted as a marker for cancer-specific alteration of biopterin metabolism (Hibiya et al., 1997).

Healthy individuals excrete via urine about $67\,\mu mol$ biopterin/mol creatinine, $142\,\mu mol$ neopterin/mol creatinine neopterin, and $257\,\mu mol$ dihydrobiopterin /mol creatinine (Tomšíková et al., 2014).

REGULATION

Biopterin status is predominantly controlled by modulation of GTP cyclohydrolase 1 expression and activity. Gamma-interferon and other cytokines increase biopterin production by inducing GTP cyclohydrolase 1 (Thöny et al., 2000). Glucocorticoids prevent cytokine-mediated induction of GTP cyclohydrolase 1 (Simmons et al., 1996), apparently via a cyclic adenosine monophosphate (cAMP)--mediated signaling cascade (Ohtsuki et al., 2002). Food deprivation of animals increases biopterin production and concentration in the blood (Koller et al., 1990).

Inflammation and infection tend to depress availability of biopterin in tissues, such as endothelia of small blood vessels, due to rapid oxidation of the reduced form (McNeill and Channon, 2012).

FUNCTION

Tyrosine synthesis: Phenylalanine can be the precursor of tyrosine through the action of the ferroenzyme phenylalanine hydroxylase (EC1.14.16.1). This hydroxylation of phenylalanine is driven by the oxidation of BH4 to 4a-hydroxytetrahydrobiopterin (4a-carbinolamine).

Catecholamine and pigment synthesis: Synthesis of the catecholamines, dopamine, noradrenaline, and adrenaline is initiated by tyrosine 3-monooxygenase (EC1.14.16.2).

Serotonin and melanotonin synthesis: Tryptophan hydroxylase (EC1.14.16.4) utilizes BH4 for the synthesis of serotonin and melatonin from L-tryptophan or tryptamin.

Nitric oxide synthesis: All isoforms (NOS1, NOS2, NOS3) of nitric oxide synthase (EC1.14.13.39) require 1 mole of BH4 and 1 mole of heme per dimer as cofactors (Rafferty et al., 1999). Diminished BH4 synthesis and the resulting decrease in nitric oxide production have been suggested to contribute importantly to impaired angiogenesis (Marinos et al., 2001), as well as epithelial dysfunction and insufficient vasodilation (Gruhn et al., 2001). It has also been suggested that reduced availability of tetrahydrobiopterin increases the production of superoxide radicals instead of the normal NOS product nitric oxide, which may contribute to vascular dysfunction (McNeill and Channon, 2012).

Cell growth and survival: Activation of neuronal Ca^{2+} channels (Koshimura et al., 1999) has been found to enhance neuronal survival. Whether biopterin actually affects the risk of Alzheimer's disease, Parkinson's disease, autism, or depression through this or another mechanism remains to be seen (Thöny et al., 2000). Dopaminergic neurons are specifically protected against free-radical damage during periods of glutathione depletion by a tetrahydrobiopterin-dependent mechanism (Nakamura et al., 2000).

Regulation of O-alkylated glycerolipids: Glyceryl-ether monooxygenase (EC1.14.16.5) is a microsomal enzyme that hydroxylates *O*-alkyl moieties in glycerolipids; the resulting hydroxyalkyl spontaneously breaks down into glycerol and a fatty aldehyde. Folic acid can serve as an electron donor instead of tetrahydropteridine.

Immune defense: Increased neopterin levels are observed in patients with acute graft rejections, viral infections, auto-immune diseases, and several malignancies (Asano et al., 1997).

Metabolic regulation: Tyrosinase activity is regulated by (6R)-L-erythro-5,6,7,8-tetrahydrobiopterin through specific allosteric inhibition (Schallreuter et al., 1999). Hydrogen peroxide inhibits the recycling of 6(R)-L-erythro-5,6,7,8-tetrahydrobiopterin by 4a-OH-tetrahydrobiopterin dehydratase and thereby might play an important role in the regulation of BH4-dependent enzymes (Schallreuter et al., 2001). Patches of skin affected by vitiligo are known to have abnormally high hydrogen peroxide concentrations, which accordingly would slow the recycling of 6(R)-L-erythro-5,6,7,8-tetrahydrobiopterin into BH4. Loss of skin pigmentation in vitiligo thus appears to occur due to the oxidized metabolite competitively, which inhibits epidermal phenylalanine hydroxylase and thus blocks the production of melanin (Schallreuter et al., 2001).

NUTRITIONAL ASSESSMENT

Deficiencies may be detected by measuring tetrahydrobiopterin concentrations in the blood or in dried blood spots (Opladen et al., 2011). Such assessment is particularly important for newborn screening because BH4 deficiency can cause PKU and responds well to treatment.

Elevated excretion of urine metabolites (biopterin, neobiopterin, and dihydrobiopterin) can be assessed as an indicator with low specificity of increased endogenous production due to infections, cancer, or progressive multiple sclerosis (Rejdak et al., 2010).

REFERENCES

Asano, T., Nakajima, F., Odajima, K., Tsuji, A., Hayakawa, M., Nakamura, H., 1997. Urinary neopterin levels in patients with genitourinary tract malignancies. Nippon Hinyokika Gakkai Zasshi Jpn. J. Urol. 88, 53–58.

Blau, N., de Klerk, J.B., Thony, B., Heizmann, C.W., Kierat, L., Smeitink, J.A., et al., 1996. Tetrahydrobiopterin loading test in xanthine dehydrogenase and molybdenum cofactor deficiencies. Biochem. Mol. Med. 58, 199–203.

Bonafé, L., Thöny, B., Penzien, J.M., Czarnecki, B., Blau, N., 2001. Mutations in the sepiapterin reductase gene cause a novel tetrahydrobiopterin-dependent monoamine-neurotransmitter deficiency without hyperphenylalaninemia. Am. J. Hum. Genet. 69, 269–277.

Fukushima, T., Shiota, T., 1972. Pterins in human urine. J. Biol. Chem. 247, 4549–4556.

Gruhn, N., Aldershvile, J., Boesgaard, S., 2001. Tetrahydrobiopterin improves endothelium-dependent vasodilation in nitroglycerin-tolerant rats. Eur. J. Pharmacol. 416, 245–249.

Hashimoto, R., Mizutani, M., Ohta, T., Nakazawa, K., 1993. On the fluctuation of plasma biopterin levels for 24 h in normal controls [in Japanese]. Yakubutsu Seishin Kodo Jpn. J. Psychopharmacol. 13, 59–63.

Hibiya, M., Teradaira, R., Shimpo, K., Matsui, T., Sugimoto, T., Nagatsu, Y., 1997. Interference of a methotrexate derivative with urinary oncopterin [N2-(3-aminopropyl)biopterin] measurement by high-performance liquid chromatography with fluorimetric detection. J. Chromatogr. B Biomed. Sci. Appl. 691, 223–227.

Hoshiga, M., Hatakeyama, K., Watanabe, M., Shimada, M., Kagamiyama, H., 1993. Autoradiographic distribution of [^{14}C]tetrahydrobiopterin and its developmental change in mice. J. Pharmacol. Exp. Ther. 267, 971–978.

Koller, M., Goldberg, M., Schramm, G., Merkenschlager, M., 1990. The influence of nutritional factors on biopterin excretion in laboratory animals. Z Ernährungsw 29, 169–177.

Koshimura, K., Tanaka, J., Murakami, Y., Kato, Y., 1999. Enhancement of neuronal survival by 6R-tetrahydrobiopterin. Neuroscience 88, 561–569.

Marinos, R.S., Zhang, W., Wu, G., Kelly, K.A., Meininger, C.J., 2001. Tetrahydrobiopterin levels regulate endothelial cell proliferation. Am. J. Physiol. Heart Circ. Physiol. 281, H482–H489.

McNeill, E., Channon, K.M., 2012. The role of tetrahydrobiopterin in inflammation and cardiovascular disease. Thromb. Haemost. 108, 832–839.

Nakamura, K., Wright, D.A., Wiatr, T., Kowlessur, D., Milstien, S., Lei, X.G., et al., 2000. Preferential resistance of dopaminergic neurons to the toxicity of glutathione depletion is independent of cellular glutathione peroxidase and is mediated by tetrahydrobiopterin. J. Neurochem. 74, 2305–2514.

Ohtsuki, M., Shiraishi, H., Kato, T., Kuroda, R., Tazawa, M., Sumi-Ichinose, C., et al., 2002. cAMP inhibits cytokine-induced biosynthesis of tetrahydrobiopterin in human umbilical vein endothelial cells. Life Sci. 70, 1–12.

Opladen, T., Abu Seda, B., Rassi, A., Thöny, B., Hoffmann, G.F., Blau, N., 2011. Diagnosis of tetrahydrobiopterin deficiency using filter paper blood spots: further development of the method and 5 years experience. J. Inherit. Metab. Dis. 34, 819–826.

Rafferty, S.E., Boyington, J.C., Kulansky, R., Sun, P.D., Malech, H.L., 1999. Stoichiometric arginine binding in the oxygenase domain of inducible nitric oxide synthase requires a single molecule of tetrahydrobiopterin per dimer. Biochem. Biophys. Res. Commun. 257, 344–347.

Rakovich, T., Boland, C., Bernstein, I., Chikwana, V.M., Iwata-Reuyl, D., Kelly, V.P., 2011. Queuosine deficiency in eukaryotes compromises tyrosine production through increased tetrahydrobiopterin oxidation. J. Biol. Chem. 286, 19354–19363.

Rejdak, K., Leary, S.M., Petzold, A., Thompson, A.J., Miller, D.H., Giovannoni, G., 2010. Urinary neopterin and nitric oxide metabolites as markers of interferon beta-1a activity in primary progressive multiple sclerosis. Mult. Scler. 16, 1066–1072.

Schallreuter, K.U., Moore, J., Tobin, D.J., Gibbons, N.J., Marshall, H.S., Jenner, T., et al., 1999. Alpha-MSH can control the essential cofactor 6-tetrahydrobiopterin in melanogenesis. Ann. NY Acad. Sci. 885, 329–341.

Schallreuter, K.U., Moore, J., Wood, J.M., Beazley, W.D., Peters, E.M., Marles, L.K., et al., 2001. Epidermal H(2)O(2) accumulation alters tetrahydrobiopterin (6BH4) recycling in vitiligo: identification of a general mechanism in regulation of all 6BH4-dependent processes? J. Invest. Dermatol. 116, 167–174.

Simmons, W.W., Ungureanu-Longrois, D., Smith, G.K., Smith, T.W., Kelly, R.A., 1996. Glucocorticoids regulate inducible nitric oxide synthase by inhibiting tetrahydrobiopterin synthesis and L-arginine transport. J. Biol. Chem. 271, 23928–23937.

Snyderman, S.E., Sansaricq, C., Pulmones, M.T., 1987. Successful long term therapy of biopterin deficiency. J. Inherit. Metab. Dis. 10, 260–266.

Thöny, B., Auerbach, G., Blau, N., 2000. Tetrahydrobiopterin biosynthesis, regeneration and functions. Biochem. J. 347, 1–16.

Tomšíková, H., Solich, P., Nováková, L., 2014. Sample preparation and UHPLC-FD analysis of pteridines in human urine. J. Pharm. Biomed. Anal. 95, 265–272.

INOSITOL

Myo-inositol (meso-inositol, i-inositol, hexahydroxycyclohexane, cyclohaxanehexol, cyclohexitol, meat sugar, inosite, mesoinosite, phaseomannite, dambose, nucite, bios 1, rat antispectacled eye factor, mouse antialopecia factor; molecular weight 180) is a cyclical glucose isomer (Figure 10.49).

ABBREVIATIONS	
IP5	inositol pentaphosphate
IP6	inositol hexaphosphate (phytate)

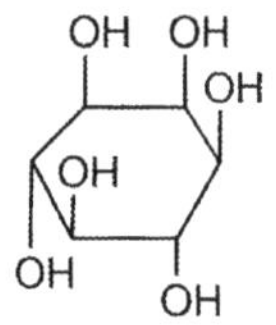

FIGURE 10.49

Myo-inositol.

NUTRITIONAL SUMMARY

Function: Inositol is a precursor for compounds that are important for membrane synthesis, the regulation of cell growth, and maintenance of intracellular osmotic pressure. It may promote lung maturation before birth, slow early cancer growth, and protect male fertility. Phytate and other inositol polyphosphates may have some anticarcinogenic activity.

> *Food sources*: Meat, poultry, fish, and dairy products are all sources of inositol.
> *Requirements*: The inositol needs of most people are met by endogenous synthesis.
> *Deficiency*: No clinical symptoms in the absence of dietary intake have been reported.
> *Excessive intake*: High inositol intake may increase the risk of neuropathy in diabetics. Phytate potently reduces intestinal absorption of iron and zinc.

ENDOGENOUS SOURCES

Inositol-3-phosphate synthase (EC5.5.1.4), an NAD-containing enzyme, catalyzes the internal cyclization of glucose-6-phosphate to inositol-3-phosphate (Figure 10.50). After removal of the phosphate group by myo-inositol-1(or 4)-monophosphatase (EC3.1.3.25) the free form is released into circulation. The kidneys produce about 4 g of myo-inositol per day. Intestinal bacteria may also contribute to inositol supplies, but the amounts available from this source are not known.

DIETARY SOURCES

Average daily intake of free inositol, inositol monophosphate, and inositol phospholipids in the United States is about 1 g, mainly from meat, poultry, fish, and dairy products. Human milk contains about 180 mg/l (Pereira et al., 1990).

Distinct from these forms with high bioavailability are the inositol polyphosphates, particularly phytate (inositol hexaphosphate, IP6). These forms are characteristic for plant-derived foods. Nuts and seeds are the richest sources, containing between 7 and 22 mg/g (poppy seeds 22 mg/g, pumpkin seeds 19 mg/g, cashew and Brazil nuts about 18 mg/g, hazelnuts and sunflower seeds 16 mg/g, pecans 15 mg/g, and peanuts and walnuts 8 mg/g). Grains and grain products are also significant sources, especially when the whole grain is consumed (white bread, cake, and other white-flour products 2 mg/g, rye and other dark bread 3 mg/g). Many fruits, vegetables, and tubers also contribute (potatoes about 1.0 mg/g, apples, tomatoes, peas, and cucumbers 0.2–0.6 mg/g). Brewed teas contain about 0.1 mg/ml (Siegenberg et al., 1991).

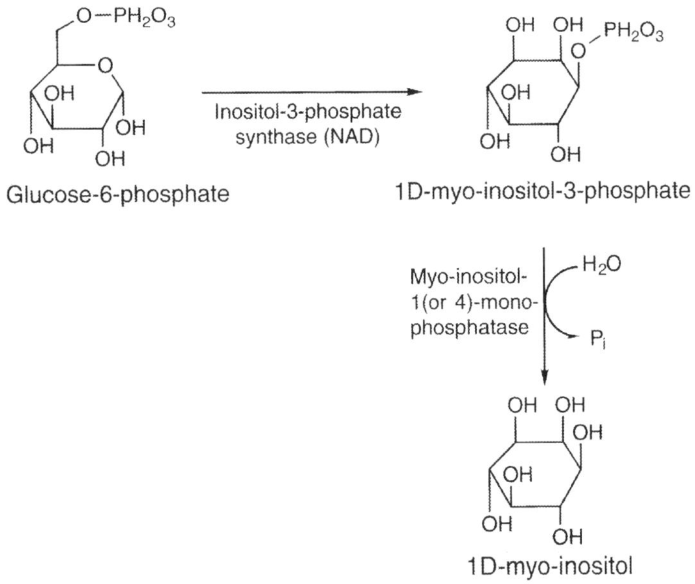

FIGURE 10.50

Inositol synthesis.

DIGESTION AND ABSORPTION

The different forms of bioavailable inositol, mainly free inositol and lipid-bound inositol, are absorbed with high efficiency from the small intestine. Free myo-inositol is taken up via the sodium-dependent myo-inositol transporter (SLC5A3) with high efficiency. Dietary inositol polyphosphates (Figure 10.51), including phytate (IP6), also reach the circulation intact (Grases et al., 2001) but are absorbed with only low efficiency, and possibly through mechanisms that do not involve the myo-inositol transporter. Phosphatidyl inositol, the main dietary source of inositol, can be hydrolyzed by pancreatic phospholipase A2. It is not clear how the lysophosphatide then enters the intestinal cell. Contrary to earlier findings (Bitar and Reinhold, 1972), release of free inositol from dietary phytate is probably minimal, and most of this is mediated by phytases already present in food. In addition, 3-phytase (EC3.1.3.8, zinc-dependent), which is present in the small intestine of rats and other mammals, does not seem to be active in the human gut. A much smaller role, if any, is played by endogenous multiple inositol-polyphosphate phosphatase (EC3.1.3.62), which is predominantly located in the liver and kidney (Craxton et al., 1997).

Some inositol from the higher inositol polyphosphates can be released by phytases in plant foods during processing or storage prior to consumption. Food processing methods that release free inositol to some (often unpredictable) degree include germination, such as is used for malting (e.g., in beer production), fermentation, exposure to baker's yeast, heating, and soaking (Sandberg et al., 1999).

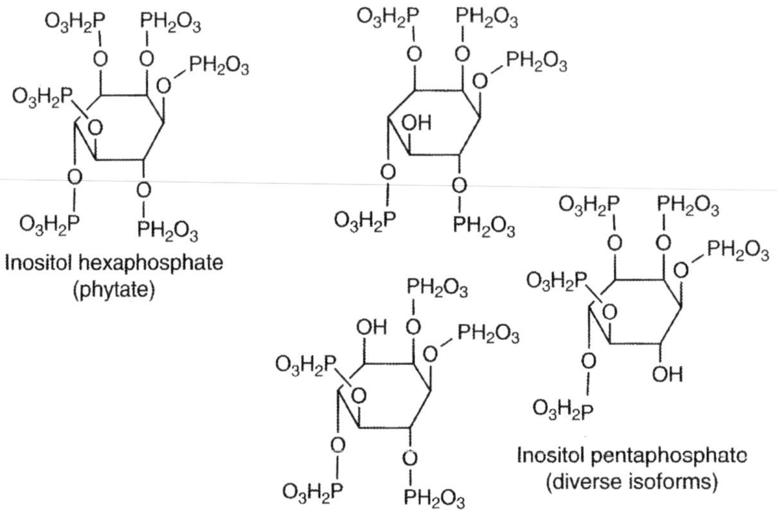

FIGURE 10.51

Inositol polyphosphates.

TRANSPORT AND CELLULAR UPTAKE

Blood circulation: Most inositol (typically about 29 μmol/l) is present in the blood in its free form; a much smaller amount is carried as phosphatidyl inositol in lipoproteins.

Inositol concentrations depend partially on intake levels and decrease in the absence of significant intake (Holub, 1986). Plasma concentrations of phytate may be as high as 0.3 mg/l in response to dietary intake (Grases et al., 2001). Its sodium-dependent transporter (SLC5A3) mediates the entry of free myo-inositol into cells in the brain, kidneys, and other tissues. Inositol polyphosphates, including IP6, are also taken up into cells (Vucenik and Shamsuddin, 1994), though the mechanism remains unclear.

METABOLISM

Inositol phosphate interconversions: Tissues contain a wide range of inositol species with 0–6 phosphates attached. The enzymes, which add phosphate groups to specific intermediates, include myo-inositol 1-kinase (EC2.7.1.64), 1-phosphatidylinositol 4-kinase (EC2.7.1.67), 1D-myo-inositol-trisphosphate 3-kinase (EC2.7.1.127), 1D-myo-inositol-trisphosphate 5-kinase (EC2.7.1.139), 1D-myo-inositol-trisphosphate 6-kinase (EC2.7.1.133), 1D-myo-inositol-tetrakisphosphate 1-kinase (EC2.7. 1.134), and 1D-myo-inositol-tetrakisphosphate 5-kinase (EC2.7.1.140).

Enzymes that remove phosphates from inositol polyphosphates include inositol-1,3,4,5-tetrakisphosphate 3-phosphatase (EC3.1.3.62), inositol-1,4,5-trisphosphate 1-phosphatase (EC3.1.3.61), inositol-1,4,5-trisphosphate 5-phosphatase (EC3.1.3.56), inositol-1,4-bisphosphate 1-phosphatase (EC3.1.3.57), inositol-1,3-bisphosphate 3-phosphatase (EC3.1.3.65), inositol-1,4-bisphosphate 1-phosphatase (EC3.1.3.57), inositol-3,4-bisphosphate 4-phosphatase (EC3.1.3.66), and myo-inositol-1(or 4)-monophosphatase (EC3.1.3.25). Expression of these enzymes in different tissues is very diverse, and in most cases, it is under the tight control of regulators such as calmodulin (Figure 10.52).

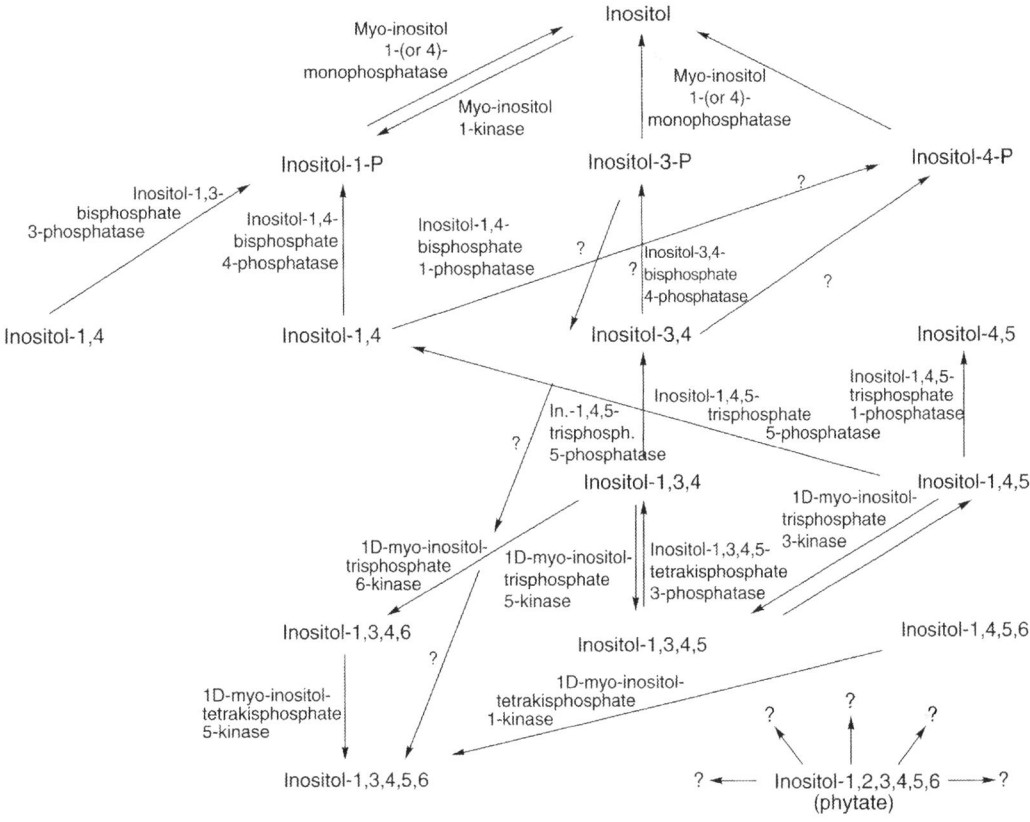

FIGURE 10.52

Inositol metabolism. The question marks indicate that the molecular nature responsible for the activity is still uncertain.

Phospholipids: CDP-diacylglycerol-inositol 3-phosphatidyltransferase (EC2.7.8.11), a manganese-dependent microsomal enzyme, uses inositol and cytidine diphosphatediacylglycerol for the synthesis of phosphatidylinositol. The complex synthesis and breakdown of the various products and intermediates of inositol-containing phospholipids utilize numerous enzymes, including phosphatidylinositol 3-kinase (EC2.7.1.137), phosphatidylinositol 4-kinase (EC2.7.1.67), 1-phosphatidylinositol-4-phosphate kinase (EC2.7.1.68), glycerophosphoinositol inositolphosphodiesterase (EC3.1.4.43), 1,2-cyclic-inositol-phosphate phosphodiesterase (EC3.1.4.36), 1-phosphatidylinositol-4,5-bisphosphate phosphodiesterase (EC3.1.4.11), 1-phosphatidylinositol phosphodiesterase (EC3.1.4.10), and others. Several of the intermediates, including inositol-1,4,5-triphosphate and phosphatidylinositol-4,5-bisphosphate, serve as second messengers.

Catabolism: Iron-dependent myo-inositol oxygenase (EC1.13.99.1) oxidizes inositol to D-glucuronate, which can be reduced by glucuronate reductase (EC1.1.1.19) to L-gulonate (Figure 10.53). Human glucuronate reductase may actually be identical with the NADP-dependent zinc-enzyme alcohol

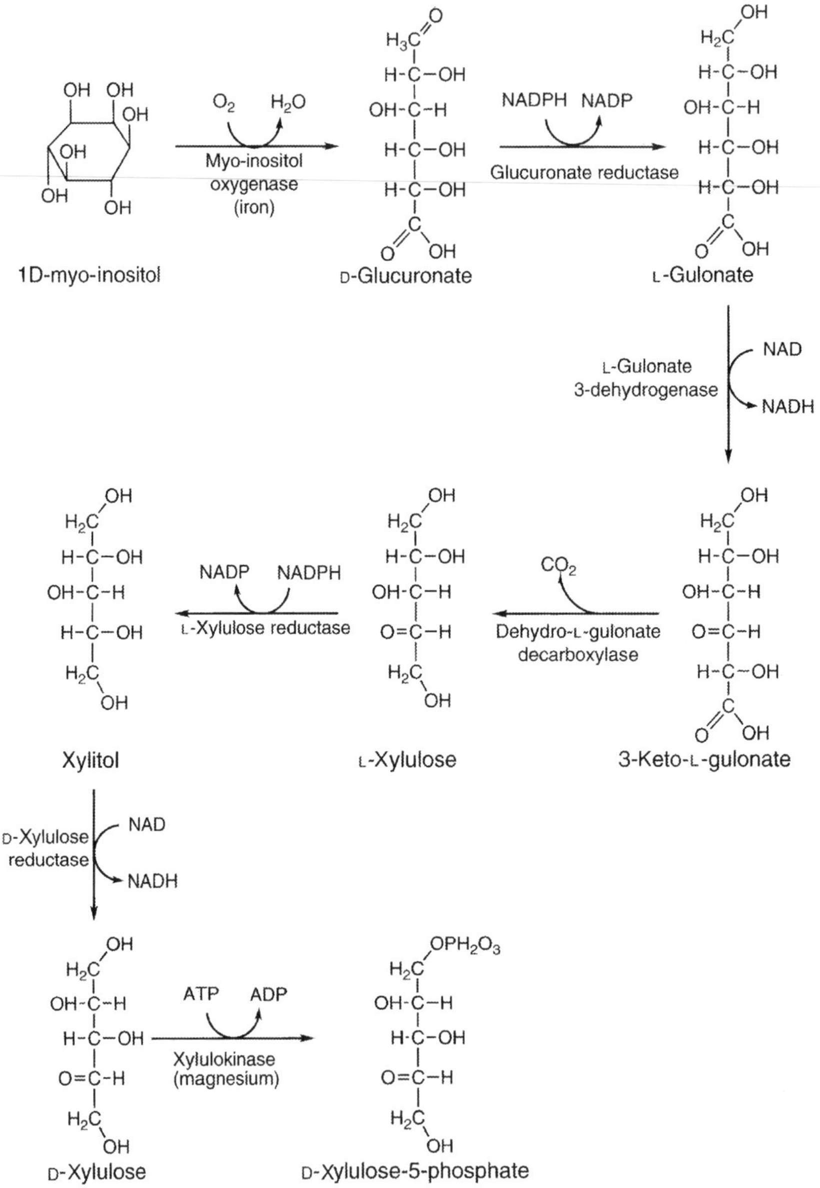

FIGURE 10.53

Inositol breakdown.

dehydrogenase (EC1.1.1.2). Further metabolism of L-gulonate proceeds to 3-keto-L-gulonate (L-gulonate 3-dehydrogenase, EC1.1.1.45), L-xylulose (magnesium- or manganese-dependent dehydro-L-gulonate decarboxylase, EC4.1.1.34), xylitol (L-xylulose reductase, EC1.1.1.10), D-xylulose (manganese requiring D-xylulose reductase, EC1.1.1.9), and xylulose 5-phosphate (magnesium-dependent xylulokinase, EC2.7.1.17). Therefore, xylulose 5-phosphate is a normal intermediate of D-glucose metabolism via the pentose phosphate pathway and readily utilized.

EXCRETION

Very little inositol (<1% of an ingested dose) is lost via urine. Filtered free inositol is reabsorbed from the proximal tubule and papillary nephron segments (Burger-Kentischer et al., 1999). High osmolarity at the basolateral side of the tubules increases uptake.

REGULATION

While activities of most enzymes for inositol phosphate and inositol phospholipids are tightly controlled, the processes that maintain overall inositol homeostasis are not well understood. Breakdown of inositol in kidney to D-glucuronic acid and D-xylulose-5-phosphate (pentose phosphate pathway) is important for the regulation of inositol homeostasis. Blood inositol concentration rises with declining kidney function (Holub, 1986).

FUNCTION

Intracellular signaling: Various agonist-triggered membrane receptors use inositol phospholipid hydrolysis as the first step of intracellular cascades that continue downstream via calcium and protein kinase C-mediated signaling. The membrane receptors activate a specific zinc-containing phospholipase C (EC3.1.4.3) that releases two signaling compounds: diacyl glycerol and inositol-1,4,5-triphosphate.

Eicosanoid synthesis: Phosphatidylinositol is a significant source of arachidonic acid and other long-chain polyunsaturated fatty acids for the intracellular synthesis of prostaglandins, thromboxans, leukotrienes, and other eicosanoids. Phosphatidylinositol deacylase (phosphatidyl inositol phospholipase A2, EC3.1.1.52) releases the precursor fatty acids in a tightly regulated fashion.

GPI anchors: Many proteins, such as the prion protein presumably involved in variant Creutzfeld–Jakob disease (more commonly known as *mad cow disease*), are specifically anchored to membranes through a GPI extension. Such GPI-anchored exoproteins can be released from membranes by HDL-associated phosphatidylinositol glycoprotein phospholipase D (EC3.1.4.50).

Diverse inositol effects: A reduction in lung cancer risk in the postinitiation period has been observed in experimental models (Wattenberg, 1999). Evidence is limited or disputed for suggested inositol effects on sperm maturation, maintenance of microtubule function and stability, and maturation of newborn lungs. It also remains to be studied whether humans, like rodents in a preliminary study (Juriloff and Harris, 2000), have an increased risk of NTDs when they have less than optimal inositol intake. Animal experiments suggest that high inositol intake may cause neuropathy, especially with an imbalance of the polyol pathway in diabetics (Williamson et al., 1986).

Osmoregulation: Inositol is one of several compounds that can protect cells from damage due to high osmotic pressure. Tubular cells in the renal papillae, where osmotic pressure is highest, promote the uptake of inositol from the intraluminal fluid via the myo-inositol/sodium cotransporter

(Burger-Kentischer et al., 1999). Brain cells also contain large amounts of inositol and increase its uptake in response to hyperosmolarity. Since the inositol transporter gene is located on chromosome 21, excessive uptake due to the 50% higher gene dose in trisomy 21 may disrupt normal neuronal function in people with Down syndrome (Berry et al., 1999).

In addition, inositol-phosphates are important regulars of water disposition through the arginine vasopressin/aquaporin-2 signaling pathway in renal collecting ducts (Pernot et al., 2011).

Mental health: One of the tissues with a very active inositol phosphate metabolism is the brain. Inositol-1,4-bisphosphate 1-phosphatase (EC3.1.3.57) and myo-inositol-1(or 4)-monophosphatase (EC3.1.3.25) are brain enzymes of particular interest because both are inhibited by lithium and may be relevant targets of lithium treatment for manic depression (Patel et al., 2002). High doses (18 g/day) of inositol intake appear to be effective for the treatment of some mental disorders, including depression, panic disorder, and obsessive-compulsive disorder (Palatnik et al., 2001).

Intracellular phytate effects: Dietary phytate has been reported to reduce carcinogenesis in various standard animal models (Shamsuddin, 1999). It may be relevant in this respect that IP6 directly up-regulates the expression of p21WAF-1/CIP1 in a dose-dependent manner.

Inhibition of intestinal cation absorption: Both phytate (IP6) and IP5 potently decrease the bioavailability of concurrently consumed nonheme iron (Sandberg et al., 1999). The minute amount of 2 mg of phytate (corresponding to 20 ml of brewed black tea) can decrease absorption by 18%, 25 mg of phytate (250 ml of tea) cuts absorption in half, and 900 mg (for instance, in 60 g of mixed nuts) reduces it to 85% (Hallberg et al., 1989; Siegenberg et al., 1991). The inhibiting effect can be counteracted by large amounts (100 mg or more) of ascorbate. Inositol derivatives with fewer phosphate groups interfere to a much lesser degree. Phytate also decreases absorption of zinc very significantly (Lönnerdal, 2000), while absorption of copper (Lönnerdal et al., 1999) is affected to a much lesser extent, and the other divalent metal cations (manganese, calcium, and magnesium) only minimally.

REFERENCES

Berry, G.T., Wang, Z.J., Dreha, S.F., Finucane, B.M., Zimmerman, R.A., 1999. *In vivo* brain myo-inositol levels in children with Down syndrome. J. Pediatr. 135, 94–97.

Bitar, K., Reinhold, J.G., 1972. Phytase and alkaline phosphatase activities in intestinal mucosae of rat, chicken, calf, and man. Biochim. Biophys. Acta. 268, 442–452.

Burger-Kentischer, A., Muller, E., März, J., Fraek, M.L., Thurau, K., Beck, F.X., 1999. Hypertonicityinduced accumulation of organic osmolytes in papillary interstitial cells. Kidney Int. 55, 1417–1425.

Craxton, A., Caffrey, J.J., Burkhart, W., Safrany, S.T., Shears, S.B., 1997. Molecular cloning and expression of a rat hepatic multiple inositol polyphosphate phosphatase. Biochem. J. 328, 75–81.

Grases, E., Simonet, B.M., Vucenik, I., Prieto, R.M., Costa-Bauza, A., March, J.G., et al., 2001. Absorption and excretion of orally administered inositol hexaphosphate (IP(6) or phytate) in humans. Biofactors 15, 53–61.

Hallberg, L., Brune, M., Rossander, L., 1989. Iron absorption in man: ascorbic acid and dose-dependent inhibition by phytate. Am. J. Clin. Nutr. 49, 140–144.

Holub, B.J., 1986. Metabolism and function of myo-inositol and inositol phospholipids. Annu. Rev. Nutr. 6, 563–597.

Juriloff, D.M., Harris, M.J., 2000. Mouse models for neural tube closure defects. Hum. Mol. Genet. 9, 993–1000.

Lönnerdal, B., 2000. Dietary factors influencing zinc absorption. J. Nutr. 130, 1378S–1383S.

Lönnerdal, B., Jayawickrama, L., Lien, E.L., 1999. Effect of reducing the phytate content and of partially hydrolyzing the protein in soy formula on zinc and copper absorption and status in infant rhesus monkeys and rat pups. Am. J. Clin. Nutr. 69, 490–496.

Palatnik, A., Frolov, K., Fux, M., Benjamin, J., 2001. Double-blind, controlled, crossover trial of inositol versus fluvoxamine for the treatment of panic disorder. J. Clin. Psychopharmacol. 21, 335–339.

Patel, S., Yenush, L., Rodriguez, P.L., Serrano, R., Blundell, T.L., 2002. Crystal structure of an enzyme displaying both inositol-polyphosphate-l-phosphatase and 3′-phosphoadenosine-5′-phosphate phosphatase activities: a novel target of lithium therapy. J. Mol. Biol. 315, 677–685.

Pereira, G.R., Baker, L., Egler, J., Corcoran, L., Chiavacci, R., 1990. Serum myoinositol concentrations in premature infants fed human milk, formula for infants, and parenteral nutrition. Am. J. Clin. Nutr. 51, 589–593.

Pernot, E., Terryn, S., Cheong, S.C., Markadieu, N., Janas, S., Blockmans, M., et al., 2011. The inositol Inpp5k 5-phosphatase affects osmoregulation through the vasopressin-aquaporin 2 pathway in the collecting system. Pflugers Arch. 462, 871–883.

Sandberg, A.S., Brune, M., Carlsson, N.G., Hallberg, L., Skoglund, E., Rossander-Hulthen, L., 1999. Inositol phosphates with different numbers of phosphate groups influence iron absorption in humans. Am. J. Clin. Nutr. 70, 240–246.

Shamsuddin, A.M., 1999. Metabolism and cellular functions of IP6: a review. Anticancer Res. 19, 3733–3736.

Siegenberg, D., Baynes, R.D., Bothwell, T.H., Macfarlane, B.J., Lamparelli, R.D., Car, N.G., et al., 1991. Ascorbic acid prevents the dose-dependent inhibitory effects of polyphenols and phytates on nonheme-iron absorption. Am. J. Clin. Nutr. 53, 537–541.

Vucenik, I., Shamsuddin, A.M., 1994. [3H] inositol hexaphosphate (phytic acid) is rapidly absorbed and metabolized by murine and human malignant cells *in vitro*. J. Nutr. 124, 861–868.

Wattenberg, L.W., 1999. Chemoprevention of pulmonary carcinogenesis by myo-inositol. Anticancer Res. 19, 3659–3661.

Williamson, J.R., Chang, K., Rowold, E., Marvel, J., Tomlinson, M., Sherman, W.R., et al., 1986. Diabetes-induced increases in vascular permeability and changes in granulation tissue levels of sorbitol, myo-inositol, chiro-inositol, and scyllo-inositol are prevented by sorbinil. Metab. Clin. Exper. 35, 41–45.

MINERALS AND TRACE ELEMENTS

CHAPTER OUTLINE

Water 673
Sodium 679
Potassium 685
Chlorine 690
Iron 697
Copper 709
Zinc 716
Manganese 725
Calcium 730
Phosphorus 737
Magnesium 744
Iodine 748
Fluorine 755
Sulfur 758
Selenium 766
Molybdenum 773
Cobalt 778
Chromium 782
Boron 785
Silicon 789
Bromine 793
Arsenic 797
Vanadium 802
Nickel 805

WATER

Water (hydrogen oxide; molecular weight 18.015) is the major liquid in humans. Less than half a percent of naturally occurring water contains one or two atoms of the stable isotope ^2H (deuterium), the stable isotopes ^{17}O or ^{18}O, or both. Water with the unstable (radioactive) isotope ^3H (tritium, half-life 12.3 years) comes almost exclusively from artificial sources (nuclear reactors or nuclear weapons). Artificially produced radioactive isotopes of oxygen (^{15}O and others) are extremely short-lived. A small portion of water dissociates into its ions H^+ and OH^-, depending on the amount of acids and bases present.

Nutrient Metabolism.

ABBREVIATION	
AQP	aquaporin

NUTRITIONAL SUMMARY

Function: Water provides the medium for metabolic reactions and for the flow of blood and other body fluids. The hydroxide and proton ions are used in various reactions for the synthesis and breakdown of body constituents. Evaporation of water from sweat protects against overheating.

Requirements: Intake must maintain adequate body hydration and be sufficient for renal excretion of solutes. Intake of 2000 ml or more is usually adequate for many people. Individuals with high sweat output or working at high altitude or intense heat need more.

Sources: Most water is consumed either straight or with very dilute beverages (such as coffee, tea, or low-calorie soda) or with nutrient-rich beverages (such as milk and other dairy products, juices, and sweetened drinks).

Deficiency: Low water intake can rapidly lead to dehydration, particularly with exertion or exposure to high temperature or dry air or when suffering from diarrhea. Dehydration impairs physical and mental performance, may cause loss of consciousness, renal failure, coma, and death.

Excessive intake: Intake of several liters more than necessary can cause abnormal dilution of body fluids and loss of essential electrolytes.

ENDOGENOUS SOURCES

Complete oxidation of all energy fuels generates water. Water yield is about 1.1 ml/g for fat and ethanol, 0.6 ml/g for carbohydrates, and about 0.4 ml/g for protein. This means that a daily intake of 260 g of carbohydrate, 60 g of protein, and 60 g of fat generates about 246 ml of water upon complete oxidation.

DIETARY SOURCES

Water is consumed with both liquids and most solid foods. The composition of the beverage matters since high electrolyte and nitrogen content may increase solvent drag and losses in the kidney, while optimal amounts of sodium and glucose (or galactose) can aid with water uptake in some situations (e.g., diarrhea).

Adequate and affordable access to clean water is becoming increasingly precarious for a growing part of the world population (Rush, 2013).

DIGESTION AND ABSORPTION

The mucosa of the proximal small intestine has a net water secretion of about 30 ml/h (Knutson et al., 1995). The water moves mainly through the gap junctions between the enterocytes, following the osmotic gradient to the more concentrated luminal contents. The higher than isoosmolar luminal concentration is due to ingested foods, glandular secretions, and sodium secreted by the enterocytes. Water is largely taken up again into small intestinal enterocytes with sodium and glucose via the sodium-glucose cotransporter (SGLT1, SLC5A1) and other sodium cotransporters. Since more than 200 water molecules are moved into the enterocyte with each glucose molecule, the absorption of 100 g of glucose pulls along about 20 times as much water (about 2000 ml).

Aquaporin 3 (AQP3), whose expression increases from the stomach and is greatest in the ileum and colon, allows water movement across the basolateral membrane (Purdy et al., 1999). Water transport across the basolateral membrane of colon enterocytes also uses AQP4 (Wang et al., 2000). In healthy people, less than 200 ml of water is lost via feces. This amount can increase to several liters in patients suffering from cholera or other types of severe diarrhea (Figure 11.1).

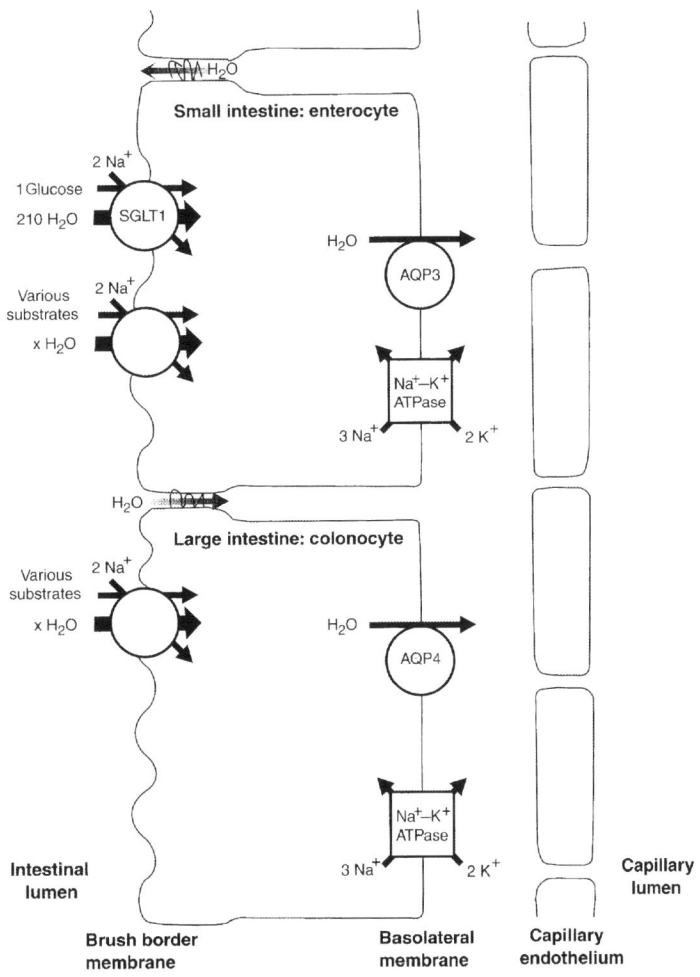

FIGURE 11.1

Mechanisms of water secretion and absorption in the small and large intestine.

TRANSPORT AND CELLULAR UPTAKE

Blood circulation: The capillary endothelium in most regions has numerous gaps and fenestrations between adjacent cells and does not form any barrier to the movement of water into or from the extra-cellular fluid space.

Blood–brain barrier (BBB): Brain capillary endothelium effectively blocks the movement of water between cells. Water can reach the brain only by transcellular transport (with a few regional exceptions). One of the exceptions is the choroid villous, where cerebrospinal fluid is produced.

Materno-fetal transfer: Net water flux to the mature fetus is 20–30 ml/day (Roberts et al., 1999) through specific water channels, including aquaporin 3 (AQ3), aquaporin 8 (AQ8), and aquaporin 9 (AQ9) (Damiano et al., 2001). Mineralocorticoid hormones participate in the control of water transport across the syntrophoblast (Hirasawa et al., 2000). The chorioamniotic membranes recover water from amniotic fluid.

EXCRETION

Water is lost via urine, feces, expired air, and sweat. The relative proportions of these losses can vary greatly depending on intake, environment, health, and other factors.

Urine: A healthy 70-kg person filters about 173 l of water per day through the glomeruli of the kidneys. Much of the reabsorption occurs from the small proximal tubule. A significant part is coupled to sodium cotransport. Another part moves across the (not-so) tight junctions by passive diffusion following the osmotic gradient due to the reabsorbed sodium chloride, bicarbonate, glucose, phosphate, amino acids, and other actively transported solutes. Toward the S3 segment of the proximal tubule, the chloride gradient becomes most important for passive water diffusion.

The blood in the capillaries that run parallel to the proximal tubules is more concentrated than normal blood because a significant amount of water has just been removed from it by filtration in the glomeruli. Water coming from the basolateral space into the capillary lumen, following the oncotic (protein concentration) gradient; thus, it comes full circle.

Water can move across the epithelium of the distal nephron without accompanying solutes by permeating through aquaporins. Aquaporin 2 (AQP2) allows water movement across the apical membrane of collecting ducts. AQP3 and aquaporin 4 (AQP4) are present in the basolateral membrane (Klussmann et al., 2000). Additional aquaporins are expressed in the kidneys.

The solute load (mainly comprising electrolytes and urea) determines the obligatory (minimal) water volume of urine. A healthy young adult can concentrate urine to about 1400 mosmol/kg (Rose, 1989). The ability to maximally concentrate urine declines with aging. Increased protein and salt consumption, therefore, increases minimal water requirements.

Exhaled water: Overly dry air damages the alveolar cells in the lungs. The oral and nasal cavity and the other segments of the air passages are set up to bring inhaled air to near 100% humidity at body temperature. Much of the moisture added during inhalation remains with exhaled air. Normal water losses with breathing are about 400 ml/day but may be considerably higher in a dry environment and with exercise (increased respiratory volume).

Sweating: Water losses from skin vary greatly depending on environmental temperature, level of exertion, fitness level, and regulatory adaptation. A trained athlete exercising in a hot environment may lose 1.6 L of water per hour or more over extended periods of time. Water loss is also typically extensive in tropical climates, especially for those who are not adapted to such climates. This does not need to be evident from profuse sweat accumulation since in a dry environment, water may evaporate as quickly as it appears on the surface of the skin.

REGULATION

Water balance is maintained through adjusting intake (thirst response) and urinary output. Thirst is perceived as a result of peripheral and central stimulatory and inhibitory signals that come together in a frontocortical brain region (Schmitt et al., 2000; Hallschmid et al., 2001). Elevated plasma sodium concentration (hypernatremia) indicates dehydration, and depressed sodium concentration (hyponatremia) points to impaired water excretion (Rose, 1989).

Thirst is a complex physiological response to water need (dehydration), but it may not always be strong enough to ensure intake. Aging tends to blunt both the mechanisms causing thirst (e.g., osmosensors) and the strength of the thirst sensation. Drinking rapidly quenches thirst even before ingested water has been absorbed. Adequate hydration also minimizes thirst sensation. Social settings and individual habits often cause intake beyond the need for adequate hydration.

Antidiuretic hormone (ADH; arginine vasopressin) is the major regulator of urinary water loss in humans, acting mainly on the luminal membranes of collecting tubules when solute concentration (primarily sodium) in the blood rises (hyperosmolality). The supraoptic and paraventricular nuclei of the hypothalamus secrete ADH in response to signals coming from osmoreceptors in the hypothalamus (vascular organ of the lamina terminalis) and other regulatory sites. ADH can bind to specific G-protein-linked receptors, increase cyclic adenosine monophosphate (cAMP) production, and thereby stimulate the translocation of AQP2 from cytoplasmic vesicles to the apical membranes of collecting ducts (Saito et al., 2000). About 3% of the AQP2 located at the luminal membrane is shed via urine and may serve as an indicator of vasopressin action (Baumgarten et al., 2000).

Immersion of the full body in water stimulates the release of ADH (Buemi et al., 2000).

FUNCTION

Solvent: Water dissolves the numerous solutes, cellular components, micellar particles and lipoproteins in the blood, cerebrospinal fluid, extravascular fluid, intracellular fluid, glandular secretions, bile, intestinal lumen content, and urine. Flow obviously would not be possible without it. Long before impaired flow becomes an issue, however, the concentration of the various nonwater constituents in body fluids is critical. The solute concentration (osmolality) of cells, extracellular fluid, blood, and cerebrospinal fluid has to be maintained within narrow ranges. Deviations of a few percentage points severely disrupt a multitude of functions and regulatory mechanisms.

Water ionization: A small fraction of water dissociates into the ions H^+ and OH^-. The concentration of protons (H^+) strongly influences chemical reaction rates. The human body and its constituent cells tolerate only minimal changes in pH from the optimum (usually around 7.4). The exception from this rule is the secretion of hydrochloric acid in the stomach for the digestion and antibacterial treatment of food. Similarly, the lysosomes are digestive cell organelles that maintain an acidic pH of around 5 to facilitate chemical processes that require this unusual milieu. For instance, the iron-regulating hormone binds to the iron exporter ferroportin (SLC40A1), and this complex is then trafficked to lysosomes, where it is broken down into its amino acid components (Preza et al., 2013).

Chemical reactions: Water is a participant in nearly all metabolic events, either as the essential solvent or as a reactant. The rates of most reactions are strongly influenced by both water ion concentration (pH) and dilution (ion strength). In many reactions, water ions (H^+ or OH^-) are substrates themselves. Common examples include all reactions involving nicotinamide adenine dinucleotide (NAD) or flavin adenine dinucleotide (FAD).

Temperature regulation: Evaporation of water from skin cools the body by dissipating about 580 kcal/l (Guyton, 1986).

REFERENCES

Baumganen, R., van de Pol, M.H., Deen, P.M., van Os, C.H., Wetzels, J.F., 2000. Dissociation between urine osmolality and urinary excretion of aquaporin-2 in healthy volunteers. Nephrol. Dial. Transplant. 15, 1155–1161.

Buemi, M., Corica, F., Di Pasquale, G., Aloisi, C., Soft, M., Casuscelli, T., et al., 2000. Water immersion increases urinary excretion of aquaporin-2 in healthy humans. Nephron 85, 20–26.

Damiano, A., Zotta, E., Goldstein, J., Reisin, I., Ibarra, C., 2001. Water channel proteins AQP3 and AQP9 are present in syncytiotrophoblast of human term placenta. Placenta 22, 776–781.

Guyton, A.C., 1986. Textbook of Medical Physiology, seventh ed. Saunders, Philadelphia, PA, Chapter 72.

Hallschmid, M., Molle, M., Wagner, U., Fehm, H.L., Born, J., 2001. Drinking related direct current positive potential shift in the human EEG depends on thirst. Neurosci. Lett. 311, 173–176.

Hirasawa, G., Takeyama, J., Sasano, H., Fukushima, K., Suzuki, T., Muramatu, Y., et al., 2000. 11Beta-hydroxysteroid dehydrogenase type II and mineralocorticoid receptor in human placenta. J. Clin. Endocrinol. Metab. 85, 1306–1309.

Klussmann, E., Maric, K., Rosenthal, W., 2000. The mechanisms of aquaporin control in the renal collecting duct. Rev. Physiol. Biochem. Pharmacol. 141, 33–95.

Knutson, T.W., Knutson, L.E., Hogan, D.L., Koss, M.A., Isenberg, J.l., 1995. Human proximal duodenal ion and water transport. Role of enteric nervous system and carbonic anhydrase. Dig. Dis. Sci. 40, 241–246.

Preza, G.C., Pinon, R., Ganz, T., Nemeth, E., 2013. Cellular catabolism of the iron-regulatory peptide hormone hepcidin. PLoS One 8, e58934.

Purdy, M.J., Cima, R.R., Doble, M.A., Klein, M.A., Zinner, M.J., Soybel, D.I., 1999. Selective decreases in levels of mRNA encoding a water channel (AQP3) in ileal mucosa after ileostomy in the rat. J. Gastroint. Surg. 3, 54–60.

Roberts, T.J., Nijland, M.J.M., Williams, L., Ross, M.G., 1999. Fetal diuretic responses to maternal hyponatremia: contribution of placental sodium gradient. J. Appl. Physiol. 87, 1440–1447.

Rose, B.D., 1989. Clinical Physiology of Acid-Base and Electrolyte Disorders. McGraw-Hill, New York, NY.

Rush, E.C., 2013. Water: neglected, unappreciated and under researched. Eur. J. Clin. Nutr. 67, 492–495.

Saito, T., Ishikawa, S.E., Ando, E., Higashiyama, M., Nagasaka, S., Sasaki, S., 2000. Vasopressin-dependent upregulation of aquaporin-2 gene expression in glucocorticoid-deficient rats. Am. J. Physiol. Renal Fluid Electrolyte Physiol. 279, F502–F508.

Schmitt, B., Molle, M., Marshall, L., Born, J., 2000. Scalp recorded direct current potential shifts associated with quenching thirst in humans. Psychophysiology 37, 766–776.

Wang, K.S., Ma, T., Filiz, E., Verkman, A.S., Bastidas, J.A., 2000. Colon water transport in transgenic mice lacking aquaporin-4 water channels. Am. J. Physiol. Gastroint. Liver Physiol. 279, G463–G470.

SODIUM

Sodium (also known as natrium, hence the abbreviation of *Na* for the atom) is an alkali metal with valence 1. The naturally occurring stable isotope ^{23}Na has an atomic weight of 22.98977. The rare isotope ^{22}Na is unstable, with a half-life of 2.6 years.

ABBREVIATIONS

GLUT2	glucose transporter 2 (SLC2A2)
MSG	monosodium glutamate
NBC1	sodium/bicarbonate cotransporter 1 (SLC4A4)
NHE1	sodium/hydrogen exchanger 1 (SLC9A1)
SGLT1	sodium-glucose cotransporter 1 (SLC5A1)

NUTRITIONAL SUMMARY

Function: Sodium is the main cationic osmolyte in the blood and extracellular fluid and mediates active transport of numerous nutrients and metabolites in the intestines, kidneys, and many other tissues.

Food sources: Meats, pickled foods, and salty snacks are the major sources of sodium.

Requirements: Intake has to match sodium losses. In most situations, a balance can be achieved with daily intake of a few hundred milligrams. Increased sweating (such as with great exertion, fever, heat, and high humidity) and diuresis can increase sodium needs to several grams per day. Very low sodium intake (such as what occurs with fasting) can cause dizziness and weakness due to hypotension.

Excessive intake: Higher than minimal sodium intake may increase blood pressure, especially in genetically susceptible individuals and when other hypertensive factors (e.g., obesity) are present.

DIETARY SOURCES

Most sodium in food comes from animal foods, added salt (sodium chloride; NaCl), and sodium salts, such as monosodium glutamate (MSG), in commercial products. Sodium content of meats tends to correlate inversely with fat content. Chicken (0.86 mg/g) and fish (0.78 mg/g) tend to be higher in sodium than beef (0.57 mg/g) and pork (0.47 mg/g). Fruits, vegetables, tubers, and grains by themselves have very low sodium content (typically <0.1 g/kg).

DIGESTION AND ABSORPTION

Sodium is absorbed nearly completely, which is facilitated by the great solubility of its salts in the aqueous environment of the digestive tract. Mechanisms of sodium uptake into enterocytes include cotransport with the sodium-glucose cotransporter (SGLT1, SLC5A1) and other sodium cotransporters. The sodium/proton exchangers NHE2 (SLC9A2) and NHE3 (SLC9A3) are equally important channels through which sodium ions can flow in the direction of the prevalent concentration gradient (Kato & Romero, 2011). There is usually net secretion of sodium into the duodenum and net absorption in the jejunum. Most of the remaining sodium of the luminal contents is absorbed from the colon. The driving force for sodium absorption is the gradient between the near-isotonic luminal contents and the low sodium concentration inside the intestinal cells. Sodium/potassium-exchanging adenosine triphosphatease (ATPase, EC3.6.3.9) pumps sodium, concurrent to the sodium flow from the lumen into the cell, across the basolateral side into the pericapillary space. Most sodium diffuses from there into capillary blood; smaller amounts leak across the tight junctions between the enterocytes back into the intestinal lumen. The main flux of sodium into small intestinal enterocytes is with bulk nutrients. About 26 g of sodium is absorbed with 100 g of glucose, for example. Since much of this sodium comes from secretions and paracellular back leakage, the few grams of ingested sodium have little influence on overall flux in health. The 2 l of water absorbed along with 100 g of glucose can become more important, particularly in diarrhea. The molecular coupling of sodium absorption with both nutrient and water absorption provides the molecular basis for the current practice of using mixed sodium-glucose solutions for rehydration in severe diarrhea.

Several channels in the basolateral membrane also mediate sodium influx to a limited extent. They are tightly controlled because otherwise they would weaken the all-important sodium concentration gradient established by the sodium/potassium-exchanging ATPase. Important examples are the sodium/bicarbonate cotransporter (NBC1, SLC4A4, Praetorius et al., 2001), the sodium-potassium-chloride transporter 1 (SLC12A2), and the sodium/proton exchanger 1 (SLC9A1). Both NHE1 and NBC1 transporters are stimulated by cAMP (Figure 11.2).

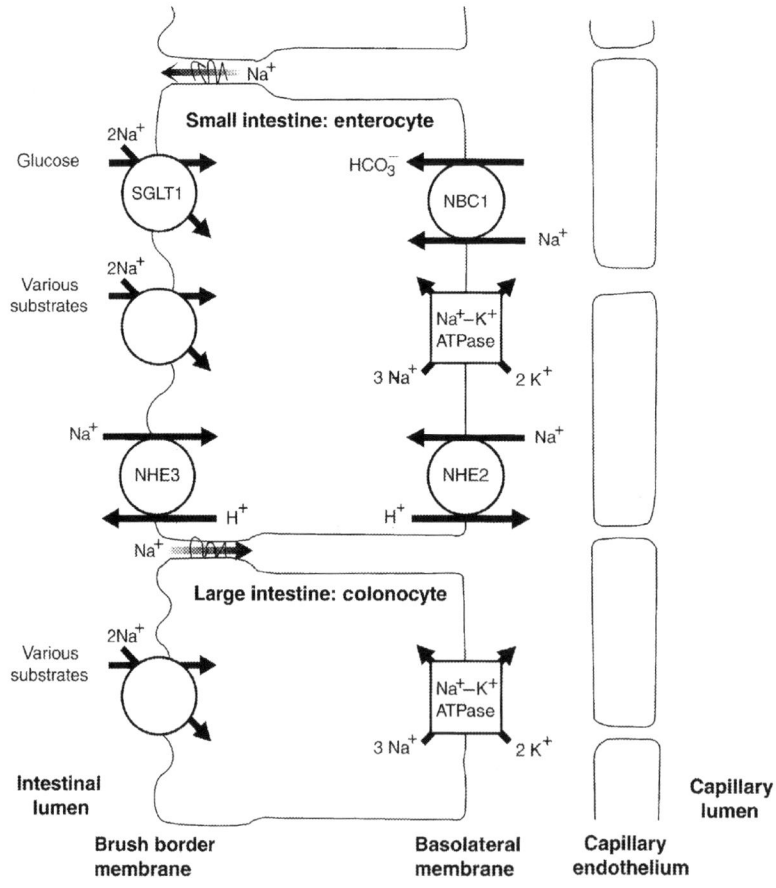

FIGURE 11.2

Intestinal sodium absorption.

TRANSPORT AND CELLULAR UPTAKE

Blood circulation: The sodium concentration in the blood is around 142 mmol/l (3.27 g/l). Sodium in blood capillaries can get into most tissues by simply diffusing through the gaps between endothelial cells. The gradient between the intracellular sodium concentration (usually around 10 mmol/l) and the concentration in extracellular fluid drives sodium entry into cells via sodium cotransporters, sodium/hydrogen exchangers (NHEs), or sodium channels. Sodium/potassium-exchanging ATPase (EC3.6.3.9) pumps sodium against this gradient out of the cell.

BBB: Highly effective tight junctions in most brain regions block the diffusion of sodium between the blood and brain in either direction across the capillary endothelium. This means that sodium flow occurs only through the epithelial cells via transporters and channels. A few brain regions are exempted from this tight enclosure. Notably, the capillary endothelium adjacent to the vascular organ of the lamina terminalis in the hypothalamus allows the relatively free exchange of sodium. Osmoreceptors in this area sense blood osmolality and thus sodium concentration (Stricker and Sved, 2000).

Materno-fetal transfer: Sodium/potassium-exchanging ATPase (EC3.6.3.9) at the maternal side maintains the low intracellular sodium concentration of the syntrophoblast layer of the placenta (Johansson et al., 2000). Since glucose, amino acids, and other nutrients and metabolites from maternal blood are transported with sodium into the syntrophoblast, large amounts have to be pumped back again. The same sodium pump is also active at the fetal side, though with less activity. Endogenous digitalis-like factor modulates ATPase activity. In addition to the pumps and cotransporters, sodium is moved in exchange for protons. The sodium/hydrogen exchanger 1 (NHE1, SLC9A1) is located on the maternal side of the placental barrier, NHE3 (SLC9A3) on the fetal side (Pepe et al., 2001). The fetal side also contains a sodium-potassium-chloride cotransporter (Zhao and Hundal, 2000).

STORAGE

A 70-kg person contains about 100 g of sodium, about half of that in bone and 40% in extracellular fluid.

EXCRETION

Sodium is lost mainly via urine and sweat, and to a much lesser extent through body secretions, skin, and feces. Losses via feces are normally less than 0.1 g/day.

Renal losses: Over 600 g/day of sodium are filtered into the renal tubules (Feraille and Doucet, 2001). Some of the filtered sodium is taken up from the proximal tubular lumen through the cotransporters for phosphate, glucose, amino acids, and other substrates (Figure 11.3).

The bulk (about 60%) of filtered sodium is absorbed from the proximal tubular lumen in exchange for protons via the sodium/hydrogen exchanger 3 (NHE3, SLC9A3) and to a lesser extent the closely related isoform NHE2 (SLC9A3). Sodium/potassium-exchanging ATPase (EC3.6.3.9) is the driving force to pump sodium across the basolateral membrane of the renal epithelial cells (Satoh et al., 1999). Additional sodium leaves the thin ascending limb of Henle's loop by passive diffusion. The kidney-specific sodium-potassium-chloride transporter 2 (SLC12A1, inhibited by the diuretic furosemide) in the TAL of Henle's loop accounts for about 30% of sodium absorption. The distal tubule reabsorbs about 5–7% of the filtered sodium via a sodium chloride transporter (SLC12A3) and basolateral ATPase-mediated transport (Hropot et al., 1985). The final sodium concentration of urine is adjusted in the distal nephron. Sodium flows into the principal cells of the cortical and medullary collecting

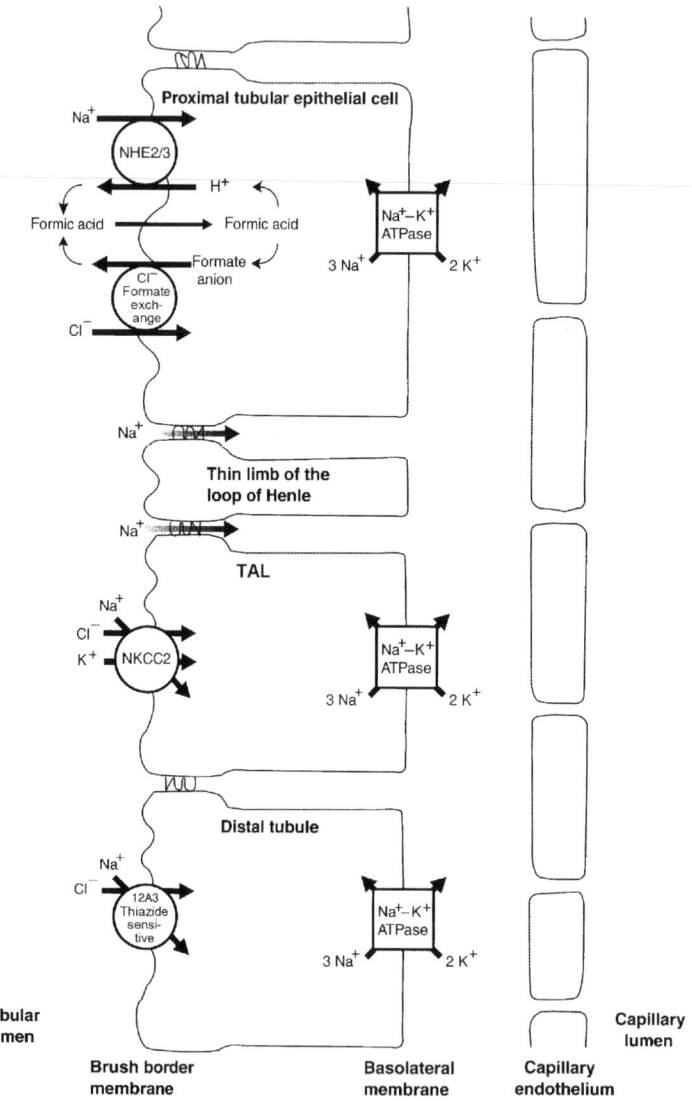

FIGURE 11.3

Renal handling of sodium.

tubules through the epithelial sodium channel (ENaC; Su and Menon, 2001) at a rate that is controlled primarily by aldosterone.

Excretion via sweat: The sodium content of sweat is about 50 mmol/l (Shirreffs and Maughan, 1998; Sanders et al., 1999). An hour of vigorous exercise at normal room temperature (22–25°C) typically induces the loss of about 0.5 l of sweat and thus about 0.6 g of sodium. Exertion near or above body temperature increases sweat production to several liters per day, with correspondingly high electrolyte losses. Ingestion of water and sugar promotes sweating (Fritzsche et al., 2000). Stimulation of the sympathetic nervous system increases sweating.

Cholecystokinin, corticotrophin-releasing hormone, enterostatin, leptin, and alphamelanocyte-stimulating hormone (melanocyte-stimulating hormone, MSH) increase sympathetic activity; neuropeptide Y (NPY), beta-endorphin, orexin, galanin, and melanin-concentrating hormone reduce it (Bray, 2000).

REGULATION

Excretory control: Since sodium is the main osmolyte (along with chloride) in the blood, its homeostasis is closely related to volume control. Factors that change renal perfusion can affect sodium homeostasis under some conditions. For example, the high estrogen concentrations during the luteal phase of the menstrual cycle in healthy women increase the glomerular filtration rate upon sodium loading (Pechère-Bertschi et al., 2002). This response to sodium loading is not seen while estrogen concentrations are lower during the follicular phase. The iuxtaglomerular cells in the kidneys secrete renin (EC3.4.23.15) in response to decreased blood flow. Renin activates angiotensin I, from which peptidyl-dipeptidase A (angiotensin 1-converting enzyme, EC3.4.15.1) generates angiotensin II. Angiotensin II promotes sodium absorption from the proximal tubule and stimulates aldosterone secretion. The steroid hormone aldosterone from the adrenal cortex increases sodium reabsorption from the connecting segment and the collecting tubule by opening the sodium channels on the luminal side.

Several hormones act on sodium/potassium-exchanging ATPase and modulate renal sodium excretion, including glucocorticoids, insulin, parathyroid hormone (PTH), vasoactive intestinal peptide (VIP), adrenaline (epinephrine), noradrenaline (norepinephrine), and dopamine (Féraille and Doucet, 2001). Myocardial cells in the atria of the heart secrete atrial natriuretic peptide (ANP) in response to volume expansion. ANP promotes sodium excretion by reducing reabsorption from the collecting tubules.

The hypothalamus releases endogenous mediator ouabain-like factor (OLF) in response to volume expansion. It is likely that OLF induces the cytoskeletal protein adducing to promote sodium/potassium-exchanging ATPase (EC3.6.3.9) activity, thereby increasing sodium reabsorption. The complete sequence of this regulatory pathway remains to be established. Glucagon, acting on its own specific receptor, has a natriuretic effect (Strazzullo et al., 2001). Insulin-resistant individuals have an impaired natriuretic response to high sodium intake (Facchini et al., 1999).

Intake control: Excessive sodium intake is partially prevented by salt aversion. Sodium channels on the apical surface of bipolar cells in the taste buds of the mouth sense the sodium concentration of ingested food and drink (Lin et al., 1999). Their sensitivity is influenced by aldosterone and other regulators of sodium homeostasis. Osmosensors in the vascular organ of the lamina terminalis in the hypothalamus sense increased blood osmolality and thus causes a likely sodium overload. The increased water intake in response to the ensuing thirst leads to the dilution of plasma sodium and provides the water volume for renal sodium excretion. The thirst sensation also suppresses salt appetite (Thunhorst and Johnson, 2001).

FUNCTION

Enzyme cofactor: Because it is the main osmolyte in the blood and the extracellular space, sodium concentration directly affects volume. All enzyme-catalyzed reactions have specific requirements for ionic strength, which can be satisfied by sodium in the extracellular milieu. Beyond that, some reactions specifically depend on sodium ions.

Cotransport: Much of the movement of nutrients and metabolites is achieved by using the power of a sodium gradient. The enterocytes of the small intestine can serve as an example. Sodium/potassium-exchanging ATPase (EC3.6.3.9) at the basolateral membrane keeps the intracellular sodium concentration low by pumping sodium from the enterocyte into the pericapillary space, from where it diffuses into the bloodstream.

The high sodium concentration in the intestinal lumen (from pancreatic and enteric secretions) creates a steep inward gradient. The sodium-glucose cotransporter 1 (SGT1, SLC5A1), like many similar cotransporters, moves sodium, water, and glucose along that sodium gradient into the cell. This transport builds up enough glucose concentration in the enterocyte that it carries this substrate via glucose transporter 2 (GLUT2, SLC2A2) right across the basolateral membrane into the pericapillary space, and eventually into the capillary bloodstream.

Signaling: Voltage-gated sodium channels are activated by the early phase of membrane polarization and allow the rapid influx of sodium into neurons and muscle cells. This flux provides the driving force for continued membrane polarization.

NUTRITIONAL ASSESSMENT

Suboptimal sodium status is indicated by low concentration in the blood (hyponatremia). Dietary intake can be estimated from the amount of excreted urinary sodium.

Blood measurement: Sodium concentration in plasma or serum is one of the most commonly measured chemistries in clinical care. Deviations of more than 10% from the normal concentration tend to indicate serious pathologies. Most often, such high concentrations are related to insufficient hydration. Low concentrations may indicate insufficient compensation of high losses, such as when some endurance athletes drink water with little or no accompanying sodium (Hoffman et al., 2013). Too many medications and medical conditions to mention here can disrupt normal sodium homeostasis and cause abnormal blood concentrations.

Urine measurement: The amount excreted with a urine sample collected over a 24-h period reflects recent sodium intake. Since the accurate collection of urine over a precise 24-h period is difficult to implement, some investigators collect a random urine sample and determine the sodium/creatinine ratio (Rhee et al., 2014). This approach is unreliable because sodium excretion varies during the course of a day and because daily creatinine excretion can vary across populations and individuals. The results can be misleading, for instance, because low creatinine excretion due to diminished muscle mass might suggest a much higher level than the actual daily sodium excretion (McQuarrie et al., 2014).

REFERENCES

Bray, G.A., 2000. Reciprocal relation of food intake and sympathetic activity: experimental observations and clinical implications. Int. J. Ob. Rel. Metab. Dis. 24, S8–S17.

Facchini, F.S., DoNascimento, C., Reaven, G.M., Yip, J.W., Ni, X.E., Humphreys, M.H., 1999. Blood pressure, sodium intake, insulin resistance, and urinary nitrate excretion. Hypertension 33, 1008–1102.

Féraille, E., Doucet, A., 2001. Sodium-potassium-adenosinetriphosphatase-dependent sodium transport in the kidney: hormonal control. Physiol. Rev. 81, 345–418.

Fritzsche, R.G., Switzer, T.W., Hodgkinson, B.J., Lee, S., Martin, J.C., Coyle, E.E., 2000. Water and carbohydrate ingestion during prolonged exercise increase maximal neuromuscular power. J. Appl. Physiol. 88, 730–737.

Hoffman, M.D., Fogard, K., Winger, J., Hew-Butler, T., Stuempfle, K.J., 2013. Characteristics of 161-km ultramarathon finishers developing exercise-associated hyponatremia. Res. Sports Med. 21, 164–175.

Hropot, M., Fowler, N., Karlmark, B., Giebisch, G., 1985. Tubular action of diuretics: distal effects on electrolyte transport and acidification. Kidney Int. 28, 477–489.

Johansson, M., Jansson, T., Powell, T.L., 2000. Na(+)-K(+)-ATPase is distributed to microvillous and basal membrane of the syncytiotrophoblast in human placenta. Am. J. Physiol. Reg. Integ. Comp. Physiol. 279, R287–R294.

Kato, A., Romero, M.F., 2011. Regulation of electroneutral NaCl absorption by the small intestine. Annu. Rev. Physiol. 73, 261–281.

Lin, W., Finger, T.E., Rossier, B.C., Kinnamon, S.C., 1999. Epithelial Na$^+$ channel subunits in rat taste cells: localization and regulation by aldosterone. J. Comp. Neurol. 405, 406–420.

McQuarrie, E.P., Traynor, J.P., Taylor, A.H., Freel, E.M., Fox, J.G., Jardine, A.G., et al., 2014. Association between urinary sodium, creatinine, albumin, and long-term survival in chronic kidney disease. Hypertension 64 (1), 111–117.

Pechère-Bertschi, A., Maillard, M., Stalder, H., Brunner, H.R., Burnier, M., 2002. Renal segmental tubular response to salt during the normal menstrual cycle. Kidney Int. 61, 425–431.

Pepe, G.J., Burch, M.G., Sibley, C.P., Davies, W.A., Albrecht, E.D., 2001. Expression of the mRNAs and Proteins for the Na(+)/H(+) exchangers and their regulatory factors in baboon and human placental syncytiotrophoblast. Endocrinology 142, 3685–3692.

Praetorius, J., Hager, H., Nielsen, S., Aaikjaer, C., Friis, U.G., Ainsworth, M.A., et al., 2001. Molecular and functional evidence for electrogenic and electroneutral Na(+)-HCO(3)(−) cotransporters in murine duodenum. Am. J. Physiol. Gastroint. Liver Physiol. 280, G332–G343.

Rhee, M.Y., Kim, J.H., Shin, S.J., Gu, N., Nah, D.Y., Hong, K.S., et al., 2014. Estimation of 24-hour urinary sodium excretion using spot urine samples. Nutrients 6, 2360–2375.

Sanders, B., Noakes, T.D., Dennis, S.C., 1999. Water and electrolyte shifts with partial fluid replacement during exercise. Eur. J. Appl. Physiol. Occ. Physiol. 80, 318–323.

Satoh, T., Owada, S., Ishida, M., 1999. Recent aspects in the genetic renal mechanisms involved in hypertension. Int. Med. 38, 919–926.

Shirreffs, S.M., Maughan, R.J., 1998. Volume repletion after exercise-induced volume depletion in humans: replacement of water and sodium losses. Am. J. Physiol. 274, F868–F875.

Strazzullo, P., Iacone, R., Siani, A., Barba, G., Russo, O., Russo, P., et al., 2001. Altered renal sodium handling and hypertension in men carrying the glucagon receptor gene (Gly40Ser) variant. J. Mol. Med. 79, 574–580.

Stricker, E.M., Sved, A.E., 2000. Thirst. Nutrition 16, 821–826.

Su, Y.R., Menon, A.G., 2001. Epithelial sodium channels and hypertension. Drug Metab. Disp. 29, 553–556.

Thunhorst, R.L., Johnson, A.K., 2001. Effects of hypotension and fluid depletion on central angiotensin-induced thirst and salt appetite. Am. J. Physiol. Reg. Integr. Comp. Physiol. 281, R1726–R1733.

Zhao, H., Hundal, H.S., 2000. Identification and biochemical localization of a Na–K–Cl cotransporter in the human placental cell line BeWo. Biochem. Biophys. Res. Commun. 274, 43–48.

POTASSIUM

Potassium (also known as kalium, hence the abbreviation *K* for the atom; atomic weight 39.098) is an alkali metal with valence 1. Two of the three naturally occurring isotopes, 39 (93.22%) and 41 (6.77%), are stable; isotope 40 (0.012%) is radioactive.

ABBREVIATION

ADH   antidiuretic hormone

NUTRITIONAL SUMMARY

Function: Potassium is the main cationic osmolyte within cells. The element plays a major role in body electricity (maintenance of cellular polarity, neuronal signaling, heart impulse transmission, and muscle contraction), nutrient and metabolite transport, and enzyme activation.

Food sources: Fruits and vegetables tend to contain large amounts of potassium, whereas animal-derived foods and grains provide smaller amounts.

Requirements: Healthy adults should consume at least 1600 mg of potassium per day. Increased losses, often due to the use of certain diuretics or laxatives, have to be balanced with higher intake. Inadequate intake leads to low plasma potassium concentration (hypokalemia), with an increased risk of heart arrhythmia, muscle weakness, paralysis, alkalosis (increased blood pH), and eventually death. Higher than minimal (but not excessive) intake is likely to be beneficial by lowering the risk of hypertension.

Excessive intake: When potassium intake exceeds the amount that can be removed (about 18 g/day), the rising plasma concentration (hyperkalemia) can cause muscle weakness, interference with electric conductance in the heart (arrhythmia), and eventually death due to cardiac arrest. The risk of hyperkalemia is particularly high in patients with renal failure.

DIETARY SOURCES

Both plant- and animal-derived foods are among the best sources. Potassium-rich foods of plant origin include avocados (6.3 mg/g), spinach (raw 5.6 g), bananas (4 mg/g), oats (3.5 mg/g), and rye flour (3.4 mg/g). Relatively potassium-rich animal products are halibut (5.8 mg), tuna (5.7 mg/g), mackerel (4.1 mg/g), and salmon (3.8 mg/g). Slightly less is available in meats, such as pork (3.2 mg/g), beef (3.0 mg/g), and chicken (2.4 mg/g). The potassium content is relatively low in wheat flour (1.5 mg/g), eggs (1.2 mg/g), and cheeses (around 1 mg/g), and very low in rice (0.4 mg/g). Orange juice (2.0 mg/g) and milk (1.5 mg/g) are important sources because of the large consumed volumes. Losses via leaching during preparation or cooking of foods can be significant due to the high solubility of potassium salts in water. Boiled and drained spinach, for example, contains 17% less potassium than the same amount of raw spinach. The difference between raw and cooked kale is nearly 50%.

Median intake in the United States has been estimated at 3.2 g/day (Green et al., 2002).

DIGESTION AND ABSORPTION

Small, but significant, amounts of potassium are secreted with saliva (15–54 mmol/day), gastric (4–12 mmol/day), and pancreatic (5–15 mmol/day) juice. Nearly all ingested and secreted potassium is absorbed from the small intestine. Net secretion occurs in the distal colon.

Significant amounts of ingested and secreted potassium follow the osmolar gradient along the paracellular route (Figure 11.4). To overcome the electric potential difference of 5 mV from lumen to basolateral fluid space, the potassium concentration in the lumen has to be at least 2 mmol/l above the concentration in the blood (Younoszai, 1989).

A hydrogen/potassium-exchanging ATPase (EC3.6.3.10) at the luminal side of the distal colonic epithelium works in conjunction with potassium channels to recycle and conserve potassium. This exchanger is distinct from the one responsible for acid production in the stomach. The export of potassium into the basolateral space is similarly balanced by active reuptake via the sodium/

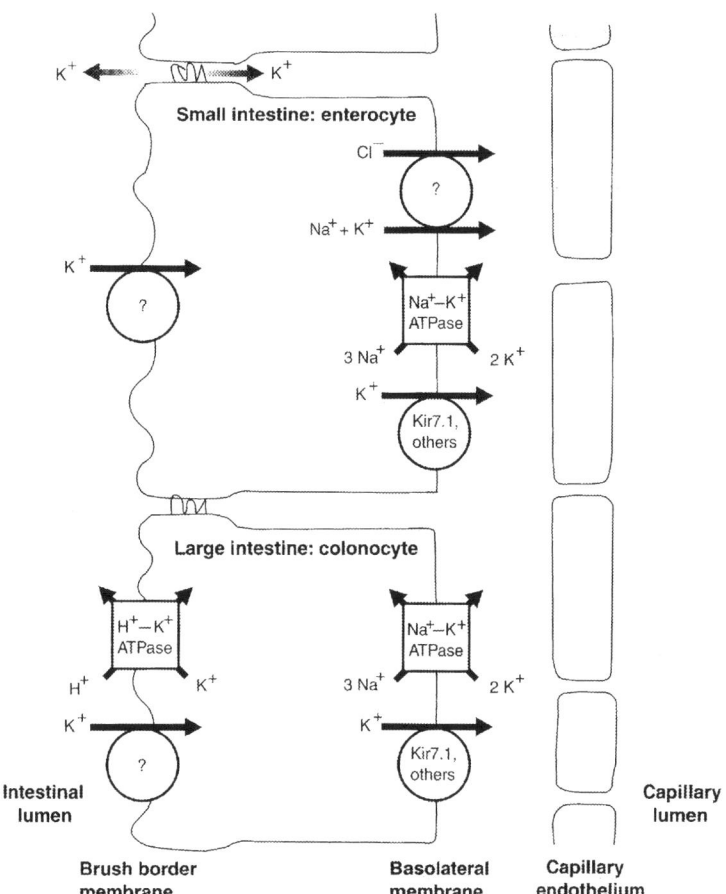

FIGURE 11.4

Intestinal absorption of potassium.

potassium-exchanging ATPase (EC3.6.3.9). These potassium-recycling systems are essential for the efficient absorption of sodium and other cations.

The colon has a hydrogen/potassium-exchanging ATPase (EC3.6.3.10), which recovers much of the large amount of potassium retained up to this stage (Sorensen et al., 2010).

TRANSPORT AND CELLULAR UPTAKE

Blood circulation: The plasma concentration of potassium is normally about 3.5–5 mmol/l. Elevated potassium concentration in the blood (hyperkalemia) can induce cardiac arrhythmia. Severe hyperkalemia will cause cardiac arrest. Low potassium levels (hypokalemia) also can induce arrhythmia.

Potassium concentration in most cells is around 160 mmol/l (Preuss, 2001). The sodium/potassium-exchanging ATPase (EC3.6.3.9) is the main mechanism that pumps potassium into cells against the steep gradient between extracellular fluid and intracellular space. Potassium exits cells via specific channels whose activity is closely regulated.

BBB: Sodium/potassium-exchanging ATPase (EC3.6.3.9) is active in endothelial capillary cells in the brain, probably both at the blood (luminal) and brain (abluminal) sides of the cells (Manoonkitiwongsa et al., 2000). Calcium-activated potassium channels also contribute to the transport into and out of the brain (Ningaraj et al., 2002). More important, these channels play an important role in the regulation of barrier permeability for other compounds (including drugs), possibly through recycling mechanisms as at other polarized epithelia.

Materno-fetal transfer: Both ATPases and potassium channels participate in the transfer of potassium to and from the fetus (Mohammed et al., 1993).

STORAGE

The potassium content of a healthy young 70-kg person is about 150 g, most (98%) of it within cells (Rose, 1989). Based on this distribution, body cell mass (BCM), and thus lean body mass, can be estimated through the measurement of total body potassium content by counting the specific radioactive emissions of ^{40}K (Andreoli et al., 2010). There are no specialized stores from which potassium could be mobilized in times of need; however, gradual depletion from optimal to dangerously low status takes several weeks, even in the absence of any intake (e.g., total fasting), because extracellular and intracellular concentrations can drop a few percentage points before undue consequences occur.

EXCRETION

The kidneys are central for the elimination of potassium excess. At least 200 mg of potassium is excreted with urine per day, more with moderate to high intake levels (Preuss, 2001). Smaller amounts are lost via feces, sweat (about 800 mg/l), and other secretions. Diarrhea can cause large fecal losses of potassium.

Since it is not associated with larger molecular complexes, all potassium (about 20–30 g/day) is filtered through the kidneys. Reabsorption occurs both in the proximal tubule and the TAL of the loop of Henle. Secretion into the connecting tubule and the collecting ducts in exchange for sodium adjusts net potassium loss. Net absorption of potassium from the proximal lumen is mainly via the paracellular pathway following the sodium and water gradient (Rose, 1989).

The sodium-potassium-chloride transporter 2 (SLC12A1) is responsible for potassium uptake across the luminal side of the TAL of the loop of Henle. This transporter carries potassium only in the presence of ADH. Some potassium is recycled back into the lumen, however, via a potassium channel (KCNJI) that is closely coupled with the sodium-potassium-chloride transporter 2. Two alternatively spliced isoforms are expressed in kidneys. Potassium back-transport confers a positive charge to the lumen, which drives the paracellular transport of other cations. Sodium/potassium-exchanging ATPase activity, which carries potassium from the basolateral fluid space into the cells, also limits net absorption. Also, some of the reabsorbed potassium is secreted again into proximal tubules. By the time the tubular fluid reaches the distal segment, however, more than 90% of the filtered potassium has been absorbed despite all these recycling pathways.

The connecting segment and the principal cells of the collecting tubules secrete potassium back into the lumen again if stimulated by aldosterone. Potassium channels are present at both the basolateral side and the luminal side. Their permeability, in conjunction with the activity of the sodium/potassium-exchanging ATPase, determines the net transfer.

REGULATION

The potassium content of the body is maintained through variation of renal excretion. Adaptation of potassium recovery from feces plays a smaller role in healthy people but becomes crucial during periods of acute intestinal infectious disease such as cholera.

Aldosterone increases the secretion of potassium from connecting segments and collecting ducts of the kidneys by acting on the mineralocorticoid receptor (NR3C2) in those segments. Receptor activation promotes expression of the sodium/potassium exchanging ATPase (EC3.6.3.9). The renin–angiotensin system links aldosterone concentration to potassium concentration in the blood. Glycorrhizic acid in licorice increases both renal and extrarenal potassium losses by inhibiting renal 11β-hydroxysteroid dehydrogenase type 2 (EC1.1.1.146) and slowing the conversion of cortisol to its metabolite cortisone (Serra et al., 2002). Cortisol, but not cortisone, strongly activates the mineralocorticoid receptor. Catecholamines and insulin promote the redistribution of potassium into the liver and skeletal muscles.

FUNCTION

Excitation: The ability of neurons, muscles, and specialized excitable tissues in the heart to depolarize relies on the change in electric charge of the cells in response to a given stimulus. These cells have a negative charge of about 70–90 mV at rest because anion concentration is slightly higher inside than outside. The polarization rapidly flips when gated potassium channels first release potassium from these cells, followed by an influx of sodium ions. The change in polarity propagates along the length of these excitable cells and can move to another excitable cell by direct electric stimulation or indirectly via a chemical transmitter (e.g., acetylcholine).

Aberrations of potassium status cause typical electrocardiographic (ECG) changes. Patients with hypokalemia show a characteristic additional wave (U-wave) just preceding and in the opposite direction of the normal T wave. Hyperkalemia flattens the P wave, lengthens the QRS interval, and sharpens the T wave.

Enzyme activation: Monovalent ions are unspecific activators for many enzymes.

Pyruvate kinase (EC2.7.1.40) and pyruvate carboxylase (EC6.4.1.1) have a slightly more specific potassium requirement. There is no indication that variation of potassium concentration within normal ranges alters enzyme activities to a biologically significant extent.

Radioactivity: The isotope ^{40}K is a γ-emitter with a half-life of 1.3×10^9 years. It has been estimated that about 90% of the body's exposure to radioactivity is due to the presence of its estimated 7 mg of radioactive potassium (Charpak & Garwin, 2001).

NUTRITIONAL ASSESSMENT

Potassium deficiency is reflected by the concentration of potassium in plasma. Dietary intake can be estimated based on potassium excretion with urine (Tasevska et al., 2006).

REFERENCES

Andreoli, A., Volpe, S.L., Ratcliffe, S.J., Di Daniele, N., Imparato, A., Gabriel, L., et al., 2010. Longitudinal study of total body potassium in healthy men. J. Am. Coll. Nutr. 29, 352–356.

Charpak, G., Garwin, R., 2001. Le DARI. Unite de mesure adaptée l'evaluation de l'effet des faibles doses d'irradiation. Bull. Acad. Natl. Med. 185, 1087–1094.

Green, D.M., Ropper, A.H., Kronmal, R.A., Psaty, B.M., Burke, G.L., 2002. Serum potassium level and dietary potassium intake as risk factors for stroke. Neurology 59, 314–320.

Manoonkitiwongsa, P.S., Schultz, R.L., Wareesangtip, W., Whitter, E.E., Nava, P.B., McMillan, P.J., 2000. Luminal localization of blood–brain barrier sodium, potassium adenosine triphosphatase is dependent on fixation. J. Histochem. Cytochem. 48, 859–865.

Mohammed, T., Stulc, J., Glazier, J.D., Boyd, R.D., Sibley, C.P., 1993. Mechanisms of potassium transfer across the dually perfused rat placenta. Am. J. Physiol. 265, R341–R347.

Ningaraj, N.S., Rao, M., Hashizume, K., Asotra, K., Black, K.L., 2002. Regulation of blood–brain tumor barrier permeability by calcium-activated potassium channels. J. Pharmacol. Exp. Ther. 301, 838–851.

Preuss, H.G., 2001. Sodium, chloride, and potassium. In: Bowman, B.A., Russell, R.M. (Eds.), Present Knowledge in Nutrition, eighth ed. ILSI Press, Washington, DC, pp. 302–310.

Rose, B.D., 1989. Clinical Physiology of Acid-Base and Electrolyte Disorders, third ed. McGraw-Hill, New York, NY.

Serra, A., Uehlinger, D.E., Ferrari, E., Dick, B., Frey, B.M., Frey, F.J., et al., 2002. Glycyrrhetinic acid decreases plasma potassium concentrations in patients with anuria. J. Am. Soc. Nephrol. 13, 191–196.

Sorensen, M.V., Matos, J.E., Praetorius, H.A., Leipziger, J., 2010. Colonic potassium handling. Pflugers Arch. Eur. J. Physiol. 459, 645–656.

Tasevska, N., Runswick, S.A., Bingham, S.A., 2006. Urinary potassium is as reliable as urinary nitrogen for use as a recovery biomarker in dietary studies of free living individuals. J. Nutr. 136, 1334–1340.

Younoszai, M.K., 1989. Physiology of mineral absorption. In: Grand, R.L., Sutphen, J.L., Dietz Jr, W.H. (Eds.), Pediatric Nutrition Butterworths, Boston, MA, pp. 163–183.

CHLORINE

Chlorine is a halogen (abbreviation Cl for the atom, atomic weight 35.4527). Isotopes 35 (75.5%) and 37 (24.5%) occur naturally. Several artificial isotopes are radioactive. Chemical valences range from 1 to 7. The chloride ion (Cl^-) is the most common chemical form in food. The hypochlorous ion (ClO^-) is produced endogenously in minute quantities for specific actions. The perchlorate ion (ClO_4^-) is a synthetic compound with distinct and largely undesirable biological activities. Organochlorine compounds (e.g., pesticides, herbicides, antibiotics, and polychlorbiphenyls/PCBs) will not be discussed here since they enter the food supply only as contaminants.

ABBREVIATIONS	
CFTR	cystic fibrosis transmembrane conductance regulator (ABCC7)
CLD	congenital diarrhea gene (DRA, SLC26A3)
NaCl	sodium chloride (table salt)
NKCC1	sodium-potassium-chloride transporter 1 (SLC12A2)
NKCC2	sodium-potassium-chloride transporter 2 (SLC12A1)

NUTRITIONAL SUMMARY

Function: Chloride is the main anionic osmolyte in the blood and extracellular fluid, and contributes to active transport of some nutrients and metabolites in the intestines, kidneys, and other tissues. Hydrochloric acid in the stomach contributes to protein digestion and inactivation of ingested microorganisms. Immune cells use directed release of hypochlorous acid to combat pathogens in the blood and tissues.

Food sources: Meats, pickled foods, and salty snacks are the major sources of chloride.

Requirements: Intake has to match chloride losses. In most situations, a balance can be achieved with daily intake of a few hundred milligrams. Increased sweating (which occurs with fever, heat, and high humidity) and diuresis can increase chloride needs to several grams per day. Very low chloride intake (for instance, with fasting), can cause dizziness and weakness due to hypotension.

Excessive intake: Higher than minimal chloride intake may increase blood pressure, especially in genetically susceptible individuals and when other hypertensive factors (such as obesity) are present.

DIETARY SOURCES

Chloride intake closely correlates with sodium intake (Al-Bander et al., 1988) since most chloride comes from animal foods and added salt (NaCl). The chloride content of meat tends to correlate inversely with fat content. Plant-derived foods are uniformly low in chloride. Some salt substitutes, such as potassium chloride, can be minor sources.

While the food composition tables in the United Kingdom contain data on chloride (UK Food Standards Agency, 2001), food composition databases in the United States lack this information.

DIGESTION AND ABSORPTION

Large amounts of chloride are secreted from the parietal cells of the stomach lining (as hydrochloric acid), pancreas, and the intestinal epithelium (Figure 11.5). Unstimulated, fasting gastric output is normally below 6 mmol/h and can increase with stimulation to 13–25 mmol/h in healthy people. Chloride from these intestinal secretions and from dietary intake is absorbed in the small and large intestine nearly to completion. Absorption occurs by both passive diffusion and active transport.

Small intestine: As gastric acid enters the duodenum, its corrosive acidity must be neutralized and the chloride recovered. The transporters SLC26A3 and SLC26A6 (putative anion transporter 1, PAT1) that secrete intracellular bicarbonate in exchange for luminal chloride achieve both at the same time (Kato & Romero, 2011). A closely related bicarbonate-chloride exchanger, SLC26A9, in crypts of the small intestine appears to have functional importance as well (Liu et al., 2014).

A much more limited amount of chloride is taken up during the absorption of some nutrients. The transport of cationic and neutral amino acids, carnitine, taurine, and beta-alanine by system $B^{0,+}$ (SLC6A14) requires the cotransport of both chloride and sodium (Nakanishi et al., 2001). The transporter IMINO (SLC6A20) for proline and hydroxyproline also has an absolute chloride requirement. Other amino acid transporters show chloride conductance that is not directly coupled to their activity.

The chloride conductance regulator (cystic fibrosis transmembrane regulator, CFTR, ABCC7) at the luminal side of the small intestine is most important for the secretion of chloride into the small intestinal lumen. CFTR is both a chloride channel and a regulator of the epithelial sodium channel and other ion transporters. Its activity is stimulated by cAMP-mediated phosphorylation. Since CFTR is an ATP-binding cassette (ABC) transporter, it is energy- and magnesium-dependent. This gene is essential

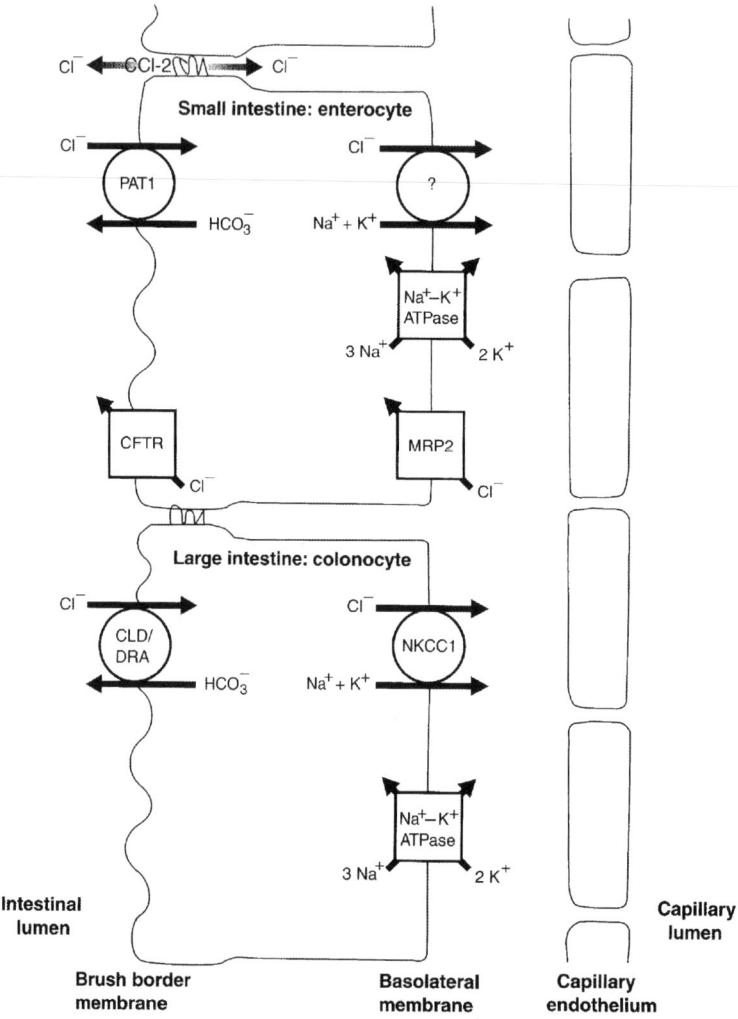

FIGURE 11.5

Intestinal chloride absorption.

for normal chloride secretion in pancreas, lung epithelium, and other tissues. Relatively common variants impair ion transport control through as-yet-unknown mechanisms. Chloride follows its gradient across the basolateral membrane mostly through the chloride channel ClC-2 (Kato & Romero, 2011).

The tight junctions between the enterocytes of the proximal intestines are relatively permeable for chloride ions. Paracellular diffusion across the tight junctions provides a chloride pathway unrelated to CFTR that can significantly contribute to chloride secretory capacity (Gyomorey et al., 2001).

Large intestine: The luminal side of the colon enterocytes contains a chloride/bicarbonate exchanger (SLC26A3; Hemminki et al., 1998), alternatively referred to as congenital diarrhea (CLD) and down-regulated in adenoma (DRA). At least two anion exchangers (including AE2 and AE3) at the basolateral membrane of the ileum and colon move chloride in exchange for bicarbonate toward the bloodstream (Alrefai et al., 2001). The sodium-potassium-chloride transporter (NKCCl, SLC12A2) mediates chloride transport across the basolateral membrane in either direction as required.

TRANSPORT AND CELLULAR UPTAKE

Blood circulation: The chloride concentration in the blood is around 103 mmol/l (3.6 g/l). Chloride enters cells mainly through specific channels following its gradient to the much lower intracellular concentration (typically around 2 mmol/l) and with sodium-potassium-chloride cotransporters. Some neurons appear to possess an additional sodium-independent, adenosine triphosphate (ATP)–driven import mechanism (Bettendorff et al., 2002).

Specific intracellular channels mediate exchange between compartments. Movement out of cells can be driven by the negative membrane potential through voltage-gated chloride channels. The ATP-energized CFTR is particularly important for transport of chloride ions across the apical membranes of epithelial cells in the pancreas, sweat glands, lungs, and other tissues. Its channel activity is up-regulated by cAMP. Potassium/chloride cotransporters are also important for transport across membranes.

Materno-fetal transfer: As the fetus grows, chloride for its expanding blood circulation and tissues must be supplied by the mother. The mechanisms are still incompletely understood. AE1 (anion exchanger 1) is important for transport across the microvillous membrane (Powell et al., 1998).

STORAGE

The body of a healthy young 70-kg person contains about 95 g of chloride. None of this is maintained in particular stores that could be mobilized in times of need. Since the chloride concentration has to be maintained within a few percentage points, there is little buffering capacity to make up for persistently insufficient or excessive intake.

EXCRETION

Chloride is lost via urine, feces, sweat, and other body secretions. Losses via sweat may be 1 g or more per day. While urine always contains at least a small amount (<200 mg), renal chloride elimination normally corresponds to recent intake.

Renal excretion: Chloride freely crosses the renal glomerular membrane, and about 630 g/day is filtered by the kidneys of a healthy 70-kg person (Figure 11.6). More than two-thirds of this amount is recovered from the proximal tubular lumen, and nearly 30% from the thick ascending limb (TAL) of the Henle loop. How much of the remainder is absorbed from the distal nephron depends on the homeostatically regulated recovery from distal tubules, connecting segments, and collecting tubules. Both active and passive transport contribute to the movement of chloride out of the proximal tubular lumen. Due to its ability to permeate the (not-so) tight junctions between proximal tubular epithelial cells, chloride readily diffuses from the highly concentrated luminal fluid directly to the basolateral and pericapillary space (i.e., paracellular transport). At the same time, active transport into the epithelial cell occurs in exchange for formate. The basis for this exchange is the fact that nonionic formate diffuses freely from the lumen into the epithelial cell, where it becomes ionized due to the higher intracellular pH (established by the apical sodium-hydrogen

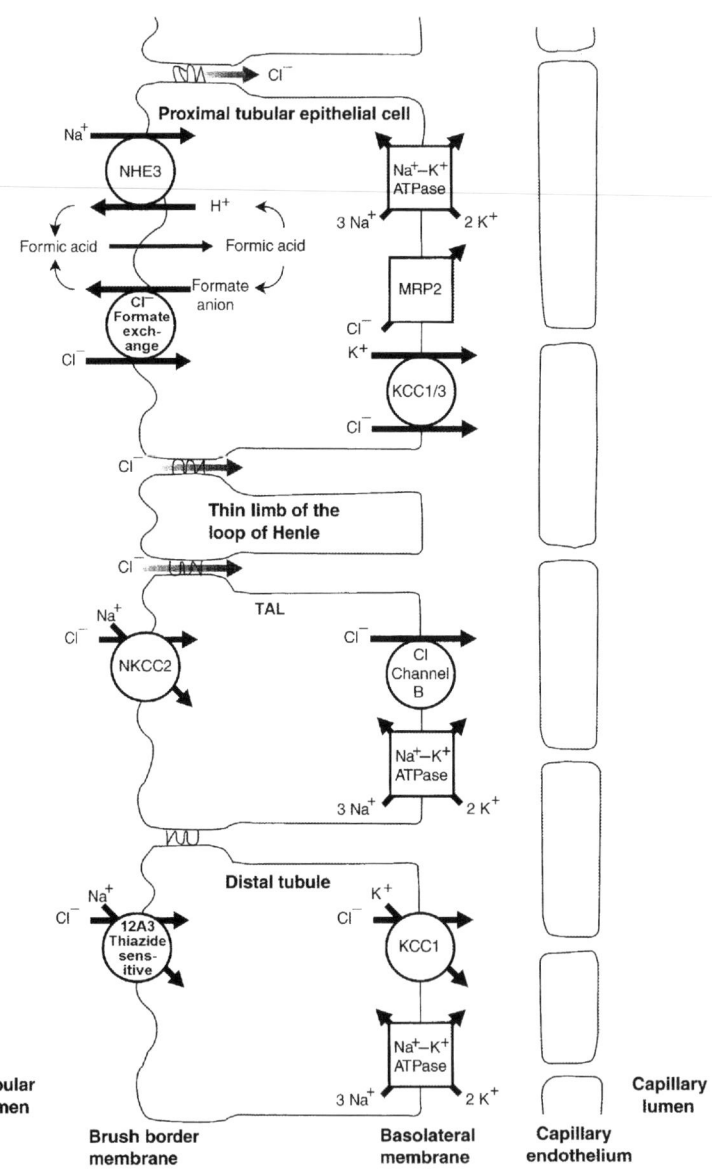

FIGURE 11.6

Reabsorption of chloride from renal tubules.

exchanger 1, NHE3, SLC9A3). The sodium/formate exchanger then transports ionized formate back into the lumen while moving chloride into the cell. Ultimately, the sodium/potassium-exchanging ATPase (EC3.6.3.9) at the basolateral membrane drives this process by maintaining low intracellular sodium concentration. Pendrin (SLC26A4; the name is derived from Pendred syndrome) explains some of the chloride transport activity, but the principal chloride/formate exchanger of the proximal tubular brush border membrane remains to be identified (Soleimani, 2001). Export across the basolateral membrane into the pericapillary space proceeds via the potassium/chloride cotransporters 3 (KCC3, SLC12A6) and 1 (SLC12A4). As in the small intestine, the cAMP-activated ABC transporter cMOAT/MRP2 (ABCC2) regulates chloride efflux across the basolateral membrane (van Kuijck et al., 1996).

Tubular chloride in the thin ascending limb of the Henle loop uses the chloride channel A (ClC-Ka/barttin) to passively follow the concentration gradient of the countercurrent system. Reabsorption of sodium chloride in the TAL of the loop of Henle is active again via the sodium-potassium-chloride transporter 2 (NKCC2, SLC12A1) on the luminal side and chloride channel B (ClC-Kb) on the basolateral side (Fahlke and Fischer, 2010).

The thiazide-sensitive sodium/chloride cotransporter (SLC12A3) provides the principal means of chloride reabsorption in the distal nephron (distal tubulus) where the losses are fine-tuned in response to chloride status. This transporter is inhibited by thiazide and related diuretics. The potassium/chloride cotransporter 1 (SLC12A4) completes the transport process on the basolateral side.

Chloride import into the intercalated cells of the cortical collecting duct and the connecting tubules via pendrin (SLC26A4), as well as the associated secretion of bicarbonate into the tubular lumen, has little relevance for overall chloride reabsorption but is important for pH regulation.

Perspiration: Chloride constitutes the main anion in sweat at a typical concentration of around 30 mmol/l. Sweat production of 1 l thus would correspond to a loss of about 1 g. High ambient temperature, physical exertion, fever, and increased sympathetic nervous system activity can induce sweating. Chloride is transported into sweat by the gene product of CLD (SLC26A3).

REGULATION

Variation of reabsorption from the distal nephron is a major mechanism for maintaining chloride balance. The control mechanisms are basically those that regulate sodium reabsorption. Renin (EC3.4.23.15) from the iuxtaglomerular cells in the kidneys cleaves angiotensin I and peptidyl-dipeptidase A (angiotensin 1-converting enzyme, EC3.4.15.1) then generates angiotensin II. Angiotensin II finally increases production and secretion of the steroid hormone aldosterone secretion from the adrenal cortex. Here, cAMP-linked intracellular signaling cascades triggered by receptor-binding of aldosterone promote chloride reabsorption from the connecting segment and the collecting tubule mainly by increasing the expression of apical chloride channels.

FUNCTION

Electrolyte balance: Chloride is the major anion in extracellular fluids and helps to maintain normal osmotic pressure.

Transport: Several transporters require chloride as a cotransported counterion or exchange chloride for another anion. The taurine transporter (TAUT, SLC6A6) and the betaine transporter (BGT1, SLC6A12) are examples that use both chloride and sodium. Chloride currents in other transporters, such as the excitatory amino acid carrier 1 (SLC1A1) responsible for terminating the neuroexcitatory action of L-glutamate, do not directly drive transport but nonstoichiometrically regulate activity.

Acid production: Large amounts of chloride are used for the production of gastric acid. The hydration of carbon dioxide by the isoforms I and II of carbonate dehydratase (carbonic anhydrase,

EC4.2.1.1, zinc-dependent) generates a prolific source of protons in parietal cells of the stomach. The hydrogen/potassium-exchanging ATPase (EC3.6.3.10, magnesium-dependent) pumps these protons across the secretory canalicular membrane in exchange for potassium ions. The parallel export of chloride ions via the chloride channel 2 (ClC2) completes gastric acid secretion (Sherry et al., 2001). The anion exchanger AE2 (SLC4A2) at the basolateral membrane stabilizes intracellular pH by removing the deprotonated bicarbonate anion in exchange for a new supply of chloride from the pericapillary fluid (Carmosino et al., 2000). The basolateral sodium-potassium-chloride cotransporter (NKCCl, SLC12A2) provides an additional pathway for chloride influx.

Chloride is also needed for acid production elsewhere, such as in endosomes for the recovery of filtered proteins in the kidneys (endocytotic pathway) or at the osteoclast ruffled membrane for bone mobilization. Distinct chloride channels are associated with the proton ATPases in the kidneys (chloride channel 5, CCL5) and bone (chloride channel 7, CCL7).

Enzyme activation: Peptidyl-dipeptidase A (angiotensin 1-converting enzyme, ACE, EC3.4.15.1) and a few other enzymes have a stringent chloride requirement. The activity of most other enzymes is strongly dependent on ionic strength, to which the chloride concentration contributes significantly in extracellular space.

Immune defense: Myeloperoxidase (EC1.11.1.7) is a lysosomal hemoprotein in neutrophils, myelocytes, macrophages, and monocytes that uses chloride and hydrogen peroxide to generate hypochlorous acid (HOCl). The highly reactive hypochlorous ion is released as part of an oxidative blast of these immune cells that is normally directed at pathogens.

NUTRITIONAL ASSESSMENT

Severe chloride depletion (for instance, due to prolonged vomiting in infants with pylorus stenosis) lowers the concentration of chloride in the blood. Less dramatic deviations from optimal status, particularly in conjunction with alkalosis, also appear to be biologically meaningful and predictive of significantly worsened health outcomes (Tani et al., 2012). The amount of chloride in 24-h urine collection reflects dietary intake with modest fidelity.

The low chloride concentration in sweat indicates defective chloride transport in people with cystic fibrosis (Accurso et al., 2014).

REFERENCES

Accurso, F.J., Van Goor, F., Zha, J., Stone, A.J., Dong, Q., Ordonez, C.L., et al., 2014. Sweat chloride as a biomarker of CFTR activity: proof of concept and ivacaftor clinical trial data. J. Cyst. Fibros. 13, 139–147.

Al-Bander, S.Y., Nix, L., Katz, R., Korn, M., Sebastian, A., 1988. Food chloride distribution in nature and its relation to sodium content. J. Am. Diet. Assoc. 88, 472–475.

Alrefai, W.A., Tyagi, S., Nazir, T.M., Barakat, J., Anwar, S.S., Hadjiagapiou, C., et al., 2001. Human intestinal anion exchanger isoforms: expression, distribution, and membrane localization. Biochim. Biophys. Acta 1511, 17–27.

Bettendorff, L., Lakaye, B., Margineanu, I., Grisar, T., Wins, E., 2002. ATP-driven, Na(+)-independent inward Cl⁻ pumping in neuroblastoma cells. J. Neurochem. 81, 792–801.

Carmosino, M., Procino, G., Casavola, V., Svelto, M., Valenti, G., 2000. The cultured human gastric cells HGT-I express the principal transporters involved in acid secretion. Pfl. Arch. Eur. J. Physiol. 440, 871–880.

Fahlke, C., Fischer, M., 2010. Physiology and pathophysiology of ClC-K/barttin channels. Front. Physiol. 1, 155.

Gyomorey, K., Garami, E., Galley, K., Rommens, J.M., Bear, C.E., 2001. Non-CFTR chloride channels likely contribute to secretion in the murine small intestine. Pfl. Arch. Eur. J. Physiol. 443, S103–S106.

Hemminki, A., Hoglund, P., Pukkala, E., Salovaara, R., Jarvinen, H., Norio, R., et al., 1998. Intestinal cancer in patients with a germline mutation in the down-regulated in adenoma (DRA) gene. Oncogene 16, 681–684.

Kato, A., Romero, M.F., 2011. Regulation of electroneutral NaCl absorption by the small intestine. Annu. Rev. Physiol. 73, 261–281.

Liu, X., Li, T., Riederer, B., Lenzen, H., Ludolph, L., Yeruva, S., et al., 2014. Loss of Slc26a9 anion transporter alters intestinal electrolyte and HCO$_3$—transport and reduces survival in CFTR-deficient mice. Pflügers Arch. June 27.

Nakanishi, T., Hatanaka, T., Huang, W., Prasad, P.D., Leibach, F.H., Ganapathy, M.E., et al., 2001. Na$^+$- and Cl$^-$-coupled active transport of carnitine by the amino acid transporter ATB(0, +) from mouse colon expressed in HRPE cells and Xenopus oocytes. J. Physiol. 532, 297–304.

Powell, T.L., Lundquist, C., Doughty, I.M., Glazier, J.D., Jansson, T., 1998. Mechanisms of chloride transport across the syncytiotrophoblast basal membrane in the human placenta. Placenta 19, 315–321.

Sherry, A.M., Malinowska, D.H., Morris, R.E., Ciraolo, G.M., Cuppoletti, J., 2001. Localization of ClC-2 Cl-channels in rabbit gastric mucosa. Am. J. Physiol. Cell Physiol. 280, C1599–C1606.

Soleimani, M., 2001. Molecular physiology of the renal chloride-formate exchanger. Curr. Opin. Nephrol. Hypertens. 10, 677–683.

Tani, M., Morimatsu, H., Takatsu, F., Morita, K., 2012. The incidence and prognostic value of hypochloremia in critically ill patients. Scientific World J. 2012, 474185.

UK Food Standards Agency, 2001. McCance and Widdowson's The Composition of Foods, sixth ed. Royal Society of Chemistry, London.

van Kuijck, M.A., van Aubel, R.A., Busch, A.E., Lang, E., Russel, F.G., Bindels, R.J., et al., 1996. Molecular cloning and expression of a cyclic AMP-activated chloride conductance regulator: a novel ATP-binding cassette transporter. Proc. Natl. Acad. Sci. USA 93, 5401–5406.

IRON

Iron (Fe, atomic weight 55.847) is a metallic divalent or trivalent transition element with naturally occurring stable isotopes 54 (5.8%), 56 (91.7%), 57 (2.2%), and 58 (0.33%). The artificial isotopes 52, 53, 55, 59, 60, and 61 are radioactive.

ABBREVIATIONS

DMT1	divalent metal ion transporter 1
HFE	hemochromatosis gene
IRE	iron-response element
IRE-BP	iron-responsive element-binding protein
IRP	iron regulatory protein
TF	transferrin
TfR	transferrin receptor

NUTRITIONAL SUMMARY

Function: Iron is essential as a cofactor of oxygen transport, respiration, amino acid, lipid, alcohol, vitamin A, and sulfite metabolism, as well as various other redox reactions.

Requirements: A dose of at least 8 mg of iron is necessary to maintain adequate stores for people consuming a mixed diet, with more for vegetarians and women between 19 and 50 (18 mg/day, 27 mg/day during pregnancy, and 9 mg/day during lactation).

Sources: All muscle foods are good iron sources. Phytate from whole grains and some vegetables interferes with iron absorption; dietary ascorbate promotes absorption.

Deficiency: Diminished stores cause loss of appetite, microcytic anemia, and impaired immune function. Deficiency slows growth and cognitive development of infants and children. This effect may be partially irreversible.

Excessive intake: Greatly expanded iron stores may damage the liver, pancreas, and heart. Iron supplement intake decreases the absorption of concomitantly ingested thyroxine, tetracycline derivatives, penicillamine, methyldopa, levodopa, carbidopa, and ciprofloxacin.

DIETARY SOURCES

Foods contain heme proteins (myoglobin, hemoglobin, cytochromes, and a few enzymes), ferrous iron (Fe^{2+}), and ferric iron (Fe^{3+}). Ferrous and ferric iron, usually referred to as *nonheme iron*, comprise about 60% of the iron in meats and most of the iron in plant-derived foods.

Iron-rich foods include all meats, such as beef (0.019 mg/g), pork (0.009 mg/g), chicken (0.012 mg/g), and fish (e.g., tuna, with 0.009 mg/g). Legumes (e.g., baked beans, with 0.003 mg/g) also are relatively iron-rich foods. Grains and grain products have to be fortified in the United States with 0.0044 mg/g.

While iron intake is very commonly inadequate worldwide, typical intake tends to be high in North America and Europe. Median daily iron consumption of American women is around 9 mg (Food and Nutrition Board, Institute of Medicine, 2001: Appendix E-5), which is well below recommendations. Median consumption of men is about 12 mg, which is above recommendations.

DIGESTION AND ABSORPTION

Uptake of iron from the intestinal lumen uses at least three distinct pathways, one for heme-bound iron through a largely uncharacterized pathway and another one for ferrous iron (Fe^{2+}) via the divalent metal ion transporter 1 (DMTI), and one for ferric iron (Fe^{3+}) via the beta3-integrin-mobilferrin pathway (Conrad et al., 2000). The relative contribution of the two nonheme iron pathways remains uncertain. Absorption is most effective in the duodenum, slightly less so in the remainder of the small intestine, and least in the colon (Figure 11.7).

Phytate and polyphenols in various commonly consumed foods and beverages strongly inhibit nonheme iron absorption (Hurrel et al., 1999). Absorption inhibitory potential is greatest for black tea (79–94%), peppermint tea (84%), pennyroyal tea (73%), cocoa (71%), and infusions from verbena (59%), lime flower (52%), and chamomile (47%). Dietary calcium reduces iron bioavailability (Fairweather-Tait et al., 2013). Ascorbate, organic acids, and heme promote the absorption of nonheme iron (Hurrel et al., 1999).

Uptake of ferrous iron: DMT1 (Nramp2, SLC11A2) transports ferrous iron (Fe^{2+}) together with a proton across the brush border membrane of the proximal small intestine (Montalbetti et al., 2013). Luminal mucin facilitates this uptake (Conrad et al., 1991). Duodenal cytochrome b at the brush border membrane helps with the reduction of Fe^{3+}.

Uptake of ferric iron: Contrary to long-held assumptions, ferric iron can be absorbed directly from the intestinal lumen by a specific pathway that depends on beta 3-integrin. A critical factor is the low solubility of Fe^{3+} at neutral to alkaline pH. Mucin and dietary chelators, such as citrate and similar organic acids, can keep this form of iron in solution and thereby greatly improve its bioavailability.

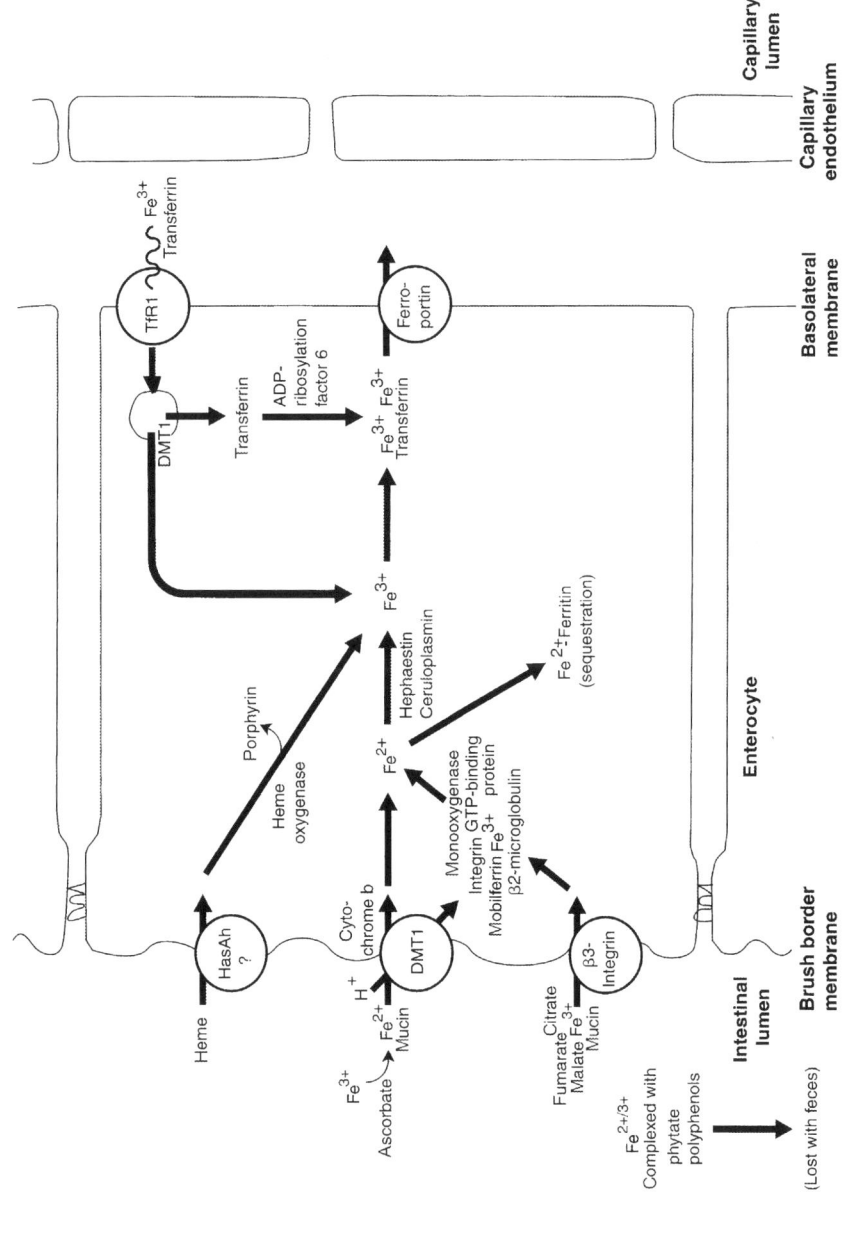

FIGURE 11.7

Intestinal absorption of iron.

Uptake of heme iron: The pathway for the uptake of heme is distinct from that for nonheme iron (Le Blanc et al., 2012). Neither iron absorption inhibitors, such as phytate, nor absorption enhancers, such as mucin or citrate, have much effect on heme uptake. Heme can be taken up by a proton-dependent carrier, folate transporter/heme carrier protein 1 (PCFT/HCP1, SLC46A1), which moves the intact heme molecule from the intestinal lumen into enterocytes (Le Blanc et al., 2012). The feline leukemia virus subgroup C receptor (FLVCR1, SLC49A1) also is a heme transporter (Khan & Quigley, 2013). Although it is better known for its critical function in erythroid bone marrow cells, it is highly expressed in the intestines. Yet another heme transporter within the enterocytes, encoded by SLC48A1 (heme responsive gene-1, HRG-1), is thought to move heme from the endosome into the cytosol (Khan & Quigley, 2013). Heme oxygenase (EC1.14.99.3), which may also aid transfer from the intestinal lumen, releases iron from its porphyrin scaffold and converts it into Fe^{3+} (Drummond et al., 1992).

Uptake of ferritin: A distinct route of iron absorption appears to be endocytosis of ingested ferritin (Theil et al., 2012). Because ferritin does not use the same mechanism for absorption, ingested heme and ferric iron will not slow its absorption.

Intracellular disposition: Iron is oxidized upon entry into the cell by as-yet-incompletely characterized oxidases, which might include ceruloplasmin and hephaestin. The iron-binding chaperone mediates cytoplasmic transport of ferric (Fe^{3+}) iron (Conrad et al., 1992). Some iron in the cytoplasma is stored temporarily with paraferritin, a large complex of at least four polypeptides, including integrin, mobilferrin (calreticulin/rho), and a flavin monooxygenase. This complex is also capable of reducing Fe^{3+} to Fe^{2+} (Umbreit et al., 1996). Of great importance for the regulation of its intestinal absorption efficiency is the storage of iron with ferritin. The spherical complexes consist of 24 identical ferritin subunits and several thousand Fe^{3+} atoms. This type of storage holds on to the iron until the enterocyte is shed at the end of its 2–3-day life span into the intestinal lumen, therefore effectively limiting iron absorption.

Some transferrin-bound iron arrives from blood circulation via transferrin receptor 1 (TfR1). The receptor-transferrin complexes concentrate in clathrin-coated pits and are directed to specific endosomal compartments. Iron is pumped out of the endosomal vesicles by DMT1 and as-yet-incompletely understood mechanisms return both transferrin and TfR1 to the basolateral membrane. ADP-ribosylation factor 6 contributes to the intracellular trafficking of the transferrin-TfR complex.

Export into blood: Only ferroportin 1 (IREG1, MTP1, and SLC40A1) can export iron from the enterocyte into blood. An important step of the transfer of iron from the enterocyte into blood is its oxidation to Fe^{3+}. Available iron-oxidizing enzymes include the ferroxidase (EC1.16.3.1) ceruloplasmin and a functionally related protein, hephaestin. The copper-containing hephaestin is expressed along the entire length of the small intestine (Frazer et al., 2001). Ferroportin finally moves iron across the basolateral membrane (Donovan et al., 2000).

TRANSPORT AND CELLULAR UPTAKE

Blood circulation: Most of the iron in the blood is contained in the hemoglobin of red blood cells amounting to about 400–600 mg/l. The much smaller amount in plasma (around 2–20 mg/l) is almost completely bound to transferrin. Ferroxidase (ceruloplasmin, EC1.16.3.1) oxidizes circulating Fe^{2+} to Fe^{3+}. Fe^{3+} can bind to two sites on transferrin, but less than half of the available sites are occupied in most healthy people. The concentration of unbound iron is normally below 10^{-8} mol/l. The maintenance of very low free iron concentration limits oxygen free radical generation and protects against the spread of iron-dependent bacteria in the blood and tissues. Even a minor increase, typical for individuals with heterozygous variants of the hemochromatosis gene *HFE*, raises the risk of deadly sepsis (Gerhard et al., 2001).

Transferrin receptors (TfRs) in clathrin-coated pits at the surface of most cells bind diferric transferrin (carrying two iron atoms) with several 100-fold preference compared to apotransferrin (carrying no iron) and mediates uptake through the endocytotic pathway, invagination of the pits leads to the formation of endosomal vesicles that become increasingly acidic by the action of a proton-transporting ATPase (Levy et al., 1999). Proton-driven DMT1 (Montalbetti et al., 2013) pumps iron from the proton-rich (acidic) vesicle into cytoplasma, where it binds to the chaperone mobilferrin (Conrad et al., 1996).

Two distinct TfRs occur with characteristic tissue distribution. TfR1 is mainly expressed in enterocytes, the precursors of most blood cells, and in hepatocytes. Transferrin receptor 2 (TfR2) is expressed in liver cells and circulating monocytes. Up to four neuraminic acid residues reside on transferrin in the blood of most healthy people. The TfR binds transferrin without neuraminic acid (asialotransferrin) less well than normally sialylated transferrin. Instead, asialotransferrin is preferentially cleared in the liver via the asialoglycoprotein receptor (Potter et al., 1992). Very high alcohol intake increases the percentage of incompletely sialylated transferrin in circulation (Walter et al., 2001). In addition to the transferrin pathway, a low-affinity transporter mediates iron uptake into cells (Garrick et al., 1999).

Macrophages, particularly in the spleen and endoplasmatic reticulum, salvage iron by the phagocytosis of old and damaged red blood cells. Iron can be released from heme inside the macrophages through the action of heme oxygenase (EC1.14.99.3). The macrophages are entirely dependent on ferroportin for the export of the accumulated iron to transferrin in circulating blood.

BBB: The mechanisms of iron transfer from blood to brain are of particular interest because inadequate supplies during infancy and early childhood hamper optimal cognitive development and because excessive iron accumulation in the brain is a hallmark of debilitating brain diseases in old age. Iron can reach the brain both by crossing the brain capillary endothelial layer and by filtration into cerebrospinal fluid and receptor-mediated uptake into neurons from there (Moos and Morgan, 2000).

Both free and protein-bound iron can also cross directly from blood circulation into brain through nonbarrier sites. The main mechanism for entry into endothelial cells of the BBB is TfR-mediated endocytosis of transferrin-bound iron. DMT1 transports iron out of endocytotic vesicles (Burdo et al., 2001). Exporting uses apotransferrin from astrocytes that is taken up via TRs, loaded with iron, and returned into the extracellular space. Another transepithelial pathway takes up lactoferrin-bound iron via low-density lipoprotein receptor-related protein (Gross et al., 1987). Axonal transport can supply iron to neurons outside the brain over considerable distances. Ferroportin 1 promotes iron export from the brain (Burdo et al., 2001).

Materno-fetal transfer: Diferric transferrin from maternal blood enters the syntrophoblast layer via TfR-mediated endocytosis (Georgieff et al., 2000). The iron is transported out of the endosomes by DMT1 and moved to the basal membrane. Iron export from the syntrophoblast into fetal blood is a complex and incompletely understood process that involves oxidation by zyklopen, a placental copper oxidase (EC1.16.3.1) similar to ceruloplasmin (Vashchenko & MacGillivray, 2013), and transfer across the basal membrane by ferroportin 1 (SLC40A1; Montalbetti et al., 2013).

By the end of a full-term pregnancy, a mother has supplied about 245 mg of iron to a single fetus, and a 75-mg dose is used for the placenta and umbilicus (Food and Nutrition Board, Institute of Medicine, 2001, p. 347). This does not take into account expansion of maternal hemoglobin mass during pregnancy (about 500 mg). Maternal iron deficiency increases the efficiency of iron transfer to the fetus, partially due to increased expression of copper oxidase and TfR (McArdle et al., 2011). This means that iron transfer to the fetus takes precedence over the mother's own needs, and maternal deficiency worsens if intake is not adequate. Blood loss at birth typically drains another 150 mg of iron from maternal stores.

Transfer into milk: The mammary glands of a breast-feeding woman secrete as much as 1 mg of iron per day into milk. The concentration of iron in mature human milk varies greatly, ranging from 0.2 to 2 mg/l, with a median value of around 0.4 mg/l (Dorea, 2000; Hannan et al., 2009). The transfer is tightly regulated and little influenced maternal iron status (Domellöf et al., 2004). The TfR mediates the uptake of iron from circulation into the mammary gland body cell. The mechanisms that control passage of iron through the cell recapitulate the events in enteral epithelia, including the export from endosomes by DMT1 and complexing with ferritin for storage. Ferroportin moves iron into vesicles, which can then secrete it into the luminal space of the mammary gland.

METABOLISM

Heme iron salvage: Heme oxygenase (EC1.14.99.3) cleaves the α-methylene bridge of heme, generates biliverdin, and releases Fe^{3+}. NADPH-ferrihemoprotein reductase (EC1.6.2.4), which is a FAD- and flavin mononucleotide (FMN)–containing component of the microsomal hydroxylating system, is needed to provide with a nicotine adenine dinucleotide phosphate (NADPH)–dependent reaction three reduced ferricytochrome molecules per cleaved heme (Figure 11.8).

The isoform 1 of heme oxygenase is highly inducible by heme, heavy metals, oxidizing agents, and endotoxin, while the genetically distinct heme oxygenase 2 is not inducible. Senescent red blood cells (i.e., those that have reached their normal life span of about 120 days) are removed from circulation mainly by the spleen and liver. The microsomal heme oxygenase in the spleen and many other tissues also breaks down heme from myoglobin, cytochromes, and heme enzymes. Biliverdin can then be metabolized further by biliverdin reductase (EC1.3.1.24) and is eventually converted into bilirubin.

About 40% of the iron from phagocytosed red blood cells is used for hemoglobin synthesis again within 12 days; the remainder goes into storage.

STORAGE

The total iron content in men is about 50 mg/kg; women have slightly less. Iron is stored within polymeric ferritin complexes composed of 24 light and heavy apoferritin monomers with 0.7-nm-wide pores. The H-chain has a region with ferroxidase activity buried inside the protein fold that promotes the reaction of one dioxygen molecule per two Fe^{2+} ions, resulting in the production of hydrogen peroxide.

Genetically distinct ferritin isoforms are expressed in the liver and hematopoietic tissues. Typically, about 2000 iron atoms (as $FeOOH^-$ ferric iron oxyhydroxide) reside in one ferritin polymer, much less than the maximum of about 4300. When the capacity of ferritin is exceeded, iron associates with the poorly soluble hemosiderin. Ferritin incorporates and releases iron rapidly; hemosiderin does so much more slowly.

Reducing substances such as ascorbate, FMNH2, and possibly sulfide facilitate the rapid release of iron from ferritin (Cassanelli & Moulis, 2001). The pores of the ferritin envelope are wide enough for glucose and other small organic molecules to diffuse into the iron-containing interior.

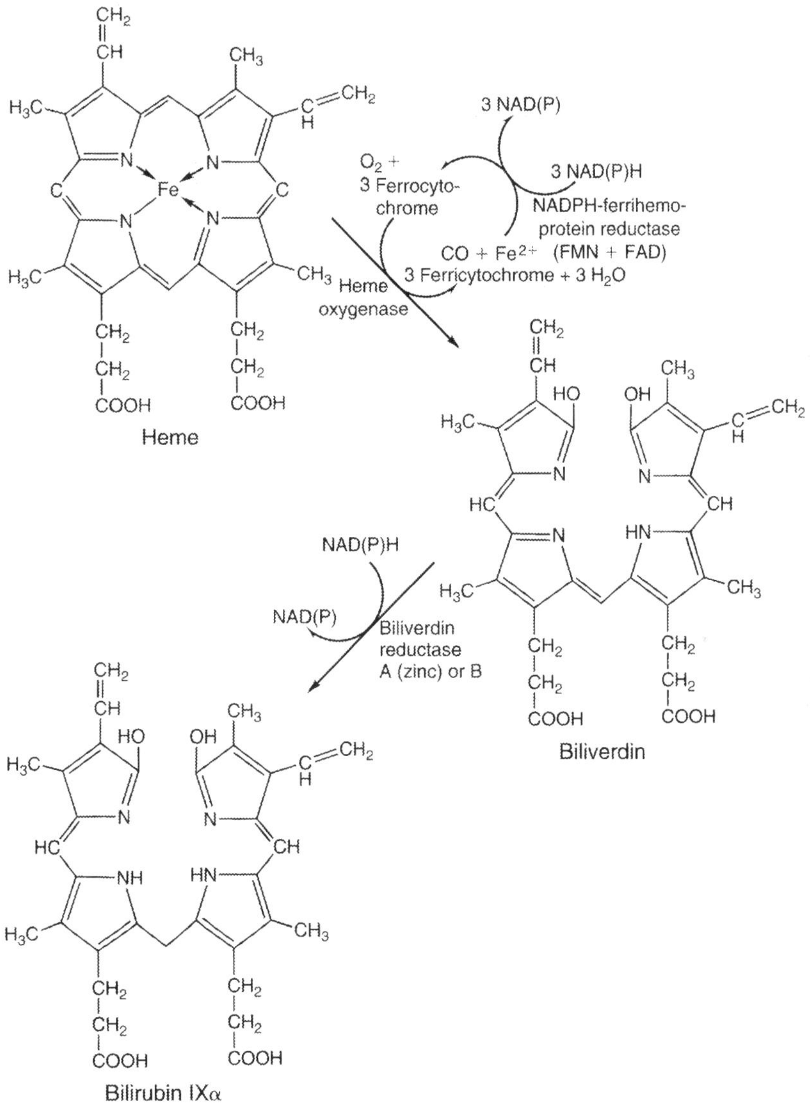

FIGURE 11.8

Heme is broken down in microsomes of the spleen and other organs.

EXCRETION

Iron is irretrievably lost with bile (84 μg/kg), skin (42 μg/kg), and urine (14 μg/kg). Typical menstrual blood loss is 30 ml per cycle. Hormonal contraception is associated with lower losses. The iron content of hemoglobin is 3.46 mg/g. At a blood hemoglobin concentration of 130 g/l, therefore, the loss of 1 ml of blood corresponds to about 0.45 mg of iron. Unrecognized low-grade blood loss (often just a few milliliters per day from the digestive tract) is a common cause of severe iron deficiency in older people.

Breast-feeding women transfer 0.3–0.6 mg/day of iron into milk (Food and Nutrition Board, Institute of Medicine, 2001). Iron losses via urine are relatively small because most iron-containing proteins are too large for glomerular filtration. Also, free hemoglobin and heme are removed very rapidly from circulation in most circumstances, and the concentration of free iron in the blood is very low. The small amounts of transferrin that are filtered nonetheless bind to cubilin at the brush border membrane of the proximal tubule and are taken up with the assistance of megalin (Kozyraki et al., 2001). The lack of either of these endocytotic receptors increases iron losses noticeably. Another mechanism of reabsorption from the TALs of Henle's loop, distal convoluted tubules, and the collecting ducts (both via principal and intercalated cells) uses DMT1 (Ferguson et al., 2001).

REGULATION

Iron status is regulated mainly at the level of intestinal absorption, but other sites, including the kidneys, also play important roles (Ganz, 2013). With full iron status, less iron can be absorbed from enterocytes. Accordingly, people with full iron stores absorb a lower percentage of ingested iron than iron-depleted people.

Hepcidin, a hormone-like protein secreted by the liver, is the master regulator of iron homeostasis. It acts by binding in the small intestine and in macrophages to ferroportin (SLC40A1), which is the exclusive exporter of iron, steering ferroportin toward lysosomes for irreversible degradation (Zhang, 2010). A cytokine from bone marrow, growth differentiation factor 15 (GDF-15), suppresses hepcidin expression and can cause iron overload in patients with myelodysplastic syndrome and other diseases (Jiang et al., 2014).

One mechanism for the control of hepcidin expression in the liver uses TfR2 to sense current iron status. When iron status is low, the limited numbers of diferric transferrin molecules in circulation bind mostly to TfR1. When the concentration of diferric transferrin is higher (usually due to larger amounts of iron in the body), TfR1 becomes saturated and more diferric transferrin can bind to TfR2. Binding of diferric transferrin to the TfR2 receptor triggers a signaling cascade that ultimately increases hepcidin expression. The *HFE* gene protein, which resembles major histocompatibility complex (MHC) class I molecules, modules binding properties of TfR2 and determines at what concentration of diferric transferrin the expression of hepcidin starts to increase (Preza et al., 2013). Individuals with certain *HFE* gene variants require a higher diferric transferrin concentration to trigger hepcidin production. These *HFE* variant carriers have greatly increased free iron concentration in the blood, tend to retain an excessive amount of iron, and often develop over time the clinical phenotype of hemochromatosis (Preza et al., 2013).

A second pathway for hepcidin regulation links circulating iron through the binding of bone morphogenic protein 6 (BMP6) to a complex involving hemojuvelin, bone morphogenic protein 6 receptor 1 (BMP6R1) and 2 (BMP6R2), and TMPRSS6. Activation of this receptor complex triggers a multistep sequence using the cytosolic signaling proteins SMAD8 (which stands for "son of mothers against decapentaplegic 8"), SMAD5, and SMAD1, and the nuclear protein SMAD4.

Infection and inflammation greatly increase hepcidin release from the liver and depress both intestinal iron absorption and mobilization of iron from macrophages (Deschemin and Vaulont, 2013). Intestinal iron absorption is regulated at additional points. One of these is the adjustment of DMT1-mediated uptake. High iron concentration in enterocytes, such as with an iron-rich meal, immediately leads to the redistribution of DMT1 from the small intestinal brush border membrane into the cytosol and only later to a decrease in DMT1 and *HFE* (hemochromatosis gene) expression (Sharp et al., 2002). Yet another mechanism is the increased sequestration in enterocytes by ferritin in response to high intracellular iron concentration. The ferritin-bound iron is blocked from proceeding out of the cell and is lost into feces when the absorptive cell is shed at the end of its 2–3-day life span (Gulec et al., 2014).

The iron enzyme aconitate hydratase (aconitase, EC4.2.1.3) and a related protein without aconitase activity act as iron regulatory proteins (IRPs) that sense the cytosolic concentration of available iron and bind to specific iron-responsive elements (IREs) in several genes. Binding of the iron-replete IRPs to the IRE of ferritin increases its expression and thereby slows intestinal absorption and promotes storage in the liver and hematopoietic (blood-forming) tissues. Binding of iron-replete IRPs to the transferrin-receptor messenger RNA (mRNA) accelerates its degradation and slows the transfer of iron from the intestine and iron-storing tissues. Analogous interactions also regulate ferroportin 1 and DMT1 expression. Much of the ferroportin translation is repressed by the iron-response miR-485-3p (Sangokoya et al., 2013).

FUNCTION

Oxygen transport: One of the longest-known functions of iron is its role as a constituent of the oxygen-binding proteins hemoglobin and myoglobin. The tetrameric hemoglobin carries oxygen from the lungs to peripheral tissues. Increased carbon dioxide concentration and associated high pH in oxygen-requiring tissues induce a conformational change (i.e., the Bohr effect) that releases the bound oxygen from hemoglobin and promotes the binding of carbon dioxide instead. Upon return of hemoglobin via blood circulation to the lungs, the carbon dioxide is exchanged for another load of oxygen. Myoglobin is the main oxygen acceptor in muscle. Cytochromes are small proteins that figure importantly in the intracellular transfer of redox equivalents for a wide range of acceptors.

Oxidative phosphorylation: Both heme and sulfur-bound iron act in the mitochondrial synthesis of ATP. The NADH dehydrogenase (EC1.6.99.3) component of the respiratory chain complex I contains iron. The cytochrome c oxidase (EC1.9.3.1) subunits of the respiratory chain complex IV use hemes to accept electrons from ferricytochrome c and use them to reduce oxygen to water.

Free radical metabolism: Free iron readily transitions between its divalent and trivalent forms, depending on its environment and potential reaction partners. It is a particularly strong catalyst for the generation of various oxygen free radicals. A typical example is the Fenton reaction, which generates hydroxyl radicals ($\cdot$OH) from superoxide anions (O_2^-). Fe^{2+} can convert hydrogen peroxide, which is an intermediate of the Fenton reaction, into hydroxyl radicals. Atherosclerosis, cancer, and other chronic diseases have been tentatively linked to excessive iron concentrations.

The iron enzyme superoxide dismutase (EC1.15.1.1) and the heme enzymes catalase (EC1.11.1.6) and peroxidase (EC1.11.1.7) keep the concentrations of superoxide, hydrogen peroxide, and other highly reactive oxygen-containing compounds in check.

DNA synthesis: Ribonucleoside-diphosphate reductase (EC1.17.4.1), which contains both iron and ATP, catalyzes the first step of DNA replication.

Nutrient metabolism: Several isoforms of the iron-containing alcohol dehydrogenase (EC1.1.1.1) catalyze the metabolism of ethanol and of various nutrient metabolites. Aldehyde oxidase (EC1.2.3.1) is one of several enzymes for the second step of ethanol catabolism. Aconitate hydratase (aconitase, EC4.2.1.3) is both an important Krebs-cycle enzyme and a cytosolic iron sensor, as described previously.

Vitamin A metabolism: Beta-carotene 15,15'-dioxygenase (EC1.13.11.21) is an iron-containing enzyme in the small intestine and liver that generates retinal from beta-carotene and a few other carotenoids. Another iron enzyme, retinal dehydrogenase (EC1.2.1.36), produces retinoic acid.

Amino acid metabolism: The iron enzymes phenylalanine hydroxylase (EC1.14.16.1) and tyrosine 3-monooxygenase (EC1.14.16.2) catalyze the initial step of phenylalanine breakdown and catecholamine and melanin synthesis. Tryptophan 5-monooxygenase (EC1.14.16.4) catalyzes the first step of serotonin synthesis. Trimethyllysine dioxygenase (EC1.14.11.8) is involved in carnitine synthesis. Gamma-butyrobetaine, 2-oxoglutarate dioxygenase (EC1.14.11.1) participates in choline metabolism. The heme enzymes tryptophan 2,3-dioxygenase (EC1.13.11.11) and indoleamine-pyrrole 2,3-dioxygenase (EC1.13.11.42) facilitate tryptophan metabolism and niacin synthesis. The latter enzyme is also important for serotonin and melatonin degradation. The final step of taurine synthesis uses iron-containing hypotaurine dehydrogenase (EC1.8.1.3).

Thyroid hormones: The heme enzyme iodide peroxidase (EC1.11.1.8) oxidizes iodide, iodinates specific tyrosines in thyroglobulin, and couples iodinated tyrosines to generate thyroxin and triiodothyronine.

Protein-modification: Specific proteins are posttranslationally modified by several iron-containing dioxygenases, including procollagen-proline 3-dioxygenase (EC1.14.11.7), procollagen-lysine 5-dioxygenase (EC1.14.11.4), procollagen-proline, 2-oxoglutarate-4-dioxygenase (EC1.14.11.2), and peptide-aspartate beta-dioxygenase (EC1.14.11.16).

Fatty acid metabolism: Two iron enzymes can add double bonds to fatty acids. Stearoyl-CoA desaturase (EC1.14.99.5) produces oleic acid. Linoleoyl-CoA desaturase (EC1.14.99.25) catalyzes the first step of arachidonic acid synthesis (precursor of prostanoids) from linoleic acid.

Sulfur metabolism: Sulfite derives directly from the diet and from the metabolism of sulfur amino acids. Sulfite oxidase (EC1.8.3.1), which contains both heme and the molybdenum cofactor, converts sulfite to sulfate.

Dementia: Patients afflicted with Alzheimer's disease were found to have onset of their symptoms 5 years earlier if they had the *HFE* variant *H63D* (Sampietro et al., 2001). This seems to indicate an increased risk of the disease in people with excessive iron stores.

Acute poisoning: Accidental ingestion of multiple adult iron supplements continues to be an all-too-common cause of poisoning in infants because the iron products are considered harmless products that do not need to be secured like other medications (Chang & Rangan, 2011).

NUTRITIONAL ASSESSMENT

The size of iron stores is reasonably well reflected by ferritin concentration in the blood (Brittenham et al., 2014). The ratio of soluble TfR over the log of ferritin concentration provides an even better estimate (Cook et al., 2003). The specificity is limited, though because molecular defects (such as thalassemia), inflammation and infection independent raise ferritin concentration (Kolnagou et al., 2013). Even direct iron measurements in the liver from biopsies or through magnetic resonance methods reach their limits in some of these cases due to uneven distribution of iron accumulations.

The concentrations of regulatory molecules in the blood also reflect iron status. Hepcidin concentration is low when iron stores are depleted. The concentration of soluble transferrin receptor (sTfR) in the circulation is thought to reflect the receptor density on iron-importing cells from which the sTfR fragments are cleaved. When cells cannot get enough iron, sTfR concentration in the blood is high

(Jankowska et al., 2014). It may be slightly more predictive of iron deficiency than ferritin, particularly in people with inflammatory diseases or malignant tumors (Braga et al., 2014).

An alternative approach for assessment in healthy populations is based on the misincorporation of zinc into protoporphyrin IX in a state of iron deficiency. The zinc protoprophyrin can be measured in red blood cells. A higher concentration indicates iron deficiency during heme synthesis for the red blood cell precursor in bone marrow (Hennig et al., 2014). The simplicity of the instrumentation is a significant advantage, particularly in resource-poor environments. The specificity, however, is limited because lead exposure has the same effect as iron deficiency on zinc protoporphyrin concentrations.

A distinct issue is the ability to limit excessive iron accumulation in the liver, heart, pancreas, and other tissues (hemochromatosis). Some individuals are prone to absorb and retain iron more than others. The commonly used measure for this predisposition is transferrin saturation. Those with very high values are at great risk of increasing iron stores excessively over time. They are also less able to sequester iron and maintain the extremely low concentration of free iron ions typical for the vast majority of humans. The elevated concentration of free iron, indicated by high transferrin saturation, greatly increases susceptibility to infections with gram-negative and other pathogenic bacteria.

REFERENCES

Braga, F., Infusino, I., Dolci, A., Panteghini, M., 2014. Soluble transferrin receptor in complicated anemia. Clin. Chim. Acta 431, 143–147.

Brittenham, G.M., Andersson, M., Egli, I., Foman, J.T., Zeder, C., Westerman, M.E., et al., 2014. Circulating non-transferrin-bound iron after oral administration of supplemental and fortification doses of iron to healthy women: a randomized study. Am. J. Clin. Nutr. 100, 813–820.

Burdo, J.R., Menzies, S.L., Simpson, I.A., Garrick, L.M., Garrick, M.D., Dolan, K.G., et al., 2001. Distribution of divalent metal transporter 1 and metal transport protein 1 in the normal and Belgrade rat. J. Neurosci. Res. 66, 1198–1207.

Cassanelli, S., Moulis, J., 2001. Sulfide is an efficient iron releasing agent for mammalian ferritins. Biochim. Biophys. Acta 1547, 174–182.

Chang, T.P., Rangan, C., 2011. Iron poisoning: a literature-based review of epidemiology, diagnosis, and management. Pediatr. Emerg. Care 27, 978–985.

Cook, J.D., Flowers, C.H., Skikne, B.S., 2003. The quantitative assessment of body iron. Blood 101, 3359–3364.

Conrad, M.E., Umbreit, J.N., 2000. Iron absorption and transport – an update. Am. J. Hematol. 64, 287–298.

Conrad, M.E., Umbreit, J.N., Moore, E.G., 1991. A role for mucin in the absorption of inorganic iron and other metal cations. A study in rats. Gastroenterology 100, 129–136.

Conrad, M.E., Umbreit, J.N., Moore, E.G., Rodning, C.R., 1992. Newly identified iron-binding protein in human duodenal mucosa. Blood 79, 244–247.

Conrad, M.E., Umbreit, J.N., Moore, E.G., Heiman, D., 1996. Mobilferrin is an intermediate in iron transport between transferrin and hemoglobin in K562 cells. J. Clin. Invest. 98, 1449–1454.

Conrad, M.E., Umbreit, J.N., Moore, E.G., Hainsworth, L.N., Porubcin, M., Simovich, M.J., et al., 2000. The effect of ceruloplasmin on iron release from placental (BeWo) cells: evidence for an endogenous Cu oxidase. Placenta 21, 805–812.

Deschemin, J.C., Vaulont, S., 2013. Role of hepcidin in the setting of hypoferremia during acute inflammation. PLoS One 8, e61050.

Domellöf, M., Lönnerdal, B., Dewey, K.G., Cohen, R.J., Hernell, O., 2004. Iron, zinc, and copper concentrations in breast milk are independent of maternal mineral status. Am. J. Clin. Nutr. 79, 111–115.

Donovan, A., Brownlie, A., Zhou, Y., Shepard, J., Pratt, S.J., Moynihan, J., et al., 2000. Positional cloning of zebrafish ferroportinl identifies a conserved vertebrate iron exporter. Nature 403, 776–781.

Dorea, J.G., 2000. Iron and copper in human milk. Nutrition 16, 209–220.

Drummond, G.S., Rosenberg, D.W., Kappas, A., 1992. Intestinal heme oxygenase inhibition and increased biliary iron excretion by metalloporphyrins. Gastroenterology 102, 1170–1175.

Fairweather-Tait, S.J., Guile, G.R., Valdes, A.M., Wawer, A.A., Hurst, R., Skinner, J., et al., 2013. The contribution of diet and genotype to iron status in women: a classical twin study. PLoS One 8 (12), e83047.

Ferguson, C.J., Wareing, M., Ward, D.T., Green, R., Smith, C.P., Riccardi, D., 2001. Cellular localization of divalent metal transporter DMT-1 in rat kidney. Am. J. Physiol. Renal Fluid Electrolyte Physiol. 280, F803–F814.

Food and Nutrition Board, Institute of Medicine, 2001. Dietary Reference Intakes for Vitamin A, Vitamin K, Arsenic, Boron, Chromium, Copper, Iodine, Iron, Manganese, Molybdenum, Nickel, Silicon, Vanadium, and Zinc. National Academy Press, Washington, DC.

Frazer, D.M., Vulpe, C.D., McKie, A.T., Wilkins, S.J., Trinder, D., Cleghorn, G.J., et al., 2001. Cloning and gastrointestinal expression of rat hephaestin: relationship to other iron transport proteins. Am. J. Physiol. Gastrointest. Liver Physiol. 281, G931–G939.

Ganz, T., 2013. Systemic iron homeostasis. Physiol. Rev. 93, 1721–1741.

Garrick, L.M., Dolan, K.G., Romano, M.A., Garrick, M.D., 1999. Non-transferrin-bound iron uptake in Belgrade and normal rat erythroid cells. J. Cell. Physiol. 178, 349–358.

Georgieff, M.K., Wobken, J.K., Welle, J., Burdo, J.R., Connor, J.R., 2000. Identification and localization of divalent metal transporter-1 (DMT-1) in term human placenta. Placenta 21, 799–804.

Gerhard, G.S., Levin, K.A., Price Goldstein, J., Wojnar, M.M., Chorney, M.J., Belchis, D.A., 2001. Vibrio vulnificus septicemia in a patient with the hemochromatosis HFE C282Y mutation. Arch. Pathol. Lab. Med. 125, 1107–1109.

Gross, P.M., Weindl, A., 1987. Peering through the windows of the brain. J. Cereb. Blood Flow Metab. 7, 663–672.

Gulec, S., Anderson, G.J., Collins, J.F., 2014. Mechanistic and regulatory aspects of intestinal iron absorption. Am. J. Physiol. Gastrointest Liver Physiol. 307, G397–G409.

Hannan, M.A., Faraji, B., Tanguma, J., Longoria, N., Rodriguez, R.C., 2009. Maternal milk concentration of zinc, iron, selenium, and iodine and its relationship to dietary intakes. Biol. Trace Elem. Res. 127, 6–15.

Hennig, G., Gruber, C., Vogeser, M., Stepp, H., Dittmar, S., Sroka, R., et al., 2014. Dual-wavelength excitation for fluorescence-based quantification of zinc protoporphyrin IX and protoporphyrin IX in whole blood. J. Biophotonics 7, 514–524.

Hurrell, R.E., Reddy, M., Cook, J.D., 1999. Inhibition of non-haem iron absorption in man by polyphenolic-containing beverages. Br. J. Nutr. 81, 289–295.

Jankowska, E.A., Kasztura, M., Sokolski, M., Bronisz, M., Nawrocka, S., Oleśkowska-Florek, W., et al., 2014. Iron deficiency defined as depleted iron stores accompanied by unmet cellular iron requirements identifies patients at the highest risk of death after an episode of acute heart failure. Eur. Heart J. 35, 2468–2476.

Jiang, F., Yu, W.J., Wang, X.H., Tang, Y.T., Guo, L., Jiao, X.Y., 2014. Regulation of hepcidin through GDF-15 in cancer-related anemia. Clin. Chim. Acta 428, 14–19.

Khan, A.A., Quigley, J.G., 2013. Heme and FLVCR-related transporter families SLC48 and SLC49. Mol. Aspects Med. 34, 669–682.

Kolnagou, A., Natsiopoulos, K., Kleanthous, M., Ioannou, A., Kontoghiorghes, G.J., 2013. Liver iron and serum ferritin levels are misleading for estimating cardiac, pancreatic, splenic and total body iron load in thalassemia patients: factors influencing the heterogenic distribution of excess storage iron in organs as identified by MRI T2*. Toxicol. Mech. Methods 23, 48–56.

Kozyraki, R., Fyfe, J., Verroust, P.J., Jacobsen, C., Dautry-Varsat, A., Gburek, J., et al., 2001. Megalin-dependent cubilin-mediated endocytosis is a major pathway for the apical uptake of transferrin in polarized epithelia. Proc. Natl. Acad. Sci. USA 98, 12491–12496.

Le Blanc, S., Garrick, M.D., Arredondo, M., 2012. Heme carrier protein 1 transports heme and is involved in heme-Fe metabolism. Am. J. Physiol. Cell Physiol. 302, C1780–C1785.

Levy, J.E., Jin, O., Fujiwara, Y., Kuo, E., Andrews, N.C., 1999. Transferrin receptor is necessary for development of erythrocytes and the nervous system. Nat. Genet. 21, 396–399.

McArdle, H.J., Lang, C., Hayes, H., 2011. Gambling L. Role of the placenta in regulation of fetal iron status. Nutr. Rev. 69 (Suppl. 1), S17–22.

Montalbetti, N., Simonin, A., Kovacs, G., Hediger, M.A., 2013. Mammalian iron transporters: families SLC11 and SLC40. Mol. Aspects Med. 34, 270–287.

Moos, T., Morgan, E.H., 2000. Transferrin and transferrin receptor function in brain barrier systems. Cell. Mol. Neurobiol. 20, 77–95.

Potter, B.J., McHugh, T.A., Beloqui, O., 1992. Iron uptake from transferrin and asialotransferrin by hepatocytes from chronically alcohol-fed rats. Alcohol. Clin. Exp. Res. 16, 810–815.

Preza, G.C., Pinon, R., Ganz, T., Nemeth, E., 2013. Cellular catabolism of the iron-regulatory peptide hormone hepcidin. PLoS One 8, e58934.

Sampietro, M., Caputo, L., Casatta, A., Meregalli, M., Pellagatti, A., Tagliabue, J., et al., 2001. The hemochromatosis gene affects the age of onset of sporadic Alzheimer's disease. Neurobiol. Aging 22, 563–568.

Sangokoya, C., Doss, J.F., Chi, J.T., 2013. Iron-responsive miR-485-3p regulates cellular iron homeostasis by targeting ferroportin. PLoS Genet. 9 (4), e1003408.

Sharp, E., Tandy, S., Yamaji, S., Tennant, J., Williams, M., Singh Srai, S.K., 2002. Rapid regulation of divalent metal transporter (DMT1) protein but not mRNA expression by non-heme iron in human intestinal Caco-2 cells. FEBS Lett. 510, 71–76.

Theil, E.C., Chen, H., Miranda, C., Janser, H., Elsenhans, B., Núñez, M.T., et al., 2012. Absorption of iron from ferritin is independent of heme iron and ferrous salts in women and rat intestinal segments. J. Nutr. 142, 478–483.

Umbreit, J.N., Conrad, M.E., Moore, E.G., Desai, M.E., Turrens, J., 1996. Paraferritin: a protein complex with ferrireductase activity is associated with iron absorption in rats. Biochemistry 35, 6460–6469.

Vashchenko, G., MacGillivray, R.T., 2013. Multi-copper oxidases and human iron metabolism. Nutrients 5, 2289–2313.

Walter, H., Hertling, I., Benda, N., Konig, B., Ramskogler, K., Riegler, A., et al., 2001. Sensitivity and specificity of carbohydrate-deficient transferrin in drinking experiments and different patients. Alcohol 25, 189–194.

Zhang, A.S., 2010. Control of systemic iron homeostasis by the hemojuvelin-hepcidin axis. Adv. Nutr. 1, 38–45.

COPPER

Copper (Cu, atomic weight 63.546) is a transition metal. Isotopes ^{63}Cu (69.1%) and ^{65}Cu (30.9%) occur naturally. Several radioactive isotopes (58–62, 64, and 66–68) can be produced artificially. The cuprous ion (Cu$^+$) is the monovalent form; the cupric ion (Cu^{2+}) is divalent.

ABBREVIATIONS

ATP7A	copper-transporting ATPase 7A (Menkes protein, EC3.6.3.4)
ATP7B	copper-transporting ATPase 7B (Wilson protein, EC3.6.3.4)
CTR1	copper transporter 1 (SLC31A1)
CTR2	copper transporter 2 (SLC31A2)

NUTRITIONAL SUMMARY

Function: Copper is essential for energy metabolism (cellular respiration), brain function (neurotransmitter regulation), soft tissues and bone (collagen synthesis), nutrient metabolism (especially iron), and antioxidant defense against free radicals (which increase the risk of cancer and cardiovascular disease).

Requirements: Adults should get 0.9 mg/day (Food and Nutrition Board, Institute of Medicine, 2001). Smoking, strenuous exercise, heat, infections, and injuries each may increase needs by 50% or more.

Food sources: The best food sources of copper include liver, shellfish, nuts, and seeds. Most other foods provide smaller amounts that in combination usually are enough to meet normal needs. People who follow the food guide pyramid recommendations ensure their adequate copper intake.

Deficiency: Deficiency is very unlikely to occur except in people with very rare genetic disorders or during prolonged starvation. Symptoms include anemia, low white cell count, accelerated bone mineral loss, and increased blood pressure and cholesterol levels. If intake is low, stores last only a few weeks.

Excessive intake: Daily doses in excess of 10 mg (typically from supplements or from contaminated stored or well water) can cause nausea, vomiting, abdominal cramps, diarrhea, and liver damage (especially in infants). Higher doses may lead to coma and death.

DIETARY SOURCES

Exceptionally rich sources of copper include liver (45 mg/kg), kidney (35–50 mg/kg), oysters (7.4 mg/kg), walnuts (16 mg/kg), and other nuts and seeds. Copper-lined pipes or vessels do not increase the copper content of their contents significantly unless exposed to acids. Average daily intake of American adults is about 1.6 mg in males and 1.2 mg in females (Food and Nutrition Board, Institute of Medicine, 2001).

DIGESTION AND ABSORPTION

Copper absorption occurs in the stomach and small intestine by saturable active transport. Bile adds about 5 mg/day to the ingested amount (Figure 11.9). More than half of a small oral dose (<1 mg) is usually absorbed. Fractional absorption decreases to about 25% with a dose of 3–4 mg, and to <15% at intake above 6–7 mg (Wapnir, 1998). Copper absorption is decreased by phytate and inositol pentaphosphate in the same meal, but the reduction is less than for iron and zinc (Lönnerdal et al., 1999). Excessive zinc intake decreases copper absorption slightly. Ceruloplasmin in human milk enhances copper absorption in the small intestines of infants. Copper is taken up from the intestinal lumen as Cu^+ through the high-affinity copper transporter CTR1 (SLC31A1). The chaperone ATOX1, together with the copper-transporting ATPase 7A (Menkes protein, ATP7A, EC3.6.3.4), then direct copper to the trans-Golgi network (de Bie et al., 2007). Transport by ATP7A is linked to the hydrolysis of ATP and is therefore magnesium-dependent.

Metallothioneins, a family of small cationic-metal-binding proteins, sequester excess copper, and limit absorption thereby. ATP7A also facilitates export across the basolateral membrane.

TRANSPORT AND CELLULAR UPTAKE

Blood circulation: Newly absorbed copper, which is mainly bound to albumin, is rapidly cleared from blood circulation by the liver, to a lesser extent by the kidneys. When copper is secreted from the liver again, it is in association with ceruloplasmin, metallothionein, and other copper-containing proteins.

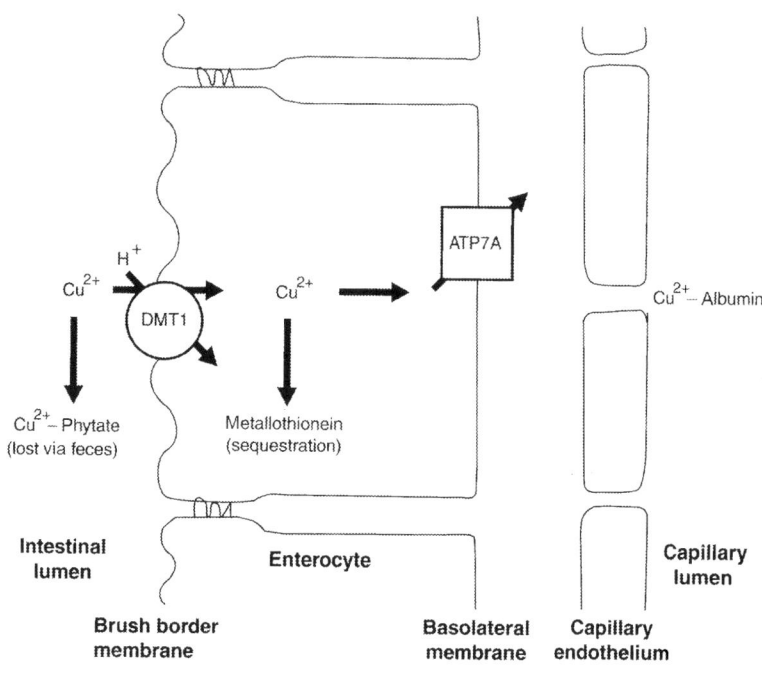

FIGURE 11.9

Intestinal copper absorption.

Because of this, most of the copper in the blood is bound to ceruloplasmin (65–90%) and albumin (5–10%). A much smaller amount is unspecifically bound to histidine. The role of transcuprein, metallothionein, and other copper-binding proteins in humans remains to be fully explored. The different copper carriers can substitute for each other, as shown by the fact that copper delivery to cells is not disrupted even in the absence of ceruloplasmin (Meyer et al., 2001). Total copper concentration in the blood of healthy adults is typically between 75 and 130 µmol/l. Copper-binding capacity in the blood normally exceeds total copper concentration by five orders of magnitude (Linder et al., 1999). Infection, tumors, pregnancy, and hormonal contraception increase the concentration of ceruloplasmin in the blood since this is an acute-phase protein. Copper separates from the various carrier systems and enters by itself. The pathway is different in liver cells, which take up ceruloplasmin through the asialoglycoprotein receptor after ceruloplasmin deglycosylation at the plasma membrane (Harris, 2000) (Figure 11.10). Uptake into other cells depends on the high-affinity copper transporters CTR1 (SLC31A1) and CTR2 (SLC31A2). Multiple distinct transcripts of both transporters occur in most tissues (Kim et al., 2013). The Cu^{2+} form from plasma has to be reduced after uptake into the cell by as-yet-incompletely characterized NADH oxidases before it can be imported through the Cu^+-specific copper transporter CTR1 (de Bie et al., 2007; Nevitt et al., 2012). Erythrocytes may also take up Cu^+ complexed with chloride and hydroxyl ions through the chloride/bicarbonate-exchanger (band 3 of the red cell membrane, SLC4A1; Bogdanova et al., 2002).

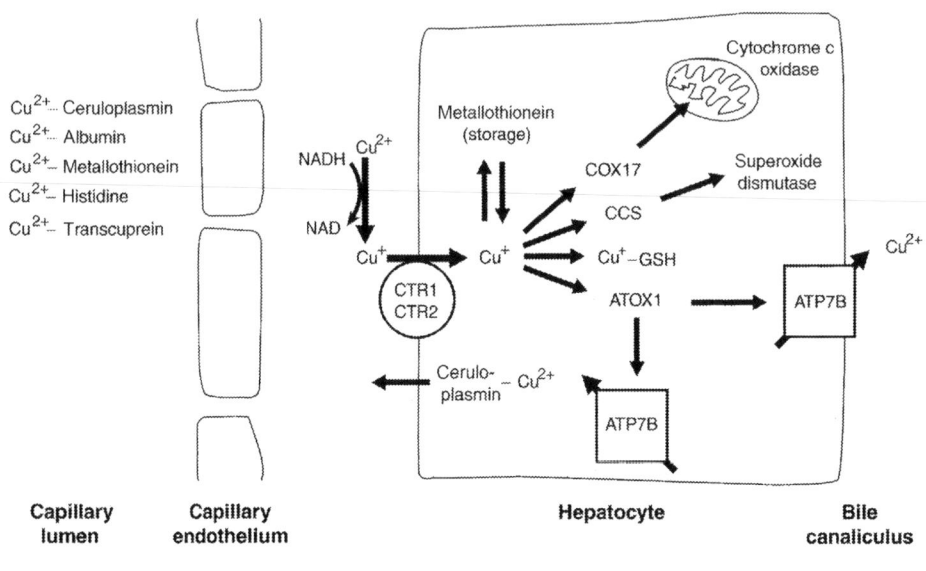

FIGURE 11.10

Intracellular copper disposition.

BBB: Movement of copper into and out of the brain of adults is very limited (Stuerenburg, 2000). Choroid plexus epithelial cells use mainly CTR1 on the side facing the brain for copper transport (Zheng et al., 2012). ATP7A moves copper toward the brain, while ATP7B mediates transfer into the blood (Nevitt et al., 2012). This arrangement supports copper transport in either direction, depending on concentration in the brain and on regulatory input.

Materno-fetal transfer: The transfer of copper across the placenta is very limited (Krachler et al., 1999). CTR1 is present at the fetal side of the syntrophoblast. ATP7A moves copper toward fetal circulation, while ATP7B mediates the transfer into maternal blood (Nevitt et al., 2012). This means that copper can move in either direction, from mother to fetus or vice versa as needed.

Transfer into milk: The concentration of copper in mature human milk is about 0.4 mg/l (Hannan et al., 2005).

The copper transporters ATP7A, ATP7B, and CTR1 are all expressed in body cells of the mammary gland, which indicates their involvement in the transfer process (Kelleher & Lönnerdal, 2006).

METABOLISM

Various copper binders keep the intracellular concentration of free copper so low ($<10^{-18}$ mol/l) that individual cells contain less than one atom on average (O'Halloran & Culotta, 2000). Reduced glutathione (GSH) avidly binds Cu^+ and facilitates its association with the metal-storage protein metallothionein. GSH is also important for the targeted intracellular transport of copper in conjunction with specific copper-binding proteins (metallochaperones), several of which direct copper to specific targets (Harris, 2000). CCS (copper chaperone for superoxide dismutase, homolog of yeast LYS7) directs copper toward newly synthesized superoxide dismutase (EC1.15.1.1). COX17 serves as a mitochondrial shuttle that delivers copper to the cytochrome oxidase c complex. Cu^+-binding proteins in the inner mitochondrial membrane, homologs of SCO1 and SCO2 in yeast, probably act as intermediaries. ATOX1 (antioxidant 1, formerly HAH1) coordinates the movement of copper to the copper-transporting ATPase 7B (ATP7B, Wilson protein, EC3.6.3.4). S-adenosyl homocysteine hydrolase (EC3.3.1.1) may participate in this transport sequence by temporarily binding Cu^{2+} (Bethin et al., 1995). The Wilson protein (ATP7B) resides mainly near the Golgi apparatus of liver cells and enables copper secretion into bile in an incompletely understood fashion (Harris, 2000). Alternative splicing of the Wilson gene product generates a truncated cytosolic form of unclear function. The closely related Menkes protein (ATP7A, EC3.6.3.54) is essential in most other tissues for the delivery of copper to newly synthesized enzymes, such as monophenol monooxygenase (tyrosinase, EC1.14.18.1), in secretory vesicles and for copper export. This copper pump is also closely associated with the trans-Golgi network. ATP7A-containing secretory vesicles cycle move rapidly to the plasma membrane, where they help to remove excess copper from the cell (Petris et al., 2000).

STORAGE

Considerable amounts of copper (50–120 mg) are bound to metallothioneins in the liver and kidneys; much smaller amounts are stored in other organs and tissues. These metallothioneins are small proteins that bind 7–12 atoms of copper, zinc, or cadmium. There are at least 10 genetically distinct metallothionein genes with differing tissue expression patterns and metal-binding properties. It has been suggested that S-adenosyl homocysteine hydrolase (EC3.3.1.1) is a bifunctional protein that provides additional copper-binding capacity in the liver (Bethin et al., 1995).

EXCRETION

Less than 10% of total losses are via skin, urine, and other secretions; the remainder (>90%) is eliminated via feces. The copper-transporting ATPase 7B (Wilson protein, EC3.6.3.4) pumps about 5 mg/day into bile. Much of this is reabsorbed from the small intestine (hepatobiliary cycling) unless concurrent dietary intake is very high or total body stores are already full.

REGULATION

Storage as a complex with metallothionein and excretion into bile are the central events for maintenance of copper homeostasis. Metallothionein transcription is induced by glucocorticoids, interleukin 6 (IL-6), copper, zinc, and cadmium. The metal-regulatory transcription factor 1, which appears to be induced by zinc, binds to and activates metal-responsive promoter elements of metallothionein genes. Copper also induces expression of the copper-transporting ATPase in the liver (ATP7B, Wilson protein). A much more rapid copper-dependent regulatory event is the phosphorylation of ATP7B, which increases its redistribution to the cell membrane (Vanderwerf et al., 2001). Intracellular distribution of the Menkes protein (ATP7A, EC3.6.3.54) similarly helps maintain copper balance in most other tissues. As the intracellular copper concentration rises, more ATP7A moves to the plasma membrane and pumps copper out of the cell (Petris et al., 2000).

FUNCTION

Iron metabolism: Ferroxidase (ceruloplasmin, iron (II):oxygen oxidoreductase, EC1.16.3.1) contains 6–7 copper atoms that can be transferred to other tissues. This multifunction protein scavenges free radicals and oxidizes iron bound to transferrin. This latter function is important but not essential for the mobilization of iron from stores (Linder et al., 1999). Excess ascorbate may interfere with ceruloplasmin activity. Hephaestin, a copper-containing metalloprotein, is present mainly in the basolateral membrane of intestinal villi and is thought to facilitate iron exit from enterocytes. The structure of this metalloprotein closely resembles that of ceruloplasmin. Zyklopen is a third iron-oxidizing enzyme. It resides in trophoblasts of the placenta, where it aids the transfer of iron to the fetus (Vashchenko & MacGillivray, 2013).

Hormone metabolism: Copper is a component of several enzymes that participate in the metabolism of catecholamines and other compounds with regulatory activity. Monophenol monooxygenase (tyrosinase, EC1.14.18.1) catalyzes the first step of dopamine and dopaquinone synthesis. Dopamine-beta-monooxygenase (dopamine hydroxylase, EC1.14.17.1, requires ascorbate and contains autocatalytically generated topaquinone) is needed for noradrenaline synthesis, particularly in the brain. Copper-containing amine oxidases (EC1.4.3.6; includes amiloride-sensitive amine oxidase, retina-specific copper amine oxidase, and membrane copper amine oxidase, which is identical to vascular adhesion protein-1) release histamine and inactivate polyamines (putrescine, spermine, and spermidine). The amine oxidases contain topaquinone (2,4,5-trihydroxyphenylalanine quinone, generated from a tyrosine-residue in the enzyme itself) as a cofactor.

Several peptide hormones, including thyroid-releasing hormone (TRH), corticosteroid-releasing hormone (CRF), gonadotropin-releasing hormone (GnRH), NPY, endorphins, gastrin, pancreatic polypeptide, atrial natriuretic factor, arginine vasopressin, alpha-melanotropin, become active only after removal of the terminal glycine residue by peptidylglycine monooxygenase (peptidyl alpha-amidating enzyme, EC1.14.17.3, also requires ascorbate).

Energy metabolism: Cytochrome c oxidase (EC1.9.3.1) catalyzes the rate-limiting terminal step of oxidative phosphorylation at the inner mitochondrial membrane. The copper in subunit 2 of the complex transfers electrons from cytochrome c to the copper and iron-containing center of the catalytic subunit 1. A vectorial Bohr mechanism (conformation shift in response to oxygenation change) moves reduced protons from the inner aqueous space through the mitochondrial membrane and the heme-copper oxidases then release them into the external space (Papa et al., 1998). This proton translocation builds up a proton gradient across the inner mitochondrial membrane that drives the synthesis of ATP.

Antioxidation: In addition to the abovementioned function as a component of the free radical scavenger ceruloplasmin, copper is also an essential constituent of several superoxide dismutases (EC1.15.1.1). Different forms require zinc, iron, or manganese for the second metal-binding site and have distinct functions. The enzymes, which are present both in cytosol and the extracellular space, dissipate highly reactive (and therefore toxic) superoxide radicals by converting them to hydrogen peroxide. It should be emphasized that free copper ions greatly promote the formation of oxygen free radicals in a variety of reactions. Excess copper, therefore, increases oxidative damage of DNA and proteins.

Connective tissue: The maturation of the structural proteins in bone, cartilage, blood vessels, and other tissues depends on the action of protein-lysine 6-oxidase (EC1.4.3.13). This copper enzyme catalyzes the hydroxylation of specific lysine residues in collagen and elastin molecules and thereby initiates the cross-linking of individual strands. Four additional lysyl oxidase-like proteins (LOXLs) have been identified so far (Csiszar, 2001; Maki et al., 2001). Functions are very diverse and include cell adhesion and motility, tumor suppression, and regulation of cellular senescence.

Blood coagulation: Copper plays a significant role in normal blood clotting through its interaction with coagulation factors V and VIII. Factor V contains a specific (type II) copper-binding site (Villoutreix & Dahlback, 1998). The interaction of the heavy- and light-chain components of factor VIII stabilizes the complex and maintains its activity (Kaufinan et al., 1997).

Gene expression: It is likely that some transcriptional events involve copper. Several copper-containing transcription factors have been identified in yeast, including Ace1p and Mac1p, but it is unclear whether there are human homologs. It is of interest in this respect that copper deficiency decreases transcription of the interleukin-2 (IL-2) gene in human mononuclear cells and lymphocytes (Hopkins & Failla, 1999).

NUTRITIONAL ASSESSMENT

Copper status is reasonably well reflected by the concentration of copper in the blood (Harvey et al., 2009; MacKay et al., 2014). The concentration of ceruloplasmin in the blood is useful only in cases of clear deficiency. Other functional biomarkers have not performed as expected.

Copper status may also be assessed by measuring its concentration in hair (Kim & Song, 2014).

REFERENCES

Bethin, K.E., Petrovic, N., Ettinger, M.J., 1995. Identification of a major hepatic copper binding protein as *S*-adenosylhomocysteine hydrolase. J. Biol. Chem. 270, 20698–20702.

Bogdanova, A.Y., Gassmann, M., Nikinmaa, M., 2002. Copper ion redox state is critical for its effects on ion transport pathways and methaemoglobin formation in trout erythrocytes. Chemico-Biol. Interact. 139, 43–59.

Csiszar, K., 2001. Lysyl oxidases: a novel multifunctional amine oxidase family. Progr. Nucl. Acid Res. Mol. Biol. 70, 1–32.

de Bie, P., Muller, P., Wijmenga, C., Klomp, L.W., 2007. Molecular pathogenesis of Wilson and Menkes disease: correlation of mutations with molecular defects and disease phenotypes. J. Med. Genet. 44, 673–688.

Food and Nutrition Board, Institute of Medicine, 2001. Dietary Reference Intakes for Vitamin A, Vitamin K, Arsenic, Boron, Chromium, Copper, Iodine, Iron, Manganese, Molybdenum, Nickel, Silicon, Vanadium, and Zinc. National Academy Press, Washington, DC, pp. 224–257.

Harris, E.D., 2000. Cellular copper transport and metabolism. Annu. Rev. Nutr. 20, 291–310.

Hopkins, R.G., Failla, M.L., 1999. Transcriptional regulation of interleukin-2 gene expression is impaired by copper deficiency in Jurkat human T lymphocytes. J. Nutr. 129, 596–601.

Hannan, M.A., Dogadkin, N.N., Ashur, I.A., Markus, W.M., 2005. Copper, selenium, and zinc concentrations in human milk during the first three weeks of lactation. Biol. Trace Elem. Res. 107, 11–20.

Harvey, L.J., Ashton, K., Hooper, L., Casgrain, A., Fairweather-Tait, S.J., 2009. Methods of assessment of copper status in humans: a systematic review. Am. J. Clin. Nutr. 89, 2009S–2024S.

Kaufinan, R.J., Pipe, S.W., Tagliavacca, L., Swaroop, M., Moussalli, M., 1997. Biosynthesis, assembly and secretion of coagulation factor VIII. Blood Coag. Fibrinolysis 8, S3–S14.

Kelleher, S.L., Lönnerdal, B., 2006. Mammary gland copper transport is stimulated by prolactin through alterations in Ctr1 and Atp7A localization. Am. J. Physiol. Regul. Integr. Comp. Physiol. 291, R1181–R1191.

Kim, H., Wu, X., Lee, J., 2013. SLC31 (CTR) family of copper transporters in health and disease. Mol. Aspects Med. 34, 561–570.

Kim, H.N., Song, S.W., 2014. Concentrations of chromium, selenium, and copper in the hair of viscerally obese adults are associated with insulin resistance. Biol. Trace Elem. Res. 158, 152–157.

Krachler, M., Rossipal, E., Micetic-Turk, D., 1999. Trace element transfer from the mother to the newborn—investigations on triplets of colostrum, maternal and umbilical cord sera. Eur. J. Clin. Nutr. 53, 486–494.

Linder, M.C., Lomeli, N.A., Donley, S., Mehrbod, E., Cerveza, E., Cotton, S., et al., 1999. Copper transport in mammals. In: Leone, F.B., Mercer, A. (Eds.), Copper Transport and its Disorder Kluwer Academic/Plenum Publishers, New York, NY, pp. 1–16.

Lönnerdal, B., Jayawickrama, L., Lien, E.L., 1999. Effect of reducing the phytate content and of partially hydrolyzing the protein in soy formula on zinc and copper absorption and status in infant rhesus monkeys and rat pups. Am. J. Clin. Nutr. 69, 490–496.

MacKay, M., Mulroy, C.W., Street, J., Stewart, C., Johnsen, J., Jackson, D., et al., 2014. Assessing copper status in pediatric patients receiving parenteral nutrition. Nutr. Clin. Pract. pii:0884533614538457.

Maki, J.M., Tikkanen, H., Kivirikko, K.I., 2001. Cloning and characterization of a fifth human lysyl oxidase isoenzyme: the third member of the lysyl oxidase-related subfamily with four scavenger receptor cysteine-rich domains. Matrix Biol. 20, 493–496.

Meyer, L.A., Durley, A.P., Prohaska, J.R., Harris, Z.L., 2001. Copper transport and metabolism are normal in aceruloplasminemic mice. J. Biol. Chem. 276, 36857–36861.

Nevitt, T., Ohrvik, H., Thiele, D.J., 2012. Charting the travels of copper in eukaryotes from yeast to mammals. Biochim. Biophys. Acta 1823, 1580–1593.

O'Halloran, T.V., Culotta, V.C., 2000. Metallochaperones, an intracellular shuttle service for metal ions. J. Biol. Chem. 275, 25057–25060.

Papa, S., Capitanio, N., Villani, G., Capitanio, G., Bizzoca, A., Palese, L.L., et al., 1998. Cooperative coupling and role of heme a in the proton pump of heme-copper oxidases. Biochimie 80, 821–836.

Petris, M.J., Strausak, D., Mercer, J.E., 2000. The Menkes copper transporter is required for the activation of tyrosinase. Hum. Mol. Genet. 9, 2845–2851.

Stuerenburg, H.J., 2000. CSF copper concentrations, blood–brain barrier function, and coeruloplasmin synthesis during the treatment of Wilson's disease. J. Neural. Transm. Gen. Sect. 107, 321–329.

Vashchenko, G., MacGillivray, R.T., 2013. Multi-copper oxidases and human iron metabolism. Nutrients 5, 2289–2313.

Vanderwerf, S.M., Cooper, M.L., Stetsenko, I.V., Lutsenko, S., 2001. Copper specifically regulates intracellular phosphorylation of the Wilson's disease protein, a human copper-transporting ATPase. J. Biol. Chem. 276, 36289–36294.

Villoutreix, B.O., Dahlback, B., 1998. Structural investigation of the A domains of human blood coagulation factor V by molecular modeling. Prot. Sci. 7, 1317–1325.

Wapnir, R.A., 1998. Copper absorption and bioavailability. Am. J. Clin. Nutr. 67, 1054S–1060S.

Zheng, G., Chen, J., Zheng, W., 2012. Relative contribution of CTR1 and DMT1 in copper transport by the blood–CSF barrier: implication in manganese-induced neurotoxicity. Toxicol. Appl. Pharmacol. 260, 285–293.

ZINC

Zinc (Zn, atomic weight 65.38) is a metallic, divalent transition metal. Naturally occurring Zn contains the stable isotopes ^{66}Zn (27.9%), ^{67}Zn (4.1%), and ^{68}Zn (318.8%), and the unstable isotopes ^{64}Zn (48.6%; with a half-life longer than 2.3×10^{18} years) and ^{70}Zn (0.6%, half-life more than 1.3×10^{16} years). The artificial isotopes ^{69}Zn, ^{69m}Zn, ^{71}Zn, ^{71m}Zn, and ^{72}Zn are radioactive. Zn can be either monovalent (Zn^{+}) or divalent (Zn^{2+}).

ABBREVIATIONS

DMT1   divalent metal transporter 1 (SLC11A2)
MTF1   metal response element-binding transcription factor 1
ZnT1   zinc transporter 1 (SLC30A1)
ZIP1   zinc- and iron-regulated protein 1 (SLC39A1)

NUTRITIONAL SUMMARY

Function: Zn is essential for the activation of numerous genes and as a cofactor for many enzyme reactions.

Requirements: At least 8 mg/day of Zn is necessary to maintain adequate stores for women consuming a mixed diet, and 11 mg for men (Food and Nutrition Board, Institute of Medicine, 2002). Vegetarians and pregnant or breast-feeding women need slightly more.

Sources: Oysters are exceptionally rich, and other shellfish and meats also are good sources. Phytate from whole grains and some vegetables interferes with Zn absorption.

Deficiency: Low intake is associated with loss of appetite, scaling skin lesions, and impaired immune function.

Excessive intake: Consumption of more than 40 mg/day may deplete copper stores, impair immune function, and lower HDL levels.

DIETARY SOURCES

Foods contain elemental Zn, much of it bound to proteins or DNA. Meats (2–3 mg/100 g) and shellfish (such as cooked clams, 2.7 mg/100 g) provide significant amounts, and oysters are an exceptionally rich source (>70 mg per serving). Plant foods are much poorer sources.

DIGESTION AND ABSORPTION

More than 70% of a small Zn dose (less than 3 mg) is absorbed from the small intestine when there is no interference from other meal constituents (Lönnerdal, 2000). Protein hydrolysates and some amino acids, particularly histidine and cysteine, increase fractional Zn absorption (Figure 11.11). Higher than minimal Zn doses, replete Zn status, and the presence of phytate in the same meal greatly decrease absorption efficiency. Concurrent high intake of iron, calcium, and other divalent metal cations does not appear to affect Zn absorption significantly (Sandström, 2001).

Significant amounts of Zn enter the intestinal lumen with secretions from the pancreas, and intraluminal digestion by proteases, DNAses, and RNAses releases Zn from the food matrix. Zn forms complexes with histidine, cysteine, and nucleotides that are absorbed better than Zn alone. Absorption is reduced by phytate (Lönnerdal, 2000). The main mechanism for Zn uptake from the intestinal lumen is through ZIP4 (SLC39A4), a member of the zinc- and iron-regulated protein (ZIP) family (Wang et al., 2002; Engelken et al., 2014). This transporter is expressed at the luminal (apical) side of enterocytes throughout the small and large intestine. A related protein, provisionally designated hORF1, may contribute to Zn uptake from the colon (Wang et al., 2002). Genetic variants of this gene are associated with the rare familial condition acrodermatitis enteropathica, but a much more common variant at the

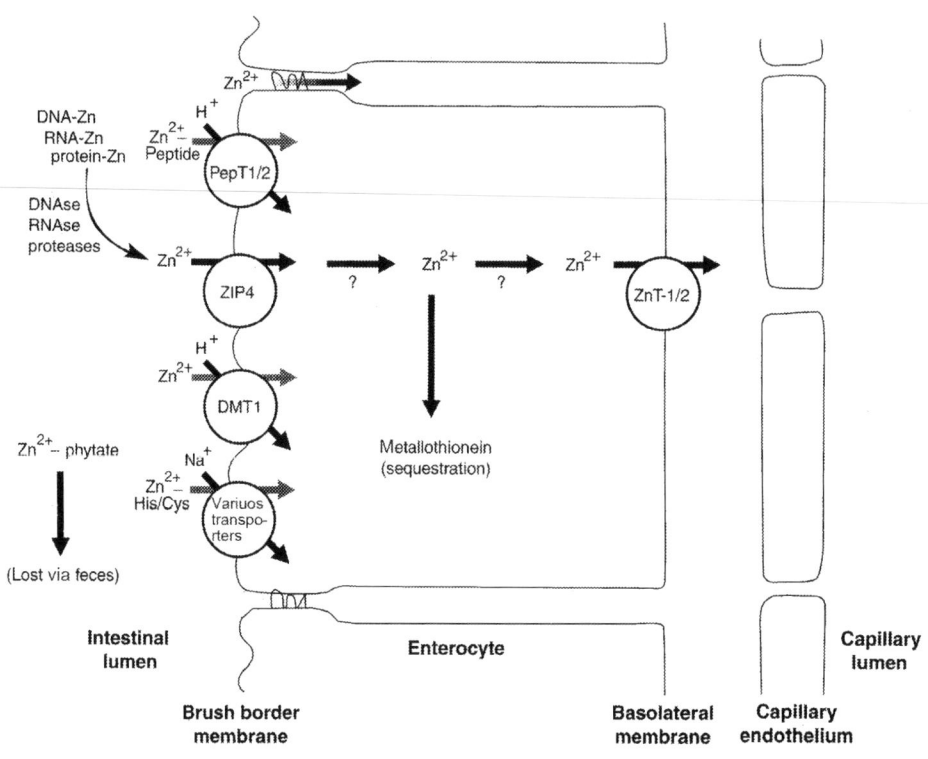

FIGURE 11.11

Intestinal zinc absorption.

same locus also constrains Zn absorption (Engelken et al., 2014). The proton-coupled divalent metal transporter 1 (DMT1, SLC11A2) appears to provide an additional minor uptake route. DMT1 also transports iron, copper, cadmium, and other divalent metal ions that may compete with Zn (McMahon and Cousins, 1998). This may be the basis for diminished Zn uptake in the presence of high amounts of calcium. Enterocyte intracellular iron concentration is the main determinant of DMT1 expression and translocation to the apical membrane; low iron status up-regulates DMT1 expression, and more so in people who are pregnant or have a familial hemochromatosis phenotype.

A possible alternative route for Zn uptake is via the hydrogen ion/peptide cotransporter (SLC15A1, PepT1) when complexed to small peptides (Tacnet et al., 1993). Uptake as a complex with individual amino acids may explain why histidine and cysteine improve intestinal Zn absorption (Lönnerdal, 2000).

Metallothioneins are a ubiquitous family of cysteine-rich small peptides that bind Zn and some other heavy metals with high capacity (up to 12 atoms per peptide). A dozen genetically distinct metallothionein isoforms occur in the liver, with additional ones in the brain, tongue, and stomach. Metallothionein regulates Zn transfer into portal blood through binding and retaining it within the enterocyte until it sheds into the intestinal lumen. High intracellular Zn concentration from luminal or endogenous increases metallothionein production, and this in turn decreases net Zn absorption (Davis et al., 1998).

Zn is exported from enterocytes into portal blood by Zn transporters 1 (ZnT-1, SLC30A1) and 2 (ZnT-2, SLC30A2), possibly functioning as Zn^{2+}/H^+ exchangers. ZnT-1 expression is up-regulated when Zn intake is high (McMahon and Cousins, 1998). Another pathway for Zn transport across the basolateral enterocyte membrane is via the Zn-regulated transporter 1 (SLC39A1).

With increasing intraluminal concentrations net Zn movement across the tight junctions of the epithelial layer (paracellular pathway) becomes more significant. Regulatory mechanisms involving DMT1 or metallothionein thus are bypassed when high-dose supplements are ingested.

TRANSPORT AND CELLULAR UPTAKE

Blood circulation: Newly absorbed Zn in portal blood is bound to albumin (Cousins, 1986) and alpha-2-macroglobulin (Osterberg & Malmensten, 1984), as is the Zn in peripheral blood. Typical venous concentrations are between 10 and 17 µmol/l, and even less in people with severe Zn deficiency.

Numerous Zn transporters mediate uptake into and efflux from specific tissues, too many to describe each one in detail here. The solute carrier families 30 (zinc transporters, ZnT, with at least 8 members) and 39 (zinc- and iron-regulated protein family, ZIP, with at least 14 members) are responsible for most of these transfer activities. ZIP usually mediate import of Zn into cells, while ZnT typically are responsible for the export from cells.

For example, the hormone-responsive Zn uptake transporter 1 (ZIP1, SLC39A1) mediates the uptake into prostate cells (Costello et al., 1999). ZIP2 (SLC39A1) is another closely related transporter that may mediate cotransport of Zn and bicarbonate into cells (Gaither & Eide, 2000), including keratinocytes (Inoue et al., 2014).

ZnT2 (SLC30A2) mediates Zn efflux or uptake into endosomal and lysosomal vesicles in the intestines, kidneys, and testis. ZnT3 (SLC30A3) is involved in Zn uptake into vesicles in neurons and in the testis. ZnT4 (SLC30A4) is another Zn exporter, which is highly expressed in mammary gland and brain. Transport of Zn across the plasma membrane of neurons is freely reversible and not ATP or ion-dependent (Colvin, 1998).

The concentration of free Zn is extremely low within the cell cytosol (Foster et al., 2014). Zn is complexed in the cell to various structures and molecules, including metalloproteins, histones, and DNA. Much of the remaining intracellular Zn is bound in zincosomes, which are distinct lysosomal compartments (Blaby-Haas & Merchant, 2014). Zn is presumably released again when Zn-containing structures are broken down by intracellular (lysosomal) digestion.

BBB: The mechanism for Zn uptake into brain capillary epithelial cells is unresolved. It is likely to involve the mediation of uptake from blood by a Zn-histidinyl complex and DMT1 (Takeda, 2000). Transport from the blood into and out of cerebrospinal fluid via the choroid plexus is a major pathway for the maintenance of brain Zn homeostasis, but the molecular mechanisms at that site are not any better understood than for brain capillary epithelium.

Materno-fetal transfer: Zn moves across the placental membrane by a slow process that equally facilitates transfer in either direction depending on the prevailing concentration gradient (Beer et al., 1992).

Transfer into milk: The mammary glands of a breast-feeding woman secrete about 1–3 mg Zn/day into milk (Picciano & Guthrie, 1976). The concentration of Zn in human milk is about 2 mg/l (Hannan et al., 2009).

Many different transporters participate in the transfer of Zn from maternal circulation into milk. The expression patterns change in characteristic manner during the transition of the nonlactating gland to the lactating state (Kelleher et al., 2012). It is thought that in the actively lactating mammary gland, ZIP8 and ZIP10 mediate most of the Zn uptake into the mammary gland body cell, ZIP7 (SLC39A7) and ZIP11 (SLC39A11) are important for release from the Golgi apparatus, and ZIP3 (SLC39A3) facilitates efflux from late endosomes. Zn is then either moved directly via ZnT4 (SLC30A4) into the mammary gland lumen or through the extrusion of Zn-containing vesicles. Zn also appears to be secreted into milk as a constitutional component of caseins and lactoferrin (Babina et al., 2004). Complexes with simultaneously secreted metallothionein greatly increase Zn bioavailability for the breastfed infant.

STORAGE

Stores in Zn-replete women are about 1.5 g, and 2.5 g for Zn-replete men (King & Keen, 1994). Muscles and bone contain most of the body's Zn. Turnover of Zn-containing proteins and DNA (which contain the bulk of Zn in muscle and bone) is very slow, with a half-life of about 300 days (Wastney et al., 2000). This means that less than 6 mg are mobilized per day as Zn-containing structures are broken down. In contrast, Zn associated with metallothionein in the liver is turning over with a half-life of about two weeks and can be readily mobilized. This much smaller pool can cover for a sudden shortfall in dietary intake, therefore. However, due to the small size of this rapidly exchangeable pool (<170 mg), Zn depletion can become functionally relevant within a week (Miller et al., 1994).

EXCRETION

About 1 mg/day appears to be lost with sweat, skin, and hair. Fecal losses (unabsorbed Zn from both diet and endogenous secretions) depend on Zn intake and status (Lee et al., 1990), and can be less than 1 mg/day (Sian et al., 1996). Ejaculate contains about 1 mg. Menstrual losses range from 0.1 to 0.5 mg. Losses via urine have been estimated to be 0.4–0.6 mg/day (2–10% of intake).

Since virtually all Zn in the blood is complexed to larger proteins (albumin, alpha-2-macroglobulin), relatively little gets into filtrate in the kidneys. The small amount that gets into ultrafiltrate is reabsorbed, mainly from the distal renal tubule via ZnT1 (Victery et al., 1981).

REGULATION

Zn concentration in cells depends on the relative activities of influx transporters (SLC39A family) and efflux transporters (SLC30A family). Specific cells may express one or more each of 14 influx transporters and 10 efflux transporters (Engelken et al., 2014).

Control of metallothionein expression is the most important mechanism for maintaining overall Zn adequacy. Increased metallothionein expression in the small intestine decreases intestinal absorption, while increased expression in the liver expands stores. Metallothionein expression is induced by the metal response element-binding transcription factor 1 (MTF1). This Zn-sensing transcription factor with its six Zn-coordinated peptide loops (zinc fingers) binds to multiple metal response elements of the metallothionein promoters when free Zn ion concentration is high (Waldron et al., 2009). MTF1 also induces expression of ZnT1. Other zinc finger proteins appear to have parallel functions, but the specifics remain to be learned. This latter change might be primarily responsible for increased excretion of Zn into bile and pancreas juice. The binding of another transcription factor, the upstream stimulatory factor family (USF), to a separate promoter sequence (antioxidant response element) is involved in the induction of metallothionein expression in response to oxygen free radical (H_2O_2) stress.

FUNCTION

Enzyme cofactor: Zn is a cofactor of several hundred enzymes and is needed for the replication and function of DNA. In fact, 16% of all enzymes need Zn for their proper function (Waldron et al., 2009). Specific metallochaperones steer Zn to the newly synthesized protein as needed.

Zn deficiency is very common, affecting possibly as much as one-third of the world's population (Das et al., 2013). Young children (Das et al., 2013), vegetarians (Foster et al., 2013), older people, and elderly patients with accelerated cognitive decline (Grønli et al., 2013) are particularly at risk. Inadequate supplies impair food digestion and absorption, synaptic signaling, gene expression, control of oxidant stress, growth and wound healing, immune function, taste and appetite, and many other functions. Less-than-adequate Zn availability appears to interfere with organ formation during early pregnancy (Hurley, 1981). It is often difficult to establish a tight link between individual Zn-dependent structures and functional status because Zn affects so many concurrent and sometimes competing processes. Only a limited selection will be touched on in the following sections.

Intestinal digestion: Zn is an essential cofactor of carbonic anhydrases, proteases, phosphatases, and other enzymes involved in food digestion and absorption. The role of carbonic anhydrases for gastric acid production is described next. The carboxypeptidases A1 and A2 (EC3.4.2.1) are Zn-dependent digestive enzymes from the pancreas. Several peptidases of the intestinal brush border membrane are Zn enzymes, including leucine aminopeptidase (LAP, EC3.4.11.1), membrane alanine aminopeptidase (aminopeptidase N, EC3.4.11.2), glutamyl aminopeptidase (aminopeptidase A, EC3.4.11.7), membrane dipeptidase (EC3.4.13.19), angiotensin I-converting enzyme (EC3.4.15.1), neprilysin (EC3.4.24.11), and meprin A (EC3.4.24.18). Another Zn-containing brush border membrane enzyme (pteroylpoly-gamma-glutamate carboxypeptidase, EC3.4.19.8) is needed to cleave off gamma-glutamyl residues from dietary folate prior to absorption. Alkaline phosphatase (EC3.1.3.1, requires both Zn and magnesium) at the intestinal brush border membrane digests complex forms of thiamin, riboflavin, and pantothenate.

pH-Regulation: The conversion of carbon dioxide to its weak acid helps cells to adjust proton concentration with a readily available and easily removable reagent. This equilibrium reaction is catalyzed by the Zn enzyme carbonate anhydrase (EC4.2.1.1). The 10 or more genetically distinct isoforms of this handy enzyme provide fine-tuned kinetic properties and regulatory characteristics for a wide range of functions that include acidification, signaling, promoting cell proliferation, bone resorption, and respiration. A long-known role of the enzymes I, II, and IV is to provide protons for hydrochloric acid production in the stomach. Gastrin, histamine, and acetylcholine activate these isoenzymes, while somatostatin and several acid-suppressing drugs inhibit them. Carbonic anhydrase VI, gustin, is a special form in salivary glands that helps to maintain taste-bud growth (possibly by acting on bud stem cells) and function (Henkin et al., 1999). Adequate Zn intake appears to promote taste acuity.

Nutrient metabolism: Zn-containing alcohol dehydrogenases (EC1.1.1.1) in the stomach wall and liver oxidize ethanol. A particular alcohol dehydrogenase isoenzyme is needed for the conversion of the transport and storage form retinol into the retinal form used for vision. Optimal Zn status appears to lower the risk of macular degeneration (Age-Related Eye Disease Study Research Group, 2001) and increase life expectancy due to reduced mortality from upper respiratory infection (Clemons et al., 2004) and cardiovascular disease (Chew et al., 2013).

Several Zn-dependent folate hydrolases in lysosomes, membranes, and cytosol are essential for folate metabolism and transport in tissues.

DNA replication and transcription: Zinc fingers are DNA-binding domains of transcription factors in which the Zn ion is tetrahedrally coordinated to residues of cysteine, histidine, or both. These Zn-complexing structures are ubiquitous features that give Zn a central role in the expression of virtually any type of protein. Zn is a cofactor of many enzymes participating in the synthesis of DNA and RNA during cell division and gene expression. Important Zn enzymes include DNA polymerase (EC2.7.7.7) and DNA-dependent RNA polymerase (EC2.7.7.6); many others are regulated through zinc-finger proteins.

RNA editing: Zn is a cofactor of the apolipoprotein B (apoB) RNA editing complex (EC3.5.4.-), which deaminates specific cytidine moieties in a few mRNA species, most prominently of apoB but also of tumor necrosis factor-a, c-myc, and other centrally important regulators of growth and differentiation (Anant & Davidson, 2000). The same complex also extensively edits the translational repressor NAT1, which is involved in postnatal heart development (Pak and Pang, 1999).

Immune function: The Zn-dependent peptide hormone thymulin comes from the thymus. Thymulin helps to maintain immune function by activating T-lymphocytes and enhancing the cytotoxicity of natural killer cells. Zn may also act directly by promoting the proliferation of lymphocytes and decrease susceptibility to programmed cell death (apoptosis). Since Zn is an essential cofactor of many enzymes involved in proliferation of any rapidly dividing cell, it becomes very difficult to differentiate this from a more specific role as a direct effector. Whatever the proximate mechanism, there is no doubt that counts of both natural killer cells and TH1 lymphocytes decline with Zn deficiency. Also related to suboptimal Zn is low production of interleukin-2, tumor necrosis factor-α, interferon-γ (Prasad, 1998), and possibly of IL-6.

Fuel metabolism: The interactions of Zn with players in carbohydrate metabolism are numerous and not yet fully understood. Only a few are to be mentioned here. Zn chelates insulin during storage and thus plays a role in control of its secretion. Glucagon rapidly lowers the intracellular concentration of the free Zn ion. Zn itself opposes the effect of cAMP on glycolysis.

Free radical metabolism: Zn contributes importantly to the defense against oxidative stress. It does so partly as a cofactor of superoxide dismutases (SOD, EC1.15.1.1) in cytoplasma (SOD1) and in the extracellular space (SOD3). Protection of sulfhydryl groups and other direct effects of Zn on redox reactions have been demonstrated (Powell, 2000).

Other functions: A link of Zn to vascular function is established through its role as a cofactor of nitric oxide synthases (EC1.14.13.39).

Observational and prospective intervention studies indicate that adequate intake is associated with favorable bone mineral density (Nielsen et al., 2011). The mechanism of this effect is uncertain.

NUTRITIONAL ASSESSMENT

Zn status is modestly well reflected by the concentration of Zn in the blood (Krebs, 2013). Challenges come from the easy contamination during sample acquisition, a large number of potential confounding factors, and the lack of clear biological end points relevant for human health.

REFERENCES

Age-Related Eye Disease Study Research Group, 2001. A randomized, placebo-controlled, clinical trial of high-dose supplementation with vitamins C and E, beta carotene, and zinc for age-related macular degeneration and vision loss: AREDS report no. 8. Arch. Ophthalmol. 119, 1417–1436.

Anant, S., Davidson, N.O., 2000. An AU-rich sequence element (UUUN[A/U]U) downstream of the edited C in apolipoprotein B mRNA is a high-affinity binding site for Apobec-1: binding of Apobec-1 to this motif in the 3′-untranslated region of c-myc increases mRNA stability. Mol. Cell. Biol. 20, 1982–1992.

Babina, S., Kanyshkova, T., Buneva, V., Nevinsky, G., 2004. Lactoferrin is the major deoxyribonuclease of human milk. Biochemistry (Moscow) 69, 1006–1015.

Beer, W.H., Johnson, R.F., Guentzel, M.N., Lozano, J., Henderson, G.I., Schenker, S., 1992. Human placental transfer of zinc: normal characteristics and role of ethanol. Alc. Clin. Exp. Res. 16, 98–105.

Blaby-Haas, C.E., Merchant, S.S., 2014. Lysosome-related organelles as mediators of metal homeostasis. J. Biol. Chem. 289, 28129–28136.

Chew, E.Y., Clemons, T.E., Agrón, E., Sperduto, R.D., Sangiovanni, J.P., Kurinij, N., et al., 2013. Long-term effects of vitamins C and E, β-carotene, and zinc on age-related macular degeneration: AREDS report no. 35. Ophthalmology 120, 1604–1611.

Clemons, T.E., Kurinij, N., Sperduto, R.D., AREDS Research Group, 2004. Associations of mortality with ocular disorders and an intervention of high-dose antioxidants and zinc in the Age-Related Eye Disease Study: AREDS Report No. 13. Arch. Ophthalmol. 122, 716–726.

Colvin, R.A., 1998. Characterization of a plasma membrane zinc transporter in rat brain. Neurosci. Lett. 247, 147–150.

Costello, L.C., Liu, Y., Zou, J., Franklin, R.B., 1999. Evidence for a zinc uptake transporter in human prostate cancer cells which is regulated by prolactin and testosterone. J. Biol. Chem. 274, 17499–17504.

Cousins, R.J., 1986. Toward a molecular understanding of zinc metabolism. Clin. Physiol. Biochem. 4, 20–30.

Das, J.K., Kumar, R., Salam, R.A., Bhutta, Z.A., 2013. Systematic review of zinc fortification trials. Ann. Nutr. Metab. 62, 44–56.

Davis, S.R., McMahon, R.J., Cousins, R.J., 1998. Metallothionein knockout and transgenic mice exhibit altered intestinal processing of zinc with uniform zinc-dependent zinc transporter-1 expression. J. Nutr. 128, 825–831.

Engelken, J., Carnero-Montoro, E., Pybus, M., Andrews, G.K., Lalueza-Fox, C., Comas, D., et al., 2014. Extreme population differences in the human zinc transporter ZIP4 (SLC39A4) are explained by positive selection in Sub-Saharan Africa. PLoS Genet. 10 (2), e1004128.

Food and Nutrition Board, Institute of Medicine, 2002. Dietary Reference Intakes for Vitamin A, Vitamin K, Arsenic, Boron, Chromium, Copper, Iodine, Iron, Manganese, Molybdenum, Nickel, Silicon, Vanadium, and Zinc. National Academy Press, Washington, DC.

Foster, A.W., Osman, D., Robinson, N.J., 2014. Metal preferences and metallation. J. Biol. Chem. 289, 28095–28103.

Foster, M., Chu, A., Petocz, P., Samman, S., 2013. Effect of vegetarian diets on zinc status: a systematic review and meta-analysis of studies in humans. J. Sci. Food Agric. 93, 2362–2371.

Gaither, L.A., Eide, D.J., 2000. Functional expression of the human hZIP2 zinc transporter. J. Biol. Chem. 275, 5560–5564.

Grønli, O., Kvamme, J.M., Friborg, O., Wynn, R., 2013. Zinc deficiency is common in several psychiatric disorders. PLoS One 8, e82793.

Hannan, M.A., Faraji, B., Tanguma, J., Longoria, N., Rodriguez, R.C., 2009. Maternal milk concentration of zinc, iron, selenium, and iodine and its relationship to dietary intakes. Biol. Trace Elem. Res. 127, 6–15.

Henkin, R.I., Martin, B.M., Agarwal, R.P., 1999. Efficacy of exogenous oral zinc in treatment of patients with carbonic anhydrase VI deficiency. Am. J. Med. Sci. 318, 392–405.

Hurley, L.S., 1981. Zinc deficiency and central nervous system malformations in humans. Am. J. Clin. Nutr. 34, 2864–2865.

Inoue, Y., Hasegawa, S., Ban, S., Yamada, T., Date, Y., Mizutani, H., et al., 2014. ZIP2 protein, a zinc transporter, is associated with keratinocyte differentiation. J. Biol. Chem. 289, 21451–21462.

Kelleher, S.L., Velasquez, V., Croxford, T.P., McCormick, N.H., Lopez, V., MacDavid, J., 2012. Mapping the zinc-transporting system in mammary cells: molecular analysis reveals a phenotype-dependent zinc-transporting network during lactation. J. Cell. Physiol. 227, 1761–1770.

King, J.C., Keen, C.L., 1994. Zinc. In: Shils, M.E., Olson, J.A., Shike, M. (Eds.), Modern Nutrition in Health and Disease, eighth ed. Lea & Febiger, Philadelphia, PA, pp. 214–230.

Krebs, N.F., 2013. Update on zinc deficiency and excess in clinical pediatric practice. Ann. Nutr. Metab. 62, 19–29.

Lee, H.H., Hill, G.M., Sikha, V.K., Brewer, G.J., Prasad, A.S., Owyang, C., 1990. Pancreaticobiliary secretion of zinc and copper in normal persons and patients with Wilson's disease. J. Lab. Clin. Med. 116, 283–288.

Lönnerdal, B., 2000. Dietary factors influencing zinc absorption. J. Nutr. 130, 1378S–1383S.

McMahon, R.J., Cousins, R.J., 1998. Mammalian zinc transporters. J. Nutr. 128, 667–670.

Miller, L.V., Hambidge, K.M., Naake, V.L., Hong, Z., Westcott, J.L., Fennessey, P.V., 1994. Size of the zinc pools that exchange rapidly with plasma zinc in humans: alternative techniques for measuring and relation to dietary zinc intake. J. Nutr. 124, 268–276.

Nielsen, F.H., Lukaski, H.C., Johnson, L.K., Roughead, Z.K., 2011. Reported zinc, but not copper, intakes influence whole-body bone density, mineral content and T score responses to zinc and copper supplementation in healthy postmenopausal women. Br. J. Nutr. 106, 1872–1879.

Osterberg, R., Malmensten, B., 1984. Methylamine-induced conformational change of alpha-2-macroglobulin and its zinc(II) binding capacity. An x-ray scattering study. Eur. J. Biochem. 143, 541–544.

Pak, B.J., Pang, S.C., 1999. Developmental regulation of the translational repressor NAT1 during cardiac development. J. Mol. Cell. Cardiol. 31, 1717–1724.

Picciano, M., Guthrie, H., 1976. Copper, iron, and zinc contents of mature humanmilk. Am. J. Clin. Nutr. 29, 242–254.

Powell, S.R., 2000. The antioxidant properties of zinc. J. Nutr. 130, 1447S–1454S.

Prasad, A.S., 1998. Zinc and immunity. Mol. Cell. Biochem. 188, 63–69.

Sandström, B., 2001. Micronutrient interactions: effects on absorption and bioavailability. Br. J. Nutr. 85, S181–S185.

Sian, L., Mingyan, X., Miller, L.V., Tong, L., Krebs, N.E., Hambidge, K.M., 1996. Zinc absorption and intestinal losses of endogenous zinc in young Chinese women with marginal zinc intakes. Am. J. Clin. Nutr. 63, 348–353.

Tacnet, F., Lauthier, E., Ripoche, P., 1993. Mechanisms of zinc transport into pig small intestine brush border membrane vesicles. J. Physiol. 465, 57–72.

Takeda, A., 2000. Movement of zinc and its functional significance in the brain. Brain Res. Rev. 34, 137–148.

Victery, W., Smith, J.M., Vander, A.J., 1981. Renal tubular handling of zinc in the dog. Am. J. Physiol. 241, F532–F539.

Wang, K., Zhou, B., Kuo, Y.M., Zemansky, J., Gitschier, J., 2002. A novel member of a zinc transporter family is defective in acrodermatitis enterpathica. Am. J. Hum. Genet. 71, 66–73.

Waldron, K.J., Rutherford, J.C., Ford, D., Robinson, N.J., 2009. Metalloproteins and metal sensing. Nature 460, 823–830.

Wastney, M.E., House, W.A., Barnes, R.M., Subramanian, K.N., 2000. Kinetics of zinc metabolism: variation with diet, genetics and disease. J. Nutr. 130, 1355S–1359S.

MANGANESE

Manganese (atomic weight 54.94) is a transition metal. Almost all of naturally occurring manganese consists of the stable isotope ^{55}Mn with small accompanying traces of the unstable isotope ^{53}Mn (half-life 3.74×10^5 years). The artificial isotope ^{52}Mn is radioactive. Manganese can exist in oxidation states 0 through $+7$.

ABBREVIATIONS

CoA	coenzyme A
DMT1	divalent metal ion transporter 1
Mn	manganese

NUTRITIONAL SUMMARY

Function: Manganese (Mn) is a cofactor for many enzymes with importance for carbohydrate metabolism, protein digestion and metabolism, biotin function, cartilage regeneration, and free radical defense.

Food sources: Black tea, nuts, grains, sweet potato, and spinach each provides at least one-sixth of adequate daily intake per serving.

Requirements: Daily intake levels of 1.8 mg for women and of 2.3 mg for men are thought to be safe and adequate (Food and Nutrition Board, Institute of Medicine, 2001).

Deficiency: Low intake causes growth retardation in children. Decreased fertility, higher susceptibility to seizures, and bone fractures are less certain consequences of deficiency.

Excessive intake: No toxic effects of ingested food-grade Mn have been reported even at many times typical intake. Very high intake or nondietary exposure induces tremor, delayed movements (akinesia), and rigidity due to extrapyramidal damage. Some derivatives used as gasoline additives and for other nonfood purposes are toxic.

DIETARY AND OTHER SOURCES

Good sources include black tea, red wine, pecans, peanuts, pineapple, oatmeal, shredded wheat, raisin bran, beans, rice, sweet potato, whole wheat, and spinach. Blueberry juice contains 21 mg/l (Karantanas et al., 2000). Water usually contributes only a few micrograms or less per day. Inhaled manganese dust particles, often due to workplace exposure, are expectorated and secondarily swallowed to some degree.

Typical daily manganese intake of American women is around 1.9 mg; men's daily intake is around 2.4 mg (Hunt and Meacham, 2001). Median intake in the Total Diet Study gave comparable results (Food and Nutrition Board, Institute of Medicine, 2001: Appendix, Table E-6). Similar average intake (2.2 mg/day) was also observed in an Indian population (Tripathi et al., 2000).

DIGESTION AND ABSORPTION

Less than 2% of ingested food Mn is absorbed throughout the small intestine (Davidsson et al., 1995, 1998). Absorption by infants from human milk (8%) and from cow's milk (2.4%) is much higher (Davidsson et al., 1989). Absorption efficiency is strongly reduced by the presence of phytate in the same meal.

Mn shares an absorption pathway with ferrous (divalent) iron (Conrad and Umbreit, 2000). The divalent metal ion transporter 1 (DMT1, SLC11A2) mediates Mn^{2+} uptake from the intestinal lumen. This transporter only functions at low pH and is under partial control of intracellular iron concentration through stabilization of its mRNA. Export across the basolateral membrane appears to occur via ferroportin (SLC40A1), which is the conduit for iron transfer into blood (Madejczyk & Ballatori, 2012).

Mn is unusual because it can be taken up by olfactory neurons and transported directly to the brain (Tjalve and Henriksson, 1999).

TRANSPORT AND CELLULAR UPTAKE

Blood circulation: Normal Mn concentrations in the blood are 0.1–0.27 μmol/l (Kristiansen et al., 1997). Mn circulates in the blood as divalent ion (Mn^{2+}). About half of this is bound to transferrin, and most of the remainder to alpha-2-macroglobulin (Scheuhammer & Cherian, 1985). Albumin carries only about 5% of total plasma Mn. Uptake by cells and exportation out of cells largely use the pathways better known for iron: DMT1 for uptake and ferroportin for efflux. SLC30A10 has been identified as an additional carrier for the export of Mn from brain cells (DeWitt et al., 2013). This newly identified carrier may have broader biological significance and be active in other tissues as well.

BBB: Free Mn^{2+} rapidly traverses from the blood into the brain, particularly via the blood–cerebrospinal fluid barrier (Bornhorst et al., 2012). While many questions remain about the mechanisms, the divalent metal transporter DMT1 on the luminal side and ferroportin on the basolateral side are likely to be involved.

Materno-fetal transfer: The placenta takes up Mn from maternal circulation through a concentrative process. Only some of the placental Mn is then transported into fetal circulation, and a significant concentration gradient from the placenta to fetal circulation remains (Miller et al., 1987). The concentrations in maternal and fetal circulation show little correlation (Krachler et al., 1999). The molecular nature of the underlying mechanisms remains uncertain. Presumably some of them are shared with iron.

STORAGE

Mn is enriched in mitochondria. The liver, bones, pancreas, and kidneys have higher concentrations than most other large organs (Gonzalez-Reimers et al., 1998). Little is known about mechanisms that could specifically mobilize Mn from tissue stores.

EXCRETION

Mn losses occur almost exclusively with bile. Newly ingested Mn reappears in bile with a considerable delay (Malecki et al., 1996). How it is transferred into bile is not well understood. Excretion of Mn with urine in people without significant industrial exposure is very low (Paschal et al., 1998) and affected little by intake level.

REGULATION

Mn uptake and disposition are partially coupled to iron status due to shared pathways (DMT1, transferrin, and possibly others). Fractional Mn absorption increases with declining iron status. The decrease of intestinal absorption efficiency with supplementation (Sandstrom et al., 1990) seems to suggest some control, but little is known about this. Strengthening the argument for the existence of regulation are the relatively constant tissue concentrations at different intake and exposure levels. Much more important for the maintenance of whole body homeostasis is the excretion of excess Mn with bile. Chronically inadequate intake nearly abolishes biliary Mn excretion (Malecki et al., 1996), but the underlying regulatory mechanism remains unknown.

FUNCTION

Enzyme cofactors: Many enzymes that use a divalent cation as a cofactor are not very specific in terms of their metal ion requirement. Specific metallochaperones are likely to steer Mn to newly synthesized apoproteins as needed, but the specifics remain to be discovered (Foster et al., 2014).

Mn often can substitute for magnesium, and sometimes for zinc. Typical examples are the cholesterol synthesis enzyme geranyltranstransferase (farnesyl pyrophosphate synthetase, EC2.5.1.10), and the key enzyme of lipid metabolism, phosphatidyicholine synthase (EC2.7.8.24), which work equally well with magnesium and Mn. The following sections will focus only on functions that specifically require Mn.

Free radical control: Mn-containing superoxide dismutase (EC1.15.1.1) is a critical component of the defenses against the free radicals that are a side product of oxidative phosphorylation and other oxidative reactions in mitochondria. Free radicals also are used by immune cells as a corrosive agent for the destruction of infectious agents. The importance of the protective function of Mn superoxide dismutase is illustrated by the fact that it is induced when monocytes become activated by bacterial endotoxin and increase free radical production.

Apoptosis: A strictly Mn-dependent enzyme, ATM protein kinase (EC2.7.11.1) phosphorylates p53, PHAS-I, RPA32, and Chk2 in response to cell damage and thereby induces cell cycle arrest at the G1/S and G2/M stages (Chan et al., 2000). The oncogene p53 participates in the control of programmed cell death (apoptosis). ATM protein kinase is defective in the genetically inherited condition ataxia teleangiectasia (hence the acronym ATM), which is characterized by high cancer incidence at young adult age, immune dysfunction, and loss of coordination of movements (ataxia). Mn may also be important for the function of DNA-protein kinases related to ATM protein kinase, possibly through modifying the type of reaction that they catalyze.

Nitrogen metabolism: Arginase (EC3.5.3.1) is the final enzyme for urea synthesis. Both the cytoplasmic and mitochondrial isoforms require manganese as a cofactor. While arginase is not involved in nitric oxide signaling, it is of note that manganese potentiates nitric oxide production by microglia (Chang et al., 1999).

Peptide hydrolysis: The ubiquitous cytosolic Xaa-Pro dipeptidase (prolidase, EC3.4.13.9) and the bradykinin-degrading Xaa-Pro aminopeptidase (EC3.4.11.9; Cottrell et al., 2000) have an absolute Mn requirement. The latter is important for the degradation of bradykinin. The protein encoded by the gene *AMPL* (cytosol aminopeptidase) exemplifies a more complex picture. This protein has two metal-binding sites, of which the first one always contains zinc. When the second binding site contains zinc, the protein has the catalytic activity of leucyl aminopeptidase (EC3.4.11.1) and preferentially cleaves an N-terminal leucine from peptides. If the second metal-binding site is occupied by Mn, on the other hand, the protein behaves as prolyl aminopeptidase (EC3.4.11.5) and cleaves the N-terminal proline from a peptide. In combination, these activities are crucial for the intracellular processing and turnover of proteins.

Glycolysis and gluconeogenesis: The bulk of energy-providing nutrients are converted into acetyl-CoA, which is then utilized through the Krebs cycle. Acetyl-CoA can only be fed into the cycle through condensation with oxaloacetate. Supplies of this indispensable intermediate are drained when significant gluconeogenesis or net-synthesis of aspartate and asparagine occurs. The Mn- and biotin-dependent enzyme pyruvate carboxylase (EC6.4.1.1) is crucial for replenishing oxaloacetate supplies because it catalyzes the only reaction that can synthesize it without drawing on preformed exogenous Krebs cycle intermediates or amino acids. Each subunit of the pyruvate carboxylase homotetramer contains one tightly bound manganese ion. The enzyme may also contain a mixture of manganese and magnesium. The mitochondrial form of the key enzyme for gluconeogenesis from lactate and proteins, phosphoenolpyruvate carboxykinase (PEPCK, EC4.1.1.32), also is Mn-dependent.

Cartilage formation: Several glycosyl transferases with a stringent Mn(II) requirement participate in the formation and maintenance of cartilage and other connective tissues.

N-acetylneuraminylgalactosylglucosylceramide beta-1,4-*N*-acetylgalactosaminyltransferase (EC2.4.1.165) and polypeptide *N*-acetylgalactosaminyltransferase (EC2.4.1.41) process fetuine, mucin, and other proteoglycans. Synthesis of the cartilage glycoprotein chondroitin is another process that relies heavily on Mn-dependent enzymes. Xylosylprotein 4-beta-galactosyltransferase (EC2.4.1.133), galactosylxylosylprotein 3-beta-galactosyltransferase (EC2.4.1.134), and galactosylgalactosylxylosylprotein 3-beta-glucuronosyltransferase (EC2.4.1.135) add the second, third, and fourth of the four sugar linkage side chain (3-beta-D-glucuronosyl-3-beta-D-galactosyl-4-beta-D-galactosyl-*O*-beta-D-xylosyl) in chondroitin sulfate and other proteoglycans, especially in fibroblasts. Glucuronylgalactosylproteoglycan beta-1,4-*N*-acetylgalactosaminyltransferase (EC2.4.1.174) adds *N*-acetyl-D-galactosamine to an acidic chondroitin sulfate side chain. Mn-dependent glycosyl transferases also are crucial for the extension of ceramide side chains. Enzymes with such activities include beta-galactosyl-*N*-acetylglucosaminylgalactosyl-glucosylceramide beta-1,3-acetylglucosaminyltransferase (EC2.4.1.163), galactosyl-*N*-acetylglucosaminylgalactosyl-glucosylceramide beta-1,6-*N*-acetylglucosaminyltransferase (EC2.4.1.164), and *N*-acetylneuraminylgalactosylglucosylceramide beta-1,4-*N*-acetylgalactosaminyltransferase (EC2.4.1.165).

Sulfate metabolism: Sulfate is released from its activated form by Mn-dependent phosphoadenylylsulfatase (EC3.6.2.2).

Toxic effects: Accumulation of Mn in the brain, often due to liver disease (impaired excretion), parenteral feeding, or industrial exposure, increases dopamine turnover. Depletion of dopaminergic brain regions may then be the cause of an Mn-induced, Parkinson's disease–like syndrome with muscle tremors, delayed movement (akinesia), and rigidity (Montes et al., 2001).

REFERENCES

Bornhorst, J., Wehe, C.A., Hüwel, S., Karst, U., Galla, H.J., Schwerdtle, T., 2012. Impact of manganese on and transfer across blood–brain and blood–cerebrospinal fluid barrier *in vitro*. J. Biol. Chem. 287, 17140–17151.

Chan, D.W., Son, S.C., Block, W., Ye, R., Khanna, K.K., Wold, M.S., et al., 2000. Purification and characterization of ATM from human placenta. A manganese-dependent, wortmannin-sensitive serine/threonine protein kinase. J. Biol. Chem. 275, 7803–7810.

Chang, J.Y., Liu, L.Z., 1999. Manganese potentiates nitric oxide production by microglia. Brain Res. Mol. Brain Res. 68, 22–28.

Conrad, M.E., Umbreit, J.N., 2000. Iron absorption and transport—an update. Am. J. Hematol. 64, 282–287.

Cottrell, G.S., Hooper, N.M., Turner, A.J., 2000. Cloning, expression, and characterization of human cytosolic aminopeptidase P: a single manganese(II)-dependent enzyme. Biochemistry 39, 15121–15128.

Davidsson, L., Almgren, A., Hurrell, R.E., 1998. Sodium iron EDTA [NaFe(III)EDTA] as a food fortificant does not influence absorption and urinary excretion of manganese in healthy adults. J. Nutr. 128, 1139–1143.

Davidsson, L., Almgren, A., Juillerat, M.A., Hurrell, R.E., 1995. Manganese absorption in humans: the effect of phytic acid and ascorbic acid in soy formula. Am. J. Clin. Nutr. 62, 984–987.

Davidsson, L., Cederblad, A., Lönnerdal, B., Sandstrom, B., 1989. Manganese absorption from human milk, cow's milk, and infant formulas in humans. Am. J. Dis. Child 143, 823–827.

DeWitt, M.R., Chen, P., Aschner, M., 2013. Manganese efflux in Parkinsonism: insights from newly characterized SLC30A10 mutations. Biochem. Biophys. Res. Commun. 432, 1–4.

Food and Nutrition Board, Institute of Medicine, 2001. Dietary Reference Intakes for Vitamin A, Vitamin K, Arsenic, Boron, Chromium, Copper, Iodine, Iron, Manganese, Molybdenum, Nickel, Silicon, Vanadium, and Zinc. National Academy Press, Washington, DC.

Foster, A.W., Osman, D., Robinson, N.J., 2014. Metal preferences and metallation. J. Biol. Chem. 289 (41), 28095–28103. [Epub ahead of print].

Gonzalez-Reimers, E., Martinez-Riera, A., Santolaria-Fernandez, E., Mas-Pascual, A., Rodriguez-Moreno, E., Galindo-Martin, L., et al., 1998. Relative and combined effects of ethanol and protein deficiency on zinc, iron, copper, and manganese contents in different organs and urinary and fecal excretion. Alcohol 16, 7–12.

Hunt, C.D., Meacham, S.L., 2001. Aluminum, boron, calcium, copper, iron, magnesium, manganese, molybdenum, phosphorus, potassium, sodium, and zinc: concentrations in common western foods and estimated daily intakes by infants; toddlers; and male and female adolescents, adults, and seniors in the United States. J. Am. Diet. Assoc. 101, 1058–1060.

Karantanas, A.H., Papanikolaou, N., Kalef-Ezra, J., Challa, A., Gourtsoyiannis, N., 2000. Blueberry juice used per os in upper abdominal MR imaging: composition and initial clinical data. Eur. Radiol. 10, 909–913.

Krachler, M., Rossipal, E., Micetic-Turk, D., 1999. Trace element transfer from the mother to the newborn—investigations on triplets of colostrum, maternal and umbilical cord sera. Eur. J. Clin. Nutr. 53, 486–494.

Kristiansen, J., Christensen, J.M., Iversen, B.S., Sabbioni, E., 1997. Toxic trace element reference levels in blood and urine: influence of gender and lifestyle factors. Sci. Total Env. 204, 147–160.

Malecki, E.A., Radzanowski, G.M., Radzanowski, T.J., Gallaher, D.D., Greger, J.L., 1996. Biliary manganese excretion in conscious rats is affected by acute and chronic manganese intake but not by dietary fat. J. Nutr. 126, 489–498.

Madejczyk, M.S., Ballatori, N., 2012. The iron transporter ferroportin can also function as a manganese exporter. BBA-Biomembranes 1818, 651–657.

Miller, R.K., Mattison, D.R., Panigel, M., Ceckler, T., Bryant, R., Thomford, P., 1987. Kinetic assessment of manganese using magnetic resonance imaging in the dually perfused human placenta *in vitro*. Environ. Health Persp. 74, 81–91.

Montes, S., Alcaraz-Zubeldia, M., Muriel, P., Rios, C., 2001. Striatal manganese accumulation induces changes in dopamine metabolism in the cirrhotic rat. Brain Res. 891, 123–129.

Paschal, D.C., Ting, B.G., Morrow, J.C., Pirkle, J.L., Jackson, R.J., Sampson, E.J., et al., 1998. Trace metals in urine of United States residents: reference range concentrations. Environ. Res. 76, 53–59.

Sandstrom, B., Davidsson, L., Erickson, R.A., Alpsten, M., 1990. Effects of long-term trace element supplementation on blood trace element levels and absorption of (75Se), (54Mn), and (65Zn). J. Trace Elem. Electrolyte Health Dis. 4, 65–72.

Scheuhammer, A.M., Cherian, M.G., 1985. Binding of manganese in human and rat plasma. Biochim. Biophys. Acta 840, 163–169.

Tjalve, H., Henriksson, J., 1999. Uptake of metals in the brain via olfactory pathways. Neurotoxicology 20, 181–195.

Tripathi, R.M., Mahapatra, S., Raghunath, R., Sastry, V.N., Krishnamoorthy, T.M., 2000. Daily intake of manganese by the adult population of Mumbai. Sci. Total Environ. 250, 43–50.

CALCIUM

Calcium is an alkaline earth metal (atomic weight 40.08) with valence 2. Naturally occurring calcium contains the stable isotopes ^{42}Ca (0.647%), ^{43}Ca (0.135%), ^{44}Ca (2.086%), ^{46}Ca (0.004%), and ^{48}Ca (0.187%), and the unstable isotopes ^{40}Ca (96.941%; half-life more than 5.9×10^{21} years) and ^{41}Ca (trace amounts; half-life 1.03×10^5 years). The synthetic isotopes ^{45}Ca and ^{47}Ca are unstable. Calcium is always divalent (Ca^{2+}).

ABBREVIATIONS

CaSR	calcium-sensing receptor
ECaC1	epithelial calcium channel 1 (TRPV5)
NCX1	Na^+/Ca^{2+} exchanger (SLC8A1)
PTH	parathyroid hormone
PTHrP	PTH-related protein

NUTRITIONAL SUMMARY

Function: Calcium is the major mineral in bone. It is needed for intracellular and hormone-like signaling, neurotransmission, muscle contraction, for the regulation of cell growth and differentiation, blood clotting, and many other functions.

Requirements: Adults should consume at least 1000 mg/day; adolescents, as well as postmenopausal and lactating women, need slightly more (Institute of Medicine (US) Committee to Review Dietary Reference Intakes for Vitamin D and Calcium, 2011).

Sources: Rich sources are all dairy foods, fortified cereals and juices, and tofu set with calcium sulfate. Kale and chard are more modest sources. Most plant-based foods contain much smaller amounts. The proportion that is absorbed from all sources strongly depends on vitamin D status (which increases calcium absorption and retention) and on phosphate, sodium, and animal protein intake (excess intake decreases calcium absorption and retention).

Deficiency: Inadequate calcium availability slows bone growth and mineralization in childhood and adolescence and causes bone mineral loss in adults. Low mineral content of bone (osteoporosis) greatly increases the risk of fractures. People with habitually low calcium intake may also have a greater than average risk of colon cancer and elevated blood pressure.

Excessive intake: Intake over 2500 mg/day increases the risk of renal stone formation in some people.

DIETARY SOURCES

Significant calcium sources include dairy products, soybean curd (tofu), a few green leafy vegetables, and calcium-fortified foods and beverages. Milk and yogurt provide about 1 mg/ml, and hard cheeses about 7 mg/g. Firm tofu contains 2 mg/g, and as much as 7 mg/g if set with calcium sulfate. Other sources are green leafy vegetables, such as Chinese mustard greens (2.5 mg/g), kale (0.7 mg/g), and chard (0.6 mg/g). Spinach also contains significant amounts of calcium (1.4 mg/g), but the high oxalate content minimizes absorption. Women in the United States get around 600 mg of calcium from foods per day, and men between 800 and 900 mg, most of this from dairy products. Older people consume even less.

Many people use dietary supplements in addition to food sources. Nonetheless, the combined intake is often well below what is considered to be adequate (Kohlmeier et al., 1997).

DIGESTION AND ABSORPTION

Calcium is absorbed with 15–70% efficiency in the small intestine (Figure 11.12). Absorption proceeds mainly by active transport at low doses, and increasingly via paracellular passive diffusion at high doses (several hundred milligrams). The efficiency of calcium absorption depends on vitamin D status, age, amount ingested, and the concentration of phosphate and other ingredients in a meal (Lemann and Favus, 1999). In addition, 1,25-dihydroxyvitamin D provides a potent stimulus for absorption. Availability for absorption is best for highly soluble complexes from which the calcium cation can dissociate easily (e.g., calcium citrate malate; Heaney et al., 1999).

Inhibition of absorption tends to be strongest at low calcium doses, when the transcellular pathway predominates. The paracellular pathway, which becomes more significant at higher doses, may accommodate some calcium complexes that are excluded from transcellular transport. Phosphate forms poorly soluble complexes with calcium and decreases its fractional absorption. Milk, meats, colas, and many processed foods are major phosphate sources. Oxalate tightly binds to calcium and strongly inhibits its absorption. This is the reason why less than 10% of the calcium in rhubarb and spinach is available for absorption (Weaver et al., 1999). Phytate also decreases calcium absorption slightly. Fractional absorption of calcium from green leafy vegetables is better than from dairy products, nonetheless, so long as oxalate content is low. About two-thirds of the modest calcium dose in broccoli is absorbed by healthy young men as compared to about a third of the calcium in milk (Weaver et al., 1999). The bioavailability of calcium from mustard greens, kale, and soybeans is still similar or slightly better than from milk. Absorption from other types of beans is slightly less effective.

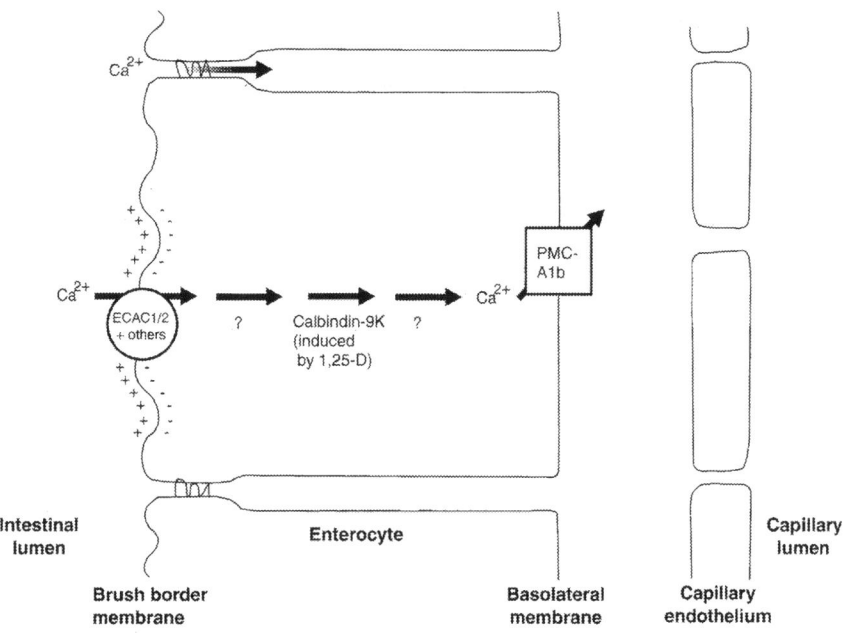

FIGURE 11.12

Intestinal calcium absorption. The question marks indicate that the molecular nature responsible for the activity is still uncertain.

The raised intraluminal concentration after a meal and the electronegative potential of the enterocyte interior drives calcium into enterocytes (Peng et al., 1999). The microvillous membrane of the proximal small intestine contains several channels through which calcium can enter, including epithelial calcium channels 1 (ECaCl, TRPV5) and 2 (ECaC2; Barley et al., 2001). The calcium ions then move across the cell with several chaperone-like proteins. Calbindin-9K is one of these translocating proteins. Its synthesis is strongly induced by 1,25-dihydroxyvitamin D (1,25-D). The driving force for calcium absorption comes from calcium-transporting ATPase Ib (plasma membrane calcium-pumping ATPase Ib, PMCAIb, EC3.6.3.8) that pumps calcium into the pericapillary space. The intestinal form hydrolyzes one ATP molecule to adenosine diphosphate (ADP) for each transported calcium ion.

TRANSPORT AND CELLULAR UPTAKE

Blood circulation: Normal blood concentration of calcium is between 2.20 and 2.65 mmol/l. About 40% of calcium in the blood is albumin-bound 10% bound to citrate, bicarbonate, phosphate, and 50% is present in the ionized form. Calcium affinity for albumin and other proteins decreases with decreasing pH (acidosis). Respiratory alkalosis induced by hyperventilation (decreased bicarbonate concentration in the blood) can rapidly lower the concentration of free ionized calcium in the blood and cause cramping and other neuromuscular symptoms.

Intracellular (cytosolic) concentration of free calcium is only a minute fraction (around 0.0001 mmol/l) of the concentration in the extracellular fluid and blood. Several calcium-transporting ATPases (EC3.6.3.8) pump calcium out of cells to maintain this steep gradient. In addition, endosomes, lysosomes, and other intracellular compartments sequester calcium with very high specificity and under tight regulation. The Na^+/Ca^{2+} exchanger (NCXI, SLC8A1) is the main transporter for calcium extrusion from heart myocytes. A related form (NCX2, SLC8A2) operates in the brain and skeletal muscle.

BBB: Calcium concentration is lower in the brain than in the blood. Calcium-transporting ATPase at the luminal side of brain capillary endothelial cells maintains this gradient by pumping calcium into the blood (Manoonkitiwongsa et al., 2000).

Materno-fetal transfer: The mismatch of the large fetal head to the narrow birth canal requires some flexibility of the skull. This may explain why mineralization of the fetal skeleton is relatively light. Nonetheless, about 140 mg of calcium have to be transferred every day from mother to fetus during the third trimester—a total of about 30 g during the entire pregnancy (Lafond et al., 2001). Information on the pathways for calcium transfer across the placenta is still incomplete. It seems clear that calcium diffuses from the maternal side into the syntrophoblast through calcium channels by passive diffusion, which is made possible by the much lower calcium concentration inside the cell layer than in maternal blood. Transport to the fetal side of the syntrophoblast depends on vitamin D–induced calbindin-9K. Calcium-transporting ATPase then pumps calcium across the basal membrane toward fetal circulation.

STORAGE

An infant at birth contains about 21 g of calcium (Crowley et al., 1998). Young mature males may carry more than 1400 g of calcium in their bones, and mature females more than 1200 g (Anderson, 2000). Turnover and net accretion are under the control of PTH, calcitonin, 1,25-dihydroxyvitamin D, sex hormones, IL-6, and many other hormonal factors and mediators.

Osteoblasts deposit hydroxyapatite $[3Ca_3(PO_4)_2](OH)_2$, the typical bone mineral. Osteoclasts release calcium from bone by breaking down the protein matrix and creating an acid microenvironment that dissolves the denuded bone surface. Tartrate-resistant alkaline phosphatase (EC3.1.3.1, requires both zinc and magnesium) also plays an important role in osteoclastic bone digestion. Calcium can then move from the extracellular space into circulation by diffusion along the concentration gradient, since the extracellular calcium concentration may be as low as 0.5 mmol/l (Baron, 1999).

EXCRETION

Obligatory endogenous losses in stool and urine are about 250 mg/day. Dermal losses are 40 mg/day or more (Jensen et al., 1983). Profuse sweating appears to increase losses significantly (Bergeron et al., 1998).

Each gram of ingested sodium increases daily losses via urine by 15 mg, and a gram of animal protein loses 1 mg (Nordin, 2000). Limiting animal protein intake to 20 g/day and sodium intake to 50 mmol/day is thought to result in a calcium requirement as low as 400 mg/day.

The kidneys filter more than 1500 mg of calcium per day into ultrafiltrate. Most of this amount is reabsorbed from the proximal renal tubules, primarily via paracellular ion-gradient driven pathways involving claudin-16 (Jeon, 2008). A smaller, but tightly regulated percentage is actively absorbed from the TAL of Henle's loop, the distal tubule, and the connecting tubule. Calcium enters the epithelial cells of those segments via the ECaCl (TRPV5). Calbindin-9K is involved in transferring calcium toward the basolateral membrane, from where it is pumped into the perivascular space by plasma membrane calcium-transporting ATPase Ib.

REGULATION

Calcium sensor: A web of hormonal and other effectors tightly regulates calcium concentration in the blood. The G-protein-coupled calcium-sensing receptor (CaSR) responds to extracellular calcium concentrations at several key sites (parathyroid glands, kidneys, intestinal epithelial cells, bone cells, keratinocytes, cervical epithelial cells, and other tissues) and sets in motion the release of various calcium-homeostatic hormones and mediators (Brown et al., 2001). The effect depends on the cell type where the receptor is located. Stimulation of the receptor by high calcium ion concentration decreases secretion of active PTH from parathyroid gland cells, increases secretion of calcitonin by C cells of the thyroid, promotes calcium reabsorption from renal epithelial cells, or up-regulates chloride channels in the intestine (Shimizu et al., 2000).

PTH: The four pea-sized parathyroid glands are located on the lateral sides of the thyroid gland. The parathyroid gland secretes PTH in very rapid response (within seconds) to decreased ionized calcium concentration in the blood (mediated by the CaSR). PTH promotes calcium mobilization from bone, calcium reabsorption from kidneys, synthesis of 1,25-dihydroxyvitamin D, and other activities that are not directly related to calcium disposition. PTH-related protein (PTHrP) is an important calcium regulator produced by the fetus, placenta and at a low level by some tissues of nonpregnant adults. This peptide hormone also influences smooth muscle relaxation, placental calcium transport, and bone development and functions as a paracrine stimulator of cellular growth and differentiation. A single G-protein-coupled receptor responds to both PTH and PTHrH and mediates most of their effects. PTH-secreting cells can adapt very quickly to different extracellular calcium concentrations by changing the intracellular degradation of PTH about to be released. If PTH synthesis rate is higher than would be appropriate for the current calcium concentration, a greater percentage is secreted as inactive carboxy-terminal fragments.

Calcium-reabsorption from the distal tubules in the kidneys is a major determinant of renal calcium losses. The bicarbonate-chloride exchanger pendrin (SLC26A4, PDS) is a key actor here (Soleimani, 2015).

Calcitonin: This calcium-regulating peptide hormone is produced in C cells of the thyroid gland, pituitary cells and a few widely dispersed neuroendocrine cells. Binding of calcitonin to the calcitonin receptor inhibits osteoclast action and indirectly calcium release.

Vitamin D: The complex of 25-hydroxyvitamin D and vitamin D–binding protein is filtered in the kidneys and taken up from the renal tubular lumen via the endocytotic receptor megalin. The production of 1,25-dihydroxyvitamin D takes place in the tubular epithelial cell and is under tight regulation by PTH. The hormone-like actions of 1,25-dihydroxyvitamin D include up-regulating the expression of calcium ATPase in the small intestine and distal renal tubules.

FUNCTION

Signaling: Calcium is a ubiquitous second messenger that links receptor-mediated hormone action to intracellular events. Binding of some effectors to their plasma membrane receptors activates associated calcium channels. The calcium-binding mediator calmodulin responds to calcium influx by activating various proteins, such as phosphorylase kinase (EC2.7.1.38), 1D-myo-inositol-trisphosphate 3-kinase (EC2.7.1.127), calcium/calmodulin-dependent protein kinase (EC2.7.1.123, phosphorylates vimentin, synapsin, glycogen synthase, myosin light-chains, and tau protein), or myosin light-chain kinase (EC2.7.1.117, in smooth muscle). Another type of receptor uses inositol triphosphate to release calcium from intracellular stores and activate protein kinase C.

Muscle contraction: Striated and other muscle cells respond to nerve impulses with a contraction. A widespread intracellular compartment, the sarcoplasmic reticulum, contains large amounts of calcium bound to calsequestrin, whereas the cytoplasma of a relaxed cell contains very little. Transmission of an action potential from a neuron end plate to the muscle cell triggers a torrential release of calcium from the sarcoplasmic reticulum into the cytoplasma and contraction begins. The contraction ends because calcium-transporting ATPase (EC3.6.3.8) pumps calcium from cytoplasma back into the sarcoplasmic reticulum. This calcium pump is particularly effective, since it moves two calcium ions with each hydrolyzed ATP. Fast-twitch skeletal muscle fibers contain a different calcium pump (ATP2A1) than slow-twitch skeletal muscle fibers and cardiac muscle fibers (ATP2A2).

Neurotransmitter release: Calcium has important roles in the release of neurotransmitters from nerve terminals, but considerable uncertainties still exist about many details of the various mechanisms (Augustine, 2001). Voltage-gated calcium channels (N-type, P-type, and Q-type) allow the rapid influx of extracellular calcium into neurons in response to depolarization and this triggers neurotransmitter release. Neurotransmitter release and neuron excitability are also regulated by calcium influx through Maxi-K potassium channels (Gribkoff et al., 2001).

Protein cofactor: Several enzymes are dependent on the presence of adequate calcium concentrations for full activity. One mode, outlined above, is the binding of calcium to calmodulin or another calcium-binding protein, which then activates specific target proteins, in other instances calcium acts directly on the protein. Numerous types of phospholipase A2 (EC3.1.1.4) require calcium as a direct cofactor, as do caldesmon kinase (EC2.7.1.120), phosphatidylinositol deacylase (EC3.1.1.52),

1,4-lactonase (EC3.1.1.25), protein-glutamine gamma-glutamyltransferase (transglutaminase, EC2.3.2.13), and calpain (EC3.4.22.17). Polypeptide *N*-acetylgalactosaminyltransferase (EC2.4.1.41) requires both calcium and manganese. A large group of metalloproteinases requires both calcium and zinc for the hydrolysis of various collagens, elastins, and other connective tissue proteins. Among these are stromelysin 1 (matrix metalloproteinase 3, EC3.4.24.17) and 2 (matrix metalloproteinase 10, EC3.4.24.22), matrilysin (matrix metalloproteinase 7, EC3.4.24.23), gelatinase A (matrix metalloproteinase 2, EC3.4.24.24) and B (EC3.4.24.35), neutrophil collagenase (matrix metalloproteinase 8, EC3.4.24.34), and macrophage elastase (matrix metalloproteinase 12, EC3.4.24.65). Apyrase (EC3.6.1.5), an ATP- and ADP-hydrolyzing ectoenzyme, requires both calcium and magnesium.

Blood clotting: A web of fibrin aggregates and platelets forms around injured blood vessel endothelia. Calcium stimulates both platelet aggregation and several steps of the blood-clotting cascade. The protein chains of four of the clotting factors (fibrinogen and factors VII, IX, and X) contain gamma-carboxylated glutamyl residues that bind calcium with high affinity. Lowering of calcium concentration, such as by the addition of ethylenediaminetedtraacetic acid (EDTA) or oxalate to blood samples, disrupts blood clotting.

Cell growth: Calcium is involved in the initiation of DNA synthesis, chromosome sorting, and regulation of cell division and differentiation. The stimulation of the calcium-sensing receptor, for example, promotes differentiation of cells such as keratinocytes (Bikle et al., 2001).

Skeletal structure: The ability of vertebrate animals to move rapidly and with precise control is very much dependent on a rigid skeleton. As in all other vertebrates, calcium is the predominant cation of bones. Inadequate bone mineralization increases the risk of fracture. Bone mineralization is promoted and controlled by several extracellular calcium-binding proteins, including osteonectin, osteocalcin, and matrix Gla protein (both vitamin K–dependent), osteopontin, and decorin. A clear understanding of the sequence of events leading to bone mineralization is still lacking, however.

Vestibular system: The inner ear contains a sensory organ that perceives balance and gives information on body position. The sensory hair cells in the vestibular system detect the minute displacement of a layer of calcium carbonate crystals (otoconia) embedded in a gelatinous layer (otolithic membrane).

NUTRITIONAL ASSESSMENT

Calcium concentrations in all body fluids and tissues are tightly regulated and deviations are due to hormonal or other regulatory deviations, not deficiency or excess. Since the largest amount of calcium by far is stored in bones, from where it can be released in times of need, bone mineral content might be considered an indicator of calcium status. However, it is subject to all the same influences of regulation and dysregulation as in other tissues and body fluids, just with a much greater delay.

Calcium intake has been estimated by some based on calcium excretion with urine. However, the reliability of that measure is disappointing, particularly with clinically relevant populations (Toren and Norman, 2005). For example, there are some individuals with disproportionally high urinary excretion of ingested calcium (Oliveira et al., 2014), and they tend to be more prone to the development of kidney stones.

REFERENCES

Anderson, J.J.B., 2000. The important role of physical activity in skeletal development: how exercise may counter low calcium intake. Am. J. Clin. Nutr. 71, 1384–1386.

Augustine, G.J., 2001. How does calcium trigger neurotransmitter release? Curr. Opin. Neurobiol. 11, 320–326.

Barley, N.E., Howard, A., O'Callaghan, D., Legon, S., Waiters, J.R., 2001. Epithelial calcium transporter expression in human duodenum. Am. J. Physiol. Gastroent. Liver Physiol. 280, G285–G290.

Baron, R., 1999. Anatomy and ultrastructure of bone. In: Favus, M.J. (Ed.), Primer on the Metaholic Bone Diseases and Disorders of Mineral Metabolism, fourth ed. Lippincott, Williams & Wilkins, Philadelphia, PA, pp. 3–10.

Bergeron, M.E., Volpe, S.L., Gelinas, Y., 1998. Cutaneous calcium losses during exercise in the heat: a regional sweat patch estimation technique. Clin. Chem. 44, A-167.

Bikle, D.D., Ng, D., Tu, C.L., Oda, Y., Xie, Z., 2001. Calcium- and vitamin D-regulated keratinocyte differentiation. Mol. Cell. Endocrinol. 177, 161–171.

Brown, E.M., MacLeod, R.J., O'Malley, B.W., 2001. Extracellular calcium sensing and extracellular calcium signaling. Physiol. Rev. 81, 239–297.

Crowley, S., Trivedi, E., Risteli, L., Risteli, J., Hindmarsh, P.C., Brook, C.G., 1998. Collagen metabolism and growth in prepubertal children with asthma treated with inhaled steroids. J. Ped. 132, 409–413.

Gribkoff, V.K., Starrett Jr, J.E., Dworetzky, S.I., 2001. Maxi-K potassium channels: form, function, and modulation of a class of endogenous regulators of intracellular calcium. Neuroscience 7, 166–177.

Heaney, R.E., Recker, R.R., Weaver, C.M., 1999. Absorbability of calcium sources: the role of limited solubility. Calcif. Tiss. Int. 46, 300–304.

Institute of Medicine (US) Committee to Review Dietary Reference Intakes for Vitamin D and Calcium, 2011.. In: Ross, A.C., Taylor, C.L., Yaktine, A.L., Del Valle, H.B. (Eds.), Dietary Reference Intakes for Calcium and Vitamin D National Academies Press (US), Washington, DC.

Jensen, F.T., Charles, P., Moskilde, L., Hansen, H.H., 1983. Calcium metabolism evaluated by 47 calcium kinetics: a physiological model with correction for faecal lag time and estimation of dermal calcium loss. Clin. Physiol. 3, 187–204.

Jeon, U.S., 2008. Kidney and calcium homeostasis. Electrolyte Blood Press. 6, 68–76.

Kohlmeier, M., Garris, S., Anderson, J.J.B., 1997. Vitamin K: a vegetarian promoter of bone health. Veg. Nutr. 1, 53–57.

Lafond, J., Goyer-O'Reilly, I., Laramee, M., Simoneau, L., 2001. Hormonal regulation and implication of cell signaling in calcium transfer by placenta. Endocrin. J. 14, 85–94.

Lemann Jr, J., Favus, M.J., 1999. The intestinal absorption of calcium, magnesium, and phosphate. In: Favus, M.J. (Ed.), Primer on the Metabolic Bone Diseases and Disorders of Mineral Metabolism, fourth ed. Lippincott, Williams & Wilkins, Philadelphia, PA, pp. 63–67.

Manoonkitiwongsa, P.S., Whitter, E.E., Wareesangtip, W., McMillan, P.J., Nava, P.B., Schultz, R.L., 2000. Calcium-dependent ATPase unlike ecto-ATPase is located primarily on the luminal surface of brain endothelial cells. Histochem. J. 32, 313–324.

Nordin, B.E.C., 2000. Calcium requirement is a sliding scale. Am. J. Clin. Nutr. 71, 1381–1383.

Oliveira, L.M., Hauschild, D.B., Leite, C.D., Baptista, D.R., Carvalho, M., 2014. Adequate dietary intake and nutritional status in patients with nephrolithiasis: new targets and objectives. J. Ren. Nutr. 24, 417–422.

Peng, J.B., Chen, X.Z., Berger, U.V., Vassilev, P.M., Ysukaguchi, H., Brown, E.M., et al., 1999. Molecular cloning and characterization of a channel-like transporter mediating intestinal calcium absorption. J. Biol. Chem. 274, 22739–22746.

Shimizu, T., Morishima, S., Okada, Y., 2000. Ca^{2+}-sensing receptor-mediated regulation of volumesensitive Cl-channels in human epithelial cells. J. Physiol. 528, 457–472.

Soleimani, M., 2015. The multiple roles of pendrin in the kidney. Nephrol. Dial. Transplant. 30, 1–10.

Toren, P.J., Norman, R.W., 2005. Is 24-hour urinary calcium a surrogate marker for dietary calcium intake? Urology 65, 459–462.

Weaver, C.M., Proulx, W.R., Heaney, R., 1999. Choices for achieving adequate dietary calcium with a vegetarian diet. Am. J. Clin. Nutr. 70, 543S–548S.

PHOSPHORUS

Phosphorus (atomic weight 29.97) is a nonmetal of the nitrogen group. The only isotope in naturally occurring phosphorus is the stable form ^{31}P. The artificial isotopes ^{32}P and ^{33}P are radioactive. Phosphorus occurs in oxidation states $+1$, $+3$, or $+5$. Phosphate (PO_4^{3-}) is the main form utilized by humans. Another form used in a small number of metabolic reactions is pyrophosphate (diphosphate, food additive E450, $P_2O_7^{4-}$).

ABBREVIATIONS

Glvr-1   gibbon ape leukemia virus receptor (Pit-1, SLC20A1)
PTH   parathyroid hormone
TmP   tubular maximum for phosphate

NUTRITIONAL SUMMARY

Function: Phosphate is the constituent anion of bone, participates in energy metabolism and storage [as ATP, guanosine triphosphate (GTP), creatine phosphate, arginine phosphate, etc.], and is an important buffer in most body compartments.

Requirements: Adults should consume at least 700 mg/day; pregnancy and lactation do not increase the need for phosphorus.

Sources: Phosphate content of foods correlates somewhat with protein content, but distinct differences exist. Milk and dairy products have the highest phosphate-to-protein ratio (near 30 mg/g). Chicken and fish are at the low end (6–7 mg/g). Processed cheese, many other processed foods, and colas and some other sodas, are also important sources. Much of the large amounts of phosphate in plant-derived foods is bound to inositol (phytate and related forms) and is not absorbed. Food tables give the misleading impression that legumes, oats, rye, and other grains are major phosphate sources, but they are not.

Deficiency: Phosphate inadequacy is usually due to low food consumption or starvation and is not uncommon in elderly people. Accelerated bone mineral loss causes osteoporosis and increases fracture risk.

Excessive intake: Phosphate intake that significantly exceeds calcium intake (on a milligram basis) induces parathyroid gland hyperplasia and PTH secretion, impairs vitamin D activation, accelerates bone mineral loss, and increases fracture risk. Extremely high intake may cause calcification of extra-osseous (nonbone) tissues, including arteries, kidneys, muscles, and tendons.

DIETARY SOURCES

It is important to recognize that foods contain three very different types of phosphate compounds (Figure 11.13). The first two, inorganic phosphate salts and most organophosphates (including phospholipids), are readily absorbed and profoundly affect human metabolism. In stark contrast, the third type, inositol polyphosphates, are hardly absorbed at all, share virtually none of the metabolic characteristics of phosphate salts or other organophosphates, and have their own distinct properties.

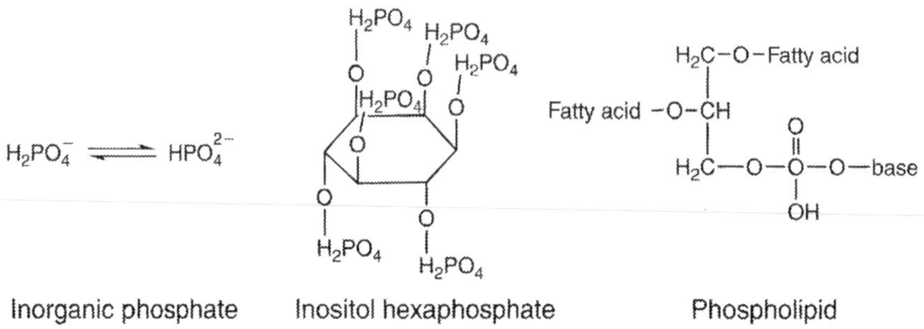

FIGURE 11.13

Different types of phosphate-containing compounds in foods.

Current food tables and other sources of food composition data routinely provide only a single value for total phosphate and become virtually meaningless in the case of grains, legumes, fruits, and vegetables. Many (but not all) of these foods provide relatively little bioavailable phosphate, currently available food composition information and practice guidelines (ADA, 1998) notwithstanding. Milk (1.0 mg/g) and dairy products are major phosphate sources. Particularly concentrated sources are hard cheeses such as cheddar (5.1 mg/g) and Swiss cheese (6.1 mg/g). Smaller amounts are in soft cheeses like cottage cheese (1.3 mg/g) and cream cheese (1.0 mg/g). An important measure is the ratio of bio-available phosphate to protein, which ranges from very high in milk (29.6), hard cheese (around 20), and eggs (14.3), to much lower in pork (11.4), beef (9.5), chicken (6.7), and fish (6.1). This ratio can be an important tool for people who need to minimize phosphate intake while obtaining adequate protein.

The phosphate content of plant-derived foods is much more difficult to assess, since much of it is bound to inositol as phytate (inositol hexaphosphate), inositol pentaphosphate, and inositol tetraphosphate.

Americans continue to increase their consumption of phosphate-containing food additives and may have average intake from this source alone that approaches 500 mg/day by now (Calvo & Park, 1996). Total daily phosphate intake of young men tends to be around 1500 mg, that of young women around 1000 mg (Food and Nutrition Board, Institute of Medicine, 1997: Appendix D). Intake varies greatly among individuals and tend to decline after young adulthood. The abovementioned caveat about the possible lack of relevance of total phosphate intake estimates, especially for groups with presumably healthful diets (rich in whole grains, legumes, fruits, and vegetables), has to be emphasized, however.

DIGESTION AND ABSORPTION

Various forms of dietary phosphates, including phosphate salts, nucleotides, and phospholipids, are absorbed with high efficiency (60–70%) from the small intestine. Many organophosphates are cleaved prior to absorption of the phosphate ion. The brush border membrane enzyme alkaline phosphatase (EC3.1.3.1), for instance, cleaves creatine phosphate. Both inorganic pyrophosphatase (EC3.6.1.1) and alkaline phosphatase (EC3.1.3.1) can hydrolyze pyrophosphate.

Type I (SLC17A1) and type IIb (NaPi3B, SLC34A2) sodium/phosphate cotransporters move inorganic phosphate and three sodium ions across the brush border membrane (Muter et al., 2001). Transport with the type I transporter is constant, while expression and transport rate with type IIb transporter depend on phosphate dose and 1,25-dihydroxyvitamin D status. An additional, sodium-independent, phosphate transporter may operate at the intestinal brush border membrane.

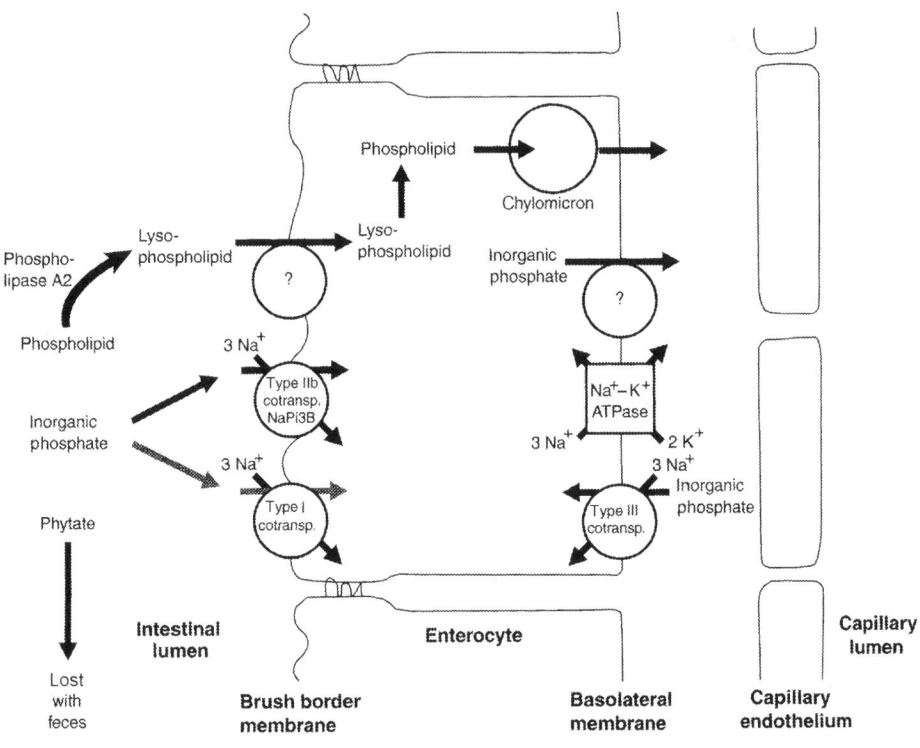

FIGURE 11.14

Intestinal absorption of phosphate. The question marks indicate that the molecular nature responsible for the activity is still uncertain.

The type III sodium cotransporters Glvr-1 (gibbon ape leukemia virus receptor, Pit-1, SLC20A1) and Pit-2 (SLC20A2) are thought to provide for phosphate influx from the basolateral side for the enterocyte's own needs (Virkki et al., 2007).

In addition to phospholipid consumed with foods, large amounts are secreted with bile and form mixed micelles with bile acids, fatty acids, cholesterol, and other minor lipids. Phospholipids are cleaved by phospholipase A2 (EC3.1.1.4) from pancreas and the resulting lysolecithin is taken up from the micelle across the brush border membrane. The liver-type fatty acid binding protein (L-FABP) helps to move lysophospholipids to intracellular compartments. A fatty acid is linked to most lysophospholipids and the phospholipid is exported with chylomicrons. Alternatively, lysophospholipase (EC3.1.1.5) can hydrolyze lysophosphatidyl choline to 3-phosphorylcholine, which is then exported and carried to the liver via the portal bloodstream (Figure 11.14).

TRANSPORT AND CELLULAR UPTAKE

Blood circulation: Most phosphate in circulation is contained in the phospholipids of lipoproteins and blood cell membranes. A smaller amount circulates as inorganic phosphate ion. Plasma concentration of inorganic phosphate is around 1 mmol/l if intake is adequate, much lower in deficiency, and slightly higher as in take increases (Food and Nutrition Board, Institute of Medicine, 1997). Phosphate enters cells via sodium cotransport, largely via the ubiquitous type III sodium cotransporters.

A slightly different use is seen in some bone where the type III sodium/phosphate cotransporter Glvr-1 (SLC20A1) pumps phosphate into the extracellular matrix during early bone mineralization (Palmer et al., 1999). Glvr-1 was also found to play a role in vascular and soft tissue calcification in response to hyperphosphatemia (Giachelli et al., 2001). Pit-2 (SLC20A2) is another type III cotransporter in many tissues. Band 3 of red cell membrane (SLC4A1) facilitates phosphate uptake in exchange for bicarbonate. Most ATP is utilized outside the mitochondria (e.g., for ion pumping), but needs to be reconstituted again by oxidative phosphorylation in the mitochondria. The mitochondrial phosphate carrier (SLC25A3), a proton-dependent symporter, and the adenine nucleotide translocators (SLC25A4, SLC25A5, and SLC25A6) ferry the precursors for ATP synthesis into mitochondria.

BBB: Phosphate transport into and from the brain is not well understood. It is likely that much of the transfer occurs through an anion antiporter that exchanges phosphate for bicarbonate (Dallaire and Bélveau, 1992).

Materno-fetal transfer: Phosphate, which has to be supplied in significant quantities to the fetus for tissue expansion and bone mineralization, crosses the maternal-facing brush border membrane of the syntrophoblast via sodium cotransport (Lajeunesse and Brunette, 1988). This process is aided by the negative charge of the syntrophoblast layer relative to its exterior.

STORAGE

The body contains 0.6–1.1% phosphorus (Aloia et al., 1984; Arunabh et al., 2002), 85% of this in bone. Osteoblasts deposit phosphate into the bone matrix, where it is part of the bone mineral hydroxyapatite. Type III sodium/phosphate cotransporters, including Glvr-1 (SLC20A1) and Pit-2 (SLC20A2), are involved in moving phosphate to its extracellular mineralization targets (Palmer et al., 1999). The release of phosphate ions from phosphoethanolamine and other phosphoesters by tissue-nonspecific (liver/bone/kidneys) form of alkaline phosphatase (EC3.1.3.1, zinc- and magnesium-dependent) is an important, though not well understood, event in bone mineralization.

Osteoclasts dissolve bone minerals by creating an acid microenvironment. Vacuolar H^+-ATPase (EC3.6.3.14) pumps protons generated by carbonic anhydrase II (EC4.2.1.1) into a restricted space between the bone surface and the ruffled border sealed off by actin fibers. An array of hydrolytic enzymes, including tartrate-resistant acid phosphatase (TRAP, ACPS, EC3.1.3.2), cysteine proteinases, and metalloproteinases, digest collagens and other bone matrix elements. As a result, the bone minerals, including phosphate, are released and can diffuse into nearby blood capillaries.

EXCRETION

Most phosphate losses occur with urine, small amounts are lost via feces (mucosal shedding) and skin. Decreased glomerular filtration in renal failure causes phosphate retention due to the inability to excrete excess.

Most of the approximately 16 g of phosphate that is filtered daily in the kidneys is reabsorbed. Renal losses depend on filtered load and thus on serum inorganic phosphate concentration. The tubular maximum for phosphate (TmE renal phosphate threshold) is a measure of the amount that can be reabsorbed.

Any filtered phosphate above this TmP threshold is excreted via urine. TmP can be estimated from a nomogram that uses measured values for plasma phosphate concentration, phosphate clearance, and creatinine clearance (Walton and Bijvoet, 1975). The type IIa (NaPi3, SLC34A1) sodium phosphate cotransporter at the brush border membrane of the proximal renal tubule provides the main mechanism of phosphate reabsorption. Much smaller amounts are recovered from the distal tubule via the type I cotransporter (NPT1, SLC17A1). Efflux across the basolateral membrane is less well understood. The type III sodium phosphate cotransporters Glvr-1 (SLC20A1) and Pit-2 (SLC20A2) are expressed at the basolateral membrane throughout the kidneys, but may be more important as "housekeeping transporters" to provide phosphate for the tubular cell's own needs. Additional transporters, including the phosphate transporter NaPi3B (SLC34A2) are present in kidneys, but their exact role in phosphate reabsorption remains to be established.

REGULATION

One master regulator of phosphate homeostasis is fibroblast growth factor 23 (FGF23), which comes mainly from osteoblasts and osteocytes (Kovesdy and Quarles, 2013). This peptide hormone binds to a complex consisting of the FGF receptor and the alpha-Klotho receptor mainly in the kidneys, decreasing expression of the type IIa sodium/phosphate cotransporter (Moz et al., 1999), increasing phosphate excretion and ultimately leading to lower phosphate and 1,25-dihydroxyvitamin D concentration in the blood, and higher PTH concentration. The response to phosphate intake appears to be delayed and indirect. Glycosylation by N-acetyl galactosamine transferase 3 (EC2.4.1.41) prevents the degradation of FGF23 and is an important part of this regulatory system (Figure 11.15).

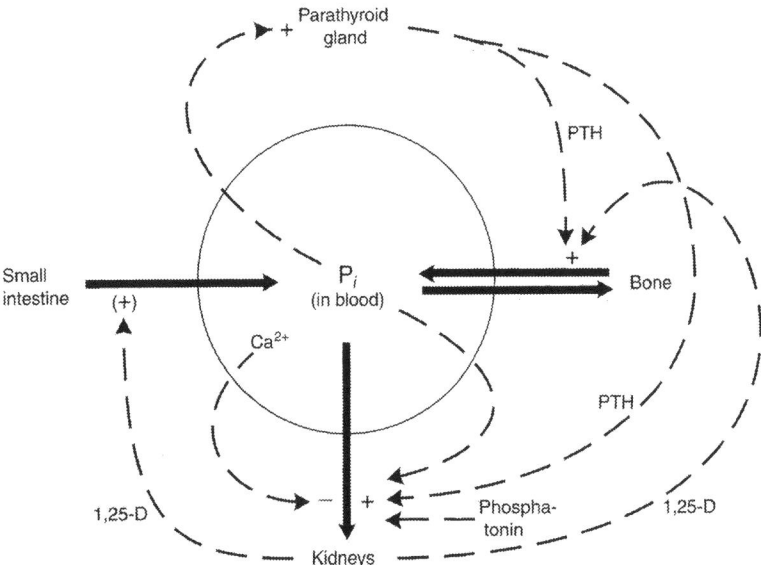

FIGURE 11.15

A web of hormones maintains phosphate balance.

The efficiency of intestinal absorption changes only little in response to variation in phosphate intake and status. Calcium, on the other hand, promotes renal tubular phosphate reabsorption. Persistently elevated blood concentrations of inorganic phosphate promote parathyroid gland proliferation and eventually cause hyperplasia. This is a typical consequence of hyperphosphatemia due to renal insufficiency (secondary hyperparathyroidism). For some time, this parathyroid gland hyperplasia is reversible if inorganic phosphate concentrations are kept lower (Barsotti et al., 1998).

FUNCTION

High-energy phosphate esters: Most transfers of chemical energy in the body involve phosphate ester bonds. This is particularly true for ATP as the main metabolic energy currency, but also applies to more specialized forms, including guanosine triphosphate (GTP), cytosine triphosphate (CTP), creatine phosphate, and arginine phosphate. These and additional nucleotides are also the essential building blocks for the synthesis of DNA and RNA. Several nucleotides, such as cAMP, are also critical for intracellular signaling.

Organophosphates: A large spectrum of endogenously synthesized compounds, aside from the ones mentioned previously, contains one or more phosphate groups as an integral part of their structure. Large amounts of various types of phospholipids are needed for the construction of cell membranes, myelin-sheathing of neurons, transport of lipids with lipoproteins, and facilitation of intestinal lipid absorption, among other functions. The phosphate-containing cholesterol precursors should also be mentioned.

Activating phosphate esters: The metabolism of many nutrients proceeds via phosphate esters at some point to provide the necessary reaction energy. Examples include the phosphates of glucose, fructose, galactose, glycerol, etc. The list of vitamins that are active only as phosphate esters includes thiamin, riboflavin, pyridoxine, niacin, and pantothenate. Phosphorylation can also serve to prevent efflux of such nutrients from cells (trapping) and aid uptake by passive diffusion. A typical example of such trapping is the intestinal absorption of vitamin B6. Free vitamin B6 diffuses across the intestinal brush border membrane via as-yet-unknown carriers and is then immediately phosphorylated by pyridoxal kinase (EC2.7.1.35). Since the phosphorylated form cannot return and intracellular concentration of the free forms is very low, diffusion into the enterocytes continues as long as there are significant amounts in the intestinal lumen.

Protein phosphorylation: The activity of many proteins is regulated through phosphorylation and dephosphorylation. The effect of an added phosphate group depends on the particular protein. Phosphorylation inactivates glycogen synthase (EC2.4.1.11), for instance, and dephosphorylation reactivates the enzyme.

Buffering: Aqueous solutions at physiological pH (around 7.4) contain about four-fifths of the inorganic phosphate as hydrogen phosphate ion (HPO_4^{2-}), and nearly one-fifth as dihydrogen phosphate ion $(H_2PO_4^-)$. This means that the significant amount of inorganic phosphate in cells, extracellular fluid, and blood acts as an effective buffer and stabilizes pH.

Polyphosphates: Some types of osteoblasts contain significant amounts of polyphosphates (Leyhausen et al., 1998). These polyphosphates can contain up to several thousand phosphates linked into a chain. The function of these structures is not well understood. Suggested roles include phosphate storage, inhibition of bone mineralization, divalent ion and basic amino acid chelation, apoptosis, pH regulation, and protection against osmotic stress. Incompletely characterized exopolyphosphatases, and in some cases pyrophosphatase (EC3.6.1.1) release phosphate ions from the polyphosphate chains.

NUTRITIONAL ASSESSMENT

Phosphate concentrations in all body fluids and tissues are tightly regulated and deviations are due to hormonal or other regulatory deviations, not deficiency or excess. Since by far the largest amount of phosphate is stored in bones, from where it can be released in times of need, bone mineral content might be considered an indicator of phosphate status, just as for calcium. However, it is subject to all the same influences of regulation and dysregulation as in other tissues and body fluids, just with a much greater delay. An initially marginal phosphate status can rapidly turn precariously low in the wake of metabolic events such as the sequestration of mobile phosphate stores by the sudden and unaccustomed influx of dietary carbohydrate and the heavy demand on phosphate for the phosphorylation of glucose. Such a potentially lethal situation can be triggered when starved individuals resume eating and is called *refeeding syndrome*. Phosphate concentrations in plasma often drop precipitously and can cause catastrophic heart muscle dysfunction (Skipper, 2012).

Phosphate absorbed from the intestines influences excretion with urine, if renal function is normal. Since many people, particularly with advancing age and with diseases such as diabetes, have diminished renal function, urine measurements have to be evaluated with caution.

REFERENCES

ADA, 1998. Pediatric manual of clinical dietetics. In: Williams, C.P. (Ed.), The American Dietetic Association, pp. 385–387.

Aloia, J.E., Vaswani, A.N.M., Yeh, J.K., Ellis, K., Cohn, S.H., 1984. Total body phosphorus in postmenopausal women. Miner. Electrolyte Metab. 10, 73–76.

Arunabh, S., Feuerman, M., Ma, R., Aloia, J.E., 2002. Total body phosphorus in healthy women and ethnic variations. Metab. Clin. Exp. 51, 180–183.

Barsotti, G., Cupisti, A., Morelli, E., Meola, M., Cozza, V., Barsotti, M., et al., 1998. Secondary hyperparathyroidism in severe chronic renal failure is corrected by very-low dietary phosphate intake and calcium carbonate supplementation. Nephron 79, 137–141.

Calvo, M.S., Park, Y.K., 1996. Changing phosphorus content of the US diet: potential for adverse effects on bone. J. Nutr. 126, 1168S–1180S.

Dallaire, L., Béliveau, R., 1992. Phosphate transport by capillaries of the blood–brain barrier. J. Biol. Chem. 267, 22323–22327.

Food and Nutrition Board, Institute of Medicine, 1997. Dietary Reference Intakes for Calcium, Phosphorus, Magnesium, Vitamin D, and Fluoride. National Academy Press, Washington, DC.

Giachelli, C.M., Jono, S., Shioi, A., Nishizawa, Y., Mori, K., Morii, H., 2001. Vascular calcification and inorganic phosphate. Am. J. Kidney Dis. 38, S34–S37.

Kovesdy, C.P., Quarles, L.D., 2013. Fibroblast growth factor-23: what we know, what we don't know, and what we need to know. Nephrol. Dial. Transplant. 28, 2228–2236.

Lajeunesse, D., Brunette, M.G., 1988. Sodium gradient-dependent phosphate transport in placental brush border membrane vesicles. Placenta 9, 117–128.

Leyhausen, G., Lorenz, B., Zhu, H., Geurtsen, W., Bohnensack, R., Muller, W.E., et al., 1998. Inorganic polyphosphate in human osteoblast-like cells. J. Bone Min. Res. 13, 803–812.

Moz, Y., Silver, J., Naveh-Many, T., 1999. Protein–RNA interactions determine the stability of the renal NaPi-2 cotransporter mRNA and its translation in hypophosphatemic rats. J. Biol. Chem. 274, 25266–25272.

Murer, H., Hernando, N., Forster, I., Biber, J., 2001. Molecular mechanisms in proximal tubular and small intestinal phosphate reabsorption. Mol. Membr. Biol. 18, 3–11.

Palmer, G., Zhao, J., Bonjour, J., Hofstetter, W., Caverzasio, J., 1999. *In vivo* expression of transcripts encoding the Glvr-1 phosphate transporter/retrovirus receptor during bone development. Bone 24, 1–7.

Virkki, L.V., Biber, J., Murer, H., Forster, I.C., 2007. Phosphate transporters: a tale of two solute carrier families. Am. J. Physiol. Renal Physiol. 293, F643–F654.

Walton, R.J., Bijvoet, O.L., 1975. Nomogram for derivation of renal threshold phosphate concentration. Lancet 2 (7929), 305–310.

MAGNESIUM

Magnesium (atomic weight 24.305) is an alkaline earth metal. Naturally occurring magnesium contains the stable isotopes ^{24}Mg (78.99%), ^{25}Mg (10.00%), and ^{26}Mg (11.01%). The oxidation state is usually +2.

ABBREVIATION

Mg   magnesium

NUTRITIONAL SUMMARY

Function: Magnesium (Mg) is an essential cofactor for a large number of reactions, including all of those involving ATP and GTP, participates in muscle and nerve depolarization, stabilizes DNA and RNA, and is a component of the mineral in bone.

Food sources: Whole grains, nuts and seeds, spinach, legumes, potatoes, and bananas are good sources. Hard tap water and some bottled mineral waters also can be significant sources.

Requirements: Men should get at least 400 mg (420 mg over age 30) from food, women at least 310 mg (320 mg over age 30, and 360 mg when pregnant).

Deficiency: Signs include confusion, disorientation, personality changes, loss of appetite, depression, muscle contractions and cramps, tingling, numbness, hypertension, abnormal heart rhythms, coronary spasm, and seizures. Deficiency is often induced by diarrhea, malabsorption, or vomiting, overuse of laxatives or diuretics, medication (cyclosporin, amphotericin, gentamycin, and cis-platinum), alcohol abuse, diabetes, or hyperparathyroidism.

Excessive intake: Intake of more than 350 mg from supplements and other nonfood sources may cause diarrhea, nausea, appetite loss, muscle weakness, mental impairment, difficulty breathing, extremely low blood pressure, and irregular heartbeat. Risk of toxicity is greater with impaired kidney function.

DIETARY SOURCES

Whole grains, nuts (66 mg/15 g), seeds (75 mg/30 g), spinach (65 mg/half a cup), legumes (35–54 mg/half a cup), potatoes (55 mg/medium-sized), and bananas (34 mg/medium-sized fruit) are good sources of Mg. Mg-rich mineral water may contain 110 mg/l or more (Sabatier et al., 2002).

Typical daily Mg intake of American women is around 220 mg; in men, it is around 300 mg (Food and Nutrition Board, Institute of Medicine, 1997).

DIGESTION AND ABSORPTION

About 30–60% of ingested Mg is absorbed (Lemann & Favus, 1999; Wauben et al., 2001; Sabatier et al., 2002). The main sites of absorption are the distal jejunum and ileum, smaller amounts in the colon. The large intestine is also capable of some absorption (Fine et al., 1991). The transient receptor potential ion channel, member 6 (TRPM6) accounts for much of the absorption capacity. Loss of TRPM6 function is associated with low Mg concentration in the blood (hypomagnesemia) and reduced calcium concentration (van der Wijst et al., 2014).

Solvent drag across the tight junctions involving claudin16 between enterocytes is an alternative pathway, particularly at high luminal concentrations and may account for more than a third of total absorption. The mechanism for transport across the basolateral membrane is not known. Phosphate and phytate, but not calcium, inhibit absorption.

TRANSPORT AND CELLULAR UPTAKE

Blood circulation: Most (55%) of the Mg in plasma (about 0.7–1.1 mmol/l) is in ionized form; smaller amounts are bound to other proteins (33%) and complexed with anions (12%). Cells take up Mg via dedicated channels including TRPM6 and TRPM7 (Ryazanova et al., 2014). Nucleotides complex most intracellular Mg, keeping free ionized Mg concentrations below 1 mmol/l. Mitochondrial adenine nucleotide translocase mobilizes Mg upon hormonal stimulation from mitochondria into cytosol (Romani et al., 1993).

BBB: Mg concentrations in cerebrospinal fluid and brain appear to be independently regulated and not related to blood concentration (Fong et al., 1995; Gee et al., 2001). The mechanisms of transport across the BBB are not well studied.

Materno-fetal transfer: Much less Mg moves from the mother to the fetus than the somewhat similar divalent cation calcium, which indicates that the transport mechanisms are not shared. How exactly Mg crosses the placental barrier remains unclear, however.

Transfer into milk: Mg concentration in human milk range widely with typical values in well-nourished women around 33 mg/l (Dórea, 2000). The divalent cation channel TRPM6 mediates Mg influx, depending on extracellular concentration (Mastrototaro et al., 2011). An as-yet-unknown sodium/Mg exchanger is thought to move Mg into the glandular lumen.

STORAGE

The body of a 70-kg adult contains about 25 g Mg: 50–60% of this is in bone, 1% in the blood and extracellular fluid (Elin, 1987; Rude, 2000). Since Mg is present mostly on the surface of the bone and is not an integral part of the bone mineral hydroxyapatite, the mechanism for deposition and mobilization is probably different from that for calcium and phosphate.

EXCRETION

Free ionized Mg and Mg complexed with small anions, which constitute the bulk (80%) of circulating Mg, are freely filtered by the renal glomerulus—normally more than 2 g/day. Healthy people lose less than 5% of the filtered Mg via urine. Some of the luminal Mg is recovered from the proximal tubules through TRPM6 (Lainez et al., 2014), a much greater proportion (70%) from the TAL of the loop of Henle, and a smaller proportion again from the distal tubule (Quamme, 1997; Cole & Quamme,

2000). To a considerable extent, Mg reabsorption occurs by equilibrative movement through the spaces between the epithelial cells. Protein complexes called tight junctions seal these intercellular spaces. Claudin 16 (paracellin-1) is a tight junction protein that controls permeation of Mg in the TAL of the loop of Henle through the paracellular pathway (Weber et al., 2001; Seo and Park, 2008).

While it is clear that uptake into the epithelial cells of the loop of Henle is driven by concentration and voltage differences, the responsible protein elements for basolateral permeation (carriers, channels) remain unknown. The sodium-chloride cotransporter (SLC12A3) may be a critical element for Mg reabsorption from the distal convoluted tubule, since its inactivity increases renal Mg losses.

REGULATION

Renal reuptake from primary filtrate is the main mechanism for the control of Mg content of the body. Urinary losses closely correlate with Mg and calcium concentrations in the blood and with dietary intake. In addition, 1,25-dihydroxyvitamin D stimulates the reabsorption of Mg from the distal renal tubule, possibly because induction of calbindin-9K promotes transcellular transport (Ritchie et al., 2001). The extracellular calcium/Mg ion sensing receptor (CaSR) at the outside face of the basolateral membrane responds to divalent ion concentration by modulating tight junction permeability and transepithelial voltage (Cole & Quamme, 2000). As Mg and calcium ion concentrations drop in the pericapillary space (and in the capillary lumen) adjacent to the basolateral membrane, the renal reabsorption of both ions increases. The Mg channels TRPM6 and TRPM7 are both tightly regulated through their integral kinase domain, which serves as a Mg sensor (Ryazanova et al., 2014).

PTH, calcitonin, antidiuretic hormone, and glucagon also may decrease renal losses, but the significance of their role in maintaining Mg homeostasis appears to be minor and is not well understood (Rude, 2000). PTH may also promote the active transport component of intestinal absorption.

FUNCTION

Nucleotide complexes: All reactions that utilize ATP and similar nucleotides work efficiently only when these nucleotides are complexes with Mg.

Enzyme cofactor: There are also some enzymes that do not involve nucleotides as cofactors but require Mg nonetheless. Manganese or other divalent cations are alternative cofactors for some of these enzymes. Their potential to substitute *in vivo* is in doubt, however, because of their low concentrations. Only a few examples of such enzymes will be given here. Phosphopyruvate hydratase (enolase, EC4.2.1.11) is both stabilized by Mg and requires the metal for catalysis. Phosphoglucomutase (EC5.4.2.2) is another enzyme in carbohydrate (galactose) metabolism that depends on Mg. The Mg-containing selenophosphate synthase (EC2.7.9.3) activates selenium prior to incorporation into proteins as selenocysteine. Choline synthesis and metabolism also depend on several Mg-requiring enzymes, including phosphatidylcholine synthase (EC2.7.8.24) and choline monooxygenase (EC1.14.15.7).

Because Mg enzymes have much higher affinity to zinc than to the Mg they need for optimal function, intracellular zinc concentration is kept extremely low (Foster et al., 2014).

Bone mineral: Mg is a normal constituent of bone, but is not a constituent of the main mineral hydroxyapatite. The exact role of Mg in bone remains unclear.

Balancing calcium: The competition of Mg with calcium for binding to transporters, signaling structures and enzymes may be of functional importance. It has been suggested that excessive calcium availability and activity with decreasing Mg concentration may be responsible for a predisposition for muscle cramps, increased blood pressure, and coronary vasospasms. Use of supplemental Mg in at-risk women to prevent eclampsia, possibly through a similar mechanism, may be less essential than believed earlier (Hall et al., 2000).

NUTRITIONAL ASSESSMENT

Mg status is modestly well reflected by the concentration of Mg in blood plasma. An initially marginal status can rapidly turn precariously low in the wake of metabolic events such as the sequestration of mobile Mg stores by the sudden and unaccustomed influx of dietary carbohydrate and the heavy demand on Mg during the phosphorylation of glucose. Such a potentially lethal situation can be triggered when starved individuals resume eating and is called *refeeding syndrome.* Mg concentrations in plasma can drop precipitously and cause catastrophic heart muscle dysfunction (Skipper, 2012).

REFERENCES

Cole, D.E.C., Quamme, G.A., 2000. Inherited disorders of renal magnesium handling. J. Am. Soc. Nephrol. 11, 1937–1947.

Dórea, J.G., 2000. Magnesium in human milk. J. Am. Coll. Nutr. 19, 210–219.

Elin, R., 1987. Assessment of magnesium status. Clin. Chem. 33, 1965–1970.

Fine, K.D., Santa Ana, C.A., Porter, J.L., Fordtran, J.S., 1991. Intestinal absorption of magnesium from food and supplements. J. Clin. Invest. 88, 396–402.

Fong, J., Gurewitsch, E.D., Volpe, L., Wagner, W.E., Gomillion, M.C., August, P., 1995. Baseline serum and cerebrospinal fluid magnesium levels in normal pregnancy and preeclampsia. Obstet. Gynecol. 85, 444–448.

Food and Nutrition Board, Institute of Medicine, 1997. Dietary Reference Intakes for Calcium, Phosphorus, Magnesium, Vitamin D, and Fluoride. National Academy Press, Washington, DC.

Foster, A.W., Osman, D., Robinson, N.J., 2014. Metal preferences and metallation. J. Biol. Chem. 289, 28095–28103.

Gee II, J.B., Corbett, R.J., Perlman, J.M., Laptook, A.R., 2001. Hypermagnesemia does not increase brain intracellular magnesium in newborn swine. Ped. Neurol. 25, 304–308.

Hall, D.R., Odendaal, H.J., Smith, M., 2000. Is the prophylactic administration of magnesium sulphate in women with pre-eclampsia indicated prior to labour? Bjog 107, 903–908.

Lainez, S., Schlingmann, K.P., van der Wijst, J., Dworniczak, B., van Zeeland, F., Konrad, M., et al., 2014. New TRPM6 missense mutations linked to hypomagnesemia with secondary hypocalcemia. Eur. J. Hum. Genet. 22 (4), 497–504.

Lemann Jr, J., Favus, M.J., 1999. The intestinal absorption of calcium, magnesium, and phosphate. In: Favus, M.J. (Ed.), Primer on the Metabolic Bone Diseases and Disorders of Mineral Metabolism, fourth ed. Lippincott, Williams & Wilkins, Philadelphia, PA, pp. 63–67.

Mastrototaro, L., Trapani, V., Boninsegna, A., Martin, H., Devaux, S., Berthelot, A., et al., 2011. Dietary Mg^{2+} regulates the epithelial Mg^{2+} channel TRPM6 in rat mammary tissue. Magn. Res. 24, S122–S129.

Quamme, G.A., 1997. Renal magnesium handling: new insights in understanding old problems. Kidney Int. 52, 1180–1195.

Ritchie, G., Kerstan, D., Dai, L.J., Kang, H.S., Canaff, L., Hendy, G.N., et al., 2001. 1,25(OH)2D 3 stimulates Mg^{2+} uptake into MDCT cells: modulation by extracellular Ca^{2+} and Mg^{2+}. Am. J. Physiol. Ren. Physiol. 280, F868–F878.

Romani, A., Marfella, C., Scarpa, A., 1993. Cell magnesium transport and homeostasis: role of intracellular compartments. Min. Electrolyte Metab. 19, 282–289.

Rude, R.K., 2000. Magnesium. In: Stipanuk, M.H. (Ed.), Biochemical and Physiological Aspects of Human Nutrition WB Saunders Company, Philadelphia, PA, pp. 671–685.

Ryazanova, L.V., Hu, Z., Suzuki, S., Chubanov, V., Fleig, A., Ryazanov, A.G., et al., 2014. Elucidating the role of the TRPM7 alpha-kinase: TRPM7 kinase inactivation leads to magnesium deprivation resistance phenotype in mice. Sci Rep 4, 7599.

Sabatier, M., Arnaud, M.J., Kastenmayer, E., Rytz, A., Barclay, D.V., 2002. Meal effect on magnesium bioavailability from mineral water in healthy women. Am. J. Clin. Nutr. 75, 65–71.

Seo, J.W., Park, T.J., 2008. Magnesium metabolism. Electrolyte Blood Press 6, 86–95.

Skipper, A., 2012. Refeeding syndrome or refeeding hypophosphatemia: a systematic review of cases. Nutr. Clin. Pract. 27, 34–40.

Wauben, I.E., Atkinson, S.A., Bradley, C., Halton, J.M., Barr, R.D., 2001. Magnesium absorption using stable isotope tracers in healthy children and children treated for leukemia. Nutrition 17, 221–224.

Weber, S., Schneider, L., Peters, M., Misselwitz, J., Ronnefarth, G., Boswald, M., et al., 2001. Novel paracellin-1 mutations in 25 families with familial hypomagnesemia with hypercalciuria and nephrocalcinosis. J. Am. Soc. Nephrol. 12, 1872–1881.

van der Wijst, J., Blanchard, M.G., Woodroof, H.I., Macartney, T.J., Gourlay, R., Hoenderop, J.G., et al., 2014. Kinase and channel activity of TRPM6 are co-ordinated by a dimerization motif and pocket interaction. Biochem. J. 460, 165–175.

IODINE

Iodine (atomic weight 126.9045) is a relatively volatile halogen. Only isotope 127 occurs naturally. The radioactive isotopes 124, 125, 128, 131, and 132 can be produced artificially. Iodine can have valences between 1 and 7. The term *iodide* refers to its ionic form, I^-, while *iodate* refers to IO_3^-.

ABBREVIATIONS

Cys	L-cysteine
TSH	thyroxine-stimulating hormone

NUTRITIONAL SUMMARY

Function: Iodine is an essential component of the thyroid hormones, which have profound impact on development, growth, and metabolism.

Food sources: Good sources are fish, shellfish, seaweed, iodized salt, and fortified foods. Other foods provide variable amounts depending on the soil iodine content of their origin.

Requirements: Adults should get 150 µg of iodine/day. Slightly more is needed in women during pregnancy and lactation.

Deficiency: A lack of iodine may cause spontaneous abortion and birth defects; it impairs irreversibly brain and physical development in the fetus and young child, causing from mild to severe mental retardation, sometimes also hearing loss or paralysis of the legs. Deficiency in older children and adults stimulates excessive growth of the thyroid gland (goiter); slows metabolic rate, mental and cardiac function; and induces a feeling of fatigue and cold intolerance.

Excessive intake: Increased iodine intake can promote thyroid hormone production in people with excessive function of the thyroid gland (autonomous hyperthyroidism). In healthy people, intake above requirements (up to 2000 µg/day) will not confer any benefit but also appear to cause no harm. Chronic intake of several milligrams per day may disrupt thyroid function.

DIETARY SOURCES

Foods contain iodine mainly as iodide (I^-), iodate (IO_3^-), or thyroxine. Best iodine sources are marine fish, shellfish, seaweed, and sea salt; iodized salt contains 76 µg iodide/g, the iodide content of seawater is 50–60 µg/l (Hetzel et al., 1990). Bread may contain iodate as a dough improver, and dairy foods may contain iodine from udder antiseptics. Unsupplemented iodine content of foods closely correlates with the concentration in soil. Plants from iodine-sufficient soil contain around 1 µg of iodine/g dry weight. Plants from areas with iodine-depleted soils (old mountain ranges and coastal areas) contain only a fraction of that amount. The median iodide content of tap water in the United States is 4.55 µg/l (Blount et al., 2010), but may be <2 µg/l in iodine-deficient areas.

Typical iodine intake of adults in the United States is around 150 µg/day, slightly more in children. As much as 10% have a very low daily intake of less than 50 µg (Caldwell et al., 2011).

DIGESTION AND ABSORPTION

Most iodine consumed as iodide in table salt is absorbed (Nath et al., 1992; Hurrell, 1997). The sodium/iodide symporter (SLC5A5) mediates active uptake of iodide from the stomach and small intestine (Nicola et al., 2009). This symporter is expressed mainly in chief and parietal cells in gastric mucosa and epithelial cells lining mucosal crypts in rectal mucosa with much lower expression levels in small intestine and colon (Ajjan et al., 1998). The mechanism of transfer into blood across the basolateral enterocyte membranes is unclear.

Fractional absorption of iodine from unsupplemented foods may be less than 20% (Wahl et al., 1995), possibly due to the presence of poorly absorbed organoiodine compounds. Only a third of ingested thyroxine appeared to be absorbed by rats (Van Middlesworth, 1985); adequately detailed investigations in humans are lacking. The bioavailability of iodate also is significantly lower compared to iodide. Nonenzymatic reduction of iodate to iodide by glutathione or other sulfhydryl compounds might be of some importance.

TRANSPORT AND CELLULAR UPTAKE

Blood circulation: Iodide is taken up by follicular cells across the basolateral membrane via the sodium/iodide symporter (SLC5A5) against the electrochemical gradient. Thiocyanate, perchlorate, and nitrate compete with iodide for uptake via the sodium/iodide symporter and reduce the production of thyroid hormones (Steinmaus et al., 2013). Thiocyanate is consumed with pungent foods of the mustard and cabbage (brassica) family, such as rutabaga, and is generated during the detoxification of cyanide (from foods such as almonds and cassava). Perchlorate is a food and water contaminants, and nitrate comes from fertilizers and food additives.

Iodide from follicular cells can then be exported across the apical membrane into the colloidal space via the chloride/iodide transporter pendrin (SLC26A4) (Pohlenz & Refetoff, 1999; Yoshida et al., 2004).

Materno-fetal transport: Uptake of iodide from the maternal side of syncytiotrophoblasts in placenta is likely to proceed via pendrin (SLC26A4) whereas transport out of the cytotrophoblasts into fetal circulation appears to use the sodium/iodide symporter (SLC5A5; Bidart et al., 2000; Mitchell et al., 2001). Transport as thyroid hormones also appears to contribute significantly to the iodine supply of the fetus.

Transfer into milk: The concentration of iodine in mature human milk is about 45 µg/l (Hannan et al., 2009). Animal experiments suggest that high bromide intake is likely to diminish iodine transfer into milk due to competition for the same transporters (Pavelka et al., 2002).

Other tissues: A combination of sodium/iodide symporter and pendrin mediates transport into the mammary gland (Riedel et al., 2001; Lacroix et al., 2001) and many other epithelial tissues. Additional mechanisms for iodide transport may exist. Evidence has been presented that an iodide/sulfate exchanger augments iodide uptake into mammary cells (Shennan, 2001).

METABOLISM

Synthesis: Thyroid hormone production takes place in the colloidal space of the thyroid and, to a much lesser extent, in the brain and possibly some other nonthyroid organs. Three distinct activities necessary for thyroid hormone synthesis are catalyzed by the heme-containing enzyme iodide peroxidase (EC1.11.1.8): (i) iodide oxidation, (ii) iodination of tyrosines in thyroglobulin (Tg), and (iii) iodothyronine coupling (phenoxy-ester formation between pairs of iodinated tyrosines to generate the thyroid hormones, thyroxine and triiodothyronine). Out of eight potential sites only four or fewer generate thyroid hormone when Tg is eventually cleaved (Dunn & Dunn, 1999), predominantly at sites Tyr^5, Tyr^{1290}, Tyr^{2553}, and Tyr^{2746} (Dunn et al., 1998). The progress of iodination is accompanied by site-specific glycation and incompletely iodinated Tg is recognized and recycled by a thyroid-specific asialoglycoprotein receptor across the basolateral membrane where the highest concentration of iodinating activity resides (Figure 11.16).

Thyroglobulin is multimerized and aggregates as a side effect of iodide peroxidation. It has also been suggested that this is actually a well-regulated process that facilitates the storage of iodinated thyroglobulin at high concentration in the colloidal space (Klein et al., 2000).

Two molecular chaperones, protein disulfide isomerase (EC5.3.4.1) and immunoglobulin heavy chain-binding protein, reduce intermolecular disulfide bridges between multimers and help with the refolding of thyroglobulin (Delom et al. 2001). This may prevent the accumulation of aggregates and thus facilitate the transport of the iodized protein back into follicular cells. Protein disulfide isomerase, a member of the thioredoxin family, contains several Cys-X-X-Cys motifs that are oxidized when interacting with disulfide bridges of suitable protein targets. The enzyme is reactivated by NADPH-dependent thioredoxin reductase (EC1.6.4.5). The thioredoxin reductase in follicle cells is a selenocystein-containing flavoprotein whose expression is tightly regulated through the calcium/phosphoinositol signaling pathway (Howie et al., 1998). Thyroglobulin containing thyroxine and triiodothyronine is taken up from the colloidal space into follicular cells by pinocytosis. Inside the follicular cells, the pinocytotic vesicles are merged with lysosomes. Proteolysis of thyroglobulin by incompletely characterized proteases then releases the active hormones. The selenium-containing 5'-thyroxine deiodinase D2 (EC1.97.1.10) produces the highly active 3,5,3'-triiodo-L-thyronine by removing iodine from thyroxine (Bianco & Kim, 2006). Removal of iodine from the 5 position (inner ring) by selenium-dependent deiodinase D3 (EC1.97.1.11) inactivates thyroxine (T4) by converting it to 3,5',3'-triiodo-L-thyronine (reverse T3). The activating deiodinase D2 is induced by overfeeding and bile acid signaling (Watanabe et al., 2006).

Some substances in cabbages and other foods have been found to interfere with normal thyroid hormone production and induce the enlargement of the thyroid gland in susceptible people, especially in conjunction with low iodine status. Many of these goitrogens (goiter-inducing compounds) contain thiocyanide or a thiocyanide moiety that can become metabolically activated. The heterocyclic compound goitrin (Figure 11.17) is a well-characterized and potent goitrogen in cabbages and Brussels sprouts.

Iodine salvage: The selenium-dependent deiodinase D3 (EC1.97.1.11) in the thyroid gland breaks down unneeded or incorrectly iodinated iodothyronines and frees up iodine for renewed incorporation into thyroglobulin. The TSH receptor regulates activity of this deiodinase via cAMP signaling (Murakami et al., 2001).

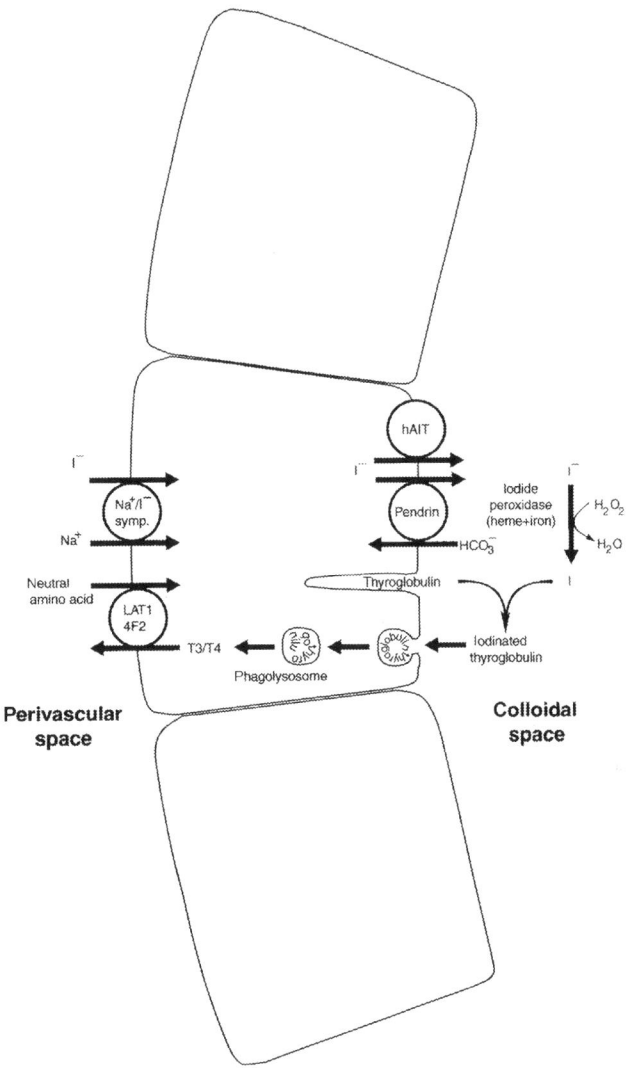

FIGURE 11.16

Synthesis of thyroid hormone.

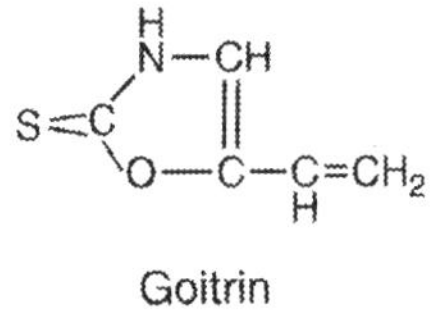

Goitrin

FIGURE 11.17

Goitrin is a thiocyanide-containing compound.

STORAGE

About 15–20 mg of iodine is stored in replete people, 70–80% of this in the thyroid. Thyroxine deiodinases (EC3.8.1.4) are a group of selenocysteine enzymes in the endoplasmic reticulum that release iodine from the thyroid hormones thyroxine and triiodothyronine, both in storage and at functional sites. At least three distinct genes code for isoforms that differ in respect to specificity and level of expression in various tissues and at different developmental stages. Thyroxine deiodinase 2 generates the most active thyroid hormone triiodothyronine in thyroid, developing brain, anterior pituitary gland, and brown adipose tissue. The isoform 1, in contrast, inactivates thyroxine in the liver and kidneys and thereby ensures that hormonal activity does not accumulate unchecked. Thyroxine deiodinase 3, which is expressed in fetal tissues and placenta, preferentially removes iodine from the inner ring of thyroid hormones, thereby inactivating both thyroxine and triiodothyronine. The enzyme thus protects fetal tissues from exposure to the much higher thyroid hormone concentrations in maternal circulation.

EXCRETION

Pendrin (SLC26A4) mediates iodide uptake across the apical membranes of proximal tubules (Soleimani et al., 2001), and intercalated cells of the cortical collecting ducts (Royaux et al., 2001). Iodide may then be pumped by the sodium/iodide symporter (SLC5A5) across the basolateral membrane into the perivascular space (Spitzweg et al., 2001) and reenter blood circulation.

REGULATION

Uptake: The sodium/iodide symporter is a phosphoprotein whose transcription, phosphorylation pattern (Riedel et al., 2001), and targeting toward the plasma membrane are controlled mainly by TSH. At low TSH levels, most transport activity is lost. Additional hormones also influence iodide uptake, though to a much lesser degree. Excess iodide directly down-regulates expression of the symporter in the gastrointestinal tract (Nicola et al., 2012).

Renal excretion: Excretion with urine depends on iodine status. It has been suggested that people who excrete more than 50 μg/g of creatinine generally have adequate iodine status. Losses via urine are regulated both through circulating levels (urine content is proportional to blood concentration) and the efficiency of tubular reabsorption.

Wolff-Chaikoff effect: Ingestion of a large dose of iodine rapidly decreases thyroid hormone production. The reason for this phenomenon is the down-regulation of the sodium/iodine symporter in response to high iodide concentration in the blood. As less iodide is taken up from circulation, the rate of thyroglobulin iodination quickly diminishes until rising TSH levels override the effect on the symporter and iodide reaches the colloidal space again.

FUNCTION

Classic role of thyroid hormones: Thyroid hormones are critically important for the fetus and young infant by affecting the development and maturation of the nervous system, skeletal muscle, and lungs through as-yet-incompletely understood mechanisms. At all life stages, thyroid hormones promote intermediary metabolism; disruption of thyroid hormone production slows oxidation, lowers body temperature, and may evolve into a life-threatening condition. About 60 μg of iodine/day is used for hormone production. If slightly less than this amount is available, goiter (enlarged thyroid gland) may develop. A reduction of iodine content of the thyroid below 0.1% usually results in hyperplastic changes (Hetzel et al., 1990). Enlargement of the thyroid can also be related to a disruption of thyroid hormone synthesis by consumption of thioisocyanates from cabbage and other cruciferous vegetables.

Nongenomic thyroid hormone actions: Thyroid hormones also act without changing gene expression (Davis and Davis, 1996). They influence the transport of sodium, calcium, glucose, and other solutes; modulate the activities of protein kinase C, cAMP-dependent protein kinase, and other kinases; affect glycolysis enzymes such as pyruvate kinase M2, oxidative phosphorylation; and regulate skeletal and vascular smooth muscle action.

NUTRITIONAL ASSESSMENT

Iodine deficiency is usually evaluated by measuring iodine in urine because low amounts mean that intake is too low to cause an overflow phenomenon (Rohner et al., 2014). Functional biomarkers with lower specificity include blood concentration of thyroid-stimulation hormone (TSH), thyroglobulin, and thyroid hormones (T4/T3), as well as assessing thyroid size for the detection of goiter.

REFERENCES

Ajjan, R.A., Watson, P.F., Findlay, C., Metcalfe, R.A., Crisp, M., Ludgate, M., et al., 1998. The sodium iodide symporter gene and its regulation by cytokines found in autoimmunity. J. Endocrinol. 158, 351–358.

Bidart, J.M., Lacroix, L., Evain-Brion, D., Caillou, B., Lazar, V., Frydman, R., et al., 2000. Expression of Na+/I−-symporter and Pendred syndrome genes in trophoblast cells. J. Clin. Endocrinol. Metab. 85, 4367–4372.

Bianco, A.C., Kim, B.W., 2006. Deiodinases: implications of the local control of thyroid hormone action. J. Clin. Invest. 116, 2571–2579.

Blount, B.C., Alwis, K.U., Jain, R.B., Solomon, B.L., Morrow, J.C., Jackson, W.A., 2010. Perchlorate, nitrate, and iodide intake through tap water. Environ. Sci. Technol. 44, 9564–9570.

Caldwell, K.L., Makhmudov, A., Ely, E., Jones, R.L., Wang, R.Y., 2011. Iodine status of the U.S. population, National Health and Nutrition Examination Survey, 2005–2006 and 2007–2008. Thyroidology 21, 419–427.

Davis, P.L., Davis, F.B., 1996. Nongenomic actions of thyroid hormone. Thyroidology 6, 497–504.

Delom, F., Mallet, B., Carayon, P., Lejeune, P.J., 2001. Role ofextracellular molecular chaperones in the folding of oxidized proteins. Refolding of colloidal thyroglobulin by protein disulfide isomerase and immunoglobulin heavy chain-binding protein. J. Biol. Chem. 276, 21337–21342.

Dunn, A.D., Corsi, C.M., Myers, H.E., Dunn, J.T., 1998. Tyrosine 130 is an important outer ring donor for thyroxine formation in thyroglobulin. J. Biol. Chem. 273, 25223–25229.

Dunn, J.T., Dunn, A.D., 1999. The importance of thyroglobulin structure for thyroid hormone biosynthesis. Biochimie 81, 505–509.

Hannan, M.A., Faraji, B., Tanguma, J., Longoria, N., Rodriguez, R.C., 2009. Maternal milk concentration of zinc, iron, selenium, and iodine and its relationship to dietary intakes. Biol. Trace Elem. Res. 127, 6–15.

Hetzel, B.S., Potter, B.J., Dulberg, E.M., 1990. The iodine deficiency disorders: nature, pathogenesis and epidemiology. World Rev. Nutr. Dieta 62, 59–119.

Howie, A.E., Arthur, J.R., Nicol, E., Walker, S.W., Beech, S.G., Beckett, G.J., 1998. Identification of a 57-kilodalton selenoprotein in human thyrocytes as thioredoxin reductase and evidence that its expression is regulated through the calcium-phosphoinositol signaling pathway. J. Clin. Endocrinol. Metab. 83, 2052–2058.

Hurrell, R.E., 1997. Bioavailability of iodine. Eur. J. Clin. Nutr. 51, S9–S12.

Klein, M., Gestmann, I., Berndorfer, U., Schmitz, A., Herzog, V., 2000. The thioredoxin boxes of thyroglobulin: possible implications for intermolecular disulfide bond formation in the follicle lumen. Biol. Chem. 381, 593–601.

Lacroix, L., Mian, C., Caillou, B., Talbot, M., Filetti, S., Schlumberger, M., et al., 2001. Na(+)/I(−) symporter and Pendred syndrome gcnc and protein expressions in human extrathyroidal tissues. Eur. J. Endocrinol. 144, 297–302.

Mitchell, A.M., Manley, S.W., Morris, J.C., Powell, K.A., Bergert, E.R., Mortimer, R.H., 2001. Sodium iodide symporter (NIS) gene expression in human placenta. Placenta 22 (2–3), 256–258.

Murakami, M., Araki, O., Hosoi, Y., Kamiya, Y., Morimura, T., Ogiwara, T., et al., 2001. Expression and regulation of type II iodothyronine deiodinase in human thyroid gland. Endocrinologica 142, 2961–2967.

Nath, S.K., Moinier, B., Thuillier, E., Rongier, M., Desjeux, J.E., 1992. Urinary excretion of iodide and fluoride from supplemented food grade salt. Int. J. Vit. Nutr. Res. 62, 66–72.

Nicola, J.P., Basquin, C., Portulano, C., Reyna Neyra, A., Paroder, M., Carrasco, N., 2009. The Na+/I− symporter mediates active iodide uptake in the intestine. Am. J. Physiol. Cell. Physiol. 296, C654–C662.

Nicola, J.P., Reyna-Neyra, A., Carrasco, N., Masini-Repiso, A.M., 2012. Dietary iodide controls its own absorption through post-transcriptional regulation of the intestinal Na+/I− symporter. J. Physiol. 590 (Pt 23), 6013–6026.

Pavelka, S., Babický, A., Lener, J., Vobecký, M., 2002. Impact of high bromide intake in the rat dam on iodine transfer to the sucklings. Food Chem. Toxicol. 40, 1041–1045.

Pohlenz, J., Refetoff, S., 1999. Mutations in the sodium/iodide symporter (NIS) gene as a cause for iodide transport defects and congenital hypothyroidism. Biochimie 81, 469–476.

Riedel, C., Levy, O., Carrasco, N., 2001. Post-transcriptional regulation of the sodium/iodide symporter by thyrotropin. J. Biol. Chem. 276, 21458–21463.

Rohner, F., Zimmermann, M., Jooste, P., Pandav, C., Caldwell, K., Raghavan, R., et al., 2014. Biomarkers of nutrition for development—iodine review. J. Nutr. 144, 1322S–1342S.

Royaux, I.E., Wall, S.M., Kamiski, L.P., Everett, L.A., Suzuki, K., Knepper, M.A., et al., 2001. Pendrin, encoded by the Pendred syndrome gene, resides in the apical region of renal intercalated cells and mediates bicarbonate secretion. Proc. Natl. Acad. Sci. USA 98, 4221–4226.

Scott, D.A., Wang, R., Kreman, T.M., Sheffield, V.C., Karniski, L.P., 1999. The Pendred syndrome gene encodes a chloride-iodide transport protein. Nat. Genet 21, 440–443.

Shennan, D.B., 2001. Iodide transport in lactating rat mammary tissue via a pathway independent from the Na+/I− cotransporter: evidence for sulfate/iodide exchange. Biochem. Biophys. Res. Commun. 280, 1359–1363.

Soleimani, M., Greeley, T., Petrovic, S., Wang, Z., Amlal, H., Kopp, E., et al., 2001. Pendrin: an apical Cl−/OH−/HCO3− exchanger in the kidney cortex. Am. J. Physiol. Renal Fluid Electrolyte Physiol. 280, F356–F364.

Spitzweg, C., Dutton, C.M., Castro, M.R., Bergert, E.R., Goellner, J.R., Heufelder, A.E., et al., 2001. Expression of the sodium iodide symporter in human kidney. Kidney Int. 59, 1013–1023.

Steinmaus, C., Miller, M.D., Cushing, L., Blount, B.C., Smith, A.H., 2013. Combined effects of perchlorate, thiocyanate, and iodine on thyroid function in the National Health and Nutrition Examination Survey 2007–08. Environ. Res. p11, S0013–S9351.

Van Middlesworth, L., 1985. Biologically available iodine in goitrogenic diets. Exp. Biol. Med. 178, 610–615.

Wahl, R., Pilz-Mittenburg, K.W., Heer, W., Kallee, E., 1995. [Iodine content in diet and excretion of iodine in urine]. [German]. Z. Ern. 34, 269–276.

Watanabe, M., Houten, S.M., Mataki, C., Christoffolete, M.A., Kim, B.W., Sato, H., et al., 2006. Bile acids induce energy expenditure by promoting intracellular thyroid hormone activation. Nature 439 (7075), 484–489.

Yoshida, A., Hisatome, I., Taniguchi, S., Sasaki, N., Yamamoto, Y., Miake, J., et al., 2004. Mechanism of iodide/chloride exchange by pendrin. Endocrinology 145, 4301–4308.

FLUORINE

Fluorine (atomic weight 18.998) is a halogen. Almost all of naturally occurring fluorine consists of the stable isotope ^{19}F with small accompanying traces of the unstable isotope ^{18}F (half-life 109.77 min). The term *fluoride* refers to the salts of the element fluorine. The body utilizes only those salts. Other fluorine-containing compounds occur as potentially harmful industrial contaminants.

ABBREVIATION

F   fluorine

NUTRITIONAL SUMMARY

Function: Fluoride increases the resistance of teeth to dental caries and may help retain bone minerals in old age.

Requirements: Fluoride is not an essential mineral, but daily intake of 0.05 mg/kg body weight greatly reduces the risk of dental caries in children and adults. Topical applications (toothpaste, tooth sealants, mouthwash, and dental floss) also promote the remineralization of teeth.

Food sources: Major dietary sources of fluoride are naturally or artificially fluoridated water, seafood, tea, and unintentional ingestion of dental products. Typical daily intake by adults in the United States has been estimated to range from 1.2 to 2.4 mg.

Deficiency: Contrary to earlier assumptions, fluoride has not been shown to be an essential trace mineral. Benefits for dental and bone health, however, exist under most conditions.

Excessive intake: Daily doses of 0.1 mg/kg during tooth development, which is only twice the recommended intake, cause mild to moderate tooth mottling (brown spotty discoloration) in children, but not in adults. Acute poisoning with nausea, vomiting, diarrhea, salivation, sweating, headache, and generalized weakness, hypotension, renal failure, severe hypocalcemia and hyperkalemia, cardiac arrhythmia, coma, and death may occur with less than 5 mg/kg in children, a dose that could result from ingesting fluoridated toothpaste. Calcium solutions, including milk, can be used to slow absorption in an emergency situation. Long-term exposure to unusually large amounts of fluoride (20 mg/day) may cause abnormal bone and ligament calcification.

DIETARY SOURCES

Major sources are fluoridated water and dietary supplements, toothpaste, seafood, and tea. Most toothpastes contain 1 mg/g, but prescription strength products may contain as much as 5 mg/g. The water in some world regions naturally has a fluoride content well in excess of 1 mg/l (Singh et al., 2001). Bottled Vichy water contains 5 mg/l; most other bottled waters contain much less. Water purification systems, with the exception of reverse osmosis and deionization systems, remove little fluoride from water (Fawell et al., 2006).

DIGESTION AND ABSORPTION

Ingested fluorides are absorbed from the oral cavity, stomach, and intestines. About half of the ingested fluoride is absorbed at both low and high intake levels (Maguire et al., 2007). High acidity promotes absorption, particularly from the stomach (Whitford and Pashley, 1984). It might be assumed that fluoride salts are absorbed with moderate efficiency, since about half of ingested fluoride is recovered via urine (Nath et al., 1992; Whitford, 2000). Fluoride in solid foods has a much lower bioavailability, especially in vegetarian diets (Goyal et al., 1998). Chloride and nitrate compete with fluoride for uptake. High calcium intake decreases fluoride absorption and may even lead to net secretion of fluoride into the intestinal lumen (Whitford, 1997). Several routes for intestinal fluoride uptake are likely to exist (He et al., 1998). So long as there is an inward proton gradient (luminal pH well below 7.4), a carrier-mediated process that involves cotransport of F with H^+ or exchange of F with OH- takes up the undissociated acid. This or a parallel carrier is an anion exchanger that is similar but not identical to SLC4A1 (band 3 protein) which otherwise mediates exchange of chloride, bicarbonate, sulfate, and glucose (Gimenez et al., 1993). Another access route may be a conductance pathway traveled in tandem with a potassium ion. Little is known about the processes that govern fluoride export across the basolateral enterocyte membrane.

TRANSPORT AND CELLULAR UPTAKE

Blood circulation: Plasma concentrations tend to be around 0.9 μmol/l, higher in people with high intake. Fluoride appears to share some transport pathways with chloride, but the exact nature of the responsible carriers remains to be resolved. Concentrations in ductal saliva are slightly lower than in plasma (Whitford et al., 1999).

Materno-fetal transfer: Concentrations in fetal blood tend to be slightly lower than in maternal blood (Malhotra et al., 1993) and respond to increases in maternal blood concentration (Forestier et al., 1990). The placenta does not constitute a significant barrier at low to moderate maternal blood concentrations, but it limits transfer at maternal concentrations above 20 μmol/l (Gupta et al., 1993).

STORAGE

Calcified tissues contain almost all of the body's fluoride. Bone typically contains between 0.6 and 1.5 g/kg of fluoride (Hać et al., 1997). About half of the ingested fluoride is deposited in mineralized tissues. Connective tissue contains fluoride bound to glycans.

EXCRETION

Nearly 50% of newly ingested fluoride is eliminated rapidly through the kidneys; another 10–20% is lost with feces (Whitford, 2000). Typical fluoride concentrations in urine from individuals with moderate intake are around 50 μmol/l. The efficiency of fluoride reabsorption from renal ultrafiltrate increases with urine acidity (Whitford 1997). The distal nephron, including the collecting ducts, is a major site of fluoride reabsorption, possibly through a pH-driven mechanism similar to the one operating in the small intestinal brush border membrane (Whitford & Pashley, 1991).

REGULATION

Fluoride concentrations in the blood or tissues are not homeostatically regulated. Fractional absorption and retention of dietary fluoride decrease with increasing plasma fluoride concentration in rats

(Whitford, 1994). However, amount or concentration of fluoride in the body or specific tissues does not seem to be maintained at a particular level.

FUNCTION

Bone and joints: Fluoride stimulates new bone growth (Farley et al., 1983). Whether low-dose fluoride intake reduces bone mineral loss in older people remains a contentiously debated issue because the newly formed bone may have a less optimal structure and be less resistant to fracture. A very high tissue concentration of fluoride (2 mmol/l) inhibits acid phosphatase in chondrocytes and bone matrix (Roach, 1999). Fluoride has also been suggested to inhibit a particular phosphotyrosine phosphatase in osteoblasts. This could promote tyrosine phosphorylation in key signaling proteins of the mitogenic transduction pathway and growth factor–dependent proliferation of bone cells (Lau & Baylink, 1998).

Teeth: Some fluoride is firmly bound to tooth minerals and forms the hard composite mineral fluorapatite. Topical application of fluorides can accelerate the remineralization of teeth. Fluoride also affects the ability of bacteria to attach through glucan-binding lectins (Cox et al., 1999) and may reduce acid production by bacteria.

High intake: Excessive intake causes a characteristically mottled appearance of teeth, but only during tooth development in children (Whitford, 1997). It has been suggested that high natural fluoride exposure (several milligrams per day) increases the risk of kidney stone formation (Singh et al., 2001).

REFERENCES

Cox, S.D., Lassiter, M.O., Miller, B.S., Doyle, R.J., 1999. A new mechanism of action of fluoride on streptococci. Biochim. Biophys. Acta 1428, 415–423.

Farley, J.R., Wergedal, J.E., Baylink, D.J., 1983. Fluoride directly stimulates proliferation and alkaline phosphatase activity of bone-forming cells. Science 222, 330–332.

Fawell, J, Bailey, K, Chilton, J, Dahi, E, Fewtrell, L, Magara, Y., 2006. Fluoride in Drinking-water. World Health Organization (WHO), IWA Publishing, London, UK, ISBN 1900222965.

Forestier, E., Daffos, F., Said, R., Brunet, C.M., Guillaume, P.N., 1990. Passage transplacentaire du fluor. Etude in utero. J. Gynecol. Obstet. Biol. Reprod. 19, 171–175.

Gimenez, I., Garay, R., Alda, J.O., 1993. Molybdenum uptake through the anion exchanger in human erythrocytes. Pfl. Arch. Eur. J. Physiol. 424, 245–249.

Goyal, A., Gauba, K., Tewari, A., 1998. Bioavailability of fluoride in humans from commonly consumed diets in India. J. Ind. Soc. Pedodont. Prev. Dent. 16, 1–6.

Gupta, S., Seth, A.K., Gupta, A., Gavane, A.G., 1993. Transplacental passage of fluorides. J. Pediatr. 123, 139–141.

Hác, E., Czarnowski, W., Gos, T., Krechniak, J., 1997. Lead and fluoride content in human bone and hair in the Gdańsk region. Sci. Total Environ. 206, 249–254.

He, H., Ganapathy, V., Isales, C.M., Whitford, G.M., 1998. pH-dependent fluoride transport in intestinal brush border membrane vesicles. Biochim. Biophys. Acta 1372, 244–254.

Lau, K.H., Baylink, D.J., 1998. Molecular mechanism of action of fluoride on bone cells. J. Bone Min. Res. 13, 1660–1667.

Maguire, A., Zohouri, F.V., Hindmarch, P.N., Hatts, J., Moynihan, P.J., 2007. Fluoride intake and urinary excretion in 6- to 7-year-old children living in optimally, sub-optimally and non-fluoridated areas. Community Dent. Oral Epidemiol. 35, 479–488.

Malhotra, A., Tewari, A., Chawla, H.S., Gauba, K., Dhall, K., 1993. Placental transfer of fluoride in pregnant women consuming optimum fluoride in drinking water. J. Ind. Soc. Pedod. Prev. Dent. 11, 1–3.

Nath, S.K., Moinier, B., Thuillier, E., Rongier, M., Desjeux, J.E., 1992. Urinary excretion of iodide and fluoride from supplemented food grade salt. Int. J. Vit. Nutr. Res. 62, 66–72.

Roach, H.I., 1999. Association of matrix acid and alkaline phosphatases with mineralization of cartilage and endochondral bone. Histochem. J. 31, 53–61.

Singh, P.R., Barjatiya, M.K., Dhing, S., Bhatnagar, R., Kothari, S., Dhar, V., 2001. Evidence suggesting that high intake of fluoride provokes nephrolithiasis in tribal populations. Urol. Res. 29, 238–244.

Whitford, G.M., 1994. Effects of plasma fluoride and dietary calcium concentrations on GI absorption and secretion of fluoride in the rat. Calcif. Tiss. Int. 54, 421–425.

Whitford, G.M., 1997. Determinants and mechanisms of enamel fluorosis. Ciba Found. Symp. 205, 226–241.

Whitford, G.M., Pashley, D.H., 1984. Fluoride absorption: the influence of gastric acidity. Calcif. Tiss. 36, 302–307.

Whitford, G.M., Pashley, D.H., 1991. Fluoride reabsorption by nonionic diffusion in the distal nephron of the dog. Proc. Soc. Exp. Biol. Med. 196, 178–183.

Whitford, G.M., Thomas, J.E., Adair, S.M., 1999. Fluoride in whole saliva, parotid ductal saliva and plasma in children. Arch. Oral Biol. 44, 785–788.

SULFUR

Sulfur (sulphur, brimstone; atomic weight 32.064) is a nonmetallic element with valences 2, 4, and 6. The natural isotopes ^{32}S (95%), ^{33}S (0.76%), ^{34}S (4.2%), and ^{36}S (0.014%) are stable. The radioactive isotopes 29, 30, 31, 35, 37, and 38 can be produced artificially.

The least oxidized simple sulfur-containing molecule is hydrogen sulfide (H_2S). Hydration of sulfur dioxide (SO_2) gives rise to sulfurous acid (H_2SO_3) and the hydrogen sulfite (bisulfite, HSO_3^-) and sulfite (HSO_3^{2-}) ions. Sulfur trioxide (SO_3) is the precursor of sulfuric acid (H_2SO_4) and the bisulfate (HSO_4^-) and sulfate (SO_4^{2-}) ions. Dithionate ($S_2O_6^{2-}$), disulfite (metabisulfite, $S_2O_5^{2-}$), dithionite ($S_2O_4^{2-}$), and thiosulfate ($S_2O_3^{2-}$) are the ions of related sulfur oxides. Thiocyanate (SCN−) is an oxygen-free mixed sulfur ion. Numerous complex organic compounds, including the amino acids methionine, cysteine, and taurine, the vitamins thiamin and biotin, and the metabolic cofactor lipoic acid contain sulfur in various configurations.

ABBREVIATIONS

APS	adenosine 5′-phosphosulfate
DTDST	diastrophic dysplasia sulfate transporter
PAPS	3′-phosphoadenosine 5′-phosphosulfate

NUTRITIONAL SUMMARY

Function: The sulfur amino acids methionine and cysteine are necessary for the synthesis of proteins and serve as precursors of important cofactors and metabolites. Development and maintenance of brain and nerves, spermatogenesis, joint repair, hormone action, and many other body functions are critically dependent on undisturbed sulfate metabolism. Sulfate is an essential constituent of many proteins, glycans, and glycolipids, and plays an important role in the activation and elimination of hormones, metabolites, phytochemicals, and xenobiotics. Various other sulfurous compounds, such as taurine, thiamin, biotin, and thiosulfate, have specific functions in the body.

Food sources: Most utilizable sulfur comes with sulfur amino acids in proteins of animal and plant origin. Much smaller amounts are consumed as sulfate. Some sulfur compounds, such as the potentially toxic hydrogen sulfide, arise when bacterial in the distal intestine act on nonabsorbed amino acids.

Requirements: Daily sulfur needs are met when adequate amounts (13 mg/kg) of methionine and cysteine are consumed (Young & Borgonha, 2000).

Deficiency: Since symptoms of methionine and cysteine deficiency will occur with inadequate intake, the consequences of isolated sulfur deficiency cannot be determined. Experience with the rare patients with defective sulfate transporters indicates a particular vulnerability of the brain and connective tissue.

Excessive intake: Excessive sulfate intake may accelerate bone mineral loss and increase the risk of osteoporosis. Exposure of sensitive individuals to large doses of sulfite (>100 mg) can trigger asthma attacks, urticaria, and related symptoms.

ENDOGENOUS SOURCES

The catabolism of the sulfur amino acids methionine, cysteine, cystine, and taurine generates significant quantities of sulfite, most of which is rapidly converted to sulfate. Typical combined intake is in excess of 40 mg/kg, and most of this is eventually broken down. Thus, more than 75% of ingested methionine is converted to sulfate within a few hours (Hamadeh & Hoffer, 2001).

Cysteine is broken down in successive steps catalyzed by cysteine dioxygenase (EC1.13.11.20, contains iron and FAD), cytosolic aspartate aminotransferase (EC2.6.1.1, requires PLP) to beta-sulfinylpyruvate. This intermediate spontaneously releases sulfite and pyruvate. Aspartate 4-decarboxylase (EC4.1.1.12) can generate sulfite from cysteinesulfinate.

Endogenous secretion of sulfate into the upper gastrointestinal tract amounts to about 0.96–2.6 mmol/day (Florin et al., 1991).

DIETARY SOURCES

Many sulfur compounds, including inhaled hydrogen sulfide, are toxic. Only a few of the numerous sulfur compounds in nature serve as significant precursors for human sulfur homeostasis. Among these the amino compounds methionine, cysteine, homocysteine, and taurine are most notable, because they contribute to the synthesis of the sulfate donor 3′-phosphoadenosine 5′-phosphosulfate (PAPS) on top of their function as precursors for proteins and other specific compounds.

Sulfites are added to some foods (especially wines and dried fruit) as preservatives, antioxidants, and bleaching agents. US regulations allow the use of sulfur dioxide, sodium sulfite, sodium bisulfite, potassium bisulfite, sodium metabisulfite, and potassium metabisulfite for food preservation. Numerous pharmaceutical products contain significant amounts of sulfites (Miyata et al., 1992). Wine and cider drinkers in France in one survey consumed an average 31.5 mg/day sulfite; those who did not drink alcoholic beverages had an average intake of 2 mg/day (Mareschi et al. 1992).

Some sulfate, including glucosamine sulfate and the food additive sodium dodecyl sulfate, is consumed directly. The sulfate content of bread is 1.5 mg/g, potatoes contain 0.3 mg/g, broccoli 0.9 mg/g, and wine 0.36 mg/g. Sulfate concentration in tap water varies widely, typically between a few milligrams to a few hundred milligrams per liter. The content of sulfate in bottled waters varies hugely. The mineral waters with especially high sulfate concentration include Eptinger (1630 mg/l), Valser (990 mg/l), Aproz (934 mg/l), Vittel, Hepar, San Pellegrino, Rietenauer, and Contrex.

Epsom salt is magnesium sulfate concentrate derived from the waters of the Epsom spa that has traditionally been used for medicinal purposes.

Glycosinolates from brassica species (cauliflower, broccoli, Brussel sprouts, cabbages, watercress, mustards, horse radish, and radishes) and from allium species (onions, garlic, leeks, and chives) release various sulfur compounds with a strong sulfurous odor. These relatively volatile odorants, such as methanethiol, dimethyl sulfide, and dimethyltrisulfide, do not contribute significantly to total sulfate intake.

While a low-sulfate diet may provide as little as 1.6 mmol/day, most people get much more. The amount of all sulfur compounds combined that reach the colon may be as much as 7 mmol/day (Florin et al., 1991).

INTESTINAL ABSORPTION

Most of the ingested sulfate up to doses of 1 g and more is absorbed form both the small and large intestine. Sulfate is taken up from the small intestinal lumen through the sodium/sulfate symporter (SLC13A1) in the brush-border membrane (Puttaparthi et al., 1999). In the ileum, sulfate can be taken up by the sodium/sulfate symporter (SLC13A1, NaSi1). A pH gradient-driven carrier-mediated SO_4^{2-} / OH^- exchange process operates in the human proximal colonic luminal membrane (Tyagi et al., 2001). Several members of the SLC26 family contribute to the transport of sulfate from the intestinal lumen. SLC26A2 (DTDST = diastrophic dysplasia sulfate transporter), SLC26A3 (CLD/DRA), and SLC26A4 (PDS) accept a wide range of small anions in addition to sulfate, including oxalate, chloride, iodine, bicarbonate, and hydroxyl anions (Lohi et al., 2000; Markovich, 2001).

The sodium dicarboxylate/sulfate transporter 2 (SDCT2, SLC13A3) exchanges sulfate against succinate or citrate plus sodium across the basolateral membrane (Chen et al., 1999).

Nearly one of two healthy adults has sulfate-reducing bacteria (SRB, *Desulfovibrio desulfuricans*) in the colon, which generate sulfide (Pitcher et al., 2000). Colonic sulfide production competes with methane production for hydrogen. High sulfide production from sulfur amino acids may explain the absence of CH_4 in the breath of many people in western populations (Pitcher et al., 1998). SAM-dependent conversion of 2-mercaptoethanol to *S*-methyl-2-mercaptoethanol is mediated by thiol methyltransferase (EC2.1.1.9, Pitcher et al., 1998).

TRANSPORT AND CELLULAR UPTAKE

Blood circulation: Sulfate is the fourth most abundant anion in human plasma, with fasting concentrations around 300 μmol/l (Hamadeh & Hoffer, 2001). Intake of sulfur amino acids or sulfate can increase plasma levels nearly twofold. Typical sulfite concentrations in serum are around 5 μmol/l with a reference range of 0–10 μmol/l (Ji et al., 1995).

Several transporters with diverse tissue expression patterns mediate the transfer of sulfate into or out of cells. Erythrocytes, leukocytes, fibroblasts, vascular smooth muscle, liver, and other tissues take up and release sulfate along with a proton in exchange for a chloride ion via SLC4A1 (band 3 protein), which otherwise mediates exchange of chloride and bicarbonate (Gimenez et al., 1993; Jennings, 1995; Chernova et al., 1997). The sulfate transporter SLC26A2 is critical for the production of sulfated proteoglycans in cartilage matrix (genetic defects cause a clinical phenotype of diastrophic dysplasia). Sulfate transporter 1 (SUT1, SLC13A4) is a sodium-dependent system in a few specialized tissues, including the heart, testes, and placenta. Its abundant presence in high endothelial venules is closely linked to the accumulation of sulfate at these vascular sites associated with lymphocyte extravasation into the extravascular space and lymphatic vessels. Sulfate is also supplied to these endothelia via the sulfate anion exchanger SLC26A7 (Vincourt et al., 2002).

Materno-fetal transfer: Both inorganic sulfate and sulfur amino acids (methionine, cysteine, cystine, and taurine) can cross the placenta in either direction, which is essential to maintain adequate sulfur supplies while avoiding excess. Sulfate anions exchange with bicarbonate at the brush-border membrane (on the maternal-facing side of the placenta; Grassl, 1996).

The sulfate transporter 1 (SUT1, SLC13A4) is abundantly expressed in the placenta, possibly at the fetal side of the syntrophoblast.

BBB: Cystine crosses into the brain with the help of the high-affinity system x(-)C (SLC7A11 plus SLC3A2). Intracerebral catabolism can then generate sulfate, though the released amounts are uncertain.

METABOLISM

Hydrogen sulfide: This small molecule is a key mediator and its tissue concentrations are tightly regulated (Kimura, 2014). The reactions catalyzed by cystathionine β-synthase (CBS, EC4.2.1.22), cystathionine γ-lyase (CSE, EC4.4.1.1), and 3-mercaptopyruvate sulfurtransferase (3MST, EC2.8.1.2) generate hydrogen sulfide. A series of other reactions in mitochondria metabolize hydrogen sulfide to a rich spectrum of compounds, including polysulfides, thiosulfate, and sulfate. Sulfide:quinone oxidoreductase (SQR, EC1.8.5.4) binds the sulfide and oxidizes it while transferring two electrons to the ubiquinone. Sulfur dioxygenase (EC1.13.11.18) then acts on the SQR-bound persulfide and oxidizes it to sulfite. Thiosulfate sulfurtransferase (rhodanese, EC2.8.1.1) finally combines the SQR-bound persulfide with sulfite and releases it as thiosulfate.

Sulfide quinone oxidoreductase (SQR, EC1.8.5.4) can generate polysulfides, which are compounds with 2–8 sulfur atoms in sequence: n HS- + n quinone $\rightleftharpoons$ polysulfide + n quinol . The polysulfide with 8 sulfur atoms forms a ring and does not accept additional sulfides.

SQR can also reduce sulfide to sulfur (sulfide + ubiquinone-1 $\rightleftharpoons$ sulfur + ubiquinol-1). The addition of sulfite to the sulfide molecule generates thiosulfate (sulfide + sulfite + ubiquinone-1 $\rightleftharpoons$ thiosulfate + ubiquinol-1). All of these SQR-mediated reactions occur in the mitochondria.

Thiocyanate: SQR catalyzes the condensation of sulfide with cyanide, which generates thiocyanate (sulfide + cyanide + ubiquinone-1 $\rightleftharpoons$ thiocyanate + ubiquinol-1). The mitochondrial enzyme thiosulfate sulfurtransferase (rhodanese, EC2.8.1.1) adds a sulfur group from thiosulfate to cyanide and thereby generates sulfite and thiocyanate.

These reactions are important for the detoxification of cyanide, especially in the case of high dietary exposure (Spencer, 1999). The zinc-enzyme 3-mercaptopyruvate sulfurtransferase (EC2.8.1.2) catalyzes the same reaction. This enzyme also accepts 3-mercaptopyruvate and cysteinesulfinate (from cysteine metabolism) as sulfur donors.

$$\text{Thiosulfate + cyanide} \quad \overset{\text{Rhodanese or 3MST}}{\rightleftharpoons} \quad \text{thiocyanate + sulfite}$$

Sulfite: The metabolism of methionine, cysteine, and a few quantitatively less significant compounds, such as thiosulfate, generates sulfite. Additional quantities are consumed directly with food.

Sulfur dioxygenase (EC1.13.11.18) can oxidize sulfur to sulfite. This iron-containing enzyme sulfite uses catalytically active glutathione, which is not oxidized or otherwise consumed in the reaction.

The flavoprotein sulfite oxidase (EC1.8.3.1) uses oxygen and water to oxidize sulfite to sulfate. The reaction releases hydrogen peroxide and relies on cytochrome c as the electron acceptor. The enzyme, which contains molybdenum cofactor and heme, resides in the mitochondrial intermembrane space and is thus readily accessible for cytosolic sulfite.

Sulfate activation: Reactions that attach sulfate to proteins or other complex molecules use PAPS as a sulfate donor. PAPS synthesis proceeds in two distinct steps. Sulfate adenylyltransferase (EC2.7.7.4) links sulfate to the adenylyl phosphate moiety of ATP and releases pyrophosphate. Adenylylsulfate kinase (EC2.7.1.25) then adds a phosphate in the 3' position. In humans, both activities are combined in the bifunctional proteins 3'-phosphoadenosine 5'-phosphosulfate synthethase 1 (PAPSS1; highly active in the brain) and 2 (PAPSS2; in the liver). PAPSS1 localizes strongly to the cell nucleus. Adenylylsulfatase (EC3.6.2.1) and phosphoadenylylsulfatase (EC3.6.2.2, manganese-dependent) reverse these synthesis steps and release sulfate again.

STORAGE

Small amounts of sulfate are stored in tissues as ascorbate sulfate. Alcohol sulfotransferase (EC2.8.2.2, requires divalent iron, manganese, or cobalt for activation) uses PAPS to modify ascorbate. Arylsulfatase A (EC3.1.6.1) releases sulfate from ascorbate sulfate.

EXCRETION

Fecal losses of sulfur compounds are small (<0.5 mmol/day) compared to urinary excretion (Florin et al., 1991). Typical total sulfur excretion via urine is around 1.3 g/day (Komarnisky et al., 2003) but may be more with high intake. About 15% of the urinary losses are in the form of organic sulfate esters (e.g., hormones, flavonoids, and xenobiotics). Most of the remainder is excreted as sulfate (Hamadeh & Hoffer, 2001).

More than 5 g of sulfate is filtered daily in the kidneys. The renal proximal tubules recover most of this amount through the sodium/sulfate symporter (SLC13A1, NaSi1) in the brush border membrane of epithelial cells of proximal tubules (Puttaparthi et al., 1999). Vitamin D status influences sulfate uptake from the proximal tubular lumen (cf. P.1522, Markovich, 2001). The sodium dicarboxylate/sulfate transporter 2 (SDCT2) then extrudes sulfate across the basolateral membrane into the perivascular space in exchange for succinate or citrate plus sodium (Chen et al., 1999).

A parallel mode of export across the basolateral membrane of epithelial cells in proximal tubules proceeds through the sodium-independent transporter Sat-1 in exchange for bicarbonate or oxalate (Beck & Silve, 2001).

Two isoforms of another sulfate/anion exchanger, SLC26A7, are also expressed in kidneys, but their precise location is not yet known (Vincourt et al., 2002).

REGULATION

Adaptation of renal reabsorption appears to be an important mechanism for the maintenance of sulfate homeostasis (Pena & Neiberger, 1997).

Several factors, including intake, aging, and pregnancy, are known to influence renal sulfate recovery via the sodium/sulfate symporter and thereby determine renal losses (Morris & Sagawa, 2000).

FUNCTION

Hydrogen sulfide: This small molecule is an important signaling compound that affects neuronal function, cellular integrity, modulation of inflammatory reactions, vascular tone, and other key homeostatic functions (Kimura, 2014).

Polysulfides have recently gained attention for their importance in the brain and other tissues, where they activate transient receptor potential channels even more potently than hydrogen sulfide. They also promote the expression of numerous genes. An important example is the sulfide-dependent sulfuration of Kelch-like ECH-associated protein 1 (Keap1), which in turn releases Nrf2 and thereby promotes the expression of antioxidant genes (Kimura, 2014).

Cyanide detoxification: Small amounts of cyanide are consumed with certain glycosides in food that are cleaved during digestion by beta-glucosidase (EC3.2.1.21). Various seeds contain amygdalin, prunasin, and other cyanogenic glycosides (Bolarinwa et al., 2015). Examples include bitter almonds and apple seeds. The only commonly consumed food with cyanogenic glycosides in significant amounts is cassava (*Manihot esculenta*). This tuber contains linamarin, from which cyanide can be released by hydroxynitrile lyase (EC4.1.2.47) from the plant itself or by bacterial beta-glucosidase (EC3.2.1.21) during food fermentation or intestinal digestion.

Sulfide quinone oxidoreductase (EC1.8.5.4) in mitochondria combines cyanide with sulfide and reduces the aggregate to thiocyanate. Thiosulfate sulfurtransferase (rhodanese, EC2.8.1.1) is another mitochondrial enzyme that chemically sequesters cyanide. In this case, thiosulfate is bound to cyanide and thiocyanate plus sulfite are then released. Yet another enzyme, 3-mercaptopyruvate sulfurtransferase (3MST, EC2.8.1.2), combines cyanide with 3-mercaptopyruvate and generates thiocyanate plus pyruvate.

Thiocyanate: Thiocyanate ($H_2S_2O_3$) selectively promotes the action of glutamate at neuronal glutamate receptors of the (RS)-alpha-amino-3-hydroxy-5-methyl-isoxazole-4-propionic acid (AMPA) subclass. Myeloperoxidase (EC1.11.2.2) and other peroxidases use thiocyanate to generate the potent free radical hypothianate ($OSCN^-$) for antimicrobial defense and in inflammatory environments (Rees et al., 2014).

Increased thiocyanate production in excess of an individual's capacity for its removal by excretion or metabolism may be responsible for persistent spastic weakness of the legs and degeneration of corresponding corticospinal pathways (konzo, which means "tied legs" in English) in people with habitual consumption of cyanogenic plants such as cassava (Boivin et al., 2013).

Antioxidant defense: The glutathione precursors *N*-acetylcysteine (NAC) and thioproline have important antioxidant properties of their own. NAC is a highly effective water-soluble antioxidant with low toxicity. The compound is commonly used to thin and break up mucus, particularly in patients with cystic fibrosis (Vasu et al., 2011).

Glycans: Sulfate groups endow various complex sugar polymers with their characteristic properties. Chondroitin (in cartilage and many other tissues) contains sulfite groups linked to the 4 or 6 position of the *N*-acetyl galactosamin (GalNAc) moieties. Chondroitin 4-sulfotransferase (EC2.9.2.5) and chondroitin 6-sulfotransferase (EC2.8.2.17) mediate sulfation at their respective positions. Similarly, GalNAc 4-sulfate, GalNAc 6-sulfate, and L-iduronic acid 2-sulfate are typical for various dermatan sulfate species with ubiquitous distribution in the body. In heparan sulfate (in skin, arterial endothelium, and other sites), the sulfur group resides at position 2 of the acylated glucose moieties and at position 2 of L-iduronic acid. The closely related heparins (in mast cells) contain an even higher proportion of sulfate-substituted sugars. Keratan sulfates (present in cartilage, cornea, and numerous other tissues) are substituted at position 6 of both their glucose and N-acetylglucose constituents by keratan sulfotransferase (EC2.8.2.21).

Arylsulfatase E (EC3.1.6.1) in bone and cartilage removes the sulfate groups again from these polymers (Daniele et al., 1998). Another important enzyme with detoxifying, antioxidant properties is triglucosylalkylacylglycerol sulfotransferase (EC2.8.2.19).

Sulfolipids: Brain and nerve tissues contain a rich assortment of sulfoglycolipids. Sulfatides (galactosylceramide 3-sulfates) and seminolipids (monogalactosylalkylglycerol 3-sulfate) are two distinct types within this class of complex compounds (Honke et al., 2002). The vitamin K–dependent enzyme galactocerebroside sulfotransferase (EC2.8.2.11) mediates the sulfate transfer from PAPS to both types, while arylsulfatase (EC3.1.6.1, also vitamin K-dependent) catalyzes the release of the sulfate group (Sundaram & Lev, 1992). An important product of such reactions is galactocerebroside 3-sulfate.

Cell adhesion proteins: Various cell surface antigens interact with their targets only when they are sulfated. Examples are 6-sulfo N-acetyllactosamine glycoconjugates such as the lymphocyte homing receptor L-selectin and the blood group compound 6-sulfo sialyl Lewis X.

Catecholamines: The activity of catecholamines is modified by sulfation.

Steroid hormones: Estrone sulfotransferase (EC2.8.2.4), steroid sulfotransferase (EC2.8.2.15), cortisol sulfotransferase (EC2.8.2.18), and to some extent less specific sulfotransferases mediate the modification of a wide range of endogenous and synthetic steroid hormones and metabolites using 3′-phosphoadenylylsulfate as a sulfate donor.

Bile acids: Both conjugated (to glucuronate, taurine, or glycine) and free bile acids can be sulfated by bile-salt sulfotransferase (EC2.8.2.14).

Modification of xenobiotics: A wide range of endobiotics, heterocyclic amines, and other complex molecules are sulfated upon uptake into the small intestinal epithelium. For instance, about 10% of the soy isoflavone genistein is sulfated and appears to be the most effective biological form. The phenol-sulfating phenol sulfotransferase 1 (SULT1A3, EC2.8.2.1) uses 3′-phosphoadenylylsulfate to convert phenols into arylsulfates.

Gastrointestinal effects: Sulfate-rich waters have been used in spas for the treatment of ailments of the digestive tract. Ingestion of sulfate-rich water appears to stimulate the emptying of the gallbladder (Gutenbrunner et al., 2001).

Renal effects: High sulfate intake with certain natural mineral waters may induce diuresis and has been suggested as a mild adjunct to the treatment of urolithiasis and pyelonephritis. Urinary sulfate may inhibit uric acid crystallization, thereby reducing the risk of kidney stone formation.

Skeletal effects: High intake of sulfur-containing amino acids and sulfate (mineral water) has been linked to accelerated bone mineral loss.

Sulfite toxicity: Exposure of sensitive individuals to several hundred milligrams of sulfite can trigger asthma attacks with airway constriction, nasal congestion, urticaria, vasculitis (Wuthrich, 1993), and other skin manifestations. Sulfites readily react with nitric oxide and related compounds (Harvey & Nelsestuen, 1995). The resulting nitric oxide depletion may interfere with its signaling function, thereby causing some of the toxic effects of sulfite.

REFERENCES

Beck, L., Silve, C., 2001. Molecular aspects of renal tubular handling and regulation of inorganic sulfate. Kidney Int. 59, 835–845.

Boivin, M.J., Okitundu, D., Makila-Mabe Bumoko, G., Sombo, M.T., Mumba, D., Tylleskar, T., et al., 2013. Neuropsychological effects of konzo: a neuromotor disease associated with poorly processed cassava. Pediatrics 131 (4), e1231–e1239.

Bolarinwa, I.F., Orfila, C., Morgan, M.R., 2015. Determination of amygdalin in apple seeds, fresh apples and processed apple juices. Food Chem. 170, 437–442.

Chen, X., Tsukaguchi, H., Chen, X.Z., Berger, U.V., Hediger, M.A., 1999. Molecular and functional analysis of SDCT2, a novel sodium-dependent dicarboxylate transporter. J. Clin. Invest. 103, 1159–1168.

Chernova, M.N., Jiang, L., Crest, M., Hand, M., Vandorpe, D.H., Strange, K., et al., 1997. Electrogenic sulfate/chloride exchange in Xenopus oocytes mediated by murine AE1 E699Q. J. Gen. Physiol. 109, 345–360.

Daniele, A., Parenti, G., d'Addio, M., Andria, G., Ballabio, A., Meroni, G., 1998. Biochemical characterization of arylsulfatase E and functional analysis of mutations found in patients with X-linked chondrodysplasia punctata. Am. J. Human Genet. 62, 562–572.

Florin, T., Neale, G., Gibson, G.R., Christl, S.U., Cummings, J.H., 1991. Metabolism of dietary sulphate: absorption and excretion in humans. Gut 32, 766–773.

Gimenez, I., Garay, R., Alda, J.O., 1993. Molybdenum uptake through the anion exchanger in human erythrocytes. Pflugers Archiv. Eur. J. Physiol. 424, 245–249.

Grassl, S.M., 1996. Sulfate transport in human placental brush-border membrane vesicles. Biochim. Biophys. Acta 1282, 115–123.

Gutenbrunner, C., El-Cherid, A., Gehrke, A., Fink, M., 2001. Circadian variations in the responsiveness of human gallbladder to sulfate mineral water. Chronobiol. Int. 18, 1029–1039.

Hamadeh, M.J., Hoffer, L.J., 2001. Use of sulfate production as a measure of short-term sulfur amino acid catabolism in humans. Am. J. Physiol. Endocrinol. Metab. 280, E857–E866.

Harvey, S.B., Nelsestuen, G.L., 1995. Reaction of nitric oxide and its derivatives with sulfite: a possible role in sulfite toxicity. Biochim. Biophys. Acta 1267, 41–44.

Honke, K., Hirahara, Y., Dupree, J., Suzuki, K., Popko, B., Fukushima, K., et al., 2002. Paranodal junction formation and spermatogenesis require sulfoglycolipids. Proc. Nat. Acad. Sci. USA 99, 422–4232.

Jennings, M.L., 1995. Rapid electrogenic sulfate-chloride exchange mediated by chemically modified band 3 in human erythrocytes. J. Gen. Physiol. 105, 21–47.

Ji, A.J., Savon, S.R., Jacobsen, D.W., 1995. Determination of serum sulfite by HPLC with fluorescence detection. Clin. Chem. 41, 897–903.

Kimura, H., 2014. Hydrogen sulfide and polysulfides as biological mediators. Molecules 19, 16146–16157.

Komarnisky, L.A., Christopherson, R.J., Basu, T.K., 2003. Sulfur: its clinical and toxicologic aspects. Nutrition 19, 54–61.

Lohi, H., Kujala, M., Kerkela, E., Saarialho-Kere, U., Kestila, M., Kere, J., 2000. Mapping of five new putative anion transporter genes in human and characterization of SLC26A6, a candidate gene for pancreatic anion exchanger. Genomics 70, 102–112.

Mareschi, J.P., Francois-Collange, M., Suschetet, M., 1992. Estimation of sulphite in food in France. Food Addit. Contam. 9, 541–549.

Markovich, D., 2001. Physiological roles and regulation of mammalian sulfate transporters. Physiol. Rev. 81, 1499–1533.

Miyata, M., Schuster, B., Schellenberg, R., 1992. Sulfite-containing Canadian pharmaceutical products available in 1991. CMAJ Can. Med. Assoc. J. 147, 1333–1338.

Morris, M.E., Sagawa, K., 2000. Molecular mechanisms of renal sulfate regulation. Clin. Rev. Clin. Lab. Sci. 37, 345–388.

Pena, D.R., Neiberger, R.E., 1997. Renal brush border sodium-sulfate cotransport in guinea pig: effect of age and diet. Ped. Nephrol. 11, 724–727.

Pitcher, M.C., Beatty, E.R., Cummings, J.H., 2000. The contribution of sulphate reducing bacteria and 5-amino-salicylic acid to faecal sulphide in patients with ulcerative colitis. Gut 46, 64–72.

Pitcher, M.C., Beatty, E.R., Harris, R.M., Waring, R.H., Cummings, J.H., 1998. Sulfur metabolism in ulcerative colitis: investigation of detoxification enzymes in peripheral blood. Dig. Dis. Sci. 43, 2080–2085.

Puttaparthi, K., Markovich, D., Halaihel, N., Wilson, P., Zajicek, H.K., Wang, H., et al., 1999. Metabolic acidosis regulates rat renal Na–Si cotransport activity. Am. J. Physiol. 276, C1398–C1404.

Rees, M.D., Maiocchi, S.L., Kettle, A.J., Thomas, S.R., 2014. Mechanism and regulation of peroxidase-catalyzed nitric oxide consumption in physiological fluids: critical protective actions of ascorbate and thiocyanate. Free Radic. Biol. Med. 72, 91–103.

Silberg, D.G., Wang, W., Moseley, R.H., Traber, P.G., 1995. The Down regulated in Adenoma (dra) gene encodes an intestine-specific membrane sulfate transport protein. J. Biol. Chem. 270, 11897–11902.

Spencer, P.S., 1999. Food toxins, ampa receptors, and motor neuron diseases. Drug Metab. Rev. 31, 561–587.

Sundaram, K.S., Lev, M., 1992. Purification and activation of brain sulfotransferase. J. Biol. Chem. 267, 24041–24044.

Tyagi, S., Kavilaveettil, R.J., Alrefai, W.A., Alsafwah, S., Ramaswamy, K., Dudeja, P.K., 2001. Evidence for the existence of a distinct SO(4)(–)-OH(–) exchange mechanism in the human proximal colonic apical membrane vesicles and its possible role in chloride transport. Exp. Biol. Med. 226, 912–918.

Vasu, V.T., de Cruz, S.J., Houghton, J.S., Hayakawa, K.A., Morrissey, B.M., Cross, C.E., et al., 2011. Evaluation of thiol-based antioxidant therapeutics in cystic fibrosis sputum: focus on myeloperoxidase. Free Radic. Res. 45, 165–176.

Vincourt, J.B., Jullien, D., Kossida, S., Amalric, F., Girard, J.P., 2002. Molecular cloning of SLC26A7, a novel member of the SLC26 sulfat/anion transporter family, from high endothelial venules and kidney. Genomics 79, 249–256.

Wuthrich, B., 1993. Adverse reactions to food additives. Ann. Allergy 71, 379–384.

Young, V.R., Borgonha, S., 2000. Nitrogen and amino acid requirements: The Massachusetts Institute of Technology amino acid requirement pattern. J. Nutr. 130, 1841S–1849S.

SELENIUM

Selenium is a nonmetallic element (atomic weight 79). Naturally occurring selenium contains the stable isotopes ^{74}Se (0.87%), ^{76}Se (9.36%), ^{77}Se (7.63%), ^{78}Se (23.78%), and ^{80}Se (49.61%), and the unstable isotopes ^{79}Se (trace amounts; half-life 3.27×10^5 years) and ^{81}Se (8.73%; half-life 1.08×10^{20} years).

Potential oxidation states are +6, +4, and −2. Important inorganic forms are sodium selenate (Na_2O_4Se), sodium selenite (Na_2O_3Se), sodium selenide (Na_2Se), and selenium chloride (Cl_2Se_2).

ABBREVIATION

Se   selenium

NUTRITIONAL SUMMARY

Function: Selenium (Se) is a cofactor for enzymes and proteins with vital importance in antioxidant defense, thyroid hormone and insulin function, regulation of cell growth, and maintenance of fertility.

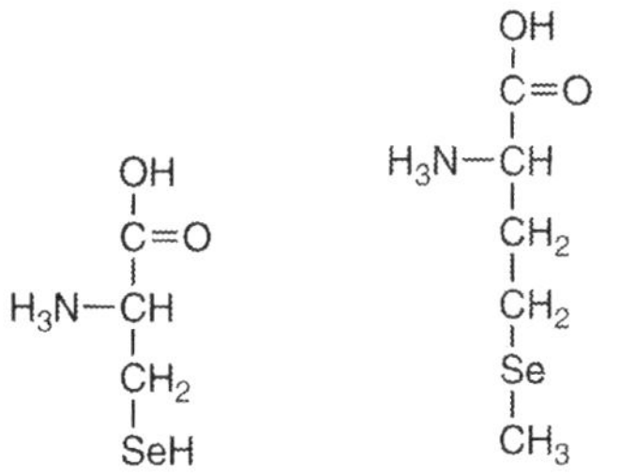

FIGURE 11.18

Se-containing amino acids.

Requirements: A daily intake of 55 µg is recommended for adults. Pregnant women should consume an additional 5 µg/day, and lactating women an additional 4 µg/day (Food and Nutrition Board, Institute of Medicine, 2000).

Food sources: Seafood and liver are the foods with the highest Se content; meat and grains contain less, and fruits and vegetables very little.

Deficiency: Deficiency increases the risk of cardiac failure, liver disease, cancer, atherosclerosis, cardiomyopathy (Keshan disease), and may cause hair loss, skin changes, and infertility.

Excessive intake: Continued consumption of 400 µg/day or more, especially with inorganic supplements, may cause peripheral neuropathy, nausea, diarrhea, dermatitis, hair loss, and nail deformities.

DIETARY AND OTHER SOURCES

Foods contain selenomethionine, selenocysteine, Se-methyl-selenomethionine, Se-methyl-selenocysteine, selenate, and selenite (Figure 11.18); selenomethionine is the main form in plants, while selenocysteine is the main form in foods of animal origin (Combs, 1984). Methylselenocysyteine is the dominant compound in broccoli grown with added Se (Davis et al., 2002). The concentration of Se in foods is greatly dependent on the Se content of the soil in which plants are grown and the Se content of animal feed.

Seafood and liver are the most Se-rich foods (400–1500 µg/kg), meat (100–400 µg/kg), grains, vegetables, and fruits usually contain much less.

Se intake varies greatly depending on region and dietary habits, with typical US consumption between 70 and 100 µg/day (Pennington et al., 1989).

Selenides, selenates, and various other inorganic Se compounds have significant toxic potential (at chronic intake levels of 1000 µg or more), mainly by causing dermatitis, hair loss, nail deformities, nausea, diarrhea, and peripheral neuropathy through unknown mechanisms.

DIGESTION AND ABSORPTION

Absorption of inorganic Se from the intestinal lumen (typically 60–70%) varies greatly, and its mechanism is not well understood. Selenate is poorly absorbed (Vendeland et al., 1994). Selenite uptake may be mediated by the formation of selenodicysteine and selenodiglutathione, which can then be transported across the apical enterocyte membrane by the corresponding amino acid and peptide transporters (Vendeland et al., 1992).

Most Se in foods, however, is present as selenocystein and selenomethionine. These modified amino acids are avidly taken up into small intestinal enterocytes, presumably through the respective amino acid and peptide transporters, amino acid transporter 1 (SLC3A1), neutral amino acid transporter (SLC1A4), and hydrogen ion/peptide cotransporter (PepT1, SLC15A1).

Further metabolism of either seleno-amino acids or inorganic Se is not well understood. While some selenide may be incorporated into selenoproteins and secreted into portal blood, another fraction, particularly following very high intake, may enter portal blood by diffusion. Selenocysteine and selenomethionine most likely are exported via the transporters of the unmodified amino acids.

TRANSPORT AND CELLULAR UPTAKE

Blood circulation: Se is transported in the blood as selenoproteins (including selenoprotein P1) and selenocysteine, and as a component of glutathione (Vendeland et al., 1992). Hepatocytes and other cells take up selenocysteine and other forms of organic Se through the same mechanisms that mediate uptake of the sulfur analogs. Inorganic Se may enter by diffusion. Selenoprotein P is essential for getting Se into bones (Pietschmann et al., 2014). This selenoprotein binds to LRP8 in bones and to LRP2 in bones and other tissues.

Genetic variants of 15-kd selenoprotein (SEP15) affect Se concentration in the blood, which indicates an important role for this gene. However, the mechanism is not known.

Materno-fetal transfer: Se as part of amino acids, particularly selenocysteine, can cross the placenta in either direction, which is essential to maintain adequate Se supplies while avoiding excess. The sulfate transporter 1 (SUT1, SLC13A4) also transports Se. It is abundantly expressed in the placenta, possibly at the fetal side of the syntrophoblast (Bergeron et al., 2013).

BBB: Selenocystine crosses into the brain with the help of the high-affinity system x(-)C (SLC7A11 plus SLC3A2). Intracerebral catabolism can then release Se, though the transferred amounts are uncertain.

Transfer into milk: The concentration of Se in mature human milk is about 36 µg/l (Hannan et al., 2009). The milk of rodents contains Se mainly in the form of selenoprotein P (Sepp1), glutathione peroxidase-3 (Gpx3), and selenomethionine (Hill et al., 2014).

METABOLISM

Selenide can be released from selenocysteine through the action of the pyridoxal phosphate-dependent selenocysteine selenide-lyase (EC4.4.1.16), which has particularly high activity in the liver. It has been proposed that the main function of this enzyme is to provide Se to selenophosphate synthetase for selenoprotein biosynthesis (Lacourciere & Stadtman, 1998).

S-adenosyl methionine-dependent enzymes (incompletely characterized) convert Se to its monomethyl, dimethyl, and trimethyl derivatives. Methylation of Se directly competes with arsenic methylation for *S*-adenosyl methionine (Davis et al., 2000).

Synthesis of selenocysteine: Selenophosphate synthase (EC2.7.9.3) phosphorylates selenide (Figure 11.19). Two genetically distinct isoenzymes exist, one of which (SPS2) is a selenoenzyme itself. In a state of Se deficiency, the decreased activity of this Se-containing isoform may be important for

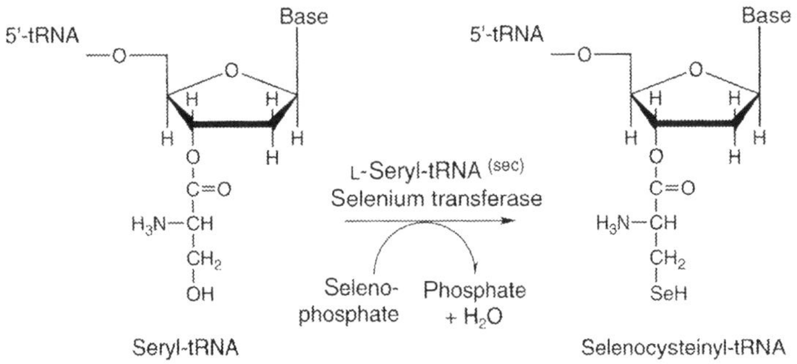

FIGURE 11.19

Selenocysteine synthesis.

down-regulation of selenide phosphorylation and sparing Se supplies. Selenophosphate and serine are the substrates for selenocysteine synthesis, which occurs only with tRNA(Ser)^{Sec} as a template (Hubert et al., 1996). First, tRNA(Ser)^{Sec} is charged by serine-tRNA ligase (EC6.1.1.11) with serine. The pyridoxal phosphate-dependent L-seryl-tRNA^{Sec} Se transferase (EC2.9.1.1) then substitutes the hydroxyl group of the serine with selenophosphate and releases the phosphate. It is not clear to what extent the newly synthesized selenocysteine can then be released directly from its tRNA. tRNA(Ser)^{Sec} recognizes the RNA-codon UGA (opal), which usually functions as a stop codon and terminates synthesis of the protein strand. The RNAs of selenoproteins, however, contain stem-loop structures (Sec insertion sequence) in 3'-untranslated regions near the UGA triplet that are recognized specifically by a Sec-tRNA^{Sec} protecting factor called *SePF* (Fujiwara et al., 1999). In an as-yet-incompletely understood way, this complex, together with a separate elongation factor, EF-1alpha, allows tRNA(Ser)^{Sec} to attach to the UGA triplet and displace the competing release factor. In this way, tRNA(Ser)^{Sec} mediates the incorporation of selenocysteine into specific proteins and protein synthesis continues beyond the stop codon. The PLP-dependent enzyme selenocysteine lyase (EC4.4.1.16) catalyzes the beta-elimination of selenocysteine, which releases hydrogen selenide and L-alanine (Soda et al., 1999).

It has been proposed that ASC reduces selenite to elemental Se, which then can form links with selenocysteine in proteins. The biological significance of such an effect is not known.

STORAGE

Se is stored mainly as selenoproteins in the liver and kidneys and less in skeletal muscle; selenoprotein P (SEPP1) is an especially Se-rich protein, which contains 10 selenocysteine moieties and is produced in the liver (Hill et al., 2012). The low-density lipoprotein receptor-related protein 8 (LRP8, apoE receptor 2) in a wide range of tissues binds SEPP1 and mediates its uptake. This means that the liver can store Se and, in times of need, provide other tissues with Se as well.

EXCRETION

Nearly half of total Se losses at average intake occur with sweat and skin cells. Selenides, particularly if ingested in excessive toxic amounts, are partially exhaled as dimethylselenide, which has a distinctive garlic odor. A significant amount is trimethylated and excreted with urine. In addition, some fecal loss (typically 50 μg/day) of elemental Se occurs.

SLC13A1 in proximal tubular cells of the kidney transports selenate (Bergeron et al., 2013).

REGULATION

Little is known about mechanisms that maintain Se homeostasis of the body. In a state of high Se availability, expression of selenoprotein P (SEPP1) is favored over that of functional selenoproteins such as glutathione peroxidases (Hill et al., 2012). This SEPP1 storage form can then be released to other tissues when Se availability declines, but the mechanisms controlling its release from liver into circulation remain to be explored. The extent of renal losses may be regulated.

FUNCTION

Selenoproteins: The number of characterized selenoproteins is still growing. The SelenoDB database (http://www.selenodb.org/) lists all genes known to code for selenoproteins, as well as genes involved in Se metabolism (Romagné et al., 2014). So far, we know about multiple genes encoding enzymes with glutathione peroxidase, thyroxine deiodinase, thioredoxin activities, and other Se-containing proteins of incompletely understood function (including selenoproteins H, I, K, M, N, O, P, R, S, T, V, and W).

Antioxidant protection: Several enzymes that contain selenocysteine as a prosthetic group are involved in the maintenance of sulfhydryl groups in reduced form and the defense against oxygen free radical. Glutathione peroxidase (EC1.11.1.9) designates four enzymes that use reduced GSH to reduce peroxides of free fatty acids and other lipids. Glutathione peroxidase activity is crucial for the protection of membranes, proteins, and DNA against damage by lipid peroxidation, and the fact that isoenzymes are encoded by at several distinct genes attests to its vital importance. The classical isoenzyme (GPX1) is ubiquitous in tissues. Gastrointestinal glutathione peroxidase (GPX2) is expressed exclusively in the gastrointestinal tract, where it may provide an important barrier against hydroxyperoxides in dietary fat (Wingler et al., 1999). Another isoenzyme (GPX3) is secreted in the blood mainly by the kidneys, as well as by tile mammary glands into milk. Each isoenzyme contains a selenocysteine residue and a FAD residue as prosthetic groups.

The hydroxyperoxides in esterified fatty acids, especially those in phospholipids, can be reduced by phospholipid-hydroperoxide glutathione peroxidase (GPX4, EC1.11.1.12) which also uses GSH as the cosubstrate. This enzyme is expressed in the inner mitochondrial membrane and cytosol of most tissues. Thioredoxin reductase (EC1.6.4.5) is a ubiquitous NADPH-dependent selenoenzyme in cytosol, which reduces both dehydroascorbate and the semidehydroascorbate radical to ascorbate (May et al., 1998).

Viral mutation: The recent recognition that low Se status of the host can increase the virulence of pathogens has thrown a new light on some manifestations of Se deficiency, such as Keshan disease. An otherwise benign strain of coxsackievirus B3 was shown to cause myocarditis in Se-deficient mice (Beck, 2007). Similarly, the rate of influenza virus mutations was related to Se status of the animal host (Nelson et al., 2001). It has been suggested that oxidative stress due to low activity of Se-containing peroxidases in the host may accelerate the emergence of mutated pathogen strains.

Thyroid hormone metabolism: Another group of selenocysteine enzymes, the thyroxine deiodinases (EC3.8.1.4), are anchored to the endoplasmic reticulum; the enzymes remove iodine from the thyroid hormones thyroxine and triiodothyronine. At least three distinct genes code for isoforms, which differ in respect to specificity and level of expression in various tissues, and at different developmental stages. Thyroxine deiodinase 2 generates the most active thyroid hormone, triiodothyronine, in the thyroid, developing brain, anterior pituitary gland, and brown adipose tissue. The isoform 1, in contrast, inactivates thyroxin in the liver and kidneys, thereby ensuring that hormonal activity does not accumulate unchecked. Thyroxine deiodinase 3, which is expressed in fetal tissues and placenta, preferentially removes iodine from the inner ring of thyroid hormones, thereby inactivating both thyroxin and triiodothyronine. The enzyme thus protects fetal tissues from exposure to the much higher thyroid hormone concentrations in maternal circulation.

Thioredoxin reductase (EC1.6.4.5) acts as a molecular chaperone which reduces intermolecular and intramolecular disulfide bonds of multimerized iodinated thyroglobulin and helps with its refolding (Delom et al., 2001). This may be important for the mobilization of stored thyroglobulin and subsequent release of active thyroid hormone (Klein et al., 2000).

Insulin metabolism: The Se-containing thioredoxin reductase (EC1.6.4.5) is an NADPH-dependent enzyme with FAD as a prosthetic group that uses thioredoxin to reduce insulin.

Cell replication: The fragile histidine triad/dinucleoside 5′,5″-P1,P3-triphosphate (Ap3A) hydrolase (EC3.6.1.29) contains four coordinated Se atoms (Supplee, 1999) and requires magnesium or manganese as cofactors. The enzyme functions as a tumor suppressor gene (Ji et al., 1999), possibly by inducing apoptosis, and through the actions of AP3A and AP4A, which are thought to regulate DNA replication and signal stress responses.

Through as-yet-unknown mechanisms, Se also appears to be important for the methylation of DNA, which is known to control gene expression and thus proliferation and differentiation. The DNA in rodents with diminished Se status appears to be distinctly less methylated than DNA from Se-replete animals. The higher cancer risk of Se deficient animals may thus be related to disrupted DNA methylation (Davis et al., 2000).

Other functions: The mitochondrial capsule selenoprotein (MCSP) maintains the crescent structure of sperm mitochondria. Selenoprotein P is an extracellular heparin-binding glycoprotein with 10 selenocysteine residues that may be important for Se transport and storage. Selenoprotein P appears to contribute to antioxidant defense in extracellular space near the endothelium. Selenoprotein W in muscle contains one selenocysteine. It appears to have antioxidant function and contribute to muscle cell differentiation (Jeon et al., 2014).

NUTRITIONAL ASSESSMENT

Se status is modestly well reflected by the concentration of Se in blood plasma or red blood cells (Costa et al., 2014). Se concentrations in hair and nail clippings, for instance, measured by neutron activation analysis (Geybels et al., 2014), also reflect long-term Se status.

REFERENCES

Beck, M.A., 2007. Selenium and vitamin E status: impact on viral pathogenicity. J. Nutr. 137, 1338–1340.

Bergeron, M.J., Clémencon, B., Hediger, M.A., Markovich, D., 2013. SLC13 family of Na(+)-coupled di- and tri-carboxylate/sulfate transporters. Mol. Aspects. Med. 34, 299–312.

Combs, G.S.B., 1984. The nutritional biochemistry of selenium. Annu. Rev. Nutr. 4, 57–280.

Costa, N.A., Gut, A.L., Pimentel, J.A., Cozzolino, S.M., Azevedo, P.S., Fernandes, A.A., et al., 2014. Erythrocyte selenium concentration predicts intensive care unit and hospital mortality in patients with septic shock: a prospective observational study. Crit. Care 18 (3), R92.

Davis, C.D., Uthus, E.O., Finley, J.W., 2000. Dietary selenium and arsenic affect DNA methylation *in vitro* in Caco-2 cells and *in vivo* in rat liver and colon. J. Nutr. 130, 2903–2909.

Davis, C.D., Zheng, H., Finley, J.W., 2002. Selenium-enriched broccoli decreases intestinal tumorigenesis in multiple intestinal neoplasia mice. J. Nutr. 132, 307–309.

Delom, E., Mallet, B., Carayon, P., Lejeune, P.J., 2001. Role of extracellular molecular chaperones in the folding of oxidized proteins. Refolding of colloidal thyroglobulin by protein disulfide isomerase and immunoglobulin heavy chain-binding protein. J. Biol. Chem. 276, 21337–21342.

Food and Nutrition Board, Institute of Medicine, 2000. Dietary Reference Intakes for Vitamin C, Vitamin E, Selenium, and Carotenoids. National Academy Press, Washington, DC.

Fujiwara, T., Busch, K., Gross, H.J., Mizutani, T.A., 1999. SECIS binding protein (SBP) is distinct from selenocysteyl-tRNA protecting factor (SePF). Biochimie 81, 213–218.

Geybels, M.S., van den Brandt, P.A., Schouten, L.J., van Schooten, F.J., van Breda, S.G., Rayman, M.P., et al., 2014. Selenoprotein gene variants, toenail selenium levels, and risk for advanced prostate cancer. J. Natl. Cancer Inst. 106 (3), dju003.

Hannan, M.A., Faraji, B., Tanguma, J., Longoria, N., Rodriguez, R.C., 2009. Maternal milk concentration of zinc, iron, selenium, and iodine and its relationship to dietary intakes. Biol. Trace Elem. Res. 127, 6–15.

Hill, K.E., Motley, A.K., Winfrey, V.P., Burk, R.F., 2014. Selenoprotein P is the major selenium transport protein in mouse milk. PLoS One 9, e103486.

Hill, K.E., Wu, S., Motley, A.K., Stevenson, T.D., Winfrey, V.P., Capecchi, M.R., et al., 2012. Production of selenoprotein P (Sepp1) by hepatocytes is central to selenium homeostasis. J. Biol. Chem. 287, 40414–40424.

Hubert, N., Walczak, R., Sturchler, C., Myslinski, E., Schuster, C., Westhof, E., et al., 1996. RNAs mediating cotranslational insertion of selenocysteine in eukaryotic selenoproteins. Biochimie 78, 590–596.

Jeon, Y.H., Park, Y.H., Lee, J.H., Hong, J.H., Kim, I.Y., 2014. Selenoprotein W enhances skeletal muscle differentiation by inhibiting TAZ binding to 14-3-3 protein. Biochim. Biophys. Acta 1843, 1356–1364.

Ji, L., Fang, B., Yen, N., Fong, K., Minna, J.D., Roth, J.A., 1999. Induction of apoptosis and inhibition of tumorigenicity and tumor growth by adenovirus vector-mediated fragile histidine triad (FHIT) gene overexpression. Cancer Res. 59, 3333–3339.

Klein, M., Gestmann, I., Berndorfer, U., Schmitz, A., Herzog, V., 2000. The thioredoxin boxes of thyroglobulin: possible implications for intermolecular disulfide bond formation in the follicle lumen. Biol. Chem. 381, 593–601.

Lacourciere, G.M., Stadtman, T.C., 1998. The NIFS protein can function as a selenide delivery protein in the biosynthesis of selenophosphate. J. Biol. Chem. 273, 30921–30926.

May, J.M., Cobb, C.E., Mendiratta, S., Hill, K.E., Burk, R.E., 1998. Reduction of the ascorbyl free radical to ascorbate by thioredoxin reductase. J. Biol. Chem. 273, 23039–23045.

Nelson, H.K., Shi, Q., Van Dael, P., Schiffrin, E.J., Blum, S., Barclay, D., et al., 2001. Host nutritional selenium status as a driving force for influenza virus mutations. FASEB J. 15, 1846–1848.

Pennington, J.A., Young, B.E., Wilson, D.B., 1989. Nutritional elements in US diets: results from the Total Diet Study, 1982 to 1986. J. Am. Diet. Assoc. 89, 659–664.

Pietschmann, N., Rijntjes, E., Hoeg, A., Stoedter, M., Schweizer, U., Seemann, P., et al., 2014. Selenoprotein P is the essential selenium transporter for bones. Metallomics 6, 1043–1049.

Romagné, F., Santesmasses, D., White, L., Sarangi, G.K., Mariotti, M., Hübler, R., et al., 2014. SelenoDB 2.0: annotation of selenoprotein genes in animals and their genetic diversity in humans. Nucleic Acids Res. 42, D437–D443.

Soda, K., Oikawa, T., Esaki, N., 1999. Vitamin B6 enzymes participating in selenium amino acid metabolism. Biofactors 10, 257–262.

Supplee, C., 1999. Physics in the 20th Century. Harry N. Abrams, New York, NY.

Vendeland, S.C., Deagen, J.T., Butler, J.A., Whanger, P.D., 1994. Uptake of selenite, selenomethionine and selenate by brush border membrane vesicles isolated from rat small intestine. Biometals 7, 305–312.

Vendeland, S.C., Deagen, J.T., Whanger, P.D., 1992. Uptake of selenotrisulfides of glutathione and cysteine by brush border membranes from rat intestines. J. Inorg. Biochem. 47, 131–140.

Wingler, K., Bocher, M., Flohé, L., Kollmus, H., Brigelius-Flohé, R., 1999. mRNA stability and selenocysteine insertion sequence efficiency rank gastrointestinal glutathione peroxidase high in the hierarchy of selenoproteins. Eur. J. Biochem. 259 (1–2), 149–157.

MOLYBDENUM

Molybdenum (atomic weight 95.94) is a transition (group 6B) element with valences 2, 3, 4, 5, or 6. Naturally occurring molybdenum contains four stable isotopes, ^{94}Mo (9.25%), ^{95}Mo (15.92%), ^{96}Mo (16.68%), and ^{97}Mo (24.13%), and two unstable isotopes, ^{92}Mo (14.92%; half-life 1.9×10^{20} years) and ^{100}Mo (9.63%; half-life 7.8×10^{18} years). Because the decay of these two isotopes is so extremely slow, there is no biologically relevant radioactivity coming from naturally occurring molybdenum. Two synthetic molybdenum isotopes, ^{93}Mo and ^{99}Mo, are radioactive.

ABBREVIATIONS

FAD	flavin adenine dinucleotide
Mo	molybdenum

NUTRITIONAL SUMMARY

Function: Humans need minute amounts of molybdenum (Mo) for the metabolism of sulfite, nucleic acids, aldehydes, and taurine. Mo may also play a role in the function of glucocorticoid hormones and immune defense.

Requirements: Recommended intake for adults is 45 μg/day (Food and Nutrition Board, Institute of Medicine, 2001). There is no indication that higher intake is beneficial.

Food sources: Grain products, beans, and leafy vegetables are the best sources, but contents vary depending on the Mo content of the soil. Typical intake in the United States has been estimated to range from 50 to 500 μg/day.

Deficiency: It is very unlikely that healthy people show the metabolic disorders seen in severe experimental deprivation, which include xanthine kidney stones and impaired conversion of toxic sulfite to harmless sulfate compounds.

Excessive intake: Daily doses in the milligram range can cause gout. A number of other adverse effects, including reproductive failure, have been observed in animals with daily intake of over 5 mg/kg body weight; none of these adverse effects has been observed in humans. The tolerable upper intake level has been set at 2 mg/day (Food and Nutrition Board, Institute of Medicine, 2001).

DIETARY SOURCES

Foods contain Mo mainly as molybdopterin and MoO_4^{2-} (molybdate). Good Mo sources include milk (50 μg/l) and dairy products, legumes (especially soybeans), cereal products, and leafy vegetables.

Estimates of mean daily intake vary widely, from 100 to 500 μg (Barceloux, 1999), 76 μg for women, and 109 μg for men (Pennington & Jones, 1987), and as low as 50 to 100 μg (Nielsen, 1994).

DIGESTION AND ABSORPTION

Absorption of ingested Mo from the stomach and proximal intestine is modestly well absorbed (as much as 85% from foods) over a wide range of intake levels (Cantone et al., 1995); only MoS_2 is poorly absorbed. The fact that sulfate inhibits absorption might suggest a shared mediated transport process, but the mechanism or involved carriers are unknown.

TRANSPORT AND CELLULAR UPTAKE

Blood circulation: Mo (typical concentration 5 nmol/l) is largely associated with alpha-2-macroglobulin as MoO_4^{-2} as well as enzyme-bound in erythrocytes (Nielsen, 1994). The uptake of MoO_3^-, and to a lesser extent MoO_4^{-2}, by erythrocytes, leukocytes, fibroblasts, vascular smooth muscle, the liver, and other tissues proceeds via SLC4A1 (band 3 protein), which otherwise mediates exchange of chloride, bicarbonate, sulfate, and glucose (Gimenez et al., 1993); several well-characterized variants of this transporter account for the Diego blood group system antigens. Tungstate is also transported by this system. An ABC transporter in bacteria has been identified that mediates Mo transport (Neubauer et al., 1999). There appears to be significant enterohepatic cycling.

Materno-fetal transfer: The Mo concentration in colostrum is eightfold higher than in maternal serum, indicating an active, concentrative transport mechanism (Krachler et al., 1999).

BBB: The mechanisms for transport across the BBB have not yet been elucidated.

Transfer into milk: About 1% of Mo ingested by a breast-feeding mother reaches the infant (Wappelhorst et al., 2002).

METABOLISM

Molybdopterin with MoO_4^{-2} coordinated to the dithiolen of the side chain is the essential cofactor of at least six human enzymes. Both the immediate precursors and the cofactor itself are unstable unless associated with an enzyme. Molybdopterin synthesis from GTP and MoO_4^{-2} requires the cooperation of at least six distinct gene products (Figure 11.20), including molybdopterin synthases 1A + 1B, 2A + 2B, molybdopterin synthase sulfurylase, and gephyrin (Stallmeyer et al., 1999; Reiss et al., 2001).

The first two steps convert GTP into the highly unstable precursor Z; these reactions are catalyzed by a bifunctional enzyme, molybdopterin synthase 1 (EC2.8.1.12), which is the fusion product of the genes MOCS1A and MOCS1B (Gray and Nicholls, 2000).

Molybdopterin synthase 2 is the bicistronic product of separately translated genes MOCS2A (small subunit) and MOCS2B (large subunit). The small subunit of the MOCS2 heterodimer becomes sulfur-charged by molybdopterin synthase sulfurylase (EC2.8.1.11) and synthesizes molybdopterin by adding a dithiolene sulfur to the pterin side chain of precursor Z (Appleyard et al., 1998). Gephyrin completes Mo cofactor synthesis (Reiss et al., 2001) by first activating (through its E domain, EC2.10.1.1) molybdate MoO_4^{-2} and then inserting it into molybdopterin (with the G domain, EC2.7.7.75). Another, distinct function of gephyrin is synaptic clustering of glycine and GABA(A) receptors. Tungstate competitively inhibits Mo cofactor synthesis.

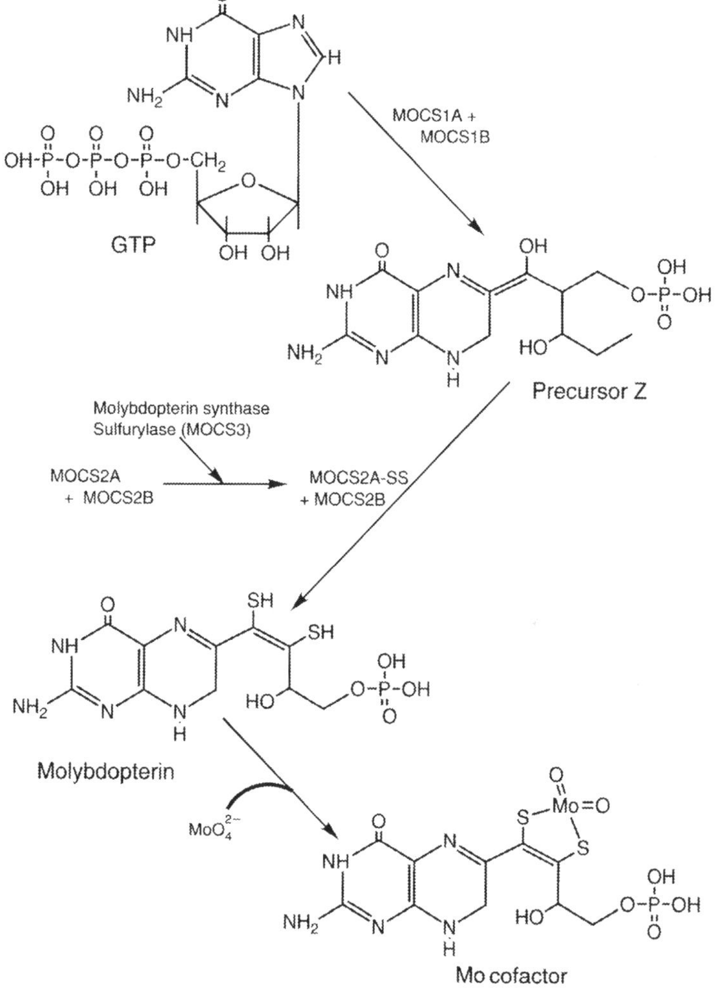

FIGURE 11.20

Mo cofactor synthesis.

STORAGE

More than half of Mo in the liver occurs as nonprotein-bound di-(carboxyaminomethyl)molybdopterin in the outer mitochondrial membrane; most of the remainder is in the form of enzyme-bound molybdopterin (Nielsen, 1994). Total Mo concentration in the human liver is about 0.9 µg/g. Total body stores depend strongly on recent intake levels and have been found to range from 900 to 5400 µg (Novotny & Turnlund, 2007).

EXCRETION

Less than 1% of systemic Mo is excreted via bile (Nielsen, 1994). Between 60% and 90% of ingested Mo is excreted via urine (Novotny & Turnlund, 2007). A significant proportion of filtered Mo is recovered, but the mechanism of renal reabsorption remains unclear (Novotny & Turnlund, 2007). Ultimately, molybdopterin is catabolized to the sulfur-containing pterin urothione and Mo is released. Urothione is mostly excreted with urine (Johnson & Rajagopalan, 1982); the amounts in urine reflect molybdopterin synthesis.

REGULATION

Healthy individuals are able to maintain stable tissue concentrations at a wide range of intake levels. Mo homeostasis appears to be regulated through adaptation of intestinal absorption and urinary excretion rates in response to intake levels (Novotny & Turnlund, 2007).

FUNCTION

Mo cofactor: Owing to its role as an essential cofactor of sulfite oxidase, aldehyde oxidases, xanthine dehydrogenase, and hypotaurine dehydrogenase, deficiency of Mo cofactor induces hypouricemia and increases the urinary excretion of sulfite, thiosulfate, *S*-sulfocysteine, taurine, hypoxanthine, and xanthine, and decreases the excretion of sulfate and urate. The deficiency may arise from a lack of Mo or from disrupted molybdopterin cofactor synthesis. Inborn lack of cofactor synthesis is usually lethal in early childhood (Reiss, 2000).

Sulfite oxidase (EC1.8.3.1) is located in the mitochondrial intermembrane space, where it converts sulfite into sulfate and generates hydrogen peroxide in the process; the enzyme contains both Mo cofactor and heme as prosthetic groups. Sulfite comes directly from ingested food (especially dried fruit and wine) and from the catabolism of cysteine (and indirectly of methionine).

The peroxisomal enzyme xanthine dehydrogenase (EC1.1.3.22), which contains Mo cofactor, iron-sulfur clusters, and FAD, converts xanthine into uric acid. The enzyme in the liver normally functions as an NAD-dependent dehydrogenase. In other tissues, the enzymes can be converted by oxidation or specific proteolysis into an oxidase, which does not require NAD but generates hydrogen peroxide. The oxidative conversion of the enzyme into the oxidase form is reversible in the presence of sulfhydryl agents. Genetic deficiency of the enzyme is associated with unusually high excretion of xanthine (xanthinuria) and formation of kidney stones rich in xanthine.

Aldehyde oxidase (EC1.2.3.1) contains Mo cofactor, heme, multiple iron/sulfur clusters, and FAD. The cytosolic enzyme is especially abundant in the liver, where it metabolizes acetaldehyde and other aldehydes to their respective acids. It may also oxidize retinal to retinoic acid (Tomita et al., 1993). Aldehyde oxidase may be as important as the mixed-function hydroxylases for the hydroxylation or oxidation of various purines, pyrimidines, pteridines, and related xenobiotics; an example is the conversion of the antiherpes agent famciclovir, a 9-substituted guanine derivative, to the active compound penciclovir (Rashidi et al., 1997).

Hypotaurine dehydrogenase (EC1.8.1.3), which contains Mo cofactor and heme, and is NAD-dependent, catalyzes the final step of taurine synthesis from cysteine (and the reverse reaction is needed for taurine catabolism). Two additional molybdo-flavoproteins, AOH1 and AOH2 (aldehyde oxidase homolog 1 and 2), have been identified based on their similarity to aldehyde oxidase and xanthine

(Terao et al., 2000). AOH1 has phenanthridine and benzaldehyde-oxidizing activity and is expressed in the liver and in spermatogonia. AOH2 also has phenanthridine-oxidizing activity and is expressed in keratinized epithelial cells and basal cells of skin and hair follicles.

Hormone-like effects: Molybdate stabilizes the glucocorticoid receptor (comprising heat-shock proteins hsp56, hsp70, and hsp90), and inhibits the dexamethason-induced transformation to the DNA-binding state; the region 595 614 of the rat glucocorticoid receptor is a molybdate-binding site (Bodine et al., 1995).

Other effects: High Mo intake may increase urinary copper losses, possibly by complexing copper. Tetrathiomolybdate is a synthetic compound that has been used as a copper chelator for the experimental treatment of Wilson disease.

NUTRITIONAL ASSESSMENT

Mo status is often assessed by measuring plasma or serum concentrations (Smorgon et al., 2004; Schultze et al., 2014). The validity of this approach is not firmly established.

Severe deficiency, including that due to genetic disorders (specifically defective synthesis of Mo cofactor), is reflected by increased excretion of sulfite, thiosulfate, and S-sulfocysteine via urine (Belaidi et al., 2012).

REFERENCES

Appleyard, M.V.C.L., Sloan, J., Kana'n, J.M., Heck, I.S., Kinghorn, J.R., Unkles, S.E., 1998. The aspergillus nidulans cnxF gene and its involvement in molybdopterin biosynthesis. J. Biol. Chem. 273, 14869–14876.

Barceloux, D.G., 1999. Molybdenum. J. Toxicol. Clin. Toxicol. 37, 231–237.

Belaidi, A.A., Arjune, S., Santamaria-Araujo, J.A., Sass, J.O., Schwarz, G., 2012. Molybdenum cofactor deficiency: a new HPLC method for fast quantification of s-sulfocysteine in urine and serum. JIMD Rep. 5, 35–43.

Bodine, P.V., Alnemri, E.S., Litwack, G., 1995. Synthetic peptides derived from the steroid binding domain block modulator and molybdate action toward the rat glucocorticoid receptor. Receptor 5, 117–122.

Cantone, M.C., de Bartolo, D., Gambarini, G., Giussani, A., Ottolenghi, A., Pirola, L., et al., 1995. Proton activation analysis of stable isotopes for a molybdenum biokinetics study in humans. Med. Phys. 22, 1293–1298.

Food and Nutrition Board, Institute of Medicine, 2001. Dietary Reference Intakes for Vitamin A, Vitamin K, Arsenic, Boron, Chromium, Copper, Iodine, Iron, Manganese, Molybdenum, Nickel, Silicon, Vanadium, and Zinc. National Academy Press, Washington, DC, pp. 333–350.

Gimenez, I., Garay, R., Alda, J.O., 1993. Molybdenum uptake through the anion exchanger in human erythrocytes. Pfl. Archiv. Eur. J. Physiol. 424, 245–249.

Gray, T.A., Nicholls, R.D., 2000. Diverse splicing mechanisms fuse the evolutionarily conserved bicistronic MOCSIA and MOCSIB open reading frames. Rna 6, 928–936.

Johnson, J.L., Rajagopalan, K.V., 1982. Structural and metabolic relationship between the molybdenum cofactor and urothione. Proc. Natl. Acad. Sci. 79, 6856–6860.

Krachler, M., Rossipal, E., Micetic-Turk, D., 1999. Trace element transfer from the mother to the newborn – investigations on triplets of colostrum, maternal and umbilical cord sera. Eur. J. Clin. Nutr. 53, 486–494.

Neubauer, H., Pantel, I., Lindgren, P.E., Gotz, E., 1999. Characterization of the molybdate transport system ModABC of *Staphylococcus carnosus*. Arch. Microbiol. 172, 109–115.

Nielsen, F.H., 1994. Ultratrace minerals. In: Shils, M.E., Olson, J.A., Shike, M. (Eds.), Modern Nutrition in Health and Disease Lea & Febiger, Philadelphia, PA, pp. 269–286.

Novotny, J.A., Turnlund, J.R., 2007. Molybdenum intake influences molybdenum kinetics in men. J. Nutr. 137, 37–42.

Pennington, J.A.T., Jones, J.W., 1987. Molybdenum, nickel, cobalt, vanadium, and strontium in total diets. J. Am. Diet. Assoc. 87, 1644–1650.

Rashidi, M.R., Smith, J.A., Clarke, S.E., Beedham, C., 1997. *In vitro* oxidation of famciclovir and 6-deoxypenciclovir by aldehyde oxidase from human, guinea pig, rabbit, and rat liver. Drug Metab. Disp. 25, 805–813.

Reiss, J., 2000. Genetics of molybdenum cofactor deficiency. Hum. Genet. 106, 157–163.

Reiss, J., Gross-Hardt, S., Christensen, E., Schmidt, E., Mendel, R.R., Schwarz, G., 2001. A mutation in the gene for the neurotransmitter receptor-clustering proetin gephyrin causes a novel form of molybdenum cofactor deficiency. Am. J. Hum. Genet. 68, 208–213.

Schultze, B., Lind, P.M., Larsson, A., Lind, L., 2014. Whole blood and serum concentrations of metals in a Swedish population-based sample. Scand. J. Clin. Lab. Invest. 74, 143–148.

Smorgon, C., Mari, E., Atti, A.R., Dalla Nora, E., Zamboni, P.F., Calzoni, F., et al., 2004. Trace elements and cognitive impairment: an elderly cohort study. Arch. Gerontol. Geriatr. Suppl. 9, 393–402.

Stallmeyer, B., Schwarz, G., Schulze, J., Nerlich, A., Reiss, J., Kirsch, J., et al., 1999. The neurotransmitter receptor-anchoring protein gephyrin reconstitutes molybdenum cofactor biosynthesis in bacteria, plants, and mammalian cells. Proc. Natl. Acad. Sci. USA 16 (96), 1333–1338.

Terao, M., Kurosaki, M., Saltini, G., Demontis, S., Marini, M., Salmona, M., et al., 2000. Cloning of the cDNAs coding for two novel molybdo-flavoproteins showing high similarity with aldehyde oxidase and xanthine oxidoreductase. J. Biol. Chem. 275, 30690–30700.

Tomita, S., Tsujita, M., Ichikawa, Y., 1993. Retinal oxidase is identical to aldehyde oxidase. FEBS Lett. 336, 272–274.

Wappelhorst, O., Kühn, I., Heidenreich, H., Markert, B., 2002. Transfer of selected elements from food into human milk. Nutrition 18, 316–322.

COBALT

Cobalt (atomic weight 58.93) is a metallic transition element. Only the stable isotope 59 occurs in nature; the radioactive isotopes ^{54}Co, ^{55}Co, ^{56}Co, ^{57}Co, ^{58}Co, ^{60}Co, ^{61}Co, ^{62}Co, ^{63}Co, and ^{64}Co can be produced artificially. Cobalt can have valences 1, 2, or 3, or (rarely) 4 or 5.

ABBREVIATION

Co   cobalt

NUTRITIONAL SUMMARY

Function: An enzyme with cobalt (Co) as an essential cofactor is involved in the regulation of translation; Co also may be a constituent of an oxygen sensor. It is an essential constituent of all forms of vitamin B12.

   Requirements: It is not known what intake is needed to maintain optimal health.

   Food sources: The Co content (other than cobalamin) of foods is not well documented.

   Deficiency: It is not known whether there are any untoward effects of low intake or what the consequences of Co deficiency in humans might be.

   Excessive intake: Pericardiomyopathy has occurred with chronic intake of 6–8 mg/day via cobalt being added to commercially sold beer. More common consequences of acute intake of 100 mg or more include hypotension, nausea and vomiting, diarrhea, loss of appetite, tinnitus, acoustic nerve damage, hypothyreosis and goiter, hyperlipidemia, and xanthomatosis. High tissue levels of divalent Co may be associated with genotoxic and carcinogenic effects.

DIETARY SOURCES

The noncorrinoid Co content of individual foods is not well documented. Investigators in the United Kingdom (Ysart et al., 1999) found significant amounts in nuts (9 µg/100 g), sugars and preserves (3 µg/100 g), potatoes (9 µg/100 g), and bread (9 µg/100 g). Estimates from studies in various countries point to total Co intake (including Co covalently bound to cobalamins and other corrinoids) of about 11 µg (Dabeka & McKenzie, 1995; Ysart et al., 1999) to about 40 µg (Health Canada, 2007). The amounts of Co that can be leached by food acids from porcelain dishes and thus become available for ingestion appear to be limited to a few micrograms (Sheets, 1998).

DIGESTION AND ABSORPTION

Some 7–37% of 8–20 µg/day intake appeared to be retained in balance studies in young girls (Engel et al., 1967). Apparent retention in healthy adults is around 40% (Vaiberg et al., 1969; Bento et al., 2013). The mechanisms that facilitate small intestinal Co absorption seem to be largely those involved in iron absorption. Binding to luminal mucin improves (Conrad et al., 1991) uptake of Co^{2+} via the divalent metal ion transporter (DMT1, DCT1, and Nramp2). This proton-dependent transmembrane transporter for divalent metals is better known for its central role as an iron transporter. Factors that modulate iron uptake via this transporter will alter Co uptake likewise. Mobilferrin is a cytoplasmic protein involved in intracellular iron transport that may also help to convey Co across the enterocyte (Conrad et al., 1992). Knowledge about the details of Co delivery to its intracellular targets or export across the basolateral membrane is still very rudimentary.

TRANSPORT AND CELLULAR UPTAKE

Blood circulation: Most Co in the blood is bound to transferrin. Little is known about the mechanisms governing the movement of Co into, within, and out of tissues. Components of the iron transport system, including TRs, are likely to be of importance.

Transfer into milk: About 12% of Co ingested by a breast-feeding mother reaches the infant (Wappelhorst et al., 2002).

METABOLISM

Little is known about the metabolism of Co in humans or other mammals. A particularly important and as-yet-unanswered question is whether Co is released *in vivo* from cobalamin (vitamin B12). Similar uncertainty exists about reactions that alter the redox state of Co.

STORAGE

Reports on Co concentrations in human fluids, tissues, and the entire body give widely differing results, partly due to persistent analytical difficulties. This makes it practically impossible to state how much

Co is stored by well-nourished people. Co can bind to ferritin, but it is not clear how stored Co might be mobilized and how it would get from storage sites to potential target tissues.

EXCRETION

Most ingested Co is excreted via feces (90%), and much of the remainder via urine (Price et al., 1970). The mechanisms involved in renal tubular reabsorption or secretion are unknown.

REGULATION

Fractional Co absorption from the small intestine increases with low iron status (Valberg et al., 1969). It is not known whether the body responds in a specific manner to the level of Co intake.

FUNCTION

Gene translation: Methionine aminopeptidase (EC3.4.11.18) may be the only enzyme in humans that has a Co requirement (Arfin et al., 1995; Hu et al., 2007), though others contend that the physiological metal cofactor is manganese (Wang et al., 2003). This enzyme plays an important role in translational regulation by preventing the phosphorylation of the alpha subunit of initiation factor-2 (Li and Chang, 1996).

Nonspecific enzyme cofactor: While there is no evidence that Co is a required or otherwise physiologically significant cofactor of enzymes other than methionine aminopeptidase (EC3.4.11.18), Co can substitute for other divalent cations in a number of enzymes. In some instances, such as mevalonate kinase (EC2.7.1.36), it may be a better activator than magnesium, manganese, or calcium (Michihara et al., 1997).

Oxygenation sensing: A heme-containing oxygen sensor (hypoxia-inducible factor 1, HIF-1), in which iron can be substituted by Co or nickel, stimulates erythropoietin production (Goldberg et al., 1988; Bunn et al., 1998). This sensor appears to be a multi-subunit flavohemoprotein complex with b5 and b5 reductase domains and an NAD(P)H oxidase capable of generating peroxide and reactive oxygen intermediates that serve as signaling molecules (Bunn et al., 1998). The physiological significance of Co for oxygen sensing remains in doubt. It may be worth noting, however, that Co has been used in the past for the controversial treatment of anemia as a stimulant of hematopoiesis. Its use for improving sports performance is particularly problematic (Lippi et al., 2005).

Toxic effects of excess: Since there is no evidence for a physiological need for added intake of inorganic Co (other than the minute trace amounts naturally present in various foods), the considerable risk of harmful effects have to be taken seriously (Simonsen et al., 2012). The concerns focus on long-term effects, even at seemingly modest intake levels, on thyroid function, immune response, and proinflammatory disposition. Another consideration is that soluble cobalt(II) salts are potential carcinogens (Koedrith and Seo, 2011).

REFERENCES

Arfin, S.M., Kendall, R.L., Hall, L., Weaver, L.H., Stewart, A.E., Matthews, B.W., et al., 1995. Eukaryotic methionyl aminopeptidases: two classes of cobalt-dependent enzymes. Proc. Natl. Acad. Sci. USA 92, 7714–7718.
Bento, J., Barros, S., Teles, P., Vaz, P., Zankl, M., 2013. Efficiency correction factors of an ACCUSCAN whole-body counter due to the biodistribution of [134]Cs, ^{137}Cs and [60]Co. Radiat. Prot. Dosimetry 155, 16–24.

Bunn, H.E., Gu, J., Huang, L.E., Park, J.W., Zhu, H., 1998. Erythropoietin: a model system for studying oxygen-dependent gene regulation. J. Exp. Biol. 201, 1197–1201.

Conrad, M.E., Umbreit, J.N., Moore, E.G., 1991. A role for mucin in the absorption of inorganic iron and other metal cations. A study in rats. Gastroenterologica 100, 129–136.

Conrad, M.E., Umbreit, J.N., Moore, E.G., Rodning, C.R., 1992. Newly identified iron-binding protein in human duodenal mucosa. Blood 79, 244–247.

Dabeka, R.W., McKenzie, A.D., 1995. Survey of lead, cadmium, fluoride, nickel, and cobalt in food composites and estimation of dietary intakes of these elements by Canadians in 1986–1988. J. AOAC Int. 78, 897–909.

Engel, R.W., Price, M.O., Miller, R.F., 1967. Copper, manganese, cobalt and molybdenum balance in pre-adolescent girls. J. Nutr. 92, 197.

Price, N.O., Bunce, G.E., Engel, R.W., 1970. Copper, manganese, cobalt, and molybdenum balance in pre-adolescence girls. Am. J. Clin. Nutr. 23, 258–260.

Goldberg, M.A., Dunning, S.R., Bunn, H.E., 1988. Regulation of the erythropoietin gene: evidence that the oxygen sensor is a heme protein. Science 242, 1412–1415.

Health Canada. 2007. Canadian total diet study. Dietary intakes of contaminants and other chemicals for different age–sex groups of Canadians. Vancouver. Health 2011;26:81–92. Available at: <http://www.hc-sc.gc.ca/fn-an/surveill/total-diet/intake-apport/chem_age-sex_chim_2007-eng.php>.

Hu, X.V., Chen, X., Han, K.C., Mildvan, A.S., Liu, J.O., 2007. Kinetic and mutational studies of the number of interacting divalent cations required by bacterial and human methionine aminopeptidases. Biochemistry 46, 12833–12843.

Koedrith, P., Seo, Y.R., 2011. Advances in carcinogenic metal toxicity and potential molecular markers. Int. J. Mol. Sci. 12, 9576–9595.

Li, X., Chang, Y.H., 1996. Evidence that the human homologue of a rat initiation factor-2 associated protein (p67) is a methionine aminopeptidase. Biochem. Biophys. Res. Commun. 227, 152–159.

Lippi, G., Franchini, M., Guidi, G.C., 2005. Cobalt chloride administration in athletes: a new perspective in blood doping? Br. J. Sports Med. 39, 872–873.

Michihara, A., Sawamura, M., Nara, Y., Ikeda, K., Yamori, Y., 1997. Purification and characterization of two mevalonate pyrophosphate decarboxylases from rat liver: a novel molecular species of 37 kDa. J. Biochem. 122, 647–654.

Sheets, R.W., 1998. Release of heavy metals from European and Asian porcelain dinnerware. Sci. Total Environ. 212, 107–113.

Simonsen, L.O., Harbak, H., Bennekou, P., 2012. Cobalt metabolism and toxicology—a brief update. Sci. Total Environ. 432, 210–215.

Valberg, L.S., Ludwig, J., Olatunbosun, D., 1969. Alteration in cobalt absorption in patients with disorders of iron metabolism. Gastroenterologica 56, 241–251.

Wang, J., Sheppard, G.S., Lou, P., Kawai, M., Park, C., Egan, D.A., et al., 2003. Physiologically relevant metal cofactor for methionine aminopeptidase-2 is manganese. Biochemistry 42, 5035–5042.

Wappelhorst, O., Kühn, I., Heidenreich, H., Markert, B., 2002. Transfer of selected elements from food into human milk. Nutrition 18, 316–322.

Ysart, G., Miller, P., Crews, H., Robb, P., Baxter, M., De UArgy, C., et al., 1999. Dietary exposure estimates of 30 elements from the UK Total Diet Study. Food Add Contain. 16, 391–403.

CHROMIUM

Chromium (Cr) is a transition metal (atomic weight 52.0) with possible valences of 2, 3, and 6, and it does not seem to be essential for humans. Naturally occurring stable isotopes are ^{52}Cr (83.789%), ^{53}Cr (9.501%), and ^{54}Cr (2.365%). Another naturally occurring isotope (^{50}Cr, 4.345%) has such a long half-life (1.3×10^{18} years) that only about 11,000 ^{50}Cr atoms decay per year in 1 g of natural chromium emitting two positrons each time, which is barely measurable with very sensitive instrumentation. Synthetic ^{51}Cr has a half-life of 27.7 days.

ABBREVIATION

Cr   chromium

NUTRITIONAL SUMMARY

Function: A peptide–chromium(III) complex increases insulin action, but this may be a pharmacological effect. Cr on its own binds to nuclear DNA and increases the number of initiation sites for RNA synthesis.

Sources: Nuts, whole grains, yeast, and cheese are good food sources of Cr.

Requirements: Recommendations for adequate intake have been set at 35 µg/day for younger healthy men, and 25 µg/day for healthy women (Food and Nutrition Board, 2002), with slightly less for men and women over the age of 50.

Deficiency: Low intake (<20 µg/day) may decrease the cellular response to insulin and slow the utilization of stored fat, but harm from low intake has not been established.

Excessive intake: Chromium(VI), but not the biologically active chromium(III), may be carcinogenic even in small amounts. Chromium picolinate may promote oxygen free radical formation.

DIETARY SOURCES

Good dietary Cr sources include yeast, whole grain and wheat germ, cereals, nuts, cheese, and organ meats (Ysart et al., 1999; Garcia et al., 2000). Human milk has been reported to contain 24 µg/l, and infant formulas have similar concentrations (Krachler et al., 2000). Synthetic chromium picolinate is a commonly used form in dietary supplements (Figure 11.21). Both natural and supplement sources contain almost exclusively chromium(III). There continues to be uncertainty about typical Cr intake of American adults, with estimates ranging from below 20 to over 40 µg/day. Investigation of double portions (i.e., all selected foods were duplicated in the same amounts for laboratory measurements) of healthy people in Spain found much higher intake (Domingo et al., 2012) suggesting that it is not hard to get enough Cr from a Mediterranean-style diet with some seafood and plenty of vegetables.

DIGESTION AND ABSORPTION

Only 0.5–2% of dietary Cr is absorbed (Anderson & Kozlovsky, 1985); fractional absorption is inversely proportional to intake. Phytate decreases bioavailability, ascorbic acid increases it. Chromium (VI) is reduced to chromium (III) by hydrochloric acid from the stomach; the chromium (III) salts of nitrate and chloride are well absorbed from the small intestine (Gammelgaard et al., 1999).

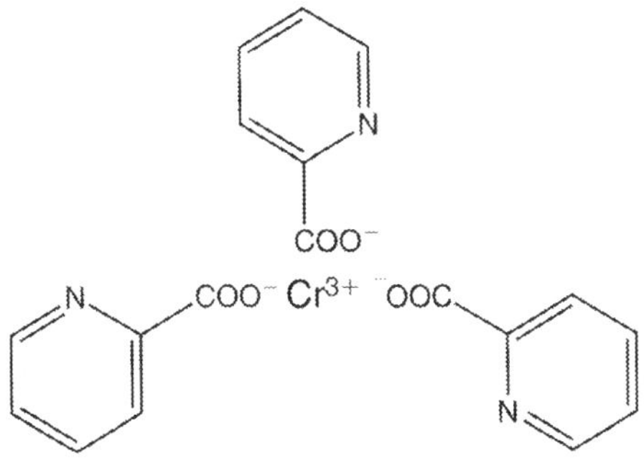

FIGURE 11.21

Chromium (III) tris (picolinate) is a commonly used dietary supplement.

Chromium picolinate tends to be better absorbed (2–5%) than the Cr in food. Absorption of this compound proceeds by a different mechanism that does not depend on the prior release of the chromic ion (Olin et al., 1994).

TRANSPORT AND CELLULAR UPTAKE

Blood circulation: Cr in the blood is bound as trivalent ion to transferrin. Typical blood concentrations are 2–3 nmol/l (Anderson, 1987). Cr enters cells with transferrin via TRs and is then transferred inside the cell to apochromodulin.

The 1500-dalton peptide chromodulin, which contains glycine, L-cysteine, L-glutamate, and L-aspartate, binds four chromium(III) ions (Vincent, 2000); it is present in the liver, the kidneys, and other tissues. Insulin, whose action is potentiated by Cr (as discussed next), has been suggested to increase Cr transport with transferrin by promoting the movement of TRs from vesicles to surface membranes (Vincent, 2000). This mechanism could facilitate Cr transport to insulin-sensitive cells, where it combines with stored apochromodulin and enhances insulin action on glucose metabolism. Patients with adult-onset diabetes appear to have diminished ability to transport Cr into leukocytes, especially when their blood glucose levels are poorly controlled (Rükgauer & Kruse-Jarres, 2001).

Transfer into milk: About 14% of Cr ingested by a breast-feeding mother reaches the infant (Wappelhorst et al., 2002).

STORAGE

Bone, spleen, muscle, skin, and the kidneys and liver have been found in rats to store most of the Cr from a large oral load (Thomann et al., 1994). The half-life of all tissue stores combined was around 100 days, and the slowest release occurred from the predominant stores in bone.

EXCRETION

Cr is excreted via urine (about 0.2 μg/day; Anderson et al., 1983), bile, and through the skin. Renal losses are increased in diabetes mellitus (Morris et al., 1999).

REGULATION

It is not clear whether and how Cr homeostasis is maintained. There is a risk of Cr accumulation with the long-term use of high doses of supplements with high bioavailability such as chromium picolinate (Stearns et al., 1995).

FUNCTION

Insulin-receptor activity: The Cr-containing peptide chromodulin binds to the insulin-activated insulin receptor and optimizes its receptor kinase activity (Vincent, 2000). Thus, chromodulin potentiates the ability of insulin to promote oxidative glucose metabolism and fatty acid synthesis from glucose in adipocytes. Raising Cr intake from low to adequate levels is likely to restore insulin receptor function if it was impaired due to Cr deficiency. Critical assessment of the available evidence does not seem to indicate that Cr is needed for the normal functioning of the insulin receptor; the observed effects may be artificially induced, druglike (i.e., pharmacological) effects (Vincent & Love, 2012). It is also doubtful whether Cr-replete diabetics will benefit from additional intake (Hellerstein, 1998).

DNA transcription: Cr binding to DNA appears to increase the number of initiation sites, thereby promoting RNA synthesis.

Other effects: Hexavalent(VI) Cr compounds have been found to induce malignant cell growth in animal models; solubility of the compounds was an important determinant of carcinogenic potential (Hayes, 1997). Most epidemiological studies have focused on the increased risk for nasal, throat, and bronchial cancer following work-related (dust) exposure; data on low-level oral exposure are lacking. An anecdotal report linking renal failure to long-term chromium picolinate ingestion (Wasser et al., 1997) has been disputed (McCarty, 1997). It has been argued that chromium picolinate may form redox-active complexes with ascorbate and thiols (Vincent, 2000). *In vitro* experiments demonstrate that mixtures of chromium picolinate and ascorbate can indeed induce the formation of hydroxyl radicals and lead to DNA damage (Speetjens et al., 1999; Hepburn & Vincent, 2002).

REFERENCES

Anderson, R.A., 1987. Chromium In: Mertz, W. (Ed.), Trace Elements in Human and Animal Nutrition, vol. 1. Academic Press, San Diego, CA, pp. 225–244.

Anderson, R.A., Kozlovsky, A.S., 1985. Chromium intake absorption and excretion of subjects consuming self-selected diets. Am. J. Clin. Nutr. 41, 768–771.

Anderson, R.A., Polansky, M.M., Btyden, N.A., Patterson, K.Y., Veillon, C., Glinsmann, W.H., 1983. Effects of chromium supplementation on urinary Cr excretion of human subjects and correlation of Cr excretion with selected clinical parameters. J. Nutr. 113, 276–281.

Domingo, J.L., Perelló, G., Giné Bordonaba, J., 2012. Dietary intake of metals by the population of Tarragona County (Catalonia, Spain): results from a duplicate diet study. Biol. Trace Elem. Res. 146, 420–425.

Food and Nutrition Board, Institute of Medicine, 2002. Dietary Reference Intakes for Vitamin A, Vitamin K, Arsenic, Boron, Chromium, Copper, Iodine, Iron, Manganese, Molybdenum, Nickel, Silicon, Vanadium, and Zinc. National Academy Press, Washington, DC, pp. 155–176.

Gammelgaard, B., Jensen, K., Steffansen, B., 1999. *In vitro* metabolism and permeation studies in rat jejunum: organic chromium compared to inorganic chromium. J. Trace Elem. Med. Biol. 13, 82–88.

Garcia, E., Cabrera, C., Lorenzo, M.L., Lopez, M.C., 2000. Chromium levels in spices and aromatic herbs. Sci. Total Environ. 247, 51–56.

Hayes, R.B., 1997. The carcinogenicity of metals in humans. Cancer Causes Control 8, 371–385.

Hellerstein, M.K., 1998. Is chromium supplementation effective in managing type II diabetes? Nutr. Rev. 56, 302–306.

Hepburn, D.D.D., Vincent, J.B., 2002. *In vivo* distribution of chromium from chromium picolinate in rats and implications for the safety of the dietary supplement. Chem. Res. Toxicol. 15, 93–100.

Krachler, M., Prohaska, T., Koellensperger, G., Rossipal, E., Stingeder, G., 2000. Concentrations of selected trace elements in human milk and in infant formulas determined by magnetic sector field inductively coupled plasma-mass spectrometry. Biol. Trace Elem. Res. 76, 97–112.

McCarty, M.E., 1997. Over-the-counter chromium and renal failure. Ann. Intern. Med. 127, 654–656.

Morris, B.W., MacNeil, S., Hardisty, C.A., Heller, S., Burgin, C., Gray, T.A., 1999. Chromium homeostasis in patients with type II (NIDDM) diabetes. J. Trace Elem. Med. Biol. 13, 57–61.

Olin, K.L., Stearns, D.M., Armstrong, W.H., Keen, C.L., 1994. Comparative retention/absorption of ^{51}chromium (^{51}Cr) from ^{51}Cr chloride, ^{51}Cr nicotinate and ^{51}Cr picolinate in a rat model. Trace Elem. Electrolytes 11, 182–186.

Rükgauer, M., Kruse-Jarres, J.D., 2001. Chromium status of patients with diabetes mellitus type 2. Clin. Lab. 47, 606.

Speetjens, J.K., Collins, R.A., Vincent, J.B., Woski, S.A., 1999. The nutritional supplement chromium(III) tris(picolinate) cleaves DNA. Chem. Res. Toxicol. 12, 483–487.

Stearns, D.M., Belbruno, J.J., Wetterhahn, K.E., 1995. A prediction of chromium(III) accumulation in humans from chromium dietary supplements. FASEB J. 9, 1650–1657.

Thomann, R.V., Snyder, C.A., Squibb, K.S., 1994. Development of a pharmacokinetic model for chromium in the rat following subchronic exposure. I. The importance of incorporating long-term storage compartment. Toxicol. Appl. Pharmacol. 128, 189–198.

Vincent, J.B., 2000. The biochemistry of chromium. J. Nutr. 130, 715–718.

Vincent, J.B., Love, S.T., 2012. The need for combined inorganic, biochemical, and nutritional studies of chromium(III). Chem. Biodivers. 9, 1923–1941.

Wappelhorst, O., Kühn, I., Heidenreich, H., Markert, B., 2002. Transfer of selected elements from food into human milk. Nutrition 18, 316–322.

Wasser, W.G., Feldman, N.S., D'Agati, V.D., 1997. Chronic renal failure after ingestion of over-the counter chromium picolinate. Ann. Intern. Med. 126, 410.

Ysart, G., Miller, P., Crews, H., Robb, P., Baxter, M., De UArgy, C., et al., 1999. Dietary exposure estimates of 30 elements from the UK Total Diet Study. Food Add Contam. 16, 391–403.

BORON

Boron (atomic weight 10.81) is a metalloid (metal-like) element. Naturally occurring boron contains two stable isotopes: ^{10}B (19–20%) and ^{11}B (80–81%). Boron has a valence of 3; its common chemical forms are boric acid (BH_3O_3, orthoboric acid, boracic acid), metaboric acid [$(BOH)_3O_3$, $B_3O_4(OH)$ (H_2O), and various polymeric forms], borate ($B_4O_7^{2-}$), boronic acid ($HB(OH)_2$), and their esters.

ABBREVIATIONS

B   boron
AI-2 autoinducer 2

NUTRITIONAL SUMMARY

Function: Boron appears to be essential in humans for fetal development, as well as energy and mineral metabolism. Beneficial or even essential effects of adequate intake on immune function, action of insulin and other hormones, and maintenance of cognitive and psychomotor functions are likely.

Food sources: Milk, vegetables, legumes, fruits, nuts, and (in some regions) water are good sources of boron.

Requirements: An intake around 1 mg/day appears to be reasonable (Food and Nutrition Board, Institute of Medicine, 2002).

Deficiency: A reduction of alertness, memory, and psychomotor skills, as well as greater vulnerability to vitamin D and mineral deficiency, may be related to low intake.

Excessive intake: Chronic intake of 1 g or more may decrease appetite, cause nausea and weight loss, and decrease sexual activity and fertility. Intake of larger amounts (known as *borism*) may result in nausea, vomiting, headache, diarrhea, hypothermia, restlessness, skin abnormalities with scaling and eruptions, kidney damage, and even death from circulatory collapse and shock.

DIETARY AND OTHER SOURCES

Boron-rich foods include prunes (20–30 μg/g), raisins (22 μg/g), peanuts (17 μg/g), peaches (5.3 μg/g), apples (3.6 μg/g), pears (2.8 μg/g), and other fruits, vegetables, and legumes (Rainey et al., 1999). Grains and meats contain very little boron. Among beverages, wine (6.1 mg/l), grape juice (3.0 mg/l), and apple juice (1.8 mg/l) are commonly consumed rich boron sources. Sodas, tea, milk, and beer contain less than 0.2 mg/l. The boron content of surface water in the United States ranges widely, with a median of about 0.08 mg/l (Coughlin, 1998).

Typical daily boron intake of American women is around 0.8 mg; men's daily intake is around 1 mg (Rainey et al., 1999; Hunt & Meacham, 2001). Individual values range widely, and both much lower and much higher intake levels are common.

Intake in Germany, Mexico, and Kenya appear to be considerably higher than in the United States (Rainey & Nyquist, 1998). Significant amounts of boron may be taken up through the skin, especially from creams and other products or work-site exposure. Uptake is greatest through damaged skin.

DIGESTION AND ABSORPTION

Most of the ingested boron from foods is absorbed. The mechanism and precise location of intestinal absorption remain unclear.

TRANSPORT AND CELLULAR UPTAKE

Blood circulation: Uncertainty about normal boron concentrations in plasma persists because of technical challenges and the impact of dietary intake levels. The average concentration in people with low intake has been reported to be about 20 μg/l (Hunt et al., 1997). Others using inductively coupled plasma mass spectrometry (ICP-MS) analytical technology have found an average concentration of 122 μg/l in pregnant women (Silberstein et al., 2014). Dietary intake from low levels to levels typical of well-nourished Americans increase concentrations only modestly. Boron appears to transfer from blood into tissues by passive diffusion (Murray, 1998). Experiments with frog eggs (*Xenopus laevis*) indicate that boric acid can directly cross lipid bilayer membranes with moderate efficiency (Dordas & Brown, 2001). It is not clear to what extent unfacilitated transmembrane diffusion contributes to the distribution of boron compounds to human tissues.

METABOLISM

Boric acid is not metabolized in human tissues to a significant extent (Murray, 1998).

EXCRETION

Only a small and relatively constant amount (about 0.04–0.16 mg/day) of boron is excreted via feces (Hunt et al., 1997). Most ingested boron is rapidly excreted via urine with a half-life of about 21 h. About half of the boron filtered in the kidneys is recovered from the tubular lumen (Pahl et al., 2001). The boron content of urine from nonexposed people closely corresponds to their intake (Naghii & Samman, 1997).

STORAGE

Bones, thyroid glands, and spleen contain higher boron concentrations than other tissues. Little is known about the deposition or release from these or other tissues.

REGULATION

Boron does not accumulate in tissues even with much higher than normal intake (Hunt et al., 1997). Similarly, blood concentrations appear to be maintained within a narrow concentration range with little impact from dietary intake levels. This has been taken to indicate regulatory mechanisms that maintain a constant boron status (Hunt, 2012). The nature of such potential homeostatic mechanisms is still not known.

FUNCTION

Adequate boron intake is likely to be important for health in many respects, but evidence for specific effects tends to be limited to findings from short-term human studies or animal experiments (Nielsen, 2000). The evaluation of potentially beneficial effects is not made any easier by the uncertainty whether any structures in humans have a prerequisite boron component. Two antibiotics and a bacterial signaling protein (Chen et al., 2002) are the only known biological compounds in any species with an absolute boron requirement. Many reports on boron needs in mammalian and nonmammalian vertebrates demonstrate detrimental (sometimes even catastrophic) health outcomes, but they did not identify specific boron-containing molecules causally linked to such effects. The recent identification of boron-containing S-adenosylmethionine, diadenosine phosphates (Ralston and Hunt, 2001), and other physiological complexes may expand the basis for understanding human boron requirements on a molecular level. NAD also binds boron with high affinity (Hunt, 2012).

Influences on hormone action: The plasma concentration of 7-β-estradiol and the ratio of 7-β-estradiol to testosterone in plasma were found to be higher with boron supplementation than with short-term boron depletion (Naghii, 1999). The health relevance of these and similar changes in steroid sex hormones remains unclear. Adequate boron status has been suggested to increase the effectiveness of several hormones, including insulin (Hunt, 2012) and 1,25-dihydroxyvitamin D.

Bone and connective tissues: Adequate boron intake appears to improve growth, mineralization, and mineral retention of bone (Hunt et al., 1997; Armstrong et al., 2000).

Uncertainties remain about the underlying mechanisms, the extent to which low boron status contributes to accelerated bone mineral loss in adults, and what optimal boron intake levels would be.

Some of the beneficial effects of adequate boron status on bone health may be directed at connective tissue. Boric acid and other boron derivatives accelerate wound healing, possibly through promoting the release of proteins and proteoglycans into the extracellular matrix and through enhancing the activity of intracellular and extracellular proteases (Benderdour et al., 2000).

Other effects in humans: Increase of intake levels from low (0.4 mg/day) to high normal (3.2 mg/day) appears to moderately increase both diastolic and systolic blood pressure (Hunt et al., 1997). Neither the mechanism nor the health significance of such effects is known.

Actions in nonmammalian species: Boron is essential for normal sexual function and offspring development in many nonmammalian vertebrates. A well-documented model is the disruption of normal organ development in boron-deficient frogs (Fort et al., 2000). Boron is also essential for the growth of many plants. The walls of growing plants in particular contain polysaccharides, such as rhamnogalacturonan-II, which are inactive without cross-linked borate esters (O'Neill et al., 2001).

Aplasmomycin and boromycin are two macrolide antibiotics produced by *Streptomyces* species. Boromycin is active against HIV (Kohno et al., 1996). The first demonstration of a compound other than antibiotics with an absolute boron requirement is the bacterial autoinducer 2 (AI-2). The active form of AI-2 in a wide variety of bacterial species contains furanosyl borate diester. Binding of active AI-2 to the sensor protein LuxP is an important mechanism for communication between bacteria (quorum sensing), even across species barriers (Chen et al., 2002).

Boron neutron capture therapy (BNCT): A form of tumor therapy known as *boron neutron capture therapy (BNCT)* takes advantage of the fact that low-energy irradiation of boron generates high-energy alpha particles ($^{10}B(n,\alpha)7Li$) that can destroy cells in its immediate vicinity. Enrichment of boron in tumor targets allows some degree of specificity (Gibson et al., 2001).

REFERENCES

Armstrong, T.A., Spears, J.W., Crenshaw, T.D., Nielsen, F.H., 2000. Boron supplernentation of a semipurified diet for weanling pigs improves feed efficiency and bone strength characteristics and alters plasma lipid metabolites. J. Nutr. 130, 2575–2581.

Benderdour, M., Van Bui, T., Hess, K., Dicko, A., Belleville, E., Dousset, B., 2000. Effects of boron derivatives on extracellular matrix formation. J. Trace Elem. Med. Biol. 14, 168–173.

Chen, X., Schauder, S., Potier, N., Van Dorsselaer, A., Pelczer, I., Bassler, B.L., et al., 2002. Structural identification of a bacterial quorum-sensing signal containing boron. Nature 415, 545–549.

Coughlin, J.R., 1998. Sources of human exposure: overview of water supplies as sources of boron. Biol. Trace Elem. Res. 66, 87–100.

Dordas, C., Brown, P.H., 2001. Permeability and the mechanism of transport of boric acid across the plasma membrane of Xenopus laevis oocytes. Biol. Trace Elem. Res. 81, 127–139.

Food and Nutrition Board, Institute of Medicine, 2002. Dietary Reference Intakes for Vitamin A, Vitamin K, Arsenic, Boron, Chromium, Copper, Iodine, Iron, Manganese, Molybdenum, Nickel, Silicon, Vanadium, and Zinc. National Academy Press, Washington, DC.

Fort, D.J., Stover, E.L., Rogers, R.L., Copley, H.E., Morgan, L.A., Foster, E.R., 2000. Chronic boron or copper deficiency induces limb teratogenesis in Xenopus. Biol. Trace Elem. Res. 77, 173–187.

Gibson, C.R., Staubus, A.E., Barth, R.E., Yang, W., Kleinholz, N.M., Jones, R.B., et al., 2001. Boron neutron capture therapy of brain tumors: investigation of urinary metabolites and oxidation products of sodium borocaptate by electrospray ionization mass spectrometry. Drug Metab. Disp. 29, 1588–1598.

Hunt, C.D., 2012. Dietary boron: progress in establishing essential roles in human physiology. J. Trace Elem. Med. Biol. 26 (2–3), 157–160.

Hunt, C.D., Meacham, S.L., 2001. Aluminum, boron, calcium, copper, iron, magnesium, manganese, molybdenum, phosphorus, potassium, sodium, and zinc: concentrations in common western foods and estimated daily intakes by infants: toddlers: and male and female adolescents, adults, and seniors in the United States. J. Am. Diet. Assoc. 101, 1058–1060.

Hunt, C.D., Herbel, J.L., Nielsen, F.H., 1997. Metabolic responses of postmenopausal women to supplemental dietary boron and aluminum during usual and low magnesium intake: boron, calcium, and rnagnesium absorption and retention and blood mineral concentrations. Am. J. Clin. Nutr. 65, 803–813.

Kohno, J., Kawahata, T., Otake, T., Morimoto, M., Mori, H., Ueba, N., et al., 1996. Boromycin, an anti-HIV antibiotic. Biosci. Biotechnol. Biochem. 60, 1036–1037.

Murray, F.J., 1998. A comparative review of the pharmacokinetics of boric acid in rodents and humans. Biol. Trace Elem. Res. 66, 331–341.

Naghii, M.R., 1999. The significance of dietary boron, with particular reference to athletes. Nutr. Hlth. 13, 31–37.

Naghii, M.R., Samman, S., 1997. The effect of boron supplementation on its urinary excretion and selected cardiovascular risk factors in healthy male subjects. Biol. Trace Elem. Res. 56, 273–286.

Nielsen, F.H., 2000. The emergence of boron as nutritionally important throughout the life cycle. Nutrition 16, 512–514.

O'Neill, M.A., Eberhard, S., Albersheim, P., Darvill, A.G., 2001. Requirement of borate cross-linking of cell wall rhamnogalacturonan II for Arabidopsis growth. Science 294, 846–849.

Pahl, M.V., Culver, B.D., Strong, P.L., Murray, F.J., Vaziri, N.D., 2001. The effect of pregnancy on renal clearance of boron in humans: a study based on normal dietary intake of boron. Toxicol. Sci. 60, 252–256.

Rainey, C., Nyquist, L., 1998. Multicountry estimation of dietary boron intake. Biol. Trace Elem. Res. 66, 79–86.

Rainey, C.J., Nyquist, L.A., Christensen, R.E., Strong, P.L., Culver, B.D., Coughlin, J.R., 1999. Daily boron intake from the American diet. J. Am. Diet Assoc. 99, 335–340.

Ralston, N.V., Hunt, C.D., 2001. Diadenosine phosphates and S-adenosylmethionine: novel boron binding biomolecules detected by capillary electrophoresis. Biochim. Biophys. Acta 1527, 20–30.

Silberstein, T., Saphier, M., Mashiach, Y., Paz-Tal, O., Saphier, O., 2015. Elements in maternal blood and amniotic fluid determined by ICP-MS. J. Matern. Fetal. Neonatal. Med. 28, 88–92. 2015.

SILICON

Silicon (Si; atomic weight 28.0855) is a metalloid element. Naturally occurring Si contains the stable isotopes ^{28}Si (92.23%), ^{29}Si (4.67%), and ^{30}Si (3.1%), and traces of the unstable isotope ^{32}Si (half-life 153 years). It occurs abundantly as silica (sand, quartz), as silicates, and in other forms, and it can have valences 2 and 4.

ABBREVIATION

Si   silicon

NUTRITIONAL SUMMARY

Function: It is unlikely that Si is essential for humans, but adequate intake may support bone growth and mineral retention and provide other health benefits.

Food sources: A typical mixed diet provides several times the estimated requirements. High-fiber plant foods, cereal products, and beer are good sources. Hard water also contains significant amounts. Another source is magnesium trisilicate, a commonly used antacid. An emerging issue is the use of nanosilica in foods.

Requirements: Intake of 5–10 mg/day is thought to be adequate for adults.

Deficiency: Very low intake in experimental animals can impair bone growth and depress collagen synthesis.

Excessive intake: Adverse effects even at many times the required intake levels have not been observed. Kidney stones in very rare instances contain silicates, but a connection with intake, especially from silicate antacids, has not been established.

DIETARY SOURCES

Most Si in food is present as silica (as in sand and quartz), an insoluble polymer that consists mostly of Si and oxygen (–O–Si–O–), and silicates. The simplest and reasonably water-soluble silanol form is orthosilicic acid, $Si(OH)_4$. Unrefined grains and other plant foods with high fiber content have the highest Si content (Pennington, 1991), mainly as amorphous $SiO_2 \cdot n(H_2O)$ (Epstein, 1994). Other cereal products and beer are good sources, too.

It may be of interest to note that even for most terrestrial plants (with the exception of *Equisitaceae*), Si may not be an essential nutrient, but it provides benefits for plant vigor and disease resistance (Epstein, 1994). Tap water may contain as little as 0.07 mmol/l silicate (corresponding to 2 mg Si/l), while mineral waters have as much as 0.4 mmol/l (Li and Chen, 2000).

The presence of silica as nanoparticulate matter in foods presents new questions, mainly because it is not known whether the particles are dissolved by digestion into plain silica, just excreted via feces, or absorbed more or less intact (Peters et al., 2012). So far, it appears that the nanoparticles persist in the context of at least some foods and make it into intestinal tissues. The long-term consequences of such exposure are not known.

Daily average Si intake has been estimated at 20–50 mg (Pennington, 1991), which is several times the recommended level of 5–10 mg. Silicates are added to some foods because of their antifoaming and anticlumping properties.

Magnesium trisilicate is commonly used as an antacid. A tablet typically contains about 6.5 mg of Si. Significant amounts of Si may also be ingested with clay (Calabrese et al., 1990). Clay-eating behavior (pica), especially by children and pregnant women, is not unusual in some rural regions of the United States.

DIGESTION AND ABSORPTION

The bioavailability of Si compounds depends foremost on their chemical form. Silica, aluminum silicate, and some other silicates are absorbed very little or not at all. In contrast, between one-third and one-half of an ingested dose of soluble monomeric silicic acid is absorbed (Bellia et al., 1996; Popplewell et al., 1998; Reffitt et al., 1999). The rapidity with which silicic acid is absorbed points to the proximal small intestine as the main site of absorption. Bioavailability of ingested Si varies greatly, ranging from more than 40% for Si from beer and green beans to just 1% for colloidal silica in dietary supplements (Sripanyakorn et al., 2009). Absorption decreases with increasing age and decreasing estrogen production.

TRANSPORT AND CELLULAR UPTAKE

Blood circulation: Silicic acid predominates in the blood, almost to the exclusion of all other chemical forms. Typical concentrations are around 5 μmol/l (Reffitt et al., 1999). Ingestion of soluble silicate in water raises plasma Si concentration moderately, but distinctly, with a peak after 1 h. Food silicate raises plasma concentrations similarly, but more slowly. It is not clear how tissues take up Si.

STORAGE

Lung, rigid connective tissue (such as the trachea and aorta), lymph nodes, and bone contain significant amounts of Si. Growth zones (epiphysis) appear to contain particularly high amounts (LeVier, 1975). Available data are limited and possibly obtained with outdated and less reliable methods. It is not known how the element gets there or how it might be mobilized again.

EXCRETION

Renal clearance is high (around 89 ml/min; Berlyne et al., 1985; Reflitt et al., 1999). This means that most of the absorbed Si is excreted via urine and little net reabsorption occurs. A decrease in renal filtration rapidly raises serum concentrations (Dobbie and Smith, 1986), as seen in patients with renal failure (Roberts and Williams, 1990).

REGULATION

High Si intake is rapidly balanced by overflow into urine, with little if any recovery. It is not clear whether additional homeostatic mechanisms exist.

FUNCTION

No Si-containing protein or other endogenously synthesized Si compound has been identified that could be linked to the suggested essential nature of Si. It has been suggested that Si affects gene expression through an as-yet-unknown mechanism. Some of the postulated benefits (Seaborn & Nielsen, 1993) could be indirect. A particularly important effect could be the ability of silicates to form insoluble complexes with aluminum, thereby hindering the absorption of this potentially harmful metal from the intestines and enhance its excretion via urine. It has even been suggested that silicate can mobilize aluminum deposits from the body (Bellia et al., 1996), but a later investigation left this in doubt (Reffitt et al., 1999).

Reports on the effects of Si inadequacy have centered on bone and the decreased collagen content in skin and connective tissue (Calomme & Vanden Berghe, 1997; Seaborn & Nielsen, 2002) of experimental animals, as well as the lower hydroxyproline content and alkaline and acid phosphatase activities in their bones. It remains to be seen whether such effects can be confirmed in humans (or in the animal models) and which molecular mechanisms might be responsible.

A large, targeted population study found evidence suggesting that long-term exposure to higher-than-average silica content of drinking water may reduce the risk of dementia in elderly people (Rondeau et al., 2009).

A modest body of evidence suggests that silica may improve the tensile strength of thin hair (Wickett et al., 2007) and improve the structure of aged skin.

The ability of inhaled silica dust (in silicosis, miners' lung, and similar occupational diseases) to induce localized collagen synthesis should not be taken to indicate a beneficial effect of dietary Si on bone and connective tissue.

REFERENCES

Bellia, J.E., Birchall, J.D., Roberts, N.B., 1996. The role of silicic acid in the renal excretion of aluminium. Ann. Clin. Lab. Sci. 26, 227–233.

Berlyne, G., Dudek, E., Adler, A.J., Rubin, J.E., Seidman, M., 1985. Silicon metabolism: the basic facts in renal failure. Kidney Int. 17, S175–S177.

Calabrese, E.J., Stanek, E.J., Gilbert, C.E., Barnes, R.M., 1990. Preliminary adult soil ingestion estimates: results of a pilot study. Reg. Toxicol. Pharmacol. 12, 88–95.

Calomme, M.R., Vanden Berghe, D.A., 1997. Supplementation of calves with stabilized orthosilicic acid. Effect on the Si, Ca, Mg, and P concentrations in serum and the collagen concentration in skin and cartilage. Biol. Trace Elem. Res. 56, 153–165.

Dobbie, J.W., Smith, M.B., 1986. Urinary and serum silicon in normal and uraemic individuals. Ciba Found. Symp. 121, 194–213.

Epstein, E., 1994. The anomaly of silicon in plant biology. Proc. Natl. Acad. Sci. USA 91, 11–17.

LeVier, R.R., 1975. Distribution of silicon in the adult rat and rhesus monkey. Bioinorg. Chem. 4, 109–115.

Li, H.B., Chen, E., 2000. Determination of silicate in water by ion exclusion chromatography with conductivity detection. J. Chromatogr. A 874, 143–147.

Pennington, J.A., 1991. Silicon in foods and diets. Food Add Cont. 8, 97–118.

Peters, R., Kramer, E., Oomen, A.G., Rivera, Z.E., Oegema, G., Tromp, P.C., et al., 2012. Presence of nano-sized silica during *in vitro* digestion of foods containing silica as a food additive. ACS Nano 6, 2441–2451.

Popplewell, J.F., King, S.J., Day, J.P., Ackrill, E., Fifield, L.K., Cresswell, R.G., et al., 1998. Kinetics of uptake and elimination of silicic acid by a human subject: a novel application of ^{32}Si and accelerator mass spectrometry. J. Inorg. Biochem. 69, 177–180.

Reffitt, D.M., Jugdaohsingh, R., Thompson, R.E., Powell, J.J., 1999. Silicic acid: its gastrointestinal uptake and urinary excretion in man and effects on aluminium excretion. J. Inorg. Biochem. 76, 141–147.

Roberts, N.B., Williams, P., 1990. Silicon measurement in serum and urine by direct current plasma emission spectrometry. Clin. Chem. 36, 1460–1465.

Rondeau, V., Jacqmin-Gadda, H., Commenges, D., Helmer, C., Dartigues, J.F., 2009. Aluminum and silica in drinking water and the risk of Alzheimer's disease or cognitive decline: findings from 15-year follow-up of the PAQUID cohort. Am. J. Epidemiol. 169, 489–496.

Seaborn, C.D., Nielsen, F.H., 1993. Silicon: a nutritional beneficence for bones, brains and blood vessels? Nutr. Today 28, 13–18.

Seaborn, C.D., Nielsen, F.H., 2002. Silicon deprivation decreases collagen formation in wounds and bone, and ornithine transaminase enzyme activity in liver. Biol. Trace Elem. Res. 89, 251–261.

Sripanyakorn, S., Jugdaohsingh, R., Dissayabutr, W., Anderson, S.H., Thompson, R.P., Powell, J.J., 2009. The comparative absorption of silicon from different foods and food supplements. Br. J. Nutr. 102, 825–834.

Wickett, R.R., Kossmann, E., Barel, A., Demeester, N., Clarys, P., Vanden Berghe, D., et al., 2007. Effect of oral intake of choline-stabilized orthosilicic acid on hair tensile strength and morphology in women with fine hair. Arch. Dermatol. Res. 299, 499–505.

BROMINE

Bromine is a halogen (atomic weight 79.9). Naturally occurring bromine contains the stable isotopes ^{79}Br (50.7%) and ^{81}Br (49.3%). Bromides (BrX) and bromates ($Br(O_3)X$) are the salts of bromine.

ABBREVIATION

Br   bromine

NUTRITIONAL SUMMARY

Function: Bromine (Br) is used by eosinophilic leukocytes for immune defense.

Food sources: Significant contributors to dietary intake grains, nuts, sea salt, seafood, and bread.

Requirements: No requirements have been established. Intake around 8 mg/day appears to be adequate.

Deficiency: The consequences of chronically low intake are uncertain; growth retardation and insomnia have been suggested. People on chronic hemodialysis may be at increased risk.

Excessive intake: High acute exposure, such as from swallowing excessively brominated pool water during swimming or from inhalation, may cause bronchospasm, headache, gastrointestinal disturbances, fatigue, reduced exercise tolerance, and myalgia.

DIETARY AND OTHER SOURCES

Grains, nuts, seafood, and sea salt are significant dietary sources. Brominated flour is sometimes used for bread and other baked goods. Brominated vegetable oil is added to some beverages at a concentration of up to 15 ppm (FDA CFR, 2013). Average Br intake in young Dutch adults has been reported to be 8 mg/day (van Dokkum et al., 1989). Ukrainians were reported to have a median daily intake of about 3 mg (Shiraishi et al., 2009).

Occasional sources may be Br-based sanitizing agents and brominated swimming pool water.

DIGESTION AND ABSORPTION

Dietary bromide is taken up into intestinal enterocytes, presumably by chloride transporters (Prat et al., 1999). Other bromide-carrying systems, such as the dual ion exchange mechanism of Na^+/H^+ and Cl^-/HCO_3^-, have been postulated for the colon (Mahajan et al., 1996).

TRANSPORT AND CELLULAR UPTAKE

Blood circulation: Br is present in the blood as bromide at a typical concentration of 4 mg/l. Transport into cells may utilize systems for the uptake of other anions, such as chloride-selective channels in neurons (Carpaneto et al., 1999) or the Na–K–2Cl cotransporter in teeth (Rajendran et al., 1995; Prostak and Skobe, 1996).

STORAGE

Tissue bromide concentrations are highest in the lungs (Hou et al., 1997) and liver (0.11 mmol/kg dry weight; Laursen et al., 1998). Radiolabel experiments in rats indicated that a large portion of ingested bromide distributes to skin due to its large volume (Pavelka et al., 2000). Preferential storage of Br in the thyroid gland was not observed by some (Pavelka et al., 2000), but not by others (Vobecky et al., 1996). It has been suggested that, especially in a state of iodine depletion, iodine atoms are replaced by Br atoms (Vobecky et al., 1996).

EXCRETION

Several hundred milligrams of bromide are filtered daily by the kidneys and are likely to be recovered via chloride transporters and other carriers, as discussed previously. Information on the precise nature of renal reabsorption of bromide is very limited. Excretion via urine is the main route of Br losses (Pavelka et al., 2000).

REGULATION

Little is known about mechanisms that might control the bromide content of the body.

FUNCTION

Immune defense: A specific peroxidase (EC1.11.1.7) in the cytoplasmic granules of eosinophils uses Br to generate a halogenating oxidant (Wu et al., 1999), which brominates tyrosines and other amino acids in proteins (Henderson et al., 2001). This heme-enzyme is structurally distinct from myeloperoxidase in neutrophils, myelocytes, and macrophages. While other peroxidases generate hypochlorous acid (HOCl) as an oxidative reactant, eosinophil peroxidase produces hypobromous acid in an analogous reaction:

$$Br^- + H_2O_2 + H^+ \rightarrow HOBr + H_2O$$

Hypobromous acid (HBrO) potently brominates the deoxycytidine moieties of DNA (Figure 11.22) and thereby disrupts the replication of invading parasites (Henderson et al., 2001). Additional corrosive reactants are generated by both eosinophils and neutrophils, including the interhalogen ClBr, as shown in the following reaction scheme:

$$HOCl + Br^- + H^+ \rightarrow BrCl + H_2O$$

This very volatile interhalogen is readily hydrolyzed:

$$BrCl + H_2O \rightarrow HOBr + Cl^- + H^+$$

Taurine can be brominated to the corrosive oxidant *N*-bromotaurine by reacting with hypobromous acid or with hypochlorous acid in the presence of bromide. *N*-bromotaurine brominates the deoxycytidine in DNA at neutral pH, while *N*-chlorotaurine can do this only in an acid environment. These

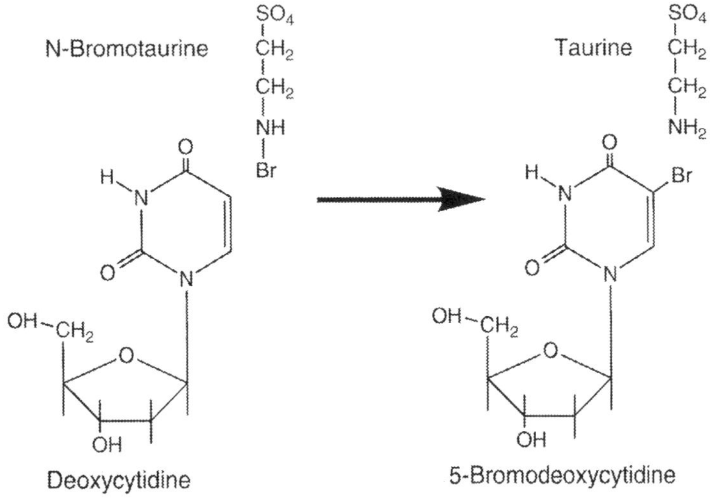

FIGURE 11.22

Bromotaurine and other brominated primary amines are corrosive reactants that attack DNA.

reactive molecules can also attack tissue proteins and DNA. Prolonged and excessive production due to inflammation has been tentatively linked to cancer and atherosclerosis.

Basement membranes: Collagen IV forms the scaffold that gives structure to basement membranes. The Br-dependent enzyme peroxidasin (EC1.11.1.7) adds sulfilimine (S=N) cross-links between individual newly formed collagen strands (McCall et al., 2014). A lack of Br, which is rare due to the ubiquitous presence of this volatile halogen, would disrupt proper three-dimensional orientation of tissues. The biological significance of this newly described connection is unknown.

Thyroid function: Br may compete with iodine for transport into the thyroid gland (Vobecky et al., 1996) and thereby mildly inhibit thyroid function (Velicky et al., 1997).

Sleep: Low Br status has been related to insomnia in patients whose hemodialysis constantly removes a significant portion of the bromide in the blood.

Before the development of newer medications, bromides were used as sedatives and anticonvulsants. Potassium bromide is still used sometimes for the treatment of epileptic seizures in veterinary medicine (Boothe et al., 2012).

DEFICIENCY

Sleep disturbances in patients depleted of Br by hemodialysis have been reported (Mahowald and Bornemann, 2006). The significance of such effects is uncertain.

EXCESS

A rare case of severe Br intoxication (bromism) due to long-term use of a nonprescription formulation (calcium bromo-galactogluconate) at a daily dose of 6 g was reported to be associated with confusion, disorientation, auditory and visual hallucinations, and loss of short-time memory (Frances et al., 2003).

REFERENCES

Boothe, D.M., Dewey, C., Carpenter, D.M., 2012. Comparison of phenobarbital with bromide as a first-choice antiepileptic drug for treatment of epilepsy in dogs. J. Am. Vet. Med. Assoc. 240, 1073–1083.

Carpaneto, A., Accardi, A., Pisciotta, M., Gambale, F., 1999. Chloride channels activated by hypotonicity in N2A neuroblastoma cell line. Exp. Brain Res. 124, 193–199.

FDA CFR—Code of Federal Regulations Title 21 (21CFR180.30, revised as of April 1, 2013). <http://www.accessdata.fda.gov/scripts/cdrh/cfdocs/cfcfr/CFRSearch.cfm?FR=180.30>.

Frances, C., Hoizey, G., Lamiable, D., Millart, H., Trenque, T., 2003. Bromism from daily over intake of bromide salt. J. Toxicol. Clin. Toxicol. 41, 181–183.

Henderson, J.E., Byun, J., Williams, M.V., Mueller, D.M., McCormick, M.L., Heinecke, J.W., 2001. Production of brominating intermediates by myeloperoxidase. A transhalogenation pathway for generating mutagenic nucleobases during inflammation. J. Biol. Chem. 276, 7867–7875.

Hou, X., Chai, Z., Chen, Q., 1997. Determination of tissue chlorine, bromine, and iodine concentrations in normal persons with neutron activation method. Chung-Hua fit Fang I Hsueh Tsa Chih [Chinese J. Prevent. Med.] 31, 288–291.

Laursen, J., Milman, N., Petersen, H.S., Mulvad, G., Jul, F., Saaby, H., et al., 1998. Elements in autopsy liver tissue samples from Greenlandic Inuit and Danes. I. Sulphur, chlorine, potassium and bromine measured by X-ray fluorescence spectrometry. J. Trace Elem. Med. Biol. 12, 109–114.

Mahajan, R.J., Baldwin, M.L., Harig, J.M., Ramaswamy, K., Dudeja, P.K., 1996. Chloride transport in human proximal colonic apical membrane vesicles. Biochim. Biophys. Acta 1280, 12–18.

Mahowald, M.W., Bornemann, M.A., 2006. Sleep and ESRD: a wake-up call. Am. J. Kidney Dis. 48, 332–334.

McCall, A.S., Cummings, C.F., Bhave, G., Vanacore, R., Page-McCaw, A., Hudson, B.G., 2014. Bromine is an essential trace element for assembly of collagen IV scaffolds in tissue development and architecture. Cell 157, 1380–1392.

Pavelka, S., Babicky, A., Vobecky, M., Lener, J., 2000. Bromide kinetics and distribution in the rat. II. Distribution of bromide in the body. Biol. Trace Elem. Res. 76, 67–74.

Prat, A.G., Cunningham, C.C., Jackson Jr, G.R., Borkan, S.C., Wang, Y., Ausiello, D.A., et al., 1999. Actin filament organization is required for proper cAMP-dependent activation of CFTR. Am. J. Physiol. 277, C1160–C1169.

Prostak, K.S., Skobe, Z., 1996. Anion translocation through the enamel organ. Adv. Dent. Res. 10, 238–244.

Rajendran, V.M., Geibel, J., Binder, H.J., 1995. Chloride-dependent Na-H exchange. A novel mechanism of sodium transport in colonic crypts. J. Biol. Chem. 270, 11051–11054.

Shiraishi, K., Ko, S., Muramatsu, Y., Zamostyan, P.V., Tsigankov, N.Y., 2009. Dietary iodine and bromine intakes in Ukrainian subjects. Health Phys. 96, 5–12.

van Dokkum, W., de Vos, R.H., Muys, T., Wesstra, J.A., 1989. Minerals and trace elements in total diets in The Netherlands. Br. J. Nutr. 61, 7–15.

Velicky, J., Titlbach, M., Duskova, J., Vobecky, M., Strbak, V., Raska, I., 1997. Potassium bromide and the thyroid gland of the rat: morphology and immunohistochernistry, RIA and INAA analysis. Anat. Anz. 179, 421–431.

Vobecky, M., Babicky, A., Lener, J., 1996. Effect of increased bromide intake on iodine excretion in rats. Biol. Trace Elem. Res. 55, 215–219.

Wu, W., Chen, Y., d'Avignon, A., Hazen, S.L., 1999. 3-Bromotyrosine and 3,5-dibromotyrosine are major products of protein oxidation by eosinophil peroxidase: potential markers for eosinophil-dependent tissue injury *in vivo*. Biochemistry 38, 3538–3548.

ARSENIC

Arsenic (As; atomic weight 74.92) can have oxidative states 3 and 5. [75]As is the only naturally occurring isotope. The synthetic isotopes [73]As and [74]As are unstable.

ABBREVIATIONS

As	arsenic
DMA(III)	dimethylarsinous acid
DMA(V)	dimethylarsinic acid
MMA(III)	monomethylarsonous acid
MMA(V)	monomethylarsonic acid
MRP1	multidrug resistance protein 1
SAM	S-adenosyl methionine
GSH	reduced glutathione

NUTRITIONAL SUMMARY

Function: Arsenic (As) may be an essential element. It has been suggested that it might play a role in the methylation of DNA, histones, or both.

Food sources: Regionally variable amounts of inorganic As come from drinking water. Seafood contains the highest amounts of organically bound As.

Requirements: If As is needed at all, the requirements are minute (much less than 50 µg/day). The molecular species (inorganic versus organic) of potentially needed As is not known.

Deficiency: As depletion in laboratory animals has been associated with damage of the central nervous and vascular systems and an increased cancer risk. Such As deficiency has not been observed in humans.

Excessive intake: All As compounds are toxic in more than miniscule amounts, causing gastrointestinal distress, skin pigmentation, keratosis (arsenicism), cancer of the skin, lung, bladder, and other sites, gangrene of the extremities (blackfoot disease), anemia, symmetric peripheral neuropathy, damage to the brain and liver, and possibly death. It is genotoxic and mutagenic at low concentrations. No As-containing supplements should be used and the As content of public water supplies has been limited by the US Environmental Protection Agency (EPA) to less than 10 µg/l due to concerns about the potential of As to cause cancer of the skin, bladder, lung, and other organs, even at very low levels of intake.

DIETARY AND OTHER SOURCES

Estimates of As intake of American adults have been as low as 12–50 µg/day (Uthus & Seaborn, 1996) and as high as 28–92 µg/day (Tao & Bolger, 1999). The highest concentration in foods consumed by Canadians was found to be in fish (1.6 µg/g in organic form, mainly arsenobetaine); much lower amounts are present in meats, baked goods, and oils (about 0.02 µg/g; Dabeka et al., 1993). Shrimp, lobster, mussels, and other types of seafood also contain As, mainly as arsenobetaine, and to a lesser extent arsenocholine. Dairy products appear to contribute nearly one-third of total intake, with much of the remainder coming from grain products (Mahaffey et al., 1975). Trivalent and pentavalent As salts are the main forms in water unless it is contaminated with As-containing compound originating from fungicide or pesticide use.

Edible seaweeds (such as nori consumed with sushi) contain arsenosugars, mainly as 5-deoxy-5-dimethylarsinyl-ribofuranoside moieties of more complex carbohydrates (Le et al., 1994).

As may also be taken up as arsine gas (ASH3), mainly during occupational exposure.

DIGESTION AND ABSORPTION

Arsenate and arsenite, the main inorganic forms present in water, are absorbed almost completely (90%) from the small intestine. Arsenobetaine, arsenocholine, and other organic forms that predominate in foods appear to be slightly less well absorbed (60–70%; Hopenhayn-Rich et al., 1993). Arsenobetaine is absorbed largely unchanged (Le et al., 1994). The mechanism of betaine uptake in humans is not well understood. Both sodium-dependent and sodium-independent components of betaine uptake appear to operate in birds (Kettunen et al., 2001). The cationic amino acid transporter 2 (SLC7A2), which mediates choline uptake, might provide the means for entry of arsenocholine. Absorbed arsenite, possibly the portion that is conjugated to glutathione, appears to be exported back into the intestinal lumen by the multidrug resistance protein 1 (MRP1; Borst et al., 2000).

TRANSPORT AND CELLULAR UPTAKE

Blood circulation: Serum typically contains about 0.8 µg/l of dimethylarsinate and about 3.5 µg/l of arsenobetaine (Zhang et al., 1996). Hemodialysis removes 68% of total As from serum and 16% from blood cells (Zhang et al., 1996).

The mechanisms responsible for the uptake of As into cells remain uncertain. Betaine transporters may play a special role, since this is the main molecular As species in the blood. Recently, the human and mouse homologs of an As-translocating plant ATPase were identified. The protein is present mainly in the kidneys and testis, and at much lower levels in the brain, liver, lungs, and skin. The precise role of this protein is not yet sufficiently understood; it may not contribute to As disposition in mammals at all (Bhattacharjee et al., 2001).

METABOLISM

Methylation is essential for the clearance of inorganic As from cells (Thomas et al., 2007). Several alternative enzymatic reactions are likely to contribute to As metabolism. A tentative outline of the quantitatively most important pathways is described in the next sections. Revisions are likely to become necessary as new research data emerge.

Inorganic As is reduced in the liver to arsenite in a glutathione-dependent reaction; the arsenate reductase (EC1.20.4.1) activity for this reaction may be supplied by glyceraldehyde 3-phosphate dehydrogenase (EC1.2.1.12) (Gregus and Németi, 2005). Transfer of a methyl group by S-adenosylmethionine-dependent arsenite methyltransferase (EC2.1.1.137) generates monomethyl arsonic acid (MMA(V)). Glutathione transferase omega 1 (EC2.5.1.18), an ubiquitous enzyme with high expression in the liver, can then reduce this intermediate (Zakharyan et al., 2001) to methylarsonous acid (MMA(III)). This is the rate-limiting reaction of As metabolism (Zakharian and Aposhian, 1999). MMA methyltransferase (EC2.1.1.XX, S-adenosylmethionine-dependent) readily generates dimethylarsinic acid (DMA(V) from MMA(III)).

This enzyme may be identical with arsenite methyltransferase. A distinct pathway catalyzed by a single As-methyltransferase may generate MMA(III) and DMA(III) directly from arsenite without intervening reduction steps (Figure 11.23).

Trimethylarsine oxide and tetramethylarsonium ion are likely to come only from exogenous sources. Arsenobetaine and arsenocholine do not appear to be significantly metabolized, but arsenosugars are (Le et al., 1994). The metabolism of different As species appears to differ greatly between individuals (Le et al., 1994). In rats, inorganic As in the form of arsine gas was shown to be converted almost

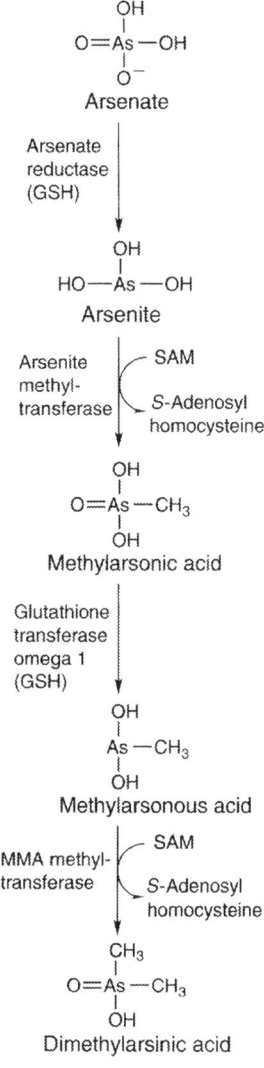

FIGURE 11.23

Arsenate becomes more toxic through *in vivo* methylation.

exclusively to its methylated forms, but not to arsenobetaine (Buchet et al., 1998). Considering very significant species differences regarding As metabolism (National Research Council, 1977), it remains to be seen whether the same applies to humans.

STORAGE

While it is clear that As is stored in the body (Aposhian et al., 1997), little is known about typical amounts, the molecular species involved, or the mechanisms of mobilization.

EXCRETION

Ingested As compounds are rapidly excreted via urine, mainly as arsenobetaine (34% of total As), DMA (28%), MMA (26%), trivalent and pentavalent As (12%), and, to a lesser extent, trimethylarsine oxide and tetramethylarsonium ions (Aposhian et al., 1997; Apostoli et al., 1999).

Whether arsenobetaine is recovered like betaine from primary filtrate via a brush border membrane transporter with high affinity to L-proline and the basolateral betaine transporter (BGT1, SLC6A12) is not known.

REGULATION

Little is known about the mechanisms that might regulate As homeostasis in the body. The activity of multidrug resistance protein 1 (MRP1/ABCC1) is likely to be important since this ABC transporter actively returns absorbed arsenite (possibly the portion that is conjugated to glutathione) into the intestinal and renal tubular lumina (Borst et al., 2000).

FUNCTION

Deficiency interfered with growth and reproductive health of a wide range of mammalian and nonmammalian species (Anke, 1986; Uthus, 1994). Rats with very low As intake were found to have unusually low *S*-adenosyl methionine concentrations in the liver (Uthus, 1994).

Arsenite competes for *S*-adenosyl methionine with selenium and cytosine DNA methyltransferase, a key enzyme for the control of gene transcription (Davis et al., 2000). It may be through this or another mechanism that arsenite influences DNA methylation, histones, or both; this interaction may modify gene expression (Meng & Meng, 1994), induce heat shock proteins and possibly affect cancer risk. Most inorganic As compounds are strongly cytotoxic and induce several types of cancer in humans. Arsenic trioxide is sometimes used in patients with malignancies (particularly multiple myeloma), resistant to other antineoplastic agents.

Other effects: Arsenic metabolites, especially MMA(III) and DMA(III) are extremely cytotoxic and can damage DNA directly; that is, they are mutagenic and genotoxic. Arsenic metabolites also inhibit many enzymes with high potency.

REFERENCES

Anke, M., 1986. Arsenic In: Mertz, W. (Ed.), Trace Elements in Human and Animal Nutrition, vol. 2 Academic Press, Orlando, FL, pp. 347–372.

Aposhian, H.V., Arroyo, A., Cebrian, M.E., Razo, L.M.D., Hurlbut, K.M., Dart, R.C., et al., 1997. DMPS-arsenic challenge test. 1: Increased urinary excretion of monomethylarsonic acid in humans given dimercaptopropane sulfonate. J. Pharmacol. Exp. Ther. 282, 192–200.

Apostoli, P., Bartoli, D., Alessio, L., Buchet, J.P., 1999. Biological monitoring of occupational exposure to inorganic arsenic. Occup. Environ. Med. 56, 825–832.

Bhattacharjee, H., Ho, Y.S., Rosen, B.E., 2001. Genomic organization and chromosomal localization of the Asna1 gene, a mouse homologue of a bacterial arsenic-translocating ATPase gene. Gene 272, 291–299.

Borst, P., Evers, R., Kool, M., Wijnholds, J., 2000. A family of drug transporters: the multidrug resistance-associated proteins. J. Natl. Cancer Inst. 92, 1295–1302.

Buchet, J.R., Apostoli, P., Lison, D., 1998. Arsenobetaine is not a major metabolite ofarsine gas in the rat. Arch. Toxicol. 72, 706–710.

Dabeka, R.W., McKenzie, A.D., Lacroix, G.M., Cleroux, C., Bowe, S., Graham, R.A., et al., 1993. Survey of arsenic in total diet food composites and estimation of the dietary intake of arsenic by Canadian adults and children. JAOAC Int. 76, 14–25.

Davis, C.D., Uthus, E.O., Finley, J.W., 2000. Dietary selenium and arsenic affect DNA methylation *in vitro* in Caco-2 cells and *in vivo* in rat liver and colon. J. Nutr. 130, 2903–2909.

Food and Nutrition Board, Institute of Medicine, 2002. Dietary Reference Intakes for Vitamin A, Vitamin K, Arsenic, Boron, Chromium, Copper, Iodine, Iron, Manganese, Molybdenum, Nickel, Silicon, Vanadium, and Zinc. National Academy Press, Washington, DC.

Gregus, Z., Németi, B., 2005. The glycolytic enzyme glyceraldehyde-3-phosphate dehydrogenase works as an arsenate reductase in human red blood cells and rat liver cytosol. Toxicol. Sci. 85, 859–869.

Hopenhayn-Rich, C., Smith, A.H., Goeden, H.M., 1993. Human studies do not support the methylation threshold hypothesis for the toxicity of inorganic arsenic. Environ. Res. 60, 161–177.

Kettunen, H., Peuranen, S., Tiihonen, K., Saarinen, M., 2001. Intestinal uptake of betaine *in vitro* and the distribution of methyl groups from betaine, choline, and methionine in the body of broiler chicks. Comp. Biochem. Physiol. A 128, 269–278.

Le, X.C., Cullen, W.R., Reimer, K.J., 1994. Human urinary arsenic excretion after one-time ingestion of seaweed, crab, and shrimp. Clin. Chem. 40, 617–624.

Mahaffey, K.R., Corneliussen, P.E., Jelinek, C.F., Fiorino, J.A., 1975. Heavy metal exposure from foods. Environ Health Persp. 12, 63–69.

Meng, Z., Meng, N., 1994. Effects of inorganic arsenicals on DNA synthesis in unsensitized human blood lymphocytes *in vitro*. Biol. Trace Elem. Res. 42, 201–208.

National Research Council, 1977. Medical and Biological Effects of Environmental Pollutants. Arsenic. National Academy Press, Washington, DC.

Tao, S.S., Bolger, P.M., 1999. Dietary arsenic intakes in the United States: FDA Total Diet Study, September 1991–December 1996. Food Add Contain 16, 465–472.

Thomas, D.J., Li, J., Waters, S.B., Xing, W., Adair, B.M., Drobna, Z., et al., 2007. Arsenic (+3 oxidation state) methyltransferase and the methylation of arsenicals. Exp. Biol. Med. (Maywood) 232, 3–13.

Uthus, E.O., 1994. Diethyl maleate, an *in vivo* chemical depletor of glutathione, affects the response of male and female rats to arsenic deprivation. Biol. Trace Elem. Res. 46, 247–259.

Uthus, E.O., Seaborn, C.D., 1996. Deliberations and evaluations of the approaches, endpoints and paradigms for dietary recommendations of the other trace elements. J. Nutr. 126, 2452S–2459S.

Zakharyan, R.A., Aposhian, H.V., 1999. Enzymatic reduction of arsenic compounds in mammalian systems: the rate-limiting enzyme of rabbit liver arsenic biotransformation is MMA(V) reductase. Chem. Res. Toxicol. 12, 1278–1283.

Zakharyan, R.A., Sampayo-Reyes, A., Healy, S.M., Tsaprailis, G., Board, P.G., Liebler, D.C., et al., 2001. Human monomethylarsonic acid (MMA(V)) reductase is a member of the glutathione-S-transferase superfamily. Chem. Res. Toxicol. 14, 1051–1057.

Zhang, X., Cornelis, R., De Kimpe, J., Mees, L., Vanderbiesen, V., De Cubber, A., et al., 1996. Accumulation of arsenic species in serum of patients with chronic renal disease. Clin. Chem. 42, 1231–1237.

VANADIUM

Vanadium is a transition element (atomic weight 50.94). Most of the naturally occurring vanadium consists of the isotope ^{51}V (99.75%) and a small amount of the stable isotope ^{50}V (0.25%). The synthetic isotopes ^{48}V and ^{49}V are unstable. Potential valences are $-1, +1, +2, +3, +4$, and $+5$. The most common biological forms are the pentavalent vanadate anion (VO_3^-) and the quadrivalent vanadyl cation (VO_2^+).

NUTRITIONAL SUMMARY

Function: Minimal amounts of vanadium appear to be needed for human health, but a specific essential function has not yet been found.

Requirements: Adults should get about 100 μg/day; only a tiny fraction (1–5%) of this will be absorbed.

Food sources: The foods containing the largest amounts of vanadium per serving include shellfish, herbs (such as parsley and dill weed), and mushrooms.

Deficiency: It is not known what the consequences of vanadium deficiency in humans are, since the requirements are always covered. Canning and other forms of processing increase the content in low-vanadium foods.

Excessive intake: Undesirable effects of very high oral doses (>10 mg/day) include gastrointestinal discomfort and green coloring of the tongue. Slowed growth, renal failure, and respiratory and cardiovascular distress might occur at very high doses.

DIETARY SOURCES

Several foods contain vanadium, especially shellfish, mushrooms, parsley, dill, and foods exposed to vanadium-containing steels during processing (dried milk) or storage (canned foods). Human milk has been reported to contain 0.18 μg/l, and infant formula an order of magnitude more than that (Krachler et al., 2000).

The estimated daily vanadium intake of the US population ranges from 10 (Uthus and Seaborn, 1996) to 60 μg/day (Barceloux, 1999).

DIGESTION AND ABSORPTION

Only a small percentage (possibly 1–5%) of ingested vanadium is absorbed. Vanadate appears to be three to five times more bioavailable than the vanadyl form. Since vanadylate is reduced in the stomach to vanadate, the efficiency of uptake can be expected to depend both on stomach conditions and residence time of an ingested dose. Vanadylate is known to bind to transferrin and ferritin, and possibly also to other iron-binding proteins; this raises the possibility that vanadium uptake proceeds at least partially via the nonheme iron absorption pathways.

TRANSPORT AND CELLULAR UPTAKE

Blood circulation: The vanadyl cation binds well to hemoglobin and transferrin. The vanadate anion, on the other hand, may be associated with phosphate-binding proteins. Thus, vanadium is likely to be transported in the blood both as part of macromolecular complexes and within erythrocytes.

METABOLISM

Vanadate can be reduced nonenzymically to vanadyl by ascorbic acid, NADH, or glutathione.

STORAGE

Vanadate predominates in extracellular body fluids, whereas vanadyl is the most common intracellular form (Barceloux, 1999). The total body content of vanadium is between 0.1 and 1 mg, with much of that in bone. Other organs with higher-than-average vanadium concentrations are the kidneys, spleen, liver, testes, and lungs. Vanadium readily forms complexes with a wide range of compounds. Vanadate competes with other transition metals for binding to metalloproteins and ATP, and complexes with cis-diols and other compounds. The vanadyl cation binds to hemoglobin and transferrin.

EXCRETION

Most ingested vanadium appears to be excreted via urine; neither the forms excreted nor the mechanisms underlying excretion have been elucidated.

REGULATION

It is not known whether and how vanadium homeostasis might be maintained.

FUNCTION

Nonmammalian organisms: Vanadium, often in conjunction with iron or other transition metals, is an important cofactor in various bacterial, fungal, and vanadate-dependent enzymes in algae, including chloroperoxidases, iodoperoxidases, and bromoperoxidases and nitrogenases. A highly specialized type of coelomic cell in *Ascidiidae*, which accumulates vanadium from seawater, is distinguished from other cells of these organisms by the functioning of the pentose-phosphate pathway (Ueki et al., 2000); whether this might indicate a special role of a vanadium species in glucose metabolism remains to be seen.

Humans: A variety of mammalian species requires trace amounts of vanadium for survival and reproductive health, but a vanadium-deficiency disease has not been identified in humans (Barceloux, 1999). Vanadium has been reported to compete with phosphate for the active sites of phosphatases, tyrosine phosphorylases, and phosphate transport proteins. Vanadium compounds (vanadyl sulfate, sodium metavanadate, and peroxovanadate-nicotinic acid) in higher than physiological doses reduce blood glucose levels independent of insulin action, possibly through the inhibition of protein phosphotyrosine phosphatases and activation of nonreceptor protein tyrosine kinases (Terziyski et al., 1999; Goldwaser et al., 2000).

Vanadium chelated to organic compounds may be less toxic than the free forms. Possibly through similar mechanisms, pharmacologic vanadium doses also may lower blood–lipid levels (Harland and Harden-Williams, 1994). Vanadium also may modulate thyroid metabolism. Vanadium deprivation was found to increase thyroid weight and decrease thyroid peroxidase activity in rats (Uthus & Nielsen, 1990).

Vanadyl-diascorbate appears to be a physiologically important inhibitor of Na-K-ATPase in a ouabain-like action. This complex may thus participate in the regulation of body fluid volume and blood pressure (Meyer-Lehnert et al., 2000). Another effect, observed in cell culture, is the promotion of NO synthesis by enhancing both endothelial (eNOS) and inducible (iNOS) isoform (EC1.14.13.39) activity (Cortizo et al., 2000).

A chemopreventive effect has been suggested on the basis of very limited animal hepatocarcinogenesis experiments (Basak and Chatterjee, 2000). Industrial exposure (oil-fired electrical generating plants, petrochemical, steel, and mining industries) to toxic doses of vanadate can induce activation of the tumor suppressor gene p53 and cell apoptosis through generation of H_2O_2 (Huang et al., 2000). The relevance of this effect for induction or prevention of cancer upon exposure to much lower dietary doses is unclear.

Vanadyl sulfate is sometimes used as a supplement by weight lifters. The efficacy of such use as an ergogenic aid has not been investigated rigorously.

REFERENCES

Barceloux, D.G., 1999. Vanadium. J. Toxicol. Clin. Toxicol. 37, 265–278.

Basak, R., Chatterjee, M., 2000. Combined supplementation of vanadium and l-alpha,25-dihydroxyvitamin D3 inhibit placental glutathione *S*-transferase positive foci in rat liver carcinogenesis. Life Sci. 68, 217–231.

Cortizo, A.M., Caporossi, M., Lettieri, G., Etcheverry, S.B., 2000. Vanadate-induced nitric oxide production: role in osteoblast growth and differentiation. Eur. J. Pharmaecol. 400, 279–285.

Goldwaser, I., Qian, S., Gershonov, E., Fridkin, M., Shechter, Y., 2000. Organic vanadium chelators potentiate vanadium-evoked glucose metabolism *in vitro* and *in vivo:* establishing criteria for optimal chelators. Mol. Pharmacol. 58, 738–746.

Harland, B.E., Harden-Williams, B.A., 1994. Is vanadium of human nutritional importance yet? J. Am. Diet Assoc. 94, 891–894.

Huang, C., Zhang, Z., Ding, M., Li, J., Ye, J., Leonard, S.S., et al., 2000. Vanadate induces p53 transactivation through hydrogen peroxide and causes apoptosis. J. Biol. Chem. 275, 32516–32522.

Krachler, M., Prohaska, T., Koellensperger, G., Rossipal, E., Stingeder, G., 2000. Concentrations of selected trace elements in human milk and in infant formulas determined by magnetic sector field inductively coupled plasma-mass spectrometry. Biol. Trace Elem. Res. 76, 97–112.

Meyer-Lehnert, H., Backer, A., Kramer, H.J., 2000. Inhibitors of Na-K-ATPase in human urine: effects of ouabain-like factors and of vanadium-diascorbate on calcium mobilization in rat vascular smooth muscle cells: comparison with the effects of ouabain, angiotensin II, and arginine-vasopressin. Am. J. Hypertens. 13, 364–369.

Terziyski, K., Tzenova, R., Milieva, E., Vladeva, S., 1999. Possible mechanism of action of vanadium ions as an antidiabetic agent. Folia Med. (Plovdiv) 41, 34–37.

Ueki, T., Uyama, T., Yamamoto, K., Kanamori, K., Michibata, H., 2000. Exclusive expression of transketolase in the vanadocytes of the vanadium-rich ascidian, *Ascidia sydneiensis* samea. Biochim. Biophys. Acta 1494, 83–90.

Uthus, E.O., Nielsen, F.H., 1990. Effect of vanadium, iodine and their interaction on growth, blood variables, liver trace elements and thyroid status indices in rats. Magn. Trace Elem. 9, 219–226.

Uthus, E.O., Seaborn, C.D., 1996. Deliberations and evaluations of the approaches, endpoints and paradigms for dietary recommendations of the other trace elements. J. Nutr. 126, 2452S–2459S.

NICKEL

Nickel (Ni; atomic weight 58.69) is an essential transition metal with potential valences 1, 2, 3, and 4. Naturally occurring nickel contains the stable isotopes ^{60}Ni (26.2%), ^{61}Ni (1.14%), ^{62}Ni (3.63%), and ^{64}Ni (0.93%), and the unstable isotopes ^{58}Ni (68.077%; half-life more than 7×10^{20} years) and ^{59}Ni (trace amounts; half-life 7×10^4 years).

ABBREVIATIONS

DMT1   divalent metal transporter (SLC11A2)
Ni   nickel

NUTRITIONAL SUMMARY

Function: Cell membrane stability and production of some hormones may be influenced by Ni availability. Ni is also a cofactor of various microbial enzymes. Ni intake, therefore, might influence microbial action in human intestines.

Food sources: Grains, chocolate, nuts, and dried legumes are good sources of Ni. Americans typically get about 300 µg/day from water and food.

Requirements: Less than 100 µg/day appears to be needed by healthy adults.

Deficiency: Low availability impairs reproduction and growth in some mammals and may impair iron, copper, and zinc metabolism.

Excessive intake: Chronic intake of several milligrams per day may cause oxidative damage to DNA and cell structures, interfere with hormonal function, promote excessive zinc and iron storage, and lead to magnesium deficiency.

DIETARY SOURCES

Significant amounts of Ni are consumed with chocolate, nuts, oatmeal and other grains, dried beans, and peas.

Americans take in between 100 (Uthus and Seaborn, 1996) and 300 (Barceloux, 1999) µg of Ni per day with food and drinking water.

DIGESTION AND ABSORPTION

Only 1% of dietary Ni is absorbed from food, but significantly more can be gotten from water (Barceloux, 1999; Amich et al., 2000). Uptake from the enteral lumen in the small intestine, particularly the jejunum, proceeds via the divalent metal transporter (DMT1), the expression of which is regulated by iron status. Ni absorption into iron-loaded cells is diminished (Tallkvist and Tjalve, 1998). Phytate, tannins, and calcium appear to decrease fractional absorption.

The Zn transporter ZIP4 (SLC39A4) on the luminal side of enterocytes can mediate the transport of Ni at micromolar concentrations (Antala & Dempski, 2012).

TRANSPORT AND CELLULAR UPTAKE

Blood circulation: Albumin binds Ni and may be the main vehicle for transport with blood (Brennan et al., 1990). Ni uptake into the liver, heart, and other tissues proceeds via the sodium/calcium exchanger (Egger et al., 1999). DMT1 in the cell membranes, as well as endosome membranes of various tissues, is known to accept Ni and thus probably contributes to its transport into erythroid cells, across the BBB, and into the placenta.

Nickel sulfate is selectively concentrated within lysosomes with an involvement of arylsulfatases (Berry, 1996). The biological significance of increased lysosomal Ni concentrations is unknown.

EXCRETION

Colonic epithelial cells may have the ability to secrete excess Ni (Tallkvist & Tjalve, 1998). DMT1 is present in the luminal membrane of the distal nephron (Ferguson et al., 2001). Whatever Ni is filtered through the kidneys may be recovered by the mechanisms best investigated for iron.

REGULATION

It is not known whether and how the body's content of Ni is regulated.

FUNCTION

A specific function of Ni in humans is not known. Ni stimulates erythropoietin production (Goldberg et al., 1988), possibly through a heme-containing sensor in which iron can be substituted by Ni or cobalt (Bunn et al., 1998). Bunn et al. suggested that the sensor is a multi-subunit assembly containing an NAD(P)H oxidase capable of generating peroxide and reactive oxygen intermediates, which serve as signaling molecules.

Ni is a cofactor for a few bacterial enzymes, mainly those involved in the utilization of hydrogen. Ni intake may thus influence human health through its importance for intestinal flora (Howlett et al., 2012). Ni-dependent bacterial enzymes include carbon-monoxide dehydrogenase (EC1.2.99.2), hydrogen dehydrogenase (EC1.12.1.2), coenzyme F420 hydrogenase (EC1.12.99.1), hydrogenase (EC1.18.99.1), urease (EC3.5.1.5), methylcoenzyme M reductase (no EC number assigned), Ni-dependent superoxide dismutase (EC1.15.1.1), lactoylglutathione lyase (glyoxalase I, EC4.4.1.5), and peptidyl-prolyl cis–trans isomerase (EC5.2.1.8).

OTHER EFFECTS

Patients with severe Ni poisoning develop intense pulmonary and gastrointestinal toxicity. Diffuse interstitial pneumonitis and cerebral edema are the main causes of death (Barceloux, 1999). The poorly water-soluble forms of Ni can be carcinogenic, causing lung and nasal cancer in humans (Hayes, 1997).

REFERENCES

Antala, S., Dempski, R.E., 2012. The human ZIP4 transporter has two distinct binding affinities and mediates transport of multiple transition metals. Biochemistry 51, 963–973.

Arnich, N., Lanhers, M.C., Cunat, L., Joyeux, M., Burnel, D., 2000. Nickel absorption and distribution from rat small intestine *in situ*. Biol. Trace Elem. Res. 74, 141–151.

Barceloux, D.G., 1999. Nickel. J. Toxicol. Clin. Toxicol. 37, 239–258.

Berry, J.P., 1996. The role of lysosomes in the selective concentration of mineral elements. A microanalytical study. Cell. Mol. Biol. 42, 395–411.

Brennan, S.O., Myles, T., Peach, R.J., Donaldson, D., George, P.M., 1990. Albumin Redhill (-1 arg, ala320-to-thr): a glycoprotein variant of human serum albumin whose precursor has an aberrant signal peptidase cleavage site. Proc. Natl. Acad. Sci. 87, 26–30.

Bunn, H.E., Gu, J., Huang, L.E., Park, J.W., Zhu, H., 1998. Erythropoietin: a model system for studying oxygen-dependent gene regulation. J. Exp. Biol. 201, 1197–1201.

Egger, M., Ruknudin, A., Niggli, E., Lederer, W.J., Schulze, D.H., 1999. Ni^{2+} transport by the human Na^+/Ca^{2+} exchanger expressed in Sf9 cells. Am. J. Physiol. 276, C1184–C1192.

Ferguson, C.J., Wareing, M., Ward, D.T., Green, R., Smith, C.P., Riccardi, D., 2001. Cellular localization of divalent metal transporter DMT-1 in rat kidney. Am. J. Physiol. Renal Fluid Electrolyte Physiol. 280, F803–F814.

Goldberg, M.A., Dunning, S.P., Bunn, H.E., 1988. Regulation of the erythropoietin gene: evidence that the oxygen sensor is a heme protein. Science 242, 1412–1415.

Hayes, R.B., 1997. The carcinogenicity of metals in humans. Cancer Causes Control 8, 371–385.

Howlett, R.M., Hughes, B.M., Hitchcock, A., Kelly, D.J., 2012. Hydrogenase activity in the foodborne pathogen *Campylobacter jejuni* depends upon a novel ABC-type nickel transporter (NikZYXWV) and is SlyD-independent. Microbiology 158 (Pt 6), 1645–1655.

Tallkvist, J., Tjalve, H., 1998. Transport of nickel across monolayers of human intestinal Caco-2 cells. Toxicol. Appl. Pharmacol. 151, 117–122.

Uthus, E.O., Seaborn, C.D., 1996. Deliberations and evaluations of the approaches, endpoints and paradigms for dietary recommendations of the other trace elements. J. Nutr. 126, 2452S–2459S.

APPLICATIONS

12

CHAPTER OUTLINE

Genetic Variation 809
Nutrient Adequacy and Supplementation 816
Nutrient Interactions 819
GRAS Database 822
Using Molecular Databases 822

GENETIC VARIATION

ABBREVIATIONS

FOLR1	folate receptor 1
HFE	hemochromatosis gene
MTHFR	5,10-methylenetetrahydrofolate reductase (EC1.7.99.5)
MTR	*S*-methyltetrahydrofolate-homocysteine *S*-methyltransferase (EC2.1.1.13)

INHERITED DIVERSITY

Humans are defined by their shared genome. It has been known for a long time, however, that people have distinguishing traits that are clearly inherited. Knowledge about genetic variation has now been extended to nutrient metabolism. Indeed, the relatively high degree of variation in this respect may reflect the ability of human populations to thrive in extremely diverse environments, in tropical rainforests, deserts, and icy northern climates. For example, variants of the hemochromatosis gene (*HFE*) may allow populations to do well on a diet with relatively low iron intake, while the now-most-common form provides some protection against the dangers of an iron-rich diet. The fact that human populations have been moving rapidly (in a long-term historical context) between very diverse regions probably prevented the equilibrium of variants from reaching a steady state appropriate to their current habitat. While musings about historic evolution may remain speculative, there cannot be much doubt about the considerable impact of genetic variance on individual nutritional needs. Many examples are well studied by now, and it is likely that some genetic variation affects availability and disposition of each nutrient (see the selection of variants affecting responses to specific nutrients in Table 12.1). Relatively common genetic variants with impacts on the individual needs of folate and iron will be described to emphasize how human health depends on the combined action of nature (genes) and nurture (nutrient intake).

Table 12.1 Examples of Genetic Variants Affecting Nutrient Metabolism

Protein	Variant	Affected Nutrient	Consequence
APOA2	T-265C	Saturated fat	Low saturated fat intake reduces obesity risk
Lactase	C-13910T	Lactose	Can usually drink two cups of milk without discomfort
Hepatic lipase	C-514T	Fat	Decreased HDL concentration at moderate to high fat intake
Aldehyde dehydrogenase 2	ALDH2*2	Ethanol	A single alcoholic drink is likely to induce facial flushing
Haptoglobin	Hp 2-2	Ascorbate	Accelerated ascorbate oxidation increases requirements
HFE	C282Y	Iron	Moderate to high habitual iron intake increases susceptibility to systemic *Vibrio vulnificans* infection
Alpha-adducin	G460W	Sodium	Added salt intake increases blood pressure
CYP1A2	*1F/*1F	Caffeine	Slow metabolism of caffeine and increased heart attack risk
F2	G20210A	Vitamin E	Extra vitamin E blunts increased risk of lung embolism

FOLATE ADEQUACY AND GENE VARIANTS

Folate provides a good example for the interaction of genetic disposition and nutrient intake. A review of 13 enzymes, receptors, and transporters directly involved in folate absorption, utilization, and regeneration demonstrates genetic variation in each. Relatively common variants in several of the genes have been firmly linked to increased risk of important health consequences. Detailed studies may also reveal risks related to polymorphisms in some of the other genes.

A lack of active folate metabolites during the first weeks of pregnancy greatly increases the risk of neural tube malformations and other birth defects. Low folate intake in conjunction with variant genes of folate metabolism is most often the underlying cause. High folate intake (0.4–4 mg/day) greatly reduces the risk associated with several, if not all, predisposing variants. It is essential that such high intake starts before conception. Increased concentration of the toxic metabolite homocysteine in people with low folate status is also associated with increased risk of cardiovascular disease, cancer, and cognitive decline.

Pteroylpoly-gamma-glutamate carboxypeptidase (GCPII, EC3.4.17.21) is a membrane-bound, zinc-dependent exopeptidase of the jejunal brush border that shortens the polyglutamate tail of folate from foods. People with the relatively common variant H475Y, which has lower than normal activity, absorb a smaller percentage of food folate; however, the bioavailability of synthetic folate is not diminished (Devlin et al., 2000). But a follow-up examination of the Framingham Offspring cohort found no significant effect on blood folate concentrations (Vargas-Martinez et al., 2002).

The high-affinity folate carrier 1 (RFC1, SLC19A1) mediates the uptake of monoglutamyl folate from the small intestine and colon. Allelic variation of RFC1 (A80G, rs1051266), with a slightly higher prevalence of the glycine (G) form in the United States, affects absorption. Among children of mothers with low folate intake, those with homozygosity for the G form are at highest risk for spina bifida (Shaw et al., 2002).

The multidrug resistance protein 2 (MRP2; ABCC2) contributes to the export of folate across the basolateral enterocyte membrane. Five functionally relevant variants of the MRP2 gene have been observed, and some of them are common (Itoda et al., 2002). Their impact on folate availability has not yet been investigated, however. The addition of multiple glutamyl residues by folylpolyglutamate synthase (EC6.3.2.17) keeps folate in the cell. Several variants of the gene are listed in the single nucleotide polymorphism database of the National Center for Biotechnology Information (NCBI; dbSNP; http://www.ncbi.nlm.nih.gov/SNP). Allele frequencies and functional significance remain to be established. Monoglutamyl folate and similar metabolites are also actively pumped out of the cell by ABC transporters, including MDR1 (ABCB1). Allelic variation of MDR1 (codon C3435T, rs1045642) affects the activity of this transporter (Kerb et al., 2001; Obermann-Borst et al., 2011).

Dihydrofolate reductase (EC1.5.1.3) catalyzes the nicotinamide adenine dinucleotide phosphate (NADP)–dependent reduction of folate to 7,8-dihydrofolate and then to 5,6,7,8-tetrahydrofolate. The Genome Database (GDB; http://www.gdb.org/) indicates four polymorphisms with maximal allele frequencies between 0.38 and 0.65. Their impact on folate metabolism is slowly becoming clearer. A common deletion in intron 1 (rs79950164, allele frequency about 0.4) reduces dihydrofolate reductase (DHFR) expression, limits the capacity of carriers to activate folic acid, and appears to be associated with a 52% increase in breast cancer risk in response to long-term multivitamin use (Xu et al., 2007).

The metabolism of serine and glycine generates large quantities of 5,10-methylenetetrahydrofolate. FADH-dependent 5,10-methylenetetrahydrofolate reductase (MTHFR, EC1.7.99.5) is essential for the recycling of folate for further reactions. The MTHFR variant A222V (codon 677C→T, rs1801133), which occurs in populations worldwide, is metabolically less active than the main form due to its decreased stability. The human gene mutation database Cardiff (HGMD; www.hgmd.org) lists another 25 mutations, some of which are common. Carriers of the A222V variant (both homozygotes and heterozygotes) have distinctly lower plasma folate and homocysteine concentrations on moderately low folate regimes (250 µg or less) than noncarriers. Daily use of additional 400 µg of synthetic folate abolishes the differences between heterozygotes and noncarriers but only narrows the gap between homozygotes and noncarriers (Ashfield-Watt et al., 2002). High folate consumption of the mothers during the first weeks of pregnancy moves their risk closer to the normal range.

The cobalamin-containing enzyme 5-methyltetrahydrofolate-homocysteine S-methyltransferase (MTR, methionine synthase, EC2.1.1.13) catalyzes the main reaction that removes the methyl group from 5-methyltetrahydrofolate and thus frees up folate again. At least 12 genetic variants have been found. The polymorphic variant D919G (codon 2756A→G; rs1805087) may increase cardiovascular risk (Wang et al., 1998). Other variants disrupt enzyme activity in homozygote carriers, but the impact on heterozygotes remains to be investigated. The cobalamin in MTR is susceptible to oxidation. The flavin adenine dinucleotide (FAD)– and flavin mononucleotide (FMN)–containing methionine synthase reductase (EC2.1.1.135) revert oxidized MTR to its active cob(I)alamin form. The polymorphism 122M (codon 66A→G, rs1801394) was found to increase the risk for birth defects several-fold in women with suboptimal vitamin B12 status (Wilson et al., 1999). The reaction uses S-adenosylmethionine as a methyl donor. Cytochrome b5, which is regenerated by nicotine adenine dinucleotide phosphate (NADPH)–dependent cytochrome P450 reductase (EC1.6.2.4), is the reductant. Genetic variants of both cytochrome b5 and its reductase are known, though the consequences for folate metabolism are not. Folate must efficiently cross the placenta to ensure adequate supplies to the fetus. Folate receptor

1 (FOLR1) binds 5-methyltetrahydrofolate at the maternally facing chorionic surface and builds up a concentration gradient that drives placental transfer.

Polymorphisms at nucleotides 762, 631, and 610 in the promoter region of FOLR1 might affect transfer efficiency, and thus the risk of a pregnancy with a birth defect (Barber et al., 2000). The closely related fetal folate receptor, FOLR2, is also expressed in the placenta. At least one variant of this gene (254A→G, rs13908) has been observed. The activity of folate-using reactions also affects folate homeostasis. An important example is genetic variation in thymidylate synthase (EC2.1.1.45), which provides the crucial dTMP precursor for DNA synthesis and is a major consumer of 5,10-methylenetetrahydrofolate. A common polymorphic variant, characterized by an additional tandem repeat (TYMS3), is associated with decreased blood folate concentration and impaired homocysteine metabolism (Trinh et al., 2002). Genetic variation of a very different kind may also influence folate status. Light with a very short wavelength (particularly ultraviolet (UV) A) can penetrate deep into fair skin and might inactivate folate in capillaries (Jablonski, 1999). Skin pigments block UV rays and minimize folate losses. A mismatch between the exposure to UV light and genetically determined pigmentation might cause enough folate inactivation to affect folate status (Table 12.2).

Table 12.2 Genes Involved in Folate Absorption and Metabolism Contain Common Variants

Gene Name	Variant	Effect
Pteroylpoly-g-glutamate carboxypeptidase (GCPII, EC3.4.17.21)	rs61886492	Decreased bioavailability of food folate
Folate carrier 1 (RFC1, SLC19A1)	rs1051266	Increased risk of birth defects at low maternal folate intakes
Multidrug resistance protein 2 (MRP2, ABCC2)	Several	Altered intracellular folate retention
Folylpolyglutamate synthase (EC6.3.2.17)	Several	Altered intracellular folate retention
MDR1 (P-glycoprotein 1, ABCB1)	rs1045642	Altered intracellular folate retention
Dihydrofolate reductase (EC1.5.1.3)	rs70991108	Slowed assimilation of folic acid; higher breast cancer risk with supplement use
5,10-Methylenetetrahydrofolate reductase (EC1.7.99.5)	rs1801133	Decreased metabolic response to folate intake; increased cardiovascular risk and risk of birth defects
5-Methyltetrahyd rofolate-homocysteine S-methyltransferase (MTR, methionine synthase) (EC2.1.1.13)	rs1805087	Increased risk of birth defects at low maternal folate intakes
Methionine synthase reductase (EC2.1.1.135)	rs1801394	Increased risk of birth defects at low maternal folate intakes
Folate receptor 1 (FOLR1)	Several	Increased risk of birth defects at low maternal folate intakes
Folate receptor 2 (FOLR2)	rs13908	Increased risk of birth defects
Folate receptor 3 (FOLR3)	Several	Increased risk of birth defects
Thymidylate synthase (EC2.1.1.45)	Several	Increased birth defect and cancer risk
Skin pigmentation	Various	Increased folate degradation upon skin exposure to sun light

THE IRON-REGULATORY GENE *HFE*

The protein produced by the *HFE* gene determines at which blood concentration of diferric transferrin the expression of the hepcidin gene in liver cells starts to increase and the release of iron from the intestine and the reticuloendothelial system is slowed. The C282Y (codon 845 G→A, rs1800562) variant of the *HFE* protein cannot interact effectively with transferrin 2 (Muckenthaler, 2014), preventing the normal reset of the signaling cascade (by ubiquitination and proteosomal degradation of the BMP receptor type 1). Thus, the signal for hepcidin expression persists without restraint (Ulvik, 2015). The increased hepcidin expression in people with the C282Y variant allows the absorption of large amounts of iron from food even when iron stores are already near or over capacity (Sangwaiya et al., 2011).

Many people of North European ancestry have one or two copies of the C282Y of the *HFE* gene on the small arm of chromosome 6 (6p21.3), while this variant is rare in people of Asian descent (Beckman et al., 1997; Merryweather-Clarke et al., 1999). In Ireland, nearly one in five newborns are heterozygous for the C282Y variant, and 1% are homozygous (Byrnes et al., 2001). The variant *HFE* gene product loses its ability to limit iron uptake when stores are filled because it is not effectively processed and moved from the Golgi compartment to the cell surface (Waheed et al., 1997). Since the *HFE* gene product favors iron uptake, the C282Y variant might confer a selective survival advantage to the offspring of iron-deficient women by enhanced iron transfer across the placenta (Parkkila et al., 1997). The downside of the C282Y variant, particularly in homozygotes, relates to the increased concentration of reactive (unbound) iron and excessive iron accumulation. Affected individuals appear to lose the ability to down-regulate iron absorption when iron stores are sufficient but retain the capacity to up-regulate in response to deficiency (Ajioka et al., 2002). Several other common variants, particularly H63D (rs1799945) and S65C (rs1800730), also compromise HFE function.

The concentration of unbound iron is normally below 10^{-8} mol/l, which limits oxygen free radical generation and protects against the spread of iron-dependent bacteria in blood and tissues. In more than 1% of the US population the concentration of unbound iron is elevated, as indicated by their very high (>60%) transferrin saturation (Looker and Johnson, 1998). C282Y homozygotes can lower their concentration of unbound iron and prevent excessive iron storage by tightly limiting their iron intake or increasing iron losses (e.g., by blood donation). Chronic excessive iron intake, on the other hand, greatly increases risk (Bell et al., 2000). The C282Y variant also increases transferrin saturation and the tendency to accumulate iron in heterozygotes (Distante et al., 1999).

The health consequences of increased unbound iron in blood can be dramatic. A single serving of raw oysters, which are very commonly contaminated with naturally occurring marine bacteria *(Vibrio vulnificus)*, has infected and killed young homozygotic carriers of the C282Y variant in good health within a few days. Even just handling contaminated seafood or swimming in water with the organisms can be dangerous (Barton and Acton, 2009). Long-term health risks of excessive iron storage include diabetes, cancer, dementia, and premature heart disease. While the risk of heterozygotes tends to be lower compared to homozygotes excessive iron intake compounds their problems. Fatal septicemia following the consumption of infected seafood has occurred in heterozygotes with iron accumulation (Gerhard et al., 2001). Expanded iron stores also increase the risk of colorectal cancer (Nelson, 2001), viral hepatitis (Fargion et al., 2001), and accelerated cognitive decline (Sampietro et al., 2001). The complexity of the issues is underscored, however, by the increased prevalence of C282Y heterozygotes among very old Sicilians (Lio et al., 2002).

Hemochromatosis responds very well to treatment when started at an early age (Islek et al., 2015). Avoidance of iron-fortified dietary supplements and foods, blood donations several times a year (Røsvik et al., 2010), and regular iron status controls are usually sufficient to maintain iron stores in the normal range.

REFERENCES

Ajioka, R.S., Levy, J.E., Andrews, N.C., Kushner, J.P., 2002. Regulation of iron absorption in Hfe mutant mice. Blood 100, 1465–1469.

Ashfield-Watt, P.A., Pullin, C.H., Whiting, J.M., Clark, Z.E., Moat, S.J., Newcombe, R.G., et al., 2002. Methylenetetrahydrofolate reductase 677C→T genotype modulates homocysteine responses to a folate-rich diet or a low-dose folic acid supplement: a randomized controlled trial. Am. J. Clin. Nutr. 76, 180–186.

Barber, R., Shalat, S., Hendricks, K., Joggerst, B., Larsen, R., Suarez, L., et al., 2000. Investigation of folate pathway gene polymorphisms and the incidence of neural tube defects in a Texas hispanic population. Mol. Genet. Metab. 70, 45–52.

Barton, J.C., Acton, R.T., 2009. Hemochromatosis and Vibrio vulnificus wound infections. J. Clin. Gastroenterol. 43, 890–893.

Beckman, L.E., Saha, N., Spitsyn, V., Van Landeghem, G., Beckman, L., 1997. Ethnic differences in the HFE codon 282 (Cys/Tyr) polymorphism. Hum. Hered. 47, 263–267.

Bell, H., Berg, J.P., Undlien, D.E., Distante, S., Raknerud, N., Heier, H.E., et al., 2000. The clinical expression of hemochromatosis in Oslo, Norway. Excessive oral iron intake may lead to secondary hemochromatosis even in HFE C282Y mutation negative subjects. Scand. J. Gastroenterol. 35, 1301–1307.

Byrnes, V., Ryan, E., Barrett, S., Kenny, P., Mayne, E., Crowe, J., 2001. Genetic hemochromatosis, a Celtic disease: is it now time for population screening? Genet. Test. 5, 127–130.

Devlin, A.M., Ling, E.H., Peerson, J.M., Fernando, S., Clarke, R., Smith, A.D., et al., 2000. Glutamate carboxypeptidase 11: a polymorphism associated with lower levels of serum folate and hyperhomocysteinemia. Hum. Mol. Gen. 9, 2837–2844.

Distante, S., Berg, J.P., Lande, K., Haug, E., Bell, H., 1999. High prevalence of the hemochromatosisassociated Cys282Tyr HFE gene mutation in a healthy Norwegian population in the city of Oslo, and its phenotypic expression. Scand. J. Gastroenterol. 34, 529–534.

Fargion, S., Stazi, M.A., Fracanzani, A.L., Mattioli, M., Sampietro, M., Tavazzi, D., et al., 2001. Mutations in the HFE gene and their interaction with exogenous risk factors in hepatocellular carcinoma. Blood Cells Mol. Dis. 27, 505–511.

Gerhard, G.S., Levin, K.A., Price Goldstein, J., Wojnar, M.M., Chorney, M.J., Belchis, D.A., 2001. Vibrio vulnificus septicemia in a patient with the hemochromatosis HFE C282Y mutation. Arch. Pathol. Lab Med. 125, 1107–1109.

Islek, A., Inci, A., Sayar, E., Yilmaz, A., Uzun, O.C., Artan, R., 2015. HFE-related hereditary hemochromatosis is not invariably a disease of adulthood: importance of early diagnosis and phlebotomy in childhood. J. Pediatr. Gastroenterol. Nutr. 60.

Itoda, M., Saito, Y., Soyama, A., Saeki, M., Murayama, N., Ishida, S., et al., 2002. Polymorphisms in the ABCC2 (cMOAT/MRP2) gene found in 72 established cell lines derived from Japanese individuals: an association between single nucleotide polymorphisms in the 5′-untranslated region and exon 28. Drug Metab. Disp. 30, 363–364.

Jablonski, N.G., 1999. A possible link between neural tube defects and ultraviolet light exposure. Med. Hypotheses 52, 581–582.

Kerb, R., Aynacioglu, A.S., Brockmoller, J., Schlagenhaufer, R., Bauer, S., Szekeres, T., et al., 2001. The predictive value of MDRI, CYP2C9, and CYP2CI9 polymorphisms for phenytoin plasma levels. Pharmacogenomics J. 1, 204–210.

Lio, D., Balistreri, C.R., Colonna-Romano, G., Motta, M., Franceschi, C., Malaguarnera, M., et al., 2002. Association between the MHC class I gene HFE polymorphisms and longevity: a study in Sicilian population. Genes Immun. 3, 20–24.

Looker, A.C., Johnson, C.J., 1998. Prevalence of elevated serum transferrin saturation in adults in the United States. Ann. Intern. Med. 129, 940–945.

Merryweather-Clarke, A.T., Simonsen, H., Shearman, J.D., Pointon, J.J., Norgaard-Pedersen, B., Robson, K.J.H., 1999. A retrospective anonymous pilot study in screening newborns for HFE mutations in Scandinavian populations. Hum. Mutat. 13, 154–159.

Muckenthaler, M.U., 2014. How mutant HFE causes hereditary hemochromatosis. Blood 124, 1212–1213.

Nelson, R.L., 2001. Iron and colorectal cancer risk: human studies. Nutr. Rev. 59, 140–148.

Obermann-Borst, SA, Isaacs, A, Younes, Z, van Schaik, RH, van der Heiden, IP, van Duyn, CM, Steegers, EA, Steegers-Theunissen, RP., 2011. General maternal medication use, folic acid, the MDR1 C3435T polymorphism, and the risk of a child with a congenital heart defect. Am. J. Obstet. Gynecol. 204 (236), e1–8.

Parkkila, S., Waheed, A., Britton, R.S., Bacon, B.R., Zhou, X.Y., Tomatsu, S., et al., 1997. Association of the transferrin receptor in human placenta with HFE, the protein defective in hereditary hemochromatosis. Proc. Natl. Acad. Sci. USA 94, 13198–13202.

Røsvik, A.S., Ulvik, R.J., Wentzel-Larsen, T., Hervig, T., 2010. Blood donors with hereditary hemochromatosis. Transfusion 50, 1787–1793.

Sampietro, M., Caputo, L., Casatta, A., Meregalli, M., Pellagatti, A., Tagliabue, J., et al., 2001. The hemochromatosis gene affects the age of onset of sporadic Alzheimer's disease. Neurobiol. Aging 22, 563–568.

Sangwaiya, A., Manglam, V., Busbridge, M., Thursz, M., Arnold, J., 2011. Blunted increase in serum hepcidin as response to oral iron in HFE-hemochromatosis. Eur. J. Gastroenterol. Hepatol. 23, 721–724.

Shaw, G.M., Lammer, E.J., Zhu, H., Baker, M.W., Neri, E., Finnell, R.H., 2002. Maternal periconceptional vitamin use, genetic variation of infant reduced folate carrier (A80G), and risk of spina bifida. Am. J. Med. Genet. 108, 1–6.

Trinh, B.N., Ong, C.N., Coetzee, G.A., Yu, M.C., Laird, P.W., 2002. Thymidylate synthase: a novel genetic determinant of plasma homocysteien and folate levels. Hum. Genet. 111, 299–302.

Ulvik, R.J., 2015. The liver in haemochromatosis. J. Trace Elem. Med. Biol. 30.

Vargas-Martinez, C., Ordovas, J.M., Wilson, P.W., Selhub, J., 2002. The glutamate carboxypeptidase gene II (C > T) polymorphism does not affect folate status in the Framingham Offspring cohort. J. Nutr. 132, 1176–1179.

Waheed, A., Parkkila, S., Zhou, X.Y., Tomatsu, S., Tsuchihashi, Z., Feder, J.N., et al., 1997. Hereditary hemochromatosis: effects of C282Y and H63D mutations on association with beta-2-microglobulin, intracellular processing, and cell surface expression of the HFE protein in COS-7 cells. Proc. Natl. Acad. Sci. USA 94, 12384–12389.

Wang, X.L., Cai, H., Cranney, G., Wilcken, D.E., 1998. The frequency of a common mutation of the methionine synthase gene in the Australian population and its relation to smoking and coronary artery disease. J. Cardiovasc. Risk 5, 289–295.

Wilson, A., Platt, R., Wu, Q., Leclerc, D., Christensen, B., Yang, H., et al., 1999. A common variant in methionine synthase reductase combined with low cobalamin (vitamin B12) increases risk for spina bifida. Mol. Genet. Metab. 67, 317–323.

Xu, X., Gammon, M.D., Wetmur, J.G., Rao, M., Gaudet, M.M., Teitelbaum, S.L., et al., 2007. A functional 19-base pair deletion polymorphism of dihydrofolate reductase (DHFR) and risk of breast cancer in multivitamin users. Am. J. Clin. Nutr. 85, 1098–1102.

NUTRIENT ADEQUACY AND SUPPLEMENTATION

ABBREVIATIONS

AI	adequate intake
DRI	Dietary Reference Intake
EAR	estimated average requirement
IOM	Institute of Medicine
RDA	recommended dietary allowance
UL	tolerable upper intake level

DIETARY REFERENCE INTAKES

Nutrients are coming to be seen more and more like medications, where the "dose makes the poison." The Dietary Reference Intake (DRI) guidelines published by the US Food and Nutrition Board (FNB) now take into consideration (at least in principle) both lower and upper desirable limits. How to determine appropriate limits for individuals or groups remains the unresolved and much-debated question. The FNB has established a basic framework for tackling this question. Needs are considered separately for each of 22 groups defined by age, gender, as well as pregnancy and lactation status. The FNB is a subdivision of the Institute of Medicine (IOM) at the National Academies of the United States and comprises panels that set guidelines for the United States and Canada. Similar institutions exist in several other countries to provide guidance on optimal nutrient intake levels.

MINIMAL NUTRIENT REQUIREMENTS

The lower limits for nutrient intakes are based on the observed consumption level of healthy populations, if data from functional investigations are lacking. This intake level is called *adequate intake (AI)* and is assumed to cover the needs of healthy people. Because of the current limitations of scientific evidence, the FNB established AIs applicable to adults for the following nutrients: total fat, omega-6 fatty acids, omega-3 fatty acids, vitamin D, vitamin K, pantothenic acid, biotin, choline, calcium, chromium, fluoride, and manganese. Because information about the requirements of infants (under 1 year old) is even more limited, only AIs were set for most nutrients.

The exceptions are more definite lower limits for protein, iron, and zinc intake of 7- to 12-month-old infants. Where functional information is found to be reasonably reliable, the FNB sets recommended dietary allowances (RDAs). This amount is thought to cover the needs of most healthy people (97–98%) in the designated group. According to this model, the RDA is determined in a three-step process. First, the intake level is sought at which the risk of inadequacy of the healthy target population (e.g., 19- to 50-year-old men) is 50%. This is called the *estimated average requirement (EAR)*—an oxymoronic expression, since it relates to the median and not the average. The second step estimates the variance of requirements. For most nutrient requirements, a normal distribution is assumed. Unless evidence to the contrary is available, the variation coefficient (standard deviation divided by the mean) is set at 10% (because this is thought to correspond to the variance of basal metabolic rate). The final step then either adds two standard deviations (usually 20%) to the EAR or determines the 97.5th percentile of requirements by a Monte Carlo simulation procedure. So far, variation coefficients of 10% were set for thiamin, riboflavin, vitamin B6, folate, vitamin B12, phosphate, magnesium, and selenium because the actual variance was thought to be unknown. The decision to set the variation coefficient for niacin requirements at 15% was based on four separate studies on a total of 29 adults with an average variation coefficient of 34%. Similarly, the variation coefficient for vitamin A requirements was set at 20%, based on a single study of the vitamin A half-life in the livers of adults that gave a 21% variation coefficient of the results. Based on a single study of adults that gave a variation coefficient of 40% the variation coefficient of iodine requirements was set to 20%. In each case, a judgment was made about the relative contributions of measurement error versus intrinsic interindividual variation. The variation coefficients for copper and molybdenum requirements were set at 15%. The panel commented that data supporting the EARs are limited, but it provided no explanation why they did not use a 10% value as for other nutrients. In all instances, the variation coefficients set for young adults were applied to children, pregnant and lactating women, and older people without the benefit of additional supporting evidence. A significant weakness of the current recommendations relates to the extremely narrow basis of supporting data for several nutrients. In the vast majority of instances, data are completely lacking for specific age and gender groups, and the recommendations are based on extrapolations from other groups. Information about children and old people is particularly sparse. When levels are set on the basis of observations in a few subjects, as is the case for most of the covered nutrients, there is little opportunity to differentiate needs by genetic disposition. Only rarely is the existence of genetic diversity acknowledged. A typical and important example pertains to niacin requirements. It is likely that many people can cover their niacin requirements through endogenous synthesis from tryptophan, while others need significant intake of preformed niacin (Food and Nutrition Board, Institute of Medicine, 1998; Fukuwatari et al., 2004).

An even-better-documented example is the greater than average susceptibility to folate deficiency (Ashfield-Watt et al., 2002; Solis et al., 2008) in people with a low-activity variant (rs1801133) of 5,10-methylenetetrahydrofolate reductase (MTHFR; EC1.7.99.5). The current intake guidelines mention such differences, but draw no consequences for guidelines (Food and Nutrition Board, Institute of Medicine, 1998). The greatest weight for setting folate requirements was given to a single study of five young women of unknown ethnicity. However, there is little hope that such a small number can capture the well-known diversity of folate requirements of women and men, young and old, with or without variants of MTHFR or other genes. We know that about 10% of Caucasians carry two copies of the low-activity MTHFR variant and need at least 50% more folate than noncarriers. It is obvious that trustworthy estimates of folate requirements must be based on data from a much larger number of individuals.

So far, the functional assessment of adequate nutrient intakes has been limited on long-known properties. Reliable knowledge about more recently recognized functions has been disregarded without good reason. Metabolic and health consequences of suboptimal or nutrient status are most likely to be detected when they are monitored by focused observation.

A major shortcoming of the current framework is the deliberate exclusion of any long-term effects, particularly chronic degenerative disease. This ignores the fact that in affluent societies, nutrition influences the main causes of morbidity and death such as atherosclerosis, cancer, and osteoporosis. It is with respect to these chronic degenerative diseases that genetic variation of nutrient metabolism is most significant today. Polymorphisms relating to metabolism of energy, glucose, lipids, folate, and iron (to name just a few) are known to be important determinants of disease risk and outcome.

EXCESSIVE INTAKES

There is little doubt that too much of any nutrient can do harm. For some nutrients, however, the amounts that might cause concerns are so high that they are not likely to be used. The DRI framework formally explores the potential for harm with high intakes and relates them to the tolerable upper intake level (UL). The aim is to find the highest level at which no adverse effect has been observed (NOAEL) or, alternatively, the lowest level at which an adverse effect has been observed (LOAEL). In either case, a judgment has to be made about how much lower the UL should be. Controversies have arisen about some specific nutrients. An important example is the UL for vitamin D, which was set at 50 μg/day based on poorly documented selective evidence (Food and Nutrition Board, Institute of Medicine, 1997; Vieth, 1999). This is much less than the approximate 250-μg dose generated in a young person lying in the summer sun for just 20 min (Vieth, 1999). In the meantime, a well-designed study of the effects of 100 μg of vitamin D in healthy adults did not find any unfavorable effects (Vieth et al., 2001). The updated 2011 recommendations recognize a higher UL (4000 IU = 100 μg), but the debate continues (Institute of Medicine (US) Committee to Review Dietary Reference Intakes for Vitamin D and Calcium, 2011; Aloia, 2011).

SUPPLEMENTATION

The practice of using concentrated sources of specific nutrients for the prevention of disease has a long history. The discovery by James Lind in 1753 (Rajakumar, 2001) that scurvy could be prevented by the judicious use of citrus fruits enabled the British navy to greatly extend the duration of their missions and build up their domination of the seas. The early ridicule (the sailors were called "limeys" because they had to eat limes) notwithstanding, the use of supplements has exploded since and now sustains a major industry. Food fortification, which might be seen as a special case of supplement use, has been well established for a long time in the United States and other countries. The United States currently requires the addition of vitamin D to milk, and thiamin, riboflavin, folic acid, and iron to grain products. Some jurisdictions have policies of adding iodine to salt or fluoride to water. Individual supplement use certainly makes sense if it balances a nutrient deficit that would be left unattended because dietary changes alone would not be sufficient or feasible. Supplementation may also be needed to cover increased needs in times of illness (Zeisel, 2000). Genetic predisposition may be another reason for supplement use—for instance, the intake of additional tyrosine by people with phenylketonuria (PKU).

Many times, however, minimal required intakes are exceeded, "to be on the safe side," and undesirable effects may occur. High nutrient intake is also often intended to achieve pharmacological effects and in this case, extensive experimental data from human studies should be as much a prerequisite as with medical drugs (Zeisel, 1999), taking into account all known nutritional and metabolic aspects of the supplemented compound.

REFERENCES

Aloia, J.F., 2011. Clinical review: the 2011 report on dietary reference intake for vitamin D: where do we go from here? J. Clin. Endocrinol. Metab. 96, 2987–2996.

Ashfield-Watt, P.A., Pullin, C.H., Whiting, J.M., Clark, Z.E., Moat, S.J., Newcombe, R.G., et al., 2002. Methylenetetrahydrofolate reductase 677C→T genotype modulates homocysteine responses to a folate-rich diet or a low-dose folic acid supplement: a randomized controlled trial. Am. J. Clin. Nutr. 76, 180–186.

Food and Nutrition Board, Institute of Medicine, 1997. Dietary Reference Intakes for Calcium, Phosphorus, Magnesium, Vitamin D, and Fluoride. National Academy Press, Washington, DC.

Food and Nutrition Board, Institute of Medicine, 1998. Dietary Reference Intakes for Thiamin, Riboflavin, Niacin, Vitamin B6, Folate, Vitamin B12, Pantothenic Acid, Biotin, and Choline. National Academic Press, Washington, DC.

Fukuwatari, T, Ohta, M, Kimtjra, N, Sasaki, R, Shibata, K., 2004. Conversion ratio of tryptophan to niacin in Japanese women fed a purified diet conforming to the Japanese Dietary Reference Intakes. J. Nutr. Sci. Vitaminol. (Tokyo) 50, 385–391.

Institute of Medicine (US) Committee to Review Dietary Reference Intakes for Vitamin D and Calcium, 2011. In: Ross, A.C., Taylor, C.L., Yaktine, A.L., Del Valle, H.B. (Eds.), Dietary Reference Intakes for Calcium and Vitamin D National Academies Press (US), Washington, DC.

Rajakumar, K., 2001. Infantile scurvy: a historical perspective. Pediatrics 108, E76.

Solis, C., Veenema, K., Ivanov, A.A., Tran, S., Li, R., Wang, W., et al., 2008. Folate intake at RDA levels is inadequate for Mexican American men with the methylenetetrahydrofolate reductase 677TT genotype. J. Nutr. 138, 67–72.

Vieth, R., 1999. Vitamin D supplementation, 25-hydroxyvitamin D concentrations, and safety. Am. J. Clin. Nutr. 69, 842–856.

Vieth, R., Chan, P.C.R., MacFarlane, G.D., 2001. Efficacy and safety of vitamin D3 intake exceeding the lowest observed adverse effect level. Am. J. Clin. Nutr. 73, 288–294.

Zeisel, S.H., 1999. Regulation of "nutraceuticals". Science 285, 1853–1855.

Zeisel, S.H., 2000. Is there a metabolic basis for dietary supplementation? Am. J. Clin. Nutr. 72, 507S–511S.

NUTRIENT INTERACTIONS

ABBREVIATIONS

DMT1 divalent metal ion transporter 1 (SLC11A2)
PLP pyridoxal 5′-phosphate

DEFICIENCIES

Some nutrients affect the metabolism of other nutrients or of medications because they are essential cofactors. A characteristic example is the disruption of amino acid, omega-3 fatty acid, carbohydrate, niacin, and selenium metabolism in vitamin B6 deficiency since pyridoxal 5′-phosphate (PLP) is such a ubiquitous cofactor. The link to endogenous amino acid synthesis (and catabolism) is obvious since all aminotransferases require PLP. Vitamin B6 is also necessary as a cofactor of 6-desaturase (EC1.14.99.25) for the endogenous synthesis of docosahexaenoic acid (DHA) from eicosapentaenoic acid (EPA) and α-linolenic acid (Tsuge and Hotta, 2000). The utilization of stored carbohydrate (glycogen) depends on the availability of PLP for glycogen phosphorylases (EC2.4.1.1). Niacin synthesis from tryptophan is diminished in people with poor vitamin B6 status (Bender et al., 1979) because the key enzyme kynureninase (EC3.7.1.3) is PLP-dependent. Vitamin B6 interfaces with selenium metabolism because it is a cofactor both for the insertion of selenium into serine by L-seryl-tRNA sec selenium transferase (EC 2.9.1.1) and the release of hydrogen selenide by selenocysteine lyase (EC4.4.1.16) for reuse after the catabolism of proteins.

Riboflavin similarly illustrates how a single nutrient affects many others because this vitamin participates in the activation or endogenous synthesis of niacin; vitamins A, K, C, B6, and B12; folate; and other nutrients, metabolites, and medical drugs. The availability of information on quantitative aspects in humans is very limited, however.

A significant proportion of niacin requirements can be covered by the catabolism of tryptophan. Activity of the FAD-containing enzyme kynurenine 3-monoxygenase (EC1.14.13.9) is decreased in riboflavin deficiency. Vitamin A use can be affected because the enzyme for retinoic acid synthesis from retinal by retinal dehydrogenase (EC1.2.1.36) requires FAD. Vitamin K reactivation proceeds by two reducing steps catalyzed by several flavoenzymes, including NADPH dehydrogenase (EC1.6.99.6) and two forms of NAD(P)H dehydrogenase (EC1.6.99.2). The reduction of oxidized vitamin C (dehydroascorbate) depends on glutathione, which needs to be maintained in a reduced state by FAD-containing thioredoxin reductase (EC1.6.4.5). Reduced glutathione is also critically important to protect sulfhydryl groups in enzymes for the metabolism of numerous nutrients. Riboflavin deficiency affects PLP availability due to impaired activity of the FMN-containing enzyme pyridoxamine-phosphate oxidase (EC1.4.3.5), which converts pyridoxine into pyridoxal, and pyridoxine 5′-phosphate (PNP) into PLP (McCormick, 1989). Vitamin B12 requires three flavoenzymes for its metabolism: cob(II)alamin reductase (EC1.6.99.9), aquacobalamin reductase/NADPH (EC1.6.99.11), and aquacobalamin reductase/NADH (EC1.6.99.8). The effect on folate metabolism is through FAD-dependent methylene tetrahydrofolate reductase (EC1.5.1.20), which ensures reutilization of the large amounts of folate metabolites generated by serine and glycine metabolism.

Such interdependencies greatly broaden the consequences of severe deficiency of a single nutrient and mimic deficiency symptoms of the affected nutrients, especially when their intake is already marginal. What needs to be investigated is the extent to which mild deficiency of many individual nutrients affects the availability of others. Meaningful human studies of such relationships have been conducted for only very few nutrients.

HIGH INTAKE

There are many ways in which high intake of a particular nutrient or group of nutrients may affect the availability of other nutrients. The most obvious way is when the intake of a precursor sustains the synthesis of a nutrient, as is the case with alpha-carotene, which is a precursor for retinol. Consumption of some nutrients furthers the bioavailability of others, illustrated by ascorbate, which enhances the absorption of nonheme iron eaten with the same meal. Of course, the opposite effect is illustrated by phytate, which strongly inhibits iron absorption.

When nutrients use the same rate-limiting transporters or metabolic pathways, competition may occur. A large dose of iron, for instance, decreases the percentage of zinc that is taken up with the same meal (Chung et al., 2002). A large dose of zinc similarly decreases iron absorption (Donangelo et al., 2002). The reason is that divalent metal cations use the same transporter, divalent metal ion transporter 1 (DMT1; SLC11A2), for entry into the intestinal cell and a large number of one kind of cation can crowd out the other. Similarly, high beta-carotene intake decreases lutein bioavailability because these fat-soluble polyisoprenoid molecules compete for the same uptake mechanism (van den Berg and van Vliet, 1998).

Even a seemingly modest nutrient dose may tip the balance and cause health problems when it coincides with marginal intake of another nutrient needed for its metabolism. An example may be the competition between branched-chain amino acid metabolism and the niacin synthesis pathway. Moderately high intake of leucin with sorghum or corn (these grains contain about 50% more leucin than other protein sources) can induce the increased expression of branched-chain amino acid transaminase (EC2.6.1.42), which will then strongly attract much of the available vitamin B6. This competition may be enough in vulnerable individuals to lower the critical activity of kynureninase (EC3.7.1.3) in the niacin synthesis pathway and induce niacin deficiency with pellagra symptoms (Krishnaswamy et al., 1976).

Regulatory effects of one nutrient can influence the absorption, metabolism, or disposition of another. Increased expression of DMT1 in response to iron deficiency will increase fractional absorption of a whole range of other metals, including manganese (Finley, 1999). Even moderately excessive zinc supplementation (20 mg/day), on the other hand, can deplete copper stores (Boukaiba et al., 1993), in part by decreasing expression of the shared storage proteins, the metallothioneins, in liver and kidneys (Santon et al., 2002).

Still another mechanism of nutrient interactions is competitive interference or the mimicking of one nutrient's action by another nutrient. Typically, such effects become significant only when a large excess of the mimicking compound is consumed because the body is clearly adapted to any cross-talk that might occur at physiological consumption levels. An important example is the occurrence of bleeding with higher-than-normal vitamin E consumption. In a long-term intervention trial, the consumption of 50 mg of all-rac-α-tocopherol was found to increase mortality from hemorrhagic stroke (ATBC Cancer Prevention Study Group, 1994), presumably due to competition of the tocoquinone metabolite with vitamin K reactivation.

REFERENCES

ATBC Cancer Prevention Study Group, 1994. The effect of vitamin E and beta carotene on the incidence of lung cancer and other cancers in male smokers. N. Engl. J. Med. 330, 1029–1035.

Bender, D.A., Earl, C.J., Lees, A.J., 1979. Niacin depletion in Parkinsonian patients treated with L-DOPA, benserazide and carbidopa. Clin. Sci. 56, 89–93.

Boukaiba, N., Flament, C., Acher, S., Chappuis, E., Piau, A., Fusselier, M., et al., 1993. A physiological amount of zinc supplementation: effects on nutritional, lipid, and thymic status in an elderly population. Am. J. Clin. Nutr. 57, 566–572.

Chung, C.S., Nagey, D.A., Veillon, C., Patterson, K.Y., Jackson, R.T., Moser-Veillon, P.B., 2002. A single 60 mg iron dose decreases zinc absorption in lactating women. J. Nutr. 132, 1903–1905.

Donangelo, C.M., Woodhouse, L.R., King, S.M., Viteri, F.E., King, J.C., 2002. Supplemental zinc lowers measures of iron status in young women with low iron reserves. J. Nutr. 132, 1860–1864.

Finley, J.W., 1999. Manganese absorption and retention by young women is associated with serum ferritin concentration. Am. J. Clin. Nutr. 70, 37–43.

Krishnaswamy, K., Rao, S.B., Raghuram, T.C., Srikantia, S.G., 1976. Effect of vitamin B6 on leucine-induced changes in human subjects. Am. J. Clin. Nutr. 29, 177–181.

McCormick, D.B., 1989. Two interconnected B vitamins: riboflavin and pyridoxine. Physiol. Rev. 69, 1170–1198.

Santon, A., Giannetto, S., Sturniolo, G.C., Medici, V., D'Inca, R., Lrato, P., et al., 2002. Interactions between Zn and Cu in LEC rats, an animal model of Wilson's disease. Histochem. Cell Biol. 117, 275–281.

Tsuge, H., Hotta, N.T., 2000. Effects of vitamin B6 on (n–3) polyunsaturated fatty acid metabolism. J. Nutr. 130, 333S–334S.

van den Berg, H., van Vliet, T., 1998. Effect of simultaneous, single oral doses of beta-carotene with lutein or lycopene on the beta-carotene and retinyl ester responses in the triacylglycerol-rich lipoprotein fraction of men. Am. J. Clin. Nutr. 68, 82–89.

GRAS DATABASE

The food industry uses a wide range of compounds to add flavor, color, texture, moisture, and other desirable properties to their products. Other compounds are used to extend shelf life or slow spoilage of foods. Such compounds can only be added when they are generally recognized as safe (GRAS). A comprehensive list of all compounds with GRAS designation can be viewed at http://www.aibmr.com/resources/GRAS-database.php.

The significance of food ingredient classifications is illustrated by the recent withdrawal of GRAS status of partially hydrogenated fats (Assaf, 2014). The previously unlimited use of such trans-fat-rich processed materials is now under scrutiny by the US Food and Drug Administration (FDA) and may be limited by administrative regulation.

REFERENCE

Assaf, R.R., 2014. Overview of local, state, and national government legislation restricting trans fats. Clin. Ther. 36, 328–332.

USING MOLECULAR DATABASES

The vast amount of data on biological systems can make it hard to find a particular piece of information. The power of current Web-based systems helps to access data with great speed. The following describes some typical scenarios and provides directions to potential resources.

ABBREVIATION

ADH alcohol dehydrogenase (EC1.1.1.1)

HOW TO GET BASIC NUTRIENT INFORMATION

Usually, a standard textbook (such as this one) will be the best starting point. The Office of Dietary Supplements has an online collection of Dietary Supplement Fact Sheets that provide detailed information on the likely benefits and risks associated with major nutrients (http://ods.od.nih.gov/).

A free online literature search system (Pubmed) is provided by the National Institutes of Health (http://www.pubmedcentral.nih.gov/).

Those with access to the electronic editions of major journals will be able to find many excellent review articles on nutrient metabolism. Publications that often carry pertinent reviews include the *Annual Review of Nutrition* (http://nutr.AnnualReviews.org/), the *American Journal of Physiology* (http://ajpcon.physiology.org/), and *American Journal of Clinical Nutrition* (http://www.ajcn.org/).

HOW TO GET INFORMATION ON NUTRIENT AND METABOLITE CONCENTRATIONS

Numerous research papers report on specific measurements in various body fluids. The Wishart metabolomics research group at the University of Alberta (Wishart et al., 2013) has collated such information into a publically accessible database (http://www.hmdb.ca/biofluids).

This resource contains qualitative and quantitative information about 15 different types of body fluids, including blood, urine, saliva, and cerebrospinal fluid. The scope extends to hundreds of nutrients and nutrient-derived metabolites and is steadily growing.

HOW TO GET INFORMATION ON ENZYMES

The Nomenclature Committee of the International Union of Biochemistry and Molecular Biology (IUBMB) has recommended a system to categorize enzymes. A four-component Enzyme Commission (EC) numbering system facilitates the reference to a specific group of enzymes. Note that the size of each of the four numeric components of the EC number is not limited. To take a particular example, alcohol dehydrogenase (ADH) has been assigned the EC number 1.1.1.1. It can be informative to review neighboring entries if searching for enzymes with similar characteristics. EC1.1.1.2, for instance, designates alcohol dehydrogenases, which use NADP as a cosubstrate.

In many instances, several human proteins correspond to the same EC number, sometimes with considerably divergent substrate specificities and other characteristics.

Humans have genes for at least seven distinct ADHs, which vary greatly in catalytic properties. While most oxidize ethanol well, ADH5 prefers long-chain primary alcohols and is actually a formaldehyde-oxidizing enzyme.

The Swiss Institute of Bioinformatics maintains extensive databases on their Expert Protein Analysis System (ExPASy) proteomics server (http://www.expasy.ch/enzyme/). This is an important, comprehensive source of information on most enzymes, including those involved in nutrient metabolism. The enzyme database can be searched by EC number, enzyme name, reactant, or cofactor, or by viewing a chart of metabolic pathways.

Brenda (the Braunschweig enzyme database) is an extensive collection of information on enzymes, substrates, products, ligands, and metabolic processes (http://www.biobase-international.com/product/brenda). Use of the database requires a paid subscription.

HOW TO GET INFORMATION ON TRANSPORTERS

Online collections of information on particular classes of transporters can be helpful to clarify the profusion of the diverse structures through which nutrients and metabolites can travel across membrane barriers.

The HUGO Gene Nomenclature Committee (HGNC) maintains a list of solute carriers (SLCs) at http://www.genenames.org/genefamilies/SLC.

The Nutrition, Metabolism, and Genomics Group at Wageningen University, the Netherlands has compiled information on all known human ABC transporters (http://nutrigene.4t.com/humanabc.htm).

HOW TO GET INFORMATION ON RECEPTORS

The International Union of Basic and Clinical Pharmacology Committee on Receptor Nomenclature and Drug Classification curates several databases on receptors and ligands, which can be accessed at http://www.guidetopharmacology.org/.

HOW TO GET INFORMATION ON GENES

One of the best sources of information on genes and genetic syndromes is OMIM, the online version of McKusick's textbook *Mendelian Inheritance in Man* (http://www3.ncbi.nlm.nih.gov), currently hosted by the NCBI at the US National Library of Medicine (NLM). The NLM also provides detailed sequence information and search engines such as BLAST (http://www.ncbi.nlm.nih.gov/BLAST/).

Sequence data can be retrieved from the Genome Database (GDB), a collaboration associated with the Human Genome Project (http://www.gdb.org/). The database also provides information on probes, primers for sequence amplification, and known polymorphisms. TRANSFAC (http://www.gene-regulation.com/) is a compilation of eukaryotic transcription factors.

The site of the 1000 Genomes Project (http://www.1000genomes.org/) provides a wealth of detailed information on sequence variants across the entire genome. Advanced bioinformatics tools are needed to mine much the available information.

Specialized Web-based databases with information on genetic variants are numerous and growing. The NCBI has a database (dbSNR; http://www.ncbi.nlm.nih.gov/SNP) with nearly 5 million polymorphisms due to single base nucleotide substitutions, short deletion, and insertions. The variants are listed with the information on each gene or protein and can be found by searching for gene name, sequence or a number of other characteristics. The Institute of Medical Genetics at the University of Wales College of Medicine in Cardiff, UK, houses the Human Gene Mutation Database (HGMD; http://www.hgmd.cf.ac.uk/ac/index.php).

Information on numerous gene variants can be found in this database (Krawczak and Cooper, 1997; Stenson et al., 2014). The Frequency of Inherited Disorders Database (FIDD; http://www.uwcm.ac.uk/uwcm/mg/hgmd0.html) at the same institution provides information on nearly 300 conditions caused by genetic variation. An example for a narrowly specialized collection is the mutation database for the human glucose-6-phosphate dehydrogenase (G6PD, EC1.1.1.49) gene (http://www.rubic.rdg.ac.uk/g6pd/).

REFERENCES

Krawczak, M., Cooper, D.N., 1997. The human gene mutation database. Trends Genet. 13, 121–122.

Stenson, P.D., Mort, M., Ball, E.V., Shaw, K., Phillips, A., Cooper, D.N., 2014. The Human Gene Mutation Database: building a comprehensive mutation repository for clinical and molecular genetics, diagnostic testing and personalized genomic medicine. Hum. Genet. 133, 1–9.

Wishart, D.S., Jewison, T., Guo, A.C., Wilson, M., Knox, C., Liu, Y., et al., 2013. HMDB 3.0—the human metabolome database in 2013. Nucleic Acids Res. 41, D801–D807.

Index

Note: Page numbers followed by "*f*" and "*t*" refer to figures and tables, respectively.

A

Absorption, 37–61
  alanine (Ala), 330–332
  amino acids, 273–275
  arginine (Arg), 416–417
  arsenic (As), 798
  asparagine (Asn), 409
  aspartate (Asp), 402
  biopterin, 658
  biotin, 643
  bromine (Br), 793
  calcium (Ca), 731–732
  carnitine, 457
  carotenoids, 491–492
  chlorine (Cl), 691–693
  chlorophyll, 180
  cholesterol (Chol), 542–544
  choline, 471
  chromium (Cr), 782–783
  citrulline, 441
  cobalt (Co), 779
  conjugated linoleic acid (CLA), 158–159
  copper (Cu), 710
  creatine, 451
  cysteine (Cys), 370
  docosahexaenoic acid (DHA), 167
  enzymes influencing, 45*t*
  ethanol, 232–233
  fatty acids, 124–125
    intestinal, 124*f*
  fluorine (F), 755–756
  folate, 622–623
  fructose (Fru), 209
  galactose (Gal), 215–216
  gastric secretions for support of, 39*t*
  glucose (Glc), 195
  glutamate, 296–297, 296*f*
  glutamine, 303
  glycine (Gly), 310–311
  heterocyclic amines (HAs), 102
  histidine (His), 432
  inositol, 665
  iodine (I), 749
  isoleucine (Ile), 397–398
  leucine (Leu), 383–384
  lipoic acid (LA), 555
  lysine (Lys), 377
  magnesium (Mg), 745
  manganese (Mn), 725–726
  melatonin, 464
  methionine (Met), 360
  molybdenum (Mo), 774
  niacin, 601–603
  nickel (Ni), 805
  nitrite/nitrate, 108
  oxalic acid, 230
  pantothenate, 648–649
  phosphorus (P), 738–739
  proline (Pro), 425
  pyruvate, 224
  queuine, 655
  riboflavin, 590–591
  selenium (Se), 767–768
  serine (Ser), 324–325
  silicon (Si), 790
  sodium, 679–680
  sulfur (S), 760
  taurine, 445
  thiamin, 581–582
  threonine (Thr), 317
  tryptophan (Trp), 351
  tyrosine (Tyr), 344
  ubiquinone (coenzyme Q), 561–563
  valine (Val), 390
  vanadium (V), 802–803
  vitamin A (vitA), 490–492, 491*f*
  vitamin B6, 612–613
  vitamin B12 (B12), 634–636
  vitamin C, 572–573
  vitamin D (vitD), 504
  vitamin E (vitE), 517
  vitamin K (vitK), 528–529
  water, 674–675
  xylitol, 220
  zinc (Zn), 717–719
Acetate, 147–153, 147*f*
  daily requirements, 147
  dietary effects, 152
  dietary sources, 147, 149
  endogenous production, 147–148, 149*f*
    carbohydrates, 147
    ethanol metabolism, 148

Acetate (*Continued*)
  fatty acids, 148
  fiber, 148
  excretion, 151–152
  functions, 147
  intestinal absorption, 149
  metabolism, 150
    ketogenesis, 150, 151*f*
  nutritional summary, 147
  regulation, 152
  risk of excessive intake, 147
  storage, 151
  transport and cellular uptake, 149–150
    via blood, 149–150
    blood–brain barrier (BBB), 150
Acetic acid, 147, 152
Acetoacetate, 134, 147–148, 340
N-Acetyl galactosamine, 308
N-Acetyl glucosamine, 212, 308
N-Acetylaspartate, 407
Acetylcholine, 475*f*
Acetyl-CoA, 147, 148*f*, 150, 154–156, 160
Acetyl-CoA C-acetyltransferase, 134, 340
Acetyl-CoA C-acyltransferase, 166, 176
Acetyl-CoA carboxylase, 120, 645
Acetyl–coenzyme A (acetyl-CoA), 120
N-Acetylmuramyl-L-alanine amidase, 330–332
N-Acetyltransferase, 281, 404
S-Acetyltransferase, 120
Acinetobacter species, 64
Aconitase, 705
Actinobacteria, 61–63
Acylcarnitine, 460
Acyl-CoA dehydrogenase, 393, 398, 400
Acyl-CoA dehydrogenase-8, 393
S-Adenosyl homocysteine hydrolase, 712–713
Adenosylhomocysteinase, 367
S-Adenosylhomocysteine, 455
Adipocytes, 135–136
Adiponectin, 144
Adriamycin, 429
Advanced glycation end products, 272
Agmatin, 421
Agouti-related protein (AgRP), 26
Alanine (Ala), 329–336
  D form, 330, 334
  daily requirements, 330
  deficiency, impact of, 330
  dietary sources, 330
  digestion and absorption, 330–332
  endogenous sources, 330
  excretion, 334

  filtered free, 334
  functions, 330, 334–335
    in energy fuel yielding, 334
    glucose–alanine cycle, 334–335
    glyoxylate metabolism, 335
    heme synthesis, 335
    protein synthesis, 334
    synthesis of nitrogen compounds, 335
  L form, 329*f*, 354
    endogenous sources, 331*f*
    intestinal absorption, 332*f*
    transamination of, 333*f*
  metabolism, 333–334
  nutritional summary, 330
  regulation, 334
  risk of excessive intake, 330
  storage, 334
  transport and cellular uptake
    via blood, 332–333
    blood–brain barrier (BBB), 333
    materno-fetal transfer, 333
Alanine aminopeptidase (AAP), 330–332
Alanine cycle, 285, 285*f*, 292
Alanine-glyoxylate aminotransferase, 310, 312, 335
Aldehyde oxidase, 776
Aldehyde reductase, 208
Aldose, 188–189
Alginate, 245
Allergies, 66
Allyl isothiocyanate, 21
Alpha-amylase, 193–195
Alpha-chymotrypsin, 272–273
Alpha-ketoglutarate, 266–271
Alpha-ketoglutarate dehydrogenase complex, 297, 354, 585
Alpha-linolenic acid, 164–166, 165*f*
Alzheimer's dementia, 2
Alzheimer's disease, 657, 706
Aminoacetone, 320
Amino acid–glyoxylate aminotransferase, 310
Amino acids, 147–148, 224
  absorption of, 273–275
    in small intestine, 275, 276*t*
    sodium gradient, role of, 273–275
    uptake of cationic amino acids, 275
  ammonium ions and synthesis of, 266–271
  branched-chain amino acids (BCAAs), 266–271
  common
  composition and structure, 266
    alpha-carbon, 266
    forms of bases and acids in, 268*f*
    isomeric forms of alanine, 267*f*
    L and D form, 266

in proteins, 267*f*
side chain, 266
specific peptides, 266
dietary sources, 271–272
heating of foods, effect of, 272
recommendations for daily intake, 271–272
in tea (*Camellia sinensis*) leaves, 272
digestion, 272–273
endogenous sources, 266–271
precursors from intermediary metabolism, 266–271
energy yield of individual, 286, 286*t*
enterohepatic circulation of, 273
excretion, 283–284
beta-alanine, 283
dipeptides, 283
fecal losses, 284
hydroxyproline, 283
proline, 283
renal salvage, 283
sodium/peptide cotransporter, role of, 283–284
taurine, 283
tripeptides, 283
with urine, 284
free amino acids, in blood, 277*t*
functions, 285–286
nonprotein mediator synthesis, 286
nucleotide synthesis, 286
protein synthesis, 285
role in neurotransmission, 286
synthesis of glucose (gluconeogenesis), 285
materno-fetal nutrient transport, 89–90
transporters, 90
metabolism, 278–281
D-alanine, 279
L-alanine, 279
arginine, 281
asparagine, 281
aspartate, 281
cysteine, 280
folate, 628
glutamate, 278–279, 281
glutamine, 281
glycine, 279
histidine, 281
iron (Fe), 706
isoleucine, 280
leucine, 280
lysine, 281
methionine, 280
ornithine synthesis, 281
phenylalanine, 280
L-proline, 280

riboflavin, 596–597
serine, 280
threonine, 280
tryptophan, 278–280
tyrosine, 280
urea synthesis, 281, 282*f*
valine, 280
vitamin B6, 617
vitamin C, 577
nonprotein, 269*t*, 270*t*
oxidation of, 286
posttranslational hydroxylation of procollagens
and elastins, role in, 271
as precursors, 271
protein, 269*t*, 270*t*
protein quality of, 293
for protein synthesis, 271
regulation, 284
nitrogen balance, 284
sodium-driven transport of, 273–275
storage, 283
structure and function of, 266–291
transport and cellular uptake of
via blood, 275–276
blood–brain barrier (BBB) and, 278
materno-fetal transfer, 278
transportation to brain, 82–83
choline, 83
glutamate, 82
insulin, 82
leptin, 82
transporters in human kidney, 73*t*
transporters in placenta, 279*t*
Aminoadipate-semialdehyde dehydrogenase, 379
2-Amino-3-carboxymuconate semialdehyde, 354
Aminomalonate, 333
2-Amino-1-methyl-6-phenylimidazo[4, 5-β]pyridine
(PhIP), 101–102, 102*f*, 105
metabolism of, 103*f*
Ammonia, 278–279
Amylin, 27
Amylose, 193–195
Anandamide, 21
Anorexia, 25–26, 580
Anorexigenic mediators, 28
Anthocyanidins, 252
Antibiotics
microbiome, effect on
ciprofloxacin, 63
clarithromycin, 63
weight gain, associated with, 63
Antidiuretic hormone (ADH), 677

Antioxidants, 479
  compounds, 485*t*
  encounter with ROS, 481
  mechanism of, 482–485
    enzyme-catalyzed reactions, 482–483, 485*t*
    metal ion chelation, 482
    regeneration of oxidized forms, 483–485
  water-soluble, 483–485
Antithiamin, 581
Aplasmomycin, 788
Apolipoprotein B (apoB), 517
Apolipoprotein B48 (apoB48), 544
Apolipoprotein E (apoE), 517
Appetite, 203–204
  central regulation, 26–27
    (appetite-decreasing) melanocortin neurons, 26
    (appetite-promoting) NPY/AgRP neurons, 26
    cannabinoids, 27
    dopamine binding receptors, 27
    hormones, 27
    melanin-concentrating hormone (MCH), 27
    neurotransmitters, 26
    serotonin (5-hydroxytryptamin), 26
  defined, 25–26
  energy balance and, 25–26
  enteral input, 28–29
    diurnal variation in ghrelin release, 28
    gastrointestinal mediators, 28
    intake inhibition, 28
    intake stimulation, 28
    neural connections, 28
    orexigenic mediators, 28
  imbalance of nutrients and, 25–26
  individual nutrients, 29–30
    calcium-containing foods, 30
    carbohydrates, 29
    fat, 29
    iron, 30
    magnesium, 30
    protein, 29
    salt, 29–30
  physical activity influence, 30
  sensory input and, 27–28
    cephalic responses, 27
    thought of a favored food, 27
  stimulation for, 25
Aquacobalamin, 633*f*, 636
Aquaporin 2 (AQP2), 676–677
Aquaporin 3 (AQP3), 675–676
Aquaporin 4 (AQP4), 676
Arachidonic acid, 114–116, 669
Aralar 1, 304, 409

Arginine (Arg), 271, 413–423
  daily requirements, 414
  decarboxylation of, 420*f*
  deficiency, impacts of, 414
  dietary sources, 414
  digestion and absorption, 416–417
  endogenous sources, 414
  excretion, 419
  functions, 414, 420–421
    of amines, 421
    in energy fuel yielding, 420
    high-energy storage, 420
    hormone stimulation, 421
    nitric oxide synthesis, 420
    protein synthesis, 420
  intestinal absorption, 416*f*
  L form, 413*f*
  metabolism, 417–419
    breakdown of arginine *N*-methyltransferases, 417–419
    catabolism, 417, 418*f*
  nitric oxide synthesis from, 421*f*
  nutritional summary, 414
  regulation, 419–420
    control of CAT-2 expression, 419–420
    control of dietary intake, 419
    glucose-induced insulin secretion, 419–420
  risk of excessive intake, 414
  storage, 419
  transport and cellular uptake, 417
    via blood, 417
    blood–brain barrier (BBB), 417
    materno-fetal transfer, 417
Argininosuccinate lyase, 281, 442
Argininosuccinate synthase, 404, 442
Arsenate, 798–799, 799*f*
Arsenic (As), 797–802
  daily requirements, 797
  deficiency, impacts of, 797, 800
  dietary sources, 797–798
  digestion and absorption, 798
  excretion, 800
  functions, 797, 800
    control of gene transcription, 800
  inorganic, 798–799
  isotopes, 797
  metabolism, 798–800
  metabolites, 800
  nutritional summary, 797
  regulation, 800
  risk of excessive intake, 797
  storage, 800
  transport and cellular uptake, 798

via blood, 798
trioxide, 800
Arsenite, 798–799
functions, 800
methyltransferase, 798–799
Arylformamidase, 352
Arylsulfatase E, 764
Ascorbate (ASC), 336, 348, 381, 481, 485, 570, 612
Ascorbate-2-sulfate (A2S), 572–573
metabolism, 575f
storage of, 574–575
Asparagine (Asn), 271, 408–413
daily requirements, 408
deficiency, impacts of, 408
depletion by chemotherapy, 412
dietary sources, 408–409
digestion and absorption, 409
endogenous sources, 409
excretion, 412
functions, 408, 412
in energy fuel yield, 412
protein synthesis, 412
L form, 408, 408f
catabolism, 411f
intestinal absorption, 410f
metabolism, 410–411
nutritional summary, 408
regulation, 412
risk of excessive intake, 408
storage, 411
synthase, 409
transport and cellular uptake, 409–410
via blood, 409
blood–brain barrier (BBB), 409–410
materno-fetal transfer, 410
Aspartate aminotransferase (AST), 340, 409, 618, 759
Aspartate (Asp), 401–408
L-aspartate, 330
daily requirements, 402
deficiency, impacts of, 402
D form, 407
dietary sources, 402
digestion and absorption, 402
endogenous sources, 402
excretion, 405
free, 402
functions, 402, 405–407
arginine synthesis, 406
asparagine synthesis, 406
in energy fuel yielding, 405
nucleotide synthesis, 406
protein synthesis, 405

vitamin metabolism, 406
L form, 401, 401f
catabolism, 405f
intestinal absorption, 403f
nucleotide synthesis, 407f
as a precursor for urea synthesis, 406f
metabolism, 403–404
Krebs-cycle reactions, 403
protein modification, 404
SAM-dependent methylation, 404
transamination, 403
urea synthesis, 404–405
nutritional summary, 402
regulation, 405
risk of excessive intake, 402
storage, 404
transport and cellular uptake, 402–403
via blood, 402
mitochondrial translocation, 403
Aspartate 4-decarboxylase, 373, 759
ATP-citrate (pro-S-)-lyase, 120
Autism, 657
Autobrewery syndrome, 66

B
Bacilli, 62
Bacteria
colonization of intestine by, 61–62
gut, in infants, 62
segment-specific density of, 61–62
Bacteroides species, 61–63
B. fragilis, 62
bariatric surgery and, 64
consumption of animal protein and cholesterol, role of, 64
excess of body fat and, 64
Balenine, 437
Basal cells, 1
Bauhinian valve, 45
Beta-aminoalanine, 330
S-Beta-aminoisobutyrate (BAIB), 393
Beta-carotene, 480
Beta-carotene 15, 15′-dioxygenase, 488
2-(Beta-D-hexopyranosyl)-1-tryptophan, 350
N(1)-(Beta-D-hexopyranosyl)-1-tryptophan, 350
Beta-glucans, 244–245
Beta-hydroxy beta-methylbutyrate, 387
Betaine transporters, 70–72, 798
Bifidobacteria species, 61–63
in infants with the HLA genotypes DQ2 or DQ8, 65
nondigestible oligosaccharides and, 65
sour taste of, 65

Bifidobacterium species, 65
  B. breve, 62
  B. infantis, 65
Bile, 43–44
  Clostridium difficile infection, effect on, 63
  digestive and absorptive activities, 46–47
  typical composition of, 43, 43*t*
  vitamin B12 (B12) excretion, 637–639
Bile acids, 310, 488, 490–491, 504, 539, 542–543, 546–547,
    549–551, 574, 652, 764
  bacterial hydrolysis of, 275
  conjugation, 315, 448
  taurine conjugates, 374, 445–446
Biliary ducts, 43
Biopterin, 336, 657–663
  daily requirements, 657
  deficiency, impacts of, 657
  dietary sources, 657–658
  digestion and absorption, 658
  endogenous sources, 658
  excretion, 661
  functions, 657, 661–662
    catecholamine and pigment synthesis, 661
    cell growth and survival, 661
    immune defense, 662
    metabolic regulation, 662
    nitric oxide synthesis, 661
    regulation of O-alkylated glycerolipids, 661–662
    serotonin and melanotonin synthesis, 661
    tyrosine synthesis, 661
  metabolism, 658–660
  nutritional assessment, 662
  nutritional summary, 657
  regulation, 661
  risk of excessive intake, 657
  storage, 660
  transport and cellular uptake, 658
Biotin, 66, 139, 280, 316, 320, 359–360, 365, 381, 383, 396,
    398, 642–647, 642*f*, 758
  daily requirements, 642
  deficiency, impacts of, 643
  dependent pyruvate carboxylase, 646*f*
  dietary sources, 642–643
  digestion and absorption, 643
  excretion, 645
  functions, 642, 645–646
    biotinylated carboxylases, 645
    catabolism of leucine, 646
    protein expression, 646
  inactive metabolites, 645*f*
  materno-fetal nutrient transport, 91
  metabolism, 643–644
    linked to carboxylases, 644*f*

  recycling of protein-bound, 644
  nutritional assessment, 646
  nutritional summary, 642–643
  regulation, 645
  risk of excessive intake, 643
  storage, 645
  transport and cellular uptake, 643
    via blood, 643
      blood–brain barrier (BBB), 643
      materno-fetal transfer, 643
    transportation to brain, 83
Bitter taste, 9–10
Blood–brain barrier (BBB), 81–87
  acetate, 150
  alanine (Ala), 333
  anatomical background, 81–82
  arginine (Arg), 417
  asparagine (Asn), 409–410
  biotin, 643
  caffeine, 96
  calcium (Ca), 732
  carnitine, 457
  cholesterol (Chol), 548
  choline, 472
  conjugated linoleic acid (CLA), 160
  copper (Cu), 712
  cysteine (Cys), 372
  docosahexaenoic acid (DHA), 167
  fatty acids, 126
  flavonoids, 254
  folate, 624
  fructose (Fru), 209
  galactose (Gal), 216
  glucose (Glc), 196–197
  glutamate, 297
  glutamine, 305
  glycine (Gly), 311
  histidine (His), 434
  iron (Fe), 701
  isoleucine (Ile), 398
  leucine (Leu), 385
  lysine (Lys), 378–379
  magnesium (Mg), 745
  manganese (Mn), 726
  methanol, 239
  methionine (Met), 361
  molybdenum (Mo), 774
  myristate (myristic acid), 154
  niacin, 603
  nitrite/nitrate, 108
  nutrient transport across, 82–85
    amino acids, 82–83
    carbohydrates, 82

lipids, 82
    minerals and trace elements, 84–85
    nonnutrient bioactives and xenobiotics, 85
    vitamins, 83–84
pantothenate, 650
passage of amino acids, 278
phenylalanine (Phe), 338
phosphorus (P), 740
phytanic acid, 180
potassium (K), 688
proline (Pro), 425
pyruvate, 225
reduction of DHA to ASC blocks, 573
riboflavin, 592
selenium (Se), 768
serine (Ser), 325
sodium, 681
sulfur (S), 761
taurine, 445
thiamin, 582–583
threonine (Thr), 318
trans-unsaturated fatty acids (trans-fatty acids), 176
tryptophan (Trp), 352
tyrosine (Tyr), 345
valine (Val), 391
vitamin A (vitA), 492
vitamin B6, 614
vitamin B12 (B12), 637
vitamin C, 573
vitamin D (vitD), 505
vitamin E (vitE), 518
vitamin K (vitK), 529
water, 676
zinc (Zn), 719
Blood–cerebrospinal fluid barrier, 81
Blood coagulation, 145
Blood hyperosmolality, 33
Boromycin, 788
Boron (B), 785–789
    common forms, 785
    daily requirements, 786
    deficiency, impacts of, 786
    dietary sources, 786
      food, 786
    digestion and absorption, 786
    excretion, 787
    functions, 785, 787–788
      actions in nonmammalian species, 788
      bone and connective tissues, 787
      boron neutron capture therapy (BNCT), 788
      influences on hormone action, 787
    half-life of, 787
    isotopes, 785

    metabolism, 787
    nutritional summary, 785–786
    regulation, 787
    risk of excessive intake, 786
    storage, 787
    transport and cellular uptake, 786
      via blood, 786
Bowman-Birk-type inhibitors, 292–293
Bowman's glands, 1
Branched-chain alpha-keto acid dehydrogenase, 585
Brenda, 823
Bromine (Br), 793–797
    daily requirements, 793
    deficiency, impacts of, 793, 795
      insomnia, 795
    dietary sources, 793
    digestion and absorption, 793
    excretion, 794
    functions, 793–795
      immune defense, 794
      inhibition of thyroid function, 795
      structuring of basement membranes, 795
    isotopes, 793
    nutritional summary, 793
    regulation, 794
    risk of excessive intake, 793, 795
    storage, 794
    transport and cellular uptake, 793
      via blood, 793
Bromotaurine, 795*f*
N-Bromotaurine, 794–795
Brunner's glands, 41
Brush border membrane, 41

C
Caffeine, 95–100, 95*f*
    abrupt caffeine withdrawal, impact of, 96
    daily requirements, 96
    dietary sources, 96
      in brewed coffees, 96
    excretion, 98–99
    functions, 96
      addictive qualities, 98
      phosphodiesterase inhibition, 98
      receptor interactions, 98
    health effects
      to assess liver function, 99
      calcium homeostasis, 98
      cognitive function, 98
      diuretic effect, 98
      ergogenic effects, 98
      lack of sleep, 98
      migraine headaches, 99

Caffeine (*Continued*)
  intestinal absorption, 96
  metabolism, 97, 97*f*
  nutritional summary, 96
  risk of excessive intake, 96, 99
  transport and cellular uptake, 96
Calbindin-9K, 732–733
Calcitroic acid, 507
Calcium (Ca), 221, 730–737
  daily requirements, 730
  deficiency, impacts of, 730
  dietary sources, 730
  digestion and absorption, 731–732
  excretion, 733
  free, 732
  functions, 730, 734–735
    in blood clotting, 735
    in cell growth, 735
    as a messenger, 734
    in muscle contraction, 734
    in neurotransmitter release and neuron
      excitability, 734
    as protein cofactor, 734–735
    for skeletal structure, 735
    for vestibular system, 735
  intestinal absorption, 731*f*
  intestinal sensing of, 19
  materno-fetal nutrient transport, 91
  nutritional assessment, 735
  nutritional summary, 730
  osteoclastic bone digestion, 733
  regulation, 733–734
    calcitonin, 734
    calcium-sensing receptor (CaSR), 733
    parathyroid hormone (PTH), 733–734
    PTH-related protein (PTHrP), 733–734
    vitamin D, 734
  renal processing of, 77
  risk of excessive intake, 730
  storage, 732–733
  taste, 12, 30
  transport and cellular uptake, 732
    via blood, 732
    blood–brain barrier (BBB), 732
    materno-fetal transfer, 732
Calcium/Mg ion sensing receptor (CaSR), 17, 746
CALHM1 ion channel, 7
Cancer, 27, 47, 112, 131–133, 143, 156, 158, 164, 172,
  178, 193, 232, 235–236, 244, 272, 291, 365, 469, 475,
  479–480, 501, 511, 515, 533–534, 567–569, 571, 591, 620,
  623, 662, 669, 705, 710, 727, 730, 767, 771, 784, 794–795,
  797, 800
Capsaicin, 21–22

Carageenan, 245
Carbamoyl synthetase II, 306
Carbohydrates, 187–191, 224–225
  daily requirements, 188
  deficiency, impacts of, 188
  dietary sources, 187–188
    animal foods, 187–188
    plant foods, 187–188
  digestible polysaccharides, 189–190
  fate of excess, 144
  functions, 187
  gustatory perception, 190
  materno-fetal nutrient transport, 88
  nutritional summary, 187–188
  oligosaccharides, 189
  risk of excessive intake, 188
  sugar units, 188–189
    sweet-tasting, 190
  transportation to brain, 82
Carbonic anhydrase VI (gustin), 6
2(2′-Carboxyethyl)-6-hydroxychromans (CEHCs), 519–522
Carboxypeptidase A, 272–273
Carboxypeptidase B, 272–273
Cardiovascular disease, 66, 143, 162, 171–172, 192, 244, 479,
  570, 710, 721–722, 810
Carnitine, 139, 271–272, 454–461
  daily requirements, 455
  deficiency, impacts of, 455
  dietary sources, 455, 457
  digestion and absorption, 457
  endogenous production, 455–457
  excretion, 458
  free, 458
  functions, 455, 458–460
    amino acid and organic acid metabolism, 460
    to boost energy, 460
    catabolism of lysine, 460
    conjugation of xenobiotics, 460
    fatty acid transport into mitochondria, 458, 459*f*
    fatty acid transport into peroxisomes, 459
    gene regulation, 460
    ketogenesis in the liver, 460
    membrane stability, 460
  L form, 454–455, 455*f*
    endogenous synthesis of, 456*f*
  metabolism, 458
    iron (Fe), role of, 706
  nutritional summary, 455
  regulation, 458
  risk of excessive intake, 455
  storage, 458
  transport and cellular uptake, 457
    via blood, 457

blood–brain barrier (BBB), 457
Carnitine acylcarnitine translocase (CACT), 160
Carnosine, 437
Catechins, 252
  metabolism, 255
Catecholamines, 2
Caveolae, 543
Cecum, 45
Cellobiose, 195
Cellulose, 244
Ceruloplasmin, 710
Chemesthesis, 20–23
Chinese restaurant syndrome, 295–296
Chinese tea (*Camellia assamica*), 96
Chlorhexidine, 13
Chlorine (Cl), 690–697
  chemical valences, 690
  conductance regulator, 691–692
  daily requirements, 691
  deficiency, impacts of, 691
  dietary sources, 691
    plant, 691
    salt, 691
  digestion and absorption, 691–693
  excretion, 693–695
    perspiration, 695
    renal, 693–695
  functions, 691, 695–696
    acid production, 695–696
    as a cotransported counterion or exchange chloride for
      another anion, 695
    electrolyte balance, 695
    enzyme activation, 696
    immune defense, 696
  intestinal absorption, 692*f*
  nutritional assessment, 696
  nutritional summary, 691
  reabsorption from renal tubules, 694*f*, 695
  regulation, 695
  risk of excessive intake, 691
  storage, 693
  transport and cellular uptake, 693
    via blood, 693
    materno-fetal transfer, 693
Chlorophyll, 179–186
  deficiency, impacts of, 180
  dietary sources, 180
  digestion and absorption, 180
  functions, 179
  nuclear hormone receptor activation, 184
  nutritional summary, 179–180
  risk of excessive intake, 180
  transport and cellular uptake, 180

Cholecystokinin (CCK), 2, 17–18, 38–40
Cholehepatic shunting, 47
Cholesterol (Chol), 539–553, 539*f*
  daily requirements, 540
  deficiency, impacts of, 540
  dietary sources, 540, 542
  digestion and absorption, 542–544, 544*f*
  endogenous synthesis, 540–542, 541*f*
  excretion, 549–550
  functions, 539, 551
    bile acid synthesis, 551
    protein lipidation, 551
    steroid hormone synthesis, 551
    vitamin D synthesis, 551
  metabolism, 548–549, 549*f*
    bile acid synthesis, 549
    modification and uptake of Chol-carrying Lps, 545*f*, 546*f*
    oxysterols, production of, 548–549, 548*f*
  nutritional summary, 539–540
  regulation, 550
    reverse Chol transport, 550
    uptake into tissues, 550
  risk of excessive intake, 540
  transport and cellular uptake, 545–548
    via blood, 545–546
    blood–brain barrier (BBB), 548
    materno-fetal transfer, 547–548
    reverse Chol transport, 546–547, 547*f*
Choline, 139, 310, 468–477, 468*f*
  daily requirements, 469
  deficiency, impacts of, 469, 475
  dietary sources, 469
  digestion and absorption, 471
  endogenous sources, 469
    by decarboxylation and demethylation of carnitine, 471*f*
    from phosphatidylcholine, 469, 470*f*, 476*f*
  excretion, 474
  free, 471
  functions, 469, 474–475
    cancer risk, 475
    cell cycle regulation, 475
    as a constituent of structural lipids, 474
    countercurrent hypertonia, 474–475
    enzyme activation, 475
    methylgroup transfer, 474
    neurotransmission, 474
  metabolism, 472
    oxidation and successive demethylation, 473*f*
  nutritional assessment, 475
  nutritional summary, 469
  phospholipids of, 472
  regulation, 474
  risk of excessive intake, 469

Choline (*Continued*)
  storage, 474
  transport and cellular uptake, 471–472
    via blood, 471–472
      blood–brain barrier (BBB), 472
      materno-fetal transfer, 472
  metabolism, 706
Cholinephosphate cytidylyltransferase, 139
Chondroitin 4-sulfotransferase, 763–764
Chondroitin 6-sulfotransferase, 763–764
Chromium (Cr), 782–785
  daily requirements, 782
  deficiency, impacts of, 782
  dietary sources, 782
    natural and supplement sources, 782
  digestion and absorption, 782–783
  excretion, 783
  functions, 782, 784
    DNA transcription, 784
    insulin-receptor activity, 784
  isotopes, 782
  nutritional summary, 782
  regulation, 784
  risk of excessive intake, 782
  storage, 783
  transport and cellular uptake, 783
    via blood, 783
    maternal circulation into milk, 783
    transport to insulin-sensitive cells, 783
    transport with transferrin, 783
Chromium (III) tris (picolinate), 783*f*
Chymotrypsin C, 272–273
Cinnamaldehyde (3-phenyl-2-propenal), 22
Citrin, 304, 409
Citrulline, 271, 440–443, 440*f*
  absorption, 441
  daily requirements, 440
  deficiency, impacts of, 440
  dietary sources, 440–441
  endogenous production, 441
  functions, 440
    arginine synthesis, 442
    hair protein modification, 442
    reverse-cholesterol transport, 442
    urea synthesis, 442
  intestinal synthesis, 441
  from L-arginine, 441
  metabolism, 442
    L-arginine synthesis, 442
    urea synthesis, 442
  nutritional summary, 440–441
  risk of excessive intake, 441

  transport and cellular uptake, 441–442
    via blood, 441
    ornithine–citrulline exchange, 441–442
    urea cycle and, 441
Claudin 16 (paracellin-1), 745–746
Clostridia, 61–62
Clostridium difficile infection
  effect on microbiomes, 63
Cobalamin, 383, 398
Cobalt (Co), 778–782
  daily requirements, 778
  deficiency, impacts of, 778
  dietary sources, 778–779
  digestion and absorption, 779
  excretion, 780
  functions, 778, 780
    as nonspecific enzyme cofactor, 780
    as oxygen sensor, 780
    role in translational regulation, 780
  isotopes, 778
  metabolism, 779
  nutritional summary, 778–779
  regulation, 780
  risk of excessive intake, 778–779
    toxic effects, 780
  storage, 779–780
  transport and cellular uptake, 779
    via blood, 779
    maternal circulation into milk, 779
Coenzyme A (CoA), 649–650, 649*f*
  synthesis, 650
Cold perception, 22
  cold-sensing sodium- and calcium-channel
    TRPM8, 22
  phenolic compounds inducing cold sensation, 22*f*
Colon, 45
  goblet cells, 45–46
  mucosa of, 45–46
Colonization of microbiome, 62
  metabolic, developmental, and health consequences, 66
Colostrum, 246
Congenital diarrhea (CLD), 230
Conjugated linoleic acid (CLA), 123, 157–164, 157*f*, 178
  deficiency, impacts of, 158
  dietary sources, 158
  digestion and absorption, 158–159
  endogenous sources, 158, 159*f*
  excretion, 162
  functions, 158, 162
    fuel metabolism, 162
    health benefits, 162
    induction of lipid peroxidation, 162

isomers in fat, 159*f*

metabolism, 160–162

  beta-oxidation, 160

  putative pathway, 161*f*

nutritional summary, 158

risk of excessive intake, 158

storage, 162

transport and cellular uptake, 160

  via blood, 160

  blood–brain barrier (BBB), 160

  materno-fetal transfer, 160

Connexin 43 (Cx43), 603

Copper (Cu), 221, 709–716

daily requirements, 710

deficiency, impacts of, 710

dietary sources, 710

digestion and absorption, 710

excretion, 713

functions, 710, 714–715

  action on connective tissue, 714–715

  antioxidation, 714

  blood coagulation, role in, 715

  energy metabolism, 714

  hormone metabolism, 714

  iron metabolism, 714

  in transcriptional events, 715

intestinal absorption, 711*f*

intracellular disposition, 712*f*

metabolism, 712–713

nutritional assessment, 715

nutritional summary, 710

regulation, 713

risk of excessive intake, 710

storage, 713

transport and cellular uptake, 710–712

  via blood, 710–711

  blood–brain barrier (BBB), 712

  materno-fetal transfer, 712

  transfer into milk, 712

transportation to brain, 84

variation coefficient for, 816–817

Cori cycle, 225

Cranial nerves V (trigeminus nerve), 21

Cranial nerves IX (glossopharyngeal nerve), 21

Cranial nerves X (vagus nerve), 21

Creatine, 450–454, 450*f*

creatinine production, 452*f*, 453

daily requirements, 450

deficiency, impacts of, 450

dietary sources, 450–451

digestion and absorption, 451

endogenous production, 450–451, 451*f*

excretion, 453

functions, 450, 453–454

metabolism, 452–453

  phosphorylation, 452, 452*f*

nutritional summary, 450

phosphate, 453–454

regulation, 453

risk of excessive intake, 450

storage, 453

transport and cellular uptake, 452

  via blood, 452

Creutzfeld–Jakob disease, 669

Crotonaldehyde, 481

Cup cells, 42

Cyanosis, 109

Cysteine (Cys), 271, 368–376, 443–445, 758–759

aminotransferase, 373

L-cysteine, 359–360

  intestinal absorption, 371*f*

  metabolism of, 372*f*

daily requirements, 369

deficiency, impacts of, 369

dietary sources, 369–370

digestion and absorption, 370

endogenous sources, 370

excretion, 373–374

functions, 369, 374–375

  conjugation of taurine, 374

  cyanide detoxification, 375

  in energy fuel yielding, 374

  glutathione metabolism, 374

  pantothenate sulfuration, 374

  protein synthesis, 374

  supply of reduced sulfur, 375

  supply of sulfate, 375

  tRNA sulfuration, 374

L form, 368–369, 369*f*

metabolism, 372–373

  oxidization of cysteine to cysteinesulfinate, 373

  single-step conversion, 373

  taurine synthesis, 373

nutritional summary, 369

regulation, 374

risk of excessive intake, 369

storage, 373

transport and cellular uptake, 370–372

  via blood, 370–371

  blood–brain barrier (BBB), 372

  materno-fetal transfer, 372

Cysteinesulfinate, 373

Cystine (CssC), 370, 373

L-Cystine, 369*f*
  intestinal absorption, 371*f*
  metabolism of, 372*f*
Cystinosin, 371
Cytidyl diphosphate (CDP), 139
Cytidyl diphosphate (CDP)–diacylglycerol, 327–328
Cytochrome b5, 120–123

D

D cells, 42
Daidzein, 254, 255*f*
Decaffeination, 96
Dehydration, 677
Dehydroalanine, 330
3-Dehydro-D-sphinganine, 321
Dehydro-L-ascorbic acid (DHA), 570, 571*f*
  blood–brain barrier (BBB) and, 573
  dietary sources, 571
  digestion and absorption, 572–573
  excretion, 576
  materno-fetal transfer, 573
  metabolism, 573–574
  storage, 576
Dehydroretinol, 486
Delta-aminolevulinate synthase, 616
Deltaproteobacteria, 64
5′-Deoxyadenosyl-cobalamin, 634*f*, 636–637
Diabetes, 50–51, 66, 109, 143, 291, 420, 567, 569, 586, 607, 743–744, 783
Diabetes mellitus, 150, 188, 192, 580, 783
Dietary fiber, 243, 246
  daily requirements, 244
  deficiency, impacts of, 244
  dietary sources, 243
  functions, 243
  nutritional summary, 243–244
  risk of excessive intake, 244–246
Dietary Reference Intake (DRI), 816
Digestible polysaccharides, 189–190
Digestion, 37–61
  alanine (Ala), 330–332
  arginine (Arg), 416–417
  arsenic (As), 798
  asparagine (Asn), 409
  aspartate (Asp), 402
  biopterin, 658
  biotin, 643
  bromine (Br), 793
  calcium (Ca), 731–732
  carnitine, 457
  chlorine (Cl), 691–693
  chlorophyll, 180

cholesterol (Chol), 542–544
choline, 471
chromium (Cr), 782–783
cobalt (Co), 779
conjugated linoleic acid (CLA), 158–159
copper (Cu), 710
creatine, 451
cysteine (Cys), 370
docosahexaenoic acid (DHA), 167
ethanol, 232–233
fatty acids, 124
fluorine (F), 755–756
folate, 622–623
fructose (Fru), 209
galactose (Gal), 215–216
gastric secretions for support of, 39*t*
glucose (Glc), 193–195
glutamate, 296–297
glutamine, 303
glycine (Gly), 310–311
heterocyclic amines (HAs), 102
histidine (His), 432
inositol, 665
iodine (I), 749
isoleucine (Ile), 397–398
leucine (Leu), 383–384
lipoic acid (LA), 555
lysine (Lys), 377
magnesium (Mg), 745
manganese (Mn), 725–726
melatonin, 464
methionine (Met), 360
molybdenum (Mo), 774
niacin, 601–603
nickel (Ni), 805
oxalic acid, 230
pantothenate, 648–649
phosphorus (P), 738–739
proline (Pro), 425
pyruvate, 224
riboflavin, 590–591
selenium (Se), 767–768
serine (Ser), 324–325
silicon (Si), 790
sodium, 679–680
taurine, 445
thiamin, 581–582
threonine (Thr), 317
tryptophan (Trp), 351
tyrosine (Tyr), 344
ubiquinone (coenzyme Q), 561–563
valine (Val), 390

vanadium (V), 802–803
vitamin A (vitA), 490–492
vitamin B6, 612–613
vitamin B12 (B12), 634–636
vitamin C, 572–573
vitamin D (vitD), 504
vitamin E (vitE), 517
vitamin K (vitK), 528–529
water, 674–675
xylitol, 220
zinc (Zn), 717–719
Digestive tract, 37
Dihydrobiopterin, 657*f*, 658
Dihydrofolate reductase, 811
Dihydrolipoic acid, 558
Dihydrovitamin K, 528*f*
Dihydroxyphenylalanine (DOPA), 341
Diketopiperazine cyclo(His-Pro), 425, 432
Dimethylargininase, 417–419
4, 8-Dimethylnonanoyl-CoA oxidation, 182–184
DNA (cytosine-5)-methyltransferase, 568–569
DNA methylation, 567, 568*f*
Docosahexaenoic acid (DHA), 123, 164–174, 164*f*, 819
   daily requirements, 164
   deficiency, impacts of, 164
   dietary sources, 164, 166–167
      fish, 166
      plant-derived foods, 166
      supplements, 166–167
   digestion and absorption, 167
   endogenous sources, 164–166
   excretion, 171
   functions, 164, 171–172
      anticancer potential, 172
      energy fuel, 171
      mental health, 172
      precursor for complex brain lipids, 171
      prostanoid synthesis, 171
      slowing development of heart disease, 171–172
   metabolism, 168–171
      energy, 168
      EPA breakdown, 170*f*
      generation of resolvins and protectin D1, 168–171
      hydroxylation pathways, 168–171
      retroconversion to EPA, 169*f*
   nutritional summary, 164
   risk of excessive intake, 164
   storage, 171
   transport and cellular uptake, 167–168
      via blood, 167
      blood–brain barrier (BBB), 167
      materno-fetal transfer, 167–168

Dopaquinone, 348
Down syndrome, 669–670
Duodenal cytochrome b561 (DcytB), 572–573

E

EC cells, 42
Eicosapentaenoic acid (EPA), 164–166
Elastase II, 272–273
Endopeptidase E, 272–273
Enoyl-CoA isomerase (ECI), 160–162
Enteral hormones, 16–17
Enterocytes, 17
Enteroendocrine cells, 16–17
Enterohepatic circulation, 46–48
   bile acids, 46–47
      intestinal conversion of cholic acid to, 47
      transport of, 47
   molybdenum, 48
   phytochemicals, 48
   vitamin B12, 47–48
Enterokinase, 272–273
Eosinophilia-myalgia syndrome (EMS), 336, 350
Epidermal growth factor (EGF), 195
Epsom salt, 760
Equol, 254, 255*f*
Ergothioneine, 437
Erucic acid, 123
Erythrulose, 188–189
Escherichia, 61–62
Ethanol, 231–238, 231*f*
   consequence of MEOS induction, 236
   daily requirements, 232
   dietary sources, 232
   digestion and absorption, 232–233
   endogenous sources, 232
   functions, 232, 235–236
      energy production, 235
      impact on intestinal function, 235
      metabolic effects, 235
   habituation and alcoholism, 236
   metabolism, 233–235
      acetate, 234–235
      compound drugs, 235
      derived amounts of acetaldehyde, 233–234
      generation of free radicals, 235
      oxidation and formation of acetyl-CoA, 233, 234*f*, 241
   nutritional summary, 232
   risk of excessive intake, 232
   syndrome related to, 232
   toxic effects, 235
   transport and cellular uptake, 233
Ethanolamine, 266–271

Ethanolamine-phosphate cytidylyltransferase, 139
Ethylene glycol, 266–271
Excessive intakes, impacts of, 818
  alanine (Ala), 330
  arginine (Arg), 414
  arsenic (As), 797
  asparagine (Asn), 408
  aspartate (Asp), 402
  biopterin, 657
  biotin, 643
  boron (B), 786
  bromine (Br), 793
  caffeine, 96, 99
  calcium (Ca), 730
  carbohydrates, 144, 188
  carnitine, 455
  chlorine (Cl), 691
  chlorophyll, 180
  cholesterol (Chol), 540
  choline, 469
  chromium (Cr), 782
  citrulline, 441
  cobalt (Co), 778–779
  copper (Cu), 710
  creatine, 450
  cysteine (Cys), 369
  dietary fiber, 244–246
  docosahexaenoic acid (DHA), 164
  ethanol, 232
  fatty acids, 112
  flavonoids, 250
  fluorine (F), 755, 757
  folate, 620
  fructose (Fru), 208
  galactose (Gal), 213
  garlic bioactives, 260
  glucose (Glc), 192
  glutamate (Glu), 295
  glutamine (Gln), 302
  glycine (Gly), 310
  heterocyclic amines (HAs), 101
  histidine (His), 431
  inositol, 664
  iodine (I), 748
  iron (Fe), 698
  isoleucine (Ile), 396
  leucine (Leu), 383
  lipoic acid (LA), 554
  lysine (Lys), 377
  magnesium (Mg), 744
  manganese (Mn), 725
  melatonin, 462
  methanol, 238
  methionine (Met), 360
  molybdenum (Mo), 773
  myristate (myristic acid), 154
  niacin, 600
  nickel (Ni), 805
  nitrite/nitrate, 107
  oxalic acid, 228
  pantothenate, 648
  phenylalanine (Phe), 336
  phosphorus (P), 737
  phytol, 180
  potassium (K), 686
  proline (Pro), 424
  pyruvate, 224
  queuine, 655
  riboflavin, 590
  selenium (Se), 767
  serine (Ser), 322
  silicon (Si), 790
  sodium (Na), 679
  sulfur (S), 759
  taurine, 443
  threonine (Thr), 317
  trans-unsaturated fatty acids (trans-fatty acids), 174
  tryptophan (Trp), 350
  tyrosine (Tyr), 343
  ubiquinone (coenzyme Q), 561
  valine (Val), 390
  vanadium (V), 802
  vitamin A (vitA), 488
  vitamin B6 (B6), 611
  vitamin B12 (B12), 632
  vitamin C, 571
  vitamin D (vitD), 502
  vitamin E (vitE), 515
  vitamin K (vitK), 527
  water, 674
  xylitol, 219
  zinc (Zn), 717
Expert Protein Analysis System (ExPASy)
  proteomics server, 823

F

Farnesyl pyrophosphate, 540
Fat taste, 11–12
Fatty acids, 167
  absorption, 124–125
    intestinal, 124f
  composition, 123
  daily requirements, 112
  deficiency, impacts of, 112

dietary, 116
dietary sources, 112, 123
digestion, 124
endogenous sources, 120–123
    chain elongation and desaturation, 120, 122*f*
    de novo synthesis, 120, 121*f*
    desaturation, 120–122
excretion, 134
functions, 112, 139–140
    complex lipid synthesis, 139
    eicosanoid synthesis, 139–140
    fuel energy, 139
    protein acylation, 139
    synthesis of cholesterol esters, 140*f*
metabolism, 127–134
    acetyl-CoA oxidation, 134
    ketogenesis, 134
    microsomal omega-oxidation, 130, 131*f*
    mitochondrial beta-oxidation, 127–129, 128*f*
    oxidative, 127
    oxidized fatty acids, 131–133, 133*f*
    peroxisomal alpha-oxidation, 129
    peroxisomal beta-oxidation, 129, 130*f*
    propionyl-CoA metabolism, 134
nutritional summary, 112
as a part of different complex lipids, 119*f*
in plants and animals, 111
regulation
    composition, 139
    fat storage, 137*f*, 138
    free fatty acid receptors, 138
    intake, 138
    peroxisomal metabolism, 138–139
risk of excessive intake, 112
storage, 135–138
structure, 111–143
    anteiso, 116, 123
    branched-chain, 118*f*
    characteristic feature, 112
    cis- and *trans*-unsaturated, 115*f*, 116
    double bond, 112–113
    iso, 116, 123
    long-chain, 116
    monounsaturated, 115*t*
    physical and metabolic properties, 112
    polyunsaturated, 117*t*
    saturated, 112, 113*t*, 114*f*
    short branched-chain, 116
transport and cellular uptake, 125–134
    via blood, 125
    blood–brain barrier (BBB), 126
    from liver, 126*f*

    materno-fetal transfer, 126–127
Fatty acid synthase, 120
Fermented milk, 65
Ferroenzyme stearyl-CoA desaturase, 120–122
Ferroxidase, 700–702, 714
FFA1 (GPR40) activation, 19
Firmicutes, 61–64
Fish histamine (*Scombroid*) poisoning, 432
Flavanones, 250
    absorption, 253
    dietary sources, 252
    metabolism, 254
Flavin adenine dinucleotide (FAD), 97, 120–122,
    154–156, 182, 569, 590–591, 811–812. *See also*
    Riboflavin
    containing cytochrome b5 reductase, 166
    dependent acyl-CoA oxidase, 166
    in flavoproteins, 595
Flavin mononucleotide (FMN), 120–122, 569,
    590–591, 811–812. *See also* Riboflavin
    in flavoproteins, 595
Flavones, 251–252
    absorption, 253
    metabolism, 254
Flavonoids, 247–260
    absorption, 252–253
    classification of, 250
    daily requirements, 250
    deficiency, impacts of, 250
    dietary sources, 250–252
    excretion, 257
    functions, 250
    health effects
        antioxidant action, 257
        drug absorption, 257
        inhibition of enzymes, 257
        sex-hormone-like actions, 257
    metabolism, 254–255
    nutritional summary, 250
    in plants, 250–251
    risk of excessive intake, 250
    storage, 255
    structures, 248*f*, 250–251, 251*f*
        oxygen heterocyclic ring in flavones, 250
    transport and cellular uptake, 253–254
        via blood, 253
        blood–brain barrier (BBB), 254
        materno-fetal transfer, 254
Flavonols, 250
    absorption, 253
    dietary sources, 252
    metabolism, 254

Fluorine (F), 755–758
  calcium intake and, 755–756
  daily requirements, 755
  deficiency, impacts of, 755
  dietary sources, 755
  digestion and absorption, 755–756
  excretion, 756
  functions, 755, 757
    bone and joints, growth of, 757
    remineralization of teeth, 757
    tooth development, 757
  nutritional summary, 755
  regulation, 756–757
  risk of excessive intake, 755, 757
  routes for intestinal fluoride uptake, 755–756
  storage, 756
  transport and cellular uptake, 756
    via blood, 756
    materno-fetal transfer, 756
Folate, 321, 381, 469, 620–632
  bioavailability, 622–623
  blood measurement, 630
  catabolism, 625–627
  daily requirements, 620
  deficiency, impacts of, 620
  dietary sources, 620, 622
  digestion and absorption, 622–623
  dihydrofolate reductase activation of, 624*f*
  endogenous sources, 620
  excretion, 627
  free, 623
  functions, 620, 627–628
    amino acid metabolism, 628
    fetal development, 628
    formate utilization, 628, 629*f*
    maintaining of circadian period length and
      rhythmicity, 628
    purine synthesis, 627
    synthesis in microorganisms, 628
    thymidylate synthesis, 627
  homocysteine concentration, 630
  interconversions, 626*f*
  intestinal absorption, 622*f*
  materno-fetal nutrient transport, 91
  metabolism, 624–627
    activation and reactivation, 624–625
    catabolism of glycine, 625
    10-formyl dihydrofolate, 625
    5-formyl-THF, 625
    10-formyl-THF, 625
    5, 10-methenyl-THF, generation of, 625
    5, 10-methylene-THF, generation of, 625

    5-methyl-THF, generation of, 625
    of polyglutamyl side chains, 624
  nutritional assessment, 630
  nutritional summary, 620
  regulation, 627
  risk of excessive intake, 620
  storage, 627
  transport and cellular uptake, 623–624
    via blood, 623
    blood–brain barrier (BBB), 624
    materno-fetal transfer, 624
  variation coefficient for, 816–817
  vitamers, 621*f*
Folate receptors, 623, 811–812
Folic acid, 620
Folylpoly-γ-glutamate carboxypeptidase, 273
Folylpolyglutamate synthase, 623–624
Food intake regulation, mediators of
  (appetite-decreasing) melanocortin neurons, 26
  (appetite-promoting) NPY/AgRP neurons, 26
  cannabinoids, 27
  cocaine- and amphetamine-regulated transcript (CART)
    peptide, 27
  corticotrophin-releasing factor (CRF), 27
  dopamine binding receptors, 27
  galanin, 27
  hormones, 27
  melanin-concentrating hormone (MCH), 27
  neurons of the lateral, dorsomedial, and perifornical
    areas, 27
  neurotransmitters, 26
  opioid peptides, 27
  orexin-producing neurons, 27
  serotonin (5-hydroxytryptamin), 26
Formaldehyde, 628
Formyl-CoA, 182
N-Formylkynurenine, 352
Foul odors, aversive reactions to, 1
Free radicals, 118*f*, 479–486, 481*t*
  composition and structure, 479–480
  oxygen, 488
  reaction with unsaturated fatty acids, 132*f*
Frequency of Inherited Disorders Database (FIDD), 824
Fructose (Fru), 207–213
  daily requirements, 208–209
  deficiency, impacts of, 208
  dietary sources, 208–209
  digestion and absorption, 209
  endogenous sources, 208
  excretion, 211
  functions, 208, 211–212
    energy fuel, 211

hexosamines, 212
as precursor, 212
intestinal absorption, 209f
metabolism, 210–211, 210f
catabolism via fructose 1-phosphate, 211
to pyruvate (glycolysis), 211
sorbitol pathway, 211
nutritional summary, 208
regulation, 211
risk of excessive intake, 208
storage, 211
structural representations, 207f
synthesis, 208f
transport and cellular uptake, 209
via blood, 209
blood–brain barrier (BBB), 209
materno-fetal transfer, 209
Fructose-bisphosphate aldolase, 197–198
Fructoselysine, 377
Fructose-6-phosphate, 308
Fumarate, 340, 346
Fumarylacetoacetase, 346

G

GABA–glutamate–glutamine cycle, 308f
Gal *N*-acetylglucosamine (GalNAc), 218
Galactomannans, 245
Galactose (Gal), 213–219, 213f
daily requirements, 213
deficiency, impacts of, 213
dietary sources, 213–215
digestion and absorption, 215–216
endogenous sources, 213
excretion, 216
functions, 213, 216–218
energy fuel, 217
glycolipid synthesis, 218
glycoprotein synthesis, 218
lactose synthesis, 216–217
as a precursor, 218
intestinal absorption, 215f
metabolism, 216, 217f
nutritional summary, 213
regulation, 216
risk of excessive intake, 213
storage, 216
synthesis, 214f
transport and cellular uptake, 216
via blood, 216
blood–brain barrier (BBB), 216
materno-fetal transfer, 216
Gamma-carboxyaspartate, 271

Gamma-carboxyglutamate, 271
Gamma-glutamyl hydrolase, 623–624
Gamma-glutamyl transpeptidase, 370–371
Gammaproteobacteria, 62
Garlic bioactives, 260–263
S-alk(en)ylcysteine sulfoxide content of, 261
biological effects, 261–262
antioxidant properties, 261
brain function, 262
energy metabolism, 262
lipoprotein metabolism, 261
spread of malodor, 261
daily requirements, 260
deficiency, impacts of, 260
dietary sources, 260–261
functions, 260
garlic powder, 261, 262f
metabolism, 261
nutritional summary, 260
risk of excessive intake, 260
Gastric lipase, 51–52
Gastricsin, 40
Gastrin, 38–40
Gastrointestinal tract, 38–46
digestion and absorption, 46
esophagus, 38
fecal matter removal, 46
large intestine, 45–46
Bauhinian valve, 45
cecum, 45
colon, 45–46
movement of food through, 46
oral cavity, 38
protein-hydrolyzing enzymes of, 274t
saliva, 38
composition, 39t
small intestine, 41–45
anatomical feature, 41
bile, 43
counteracting stomach acidity, 42–43
cup cells, 42
digestion, 44
duodenum, 41, 43–44
endocrine products, 43–44
enterocytes, 41
enteroendocrine cells of, 42
epithelial cells, 41
exocrine secretions, 43–44
goblet cells, 41
gross anatomical structure, 41
intestinal mucosa, 41
jejunum, 41

Gastrointestinal tract (*Continued*)
   luminal surface, 41
   main function of, 41
   major cell types of, 41
   mesentery, 41
   microanatomical features, 41
   microfold or membraneous cells (M cells), 42
   pancreatic secretions, 43–44
   Paneth cells, 42
   pH adjustment, 40, 42–43
   tuft cells, 42
  stomach, 38–40
   acid-stimulating effect, 38–40
   anatomical structure, 38
   calcium-sensing receptor (SCAR), 38–40
   hydrochloric acid, role of, 40
   mixing and grinding of ingested foods, 40
   motility, 40
   parietal cells, 40
   release of gastric contents, 40, 40*f*
   secretory function, 38–40
Generally recognized as safe (GRAS) database, 822
Genetic variation
  affecting nutrient metabolism, 810*t*
  folate adequacy and, 810–812
   genes involved in folate absorption and metabolism, 812*t*
  of hemochromatosis (HFE) gene, 809
  impacts on folate and iron, 809
  iron-regulatory gene HFE, 813–814
Genistein, 254, 256*f*
Genome Database (GDB), 824
Ghrelin *O*-acyltransferase (GOAT), 28
Ghrelin-secreting cells, 16–18
GHRL gene, 17
Gla-rich protein (GRP), 533
Glucagon-like peptides, 6, 17
Glucogenic, 279
Gluconeogenesis, 192–193
  from endogenous and exogenous precursors, 194*f*
Glucosamine-fructose-6-phosphate aminotransferase, 616
Glucosamine sulfate, 759
Glucose–alanine cycle, 334–335
Glucose (Glc), 191–207
  appetite and satiety, 203–204
  bioavailability, 193
  daily requirements, 191
  deficiency, impacts of, 192
  D form, 191, 191*f*
  dietary sources, 191, 193
  digestion and absorption, 193–195
  endogenous sources, 192–193
  excretion, 202
  functions, 191, 204–205

   as carbon source, 205
   fuel energy, 204
   hexosamines, 205
   reducing equivalents, 204
   synthesis of fructose, 205
   as UDP-galactose precursor, 205
  insulin and, 203
   satiety-inducing effect of, 203–204
  intestinal absorption, 196*f*
  L form, 191*f*
  metabolism, 197–201
   activation of glycogen synthase, 201–202
   aerobic, 199, 199*f*
   anaerobic, 198–199
   cAMP-induced actions, 202
   fasting and starvation, 204
   glycolysis, 197–198, 198*f*
   glycolytic pathway, 197
   for mode of fuel utilization, 204
   pentose-phosphate pathway, 199–201, 200*f*
   postprandial, 204
   to pyruvate, 197–198
  nutritional summary, 191–192
  in plant oligosaccharides and polysaccharides, 193
  regulation, 203–204
   hormonal, 203
  risk of excessive intake, 192
  storage, 201–202
  transport and cellular uptake, 196–197
   via blood, 196
   blood–brain barrier (BBB), 196–197
   materno-fetal transfer, 197
  transport into brain, 82
Glucose 6-phosphate, 197–198
Glucose transporters, 195–196, 203, 209
Glucuronides, 104
Glutamate (Glu), 278–279, 281, 294–302, 333
  deficiency impacts, 295
  dietary intake of, 295
  dietary sources, 295–296
   free glutamate sources, 295–296
   Glu-rich proteins, 295
  digestion and absorption, 296–297, 296*f*
  endogenous sources, 295
  excretion, 299
  food sources, 295
  functions, 299–300
   amino acid synthesis, 300
   in energy fuel yielding, 299
   glutamylation, 299–300
   glutathione degradation, 300
   health effects, 300
   Krebs cycle anapleurosis, 300

metabolic effects, 300
neurotransmission, 300
protein synthesis, 299
L-glutamate, 18, 305, 402, 405
control of Gln in and out of brain, 305
ornithine synthesis from, 281
health risks of excessive intake, 295
L form, 414
metabolism, 297, 298f
nutritional summary, 295
as a precursor for intestinal arginine synthesis, 415f
regulation, 299
storage, 299
taste, 296
transport and cellular uptake, 297
via blood, 297
blood–brain barrier (BBB), 297
materno-fetal transfer, 297
mitochondrial transportation, 297
Glutamate dehydrogenase NAD(P) 1, 308
Glutaminase, 305–306
Glutamine (Gln), 271, 302–309, 333
daily intake, 302
deficiency impact, 302
dietary sources, 303
digestion and absorption, 303
endogenous sources, 303
excretion, 306–307
food sources, 302
functions, 302, 307–308
in energy fuel yielding, 307
GABA–glutamate–glutamine cycle, 308f
hexosamine synthesis, 308
Krebs cycle, 307
neurotransmitter cycling, 307
nucleotide synthesis, 307–308
pH regulation, 308
protein synthesis, 307
pyrimidine nucleotide synthesis, 307–308
L-glutamine (Gln), 299, 340
γ-amido group of, 305–306
endogenous synthesis of, 303f
intestinal absorption, 304f
metabolism, 306f
metabolism, 305–306
in muscle cells, 304
nutritional summary, 302
regulation, 307
stress and infection, effect of, 307
risk of excessive intake, 302
storage, 306
transport and cellular uptake, 303–305
via blood, 303–304

blood–brain barrier (BBB), 305
materno-fetal transfer, 305
Glutamine N-phenylacetyltransferase, 340
Glutamine-phenylpyruvate aminotransferase, 340
Glutaryl-CoA dehydrogenase, 379
Glutathione-cystine transhydrogenase, 371
Glutathione (GSH), 483
Glutathione S-transferase, 340, 346
Glutaurine, 446, 448
Glycans, 763–764
Glyceraldehyde, 188–189
Glycerol-3-phosphate O-acyltransferase, 136
Glycine amidinotransferase, 453
Glycine aminotransferase, 310
Glycine (Gly), 309–316, 309f, 333, 810–811
as a biomarker of dietary availability, 315
as a component of dipeptides or tripeptides, 310
deficiency, impact of, 310
dietary intake, 310
dietary sources, 310
digestion and absorption, 310–311, 312f
endogenous sources, 310
synthesis of, 311f
excretion, 314
food sources, 310
free, 310
functions, 310
bile acid conjugation, 315
in brain function, 315
creatine synthesis, 314–315
detoxification, 315
in energy fuel yielding, 314
glutathione synthesis, 314
methionine metabolism, 314
porphyrine synthesis, 314
protein synthesis, 314
purine synthesis, 314
metabolism, 312, 313f
cleavage system, 312
conversion to serine, 312
pathways, 312
nutritional summary, 310
as a precursor of purine nucleotides, 315f
regulation, 314
Gly catabolism, 314
risk of excessive intake, 310
storage, 314
transport and cellular uptake, 311
via blood, 311
blood–brain barrier (BBB), 311
materno-fetal transfer, 311
uptake from the intestinal lumen, 311
Glycolate, 266–271

Glycosides, 250–251
Glycosinolates, 760
Glyoxylate, 266–271
Goitrin, 751*f*
Goitrogens, 750
GPR81, 17
GPR120, 17
G-protein-coupled receptor 120 (GPR120), 6, 18–19

H

Hartnup disease, 350–351
Hartnup syndrome, 273–275, 600
Hearty taste (*kokumi*), 11
Helicobacter pylori, 63
Hemicelluoses, 244
Hemochromatosis (HFE) gene, 809
Henle's loop, 681–682, 688, 695, 704, 733
Hepcidin, 704–707
Heptadecanoic acid, 123
Heterocyclic amines (HAs), 100–107
    daily requirements, 101
    dietary sources, 101–102, 101*f*, 102*f*
    digestion and absorption, 102
    excretion, 105
        fractional, 105
    functions, 101
    health effects, 105
    metabolism, 102–104
        phase I reactions, 102–104
        phase II reactions, 104
    nutritional summary, 101
    risk of excessive intake, 101
Hexavalent(VI) Cr compounds, 784
Hexosamine metabolites, 145
Hexose, 188–189
HFE gene protein, 704
High-density lipoprotein (HDL), 133–134, 483
    cholesterol, 608
    transportation to brain, 84
Histamine, 438, 438*f*
Histidine (His), 271–272, 430–440
    as biomarkers of dietary intake, 439
    containing dipeptides, 437
    daily requirements, 431
    deficiency, impacts of, 431
    dietary sources, 431–432
    digestion and absorption, 432
    excretion, 437
        normal daily excretion with urine, 432
    functions, 431, 437–438
        effect on tissues, 438
        in energy fuel yielding, 437
        intestinal zinc absorption, role in, 438

    protein synthesis, 437
    synthesis of GABA, 437
    in hemoglobin, 436
    L forms, 430, 431*f*
        intestinal absorption, 434*f*
        metabolic fates of, 435*f*
        rich peptides, 436*f*
    metabolism, 434–436
        alternative pathways, 434–436
        main pathway, 434
        protein histidyl methylation, 436
    nutritional summary, 431
    regulation, 437
    risk of excessive intake, 431
    storage, 436
    transport and cellular uptake, 432–434
        via blood, 432
        blood–brain barrier (BBB), 434
        materno-fetal transfer, 432–434
Homocarnosin, 437
Homocysteine, 759
L-Homocysteine, 370
Homogentisate 1, 2-dioxygenase, 346
Hot, spicy taste, 21–22
HUGO Gene Nomenclature Committee
    (HGNC), 823
Human Genome Project, 824
Hydrochloric acid, 691
Hydrogen peroxide, 480, 482–483
Hydrogen sulfide, 763
3-Hydroxyanthralinate, 354
4-Hydroxybenzoate, 341, 348
3-Hydroxy-kynurenine, 354
3-Hydroxykynurenine glycoside, 357
Hydroxylysine, 271
Hydroxymethylglutaryl-CoA synthase, 540
4-Hydroxyphenylpyruvate dioxygenase, 340, 346
Hydroxyproline, 266–271
4-Hydroxyproline (Hypro), 423–424, 423*f*
    breakdown of, 428*f*
    catabolism of, 426
4-Hydroxy-2, 3-trans-nonenal (HNE), 481
5-Hydroxytryptamine (5-HT), 350
5-Hydroxytryptophan, 350, 352
5-Hydroxytryptophol (5-HTOL), 354
Hyperkeratosis, 487–488
Hyperuricemia, 145–146
Hypobromous acid (HBrO), 794
Hypodipsic hypernatremia, 34
Hypokalemia, 686–689
Hypomethylation, 568–569
Hypotaurine, 446, 448
Hypotaurine dehydrogenase, 445–446, 776–777

I

I cells, 17, 42
Ileal bile acid–binding protein (IBABP), 47
Ileal brake, 38–40
Imidazolonepropionase, 434
Immunoglobulin A (IgA), 17
Indigestible carbohydrates, 243–247
    daily requirements, 244
    deficiency, impacts of, 244
    dietary sources, 243
    distinction between insoluble and soluble fiber, 243
    functions, 243
    human digestive enzymes, 244t
    nutritional summary, 243–244
    risk of excessive intake, 244
Indigestible oligosaccharides, 246–247
    milk, 246
        beneficial probiotic effects, 246
        protective effects, 246
    plant-based, 246
Indigestible polysaccharides, 244
Indoleamine-pyrrole 2, 3-dioxygenase, 352
Infants, microbiome in, 62
    with HLA-DQ2, 63
        feeding cultured *Bifidobacteria*, impact of, 65
Information
    on enzymes, 823
    on genes and genetic syndromes, 824
    getting, 823
    metabolite concentrations, 823
    nutrient, 822
    on receptors and ligands, 823
    on transporters, 823
Inherited diversity, 809
Inositol, 663–671
    bioavailability of, 664–665
    breakdown, 668f
    daily requirements, 664
    deficiency, impacts of, 664
    dietary sources, 664
    digestion and absorption, 665
    endogenous sources, 664
    excretion, 669
    functions, 664, 669–670
        diverse inositol effects, 669
        eicosanoid synthesis, 669
        GPI-anchored exoproteins, 669
        inhibition of intestinal cation absorption, 670
        intracellular phytate effects, 670
        intracellular signaling, 669
        for mental health, 670
        osmoregulation, 669–670
    metabolism, 666–669, 667f

    catabolism, 667–669
        inositol phosphate interconversions, 666
        phospholipids, generation of, 667
    nutritional summary, 664
    polyphosphates, 665, 666f
    regulation, 669
    risk of excessive intake, 664
    synthesis, 665f
    transport and cellular uptake, 666
        via blood, 666
Inositol-3-phosphate synthase, 664
Insoluble fiber, 244
    health benefits, 244
Insulin resistance, 144–145
Intake behavior, regulation of, 26
International Union of Basic and Clinical
    Pharmacology Committee on Receptor Nomenclature
    and Drug Classification, 823
International Union of Biochemistry and Molecular
    Biology (IUBMB), 823
Intestinal absorption
    acetate, 149
    alanine (Ala), 332f
    anatomical considerations, 48
        lymph system, 48
        portal vein, 48
    arginine (Arg), 416f
    asparagine (Asn), 410f
    aspartate (Asp), 403f
    caffeine, 96
    calcium (Ca), 731f
    chlorine (Cl), 692f
    copper (Cu), 711f
    L-cysteine (Cys), 371f
    L-cystine, 371f
    of fat-soluble vitamins
        vitamin A, 53
        vitamin D, 54
        vitamin E, 54
        vitamin K, 54
    folate, 622f
    fructose (Fru), 209f
    galactose (Gal), 215f
    glucose (Glc), 196f
    L-glutamine (Gln), 304f
    histidine (His), 434f
    iron (Fe), 699f
    isoleucine (Ile), 397f
    leucine (Leu), 384f
    lysine (Lys), 378f
    of macronutrients, 51–52
        amino acids, 51
        carbohydrates, 51

Intestinal absorption (*Continued*)
   diglycerides, 51–52
   lipids, 51–52
   peptides, 51
   proteins, 51
   triglycerides, 51–52
  methanol, 239
  methionine (Met), 361*f*
  of minerals and trace elements
   arsenic, 57
   bromide, 56
   calcium, 55
   chromium, 56
   copper, 56
   fluoride, 56
   iodine, 56
   iron, 55–56
   magnesium, 55
   manganese, 56
   molybdenum, 56
   nickel, 57
   phosphate, 55
   selenium, 56
   sulfur, 55
   transferrin, 56
   vanadium, 57
   zinc, 56
  molecular transport mechanisms, 48–51
   active ATP-driven transport, 49–50
   chloride cotransport, 50
   exchangers, 50
   facilitated diffusion, 50
   intracellular transformation, 50
   paracellular diffusion, 48–49
   proton cotransport, 49*f*, 50
   sodium cotransport, 50
   transcytosis, 51
   unmediated transcellular diffusion, 51
  phenylalanine (Phe), 337*f*
  phosphorus (P), 739*f*
  of phytochemicals, 57
   polyphenols, 57
  potassium (K), 687*f*
  proline (Pro), 426*f*
  retinol, 490–492
  serine (Ser), 324*f*
  sulfur (S), 760
  thiamin, 582*f*
  threonine (Thr), 318*f*
  tryptophan (Trp), 351*f*
  tyrosine (Tyr), 345*f*
  valine (Val), 391*f*
  vitamin A (vitA), 490–492
  vitamin B6 (B6), 613*f*
  vitamin B12 (B12), 635*f*
  vitamin D (vitD), 505*f*
  vitamin K (vitK), 528*f*
  of water and electrolytes
   chloride, 55
   pH adjustments, 54
   sodium, 54
   water secretion, 54
  of water-soluble vitamins
   ascorbate, 52
   biotin, 53
   folate, 53
   lipoic acid, 53
   niacin, 53
   pantothenic acid, 53
   riboflavin, 52
   thiamine, 52
   vitamin B6, 53
   vitamin B12, 53
  zinc (Zn), 718*f*
Intestinal fermentation, 245
Intestinal infections
  effect on microbiomes, 63
Intestinal sensing, 16–20
  amino acids, 18
  fat, 18–19
  of ingested calcium, 19
  sweet taste, 18
Inulin, 65, 246
Iodine (I), 748–755
  daily requirements, 748
  deficiency, impacts of, 748, 753
  dietary sources, 748–749
   plants, 748–749
   seawater, 748–749
  digestion and absorption, 749
   fractional absorption, 749
   nonenzymatic reduction of iodate, 749
  excretion, 752
   renal, 752
  functions, 748
   nongenomic thyroid hormone actions, 753
   thyroid hormone production, 752–753
  metabolism, 750
   iodine salvage, 750
   synthesis, 750
  nutritional assessment, 753
  nutritional summary, 748

regulation, 752
  iodide uptake, 752
risk of excessive intake, 748
role in synthesis of thyroid hormone, 751*f*
storage, 752
transport and cellular uptake, 749–750
  via blood, 749
  maternal circulation into milk, 749
  materno-fetal transfer, 749
Iron (Fe), 139, 221, 279–281, 299, 302, 316, 320–321, 334,
  336, 348, 350, 352, 359–360, 365, 381, 383, 396, 402, 405,
  417, 420, 423, 429, 443–445, 487, 697–709
breast-feeding transfer, 704
daily requirements, 697
deficiency, and cravings for nonfood items, 30
deficiency, impacts of, 698
dietary sources, 698
digestion and absorption, 698–700
  export into blood, 700
  ferric form, 698
  ferritin, 700
  ferrous form, 698
  heme form, 700
  intracellular disposition, 700
  transferrin-bound, 700
excretion, 703–704
from ferritin, 702, 703*f*
functions, 697, 705–706
  amino acid metabolism, 706
  carnitine synthesis, 706
  catecholamine and melanin synthesis, 706
  cause of poisoning in infants, 706
  choline metabolism, 706
  dementia risk, 706
  DNA synthesis, 705
  fatty acid metabolism, 706
  free radical metabolism, 705
  generation of thyroxin and triiodothyronine, 706
  nutrient metabolism, 705
  oxidative phosphorylation, 705
  oxygen transport, 705
  protein modification, 706
  serotonin and melatonin degradation, 706
  sulfur metabolism, 706
  vitamin A metabolism, 706
heme iron salvage, 702
in hemoglobin, 703–704
intestinal absorption, 699*f*
materno-fetal nutrient transport, 91
metabolism, 702
  copper, role of, 714

nonheme, 698
nutritional assessment, 706–707
nutritional summary, 697–698
regulation, 704–705
risk of excessive intake, 698
storage, 702
transport and cellular uptake, 700–702
  via blood, 700–701
  blood–brain barrier (BBB), 701
  materno-fetal transfer, 701–702
  transfer into milk, 702
  transferrin receptors (TfRs), 701
  transportation to brain, 84
Iron regulatory proteins (IRPs), 705
Iron-regulatory gene HFE, 813–814
Iron-responsive elements (IREs), 705
Isoflavones, 247–260
  absorption, 253
  B-ring in, 250
  dietary sources, 251
  excretion, 257
  metabolism, 254
Isoleucine (Ile), 271–272, 396–401
  daily requirements, 396
  deficiency, impacts of, 396
  dietary sources, 396–397
  digestion and absorption, 397–398
  excretion, 400
  functions, 396
    in energy fuel yielding, 400
    protein synthesis, 400
  L form, 396, 396*f*
    intestinal absorption, 397*f*
    metabolism of, 399*f*
  metabolism, 398–400
  nutritional summary, 396
  regulation, 400
  risk of excessive intake, 396
  storage, 400
  transport and cellular uptake, 398
    via blood, 398
    blood–brain barrier (BBB), 398
    materno-fetal transfer, 398
lysinoalanine, 377

K

K cells, 17, 42
Kallikrein, 272–273
Keratan sulfates, 763–764
Ketone bodies, 134
Ketose, 188–189

Kidneys
  aspartate metabolism, 410–411
  Asp functions for urea synthesis, 404
  calcium (Ca) excretion, 733
  L-carnitine synthesis in, 455
  choline breakdown, 472
  decomposition of S-nitrosoglutathione in, 437
  free Cys, 371
  glutamate metabolism in, 297
  glutamine excretion, 306–307
  Met remethylation, 363
  phosphorus (P) excretion, 740–741
  potassium (K) excretion, 688
  proline (Pro) breakdown, 425
  retinol excretion, 495
  taurine, role of, 443
  taurine and, 448
  vitamin B12 (B12) excretion, 639
Krebs cycle, 134, 152, 162, 192–193, 197, 234–235,
    292, 297, 340, 727–728
  intermediate succinyl-CoA, 392–393
Kynurenic acid, 352, 354
Kynurenine, 352
Kynurenine-glyoxylate aminotransferase, 310, 340, 352
Kynurenine 3-monoxygenase, 596
Kynurenine-oxoglutarate aminotransferase (KAT), 352

L

L cells, 17, 42
D-Lactate, 320
Lactobacilli species, 61–62, 64
  L. acidophilus, 61
  nondigestible oligosaccharides and, 65
  sour taste of, 65
Lactobacillus species, 65
Lactococcus species, 65
Lactoferrin, transportation to brain, 84
Lactose, 195, 214–215
Large intestine, 45–46
  Bauhinian valve, 45
  cecum, 45
  chlorine (Cl) absorption, 692–693
  colon, 45–46
  water secretion, 675f
Lectins, 293
Legumes, 292
Leucine (Leu), 271–272, 382–389
  daily requirements, 383
  deficiency, impacts of, 383
  dietary sources, 383
  digestion and absorption, 383–384
  excretion, 387

functions, 383, 388
  cell proliferation, 388
  in energy fuel yielding, 388
  ketogenesis, 388
  neurotransmitter metabolism, 388
  protein synthesis, 388
  toxicity, 388
L form, 382, 382f
  intestinal absorption, 384f
  metabolism of, 386f
metabolism, 385–387
  beta-keto pathway, 387
  conversions, 385, 387
  main catabolic pathway, 385
nutritional summary, 383
regulation, 387–388
risk of excessive intake, 383
storage, 387
transport and cellular uptake, 384–385
  via blood, 384
  blood–brain barrier (BBB), 385
  materno-fetal transfer, 384–385
Lignin, 245
Linoleic acid, 123
Linolenic acid, 123
Linoleoyl-CoA desaturase, 123
Lipids
  materno-fetal nutrient transport, 88–89
  transportation to brain, 82
    fatty acid–derived metabolites, 82
    long-chain polyunsaturated fatty acids, 82
Lipoamide, 557f
Lipoate, 279–281, 299, 302, 316, 320–321, 334, 336,
    348, 350, 352, 359–360, 365, 383, 396, 402, 405, 417,
    420, 423, 429, 483–485, 554f. *See also* Lipoic acid (LA)
  enantiomers, 555f
  endogenous synthesis, 554f
  oxidized, 556
  storage and release of, 556
  transportation to brain, 83
Lipoic acid (LA), 553–560
  daily requirements, 554
  deficiency, impacts of, 554
  dietary sources, 554–555
  digestion and absorption, 555
  endogenous synthesis, 554
  excretion, 556
  functions, 554, 556–558
    antioxidant potential of, 558
    branched-chain alpha-keto acid dehydrogenase,
      development of, 557–558
    glucose metabolism, 558

glycine dehydrogenase, development of, 558
impairment of glycine, 558
liver protection, 558
2-oxoglutarate dehydrogenase, development of, 557
pyruvate dehydrogenase, development of, 556–557
reactivation of oxidized forms of vitamins C and E, 558
scavenging of free radicals and protein stabilization, 558
synthesis of acetylcholine, 558
materno-fetal nutrient transport, 91
metabolism, 555–556
from normal intestinal bacteria, 554
nutritional summary, 554
as a potent redox reagent, 556*f*
regulation, 556
risk of excessive intake, 554
sodium-dependent carrier for, 555
storage and release of, 556
transport and cellular uptake, 555
via blood, 555
materno-fetal transfer, 555
Lipoprotein lipase (LPL), 167
Lipoyltransferase, 555–556
Liver
arginine (Arg) breakdown in, 417, 418*f*
aspartate metabolism, 410–411
Asp functions for urea synthesis, 404
L-carnitine synthesis in, 455
choline breakdown, 472
decomposition of S-nitrosoglutathione in, 437
glutamate metabolism in, 297
glycogen content in, 201–202
HA metabolism in, 104
histidine (His) breakdown, 434–436
hypro breakdown in, 426
Ile utilization, 400
inorganic As, reduction of, 798–799
Met remethylation, 363
proline (Pro) breakdown, 425
RBP4 production, 495
risk associated with niacin intake, 600
taurine breakdown, 445–446
transamination of Val in, 393
triglyceride-depleted Chylos, uptake of, 504
vitA in, 494–495
vitD synthesis, 505
vitE uptake, 517–518
Liver-type fatty acid binding protein (L-FABP), 739
Low-density lipoprotein (LDL), 144–145, 482
cholesterol, 608
Lumiflavin, 590, 590*f*
Luminal amino acids, 18
Lysine (Lys), 271–272, 376–382

daily requirements, 377
deficiency, impacts of, 377
dietary sources, 377
digestion and absorption, 377
excretion, 379
free, 377
functions, 377, 381
carnitine synthesis, 381
in energy fuel yielding, 381
polyamine synthesis, 381
protein and peptide synthesis, 381
L form, 376*f*
intestinal absorption, 378*f*
metabolism, 379
breakdown, steps of, 379
catabolism, 379
oxidation processes, 379
pathways of degradion for L-lysine via
saccharopine or pipecolate, 380*f*
nutritional summary, 377
regulation, 381
risk of excessive intake, 377
storage, 379
transport and cellular uptake, 378–379
via blood, 378
blood–brain barrier (BBB), 378–379
materno-fetal transfer, 378
Lysophospholipase, 471

M

Macroglycogen (MG), 201–202
Magnesium (Mg), 139, 221, 279–281, 299, 302, 316,
320–321, 334, 336, 348, 350, 352, 359–360, 365, 383,
396, 402, 405, 417, 420, 423, 429, 469, 744–748
daily requirements, 744
deficiency, impacts of, 744
dietary sources, 744
digestion and absorption, 745
excretion, 745–746
free ionized, 745–746
functions, 744, 746–747
in balancing calcium, 746–747
as a constituent of bone, 746
as enzyme cofactor, 746
nucleotide complexes, 746
luminal, 745–746
maintaining Mg homeostasis, 746
nutritional assessment, 747
nutritional summary, 744
regulation, 746
risk of excessive intake, 744
storage, 745

Magnesium (Mg) (*Continued*)
  transport and cellular uptake, 745
    via blood, 745
    blood–brain barrier (BBB), 745
    maternal circulation into milk, 745
    materno-fetal transfer, 745
    TRPM6 function and, 745
    variation coefficient for, 816–817
Magnesium trisilicate, 790
Maillard products, 272
Maleylacetoacetate isomerase, 340, 346
S-Malonyltransferase, 120
Malondialdehyde, 481
Malonyl-CoA, 145
Maltase-glucoamylase, 193–195
Maltose, 189
Manganese (Mn), 221, 479, 724 730
  daily requirements, 725
  deficiency, impacts of, 725
  dietary sources, 725
  digestion and absorption, 725–726
  excretion, 726
  functions, 725, 727–728
    in apoptosis, 727
    cartilage formation, 728
    as enzyme cofactors, 727
    for free radical control, 727
    gluconeogenesis, 727–728
    glycolysis, 727–728
    nitrogen metabolism, 727
    peptide hydrolysis, 727
    sulfate metabolism, 728
  nutritional summary, 725
  regulation, 726
  risk of excessive intake, 725
  storage, 726
  toxic effects of, 728
  transport and cellular uptake, 726
    via blood, 726
    blood–brain barrier (BBB), 726
    maternal circulation into milk, 726
  transportation to brain, 85
Maternal circulation into milk
  calcium (Ca), 732
  chromium (Cr), 783
  cobalt (Co), 779
  copper (Cu), 712
  iodine (I), 749
  iron (Fe), 702
  magnesium (Mg), 745
  manganese (Mn), 726
  molybdenum (Mo), 774

  selenium (Se), 768
  zinc (Zn), 719–720
Materno-fetal barrier, 88–92
Materno-fetal nutrient transport, 87–93
  amino acids, 89–90
  carbohydrates, 88
  glucose, 88
  lipids, 88–89
    arachidonic acid, 88–89
    docosahexaenoic acid (DHA), 88–89
    essential polyunsaturated fatty acids, 88–89
    maternal HDL, 89
  minerals and trace metals, 91
    calcium, 91
    iron, 91
    xenobiotics, 91
  vitamins, 90–91
Materno-fetal transfer
  alanine (Ala), 333
  amino acids, 278
  arginine (Arg), 417
  asparagine (Asn), 410
  biotin, 643
  caffeine, 96
  calcium (Ca), 732
  chlorine (Cl), 693
  cholesterol (Chol), 547–548
  choline, 472
  conjugated linoleic acid (CLA), 160
  copper (Cu), 712
  cysteine (Cys), 372
  docosahexaenoic acid (DHA), 167–168
  fatty acids, 126–127
  flavonoids, 254
  fluorine (F), 756
  folate, 624
  fructose (Fru), 209
  galactose (Gal), 216
  glucose (Glc), 197
  glutamate, 297
  glutamine, 305
  glycine (Gly), 311
  histidine (His), 432–434
  iodine (I), 749
  iron (Fe), 701–702
  isoleucine (Ile), 398
  leucine (Leu), 384–385
  lipoic acid (LA), 555
  lysine (Lys), 378
  magnesium (Mg), 745
  methanol, 239
  methionine (Met), 361

molybdenum (Mo), 774
myristate (myristic acid), 154
niacin, 603
nitrite/nitrate, 108
pantothenate, 650
phenylalanine (Phe), 338
phosphorus (P), 740
phytanic acid, 180
potassium (K), 688
proline (Pro), 425
pyruvate, 225
riboflavin, 591–592
selenium (Se), 768
serine (Ser), 325
sodium, 681
sulfur (S), 761
taurine, 445
thiamin, 583
threonine (Thr), 318
trans-unsaturated fatty acids (trans-fatty acids), 176
tryptophan (Trp), 352
tyrosine (Tyr), 345
valine (Val), 391
vitamin A (vitA), 492
vitamin B6, 614
vitamin B12 (B12), 637
vitamin C, 573
vitamin D (vitD), 505
vitamin E (vitE), 518
vitamin K (vitK), 529
water, 676
zinc (Zn), 719
Mature placenta, 88
    amino acid metabolism in, 90
    amino acid transporters in, 279t
    fetal capillaries, 88
    maternal iron deficiency and, 91
    placental villi, 88
    structural organization of, 89f
Meaty taste (*umami*), 11
Mechanoreceptors, 21
Megalin, 507
    transportation to brain, 84
Megaloblastic anemia, 580
Mei-Tei-Sho syndrome, 66
Melanocortin receptors, 26
Melatonin, 350, 352, 357, 461–468, 462f
    catabolism, 465f
    daily requirements, 462
    deficiency, impacts of, 462
    dietary sources, 462–464
    digestion and absorption, 464

endogenous sources, 462
excretion, 466
functions, 461, 466–467
    circadian rhythm, role in, 466
    fertility, effect on, 467
    free radical scavenging, 466
    hormonal function, 467
    immune response, 466–467
    influences on mood, 466
    receptor binding, 466
    seasonal rhythm, role in, 466
    skin pigmentation, 466
metabolism, 464
    synthesis in pineal gland, 463f
nutritional summary, 461–462
regulation, 466
risk of excessive intake, 462
storage, 464
synthesis
    Trp, 357
transport and cellular uptake, 464
    via blood, 464
Menaquinone-4, 531, 531f
Menaquinones, 526–529
Menkes protein (ATP7A), 712–713
Menthol ([1α, 2β, 5a]-5-methyl-2-[1-methylethyl]
    cyclohexanol), 22
1-Menthone ([2β, 5α]-5-methyl-2-[1-methylethyl]
    cyclohexanone), 22
Mercaptopyruvate sulfurtransferase, 373
Metabotropic glutamate receptor 5 (mGluR5, GRM5), 18
Metallic tastes, 12
Metalloproteinases, 734–735
Metallothioneins, 710, 713, 718
Methanol, 238–242, 238f
    daily requirements, 238
    dietary sources, 238–239
    endogenous sources, 238–239
    functions, 238, 241
    intestinal absorption, 239
    metabolism, 240f
        formaldehyde, generation of, 240
        formic acid, generation of, 241
        oxidation, 239–240
    nutritional summary, 238
    risk of excessive intake, 238
    toxic effects, 241
    transport and cellular uptake, 239
        via blood, 239
        blood–brain barrier (BBB), 239
        materno-fetal transfer, 239
Methanomassiliicoccus luminyensis B10, 64

Methionine (Met), 271–272, 359–368, 469, 758–759
daily requirements, 360
deficiency, impacts of, 360
dietary sources, 360
digestion and absorption, 360
excretion, 365
functions, 359–360
cysteine synthesis, 366–367
in energy fuel yielding, 365
homocysteine remethylation, 363f, 365–366
methyl-group transfer, 365
polyamine synthesis, 367
production of acids, 367
protein synthesis, 365
L form, 359–360, 359f, 369
intestinal absorption, 361f
metabolic fates of, 362f
metabolism, 361–364
remethylation, 363
repair of oxidative damage, 364
SAM synthesis, 362–363
transamination method, 361–362
transsulfuration method, 361–364
nutritional summary, 359–360
regulation, 365
risk of excessive intake, 360
storage, 365
transport and cellular uptake, 361
via blood, 361
blood–brain barrier (BBB), 361
materno-fetal transfer, 361
Methylation, 567–570, 798
DNA, 567, 568f
Hcys remethylation, 569
of parental DNA, 569
reductive, 569
SAM-dependent, 568–569
sources of methyl groups, 569
Methylcrotonyl-CoA carboxylase, 387
Methyldehydroalanine, 330
Methylenetetrahydrofolate reductase (MTHFR), 569, 811, 817
Methylglyoxal, 319–320
Methylhistidine, 271
N-Methylnicotinamide (MNA), 605–606
Methyl-S-adenosylthiopropylamine, 367
5-Methyltetrahydrofolate, 623
5-Methyl-tetrahydrofolate-homocysteine methyltransferase, 362–363
5-Methyltetrahydrofolate-homocysteine S-methyltransferase (MTR), 811–812
5′-Methyl-thioadenosine phosphorylase, 367
N-Methyltransferases, 417–419, 436, 455
Methylxantines, 96

Microbiome, 61–69
colonization of, 62
disruption of gut, 63–64
antibiotics, 63
chronic intestinal infections, 63
in infants, 62
life forms referred as, 61
in luminal surface of gastrointestinal walls, 63
metabolic, developmental, and health consequences, 66
microbial metabolites, 66
nutrients, 66
toxic, 66
nomenclature and classification, 61
nutritional factors and influence on
high fat intake and low fiber consumption, 64
high fiber intake and low fat consumption, 64
meat consumption, 64
milk consumption, 64
obesity, 64
trimethylamine (TMA) production, 64
prebiotics, 65
probiotics, 65
segment-specific density of, 61–62
bacteria, 61–62
viruses (virome), 62
yeast species, 62
Microbiota, 61
Microfold or membraneous cells (M cells), 42
Microsomal ethanol oxidizing system (MEOS), 233–234
Microsomal triglyceride transfer protein (MTP), 158–159
Microvilli, 41
Milk
indigestible oligosaccharides of, 65
microbiome composition, 64
Mitochondrial capsule selenoprotein (MCSP), 771
Mitochondrial enoyl-CoA isomerase (MECI), 160–162
Molecular databases, 822–824
Molybdenum (Mo), 443–445, 773–778
daily requirements, 773
deficiency, impacts of, 773
dietary sources, 773–774
digestion and absorption, 774
enterohepatic circulation of, 48
excretion, 776
functions, 773, 776–777
as a cofactor, 775f, 776
hormone-like effects, 777
isotopes, 773
metabolism, 774
nutritional assessment, 777
nutritional summary, 773
regulation, 776
risk of excessive intake, 773

storage, 775
transport and cellular uptake, 774
  via blood, 774
  blood–brain barrier (BBB), 774
  maternal circulation into milk, 774
  materno-fetal transfer, 774
variation coefficient for, 816–817
Molybdopterin synthesis, 774
Monophenol monooxygenase, 341
Monosodium glutamate (MSG), 295–296. *See also* Glutamate
Multidrug resistance protein 2 (MRP2), 811
Muscle, 127, 138, 225
acetate transfer, 234–235
acetate utilization, 150
acidification of cells, 204
Ala content, 330, 334
alpha enzyme in, 120
amino acids, 266, 284
anaerobic metabolism of, 198–199
bile, 43
carnosine content of, 436
cells of the heart, 125
CLA uptake, 160
fatty acids uptake, 167, 176
Glc entry, 196
Glu content in, 299
glucose oxidizing reaction, 197–198
glycogen synthesis, 201–202, 204, 211
methylation of actin and myosin in, 271
Met uptake, 361
movement of food through the mouth and, 46
phosphate storage in, 414
proteins, 283, 285f, 306, 340, 404, 411, 450
protein synthesis in, 284
pyruvate transfer, 224
skeletal, 144, 196, 285, 292, 388, 394, 396, 400, 460, 545–546, 556, 573, 605, 720
  Arg content of, 419
stomach, 40
wasting, 580
Zn content in, 720
Myo-inositol, 663, 664f
Myristate (myristic acid), 153–157, 153f
breakdown of, 155f
daily requirements, 153
deficiency, impacts of, 153
dietary sources, 153–154
digestion and absorption, 154
endogenous sources, 154
excretion, 156
functions, 153, 156–157
  fuel energy, 156
  hyperlipidemic potential, 157

membrane anchor for proteins, 156
metabolism, 154–156
  chain elongation and desaturation, 154
  mitochondrial catabolism, 154
  peroxisomal catabolism, 156
nutritional summary, 153–154
regulation, 156
risk of excessive intake, 154
storage, 156
transport and cellular uptake, 154
  via blood, 154
  blood–brain barrier (BBB), 154
  materno-fetal transfer, 154

N

N epsilon-carboxymethyllysine, 377
Neopterin, 658
Neuropeptide Y (NPY) neurons, 26
Neurotransmitters, role in regulation of appetite, 26
Niacin, 139, 221, 279–281, 299, 302, 316, 320–321, 334, 336, 348, 350, 352, 359–360, 365, 381, 383, 396, 402, 405, 414, 417, 420, 423, 429, 443–445, 469, 487, 599–610, 599f
blood measurements, 608
daily requirements, 600
  equivalents (NEs)/day, 600
deficiency, impacts of, 600
dietary sources, 600–601
digestion and absorption, 601–603
dinucleotides of, 601f
endogenous sources, 600
excretion, 605–606
functions, 599, 607–608
  ADP ribosylation of proteins, 607
  calcium signaling, 608
  DNA deacetylation, 607
  DNA repair, 607
  glucose signaling, 608
  pharmacological uses, 608
  as receptors, 607
materno-fetal nutrient transport, 91
metabolism, 603–605, 604f
  ADP-ribose synthesis, 605
  breakdown, 605
  catabolism, 606–607, 606f
  nucleotide synthesis, 603
  salvage pathways, 605
nutritional assessments, 608
nutritional summary, 599–600
regulation, 606–607
risk of excessive intake, 600
storage, 605
synthesis from tryptophan, 819–820
transport and cellular uptake, 603

Niacin (*Continued*)
    via blood, 603
        blood–brain barrier (BBB), 603
        materno-fetal transfer, 603
    transportation to brain, 83
    urine measurements, 608
    water-soluble forms, 599
Nickel (Ni), 805–807
    daily requirements, 805
    deficiency, impacts of, 805
    dietary sources, 805
    digestion and absorption, 805
    excretion, 806
    functions, 805–806
        as a cofactor, 806
    isotopes, 805
    nutritional summary, 805
    poisoning, 806
    regulation, 806
    risk of excessive intake, 805
    transport and cellular uptake, 806
        via blood, 806
Nickel sulfate, 806
Nicotinamide, 605
    storage, 605
Nicotinamide adenine dinucleotide (NAD), 154–156,
    188–189, 221, 600, 605
    endogenous synthesis of precursors, 602*f*
    functions, 607
    as a prosthetic group, 607
Nicotinamide adenine dinucleotide phosphate (NADP), 120,
    600, 605
    functions, 607
Nicotinamide mononucleotide (NMN), 605, 607
Nicotinate, 603, 603*f*
Nicotinate mononucleotide, 603
Nicotinate-nucleotide pyrophosphorylase, 600
Nicotinate phosphoribosyltransferase, 605
Nicotine adenine dinucleotide phosphate (NADPH), 120,
    188–189, 199–201, 221, 261, 811–812
Nicotinic acid, 599–601
Nicotinuric acid, 605
Niemann–Pick-C-protein-like 1 (NPC1L1), 491
Nitric oxide (NO), 481–482
Nitrite/nitrate, 107–110, 107*f*
    absorption, 108
    daily requirements, 107
    deficiency, impacts of, 107
    dietary sources, 107–108
    endogenous production, 107
    excretion, 109
    functions, 107, 109
        antimicrobial action, 109

        energy metabolism, 109
        methemoglobin formation, 109
        nitrosamine formation, 109
        vasodilation, 109
    health impacts
        diabetes, 109
        thyroid function, 109
    metabolism, 108–110
    nutritional summary, 107
    risk of excessive intake, 107
    transport and cellular uptake, 108
        via blood, 108
        blood–brain barrier (BBB), 108
        materno-fetal transfer, 108
Nitrogen balance, 284
Nociceptors, 21
    autonomous responses to, 21
    irritants that trigger nociceptor neurons, 21
Noro viruses, 63
Nutrient adequacy and supplementation, 816–819
    adequate intake (AI), 816
        functional assessment, 817
    Dietary Reference Intake (DRI), 816
    estimated average requirement (EAR), 816–817
    excessive intakes, 818
    minimal nutrient requirements, 816–817
    supplementation, 818
Nutrient homeostasis, 37
Nutrient information, 822
Nutrient interactions, 819–822
    competitive interference, 821
    deficiencies, 819–820
    high intake, 820–821
Nutrient sensing, 145
Nutrient-sensing cell types, 16–17
    active in feeding state, 17
    active nutrient sensors, 16–17
    properties of individual, 16–17
Nutrition, Metabolism, and Genomics Group, 823
Nutritional assessment
    amino acids, 287
    biopterin, 662
    biotin, 646
    calcium (Ca), 735
    chlorine (Cl), 696
    choline, 475
    copper (Cu), 715
    folate, 630
    iodine (I), 753
    magnesium (Mg), 747
    niacin, 608
    phytanic acid, 185
    selenium (Se), 771

sodium, 684
thiamin, 586
ubiquinone (coenzyme Q), 564
vitamin A (vitA), 498
vitamin B6, 618–619
vitamin B12 (B12), 640
vitamin C, 578
vitamin D (vitD), 511
vitamin E (vitE), 523–524
vitamin K (vitK), 535
zinc (Zn), 722
Nutritional wisdom, 25–26
Nutriture of the early embryo, 87–88

O

Obesity, 66
blood coagulation in, 145
cardiovascular risk, 145–146
de novo fat synthesis, 144
hyperlipidemia in, 144–145
hyperuricemia, risk of, 145–146
insulin resistance in, 144–145
issue of excess carbohydrates, 144
nutrient-sensing mechanism, 145
turnover of fat stores, 144–145
Octadecanoic acid, 112
Odorant-binding protein, 2
Odorants, 1
Oleic acid, 18–19, 123
Olfaction, 2
detection of specific odors, 2
olfactory signaling cascade, 2
Olfactory nerve, 2
Olfactory system, 1–2
detecting units, 1
impairment of sense of smell, 2
odor discrimination, 2
odor-specific coordination, 2
Oligosaccharides, 189
Omega-3 fatty acids, 114–116, 116*f*, 123, 166–167
mental health, 172
Omega-6 polyunsaturated fatty acids, 117*f*
Oral and pharyngeal cavities, 20
Oral cavity, 38
Orexigenic mediators, 25–26, 28
Ornithine, 271–272
Ornithine-delta-aminotransferase, 281
Oro-cecal transit time, 46
Osmosensors, 683
Otitis media, 66
Ouabain-like factor (OLF), 683
Overfeeding, 143–147
blood coagulation, 145

fate of excess carbohydrate, 144
hyperlipidemia, 144–145
hyperuricemia, 145–146
insulin resistance, 144–145
nature of, 143–144
nutrient sensing, 145
Oxalate, 228–229, 229*f*
in blood, 230
metabolism, 230
transport of, 230
urinary excretion of, 229–230
Oxalic acid, 228–231, 228*f*
daily requirements, 228
deficiency, impacts of, 228
dietary sources, 228–229
digestion and absorption, 230
endogenous sources, 228–229
excretion, 230
functions, 228
metabolism, 230
nutritional summary, 228
risk of excessive intake, 228
transport and cellular uptake, 230
via blood, 230
Oxaloacetate, 266–271
3-Oxoacid CoA-transferase, 340
2-Oxobutanoate, 370
4-Oxoretinol, 497, 497*f*
Oxygen free radicals, 488
Ozone (O$_3$), 480

P

Pan, 350
Pancreatic endopeptidase E, 330–332
Pancreatic lipase, 51–52
Pancreatic secretions, 43–44
typical composition of, 44*t*
Paneth cells, 17–18, 42
Pantothenate, 139, 280, 299, 302, 316, 320–321, 334, 336, 348, 350, 352, 359–360, 365, 383, 396, 402, 405, 417, 420, 423, 429, 469, 647–654, 648*f*
daily requirements, 648
deficiency, impacts of, 648
dietary sources, 648
digestion and absorption, 648–649
excretion, 650–652
functions, 648, 652–653
of acyl carrier protein, 653
anti-aggregation activity, 653
intermediary metabolism, 652
lipid metabolism, 652
protein acylation, 652
xenobiotic detoxification, 652–653

Pantothenate (*Continued*)
  materno-fetal nutrient transport, 91
  metabolism, 650, 651*f*
    breakdown, 650
    CoA synthesis, 650, 652
    phosphorylation, 650
  nutritional summary, 648
  regulation, 652
  risk of excessive intake, 648
  storage, 650
  transport and cellular uptake, 649–650
    via blood, 649–650
    blood–brain barrier (BBB), 650
    materno-fetal transfer, 650
  transportation to brain, 83
Papilla of Vater (papilla Vateri), 43–44
Parkinson's disease, 84, 657, 728
Pectins, 245
Pendred syndrome, 693–695
Pendrin, 752
Pentapeptide enterostatin (Val–Pro–Asp–Pro–Arg), 29
Pentose, 188–189
Pepsinogen A, 40
Peptide–chromium(III) complex, 782
Peptide YY (PYY), 7, 38–40
Peptidylarginine deiminase, 442
Peripheral mediators of food intake regulation, 27
Peroxiredoxins (Prx), 482–485
Peyer's plaques, 41, 61–62
Phenylalanine (histidine) aminotransferase, 340
Phenylalanine (Phe), 271–272, 336–342, 661
  daily requirements, 336
  deficiency, impacts of, 336
  dietary sources, 336–337
  digestion and absorption, 337
  excretion, 341
  functions, 336, 341
    catecholamine synthesis, 341
    3, 4-dihydroxyphenylalanine (DOPA) synthesis, 341
    in energy fuel yielding, 341
    melanin synthesis, 341
    protein synthesis, 341
    thyroid hormone synthesis, 341
    ubiquinone synthesis, 341
  L form, 336, 336*f*
    intestinal absorption, 337*f*
  metabolism, 338–340
    conversion to L-tyrosine, 340
    direct transamination, 340
    tyrosine catabolism, 340
  nutritional summary, 336
  L-phenylalanine, 333
  regulation, 341

risk of excessive intake, 336
storage, 340
transport and cellular uptake, 338
  via blood, 338
  blood–brain barrier (BBB), 338
  materno-fetal transfer, 338
  transport of free, 337
Phenylketonuria (PKU), 336, 561
Phosphate, variation coefficient for, 816–817
Phosphatidylcholine, 139, 327–328
Phosphatidylcholine-retinol *O*-acyltransferase, 490–491
Phosphatidylcholine-sterol *O*-acyltransferase, 139
Phosphatidylethanolamine, 327–328
Phosphatidylserine, 327–328
Phosphatidylserine synthase (PSS), 139
3′-Phosphoadenosine 5′-phosphosulfate, 762
Phosphofructokinase, 197–198
Phospholipase C, 472
Phospholipids, 116
Phosphoribosyl formylglycinamidine synthase (FGAM
  synthase), 307–308
Phosphorus (P), 737–744
  daily requirements, 737
  deficiency, impacts of, 737
  dietary sources, 737–738
    foods, 738
    phosphate-containing compounds, 738*f*
    plant-derived foods, 738
  digestion and absorption, 738–739
  excretion, 740–741
  functions, 737, 742
    activating phosphate esters, 742
    as an effective buffer, 742
    high-energy phosphate esters, 742
    organophosphates, 742
    polyphosphates, role of, 742
    protein phosphorylation, 742
  hormones maintaining phosphate balance, 741*f*
  intestinal absorption, 739*f*, 742
  nutritional assessment, 743
  nutritional summary, 737
  regulation, 741–742
  risk of excessive intake, 737
  storage, 740
  transport and cellular uptake, 739–740
    via blood, 739–740
    blood–brain barrier (BBB), 740
    materno-fetal transfer, 740
Phosphorylated erythrose, 188–189
Physical sensing, 20–23
Phytanic acid, 116, 129, 179–186, 179*f*
  alpha-oxidation, 181–182, 181*f*
  breakdown, 182

consequences of impaired, 184
    digestion and absorption, 180
    nutritional assessment, 185
    regulation, 184
    storage, 184
Phytanoyl-CoA, 182
Phytanoyl-CoA dioxygenase, 182
Phytochemicals, enterohepatic circulation of, 48
Phytol, 179–186, 179f
    deficiency, impacts of, 180
    dietary sources, 180
    digestion and absorption, 180
    excretion, 184
    functions, 179
    nutritional summary, 179–180
    oxidation, 181
    regulation, 184
    risk of excessive intake, 180
    transport and cellular uptake, 180
Pica behavior, 30
Piperine, 21
PLP-dependent 2-amino-3-ketobutyrate coenzyme A (CoA)
    ligase, 310
PLP-dependent BCAA aminotransferase 2, 398
Potassium (K), 685–690
    daily requirements, 686
    deficiency, impacts of, 686–689
    dietary sources, 686
        animal products, 686
        plants, 686
    digestion and absorption, 686–687
    excretion, 688–689
        glycorrhizic acid and, 689
        urine, 688
    functions, 686, 689
        enzyme activation, 689
        excitation, 689
        radioactivity, 689
    hydrogen/potassium-exchanging ATPase, 686–687
    intestinal absorption, 687f
    nutritional assessment, 689
    nutritional summary, 686
    regulation, 689
    risk of excessive intake, 686
    storage, 688
    transport and cellular uptake, 687–688
        via blood, 687–688
        blood–brain barrier (BBB), 688
        materno-fetal transfer, 688
Prebiotics, 65
8-Prenylquercetin, 48
Presynaptic (type III) cells, 6
Prevotella species, 61–62, 64

Pristanic acid, 116, 180
    beta-oxidation, 182, 183f
    storage, 184
Probiotics, 65
    bacteria species in, 65
Proglycogen (PG), 201–202
Proline (Pro), 271, 423–430
    daily requirements, 423
    deficiency, impacts of, 423
    dehydrogenase, 429
    de novo synthesis of, 424
    dietary sources, 423, 425
    digestion and absorption, 425
    dipeptides and tripeptides of, 425
    endogenous sources, 424
    excretion, 429
    functions, 423, 429
        arginine synthesis, 429
        of brain, 429
        in energy fuel yielding, 429
        oxygen sensing, 429
        protein synthesis, 429
        redox shuttle, 429
        regulation of p53-dependent apoptosis, 429
        urea cycle, 429
    L form, 423, 423f
        breakdown of, 427f
        endogenous synthesis of, 424f
        intestinal absorption, 426f
    metabolism, 425–426
        conversion to delta-1-pyrroline-5-carboxylate (P5C), 425
        conversion to glutamate, 426
    nutritional summary, 423–424
    regulation, 429
    risk of excessive intake, 424
    storage, 429
    transport and cellular uptake, 425
        via blood, 425
        blood–brain barrier (BBB), 425
        materno-fetal transfer, 425
Proline-rich Gla protein 2 (PRGP2), 534
Propionyl-CoA, 136f
    metabolism, 182
Protein availability, 292–293
    browning and oxidation, effect of, 293
    cooking practices and, 293
    poor chewing and inadequate predigestion, effect of, 293
Protein biomarkers, 287
Protein deficiency (protein malnutrition), 287
Protein digestion, 272–273
    protein-digesting enzymes, 272–273
Protein quality, 293

Protein turnover, 291–292
Protein-energy malnutrition (PEM), 291
Proteobacteria, 61–62
Protobacteria, 63
Proton-coupled folate transporter (PCFT), 623
Protoporphyrin IX, 707
Provitamin A, 486, 492
   carotenoids, 488
Pteroylpoly-gamma-glutamate carboxypeptidase (GCPII), 810
Pyridoxal-5′-phosphate (PLP), 611, 614, 616–617, 616*f*
Pyridoxamine-phosphate oxidase, 596
Pyridoxamine-5′-phosphate (PMP), 611, 614
Pyridoxine, 302, 383, 443–445
Pyruvate, 223–228, 266–271
   carboxylase, 226, 645
   daily requirements, 224
   deficiency, impacts of, 224
   dehydrogenase, 225–226, 585
   dietary sources, 223–224
   digestion and absorption, 224
   endogenous sources, 224
   excretion, 225
   functions, 223, 226–227
      amino acid synthesis, 227
      as enzyme cofactor, 227
      fuel metabolism, 226
      for performance enhancement, 227
   metabolism, 225
      anaplerotic, 225
      carboxylation, 225
      metabolic fate of, 225
      oxidative decarboxylation, 225
   multisubunit enzyme complex, 226*f*
   nutritional summary, 223–224
   paradox, 226
   regulation, 225–226
   risk of excessive intake, 224
   transport and cellular uptake, 224–225
      via blood, 224
      blood–brain barrier (BBB), 225
      materno-fetal transfer, 225

Q

Quercetin, 254, 256*f*, 257
Queuine, 66, 654–657, 654*f*, 656*f*
   absorption and transport, 655
   daily requirements, 655
   deficiency, impacts of, 655
   dietary sources, 655
   functions, 654–655
   metabolism, 655
   nutritional summary, 654–655
   risk of excessive intake, 655
Queuosine (Q), 655

R

Reactive oxygen species (ROS), 479–481
   induced damages, 482
      DNA modifications, 482
      oxidative damages, 482, 483*f*, 484*f*
   during oxidative phosphorylation, 480
   ozone (O_3), 480
   physiological functions, 481–482
      in oxidative blast, 481–482
      programmed cell death (apoptosis), 481–482
      removal of pathogens, 481–482
   sources of, 480
      sunlight and radiation, 480
Receptor-mediated pinocytosis, 72
Refeeding syndrome, 743, 747
Refsum's disease, 184–185
Relaxin, 33
Renal anatomy, 69–76
   Bowman's space, 69
   distal tubule, 72
   glomerular filtration rate (GFR), 69–70
   Henle loop, 70–72, 76–77
   intercalated cells, 72
   kidneys, 69
   medullary-collecting tubule, 72
   nephrons, 69, 70*f*
   podocytes, 69
   proximal tubule, 69–70
Renal processing, 69–81
   active secretion of food compounds, 77
      acidic xenobiotics, 77
      N-methyl-nicotinamide (NMN), 77
      xenobiotic–cysteine conjugates, 77
   aquaporins (AQPs), 76
   calcium, 77
   chloride, 76
   complex nutrients, 72–75
      amino acids, 73–74, 73*t*, 283
      beta-alanine, 74
      carbohydrates, 72–73
      carnitine, 74
      citrate, 73
      Clara cell secretory protein (CCSP), 75
      cobalamin, 75
      glycine, 74
      hydroxyproline, 74
      insulin, 73
      lysozyme, 73
      major driving force for uptake, 72

D-mannose homeostasis, 72–73
  pantothenate, 75
  proline, 74
  proteins, 73
  recovery of D-glucose and D-galactose from the lumen,
    72–73
  recovery of nutrients from the proximal tubular lumen, 71*f*
  retinol-binding protein, 73, 75
  taurine, 74
  thiamin pyrophosphate, 75
  transcobalamin II, 73, 75
  transport, 74–75
  uptake of dipeptides and tripeptides, 73–74
  urea, 75
  vitamin D–binding protein (VBP), 73
  vitamins, 75
hormones affecting, 78–79
  ADH, 78
  aldosterone, 78
  angiotensin II, 78
  atrial natriuretic peptide (ANP), 78
  prostaglandins, 78
  PTH, 79
iodide, 77
magnesium, 77
phosphate, 77
potassium, 76
sodium, 76
sodium chloride, 76–77
water, 76
Resistant starch, 189–190
Retinoic acid, 488, 490–491
  excretion, 495
  regulation, 495
Retinoid X receptor (RXR) group, 497
Retinol. *See also* Vitamin A (vitA)
  blood measurements, 498
  catabolism, 236
  dietary intake, 488–490
  digestion and intestinal absorption, 490–492
  endogenous synthesis, 488
  excretion, 494–495
  free, 494–495
  functions, 496–498
  metabolism
    conversion of all-trans-retinol to 9-cis-retinol, 493–494
    conversion of retinol via retinal to retinoic acid, 492–493
    conversion to hydroxylated retinoids, 494
    retinoic acid breakdown, 494
  metabolite 3, 4 didehydroretinoic acid, 494*f*
  nutritional summary, 487–488
  as a precursor of retinoic acid isomers, 493*f*

regulation, 495
transport and cellular uptake
  via blood, 492
  blood–brain barrier (BBB), 492
  materno-fetal transfer, 492
Retinol-binding protein 2 (RBP2), 490–491
Retinol-binding protein 4 (RBP4), 492, 495
Riboflavin, 139, 279–281, 299, 302, 316, 320–321, 334, 336,
  348, 350, 352, 359–360, 365, 383, 396, 398, 402, 405, 417,
  420, 423, 429, 487, 589–599, 589*f*, 820
  blood measurement, 597
  daily requirements, 590
  deficiency, impacts of, 590
  dietary sources, 590
  digestion and absorption, 590–591
  excretion, 593–594
  free, 591
  functions, 590, 595–597
    amino acid and amine metabolism, 596–597
    compliance monitoring, 597
    detoxification, 597
    fatty acid beta-oxidation, 596
    glutathione-linked reactions, 595
    heme metabolism, 596
    hormones and cell signaling, 597
    intermediary metabolism, 595
    lipid metabolism, 596
    nucleotide metabolism, 596
    oxidative phosphorylation, 595
    redox reactions, 595
    vitamin metabolism, 596
  materno-fetal nutrient transport, 91
  metabolism, 592–593, 592*f*
    catabolism, 593, 593*f*
    flavin cofactor synthesis, 592–593
    intracellular degradation of proteins, 593
  nutritional assessment, 597
  nutritional summary, 590
  regulation, 595
  risk of excessive intake, 590
  storage, 593
  transport and cellular uptake, 591–592
    via blood, 591
    blood–brain barrier (BBB), 592
    materno-fetal transfer, 591–592
  transportation to brain, 83
  urinary metabolites related to ingestion, 594*f*
  urine measurement, 597
  variation coefficient for, 816–817
Riboflavin carrier protein (RCP), 591
Ribose, 188–189
Ruminococcus, 61–62

S

S cells, 17, 42–44
S-adenosyll-homocysteine (SAH), 605
S-adenosylmethione (SAM), 359–360, 381, 450–451, 561, 605
  cycle, 362–363
  dependent creatine synthesis, 365
  dependent methylation, 568–569
    methyl group for, 569
  as a precursor for polyamine synthesis, 366*f*
Saliva, 38
  composition, 39*t*
Salt-tasting sensitivity, 8
Salty taste, 8
Sarcosin oxidase, 425
Schizophrenia, and excessive water intake, 34
Scurvy, 571
Secretion, 37
Selenium (Se), 479, 766–773
  containing amino acids, 767*f*
  containing glutathione peroxidases, 483
  daily requirements, 767
  deficiency, impacts of, 767
  dietary sources, 767
  digestion and absorption, 767–768
  excretion, 769–770
  functions, 766, 770–771
    antioxidant protection, 770
    cell replication, 771
    code for selenoproteins, 770
    insulin metabolism, 771
    thyroid hormone metabolism, 770–771
    virus mutations, 770
  inorganic forms, 766
  isotopes, 766
  metabolism, 768–769, 819
    synthesis of selenocysteine, 768–769, 769*f*
  nutritional assessment, 771
  nutritional summary, 766–767
  regulation, 770
  risk of excessive intake, 767
  storage, 769
  transport and cellular uptake, 768
    via blood, 768
    blood–brain barrier (BBB), 768
    maternal circulation into milk, 768
    materno-fetal transfer, 768
  variation coefficient for, 816–817
Selenocysteine, 56, 271, 321, 325, 330
Selenomethionine, 56
Selenoprotein P (SEPP1), 770–771
Semidehydroascorbate (SDA), 570, 571*f*
  dietary sources, 571

metabolism, 573–574
Sensitivity to smell, 3
  androstenone, 3
  asparagus consumption, 3
  musk, 3
Sensorineural deafness, 580
SePF, 768–769
Serine (Ser), 139, 271, 321–329, 469
  daily requirements, 322
  deficiency, impact of, 322
  dietary sources, 322–324
  digestion and absorption, 324–325
  dipeptides and tripeptides of, 324
  enantiomers, 322*f*
  endogenous sources, 322
    D-serine, synthesis from, 322
    glycine intermediates, synthesis from, 322
    glycolysis intermediates, synthesis from, 322
  excretion, 327
  export across basolateral membrane, 324–325
  functions
    connective tissue, 328
    cysteine synthesis, 327
    in energy fuel yielding, 327
    glycine synthesis, 327
    neuronal function, 328
    phospholipid synthesis, 327–328
    posttranslational protein modification, 328
    protein synthesis, 327
    selenocysteine synthesis, 327
  L form, 321, 325
    endogenous synthesis of, 323*f*
  metabolism, 325–326, 326*f*
    conversion to glycine, 325
    intestinal absorption, 324*f*
    L-serine-sulfate, cleaving, 326
    major catabolic pathway, 325
    minor catabolic pathway, 326
  nutritional summary, 321–322
    functions, 321
  regulation, 327
  risk of excessive intake, 322
  L-serine, 333, 370
  storage, 327
  transport and cellular uptake, 325
    via blood, 325
    blood–brain barrier (BBB), 325
    materno-fetal transfer, 325
Serine-sulfate ammonia-lyase, 326
Serotonin (5-HT), 26, 38, 42, 235, 280, 286, 350, 352, 354–357, 462, 464, 597, 600, 614, 616–617, 657, 661, 706
Serratia marcescens, 64

L-Seryl-tRNA sec selenium transferase, 327
Sibutramine, 27
Silicon (Si), 789–793
    beneficial effect of dietary, 792
    bioavailability of, 790
    daily requirements, 790
    deficiency, impacts of, 790–791
    dietary sources, 790
    digestion and absorption, 790
    excretion, 791
    forms, 789
    functions, 789, 791–792
    isotopes, 789
    nutritional summary, 789–790
    regulation, 791
    risk of excessive intake, 790
    storage, 791
    transport and cellular uptake, 791
        via blood, 791
SLC6A19 gene, 273–275
Small intestine, 41–45
    absorption of amino acids, 275, 276*t*
    alanine (Ala), uptake of, 332
    anatomical feature, 41
    arginine (Arg) synthesis in, 414
    bile, 43
    chlorine (Cl) absorption, 691
    citrulline synthesis, 414
    counteracting stomach acidity, 42–43
    cup cells, 42
    digestion, 44
    duodenum, 41, 43–44
    endocrine products, 43–44
    enterocytes, 41
    enteroendocrine cells of, 42
    epithelial cells, 41
    exocrine secretions, 43–44
    Gln uptake, 303
    goblet cells, 41
    gross anatomical structure, 41
    intestinal mucosa, 41
    jejunum, 41
    luminal surface, 41
    main function of, 41
    major cell types of, 41
    mesentery, 41
    microanatomical features, 41
    microfold or membraneous cells (M cells), 42
    pancreatic secretions, 43–44
    Paneth cells, 42
    pH adjustment, 40, 42–43
    threonine (Thr), uptake of, 317

    tuft cells, 42
    water secretion, 675*f*
Smell, 1–4
    definition, 1
    impaired ability to, 1–2, 28
    odors as appetite stimulant, 1
    sensitivity, variation in, 3
Smell-blindness, 3
Sodium (Na), 678–685
    blood measurement, 684
    daily requirements, 679
    deficiency, impacts of, 679
    dietary sources, 679
    digestion and absorption, 679–680, 680*f*
    excretion, 681–683
        renal losses, 681, 682*f*
        sweating, 682–683
    functions, 679, 683–684
        cotransporters, 684
        enzyme-catalyzed reactions, 683
        signaling, 684
    net secretion, 679–680
    nutritional assessment, 684
    nutritional summary, 679
    regulation, 683
        excretory control, 683
        intake control, 683
    risk of excessive intake, 679
    storage, 681
    transport and cellular uptake, 681
        via blood, 681
        blood–brain barrier (BBB), 681
        materno-fetal transfer, 681
    urine measurement, 684
Sodium–amino acid cotransport systems and uptake
    mechanism, 303, 679–680
    Ala, 325, 332–334
    Arg, 416
    Asn, 409–410, 412
    biotin, 643
    exchange of Cys and CssC, 370
    free Trp, 351
    Gln, 303, 305–307
    Gly, 311, 314
    Met, 361
    Phe, 337, 341
    Ser, 324–325, 327
    Thr, 317–318, 320
    Trp, 351
Sodium-dependent myo-inositol transporter (SLC5A3), 70–72
Sodium dicarboxylate/sulfate transporter 2 (SDCT2, SLC13A3), 760

Sodium dodecyl sulfate, 759
Sodium/glucose cotransporters, 195, 215
Sodium/iodide symporter (SLC5A5), 749, 752
Sodium/potassium-exchanging ATPase, 679–681, 683, 687–688, 693–695, 738
Soluble fiber, 244–245
    alginate, 245
    beta-glucans, 244
    carageenan, 245
    galactomannans, 245
    lignin, 245
    long-term benefits, 245
    pectins, 245
    xanthan gum, 245
Sour taste, 10–11
Spermidine synthase, 367
Sphingolipids, 327–328
Starches, 189–190
Starvation, 303
    alanine (Ala) synthesis, 332, 334
    asparagine (Asn) transport and, 412
    protein availability, 292–293
    protein quality, 293
    protein turnover, 291–292
    Ser absorption and, 324–325
    as a serious health threat, 291
        changes in protein and amino acid metabolism, 292
        oxidation of amino acids, 292
    Val metabolism and, 394
Stearidonic acid, 123
Stomach, 38–40
    acid-stimulating effect, 38–40
    anatomical structure, 38
    calcium-sensing receptor (SCAR), 38–40
    hydrochloric acid, role of, 40
    mixing and grinding of ingested foods, 40
    motility, 40
    parietal cells, 40
    release of gastric contents, 40, 40f
    secretory function, 38–40
Streptococcus species, 65
Sucralose, 18
Sucrose, 189
Sugars, 18
Sulfate-reducing bacteria (SRB), 760
Sulfidequinone oxidoreductase (SQR), 761
Sulfinoalanine decarboxylase, 443–445, 616–617
3-Sulfino-L-alanine, 330
Sulfite oxidase, 776
Sulfur (S), 758–766
    catabolism, 759
    daily requirements, 759

    deficiency, impacts of, 759
    dietary sources, 759–760
        concentration in tap water, 759
        pharmaceutical products, 759
    endogenous sources, 759
    excretion, 762
    functions, 758, 763–764
        antioxidant defense, 763
        in bile acids, 764
        catecholamines, 764
        cell adhesion proteins, 764
        cyanide detoxification, 763
        gastrointestinal effects, 764
        glycans, 763–764
        homeostatic, 763
        modification of xenobiotics, 764
        renal effects, 764
        skeletal effects, 764
        steroid hormones, 764
        sulfolipids, 764
        thiocyanate ($H_2 S_2 O_3$), 763
    intestinal absorption, 760
    metabolism, 761–762
        hydrogen sulfide, generation of, 761
        sulfate activation, 762
        sulfite, generation of, 761
        thiocyanate, generation of, 761
    natural isotopes, 758
    nutritional summary, 758–759
    oxidized forms, 758
    precursors, 759
    regulation, 762–763
    risk of excessive intake, 759
    storage, 762
    toxic effects of sulfite, 764
    transport and cellular uptake, 760–761
        via blood, 760
        blood–brain barrier (BBB), 761
        materno-fetal transfer, 761
Sulfur dioxygenase, 762
Superoxide anion, 480–481
Supertasters, 12–13
Sweet taste, 8–9
    interaction of ingested molecules, 9
    proteins with, 8–9
    spectrum of sweet-tasting molecules, 8
Sweet-tasting carbohydrates, 8

T

Tagatose, 188–189
Tannins, 38
TAS1R1–TAS1R3 dimer, 18

Taste (gustation), 4–16
  basic taste qualities, 4, 12
  bitter, 9–10
  calcium, 12
  cold perception, 22
  digestion and, 4
  enhancing secretions, 7
  fat, 11–12
  hearty (*kokumi*), 11
  hot, spicy, 21–22
  meaty (*umami*), 11
    Glu flavor, 296
  metallic, 12
  molecular basis of signaling in, 7
  salty, 8
  sensitivity to, 12–13
  sour, 10–11
  supertasters, 12–13
  sweet, 8–9, 18
Taste buds, 4–6
  cells in, 6
  circumvallate papillae, 4, 5*f*
    trenches of, 4, 5*f*
  detection of nutrients, 6
  foliate papillae, 4, 5*f*
  fungiform papillae, 4, 5*f*
  innervation of, 6–7
  receptor cells of, 6
  receptors, 4
  sensors attached to, 6
Taste cell–associated cation-channel TRPM5, 17
Taste PROP, 12–13
Taste receptor–associated G-protein alpha-gustducin, 17
Taste sensitivity
  age and, 13
  physical status and medications, role of, 13
Taurine, 46–47, 271–272, 443–450, 443*f*, 616–617, 758–759, 794–795
  as an ampholyte, 448*f*
  daily requirements, 443
  deficiency, impacts of, 443
  dietary sources, 443, 445
  digestion and absorption, 445
  endogenous sources, 443–445, 444*f*
  excretion, 446
  functions, 443, 448
    as an antioxidant, 448
    bile acid conjugation, 448
    osmolar buffering, 448
  metabolism, 445–446, 447*f*
  nutritional summary, 443
  regulation, 446

  risk of excessive intake, 443
  storage, 446
  transport and cellular uptake, 445
    via blood, 445
    blood–brain barrier (BBB), 445
    materno-fetal transfer, 445
Tetrahydrobiopterin, 485, 657–658, 657*f*
  reactivation, 660*f*
  synthesis, 659*f*
5, 6, 7, 8-Tetrahydrobiopterin (BH4), 343
Tetrathiomolybdate, 777
L-Theanine (5-*N*-ethylglutmine), 272
Thiamin, 139, 221, 279–281, 299, 302, 316, 320–321, 334, 336, 348, 352, 359–360, 365, 383, 396, 398, 402, 405, 420, 423, 429, 580–589, 580*f*, 758
  anaphylactic reactions, 580
  creatinine/thiamin renal clearance ratio, 584
  daily requirements, 580
  deficiency, impacts of, 580
  dietary sources, 580–581
  digestion and absorption, 581–582
  excretion, 584
  functions, 580, 585–586
    brain function, 585–586
    cardiovascular function, 586
    catabolism of the branched-chain amino acids, 585
    detoxification program, 585–586
    glucose metabolism, 585
    mitochondrial function, 586
    phytanic acid metabolism, 585
  homeostasis, 584–585
  intestinal absorption, 582*f*
  materno-fetal nutrient transport, 91
  metabolism, 583–584, 583*f*
  nutritional assessment, 586
  nutritional summary, 580
  regulation, 584–585
  storage, 584
  transport and cellular uptake, 582–583
    via blood, 582–583
    blood–brain barrier (BBB), 582–583
  transportation to brain, 83
  variation coefficient for, 816–817
Thiaminase, 581
Thiamin pyridinylase, 581
Thiamin pyrophosphate (TPP), 580
  in blood plasma, 582–583, 586
  dietary sources, 581
  metabolism, 583–584
Thiazide-sensitive sodium/chloride cotransporter, 695
Thioredoxin, 572–573
Thioredoxin reductase, 483–485, 596

Thiosulfate, 758, 763
Thirst, 677. *See also* Water
  brain sensors of, 33
    angiotensin II, 33
    relaxin, 33
    sodium-sensitive and solute-density (osmolality-) sensitive
      neurons, 33
  complex central nervous system input, 34
  peripheral sensors of, 32–33
    blood pressure variation and, 32–33
    otopharyngeal mechanoreceptors, 33
Threonine (Thr), 271–272, 316–321
  adequate daily intake, 317
  deficiency, effects of, 317
  dietary sources, 316–317
  digestion and absorption, 317
  endogenous sources, 317
  excretion, 320
  functions, 316
    in energy fuel yielding, 320
    glycine synthesis, 320
    protein synthesis, 320
  L form, 316, 316*f*
    intestinal absorption, 318*f*
    metabolism of, 319*f*
  metabolism
    in cytosol, 318–319
    in mitochondria, 319–320
  nutritional summary, 316–317
  regulation, 320
  requirements of adults, 317
  risk of excessive intake, 317
  storage, 320
  transport and cellular uptake, 317–318
    via blood, 317–318
    blood–brain barrier (BBB), 318
    materno-fetal transfer, 318
L-Threonine dehydrogenase, 319–320
Thrombinogen, 532–533
Thymine, 330
Thyroglobulin, 750
Toxic metabolites, 66
  Candida albicans and, 66
  Deltaproteobacteria, 66
  production of, 66
Transaldolase, 199–201, 221
Transcobalamins I and III, 635–636
TRANSFAC, 824
Transient receptor potential ankyrin 1 (TRPA1), 22
Transient receptor potential vanilloid 1 (TRPV1), 21–22
Transketolase, 199–201, 221, 585
Transport and cellular uptake

acetate, 149–150
alanine (Ala), 332–333
amino acids, 275–276, 278
arginine (Arg), 417
arsenic (As), 798
asparagine (Asn), 409–410
aspartate (Asp), 402–403
biopterin, 658
biotin, 643
boron (B), 786
bromine (Br), 793
calcium (Ca), 732
carnitine, 457
chlorine (Cl), 693
chlorophyll, 180
cholesterol (Chol), 545–548
choline, 471–472
chromium (Cr), 783
citrulline, 441–442
cobalt (Co), 779
conjugated linoleic acid (CLA), 160
copper (Cu), 710–712
creatine, 452
cysteine (Cys), 370–372
docosahexaenoic acid (DHA), 167–168
ethanol, 233
fatty acids, 125–134
flavonoids, 253–254
fluorine (F), 756
fructose (Fru), 209
galactose (Gal), 216
glucose (Glc), 196–197
glutamate (Glu), 297
glutamine (Gln), 303–305
glycine (Gly), 311
histidine (His), 432–434
inositol, 666
iodine (I), 749–750
iron (Fe), 700–702
isoleucine (Ile), 398
leucine (Leu), 384–385
lipoic acid (LA), 555
lysine (Lys), 378–379
magnesium (Mg), 745
manganese (Mn), 726
melatonin, 464
methanol, 239
methionine (Met), 361
molybdenum (Mo), 774
niacin, 603
nickel (Ni), 806
nitrite/nitrate, 108

pantothenate, 649–650

phenylalanine (Phe), 338

phosphorus (P), 739–740

potassium (K), 687–688

proline (Pro), 425

pyruvate, 224–225

retinol, 492

riboflavin, 591–592

selenium (Se), 768

serine (Ser), 325

silicon (Si), 791

sodium, 681

sulfur (S), 760–761

taurine, 445

thiamin, 582–583

threonine (Thr), 317–318

trans-unsaturated fatty acids (trans-fatty acids), 176

tryptophan (Trp), 351–352

tyrosine (Tyr), 345

ubiquinone (coenzyme Q), 563

valine (Val), 391

vanadium (V), 803

vitamin A (vitA), 492

vitamin B6, 613–614

vitamin B12 (B12), 636–637

vitamin C, 573

vitamin D (vitD), 504–505

vitamin E (vitE), 517–518

vitamin K (vitK), 529

water, 676

xylitol, 221

zinc (Zn), 719–720

Trans-unsaturated fatty acids (trans-fatty acids), 174–179, 175f

daily requirements, 174

deficiency, impacts of, 174

dietary sources, 174–176

endogenous sources, 174

excretion, 177

functions, 174, 178

fuel energy, 178

on glucose homeostasis, 178

GRAS status, 174–175

health impacts

atherosclerosis and risk of myocardial infarction, 178

systemic inflammation, 178

hydrogenation and, 174–176

metabolism, 176

beta-oxidation cycles, 176

mitochondrial catabolism, 176, 177f

peroxisomal catabolism, 176

nutritional summary, 174

regulation, 178

risk of excessive intake, 174

storage, 176–177

in adipose tissue, 176–177

transport and cellular uptake, 176

via blood, 176

blood–brain barrier (BBB), 176

materno-fetal transfer, 176

Trans-vaccenic acid (TVA), 123, 158, 159f

Trehalase deficiency, 195

Trehalose, 189, 195

Triglycerides, 116, 123, 136, 138

Trigonelline, 603f

Trimethoprim, 657

Trimethyllysine, 457

Trimethyllysine dioxygenase, 706

Triose isomerase, 197–198

TRNA(Ser) sec, 327

Trypsin, 272–273, 635–636

Tryptamine, 350, 352

L-Tryptophan (Trp), 330, 600

Tryptophan 2, 3-dioxygenase, 352

Tryptophan 5-monooxygenase, 706

Tryptophan (Trp), 271–272, 343, 349–359, 600–601, 606–607

daily requirements, 350

deficiency, impacts of, 350, 356–357

dietary sources, 350

digestion and absorption, 351

dipeptides and tripeptides of, 351

excretion, 355

free, 351

functions, 350, 355–357

in energy fuel yielding, 355

melatonin synthesis, 357

NAD synthesis, 355

photoprotection, 357

protein synthesis, 355

serotonin synthesis, 355–357

L form, 349, 349f

intestinal absorption, 351f

pathways of breakdown, 353f

uses, 356f

mechanism of loss during heating, 350

metabolism, 352–354

catabolism, 352, 354

nutritional summary, 350

regulation, 355

risk of excessive intake, 350

storage, 354

transport and cellular uptake, 351–352

via blood, 351–352

blood–brain barrier (BBB), 352

materno-fetal transfer, 352

Tuft cells, 17, 42
Tyramine, 344
Tyrosinemia, 340
Tyrosine (Tyr), 271, 342–349
  daily requirements, 343
  deficiency, impacts of, 343
  dietary sources, 343–344
  digestion and absorption, 344
  endogenous sources, 343
  excretion, 346
  functions, 343
    catecholamine synthesis, 348
    in energy fuel yielding, 348
    melanin synthesis, 348
    protein synthesis, 348
    thyroid hormone synthesis, 348
    ubiquinone synthesis, 348
  L form, 342–343, 342f
    intestinal absorption, 345f
    from L-phenylalanine, 344f
    metabolism of, 347f
  metabolism of, 346
    mitochondrial and cytosolic synthesis, 346
  nutritional summary, 343
  regulation, 346
  risk of excessive intake, 343
  storage, 346
  transport and cellular uptake, 345
    via blood, 345
    blood–brain barrier (BBB), 345
    materno-fetal transfer, 345
  L-tyrosine, 18
    fumerate and acetoacetate of, 337f
Tyrosine 3-monooxygenase, 341

U

Ubiquinone (coenzyme Q), 139, 279–281, 299, 302, 316,
    320–321, 334, 336, 341, 348, 350, 352, 359–360, 365, 383,
    396, 398, 402, 405, 417, 420, 423, 429, 485, 558, 560–565
  catabolism, 563
  daily requirements, 560
  deficiency, impacts, 560–561
  de novo synthesis of, 561
  dietary sources, 560–561
  digestion and absorption, 561–563
  endogenous sources, 561
  excretion, 563
  functions, 560, 563–564
  nutritional assessment, 564
  nutritional summary, 560–561
  regulation, 563
  risk of excessive intake, 561
  transport and cellular uptake, 563
Ubiquinone-10 (Q10), 560, 560f, 562f

V

Valine (Val), 271–272, 389–396
  daily requirements, 390
  deficiency, impacts of, 390
  dietary sources, 390
  digestion and absorption, 390
  excretion, 394
  functions, 390
    anabolic effects on skeletal muscle, 394
    in energy fuel yielding, 394
    neurotransmitter metabolism, 394
    protein synthesis, 394
  L form, 389, 389f
    intestinal absorption, 391f
    metabolism of, 392f
  metabolism, 392–393
    catalyzation, 393
    pathway of Val breakdown to propionyl-CoA, 392–393
    transamination products, 393
  nutritional summary, 390
  regulation, 394
  risk of excessive intake, 390
  storage, 394
  transport and cellular uptake, 391
    via blood, 391
    blood–brain barrier (BBB), 391
    materno-fetal transfer, 391
Vanadate, 802–803
Vanadium (V), 802–805
  daily requirements, 802
  deficiency, impacts of, 802
  dietary sources, 802
  digestion and absorption, 802–803
  excretion, 803
  functions, 802–804
    as an inhibitor of Na-K-ATPase, 804
    chemopreventive effects, 804
    as a cofactor, 803
    for lowering blood–lipid levels, 804
    for survival and reproductive health, 803
  isotopes, 802
  metabolism, 803
  nutritional summary, 802
  regulation, 803
  risk of excessive intake, 802
  storage, 803
  transport and cellular uptake, 803
    via blood, 803
Vanadylate, 802–803
Vanadyl sulfate, 804
Viral diarrhea, 66
Vitamin A (vitA), 486–501, 820
  daily requirements, 487
    women's needs, 487

deficiency, impacts of, 487–488
dietary sources, 487–490
    in animal-derived foods, 489
    from beta-carotene, 489*f*
    from carotenoids, 490*f*
    in natural foods, 488
    from plants, 489–490
    synthetic retinol, 488
digestion and intestinal absorption, 490–492
endogenous synthesis, 488
excretion, 494–495
functions, 487, 496–498
    cell cycle and apoptosis, 497
    cell signaling, 498
    nuclear actions, 497
    vision, 496
metabolism, 492–494
    all-trans-retinoic acid synthesis, 492–493
    conversion of all-trans-retinol to 9-cis-retinol, 493–494
    3, 4-didehydroretinoic acid, formation of, 494
    hydroxylated retinoids, formation of, 494
    in the retina, 496*f*
    retinoic acid breakdown, 494
metabolites, 486
nutritional assessment, 498
nutritional summary, 487–488
regulation, 495
retinol activity equivalents (RAE), 489
risk of excessive intake, 488
storage, 494–495
structurally related compounds with, 487*f*
transport and cellular uptake, 492
    via blood, 492
    blood–brain barrier (BBB), 492
    materno-fetal transfer, 492
variation coefficient for, 816–817
Vitamin B6 (B6), 279–281, 299, 316, 320–321, 334, 336, 348, 350, 352, 359–360, 365, 381, 396, 402, 405, 409, 414, 417, 420, 423, 429, 610–620
in animal-derived foods, 611, 611*f*
bioavailability of dietary B6 glucoside, 612–613
blood measurement, 618–619
daily requirements, 610
deficiency, impacts of, 611, 819
dietary sources, 610–612
digestion and absorption, 612–613
effect of cooking, 612
excretion, 614
functions, 610, 616–618
    amino acid synthesis and catabolism, 617
    arginine synthesis, 618
    carbohydrate metabolism, 616
    carnitine synthesis, 618
    glutamate/glutamine/aspartate/asparagine synthesis, 618

glycine synthesis, 617
heme synthesis, 616
histidine transamination, 618
hormone metabolism, 618
L-alanine synthesis, 617
lipid metabolism, 618
lysine conversion into cadaverine, 618
metabolic regulation, 618
methionine/cysteine transamination reaction, 617
neurotransmitter synthesis, 616–617
phenylalanine/tyrosine synthesis, 617
proline synthesis, 617
selenium metabolism, 618
serine conversion, 617
taurine synthesis, 618
threonine catabolism, 617
tryptophan metabolism, 617
urea cycle, 617
valine/leucine/isoleucine breakdown, 617
vitamin metabolism, 618
xenobiotic metabolism, 618
inactive derivatives of, 612*f*
intestinal absorption, 613*f*
materno-fetal nutrient transport, 91
metabolism, 614, 615*f*
nutritional assessment, 618–619
nutritional summary, 610–611
in plant-derived foods, 611*f*, 612
regulation, 614–616
risk of excessive intake, 611
selenium metabolism and, 819
storage, 614
transport and cellular uptake, 613–614
    via blood, 613
    blood–brain barrier (BBB), 614
    materno-fetal transfer, 614
transportation to brain, 83
urine measurement, 619
variation coefficient for, 816–817
Vitamin B12 (B12), 139, 280, 316, 320, 359–360, 365, 396, 469, 632–642, 820
absorption, 53
blood measurement, 640
daily requirements, 632
deficiency, impacts of, 632
dietary sources, 632–633
    plant-derived foods, 633
digestion and absorption, 634–636
    by pepsin, 635–636
    transcellular transport, 636
    transcobalamin-II-mediated transport, 636
    uptake into ileal enterocytes, 636
    using endogenous B12-binders, 635
enterohepatic circulation of, 47–48

Vitamin B12 (B12) (*Continued*)
  excretion, 637–639
  functions, 632
    as cyanide antidote, 640
    homocysteine remethylation, 639
    leucine metabolism, 639–640
    propionyl-CoA metabolism, 639
  intestinal absorption, 635*f*
  metabolic activation of, 638*f*
  metabolism, 637
    synthesis in mitochondria, 637
  nutritional assessment, 640
  nutritional summary, 632–634
  regulation, 639
  risk of excessive intake, 632
  storage, 637
  transport and cellular uptake, 636–637
    via blood, 636
    blood–brain barrier (BBB), 637
    materno-fetal transfer, 637
    via transcobalamin-II/B12 complex, 636–637
  transportation to brain, 83
  urine measurement, 640
  variation coefficient for, 816–817
Vitamin C, 479, 488, 570–580
  daily requirements, 571
  deficiency, impacts of, 571
  dietary sources, 571–572
  digestion and absorption, 572–573
  endogenous sources, 571
  excretion, 576
  functions, 570, 576–578
    aldosterone synthesis, 577
    amino acid metabolism, 577
    antioxidant defense against free radicals, 570
    bile acid synthesis, 577
    decreasing carcinogen exposure, role in, 578
    DNA modification, 577–578
    free radical scavenging, 576
    iron metabolism, 578
    protein modification, 577
    sulfate transfer, 578
  intestinal uptake of, 572*f*
  materno-fetal nutrient transport, 90–91
  metabolism, 573–574
  nonenzymic degradation of, 576*f*
  nutritional assessment, 578
  nutritional summary, 570–571
  redox equivalents, 575*f*
  redox states, 570, 571*f*
  reduction of DHA to ASC, 572–573
  regulation, 576
  risk of excessive intake, 571

  storage, 574–575
  transport and cellular uptake, 573
    via blood, 573
    blood–brain barrier (BBB), 573
    materno-fetal transfer, 573
  transportation to brain, 83
Vitamin D (vitD), 166, 501–514, 731–732, 741
  blood measurements, 511
  common form D₃ (D3, cholecalciferol), 501–502
    light-induced synthesis of, 503*f*
  daily requirements, 502
  deficiency, impacts of, 501*f*, 502
  dietary sources, 502–504
    in fish, 502
    in milk, 504
  digestion and absorption, 504
    intestinal absorption, 505*f*
  endogenous sources, 502
    UV light, 502
  excretion, 507
  functions, 501, 510–511
    calcium regulation, 734
    cell differentiation, 511
    cellular effects, 511
    effects on bone, 511
    intestinal calcium absorption, 510–511
    nuclear effects, 510
  metabolism, 505–507
    catabolic pathways, 507, 508*f*, 509*f*
  metabolites derived from vitamin D₃ –epi-intermediates, 510*f*
  nutritional assessment, 511
  nutritional summary, 501–502
  regulation, 508–509, 510*f*
  renal filtration and reabsorption, 506*f*
  renal processing of, 75
  renal production of 1, 25-D is PTH, 509
  response elements (VDREs), 510
  risk of excessive intake, 502
  storage, 507
  transport and cellular uptake, 504–505
    via blood, 504
    blood–brain barrier (BBB), 505
    materno-fetal transfer, 505
  transportation to brain, 84
  vitamin D₂ (ergocalciferol), 501
Vitamin E (vitE), 479–481, 485, 488, 514–526
  biological potency, 516
  blood measurements, 523
  daily requirements, 514
  deficiency, impacts of, 515
  dietary sources, 514–517
    synthetic all-rac alpha-tocopherol, 516*f*

synthetic production of, 516
digestion and absorption, 517
excretion, 519–522
functions, 514, 522–523
   antioxidant protection, 522–523
   fertility, 523
   inhibiting effects on platelet aggregation, 523
   nonantioxidant functions, 523
   pathogen mutations, 523
   quenching of free radicals, 522*f*
metabolism, 519, 520*f*
   reaction with free radicals, 521*f*
   reaction with one-electron oxidants, 519
   ring modification, 519
   side-chain breakdown, 519
naturally occurring forms of, 514, 515*f*
nutritional assessment, 523–524
nutritional summary, 514–515
properties, 515–516
regulation, 522
risk of excessive intake, 515
RRR-alpha-tocopherol, 515–518, 518*f*
transport and cellular uptake, 517–518
   via blood, 517–518
   blood–brain barrier (BBB), 518
   from extrahepatic tissues, 518
   materno-fetal transfer, 518
transportation to brain, 84
urine measurements, 524
Vitamin K (vitK), 66, 271, 526–539, 764, 820
blood measurements, 535
daily requirements, 527
deficiency, impacts of, 527
dependent proteins, 534*t*
dietary sources, 527
digestion and absorption, 528–529
endogenous sources, 527
excretion, 531
functions, 527, 532–535
   blood coagulation, 532–533
   bone mineralization, 533
   embryonic heart valve development, 534
   fertility, 533
   of Gla-proteins, 534
   glucose homeostasis, 533
   GRP carboxylation, 533
   microtubulin stability, 534
   oxidative activity in mitochondria, 535
   prostaglandin metabolism, 535
   remodeling of extracellular matrix, 534
   signaling, 533–534
   sulfur metabolism, 534–535
   of tissue mineralization, 533

intestinal absorption, 528*f*
metabolism, 529–531
   carboxylation, 529
   catabolism, 531
   gamma-carboxylation of specific glutamyl
      residues, 529
   by omega-oxidation, 532*f*
   side-chain resynthesis, 530
   vitK cycle, 529–530
natural form, 526, 526*f*
nutritional assessment, 535
nutritional summary, 527
regulation, 532
risk of excessive intake, 527
storage, 531
transport and cellular uptake, 529
   via blood, 529
   blood–brain barrier (BBB), 529
   materno-fetal transfer, 529
urine measurements, 535
Vitamins, transportation to brain, 83–84
Vomeronasal organ (VNO, Jacobson's organ), 2
Von Ebner's glands, 7, 38

W

Water, 673–678
access to clean, 674
daily requirements, 674
deficiency, impacts of, 674
deprivation, 32
dietary sources, 674
digestion and absorption, 674–675
endogenous sources, 674
excretion, 676–677
   exhaled, 676
   sweating, 676–677
   urinary, 676
functions, 674
   in chemical reactions, 677–678
   ionization, 677
   as solvent, 677
   in temperature regulation, 678
mechanisms of secretion and absorption in small and
   large intestine, 675*f*
net secretion, 674–675
nutritional summary, 674
regulation, 677
risk of excessive intake, 674
transport and cellular uptake, 676
   via blood, 676
   blood–brain barrier (BBB), 676
   across the epithelium, 676
   materno-fetal transfer, 676

Water intake. *See* Thirst
Wernicke–Korsakoff syndrome (WKS), 199–201, 221, 580, 585–586
Wilson disease, 777
Wilson protein (ATP7B), 712–713
Wolff–Chaikoff effect, 752

X

Xaa-Pro aminopeptidase, 727
Xaa-Pro dipeptidase, 727
Xanthan gum, 245
Xenobiotics, 104
   materno-fetal nutrient transport, 91
   renal processing, 77
   transportation to brain, 85
Xylitol, 219–223, 219*f*
   daily requirements, 219
   dietary sources, 219–220
   digestion and absorption, 220
   endogenous production, 219–220, 220*f*
   functions, 219, 221
   metabolism, 221, 222*f*
      pentitol pathway, 221
   nutritional summary, 219
   risk of excessive intake, 219
   transport and cellular uptake, 221
      via blood, 221

Z

Zellweger syndrome, 185
Zinc (Zn), 479, 487, 716–724
   artificial isotopes, 716
   brain Zn homeostasis, 719
   daily requirements, 717
   deficiency, impacts of, 717, 721
   dietary sources, 717
   digestion and absorption, 717–719
      export to blood, 719
      via hydrogen ion/peptide cotransporter, 718
      movement across epithelial layer, 719
   excretion, 720
   free, 719
   functions, 717, 721–722
      DNA replication and transcription, 722
      as enzyme cofactor, 721
      free radical metabolism, 722
      fuel metabolism, 722
      intestinal digestion, 721
      maintaining immune function, 722
      nutrient metabolism, 721–722
      pH-regulation, 721
      RNA editing, 722
      in vascular function, 722
   intestinal absorption, 718*f*
   nutritional assessment, 722
   nutritional summary, 717
   regulation, 720–721
      control of metallothionein expression, 720–721
   risk of excessive intake, 717
   stable isotopes, 716
   storage, 720
   transport and cellular uptake, 719–720
      via blood, 719
      blood–brain barrier (BBB), 719
      maternal circulation into milk, 719–720
      materno-fetal transfer, 719
   transportation to brain, 84–85